Madan Lal Puri Selected Collected Works
Volume 2

Madan Lal Puri

Dedication

These three volumes of "Selected Collected Works of Madan Lal Puri" will serve to preserve, in a unified and easily accessible form, the knowledge and wisdom conveyed in his many research papers, so as to aid its dissemination to future generations.

The Editors

Madan Lal Puri
Selected Collected Works

Volume 2

Probability Theory and Extreme Value Theory

Editors
Peter G. Hall, Marc Hallin and George G. Roussas

///VSP///
Utrecht • Boston
2003

VSP
P.O. Box 346
3700 AH Zeist
The Netherlands

Tel: +31 30 692 5790
Fax: +31 30 693 2081
vsppub@compuserve.com
www.vsppub.com

First published in 2003

ISBN 90-6764-385-8 (Volume 2)
ISBN 90-6764-387-4 (Set)

Printed in The Netherlands by Ridderprint bv, Ridderkerk

CONTENTS

Preface xii

Part I. LIMIT THEOREMS, RATES OF CONVERGENCE, AND RELATED TOPICS (INDEPENDENT CASE)

Order of normal approximation for rank test statistics distribution
(with Jana Jurečková)
Annals of Probability, 3 (1975), 526-533. 3

Convergence and remainder terms in linear rank statistics
(with Harald Bergström)
Annals of Statistics, 5 (1977), 671-680. 11

Invariance principles for rank statistics for testing independence
(with Lanh T. Tran)
Contributions to Probability Theory. Academic Press, New York (1981), 267-282. 21

On the degeneration of the variance in the asymptotic normality of signed rank statistics
(with Stefan S. Ralescu)
Statistics and Probability: Essays in Honor of C.R. Rao. North Holland Publishing Company (1982), 591-608. 37

On the order of magnitude of cumulants of von Mises functionals and related statistics
(with R.N. Bhattacharya)
Annals of Probability, 11 (1983), 346-354. 54

On Berry–Esséen rates, a law of the iterated logarithm and an invariance principle for the proportion of the sample below the sample mean
(with Stefan Ralescu)
Journal of Multivariate Analysis, 14 (1984), 231-247. 63

Cramér type large deviations for generalized rank statistics
(with Munsup Seoh and Stefan S. Ralescu)
Annals of Probability, 13 (1985), 115-125. 80

On the rate of convergence in the central limit theorem for signed rank statistics
(with Stefan S. Ralescu)
Advances in Applied Mathematics, 6 (1985), 23-51. 91

A sharpening of the remainder term in the higher-dimensional central limit theorem for multilinear rank statistics
(with Manfred Denker and Uwe Rösler)
Journal of Multivariate Analysis, 17 (1985), 148-167. 120

The order of normal approximation for signed linear rank statistics
(with T.-J. Wu)
Teoria Veroyatnostei i ee Primeneniya (also in Theory of Probability and Its Applications, Translation by SIAM), 31 (1986), 156-163. 140

Central limit theorem for perturbed empirical distribution functions evaluated at a random point
(with Stefan S. Ralescu)
Journal of Multivariate Analysis, 19 (1986), 273-279. 147

Limit theorems for random central order statistics
(with Stefan S. Ralescu)
Adaptive Statistical Procedures and Related Topics, IMS Lecture Notes and Monograph Series, 8 (1987), 447-475. 154

Asymptotic expansions for sums of nonidentically distributed Bernoulli random variables
(with Paul Deheuvels and Stefan S. Ralescu)
Journal of Multivariate Analysis, 28 (1989), 282-303. 183

On the rate of convergence in normal approximation and large deviation probabilities for a class of statistics
(with M. Seoh)
Teoria Veroyatnostei i ee Primeneniya (also in Theory of Probability and its Applications, Translation by SIAM), 33 (1989), 735-750. 205

On Hilbert-space-valued U-statistics
(with V.V. Sazonov)
Teoria Veroyatnostei i ee Primeneniya (also in Theory of Probability and Its Applications, Translation by SIAM), 36 (1991), 604-605. 221

On the central limit theorem in Hilbert space with application to U-statistics
(with V.V. Sazonov)
In: Statistical Sciences and Data Analysis: Proceedings of the Third Pacific Area Statistical Conference. VSP International Science Publishers, Holland. (Editors: Kameo Matusita, Madan L. Puri and Takesi Hayakawa) (1993), 407-413. 224

Asymptotic expansions in statistics: A review of methods and applications
(with Rabi N. Bhattacharya)
In: Advances in Econometrics and Quantitative Economics: Essays in honor of C.R. Rao. Blackwell Publishers, Oxford, United Kingdom. (Editors: G.S. Maddala, Peter C.B. Phillips and T.N. Srinavasan) (1995), 88-122. 231

Normal approximation of U-statistics in Hilbert space
(with Yu.V. Borovskikh and V.V. Sazonov)
Teoria Veroyatnostei i ee Primeneniya (also in Theory of Probability and its Applications, Translation by SIAM), 41 (1997), 481-504. 266

Part II. LIMIT THEOREMS (DEPENDENT CASE)

Empirical distribution functions and functions of order statistics for mixing random variables
(with Lanh T. Tran)
Journal of Multivariate Analysis, 10 (1980), 405-425. 289

An invariance principle for processes indexed by two parameters and some statistical applications
(with Manfred Denker)
Probability and Mathematical Statistics, 9 (1988), 25-76. 310

Limiting behavior of U-statistics, V-statistics, and one sample rank order statistics for nonstationary absolutely regular processes
(with Michel Harel)
Journal of Multivariate Analysis, 30 (1989), 181-204. 362

Weak invariance of generalized U-statistics for nonstationary absolutely regular processes
(with Michel Harel)
Stochastic Processes and Their Applications, 34 (1990), 341-360. 386

The space $\tilde{D}_k$ and weak convergence for the rectangle-indexed processes under mixing
(with Michel Harel)
Advances in Applied Mathematics, 11 (1990), 443-474. 406

Weak invariance of the multidimensional rank statistic for nonstationary absolutely regular processe
(with Michel Harel)
Journal of Multivariate Analysis, 36 (1991), 204-221. 438

Weak convergence of the simple linear rank statistic under mixing conditions in the nonstationary case
(with M. Harel)
Teoria Veroyatnostei i ee Primeneniya (also in Theory of Probability and Its Applications, Translation by SIAM) 38 (1993), 579-599. 456

Weak convergence of weighted empirical U-statistics processes for dependent random variables
(with Michel Harel and Colm Art O'Cinneide)
Journal of Multivariate Analysis, 48 (1994), 297-314. 474

Law of the iterated logarithm for perturbed empirical distribution functions evaluated at a random point for nonstationary random variables
(with Michel Harel)
Journal of Theoretical Probability, 7 (1994), 831-855. 492

Valid Edgeworth expansions of M-estimators in regression models with weakly dependent residuals
(with Masanobu Taniguchi)
Econometric Theory, 12 (1996), 331-346. 517

Conditional U-statistics for dependent random variables
(with Michel Harel)
Journal of Multivariate Analysis, 57 (1996), 84-100. 533

Weak convergence of sequences of first passage processes and applications
(with Stefan S. Ralescu)
Stochastic Processes and Their Applications, 62 (1996), 327-345. 550

Conditional empirical processes defined by nonstationary absolutely regular sequences
(with Michel Harel)
Journal of Multivariate Analysis, 70 (1999), 250-285. 569

Part III. EXTREME VALUE THEORY

A strong invariance principle concerning the J-upper order statistics for stationary m-dependent sequences
(with George Haiman)
Journal of Statistical Planning and Inference, 25 (1990), 43-51. 607

A strong invariance principle concerning the J-upper order statistics for stationary Gaussian sequences
(with George Haiman)
Annals of Probability, 21 (1993), 86-135. 616

Extremes of Markov sequences
(with George Haiman and Maxime Kiki)
Journal of Statistical Planning and Inference, 45 (1995), 185-201. 666

Records and 2-block records of 1-dependent stationary sequences under local dependence
(with George Haiman, Nelly Mayeur and Valéry Nevzorov)
Annales de l'institut Henri Poincaré, 34 (1998), 481-503. 683

Part IV. APPENDICES

A. Volume 1.
Nonparametric Methods in Statistics and Related Topics
Contents 709

B. Volume 3.
Time Series, Fuzzy Analysis, and Miscellaneous Topics
Contents 715

C. The Publications of Madan Lal Puri 721

Part I

Limit Theorems, Rates of Convergence, and Related Topics

(Independent Case)

ORDER OF NORMAL APPROXIMATION FOR RANK TEST STATISTICS DISTRIBUTION

JANA JUREČKOVÁ and MADAN L. PURI[1]
Charles University and Indiana University

0. Summary. Under suitable assumptions, it is established that the rate of convergence of the cdf (cumulative distribution function) of the simple linear rank statistics

$$S_N = \sum_{i=1}^N C_{Ni}\varphi\left(\frac{R_{Ni}}{N+1}\right)$$

to the normal one is $O(N^{-\frac{1}{2}+\delta})$ for any $\delta > 0$. Here $C_{N1}, \cdots, C_{NN}$ are known constants, $R_{N1}, \cdots, R_{NN}$ are the ranks of independent observations $X_{N1}, \cdots, X_{NN}$, and φ is a score generating function defined in Section 1.

1. Introduction. Let X_{Ni}, $i = 1, \cdots, N$ be independent rvs distributed according to the cdf $F_i(x) = F(x - \Delta d_{Ni})$, $i = 1, \cdots, N$. We assumed that $F(x)$ is absolutely continuous having the density function $f(x)$ whose derivative $f'(x)$ exists. Furthermore, $F(x)$ is assumed to have the finite Fisher information, that is,

$$I(f) = \int_{-\infty}^{\infty} [f'(x)/f(x)]^2 f(x)\,dx < \infty\,. \tag{1.1}$$

Δ is an unknown parameter, and d_{Ni}, $i = 1, \cdots, N$ are known constants. Let R_{Ni} be the rank of X_{Ni} among $X_{N1}, \cdots, X_{NN}$. Setting $u(x) = 1$ if $x \geqq 0$, and $u(x) = 0$ otherwise, we can write

$$R_{Ni} = \sum_{j=1}^N u(X_{Ni} - X_{Nj})\,, \qquad i = 1, \cdots, N\,. \tag{1.2}$$

Consider now the simple linear rank statistics

$$S_N = \sum_{i=1}^N C_{Ni}\, a_N(R_{Ni}) \tag{1.3}$$

where $C_{N1}, \cdots, C_{NN}$ are known constants, and $a_N(i)$, $i = 1, \cdots, N$ are "scores" generated by a function $\varphi(t)$ in the following manner:

$$a_N(i) = \varphi\left(\frac{i}{N+1}\right), \qquad 1 \leqq i \leqq N\,. \tag{1.4}$$

Statistics of the type (1.3) play an important role in the theory of nonparametric inference. For example, in the two sample problem where $F_1 = \cdots = F_m \equiv F$, and

$$F_{m+1} = \cdots = F_N \equiv G\,,$$

Received March 26, 1974; revised October 1, 1974.

[1] Work supported by the Air Force Office of Scientific Research, AFSC, USAF, under Grant No. AFOSR 71-2009. Reproduction in whole or in part is permitted for any purpose of the U.S. Government. Part of this work was done while the author was the recipient of a Senior U.S. Scientist Award from the Alexander-von-Humboldt-Foundation.

AMS 1970 *subject classifications.* Primary 62E20; Secondary 60F05, 60F99.

Key words and phrases. Linear rank statistics, score generating function, rate of convergence.

for testing the hypothesis H_0: $F \equiv G$, many rank tests are based on the statistic

$$S_N' = \sum_{i=1}^{m} a_N(R_{Ni})$$

which is a special case of (1.3) when $C_{N1} = \cdots = C_{Nm} = 1$ and $C_{Nm+1} = \cdots = C_{NN} = 0$. It is well known (see e.g., Capon (1961)) that the statistics of the form (1.3) for different score functions yield locally most powerful rank tests. Under suitable assumptions on the C's and the score generating function φ, Hájek (1962) [see also Hájek–Šidák (1967)] established the asymptotic normality of S_N. However, the problem of determining the rate of convergence of the cdf of S_N to the limiting normal distribution has remained open. This problem is investigated in this paper for the case $\Delta = 0$ as well as for $\Delta \neq 0$. In both cases, the rate of convergence is proved to be $O(N^{-\frac{1}{2}+\delta})$ for $\delta > 0$. For the case $\Delta = 0$, the result is valid for the φ functions having the bounded first derivative, and for the case $\Delta \neq 0$, it is necessary to assume the boundedness of the fourth derivative of φ.

Throughout the paper, we shall make the following assumptions on C's and d's.

(1.5) $$\sum_{i=1}^{N} C_{Ni} = \sum_{i=1}^{N} d_{Ni} = 0\,, \qquad \sum_{i=1}^{N} C_{Ni}^2 = \sum_{i=1}^{N} d_{Ni}^2 = 1\,,$$

(1.6) $$\max_{1 \le i \le N} C_{Ni}^2 = O(N^{-1} \log N)\,, \qquad \max_{1 \le i \le N} d_{Ni}^2 = O(N^{-1} \log N)\,.$$

It may be noted that the assumption (1.5) can be made without any loss of generality. Furthermore, it may be noted [cf. Hájek–Šidák (1967)] that if φ is the difference of two non-decreasing, square integrable functions in (0, 1), then S_N has asymptotically $\eta(0, \sigma^2)$ distribution under $\Delta = 0$, and $\eta(ES_N, \sigma^2)$ or

$$\eta\left(\Delta \sum_{i=1}^{N} C_{Ni} d_{Ni} \int_0^1 \varphi(t)\varphi(t, f)\, dt, \sigma^2\right)$$

distribution under $\Delta \neq 0$. Here

$$\sigma^2 = \int_0^1 (\varphi(t) - \bar{\varphi})^2\, dt\,, \qquad \bar{\varphi} = \int_0^1 \varphi(t)\, dt\,, \qquad \varphi(t, f) = \frac{-f'(F^{-1}(t))}{f(F^{-1}(t))}$$

and $\eta(\xi, \sigma^2)$ stands for the normal distribution with mean ξ and variance σ^2.

2. Rate of convergence for $\Delta = 0$. The main result of this section is the following theorem.

THEOREM 2.1. *Let $\Delta = 0$ and the first derivative of $\varphi(t)$ exist and be bounded in* (0, 1). *Then, under the assumptions of Section* 1, *corresponding to any $\delta > 0$, there exists a constant $A(\delta) > 0$, and a positive integer N_δ such that for all $N > N_\delta$,*

(2.1) $$\sup_{-\infty < x < \infty} |F_N(x) - \Phi(x)| \le A(\delta) N^{(-\frac{1}{2}+\delta)}$$

where $F_N(x)$ is the cdf *of $\sigma^{-1} S_N$ and $\Phi(x)$ is the standard normal* cdf.

The proof of this theorem is based on the following two lemmas, the second of which is a consequence of Theorem 6, Chapter 5 of Petrov (1972).

LEMMA 2.1. *Under the assumptions of Theorem* 2.1, *corresponding to any positive integer k, where $2k + 1 < N$, there exists a constant $B(k) > 0$ and a positive integer N_k such that for all $N > N_k$,*

(2.2) $$E(S_N - T_N)^{2k} \le B(k) N^{-k}$$

where

$$T_N = \sum_{i=1}^N C_i \varphi(F(X_i)) \,. \tag{2.3}$$

LEMMA 2.2. *Under assumptions of Section 2 and Theorem 2.1, for any positive integer N,*

$$\sup_{-\infty<x<\infty} |F_N{}^*(x) - \Phi(x)| \leqq A \int_0^1 |\varphi(t) - \bar{\varphi}|^3 \, dt \cdot \sum_{i=1}^N |C_{Ni}|^3 \tag{2.4}$$

where $A > 0$ is a constant independent of N, and $F_N{}^*$ is the cdf *of $\sigma^{-1}T_N$ under $\Delta = 0$.*

In what follows, we shall suppress the subscript N in C_{Ni}, d_{Ni}, R_{Ni}, etc. whenever there is no confusion.

PROOF OF LEMMA 2.1. Set $U_i = F(X_i)$, $i = 1, \cdots, N$. Denoting $Y_i = a_N(R_i) - \varphi(U_i)$, $i = 1, \cdots, N$, we get

$$\begin{aligned} E[(S_N - T_N)^{2k}] &= E\{(\textstyle\sum_{i=1}^N c_i Y_i)^{2k}\} \\ &= \sum \frac{(2k)!}{p_1! \cdots p_N!} c_1{}^{p_1} \cdots c_N{}^{p_N} E(\textstyle\prod_{i=1}^N Y_i{}^{p_i}) \end{aligned} \tag{2.5}$$

where the sum extends over the set A of vectors $(p_1, \cdots, p_N)$ of integers such that $0 \leqq p_i \leqq 2k$, $i = 1, \cdots, N$, $\sum_{i=1}^N p_i = 2k$.

Each point of A could have at most $2k$ positive components. Noting this fact, we may decompose A into $2k$ disjoint parts such that the jth part consists of those points which have just j positive components. Thus we may rewrite (2.5) as

$$\begin{aligned} E[(S_N - T_N)^{2k}] &= \textstyle\sum_{i=1}^N c_i^{2k} E Y_i^{2k} + \cdots \\ &\quad + \sum_{1 \leqq p_1, \cdots, p_m < 2k, p_1 + \cdots + p_m = 2k} \frac{(2k)!}{p_1! \cdots p^m!} \\ &\quad \times \textstyle\sum_{i_1, \cdots, i_m = 1, \text{different}}^N c_{i_1}^{p_1} \cdots c_{i_m}^{p_m} E(Y_{i_1}^{p_1} \cdots Y_{i_m}^{p_m}) + \cdots \\ &\quad + \textstyle\sum_{i_1, \cdots, {}_{2k} = 1, \text{different}}^N c_{i_1} \cdots c_{i_{2k}} E(Y_{i_1} \cdots Y_{i_{2k}}) \,. \end{aligned} \tag{2.6}$$

In view of (1.5) and (1.6), it follows that

$$|\textstyle\sum_{i_1, \cdots, i_m = 1, \text{different}}^N c_{i_1}^{p_1} \cdots c_{i_m}^{p_m}| \leqq K \qquad \text{for } N > N_k \tag{2.7}$$

for any $m = 1, \cdots, 2k$ and any p_i, $0 < p_i < 2k$, $i = 1, \cdots, m$, $\sum_{i=1}^m p_i = 2k$, $K > 0$ is a constant dependent only on k. Actually, if $p_i \geqq 2$ for $i = 1, \cdots, m$, then

$$|\textstyle\sum_{i_1, \cdots, i_m = 1, \text{different}}^N c_{i_1}^{p_1} \cdots c_{i_m}^{p_m} \leqq |\prod_{j=1}^m (\sum_{i=1}^N |c_i|^{p_j}) \leqq \max_{1 \leqq i \leqq N} |c_i|^{2(k-m)} \,.$$

On the other hand, suppose that some of p_i's are equal to one, say $p_m = 1$. Then in view of (1.5)

$$\begin{aligned} &\textstyle\sum_{i_1, \cdots, i_m = 1, \text{different}}^N c_{i_1}^{p_1} \cdots c_{i_1}^{p_m} \\ &\quad = \textstyle\sum_{i_1, \cdots, i_{m-1} = 1, \text{different}}^N c_{i_1}^{p_1} \cdots c_{i_{m-1}}^{p_{m-1}} (-c_{i_1} - \cdots - c_{i_{m-1}}) \end{aligned} \tag{2.8}$$

so that we get $m - 1$ sums of similar type; each of them sums the products of $(m - 1)$ factors. Considering any of these sums, we may have again two cases:

either all exponents are at least two, so that we are in the first case; or some of them equal one and we may write an equality analogous to (2.8). We continue in this way until after a finite number of steps (in which we decompose the original expression into at most $m!$ sums) we get only the sums with exponents greater than or equal to two. Actually, the extreme case is the sum of the type

$$\sum_{i_1,i_2=1,i_1\neq i_2}^{N} c_{i_1}^{2k-1}c_{i_2} = -\sum_{i_1=1}^{N} c_{i_1}^{2k},$$

so that (2.7) is proved.

Further, using the generalized Cauchy–Schwarz inequality

$$E|\prod_{i=1}^{n} Z_i| \leqq (\prod_{i=1}^{n} E|Z_i^n|)^{1/n}, \qquad n = 2, 3, \cdots \tag{2.9}$$

we see that

$$\begin{aligned} E|Y_{i_1}^{p_1}\cdots Y_{i_m}^{p_m}| &\leqq (\prod_{j=1}^{m} E|Y_{i_j}^{mp_j}|)^{1/m} < (\prod_{j=1}^{m} E|Y_{i_j}^{2kp_j}|)^{1/2k} \\ &= (\prod_{j=1}^{m} E|a_N(R_1) - \varphi(U_1)|^{2kp_j})^{1/2k} \end{aligned} \tag{2.10}$$

holds for any $m = 1, \cdots, 2k$ and any p_i, $0 < p_i \leqq 2k$, $\sum_{i=1}^{m} p_i = 2k$. Finally, the expression

$$\sum_{m=1}^{2k} \sum_{1\leq p_1,\cdots,p_m\leq 2k,\, p_1+\cdots+p_m=2k} \frac{(2k)!}{p_1!\cdots p_m!} \tag{2.11}$$

depends only on k.

Now, if $a_N(i) = \varphi(i/(N+1))$, $i = 1, \cdots, N$, where φ has a bounded derivative we get the inequality

$$E[a_N(R_1) - \varphi(U_1)]^{2kp_j} \leqq B_2(k)E\left[\frac{R_1}{N+1} - U_1\right]^{2kp_j} \tag{2.12}$$

which is varied for $j = 1, \cdots, m$; $m = 1, \cdots, 2k$.

U_1 being fixed, R_1 is the sum of independent zero-one random variables (see (1.2)) so that

$$E\left(\frac{R_{N1}}{N+1} - U_{N1}\right)^{2kp_j} \leqq B_3(k)N^{-kp_j}. \tag{2.13}$$

(2.6), (2.7), (2.10), (2.11), (2.12) and (2.13) then prove the lemma.

Proof of Theorem 2.1. Since for any $\varepsilon > 0$ and any N, we have

$$P\{\sigma^{-1}S_N \leqq x\} \leqq P\{\sigma^{-1}T_N \leqq x + \varepsilon\} + P\{\sigma^{-1}|S_N - T_N| \geqq \varepsilon\} \tag{2.14}$$

and analogously

$$P\{\sigma^{-1}S_N \leqq x\} \geqq P\{\sigma^{-1}T_N \leqq x - \varepsilon\} - P\{\sigma^{-1}|S_N - T_N| \geqq \varepsilon\}, \tag{2.15}$$

it follows using Lemmas 2.1 and 2.2, that

$$\sup_{-\infty<x<\infty} |F_N(x) - \Phi(x)| \leqq (\varepsilon\sigma)^{-2k}B(k)N^{-k} + c_2\sum_{i=1}^{N} |c_{Ni}|^3 + O(\varepsilon) \tag{2.16}$$

holds for any $\varepsilon > 0$, any k and for $N > N_k$.

For $\delta > 0$ being fixed, take k such that $2k + 1 > 1/2\delta \geqq 2k$ and put $\varepsilon = N^{-\frac{1}{2}(1-1/(2k+1))}$. The theorem then follows from (2.13) and from the assumption (1.6).

3. Rate of convergence for $\Delta \neq 0$. Without loss of generality, we assume that $\Delta > 0$. For convenience we shall use the following representation in this section. Let X_{Ni}, $i = 1, \cdots, N$ be independent and identically distributed rvs each having the cdf $F(x)$ such that $I(f) < \infty$. Let R_{Ni}^{Δ} be the rank of $X_{Ni} + \Delta d_{Ni}$, that is

$$R_{Ni}^{\Delta} = \sum_{j=1}^{N} u(X_{Ni} - X_{Nj} + \Delta(d_{Ni} - d_{Nj}))\,.$$

Consider now the statistics

$$S_{\Delta N} = \sum_{i=1}^{N} c_{Ni}\,\varphi\left(\frac{R_{Ni}^{\Delta}}{N+1}\right).$$

The asymptotic distribution of $S_{\Delta N} - S_{0N}$ was investigated by Jurečková for Wilcoxon scores in (1973a) and for general score function φ in (1973b). In the case of general scores function φ, it was assumed that the φ function has the four bounded derivatives in (0, 1).

Suppose now that the vectors $(c_{N1}, \cdots, c_{NN})$ and $(d_{N1}, \cdots, d_{NN})$ satisfy (1.5), (1.6) and the following:

$$\lim_{N\to\infty} \sum_{i=1}^{N} c_{Ni}\,d_{Ni} = a^2\,, \qquad 0 < a^2 < \infty\,, \tag{3.1}$$

$$\lim_{N\to\infty} [\max_{1\le i\le N} (c_{Ni}^2 d_{Ni}^2)(\textstyle\sum_{i=1}^{N} c_{Ni}^2 d_{Ni}^2)^{-1}] = 0\,, \tag{3.2}$$

and

$$\lim_{N\to\infty} [N^{-1}(\textstyle\sum_{i=1}^{N} c_{Ni}\,d_{Ni})^2(\sum_{i=1}^{N} c_{Ni}^2 d_{Ni}^2)^{-1}] = \gamma \geqq 0\,. \tag{3.3}$$

Then, [cf. Jurečková (1973b)] for φ having four bounded derivatives in (0, 1), the asymptotic distribution of

$$A_N^{-1}(S_{\Delta N} - S_{0N} - \Delta a_N - \Delta^2 b_N) \tag{3.4}$$

is $\eta(0, \Delta^2\rho^2)$ where

$$A_N^2 = \sum_{i=1}^{N} c_{Ni}^2 d_{Ni}^2 + 3N^{-1}(\textstyle\sum_{i=1}^{N} c_{Ni}\,d_{Ni})^2 \tag{3.5}$$

$$a_N = \sum_{i=1}^{N} c_{Ni}\,d_{Ni} \int \varphi'(F(x))f^2(x)\,dx = \sum_{i=1}^{N} c_{Ni}\,d_{Ni} \int_0^1 \varphi(t)\varphi(t, f)\,dt \tag{3.6}$$

$$b_N = \tfrac{1}{2} \sum_{i=1}^{N} c_{Ni}\,d_{Ni}^2 \int \varphi''(F(x))f^3(x)\,dx \tag{3.7}$$

and

$$\begin{aligned}\rho^2 = {} & \int [\varphi'(F(x))]^2 f^3(x)\,dx - (\int [\varphi'(F(x))]^2 f^2(x)\,dx)^2 + 2\gamma(1 + 3\gamma)^{-1} \\ & \times [\textstyle\iint_{x<y} F(x)(1 - F(y))\varphi''(F(x))\varphi''(F(y))f^2(x)f^2(y)\,dx\,dy \\ & + \textstyle\iint_{x<y} \varphi'(F(x))\varphi''(F(y))f^2(x)f^2(y)\,dx\,dy \\ & - \int \varphi'(F(x))f(x)\,dx \cdot \int \varphi''(F(x))F(x)f^2(x)\,dx\,.\end{aligned} \tag{3.8}$$

Let $F_{N\Delta}$ denote the cdf of $\sigma^{-1}(S_{\Delta N} - \Delta a_N)$. Then we have the following theorem.

THEOREM 3.1. *Suppose that c_{Ni}, d_{Ni}, $i = 1, \cdots, N$ satisfy* (1.5), (1.6), (3.1)—(3.3) *and that the score-generating function has four bounded derivatives on* (0, 1).

Then

$$\sup_x |F_{N\Delta}(x) - \Phi(x)| = O(N^{-\frac{1}{2}+\delta}) \tag{3.9}$$

holds for any $\delta > 0$ and any fixed Δ.

PROOF. We may write for any $\varepsilon > 0$ and for any x

$$\begin{aligned} P\{\sigma^{-1}(S_{\Delta N} - \Delta a_N - \Delta^2 b_N) \leqq x\} & \\ \leqq P\{\sigma^{-1}S_{0N} \leqq x + \varepsilon\} & \\ + P\{\sigma^{-1}|S_{\Delta N} - S_{0N} - \Delta a_N - \Delta^2 b_N| \geqq \varepsilon\} & \end{aligned} \tag{3.10}$$

and analogously

$$\begin{aligned} P\{\sigma^{-1}(S_{\Delta N} - \Delta a_N - \Delta^2 b_N) \leqq x\} & \\ \geqq P\{\sigma^{-1}S_{0N} \leqq x - \varepsilon\} - P\{\sigma^{-1}|S_{\Delta N} - S_{0N} - \Delta a_N - \Delta^2 b_N| \geqq \varepsilon\} \,. & \end{aligned}$$

Then by Theorem 2.1,

$$\begin{aligned} \sup_x |F_{N\Delta}(x + \sigma^{-1}\Delta^2 b_N) - \Phi(x)| & \\ \leqq C \cdot \varepsilon + P\{\sigma^{-1}|S_{\Delta N} - S_{0N} - \Delta a_N - \Delta^2 b_N| \geqq \varepsilon\} & \\ + A(\delta)N^{-\frac{1}{2}+\delta} & \end{aligned} \tag{3.11}$$

holds for any $\delta > 0$ and $N > N_\delta$.

Let us consider the third member of the right-hand side of (3.11). We shall use the following theorem:

THEOREM 3.2 (Petrov). *Let $H(x)$ be any* cdf *and $\Phi(x)$* cdf *of the normal* (0, 1) *distribution.*

Let

$$\nu = \sup_{-\infty<x<\infty} |H(x) - \Phi(x)|$$

and let M_p denote the set of distribution functions possessing the finite absolute moment of order $p > 0$. Then, if $0 < \nu \leqq e^{-\frac{1}{2}}$ and $H(x) \in M_p$, there exists a constant C_p depending on p only such that

$$|H(x) - \Phi(x)| \leqq \frac{C_p \nu \left(\log \dfrac{1}{\nu}\right)^{p/2} + \lambda_p}{1 + |x|^p} \tag{3.12}$$

holds for all real x; here

$$\lambda_p = |\textstyle\int |x|^p \, dH(x) - \int |x|^p \, d\Phi(x)| \,.$$

For the proof, see Petrov (1972).

Let us denote by $G_{N\Delta}$ the cdf of $\Delta^{-1}A_N^{-1}\rho^{-1}(S_{\Delta N} - S_{0N} - \Delta a_N - \Delta^2 b_N)$. On account of the boundedness of φ, $G_{N\Delta}$ has finite absolute moments of any order for any fixed N and any fixed Δ. On the other hand, it follows from Theorem 2.1 of [6] (see (3.1)—(3.8) of the present paper) that $\lim_{N\to\infty} \sup_x |G_{N\Delta}(x) - \Phi(x)| = 0$ for any fixed Δ and that for $N > N_\Delta$

$$\sup_x |G_{N\Delta}(x) - \Phi(x)| < e^{-\frac{1}{2}} \,.$$

The assumptions of Theorem 3.2 are satisfied for any $p = k = 1, 2, \cdots$, so that there exists a constant C_k^* to any k such that

$$(3.13) \qquad |G_{N\Delta}(x) - \Phi(x)| \leqq C_k^*(1 + |x|^k)^{-1}$$

holds for all $x \in (-\infty, \infty)$.

We have

$$(3.14) \qquad P\{\sigma^{-1}|S_{\Delta N} - S_{0N} - \Delta a_N - \Delta^2 b_N| \geqq \varepsilon\} = 2[1 - G_{N\Delta}(\Delta^{-1}\rho^{-1}\sigma A_N^{-1}\varepsilon)]$$

so that (3.13) implies that

$$(3.15) \qquad \begin{aligned} &P\{\sigma^{-1}|S_{\Delta N} - S_{0N} - \Delta a_N - \Delta^2 b_N| \geqq \varepsilon\} \\ &\leqq 2[1 - \Phi(\Delta^{-1}\rho^{-1}\sigma A_N^{-1}\varepsilon)] + 2C_k^*[1 + (\Delta^{-1}\rho^{-1}\sigma)^k A_N^{-k}\varepsilon^k]^{-1} \end{aligned}$$

holds for any $\varepsilon > 0$, any $k = 1, 2, \cdots$ and for $N > N_\Delta$.

Let us fix δ, $\delta > 0$ and put $\varepsilon = A_N \cdot N^{\delta/2}$. Then in view of (3.15) and Lemma 2, Chapter VII of Feller (1957) we have that for any $N > N_\Delta$ and sufficiently large k

$$(3.16) \qquad \sup |F_{N\Delta}(x + \sigma^{-1}\Delta^2 b_N) - \Phi(x)| \leqq C_\delta'' N^{-\frac{1}{4}+\delta} + O(N^{-1+2\delta}) .$$

Thus

$$(3.17) \qquad \begin{aligned} &\sup_{-\infty<x<\infty} |F_{N\Delta}(x) - \Phi(x)| \\ &\leqq \sup_x |F_{N\Delta}(x) - \Phi(x + \sigma^{-1}\Delta^2 b_N)| \\ &\quad + \sup_x |\Phi(x + \sigma^{-1}\Delta^2 b_N) - \Phi(x)| \\ &\leqq \sup_x |F_{N\Delta}(x - \sigma^{-1}\Delta^2 b_N) - \Phi(x)| + K \cdot \sigma^{-1}\Delta^2 b_N . \end{aligned}$$

(3.16) and (3.17) together with assumption (1.5) complete the proof of the Theorem.

Acknowledgment. It is a pleasure to express our appreciation to Dr. Søren Johansen and Martin Jacobsen for some helpful discussions, and to the referee for some very valuable comments. We would also like to thank the referee for bringing to our attention a paper of Bickel (1974) where he established that a Berry–Esseén bound of order $N^{-\frac{1}{2}}$ was valid for the Wilcoxon statistic. This suggests that the bound in the present paper may be improved upon. The reader is also referred to Vizková, (1974) for a related problem.

REFERENCES

[1] Bickel, P. J. (1974). Edgeworth expansions in nonparametric statistics. *Ann. Statist.* **2** 1-20.

[2] Capon, J. (1961). Asymptotic efficiency of certain locally most powerful rank tests. *Ann. Math. Statist.* **32** 88-100.

[3] Feller, W. (1957). *An Introduction to Probability Theory and its Applications.* Wiley, New York.

[4] Hájek, J. (1962). Asymptotically most powerful rank order tests. *Ann. Math. Statist.* **33** 1124-1147.

[5] Hájek, J. and Sidák, Z. (1967). *Theory of Rank Tests.* Academic Press, New York.

[6] Jurečková, J. (1973a). Central limit theorem for Wilcoxon rank statistics process. *Ann. Statist.* **1** 1046-1060.

[7] Jurečková, J. (1973 b). Central limit theorem for rank statistics process. (Submitted).

[8] Petrov, V. V. (1972). *The Sums of Independent Random Variables* (in Russian). Nauka, Moscow.

[9] Vízková, Z. (1974). Local limit theorems for rank statistics. Ph. D. dissertation, Charles Univ., Prague.

Jana Jurečková
Charles University
Department of Statistics
Sokolovska Ul. 83
Prague 8, Czechoslovakia

Madan L. Puri
Institut fur Mathematische Statistick
und Wirtschaftsmatik
Universitat Gottingen
34 Gottingen
Lotzestrabe 13, West Germany

CONVERGENCE AND REMAINDER TERMS IN LINEAR RANK STATISTICS[1]

HARALD BERGSTRÖM and MADAN L. PURI

University og Göteborg, Chalmers University of Technology, and Indiana University

A new approach to the asymptotic normality of simple linear rank statistics for the regression case studied earlier by Hájek (1968) is provided along with the estimation of the remainder term in the approximation to normality.

1. Introduction and summary. Let $X_1, \cdots, X_n$ be independent random variables having continuous cdf's (cumulative distribution functions) $F_1(x), \cdots, F_n(x)$ respectively. Consider a statistic $S_n = s(X_1, \cdots, X_n)$ with $ES_n = 0$ and $ES_n^2 < \infty$. Then, to prove the asymptotic normality of S_n (as $n \to \infty$), Hájek (1968) uses the method of projection which gives to the statistic S_n, the approximation of the form

$$\hat{S}_n = \sum_{j=1}^{n} E[S_n \mid X_j] . \tag{1.1}$$

Consider now the simple linear rank statistic S_n introduced by Hájek (1962, 1968)

$$S_n = \sum_{j=1}^{n} c_j\{\psi(R_j/n) - E[\psi(R_j/n)]\} \tag{1.2}$$

where the c's are known constants, R_j is the rank of X_j among $(X_1, \cdots, X_n)$ and $\psi(\cdot)$ is a score generating function defined on (0, 1). Hájek (1962) [see also Hájek–Šidák (1967)] established the asymptotic normality of S_n in (1.2) under the assumption that the F_i are contiguous, e.g., when $F_i(x) = F(x - \Delta d_{ni})$ where Δ is the unknown parameter and the d's are the known constants. Later on Hájek (1968) studied the asymptotic normality of S_n for the general $F_i(x)$ (the noncontiguous case). Under the setup of Hájek (1962), Jurečková and Puri (1975), referred to hereafter as JP, studied the problem of determining the rate of convergence of the cdf of S_n to the limiting normal cdf and established it of order $O(N^{-\frac{1}{2}+\delta})$ for $\delta > 0$. In this paper we not only give a new approach to the asymptotic normality of S_n for the general F_i (i.e., not necessarily contiguous) but improve the results of JP in providing a sharper bound (for the general F_i's). In the passing, we may also mention that whereas JP requires ψ to have a bounded fourth derivative, here we only require the boundedness of the second

Received May 1975; revised January 1977.

[1] Work supported by the Air Force Office of Scientific Research, AFSC, USAF, under Grant No. AFOSR 76-2927. Reproduction in whole or in part is permitted for any purpose of the U.S. Government. Part of this work was done while the author was the recipient of a Senior U.S. Scientist Award from the Alexander-von-Humboldt-Foundation.

AMS 1970 *subject classifications.* Primary 62E20; Secondary 60F05, 60F99.

Key words and phrases. Linear rank statistics, score generating functions, rate of convergence.

derivative. Furthermore whereas this paper gives more explicit error bounds than the JP paper, the latter gives more information on the limiting behavior of ES_n and Var S_n.

We now introduce some notations. We define $\psi(\cdot) = 0$ outside (0, 1). Then, we can use the supremum norm

(1.3) $$||\psi|| = \sup_{t\in(-\infty,\infty)} |\psi(t)| .$$

Set

(1.4) $$\rho_i = R_i/n , \qquad \rho_{ii} = E[\rho_i | X_i] , \qquad u(x) = 1 \quad \text{if} \quad x \geqq 0$$
$$\text{and} \quad u(x) = 0 \quad \text{otherwise.}$$

Then

(1.5) $$R_i = \sum_{j=1}^n u(X_i - X_j) .$$

In this paper, we shall deal with the following approximation of S_n.

(1.6) $$T_n = \sum_{i=1}^n c_i\{\psi(\rho_{ii}) - E[\psi(\rho_{ii})] + (\rho_i - \rho_{ii})\psi'(\rho_{ii})\} ,$$

assuming that ψ' exists on (0, 1) and

(1.7) $$\hat{T}_n = \sum_{j=1}^n E[T_n | X_j] .$$

Since $E[(\rho_i - \rho_{ii})\psi'(\rho_{ii})] = 0$, it follows that

(1.8) $$\hat{T}_n = \sum_{i=1}^n c_i\{\psi(\rho_{ii}) - E[\psi(\rho_{ii})] + \sum_{j\neq i}^n E[(\rho_i - \rho_{ii})\psi'(\rho_{ii}) | X_j]\} .$$

Let H_n, G_n and $\hat{G}_n$ be the cdf's of S_n, T_n and $\hat{T}_n$ respectively, and put

(1.9) $$\sigma_n^2 = E[S_n^2] , \qquad \hat{\delta}_n^2 = E[\hat{T}_n^2] , \qquad \Gamma_{nr}^{2r} = \frac{1}{n} \sum_{i=1}^n c_i^{2r} , \qquad \Gamma_{nr} > 0 .$$

Then our theorems are the following:

THEOREM 1.1. *If ψ has a derivative on* (0, 1) *then*

(1.10) $$||\hat{G}_n(\hat{\delta}_n \cdot) - \Phi(\cdot)|| \leqq 4C[2||\psi||^3 + ||\psi'||^3] \sum_{i=1}^n |c_i|^3 \hat{\delta}_n^{-3} ;$$
$$\Phi(x) = (2\pi)^{-\frac{1}{2}} \int_{-\infty}^x e^{-t^2/2}\, dt$$

where C is the constant in Berry–Esseen's inequality (Zolotarev (1967) gives the approximation 0.9051). *Further,*

(1.11) $$|\hat{\delta}_n - \sigma_n| \leqq C_1(||\psi'|| + ||\psi''||)\Gamma_{n,1}$$

with an absolute constant C_1, provided ψ'' exists on (0, 1).

THEOREM 1.2. *If ψ has a second order derivative on* (0, 1), *then for any positive integers n and r such that $n^{-1}r^3 \leqq \frac{3}{8}$,*

(1.12) $$||H_n(\hat{\delta}_n \cdot) - \Phi(\cdot)|| \leqq 4C(2||\psi||^3 + ||\psi'||^3) \sum_{i=1}^n |c_i|^3 \hat{\delta}_n^{-3}$$
$$+ C_2[\hat{\delta}_n^{-1}(||\psi'|| + ||\psi''||) r\Gamma_{nr}]^{2r/(2r+1)} ,$$

where C_2 is an absolute constant.

REMARK. If the c_i are chosen such that $|c_i| \leqq a/n^{\frac{1}{2}}$ with constant a for all

i and n, then

$$\Gamma_{nr} \leqq a/n^{\frac{1}{2}} ,$$

and for $r = [\log n]$, $[r\Gamma_{nr}]^{2r/(2r+1)} \leqq a\sqrt[4]{e}\,(\log n)\, n^{-\frac{1}{2}}(1 + O(1/\log n))$.

Note that $\hat{\delta}_n^{-1}c_i$ is invariant and thus also $\hat{\delta}_n^{-1}\Gamma_{nr}$ is invariant under the transformation $c_i \to \gamma c_i$, $i = 1, 2, \cdots$.

2. Some lemmas.

LEMMA 2.1. *For any positive integers r and n, $2r \leqq n$, we have*

$$E[(\rho_i - \rho_{ii})^{2r}] \leqq b(r)n^{-r} \tag{2.1}$$

with

$$b(r) \leqq n^{-r} \textstyle\sum_{t=1}^{r} \binom{n-1}{t} \dfrac{(2r)!}{(2r-2t)!} t^{2r-2t} \cdot 2^{-3t} \tag{2.2}$$

and for $n^{-1}r^3 \leqq \frac{3}{4}$

$$b(r) \leqq 2^{-3r} \frac{(2r)!}{r!} [1 + 8n^{-1}r^3] . \tag{2.3}$$

PROOF. By (1.4) we obtain

$$\rho_i - \rho_{ii} = \frac{1}{n} \textstyle\sum_{j\neq i}^{n} [u(X_i - X_j)F_j(X_i)] .$$

By the polynomial theorem we then get

$$E[(\rho_i - \rho_{ii})^{2r}] = n^{-2r} \sum \frac{(2r)!}{s_1! \cdots s_n!} E \textstyle\prod_{j\neq i}^{n} [u(X_i - X_j) - F_j(X_i)]^{s_j} , \tag{2.4}$$
$$s_1 + \cdots + s_n = 2r .$$

We claim that any term in this sum is equal to zero if $s_{j_0} = 1$ for some j_0. Indeed we find that the conditional expection of the product with respect to all X_j, $j \neq j_0$ is equal to 0 if $s_{j_0} = 1$. Hence we have only to regard terms with $s_j = 0$ or $\geqq 2$ for any j, and there can be at most $t \leqq r$ exponents s_j different from 0. If $s_j \geqq 2$, $j = 1, 2, \cdots, t$, $s_j = 0$ for $j > t$, $i > t$ we obtain, observing that

$$|u(X_i - X_j) - F_j(X_i)| \leqq 1$$

$$\begin{aligned} E[\textstyle\prod_{j=1}^{t} [u(X_i - X_j) - F_j(X_i)]^{s_j} &\leqq E \textstyle\prod_{j=1}^{t} [u(X_i - X_j) - F_j(X_i)]^2 \\ &= E[\textstyle\prod_{j=1}^{t} [F_j(X_i) - F_j^2(X_i)]] \leqq 4^{-t} . \end{aligned} \tag{2.5}$$

This inequality remains true for all permutations of the indices $1, \cdots, n$. Put

$$\gamma(t) = \textstyle\sum_{s_1+\cdots+s_t=2r;\, s_j\geqq 2,\, j=1,\cdots,t} \dfrac{(2r)!}{s_1! \cdots s_t!} . \tag{2.6}$$

Since t indices out of $n - 1$ indices can be chosen in $\binom{n-1}{t}$ different ways we obtain from (2.4) through (2.6),

$$E[(\rho_i - \rho_{ii})^{2r}] \leqq n^{-2r} \textstyle\sum_{t=1}^{r} \binom{n-1}{t}\gamma(t)4^{-t} . \tag{2.7}$$

We claim that

$$\gamma(t) \leqq \frac{(2r)!}{(2r-2t)!} 2^{-t} t^{2r-2t} . \tag{2.8}$$

Indeed, differentiating the identity

$$(\sum_{j=1}^{t} y_j)^{2r} = \sum_{s_1+\cdots+s_t=2r} \frac{(2r)!}{s_1! \cdots s_t!} \prod_{j=1}^{t} y_j^{s_j}$$

twice with respect to all y_j and then putting all y_j equal to 1, we obtain

$$\frac{(2r)!}{(2r-2t)!} t^{(2r-2t)} = \sum_{s_1+\cdots+s_t=2r; s_j \geqq 2, j=1\cdots t} \prod_{j=1}^{t} s_j(s_j-1) \frac{(2r)!}{s_1! \cdots s_t!} .$$

Now using (2.7) and (2.8), we get (2.1) and (2.2). We now estimate $b(r)$ further, mainly for use when n and r are large. Put $r - t = u$. Then we can write

$$b(r) \leqq 2^{-3r} \sum_{u=0}^{r-1} k(u) \tag{2.9}$$

with

$$k(u) = \frac{n^{-u}(2r)!\,(r-u)^{2u} 2^{3u}}{(r-u)!\,(2u)!} .$$

Particularly

$$k(0) = \frac{(2r)!}{r!}, \qquad k(1) < 4n^{-1} r^3 \cdot \frac{(2r)!}{r!}$$

and for $u \geqq 1$

$$\frac{k(u+1)}{k(u)} = n^{-1}\left(1 - \frac{1}{r-u}\right)^{2u} \cdot 2^3 \cdot (r-u) \frac{(r-u-1)^2}{(2u+1)(2u+2)}$$
$$< \tfrac{2}{3} n^{-1} r^3 \leqq \tfrac{1}{2} \quad \text{for} \quad n^{-1} r^3 \leqq \tfrac{3}{4} .$$

Hence

$$b(r) \leqq 2^{-3r} \cdot \frac{(2r)!}{r!} [1 + 8n^{-1} r^3]$$

for $n^{-1} r^3 \leqq \frac{3}{4}$.

LEMMA 2.2. *For any positive integers r and n, $2r \leqq n$, we have*

$$E(T_n - \hat{T}_n)^{2r} \leqq c(r) \|\psi\|^{2r} \Gamma_{n,r}^{2r} \tag{2.10}$$

if ψ' exists on (0, 1), *and if ψ'' exists on* (0, 1)

$$E[(S_n - T_n)^{2r}] \leqq b(2r) \|\psi''\|^{2r} \Gamma_{n,r}^{2r} , \tag{2.11}$$

$$E[(S_n - \hat{T}_n)^{2r}] \leqq d(r, \psi) \Gamma_{n,r}^{2r} \tag{2.12}$$

with

$$b(2r) \leqq n^{-2r} \sum_{t=1}^{2r} \binom{n-1}{t} \frac{(4r)!}{(4r-2t)!} t^{4r-2t} \cdot 2^{-3t}$$

$$c(r) \leqq 2^{2r} n^{-2r} \sum_{t=1}^{2r} \binom{n}{t} \frac{(4r)!}{(4r-2t)!} t^{4r-2t} \cdot 2^{-t}$$

$$d(r, \psi) \leqq [[b(2r)]^{1/2r} \|\psi''\| + [c(r)]^{1/2r} \|\psi'\|]^{2r} .$$

Further we have the estimates

$$b(2r) \leqq 2^{-6r}\frac{(4r)!}{(2r)!}[1 + 2^6 n^{-1} r^3] \tag{2.13}$$

for $2^3 n^{-1} r^3 \lneqq \frac{3}{4}$,

$$c(r) \leqq \frac{(4r)!}{(2r)!}[1 + 2^3 n^{-1} r^3] \qquad \textit{for} \quad n^{-1} r^3 \leqq \tfrac{3}{8}\,. \tag{2.14}$$

Remark. By Stirling's approximation of the Γ-function we have

$$\frac{(4r)!}{(2r)!} \leqq 2^{6r+\frac{1}{2}} r^{2r} (\exp -2r) \exp \frac{1}{48r}\,.$$

Proof. By (1.6) and (1.8) we get

$$T_n - \hat{T}_n = \textstyle\sum_{i=1}^n c_i\{(\rho_i - \rho_{ii})\psi'(\rho_{ii}) - \sum_{j=1; j\neq i}^n E[(\rho_i - \rho_{ii})\psi'(\rho_{ii}) \,|\, X_j]\} \tag{2.15}$$

and for $j \neq i$

$$\begin{aligned} E[(\rho_i - \rho_{ii})\psi'(\rho_{ii}) \,|\, X_j] &= \frac{1}{n} \textstyle\sum_{k\neq i}^n E\{[u(X_i - X_k) - F_k(X_i)]\psi'(\rho_{ii}) \,|\, X_j\} \\ &= \frac{1}{n} E[u(X_i - X_j) - F_j(X_i)]\psi'(\rho_{ii}) \,|\, X_j]\,, \end{aligned} \tag{2.16}$$

since the conditional expectations in the sum are zero for $j \neq k, i$. Now using the relation

$$(\rho_i - \rho_{ii})\psi'(\rho_{ii}) = \frac{1}{n} \textstyle\sum_{j\neq i}^n [u(X_i - X_j) - F_j(X_i)]\psi'(\rho_{ii})\,,$$

and noting that

$$E[(\rho_i - \rho_{ii})\psi'(\rho_{ii}) \,|\, X_i] = 0$$

we obtain from (2.15)

$$T_n - \hat{T}_n = \frac{1}{n} \textstyle\sum_{i=1}^n \sum_{j\neq i}^n c_i V_{ij} \tag{2.17}$$

with

$$\begin{aligned} V_{ij} = {} & [u(X_i - X_j) - F_j(X_i)]\psi'(\rho_{ii}) \\ & - E\{[u(X_i - X_j) - F_j(X_i)]\psi'(\rho_{ii}) \,|\, X_j\}\,. \end{aligned} \tag{2.18}$$

Clearly

$$E[V_{ij} \,|\, X_j] = 0\,, \qquad E[V_{ij} \,|\, X_i] = 0\,. \tag{2.19}$$

By the polynomial theorem we get

$$\begin{aligned} E[(T_n - \hat{T}_n)^{2r}] &= n^{-2r} E[\textstyle\sum_{i=1}^n \sum_{j\neq i}^n c_i V_{ij}]^{2r} \\ &= n^{-2r} \sum \frac{(2r)!}{\prod_{i=1}^n \prod_{j\neq i}^n (s_{ij}!)} E\{\textstyle\prod_{i=1}^n \prod_{j\neq i}^n (c_i V_{ij})^{s_{ij}}\} \end{aligned} \tag{2.20}$$

where the sum should be taken over terms corresponding to different vector solutions $\{s_{ij}\}$, $i, j = 1, \cdots n$, $j \neq i$ of the equation

$$\textstyle\sum_{i=1}^n \sum_{j\neq i}^n s_{ij} = 2r\,. \tag{2.21}$$

(2.22) $$E[\prod_{i=1}^{n}\prod_{j\neq i}^{n} V_{ij}^{s_{ij}}]$$

is equal to 0 for some vector solutions of (2.21) since (2.19) holds, and we have only to regard those solutions for which the expectation (2.22) is not equal to 0.

We say that s_{ij} gives the contribution $\frac{1}{2}s_{ij}$ to the sum (2.21) from each of the indices i and j. Hence according to this notation an index k gives the contribution

(2.23) $$g(k) = \tfrac{1}{2}\sum_{j\neq k}^{n} s_{kj} + \tfrac{1}{2}\sum_{j\neq k}^{n} s_{jk}$$

to the sum (2.21). By conditioning with respect to all X_j, $j \neq k$ we easily find that the expectation (2.22) is equal to 0 if k gives the contribution $\frac{1}{2}$ to the sum (2.21), i.e., if $s_{kj} = 1$ for exactly one index $j \neq k$, and $s_{jk} = 0$ for $j \neq k$ or if $s_{jk} = 1$ for exactly one j and $s_{kj} = 0$ for $j \neq k$.

The sum $\sum$ on the right-hand side of (2.20) can be divided into partial sums as follows. Let C be a collection of different positive integers belonging to the set $1, \cdots, 2r$, say $C = (1, 2, \cdots, t)$. Let $\sum_C$ consist of all terms in (2.20) corresponding to the vector solutions of (2.21) such that

(a) $s_{ij} = 0$ if not both i and j belong to C;

(b) for any $k \in C$ the contribution to the sum (2.21) is larger than $\frac{1}{2}$. Note that C can contain at most 2r different integers since every $k \in C$ gives at least the contribution 1 to the sum (2.21). Clearly partial sums $\sum_{C_1}$ and $\sum_{C_2}$ contain no common terms if $C_1 \neq C_2$. Consider now the expectation

$$E[\prod_{i=1}^{t}\prod_{j\neq i}^{t} (c_i V_{ij})^{s_{ij}}]$$

where the i and j belong to the collection C. Note that s_{ij} may be equal to 0 for some pairs (i, j). By Hölder's inequality we get, using the fact that $|V_{ij}| \leqq 2||\psi'||$,

(2.24) $$\begin{aligned} |E\prod_{i=1}^{t}\prod_{j\neq i}^{t} (c_i V_{ij})^{s_{ij}}| &\leqq \prod_{i=1}^{t}\prod_{j\neq i}^{t} |c_i|^{s_{ij}}\{E[(V_{ij})^{2r}]\}^{s_{ij}/2r} \\ &\leqq 2^{2r}||\psi'||^{2r}\prod_{i=1}^{t}|c_i|^{s_i} \end{aligned}$$

where

(2.25) $$s_i = \sum_{j=1}^{t} s_{ij}\,, \qquad \sum_{i=1}^{t} s_i = 2r\,.$$

The partial sum corresponding to C is then estimated by

(2.26) $$\sum_C' \frac{(2r)!}{\prod_{i=1}^{t}\prod_{i\neq j}^{t}(s_{ij})!}\,(2^{2r}||\psi'||^{2r}\prod_{i=1}^{t}|c_i|^{s_i})\,.$$

Note that $(2r)!/\prod_{i=1}^{t}\prod_{i\neq j}^{t}(s_{ij})!$ is an integer. Hence we have

$$N(t) = \sum_C' \frac{(2r)!}{\prod_{i=1}^{t}\prod_{j\neq i}^{t}(s_{ij})!}$$

terms in the class C which are estimated by (2.24). Let $\mathscr{C}_t$ be the set of all terms

$$\sum \prod_{i=1}^{n}\prod_{j\neq 1}^{n}(c_i V_{ij})^{s_{ij}}$$

$s_2, \cdots, s_t)$ in (2.26) be given, $0 \leqq s_1 \leqq s_2 < \cdots \leqq s_t$, $\sum_{i=1}^t s_i = 2r$. Then according to the symmetry the set $\mathscr{C}_t$ contains a sum of terms, each estimated by

(2.27) $$2^{2r}||\psi'||^{2r} \prod_{i=1}^t |c_{k_i}|^{s_i}$$

where $(k_1 \cdots k_t)$ is any combination of numbers $1, 2, \cdots, n$ to the tth class and in any order within this class. Let the number of terms in C_t for a fixed vector $(s_1, s_2, \cdots, s_t)$ as above be $n(t)$ and the sum of terms (2.27) belonging to $(s_1, s_2, \cdots, s_t)$ be $A(s_1, s_2, \cdots, s_t)$. (Note that $n(t)$ depends on $s_1, \cdots, s_t$.) Then, since $A(s_1, \cdots, s_t)$ is a symmetrical function

(2.28) $$A(s_1, s_2, \cdots, s_t) = \frac{n(t)}{n!} \sum{}' 2^{2r}||\psi'||^{2r} \prod_{i=1}^t |c_{k_i}|^{s_i}$$

where $\sum'$ is the sum all terms belonging to all permutations of the numbers $1, 2, \cdots, n$. By Hölder's inequality we get, observing that

$$|c_{k_i}|^{s_i} = [c_{k_i}^{2r}]^{s_i/2r}, \qquad \sum_{i=1}^t \frac{s_i}{2r} = 1,$$

(2.29) $$\sum{}' \prod_{i=1}^t |c_{k_i}|^{s_i} \leqq \prod_{i=1}^t \left(\sum{}' c_{k_i}^{2r}\right)^{s_i/2r}$$

and here

$$\sum{}' c_{k_i}^{2r} = \frac{n!}{n} \sum_{i=1}^n c_i^{2r}.$$

Hence we obtain by (2.28) and (2.29)

$$A(s_1, s_2, \cdots, s_t) \leqq 2^{2r}||\psi'||^{2r} \cdot n(t) \cdot \frac{1}{n} \sum_{i=1}^n c_i^{2r}.$$

Since $\mathscr{C}_t$ contains $\binom{n}{t}N(t)$ terms we then find that $\mathscr{C}_t$ gives at most the contribution

$$n^{-2r}2^{2r}||\psi'||^{2r}\binom{n}{t}N(t) \cdot \frac{1}{n} \sum_{i=1}^n c_i^{2r}$$

to the right-hand side of (2.20). Putting

$$\Gamma_{nr}^{2r} = \frac{1}{n} \sum_{i=1}^n c_i^{2r}, \qquad \Gamma_{nr} \geqq 0,$$

and regarding the sets $\mathscr{C}_t$ for $t = 1, 2, \cdots, 2r$, we obtain from (2.20) that

(2.30) $$E[(T_n - \hat{T}_n)^{2r}] \leqq 2^{2r}n^{-2r}||\psi'||^{2r}\Gamma_{nr}^{2r} \sum_{t=1}^{2r}\binom{n}{t}N(t).$$

We estimate $N(t)$ in the following way. Consider the identity

(2.31) $$\left(\sum_{i=1}^t \sum_{j=i}^t x_i x_j\right)^{2r} = \sum \frac{(2r)!}{\prod_{i=1}^t \prod_{j=1}^t (s_{ij})!} \prod_{i=1}^t \prod_{j\neq i}^t (x_i x_j)^{s_{ij}}.$$

If an index k gives the contribution $\geqq 1$ to the sum (2.21), i.e., to the sum

$$\sum_{i=1}^t \sum_{j\neq i}^t s_{ij} = 2r,$$

then the double product

$$\prod_{j=1}^{t} \prod_{j\neq i}^{t} (x_i x_j)^{s_{ij}}$$

contains x_k as factor at least in the power 2. Hence differentiating the identity twice with respect to each x_k, $k = 1, 2, \cdots, t$ and then putting all x_n equal to 1 we get the inequality

$$2^t N(t) \leqq \left\{\prod_{k=1}^{t} \frac{\partial^2}{\partial_{x_k}} (\textstyle\sum_{i=1}^{t} \sum_{j\neq i}^{t} x_i x_j)^{2r}\right\}_{x_k=1, k=1,2,\cdots,t} . \tag{2.32}$$

The right-hand side, however, is at most equal to

$$\left\{\prod_{k=1}^{t} \frac{\partial^2}{\partial x_k} ((\textstyle\sum_{i=1}^{t} x_i)^{4r})\right\}_{x_k=1,k=1,\cdots,t} = \frac{(4r)!}{(4r-2t)!} t^{4r-2t} . \tag{2.33}$$

Combining (2.30), (2.32) and (2.33), we get

$$E[(T_n - \hat{T}_n)^{2r}] \leqq c(r) ||\psi'||^{2r} \Gamma_{nr}^{2r}$$

with

$$c(r) = 2^{2r} n^{-2r} \textstyle\sum_{t=1}^{2r} \binom{n}{t} \dfrac{(4r)!}{(4r-2t)!} t^{4r-2t} \cdot 2^{-t}$$

$$\Gamma_{nr}^{2r} = \frac{1}{n} \textstyle\sum_{i=1}^{n} |c_i|^{2r} .$$

We estimate $c(r)$ exactly in the same way as we have estimated $b(r)$ in Lemma 2.1 and then obtain for $u = 2r - t$

$$c(r) \leqq \textstyle\sum_{u=0}^{r-1} k(u)$$

with

$$k(u) = n^{-u} \frac{(4r)!}{(2u)!\,(2r-u)!} (2r-u)^{2u} \cdot 2^u .$$

Hence

$$k(0) = \frac{(4r)!}{(2r)!} , \qquad k(1) < n^{-1} \cdot (2r)^3 \frac{(4r)!}{(2r)!}$$

and for $u \geqq 1$

$$\frac{k(u+1)}{k(u)} \leqq \tfrac{4}{3} n^{-1} r^3 \leqq \tfrac{1}{2} \qquad \text{for} \quad n^{-1} r^3 \leqq \tfrac{3}{8} .$$

Hence for $n^{-1} r^3 \leqq \frac{3}{8}$

$$c(r) \leqq \frac{(4r)!}{(2r)!} [1 + 8n^{-1} r^3] .$$

Thus we have proved (2.13) and (2.14) of the lemma.

It follows by the definition of T_n that

$$S_n - T_n = \textstyle\sum_{i=1}^{n} c_i [\xi_i - E(\xi_i)]$$

with

$$|\xi_i| \leqq \tfrac{1}{2} (\rho_i - \rho_{ii})^2 ||\psi''|| .$$

Hence

$$E[(S_n - T_n)^{2r}] \leqq n^{2r-1} \sum_{i=1}^{n} c_i^{2r} E[(\xi_i - E\xi_i)^{2r}]$$

and by Lemma 2.1

$$E[(\xi_i - E(\xi_i))^{2r}] \leqq 2^{2r} E[\xi_i^{2r}] \leqq \|\psi''\|^{2r} E[(\rho_i - \rho_{ii})^{4r}] \leqq n^{-2r} b(2r) \|\psi''\|^{2r} .$$

Thus we get (2.11)

$$E[(S_n - T_n)^{2r}] \leqq b(2r)\Gamma_{nr} .$$

By Minkovski's inequality we obtain (2.12) from (2.10) and (2.11)

$$E^{1/2r}[(S_n - \hat{T}_n)^{2r}] \leqq E^{1/2r}[(S_n - T_n)^{2r}] + E^{1/2r}[(T_n - \hat{T}_n]^{2r}] .$$

LEMMA 2.3. $\hat{T}_n = \sum_{j=1}^{n} \hat{T}_n^{(j)}$ *with independent random variables*

$$\text{(i)} \qquad \hat{T}_n^{(j)} = c_j\{\psi(\rho_{jj}) - E[\psi(\rho_{jj})]\} + \frac{1}{n} \sum_{i \neq j}^{n} c_i [E(u(X_i - X_j) - F_j(X_i))\psi'(\rho_{ii}) \,|\, X_j] .$$

Further,

$$\text{(ii)} \qquad \sum_{j=1}^{n} [E|\hat{T}_n^{(j)}|^3] \leqq 4[2\|\psi\|^3 + \|\psi'\|^3\} \sum_{j=1}^{n} |c_i|^3 .$$

PROOF. We get the representation (i) by (2.16). Using well-known inequalities

$$|(a + b)^3| \leqq 4[|a|^3 + |b|^3] , \qquad |(\textstyle\sum_{i=1}^{n} a_i)^3| \leqq n^2 \sum_{i=1}^{n} |a_i|^3$$

we obtain

$$E[|\hat{T}_n^{(j)}|^3] \leqq 4|c_j|^3 E[|[\psi(\rho_{jj})] - E\psi(\rho_{jj})|^3] + \frac{4}{n} \sum_{i \neq j}^{n} |c_i|^3 \|\psi'\|^3 .$$

Here

$$E[|\psi(\rho_{jj}) - E[\psi(\rho_{jj})]|^3] \leqq 2\|\psi\| E(\psi(\rho_{jj}) - E(\psi(\rho_{jj}))^2 .$$

Thus we get (ii).

3. Proofs of the theorems.

(a) PROOF OF THEOREM 1.1. (1.10) follows from Berry–Esseen's inequality and Lemma 2.3 and (1.11) from Lemma 2.2 (2.12).

(b) PROOF OF THEOREM 1.2. For $h > 0$ we get

$$\text{(3.1)} \qquad P[S_n \leqq \hat{\delta}_n x] \leqq P(S_n \leqq \hat{\delta}_n x, |S_n - \hat{T}_n| < h\hat{\delta}_n) + P[|S_n - \hat{T}_n| \geqq h\delta_n] \leqq P[\hat{T}_n \leqq \hat{\delta}_n(x + h)] + P[|S_n - \hat{T}_n| \geqq h\hat{\delta}_n] .$$

Applying Theorem 1.1 we get

$$\text{(3.2)} \qquad P[\hat{T}_n \leqq \hat{\delta}_n(x + h)] \leqq \Phi(x + h) + 4C(2\|\psi\|^3 + \|\psi'\|^3) \cdot \sum_{i=1}^{n} |c_i|^3 \hat{\delta}_n^{-3} .$$

Here

$$\text{(3.3)} \qquad \Phi(x + h) \leqq \Phi(x) + \|\Phi'(x)\| = \Phi(x) + \frac{h}{(2\pi)^{\frac{1}{2}}} .$$

By Chebyshev's inequality and the inequality (2.12) of Lemma 2.2 we get

$$\text{(3.4)} \qquad P[|S_n - \hat{T}_n| \geqq h\hat{\delta}_n] \leqq d(r, \psi)\Gamma_{nr}^{2r}(h\hat{\delta}_n)^{-2r} .$$

Now we choose n such that

$$\frac{h}{(2\pi)^{\frac{1}{2}}} = d(r, \psi)\Gamma_{nr}^{2r}(h\hat{\delta}_n)^{-2r} ,$$

i.e.,

(3.5) $$h = [(2\pi)^{\frac{1}{2}} d(r, \psi)\hat{\delta}_n^{-2r}\Gamma_{nr}^{2r}]^{1/(2r+1)} .$$

It follows by Lemma 2.2, (2.12), (2.13) and (2.14), and the remark made after Lemma 2.2 that for $n^{-1}r^3 \leqq \frac{3}{8}$

$$[d(r, \psi)]^{1/2r} \leqq C'r(||\psi'|| + ||\psi''||)$$

with an absolute constant C'. Then it follows by (3.4) and (3.5) that

$$\frac{h}{(2\pi)^{\frac{1}{2}}} + d(r, \psi)\Gamma_{nr}^{2r}(h\hat{\delta}_n)^{-2r} \leqq C_2[\hat{\delta}_n^{-1}(||\psi'|| + ||\psi''||)r\Gamma_{nr}]^{2r/(2r+1)} .$$

By (3.1)—(3.6) we get the inequality (1.12) in one direction. It follows for the other direction in the same way.

Acknowledgment. The authors would like to express their sincere appreciation to Professor I. R. Savage, the associate editor, and the referees for the critical examination of the original draft. Their constructive criticism, detailed comments and suggestions for improvements are gratefully acknowledged.

REFERENCES

[1] Hájek, J. (1962). Asymptotically most powerful rank order tests. *Ann. Math. Statist.* **33** 1124–1147.

[2] Hájek, J. (1968). Asymptotic normality of simple linear rank statistics under alternatives. *Ann. Math Statist.* **39** 325–346.

[3] Hájek, J. and Šidák, Z. (1967). *Theory of Rank Tests.* Academic Press, New York.

[4] Jurečková, J. and Puri, Madan L. (1975). Order of normal approximation for rank test statistics distribution. *Ann. Probability* **3** 526–533.

[5] Zolotarev, V. M. (1967). A sharpening of the inequality of Berry–Esseen. *Z. Wahrscheinlichkeitstheorie und Verw. Gebiete* **8** 332–342.

Department of Mathematics
Indiana University
Swain Hall, East
Bloomington, IN 47401

INVARIANCE PRINCIPLES FOR RANK STATISTICS FOR TESTING INDEPENDENCE

MADAN L. PURI* and LANH T. TRAN†

ABSTRACT

In this paper we consider a general class of rank order statistics for testing independence in bivariate populations. Each statistic is represented as a sum of independent and identically distributed (i.i.d) random variables and a remainder term. Suitable order (a.s.) of the remainder term is found and then some invariance principles are obtained. The results obtained are extensions of the results of Chernoff and Savage (1958), Bhuchongkul (1964), Bahadur (1966), Ruymgaart *et al.* (1972), Sen and Ghosh (1974) and Lai (1975).

Key words and phrases: Invariance principles, linear rank statistics, score functions.

* Research supported by the Air Force Office of Scientific Research AFSC, USAF, under Grant No. AFOSR76-2927. This research was done at the University of Washington. during my leave of absence from Indiana University. I appreciate very much the use of facilities that were made available to me by the Department of Biostatistics at the University of Washington.

† Research supported by National Science Foundation Grant No. MCS76-00951. Reproduction in whole or part is permitted for any purpose of the U.S. Government.

1. Introduction

Let $\{(X_i, Y_i), 1 \le i \le N\}$ be N independent and identically distributed i.i.d. random vectors, each having a continuous cumulative distribution function (c.d.f.) $H(x, y)$. Let $F(x)$ and $G(y)$ denote the marginal c.d.f.'s of X_i and Y_i, respectively. Denote by $F_N(x)$, $G_N(y)$, and $H_N(x, y)$, the empirical c.d.f.'s of $\{X_i, 1 \le i \le N\}$, $\{Y_i, 1 \le i \le N\}$, and $\{(X_i, Y_i), 1 \le i \le N\}$, respectively. Finally, let R_{Ni} (and S_{Ni}) denote the rank of X_i (and Y_i) among $X_i, 1 \le i \le N$ (and Y_i, $1 \le i \le N$). Then many rank tests for the hypothesis of independence

$$H_0: H(x, y) = F(x)G(y) \tag{1}$$

are based on the statistic

$$T_N = N^{-1} \sum_{i=1}^{N} J_N\left(\frac{R_{Ni}}{N}\right) L_N\left(\frac{S_{Ni}}{N}\right) = \int J_N[F_N(x)] L_N[G_N(y)]\, dH_N(x, y) \tag{2}$$

where $J_N(i/N) = EJ(U_{Ni})$ or $J(i/N + 1)$, $L_N(i/N) = EL(U_{Ni})$ or $L(i/(N + 1))$, U_{Ni}, $(1 \le i \le N)$ is the ith order statistic in a sample of size N from the uniform distribution over $(0, 1)$, and $J(u)$, $L(u)$, $0 < u < 1$, are nondecreasing, twice differentiable score functions [cf. Bhuchongkul (1964)].

For the case $J_N(i/N) = EJ(U_{Ni})$ and $L_N(i/N) = EL(U_{Ni})$, Sen and Ghosh (1974) have obtained some invariance principles for $\{T_N\}$ when the null hypothesis H_0 in (1) holds. Their results are based on a fundamental martingale property possessed by $\{T_N, \mathscr{F}_N\}$ when $\{R_{Ni}, 1 \le i \le N\}$ and $\{S_{Ni}, 1 \le i \le N\}$ are stochastically independent. Here $\mathscr{F}_N$ denotes the σ field generated by $\{R_{Ni}, S_{Ni}; 1 \le i \le N\}$.

In this paper the invariance principles are established for $\{T_N\}$ *under alternatives*. When $H(x, y) \ne F(x)G(y)$, the techniques of Sen and Ghosh (1974) are not applicable since $\{T_N, \mathscr{F}_N\}$ is not a martingale. Our methods are related to those of Chernoff and Savage (1958), Bhuchongkul (1964), Bahadur (1966), Sen and Ghosh (1973), Ruymgaart *et al.* (1972), and Lai (1975). The main argument is based on a representation of T_N as the sum of i.i.d. random variables and a remainder term which is shown to converge a.s. to zero at an appropriate rate. The contents of this paper are as follows:

In Sec. 2, assumptions on the score functions are stated and preliminary lemmas are presented. Section 3 deals with the order of magnitude of the remainder term. Some invariance principles are then established in Sec. 4. In what follows K is used as a generic constant whose values may differ from line to line.

2. Assumptions and Some Preliminary Lemmas

ASSUMPTION 2.1. $J(u)$ and $L(u)$, $0 < u < 1$, are absolutely continuous, twice differentiable score functions, with

$$|J^{(i)}(u)| \le K[u(1-u)]^{-\alpha-i}, \; |L^{(i)}(u)| \le K[u(1-u)]^{-\beta-i}, \qquad i = 0, 1, 2, \tag{3}$$

where

$$\alpha = (1 - 2\delta)/2p, \qquad \beta = (1 - 2\delta)/2q \tag{4}$$

for some $0 < \delta < \frac{1}{2}$ and some $p, q > 1$ with $p^{-1} + q^{-1} = 1$.

We shall start with the following lemmas which are slight variations or generalizations of some of the results of Bahadur (1966), Sen (1972), Sen and Ghosh (1974), and which we shall need in sequel.

LEMMA 2.1. *Let X_i, $1 \leq i \leq N$ be i.i.d. random variables, each having a continuous c.d.f. $F(x)$. Let $U_N(u)$ be the empirical c.d.f. of $\{F(X_i), 1 \leq i \leq N\}$. Then for every $\varepsilon > 0$,*

$$\sup N^{1/2}\{u(1-u)\}^{\varepsilon - 1/2}|U_N(u) - u| = o(\log N) \qquad \text{a.s.} \tag{5}$$

Proof. Follows from Lemma 2.1 of Sen and Ghosh (1974).

LEMMA 2.2. *Let* $F_{N,\theta}(x) = \theta N F_N(x)/(1+N) + (1-\theta)F(x), \quad 0 \leq \theta \leq 1.$ *Then*

$$1 - F_{N,\theta}(x) \geq \{1 - F(x)\}\{1 - 0(1)\} \text{ a.s.} = \{1 - F_N(x)\}\{1 - 0(1)\} \text{ a.s.} \tag{6}$$

as $N \to \infty$ for $F^{-1}(N^{-1+\lambda}) \leq x \leq F^{-1}(1 - N^{-1+\lambda})$, where λ is an arbitrary positive number < 1.

Proof. Follows from Sen and Ghosh (1974, p. 164).

LEMMA 2.3. *Let $D = (0,1) \times (0,1)$ and let $\bar{w} = (u, v) \in D$. If $H(\bar{w})$ is a continuous c.d.f. with uniform $(0,1)$ marginal c.d.f.'s, then*

$$\sup_{\bar{w} \in D} \sup_{\bar{w}' \in D} \{|H_N(\bar{w}') - H_N(\bar{w}) - H(\bar{w}') + H(\bar{w})| : |\bar{w}' - \bar{w}| \leq N^{-1/2}\}$$
$$= O(N^{-3/4} \log N) \text{ a.s.} \tag{7}$$

Proof. By a straightforward generalization of Lemma 1 of Bahadur (1966), we have

$$\sup_{\bar{w}' \in D} \{|H_N(\bar{w}') - H_N(\bar{w}) - H(\bar{w}') + H(\bar{w})| : |\bar{w}' - \bar{w}| \leq N^{-1/2}\}$$
$$= O(N^{-3/4} \log N) \text{ a.s.,}$$

where $\bar{w} \in (0,1) \times (0,1)$ is an arbitrary fixed point. Then (7) can be obtained by the same line of argument as in Theorem 4.2 of Sen and Ghosh (1971).

LEMMA 2.4. *Let $0 < \lambda < 1$. Then*

$$|F_N(F^{-1}(1 - N^{-1+\lambda})) - (1 - N^{-1+\lambda})| = O(N^{-1+\lambda/2} \log N) \text{ a.s. as } N \to \infty.$$

Proof. Follows from a slight variation of Lemma 4.1 of Sen (1972).

3. Order of Magnitude of the Remainder Term

THEOREM 3.1. *Let T_N defined in (2) be written as*

$$T_N = \sum_{i=1}^{3} A_{iN} + R_N, \tag{8}$$

where

$$A_{1N} = \int_{-\infty}^{\infty} \int_{-\infty}^{\infty} J[F(x)]L[G(y)]\, dH_N(x, y), \tag{9}$$

$$A_{2N} = \int_{-\infty}^{\infty} \int_{-\infty}^{\infty} [F_N(x) - F(x)]J'[F(x)]L(G(y)]\, dH(x, y), \tag{10}$$

$$A_{3N} = \int_{-\infty}^{\infty} \int_{-\infty}^{\infty} [G_N(y) - G(y)]L'[G(y)]J[F(x)]\, dH(x, y), \tag{11}$$

and R_N is defined below (13). Then, under the assumption 2.1, $\lim_{N\to\infty} N^{1/2}R_N = 0$ a.s.

Proof. Define a_{1N}, b_{1N}, c_{1N}, and d_{1N} by

$$F(a_{1N}) = N^{-1+\delta_1};\ F(b_{1N}) = 1 - N^{-1+\delta_1};$$
$$G(c_{1N}) = N^{-1+\delta_2},\ G(d_{1N}) = 1 - N^{-1+\delta_2},$$

where $\delta_1 = \delta p^{-1}$ and $\delta_2 = \delta q^{-1}$.

Let ξ be any positive number smaller than $\frac{1}{4}$. Pick γ_1 and γ_2 with $0 < \gamma_1 < \min(\xi(1+\alpha)^{-1}, (1-2\beta)(8\alpha)^{-1})$ and $0 < \gamma_2 < \min(\xi(1+\beta)^{-1}, (1-2\alpha)(8\beta)^{-1})$. Define a_{2N}, b_{2N}, c_{2N}, and d_{2N} by $F(a_{2N}) = N^{-\gamma_1}$, $F(b_{2N}) = 1 - N^{-\gamma_1}$, $G(c_{2N}) = N^{-\gamma_2}$ and $G(d_{2N}) = 1 - N^{-\gamma_2}$. Let

$$I_{1,N} = [a_{1N}, b_{1N}] \times [c_{1N}, d_{1N}],\ I_{2N} = [a_{2N}, b_{2N}] \times [c_{2N}, d_{2N}]. \tag{12}$$

and denote the complements of I_{1N} and I_{2N} by I_{1N}^c and I_{2N}^c, respectively. Then from the decomposition (8) we have

$$R_N = \sum_{i=1}^{17} B_{i,N}, \tag{13}$$

where

$$B_{1N} = \int_{-\infty}^{\infty} \int_{-\infty}^{\infty} \{L_N[G_N(y)] - L[NG_N(y)/(N+1)]\}\{J_N[F_N(x)] - J[NF_N(x)/(N+1)]\}\, dH_N(x, y),$$

$$B_{2N} = \int_{-\infty}^{\infty} \int_{-\infty}^{\infty} L[NG_N(y)/(N+1)][J_N[F_N(x)] - J(NF_N(x)/(N+1))]\, dH_N(x, y),$$

$$B_{3N} = \int_{-\infty}^{\infty} \int_{-\infty}^{\infty} J[NF_N(x)/(N+1)]\{L_N[G_N(y)] - L[NG_N(y)/(N+1)]\}\, dH_N(x, y),$$

$$B_{4N} = \iint_{I_{1,N}} \{L[NG_N(y)/(N+1)] - L[G(y)]\}\{J[NF_N(x)/(N+1)] - J[F(x)]\}\, dH_N(x, y),$$

$$B_{5N} = \iint_{I_{1,N}^c} \{L[NG_N(y)/(N+1)] - L[G(y)]\}\{J[NF_N(x)/(N+1)] - J[F(x)]\}\, dH_N(x, y),$$

$$B_{6N} = \iint_{I_{1,N}} L[G(y)]\{J[NF_N(x)/(N+1)] - J[F(x)] - J'[F(x)]\}\{NF_N(x)/(N+1) - F(x)\}\, dH_N(x, y),$$

$$B_{7N} = \iint_{I_{1,N}} J[F(x)]\{L[NG_N(y)/(N+1)] - L[G(y)] - L'[G(y)][NG_N(y)/(N+1) - G(y)]\}\, dH_N(x, y),$$

$$B_{8N} = \iint_{I_{1,N}^c} L[G(y)]\{J[NF_N(x)/(N+1)] - J[F(x)]\}\, dH_N(x, y),$$

$$B_{9N} = \iint_{I_{1,N}^c} J[F(x)]\{L[NG_N(y)/(N+1)] - L[G(y)]\}\, dH_N(x, y),$$

$$B_{10N} = -\iint_{I_{2,N}^c} L[G(y)]J'[F(x)]\{F_N(x) - F(x)\}\, dH(x, y),$$

$$B_{11N} = -\iint_{I_{2,N}^c} J[F(x)]L'[G(y)]\{G_N(y) - G(y)\}\, dH(x, y),$$

$$B_{12N} = \iint_{I_{1,N} \cap I_{2,N}^c} L[G(y)]J'[F(x)]\{F_N(x) - F(x)\}\, dH_N(x, y),$$

$$B_{13N} = \iint_{I_{1,N} \cap I_{2,N}^c} J[F(x)]L'[G(y)]\{G_N(y) - G(y)\}\, dH_N(x, y),$$

$$B_{14N} = \iint_{I_{2,N}} L[G(y)]J'[F(x)][F_N(x) - F(x)]\, d[H_N(x, y) - H(x, y)],$$

$$B_{15N} = \iint_{I_{2,N}} J[F(x)]L'[G(y)][G_N(y) - G(y)]\, d[H_N(x, y) - H(x, y)],$$

$$B_{16N} = -(N+1)^{-1} \iint_{I_{1,N}} L[G(y)]J'[F(x)]F_N(x)\, dH_N(x, y),$$

$$B_{17N} = -(N+1)^{-1} \iint_{I_{1,N}} J[F(x)]L'[G(y)]G_N(y)\, dH_N(x, y).$$

For each $1 \le i \le 17$, we shall show that $|B_{i,N}| = O(N^{-1/2-\eta})$ for some $\eta > 0$.

For reasons of symmetry, we do not need to treat $B_{3N}, B_{7N}, B_{9N}, B_{11N}, B_{15N}, B_{17N}$.

First, consider B_{1N}. By Holder's inequality, $|B_{1N}|$ is bounded by

$$\left(\int_{-\infty}^{\infty} |J_N[F_N(x)] - J[NF_N(x)/(N+1)]|^p \, dF_N(x)\right)^{1/p}$$
$$\times \left(\int_{-\infty}^{\infty} |L_N[G_N(y)] - L[NG_N(y)/(N+1)]|^q \, dG_N(y)\right)^{1/q}. \tag{14}$$

From Chernoff and Savage (1958, Theorem 2), it follows that

$$|J_N(1/N) - J(1/(N+1))| < KN^{\alpha}; \qquad |J_N(i/N) - J(i/(N+1))| < KN^{\alpha}[\Phi(-\sqrt{i}/K) + i^{-1-\alpha}], \qquad 1 < i \leq N/2. \tag{15}$$

Thus

$$\int_{1 \leq NF_N(x) \leq N/2} |J_N[F_N(x)] - J[NF_N(x)/(N+1)]|^p \, dF_N(x)$$

is bounded by

$$N^{-1}\left(KN^{\alpha p} + \sum_{i=2}^{N/2} KN^{\alpha p}[\Phi(-\sqrt{i}/K) + i^{-1-\alpha}]^p\right) \leq KN^{\alpha p - 1}.$$

By a symmetric argument we can cover the range $N/2 \leq NF_N(x) \leq N$. Hence the first factor of (14) is bounded by $KN^{(\alpha p - 1)p^{-1}}$. Similarly the second factor of (14) is bounded by $KN^{(\beta q - 1)q^{-1}}$. Thus

$$|B_{1N}| = O(N^{(\alpha p - 1)p^{-1} + (\beta q - 1)q^{-1}}) = O(N^{-\delta_1 - \delta_2 - 1/2}) = O(N^{-\eta - 1/2}) \qquad \text{for} \quad \eta \leq \delta_1 + \delta_2. \tag{16}$$

We now consider B_{2N}. By (3), it is clear that $L[NG_N(y)/(N+1)] \leq KN^{\beta}$ for $N^{-1} \leq G_N(y) \leq 1$. Hence $|B_{2N}|$ is bounded by

$$KN^{\beta} \int_{-\infty}^{\infty} |J_N(F_N(x)] - J[NF_N(x)/(N+1)]| \, dF_N(x). \tag{17}$$

Using (15), it follows that (17) is bounded by

$$KN^{\beta}N^{-1}\left(KN^{\alpha} + \sum_{i=2}^{N/2} KN^{\alpha}[\Phi(-\sqrt{i}/K) + i^{-1-\alpha}]\right) \leq KN^{-\delta_1 - \delta_2 - 1/2}.$$

Consequently

$$|B_{2N}| = O(N^{-\eta - 1/2}) \qquad \text{for} \quad \eta < \delta_1 + \delta_2. \tag{18}$$

We now consider B_{4N}. By Holder's inequality, $|B_{4N}|$ is bounded by

$$\left(\int_{a_{1N}}^{b_{1N}} |J[NF_N(x)/(N+1)] - J[F(x)]|^p \, dF_N(x)\right)^{p^{-1}}$$
$$\times \left(\int_{c_{1N}}^{d_{1N}} |L[NG_N(y)/(N+1)] - L[G(y)]|^q \, dG_N(y)\right)^{q^{-1}} \tag{19}$$

By the mean value theorem, the integral in the first factor of (19) is bounded by

$$\int_{a_{1N}}^{b_{1N}} |NF_N(x)/(N+1) - F(x)|^p \, |J'[F_{N,\theta}(x)]|^p \, dF_N(x), \tag{20}$$

where $F_{N,\theta}(x)$ is defined in Lemma 2.2.

Let ε be a positive number to be specified later. By (3) and Lemma 2.1, (20) is bounded by

$$O(N^{-p/2}(\log N)^p) \times \int_{a_{1N}}^{b_{1N}} [F(x)(1-F(x)]^{p(-\varepsilon+1/2)}[F_{N,\theta}(x)(1-F_{N,\theta}(x))]^{-p(1+\alpha)} \, dF_N(x). \tag{21}$$

Pick a with $F(a) = \frac{1}{2}$. Then (21) is bounded by

$$O(N^{-p/2}(\log N)^p) \int_{a}^{b_{1N}} [1-F(x)]^{p(-\varepsilon+1/2)}[F_{N,\theta}(x)(1-F_{N,\theta}(x)]^{-p(1+\alpha)} \, dF_N(x)$$
$$+ O(N^{-p/2}(\log N)^p) \int_{a_{1N}}^{a} [F(x)]^{p(-\varepsilon+1/2)}[F_{N,\theta}(x)]^{-p(1+\alpha)} \, dF_N(x). \tag{22}$$

By Lemma 2.2, the first term of (22) is bounded by

$$O(N^{-p/2}(\log N)^p) \int_{a}^{b_{1N}} [1-F_N(x)]^{-p(\varepsilon+\alpha+1/2)} \, dF_N(x). \tag{23}$$

Observe that the integral in (23) tends to infinity as $N \to \infty$ if and only if $(\frac{1}{2}+\alpha+\varepsilon)p \geq 1$. For simplicity, we pick $\varepsilon > \delta_1$. Then $(\frac{1}{2}+\alpha+\varepsilon)p > 1$ and only the case where this integral tends to infinity has to be considered. By a simple integration, we note that (23) is bounded by

$$O(N^{-p/2}(\log N)^p)(1-F_N(b_{1N}))^{-(1/2+\alpha+\varepsilon)p+1}. \tag{24}$$

By Lemma 2.4 it can easily be seen that (24) is bounded by

$$\begin{aligned} &O(N^{-p/2}(\log N)^p)[N^{-1+\delta_1} \pm 0\,(N^{-1+\delta_1/2}\log N)]^{-(1/2+\alpha+\varepsilon)p+1} \\ &\quad = O(N^{-p/2}(\log N)^p)N^{(-1+\delta_1)[-(1/2+\alpha+\varepsilon)p+1]} \\ &\quad = O(N^{(-1+\delta_1)(-p\alpha+1)}(\log N)^p)N^{-p[2^{-1}\delta_1-\varepsilon(1-\delta_1)]} \\ &\quad = O(N^{(-1+\delta_1)(-p\alpha+1)}) \end{aligned}$$

by choosing $\varepsilon < \delta_1\,(1-\delta_1)^{-1}2^{-1}$. Clearly, the second term of (22) is also bounded by $O(N^{(-1+\delta_1)(-p\alpha+1)})$. Hence, the first factor of (19) is bounded by $O(N^{(-1+\delta_1)(-p\alpha+1)p^{-1}})$. Similarly, the second factor of (19) is bounded by $O(N^{(-1+\delta_2)(-q\beta+1)q^{-1}})$. Finally,

$$\begin{aligned} |B_{4N}| &= O(N^{-1/2-\delta_1(1-2^{-1}p^{-1}-\delta_1)-\delta_2(1+2^{-1}q^{-1}+\delta_2)}) \\ &= O(N^{-\eta-1/2}) \qquad \text{for some } \eta > 0. \end{aligned}$$

Next, consider B_{5N}. Observe that $I_{1N}^c = \bigcup_{i=1}^8 I_{1,iN}$, where

$$I_{1,1N} = (b_{1N}, \infty) \times (d_{1N}, \infty), \qquad I_{1,2N} = (-\infty, a_{1N}) \times (-\infty, c_{1N}),$$
$$I_{1,3N} = (b_{1N}, \infty) \times (-\infty, c_{1N}), \qquad I_{1,4N} = (-\infty, a_{1N}) \times (d_{1N}, \infty),$$
$$I_{1,5N} = (a_{1N}, b_{1N}) \times (d_{1N}, \infty), \qquad I_{1,6N} = (a_{1N}, b_{1N}) \times (-\infty, c_{1N}),$$
$$I_{1,7N} = (b_{1N}, \infty) \times (c_{1N}, d_{1N}), \qquad I_{1,8N} = (-\infty, a_{1N}) \times (c_{1N}, d_{1N}).$$

Define

$$B_{5,iN} = \iint_{I_{1,iN}} \left\{ L\left[\frac{NG_N(y)}{(N+1)}\right] - L[G(y)] \right\} \left\{ J\left[\frac{NF_N(x)}{(N+1)}\right] - J[F(x)] \right\} dH_N(x, y).$$

Then $B_{5N} = \sum_{i=1}^8 B_{5,iN}$.

First, consider $B_{5,1N}$. By Holder's inequality, $|B_{5,1N}|$ is bounded by

$$\left(\int_{b_{1N}}^{\infty} |J[NF_N(x)/(N+1)] - J[F(x)]|^p \, dF_N(x) \right)^{1/p} \times \left(\int_{d_{1N}}^{\infty} |L[NG_N(y)/(N+1)] - L[G(y)]|^q \, dG_N(y) \right)^{1/q}. \tag{25}$$

The integral in the first factor of (25) is bounded by

$$K \int_{b_{1N}}^{\infty} |J[NF_N(x)/(N+1)]|^p \, dF_N(x) + K \int_{b_{1N}}^{\infty} |J[F(x)]|^p \, dF_N(x). \tag{26}$$

By (3), the first term of (26) is smaller than

$$K \int_{b_{1N}}^{\infty} [1 - NF_N(x)/(N+1)]^{-p\alpha} \, dF_N(x),$$

which upon integration equals

$$K(1 - N(N+1)^{-1} F_N(b_N))^{-p\alpha+1}. \tag{27}$$

Lemma 2.4 now implies that (27) equals

$$K(1 - N(N+1)^{-1}(F(b_N) \pm O(N^{-1+\delta_1/2} \log N))^{-p\alpha+1} = O(N^{(-1+\delta_1)(-p\alpha+1)}).$$

Next the second term of (26) which can be written as $-K \int_{b_{1N}}^{\infty} |J[F(x)]|^p \, d[1 - F_N(x)]$ is bounded by

$$K|J[F(b_N)]|^p [1 - F_N(b_N)] + K \int_{b_{1N}}^{\infty} [1 - F_N(x)] |J[F(x)]|^{p-1} |J'[F(x)]| \, dF(x) \tag{28}$$

upon integration by parts.

Now using (3) and some routine computations, it follows that the first term of (28) is bounded by $K(1 - F(b_N)]^{-p\alpha}[1 - F(b_N) + O(N^{-1+\delta_1/2} \log N)] = O(N^{(-1+\delta_1)(-p\alpha+1)})$.

By Lemma 2.1, the second term of (28) is bounded by

$$K \int_{b_{1N}}^{\infty} [1 - F(x)]|J(F(x))|^{p-1}|J'(F(x))|\, dF(x) + O(N^{-1/2} \log N)$$
$$\times \int_{b_{1N}}^{\infty} [1 - F(x)]^{-\varepsilon - 1/2}|J(F(x))|^{p-1}|J'(F(x))|\, dF(x), \tag{29}$$

where ε is any arbitrary positive number.

Now using (3), it is easy to verify that (29) is bounded by

$$K \int_{b_{1N}}^{\infty} [1 - F(x)]^{-p\alpha}\, dF(x) + O(N^{-1/2} \log N) \int_{b_{1N}}^{\infty} [1 - F(x)]^{-1/2 - \alpha p - \varepsilon}\, dF(x)$$
$$\leq O(N^{(-1+\delta_1)(-p\alpha+1)}) + O(N^{-1/2} \log N)(N^{(-1+\delta_1)(1/2 - p\alpha - \varepsilon)})$$
$$= O(N^{(-1+\delta_1)(-p\alpha+1)}) \qquad \text{if} \quad \varepsilon < \delta_1(1 - \delta_1)^{-1}2^{-1}.$$

It is now clear that the first factor of (25) equals $O(N^{(-1+\delta_1)(-p\alpha+1)p^{-1}})$. Similarly, the second factor equals $O(N^{(-1+\delta_2)(-q\beta+1)q^{-1}})$. Hence $|B_{5,1N}| = O(N^{-\eta-1/2})$ for some $\eta > 0$.

Using the same arguments, $|B_{5,iN}|$ can be shown to be equal to $O(N^{-\eta-1/2})$ for some $\eta > 0$, $i = 1, \ldots, 8$.

Next, consider $B_{6,N}$. By (3), it is easy to verify that $L[G(y)]$ is bounded by $KN^{\beta(1-\delta_2)}$ for $y \in [c_{1N}, d_{1N}]$. Hence, $|B_{6N}|$ is smaller than or equal to

$$KN^{\beta(1-\delta_2)} \int_{a_{1N}}^{b_{1N}} |J[NF_N(x)/(N+1)] - J[F(x)]$$
$$- J'[F(x)][NF_N(x)/(N+1) - F(x)]|\, dF_N(x). \tag{30}$$

Using (30) and the mean value theorem,

$$|B_{6N}| \leq KN^{\beta(1-\delta_2)} \int_{a_{1N}}^{b_{1N}} \{[NF_N(x)/(N+1)] - F(x)\}^2 |J''(F_{N,\theta}(x)]|\, dF_N(x), \tag{31}$$

where $F_{N,\theta}$ is defined in Lemma 2.2. Next, (3) and Lemma 2.2 imply that (31) is bounded by

$$N^{\beta(1-\delta_2)} O(N^{-1}(\log N)^2) \int_{a}^{b_{1N}} [1 - F(x)]^{1-2\varepsilon}[1 - F_{N,\theta}(x)]^{-2-\alpha}\, dF_N(x)$$
$$+ N^{\beta(1-\delta_2)} O(N^{-1}(\log N)^2) \int_{a_{1N}}^{a} [F(x)]^{1-2\varepsilon}[F_{N,\theta}(x)]^{-2-\alpha}\, dF_N(x) \tag{32}$$

for all $\varepsilon > 0$. By Lemma 2.2, the first term of (32) is bounded by

$$N^{\beta(1-\delta_2)} O(N^{-1}(\log N)^2) \int_{a}^{b_{1N}} [1 - F_N(x)]^{-1-\alpha-2\varepsilon}\, dF_N(x) \tag{33}$$

which, on integration, equals

$$N^{\beta(1-\delta_2)} O(N^{-1}(\log N)^2) O(N^{(1-\delta_1)(\alpha+2\varepsilon)}) = O(N^{-\eta-1/2})$$

for some $\eta > 0$ since ε can be chosen to be smaller than $(\delta_1 + \delta_2 + \alpha\delta_1 + \beta\delta_2)2^{-1}(1-\delta_1)^{-1}$. Similarly, the second term of (32) equals $O(N^{-\eta-1/2})$ for some $\eta > 0$. Hence $|B_{6N}| = O(N^{-\eta-1/2})$ for some $\eta > 0$.

Consider B_{8N}. Define

$$B_{8,iN} = \iint_{I_{1,iN}} L[G(y)]\{J[NF_N(x)/(N+1)] - J[F(x)]\}\, dH_N(x,y).$$

Then $B_{8N} = \sum_{i=1}^{8} B_{8,iN}$.

By Holder's inequality,

$$|B_{8,1N}| \le \left(\int_{b_{1N}}^{\infty} |J[NF_N(x)/(N+1)] - J[F(x)]|^p\, dF_N(x)\right)^{1/p} \times \left(\int_{d_{1N}}^{\infty} |L[G(y)]|^q\, dG_N(y)\right)^{1/q}. \tag{34}$$

The first factor in (34) is identical to the first factor of (25), which has been shown to be equal to $O(N^{(-1+\delta_1)(-p\alpha+1)p^{-1}})$. Since the second term of (26) is bounded by $O(N^{(-1+\delta_1)(-p\alpha+1)})$, it follows easily that the second term of (34) is bounded by $O(N^{(-1+\delta_2)(-q\beta+1)q^{-1}})$. Thus $|B_{8,1N}| = O(N^{-\eta-1/2})$ for some $\eta > 0$. By similar treatments, we see that $|B_{8,2N}|$, $|B_{8,3N}|$, and $|B_{8,4N}|$ are all equal to $O(N^{-\eta-1/2})$ for sufficiently small η.

By Holder's inequality, $|B_{8,5N}|$ is bounded by

$$\left(\int_{a_{1N}}^{b_{1N}} |J[NF_N(x)/(N+1)] - J[F(x)]|^p\, dF_N(x)\right)^{1/p} \left(\int_{d_{1N}}^{\infty} |L[G(y)]|^q\, dG_N(y)\right)^{1/q}. \tag{35}$$

The first and second factor of (35) are respectively equal to the first factor of (19) and the second factor of (34). Therefore, $|B_{8,5N}| = O(N^{-\eta-1/2})$ for some $\eta > 0$. Similarly, $|B_{8,6N}| = O(N^{-\eta-1/2})$ for sufficiently small η.

Consider $B_{8,7N}$. By (3), $L[G(y)] \le KN^{\beta(1-\delta_2)}$ for $y \in [c_{1N}, d_{1N}]$. Hence $|B_{8,7N}|$ is bounded by

$$KN^{\beta(1-\delta_2)} \int_{b_{1N}}^{\infty} |J[NF_N(x)/(N+1)] - J[F(x)]|\, dF_N(x). \tag{36}$$

Recall that the first factor of (25) is equal to $O(N^{(-1+\delta_1)(-p\alpha+1)p^{-1}})$. It is easy to deduce that the integral in (26) equals $O(N^{(-1+\delta_1)(-\alpha p+1)})$. Hence $|B_{8,7N}| = O(N^{\beta(1-\delta_2)+(-1+\delta_1)(-\alpha p+1)}) = O(N^{-\eta-1/2})$ for some $\eta > 0$. One can treat $B_{8,8N}$ by using the same line of arguments. Finally, $|B_{8,N}| = O(N^{-\eta-1/2})$ for sufficiently small η.

Consider $B_{10,N}$. Write $I_{2,N}^c = \bigcup_{i=1}^{8} I_{2,iN}$, where $I_{2,1N} = (b_{2N,\infty}) \times (d_{2N,\infty})$, $I_{2,2N} = (-\infty, a_{2N}) \times (-\infty, c_{2N})$, $I_{2,3N} = (b_{2N,\infty}) \times (-\infty, c_{2N})$, $I_{2,4N} = (-\infty, a_{2N}) \times (d_{2N,\infty})$, $I_{2,5N} = (a_{2N}, b_{2N}) \times (d_{2N,\infty})$, $I_{2,6N} = (a_{2N}, b_{2N}) \times (-\infty, c_{2N})$, $I_{2,7N} = (b_{2N,\infty}) \times (c_{2N}, d_{2N})$, $I_{2,8N} = (-\infty, a_{2N}) \times (c_{2N}, d_{2N})$.

Define

$$B_{10,iN} = -\iint_{I_{2,iN}} L[G(y)]J'[F(x)][F_N(x) - F(X)]\,dH(x, y).$$

Then

$$B_{10,N} = \sum_{i=1}^{8} B_{10,iN}.$$

By Holder's inequality, $|B_{10,1N}|$ is bounded by

$$\left(\int_{b_{2N}}^{\infty} |J'[F(x)][F_N(x) - F(x)]|^{2p/(p+1)}\,dF(x)\right)^{(p+1)/2p} \times \left(\int_{d_{2N}}^{\infty} |L[G(y)]|^{2q}\,dG(y)\right)^{1/2q} \tag{37}$$

Let $v = (\alpha + \varepsilon + \frac{1}{2})2p/(p+1) = (\alpha + \varepsilon + \frac{1}{2})/(\alpha + \delta_1 + \frac{1}{2})$. Pick $0 < \varepsilon < \delta_1$ and note that $\frac{1}{2} < v < 1$. Lemma 2.1 and (3) imply that the first factor of (37) is bounded by

$$O(N^{-1/2}\log N)\left(\int_{b_{2N}}^{\infty} [1 - F(x)]^{-v}\,dF(x)\right)^{(p+1)/2p}, \tag{38}$$

which equals

$$O(N^{-1/2}\log N)(N^{-\gamma_1(1-v)(p+1)/2p}) = O(N^{-\eta-1/2}) \qquad \text{for some} \quad \eta > 0.$$

Again by (3), the second factor of (37) is bounded by

$$K\left(\int_{d_{2N}}^{\infty} [1 - G(y)]^{-2\beta q}\,dG(y)\right)^{1/2q},$$

which, on integration, equals $O(N^{-\gamma_2(1-2\beta q)/2q}) = O(N^{-\eta})$ for some $\eta > 0$. Hence $|B_{10,1N}| = O(N^{-\eta-1/2})$. By similar arguments, $|B_{10,iN}| = O(N^{-\eta-1/2})$ for $i = \ldots, 11$. We now consider $B_{12,N}$, which can be decomposed as

$$B_{12,N} = \sum_{i=1}^{8} B_{12,iN}, \tag{39}$$

where

$$B_{12,iN} = \iint_{I_{1N}\cap I_{2,iN}} L[G(y)]J'[F(x)][F_N(x) - F(x)]\,dH_N(x, y). \tag{40}$$

It is easy to check (by Holder's inequality) that

$$|B_{12,1N}| \le \left(\int_{b_{2N}}^{b_{1N}} |J'[F(x)][F_N(x) - F(x)]|^{2p/(p+1)}\right)^{(p+1)/2p} \times \left(\int_{d_{2N}}^{d_{1N}} L[G(y)]^{2q}\,dG_N(y)\right)^{1/2q}. \tag{41}$$

Using (3) and Lemma 2.1, the first factor of (41) equals

$$O(N^{-1/2}\log N)\left(\int_{b_{2N}}^{b_{1N}}[1-F(x)]^{-\nu}\,dF_N(x)\right)^{(p+1)/2p}, \tag{42}$$

i.e.,

$$O(N^{-1/2}\log N)\left(-\int_{b_{2N}}^{b_{1N}}[1-F(x)]^{-\nu}\,d[1-F_N(x)]\right)^{(p+1)/2p}. \tag{43}$$

Now integrating by parts and using Lemmas 2.1 and 2.4, we obtain, after routine computations, that (43) equals $O(N^{-\eta})$ for some η. Similarly the second factor on the right-hand side of (41) equals $O(N^{-\eta})$ for sufficiently small η. Thus $|B_{12,1N}| = O(N^{-\eta-1/2})$. The proof that the other terms in (39) equal $O(N^{-\eta-1/2})$ is similar and therefore omitted.

Next we consider $B_{14,N}$. First, observe that the rank of X_i (and Y_i) among $X_1,\ldots,X_N$ (and $Y_1,\ldots,Y_N$) is the same as the rank of $F(X_i)$ [and $F(Y_i)$] among $F(X_1),\ldots,F(x_N)$ [and $F(Y_1),\ldots,F(Y_N)$]. Following Groeneboom *et al.* (1976), we define $\bar{H}(u,v) = H(F^{-1}(u), G^{-1}(v))$ for $(u,v)\in(0,1)\times(0,1)$. Clearly, $\bar{H}(u,v) = P(\{F(X)\le u,\ G(Y)\le v\})$ so that it assigns mass 1 to the unit square and has uniform (0, 1) marginal distribution functions. Without loss of generality, we can then assume that $H(u,v)$ has uniform (0, 1) marginal distribution functions.

For reasons of symmetry, it is enough to consider

$$\left|\int_{1/2}^{1-N^{-\gamma_2}}\int_{1/2}^{1-N^{-\gamma_1}} L(v)J'(u)[F_N(u)-u]\,d[H_N(u,v)-H(u,v)]\right|. \tag{44}$$

We now show that (44) is equal to $O(N^{-\eta-1/2})$ for some $\eta>0$ by extending a method used in Sen and Ghosh (1974).

Define $\bar{I}_{1,Ni} = [\frac{1}{2}+(i-1)N^{-1/2}, \frac{1}{2}+iN^{-1/2}]$, $i=1,2,\ldots,N_1^*-1$, where N_1^* is the largest positive integer such that $\frac{1}{2}+(N_1^*-1)N^{-1/2} < 1-N^{-\gamma_1}$. Define $\bar{I}_{1,NN_1^*} = [\frac{1}{2}+(N_1^*-1)N^{-1/2}, 1-N^{-\gamma_1}]$, $u_{Ni} = \frac{1}{2}+iN^{-1/2}$, $0\le i\le N_1^*-1$, $u_{NN_1^*} = 1-N^{-\gamma_1}$. Define $\bar{I}_{2,Nj} = [\frac{1}{2}+(j-1)N^{-\xi}, \frac{1}{2}+jN^{-\xi}]$, $j=1,2,\ldots,N_2^*-1$, where N_2^* is the largest positive integer such that $\frac{1}{2}+(N_2^*-1)N^{-\xi}<1-N^{-\gamma_2}$. Define $\bar{I}_{2,NN_2^*}=[\frac{1}{2}+(N_2^*-1)N^{-\xi}, 1-N^{-\gamma_2}]$, $v_{Nj} = \frac{1}{2}+jN^{-\xi}$, $0\le j\le N_2^*-1$, $v_{NN_2^*} = 1-N^{-\gamma_2}$. Observe that $N_1^* = O(N^{1/2})$, $N_2^* = O(N^{\xi})$. Let $\bar{I}_{Nij} = \bar{I}_{1,Ni}\times\bar{I}_{2,Nj}$, $i=1,\ldots,N_1^*; j=1,\ldots,N_2^*$. If $u\in\bar{I}_{1,Ni}$, then $J'(u) = J'(u_{Ni}) + O(N^{-1/2})O(\{1-u_{Ni}\}^{-2-\alpha})$ and by Lemma 2.3,

$$F_N(u) - u = F_N(u_{Ni}) + O(N^{-3/4}\log N) \text{ a.s.}$$

If $v\in\bar{I}_{2,Nj}$, then $L(v) = L(v_{Nj}) + O(N^{-\xi})O(\{1-v_{Nj}\}^{-1-\beta})$. Note that for v, $v_{Nj}\in[\frac{1}{2}, 1-N^{-\gamma_2}]$, $|L(v)|\le KN^{\gamma_2\beta}$ and $\{1-v_{Nj}\}^{-1-\beta}\le N^{\gamma_2(1+\beta)}$. It is easy to see that (44) is bounded by

$$\sum_{j=1}^{N_2^*}\sum_{i=1}^{N_1^*}|F_N(u_{Ni})-u_{Ni}||J'(u_{Ni})||L(v_{Nj})|\left|\iint_{I_{Nij}} d[H_N(u,v)-H(u,v)]\right|$$
$$+\sum_{i=1}^{N_1^*}|F_N(u_{Ni})-u_{Ni}||J'(u_{Ni})|O(N^{-\xi})KN^{\gamma_2(1+\beta)}$$
$$\times\left|\int_{1/2}^{d_{2N}}\int_{\bar{I}_{1,Ni}} d[H_N(u,v)-H(u,v)]\right|$$
$$+\sum_{i=1}^{N_1^*}|F_N(u_{Ni})-u_{Ni}|O(N^{-1/2})O(\{1-u_{Ni}\}^{-2-\alpha})KN^{\gamma_2\beta}$$
$$\times\left|\int_{1/2}^{d_{2N}}\int_{\bar{I}_{1,Ni}} d[H_N(u,v)-H(u,v)]\right|$$
$$+\sum_{i=1}^{N_1^*}O(N^{-3/4}\log N)|J'(u_{Ni})|KN^{\gamma_2\beta}\int_{1/2}^{d_{2N}}\int_{\bar{I}_{1,Ni}}|d[H_N(u,v)-H(u,v)]|$$
$$+\sum_{i=1}^{N_1^*}O(N^{-5/4}\log N)O(\{1-u_{Ni}\}^{-2-\alpha})KN^{\gamma_2\beta}$$
$$\times\int_{1/2}^{d_{2N}}\int_{\bar{I}_{1,Ni}}|d[H_N(u,v)-H(u,v)]|. \tag{45}$$

Consider the first term of (45). Observe that

$$\left|\iint_{I_{Nij}} d[H_N(u,v)-H(u,v)]\right|$$
$$\leq |H_N(u_{Ni},v_{Nj})-H_N(u_{Ni-1'},v_{Nj})-H(u_{Ni},v_{Nj})+H(u_{Ni-1'},v_{Nj})|$$
$$+|H_N(u_{Ni-1'}v_{Nj-1})-H_N(u_{Ni'}v_{Nj-1})-H(u_{Ni-1'}v_{Nj-1})+H(u_{Ni'}v_{Nj-1})|. \tag{46}$$

Lemma 2.3 implies that (46) is bounded by $O(N^{-3/4}\log N)$.

Using (3) and Lemma 2.1, it is easily seen that the first term of (45) is bounded by

$$O(N^{-5/4}(\log N)^2)\left(\sum_{i=1}^{N_1^*}[1-u_{Ni}]^{-1/2-\alpha-\varepsilon}\right)\left(\sum_{j=1}^{N_2^*}[1-v_{Ni}]^{-\beta}\right) \tag{47}$$

for any $\varepsilon > 0$. Note that if ε is small enough, then $\frac{1}{2}+\alpha+\varepsilon<1$. Hence

$$\sum_{i=1}^{N_1^*}[1-u_{Ni}]^{-1/2-\alpha-\varepsilon}\leq\sum_{i=1}^{N_1^*}(iN^{-1/2})^{-1/2-\alpha-\varepsilon}=O(N^{-1/2}). \tag{48}$$

Also note that

$$\sum_{j=1}^{N_2^*} [1 - v_{Nj}]^{-\beta} \leq \sum_{j=1}^{N_2^*} (jN^{-\xi})^{-\beta} = O(N^{\xi}).$$

Since $\xi < \frac{1}{4}$, we see that (47) is equal to $O(N^{-\eta-1/2})$ for some $\eta > 0$. It can similarly be shown that the other terms in (45) also equal $O(N^{-\eta-1/2})$. Hence $|B_{14,N}| = O(N^{-\eta-1/2})$ for sufficiently small η.

Next consider $B_{16,N}$. For $y \in [c_{1N}, d_{1N}]$, $L[G(y)] \leq KN^{\beta(1-\delta_2)}$,

$$|B_{16,N}| \leq (N+1)^{-1} KN^{\beta(1-\delta_2)} \int_a^{b_{1N}} [1 - F(x)]^{-1-\alpha} \, dF_N(x)$$
$$+ (N+1)^{-1} KN^{\beta(1-\delta_2)} \int_{a_{1N}}^{a} [F(x)]^{-1-\alpha} \, dF_N(x). \tag{49}$$

Lemma 2.2 implies that the first term of (49) is bounded by

$$O(N^{-1+\beta(1-\delta_2)}) \int_a^{b_{1N}} [1 - F_N(x)]^{-1-\alpha} \, dF_N(x). \tag{50}$$

Also the integral inside (50) equals $O(N^{\alpha(1-\delta_1)})$. Thus Eq. (50) equals

$$O(N^{-1/2-\delta_1-\delta_2-\beta\delta_2-\alpha\delta_1}) = O(N^{-\eta-1/2}) \qquad \text{for} \quad \eta < \delta_1 + \delta_2 + \beta\delta_2 + \alpha\delta_1.$$

Similarly, the second term of (49) equals $O(N^{-\eta-1/2})$ for sufficiently small η. Hence $|B_{16,N}| = O(N^{-\eta-1/2})$. The proof of the theorem is now complete.

Now set

$$\mu = \int_{-\infty}^{\infty} \int_{-\infty}^{\infty} J[F(x)]L[G(y)] \, dH(x, y) \tag{51}$$

and

$$\sigma^2 = \operatorname{var}\Bigg(J[F(X)]L[G(Y)] + \int_{-\infty}^{\infty} \int_{-\infty}^{\infty} [\phi_X(x) - F(x)]J'[F(x)]L[G(y)] \, dH(x, y) + \int_{-\infty}^{\infty} \int_{-\infty}^{\infty} [\phi_Y(y) - G(y)]J[F(x)]L'[G(y)] \, dH(x, y) \Bigg),$$

where $\phi_X(x) = 1$ ($\phi_Y(y) = 1$) if $X \leq x$ ($Y \leq y$) and is zero otherwise. Then we have the following result.

THEOREM 3.2. *Under the assumption* 2.1, $N^{1/2}(T_N - \mu) \xrightarrow{d} N(0, \sigma^2)$ *as* $N \to \infty$, *uniformly with respect to* $H(x, y)$.

Proof. From Ruymgaart *et al.* (1972), $N^{1/2}(\sum_{i=1}^{N} A_{iN} - \mu) \xrightarrow{d} N(0, \sigma^2)$ uniformly with respect to $H(x, y)$ as $N \to \infty$. Since (by Theorem 2.1), $N^{1/2}R_N \to 0$ a.s., the proof follows.

4. Invariance Principles

As a consequence of Theorem 3.1, we have the following law of the iterated logarithm.

THEOREM 4.1. *Under Assumption* 2.1,

$$\limsup_{N\to\infty} N^{1/2}(T_N - \mu)/(2\log\log N)^{1/2} = \sigma \text{ a.s.},$$
$$\liminf_{N\to\infty} N^{1/2}(T_N - \mu)/(2\log\log N)^{1/2} = -\sigma \text{ a.s.}$$

Proof: It can be shown that $\sum_{i=1}^{3} A_{i,N}$ is the average of N independent and identically distributed random variables, each having mean μ and variance σ^2. Hence

$$\limsup_{N\to\infty} N^{1/2}\left[\left(\sum_{i=1}^{3} A_{i,N}\right) - \mu\right]\Big/(2\log\log N)^{1/2} = \sigma \text{ a.s.},$$
$$\liminf_{N\to\infty} N^{1/2}\left[\left(\sum_{i=1}^{3} A_{i,N}\right) - \mu\right]\Big/(2\log\log N)^{1/2} = -\sigma \text{ a.s.}$$

Also by Theorem 3.1,

$$\limsup_{N\to\infty} N^{1/2}R_N = 0 \text{ a.s.}$$

The proof follows.

It is also easy to establish the following.

THEOREM 4.2. *Let* $W_N = N^{1/2}(T_N - \mu)$. *If Assumption 2.1 holds, then* $N^{-1/2}W_{[Nt]}/\sigma$, $0 \le t \le 1$, *converges weakly to the standard Wiener process.*

Both Theorems 4.1 and 4.2 have useful applications to sequential tests for independence. Along the lines of Sen and Ghosh (1973), a class of sequential tests having power 1 and arbitrary small Type 1 error can be constructed for testing $H(x, y) = F(x)G(y)$. The invariance principles obtained here are useful for the study of the asymptotic properties of these tests, especially when the null hypothesis does not hold.

References

Bahadur, R. R. (1966). A note on quantiles in large samples. *Ann. Math. Statist.* **37**, 577–580.

Bhuchongkul, S. (1964). A class of nonparametric tests for independence in bivariate populations. *Ann. Math. Statist.* **35**, 138–149.

Chernoff, H., and Savage, I. R. (1958). Asymptotic normality and efficiency of certain nonparametric test statistics. *Ann. Math. Statist.* **29**, 972–944.

Groeneboom, P., Lepage, Y., and Ruymgaart, F. H. (1976). Rank tests for independence with best strong exact Bahadur slope. *Z. Wahrsch. Verw. Gebiete* **36**, 119–127.

Lai, T. L. (1975). On Chernoff-Savage statistics and sequential rank tests. *Ann. Statist.* **3**, 825–845.

Ruymgaart, F. H., Shorack, G. R., and van Zwet, W. R. (1972). Asymptotic normality of nonparametric tests for independence. *Ann. Math. Statist.* **43**, 1122–1135.

Sen, P. K. (1972). On the Bahadur representation of sample quantities for sequences of ϕ-mixing random variables. *J. Multivariate Anal.* **2**, 77–95.

Sen, P. K., and Ghosh, M. (1971). On bounded length sequential confidence intervals based on one sample rank order statistics. *Ann. Math. Statist.* **42**, 189–203.

Sen, P. K., and Ghosh, M. (1973). Some invariance principles for rank statistics for testing independence. *Z. Wahrsch. Verw. Gebiete* **29**, 93–107.

Sen, P. K., and Ghosh, M. (1974). A Chernoff-Savage representation of rank order statistics for stationary ϕ-mixing processes. *Sankhyà, Ser. A* **35**, 153–172.

Department of Mathematics
Indiana University
Bloomington, Indiana

AMS 1970 Subject Classification:
Primary 62E20, 60B10;
Secondary 62G30.

Reprinted from
Contributions to Probability: A Collection of Papers
Dedicated to Eugene Lukacs (1981), pages 267-282,
edited by J. Gani and V.K. Rohatgi,
by permission of the publisher,

ON THE DEGENERATION OF THE VARIANCE IN THE ASYMPTOTIC NORMALITY OF SIGNED RANK STATISTICS

MADAN L. PURI and STEFAN S. RALESCU

Indiana University, Bloomington, IN, and Brown University, Providence, RI, U.S.A.

The purpose of this paper is to establish the asymptotic normality of simple linear signed rank statistics S_N^+ considered by Hušková (1970), Koul and Staudte (1972), and Puri and Ralescu (1980) for the case when the score-generating function is discontinuous and $\operatorname{Var}(S_N^+)$ compared with the variance of S_N^+ under the hypothesis of symmetry is allowed some degree of degeneracy.

The results obtained are extensions of those by Hájek (1968), Dupač and Hájek (1969), Dupač (1970), Koul and Staudte (1972) and Puri and Ralescu (1980).

1. Preliminaries

Let $X_{N1},\ldots, X_{NN}$, $N\geqslant 1$ be independent random variables, with continuous distribution function $F_{N1},\ldots, F_{NN}$ respectively, and let R_{Ni}^+ be the rank of $|X_{Ni}|$ among $|X_{N1}|,\ldots,|X_{NN}|$. Consider the statistic

$$S_N^+ = \sum_{i=1}^{N} c_{Ni} a_N(R_{Ni}^+) \operatorname{sgn} X_{Ni} \tag{1.1}$$

where $c_{N1},\ldots, c_{NN}$ are known regression constants, $a_N(1),\ldots, a_N(N)$ are scores and $\operatorname{sgn} x=1$ if $x\geqslant 0$, $\operatorname{sgn} x=-1$ if $x<0$.

For simplicity of notation, we shall drop the subscript N in X_{Ni}, c_{Ni} and R_{Ni}^+ in the sequel.

In order to study the asymptotic behavior of S_N^+, the ratio $\operatorname{Var}(S_N^+)/\Sigma_{i=1}^N c_i^2$ plays an important role [see Hájek (1968), Dupač and Hájek (1969), Dupač (1970) and Koul and Staudte (1972)]. For the case of the unit step score-generating function $\psi(t)=1$ for $t\geqslant v$, $\psi(t)=0$ for $t<v$ $(0<v<1)$, under suitable conditions on the distribution functions and regression constants, we shall prove that if the ratio $\operatorname{Var}(S_N^+)/\Sigma_{i=1}^N c_i^2$ goes to zero at most at the rate $N^{-\alpha}$ for some $0<\alpha<\frac{1}{2}$, then S_N^+ is asymptotically normal with natural parameters $(E(S_N^+), \operatorname{Var}(S_N^+))$ as well as with some other simpler parameters (μ_N^+, σ_N^2).

We assume that the c_i's satisfy the condition

$$\max_{1\leqslant i\leqslant N} c_i^2 / \sum_{i=1}^{N} c_i^2 = O(N^{-1/2}). \tag{1.2}$$

Let $F_i^*(x)$ be the distribution function of $|X_i|$ and define

$$H_N^*(x)=\frac{1}{N}\sum_{i=1}^{N} F_i^*(x),$$

$$H_N^{*-1}(t)=\inf\{x\colon H_N^*(x)\geq t\},\quad 0<t<1,$$

$$L_i(t)=F_i\big(H^{*-1}(t)\big),\quad 0<t<1,$$

$$M_i(t)=-F_i\big(-H^{*-1}(t)\big),\quad 0<t<1,$$

$$G_i(t)=F_i^*\big(H^{*-1}(t)\big)=L_i(t)+M_i(t),\quad 0<t<1. \tag{1.3}$$

Assume that the scores are generated by a function $\psi(t)$, $0<t<1$, either by interpolation

$$a_N(i)=\psi(i/(N+1)),\quad 1\leq i\leq N \tag{1.4}$$

or by a procedure satisfying

$$\sum_{i=1}^{N} |a_N(i)-\psi(i/(N+1))|=O(1). \tag{1.5}$$

If $v\in(0,1)$ represents a jump point of the score-generating function ψ, then for every $K>0$ we assume the existence of the derivatives $L_i'(t)$ and $M_i'(t)$ in the interval $|t-v|\leq KN^{-1/2}Lg^{1/2}N$ and the satisfaction of the following conditions.

$$\max_{1\leq i\leq N}|L_i'(t)|=O(1), \tag{1.6}$$

$$\max_{1\leq i\leq N}|M_i'(t)|=O(1), \tag{1.7}$$

$$\max_{1\leq i\leq N}\ \sup_{|t-v|\leq KN^{-1/2}Lg^{1/2}N}|L_i'(t)-L_i'(v)|=O(N^{-1/2}Lg^{1/2}N), \tag{1.8}$$

$$\max_{1\leq i\leq N}\ \sup_{|t-v|\leq KN^{-1/2}Lg^{1/2}N}|M_i'(t)-M_i'(v)|=O(N^{-1/2}Lg^{1/2}N). \tag{1.9}$$

Another condition concerning the G_i's that we use is:

$$\liminf_{N\to\infty} N^{-1}\sum_{i=1}^{N} G_i(v)(1-G_i(v))>0. \tag{1.10}$$

Sometimes, mainly for purposes of applications, we replace (1.6)–(1.10) by the following condition which is easier to verify:

Suppose that each F_i has a density f_i. For each $\varepsilon>0$ denote $I_\varepsilon=(H^{*-1}(v)-\varepsilon, H^{*-1}(v)+\varepsilon)$.

Suppose that there exist $\varepsilon_1, \varepsilon_2, \varepsilon_3>0$ such that:

(a)

$$\liminf_{N\to\infty} H_N^{*-1}(v)>0, \tag{1.11}$$

(b) $f_i'(x)$ are uniformly bonded (in x, i, N) on $I_{\varepsilon_1} \cup (-I_{\varepsilon_1})$,
(c) for all $N \geqslant 1$,

$$\frac{1}{N}\operatorname{Card}\left\{1 \leqslant i \leqslant N:\ \inf_{x \in I_{\varepsilon_1}} f_i^*(x) > \varepsilon_2\right\} > \varepsilon_3$$

where f_i^* is the density of F_i^*.

The last condition that we require concerns some possible degeneration of $\operatorname{Var}(S_N^+)$ in the form

$$\liminf_{N\to\infty} \operatorname{Var}(S_N^+) \Big/ \left(N^{-\alpha} \sum_{i=1}^{N} c_i^2 \right) > 0 \tag{1.12}$$

for some $0 < \alpha < \frac{1}{2}$.

Alternatively we shall assume that (1.12) holds with $\operatorname{Var}(S_N^+)$ replaced by some approximate variance σ_N^2:

$$\liminf_{N\to\infty} \sigma_N^2 \Big/ \left(N^{-\alpha} \sum_{i=1}^{N} c_i^2 \right) > 0 \quad (0 < \alpha < \tfrac{1}{2}). \tag{1.13}$$

2. Main theorems

Let $u(t)$ be 1 or 0 according to $t \geqslant 0$ or $t < 0$.

The main result of this paper is the following theorem:

Theorem 2.1. *Let S_N^+ given by* (1.1) *have scores given by* (1.5) *where $\psi(t) = u(t - v)$*, $0 < v < 1$.

Then S_N^+ is asymptotically normal with natural parameters $(E(S_N^+), \operatorname{Var}(S_N^+))$ if any of the following sets of conditions is satisfied:

$$(\tilde{C}_1^+): (1.2), (1.6), (1.7), (1.8), (1.9), (1.10), (1.12),$$

$$(\tilde{C}_2^+): (1.2), (1.11), (1.12).$$

Proof. We show that S_N^+ is asymptotically equivalent to its projection $\hat{S}_N^+$ onto the space of linear statistics and then that $\hat{S}_N^+$ is asymptotically equivalent to a sum of independent random variables to which the Lindeberg central limit theorem applies.

Let us begin by assuming that scores are given by (1.4) and that $(\tilde{C}_1^+)$ holds.

First we would like to derive an upper bound for the residual variance $E(S_N^+ - \hat{S}_N^+)^2$, where:

$$\hat{S}_N^+ = \sum_{i=1}^{N} E(S_N^+ \mid X_i) - (N-1)E(S_N^+).$$

This will be accomplished by using the Residual variance inequality [see Hájek (1968) and Koul and Staudte (1972)]:

$$E(S_N^+ - \hat{S}_N^+)^2 \leqslant \sum_{i=1}^{N} c_i^2 E\big(a(R_i^+) - E(a(R_i^+)|X_i)\big)^2$$

$$+ \sum_{i\neq j} c_i c_j \Big\{ E\big(\text{sgn } X_i \text{sgn } X_j \text{Cov}\big(a(R_i^+), a(R_j^+)|X_i, X_j\big)\big)$$

$$+ E\{\text{sgn } X_i \text{sgn } X_j \big[E\big(a(R_i^+)|X_i, X_j\big) - E\big(a(R_i^+)|X_i\big)\big]$$

$$\times \big[E\big(a(R_j^+)|X_i, X_j\big) - E\big(a(R_j^+)|X_j\big)\big]\}$$

$$- \sum_{k\neq i,j} \text{Cov}\{E\big(\text{sgn } X_i a(R_i^+)|X_k\big), E\big(\text{sgn } X_j a(R_j^+)|X_k\big)\}\Big\}.$$

We investigate each term in the above inequality. The proof is divided in several steps:

Lemma 2.1. *Let $x, y \in \mathbb{R}$. Then for each $K_1 > \sqrt{6}$ there exists a $K_2 \geqslant \frac{3}{2}$ such that for all $N > N_0(K_1)$ we have*

(i) $v - H^*(|x|) > K_1 N^{-1/2} Lg^{1/2} N \Rightarrow P(R_i^+ \geqslant V | X_i = x,\ X_j = y) < N^{-K_2}$,

(ii) $v - H^*(|x|) < -K_1 N^{-1/2} Lg^{1/2} N \Rightarrow P(R_i^+ \leqslant V | X_i = x,\ X_j = y) < N^{-K_2}$ *where* $V = [(N+1)v]$. ($[\cdot]$ = *integer part*).

Furthermore, (i) *and* (ii) *remain true even when the condition $X_j = y$ is omitted.*

Let

$$D^2 = N^{-1} \sum_{i=1}^{N} G_i(v)(1 - G_i(v)).$$

Lemma 2.2. *Suppose that $|v - H^*(|x|)| \leqslant K_3 N^{-1/2} Lg^{1/2} N$. Then for sufficiently large N, we have*

(i) $$\left| \sum_{i=1}^{N} F_i^*(|x|)(1 - F_i^*(|x|)) - ND^2 \right| \leqslant K_4 N^{1/2} Lg^{1/2} N,$$

(ii) $$\left| \phi\left(V;\ \sum_{i=1}^{N} F_i^*(|x|),\ \sum_{i=1}^{N} F_i^*(|x|)(1 - F_i^*(|x|)) \right) - \phi\left(Nv;\ \sum_{i=1}^{N} F_i^*(|x|),\ ND^2 \right) \right| \leqslant K_5 N^{-1} Lg^{1/2} N,$$

(iii)
$$\left|\Phi\left(V;\ \sum_{i=1}^{N} F_i^*(|x|),\ \sum_{i=1}^{N} F_i^*(|x|)(1-F_i^*(|x|))\right)\right.$$
$$\left.-\Phi\left(Nv;\ \sum_{i=1}^{N} F_i^*(|x|),\ ND^2\right)\right| \leqslant K_6 N^{-1/2} Lg^{1/2} N$$

where $\phi(x;\mu,\sigma^2)$, $\Phi(x;\mu,\sigma^2)$ *denote the normal density, respectively the normal distribution function with parameters* (μ,σ^2).

The proofs of Lemmas 2.1 and 2.2 are analogous to those of Lemmas 5 and 6 of Dupač and Hájek (1969) and are therefore omitted.

Lemma 2.3. *For* $N\to\infty$, *we have*

$$E\big(a(R_i^+)-E(a(R_i^+)|X_i)\big)^2 = o(N^{-\alpha})$$

uniformly in $1\leqslant i\leqslant N$.

Proof. Let $\nabla^+(X_i)=E(a(R_i^+)|X_i)-[E(a(R_i^+)|X_i)]^2$. Then, by conditioning, we obtain

$$E\big[a(R_i^+)-E(a(R_i^+)|X_i)\big]^2 = E(\nabla^+(X_i)).$$

Now, by definition

$$E(a(R_i^+)|X_i=x)=P(R_i^+>V|X_i=x).$$

Thus

$$\nabla^+(X_i=x)=P(R_i^+>V|X_i=x)\cdot P(R_i^+\leqslant V|X_i=x).$$

Let $I=\{x:\ |H^*(|x|)-v|\leqslant K_1 N^{-1/2} Lg^{1/2} N\}$ with $K_1>\sqrt{6}$. By Lemma 2.1, if $x\notin I$ we have:

$$\nabla^+(X_i=x)<N^{-K_2} \quad \text{for every } N>N_0(K_1),\ K_2\geqslant\tfrac{3}{2}.$$

On the other hand, if $x\in I$, then since $P(R_i^+=k\,|\,X_i=x)=B^i(k, F_1^*(|x|),\ldots,F_N^*(|x|))$ (in the notation used by Dupač and Hájek (1969)) we obtain, using Lemmas 2.1 and 2.2, that

$$\nabla^+(X_i=x)=\left\{\sum_{k>V} B^i(k, F_1^*(|x|),\ldots,F_N^*(|x|))\right\}$$
$$\times\left\{\sum_{l\leqslant V} B^i(l, F_1^*(|x|),\ldots,F_N^*(|x|))\right\}$$
$$=\cdots=\Phi\left(\frac{H^*(|x|)-v}{DN^{-1/2}}\right)\left\{1-\Phi\left(\frac{H^*(|x|)-v}{DN^{-1/2}}\right)\right\}+\theta_1 N^{-1/2} Lg^{1/2} N$$

for sufficiently large N, $|\theta_1|\leqslant K_7$. Here Φ denotes the standard normal distribution function. We use ϕ for its density function in the sequel.

We observe that the last equality remains true even if we enlarge I to

$$I' = \{x : |H^*(|x|) - v| \leqslant K_9 D N^{-1/2} Lg^{1/2} N\},$$

where K_9 is such that $K_9 = {}^{K}1/2K_8$ with $K_8 \leqslant D \leqslant \frac{1}{2}$.

Now, using (1.6)–(1.9) it is easy to show that

$$N^{-1/2} Lg^{1/2} N \int_{I'} \theta_1 \, \mathrm{d}F_i(x) = \mathrm{o}(N^{-\alpha})$$

and

$$\int_{I'} \Phi\left(\frac{H^*(|x|) - v}{DN^{-1/2}}\right)\left\{1 - \Phi\left(\frac{H^*(|x|) - v}{DN^{-1/2}}\right)\right\} \mathrm{d}F_i(x) = \mathrm{o}(N^{-\alpha})$$

uniformly in $1 \leqslant i \leqslant N$. Hence

$$E(\nabla^+(x_i)) = \mathrm{o}(N^{-\alpha}) \text{ uniformly in } 1 \leqslant i \leqslant N$$

and the proof follows.

Lemma 2.4. *For $N \to \infty$ we have*

$$E\{\operatorname{sgn} X_i \operatorname{sgn} X_j [E(a(R_i^+)|X_i, X_j) - E(a(R_i^+)|X_i)]$$
$$\times [E(a(R_j^+)|X_i, X_j) - E(a(R_j^+)|X_j)]\} = \mathrm{o}(N^{-1-\alpha}),$$

uniformly in $1 \leqslant i, j \leqslant N$.

The proof of this lemma is similar to that of Lemma 2.3 and is therefore omitted.

Lemma 2.5. *For $N \to \infty$ we have*

$$E[\operatorname{sgn} X_i \operatorname{sgn} X_j \operatorname{Cov}(a(R_i^+), a(R_j^+)|X_i, X_j)]$$
$$= N^{-1} D^2 (L_i'(v) - M_i'(v))(L_j'(v) - M_j'(v)) + \mathrm{o}(N^{-1-\alpha})$$

uniformly in $1 \leqslant i, j \leqslant N$.

Proof. We have

$$\Delta^+ = \operatorname{Cov}(a(R_i^+), a(R_j^+)|X_i = x, X_j = y)$$
$$= \begin{cases} P(R_i^+ > V|X_i = x, X_j = y) P(R_j^+ \leqslant V|X_i = x, X_j = y), \\ \quad \text{if } |x| < |y|, \\ P(R_j^+ > V|X_i = x, X_j = y) P(R_i^+ \leqslant V|X_i = x, X_j = y), \\ \quad \text{if } |x| \geqslant |y|. \end{cases}$$

Let $K_1 > \sqrt{6}$. Denote

$$I = \{(x, y) : \max(|H^*(|x|) - v|, |H^*(|y|) - v|) \leqslant K_1 N^{-1/2} Lg^{1/2} N\}.$$

By considerations as used in the derivation of (4.11) and (4.12) in Dupač and Hájek (1969) we obtain:

$$\Delta^+(x,y)\begin{cases} <N^{-K_2} \quad \text{for } (x,y)\notin I,\ N>N_0(K_2) \\ =\Phi\left(\dfrac{H^*(|x|)-v}{DN^{-1/2}}\right)\left\{1-\Phi\left(\dfrac{H^*(|y|)-v}{DN^{-1/2}}\right)\right\}+\theta_2 N^{-1/2}Lg^{1/2}N \\ \quad \text{for } N \text{ sufficiently large, } (x,y)\in I,\ |x|<|y| \\ \quad \text{and } |\theta_2|\leqslant K_{10} \\ =\Phi\left(\dfrac{H^*(|y|)-v}{DN^{-1/2}}\right)\left\{1-\Phi\left(\dfrac{H^*(|x|)-v}{DN^{-1/2}}\right)\right\}+\theta_3 N^{-1/2}Lg^{1/2}N \\ \quad \text{for } N \text{ sufficiently large, } (x,y)\in I,\ |x|\geqslant|y| \\ \quad \text{and } |\theta_3|\leqslant K_{11} \end{cases} \tag{2.1}$$

where $K_2 \geqslant \frac{3}{2}$. We note that the *equality* in (2.1) remains true even if we enlarge I to I' where

$$I'=\left\{(x,y): \max(|H^*(|x|)-v|,|H^*(|y|)-v|)\leqslant K_9 DN^{-1/2}Lg^{1/2}N\right\}$$

where $K_9 = {}^{K}1/2K_8$, $K_8 \leqslant D \leqslant \frac{1}{2}$.

We have, using (2.1) that

$$E(\operatorname{sgn} X_i \operatorname{sgn} X_j \operatorname{Cov}(a(R_i^+), a(R_j^+))|X_i, X_j)=$$

$$=\iint_{I'\cap\{|x|<|y|\}} \operatorname{sgn} x \operatorname{sgn} y\, \Phi\left(\frac{H^*(|x|)-v}{DN^{-1/2}}\right)\left\{1-\Phi\left(\frac{H^*(|y|)-v}{DN^{-1/2}}\right)\right\} \mathrm{d}F_i(x)\,\mathrm{d}F_j(y)$$

$$+\iint_{I'\cap\{|x|\geqslant|y|\}} \operatorname{sgn} x \operatorname{sgn} y\, \Phi\left(\frac{H^*(|y|)-v}{DN^{-1/2}}\right)\left\{1-\Phi\left(\frac{H^*(|x|)-v}{DN^{-1/2}}\right)\right\} \mathrm{d}F_i(x)\,\mathrm{d}F_j(y) \tag{2.2}$$

$$+N^{-1/2}Lg^{1/2}N\iint_{I'} \operatorname{sgn} x \operatorname{sgn} y\, \theta_4(x,y)\,\mathrm{d}F_i(x)\,\mathrm{d}F_j(y)+\theta_5 N^{-K_2}$$

with $|\theta_4| \leqslant K_{12}$, $|\theta_5| \leqslant 1$.

The last two terms are $o(N^{-1-\alpha})$ uniformly in i, j as follows by using (1.6)–(1.9) and $K_2 \geqslant \frac{3}{2}$. It remains to estimate the first two terms.

Denote the first term by $\mathfrak{T}$. Consider

$$\mathfrak{T}_1 = \iint_{A_{xy}} \Phi\left(\frac{H^*(x)-v}{DN^{-1/2}}\right)\left\{1-\Phi\left(\frac{H^*(y)-v}{DN^{-1/2}}\right)\right\} \mathrm{d}F_i(x)\,\mathrm{d}F_j(y)$$

where

$$A_{xy}=\left\{(x,y): \begin{matrix} x>0 \\ y>0 \\ x<y \end{matrix},\ \max(|H^*(x)-v|,|H^*(y)-v|)\leqslant K_9 DN^{-1/2}Lg^{1/2}N\right\}.$$

Set

$$p=\frac{H^*(x)-v}{DN^{-1/2}}, \qquad q=\frac{H^*(y)-v}{DN^{-1/2}},$$

and

$$I''=\{(p,q)\colon \max(|p|,|q|)\leqslant K_9 Lg^{1/2}N\}.$$

Then

$$\mathfrak{T}_1=\iint_{I''\cap\{p<q\}}\Phi(p)(1-\Phi(q))\,\mathrm{d}L_i(v+DN^{-1/2}p)\,\mathrm{d}L_j(v+DN^{-1/2}q).$$

Let $\Omega=I''\cap\{p<q\}$ and $\Omega^*=\{p<q\}\setminus\Omega$. Then using (1.6)–(1.9), one can easily show that

$$\mathfrak{T}_1=D^2N^{-1}L_i'(v)L_j'(v)\iint_{\Omega}\Phi(p)(1-\Phi(q))\,\mathrm{d}p\,\mathrm{d}q+\mathrm{o}(N^{-1-\alpha}) \tag{2.3}$$

uniformly in $1\leqslant i, j\leqslant N$.

Let $\Omega_1^*=\Omega^*\cap\{p>-q\}$. Then, by Fubini's theorem we have:

$$N^{-1}\iint_{\Omega_1^*}\Phi(p)(1-\Phi(q))\,\mathrm{d}p\,\mathrm{d}q=N^{-1}\int_{K_9Lg^{1/2}N}^{\infty}(1-\Phi(q))\left(\int_{-q}^{q}\Phi(p)\,\mathrm{d}p\right)\mathrm{d}q$$

$$\leqslant 2N^{-1}\int_{K_9Lg^{1/2}N}^{\infty}q\Phi(-q)\,\mathrm{d}q.$$

Using integration by parts it follows that:

$$N^{-1}\iint_{\Omega_1^*}\Phi(p)(1-\Phi(q))\,\mathrm{d}p\,\mathrm{d}q=\mathrm{o}(N^{-1-\alpha}). \tag{2.4}$$

Similarly we can show that:

$$N^{-1}\iint_{\Omega_2^*}\Phi(p)(1-\Phi(q))\,\mathrm{d}p\,\mathrm{d}q=\mathrm{o}(N^{-1-\alpha}) \tag{2.5}$$

where $\Omega_2^*=\Omega^*\setminus\Omega_1^*$.

By (2.4) and (2.5)

$$D^2N^{-1}L_i'(v)L_j'(v)\iint_{\Omega^*}\Phi(p)(1-\Phi(q))\,\mathrm{d}p\,\mathrm{d}q=\mathrm{o}(N^{-1-\alpha}). \tag{2.6}$$

From (2.3), (2.6) and the fact that

$$\iint_{\{p<q\}}\Phi(p)(1-\Phi(q))\,\mathrm{d}p\,\mathrm{d}q=\tfrac{1}{2}$$

we obtain:

$$\mathfrak{T}_1=\tfrac{1}{2}D^2N^{-1}L_i'(v)L_j'(v)+\mathrm{o}(N^{-1-\alpha}) \tag{2.7}$$

uniformly in $1\leqslant i, j\leqslant N$.

Let

$$\mathfrak{T}_2=\iint_{B_{xy}} -\Phi\left(\frac{H^*(-x)-v}{DN^{-1/2}}\right)\left\{1-\Phi\left(\frac{H^*(y)-v}{DN^{-1/2}}\right)\right\}\mathrm{d}F_i(x)\mathrm{d}F_j(y)$$

$$=\iint_{I''\cap\{p<q\}} -\Phi(p)(1-\Phi(q))\,\mathrm{d}M_i(v+DN^{-1/2}p)\,\mathrm{d}L_j(v+DN^{-1/2}q)$$

where

$$B_{xy}=\left\{(x,y):\begin{array}{l} x<0\\ y>0\\ -x<y\end{array},\max(|H^*(-x)-v|,|H^*(y)-v|)\leqslant K_9DN^{-1/2}Lg^{1/2}N\right\},$$

$$\mathfrak{T}_3=\iint_{C_{xy}} \Phi\left(\frac{H^*(-x)-y}{DN^{-1/2}}\right)\left\{1-\Phi\left(\frac{H^*(-y)-v}{DN^{-1/2}}\right)\right\}\mathrm{d}F_i(x)\mathrm{d}F_j(y)$$

and

$$\mathfrak{T}_4=\iint_{D_{xy}} -\Phi\left(\frac{H^*(x)-y}{DN^{-1/2}}\right)\left\{1-\Phi\left(\frac{H^*(-y)-v}{DN^{-1/2}}\right)\right\}\mathrm{d}F_i(x)\mathrm{d}F_j(y)$$

where

$$C_{xy}=\left\{(x,y):\begin{array}{l} x<0\\ y<0\\ -x<-y\end{array},\max(|H^*(-x)-v|,|H^*(-y)-v|)\leqslant K_9DN^{-1/2}Lg^{1/2}N\right\}$$

and

$$D_{xy}=\left\{(x,y):\begin{array}{l} x>0\\ y<0\\ x<-y\end{array},\max(|H^*(x)-v|,|H^*(-y)-v|)\leqslant K_9DN^{-1/2}Lg^{1/2}N\right\}.$$

We now repeat the steps used in the derivation of (2.7), this time applying them to $\mathfrak{T}_2$, $\mathfrak{T}_3$ and $\mathfrak{T}_4$ to obtain

$$\mathfrak{T}_2=-\tfrac{1}{2}D^2N^{-1}M_i'(v)L_j'(v)+o(N^{-1-\alpha})\quad\text{uniformly in } 1\leqslant i,\ j\leqslant N,$$

$$\mathfrak{T}_3=\tfrac{1}{2}D^2N^{-1}M_i'(v)M_j'(v)+o(N^{-1-\alpha})\quad\text{uniformly in } 1\leqslant i,\ j\leqslant N$$

and

$$\mathfrak{T}_4=-\tfrac{1}{2}D^2N^{-1}L_i'(v)M_j'(v)+o(N^{-1-\alpha})\quad\text{uniformly in } 1\leqslant i,\ j\leqslant N.$$

Thus

$$\mathfrak{T}=\mathfrak{T}_1+\mathfrak{T}_2+\mathfrak{T}_3+\mathfrak{T}_4=\tfrac{1}{2}D^2N^{-1}(L_i'(v)-M_i'(v))(L_j'(v)-M_j'(v))+o(N^{-1-\alpha})$$

uniformly in $1\leqslant i,\ j\leqslant N$. (2.8)

Proceeding as above, it can be shown that the second term of (2.2) is the same as (2.8). The proof follows.

Lemma 2.6. *For $N \to \infty$ we have (for $i \neq j$)*

$$\sum_{k \neq i,j} \operatorname{Cov}\{E(\operatorname{sgn} X_i a(R_i^+)|X_k),\, E(\operatorname{sgn} X_j a(R_j^+)|X_k)\} =$$

$$= D^2 N^{-1}(L_i'(v) - M_i'(v))(L_j'(v) - M_j'(v)) + o(N^{-1-\alpha})$$

uniformly in $1 \leq i, j \leq N$.

Proof. By Lemma 3.2 in Hájek (1968) we have

$$E(a(R_i^+)\operatorname{sgn} X_i | X_i = x, X_k = z) - E(a(R_i^+)\operatorname{sgn} X_i | X_i = x) =$$
$$= \operatorname{sgn} x[u(|x| - |z|) - F_k^*(|x|)] \cdot P(R_i^+ = V + 1 | X_i = x, |X_k| = |x| - 1).$$

From Lemma 2.1, we have

$$P(R_i^+ = V + 1 | X_i = x, |X_k| = |x| - 1) < N^{-K_2} \tag{2.9}$$

for some $K_2 \geq \frac{3}{2}$ and all $|H^*(|x|) - v| \geq K_1 N^{-1/2} Lg^{1/2} N$.

Furthermore Lemmas 2.1 and 2.2 imply:

$$P(R_i^+ = V + 1 | X_i = x, |X_k| = |x| - 1) = \phi\left(Nv;\ \sum_{j=1}^N F_j^*(|x|),\, ND^2\right) + \theta_6 N^{-1} Lg^{1/2} N$$

for some $|\theta_6| \leq K_{13}$ and all $|H^*(|x|) - v| \leq K_1 N^{-1/2} Lg^{1/2} N$.

As before, the last equality remains true even if $|H^*(|x|) - v| \leq K_9 D N^{-1/2} Lg^{1/2} N$. Let

$$I' = \{x:\ |H^*(|x|) - v| \leq K_9 D N^{-1/2} Lg^{1/2} N\}.$$

Then

$$E(a(R_i^+)\operatorname{sgn} X_i | X_k = z) - E(a(R_i^+)\operatorname{sgn} X_i) =$$
$$= \int \operatorname{sgn} x[u(|x| - |z|) - F_k^*(|x|)]$$
$$\times P(R_i^+ = V + 1 | X_i = x, |X_k| = |x| - 1)\, dF_i(x)$$
$$= \int_{I'} (\cdots)\, dF_i(x) + \int_{\mathbb{R} \setminus I'} (\cdots)\, dF_i(x).$$

The second integral is $o(N^{-1-\alpha})$ by (2.9), while the first is equal to

$$\int_{I'} \operatorname{sgn} x[u(|x| - |z|) - F_k^*(|x|)]\, \phi\left(Nv;\ \sum_{j=1}^N F_j^*(|x|),\, ND^2\right) dF_i(x)$$
$$+ \int_{I'} \operatorname{sgn} x[u(|x| - |z|) - F_k^*(|x|)]\, \theta_6 N^{-1} Lg^{1/2} N\, dF_i(x).$$

In the last expression, let us denote by $\mathfrak{T}_5$ the first term and $\mathfrak{T}_6$ the second. From (1.6)–(1.9) and the Mean Value Theorem, it follows that:

$$\max_{1 \leq k \leq N}\ \sup_{|p| \leq K_9 Lg^{1/2} N} |G_k(v + D N^{-1/2} p) - G_k(v)| = O(N^{-1/2} Lg^{1/2} N). \tag{2.10}$$

Then it is easy to show that

$$D^{-1}N^{-1/2}\int_{\{|p|\leqslant K_9Lg^{1/2}N\}} G_k(v+DN^{-1/2}p)\phi(p)\,\mathrm{d}L_i(v+DN^{-1/2}p)=$$

$$=D^{-1}N^{-1/2}\int_{\{|p|\leqslant K_9Lg^{1/2}N\}} G_k(v)\phi(p)\,\mathrm{d}L_i(v+DN^{-1/2}p)+\mathrm{o}(N^{-1-\alpha})$$

uniformly in i and k.

We write

$$\mathfrak{T}_5=D^{-1}N^{-1/2}\int_{\{x>0:|H^*(x)-v|\leqslant K_9DN^{-1/2}Lg^{1/2}N\}}(\cdots)\,\mathrm{d}F_i(x)$$

$$+D^{-1}N^{-1/2}\int_{\{x<0:\,|H^*(-x)-v|\leqslant K_9DN^{-1/2}Lg^{1/2}N\}}(\cdots)\,\mathrm{d}F_i(x)$$

$$=\mathfrak{T}_5'+\mathfrak{T}_5''.$$

We have:

$$\mathfrak{T}_5'=N^{-1}L_i'(v)\int_{\{|p|\leqslant K_9Lg^{1/2}N\}}[u(p-q)-G_k(v)]\phi(p)\,\mathrm{d}p$$

$$+N^{-1}\mathrm{O}(N^{-1/2}Lg^{1/2}N)\int_{\{|p|\leqslant K_9Lg^{1/2}N\}}[u(p-q)-G_k(v)]\phi(p)\,\mathrm{d}p$$

$$+\mathrm{o}(N^{-1-\alpha}).$$

In the last expression, the second term is $\mathrm{o}(N^{-1-\alpha})$ while the first is equal to

$$N^{-1}L_i'(v)[1-\Phi(q)-G_k(v)]$$

$$-N^{-1}L_i'(v)\int_{\{|p|>K_9Lg^{1/2}N\}}[u(p-q)-G_k(v)]\phi(p)\,\mathrm{d}p.$$

But

$$\left|N^{\alpha}\int_{K_9Lg^{1/2}N}^{\infty}[u(p-q)-G_k(v)]\phi(p)\,\mathrm{d}p\right|\leqslant$$

$$\leqslant N^{\alpha}\int_{K_9Lg^{1/2}N}^{\infty}\phi(p)\,\mathrm{d}p=N^{\alpha}\Phi(-K_9Lg^{1/2}N)\to 0$$

and we obtain:

$$\mathfrak{T}_5'=N^{-1}L_i'(v)[1-\Phi(q)-G_k(v)]+\mathrm{o}(N^{-1-\alpha})$$

uniformly in z, $1\leqslant i\leqslant N$, where $q=(H^*(|z|)-v)/DN^{-1/2}$.

Similarly

$$\mathfrak{T}_5''=-N^{-1}M_i'(v)[1-\Phi(q)-G_k(v)]+\mathrm{o}(N^{-1-\alpha}).$$

Also, it is easy to check that $\mathfrak{T}_6 = o(N^{-1-\alpha})$. Hence

$$\begin{aligned} &E(a(R_i^+)\operatorname{sgn} X_i | X_k = z) - E(a(R_i^+)\operatorname{sgn} X_i) \\ &= \mathfrak{T}_5' + \mathfrak{T}_5'' + o(N^{-1-\alpha}) \\ &= N^{-1}(L_i'(v) - M_i'(v))[1 - \Phi(q) - G_k(v)] + o(N^{-1-\alpha}) \end{aligned} \tag{2.11}$$

uniformly in $-\infty < z < \infty$.

We now show that under (1.6)–(1.9)

$$\int_{-v/DN^{-1/2}}^{(1-v)/DN^{-1/2}} (1 - \Phi(q))\, \mathrm{d}G_k(v + DN^{-1/2}q) = G_k(v) + o(N^{-\alpha}). \tag{2.12}$$

Indeed, using integration by parts

$$\int_{-v/DN^{-1/2}}^{(1-v)/DN^{-1/2}} (1 - \Phi(q))\, \mathrm{d}G_k(v + DN^{-1/2}q) - G_k(v) =$$

$$= \left[1 - \Phi\left(\frac{1-v}{DN^{-1/2}}\right)\right]$$

$$+ \int_{-\infty}^{\infty} \left[I_{(-(v/DN^{-1/2}),((1-v)/DN^{-1/2}))}(q) G_k(v + DN^{-1/2}q) - G_k(v)\right] \phi(q)\, \mathrm{d}q.$$

Let $\mathcal{Q}$ denote the last integral in the above relation. Then:

$$\mathcal{Q} = \int_{\{|q| \geqslant Lg^{1/2}N\}} (\cdots)\phi(q)\, \mathrm{d}q + \int_{\{|q| \leqslant Lg^{1/2}N\}} (\cdots)\phi(q)\, \mathrm{d}q = \mathcal{Q}_1 + \mathcal{Q}_2.$$

Since

$$N^\alpha \int_{Lg^{1/2}N}^{\infty} \phi(q)\, \mathrm{d}q = N^\alpha \Phi(-Lg^{1/2}N) \to 0$$

and

$$|\mathcal{Q}_1| \leqslant \int_{\{|q| \geqslant Lg^{1/2}N\}} \phi(q)\, \mathrm{d}q$$

it follows that $\mathcal{Q}_1 = o(N^{-\alpha})$. On the other hand, (2.10) entails for sufficiently large N that:

$$\begin{aligned} |N^\alpha \mathcal{Q}_2| &\leqslant N^\alpha \int_{\{|q| \leqslant Lg^{1/2}N\}} |G_k(v + DN^{-1/2}q) - G_k(v)| \phi(q)\, \mathrm{d}q \\ &= O(N^{\alpha - 1/2} Lg^{1/2}N). \end{aligned}$$

Thus $\mathcal{Q}_2 = o(N^{-\alpha})$ and we get

$$\mathcal{Q} = \mathcal{Q}_1 + \mathcal{Q}_2 = o(N^{-\alpha}).$$

This, together with the fact that

$$1 - \Phi\left(\frac{1-v}{DN^{-1/2}}\right) = o(N^{-\alpha})$$

proves (2.12).

Similarly we can show that

$$\int_{-v/DN^{-1/2}}^{(1-v)/DN^{-1/2}} (1-\Phi(q))^2 \mathrm{d}G_k(v+DN^{-1/2}q)=G_k(v)+\mathrm{o}(N^{-\alpha}). \tag{2.13}$$

Finally, using (2.11), (2.12) and (2.13) and proceeding as in Dupač and Hájek (1969), the proof follows.

By Lemmas 2.3–3.6 and the Residual variance inequality

$$E(S_N^+ - \hat{S}_N^+)^2 = \mathrm{o}\left(N^{-\alpha} \sum_{i=1}^{N} c_i^2\right). \tag{2.14}$$

Let us show now that

$$E\left\{\Phi\left(\frac{v-H^*(|X_i|)}{DN^{-1/2}}\right)-u(v-H^*(|X_i|))\right\}^2=\mathrm{o}(N^{-\alpha}). \tag{2.15}$$

The left-hand side of (2.15) equals $I_1 + I_2$ where

$$\begin{aligned} I_1 &= \int_{-v/DN^{-1/2}}^{0} \Phi^2(p)\,\mathrm{d}G_i(v+DN^{-1/2}p) \\ &= -\int_{-\infty}^{0} \left[I_{(-v/DN^{-1/2},0)}(p)G_i(v+DN^{-1/2}p)-G_i(v)\right]2\Phi(p)\phi(p)\,\mathrm{d}p \end{aligned}$$

and

$$\begin{aligned} I_2 &= \int_{0}^{(1-v)/DN^{-1/2}} [1-\Phi(p)]^2\,\mathrm{d}G_i(v+DN^{-1/2}p) \\ &= \Phi^2\left(-\frac{(1-v)}{DN^{-1/2}}\right)+\int_0^{\infty}\left[I_{(0,((1-v')/DN^{-1/2}))}(p)G_i(v+DN^{-1/2}p)-G_i(v)\right] \\ &\qquad \times 2\Phi(-p)\phi(p)\,\mathrm{d}p. \end{aligned}$$

Then, proceeding as in the derivation of (2.12) and (2.13) it follows that $I_1 = \mathrm{o}(N^{-\alpha})$ and $I_2 = \mathrm{o}(N^{-\alpha})$ and hence (2.15) holds true.

Now, using (1.6)–(1.9) and (2.15) we obtain

$$E(Y_i - Z_i)^2 = \mathrm{o}\left(N^{-1-\alpha} \sum_{i=1}^{N} c_i^2\right) \tag{2.16}$$

where

$$Y_i = \sum_{j=1}^{N} c_j\{E(a(R_j^+)\operatorname{sgn} X_j|X_i)-E(a(R_j^+)\operatorname{sgn} X_j)\}, \quad 1\leqslant i\leqslant N \tag{2.17}$$

and

$$\begin{aligned} Z_i &= N^{-1}\left[\sum_{\substack{j=1\\ j\neq i}}^{N} c_j(L_j'(v)-M_j'(v))\right][u(v-H^*(|X_i|))-G_i(v)] \\ &\quad + c_i[E(\operatorname{sgn} X_i a(R_i^+)|X_i)-E(\operatorname{sgn} X_i a(R_i^+))], \quad 1\leqslant i\leqslant N. \end{aligned} \tag{2.18}$$

Then, since $\hat{S}_N^+ - E(\hat{S}_N^+) = \Sigma_{i=1}^N Y_i$, (2.14) and (2.16) entail

$$\operatorname{Var}\left(S_N^+ - \sum_{i=1}^N Z_i \right) = o\left(N^{-\alpha} \sum_{i=1}^N c_i^2 \right). \tag{2.19}$$

Proceeding as in Lemma 13 of Dupač and Hájek (1969), it is easy to show that (1.12) holds if and only if (1.13) holds with $\sigma_N^2 = \Sigma_{i=1}^N \operatorname{Var}(Z_i)$ and in this case

$$\lim_{N\to\infty} \operatorname{Var}(S_N^+)/\sigma_N^2 = 1.$$

Finally, the asymptotic normality of $\Sigma_{i=1}^N Z_i$ with parameters $(0, \sigma_N^2)$ follows as in Lemma 14 of Dupač and Hájek (1969) with the help of (1.2), (1.6), (1.7), (1.13) and the Lindeberg central limit theorem.

Hence, since we have proved that

$$\sum_{i=1}^N Z_i/\sigma_N \xrightarrow{\mathfrak{D}} N(0,1), \qquad \left(S_N^+ - E(S_N^+) - \sum_{i=1}^N Z_i \right) \Big/ \sigma_N \xrightarrow{\mathfrak{L}^2} 0$$

and

$$\operatorname{Var}(S_N^+)/\sigma_N^2 \to 1,$$

we obtain

$$(S_N^+ - E(S_N^+))/(\operatorname{Var} S_N^+)^{1/2} \xrightarrow{\mathfrak{D}} N(0,1).$$

Suppose we want to relax condition (1.4) to (1.5). Let us denote the statistic corresponding to (1.4) by S_N^+ and the statistic corresponding to (1.5) by S_N^{+*}. Then, using (1.2) and (1.5)

$$\operatorname{Var}(S_N^+ - S_N^{+*}) = o\left(N^{-\alpha} \sum_{i=1}^N c_i^2 \right).$$

Consequently, the asymptotic normality of S_N^{+*} easily follows from the last relation and the asymptotic normality of S_N^+.

We have proved Theorem 2.1 under condition $(\tilde{C}_1^+)$. It remains to show that this set of conditions is implied by the conditions $(\tilde{C}_2^+)$. The proof of this fact is similar to the implications $(C_3) \Rightarrow (C_1)$ and $(C_2) \Rightarrow (C_1)$ in Dupač and Hájek (1969, Section 5) and is therefore omitted.

The following theorem shows that under the same conditions $(\tilde{C}_1^+)$ or $(\tilde{C}_2^+)$, S_N^+ is asymptotically normal with (simpler) parameters (μ_N^+, σ_N^2). This problem is of practical interest since μ_N^+ and σ_N^2 are easier to evaluate:

Let us define:

$$\mu_N^+ = \sum_{i=1}^N c_i E[\operatorname{sgn} X_i \psi(H^*(|X_i|))] \tag{2.20}$$

and

$$\sigma_N^2 = \sum_{i=1}^N \operatorname{Var} Z_i' \tag{2.21}$$

where

$$Z_i' = N^{-1}\left[\sum_{\substack{j=1\\j\neq i}}^{N} c_j\big(L_j'(v)-M_j'(v)\big)\right]\big[u(v-H^*(|X_i|))-G_i(v)\big]$$

$$+c_i\big[\operatorname{sgn} X_i\psi(H^*(|X_i|))-E(\operatorname{sgn} X_i\psi(H^*(|X_i|)))\big]. \tag{2.22}$$

Theorem 2.1. *Let S_N^+ be given by* (1.1) *with scores satisfying* (1.5) *where $\psi(t)=u(t-v)$.*

Assume that $(\tilde{C}_1^+)$ *or* $(\tilde{C}_2^+)$ *holds, with* (1.12) *replaced by* (1.13), *where σ_N^2 is given by* (2.21).

Then S_N^+ is asymptotically normal with parameters (μ_N^+, σ_N^2) defined in (2.20) *and* (2.21).

Proof. We shall follow the proof of Theorem 2.1 (where it is first assumed that $a(i)=\psi(i/(N+1))$ and that $(\tilde{C}_i^+)$ holds).

With Y_i and Z_i defined by (2.17) and (2.18) respectively, we have:

$$E(Y_i-Z_i)^2=\mathrm{o}\left(N^{-1-\alpha}\sum_{j=1}^{N}c_j^2\right) \tag{2.23}$$

uniformly in $1\leqslant i\leqslant N$, as follows from (2.11) and (2.15).

Define

$$\Delta_i(X_i)=\big\{E(\operatorname{sgn} X_i a(R_i^+)|X_i)-E(\operatorname{sgn} X_i a(R_i^+))\big\}$$

$$-\big\{\operatorname{sgn} X_i u(H^*(|X_i|)-v)-E(\operatorname{sgn} X_i u(H^*(|X_i|)-v))\big\}.$$

Proceeding as in Dupač (1970) it can be shown (omitting the details of computation) that

$$\operatorname{Var}(\Delta_i)=E(\Delta_i^2)=\mathrm{O}(N^{-1/2})$$

where $\Delta_i=\Delta_i(X_i)$.

Then, since $Z_i=Z_i'+c_i\Delta_i$, we have

$$E(Z_i-Z_i')^2=c_i^2\mathrm{O}(N^{-1/2}) \tag{2.24}$$

uniformly in $1\leqslant i\leqslant N$.

But $\hat{S}_N^+ - E(\hat{S}_N^+)=\Sigma_{i=1}^N Y_i$ and from (2.23) and (2.24) we obtain:

$$\operatorname{Var}\left(\hat{S}_N^+ - \sum_{i=1}^{N} Z_i'\right)=\mathrm{o}\left(N^{-\alpha}\sum_{i=1}^{N}c_i^2\right).$$

This together with (2.14) entails

$$\operatorname{Var}\left(S_N^+ - \sum_{i=1}^{N} Z_i'\right)=\mathrm{o}\left(N^{-\alpha}\sum_{i=1}^{N}c_i^2\right).$$

Then, proceeding precisely as in the proof of Theorem 2.1, it follows that:

$$\sum_{i=1}^{N} Z_i'/\sigma_N \xrightarrow{\mathfrak{D}} N(0,1),$$

$$\left(S_N^+ - E(S_N^+) - \sum_{i=1}^{N} Z_i'\right)/\sigma_N \xrightarrow{\mathfrak{L}^2} 0$$

and

$$\operatorname{Var}(S_N^+)/\sigma_N^2 \to 1. \tag{2.25}$$

Further set

$$\rho_i = E(\operatorname{sgn} X_i a(R_i^+) - E(\operatorname{sgn} X_i u(H^*(|X_i|) - v)), \quad 1 \leqslant i \leqslant N.$$

It can be shown (again omitting the details of computation) that:

$$\rho_i = o(N^{-\alpha/2 - 1/2}) \quad \text{uniformly in } 1 \leqslant i \leqslant N. \tag{2.26}$$

Now, using the inequality:

$$\left(E(S_N^+) - \mu_N^+\right)^2 \leqslant \left(\sum_{i=1}^{N} c_i^2\right) \cdot \left(\sum_{i=1}^{N} \rho_i^2\right)$$

together with (2.26) we get

$$\left(E(S_N^+) - \mu_N^+\right)^2 = o\left(N^{-\alpha} \sum_{i=1}^{N} c_i^2\right). \tag{2.27}$$

Finally, making use of (1.13), (2.25) and (2.27), the proof follows.

3. An example

Assume that $X_1, \ldots, X_N$ are i.i.d. with common density function f, and consider the problem of testing the hypothesis of symmetry with normal underlying density H_0: $f(x) = \phi(x)$ against the sequence of shift alternatives

$$H_1: f(x) = \phi(x - \Delta) \quad (\Delta = \Delta_N > 0).$$

Assume that $\Delta \to \infty$ sufficiently slow such that:

$$\limsup_{N \to \infty} \Delta \cdot Lg^{-1/2} N < \tfrac{1}{2}. \tag{3.1}$$

We shall prove that under H_1, (3.1) implies the asymptotic normality of S_N^+ where

$$S_N^+ = \sum_{i=1}^{N} u\left(\frac{R_i^+}{N+1} - v\right) \operatorname{sgn} X_i.$$

First we note that since $\Delta \to \infty$, $H^{*-1}(v) \to \infty$, in such a way that:

$$\lim_{N \to \infty} \left(H^{*-1}(v) - \Delta\right) = 0. \tag{3.2}$$

Using (3.2) it is easy to see that condition (1.11) is satisfied. Also, since $c_1 = \cdots = c_N = 1$, it is easy to check that $\sigma_N^2/\Sigma_{i=1}^N c_i^2 \to 0$, where σ_N^2 is defined by (2.21).

Now, (3.1) implies the existence of a constant $0 < C < \frac{1}{2}$ such that, for sufficiently large N

$$\Delta^2 \leqslant C^2 LgN. \tag{3.3}$$

It can be shown (omitting the details of computation) that for sufficiently large N,

$$\sigma_N^2 / \sum_{i=1}^N c_i^2 \geqslant C'\Phi(-2\Delta - 1) \tag{3.4}$$

for some constant $C' > 0$.

Let α and C'' satisfy $\frac{1}{2} < C'' < (1/8C^2)$ and $4C''C^2 < \alpha < \frac{1}{2}$. Thus from (3.3) and (3.4) we have:

$$\sigma_N^2 \Bigg/ \left(N^{-\alpha} \sum_{i=1}^N c_i^2 \right) > C' N^{\alpha - 4C''C^2}.$$

The last relation clearly implies the satisfaction of (1.13). The result follows by an application of Theorem 2.2.

References

Dupač, V. (1970). *A Contribution to the Asymptotic Normality of Simple Linear Rank Statistics. Nonparametric Techniques in Statistical Inference*. Cambridge Univ. Press, London, pp. 75–88.

Dupač, V. and Hájek, J. (1969). Asymptotic normality of simple linear rank statistics under alternatives II. *Ann. Math. Statist.* **40**, 1992, 2017.

Hájek, J. (1968). Asymptotic normality of simple linear rank statistics under alternatives. *Ann. Math. Statist.* **39**, 325–346.

Husková, M. (1970). Asymptotic distribution of simple linear rank statistics for testing symmetry. *Z. Wahrscheinlichkeitstheorie verw. Geb.* **14**, 308–322.

Koul, H.L. and Staudte, R.G. (1972). Asymptotic normality of signed rank statistics. *Z Wahrscheinlichkeitstheorie verw. Geb.* **22**, 295–300.

Puri, M.L. and Ralescu, S.S. (1980). Asymptotic normality of signed rank statistic with discontinuous score generating function. Preprint.

Reprinted from
Statistics and Probability: Essays in Honor of C.R. Rao,
dited by G. Kallianpur, P.R. Krishnaiah and J.K. Ghosh,
by permission of M.L. Puri's co-author

ON THE ORDER OF MAGNITUDE OF CUMULANTS OF VON MISES FUNCTIONALS AND RELATED STATISTICS

R. N. BHATTACHARYA[1] and M. L. PURI[2]
Indiana University

It is shown that under appropriate conditions the sth cumulant of a von Mises statistic or a U (or V) statistic is $O(n^{-s+1})$, $s \geq 2$, as the sample size n goes to infinity. A possible route toward the derivation of an asymptotic expansion of the characteristic function is indicated.

1. Introduction. The Edgeworth expansion of the characteristic function of a normalized sum of n independent and identically distributed (i.i.d) random variables derives from the order of magnitude $O(n^{-(s-2)/2})$ of the sth *cumulant* ($s \geq 2$) (See, e.g., Bhattacharya, 1977). For statistics which may be expressed as or approximated by polynomials in several average sample characteristics (e.g., (i) polynomials in sample moments and (ii) maximum likelihood estimators in the regular case), the validity of the so-called "formal Edgeworth expansion" depends crucially on the above order of magnitude of the sth cumulant ($s \geq 2$) of the normalized statistic (see Bhattacharya and Ghosh, 1978). In this note it is shown that cumulants of normalized *U-statistics* and *von Mises functionals* have the above order of magnitude, if certain conditions are satisfied. For general background on these statistics we refer to von Mises (1947) and Serfling (1980). Assuming the validity of (a) the above order of magnitude of the cumulants and (b) the Edgeworth expansion of the distribution function of a von Mises functional, Withers (1980) has given an algorithm for computing the coefficients in the asymptotic expansion. Some of the moment computations in Section 2 are similar to those in Withers (loc. cit). In Section 3 a new method of derivation of Cramér-Edgeworth expansions of characteristic functions of a class of statistics is provided.

2. Moments and cumulants. Let χ be a separable metric space (e.g., a subset of R^d), $\mathscr{B}_\chi$ its Borel sigma field, and P a given probability measure on $\mathscr{B}_\chi$, whose *support* is S. Let $\mathscr{P}_f$ denote the set of all probability measures on $\mathscr{B}_\chi \cap S$ having *finite supports*. Endow $\mathscr{P}_f \cup \{P\}$ with the weak-star topology. Consider for each n the product space $(\chi^n, \mathscr{B}_{\chi^n})$, and let $X_1, \cdots, X_n$ be the n coordinate random variables. Let $G^{\otimes n} = G \times G \times \cdots \times G$ denote the product probability measure on $\mathscr{B}_{\chi^n}$, where G is a probability measure on $\mathscr{B}_\chi$. Under $G^{\otimes n}$ the random variables $X_1, \cdots, X_n$ are i.i.d. with common distribution G. We shall write E_G to denote expectation under $G^{\otimes n}$. Denote the *empirical distribution* of the "observations" $X_1, \cdots, X_n$ by F_n, i.e., $F_n = n^{-1}\sum_{i=1}^n \delta_{X_i}$, where δ_x is the *Dirac measure* with point mass at x.

Let $h(x_1, x_2, \cdots, x_r)$ be a real-valued, Borel measurable, symmetric function on χ^r, for some $r \geq 2$. Define the *V-statistic* (with *kernel h*)

$$V_n = n^{-r} \sum_{i_1=1}^n \cdots \sum_{i_r=1}^n h(X_{i_1}, X_{i_2}, \cdots, X_{i_r}), \tag{2.1}$$

and the *U-statistic* (with *kernel h*)

$$U_n = \binom{n}{r}^{-1} \sum h(X_{i_1}, X_{i_2}, \cdots, X_{i_r}) \tag{2.2}$$

where the summation is over $1 \leq i_1 < i_2 < \cdots < i_r \leq n$.

Received February 1982; revised August 1982.
[1] Research supported by NSF Grant MCS 8201628.
[2] Research supported by Office of Naval Research under Contract N00014-77-C-0659.
AMS 1970 *subject classifications.* Primary 62E20; secondary 62G05, 62G10.
Key words and phrases. V-Statistics, U-statistics, Edgeworth expansion.

Theorem 2.1. (a) *If for some integer $s \geq 3$ one has*

$$E_P|h(X_{j_1}, X_{j_2}, \cdots, X_{j_r})|^s < \infty \tag{2.3}$$

for all choices of $j_1, j_2, \cdots, j_r (1 \leq j_1, j_2, \cdots, j_r \leq r)$, then the pth *cumulant $k_{p,n}(P)$ of V_n under P is of the form*

$$k_{p,n}(P) = \textstyle\sum_{m=p-1}^{s-1} n^{-m}\lambda_{m,p}(P) + o(n^{-s+1}), \qquad (2 \leq p \leq s). \tag{2.4}$$

The quantities $\lambda_{m,p}(P)$ are independent of n.

(b) *Suppose that, for some integer $s \geq 3$ one has*

$$E_P|h(X_1, X_2, \cdots, X_r)|^s < \infty. \tag{2.5}$$

Then the cumulants of the statistic U_n also are of the form (2.4).

Proof. (a) Write

$$V_n = \int \cdots \int h(x_1, \cdots, x_r)F_n(dx_1) \cdots F_n(dx_r). \tag{2.6}$$

For $G = \sum_{i=1}^{q} \alpha_i\delta_{y_i}$ in $\mathscr{P}_f$, F_n may be expressed as $\sum_{i=1}^{q} \hat{\alpha}_i\delta_{y_i}$ (with $G^{\otimes n}$-probability one), where $\hat{\alpha}_i$ is the proportion of y_i's in the "sample" $\{X_1, \cdots, X_n\}$. Thus V_n becomes a polynomial in the q-variables $\hat{\alpha}_i$, $1 \leq i \leq q$. Hence by a result of James and Mayne (1962) (this may also be derived from the results of Leonov and Shiryaev, 1959), the pth cumulant of V_n under G is of the form

$$k_{p,n}(G) = \textstyle\sum_{m=p-1}^{rp-1} n^{-m}\lambda_{m,p}(G), \qquad (p \geq 2). \tag{2.7}$$

On the other hand, for all G such that $E_G|V_n|^p < \infty$, one has (for all $n > rp$)

$$\begin{aligned} E_G V_n^p &= n^{-rp}E_G \int \cdots \int (\textstyle\prod_{t=1}^{p} h(x_{r(t-1)+1}, \cdots, x_{rt})) \\ &\quad \cdot (\delta_{X_1} + \delta_{X_2} + \cdots + \delta_{X_n})(dx_1) \cdots (\delta_{X_1} + \delta_{X_2} + \cdots + \delta_{X_n})(dx_{rp}) \\ &= n^{-rp} \int \cdots \int (\textstyle\prod_{t=1}^{p} h(x_{r(t-1)+1}, \cdots, x_{rt})) \\ &\quad \Big[\textstyle\sum_{m=1}^{rp} \dfrac{n!}{(n-m)!} \sum_2 \sum_1 E_G\{\delta_{X_1}(dx_{j_{11}})\delta_{X_1}(dx_{j_{12}}) \cdots \delta_{X_1}(dx_{j_{1s_1}}) \\ &\quad \delta_{X_2}(dx_{j_{21}}) \cdots \delta_{X_2}(dx_{j_{2s_2}}) \cdots \delta_{X_m}(dx_{j_{m1}}) \cdots \delta_{X_m}(dx_{j_{ms_m}})\}\Big]. \end{aligned} \tag{2.8}$$

Here, for a given m, $\sum_2$ denotes summation over all collections of m positive integers $\{s_1, s_2, \cdots, s_m\}$ satisfying $\sum s_i = rp$; for a given collection $\{s_1, s_2, \cdots, s_m\}$, $\sum_1$ denotes summation over all partitions of $\{1, 2, \cdots, rp\}$ into m groups of $s_1, s_2, \cdots, s_m$ elements, a typical partition being $(\{j_{11}, j_{12}, \cdots, j_{1s_1}\}, \{j_{21}, j_{22}, \cdots, j_{2s_2}\}, \{j_{m1}, j_{m2}, \cdots, j_{ms_m}\})$. Denote by $H_s(G)$ the distribution of the s-dimensional random vector $(X_1, X_1, \cdots, X_1)$ under G, and let $H_{s_1,s_2,\cdots,s_m}(G)$ stand for the measure

$$\begin{aligned} H_{s_1,s_2,\cdots,s_m}(G)(dx_1dx_2 \cdots dx_{rp}) &= \textstyle\sum_1 H_{s_1}(G)(dx_{j_{11}}dx_{j_{12}} \cdots dx_{j_{1s_1}})H_{s_2}(G) \\ &\quad \cdot (dx_{j_{21}}dx_{j_{22}} \cdots dx_{j_{2s_2}}) \cdots H_{s_m}(dx_{j_{m1}}dx_{j_{m2}} \cdots dx_{j_{ms_m}}). \end{aligned} \tag{2.9}$$

Also note that

$$\begin{aligned} n^{-rp}\frac{n!}{(n-m)!} &= n^{-rp}n(n-1) \cdots (n-m+1) \\ &= \textstyle\sum_{m'=1}^{m} (-1)^{m-m'} n^{-rp+m'}\theta(m-m'; m-1), \end{aligned} \tag{2.10}$$

where $\theta(i; N)$ is the sum of all products of i distinct integers taken from $\{1, 2, \cdots, N\}$,

$\theta(0; N) = 1$. From (2.8)–(2.10) one obtains

$$E_G V_n^p = \sum_{m=1}^{rp} \sum_{m'=1}^{m} (-1)^{m-m'} n^{-rp+m'} \theta(m-m'; m-1)$$

$$\cdot \int \cdots \int (\prod_{t=1}^{p} h(x_{r(t-1)+1}, \cdots, x_{rt})) \sum_2 H_{s_1, s_2, \cdots, s_m}(G)(dx_1 \cdots dx_{rp})$$

(2.11) $$= \sum_{j=0}^{rp-1} n^{-j} \{ \sum_{m=rp-j}^{rp} (-1)^{m-rp+j} \theta(m-rp+j; m-1)$$

$$\cdot \int \cdots \int (\prod_{t=1}^{p} h(x_{r(t-1)+1}, \cdots, x_{rt})) \sum_2 H_{s_1, s_2, \cdots, s_m}(G)(dx_1 \cdots dx_{rp}) \}$$

$$= \sum_{j=0}^{rp-1} n^{-j} \mu_{j,p}(G), \qquad (1 \le p \le s),$$

say. Here $\mu_{j,p}(G)$ is a linear combination (with coefficients not depending on n, G or h) of terms like

(2.12) $$\int \cdots \int (\prod_{t=1}^{p} h(x_{r(t-1)+1}, \cdots, x_{rt})) H_{s_1, s_2, \cdots, s_m}(G)(dx_1 \cdots dx_{rp}).$$

Using the familiar relations between moments and cumulants one has

(2.13) $$k_{p,n}(G) = \sum_{j=0}^{rp-1} n^{-j} \bar{\lambda}_{j,p}(G),$$

where $\bar{\lambda}_{j,p}(G)$ is a polynomial in $\mu_{j,p'} (1 \le p' \le p)$, whose coefficients are absolute constants. Since the map $G \to H_{s_1}(G)$ is continuous in the weak-star topology, so is the map $G \to H_{s_1, s_2, \cdots, s_m}(G)$. It follows that for a bounded continuous h the integral (2.12) is a weak-star continuous function of G; this implies that the maps $G \to \mu_{j,p}$ and, therefore, $G \to \bar{\lambda}_{m,p}(G)$ are continuous. If $p \ge 2$, then $\bar{\lambda}_{m,p}(G) = 0$ for $2 \le m < p-1$ and $G \in \mathscr{P}_f$. Also there exists $G_N \in \mathscr{P}_f (N = 1, 2, \cdots)$ such that G_N converges to P (This is where the separability of χ is made use of; see, e.g., Parthasarathy (1967), Theorem 6.3). Therefore, one must have $\bar{\lambda}_{m,p}(P) = 0$ for $1 \le m < p-1$. This completes the proof of (a) for bounded continuous h. Since functions of the form $\prod_{t=1}^{p} h(x_{r(t-1)+1}, \cdots, x_{rt})$ belonging to $L^1(\chi^{rp}, H_{s_1, s_2, \cdots, s_m}(P))$ may be approximated (in L^1) by continuous bounded functions of the same form, the proof is complete. Note that for this last argument (2.3) is needed.

(b) First assume (i) $h(x_1, x_2, \cdots, x_r) = 0$ *if* $x_i = x_j$ *for some* $i, j (i \ne j)$. Then the cumulants of U_n satisfy (2.4), since

$$U_n = \left(\frac{n!}{(n-r)!} \right)^{-1} n^r V_n = \left(\left(1 - \frac{1}{n}\right)\left(1 - \frac{2}{n}\right) \cdots \left(1 - \frac{r-1}{n}\right) \right)^{-1} V_n = (1 + o(1)) V_n.$$

Next, instead of (i) assume (ii) *P has no atoms*. Then modify h so as to satisfy (i); this does not change U_n, except on a set of probability zero. Finally, consider an arbitrary P. Let D be its set of atoms. Let D' be a subset of reals in one-one correspondence with D. Consider the space $\chi' = (\chi \backslash D) \cup R$, with $\chi \backslash D$ and R each carrying its own topology but their union is topologically disconnected. Then P lifted to this space χ' (by placing the discrete mass on D') is a weak-star limit of nonatomic probability measures. Extend h to $(\chi')^r$ by setting it zero if any coordinate is in $R \backslash D'$. Now apply an argument entirely analogous to that in the preceding paragraph.

Remark 2.1.1. The U_n and V_n defined above are not centered around their expectations (under P). Centering has been avoided deliberately to ensure that h does not depend on P. For the general von Mises functional considered below centering seems unavoidable; this causes some technical problems.

Remark 2.1.2. Under the hypotheses of Theorem 2.1 the pth cumulants of the normalized statistics $\sqrt{n}(V_n - EV_n)$, $\sqrt{n}(U_n - EU_n)$ are of the order $O(n^{-(p-2)/2})$, $2 \le p \le s$.

Let T be a *von Mises functional* defined on $\mathscr{P}_f \cup \{P\}$, and let the statistic $T(F_n)$ have the expansion

$$T(F_n) - T(G) = \sum_{i=1}^{r} \int \cdots \int T^{(i)}(G; x_1, x_2, \cdots, x_i) \prod_{j=1}^{i} (F_n - G)(dx_j) + R_n \tag{2.14}$$
$$= \sum_{i=1}^{r} V_{n,i}(G) + R_n \qquad (G \in \mathscr{P}_f \cup \{P\}),$$

where $T^{(i)}$ is a real-valued, symmetric (in the arguments $x_1, \cdots, x_i$), Borel measurable function on $\mathscr{P}_f \cup \{P\} \times \chi^i$ satisfying

$$E_P|T^{(i)}(P; X_{j_1}, X_{j_2}, \cdots, X_{j_i})|^s < \infty, \qquad (1 \leq i \leq r), \tag{2.15}$$

for all $1 \leq j_1, j_2, \cdots, j_i \leq r$, and the "remainder term" R_n satisfies

$$E_P|R_n|^p = o(n^{-s+1}), \qquad (1 \leq p \leq s). \tag{2.16}$$

Write

$$V_n(G) = \sum_{i=1}^{r} V_{n,i}(G). \tag{2.17}$$

Then

$$E_G V_n^p(G) = \sum\nolimits_3 V_{n,i_1}(G) V_{n,i_2}(G) \cdots V_{n,i_p}(G), \tag{2.18}$$

where $\sum_3$ denotes summation over all p-tuples $(i_1, i_2, \cdots, i_p)$ such that $1 \leq i_1, \cdots, i_p \leq r$. Now let $I_p = i_1 + \cdots i_p$ and write, as in (2.8),

$$E_G V_n^p(G) = \sum\nolimits_3 E_G \int \cdots \int \left(\prod_{t=1}^{p} T^{(i_t)}(G; x_{I_{t-1}+1}, \cdots, x_{I_t})\right)$$
$$\cdot \left[n^{-I_p} \sum_{m=1}^{[I_p/2]} \frac{n!}{(n-m)!} \sum\nolimits_2' \sum\nolimits_1' (\delta_{X_1} - G)(dx_{j_{11}})(\delta_{X_1} - G)(dx_{j_{12}}) \right. \tag{2.19}$$
$$\left. \cdots (\delta_{X_1} - G)(dx_{j_{1s_1}}) \cdots (\delta_{X_m} - G)(dx_{j_{m1}}) \cdots (\delta_{X_m} - G)(dx_{j_{ms_m}}) \right].$$

Here, for a given m, $\sum_2'$ denotes summation over all collections of m integers $\{s_1, s_2, \cdots, s_m\}$ satisfying $s_i \geq 2$ and $\sum s_i = I_p$; and $\sum_1'$ denotes, for each collection $\{s_1, s_2, \cdots, s_m\}$, summation over all partitions of $\{1, 2, \cdots, I_p\}$ into m subgroups of $s_1, s_2, \cdots, s_m$ elements such as $(\{j_{11}, j_{12}, \cdots, j_{1s_1}\}, \cdots, \{j_{m1}, j_{m2}, \cdots, j_{ms_m}\})$. Note that expectations of terms involving $s_i = 1$ for some i vanish. Next let $H_q(i_1, i_2, \cdots, i_\ell; G)$ denote the distribution of a q-dimensional random vector whose i_1th, $\cdots$, i_ℓth coordinates are X_1, while the remaining coordinates are i.i.d with distribution G and independent of X_1. Write $\tilde{H}_q(G)$ for the *signed* measure

$$\tilde{H}_q(G) = \sum_{\ell=0}^{q} (-1)^{q-\ell} \sum\nolimits_4 H_q(i_1, \cdots, i_\ell; G), \tag{2.20}$$

where $\sum_4$ denotes summation over all choices $\{i_1, i_2, \cdots, i_\ell\}$ of ℓ distinct integers from $\{1, 2, \cdots, q\}$. Now define

$$\tilde{H}_{s_1, s_2, \cdots, s_m}(G; dx_1 dx_2 \cdots dx_{I_p}) \tag{2.21}$$
$$= \sum\nolimits_1' \tilde{H}_{s_1}(G)(dx_{j_{11}} dx_{j_{12}} \cdots dx_{j_{1s_1}}) \cdots \tilde{H}_{s_m}(G)(dx_{j_{m1}} dx_{j_{m2}} \cdots dx_{j_{ms_m}}).$$

Then, as in (2.10),

$$E_G V_n^p(G) = \sum\nolimits_3 \left[\sum_{j=\left[\frac{I_p+1}{2}\right]}^{I_p-1} n^{-j} \left\{ \sum_{m=I_p-j}^{[I_p/2]} (-1)^{m-I_p+j} \theta(m - I_p + j; m-1) \right. \right. \tag{2.22}$$
$$\left. \left. \cdot \int \cdots \int \left(\prod_{t=1}^{p} T^{(i_t)}(G; x_{I_{t-1}+1}, \cdots, x_{I_t})\right) \sum\nolimits_2' \tilde{H}_{s_1, \cdots, s_m}(G; dx_1 \cdots dx_{I_p}) \right\} \right].$$

For $G = \sum_{i=1}^{q} \alpha_i \delta_{y_i}$, $V_n(G)$ is a polynomial in $\hat{\alpha}_i - \alpha_i$, so that the pth cumulant of $V_n(G)$ is of the order $O(n^{-p+1})$ under $G(2 \leq p \leq s)$. In view of (2.16) and (2.22), the proof of the following theorem is now complete.

THEOREM 2.2 *Suppose that* (2.14)–(2.16) *hold. Assume, in addition, that there exists a sequence* $\{G_N : N \geq 1\}$ *having finite support such that*

$$\lim_{N\to\infty} E_{G_N}\left(\prod_{t=1}^{p} T^{(i_t)}(G_N; X_{t_1}, \cdots, X_{t_{i_t}})\right) = E_P\left(\prod_{t=1}^{p} T^{(i_t)}(P; X_{t_1}, \cdots, X_{t_{i_t}})\right) \tag{2.23}$$

for all $1 \leq i_1, i_2, \cdots, i_p \leq r$, *and all* $1 \leq t_1, t_2, \cdots, t_{i_t} \leq rp$ $(1 \leq t \leq p)$. *Then the* pth *cumulant of* $T(F_n)$ *under* P *is of the order* $O(n^{-p+1})$ *for* $2 \leq p \leq s$.

REMARK 2.2.1. Notice that the statement "condition (2.23) holds for *some* $\{G_N;\ N \geq 1\} \subset \mathscr{P}_f$" is much weaker than the statement "condition (2.23) holds for all sequences $\{G_N\colon N \geq 1\}$ converging to P (weak-star)", the latter being equivalent to saying that the integral is weak-star continuous at P (on $\mathscr{P}_f \cup \{P\}$). To illustrate this point, note that even such functionals as $T(G) = \int x^k G(dx)$, $k \geq 1$, are *not* weak-star continuous on $\mathscr{P}_f \cup \{P\}$, where P is a probability measure on the line having a finite kth moment. The difficulty is that one may place a mass $O(N^{-k/2})$ at $x = N$ which goes to zero to ensure weak-star convergence, but is large enough to blow up the integral as $N \to \infty$. On the other hand, one may integrate (with respect to P) a step-function approximation, $f_N(x)$ to x^k, which amounts to integrating x^k with respect to an appropriate $G_N \in \mathscr{P}_f$; and the latter integral $\int x^k G_N(dx)$ will converge to $\int x^k P(dx)$, as the intervals of constancy decrease to zero in width. These considerations apply to more general functions (see, Serfling (1980), pages 214–216, for examples).

REMARK 2.2.2. The fact that the sth cumulant of V_n (or T_n) is $O(n^{-s+1})$ when G has finite support means the vanishing of a number of polynomials in the variables $\mu_p(G)$. One should be able to prove that these polynomials are identically zero by showing that the $\mu_p(G)$'s assume a broad enough spectrum of values as G ranges over the set of all probability measures having finite support. This would enable one to dispense with the condition (2.23) in Theorem 2.2. However, we are unable to make this algebraic argument firm.

Finally, the method used here should be more widely applicable in deriving orders of magnitudes of cumulants.

3. A method of derivation of Edgeworth expansions of characteristic functions, and an unsolved problem. In the present section we provide a method (which appears to be new) for the derivation of Cramér-Edgeworth expansions of characteristic functions of a class of statistics T_n having zero means, finite moment generating functions (m.g.f.'s), and cumulants $\chi_{p,n}$ satisfying

$$\chi_{p,n} = n^{-(p-2)/2}\lambda_p + o(n^{-(p-2)/2}), \qquad (p \geq 2), \qquad \lambda_2 > 0. \tag{3.1}$$

Let

$$f_n(\xi) = E\exp\{i\xi T_n\} \tag{3.2}$$

denote the characteristic function of T_n. One may write

$$f_n(\xi) = f(i\xi, \varepsilon), \tag{3.3}$$

with $\varepsilon = n^{-1/2}$. Under the additional assumption that $f(i\xi, \varepsilon)$ has an absolutely convergent power series expansion in ξ and ε in a neighborhood of the origin (0, 0), it is shown in Theorem 3.1 that $f_n(\xi)$ and its derivatives have a proper asymptotic expansion of the Cramér-Edgeworth type. The *unsolved problem* is to identify a large enough class of von Mises functionals for which this analyticity holds. In particular, we do not know if the analyticity property holds for U-statistics (see (2.2)) with kernels h satisfying:

$$E\exp\{th(X_1, X_2, \cdots, X_r)\} < \infty, \qquad (-\infty < t < \infty). \tag{3.4}$$

In remarks following the corollaries to Theorem 3.1 it is shown that the assumption of analyticity does hold for some special classes. We expect the moment computations of Section 2 to be crucial in resolving the problem of analyticity in the general case.

THEOREM 3.1. *Let $T_n (n = 1, 2, \cdots)$ be a sequence of random variables having zero means. Assume that* (i) $E \exp\{tT_n\} < \infty$ *for all* $t(-\infty < t < \infty)$ *and* n, (ii) $f(i\xi, \varepsilon)$ *can be extended as an analytic function* $f(z, \eta)$ *of the complex variables* z *and* η *in a neighborhood of the origin* (0, 0) *in* $\mathbb{C}^2$, *and* (iii) *the cumulants* $\chi_{p,n}$ *of* T_n *satisfy* (3.1). *Then the following results hold:*

(a) *There exist a positive constant* δ_0 *and polynomials* P_j, *whose coefficients do not depend on* n, *such that for all* ξ, $-\delta_0\sqrt{n} < \xi < \delta_0\sqrt{n}$, *one has*

$$f_n(\xi) = \exp\left\{-\frac{\lambda_2}{2}\xi^2\right\}(1 + \textstyle\sum_{j=1}^{\infty} n^{-j/2}P_j(i\xi)).$$

(b) *For every pair of integers* m *and* p *satisfying* $p \geq 2$, $0 \leq m \leq p$, *there exist positive constants* δ_0, c_1, c_2 *such that*

$$\left|\frac{d^m}{d\xi^m}\left[f_n(\xi) - \exp\left\{-\frac{\lambda_2}{2}\xi^2\right\}(1 + \textstyle\sum_{j=1}^{p-2} n^{-j/2}P_j(i\xi))\right]\right|$$

$$\leq \frac{c_1}{n^{(p-1)/2}}[|\xi|^{p+1-m} + |\xi|^{3(p-1)+m}]\exp\{-c_2\xi^2\}, \qquad (|\xi| < \delta_0\sqrt{n}).$$

PROOF. Since $f(z, \eta)$ is analytic in a neighborhood of (0, 0), and $f(0, 0) = 1$, $\phi(z, \eta) = \log f(z, \eta)$ (we take the principal branch of the logarithm) is defined and analytic in a neighborhood of (0, 0). In view of (3.1) and the fact that $ET_n = 0$, one may express $\phi(z, \eta)$ as

(3.5)
$$\begin{aligned}\phi(z, \eta) &= \frac{z^2}{2!}(\lambda_2 + \textstyle\sum_{j=1}^{\infty}\lambda_{2,j}\eta^j) + \cdots + \frac{z^k}{k!}\eta^{k-2}(\sum_{j=0}^{\infty}\lambda_{k,j}\eta^j) + \cdots\\ &= z^2\left[\textstyle\sum_{k=2}^{\infty}\frac{(\eta z)^{k-2}}{k!}(\sum_{j=0}^{\infty}\lambda_{k,j}\eta^j)\right], \qquad (\lambda_{k,0} = \lambda_k).\end{aligned}$$

Since this last series is absolutely convergent in a neighborhood of (0, 0), so is the series within square brackets. Let δ_1, δ_2 be two positive numbers such that this last series is absolutely convergent for $|z| = \delta_1$, $|\eta| = \delta_2$. Then

(3.6)
$$\textstyle\sum_{k=2}^{\infty}\frac{(\delta_1\delta_2)^{k-2}}{k!}\sum_{j=0}^{\infty}|\lambda_{k,j}|\delta_2^j < \infty.$$

It follows that (3.5) is absolutely convergent for $|z\eta| \leq \delta_1\delta_2$ and $|\eta| \leq \delta_2$. Therefore, the last expression in (3.5) defines an analytic function in the region $D = \{(z, \eta) \in \mathbb{C}^2\colon |z| < \delta_1\delta_2/|\eta|, |\eta| < \delta_2\}$, and over this region $\exp\{\phi(z, \eta)\}$ defines an analytic continuation of $f(z, \eta)$. We shall refer to this extension also by $f(z, \eta)$. Since the characteristic function $\xi \to f_n(\xi)$ is entire (by assumption (i)) and since analytic continuations are unique, $f_n(\xi) = f(i\xi, n^{-1/2})$ for $-\infty < \xi < \infty$ (note that one could not assume a priori that this equality holds between f_n and the analytically extended f). In addition, on D one has

(3.7)
$$|f(z, \eta) - 1| < c' < 1,$$

for some constant c', and $\phi(z, \eta)$ is the principal branch of the logarithm of $f(z, \eta)$ on D. The relations (3.5) now hold on D and one may rewrite the first relation in (3.5) as

(3.8)
$$\log f(z, \eta) - \frac{\lambda_2}{2!}z^2 = \textstyle\sum_{j=1}^{\infty}\eta^j Q_j(z), \qquad (z, \eta) \in D,$$

where Q_j is a polynomial of degree $j + 2$. Thus

(3.9) $$f(z, \eta) \exp\left\{-\frac{\lambda_2}{2} z^2\right\} = \exp\left\{\sum_{j=1}^{\infty} \eta^j Q_j(z)\right\} = 1 + \sum_{j=1}^{\infty} \eta^j P_j(z), \qquad (z, \eta) \in D,$$

where P_j's are appropriate polynomials. From (3.9) one gets

(3.10) $$f(z, \eta) = \exp\left\{\frac{\lambda_2}{2} z^2\right\} (1 + \sum_{j=1}^{\infty} \eta^j P_j(z)), \qquad (z, \eta) \in D,$$

and, in particular (with $z = i\xi$, $\eta = n^{-1/2}$),

(3.11) $$f_n(\xi) = \exp\left\{-\frac{\lambda_2}{2} \xi^2\right\} (1 + \sum_{j=1}^{\infty} n^{-j/2} P_j(i\xi)), \qquad (-\delta_1\delta_2 \sqrt{n} < \xi < \delta_1\delta_2 \sqrt{n}).$$

This proves part (a). To prove part (b) one may first approximate $\log f(z, \eta)$ by

(3.12) $$\phi_p(z, \eta) = z^2 \sum_{k=2}^{p+2} \frac{(\eta z)^{k-2}}{k!} (\sum_{j=0}^{\infty} \lambda_{k,j} \eta^j).$$

Writing

(3.13) $$\psi(z, \eta) = \phi(z, \eta) - \frac{\lambda_2}{2} z^2, \qquad \psi_p(z, \eta) = \phi_p(z, \eta) - \frac{\lambda_2}{2} z^2,$$

one has (using (3.6), or analyticity on D)

(3.14) $$|\phi(z, \eta) - \phi_p(z, \eta)| = |\psi(z, \eta) - \psi_p(z, \eta)| \leq c_3 |\eta|^{p-1} |z|^{p+1}, \qquad (z, \eta) \in D,$$

for an appropriate constant c_3. By (3.6) and (3.14), if δ_1 is small, then

(3.15) $$|\exp\{\psi(z, \eta)\} - \exp\{\psi_p(z, \eta)\}| \leq c_4 |\eta|^{p-1} |z|^{p+1} \exp\left\{\frac{\lambda_2 |z|^2}{4}\right\},$$

for some constant c_4; this may be written as

(3.16) $$\left| e^{-\frac{\lambda_2}{2} z^2} [f(z, \eta) - \exp\{\phi_p(z, \eta)\}] \right| \leq c_4 |\eta|^{p-1} |z|^{p+1} \exp\left\{\frac{\lambda_2 |z|^2}{4}\right\}.$$

Letting $z = i\xi$, $\eta = n^{-1/2}$, (3.16) becomes

(3.17) $$|f_n(\xi) - \exp\{\phi_p(i\xi, n^{-1/2})\}| \leq c_4 n^{-(p-1)/2} |\xi|^{p+1} \exp\left\{-\frac{\lambda_2}{4} \xi^2\right\}, \qquad (|\xi| < \delta_1\delta_2 \sqrt{n}).$$

The comparison of $\exp\{\phi_p(i\xi, n^{-1/2})\}$ with $\exp\left\{-\frac{\lambda_2}{2} \xi^2\right\} (1 + \sum_{j=1}^{p-2} n^{-j/2} P_j(i\xi))$ is carried out exactly as in Lemmas 9.7, 9.8 in Bhattacharya and Ranga Rao (1976). □

COROLLARY 3.1.1. *Under the hypothesis of Theorem* 3.1 *one has the Berry-Esseen bound*

(3.18) $$\sup_x |P(T_n \leq x) - \Phi_{\lambda_2}(x)| \leq c n^{-1/2},$$

for some constant $c > 0$. *Here* Φ_{λ_2} *is the normal distribution function with mean zero and variance* λ_2.

PROOF. Use Theorem 3.1 (b) and Esseen's inequality (see Lemmas 12.1, 12.2 in Bhattacharya and Ranga Rao (1976)). □

COROLLARY 3.1.2. *Assume the hypothesis of Theorem* 3.1. *If, for some* $p \geq 2$, g *is a* p*-times continuously differentiable function on* $\mathbb{R}^1$ *such that* $\sup\{(1 + |x|^p)|g^{(m)}(x)|:$

$x \in I\!R^1\} < \infty$ *for* $0 \leq m \leq p$, *then*

$$(3.19) \qquad |Eg(T_n) - \int_{I\!R^1} g(x) \left[1 + \sum_{j=1}^{p-2} n^{-j/2} P_j \left(-\frac{d}{dx} \right) \right] \phi_{\lambda_2}(x) dx | \leq dn^{-(p-1)/2}$$

for some positive constant d.

PROOF. One may apply the method of Götze and Hipp (1978) to the estimate in Theorem 3.1 (b) to derive (3.19) directly. Alternatively, first establish (3.19) for the class of all Schwartz functions as in Bhattacharya and Ranga Rao (1976), Theorem 20.7, expressing the error estimate in terms of a Sobolev norm; then extend the result to a wider class by completion in the Sobolev norm. □

REMARK 3.1.3. Let $X_1, X_2, \cdots$ be an i.i.d. sequence having mean zero and a positive variance. The hypothesis of Theorem 3.1 is satisfied for the statistics $T_n = n^{-1/2}(X_1 + \cdots + X_n)$ if the m.g.f. of X_1 is finite everywhere. Of course, in this classical case Theorem 3.1 (b) holds under less stringent assumptions (see, e.g., Bhattacharya and Ranga Rao (1976), Chapter 2). Note, however, the conclusion of part (a) of Theorem 3.1 requires stronger assumptions than finiteness of moments.

REMARK 3.1.4 Let U_n be a U-statistic with kernel h (see (2.2)). Assume, without loss of generality, that $Eh(X_1, X_2, \cdots, X_r) = 0$. If $E \exp\{th(X_1, \cdots, X_r)\} < \infty$ for all t, $-\infty < t < \infty$, then hypothesis (i) of Theorem 3.1 is satisfied for the statistic $T_n = \sqrt{n}\ U_n$ (see Serfling (1980), Lemma C, page 200). In addition, assume $E\phi^2(X_1) = \lambda_2 > 0$, where $\phi(x) = Eh(x, X_2, \cdots, X_r)$. Then T_n is asymptotically normal (see Serfling (1980), Theorem A, page 192) and, by Theorem 2.1 (b), hypothesis (iii) of Theorem 3.1 also holds. It would be of great interest to see if hypothesis (ii) of Theorem 3.1 is a consequence of the above assumptions. We emphasize that this is the main unresolved problem in the context of the present article. For kernels h which are sums of products of functions of single variables, analyticity of $f(z, \eta)$ in a neighborhood of the origin in $\mathbb{C}^2$ has been proved by methods of statistical mechanics (see, e.g., Ruelle, 1969). However, for these special kernels an adequate theory of Edgeworth expansions has been derived in Bhattacharya and Ghosh (1978) under less stringent assumptions.

REMARK 3.1.5. Some partial expansions of characteristic functions of U-statistics have been obtained by Callaert, Janssen and Veraverbeke (1980).

Acknowledgement. The authors wish to thank the referee for suggesting that some terse comments in an earlier draft be expanded; the present Section 3 constitutes the resulting expansion. Thanks are also due to the Associate Editor for indicating that asymptotic expansions for von Mises functionals are of potential use in robust estimation. Finally, we are indebted to Charles M. Newman for pointing out that the analyticity hypothesis in Theorem 3.1 has been verified in the context of statistical mechanics for the special class of kernels mentioned at the end of Remark 3.1.4.

REFERENCES

BHATTACHARYA, R. N. (1977). Refinements of the multidimensional central limit theorem and applications. *Ann. Probab.* **5** 1–27.

BHATTACHARYA, R. N. and GHOSH, J. K. (1978). On the validity of the formal Edgeworth expansion. *Ann. Statist.* **6** 434–451.

BHATTACHARYA, R. N. and RANGA RAO, R. (1976). *Normal Approximation and Asymptotic Expansions*. Wiley, New York.

CALLAERT, H., JANSSEN, P. and VERAVERBEKE, N. (1980). An Edgeworth expansion for *U*-statistics. *Ann. Probab.* **8** 299–312.

GÖTZE, F. and HIPP, C. (1978). Asymptotic expansions in the central limit theorem under moment conditions. *Z. Wahrsch. verw. Gebiete* **42** 67–87.

JAMES, G. S. and MAYNE ALAN, J. (1962). Cumulants of functions of random variables. *Sankhya Ser. A* **24** 47–54.

LEONOV, V. P. and SHIRYAEV, A. N. (1959). On a method of calculation of semi-invariants. *Theor. Probab. Appl.*, 4 319–329.

PARTHASARATHY, K. R. (1967). *Probability Measures on Metric Spaces.* Academic, New York.

RUELLE, D. (1969). *Statistical Mechanics: Rigorous Results.* Benjamin, Reading.

SERFLING, ROBERT J. (1980). *Approximation Theorems of Mathematical Statistics.* Wiley, New York.

VON MISES, R. (1947). On the asymptotic distribution of differentiable statistical functions. *Ann. Math. Statist.* **18** 309–348.

WITHERS, C. S. (1980). The distribution and quantiles of a regular functional of the empirical distribution, Report No. 96, D.S.I.R., Wellington, New Zealand.

DEPARTMENT OF MATHEMATICS
SWAIN HALL EAST
INDIANA UNIVERSITY
BLOOMINGTON, INDIANA 47405

ON BERRY–ESSÉEN RATES, A LAW OF THE ITERATED LOGARITHM AND AN INVARIANCE PRINCIPLE FOR THE PROPORTION OF THE SAMPLE BELOW THE SAMPLE MEAN*

STEFAN RALESCU and MADAN L. PURI

Brown University and Indiana University Communicates by M. Rosenblatt

Let $F_n(x)$ be the empirical distribution function based on n independent random variables $X_1, ..., X_n$ from a common distribution function $F(x)$, and let $\bar{X} = \sum_{i=1}^{n} X_i/n$ be the sample mean. We derive the rate of convergence of $F_n(\bar{X})$ to normality (for the regular as well as nonregular cases), a law of iterated logarithm, and an invariance principle for $F_n(\bar{X})$.

1. INTRODUCTION

Let $X_1, ..., X_n$ be independent real valued *rv*'s with common distribution function (*df*) $F(x)$, and let $F_n(x)$ be the corresponding empirical *df*, i.e., $nF_n(x) =$ number of $X_i \leqslant x$, $1 \leqslant i \leqslant n$. Let $\bar{X} = \sum_{i=1}^{n} X_i/n$, and consider the statistic

$$T_n = F_n(\bar{X}) \tag{1.1}$$

which represents the proportion of the sample below the sample mean. Such a statistic is often used in estimating a functional $\theta = F(\mu)$, where $\mu = EX_1$ if both F and μ are unknown or in testing the hypothesis that F is symmetric about an unknown location μ against certain classes of alternatives (see Gastwirth (1971)). The asymptotic normality of T_n was first derived by David (1962) under the assumption that F is normal. Later Ghosh (1971) derived this result under weaker assumptions that $0 < \text{Var}\, X_1 < \infty$ and F is differentiable at μ with $0 < F'(\mu) < \infty$. (See also Sarkadi, Schnell and Vincze

Received July 26, 1982; revised November 10, 1982.

AMS 1980 subject classification: Primary 60F05, 60F15; secondary 62E20.

Key words and phrases: Berry–Esséen rates, law of iterated logarithm, invariance principle, Gâteaux-differential.

* Research supported by the National Science Foundation Grant MCS-8301409.

(1962) for the connection between the limit law of T_n and the occupancy problem).

Quite often, however, one needs more precise information than the asymptotic normality can provide. On the one hand, in view of applications, one may need bounds on the rate of convergence of T_n to normality, and, on the other hand, one may be interested in deriving the rates of strong convergence of T_n to $F(\mu)$ or certain invariance principles for T_n and incorporate them in the study of the asymptotic properties of the procedures (testing and estimation etc.) based on this statistic. The present note addresses these problems. Under different assumptions on F we derive (i) the Berry–Esséen rate $O(n^{-1/2})$ for the convergence of T_n to normality in the nonregular cases (i.e., when $F'(\mu)=0$), (ii) a law of the iterated logarithm, and (iii) an invariance principle for T_n. We also obtain a bound on the rate at which T_n converges to normality in the general case when $F'(\mu)$ does not necessarily vanish.

2. The Berry–Esséen Theorem for T_n in a Nonregular Case

Here we consider the question: What happens to the asymptotic law of T_n when $F'(\mu)=0$? Note that in such a case the asymptotic variance of the modified sign test equals that of the regular sign test (Gastwirth (1971)). Ghosh's (1971) method fails when $F'(\mu)=0$, while Gastwirth (1971) provides a heuristic argument. However, it will become clear from our Lemma 4.1 that if one is just interested in the asymptotic law of T_n under the assumptions that $F'(\mu)=0$, $0<\operatorname{Var} X_1<\infty$ and $0<F(\mu)<1$, then one may derive the representation

$$F_n(\bar{X})=F_n(\mu)+R_n \tag{2.1}$$

and show that $n^{1/2}R_n \to^P 0$. The asymptotic normality of T_n then follows immediately.

To motivate our study, consider the following example (cf. Chandra (1975)). Let

$$\begin{aligned} F(x) &= 0 && \text{if} \quad x\leqslant \mu-1 \\ &= \frac{(x-\mu)^{2m+1}+1}{2} && \text{if} \quad |x-\mu|\leqslant 1 \\ &= 1 && \text{if} \quad x\geqslant \mu+1 \end{aligned} \tag{2.2}$$

where $m > 0$. Then $EX_1 = \mu$ and $F'(\mu) = 0$. Note that if m is an integer, we have $F^{(j)}(\mu) = 0$, $1 \leqslant j \leqslant 2m$, $F^{(2m+1)}(\mu) = (2m+1)!/2$, and, in fact,

$$\lim_{h \to 0} \frac{|F(\mu + h) - F(\mu)|}{|h|^{2m+1}} = \frac{1}{2}. \tag{2.3}$$

To study the rate of convergence of T_n to normality in a nonregular case (i.e., $F'(\mu) = 0$) we shall need conditions similar to (2.3) regarding the variation of F at μ.

Precisely, we assume

$$E|X_1|^3 < \infty, \qquad \sigma^2 = \operatorname{Var} X_1 > 0, \qquad 0 < F(\mu) < 1. \tag{2.4}$$

For $h \in \mathbb{R}$, define $D(h) = F(\mu + h) - F(\mu)$, and assume that

$$D(h) = O(|h|^\rho) \qquad \text{as} \quad h \to 0, \text{ for some } \rho > 2. \tag{2.5}$$

Then we have the following main theorem of this section.

THEOREM 2.1. *Under* (2.4) *and* (2.5),

$$\sup_x |P\{n^{1/2}(T_n - \theta) \leqslant x[\theta(1-\theta)]^{1/2}\} - \Phi(x)| = O(n^{-1/2}), \qquad \text{as} \quad n \to \infty, \tag{2.6}$$

where $\theta = F(\mu)$, *and* $\Phi(x)$ *is the standard normal df.*

Proof. Set $s = [\theta(1-\theta)]^{1/2}$, $G_n(x) = P\{n^{1/2}(T_n - \theta) \leqslant xs\}$ and $c_n = C(\log n)^{1/2}$, $C > 0$. Note that

$$\begin{aligned} \sup_{|x| > c_n} |G_n(x) - \Phi(x)| &\leqslant \max\{G_n(-c_n) + \Phi(-c_n), 1 - G_n(c_n) + \Phi(-c_n)\} \\ &\leqslant [G_n(-c_n) + 1 - G_n(c_n)] + [1 - \Phi(c_n)] \qquad (2.7) \\ &= P\{|F_n(\bar{X}) - F(\mu)| \geqslant sc_n n^{-1/2}\} + [1 - \Phi(c_n)]. \end{aligned}$$

Using the well known inequality $1 - \Phi(x) \leqslant \phi(x)/x$ for $x > 0$, where $\phi(x)$ denotes the density of Φ, we obtain by choosing $C^2 \geqslant 1$,

$$1 - \Phi(c_n) \leqslant n^{-C^2/2}/(2\pi)^{1/2}\, c_n = O(n^{-1/2}), \qquad \text{as} \quad n \to \infty. \tag{2.8}$$

Now, from the classical Berry–Esséen theorem, and the fact that $1 - \Phi((\log n)^{1/2}) = O(n^{-1/2})$, as $n \to \infty$, it follows that

$$r_n = P\{|\bar{X} - \mu| > \sigma n^{-1/2}(\log n)^{1/2}\} = O(n^{-1/2}), \qquad \text{as} \quad n \to \infty. \tag{2.9}$$

Now consider $P\{F_n(\bar{X}) - F(\mu) \geqslant sc_n n^{-1/2}\}$ and write it as

$$\begin{aligned} P\{F_n(\bar{X}) - F(\mu) \geqslant sc_n n^{1/2}\} \\ &= P\{F_n(\bar{X}) - F(\mu) \geqslant sc_n n^{-1/2}, |\bar{X} - \mu| \leqslant \sigma n^{-1/2}(\log n)^{1/2}\} \\ &\quad + P\{F_n(\bar{X}) - F(\mu) \geqslant sc_n n^{-1/2}, |\bar{X} - \mu| > \sigma n^{-1/2}(\log n)^{1/2}\} \\ &\leqslant P\{F_n(\mu + \sigma n^{-1/2}(\log n)^{1/2}) - F(\mu) > sc_n n^{-1/2}\} + r_n \\ &= P\left\{\sum_{i=1}^{n} [u(a_n - X_i) - F(a_n)] > nt_n\right\} + r_n, \end{aligned} \tag{2.10}$$

where $u(t) = 1$ or 0 according as $t \geqslant 0$ or < 0, $a_n = \mu + \sigma n^{-1/2}(\log n)^{1/2}$, and $t_n = sCn^{-1/2}(\log n)^{1/2} + F(\mu) - F(a_n)$.

Now let $\xi_n = [F(a_n) - F(\mu)]/(a_n - \mu)^\rho$. Noting that (on account of (2.5)), $\xi_n = O(1)$, as $n \to \infty$, we have for sufficiently large n, $t_n = n^{-1/2}(\log n)^{1/2}$ $\{sC + \sigma^\rho n^{(1-\rho)/2}(\log n)^{(\rho-1)/2}\,\xi_n\} > 0$. Then since $\sum_{i=1}^{n} u(a_n - X_i)$ has a binomial distribution with parameters n and $F(a_n)$, we obtain by using the Bernstein inequality (cf. Uspensky (1937), pp. 204–205)

$$P\left\{\sum_{i=1}^{n} [u(a_n - X_i) - F(a_n)] > nt_n\right\} \leqslant 2\exp(-\gamma_n), \tag{2.11}$$

where

$$\begin{aligned} \gamma_n &= \frac{nt_n^2}{2[F(a_n) + t_n]} \\ &= \frac{n\{s^2C^2n^{-1}\log n + [F(\mu) - F(a_n)]^2 + 2sCn^{-1/2}(\log n)^{1/2}\,[F(\mu) - F(a_n)]\}}{2[sCn^{-1/2}(\log n)^{1/2} + F(\mu)]} \\ &\geqslant \frac{s^2C^2\log n + 2sCn^{1/2}(\log n)^{1/2}\,[F(\mu) - F(a_n)]}{2[sCn^{-1/2}(\log n)^{1/2} + F(\mu)]} \\ &= (\log n)\left\{\frac{s^2C^2 + 2sC\sigma^\rho n^{1-\rho/2}(\log n)^{(\rho-1)/2}\,\xi_n}{2[sCn^{-1/2}(\log n)^{1/2} + F(\mu)]}\right\} = (\log n)\,\delta_n, \end{aligned} \tag{2.12}$$

say. Now, since $\delta_n \to s^2C^2/2F(\mu)$ as $n \to \infty$, by choosing

$$C^2 > F(\mu)/s^2 \tag{2.13}$$

and using (2.11) and (2.12) we obtain, for sufficiently large n

$$P\left\{\sum_{i=1}^{n} [u(a_n - X_i) - F(a_n)] > nt_n\right\} \leqslant 2n^{-1/2}. \tag{2.14}$$

Thus, combining (2.9), (2.10) and (2.14)

$$P\{F_n(\bar{X}) - F(\mu) \geqslant sc_n n^{-1/2}\} = O(n^{-1/2}), \qquad \text{as} \quad n \to \infty. \tag{2.15}$$

In the same way it may be shown that

$$P\{F(\mu) - F_n(\bar{X}) \geqslant sc_n n^{-1/2}\} = O(n^{-1/2}), \qquad \text{as} \quad n \to \infty. \tag{2.16}$$

Hence, from (2.6), (2.7), (2.15) and (2.16) it follows that with $C^2 > 1/(1 - F(\mu))$,

$$\sup_{|x| > c_n} |G_n(x) - \Phi(x)| = O(n^{-1/2}), \qquad \text{as} \quad n \to \infty. \tag{2.17}$$

We now consider $\sup_{|x| \leqslant c_n} |G_n(x) - \Phi(x)|$. To do so, observe that

$$\begin{aligned} G_n(x) - \Phi(x) &\leqslant [P\{F_n(\bar{X}) \leqslant F(\mu) + xsn^{-1/2}, |\bar{X} - \mu| \\ &\qquad \leqslant \sigma n^{-1/2}(\log n)^{1/2}\} - \Phi(x)] + r_n \qquad (2.18) \\ &\leqslant [P\{F_n(\mu - \sigma n^{-1/2}(\log n)^{1/2}) \leqslant F(\mu) + xsn^{-1/2}\} - \Phi(x)] + r_n. \end{aligned}$$

Now, applying the Barry–Esséen theorem for a double sequence of (row-wise) i.i.d. random variables, we have

$$\begin{aligned} \sup_t &\left| P\left\{ \frac{\sum_{i=1}^n [u(b_n - X_i) - F(b_n)]}{n^{1/2}[F(b_n)(1 - F(b_n)]^{1/2}} \leqslant t \right\} - \Phi(t) \right| \\ &\leqslant K \frac{E\,|u(b_n - X_1) - F(b_n)|^3}{\{F(b_n)(1 - F(b_n))\}^{3/2}} n^{-1/2}, \qquad n \geqslant 1, \end{aligned} \tag{2.19}$$

where K is a positive constant and $b_n = \mu - \sigma n^{-1/2}(\log n)^{1/2}$.

Thus, with

$$b_n(x) = \frac{n^{1/2}[F(\mu) - F(b_n) + xsn^{-1/2}]}{[F(b_n)(1 - F(b_n))]^{1/2}},$$

from (2.18) and (2.19) we obtain

$$\begin{aligned} G_n(x) - \Phi(x) \leqslant K \frac{F^2(b_n) + (1 - F(b_n))^2}{[F(b_n)(1 - F(b_n))]^{1/2}} n^{-1/2} \\ + [\Phi(b_n(x)) - \Phi(x)] + r_n. \end{aligned} \tag{2.20}$$

Similarly, with $a_n = \mu + \sigma n^{-1/2}(\log n)^{1/2}$ as before and

$$a_n(x) = \frac{n^{1/2}[F(\mu) - F(a_n) + xsn^{-1/2}]}{[F(a_n)(1 - F(a_n))]^{1/2}},$$

we have

$$\begin{aligned}\Phi(x) - G_n(x) &= P\{F_n(\bar{X}) > sxn^{-1/2} + F(\mu)\} - \Phi(-x) \\ &\leqslant P\{F_n(a_n) > xsn^{-1/2} + F(\mu)\} - \Phi(-x) + r_n \qquad (2.21) \\ &\leqslant K\frac{F^2(a_n) + (1 - F(a_n))^2}{[F(a_n)(1 - F(a_n))]^{1/2}}n^{-1/2} \\ &\quad + [\Phi(x) - \Phi(a_n(x))] + r_n.\end{aligned}$$

Now, using (2.20) and (2.21)

$$\sup_{|x| \leqslant c_n} |G_n(x) - \Phi(x)| \leqslant K\alpha_n n^{-1/2} + \sup_{|x| \leqslant c_n} \beta_n(x) + r_n, \qquad (2.22)$$

where

$$\alpha_n = \max\left\{\frac{F^2(a_n) + (1 - F(a_n))^2}{[F(a_n)(1 - F(a_n))]^{1/2}}, \frac{F^2(b_n) + (1 - F(b_n))^2}{[F(b_n)(1 - F(b_n))]^{1/2}}\right\}$$

and

$$\beta_n(x) = \max\{\Phi(x) - \Phi(a_n(x)), \Phi(b_n(x)) - \Phi(x)\}.$$

Observe that α_n is bounded as

$$\alpha_n \to \frac{F^2(\mu) + (1 - F(\mu))^2}{[F(\mu)(1 - F(\mu))]^{1/2}}, \qquad n \to \infty,$$

and from (2.9), $r_n = O(n^{-1/2})$, $n \to \infty$.

Finally, by using (2.5) and the mean value theorem it may easily be shown that

$$\sup_{|x| \leqslant c_n} \beta_n(x) = O(n^{-\rho/2}(\log n)^{(1+\rho)/2}) + O(n^{(1-\rho)/2}(\log n)^{\rho/2}) \qquad \text{as} \quad n \to \infty \qquad (2.23)$$

and since $\rho > 2$ one gets R.H.S. of (2.23) as $O(n^{-1/2})$, $n \to \infty$. The proof of the theorem is then completed by (2.17), (2.22) and (2.23).

Remark 2.1. It is not hard to show (cf. Chandra (1975)) that if $F^{(2m+1)}$ exists and is continuous in a neighborhood of μ for some integer $m \geqslant 1$, with $F^{(2j-1)}(\mu) = 0$, $1 \leqslant j \leqslant m$, and $F^{(2m+1)}(\mu) > 0$, then (2.5) holds with $\rho = 2m + 1$, and, in fact,

$$\lim_{h \to 0} \frac{|D(h)|}{|h|^{2m+1}} = \frac{F^{(2m+1)}(\mu)}{(2m+1)!}.$$

However, Theorem 2.1 shows that in order to obtain the best Berry–Esséen rate $O(n^{-1/2})$ one needs sharp estimates of $D(h) = F(\mu + h) - F(\mu)$ for small h and not necessarily existence of derivatives of F around μ.

Remark 2.2. If $F'(\mu) = 0$ and F'' exists and is bounded in a neighborhood of μ (this situation corresponds to the case $\rho = 2$ in (2.5)), by the proof of Theorem 2.1 (see (2.23)) one gets the Berry–Esséen rate $O(n^{-1/2} \log n)$.

3. An Almost Sure Representation of T_n

In what follows we shall show that under appropriate conditions it is possible to express the statistic $T_n = F_n(\bar{X})$ asymptotically as an average of i.i.d. r.v.'s, via an almost sure representation having a number of important implications.

Note that T_n can be conveniently expressed as a functional of the empirical d.f. F_n, namely, $T_n = T(F_n)$, where for any *df* G, $T(G) = G(\mu_G)$, $\mu_G =$ mean of G. And if F is differentiable at μ, by using the Gâteaux-differential approach for differentiable statistical functions (cf. Serfling (1980)) we get $d_1 T(F; F_n - F) = F_n(\mu) - F(\mu) + (\bar{X} - \mu) F'(\mu)$ which represents the principal part in the decomposition of $T(F_n) - T(F)$. The reduction step concerning the remainder term $R_n = T(F_n) - T(F) - d_1 T(F; F_n - F)$ was carried out by Ghosh (1971) who showed that if $F'(\mu) > 0$, $R_n = o_P(n^{-1/2})$. In the following theorem we are able to establish, under stronger assumptions, the improved stronger rate, namely, $O(n^{-3/4}(\log \log n)^{3/4})$, and we conjecture that this is the sharpest possible order.

Theorem 3.1. *If $F''(\mu)$ exists (finite), then*

$$F_n(\bar{X}) = F_n(\mu) + (\bar{X} - \mu) F'(\mu) + R_n, \tag{3.1}$$

where

$$R_n = O(n^{-3/4}(\log \log n)^{3/4}) \qquad \text{a.s. } as\ n \to \infty.$$

In proving Theorem 3.1, the main development is contained in the following result.

Lemma 3.1. *Let $c_0 > 0$ be a constant and $\alpha_n = n^{-1/2}(\log \log n)^{1/2}$. Then*

$$\sup_{|x-\mu| \leqslant c_0 \alpha_n} |F_n(x) - F_n(\mu) - F(x) + F(\mu)|$$
$$= O(n^{-3/4}(\log \log n)^{3/4}) \qquad \text{a.s. } as\ n \to \infty. \tag{3.2}$$

Proof of Lemma 3.1. Set $k_n = \sup\{F'(\mu + x): |x| \leqslant n^{-1}(1 + [c_0 n\alpha_n])\}$ ($[\cdot]$ = integer part). Then, using the monotonicity of F_n and F, we have by the mean value theorem

$$\sup_{|x-\mu| \leqslant c_0\alpha_n} |F_n(x) - F_n(\mu) - F(x) + F(\mu)| \leqslant G_n + n^{-1}k_n, \tag{3.3}$$

where

$$G_n = \max_{|j| \leqslant 1 + [c_0 n\alpha_n]} \left| F_n\left(\mu + \frac{j}{n}\right) - F_n(\mu) - F\left(\mu + \frac{j}{n}\right) + F(\mu) \right|.$$

Since $k_n \leqslant c_1$, for some $c_1 > 0$, and all n sufficiently large, it is seen from (3.3) that (3.2) will follow if we show

$$G_n = O(g(n)), \qquad \text{a.s. as } n \to \infty, \tag{3.4}$$

where $g(n) = n^{-3/4}(\log\log n)^{3/4}$. To this end, define for $i \geqslant 1$, $-\infty < x$, $y < \infty$

$$B_i(x, y) = u(\max(x, y) - X_i) - u(\min(x, y) - X_i)$$
$$- F(\max(x, y)) + F(\min(x, y)).$$

Then, if $T_{n,j} = F_n(\mu + j/n) - F_n(\mu) - F(\mu + j/n) + F(\mu)$, we have

$$G_n = \max_{|j| \leqslant 1 + [c_0 n\alpha_n]} |T_{n,j}| = n^{-1} \max_{|j| \leqslant 1 + [c_0 n\alpha_n]} \left| \sum_{i=1}^{n} B_i(\mu, \mu + jn^{-1}) \right|. \tag{3.5}$$

Let $\{n_k, k = 1, 2, \ldots\}$ be an increasing sequence of positive integers for which, as $k \to \infty$, $n_k \sim e^{\sqrt{k}}$. In the sequel, it will be convenient to take $n_k = [e^{\sqrt{k}}] + 1$ and $I_k = \{n \text{ integer: } n_k \leqslant n < n_{k+1}\}$. Let for $k \geqslant 1$, $a_k = n_k^{1/4}(\log\log n_k)^{3/4}$ and $b_k = 2c_0 k^{-1}\alpha_{n_k}$. For $k \geqslant 1$, $n \in I_k$ and $M > 0$ (to be specified later) define events

$$A_k = \left\{ \max_{|j| \leqslant k} \left| \sum_{i=1}^{n_k} B_i(\mu, \mu + jb_k) \right| > Ma_k \right\},$$

$$B_n = \left\{ \max_{|j| \leqslant k} \left| \sum_{i=n_k+1}^{n} B_i(\mu, \mu + jb_k) \right| > Ma_k \right\},$$

$$C_n = \left\{ \max_{|j| \leqslant 1 + [c_0 n\alpha_n]} \left| \sum_{i=1}^{n} B_i(\mu + [jn^{-1}b_k^{-1}]\, b_k, \mu + jn^{-1}) \right| > Mng(n) \right\}.$$

Using the relations $B_i(\mu + a,\ \mu + b) = B_i(\mu, \mu + b) - B_i(\mu, \mu + a)$, for

$-\infty < a < b < \infty$, we have on the complement of C_n, $n \in I_k$, that for all $|j| \leqslant 1 + [c_0 n\alpha_n]$

$$\left| \sum_{i=1}^{n} B_i(\mu, \mu + jn^{-1}) \right| \leqslant Mng(n) + \left| \sum_{i=1}^{n} B_i(\mu, \mu + [jn^{-1}b_k^{-1}]\, b_k) \right|. \quad (3.6)$$

Now, if $J = [jn^{-1}b_k^{-1}]$ and $|j| \leqslant 1 + [c_0 n\alpha_n]$, the admissible range of J is included in $\{J \text{ integer: } |J| \leqslant k\}$, and using (3.6), we obtain on $(A_k \cup B_n \cup C_n)^c$, $n \in I_k$,

$$G_n \leqslant 3Mg(n). \quad (3.7)$$

Thus, using (3.7) and the Borel–Cantelli lemma, to prove (3.4) it suffices to show that for some $M > 0$ (sufficiently large)

$$\sum_{k=1}^{\infty} P\left\{A_k \cup \bigcup_{n \in I_k} (B_n \cup C_n)\right\} < \infty. \quad (3.8)$$

For $|j| \leqslant k$, we obtain by using Berstein's inequality

$$P\left\{\left| \sum_{i=1}^{n_k} B_i(\mu, \mu + jb_k) \right| > Ma_k\right\} \leqslant 2\exp(-\gamma_k), \quad (3.9)$$

where

$$\gamma_k = \frac{M^2 a_k^2}{2v_{n_k} + \frac{2}{3} M a_k}, \qquad v_{n_k} = \sum_{i=1}^{n_k} \operatorname{Var} B_i(\mu, \mu + jb_k). \quad (3.10)$$

Using the mean value theorem, we have $v_{n_k} \leqslant c_1 k n_k b_k$, implying

$$\gamma_k \geqslant \frac{M^2 \log\log n_k}{4c_0c_1 + \frac{2}{3} M n_k^{-1/4} (\log\log n_k)^{1/4}} \geqslant \frac{M^2}{5c_0c_1} (\log\log n_k), \quad (3.11)$$

where the last inequality holds for all k sufficiently large.

From (3.9)–(3.11) we get

$$P(A_k) \leqslant 2(2k+1) \exp\left\{-\frac{M^2}{5c_0c_1} (\log\log n_k)\right\}. \quad (3.12)$$

Since $\log\log n_k > \frac{1}{2}\log k$, by choosing M such that $M^2 > 30c_0c_1$ it is seen that the right hand side of (3.12) is bounded by k^{-2}, for sufficiently large k. Thus

$$\sum_{k=1}^{\infty} P(A_k) < \infty. \quad (3.13)$$

Now, for $|j| \leqslant k$ and $n \in I_k$, another application of Bernstein's inequality gives

$$P\left\{\left|\sum_{i=n_k+1}^{n} B_i(\mu, \mu + jb_k)\right| > Ma_k\right\} \leqslant 2\exp(-\delta_n), \tag{3.14}$$

where

$$\delta_n = \frac{M^2 a_k^2}{2v_{n-n_k} + \frac{2}{3}Ma_k}, \qquad v_{n-n_k} = \sum_{i=n_{k+1}}^{n} \operatorname{Var} B_i(\mu, \mu + jb_k). \tag{3.15}$$

As before, $v_{n-n_k} \leqslant c_1 k(n - n_k)\, b_k$ entailing

$$\delta_n \geqslant \frac{M^2 k^{1/2}(\log\log n_k)}{4c_0 c_1 + \frac{2}{3}Mk^{1/2} n_k^{-1/4}(\log\log n_k)^{1/4}} \geqslant \frac{M^2 k^{1/2}(\log\log n_k)}{5c_0 c_1}, \tag{3.16}$$

where the last inequality holds for all k sufficiently large and we used the fact that $\lim_{k\to\infty} k^{1/2} n_k^{-1/4}(\log\log n_k)^{1/4} = 0$.

From (3.14)–(3.16) we obtain for $n \in I_k$ and k sufficiently large

$$P(B_n) \leqslant 2(2k+1)\exp\left\{-\frac{M^2}{5c_0 c_1} k^{1/2}(\log\log n_k)\right\}. \tag{3.17}$$

If $M^2 > 30c_0 c_1$, it is easily seen that for k sufficiently large

$$\begin{aligned} &6kn_k^2 \exp\left\{-\frac{M^2}{5c_0 c_1} k^{1/2}(\log\log n_k)\right\} \\ &\qquad \leqslant 24k\exp\left\{k^{1/2}\left(2 - \frac{M^2}{10c_0 c_1}\log k\right)\right\} \leqslant 1 \end{aligned} \tag{3.18}$$

from which it follows that the right hand side of (3.17) is bounded by n_k^{-2} for $k \geqslant k_0$. Therefore,

$$\sum_{k=k_0}^{\infty}\left\{\sum_{n\in I_k} P(B_n)\right\} \leqslant \sum_{k=k_0}^{\infty} \frac{(n_{k+1} - n_k)}{n_k^2} \leqslant \sum_{k=k_0}^{\infty} \frac{1}{k^{1/2} n_k} < \infty. \tag{3.19}$$

Finally, again by Bernstein's inequality we have for $|j| \leqslant 1 + [c_0 n\alpha_n]$ and $n \in I_k$

$$P\left\{\left|\sum_{i=1}^{n} B_i(\mu + [jn^{-1}b_k^{-1}]\, b_k, \mu + jn^{-1})\right| > Mng(n)\right\} \leqslant 2\exp(-\rho_n), \tag{3.20}$$

where

$$\rho_n = \frac{M^2 n^2 g^2(n)}{2\omega_n + \frac{2}{3}Mng(n)},$$

$$\omega_n = \sum_{i=1}^{n} \operatorname{Var} B_i(\mu + [jn^{-1}b_k^{-1}]\, b_k, \mu + jn^{-1}). \tag{3.21}$$

Now $0 \leqslant jn^{-1} - [jn^{-1}b_k^{-1}]\, b_k \leqslant b_k$ which implies $\omega_n \leqslant c_1 n b_k$ and we have for k sufficiently large

$$\rho_n \geqslant \frac{M^2(\log\log n)}{4c_0c_1(n/n_k)^{1/2}\,(\log\log n_k/\log\log n)^{1/2}\, k^{-1} + \frac{2}{3}Mn^{-1/4}(\log\log n)^{1/4}}$$

$$\geqslant \frac{M^2k(\log\log n_k)}{8c_0c_1 + \frac{2}{3}Mkn_k^{-1/4}(\log\log n_{k+1})^{1/4}} \geqslant \frac{M^2k(\log\log n_k)}{9c_0c_1}. \tag{3.22}$$

Thus, using (3.20)–(3.22)

$$P(C_n) \leqslant 2[2(1 + [c_0 n\alpha_n]) + 1] \exp\left\{-\frac{M^2k(\log\log n_k)}{9c_0c_1}\right\}$$

$$\leqslant 10c_0 n\alpha_n \exp\left\{-\frac{M^2k\log k}{18c_0c_1}\right\} \tag{3.23}$$

and if $M^2 > 30c_0c_1$, arguing as before we get $P(C_n) \leqslant n_k^{-2}$ for all k sufficiently large, say $k \geqslant k_1$. Then

$$\sum_{k=k_1}^{\infty} \left\{\sum_{n\in I_k} P(C_n)\right\} < \infty. \tag{3.24}$$

From (3.13), (3.19) and (3.24) we obtain (3.8). This completes the proof of the lemma.

Remark 3.1. Using Lemma 3.1 in connection with the results of Bahadur (1966) and Kiefer (1967) it follows that the order bound in (3.2) can not be improved in general and thus is the best possible.

Proof of Theorem 3.1. We shall only sketch the proof since the line of argument is the same as that of Bahadur (1966).

First, note that form the classical law of the iterated logarithm

$$|\bar{X} - \mu| \leqslant 2\sigma n^{-1/2}(\log\log n)^{1/2}, \qquad \text{a.s. as } n \to \infty. \tag{3.25}$$

Next, by using Young's form of Taylor's theorem (Serfling (1980), p. 45), since $\bar{X} \to \mu$ a.s. as $n \to \infty$, we get

$$F(\bar{X}) = F(\mu) + (\bar{X} - \mu) F'(\mu) + O(n^{-1} \log \log n), \qquad \text{a.s. as } n \to \infty, \quad (3.26)$$

and, as a result, (3.1) follows from (3.2), (3.25) and (3.26).

From the representation (3.1) may be obtained important properties of T_n, including asymptotic normality and the following law of the interated logarithm (LIL).

COROLLARY 3.1. (A LIL for T_n) *Under the conditions of Theorem* 3.1, *if* $v^2 > 0$, *then*

$$\limsup_{n \to \infty} \frac{n^{1/2}(T_n - \theta)}{(2v^2 \log \log n)^{1/2}} = 1, \qquad \text{a.s.}$$

where $\theta = F(\mu)$ *and* $v^2 = \mathrm{Var}[u(\mu - X_1) + (X_1 - \mu) F'(\mu)]$.

Remark 3.2. As another consequence of Theorem 3.1, the strong consistency of T_n in estimating the parameter $\theta = F(\mu)$ when both F and μ are unknown, is obtained. Moreover, the representation (3.1) reveals that the difference of the r.v.'s $n^{1/2}(T_n - \theta)$ and $n^{1/2}[F_n(\mu) - F(\mu) + (\bar{X} - \mu) F'(\mu)]$ (both having a normal limit distribution) tends to zero a.s. at the rate $O(n^{-1/4}(\log \log n)^{3/4})$.

4. An Invariance Principle for T_N

In this section our main concern is the study of the weak convergence of a suitable stochastic process constructed from the sequence $\{T_i, i \geqslant 1\}$ and summarizing the past history of $\{T_i, 1 \leqslant i \leqslant n\}$. We assume

$$E|X_i|^{4+\delta} < \infty \text{ for some } \delta > 0 \qquad \text{and} \qquad \sigma^2 = \mathrm{Var}\, X_1 > 0, \quad (4.1)$$

and

$$F'' \text{ exists and is bounded in the neighborhood of } \mu = EX_1. \quad (4.2)$$

Again, let $\theta = F(\mu)$, $v^2 = \mathrm{Var}[u(\mu - X_1) + (X_1 - \mu) F'(\mu)] > 0$.

THEOREM 4.1. *For each* $n \geqslant 1$, *let* W_n *be a random function on* $[0, 1]$ *defined as*

$$W_n(0) = 0, \qquad W_n\left(\frac{i}{n}\right) = \frac{i(T_i - \theta)}{n^{1/2} v}, \qquad 1 \leqslant i \leqslant n, \quad (4.3)$$

and $W_n(t)$, $0 \leqslant t \leqslant 1$, defined elsewhere, by linear interpolation. Then, under (4.1) and (4.2), we have $W_n \to^{\mathscr{D}} W$ on $(C[0, 1], d)$, where $C[0, 1]$ is the space of all continuous functions on $[0, 1]$, d is the uniform metric, and W is a standard Wiener process on $[0, 1]$.

The proof of this theorem is based on the following lemma.

LEMMA 4.1. *Under (4.1) and (4.2), there exist constants $\varepsilon > 0$, $M > 0$, $K > 0$ and $n_0 \geqslant 1$ such that for $n \geqslant n_0$*

$$P\{|Q_n| > Mn^{-1/4} \log n\} \leqslant Kn^{-1-\varepsilon}, \tag{4.4}$$

where

$$Q_n = n^{1/2}[F_n(\bar{X}) - F_n(\mu) - (\bar{X} - \mu) F'(\mu)].$$

Proof of Lemma 4.1. Let $q = 4 + \delta$ and select a constant c^2 such that $2 < c^2 < 2 + \delta$. We apply the following well-known moderate deviation result (cf. Petrov (1975), p. 254): If $E|X_1|^q < \infty$ for some $q > c^2 + 2$, then

$$P\{|\bar{X} - \mu| > c\sigma n^{-1/2}(\log n)^{1/2}\} = \frac{2n^{-c^2/2}}{c(2\pi \log n)^{1/2}} \left\{1 + O\left(\frac{1}{\log n}\right)\right\}.$$

Using this result, we note that if $0 < \varepsilon_1 < c^2/2 - 1$, there exist constants $K_1 > 0$ and $n_1 \geqslant 1$ such that for $n \geqslant n_1$,

$$P\{|\bar{X} - \mu| > c\sigma n^{-1/2}(\log n)^{1/2}\} \leqslant K_1 n^{-1-\varepsilon}. \tag{4.5}$$

Then, with $a_n = Mn^{-1/4} \log n$, we get for $n \geqslant n_1$

$$P\{Q_n > a_n\} \leqslant P\{Q_n > a_n, |\bar{X} - \mu| \leqslant c\sigma n^{-1/2}(\log n)^{1/2}\} + K_1 n^{-1-\varepsilon_1}. \tag{4.6}$$

Now partition the interval $[-c\sigma(\log n)^{1/2}, c\sigma(\log n)^{1/2}]$ into $2m_n$ intervals with end points $-c\sigma(\log n)^{1/2} = \Delta_{-m_n} < \cdots < \Delta_0 = 0 < \cdots < \Delta_{m_n} = c\sigma(\log n)^{1/2}$ such that for all $i = -m_n + 1, \ldots, m_n$

$$(\Delta_i - \Delta_{i-1}) F'(\mu) < \frac{a_n}{2}. \tag{4.7}$$

Then, by the monotonicity of F_n and (4.7), it is easily seen that

$$P\{Q_n > a_n, |\bar{X} - \mu| \leqslant c\sigma n^{-1/2}(\log n)^{1/2}\}$$
$$\leqslant \sum_{i=-m_n+1}^{m_n} P\left\{n^{1/2}[F_n(\mu + \Delta_i n^{-1/2}) - F_n(\mu)] - \Delta_i F'(\mu) > \frac{a_n}{2}\right\}. \tag{4.8}$$

Fix $-m_n+1 \leqslant i \leqslant m_n$ and write

$$P\left\{n^{1/2}[F_n(\mu+\Delta_i n^{-1/2})-F_n(\mu)]-\Delta_i F'(\mu)>\frac{a_n}{2}\right\}$$

$$=P\left\{\sum_{j=1}^{n}[Z_j-E(Z_j)]\geqslant\omega\right\}, \tag{4.9}$$

where

$$Z_j=u(\mu+\Delta_i n^{-1/2}-X_j)-u(\mu-X_j)$$

and

$$\omega=n^{1/2}\left[\Delta_i F'(\mu)+\frac{a_n}{2}\right]-n[F(\mu+\Delta_i n^{-1/2})-F(\mu)].$$

Using (4.2) and Taylor's expansion, we have $\omega \geqslant (M/4)\, n^{1/4}\log n$, for sufficiently large n.

Now, if $\Delta_i>0$, using Bernstein's inequality, we note that the right side of (4.9) $\leqslant\exp(-\omega^2/(2pn+2\omega))$, where $p=F(\mu+\Delta_i n^{-1/2})-F(\mu)$, and $\omega^2/(2pn+2\omega)\geqslant(M^2/16)(\log^2 n/(Mn^{-1/4}\log n+2\Delta_i F'(\mu)))\geqslant\alpha\log n$, for some $\alpha>1$ and n large, if M is choosen sufficiently large. On the other hand, if $\Delta_i<0$, then again by Bernstein's inequality, the right side of (4.9) $\leqslant\exp(-\omega^2/(2nq(1-q)+2\omega))$, where $q=F(\mu)-F(\mu+\Delta_i n^{-1/2})$, and for sufficiently large n, and M choosen appropriately, we obtain

$$\frac{\omega^2}{2nq(1-q)+2\omega}\geqslant\frac{M^2}{16}\,\frac{\log^2 n}{Mn^{-1/2}\log n-2F'(\mu)\Delta_i+2\Delta_i^2 n^{-1/2}c_1}\geqslant\beta\log n,$$

where $c_1>0$ and $\beta>1$.

Using the above results, (4.6), (4.8), (4.9) and the fact that $m_n\sim c_0(\log n)^{-1/2}n^{1/4}$, as $n\to\infty$, for some $c_0>0$ and $\Delta_j=O((\log n)^{1/2})$ uniformly in $-m_n\leqslant j\leqslant m_n$, $n\geqslant 1$, we can find $M>0$ and $\varepsilon>0$ such that

$$P\{Q_n>a_n\}=O(n^{-1-\varepsilon}),\qquad\text{as}\quad n\to\infty.$$

Since a similar statement holds for $p\{Q_n<-a_n\}$, the proof of the Lemma 4.1 follows.

Remark 4.1. Corresponding to the asymptotic normality of the T_n in the general case when $F'(\mu)$ does not necessarily vanish, we may obtain an associated Berry–Esséen rate by using the following device: For any sequence of random variables $\{U_n\}$ and $\{V_n\}$ and a sequence of positive constants $\{c_n\}$,

$$\sup_x |P\{U_n \leqslant x\} - \Phi(x)| \leqslant \sup_x |P\{V_n \leqslant x\} - \Phi(x)| + P\{|U_n - V_n| \geqslant c_n\} + O(c_n).$$

By this inequality, the classical Berry–Esséen theorem and Lemma 4.1, we get under (4.1), (4.2) and $\nu^2 > 0$ that

$$\sup_x |P\{n^{1/2}(T_n - \theta) \leqslant \nu x\} - \Phi(x)| = O(n^{-1/4} \log n), \qquad n \to \infty. \tag{4.10}$$

We are now in a position to prove the main theorem of this section.

Proof of Theorem 4.1. Define the random function W_n^*, $n \geqslant 1$, on $[0, 1]$ by

$$W_n^*(0) = 0,$$

$$W_n^*\left(\frac{i}{n}\right) = \frac{\sum_{j=1}^{i} [u(\mu - X_j) - F(\mu) + (X_j - \mu) F'(\mu)]}{\nu n^{1/2}}, \qquad 1 \leqslant i \leqslant n, \tag{4.11}$$

and linearly interpolated on $[(i-1)/n, i/n]$, $1 \leqslant i \leqslant n$.

By Donsker's theorem (cf. Billingsley (1968)) we obtain

$$W_n^* \xrightarrow{\mathscr{D}} W \qquad \text{in} \quad (C[0, 1], d), \tag{4.12}$$

where W is a standard Wiener process on $[0, 1]$.

The proof will be completed if we show that

$$d(W_n, W_n^*) \xrightarrow{P} 0. \tag{4.13}$$

But

$$d(W_n, W_n^*) \leqslant \max_{1 \leqslant i \leqslant n} \frac{i^{1/2} |Q_i|}{\nu n^{1/2}}, \tag{4.14}$$

where Q_i is defined in (4.4).

Now let ε_0, $\delta > 0$ be given and pick $n_1 \geqslant n_0$ such that

$$K \sum_{i=n_1+1}^{n} i^{-1-\varepsilon} < \delta, \tag{4.15}$$

where n_0, K and ε are given by Lemma 4.1.

Then, for $n \geqslant n_1$,

$$P\left\{\max_{1 \leqslant i \leqslant n} \frac{i^{1/2}|Q_i|}{vn^{1/2}} > \varepsilon_0\right\} \leqslant \sum_{i=1}^{n_1} P\{i^{1/2}|Q_i| > \varepsilon_0 vn^{1/2}\}$$

$$+ \sum_{i=n_1+1}^{n} P\{i^{1/2}|Q_i| > \varepsilon_0 vn^{1/2}\}. \quad (4.16)$$

Now clearly the first sum in the right side of (4.16) tends to 0 as $n \to \infty$. Also, for n_1 chosen sufficiently large, since for $n > i > n_1$, $\{i^{1/2}|Q_i| > \varepsilon_0 vn^{1/2}\} \subset \{|Q_i| > Mi^{-1/4} \log i\}$, we find by using Lemma 4.1 and (4.15) that the second sum in the right side of (4.16) is $\leqslant \delta$ for $n \geqslant n_1$. Consequently (4.14) and (4.16) entail (4.13). The proof of the theorem now follows from (4.12) and (4.13).

Remark 4.2. Theorem 4.1 clearly contains the central limit theorem for T_n and likewise provides additional information such as for $x > 0$, $P\{\max_{1 \leqslant i \leqslant n} i(T_i - \theta) \geqslant xvn^{1/2}\} \to 2[1 - \Phi(x)]$, as $n \to \infty$.

Acknowledgment

The authors would like to express their sincere thanks to the referee for raising interesting questions on their original draft which led to considerable improvement.

References

[1] Antille, A., and Kersting, G. (1977). Tests for Symmetry. *Z. Wahrsch. Verw. Gebiete* **39**, 235–255.
[2] Bahadur, R. R. (1966). A note on quantiles in large samples. *Ann. Math. Statist.* **37**, 577–580.
[3] Billingsley, P. (1968). *Convergence of Probability Measures*. Wiley, New York.
[4] Chandra, K. C. (1975). Some comments on the asymptotic probability laws of sample quantiles. *Calcutta Stat. Assoc. Bulletin*, 123–126.
[5] David, H. T. (1962). The sample mean among the moderate order statistics. *Ann. Math. Statist.* **33**, 1160–1166.
[6] Gastwirth, J. L. (1971). On the sign test for symmetry. *J. Amer. Statist. Assoc.* **66**, 821–823.
[7] Ghosh, J. K. (1971). A new proof of the Bahadur representation of quantiles and an application. *Ann. Math. Statist.* **42**, 1957–1961.
[8] Kiefer, J. (1967). On Bahadur's representation of sample quantiles. *Ann. Math. Statist.* **38**, 1323–1342.
[9] Petrov, V. V. (1975). *Sums of Independent Random Variables*. Springer-Verlag, New York.
[10] Ralescu, S., and Puri, M. L. (1984). On the rate of convergence for central limit theorem for signed rank statistics. *Adv. in Appl. Math.* To appear.

[11] SARKADI, K., SCHNELL, E., AND VINCZE, I. (1962). On the position of the sample mean among the ordered sample elements. *Publ. Math. Inst. Hung. Acad. Sci.* **7A**, 239–254.
[12] SERFLING, R. J. (1980). *Approximation Theorems of Mathematical Statistics.* Wiley, New York.
[13] USPENSKY, J. V. (1937). *Introduction to Mathematical Probability.* McGraw-Hill, New York.

CRAMÉR TYPE LARGE DEVIATIONS FOR GENERALIZED RANK STATISTICS[1]

MUNSUP SEOH, STEFAN S. RALESCU and MADAN L. PURI

Wright State University, Brown University, and Indiana University

A Cramér type large deviation theorem is proved under alternatives as well as under hypothesis for the generalized linear rank statistic which includes as special cases (unsigned) linear rank statistics, signed linear rank statistics, linear combination of functions of order statistics, and a rank combinatorial statistic.

1. Introduction. Let $X_{N1}, X_{N2}, \cdots, X_{NN}$ be independent r.v.'s (random variables) and let g be a real valued measurable function such that $X^g_{Nj} \equiv g(X_{Nj})$, $1 \le j \le N$, has a c.d.f. (cumulative distribution function) G_{Nj}. We introduce the generalized linear rank statistic

$$T_N = \textstyle\sum_{j=1}^N c_{Nj} a_{NR^g_{Nj}}(X_{Nj}) \tag{1.1}$$

where $\{c_{Nj}: 1 \le j \le N\}$ is an array of regression constants; $\{a_{Nj}(\cdot): 1 \le j \le N\}$ is an array of known real functions (called scores); and $R^g_{Nj} = \sum_{k=1}^N u(X^g_{Nj} - X^g_{Nk})$ is a generalized rank of X^g_{Nj} among $\{X^g_{Nk}: 1 \le k \le N\}$ where $u(x) = 1$ if $x \ge 0$ and $u(x) = 0$ otherwise. We assume that the scores $a_{Nj}(\cdot)$, $1 \le j \le N$ are generated by a nonconstant score generating function $\phi(s, t)$, $0 < s < 1$, $-\infty < t < \infty$, in either of the following two ways;

$$a_{Nj}(t) = \phi(EU_{N:j}, t), \quad j = 1, 2, \cdots, N \text{ (approximate scores)} \tag{1.2}$$

$$a_{Nj}(t) = E\phi(U_{N:j}, t) \quad j = 1, 2, \cdots, N \text{ (exact scores)} \tag{1.3}$$

where $U_{N:j}$ is the jth order statistic in a random sample of size N from the uniform distribution over (0, 1). To avoid the trivialities, we assume that $\sum_{j=1}^N |c_{Nj}| > 0$.

When $g(x) = x$ and $\phi(x, y) = \tilde{\phi}(x)$, the statistic (1.1) reduces to the (unsigned) linear rank statistic

$$T_N = \textstyle\sum_{j=1}^N c_{Nj} \tilde{a}_{NR_{Nj}} \tag{1.4}$$

where R_{Nj}, $1 \le j \le N$, is the rank of X_{Nj} among $\{X_{Nk}: 1 \le k \le N\}$ and $\tilde{a}_{Nj}$'s are usual scores of constants. On the other hand, when $g(x) = |x|$ and $\phi(x, y) = \tilde{\phi}(x)\operatorname{sgn} y$, where $\operatorname{sgn} y = 2u(y) - 1$, the statistic (1.1) reduces to the signed linear

Received July 1983; revised March 1984.

Research supported by NSF Grant MCS-8301409.

AMS 1980 *subject classifications.* Primary 60F10; secondary 62E20.

Key words and phrases. Linear rank statistics, order statistics, rank combinatorial statistic, large deviation probabilities.

rank statistic

$$T_N = \sum_{j=1}^{N} c_{Nj}\tilde{a}_{NR_{Nj}^+}\text{sgn } X_{Nj} \tag{1.5}$$

where R_{Nj}^+, $1 \le j \le N$, is the rank of $|X_{Nj}|$ among $\{|X_{Nk}|: 1 \le k \le N\}$.

Also note that when $c_{N1} = \cdots = c_{NN} = N^{-1}$, $g(x) = x$, $\phi(x, y) = \tilde{\phi}(x)\psi(y)$, and the underlying c.d.f.'s G_{Nj} are continuous, the statistic (1.1) reduces to the linear combination of functions of order statistics

$$T_N = (1/N) \sum_{j=1}^{N} \tilde{a}_{NR_{Nj}}\psi(X_{Nj}) = (1/N) \sum_{j=1}^{N} \tilde{a}_{Nj}\psi(X_{N:j}) \tag{1.6}$$

where $X_{N:j}$ is the jth order statistic among $\{X_{Nk}, 1 \le k \le N\}$. Furthermore, denoting $Y_{jk} = c_{Nj}a_{Nk}(X_{Nj})$, $1 \le j, k \le N$, we can rewrite (1.1) as

$$T_N = \sum_{j=1}^{N} Y_{jR_{Nj}} \tag{1.7}$$

which is a rank combinatorial statistic. For these four different types of statistics, several authors have investigated problems concerning asymptotic normality, rate of convergence to normality and higher order expansions. For a review, the reader is referred to Hájek (1962, 1968), Hušková (1970, 1977, 1979), Jurečková and Puri (1975), Bergström and Puri (1977), Puri and Seoh (1984a, b, c), Does (1982), Puri and Wu (1983) (for statistics of the type (1.4) and (1.5)); Shorack (1969, 1972), Stigler (1974), Bjerve (1977) and Helmers (1977, 1980, 1981) (for statistics of the type (1.6)); Hoeffding (1951), Motoo (1956), von Bahr (1976) and Ho and Chen (1978) (for the statistics of the type (1.7)) and the papers cited therein, among others.

In recent years, there has been a great upsurge of activity in the theory of large deviations initiated by Cramér (1938) and studied in detail by Petrov (1975) for the case of independent summands. Large deviation probabilities for U-statistics were obtained by Malevich and Abdalimov (1979) while the corresponding results for the case of the statistic (1.4) were studied by Kallenberg (1982). But these results are restrictive in the sense that the observations $X_{N1}, X_{N2}, \cdots, X_{NN}$ are identically distributed and the underlying distribution function is *continuous* (Kallenberg, 1982). In addition, Robinson (1977) has dealt with the case of large deviations for samples from finite populations.

In this paper we shall be concerned with the relative error of the normal approximation to the distribution of the (properly normalized) generalized linear rank statistic (1.1) under general alternatives, i.e., assuming only that the observations are independent (*not necessarily identically distributed*) *and without assuming the continuity of the underlying distribution functions.* The results obtained not only include the results of Kallenberg (1982) as a special case, but also provide the large deviation theory for the statistics of the type (1.5), (1.6), and (1.7) which to the best of our knowledge has not been considered in the generality of the present paper.

2. Assumptions and main theorem. Throughout this paper we make the following assumptions.

ASSUMPTION (A). The variance of T_N satisfy

$$\liminf \tau_N^2 = \liminf \operatorname{Var} T_N > 0.$$

ASSUMPTION (B). The regression constants $c_{N1}, c_{N2}, \cdots, c_{NN}$ satisfy

$$\textstyle\sum_{j=1}^N c_{Nj}^2 = 1, \quad \max_{1\le j\le N} |c_{Nj}| \le A_1 N^{-1/3}, \qquad \sum_{j=1}^N |c_{Nj}|^3 \le A_2 N^{-1/2},$$

where A_1 and A_2 are absolute constants.

ASSUMPTION (C). The score generating function $\phi(s, t)$ is differentiable with respect to its first argument s such that its first partial derivatives $\phi_1(s, t) = \partial\phi(s, t)/\partial s$ satisfy Lipschitz's condition of order one with respect to s, i.e., there is a constant Δ such that for any $x, y \in (0, 1)$

$$\sup_{-\infty<t<\infty} |\phi_1(x, t) - \phi_1(y, t)| \le \Delta |x - y|. \tag{2.1}$$

(Note that the normal scores statistic does not satisfy this assumption).

We now introduce some notations. Let $S°$ denote a r.v. S centered at its expectation, i.e. $S° = S - ES$. Denote

$$\rho_{Nj} = R_{Nj}^g/(N + 1), \qquad \rho_{Njj} = E(\rho_{Nj} | X_j). \tag{2.2}$$

Furthermore, we shall use the r.v. $\hat{S}_N$ (Hájek's projection), as an approximation of the statistic T_N, defined by

$$\hat{S}_N = \textstyle\sum_{\ell=1}^N E(S_N | X_\ell) - (N - 1)ES_N \tag{2.3}$$

where S_N is the first two terms of Taylor's expansion of T_N with approximate scores, i.e.,

$$S_N = \textstyle\sum_{j=1}^N c_{Nj}\{\phi(\rho_{Njj}, X_{Nj}) + (\rho_{Nj} - \rho_{Njj})\phi_1(\rho_{Njj}, X_{Nj})\}. \tag{2.4}$$

Let $\Phi(\cdot)$ denote the standard normal c.d.f. and put $\hat{\sigma}_N^2 = \operatorname{var} \hat{S}_N$. Then

THEOREM 2.1. *Under assumptions* (A), (B) *and* (C), *uniformly in the region* $0 < x \le \rho_N N^{1/6}$, $\rho_N = o(1)$, *we have as* $N \to \infty$

$$P(T_N - ET_N > \tau_N x)[1 - \Phi(x)]^{-1} = 1 + o(1) \tag{2.5}$$

which remains true if we replace τ_N by $\hat{\sigma}_N$.

REMARK 2.1. Note that the result of Kallenberg (1982) which deals with the statistic (1.4) and holds for the case of iid r.v.'s is a special case of our result, but we impose somewhat stronger assumption on the score generating function. This is due to our generalized statistic (1.1) and weaker assumptions on underlying distributions. However, for the case of iid r.v.'s, the results of the above theorem hold under relatively weaker assumptions (see Remark 4.1).

From now on, we shall suppress the subscript N in c_{Nj}, a_{Nj}, R^g_{Nj}, X_{Nj}, ρ_{Nj}, ρ_{Njj}, etc., whenever it causes no confusion.

3. Preliminaries.

In this section, we derive bounds on the $2r$th moments (r is any positive integer) for statistics defined in Lemma 3.1–3.5. These bounds play an important role in this paper and, since the method of their derivation depends heavily on that of Bickel (1974), we will give only the brief outline of the proofs.

Let $\{Y_j\}_{j=1}^{\infty}$ be a sequence of r.v.'s and $\{d_j\}_{j=1}^{\infty}$ a sequence of real numbers. Then we have the following lemmas.

LEMMA 3.1. *Let Z_j, $j \geq 1$, be r.v.'s of the form $Z_j = g_j(Y_1, Y_2, \cdots, Y_j)$ such that for $j \geq 2$, $E(Z_j \mid Y_1, Y_2, \cdots, Y_{j-1}) = 0$. If the sequence $\{d_j\}$ is nonincreasing in absolute values, then for any positive integers r and ℓ,*

$$E(\textstyle\sum_{j=1}^{\ell} d_j Z_j)^{2r} \leq (4e)^r \, (\sum_{j=1}^{\ell} d_j^2)^r \, r^r \max_{i \leq j \leq \ell} EZ_j^{2r}.$$

LEMMA 3.2. *Let Y_j, $j \geq 1$, be independent r.v.'s and let $\tilde{V}_{jk}$ be r.v.'s of the form $\tilde{V}_{jk} = g_{jk}(Y_j, Y_k)$, $1 \leq j, k < \infty$, such that for any j and k, $j \neq k$, $E(\tilde{V}_{jk} \mid Y_j) = E(\tilde{V}_{jk} \mid Y_k) = 0$. Then, for any positive integers ℓ and r,*

$$E[(1/\ell) \textstyle\sum_{j=1}^{\ell} \sum_{k\neq j}^{\ell} d_j \tilde{V}_{jk}]^{2r} \leq (4e)^{2r} (\sum_{j=1}^{\ell} d_j^2)^r (2r)^{2r} \ell^{-r} \max_{1 \leq j,k \leq \ell; j \neq k} E\tilde{V}_{jk}^{2r}.$$

REMARK 3.1. Lemmas 3.1 and 3.2 are generalizations of Lemmas 1 and 4, respectively, of Hušková (1979). (We may point out that, in Hušková (1979), the proof of Lemma 4 is incorrect in its application of Lemma 1, especially in deriving (23) and (24) of her paper). Lemma 3.2 is also a generalization, as well as an improvement, of Lemma 2.2 of Bergström and Puri (1977).

PROOF OF LEMMA 3.1. For $r \geq \ell$, the proof follows by applying Hölder's inequality to $(\sum_{j=1}^{\ell} d_j Z_j)^{2r}$ and some routine computations. For $r \leq \ell$, the proof follows by induction on ℓ with r fixed.

PROOF OF LEMMA 3.2. Since the assumptions and the conclusion of this lemma are invariant under simultaneous permutation of d_j's and Y_j's, we may, without loss of generality, assume that $|d_1| \geq |d_2| \geq \cdots \geq |d_\ell|$.

Define $Z_1 = \tilde{Z}_1 = 0$, $Z_j = \sum_{k=1}^{j-1} \tilde{V}_{jk}$ and $\tilde{Z}_j = \sum_{k=1}^{j-1} d_k \tilde{V}_{jk}$, $2 \leq j \leq \ell$, so that $(1/\ell) \sum_{j=1}^{\ell} \sum_{k\neq j}^{\ell} d_j \tilde{V}_{jk} = (1/\ell)(\sum_{j=1}^{\ell} d_j Z_j + \sum_{j=1}^{\ell} \tilde{Z}_j) = \hat{V}_\ell$, say. The proof then follows by using the following facts:

(a) $E\hat{V}_\ell^{2r} \leq \ell^{-2r} 2^{2r-1} (4e)^r r^r \{(\sum_{j=1}^{\ell} d_j^2)^r \max_{1 \leq j \leq \ell} EZ_j^{2r} + \ell^r \max_{1 \leq j \leq \ell} E\tilde{Z}_j^{2r}\}$.

(b) $EZ_j^{2r} \leq (4e)^r (j-1)^r r^r \max_{1 \leq k \leq \ell; k \neq j} EV_{jk}^{2r}$ and

(c) $E\tilde{Z}_j^{2r} \leq (4e)^r (\sum_{k=1}^{j-1} d_k^2)^r r^r \max_{1 \leq k \leq \ell; k \neq j} EV_{kj}^{2r}$.

LEMMA 3.3. *Let T_N and S_N be defined by* (1.1) *and* (2.4). *Then for any positive* *integer* $r \geq 1$, $E(T_N^0 - S_N^0)^{2r} \leq (16e\Delta)^{2r}(2r)^{2r}N^{-r}$.

PROOF. Denote $H_{jk} = u(X_j^g - X_k^g) - G_k(X_j^g)$, $1 \leq j, k \leq N$. Then, we have

$$\rho_j - \rho_{jj} = (N+1)^{-1} \textstyle\sum_{k\neq j}^N H_{jk} \tag{3.1}$$

and

$$S_N = \textstyle\sum_{j=1}^N c_j\{\phi(\rho_{jj}, X_j) + (N+1)^{-1} \sum_{k\neq j}^N H_{jk}\phi_1(\rho_{jj}, X_j)\}. \tag{3.2}$$

First consider the statistic T_N with approximate scores. Then, by Taylor's expansion, for some $0 \leq \lambda \leq 1$,

$$T_N = \textstyle\sum_{j=1}^N c_j\{\phi(\rho_{jj}, X_j) + (\rho_j - \rho_{jj})\phi_1(\lambda\rho_{jj} + (1-\lambda)\rho_j, X_j)\}. \tag{3.3}$$

Using (3.2) and (3.3) along with assumptions (B) and (C), we obtain

$$E(T_N - S_N)^{2r} \leq \Delta^{2r}N^{r-1} \textstyle\sum_{j=1}^N E(\rho_j - \rho_{jj})^{4r}. \tag{3.4}$$

For each j, conditionally given X_j^g, $\rho_j - \rho_{jj}$ is the sum of independent r.v.'s with zero means and thus we may apply Lemma 3.1 to obtain that for any integer $r \geq 1$,

$$E(\rho_j - \rho_{jj})^{4r} \leq (4e)^{2r}(2r)^{2r}N^{-2r}. \tag{3.5}$$

The proof follows using (3.4) and (3.5).

We now consider the exact scores. Let $\tilde{T}_N$ be the statistic (1.1) with exact scores and put for $1 \leq j \leq N$, $\tilde{a}_j(t) = E\phi(U_{N:j}, t)$ to distinguish it from the statistic T_N with approximate scores. The proof then follows by using $|E\phi(U_{N:j}, t) - \phi(j/(N+1), t)| \leq \Delta N^{-1}$, $E(\tilde{T}_N^0 - T_N^0)^{2r} \leq (2\Delta)^{2r}N^{-r}$, $E(T_N^0 - S_N^0)^{2r} \leq 2^{2r}E(T_N - S_N)^{2r} \leq (8e\Delta)^{2r}(2r)^{2r}N^{-r}$ and routine computations.

LEMMA 3.4. *Let* $\hat{S}_N$ *and* S_N *be defined by* (2.3) *and* (2.4). *Then for any integer* $r \geq 1$, $E(S_N^0 - \hat{S}_N^0)^{2r} \leq (8e\,\|\phi_1\|)^{2r}(2r)^{2r}N^{-r}$ *where*

$$\|\phi_1\| = \sup_{0<s<1}\sup_{-\infty<t<\infty}|\phi_1(s, t)|.$$

PROOF. The proof follows by using Lemma 3.2 and noting that

$$\begin{aligned} E(S_N|X_\ell) &= c_\ell\phi(\rho_{\ell\ell}, X_\ell) + ES_N - c_\ell E\phi(\rho_{\ell\ell}, X_\ell) \\ &\quad + (N+1)^{-1} \textstyle\sum_{j=1}^N c_jE\{H_{j\ell}\phi_1(\rho_{jj}, X_j)|X_\ell\}, \end{aligned} \tag{3.6}$$

and

$$S_N^0 - \hat{S}_N^0 = (N+1)^{-1} \textstyle\sum_{j=1}^N \sum_{k\neq j}^N c_j[H_{jk}\phi_1(\rho_{jj}, X_j) - E\{H_{jk}\phi_1(\rho_{jj}, X_j)|X_k\}].$$

LEMMA 3.5. *Let* T_N *and* $\hat{S}_N$ *be defined by* (1.1) *and* (2.3). *Then, for any real* $r \geq 1/2$,

$$E(T_N^0 - \hat{S}_N^0)^{2r} \leq (64e(\Delta + \|\phi_1\|))^{2r}(2r)^{2r}N^{-r}. \tag{3.7}$$

PROOF. Denote by $[x]$, the smallest integer $\geq x$. Then, applying Hölder's inequality,

$$[E(T_N^0 - \hat{S}_N^0)^{2r}]^{[r]/r} \le E(T_N^0 - \hat{S}_N^0)^{2[r]}.$$

Since, by the C_r-inequality,

$$\text{the right-hand side term} \le 2^{2[r]-1}[E(T_N^0 - S_N^0)^{2[r]} + E(S_N^0 - \hat{S}_N^0)^{2[r]}],$$

we have

$$[E(T_N^0 - \hat{S}_N^0)^{2r}]^{[r]/r} \le 2^{2[r]-1}[E(T_N^0 - S_N^0)^{2[r]} + E(S_N^0 - \hat{S}_N^0)^{2[r]}].$$

Now using Lemmas 3.3 and 3.4 to the terms on the right-hand side, we obtain, after omitting some details of computations, that

$$[E(T_N^0 - \hat{S}_N^0)^{2r}]^{[r]/r} \le \{32e(\Delta + \| \phi_1 \|)\}^{2[r]}(2[r])^{2[r]}N^{-[r]}.$$

Thus,

$$E(T_N^0 - \hat{S}_N^0)^{2r}$$
$$\le \{32([r]/r)e(\Delta + \| \phi_1 \|)\}^{2r}(2r)^{2r}N^{-r} \le \{64e(\Delta + \| \phi_1 \|)\}^{2r}(2r)^{2r}N^{-r}$$

where the last inequality follows from the fact that $1 \le [r]/r \le 2$ for any real $r \ge 1/2$.

4. Proof of Theorem 2.1. By assumption (A) and Lemma 3.5, we have

$$| \hat{\sigma}_N^2 - \tau_N^2 | = | 2\,\mathrm{Cov}(T_N, \hat{S}_N - T_N) + \mathrm{Var}(\hat{S}_N - T_N) |$$
$$\le 2\tau_N(\mathrm{Var}(\hat{S}_N - T_N))^{1/2} + \mathrm{Var}(\hat{S}_N - T_N)$$

which implies that $\hat{\sigma}_N^2/\tau_N^2 = 1 + O(N^{-1/2})$ and that there is a positive constant σ such that for all $N \ge N_0$,

$$\hat{\sigma}_N^2 \ge \sigma^2. \tag{4.1}$$

Hence, in order to prove Theorem 2.1, it is sufficient to show that

$$P(T_N - ET_N > \hat{\sigma}_N x)[1 - \Phi(x)]^{-1} = 1 + o(1) \tag{4.2}$$

uniformly in the region $0 < x < \rho_N N^{1/6}$.

By standard arguments we have

$$\begin{aligned} &P(T_N^0 > \hat{\sigma}_N x) \le P(\hat{S}_N^0 > (x - N^{-1/6})\hat{\sigma}_N) + P(| T_N^0 - \hat{S}_N^0 | > N^{-1/6}\hat{\sigma}_N),\\ &P(T_N^0 > \hat{\sigma}_N x) > P(\hat{S}_N^0 > (x + N^{-1/6})\hat{\sigma}_N) - P(| T_N^0 - \hat{S}_N^0 | > N^{-1/6}\hat{\sigma}_N). \end{aligned} \tag{4.3}$$

Using Chebyshev's inequality and applying Lemma 3.5 with $r = 1/2\delta N^{1/3}$, $\delta = \sigma(64(\Delta + \| \phi_1 \|)e^2)^{-1}$, we get

$$P(| T_N^0 - \hat{S}_N^0 | > N^{-1/6}\hat{\sigma}_N) \le E(T_N^0 - \hat{S}_N^0)^{2r} N^{r/3}\hat{\sigma}_N^{-2r} \le \exp(-\delta N^{1/3}) \tag{4.4}$$

which implies that uniformly for $0 < x \le \rho_N N^{1/6}$,

$$\begin{aligned} &P(| T_N^0 - \hat{S}_N^0 | > N^{-1/6}\hat{\sigma}_N)[1 - \Phi(x)]^{-1}\\ &\qquad \le \exp(-\delta N^{1/3})[1 - \Phi(\rho_N N^{1/6})]^{-1} = o(1) \end{aligned} \tag{4.5}$$

where the last equality follows by Lemma VII.1.2 of Feller (1968).

Now, in view of (4.3) and (4.5), to prove (4.2) it suffices to show that as $N \to \infty$

$$P(\hat{S}_N^0 > x_N \hat{\sigma}_N)[1 - \Phi(x)]^{-1} = 1 + o(1) \tag{4.6}$$

uniformly in $|x_N - x| = N^{-1/6}$, $0 \le x \le \rho_N N^{1/6}$.

Using (2.3) and (3.6) we can write $\hat{S}_N^0 = \sum_{j=1}^N \hat{S}_N^{(j)}$ where

$$\hat{S}_N^{(j)} = c_j[\phi(\rho_{jj}, X_j) - E\phi(\rho_{jj}, X_j)]$$
$$+ (N+1)^{-1} \textstyle\sum_{k\ne j}^N c_k E\{[u(X_k^g - X_j^g) - G_j(X_k^g)]\phi_1(\rho_{kk}, X_k) \mid X_j\}.$$

Thus $\hat{S}_N^0$ is a sum of independent r.v.'s with means zero and

$$|S_N^{(j)}| \le 2|c_j| \, \|\phi\| + (N+1)^{-1} \textstyle\sum_{k\ne j}^N |c_k| \, \|\phi_1\|$$
$$\le (2\|\phi\| + \|\phi_1\|)A_1 N^{-1/3}, \quad 1 \le j \le N,$$

where $\|\phi\| = \sup_{0<s<1}\sup_{-\infty<t<\infty}|\phi(s,t)|$. Furthermore, there is an integer N_0 such that for all $N \ge N_0$ and $0 < x < \rho_N N^{1/6}$, $|x_N - x| \le N^{-1/6}$,

$$0 < (2\|\phi\| + \|\phi_1\|)A_1 N^{-1/3}\hat{\sigma}_N^{-1} x_N \le 1/12 \tag{4.7}$$

in view of (4.1). Thus we can use Theorem 1 of Feller (1943) (cf. also Petrov, 1975, page 253) to obtain that for all $N \ge N_0$,

$$P(\hat{S}_N^0 > x_N \hat{\sigma}_N) = \exp\{-\tfrac{1}{2} x_N^2 Q_N(x_N)\}[1 - \Phi(x_N) + \theta_N \lambda_N \exp(-\tfrac{1}{2} x_N^2)] \tag{4.8}$$

where

$$\lambda_N = (2\|\phi\| + \|\phi_1\|)A_1 N^{-1/3}\hat{\sigma}_N^{-1}, \quad \theta_N < 7.465,$$
$$Q_N(x) = \textstyle\sum_{j=1}^\infty q_{Nj} x^j, \qquad q_{N1} = 3^{-1}\hat{\sigma}_N^{-3} \sum_{j=1}^N E(\hat{S}_N^{(j)})^3, \tag{4.9}$$
$$|q_{Nj}| < 8^{-1}(12\lambda_N)^j, \qquad j \ge 2.$$

Note that $|x_N| \le \rho_N N^{1/6} + N^{-1/6}$. Since $\rho_N = o(1)$ we have $x_N = o(N^{1/6})$ as $N \to \infty$. Let $K > 0$ be such that

$$|x_N| \le \frac{K\hat{\sigma}_N N^{1/6}}{12A_1(2\|\phi\| + \|\phi_1\|)}, \qquad N \ge 1. \tag{4.10}$$

Also it follows by assumption (B) and (4.1) that

$$|q_{N1} x_N| \le \hat{\sigma}_N^{-3}\, 4/3\, (2\|\phi\|^3 + \|\phi_1\|^3) \textstyle\sum_{j=1}^N |c_j|^3 x_N \le A_3 N^{-1/3} \tag{4.11}$$

where A_3 is an absolute constant. Hence combining (4.9), (4.10) and (4.11) we obtain that

$$|Q_N(x_N)| \le A_3 N^{-1/3} + \textstyle\sum_{j=2}^\infty |q_{Nj} x_N^j| \le A_3 N^{-1/3} + 8^{-1} \sum_{j=2}^\infty (KN^{-1/6})^j$$

which implies that as $N \to \infty$

$$x_N^2 Q_N(x_N) = x_N^2 O(N^{-1/3}) = o(N^{1/3})O(N^{-1/3}) = o(1) \tag{4.12}$$

uniformly in $0 < x \le \rho_N N^{1/6}$. Moreover Lemma VII.1.2 of Feller (1968) ensures that

$$\theta_N \lambda_N \exp(-\tfrac{1}{2} x_N^2)[1 - \Phi(x_N)]^{-1} = O(\lambda_N x_N) = o(1),$$
(4.13)
$$[1 - \Phi(x_N)]^{-1}[1 - \Phi(x)] = 1 + o(1).$$

(4.2) now follows uniformly in $0 < x \leq \rho_N N^{1/6}$ by using (4.8), (4.12) and (4.13).

REMARK 4.1. The above theorem obviously holds for the case of the iid r.v.'s. However, for such a case, and for the statistics (1.4) and (1.5), the theorem holds under somewhat weaker assumptions on the score generating function, if we assume somewhat stronger assumption on the underlying distribution function. For the case of the statistic (1.4), we refer to Kallenberg (1982). For the case of the statistic (1.5), we have the following theorem:

THEOREM 4.1. *Let* X_{Nj}, $1 < j < N$, $N \geq 1$ *be iid r.v.'s with a continuous c.d.f.* $F_N(x)$ *symmetric about* 0. *Assume the following*:

ASSUMPTION (D). $\sum_{j=1}^N c_{Nj}^2 = 1$, $\max_{1 \leq j \leq N} | c_{Nj} | \leq A_1 N^{-1/3}$ where A_1 is an absolute constant.

ASSUMPTION (E). *The score generating function* $\tilde{\phi}$ *(defined in Section* 1) *is not identically zero and satisfies a Lipschitz condition of order* 1 *on* (0, 1), *that is, there exists a constant C such that* $| \tilde{\phi}(t) - \tilde{\phi}(s) | \leq C | t - s |$ *for all* $t, s \in (0, 1)$.

Then, the conclusion of the Theorem 2.1 *holds.*

To prove this theorem we use the following two lemmas and proceed essentially as in the proof of Theorem 2.1. (As before, we suppress the subscript N in c_{Nj}, X_{Nj} etc. whenever it causes no confusion).

LEMMA 4.1. *Let* $Z_1, \cdots, Z_N$ *be random variables such that for any permutation* $(i_1, \cdots, i_N)$ *of* $(1, \cdots, N)$

(4.14)
$$E \prod_{j=1}^N Z_j^{\alpha_j} = E \prod_{j=1}^N Z_{i_j}^{\alpha_j}$$

where α_j*'s are nonnegative integers such that* $\sum_{j=1}^N \alpha_j = 2k$, $k \geq 1$ *integer. Furthermore, assume that*

(4.15)
$$E \prod_{j=1}^N Z_j^{\alpha_j} = 0$$

if at least one of the α_j's *is odd. Then if* $\sum_{j=1}^N c_j^2 = 1$, *for any integer* $k \leq d(\max_{1 \leq j \leq N} | c_j |)^{-1}$, $d > 0$

(4.16)
$$E(\textstyle\sum_{j=1}^N c_j Z_j)^{2k} \leq 2^{2k+1}(\max\{1, d\})^{2k} k^k E Z_1^{2k}.$$

Consider now the statistic $S_N = \sum_{j=1}^N c_j \tilde{\phi}(F_N^*(| X_j |)) \operatorname{sgn} X_j$ where $F_N^*(x) = P(| X_1 | \leq x)$, $0 \leq x < \infty$.

LEMMA 4.2. *For all real p,* $1 \le p \le N$,

$$E(T_N - S_N)^{2p} \le A_2^{2p}(2p)^{2p}N^{-p}(\max\{1, p \max_{1\le j\le N} | c_j |\})^{2p}$$

where A_2 *is a constant independent of N and p.*

PROOF OF LEMMA 4.1. Because of (4.14) and (4.15), the multinomial expansion yields

$$\begin{aligned} &E(\textstyle\sum_{j=1}^N c_j Z_j)^{2k} \\ (4.17)\qquad &= \textstyle\sum_{\alpha=1}^k \sum_{(k_1,\cdots,k_\alpha)\in A_\alpha} \dfrac{(2k)!}{\prod_{\nu=1}^\alpha [(2\nu)!]^{k_\nu} \prod_{\nu=1}^\alpha (k_\nu)!} \\ &\cdot \textstyle\sum_{(i_1,\cdots,i_{K_\alpha})\neq} E \prod_{j=1}^{K_1} (c_{i_j} Z_{i_j})^2 \prod_{j=K_1+1}^{K_2} (c_{i_j} Z_{i_j})^4 \cdots \prod_{j=K_{\alpha-1}}^{K_\alpha} (c_{i_j} Z_{i_j})^{2\alpha} \end{aligned}$$

where k_j, $1 \le j \le \alpha$ are nonnegative integers, $K_\beta = \sum_{\nu=1}^\beta k_\nu$, $1 \le \beta \le \alpha$, $A_\alpha = \{(k_1, k_2, \cdots, k_\alpha): \sum_{\nu=1}^\alpha \nu k_\nu = k\}$ and $\sum_{(i_1,\cdots,i_{K_\alpha})\neq}$ means that the sum is taken over mutually different indices $1 \le i_1, i_2, \cdots i_{K_\alpha} \le N$.

From (4.14) and generalized Hölder's inequality, it follows that

$$\begin{aligned} &\textstyle\sum_{(i_1,\cdots,i_{K_\alpha})\neq} E \prod_{j=1}^{K_1} (c_{i_j} Z_{i_j})^2 \prod_{j=K_1+1}^{K_2} (c_{i_j} Z_{i_j})^4 \cdots \prod_{j=K_{\alpha-1}}^{K_\alpha} (c_{i_j} Z_{i_j})^{2\alpha} \\ (4.18)\qquad &\le \textstyle\sum_{(i_1,\cdots,i_{K_\alpha})\neq} \pi(k_1, k_2, \cdots, k_\alpha) E Z_1^{2k} \end{aligned}$$

where

$$\pi(k_1, k_2, \cdots, k_\alpha) = \prod_{j=1}^{K_1} c_{i_j}^2 \prod_{j=K_1+1}^{K_2} c_{i_j}^4 \cdots \prod_{j=K_{\alpha-1}+1}^{K_\alpha} c_{i_j}^{2\alpha}.$$

Now, if we let $p_1 = \cdots = p_{K_1} = 2$,

$$p_{K_1+1} = \cdots = p_{K_2} = 4, \cdots, p_{K_{\alpha-1}+1} = \cdots = p_{K_\alpha} = 2\alpha,$$

by using the conditions $k \le d(\max_{1\le j\le N} | c_j |)^{-1}$ and $\sum_{j=1}^N c_j^2 = 1$; we get

$$\begin{aligned} &\textstyle\sum_{(i_1,\cdots,i_{k_\alpha})\neq} \pi(k_1, \cdots, k_\alpha) \le \sum_{\ell_1,\cdots,\ell_{K_\alpha}=1}^N c_{\ell_1}^{p_1} \cdots c_{\ell_{K_\alpha}}^{p_{K_\alpha}} \le \prod_{j=1}^{K_\alpha} (\sum_{i=1}^N c_i^{p_j}) \\ (4.19)\qquad &\le (\max | c_i |)^{2k-2K_\alpha} \le (d/k)^{2k-2K_\alpha}. \end{aligned}$$

Thus, relations (4.17)–(4.19) yield the estimate

$$\begin{aligned} &E(\textstyle\sum_{j=1}^N c_j Z_j)^{2k} \\ (4.20)\qquad &\le \textstyle\sum_{\alpha=1}^k \sum_{(k_1,\cdots,k_\alpha)\in A_\alpha} \dfrac{(2k)!}{\prod_{\nu=1}^\alpha [(2\nu)!]^{k_\nu} \prod_{\nu=1}^\alpha (k_\nu)!} \left(\dfrac{d}{k}\right)^{2k-2K_\alpha} E Z_1^{2k}. \end{aligned}$$

Further, by Stirling's formula (see Feller, 1968)

$$(2\pi)^{1/2} n^{n+1/2} e^{-n} < n! \le (2\pi)^{1/2} n^{n+1/2} e^{-n} (1 + (1/4n))$$

we obtain, after some computations, that

$$(4.21)\qquad \frac{\prod_{\nu=1}^\alpha (\nu!)^{k_\nu} (2k)!(k - K_\alpha)!}{\prod_{\nu=1}^\alpha [(2\nu)!]^{k_\nu} (k!)^2} \left(\frac{d}{k}\right)^{2k-2K_\alpha} \le 2^{2k+1} (\max\{1, d\})^{2k}.$$

Finally, on the basis of the expansion

$$k^k = \sum_{\alpha=1}^{k} \sum_{(k_1,\cdots,k_\alpha)\in A_\alpha} (k!)^2/\{\prod_{\nu=1}^{\alpha} (\nu!)^{k_\nu} \prod_{\nu=1}^{\alpha} (k_\nu)!(k - K_\alpha)!\}$$

inequalities (4.20) and (4.21) lead to the assertion (4.16) in Lemma 4.1.

PROOF OF LEMMA 4.2. Let $1 \leq p \leq N$ be any given integer and let $Z_j = \{\tilde{\phi}(R_j^+/(N+1)) - \tilde{\phi}(F^*(|X_j|))\} \operatorname{sgn} X_j, \quad 1 \leq j \leq N$. Then, using Lemma 4.1 with $d = p \max_{1\leq j\leq N} |c_j|$, it follows that

$$E(T_N - S_N)^{2p} \leq 2^{2p+1}(\max\{1, p \max_{1\leq j\leq N} |c_j|\})^{2p} p^p E Z_1^{2p}. \tag{4.22}$$

Now, using Hölder's inequality and proceeding as in Kallenberg (1982), we obtain the desired result for any real $p, \quad 1 \leq p \leq N$. The details are omitted.

REFERENCES

BERGSTRÖM, H. and PURI, M. L. (1977). Convergence and remainder terms in linear rank statistic. *Ann. Statist.* **5** 671–680.

BICKEL, P. J. (1974). Edgeworth expansions in nonparametric statistics. *Ann. Statist.* **2** 1–20.

BJERVE, S. (1977). Error bounds for linear combinations of order statistics. *Ann. Statist.* **5** 357–369.

CRAMÉR, H. (1938). Sur un nouveau theoreme-limite de la theorie des probabilities. *Actualites Sci. Ind.* **736** 5–23.

DOES, R. J. M. M. (1982). *Higher Order Asymptotics for Simple Linear Rank Statistics.* Mathematisch Centrum, Amsterdam.

FELLER, W. (1943). Generalization of a probability limit theorem of Cramér. *Trans. Amer. Math. Soc.* **54** 361–372.

FELLER, W. (1968). *An Introduction to Probability Theorem and its Applications*, Vol. I, 3rd edition. Wiley, New York.

HÁJEK, J. (1962). Asymptotically most powerful rank order tests. *Ann. Math. Statist.* **33** 1129–1147.

HÁJEK, J. (1968). Asymptotic normality of simple linear rank statistics under alternatives. *Ann. Math. Statist.* **39** 325–346.

HELMERS, R. (1977). The order of the normal approximation for linear combinations of order statistics with smooth weight functions. *Ann. Probab.* **5** 940–953.

HELMERS, R. (1980). A Berry-Esséen theorem for linear combinations of order statistics with smooth weight functions. *Ann. Statist.* **8** 1361–1374.

HELMERS, R. (1981). A Berry-Esséen theorem for linear combinations of order statistics. *Ann. Probab.* **9** 342–347.

HO, S. T. and CHEN, L. Y. H. (1978). An L_p bound for the remainder in a combinatorial central limit theorem. *Ann. Probab.* **6** 231–249.

HOEFFDING, W. (1951). A combinatorial central limit theorem. *Ann. Math. Statist.* **22** 558–566.

HUŠKOVÁ, M. (1970). Asymptotic distribution of simple linear rank statistics for testing symmetry. *Z. Wahrsch. verw. Gebiete.* **14** 308–322.

HUŠKOVÁ, M. (1977). The rate of convergence of simple linear rank statistics under hypothesis and alternatives. *Ann. Statist.* **5** 658–670.

HUŠKOVÁ, M. (1979). The rate of convergence of simple linear rank statistics under alternatives. *Contributions to Statistics*, 99–108. Ed. by J. Jurečková. Academic Publishing House, Prague.

JUREČKOVÁ, J. AND PURI, MADAN L. (1975). Order of normal approximation for rank test statistics distribution. *Ann. Probab.* **3** 526–533.

KALLENBERG, W. C. M. (1982). Cramér type large deviations for simple linear rank statistics. *Z. Wahrsch. verw. Gebiete.* **60** 403–409.

MALEVICH, T. L. and ABDALIMOV, B. (1979). Large deviation probabilities for *U*-statistics. *Theory Probab. Appl.* **24** 215–219.

MOTOO, M. (1956). On the Hoeffding's combinatorial central limit theorem. *Ann. Inst. Statist. Math.* **8** 145–154.

PETROV, V. V. (1975). *Sums of Independent Random Variables.* Springer-Verlag, Berlin.

PURI, M. L. and RALESCU, S. (1984). Centering of signed rank statistics with continuous score generating function. *Teor. Veroyatnost. i Primenen.* **29** (to appear).

PURI, M. L. and SEOH, M. (1984a). Berry-Esséen theorems for signed linear rank statistics with regression constants. *Colloquium on Limit Theorems in Probability and Statistics.* Veszprém (Hungary). To appear.

PURI, M. L. and SEOH, M. (1984b). Edgeworth expansions for signed linear rank statistics with regression constants. *J. Statist. Plann. Inference* **10** 137–149.

PURI, M. L. and SEOH, M. (1984c). Edgeworth expansions for signed linear rank statistics under near location alternatives. *J. Statist. Plann. Inference* **10** 289–309.

PURI, M. L. and SEOH, M. (1984d). On the rate of convergence in normal approximation for a class of statistics. Tech. report no. 23, Dept. of Mathematics, Indiana University, Bloomington.

PURI, M. L. and WU, T. J. (1985). The order of normal approximation for signed linear rank statistics. *Teor. Verojatnost. i Primenen.* (To appear).

ROBINSON, J. (1977). Large deviation probabilities for samples from a finite population. *Ann. Probab.* **5** 913–925.

SHORACK, G. R. (1969). Asymptotic normality of linear combinations of functions of order statistics. *Ann. Math. Statist.* **40** 2041–2050.

SHORACK, G. R. (1972). Functions of order statistics. *Ann. Math. Statist.* **43** 412–427.

STIGLER, S. M. (1974). Linear functions of order statistics with smooth weight functions. *Ann. Statist.* **2** 676–693.

VON BAHR, B. (1976). Remainder term estimate in a combinatorial limit theorem. *Z. Wahrsch. verw. Gebiete* **35** 131–139.

MUNSUP SEOH
DEPARTMENT OF MATHEMATICS
AND STATISTICS
WRIGHT STATE UNIVERSITY
DAYTON, OHIO 45435

STEFAN S. RALESCU
DIVISION OF APPLIED MATHEMATICS
BROWN UNIVERSITY
PROVIDENCE, R.I. 02912

MADAN L. PURI
DEPARTMENT OF MATHEMATICS
INDIANA UNIVERSITY
BLOOMINGTON, INDIANA 47405

ON THE RATE OF CONVERGENCE IN THE CENTRAL LIMIT THEOREM FOR SIGNED RANK STATISTICS

STEFAN S. RALESCU and MADAN L. PURI*

Division of Applied Mathematics, Brown University, Providence, Rhode Island 02912,
Department of Mathematics, Indiana University, Bloomington, Indiana 47405

DEDICATED TO THE MEMORY OF THE LATE PROFESSOR JACK KIEFER

The problem of the rate of convergence of the cumulative distribution function of the one-sample rank order statistics S_N^+ to the limiting normal one is studied for the case when the underlying observations are independent but not necessarily identically distributed. The results obtained are then used to derive a strong law and functional form of a central limit theorem for S_N^+.

1. Introduction

In recent years, the problem of determining the rate of convergence of the d.f.'s (distribution functions) of different rank statistics to the limiting normal d.f. has been the subject of several papers. In the regression setup, this problem has been studied by Jurečková and Puri [14], Bergström and Puri [3], and Hušková [12, 13] among others (see also Albers, Bickel, and van Zwet [1] and the references cited therein). However, in all these papers the score generating function is assumed to be smooth and bounded. Recently, Müller-Funk and Witting [17] considered this problem for the one sample case and were able to cover also the unbounded score generating function which includes, among others, the important normal scores statis-

*This work was partially supported by the Office of Naval Research under Contract N00014-77-C-0659. The U.S. Government's right to retain a nonexclusive royalty-free license in and to the copyright covering this paper, for governmental purposes, is acknowledged.

tics. The main restriction in their paper was, however, that the observations were all assumed to be independent and identically distributed. Also they had made a strong assumption regarding the monotonicity of the score generating function.

The present paper is an extended study of the problem considered by Müller-Funk and Witting [17] for the case when the score generating function is unbounded (and also not necessarily monotone), and when the underlying r.v.'s are independent but not necessarily identically distributed. In dealing with this problem we use the Chernoff–Savage approach which relies on the possibility of expressing the one sample rank statistic as integrals in terms of empirical d.f.'s. In this approach, the empirical d.f.'s and their properties serve as a probabilistic tool to arrive at results for the rank statistics. Hitherto, these properties were known in the i.i.d. case only (see Dvoretzky, Kiefer, and Wolfowitz [8]), and our problem requires knowledge of the behavior of the empirical d.f. in the non-i.i.d. case. These fundamental properties of the empirical d.f.'s in the non-i.i.d. case are derived in Section 3, and they generalize the results of Dvoretzky, Kiefer, and Wolfowitz [8].

The main result of the present study is that if the score generating function increases to infinity (near one) in a suitable manner (see (2.3)), then the rate of convergence of the d.f. of the one-sample rank statistic to the limiting normal distribution is $O(N^{-\alpha})$ for some $0 < \alpha < \frac{1}{2}$.

Section 5 deals with applications; it contains a functional limit theorem (Donsker-type invariance principle) and a strong law of large numbers for the one sample rank statistics in the non-i.i.d. case. Some of these applications are useful in sequential analysis and in this connection we refer to Lai [15] and Müller-Funk [16].

2. Notations and the Main Result

Let $X_{N1}, \ldots, X_{NN}$, $N \geq 1$, be independent random variables with continuous d.f.'s $F_{N1}, \ldots, F_{NN}$, $N \geq 1$, respectively. Consider the one-sample rank order statistic

$$S_N^+ = \frac{1}{N} \sum_{i=1}^{N} u(X_{Ni}) a_N(R_{Ni}^+) \tag{2.1}$$

where R_{Ni}^+ is the rank of $|X_{Ni}|$ among $|X_{N1}|, \ldots, |X_{NN}|$ respectively, $u(t) = 1$ or 0 according as $t \geq 0$ or < 0 and $a_N(i)$, $1 \leq i \leq N$, are scores generated by a known function $\psi(t)$, $0 < t < 1$, by

$$a_N(i) = \psi(i/(N+1)), \qquad 1 \leq i \leq N. \tag{2.2}$$

We shall require the following Chernoff–Savage assumption concerning the smoothness and the order of magnitude of the score generating function near the boundary of the unit interval. We assume that ψ is twice differentiable on $(0,1)$ and for all $0 < t < 1$,

$$|\psi(t)| \le K(1-t)^{-1/2+\lambda}, \qquad |\psi^{(i)}(t)| \le K(1-t)^{-i-1/2+\delta} \tag{2.3}$$

for $i = 1, 2$ and some $\frac{1}{4} < \delta < \lambda < \frac{1}{2}$, $2\lambda - \delta > \frac{1}{2}$, where K is a positive constant.

We remark that (2.3) holds true, for example, if $\psi(t) = G^{-1}((1+t)/2)$, $0 \le t < 1$, where G is a symmetric d.f. having a smooth density and whose tail has an increasing failure rate. In particular, if $G = \Phi$ (standard normal d.f.) S_N^+ is termed the *normal scores statistic.*

For the ease of convenience, we shall suppress, from now on, the subscript N from F_{Ni}, R_{Ni}^+, etc.

We now set

$$\begin{aligned} \hat{F}_N(x) &= \frac{1}{N}\sum_{i=1}^{N} u(x - X_i), &\qquad \bar{F}_N(x) &= \frac{1}{N}\sum_{i=1}^{N} F_i(x) \\ \hat{H}_N(x) &= \frac{1}{N}\sum_{i=1}^{N} u(x - |X_i|), &\qquad \bar{H}_N(x) &= \frac{1}{N}\sum_{i=1}^{N} F_i^*(x) \end{aligned} \tag{2.4}$$

where $F_i^*(x) = P\{|X_i| \le x\}$, $1 \le i \le N$.

Now S_N^+ in (2.1) may be written as

$$S_N^+ = \int_0^\infty \psi\left(\frac{N}{N+1}\hat{H}_N(x)\right) d\hat{F}_N(x). \tag{2.5}$$

Set

$$\mu_N^+ = \int_0^\infty \psi(\bar{H}_N(x))\, d\bar{F}_N(x) \tag{2.6}$$

and note that under the assumption (2.3), μ_N^+ is finite.

Then, with probability one, we have the decomposition

$$S_N^+ - \mu_N^+ = A_N + B_N + Q_N \tag{2.7}$$

where

$$A_N = \int_0^\infty \psi(\bar{H}_N)\, d(\hat{F}_N - \bar{F}_N), \qquad B_N = \int_0^\infty (\hat{H}_N - \bar{H}_N)\psi'(\bar{H}_N)\, d\bar{F}_N, \tag{2.8}$$

and

$$Q_N = \int_0^\infty \left(\psi\left(\frac{N}{N+1} \hat{H}_N \right) - \psi(\overline{H}_N) \right) d\hat{F}_N - \int_0^\infty (\hat{H}_N - \overline{H}_N) \psi'(\overline{H}_N)\, d\overline{F}_N. \tag{2.9}$$

As in Chernoff and Savage [6] we term $A_N + B_N$ and Q_N respectively the principal and remainder terms of the (centered) S_N^+.

Denote

$$\sigma_N^2 = \operatorname{Var}(A_N + B_N) \tag{2.10}$$

and note that, under (2.3), this quantity is well defined.

We shall require the nondegeneration of σ_N^2 in the form

$$\liminf_{N \to \infty} N\sigma_N^2 > 0. \tag{2.11}$$

The main result of this paper is the following.

THEOREM 2.1. *Consider the statistic S_N^+ defined by* (2.1) *with scores given by* (2.2). *Then if* (2.3) *and* (2.11) *are satisfied, there exist $C > 0$ and $0 < \alpha < \frac{1}{2}$ (both depending on our assumptions only) such that*

$$\sup_{-\infty < x < \infty} |P(S_N^+ - \mu_N^+ \leq \sigma_N x) - \Phi(x)| \leq CN^{-\alpha}$$

where μ_N^+ and σ_N^2 are defined respectively by (2.6) *and* (2.10).

Remark. It may be noted that we do not assume the monotonicity of the score generating function ψ.

3. PRELIMINARY LEMMAS ON EMPIRICAL d.f.'s.

Let $U_1, \ldots, U_N$, $N \geq 1$, be i.i.d. r.v.'s from the uniform $(0,1)$ d.f. and denote by $\hat{F}_N$ the corresponding empirical d.f. The following basic result concerning a certain asymptotic order for the tail probabilities was obtained by Dvoretzki, Kiefer, and Wolfowitz [8]. For every $\epsilon > 0$, there exist $C_1, C_2 > 0$ and N_0 (all depending on ϵ) such that

$$P\left\{ \sup_{0<t<1} |\hat{F}_N(t) - t| \geq C_1 N^{-1/2} (\operatorname{Log} N)^{1/2} \right\} \leq C_2 N^{-\epsilon}, \qquad N \geq N_0. \tag{3.1}$$

Our main goal in this section is to derive statements similar to (3.1) concerning fluctuations of the empirical d.f. $\hat{F}_N$ from the averaged theoretical d.f. $\bar{F}_N$ when the N sample elements are assumed to be independent but not necessarily identically distributed. These results are then used in the next section for obtaining estimates for the speed of convergence in the C.L.T. for signed rank statistics.

One basic tool for our study is the following inequality due to S. Bernstein.

PROPOSITION 3.1. *Suppose $Y_1, \ldots, Y_N$ are independent random variables such that $E(Y_i) = 0$ and $|Y_i| \leq c$ for some $0 < c < \infty$, $1 \leq i \leq N$. Then for every $\omega > 0$,*

$$P\left\{\left|\sum_{i=1}^{N} Y_i\right| > \omega\right\} < 2\exp(-h_N) \tag{3.2}$$

where

$$h_N = \omega^2/(2B_N + 2c\omega), \qquad B_N = \operatorname{Var}\left(\sum_{i=1}^{N} Y_i\right). \tag{3.3}$$

For a proof of this inequality we refer to Uspensky [21, p. 205].

Our first result is a generalization to the non-i.i.d. case of a lemma of Ghosh [10] which has been applied to the theory of linear rank statistics by Müller-Funk [16] and Müller-Funk and Witting [17].

LEMMA 3.1. *For every $\epsilon > 0$, there exist $M > 0$ and N_0 (both depending on ϵ) such that for $N \geq N_0$ we have*

$$P\left\{\sup_{N^{-1} \leq \bar{F}(x) \leq 1 - N^{-1}} \frac{|\hat{F}_N(x) - \bar{F}(x)|}{\{\bar{F}(x)(1 - \bar{F}(x))\}^{1/2}} \geq MN^{-1/2}\operatorname{Log} N\right\} \leq 2N^{-1-\epsilon} \tag{3.4}$$

and

$$P\left\{\sup_{R} \frac{|\hat{F}_N(x) - \bar{F}(x)|}{\{\bar{F}(x)(1 - \bar{F}(x))\}^{1/2-\epsilon_0}} \geq MN^{-1/2}\operatorname{Log} N\right\} \leq 2N^{-1-\epsilon} \tag{3.5}$$

where $\epsilon_0 = (1 + \epsilon)/2(2 + \epsilon)$.

Proof. First we note that since

$$P\{\omega : \hat{F}_N(x) = \hat{F}_N(\bar{F}_N^{-1}(\bar{F}_N(x))), \text{ all } x \in R, \text{ all } N \geq 1\} = 1 \tag{3.6}$$

we have, with probability 1,

$$\sup_{N^{-1}\le \bar{F}(x)\le 1-N^{-1}} \frac{|\hat{F}_N(x)-\bar{F}(x)|}{\{\bar{F}(x)(1-\bar{F}(x))\}^{1/2}} = \sup_{N^{-1}\le t\le 1-N^{-1}} \frac{|\hat{F}_N(\bar{F}^{-1}(t))-t|}{\{t(1-t)\}^{1/2}}. \tag{3.7}$$

Set $G_N(t)=\{t(1-t)\}^{-1/2}[\hat{F}_N(\bar{F}^{-1}(t))-t]$, $0<t<1$ and $\alpha_{j,N}=j/N$, $1\le j\le N-1$.

Then, for $t\in[\alpha_{j-1,N},\alpha_{j,N}]$, $2\le j\le N-1$,

$$\left(\frac{j-1}{j}\right)^{1/2} G_N(\alpha_{j-1,N})-[j(N-j+1)]^{-1/2} \le G_N(t)\le \left(\frac{j}{j-1}\right)^{1/2} G_N(\alpha_{j,N})+[(j-1)(N-j)]^{-1/2} \tag{3.8}$$

and, since for $2\le j\le N-1$, $N\ge 4$, $\max\{[(j-1)(N-j)]^{-1/2}, [j(N-j+1)]^{-1/2}\}\le \sqrt{2}N^{-1/2}$ and $(j/(j-1))^{1/2}\le\sqrt{2}$, we get from (3.8) that

$$\sup_{N^{-1}\le t\le 1-N^{-1}} |G_N(t)|\le \sqrt{2}\max_{1\le j\le N-1}|G_N(\alpha_{j,N})|+\sqrt{2}N^{-1/2}. \tag{3.9}$$

Now, with $1\le j\le N-1$ fixed, note that

$$N\hat{F}_N(\bar{F}^{-1}(\alpha_{j,N}))=\sum_{i=1}^{N} u(\bar{F}^{-1}(\alpha_{j,N})-X_i)=\sum_{i=1}^{N} Z_i$$

where Z_i, $1\le i\le N$, are independent Bernoulli r.v.'s with $P(Z_i=1)=F_i(\bar{F}^{-1}(\alpha_{j,N}))=p_i$, $1\le i\le N$.

Consequently, with $M_1>0$ to be specified later, by Bernstein's inequality (3.2) with $Y_i=Z_i-p_i$, $1\le i\le N$, and $\omega=M_1\{N\bar{p}(1-\bar{p})\}^{1/2}\operatorname{Log} N$ where $\bar{p}=(1/N)\Sigma_{i=1}^{N} p_i=\alpha_{j,N}$, we obtain for $1\le j\le N-1$

$$P\left\{|G_N(\alpha_{j,N})|\ge M_1 N^{-1/2}\operatorname{Log} N\right\}\le 2\exp(-h_{j,N}) \tag{3.10}$$

where

$$h_{j,N}\ge \frac{M_1^2}{2}\cdot\frac{\operatorname{Log}^2 N}{1+M_1[N\alpha_{j,N}(1-\alpha_{j,N})]^{-1/2}\operatorname{Log} N}. \tag{3.11}$$

The last inequality follows from (3.3) and the fact that

$$B_N = \sum_{i=1}^{N} \operatorname{Var} Z_i = \sum_{i=1}^{N} p_i(1-p_i) \le N\bar{p}(1-\bar{p}).$$

Now using (3.11) and the fact that $[\alpha_{j,N}(1-\alpha_{j,N})]^{-1/2} \le (2N)^{1/2}$ for all $1 \le j \le N-1$, it is easily seen that if we choose $M_1 \ge 2^{5/2}(2+\epsilon)$, then

$$\sum_{j=1}^{N-1} \exp(-h_{j,N}) \le N^{-1-\epsilon}, \qquad N \ge N_1. \tag{3.12}$$

Finally, if we let $M = \sqrt{2}(1 + M_1)$, the assertion (3.4) follows from (3.7), (3.9), (3.10), and (3.12).

To prove (3.5) we first show that with probability $\ge 1 - N^{-1-\epsilon}$

$$\sup_{0<t<N^{-2-\epsilon}} |G_N(t)| = O(N^{-1-\epsilon/2}). \tag{3.13}$$

Indeed, using Bernoulli's inequality we have

$$P\left\{\sup_{0<t<N^{-2-\epsilon}} \hat{F}_N(\bar{F}^{-1}(t)) = 0\right\} \ge 1 - \sum_{i=1}^{N} P\{X_i \le \bar{F}^{-1}(N^{-2-\epsilon})\}$$
$$= 1 - N^{-1-\epsilon}$$

which implies that with probability $\ge 1 - N^{-1-\epsilon}$

$$\sup_{0<t<N^{-2-\epsilon}} |G_N(t)| \le \sup_{0<t<N^{-2-\epsilon}} t^{1/2}(1-t)^{-1/2} = O(N^{-1-\epsilon/2}).$$

Similarly one proves that with probability $\ge 1 - N^{-1-\epsilon}$

$$\sup_{1-N^{-2-\epsilon}<t<1} |G_N(t)| = O(N^{-1-\epsilon/2}). \tag{3.14}$$

We now show that for every $\epsilon > 0$, there exist $M' > 0$ and N_1' such that

$$P\left\{\sup_{N^{-2-\epsilon} \le t < N^{-1}} |K_N(t)| \ge M' N^{-1/2} \operatorname{Log} N\right\} \le 2N^{-1-\epsilon}, \qquad N \ge N_1' \tag{3.15}$$

where $K_N(t) = \{t(1-t)\}^{\epsilon_0} G_N(t)$, $0 < t < 1$.

By considerations similar to those used in the derivation of (3.9) we obtain

$$\sup_{N^{-2-\epsilon} \le t < N^{-1}} |K_N(t)| \le \sqrt{2} \max_{1 \le j \le [N^{1+\epsilon}]} \left| K_N(\beta_{j,N}) \right| + O(N^{-3/2-\epsilon}) \tag{3.16}$$

where $\beta_{j,N} = j/N^{2+\epsilon}$, $1 \le j \le [N^{1+\epsilon}]$ ($[\cdot]$ = *integer part*).

Therefore, to prove (3.15) it suffices to show that there exist M_1' and N_1' such that for $N \ge N_1'$

$$P\left\{ \max_{1 \le j \le [N^{1+\epsilon}]} \left| K_N(\beta_{j,N}) \right| \ge M_1' N^{-1/2} \operatorname{Log} N \right\} \le 2N^{-1-\epsilon}. \tag{3.17}$$

Now, using Bernstein's inequality and the relations

$$N^{1/2} \operatorname{Log}^{-1} N \left\{ \beta_{j,N}(1 - \beta_{j,N}) \right\}^{1/2+\epsilon_0} \le N^{-\epsilon_0} \operatorname{Log}^{-1} N,$$

$$N^{1/2} \left\{ \beta_{j,N}(1 - \beta_{j,N}) \right\}^{1/2-\epsilon_0} \ge 2^{-1/2(2+\epsilon)}, \qquad N \ge 2,$$

we get, after some calculation, that for $N \ge N_1' = N_1'(M_1', \epsilon)$

$$P\left\{ \left| K_N(\beta_{j,N}) \right| \ge M_1' N^{-1/2} \operatorname{Log} N \right\} \le 2\exp(-l_{j,N}) \tag{3.18}$$

where $l_{j,N} \ge (M_1'/2)(\operatorname{Log} N) 2^{-1/2(2+\epsilon)}/[1 + (M_1' \operatorname{Log} N)^{-1} N^{-\epsilon_0}] \ge M_2' \operatorname{Log} N^{1/4}$ with $M_2' = M_1' 2^{-1/2(2+\epsilon)-2}$.

Hence, if we choose $M_2' > 2 + 2\epsilon$, we obtain $2\Sigma_{j=1}^{[N^{1+\epsilon}]} \exp(-l_{j,N}) \le 2N^{1+\epsilon-M_2} \le 2N^{-1-\epsilon}$ which together with (3.18) entails the satisfaction of (3.17). Thus (3.15) holds true.

Similarly one proves that for every $\epsilon > 0$, there exist $M'' > 0$ and N_2' such that

$$P\left\{ \sup_{1-N^{-1} < t \le 1 - N^{-2-\epsilon}} |K_N(t)| \ge M'' N^{-1/2} \operatorname{Log} N \right\} \le 2N^{-1-\epsilon},$$

$$N \ge N_2'. \tag{3.19}$$

Finally (3.4), (3.13), (3.14), (3.15), and (3.19) entail the satisfaction of (3.5). The proof of Lemma 3.1 is completed.

Remark. Using the inequality

$$\left| \hat{F}_N^*(\bar{F}^{-1}(t)) - t \right| \le \frac{N}{N+1} \left\{ \left| \hat{F}_N(\bar{F}^{-1}(t)) - t \right| + \frac{t}{N} \right\},$$

it is easy to see that (3.4) and (3.5) remain true if we replace $\hat{F}_N$ by $\hat{F}_N^* = (N/(N+1))\hat{F}_N$.

Lemma 3.2. *For every $\epsilon > 0$ and every $\delta \in (0, \frac{1}{2})$ there exist $M > 0$ and N_0 (both depending on ϵ only) such that for $N \geq N_0$ we have*

$$P\left\{ \sup_{N^{-1+\delta} \leq \bar{F}(x) \leq 1-N^{-1+\delta}} \left| \frac{\hat{F}_N(x)}{\bar{F}(x)} - 1 \right| \geq MN^{-\delta/2} \operatorname{Log} N \right\} \leq 2N^{-1-\epsilon} \tag{3.20}$$

$$P\left\{ \sup_{N^{-1+\delta} \leq \bar{F}(x) \leq 1-N^{-1+\delta}} \left| \frac{1-\hat{F}_N(x)}{1-\bar{F}(x)} - 1 \right| \geq MN^{-\delta/2} \operatorname{Log} N \right\} \leq 2N^{-1-\epsilon} \tag{3.21}$$

$$P\left\{ \frac{N^{-1+\delta}}{2} \leq \hat{F}_N\left(\bar{F}^{-1}(N^{-1+\delta})\right) \leq \frac{3N^{-1+\delta}}{2}, \right.$$
$$\left. \frac{N^{-1+\delta}}{2} \leq 1 - \hat{F}_N\left(\bar{F}^{-1}(1-N^{-1+\delta})\right) \leq \frac{3N^{-1+\delta}}{2} \right\} \geq 1 - 2N^{-1-\epsilon}. \tag{3.22}$$

Moreover, the above inequalities remain true if we replace $\hat{F}_N$ by $\hat{F}_N^*$.

Proof. The proof of (3.20) follows by first noting that for $N^{-1+\delta} \leq \bar{F}(x) \leq 1 - N^{-1+\delta}$,

$$\left| \frac{\hat{F}_N(x)}{\bar{F}(x)} - 1 \right| \leq N^{(1-\delta)/2} \sup_{N^{-1+\delta} \leq \bar{F}(x) \leq 1-N^{-1+\delta}} \frac{|\hat{F}_N(x) - \bar{F}(x)|}{\{\bar{F}(x)(1-\bar{F}(x))\}^{1/2}}$$

and then applying (3.4).

The proof of (3.21) is similar.

Finally, from (3.4) and the inequalities

$$\left|\hat{F}_N\left(\bar{F}^{-1}(N^{-1+\delta})\right) - N^{-1+\delta}\right|$$
$$\leq N^{(-1+\delta)/2} \sup_{N^{-1+\delta} \leq \bar{F}(x) \leq 1-N^{-1+\delta}} \frac{|\hat{F}_N(x) - \bar{F}(x)|}{\{\bar{F}(x)(1-\bar{F}(x)\}^{1/2}}$$

and

$$\left|\left[1 - \hat{F}_N\left(\bar{F}^{-1}(1-N^{-1+\delta})\right)\right] - N^{-1+\delta}\right|$$
$$\leq N^{(-1+\delta)/2} \sup_{N^{-1+\delta} \leq \bar{F}(x) \leq 1-N^{-1+\delta}} \frac{|\hat{F}_N(x) - \bar{F}(x)|}{\{\bar{F}(x)(1-\bar{F}(x))\}^{1/2}},$$

we obtain (3.22). The proof of Lemma 3.2 follows.

Lemma 3.3. *For every $k > 1$ and every $\delta \in (0, \frac{1}{2})$, there exist $M > 0$ and N_0 (both depending on k only) such that for $N \geq N_0$ we have*

$$P\left\{ \sup_{|\bar{F}(x) - \bar{F}(y)| \leq N^{-1+\delta} \operatorname{Log} N} |\hat{F}_N(x) - \hat{F}_N(y) - \bar{F}(x) + \bar{F}(y)| \right.$$

$$\left. \geq MN^{-1+\delta/2} \operatorname{Log} N \right\} \leq N^{-k}. \tag{3.23}$$

Proof. As in the proof of Lemma 3.1 it suffices to show that there exist $M > 0$ and N_0 such that for $N \geq N_0$

$$P\left\{ \sup_{|s-t| \leq N^{-1+\delta} \operatorname{Log} N} \left| \hat{F}_N(\bar{F}^{-1}(s)) - \hat{F}_N(\bar{F}^{-1}(t)) - s + t \right| \right.$$

$$\left. \geq MN^{-1+\delta/2} \operatorname{Log} N \right\} \leq N^{-k}. \tag{3.24}$$

With this end in view, set

$$D_N(t) = \sup_{|s-t| \leq N^{-1+\delta} \operatorname{Log} N} \left| \hat{F}_N(\bar{F}^{-1}(s)) - \hat{F}_N(\bar{F}^{-1}(t)) - s + t \right|,$$

$$0 < t < 1 \tag{3.25}$$

and

$$\theta_{N,j} = j/[N^{1-\delta}], \qquad j = 0, 1, \ldots, [N^{1-\delta}] \ ([\cdot] = \text{integer part}). \tag{3.26}$$

Let us show first that

$$\sup_{|s-t| \leq N^{-1+\delta} \operatorname{Log} N} \left| \hat{F}_N(\bar{F}^{-1}(s)) - \hat{F}_N(\bar{F}^{-1}(t)) - s + t \right|$$

$$\leq 3 \max_{1 \leq j \leq [N^{1-\delta}]} D_N(\theta_{N,j}). \tag{3.27}$$

Indeed, if $s, t \in [\theta_{N,j}, \theta_{N,j+1}]$, $|s - t| \leq N^{-1+\delta} \operatorname{Log} N$, then since $\max\{|s - \theta_{N,j+1}|, |t - \theta_{N,j}|\} \leq \theta_{N,j+1} - \theta_{N,j} \leq N^{-1+\delta} \operatorname{Log} N$, we get

$$\left| \hat{F}_N(\bar{F}^{-1}(s)) - \hat{F}_N(\bar{F}^{-1}(t)) - s + t \right| \leq 2 D_N(\theta_{N,j+1}). \tag{3.28}$$

On the other hand, if $s \in [\theta_{N,j}, \theta_{N,j+1}]$ and $t \in [\theta_{N,r}, \theta_{N,r+1}]$ with $j+1 \le r$, $|s-t| \le N^{-1+\delta}\mathrm{Log}\, N$, then since $\theta_{N,r} - \theta_{N,j+1} \le t - s \le N^{-1+\delta}\mathrm{Log}\, N$ we find

$$\left|\hat{F}_N(\bar{F}^{-1}(s)) - \hat{F}_N(\bar{F}^{-1}(t)) - s + t\right| \le D_N(\theta_{N,j+1}) + 2D_N(\theta_{N,r}). \tag{3.29}$$

Thus (3.28) and (3.29) entail (3.27).

Further, set

$$T_{Nj}(s) = \hat{F}_N(\bar{F}^{-1}(s)) - \hat{F}_N(\bar{F}^{-1}(\theta_{N,j})) - s + \theta_{N,j}, \qquad 1 \le j \le [N^{1-\delta}] \tag{3.30}$$

and

$$\xi_{r,N} = \theta_{N,j} + \frac{N^{-1+\delta}\mathrm{Log}\, N}{[N^{\delta/2}]} r, \qquad r \in Z. \tag{3.31}$$

Now, if $s \in [\xi_{r,N}, \xi_{r+1,N}]$ for some $-[N^{\delta/2}] \le r \le [N^{\delta/2}] - 1$, it is easy to see that

$$T_{Nj}(s) \le T_{Nj}(\xi_{r+1,N}) + (\xi_{r+1,N} - \xi_{r,N})$$

and

$$T_{Nj}(s) \ge T_{Nj}(\xi_{r,N}) - (\xi_{r+1,N} - \xi_{r,N}).$$

Consequently,

$$\sup_{\xi_{r,N} \le s \le \xi_{r+1,N}} |T_{Nj}(s)| \le \max_{l=r,r+1} |T_{Nj}(\xi_{l,N})| + (\xi_{r+1,N} - \xi_{r,N}) \tag{3.32}$$

and since $\xi_{r+1,N} - \xi_{r,N} \le 2N^{-1+\delta/2}\mathrm{Log}\, N$

$$|D_N(\theta_{N,j})| \le \max_{-[N^{\delta/2}] \le r \le [N^{\delta/2}]-1} |T_{Nj}(\xi_{r,N})| + 2N^{-1+\delta/2}\mathrm{Log}\, N. \tag{3.33}$$

Let us now evaluate $P\{|T_{Nj}(\xi_{r,N})| \ge MN^{-1+\delta/2}\mathrm{Log}\, N\}$. Note that

$$NT_{Nj}(\xi_{r,N}) = \sum_{i=1}^{N} \Big[u(\bar{F}^{-1}(\xi_{r,N}) - X_i) - u(\bar{F}^{-1}(\theta_{N,j}) - X_i)$$

$$- F_i(\bar{F}^{-1}(\xi_{r,N})) + F_i(\bar{F}^{-1}(\theta_{N,j}))\Big] = \sum_{i=1}^{N} W_i,$$

$$|W_i| \le 1, 1 \le i \le N.$$

Assume for the moment that $r \geq 0$. Set $u_i = F_i(\bar{F}^{-1}(\xi_{r,N}))$ and $v_i = F_i(\bar{F}^{-1}(\theta_{N,j}))$. Then $u_i > v_i$ and $\operatorname{Var} W_i$, after some calculations, equals $(u_i - v_i)[1 - (u_i - v_i)] = p_i(1 - p_i)$ with $p_i = u_i - v_i$, $0 < p_i < 1$, $1 \leq i \leq N$. Now, with $\bar{p} = \sum_{i=1}^N p_i/N$, we have

$$B_N = \sum_{i=1}^{N} \operatorname{Var} W_i = \sum_{i=1}^{N} p_i(1 - p_i) \leq N\bar{p}(1 - \bar{p}) \leq N^{\delta} \operatorname{Log} N$$

where the last inequality holds since

$$\bar{p} = \xi_{r,N} - \theta_{N,j} = \frac{N^{-1+\delta}\operatorname{Log} N}{[N^{\delta/2}]} r \leq N^{-1+\delta}\operatorname{Log} N.$$

Thus, using Bernstein's inequality (3.2), we get

$$P\left\{ \left| NT_{Nj}(\xi_{r,N}) \right| > MN^{\delta/2}\operatorname{Log} N \right\} \leq 2\exp(-h_N) \tag{3.34}$$

where

$$h_N = \frac{M^2 N^{\delta} \operatorname{Log}^2 N}{2B_N + 2MN^{\delta/2}\operatorname{Log} N} \geq \frac{M^2}{2} \cdot \frac{\operatorname{Log} N}{1 + MN^{-\delta/2}} = \lambda_N.$$

Thus

$$P\left\{ \left| T_{Nj}(\xi_{r,N}) \right| > MN^{-1+\delta/2}\operatorname{Log} N \right\} \leq 2\exp(-\lambda_N). \tag{3.35}$$

A similar bound holds also for $r < 0$. Hence

$$P\left\{ \max_{-[N^{\delta/2}] \leq r \leq [N^{\delta/2}]-1} \left| T_{Nj}(\xi_{r,N}) \right| > MN^{-1+\delta/2}\operatorname{Log} N \right\}$$
$$\leq 4N^{\delta/2}\exp(-\lambda_N). \tag{3.36}$$

Now, from (3.33) and (3.36), we obtain

$$P\left\{ \left| D_N(\theta_{N,j}) \right| \geq MN^{-1+\delta/2}\operatorname{Log} N \right\}$$
$$\leq 4N^{\delta/2}\exp\left\{ -\frac{(M-2)^2}{2} \cdot \frac{\operatorname{Log} N}{1 + (M-2)N^{-\delta/2}} \right\} = 4N^{\delta/2}\exp(-\delta_N) \tag{3.37}$$

which, together with (3.27), entails

$$P\left\{ \sup_{|s-t|\le N^{-1+\delta}\operatorname{Log} N} \left|F_N\left(\bar{F}^{-1}(s)\right) - F_N\left(\bar{F}^{-1}(t)\right) - s + t\right| \right.$$
$$\left. \ge 3MN^{-1+\delta/2}\operatorname{Log} N \right\}$$
$$\le P\left\{ \max_{1\le j\le [N^{1-\delta}]} D_N\left(\theta_{N,j}\right) \ge MN^{-1+\delta/2}\operatorname{Log} N \right\}$$
$$\le 4N^{1-\delta/2}\exp(-\delta_N) = \rho_N. \tag{3.38}$$

It follows easily from the definition of δ_N and ρ_N that

$$\frac{\log \rho_N}{\log N} \to \left(1 - \frac{\delta}{2}\right) - \frac{(M-2)^2}{2} < 1 - \frac{(M-2)^2}{2} \qquad \text{as } N \to \infty.$$

Thus, if M is chosen sufficiently large (depending on $k > 1$) we obtain for all $N \ge N_1(k)$ that

$$\log \rho_N/\log N \le -k, \qquad \text{i.e., } \rho_N \le N^{-k}.$$

Therefore (3.23) is entailed by (3.38) and the proof of Lemma 3.3 is terminated.

LEMMA 3.4. *For every $k > 1$ and every $\delta \in (0, \frac{1}{2})$, there exist $M > 0$ and N_0 (both depending on k only) such that for $N \ge N_0$ we have*

$$P\left\{|\hat{F}_N - \bar{F}|(I) \ge MN^{-1+\delta}\right\} \le N^{-k} \tag{3.39}$$

where $|\hat{F}_N - \bar{F}|$ denotes the total variation of the measure $(\hat{F}_N - \bar{F})(B) = \int_B d(\hat{F}_N - \bar{F})$, $B \subset R$ is a Borel set, and

$$I = \left(\bar{H}^{-1}(s), \bar{H}^{-1}(t)\right], \text{ with } N^{-1+\delta} \le s < t \le 1 - N^{-1+\delta},$$
$$t - s \le N^{-1+\delta}.$$

Proof. With I as defined above, we have

$$|\hat{F}_N - \bar{F}|(I) \le |\hat{H}_N(I) - \bar{H}(I)| + 2\bar{H}(I)$$
$$\le \sup_{|\bar{H}(x)-\bar{H}(y)|\le N^{-1+\delta}\operatorname{Log} N} |\hat{H}_N(x) - \hat{H}_N(y) - \bar{H}(x) + \bar{H}(y)| + 2N^{-1+\delta}.$$

The last relation, together with Lemma 3.3 (applied with $\hat{F}_N$ replaced by $\hat{H}_N$ and $\bar{F}$ replaced by $\bar{H}$), clearly implies (3.39).

Finally, we shall need the following result.

LEMMA 3.5. *For every $k > 1$ we have*

$$P\Big\{\max_{1\le i\le N}|X_i| \le \overline{H}^{-1}(1-N^{-k-1})\Big\} \ge 1 - N^{-k}. \tag{3.40}$$

Proof. Applying Bernoulli's inequality we have

$$P\Big\{\max_{1\le i\le N}|X_i| \le \overline{H}^{-1}(1-N^{-k-1})\Big\}$$
$$\ge 1 - \sum_{i=1}^{N} P\big\{|X_i| > \overline{H}^{-1}(1-N^{-k-1})\big\} = 1 - N^{-k}$$

as was to be proved.

4. PROOF OF THEOREM 2.1.

Considering the decomposition (2.7), we shall investigate the rate of convergence for $(A_N + B_N)$ and Q_N separately and finally use an argument which is based on the following lemma (cf. Petrov [18, p. 16]).

LEMMA 4.1. *Consider random variables U_N, V_N, and Q_N such that $U_N = V_N + Q_N$ and let G_N and K_N be the d.f.'s of V_N and U_N respectively. Assume that*

$$\sup_{-\infty<x<\infty} |G_N(x) - \Phi(x)| \le d_N.$$

Then if $P\{|Q_N| \ge \epsilon_N\} \le \beta_N$, we have

$$\sup_{-\infty<x<\infty} |K_N(x) - \Phi(x)| \le d_N + \frac{\epsilon_N}{\sqrt{2\pi}} + \beta_N. \tag{4.1}$$

For convenience, we shall divide the proof into two parts as follows:

Part 1: *Rate of Convergence for the Principal Term*

Our intention is to apply the following theorem due to Esseen (cf. Petrov [18, p. 115]).

THEOREM 4.1. *Let $Y_{N1},\dots,Y_{NN}$, $N \ge 1$ be a double sequence of (row-wise) independent r.v.'s such that $\sum_{i=1}^{N} E(Y_{Ni}) = 0$ and for some $0 < \eta \le 1$, $E|Y_{Ni}|^{2+\eta}$ exists for all $1 \le i \le N$, $N \ge 1$. Then there exists a finite constant*

$C_\eta > 0$ *(depending on* η *only) such that*

$$\sup_{-\infty<x<\infty}\left|P\left(N^{-1}\sum_{i=1}^{N} Y_{Ni} \le \tau_N x\right) - \Phi(x)\right| \le C_\eta \sigma_N^{-(2+\eta)}\beta_N^{(\eta)}N^{-\eta/2} \tag{4.2}$$

where $\tau_N^2 = \mathrm{Var}(N^{-1}\Sigma_{i=1}^N Y_{Ni})$, $\sigma_N^2 = N\tau_N^2$, *and* $\beta_N^{(\eta)} = N^{-1}\Sigma_{i=1}^N E|Y_{Ni}|^{2+\eta}$.

Now, with probability 1,

$$A_N + B_N = N^{-1}\sum_{i=1}^{N} Y_i \tag{4.3}$$

where $Y_i = W_i + T_i$ with

$$W_i = u(X_i)\psi(\bar{H}(|X_i|)) - E(u(X_i)\psi(\bar{H}(|X_i|))), \qquad 1 \le i \le N \tag{4.4}$$

and

$$T_i = \int_0^\infty [u(\bar{H}(x) - \bar{H}(|X_i|)) - \bar{H}(x)]\psi'(\bar{H}(x)\,d\bar{F}(x), \qquad 1 \le i \le N. \tag{4.5}$$

It is clear that Y_i, $1 \le i \le N$, are independent r.v.'s with $\Sigma_{i=1}^N E(Y_i) = 0$ and $\sigma_N^2 = N^{-1}\sigma_N^2$ where $\sigma_N^2 = N^{-1}\mathrm{Var}(\Sigma_{i=1}^N Y_i)$. Thus, by (2.11)

$$\sigma_N^2 = N\sigma_N^2 \ge K_1, \qquad \text{for some } K_1 > 0, \text{ all } N \ge 1. \tag{4.6}$$

In order to apply Esseen's theorem, we will prove that there exists $0 < \eta \le 1$ such that

$$\limsup_{N\to\infty} N^{-1}\sum_{i=1}^{N} E|W_i|^{2+\eta} < \infty \tag{4.7}$$

and

$$\limsup_{N\to\infty} N^{-1}\sum_{i=1}^{N} E|T_i|^{2+\eta} < \infty. \tag{4.8}$$

To this end, pick $0 < \eta < \delta$, and first note that from (2.3) we have

$$N^{-1}\sum_{i=1}^{N} E\left|u(X_i)\psi(\bar{H}(|X_i|))\right|^{2+\eta} \le K\int_0^1 (1-t)^{(-1/2+\lambda)(2+\eta)}\,dt < \infty \tag{4.9}$$

where the finiteness of the last integral follows from our assumption on λ.

We now make use of the following inequality (cf. Ruymgaart [20, p. 27].

LEMMA 4.2. *For every* $\eta \in (0, \frac{1}{2}]$, *there exists a positive constant* M (*depending on* η *only*) *such that for all* $s, t \in (0,1)$ *we have*

$$|u(s-t) - s| \leq M[r(t)]^{1/2-\eta}[r(s)]^{-1/2+\eta} \tag{4.10}$$

where $r(x) = \{x(1-x)\}^{-1}$, $x \in (0,1)$.

Now, using (2.3) and (4.10) we have

$$N^{-1}\sum_{i=1}^{N} E|T_i|^{2+\eta} \leq (MK)^{2+\eta}\int_0^1 [r(t)]^{(1/2-\eta)(2+\eta)} dt\left(\int_0^1 t^{1/2-\eta}(1-t)^{-1-\eta+\delta}\,dt\right)^{2+\eta} < \infty \tag{4.11}$$

since $0 < \eta < \delta$ and $(\frac{1}{2} - \eta)(2+\eta) < 1$.

The bounds in (4.9) and (4.11) being independent of N, the proofs of (4.7) and (4.8) follow.

Therefore, using Esseen's theorem and (4.6), there exist $0 < \eta < \frac{1}{2}$ and $C = C(\lambda, \delta, \eta, K, K_1) > 0$ such that

$$\sup_{-\infty < x < \infty} |P(A_N + B_N \leq \sigma_N x) - \Phi(x)| \leq CN^{-\eta}, \qquad N \geq 1. \tag{4.12}$$

Part 2: *Remainder Terms*

According to Lemma 4.1 we have to obtain an inequality of the form

$$P\left\{\left|\frac{Q_N}{\sigma_N}\right| \geq C_1 N^{-\alpha}\right\} \leq C_2 N^{-\alpha} \tag{4.13}$$

for some $C_1, C_2 > 0$ and some $0 < \alpha < \frac{1}{2}$.

Now (4.13) will be entailed by (4.6) and the following stronger assertion which will be needed in the next section.

There exist $0 < \alpha < \frac{1}{2}$ and $k > 1$ such that

$$P\{|Q_N| \geq C_1 N^{-1/2-\alpha}\} \leq C_2 N^{-k} \tag{4.14}$$

for some $C_1, C_2 > 0$.

With this end in view, we first choose $k > 1$ such that

$$\frac{k}{2k+2} < \delta \qquad \text{and} \qquad k < \frac{2(\lambda - \delta)}{1 - 2\lambda}. \tag{4.15}$$

Now, with δ and λ given by (2.3) and $k > 1$ chosen to satisfy (4.15), we consider for each $M > 0$ the sets

$$\Omega_{N1} = \left\{ \sup_{N^{-1} \leq \overline{H}(x) \leq 1-N^{-1}} \frac{|\hat{H}_N(x) - \overline{H}(x)|}{\{\overline{H}(x)(1-\overline{H}(x))\}^{1/2}} \leq MN^{-1/2}\operatorname{Log} N \right\}$$

$$\Omega_{N2} = \left\{ \sup_{x \in R} \frac{|\hat{H}_N(x) - \overline{H}(x)|}{\{\overline{H}(x)(1-\overline{H}(x))\}^{1/2-\epsilon_0}} \leq MN^{-1/2}\operatorname{Log} N \right\}$$

$$\Omega_{N3} = \left\{ \sup_{N^{-1+\delta} \leq \overline{H}(x) \leq 1-N^{-1+\delta}} \left| \frac{\hat{H}_N(x)}{\overline{H}(x)} - 1 \right| \leq MN^{-\delta/2}\operatorname{Log} N \right\}$$

$$\Omega_{N4} = \left\{ \sup_{N^{-1+\delta} \leq \overline{H}(x) \leq 1-N^{-1+\delta}} \left| \frac{1-\hat{H}_N(x)}{1-\overline{H}(x)} - 1 \right| \leq MN^{-\delta/2}\operatorname{Log} N \right\}$$

$$\Omega_{N5} = \left\{ \frac{N^{-1+\delta}}{2} \leq \hat{H}_N\big(\overline{H}^{-1}(N^{-1+\delta})\big) \leq \frac{3N^{-1+\delta}}{2}, \right.$$
$$\left. \frac{N^{-1+\delta}}{2} \leq 1 - \hat{H}_N\big(\overline{H}^{-1}(1-N^{-1+\delta})\big) \leq \frac{3N^{-1+\delta}}{2} \right\}$$

$$\Omega_{N6} = \left\{ \sup_{|\overline{F}(x)-\overline{F}(y)| \leq N^{-1+\delta}\operatorname{Log} N} |\hat{F}_N(x) - \hat{F}_N(y) - \overline{F}(x) + \overline{F}(y)| \right.$$
$$\left. \leq MN^{-1+\delta/2}\operatorname{Log} N \right\}$$

$$\Omega_{N7} = \left\{ |\hat{F}_N - \overline{F}|(I) \leq MN^{-1+\delta} \right\},$$

$$\Omega_{N8} = \left\{ \max_{1 \leq i \leq N} |X_i| \leq \overline{H}^{-1}(1 - N^{-k-1}) \right\}$$

where $\epsilon_0 = k/(2k+2)$ and $I = (\overline{H}^{-1}(s), \overline{H}^{-1}(t)]$, $N^{-1+\delta} \leq s < t \leq 1 - N^{-1+\delta}$, $t - s \leq N^{-1+\delta}$.

Also define Ω^*_{Ni}, $1 \leq i \leq 5$, to be the set Ω_{Ni} where we replace $\hat{H}_N$ by $\hat{H}^*_N = (N/(N+1))\hat{H}_N$ and finally let Ω^*_{N6} be the set Ω_{N6} where we replace $\hat{F}_N$ by $\hat{H}_N$ and $\overline{F}$ by $\overline{H}$.

Then, Lemmas 3.1 through 3.5 entail the existence of a constant $M > 0$ and N_0 such that for $N \geq N_0$

$$P(\Omega_N) \geq 1 - \text{const} \cdot N^{-k} \tag{4.16}$$

where $\Omega_N = [\cap_{i=1}^{6}(\Omega_{Ni} \cap \Omega^*_{Ni})] \cap \Omega_{N7} \cap \Omega_{N8}$. (We may mention that the const (constant) appearing in (4.16) and in the sequel is a generic positive constant which does not depend on N.)

The set Ω_N will play a major role in the following estimation without explicit reference.

Set $a_N = \overline{H}^{-1}(N^{-1+\delta})$, $b_N = \overline{H}^{-1}(1 - N^{-1+\delta})$, and $I_N = (a_N, b_N]$.

It is easy to see that the remainder term Q_N may be decomposed as $Q_N = \Sigma_{i=1}^{6} Q_{iN}$, where

$$Q_{1N} = \int_{I_N} (\psi(\hat{H}_N^*) - \psi(\overline{H}) - (\hat{H}_N^* - \overline{H})\psi'(\overline{H}))\, d\hat{F}_N,$$

$$Q_{2N} = \int_{I_N} -\frac{1}{N+1}\hat{H}_N \psi'(\overline{H})\, d\hat{F}_N,$$

$$Q_{3N} = \int_{I_N} (\hat{H}_N - \overline{H})\psi'(\overline{H})\, d(\hat{F}_N - \overline{F}),$$

$$Q_{4N} = \int_{I_N^c} -(\hat{H}_N - \overline{H})\psi'(\overline{H})\, d\overline{F}$$

$$Q_{5N} = \int_{I_N^c} -\psi(\overline{H})\, d\hat{F}_N, \qquad \text{and} \qquad Q_{6N} = \int_{I_N^c} \psi(\hat{H}_N^*)\, d\hat{F}_N$$

with $I_N^c = [0, \infty) \setminus I_N$.

On the set Ω_N we shall find upper bounds for each term Q_{iN}, $1 \le i \le 6$.

The term Q_{1N}: By Taylor's formula, the integrand of Q_{1N} is equal to $\frac{1}{2}(\hat{H}_N^* - \overline{H})^2 \psi''((1-\theta)\overline{H} + \theta \hat{H}_N^*)$ where $0 \le \theta \le 1$.

$$|\psi''((1-\theta)\overline{H} + \theta\hat{H}_N^*)| \le K(1-\overline{H})^{-5/2+\delta} \cdot \left(1 + \theta\frac{(\overline{H} - \hat{H}_N^*)}{(1-\overline{H})}\right)^{-5/2+\delta} \tag{4.17}$$

On the other hand, for $x \in I_N$

$$1 + \theta\frac{(\overline{H} - \hat{H}_N^*)}{(1-\overline{H})} \ge 1 - N^{(1-\delta)/2} \sup_{I_N} \frac{|\hat{H}_N^* - \overline{H}|}{\{\overline{H}(1-\overline{H})\}^{1/2}}$$

$$\ge 1 - MN^{-\delta/2}\operatorname{Log} N. \tag{4.18}$$

Now, clearly $\sup_{I_N}((1 - \hat{H}_N^*)/(1 - \overline{H})) \le$ const on Ω_N, and using (4.17) and (4.18), we have

$$|Q_{1N}| \le \text{const} \cdot N^{-1}\operatorname{Log}^2 N \int_{I_N} (1 - \hat{H}_N^*)^{-3/2+\delta}\, d\hat{H}_N. \tag{4.19}$$

Now

$$\int_{I_N} (1 - \hat{H}_N^*)^{-3/2+\delta} \, d\hat{H}_N \le \frac{1}{N} \sum_{j=1}^{N\hat{H}_N(b_N)} \left(1 - \frac{j}{N+1}\right)^{-3/2+\delta}$$

$$\le 2 \int_0^{(N\hat{H}_N(b_N)+1)/(N+1)} (1-s)^{-3/2+\delta} \, ds \le \text{const} \cdot N^{(\delta-1)(-1/2+\delta)}. \tag{4.20}$$

Using (4.19) and (4.20) we finally obtain

$$|Q_{1N}| \le \text{const} \cdot N^{-1+(\delta-1)(\delta-1/2)} \cdot \text{Log}^2 N. \tag{4.21}$$

The term Q_{2N}: Proceeding as in the derivation of (4.19) and (4.20) we find

$$|Q_{2N}| \le KN^{-1} \left\{ \sup_{I_N} \left(\frac{1 - \hat{H}_N^*}{1 - \overline{H}} \right)^{3/2-\delta} \right\} \int_{I_N} (1 - \hat{H}_N^*)^{-3/2+\delta} \, d\hat{H}_N$$

$$\le (\text{const} \cdot N^{-1}) \left\{ \frac{1}{N} \sum_{j=1}^{N\hat{H}_N(b_N)} \left(1 - \frac{j}{N+1}\right)^{-3/2+\delta} \right\}$$

$$\le \text{const} \cdot N^{-1+(\delta-1)(\delta-1/2)}. \tag{4.22}$$

The term Q_{3N}: Set $m_N = [N^{1-\delta}] + 1$, $c_i = N^{-1+\delta} + i/m_N(1 - 2N^{-1+\delta})$, $d_i = \overline{H}^{-1}(c_i)$, and $J_{Ni} = (d_{i-1}, d_i]$, $0 \le i \le m_N$.

Then it is easy to check that

$$Q_{3N} = \sum_{i=1}^{4} \sum_{j=1}^{m_N} Q_{3Nij}$$

where

$$Q_{3N1j} = \left[\hat{H}_N(d_{j-1}) - \overline{H}(d_{j-1})\right] \psi'\left(\overline{H}(d_{j-1})\right)\left(\hat{F}_N - \overline{F}\right)(J_{Nj})$$

$$Q_{3N2j} = \int_{J_{Nj}} \left[\hat{H}_N(x) - \overline{H}(x) - \hat{H}_N(d_{j-1}) + \overline{H}(d_{j-1})\right] \psi'\left(\overline{H}(d_{j-1})\right) d\left(\hat{F}_N - \overline{F}\right)(x)$$

$$Q_{3N3j} = \left[\hat{H}_N(d_{j-1}) - \overline{H}(d_{j-1})\right] \times \int_{J_{Nj}} \left[\psi'\left(\overline{H}(x)\right) - \psi'\left(\overline{H}(d_{j-1})\right)\right] d\left(\hat{F}_N - \overline{F}\right)(x)$$

and

$$Q_{3N4j} = \int_{J_{Nj}} \left[\hat{H}_N(x) - \overline{H}(x) - \hat{H}_N(d_{j-1}) + \overline{H}(d_{j-1})\right]$$
$$\cdot\left[\psi'(\overline{H}(x)) - \psi'(\overline{H}(d_{j-1}))\right] d(\hat{F}_N - F)(x).$$

Using (2.3), we have on Ω_N

$$|Q_{3N1j}| \le K\left(\sup_{I_N} \frac{|\hat{H}_N - \overline{H}|}{\{\overline{H}(1-\overline{H})\}^{1/2}}\right)(1 - c_{j-1})^{-1+\delta}$$
$$\cdot \sup_{|\overline{F}(x)-\overline{F}(y)| \le N^{-1+\delta}\operatorname{Log} N} |\hat{F}_N(x) - \hat{F}_N(y) - \overline{F}(x) + \overline{F}(y)|$$
$$\le \text{const} \cdot N^{(-3+\delta)/2}\operatorname{Log}^2 N(1 - c_{j-1})^{-1+\delta}.$$

Therefore, since $(1/m_N)\Sigma_{j=1}^{m_N}(1 - c_{j-1})^{-1+\delta}$ is bounded, we get

$$\left|\sum_{j=1}^{m_N} Q_{3N1j}\right| \le \text{const} \cdot N^{(-1-\delta)/2} \cdot \operatorname{Log}^2 N \cdot \frac{1}{m_N}\sum_{j=1}^{m_N}(1 - c_{j-1})^{-1+\delta}$$
$$\le \text{const} \cdot N^{(-1-\delta)/2}\operatorname{Log}^2 N. \tag{4.23}$$

Similarly, using the fact that

$$\frac{1}{m_N}\sum_{j=1}^{m_N}(1 - c_{j-1})^{-3/2+\delta} \le \frac{1}{1 - 2N^{-1+\delta}}\int_0^{c_{m_N}}(1 - s)^{-3/2+\delta}\,ds$$
$$\le \text{const} \cdot N^{(\delta-1)(\delta-1/2)}$$

we find on Ω_N

$$\left|\sum_{j=1}^{m_N} Q_{3N2j}\right| \le K \sup_{|\overline{H}(x)-\overline{H}(y)| \le N^{-1+\delta}\operatorname{Log} N} \cdot|\hat{H}_N(x) - \hat{H}_N(y) - \overline{H}(x)$$
$$+ \overline{H}(y)| \sum_{j=1}^{m_N}(1 - c_{j-1})^{-3/2+\delta}|\hat{F}_N - \overline{F}|(J_{Nj})$$
$$\le \text{const} \cdot N^{-1+\delta/2+(\delta-1)(\delta-1/2)} \cdot \operatorname{Log} N \qquad (4.24).$$

Further, using Taylor's formula, (2.3), and the fact that for $x \in J_{Nj}$ and sufficiently large N

$$1 - \theta\frac{\overline{H}(x) - \overline{H}(d_{j-1})}{1 - \overline{H}(d_{j-1})} \ge \text{const} > 0$$

we obtain

$$\left|\psi'(\overline{H}(x)) - \psi'(\overline{H}(d_{j-1}))\right| \leq \text{const} \cdot N^{-1+\delta}(1 - c_{j-1})^{-5/2+\delta},$$
$$x \in J_{Nj}. \quad (4.25)$$

Then, since

$$\frac{1}{m_N}\sum_{j=1}^{m_N}(1 - c_{j-1})^{-2+\delta} \leq \frac{1}{1 - 2N^{-1+\delta}}\int_0^{c_{m_N}}(1 - s)^{-2+\delta}\,ds$$
$$\leq \text{const} \cdot N^{(1-\delta)^2}$$

we find that

$$\left|\sum_{j=1}^{m_N} Q_{3N3j}\right| \leq KN^{-1+\delta}\left[\sup_{I_N}\frac{\hat{H}_N - \overline{H}}{\{\overline{H}(1 - \overline{H})\}^{1/2}}\right]$$
$$\times \sum_{j=1}^{m_N}(1 - c_{j-1})^{-2+\delta}|\hat{F}_N - \overline{F}|(J_{Nj})$$
$$\leq \text{const} \cdot N^{-3/2+\delta+(\delta-1)^2} \cdot \operatorname{Log} N. \quad (4.26)$$

Similarly, we obtain the estimate

$$\left|\sum_{j=1}^{m_N} Q_{3N4j}\right| \leq \text{const} \cdot N^{-2+3\delta/2+(1-\delta)(3/2-\delta)}\operatorname{Log} N. \quad (4.27)$$

The term Q_{4N}: With $\epsilon_0 = k/(2k + 2) < 1/2$ we have

$$|Q_{4N}| \leq K\left[\sup_R \frac{|\hat{H}_N - \overline{H}|}{\{\overline{H}(1 - \overline{H})\}^{1/2-\epsilon_0}}\right]\int_{I_N^c}(1 - \overline{H})^{-1-\epsilon_0+\delta}\,d\overline{H}$$
$$\leq KMN^{-1/2}\operatorname{Log} N\left\{\int_0^{a_N}(1 - \overline{H})^{-1-\epsilon_0+\delta}\,d\overline{H} + \int_{b_N}^{\infty}(1 - \overline{H})^{-1-\epsilon_0+\delta}\,d\overline{H}\right\}.$$

The first integral in the second line of the above inequality is clearly bounded by (const) $\cdot N^{-1+\delta}$, while the second is bounded by

$$\int_{1-N^{-1+\delta}}^{1}(1 - t)^{-1-\epsilon_0+\delta}\,dt = \frac{1}{\delta - \epsilon_0}N^{(-1+\delta)(\delta-\epsilon_0)}.$$

Thus

$$|Q_{4N}| \leq \text{const} \cdot N^{-3/2+\delta} \cdot \operatorname{Log} N + \text{const} \cdot N^{-1/2+(\delta-1)(\delta-\epsilon_0)}\operatorname{Log} N. \quad (4.28)$$

The term Q_{5N}: Denote

$$\mathscr{L}_1 = \int_0^{a_N} - \psi(\overline{H})\, d\hat{F}_N \qquad \text{and} \qquad \mathscr{L}_2 = \int_{b_N}^{\infty} - \psi(\overline{H})\, d\hat{F}_N.$$

Using (2.3) and the fact that for $x \le a_N$, $1 - \overline{H}(x) \ge \frac{1}{2}$ for N sufficiently large, we find

$$|\mathscr{L}_1| \le K\left(\tfrac{1}{2}\right)^{-1/2+\lambda} \hat{H}_N(a_N) \le K\left(\tfrac{1}{2}\right)^{-1/2+\lambda} \tfrac{3}{2} N^{-1+\delta}. \tag{4.29}$$

On the other hand, since on Ω_{N8}

$$\left[1 - \overline{H}(|X_i|)\right]^{-1/2+\lambda} \le N^{(k+1)(1/2-\lambda)}, \qquad 1 \le i \le N$$

we get

$$\begin{aligned} |\mathscr{L}_2| &\le K \int_{b_N}^{\infty} (1 - \overline{H})^{-1/2+\lambda}\, d\hat{H}_N \\ &= \frac{K}{N} \sum_{i=1}^{N} \chi_{[b_N, \infty)}(|X_i|)(1 - \overline{H}(|X_i|))^{-1/2+\lambda} \\ &\le K N^{(k+1)(1/2-\lambda)} (1 - \hat{H}_N(b_N)) \\ &\le \frac{3K}{2} N^{-1+\delta+(k+1)(1/2-\lambda)}. \end{aligned} \tag{4.30}$$

where χ denotes the indicator function.

From (4.29) and (4.30), since $Q_{5N} = \mathscr{L}_1 + \mathscr{L}_2$, we get

$$|Q_{5N}| \le \text{const} \cdot N^{-1+\delta} + \text{const} \cdot N^{-1+\delta+(k+1)(1/2-\lambda)}. \tag{4.31}$$

The term Q_{6N}: Set

$$\mathscr{M}_1 = \int_0^{a_N} \psi(\hat{H}_N^*)\, d\hat{F}_N, \qquad \mathscr{M}_2 = \int_{b_N}^{\infty} \psi(\hat{H}_N^*)\, d\hat{F}_N.$$

Then, since for all $x \le a_N : 1 - \hat{H}_N(x) \ge 1 - \frac{3}{2}N^{-1+\delta} \ge \frac{1}{2}$ on Ω_N for N sufficiently large, we have

$$|\mathscr{M}_1| \le K\left(\tfrac{1}{2}\right)^{-1/2+\lambda} \hat{H}_N(a_N) \le \text{const} \cdot N^{-1+\delta}. \tag{4.32}$$

On the other hand, using (2.3) we find

$$|\mathcal{M}_2| \le K \int_{b_N}^{\infty} (1 - \hat{H}_N^*)^{-1/2+\lambda}\, d\hat{H}_N \le \frac{K}{N} \sum_{j=N\hat{H}_N(b_N)}^{N} \left(1 - \frac{j}{N+1}\right)^{-1/2+\lambda}$$

$$\le 2K \int_{N\hat{H}_N(b_N)/(N+1)}^{1} (1-t)^{-1/2+\lambda}\, dt = \frac{4K}{1+2\lambda}\left(1 - \frac{N\hat{H}_N(b_N)}{N+1}\right)^{1/2+\lambda}$$

$$\le \text{const} \cdot N^{(-1+\delta)(1/2+\lambda)}. \tag{4.33}$$

From (4.32), (4.33), and the fact that $Q_{6N} = \mathcal{M}_1 + \mathcal{M}_2$, we infer that

$$|Q_{6N}| \le \text{const} \cdot N^{-1+\delta} + \text{const} \cdot N^{(-1+\delta)(1/2+\lambda)}. \tag{4.34}$$

Finally, straightforward combination of relations (4.21)–(4.24), (4.26)–(4.28), (4.31), and (4.34) together with (4.15) and the fact that $\frac{1}{4} < \delta < \lambda < \frac{1}{2}$, $2\lambda - \delta > \frac{1}{2}$ lead to the satisfaction of (4.14).

The proof follows from (4.12), (4.13), and Lemma 4.1.

5. Applications

In this section we shall obtain, as applications of our main result, (i) a functional Central Limit Theorem (Donsker-type invariance principle) for S_N^+ and (ii) a strong law for S_N^+.

These results are useful for the study of the properties of sequential tests and estimates based on rank order statistics (see, e.g., Lai [15] and Müller-Funk [16]).

(i) *A Functional C.L.T. for S_N^+ in the Non-i.i.d. Case*

Denote by $C[0,1]$ the space of real, continuous functions on $[0,1]$ with the uniform metric $d(x, y) = \sup_{t \in [0,1]} |x(t) - y(t)|$, $x, y \in C[0,1]$. We adopt the notation used in Billingsley [4], $U_N \xrightarrow{\mathcal{D}} U$ for convergence in distribution of a sequence $\{U_N\}$ of random elements of $C[0,1]$ to a random element U.

Now, for the principal term $A_N + B_N$, we consider the following decomposition similar to (4.3),

$$A_N + B_N = N^{-1} \sum_{i=1}^{N} Z_{Ni} \tag{5.1}$$

where

$$Z_{Ni} = W_i + \int_0^{\infty} [u(x - |X_i|) - F_i^*(x)] \psi'(\overline{H}(x))\, d\overline{F}(x) \qquad (5.2)$$

with W_i defined by (4.4) ($W_i = W_{Ni}$).

Define the combined (N, i)-rank order statistic T_{Ni}^+ by

$$T_{Ni}^+ = \frac{1}{i} \sum_{j=1}^{i} (Z_{Nj} - Z_{ij}) + S_i^+, \qquad 1 \le i \le N,\ N \ge 1. \qquad (5.3)$$

Set $t_{Ni}^2 = \Sigma_{j=1}^{i} \operatorname{Var} Z_{Nj}$, $1 \le i \le N$ and $t_N^2 = t_{NN}^2$.

Now, for every $N \ge 1$, define the stochastic process W_N by

$$\begin{aligned} W_N(t) &= 0 && \text{for } t = 0 \\ &= i(T_{Ni}^+ - \mu_i^+)/t_N && \text{for } t = t_{Ni}^2/t_N^2,\ 1 \le i \le N \end{aligned} \qquad (5.4)$$

linearly interpolated for $t \in [t_{Ni}^2/t_N^2, t_{N,i+1}^2/t_N^2]$, $0 \le i \le N-1$, $(t_{N0}^2 = 0)$.

Then we have the following theorem.

THEOREM 5.1. *Consider the statistic S_N^+ defined by* (2.1) *with scores given by* (2.2) *and assume that* (2.3) *and* (2.11) *are satisfied.*

Then the process W_N, defined by (5.4), *converges in distribution in the uniform topology on* $C[0,1]$ *to a standard Brownian motion process* $W = \{W(t) : 0 \le t \le 1\}$, *i.e.*,

$$W_N \xrightarrow{\mathcal{D}} W. \qquad (5.5)$$

Remark. We may mention first that the above theorem contains as a particular case the asymptotic normality of S_N^+ with parameters (μ_N^+, σ_N^2) defined respectively by (2.6) and (2.10).

Indeed, for every $v \in R$, define $\Delta_v = \{x \in C[0,1] : x(1) \le v\}$. Let P_N denote the probability distribution of W_N, $N \ge 1$. Then, since $T_{NN}^+ = S_N^+$, we have $W_N(1) = (S_N^+ - \mu_N^+)/\sigma_N$. Now, it is well known that $W(\partial \Delta_v) = 0$ (where W denotes also the Wiener measures on $C[0,1]$) and since by (5.5), P_N converges weakly to W, we have

$$P\left\{ \frac{S_N^+ - \mu_N^+}{\sigma_N} \le v \right\} \to W\{w : x(1) \le v\} = \Phi(v).$$

To prove Theorem 5.1 we require the following lemma which shows that the sequence $\{Q_N\}$, $N \geq 1$ is bounded in L^1-norm.

LEMMA 5.1. *Let Q_N be defined by* (2.9). *Then if* (2.3) *is satisfied, we have*

$$\limsup_{N\to\infty} E|Q_N| < \infty. \tag{5.6}$$

Proof. Denote the first integral in (2.9) by Q'_N, and note that $Q_N = Q'_N - B_N$.

Then, using (2.3) we have

$$E\left|\int_0^\infty \psi(\overline{H})\, d\hat{F}_N\right| \leq K\int_0^1 (1-s)^{-1/2+\lambda}\, ds = \frac{K}{1/2+\lambda}, \tag{5.7}$$

and

$$\begin{aligned}\left|\int_0^\infty \psi\left(\frac{N}{N+1}\hat{H}_N\right) d\hat{F}_N\right| &\leq K\int_0^\infty \left(1 - \frac{N}{N+1}\hat{H}_N\right)^{-1/2+\lambda} d\hat{H}_N \\ &= KN^{-1}\sum_{i=1}^N \left(1 - \frac{i}{N+1}\right)^{-1/2+\lambda} \\ &\leq 2K\int_{1/(N+1)}^1 (1-s)^{-1/2+\lambda}\, ds \leq \frac{2K}{1/2+\lambda},\end{aligned} \tag{5.8}$$

with probability 1. Therefore, by (5.7) and (5.8),

$$E|Q_N| \leq \frac{3K}{1/2+\lambda}. \tag{5.9}$$

On the other hand, using the fact that $B_N = N^{-1}\sum_{i=1}^N \int_0^\infty [u(\overline{H}(x) - \overline{H}(|X_i|)) - \overline{H}(x)]\psi'(\overline{H}(x))\, d\overline{F}(x)$, (2.3), and Lemma 2.4 (with $0 < \eta < \delta$) we obtain

$$\begin{aligned}|B_N| \leq KM&\left\{\int_0^\infty [r(\overline{H})]^{-1/2+\eta}(1-\overline{H})^{-3/2+\delta}\, d\overline{H}\right\} \\ &\times\left\{N^{-1}\sum_{i=1}^N [r(\overline{H}(|X_i|))]^{1/2-\eta}\right\}\end{aligned}$$

which implies

$$E|B_N| \leq KM\left\{\int_0^1 [r(t)]^{-1/2+\eta}(1-t)^{-3/2+\delta}\, dt\right\}\int_0^1 [r(t)]^{1/2-\eta}\, dt < \infty. \tag{5.10}$$

The bound in (5.10) being independent of N, the proof follows from (5.9) and (5.10).

Proof of Theorem 5.1. This theorem is an application of Prohorov's [19] generalization of Donsker's theorem together with (4.14) and Lemma 5.1. We will approximate W_N by a process W_N^* in "regular position" (converging in distribution to a standard Brownian motion W) and then show that the distance between W_N and W_N^* is negligible in probability as $N \to \infty$.

To this end, define the random process $W_N^* = \{W_N^*(t) : 0 \le t \le 1\}$ by

$$\begin{aligned} W_N^*(t) &= 0 && \text{for } t = 0 \\ &= \left(\textstyle\sum_{j=1}^{i} Z_{Nj}\right)/t_N && \text{for } t = t_{Ni}^2/t_N^2,\ 1 \le i \le N \\ &\quad \text{linearly interpolated} && \text{for } t \in [t_{Ni}^2/t_N^2,\ t_{N,i+1}^2/t_N^2], \\ & && 0 \le i \le N-1,\ (t_{N0}^2 = 0). \end{aligned} \tag{5.11}$$

Then, proceeding as in Part 1 of the proof of Theorem 2.1, it can be shown that the sequence $\{Z_{Ni}\}$, $1 \le i \le N$, $N \ge 1$ satisfies the Liapounov condition for asymptotic normality. Hence, by Prohorov's Theorem (cf. also Billingsley [4, p. 77])

$$W_N^* \xrightarrow{\mathcal{D}} W \tag{5.12}$$

where W is a standard Brownian motion process.

Now, it is readily seen that

$$d(W_N, W_N^*) \le \frac{\max_{1 \le i \le N} i|Q_i|}{t_N} \tag{5.13}$$

and it remains to show that the right hand side of (5.13) converges to zero in probability.

Now, since $\sigma_N = t_N/N$, it follows from (4.6) that

$$t_N \ge K'N^{1/2}, \qquad \text{all } N \ge 1 \tag{5.14}$$

for some $K' > 0$. (In particular, $\lim_{N\to\infty} t_N = \infty$.)

Let ϵ_0, $\epsilon > 0$ be given. Let $C_1, C_2 > 0$, $0 < \alpha < \frac{1}{2}$, and $k > 1$ be defined by (4.14). Pick N_0 such that for $N \ge N_0$

$$\epsilon_0 K'N^{1/2} \ge C_1 N^{1/2-\alpha}. \tag{5.15}$$

Also, choose $N_1 \geq N_0$ such that

$$C_2 \sum_{i=N_1}^{\infty} i^{-k} < \epsilon/2 \tag{5.16}$$

Finally, let $N_2 \geq N_1$ be such that for all $N \geq N_2$

$$t_N \geq \frac{2(N_1 - 1)^2 K''}{\epsilon_0 \epsilon} \tag{5.17}$$

where, by Lemma 5.1,

$$E|Q_N| \leq K'', \qquad \text{all } N \geq 1, \text{ some } K'' > 0. \tag{5.18}$$

Let $N \geq N_2$. Then, by Chebyshev's inequality, (5.17), and (5.18) we have

$$\sum_{i=1}^{N_1 - 1} P\{i|Q_i| \geq \epsilon_0 t_N\} \leq \epsilon/2. \tag{5.19}$$

On the other hand, from (4.14), (5.15), and (5.16), we obtain

$$\sum_{i=N_1}^{N} P\{i|Q_i| \geq \epsilon_0 t_N\} \leq \sum_{i=N_1}^{N} P\{|Q_i| \geq C_1 i^{-1/2-\alpha}\} \leq C_2 \sum_{i=N_1}^{\infty} i^{-k} < \epsilon/2. \tag{5.20}$$

Consequently, by (5.19) and (5.20), $(\max_{1 \leq i \leq N} i|Q_i|)/t_N \xrightarrow{\mathscr{P}} 0$, which together with (5.12) and (5.13) yields the desired result (5.5). The proof follows.

(ii) *Strong Law for* S_N^+

We now prove that under suitable conditions on the score-generating function ψ, $S_N^+ - \mu_N^+ \to 0$ a.s.

Here we provide a simple proof of this fact, utilizing the decomposition (2.7) together with the results of the previous section.

THEOREM 5.2. *Consider the statistic* S_N^+ *defined by* (2.1) *with scores given by* (2.2). *Then, under the assumption* (2.3), $S_N^+ - \mu_N^+ \to 0$ a.s., *where* μ_N^+ *is given by* (2.6).

Proof. Considering the decomposition (2.7), it suffices to show that $A_N + B_N \to 0$ a.s. and $Q_N \to 0$ a.s.

In dealing with the principal term we shall need the following lemma, the proof of which follows from the standard arguments, and is therefore omitted.

LEMMA 5.2. *Consider a double sequence* $\{Y_{Ni}\}$, $1 \le i \le N$, $N \ge 1$, *of row-wise independent* r.v.'s *and assume that*

$$\limsup_{N \to \infty} N^{-1} \sum_{i=1}^{N} E|Y_{Ni}|^{2+\eta} < \infty, \qquad \text{for some } \eta > 0. \tag{5.21}$$

Then, there exists $l > 1$ *such that for every* $\epsilon > 0$ *there exist* $C_\epsilon > 0$ *and* $N_0(\epsilon)$ *such that for all* $N \ge N_0(\epsilon)$

$$P\{|\overline{Y}_N - E(\overline{Y}_N)| > \epsilon\} \le C_\epsilon N^{-l} \tag{5.22}$$

where $\overline{Y}_N = N^{-1}\sum_{i=1}^{N} Y_{Ni}$. *In particular, it follows from* (5.22) *that* $\overline{Y}_N - E(\overline{Y}_N) \to 0$ a.s.

Now, from (4.3), (4.7), (4.8), and Lemma 5.2, we have $A_N + B_N \to 0$ a.s. On the other hand, (4.14) clearly implies that $Q_N \to 0$ a.s. The proof follows.

ACKNOWLEDGMENTS

The authors are grateful to the late Professor Jack Kiefer for helpful discussions and some highly valuable suggestions especially concerning the contents of Sections 3 and 5.

REFERENCES

1. W. ALBERS, P. J. BICKEL AND W. R. VAN ZWET, Asymptotic expansions for the power of distribution free tests in the one sample problem, *Ann. Statist.* **4** (1976), 108–156.
2. R. R. BAHADUR, A note on quantiles in large samples, *Ann. Math. Statist.* **37** (1966), 577–580.
3. H. BERGSTRÖM AND M. L. PURI, Convergence and remainder terms in linear rank statistics, *Ann. Statist.* **5** (1977), 671–680.
4. P. BILLINGSLEY, "Convergence of Probability Measures," Wiley, New York, 1968.
5. D. R. BRILLINGER, A note on the rate of convergence of a mean, *Biometrika* **49** (1962), 574–576.
6. H. CHERNOFF AND I. R. SAVAGE, Asymptotic normality and efficiency of certain nonparametric test statistics, *Ann. Math. Statist.* **29** (1958) 972–994.
7. V. DUPAČ AND J. HÁJEK, Asymptotic normality of simple linear rank statistics under alternatives, II, *Ann. Math. Statist.* **40** (1969), 1992–2017.
8. A. DVORETZKY, J. KIEFER, AND J. WOLFOWITZ, Asymptotic minimax character of the sample distribution functions and of the classical multinomial estimator, *Ann. Math. Statist.* **27** (1956), 642–669.
9. C. G. ESSEEN, Fourier analysis of distribution functions. A mathematical study of the Laplace–Gaussian law, *Acta Math.* **77** (1945), 1–125.
10. M. GHOSH, On the representation of linear functions of order statistics, *Sankhyā Ser. A* **34** (1972), 349–356.

11. J. HÁJEK, Asymptotic normality of simple linear rank statistics under alternatives, *Ann. Math. Statist.* **39** (1968), 325–346.
12. M. HUŠKOVÁ, The rate of convergence of simple linear rank statistics under hypothesis and alternatives. *Ann. Statist.* **4** (1977), 658–670.
13. M. HUŠKOVÁ, The rate of convergence of simple linear rank statistics under alternatives, *in* "Contributions to Statistics" (J. Jureckova, Ed.), pp. 99–108, Academic Publishing House, Prague, 1979.
14. J. JUREČKOVÁ AND M. L. PURI, Order of normal approximation for rank test statistics distribution, *Ann. Probab.* **3** (1977), 526–533.
15. T. L. LAI, On Chernoff–Savage statistics and sequential rank tests, *Ann. Statist.* **3** (1975), 825–845.
16. U. MÜLLER-FUNK, Nonparametric sequential tests for symmetry, *Z. Wahrsch. Verw. Gebiete* **46** (1979), 325–342.
17. U. MÜLLER-FUNK AND H. WITTING, On the rate of convergence in the C.L.T. for signed linear rank statistics, preprint, 1980.
18. V. V. PETROV, "Sums of Independent Random Variables," Springer-Verlag, Berling/Heidelberg/New York, 1975.
19. Y. V. PROHOROV, Convergence of random processes and limit theorems in probability theory, *Teor. Veroyatnost. i Primenen.* **1** (1956), 177–238.
20. F. H. RUYMGAART, "Asymptotic Theory of Rank Tests for Independence," Mathematical Centre Tracts No. 43, Math. Centrum, Amsterdam, 1973.
21. J. V. USPENSKY, "Introduction to Mathematical Probability," McGraw–Hill, New York, 1937.
22. M. C. A. VAN ZUIJLEN, "Empirical distribution and rank statistics," Mathematical Centre Tracts No. 79 Math. Centrum, Amsterdam, 1977.

A SHARPENING OF THE REMAINDER TERM IN THE HIGHER-DIMENSIONAL CENTRAL LIMIT THEOREM FOR MULTILINEAR RANK STATISTICS

MANFRED DENKER, MADAN L. PURI* and UWE RÖSLER

Universität Göttingen, Göttingen, West Germany
Indiana University
Universität Göttingen, Göttingen, West Germany
communicated by J. Pflanzagl

A new approach to the asymptotic normality of the multivariate linear rank statistics is provided along with the Berry–Esséen and the Prohorov distance estimates for the remainder term in the convergence to normality.

1. Introduction

Let $X_{Ni} = (X_{Ni}^{(1)}, \ldots, X_{Ni}^{(p)})'$, $i = 1, \ldots, N$, be a sequence of independent p-variate $(p \geqslant 1)$ random vectors having continuous cumulative distribution functions (cdf) F_{Ni}, $i = 1, \ldots, N$, respectively. Consider the $p \times N$ random matrix X_N corresponding to $(X_{N1}, \ldots, X_{NN})$, i.e.,

$$X_N = (X_{Ni}^{(\nu)})_{\substack{i=1,\ldots,N \\ \nu=1,\ldots,p}} \tag{1.1}$$

Received December 9, 1982; revised January 17, 1984.

AMS 1980 subject classifications: primary 62E20, secondary 60F05.

Key words and phrases: linear rank statistics, Berry–Esséen estimates, Prohorov distance estimates.

* Research supported by the National Science Foundation Grant MCS-8301409.

and observe that each row of X_N is composed of N independent univariate random variables. Let $R_{Ni}^{(v)}$ be the rank of $X_{Ni}^{(v)}$ among $X_{N1}^{(v)},..., X_{NN}^{(v)}$ for each $v = 1,..., p$. Then, corresponding to the observation matrix X_N, we have a rank collection matrix

$$R_N = (R_{Ni}^{(v)})_{\substack{i=1,...,N \\ v=1,...,p}}. \tag{1.2}$$

Consider now p sets of scores $(a_N^{(v)}(i), 1 \leqslant i \leqslant N)$, $v = 1,..., p$, generated by known functions $\phi_v: (0, 1) \to \mathbb{R}$ in either of the following ways:

$$a_N^{(v)}(i) = \phi_v\left(\frac{i}{N+1}\right), \qquad i = 1,..., N;\ v = 1,..., p \tag{1.3}$$

$$a_N^{(v)}(i) = E\phi_v(U_N^{(i)}), \qquad i = 1,..., N;\ v = 1,..., p, \tag{1.4}$$

where $U_N^{(i)}$ is the ith-order statistic in a random sample of size N from the uniform distribution over $(0, 1)$. Consider now the simple linear rank vector S_N corresponding to X_N (or R_N) and $\phi = (\phi_1,..., \phi_p)$, i.e.,

$$\begin{aligned} S_N &= (S_N^{(1)},..., S_N^{(p)}), \\ S_N^{(v)} &= \sum_{i=1}^{N} C_{Ni}^{(v)} a_N^{(v)}(R_{Ni}^{(v)}), \qquad 1 \leqslant v \leqslant p, \end{aligned} \tag{1.5}$$

where $(C_{Ni}^{(v)}, 1 \leqslant i \leqslant N)$, $1 \leqslant v \leqslant p$, are p sets of known (regression) constants.

In Section 3 we obtain estimates for the order of normal approximation for S_N when each ϕ_v has a bounded second derivative. The Berry–Esséen-type theorem is of order $N^{-1/2}(\log N)^2$, and the estimate of the Prohorov distance has order $N^{-1/8}p^{1/4} + p^{1/2}N^{-1/2}(\log N)^2$. Using the approximation method in Denker and Rösler [5] (restated in Section 2 among other things) we also establish as a corollary the asymptotic normality of S_N in Section 3. For a different approach to our results on asymptotic normality, see Hájek [6] and Puri and Sen [12]. In Section 4 we derive estimates for the order of normal approximation for S_N for a more general score generating function. In particular, if each ϕ_v has a locally Hölder-continuous derivative and if ϕ_v' is bounded by some function $t \to (t(1-t))^{-3/2+\eta}$, then the approximation is still of order $N^{-\lambda}$ for some specified $\lambda > 0$, both in the Prohorov distance and the Berry–Esséen type. This constitutes the extensions as well as generalizations of the results of Jurečková and Puri [10], Bergström and Puri [1], and Hušková [9], where the problems are treated in the univariate setups. The multivariate extensions in the generality of our paper do not appear to exist in the literature so far. (For a rather special multivariate case, the reader is referred to Hušková [8].)

2. Some Approximate Lemmas

In this section we shall give the basic lemmas of an approximation method in the univariate case, which later will be applied coordinatewise to S_N.

Let us denote by μ the measure on $(0, 1)$ given by its density function $(t(1-t))^{-1/2}$ with respect to the Lebesgue measure and throughout this paper we consider functions ϕ given by

$$\phi(z) = \int_{1/2}^{z} dv^*. \tag{2.1}$$

(In the multivariate case each ϕ_ν, $\nu = 1,\ldots, p$, will have this form.) v^* is a signed measure on $(0, 1)$ with finite total variation norm $\| \ \|_T$ on every compact set in $(0, 1)$. In order to make (2.1) precise, we use the convention

$$\int_a^b dv^* = \int I_{(a,b]}\, dv^* \qquad \text{for} \quad a \leqslant b$$

and

$$\int_a^b dv^* = -\int I_{(b,a]}\, dv^* \qquad \text{for} \quad b \leqslant a,$$

where I denotes the indicator function. Note that $\phi(\frac{1}{2}+) = 0$ is no restriction. Let $v^* = v_1 - v_2$ be the Jordan–Hahn decomposition of v^* into its positive and negative parts. Then we can write $\phi = \phi^1 - \phi^2$, where $\phi^i(z) = \int_{1/2}^{z} dv_i$, $i = 1, 2$. This representation of ϕ by monotone functions is unique. The norm of ϕ is defined by

$$\|\phi\| = \sum_{i=1}^{2} \int |\phi^i|\, d\mu. \tag{2.2}$$

Denote by $\mathscr{H}$ the set of all functions ϕ with finite norm and by $\mathscr{G}$ the set of all functions ϕ with finite norm for which the corresponding v^* is absolutely continuous with respect to Lebesgue measure. Clearly, $\mathscr{G} \subset \mathscr{H}$ and $\mathscr{G}$ is the $\| \ \|$-closure of C_2-functions with bounded second derivative. Moreover, we shall need for $\phi \in \mathscr{H}$

$$\int (t(1-t))^{1/2}\, d|\phi|(t) \leqslant \|\phi\| \ll \int (t(1-t))^{1/2}\, d|\phi|(t), \tag{2.3}$$

where $d|\phi|$ denotes the total variation measure associated with $d\phi$. (Here and in the following $a \ll b$ means $a \leqslant \gamma b$ for some constant γ.)

$$\int \phi^2(t)\,dt \ll \|\phi\|^2 \tag{2.4}$$

$$\lim_{t \to \{0,1\}} \phi(t)(t(1-t))^{1/2} = 0. \tag{2.5}$$

(2.3)–(2.5) can be found in Denker and Rösler [5, Proposition 1].

We shall keep the notations of Section 1 but omit the upper index v for simplicity. In addition, define

$$H_N = \frac{1}{N} \sum_{i=1}^{N} F_{Ni}, \qquad G_N = \sum_{i=1}^{N} C_{Ni} F_{Ni} \tag{2.6}$$

$$\hat{H}_N = \frac{1}{N} \sum_{i=1}^{N} \hat{F}_{Ni}, \qquad \hat{G}_N = \sum_{i=1}^{N} C_{Ni} \hat{F}_{Ni}, \tag{2.7}$$

where $\hat{F}_{Ni}$ is the empirical measure defined by X_{Ni}, $i = 1, \ldots, N$. For a function ϕ and for scores $a_N(i)$ satisfying (1.3), the statistic $S_N = S_N(\phi)$ can be written as

$$S_N(\phi) = \int \phi\left(\frac{N}{N+1} \hat{H}_N(t)\right) d\hat{G}_N(t).$$

We shall make the following assumption throughout this paper:

$$\max_{1 \leqslant i \leqslant N} |C_{Ni}^{(v)}| \ll N^{-1}, \qquad 1 \leqslant v \leqslant p. \tag{2.8}$$

It follows immediately that $|G_N(x)| \ll H_N(x)$, $|\hat{G}_N(x)| \ll \hat{H}_N(x)$ and in the total variation norm one has also $\|G_N\|_T \ll \|H_N\|_T$ and $\|\hat{G}_N\|_T \ll \|\hat{H}_N\|_T$. Also it is easily verified that

$$E(\sqrt{N}(\hat{H}_N(x) - H_N(x)))^2$$
$$= N^{-1} \sum_{i=1}^{N} E(\hat{F}_{Ni}(x) - F_{Ni}(x))^2 \leqslant H_N(x)(1 - H_N(x))$$

and

$$E(\sqrt{N}(\hat{G}_N(x) - G_N(x)))^2 \ll E(\sqrt{N}(\hat{H}_N(x) - H_N(x)))^2.$$

The following two lemmas are contained in Proposition 2 of Denker and Rösler [5] and are restated here for completeness.

LEMMA 2.1. *For every $\phi \in \mathscr{H}$ we have*

$$E\left|\sqrt{N}\int_{-\infty}^{\infty}\phi\left(\frac{N}{N+1}\hat{H}_N(t)\right)d\hat{G}_N(t)\right.$$

$$\left.-\sqrt{N}\int_{-\infty}^{\infty}\phi\left(\frac{N}{N+1}H_N(t)\right)d\hat{G}_N(t)\right|^2 \leqslant C\|\phi\|^2 \tag{2.9}$$

$$\sqrt{N}\left|\int_{-\infty}^{\infty}\left(\phi(H_N(t))-\phi\left(\frac{N}{N+1}H_N(t)\right)\right)dG_N(t)\right| \leqslant C\|\phi\|. \tag{2.10}$$

Note that the constant C does not depend on ϕ.

LEMMA 2.2. *There exists a constant C such that for any $\phi \in \mathscr{H}$,*

$$E\sqrt{N}\left|\int_{-\infty}^{\infty}\phi\left(\frac{N}{N+1}\hat{H}_N(t)\right)d\hat{G}_N(t)-\int_{-\infty}^{\infty}\phi(H_N(t))\,dG_N(t)\right|^2$$

$$\leqslant C\|\phi\|^2.$$

If $\phi \in \mathscr{G}$, let g denote the function defined by $g(s)=\int_s^{\infty}\phi'(H_N(t))\,dG_N(t)$ if ϕ' is integrable. Then define

$$T_N(\phi)=\sqrt{N}\int_{-\infty}^{\infty}g(s)\,d(\hat{H}_N-H_N)(s)$$

$$+\sqrt{N}\int_{-\infty}^{\infty}\phi(H_N(t))\,d(\hat{G}_N-G_N)(t). \tag{2.11}$$

LEMMA 2.3. *There exists a constant C, such that for every $\phi \in \mathscr{G}$ with integrable derivative,*

$$E(T_N(\phi))^2 \leqslant C\|\phi\|^2. \tag{2.12}$$

Proof. It is easy to see that the variance of the second summand in (2.11) is bounded by $C\|\phi\|^2$. Hence we consider the first summand:

$$E\left|\sqrt{N}\int_{-\infty}^{\infty}g(s)\,d(\hat{H}_N-H_N)(s)\right|^2$$

$$=\frac{1}{N}\sum_{i=1}^{N}\iint_s^{\infty}\phi'(H_N(t))\,dG_N(t)\int_s^{\infty}\phi'(H_N(r))\,dG_N(r)\,dF_{Ni}(s)$$

$$-\iint_s^{\infty}\phi'(H_N(t))\,dG_N(t)\,dF_{Ni}(s)\iint_u^{\infty}\phi'(H_N(r))\,dG_N(r)\,dF_{Ni}(u)$$

$$=\frac{2}{N}\sum_{i=1}^{N}\iint_{t<r}\int_{-\infty}^{t} dF_{Ni}(s)\,\phi'(H_N(t))\,\phi'(H_N(r))\,dG_N(t)\,dG_N(r)$$

$$-\iint_{t<r}\int_{-\infty}^{t} dF_{Ni}(s)\int_{0}^{r} dF_{Ni}(u)\,\phi'(H_N(t))\,\phi'(H_N(r))\,dG_N(t)\,dG_N(r)$$

$$=\frac{2}{N}\sum_{i=1}^{N}\iint_{t<r} F_{Ni}(t)(1-F_{Ni}(r))\,\phi'(H_N(t))\,\phi'(H_N(r))\,dG_N(t)\,dG_N(r)$$

$$\ll\frac{2}{N}\sum_{i=1}^{N}\iint_{t<r}\sqrt{F_{Ni}(t)(1-F_{Ni}(t))}\sqrt{F_{Ni}(r)(1-F_{Ni}(r))}$$

$$\times|\phi'(H_N(t))|\,|\phi'(H_N(r))|\,dH_N(t)\,dH_N(r)$$

$$\ll\left(\int\sqrt{H_N(t)(1-H_N(t))}\,|\phi'(H_N(t))|\,dH_N(t)\right)^2$$

$$=\left(\int_0^1\sqrt{z(1-z)}\,|\phi'(z)|\,dz\right)^2\ll\|\phi\|^2.$$

Remark. The statistic $T_N(\phi)$ can be written in a different form as follows: Let

$$A_{Ni}(x)=\frac{1}{N}\sum_{j=1}^{N}(C_{Nj}-C_{Ni})\int_{-\infty}^{\infty}(I_{\{x\leqslant t\}}-F_{Ni}(t))\,\phi'(H_N(t))\,dF_{Nj}(t). \quad (2.13)$$

Then

$$\frac{1}{N}\int_{-\infty}^{\infty} g(s)\,d(\hat{F}_{Ni}-F_{Ni})(s)+C_{Ni}\int_{-\infty}^{\infty}\phi(H_N(t))\,d(\hat{F}_{Ni}-F_{Ni})(t)$$

$$=\frac{1}{N}\int_{X_{Ni}}^{\infty}\phi'(H_N(t))\sum_{j=1}^{N}C_{Nj}\,dF_{Nj}(t)$$

$$-\frac{1}{N}\int_{-\infty}^{\infty}\int_{s}^{\infty}\phi'(H_N(t))\sum_{j=1}^{N}C_{Nj}\,dF_{Nj}(t)\,dF_{Ni}(s)$$

$$+C_{Ni}\int_{-\infty}^{H_N(X_{Ni})}\phi'(s)\,ds-C_{Ni}\int_{-\infty}^{\infty}\int_{-\infty}^{H_N(t)}\phi'(s)\,ds\,dF_{Ni}(t)$$

$$=A_{Ni}(X_{Ni}).$$

Consequently

$$T_N(\phi)=\sqrt{N}\sum_{i=1}^{N}A_{Ni}(X_{Ni}),$$

and by Lemma 2.3 and the density of C_2-functions with bounded derivatives, we obtain (2.12) *for arbitrary* $\phi \in \mathcal{G}$ (not necessarily with integrable derivative). In particular (2.12) implies the continuity of the variances of $T_N(\phi)$, and in the next section the continuity of the covariance matrices of the vector $(T_N(\phi_\nu))_{\nu=1,\dots,p}$.

LEMMA 2.4. *Let g^r be a square integrable function with respect to the Lebesgue measure, where r is a positive integer. Then for every $m \in \mathbb{N}$ there exists a constant c independent of r such that*

$$E\left(\int g(H_N(t)) \left|\frac{N}{N+1} \hat{H}_N(t) - H_N(t)\right|^m d\hat{H}_N(t)\right)^{2r}$$

$$\leqslant c^r r^{4rm} \sum_{l=0}^{mr} N^{-2mr+l} \int g^{2r}(t)(t(1-t))^l \, dt \tag{2.14}$$

and

$$E\left(\int g(H_N(t)) \left(\frac{N}{N+1} \hat{H}_N(t) - H_N(t)\right) d(\hat{G}_N(t) - G_N(t))\right)^{2r}$$

$$\leqslant c^r r^{4r} N^{-2r} \int g^{2r}(t) \, dt. \tag{2.15}$$

Proof. Omitting the index N in X_{Ni} and F_{Ni}, first note that

$$E(\hat{H}_N(t) - H_N(t))^{2mr} = \sum_{i_1=1}^{N} \cdots \sum_{i_{2mr}=1}^{N} N^{-2mr} E \prod_{j=1}^{2mr} (1_{\{X_{i_j} \leqslant t\}} - F_{i_j}(t))$$

$$= \sum_{l=1}^{mr} \sum_{1 < i_1 < \cdots < i_l \leqslant N} N^{-2mr} \sum_{\substack{\alpha_j \geqslant 2 \\ j=1,\dots,l \\ \sum \alpha_j = 2mr}} \frac{(2mr)!}{\prod \alpha_j!}$$

$$\times E \prod_{k=1}^{l} (1_{\{X_{i_k} \leqslant t\}} - F_{i_k}(t))^{\alpha_k}$$

$$\ll \sum_{l=1}^{mr} \sum_{1 < i_1 < \cdots < i_l \leqslant N} \frac{(2mr)!}{(2mr-2l)!}$$

$$\times l^{2mr-2l} N^{-2mr} \prod_{k=1}^{l} F_{i_k}(t)(1 - F_{i_k}(t))$$

$$\leqslant c^r r^{4mr} \sum_{l=1}^{mr} N^{-2mr+l} (H_N(t)(1 - H_N(t)))^l.$$

Furthermore, applying the last estimate for each X_i fixed by taking expectation with respect to all X_j $(j \neq i)$ first, we obtain

$$E\left(\int g(H_N(t))\,|\hat{H}_N(t) - H_N(t)|^m \, d\hat{H}_N(t)\right)^{2r}$$

$$= E\left(\sum_{1 \leqslant i \leqslant N} N^{-1} g(H_N(X_i))\,|\hat{H}_N(X_i) - H_N(X_i)|^m\right)^{2r}$$

$$\leqslant N^{-1} \sum_{1 \leqslant i \leqslant N} E(g^{2r}(H_N(X_i))(\hat{H}_N(X_i) - H_N(X_i))^{2mr})$$

$$\leqslant c^r N^{-2mr} \frac{1}{N} \sum_{i=1}^{N} Eg^{2r}(H_N(X_i)) + \frac{1}{N} \sum_{i=1}^{N}$$

$$\times E\left(c^r r^{4mr} \sum_{l=1}^{mr} g^{2r}(H_N(X_i))[H_N(X_i)(1 - H_N(X_i)) + O(N^{-1})]^l N^{-2mr+l}\right)$$

$$\ll c^r r^{4mr} \sum_{l=0}^{mr} N^{-2mr+l} \int g^{2r}(t)(t(1-t))^l \, dt.$$

This proves (2.14).

To prove (2.15), we proceed as follows: omitting the index N as before, let

$$U_{(i,j)} = g(H_N(X_i))\, I_{\{X_j \leqslant X_i\}} - g(H_N(X_i))\, F_j(X_i)$$
$$- \int g(H_N(t))\, I_{\{X_j \leqslant t\}}\, dF_i(t) + \int g(H_N(t))\, F_j(t)\, dF_i(t),$$
$$1 \leqslant i, j \leqslant N.$$

Then

$$E\left(\int g(H_N(t))\left(\frac{N}{N+1}\hat{H}_N(t) - H_N(t)\right) d(\hat{G}_N(t) - G_N(t))\right)^{2r}$$

$$= E\left(\sum_{i=1}^{N} \sum_{j=1}^{N} \frac{C_i}{N} U_{(i,j)} - \frac{1}{N+1} \int g(H_N(t))\, \hat{H}_N(t)\, d(\hat{G}_N(t) - G_N(t))\right)^{2r}$$

$$\leqslant 4^r \left(E\left(\sum_{i,j=1}^{N} \frac{C_i}{N} U_{(i,j)}\right)^{2r} + N^{-2r} \int g^{2r}(t)\, dt\right).$$

If $1 \leqslant i_1, \ldots, i_{4r} \leqslant N$, and if one of the indices i_l is different from all the others, then it is easy to check that

$$EU_{(i_1,i_2)} \cdots U_{(i_{4r-1}, i_{4r})} = 0.$$

Consequently,

$$E\left(\sum_{i,j=1}^{N} \frac{C_i}{N} U_{(i,j)}\right)^{2r} \leqslant N^{-4r} \sum E|U_{(i_1,i_2)} \cdots U_{(i_{4r-1},i_{4r})}|,$$

where the summation extends over all choices of $i_1,\ldots, i_{4r}$ not satisfying the above condition. Using the Cauchy–Schwarz inequality and a simple combinatorial argument as before, and observing that $E|U_{(i_1,i_2)} \cdots U_{(i_{4r-1},i_{4r})}| \leqslant c^r \int g^{2r}(t)\, dt$, we obtain

$$E\left(\sum_{i,j=1}^{N} \frac{C_i}{N} U_{(i,j)}\right)^{2r} \leqslant c^r N^{-2r} r^{4r} \int g^{2r}(t)\, dt.$$

The proof follows.

3. Asymptotic Normality of S_N

Consider a function $\phi = (\phi_1,\ldots, \phi_p)$ such that $\phi_\nu \in \mathscr{G}$, $\nu = 1,\ldots, p$. Applying (2.13) to each coordinate, we can define $A_{Ni}^{(\nu)}$ and write $T_N^\nu(\phi)$ as

$$T_N^{(\nu)}(\phi) = \sqrt{N} \sum_{i=1}^{N} A_{Ni}^{(\nu)}(X_{Ni}^{(\nu)}), \qquad \nu = 1,\ldots, p.$$

Let us denote by $V = V_N(\phi)$ the covariance matrix of $T_N = (T_N^{(1)}(\phi),\ldots, T_N^{(p)}(\phi))$. Then, by the remark after Lemma 2.3, we have

$$\|V\| \leqslant \|\phi\|^2 := \max_{1 \leqslant \nu \leqslant p} \|\phi_\nu\|^2.$$

Since $T_N(\phi)$ is linear in $\phi \in \mathscr{G}$, it follows (using the Cauchy–Schwarz inequality) that for $\phi, \psi \in \mathscr{G}$,

$$\|V_N(\phi) - V_N(\psi)\| \leqslant (\|\phi\| + \|\psi\|)\, \|\phi - \psi\|. \tag{3.1}$$

The matrix $V = V_N(\phi)$, $\phi \in \mathscr{G}$, is symmetric and positive definite. Denote by $\mathscr{N}(V)$ the distribution function of the $\mathbb{R}^p$-valued centered Gaussian measure defined by V. It is then straightforward to see that for $\phi, \psi \in \mathscr{G}$,

$$\begin{aligned} |\mathscr{N}(V_N(\phi))(C) - \mathscr{N}(V_N(\psi))(C)| &\leqslant \lambda^{-p} \|V_N(\phi) - V_N(\psi)\|^p \\ &\leqslant \lambda^{-p}[(\|\phi\| + \|\psi\|)\, \|\phi - \psi\|]^p, \quad (3.2) \end{aligned}$$

where λ denotes the smallest eigenvalue of $V_N(\phi)$. Indeed for two normal distributions P and Q with covariance matrices A and B, we have

$$|P(C)-Q(C)| = \left(\frac{1}{2\pi}\right)^{p/2} \left| \int_C \left(\frac{1}{\sqrt{\det A}} \exp\left(-\frac{1}{2} x' A^{-1} x\right) - \frac{1}{\sqrt{\det B}} \exp\left(-\frac{1}{2} x' B^{-1} x\right)\right) dx \right|$$

$$\leqslant \left(\frac{1}{2\pi}\right)^{p/2} \left| \int_{O^{-1}B^{-1/2}C} \left(\frac{1}{\sqrt{\det \Lambda}} \exp\left(-\frac{1}{2} y' \Lambda^{-1} y\right) - \exp\left(-\frac{1}{2} y' y\right)\right) dy \right|$$

$$\leqslant \prod_{i=1}^{p} |1-\sigma_i| = \det(\Lambda - I) \ll \lambda^{-p} \|A-B\|^p,$$

where O is a matrix with

$$\Lambda = OB^{1/2}A^{-1}B^{1/2}O^{-1} = \begin{pmatrix} \sigma_1 & & O \\ & \sigma_2 \ \ddots & \\ O & & \sigma_p \end{pmatrix}.$$

Denote by $Q_N(\phi)$ the distribution of the centered statistic $\tilde{S}_N(\phi)$, where

$$\tilde{S}_N(\phi) = \sqrt{N}\left(S_N(\phi) - \left(\int \phi_\nu(H_N^{(\nu)}(t))\, dG_N^{(\nu)}(t) \right)_{1 \leqslant \nu \leqslant p} \right)$$

and where $H_N^{(\nu)}$ and $G_N^{(\nu)}$ are defined as in (2.7). We first obtain an estimate for the remainder in the normal approximation to $Q_N(\phi)$ where ϕ has a bounded second derivative.

THEOREM 3.1. *Let the scores $a_N^{(\nu)}(i)$ $(1 \leqslant \nu \leqslant p)$ be defined by (1.3) for functions ϕ_ν which have a bounded, continuous second derivative. Assume that the regression constants $C_{Ni}^{(\nu)}$ $(1 \leqslant \nu \leqslant p)$ satisfy (2.8). Then there exists a universal constant $C(p)$, independent of N, ϕ, and $C_{Ni}^{(\nu)}$, such that*

$$\sup_{x \in \mathbb{R}^p} |Q_N(\phi)(x) - \mathcal{N}(V_N(\phi))(x)|$$

$$\leqslant C(p)\, \lambda^{-3/2} \rho_3 N^{-1/2} + C(p)\, N^{-1/2} (\log N)^2 \left(\|\phi''\|_\infty^2 / \lambda\right)^{\operatorname{Log} N/(1+2\operatorname{Log} N)}, \quad (3.3)$$

where $\|\phi''\|_\infty = \max_{1 \leqslant \nu \leqslant p} \operatorname{ess\,sup} |\phi_\nu''|$, λ is the smallest eigenvalue of $V_N(\phi)$, and $\rho_3 = (1/N) \sum_{i=1}^N E\|N(A_{Ni}^{(\nu)}(X_{Ni})_{1 \leqslant \nu \leqslant p})\|^3$.

Remark. If the X_{Ni} have identical distributions, the approximation (3.3) is of order $N^{-1/2+\varepsilon}$ for any $\varepsilon > 0$.

Proof. We write

$$\begin{aligned}\tilde{S}_N^{(\nu)}(\phi) = T_N^{(\nu)}(\phi_\nu) + \sqrt{N} \int_{-\infty}^{\infty} & \phi_\nu'(H_N^{(\nu)}(t)) \\ & \times \left(\frac{N}{N+1} \hat{H}_N^{(\nu)}(t) - H_N^{(\nu)}(t)\right) d(\hat{G}_N^{(\nu)} - G_N^{(\nu)})(t) \\ & - \frac{\sqrt{N}}{N+1} \int_{-\infty}^{\infty} \phi_\nu'(H_N^{(\nu)}(t))\, \hat{H}_N^{(\nu)}(t)\, dG_N^{(\nu)}(t) \\ & + \frac{1}{2} \sqrt{N} \int_{-\infty}^{\infty} \phi_\nu'' \Big(\theta(H_N^{(\nu)}(t)) \\ & \times \left(\frac{N}{N+1} \hat{H}_N^{(\nu)}(t) - H_N^{(\nu)}(t)\right)^2 d\hat{G}_N^{(\nu)}(t)\Big)\end{aligned}$$

and calculate (using Lemma 2.4) with $q \geqslant 1$,

$$\begin{aligned} & E\left(\sqrt{N} \int_{-\infty}^{\infty} \phi_\nu'(H_N^{(\nu)}(t)) \left(\frac{N}{N+1} \hat{H}_N^{(\nu)}(t) - H_N^{(\nu)}(t)\right) d(\hat{G}_N^{(\nu)} - G_N^{(\nu)})(t)\right)^{2q} \\ & \qquad \ll N^{-q} \|\phi''\|_\infty^{2q} q^{4q} c^q, \\ & E\left(\frac{\sqrt{N}}{N+1} \int_{-\infty}^{\infty} \phi_\nu''(H_N^{(\nu)}(t))\, \hat{H}_N^{(\nu)}(t)\, dG_N^{(\nu)}(t)\right)^{2q} \\ & \qquad \ll N^{-q} \|\phi''\|_\infty^{2q}, \\ & E\left(\sqrt{N} \int_{-\infty}^{\infty} \phi_\nu'' \left(\theta(H_N^{(\nu)}(t))\right) \left(\frac{N}{N+1} \hat{H}_N^{(\nu)}(t) - H_N^{(\nu)}(t)\right)^2 d\hat{G}_N^{(\nu)}(t)\right)^{2q} \\ & \qquad \ll N^{-q} \|\phi''\|_\infty^{2q} q^{4q} c^q,\end{aligned}$$

where $\theta(\cdot)$ arises from the Taylor expansion. First, an application of Corollary 17.2 (page 165) of Bhattacharya and Ranga Rao [2] yields

$$\sup_{x \in \mathbb{R}^p} |P(T_N^{(\nu)}(\phi) \leqslant x_\nu (1 \leqslant \nu \leqslant p)) - \mathcal{N}(V_N(\phi))(x)| \leqslant C(p)\, \lambda^{-3/2} \rho_3 N^{-1/2}.$$

Since

$$\begin{aligned} & E\left|\sqrt{N}\left(S_N^{(\nu)}(\phi) - \int \phi_\nu(H_N^{(\nu)}(t))\, dG_N^{(\nu)}(t)\right) - T_N^{(\nu)}(\phi)\right|^{2q} \\ & \qquad \ll N^{-q} \|\phi''\|_\infty^{2q} q^{4q} c^q,\end{aligned}$$

(3.3) follows from the Markov inequality with $q = \log N$. The proof follows.

Denote by $\langle \cdot, \cdot \rangle$ the usual inner product on $\mathbb{R}^p$ and by $|x|^2 = \sum_{\nu=1}^{p} x_\nu^2$ the Euclidean norm on $\mathbb{R}^p$. The Prohorov distance $\Pi(F, G)$ of two probability distributions F and G on $\mathbb{R}^p$ is defined by

$$\Pi(F, G) = \inf\{\varepsilon > 0 \mid F(A) \leqslant G(A^\varepsilon) + \varepsilon \text{ for all measurable } A\}, \tag{3.4}$$

where $A^\varepsilon = \{x \in \mathbb{R}^p \mid \inf_{y \in A} |x - y| < \varepsilon\}$.

THEOREM 3.2. *Under the assumptions of Theorem* 3.1, *there exists some universal constant C (independent of p, N, ϕ, and $C_{Ni}^{(\nu)}$) such that*

$$\Pi(\tilde{Q}_N(\phi), \mathcal{N}(\sigma^{-2} V_N(\phi))) \leqslant CL_N + CN^{-1/2}(\log N)^2\, p^{1/2}(\|\phi''\|_\infty\, \sigma^{-1})^{1-1/(1+2\log N)}, \tag{3.5}$$

where

$$L_N = \sum_{\nu=1}^{p} \frac{\sum_{j=1}^{N} E|\langle e_\nu, \sqrt{N}(A_{Nj}^{(s)}(X_{Nj}^{(s)}))_{1 \leqslant s \leqslant p}\rangle|^3}{(E\langle e_\nu, T_N(\phi)\rangle^2)^{3/2}},$$

where $\tilde{Q}_N(\phi)$ denotes the distribution of $\sigma^{-1}\tilde{S}_N(\phi)$, σ^2 is the largest eigenvalue of $V_N(\phi)$, and e_ν are the eigenvectors of $\sigma^{-1}V_N(\phi)$.

Furthermore, we have

$$\Pi(Q_N(\phi), \mathcal{N}(V_N(\phi))) \leqslant Cp^{1/4}\rho_3^{1/4}N^{-1/8}(|\log \rho_3 N^{-1/2}|^{1/2} + (\log p)^{1/2}) + Cp^{1/2}N^{-1/2}(\log N)^2\,\|\phi''\|_\infty^{1-1/(2+\log N)}, \tag{3.6}$$

where ρ_3 is defined in Theorem 3.1.

Remark. If all X_{Ni} are identically distributed, then L_N is of order $N^{-1/2}$.

Proof. The proof is carried out as in Theorem 3.1. To prove (3.5), one can use the theorem on page 9 of Yurinskii [14] and for (3.6) the result follows from Theorem 1 (page 236) of Yurinskii [15] (along with the fact that $P(Y \in C) \leqslant P(Z \in C^\varepsilon) + P(\|Y - Z\| \geqslant \varepsilon)$ for two random vectors Y and Z to estimate the Prohorov distance of their distributions).

We continue this section proving asymptotic normality for S_N. This will be done using the results of Section 2 and the fact that functions with bounded second derivative are dense in $\mathcal{G}$.

THEOREM 3.3. *Let the scores $a_N^{(\nu)}(i)$ $(1 \leqslant \nu \leqslant p)$ be defined by* (1.3) *with respect to functions $\phi_\nu \in \mathcal{G}$, and assume that the $C_{Ni}^{(\nu)}$ $(1 \leqslant \nu \leqslant p)$ satisfy* (2.8). *Then, for every vector $\lambda \in \mathbb{R}^p$,*

$$\lambda' \tilde{S}_N(\phi)(\lambda' V_N(\phi)\, \lambda)^{-1/2} \tag{3.7}$$

has asymptotically (as $N \to \infty$*) the* $\mathcal{N}(0,1)$ *distribution, provided* lim inf $\lambda' V_N(\phi)\,\lambda > 0$.

Proof. By Theorem 3.2, for $\phi_\nu \in \mathcal{G}$ with bounded second derivative, (3.7) converges to the standard normal distribution.

Now let $\phi_\nu \in \mathcal{G}$ $(1 \leqslant \nu \leqslant p)$ be arbitrary. Let $\psi_\nu \in \mathcal{G}$ have a bounded continuous second derivative $(1 \leqslant \nu \leqslant p)$. Then, by Lemma 2.2 (applied for each $\nu = 1, \dots, p$) we have, uniformly in N,

$$E\,\|\tilde{S}_N(\phi) - \tilde{S}_N(\psi)\|^2 \ll \|\phi - \psi\|^2$$

and therefore

$$E\,\|\lambda'\tilde{S}_N(\phi - \psi)\|^2 \ll \|\phi - \psi\|^2.$$

Since $V_N(\phi)$ is a continuous function on $\mathcal{G}$ (cf. (3.1)), and since every $\phi_\nu \in \mathcal{G}$ can be approximated arbitrarily well by C_2-functions with bounded second derivative, the theorem follows immediately.

We next give a corollary which extends these results to the case when the scores are generated according to (1.4).

COROLLARY 3.4. *Let* $\phi_\nu \in \mathcal{G}$ *be the inverse of a distribution function and assume that the scores* $a_N^{(\nu)}(i)$ *are defined by* (1.4) *with respect to* ϕ_ν $(1 \leqslant \nu \leqslant p)$. *If also* (2.8) *is satisfied, then the statistic*

$$\lambda' \sqrt{N}\left(S_N - \left(\left(\int \phi_\nu(H_N^{(\nu)}(t))\, dG_N^{(\nu)}(t)\right)_{1 \leqslant \nu \leqslant p}\right)\right)(\lambda' V_N(\phi)\,\lambda)^{-1/2} \qquad (3.8)$$

has asymptotically the $N(0,1)$ *distribution, provided* lim inf $\lambda' V_N(\phi)\,\lambda > 0$.

Proof. Note that S_N and $S_N(\phi)$ are different, but it suffices to show that $\sqrt{N}\,(S_N - S_N(\phi))(\lambda' V_N(\phi)\,\lambda)^{-1/2} \to 0$ in probability.

Define $\phi_{N,\nu}(t) = \sum_{i=1}^{[Nt]} (a_N^{(\nu)}(i) - a_N^{(\nu)}(i-1))$, $a_N^{(\nu)}(0) = 0$, $0 < t < 1$, $\nu = 1, \dots, p$, where $[\alpha]$ denotes the largest integer not exceeding α. It is easy to check that

$$S_N^\nu = \int \phi_{N,\nu}\left(\frac{N}{N+1}\,\hat{H}_N^{(\nu)}(t)\right) d\hat{G}_N^{(\nu)}(t) = S_N^{(\nu)}(\phi_{N,\nu}).$$

The proof of Corollary 3.4 will be accomplished if we prove the following:

(i) $\lim \phi_{N,\nu}(t) = \phi_\nu(t)$

(ii) $$\sqrt{N}\left|\int_{-\infty}^{\infty}\left\{\phi_{N,\nu}\left(\frac{N}{N+1}\,\hat{H}_N^{(\nu)}(t)\right) - \phi_\nu\left(\frac{N}{N+1}\,\hat{H}_N^{(\nu)}(t)\right)\right\} d\hat{G}_N(t)\right| \to 0.$$

We shall only prove (ii) since the proof of (i) can be found on pages 408–409 of Puri and Sen [13]. By inequality (2.8), we have

$$\sqrt{N}\left|\int_{-\infty}^{\infty}\left\{\phi_{N,\nu}\left(\frac{N}{N+1}\hat{H}_N^{(\nu)}(t)\right)-\phi_\nu\left(\frac{N}{N+1}\hat{H}_N^{(\nu)}(t)\right)\right\}d\hat{G}_N^{(\nu)}(t)\right|$$

$$\ll N^{-1/2}\sum_{i=1}^{N}\left|\phi_{N,\nu}\left(\frac{i}{N+1}\right)-\phi_\nu\left(\frac{i}{N+1}\right)\right|.$$

By Puri and Sen [13, pp. 409–411], this last term tends to zero. The proof follows.

A consequence of Corollary 3.4 is that in many cases of practical interest (such as the normal scores), asymptotic normality holds with the same centering sequence, whether the scores are given by (1.3) or (1.4). Thus Theorem 3.3 and Corollary 3.4 serve to unify the results of Hájek [6] and Hoeffding [7], and also to some extent simplify the results of the latter paper.

4. Remainder Term of S_N

In this section we extend the estimate in Theorem 3.1 for the remainder in the normal approximation of $S_N(\phi)$ to more general $\phi_\nu \in \mathscr{G}$ $(\nu = 1,..., p)$. We keep the notations of the first three sections.

THEOREM 4.1. *Let $\phi_\nu \in \mathscr{G}$ $(\nu = 1,..., p)$ be C_2-functions satisfying*

$$\max(|\phi''(t)|, |\phi'(t)|) \leqslant \Delta^{-1}(t(1-t))^{-3/2+\eta} \tag{4.1}$$

for some $\eta, \Delta > 0$. Assume that the scores $a_N^{(\nu)}(i)$ are given by (1.3) *with respect to ϕ_ν and let* (2.8) *be satisfied. Then, for every $\varepsilon > 0$, there exists a constant $C = C(p, \varepsilon, \eta)$ (independent of N, ϕ and $C_{Ni}^{(\nu)}$) such that*

$$\sup_{x\in\mathbb{R}^p} |Q_N(\phi)(x) - \mathscr{N}(V_N(\phi))(x)|$$

$$\leqslant C\lambda^{-3/2}\rho_3 N^{-1/2} + C(p/N\lambda\,\Delta^2)^{1/3}\,N^{\varepsilon}. \tag{4.2}$$

Proof. Proceeding as in the proof of Theorem 3.1 and using Taylor's expansion, we obtain

$$\tilde{S}_N^{(\nu)}(\phi_\nu) = T_N^{(\nu)}(\phi_\nu) + \sqrt{N}\int_{-\infty}^{\infty}\phi_\nu'(H_N^{(\nu)}(t))\left(\frac{N}{N+1}\hat{H}_N^{(\nu)}(t) - H_N^{(\nu)}(t)\right)$$

$$\times\, d(\hat{G}_N^{(\nu)} - G_N^{(\nu)})(t) - \frac{\sqrt{N}}{N+1}\int_{-\infty}^{\infty}\phi_\nu'(H_N^{(\nu)}(t))\,\hat{H}_N^{(\nu)}(t)\,dG_N^{(\nu)}(t)$$

$$+\sqrt{N}\int_{-\infty}^{\infty}\left(\int_{H_N^{(\nu)}(t)}^{(N/(N+1))\hat{H}_N^{(\nu)}(t)}\left(\frac{N}{N+1}\hat{H}_N^{(\nu)}(t)-s\right)\phi_\nu''(s)\,ds\right)$$

$$\times d\hat{G}_N^{(\nu)}(t).$$

By Lemma 2.4 (omitting the index ν)

$$E\left(\sqrt{N}\int_{-\infty}^{\infty}\phi'(H_N(t))\left(\frac{N}{N+1}\hat{H}_N(t)-H_N(t)\right)d(\hat{G}_N-G_N)(t)\right)^2$$

$$\ll N^{-1}\int_0^1\phi'^2(t)\,dt \ll \Delta^{-2}N^{-1}\int_0^1(t(1-t))^{-1+2\eta}\,dt \ll \Delta^{-2}N^{-1}.$$

Also it follows immediately that

$$E\left|\frac{\sqrt{N}}{N+1}\int_{-\infty}^{\infty}\phi'(H_N(t))\,\hat{H}_N(t)\,dG_N(t)\right|^2 \ll N^{-1}\Delta^{-2}.$$

Let $a=H_N^{-1}(1/N)$ and $b=H_N^{-1}(1-1/N)$. Then

$$E\left(\sqrt{N}\int_a^b\left(\int_{H_N(t)}^{(N/(N+1))\hat{H}_N(t)}\left(\frac{N}{N+1}\hat{H}_N(t)-s\right)\phi''(s)\,ds\right)d\hat{G}_N(t)\right)^2$$

$$\ll E\left(\int_a^b\left|\frac{N}{N+1}\hat{H}_N(t)-H_N(t)\right|\right.$$

$$\left.\times\left|\int_{H_N(t)}^{(N/(N+1))\hat{H}_N(t)}\Delta^{-1}(s(1-s))^{-3/2+\eta}\,ds\right|\,d\hat{H}_N(t)\right)^2$$

$$\ll \Delta^{-2}NE\left(\int_a^b\left|\frac{N}{N+1}\hat{H}_N(t)-H_N(t)\right|\right.$$

$$\times\left|\left(\frac{N}{N+1}\hat{H}_N(t)\left(1-\frac{N}{N+1}\hat{H}_N(t)\right)\right)^{-1/2+\eta}\right.$$

$$\left.\left.-(H_N(t)(1-H_N(t)))^{-1/2+\eta}\right|\,d\hat{H}_N(t)\right)^2$$

$$\ll \Delta^{-2}NE\left(\int_a^b\left|\frac{N}{N+1}\hat{H}_N(t)-H_N(t)\right|^2\right.$$

$$\times\left[\frac{N}{N+1}\hat{H}_N(t)\left(1-\frac{N}{N+1}\hat{H}_N(t)\right)H_N(t)(1-H_N(t))\right]^{-1/2+\eta}$$

$$\left.\times\sup_{x\in A}(x(1-x))^{-1/2+\eta}\,d\hat{H}_N(t)\right)^2$$

with

$$A=\left[H_N(t),\frac{N}{N+1}\hat{H}_N(t)\right]\cup\left[\frac{N}{N+1}\hat{H}_N(t),H_N(t)\right].$$

(The supremum is taken either at $H_N(t)$ or at $(N/(N+1))\,\hat{H}_N(t)$. Let $p^{-1}+q^{-1}=1$ and $p\geqslant 2$ sufficiently large.)

$$\begin{aligned}
&\ll \Delta^{-2}NE\left[\left(\int_a^b\left|\frac{N}{N+1}\hat{H}_N(t)-H_N(t)\right|^{2p}(H_N(t)(1-H_N(t)))^{-p/2+p\eta}\right)^{2/p}\right.\\
&\quad\left.\times\left(\int\left(\frac{N}{N+1}\hat{H}_N(t)\left(1-\frac{N}{N+1}\hat{H}_N(t)\right)\right)^{-q}d\hat{H}_N(t)\right)^{2/q}\right]\\
&\quad+\Delta^{-2}NE\left[\left(\int_a^b\left|\frac{N}{N+1}\hat{H}_N(t)-H_N(t)\right|^{2p}(H_N(t)(1-H_N(t)))^{-p}d\hat{H}_N(t)\right)^{2/p}\right.\\
&\quad\left.\times\left(\int\left(\frac{N}{N+1}\hat{H}_N(t)\left(1-\frac{N}{N-1}\hat{H}_N(t)\right)\right)^{(-1/2+\eta)q}d\hat{H}_N(t)\right)^{2/q}\right]\\
&\ll \Delta^{-2}N\left(\sum_{l=0}^{2p}N^{-4p+l}\int_{N^{-1}}^{1-N^{-1}}(t(1-t))^{(-1/2+\eta)2p+l}\,dt\right)^{1/p}N^{2-2/q}\\
&\quad+\Delta^{-2}N\left(\sum_{l=0}^{2p}N^{-4p+l}\int_{N^{-1}}^{1-N^{-1}}(t(1-t))^{-2p+l}\,dt\right)^{1/p}\\
&\ll \Delta^{-2}N^{1-2/q}+\Delta^{-2}N^{-1}\ll \Delta^{-2}N^{-1+\varepsilon}.
\end{aligned}$$

On the other hand, if $H_N(t)\leqslant 1/N$ or $H_N(t)\geqslant 1-1/N$, then

$$\begin{aligned}
&E\left(\sqrt{N}\int_{[a,b]^c}\left(\int_{H_N(t)}^{(N/(N+1))\hat{H}_N(t)}\left(\frac{N}{N+1}\hat{H}_N(t)-s\right)\phi''(s)\,ds\right)d\hat{G}_N(t)\right)^2\\
&\qquad\ll \Delta^{-2}NE\left[\int_{[a,b]^c}\left|\frac{N}{N+1}\hat{H}_N(t)-H_N(t)\right|\right.\\
&\qquad\quad\left.\times(H_N(t)(1-H_N(t)))^{-1/2+\eta}\,d\hat{H}_N(t)\right]^2\\
&\qquad\ll \Delta^{-2}N\left(\sum_{l=0}^{1}N^{-2+l}\int_0^{N^{-1}}(t(1-t))^{-1+2\eta+l}\,dt\right)\\
&\qquad\ll \Delta^{-2}N(N^{-2}+N^{-1}N^{-1})\ll \Delta^{-2}N^{-1}.
\end{aligned}$$

(4.2) follows the same way as the proof of (3.3).

COROLLARY 4.2. *There exist constants C, and for every* $\varepsilon > 0$ $C(\varepsilon, \eta)$, *such that under the assumptions of Theorem* 4.1, *one has*

$$\Pi(\tilde{Q}_N(\phi), \mathcal{N}(\sigma^{-2}V_N(\phi))) \leqslant CL_N + C(\varepsilon, \eta)(p/N\sigma^2\Delta^2)^{1/3}N^{\varepsilon} \tag{4.3}$$

and

$$\Pi(Q_N(\phi), \mathcal{N}(V_N(\phi))) \leqslant Cp^{1/4}\rho_3^{1/4}N^{-1/8}(|\log \rho_3 N^{-1/2}|^{1/2} + (\log p)^{1/2}) + C(\varepsilon, \eta)(p/N\Delta^2)^{1/3}N^{\varepsilon}. \tag{4.4}$$

LEMMA 4.3. *Let* $\phi \in \mathcal{G}$ *have a continuous derivative* ϕ' *satisfying*

$$|\phi'(t)| \leqslant C_0(t(1-t))^{-3/2+\eta}$$

for some $\eta > 0$ *and some constant* $C_0 > 0$. *Define a* C_2*-function* $g = g_\Delta$ ($\Delta > 0$) *by its second derivative and* $g'_\Delta(\frac{1}{2}) = \phi'(\frac{1}{2})$, $g_\Delta(\frac{1}{2}) = \phi(\frac{1}{2})$,

$$\begin{aligned} g''_\Delta(t) &= \frac{\phi'(t) - \phi'(t-\Delta)}{\Delta} \qquad \text{if} \quad t \geqslant \tfrac{1}{2} \\ &= \frac{\phi'(t) - \phi'(t+\Delta)}{\Delta} \qquad \text{if} \quad t < \tfrac{1}{2}. \end{aligned} \tag{4.5}$$

Then

$$\|g_\Delta - \phi\| \to 0 \qquad \text{as} \quad \Delta \to 0$$

and

$$|g''_\Delta(t)| \leqslant C_0\,\Delta^{-1}(t(1-t))^{-3/2+\eta}.$$

Moreover, if ϕ' *is locally Hölder-continuous with exponent* γ, *i.e.*, $|\phi'(t) - \phi'(s)| \ll |t-s|^\gamma \max(|\phi'(t)|, |\phi'(s)|)$, *then* $\|g_\Delta - \phi\| \ll \Delta^\gamma$.

Proof. Without less of generality $\phi'(\frac{1}{2}) = 0 = \phi(\frac{1}{2})$. Let $t \geqslant \frac{1}{2}$. Then

$$g'_\Delta(t) = \int_{1/2}^{t} g''_\Delta(s)\,ds = \Delta^{-1}(\phi(t) - \phi(t-\Delta) + \phi(\tfrac{1}{2}-\Delta) - \phi(\tfrac{1}{2})).$$

If $t < \frac{1}{2}$, then $g'_\Delta(t) = \Delta^{-1}(\phi(t) - \phi(t+\Delta) + \phi(\frac{1}{2}+\Delta) - \phi(\frac{1}{2}))$. If ϕ' is Hölder-continuous with exponent γ, then, by the Mean Value Theorem, for $t \geqslant \frac{1}{2}$,

$$\begin{aligned} &\sqrt{t(1-t)}\,|\phi'(t) - g'(t)| \\ &\leqslant \left(\left|\frac{\phi(\frac{1}{2}-\Delta) - \phi(\frac{1}{2})}{\Delta}\right| + \left|\phi'(t) - \frac{\phi(t) - \phi(t-\Delta)}{\Delta}\right|\right)\sqrt{t(1-t)} \\ &\ll \Delta^\gamma(t(1-t))^{-1+\eta}, \end{aligned}$$

and similarly if $t<\frac{1}{2}$. It follows from (2.3) that

$$\|\phi-g\| \ll \int \sqrt{t(1-t)}\, |\phi'(t)-g'(t)|\, dt \ll \Delta^{\gamma}.$$

In general, if $\Delta \to 0$, $|\phi'(t)-g'(t)| \to 0$ and is bounded by $t \to (t(1-t))^{-3/2+\eta}$, since

$$\begin{aligned} |g'(t)| &\leqslant \left|\frac{\phi(t)-\phi(t-\Delta)}{\Delta}\right| + \left|\frac{\phi(\frac{1}{2}-\Delta)-\phi(\frac{1}{2})}{\Delta}\right| \\ &= |\phi'(\zeta)| + |\phi'(\zeta')| \leqslant C_0(t(1-t))^{-3/2}, \end{aligned}$$

where $t-\Delta \leqslant \zeta \leqslant t$, $\frac{1}{2}-\Delta \leqslant \zeta' \leqslant \frac{1}{2}$, and $t \geqslant \frac{1}{2}+\Delta$; the other cases follow similarly.

Hence $\sqrt{t(1-t)}\, |\phi'(t)-g'(t)| \to 0$ for every t, and this sequence of functions is bounded by the integrable function $t \to (t(1-t))^{-1+\eta}$. Using (2.3) and the Dominated Convergence Theorem we obtain $\|\phi-g\| \to 0$ as $\Delta \to 0$. As in the proof of Theorem 3.3 we can use the results of Section 2 (Lemma 2.2) together with (3.1), (3.2), and Theorem 4.1 to obtain

THEOREM 4.4. *Let $\phi_\nu \in \mathscr{G}$ be continuously differentiable with derivative ϕ'_ν satisfying $|\phi'_\nu(t)| \leqslant C_0(t(1-t))^{-3/2+\eta}$ for some constants $\eta, C_0>0$ $(\nu=1,\dots,p)$. Let the scores be defined by* (1.3) *and the regression constants satisfy* (2.8). *Then there exist universal constants C, $C(p)$ (depending on C_0 and η) and for every $\varepsilon>0$ constants $C(\varepsilon,\eta)$ and $C(p,\varepsilon,\eta)$ such that*

$$\begin{aligned} &\sup_{x\in\mathbb{R}^p} |Q_N(\phi)(x)-(V_N(\phi))(x)| \\ &\qquad \leqslant C(p)\,\lambda^{-3/2}\rho_3 N^{-1/2} + C(p,\varepsilon,\eta)(N\lambda\,\Delta^2)^{-1/3} N^{\varepsilon} \\ &\qquad\quad + C(p)(\|\phi-g\|^2\lambda^{-1})^{1/3}, \end{aligned} \tag{4.6}$$

$$\begin{aligned} &\Pi(\tilde{Q}_N(\phi), \mathscr{N}(\sigma^{-2}V_N(g))) \\ &\qquad \leqslant CL_N + C(\varepsilon,\eta)(p/N\sigma^2\Delta^2)^{1/3}N^{\varepsilon} + C(p\|\phi-g\|^2/\sigma^2)^{1/3}, \end{aligned} \tag{4.7}$$

$$\begin{aligned} &\Pi(Q_N(\phi), \mathscr{N}(V_N(g))) \\ &\quad \leqslant Cp^{1/4}\rho_3^{1/4}N^{-1/8}((\log p)^{1/2} + |\log \rho_3 N^{-1/2}|^{1/2}) + C(\varepsilon,\eta)(p/N\Delta^2)^{1/3}N^{\varepsilon} \\ &\qquad + C(p\|\phi-g\|^2)^{1/3}, \end{aligned} \tag{4.8}$$

where σ^2, ρ_3, λ, and L_N are defined as in Theorems 3.1 *and* 3.2 *for g, where $\Delta>0$ and where g is the function defined in* (4.5) *by Δ. In particular, if ϕ'_ν is locally Hölder-continuous with exponent γ, then we can choose Δ as some power of N and obtain in* (4.6)–(4.8) *a rate of convergence of order $N^{-\alpha}$ for* *some $\alpha > 0$. In general, Δ can be chosen in such a way that the right-hand side tends to zero.*

Proof. Let $\Delta > 0$ be given and let g be defined by (4.5). According to (3.2) the error we make when replacing $\mathcal{N}(V_N(\phi))$ by $\mathcal{N}(V_N(g))$ is of order $\|\phi - g\|^p \lambda^{-p} + \|\phi - g\|^{2p} \lambda^{-p}$ (for the estimate (4.6)). When passing from $\tilde{S}_N^{(v)}(\phi_v)$ to $\tilde{S}_N^{(v)}(g_v)$, Lemma 2.2 shows that the error is of order $\lambda^{-1/3} \|\phi - g\|^{2/3}$, resp. $p^{1/3} \|\phi - g\|^{2/3}$ or $p^{1/3}(\sigma^{-1} \|\phi - g\|)^{2/3}$. (4.6), (4.7), and (4.8) follow now from Theorem 4.1 and its Corollary; the Hölder-continuous case is already covered in Lemma 4.3.

Remarks. 1. In case $\eta > 1/6$ we have $|g'(t)|^3 \ll \Delta^{-3}(t(1-t)^{-1+\delta})$ (for some $\delta > 0$), and therefore we can replace ρ_3 by Δ^{-3} in the last theorem. Similarly L_N may be replaced by $N^{-1/3} \Delta^{-3}$. Concerning the Prohorov distance of $N(V_N(\phi))$ and $\mathcal{N}(V_N(g))$, it is easy to check (using the Euclidean norm) that $\Pi(N(V_N(\phi)); \mathcal{N}(V_N(g))) \leqslant \|V_N^{1/2}(\phi) - V_N^{1/2}(g)\|^{2/3}$ or $\leqslant \lambda^{-p}(\|\phi - g\|^p)$ by (3.2).

2. From the expressions for $V_{vv'} := \mathrm{Cov}(T_N^{(v)}(\phi_v), T_N^{(v')}(\phi_{v'}))$ $(1 \leqslant v, v' \leqslant p)$, it is clear that the smallest eigenvalue λ of $V_N(\phi)$ is extremely hard to compute. However, if the components $X_{Ni}^{(v)}$ of the vectors X_{Ni} satisfy the condition of "weak dependence" in the sense of $V_{vv}^2 - \sum_{v' \neq v''} |V_{v'v''}| > 0$ $(v = 1,..., p)$ the classical Grischgorin Theorem (cf. Dahlquist and Björek [4]), permits us to replace λ by a much simpler expression, assuming (without loss of generality) that $\lambda > 0$. If one is prepared to work numerically, much better estimates for λ can be obtained (op. cit. Dahlquist and Björek [4]).

References

[1] Bergström, H. and Puri, M. L. (1977). Convergence and remainder terms in linear rank statistics. *Ann. Statist.* **5** 671–680.

[2] Bhattacharya, R. N. and Ranga Rao, R. (1976). *Normal Approximation and Asymptotic Expansions.* Wiley, New York.

[3] Chernoff, H. and Savage, I. R. (1958). Asymptotic normality and efficiency of certain nonparametric test statistics. *Ann. Math. Statist.* **29** 972–996.

[4] Dahlquist, G. and Björek, H. L. (1974). *Numerical Methods* (English transl. by N. Anderson). Prentice-Hall, Englewood Cliffs, N.J.

[5] Denker, M. and Rösler, U. (1982). Some contributions to Chernoff-Savage theorems. *Statist. Decisions,* in press.

[6] Hájek, J. (1968). Asymptotic normality of simple linear rank statistics under alternatives. *Ann. Math. Statist.* **39** 325–346.

[7] Hoeffding, W. (1973). On the centering of a simple linear rank statistic. *Ann. Math. Statist.* **1** 56–66.

[8] Hušková, M. (1977). Rates of convergence of quadratic rank statistics. *J. Multivar. Anal.* **1** 63–73.

[9] HUŠKOVÁ, M. (1977). The rate of convergence of simple linear rank statistics under hypothesis and alternatives. *Ann. Math. Statist.* **5** 658–670.

[10] JUREČKOVÁ, J. AND PURI, M. L. (1975). Order of normal approximation for rank test statistics distribution. *Ann. Probab.* **3** 526–533.

[11] LOÈVE, M. (1963). *Probability Theory.* Van Nostrand, New York.

[12] PURI, M. L. AND SEN, P. K. (1969). A class of rank order tests for a general linear hypothesis. *Ann. Math. Statist.* **40** 1325–1343.

[13] PURI, M. L. AND SEN, P. K. (1971). *Nonparametric Methods in Multivariate Analysis.* Wiley, New York.

[14] YURINSKII, V. V. (1975). A smoothing inequality for estimates of the Lévy–Prokhorov distance. *Theory Probab. Appl.* **20** 1–10.

[15] YURINSKII, V. V. (1977). On the error of the Gaussian approximation for convolutions. *Theory Probab. Appl.* **22** 236–247.

THE ORDER OF NORMAL APPROXIMATION FOR SIGNED LINEAR RANK STATISTICS*

M. L. PURI[1] and T.-J. WU

Summary. The rate of convergence of the cdf (cumulative distribution function) of the signed linear rank statistics to the normal one is investigated. Under suitable assumptions, it is shown that the convergence rate is of order $O(N^{-1/2+\delta})$ for any $\delta > 0$.

1. Introduction. Let Y_{Ni}, $1 \leqq i \leqq N$, $N \geqq 1$, be independent random variables distributed according to the cdf's $F_{Ni}(y) = F(y - \Delta d_{Ni})$, $1 \leqq i \leqq N$, $N \geqq 1$, where Δ is an unknown real parameter, and d_{Ni}, $1 \leqq i \leqq N$, $N \geqq 1$, are known real constants. Consider the signed linear rank statistics

$$S_{\Delta N} = \sum_{i=1}^{N} c_{Ni}\varphi\left(\frac{R^{\Delta}_{Ni}}{N+1}\right) \operatorname{sgn}(Y_{Ni}), \tag{1.1}$$

where R^{Δ}_{Ni} is the rank of $|Y_{Ni}|$ among $|Y_{N1}|, \cdots, |Y_{NN}|$, c_{Ni}'s are known real constants, $\operatorname{sgn}(y) = 1$ or -1 according as $y \geqq 0$ or <0, and φ is a score generating function defined over $(0, 1)$. Under suitable assumptions on the c's and the score generating function φ, the asymptotic normality of $S_{\Delta N}$ has been established by Hušková [3]. However, the problem of determining the rate of convergence of the cdf of $S_{\Delta N}$ to the limiting normal one has not been investigated so far. The purpose of this paper is to study this problem. It is shown that the rate of convergence is of order $O(N^{-1/2+\delta})$ for any $\delta > 0$ for the case $\Delta = 0$ as well as $\Delta \neq 0$. For the corresponding study of the similar problem in the case of (unsigned) linear rank statistics of the form

$$S_N = \sum_{i=1}^{N} c_{Ni}\varphi\left(\frac{R_{Ni}}{N+1}\right),$$

where R_{Ni} is the rank of Y_{Ni} among $Y_{N1}, \cdots, Y_{NN}$, the reader is referred to Jurečková and Puri [5], Bergström and Puri [1], and Hušková [4], among others. Our methods are mainly the adaptations of the ideas of Jurečková and Puri [5], Bergström and Puri [1] and Hušková [4]. Throughout the paper we make the following assumptions:

A1) $F(x)$ is continuous and symmetric about zero,

A2) $\sum_{i=1}^{N} c^2_{Ni} = 1$, $\max_{1 \leqq i \leqq N} |c_{Ni}| = O(N^{-1/2}\sqrt{\log N})$,

A3) φ is not identically zero and satisfies the Lipschitz condition of order one on $(0, 1)$, that is, there exists a constant C such that

$$|\varphi(u) - \varphi(v)| \leqq C|u - v| \text{ for all } u, v \in (0, 1).$$

Remark. Unlike in earlier papers referred to above where the derivatives of different orders on φ are assumed, here the Lipschitz condition on φ is the only assumption we are making.

For simplicity of notation, we shall suppress the subscript N in c_{Ni}, d_{Ni}, R^{Δ}_{Ni} etc., whenever this causes no confusion.

2. Rate of convergence for $\Delta = 0$. Here we assume that $\Delta = 0$, so that $Y_{Ni}, \cdots, Y_{NN}$, $N \geqq 1$, are i.i.d. random variables distributed according to a continuous and symmetric cdf $F(y)$. We shall need the following theorem (cf. Petrov [7, Chap. V, Thm. 6]).

THEOREM 2.1. *Let $V_1, \cdots, V_N$ be independent random variables with $\mathbf{E}V_j = 0$, $\mathbf{E}|V_j|^{2+\delta} < \infty$ for some positive $\delta \leqq 1$ $(j = 1, \cdots, N)$. Then*

$$\sup_{x \in (-\infty, \infty)} |G_N(x) - \Phi(x)| \leqq AB_N^{-1-\delta/2} \sum_{j=1}^{N} \mathbf{E}|V_j|^{2+\delta},$$

where $A > 0$ is a constant independent of N, $B_N = \sum_{j=1}^{N} \mathbf{D}V_j$, $G_N(x) = \mathbf{P}[B_N^{-1/2} \sum_{j=1}^{N} V_j \leqq x]$, and $\Phi(x)$ is the standard cumulative normal distribution function.

* Received by the editors July 13, 1982; in revised form May 20, 1985.

[1] Work supported by the Office of Naval Research under contract No. N00014-77-C-0659. Reproduction in whole or part is permitted for any purpose of the United States Government.

AMS 1970 subject classifications. Primary 62E20, Secondary 60F99.

Now let us denote

$$T_{\Delta N}=\sum_{i=1}^{N} c_{Ni}\varphi(F^*(|Y_{Ni}|))\,\mathrm{sgn}\,(Y_{Ni}),\qquad F^*(x)=P_0[|Y_{11}|\leqq x], \tag{2.1}$$

where P_0 means that the probability is computed under $\Delta=0$, and note that $T_{\Delta N}$ is a sum of independent random variables. We shall use $T_{\Delta N}$ to approximate $S_{\Delta N}$.

LEMMA 2.1. *Under assumptions* A1–A3, *we have*

$$\sup_{x\in(-\infty,\infty)}|G_{0N}(x)-\Phi(x)|=O(N^{-1/2}\sqrt{\log^3 N}),$$

where G_{0N} is the cdf of $T_{0N}/\|\varphi\|_2$ and T_{0N} is given by (2.1).

Proof. Since $\mathbf{E}[c_i\varphi(F^*(|Y_i|)\,\mathrm{sgn}\,(Y_i)]=0$, and $\mathbf{E}|c_i\varphi(F^*(|Y_i|))\,\mathrm{sgn}\,(Y_i)|^3=|c_i|^3\int_0^1|\varphi^3(u)|\,du<\infty$ for $1\leqq i\leqq N$, it follows from Theorem 2.1 that

$$\sup_{x\in(-\infty,\infty)}|G_{0N}(x)-\Phi(x)|\leqq AB_N^{-3/2}\sum_{i=1}^{N}|c_i|^3\int_0^1|\varphi^3(u)|\,du, \tag{2.2}$$

where $A>0$ is a constant independent of N, and $B_N=\sum_{i=1}^N c_i^2\int_0^1\varphi^2(u)\,du$. The proof now follows from A2, A3 and (2.2).

LEMMA 2.2. *Under assumptions* A1–A3,

$$\mathbf{E}(S_{0N}-T_{0N})^{2K}=O[N^{-K}(\log N)^K]$$

for any fixed positive integer K, where S_{0N} and T_{0N} are given by (1.1) *and* (2.1) *respectively.*

Proof. Put $Z_i=[\varphi(R_i^0/(N+1))-\varphi(F^*(|Y_i|))]\,\mathrm{sgn}\,(Y_i)$. Then, for arbitrary nonnegative integers $p_1,p_2,\cdots,p_N$ satisfying $\sum_{i=1}^N p_i=2K$ we have

$$\mathbf{E}\left(\prod_{i=1}^{N}Z_i^{p_i}\right)=0, \tag{2.3}$$

if at least one of the p_i's is an odd number. Then using (2.3) we obtain

$$\begin{aligned}\mathbf{E}[(S_{0N}-T_{0N})]^{2K}&=\mathbf{E}\left[\left(\sum_{i=1}^{N}c_iZ_i\right)^{2K}\right]\\
&=\sum_{\substack{l_1=0\\ l_1+l_2+\cdots+l_N=K}}^{K}\sum_{l_2=0}^{K}\cdots\sum_{l_N=0}^{K}C_{2\mathbf{l}_N}^{2K}c_1^{2l_1}c_2^{2l_2}\cdots c_N^{2l_N}\mathbf{E}\left(\prod_{i=1}^{N}Z_i^{2l_i}\right)\\
&=\sum_{m=1}^{K}\Bigg[\sum_{\substack{i_1=1\\ i_1<i_2<\cdots<i_m}}^{N}\sum_{i_2=1}^{N}\cdots\sum_{i_m=1}^{N}\\
&\qquad\cdot\Bigg[\sum_{\substack{l_1=1\\ l_1+l_2+\cdots+l_m=K}}^{K}\sum_{l_2=1}^{K}\cdots\sum_{l_m=1}^{K}C_{2\mathbf{l}_m}^{2K}c_{i_1}^{2l_1}\cdots c_{i_m}^{2l_m}\mathbf{E}(Z_{i_1}^{2l_1}Z_{i_2}^{2l_2}\cdots Z_{i_m}^{2l_m})\Bigg]\Bigg],\end{aligned} \tag{2.4}$$

where

$$\mathbf{l}_j=(l_1,l_2,\cdots,l_j)\quad\text{and}\quad C_{2\mathbf{l}_j}^{2K}=\frac{(2K)!}{(2l_1)!(2l_2)!\cdots(2l_j)!}\quad\text{for } j=N \text{ or } m.$$

Now, by the generalized Cauchy–Schwarz inequality and A3, we have for any $i_1<i_2<\cdots<i_m$, m; $l_j=1,\cdots,K$; $j=1,\cdots,m$, $\sum_{j=1}^m l_j=K$, that

$$\begin{aligned}\mathbf{E}(Z_{i_1}^{2l_1}Z_{i_2}^{2l_2}\cdots Z_{i_m}^{2l_m})&\leqq\left(\prod_{j=1}^{m}\mathbf{E}(Z_j^{2ml_j})\right)^{1/m}\\
&=\left\{\prod_{j=1}^{m}\mathbf{E}[\varphi(R_1^0/(N+1))-\varphi(F^*(|Y_1|))]^{2ml_j}\right\}^{1/m}\\
&\leqq C^{2K}\left\{\prod_{j=1}^{m}\mathbf{E}\left(\frac{R_1^0}{N+1}-F^*(|Y_1|)\right)^{2ml_j}\right\}^{1/m}\leqq M_kN^{-K},\end{aligned} \tag{2.5}$$

where M_K is a constant depending only on K. To obtain the last inequality in (2.5) we have used the well-known fact (see e.g. Kallenberg [6, p. 406]) that

$$\mathbf{E}\left(\frac{R_1^0}{N+1}-F^*(|Y_1|)\right)^{2p}=O(N^{-p})$$

for any positive integer p. Also, in view of A2, there exists a constant M'_K (depending only on K) such that

$$|c_{i_1}^{2l_1}c_{i_2}^{2l_2}\cdots c_{i_m}^{2l_m}|\leqq M'_K N^{-K}(\log N)^K \tag{2.6}$$

for any $i_1<\cdots<i_m$, $m=1,\cdots,K$, and any $l_j=1,\cdots,K$; $j=1,\cdots,m$, $\sum_{j=1}^m l_j=K$. The proof now follows from (2.4), (2.5) and (2.6).

The main theorem of this section is the following.

THEOREM 2.2. *Under assumptions* A1–A3, *we have for any* $\delta>0$ *that*

$$\sup_{x\in(-\infty,\infty)}|E_{0N}(x)-\varphi(x)|=O(N^{-1/2+\delta}),$$

where F_{0N} *is the* cdf *of* $S_{0N}/|\varphi\|_2$.

Proof. For any $\varepsilon>0$ and any N, we have

$$\mathbf{P}[\|\Phi\|_2^{-1}S_{0N}\leqq x]\leqq\mathbf{P}[\|\Phi\|_2^{-1}T_{0N}\leqq x+\varepsilon]+\mathbf{P}[\|\varphi\|_2^{-1}|S_{0N}-T_{0N}|\geqq\varepsilon] \tag{2.7}$$

and

$$\mathbf{P}[\|\Phi\|_2^{-1}S_{0N}\leqq x]\geqq\mathbf{P}[\|\Phi\|_2^{-1}T_{0N}\leqq x-\varepsilon]-\mathbf{P}[\|\Phi\|_2^{-1}|S_{0N}-T_{0N}|\geqq\varepsilon]. \tag{2.8}$$

It then follows from (2.7), (2.8), Lemma 2.1, Lemma 2.2 and Markov's inequality that for any $\delta>0$

$$\sup_{x\in(-\infty,\infty)}|F_{0N}(x)-\Phi(x)|\leqq O(N^{-1/2}\sqrt{\log^3 N})+O(\varepsilon)+\frac{O(N^{-K}(\log N)^K)}{(\varepsilon\|\Phi\|_2)^{2K}} \tag{2.9}$$

holds for any $\varepsilon>0$ and any $K\geqq 1$. Now, pick K large enough such that $(2K+1)^{-1}<2\delta$ and put $\varepsilon=N^{-1/2+1/(2(2K+1))}$. Then using (2.9), the proof follows immediately.

3. Rate of convergence under near alternatives. In this section, we assume $\Delta\neq 0$ so that $Y_{N1},\cdots,Y_{NN}$, $N\geqq 1$, are independently distributed with cdf's $F(y-\Delta d_{N_1}),\cdots,F(y-\Delta d_{NN})$, $N\geqq 1$, respectively. Furthermore, besides the assumptions A1–A3, we assume:

B1) F is absolutely continuous, $f(x)=F'(x)$ is absolutely continuous, and there exists a constant $\theta_0>0$ such that

$$\sup_{\theta\in[-\theta_0,\theta_0]}\int_{-\infty}^{\infty}\frac{[f'(x-\theta)]^2}{f(x)}\,dx<\infty,$$

and

B2) $$\sum_{i=1}^N d_{Ni}^2=1,\quad \sum_{i=1}^N d_{Ni}=0,\ N=2,3,\cdots,\ \max_{1\leqq i\leqq N}|d_{Ni}|=O(N^{-1/2}\sqrt{\log N}).$$

In what follows $\mathbf{E}_0$ and $\mathbf{E}_\Delta$ will denote the expectations computed under $\Delta=0$ and $\Delta\neq 0$ respectively. Furthermore, $P_{\Delta r}$ will denote the restriction of P_Δ to the set $D_N=\{\prod_{i=1}^N f(Y_{Ni})\neq 0\}$ so that for any measurable function h of $(Y_1,Y_2,\cdots,Y_N)$

$$\mathbf{E}_{\Delta r}h(Y_1,Y_2,\cdots,Y_N)=\int_{\{\prod_{i=1}^N f(y_i)\neq 0\}}h(y_1,y_2,\cdots,y_N)\cdot\prod_{i=1}^N f(y_i-\Delta d_i)\,dy_1\,dy_2\cdots dy_N, \tag{3.1}$$

i.e., $\mathbf{E}_{\Delta r}h(Y_1,Y_2,\cdots,Y_N)=\mathbf{E}_\Delta\{h(Y_1,Y_2,\cdots,Y_n)I_{D_N}(Y_1,Y_2,\cdots,Y_N)\}$, where I denotes the indicator function.

LEMMA 3.1. *Under assumptions* A1, B1 *and* B2,

$$P_\Delta(D_N^c)=O(N^{-1/2}\sqrt{\log^3 N})$$

holds for any $\Delta\neq 0$, *where* D_N^c *is the complement of* D_N.

Remark. Our assumption B1 is equivalent to assumption V in Hušková [4]. In fact, Hušková's assumption V implies our assumption B1 and her Lemma 3.5 implies our Lemma 3.1.

Proof. By Lemma 3.5 of Hušková [4] and B2, there exists a constant C_1 (independent of N) and a positive integer N_Δ such that for all $N \geqq N_\Delta$ we have

$$P_\Delta\left[\prod_{i=1}^{N} f(Y_{Ni})=0\right] \leqq C_1|\Delta|^3 \sum_{i=1}^{N} |d_i|^3 = O(N^{-1/2}\sqrt{\log^3 N}).$$

The proof follows.

LEMMA 3.2. *Under assumptions* A1 *to* A3, B1 *and* B2, *it holds true for any positive integer* K *and any* $\Delta \neq 0$ *that*

$$\{\mathbf{E}_\Delta[(S_{\Delta N}-T_{\Delta N})I_{D_N^c}(Y_1,\cdots,Y_N)]\}^{2K} = O[(N^{-K}(\log N)^{3K})^{1-1/(2p)}(\log N)^{2K}] \tag{3.2}$$

for all $p=1,2,\cdots$.

Proof. By the Hölder inequality, the left-hand side of (3.2) is bounded by

$$[\mathbf{E}_\Delta(S_{\Delta N}-T_{\Delta N})^{2p}]^{K/p}[P_\Delta(D_N^c)]^{2K(1-1/(2p))} \tag{3.3}$$

for any $p=1,2,\cdots$. Thus the assertion (3.2) will be proved in view of (3.3) and Lemma 3.1, if we can show that $\mathbf{E}_\Delta(S_{\Delta N}-T_{\Delta N})^{2p} = O[(\log N)^{2p}]$ for any $p=1,2,\cdots$. Now, put

$$Z_{i\Delta} = \left[\varphi\left(\frac{R_i^\Delta}{N+1}\right) - \varphi(F^*(|Y_i|))\right] \operatorname{sgn}(Y_i).$$

Then from (1.1), (2.1) and (2.4) we have

$$\begin{aligned}\mathbf{E}_\Delta(S_{\Delta N}-T_{\Delta N})^{2p} &= \mathbf{E}_\Delta\left[\left(\sum_{i=1}^{N} c_i Z_{i\Delta}\right)^{2p}\right]\\ &= \sum_{m=1}^{2p}\left[\sum_{\substack{i_1=1\\ i_1<\cdots<i_m}}^{N}\cdots\sum_{i_m=1}^{N}\left[\sum_{\substack{l_1=1\\ l_1+\cdots+l_m=2p}}^{2p}\cdots\sum_{l_m=1}^{2p} C_{l_1 \cdots l_m}^{2p} c_{i_1}^{l_1}\cdots c_{i_m}^{l_m}\mathbf{E}_\Delta(Z_{i_1\Delta}^{l_1}\cdots Z_{i_m\Delta}^{l_m})\right]\right],\end{aligned} \tag{3.4}$$

where

$$C_{l_1 \cdots l_m}^{2p} = \frac{(2p)!}{(l_1!)\cdots(l_m!)}.$$

We now consider the expectation $\mathbf{E}_\Delta Z_{i\Delta}^{2q}$ for any $q=1,2,\cdots$ and any $i=1,\cdots,N$.

From A3 it follows that

$$\begin{aligned}\mathbf{E}_\Delta Z_{i\Delta}^{2q} &\leqq C^{2q}\mathbf{E}_\Delta\left(\frac{R_i^\Delta}{N+1} - F^*(|Y_i|)\right)^{2q}\\ &\leqq 2^{2q-1}C^{2q}\left[\mathbf{E}_\Delta\left(\frac{R_i^\Delta}{N+1} - V_{ii\Delta}\right)^{2q} + \mathbf{E}_\Delta(V_{ii\Delta}-F^*(|Y_i|))^{2q}\right],\end{aligned} \tag{3.5}$$

where $V_{ii\Delta} = \mathbf{E}_\Delta(R_i^\Delta/(N+1)|Y_i)$. By Lemma 2.1 (after slight and trivial modification) of Bergström and Puri [1] we obtain

$$\mathbf{E}_\Delta\left(\frac{R_i^\Delta}{N+1} - V_{ii\Delta}\right)^{2q} = O(N^{-q}), \quad q=1,2,\cdots, \tag{3.6}$$

uniformly in $i=1,2,\cdots,N$. Using the Mean Value Theorem and B1, we can easily show that

$$\mathbf{E}_\Delta(V_{ii\Delta}-F^*(|Y_i|))^{2q} = O(\max_{i\leqq j\leqq N}|d_j|^{2q}), \quad q=1,2,\cdots, \tag{3.7}$$

uniformly in $i=1,2,\cdots,N$. It now follows from B2, (3.5), (3.6) and (3.7) that

$$\mathbf{E}_\Delta Z_{i\Delta}^{2q} = O[N^{-q}(\log N)^q], \quad q=1,2,\cdots, \tag{3.8}$$

holds uniformly in $i=1,\cdots,N$. In view of (3.4) and (3.8), it follows, by using arguments analogous to those in the proof of Lemma 2.2, that

$$\mathbf{E}_\Delta(S_{\Delta N}-T_{\Delta N})^{2p} = O[(\log N)^{2p}], \quad p=1,2,\cdots. \tag{3.9}$$

The proof follows.

LEMMA 3.3. *Under assumptions* A1-A3, B1 *and* B2, *it holds for any* $K=1,2,\cdots$ *and any* $\Delta \neq 0$ *that*

$$\mathbf{E}_{\Delta_r}[(S_{\Delta N}-\mathbf{E}_\Delta S_{\Delta N})-(T_{\Delta N}-\mathbf{E}_\Delta T_{\Delta N})]^{2K} = O[N^{-K(1-1/(2p))}(\log N)^{q_{Kp}}] \tag{3.10}$$

for all $p=1,2,\cdots$, *where* $q_{Kp}=(5-3/(2p))K$.

Proof. The left-hand side of (3.10) is bounded by

$$2^{2K-1}\mathbf{E}_{\Delta r}(S_{\Delta N}-T_{\Delta N})^{2K}+2^{2K-1}[\mathbf{E}_{\Delta}(S_{\Delta N}-T_{\Delta N})]^2K. \tag{3.11}$$

In view of Lemma 3.6 of Hušková [4], there exist a constant C_2 (independent of N) and a positive integer N'_Δ such that for all $N \geqq N'_\Delta$

$$\mathbf{E}_{\Delta r}(S_{\Delta N}-T_{\Delta N})^{2K} \leqq C_2\{\mathbf{E}_0(S_{0N}-T_{0N})^{4K}\}^{1/2}=O[N^{-K}(\log N)^K], \tag{3.12}$$

where the last equality is by Lemma 2.2. Next, we have

$$\begin{aligned}[\mathbf{E}_\Delta(S_{\Delta N}-T_{\Delta N})]^{2K}&=\{\mathbf{E}_{\Delta r}(S_{\Delta N}-T_{\Delta N})+\mathbf{E}_\Delta[(S_{\Delta N}-T_{\Delta N})I_{D_N^c}(Y_1,\cdots,Y_N)]\}^{2K}\\ &\leqq 2^{2K-1}[\mathbf{E}_{\Delta r}(S_{\Delta N}-T_{\Delta N})]^{2K}+2^{2K-1}\{\mathbf{E}_\Delta[(S_{\Delta N}-T_{\Delta N})I_{D_N^c}(Y_1,\cdots,Y_N)]\}^{2K}.\end{aligned} \tag{3.13}$$

Using (3.1) and Jensen's inequality, it follows that

$$[\mathbf{E}_{\Delta r}(S_{\Delta N}-T_{\Delta N})]^{2K} \leqq \mathbf{E}_{\Delta r}(S_{\Delta N}-T_{\Delta N})^{2K}. \tag{3.14}$$

The proof now follows from (3.11)-(3.14) and Lemma 3.2.

The following lemma is an immediate consequence of the previous lemmas and the Markov inequality.

LEMMA 3.4. *Under assumptions* A1-A3, B1 *and* B2, *it holds true for any* $\Delta \neq 0$ *and* $\varepsilon > 0$ *that*

$$\begin{aligned}&P_\Delta[|(S_{\Delta N}-\mathbf{E}_\Delta S_{\Delta N})-(T_{\Delta N}-\mathbf{E}_\Delta T_{\Delta N})|\,\|\varphi\|_2^{-1}>\varepsilon]\\ &\qquad\leqq M_{Kp}[N^{-1/2}\sqrt{\log^3 N}+(\|\varphi\|_2\varepsilon)^{-2K}N^{-K(1-1/(2p))}(\log N)^{q_{Kp}}]\end{aligned} \tag{3.15}$$

for all $K, p=1,2,\cdots$ *and for all* $N \geqq N_{\Delta Kp}$ *where* $N_{\Delta Kp}$ *is a positive integer,* M_{Kp} *is a constant independent of* N *and* $q_{Kp}=(5-3/(2p))K$.

Proof. For any $\varepsilon>0$, the left-hand of (3.15) is bounded by

$$P_\Delta(D_N^c)+P_{\Delta r}[|(S_{\Delta N}-\mathbf{E}_\Delta S_{\Delta N})-(T_{\Delta N}-\mathbf{E}_\Delta T_{\Delta N})|\,\|\varphi\|_2^{-1}>\varepsilon]. \tag{3.16}$$

(Recall from (3.1) that $P_{\Delta r}(A)=P_\Delta(A\cap D_N)$ for any measurable set A.)

The first term in (3.16) is of order $O(N^{-1/2}\sqrt{\log^3 N})$ as can be seen from Lemma 3.1. The second term in (3.16), by Markov's inequality, is bounded by

$$\mathbf{E}_{\Delta r}[(S_{\Delta N}-\mathbf{E}_\Delta S_{\Delta N})-(T_{\Delta N}-\mathbf{E}_\Delta T_{\Delta N})]^{2K}(\|\varphi\|_2\varepsilon)^{-2K} \text{ for any } K=1,2,\cdots.$$

The lemma now follows from Lemma 3.3.

THEOREM 3.1. *Under assumptions* A1-A3, B1 *and* B2, *it holds true that*

$$\sup_{x\in(-\infty,\infty)}|G_{\Delta N}(x)-\Phi(x)|=O(N^{-1/2}\sqrt{\log^3 N}),$$

where $G_{\Delta N}$ *is the cdf of* $(T_{\Delta N}-\mathbf{E}_\Delta T_{\Delta N})\|\varphi\|_2^{-1}$ *and* $T_{\Delta N}$ *is given by* (2.1).

Proof. Let us denote $W_{i\Delta}=\varphi(F^*(|Y_i|)\operatorname{sgn}(Y_i)$. Then using (2.1), we have $T_{\Delta N}-\mathbf{E}_\Delta T_{\Delta N}=\sum_{i=1}^N c_i(W_{i\Delta}-\mathbf{E}_\Delta W_{i\Delta})$ which is the sum of independent random variables, and by A2 we have

$$\begin{aligned}|\mathbf{D}_\Delta T_{\Delta N}-\|\varphi\|_2^2|&=\left|\sum_{i=1}^N c_i^2[\mathbf{E}_\Delta W_{i\Delta}^2-\|\varphi\|_2^2-(\mathbf{E}_\Delta W_{i\Delta})^2]\right|\\ &\leqq \max_{1\leqq i\leqq N}|\mathbf{E}_\Delta W_{i\Delta}^2-\|\varphi\|_2^2|+\max_{1\leqq i\leqq N}|\mathbf{E}_\Delta W_{i\Delta}|^2.\end{aligned} \tag{3.17}$$

Now, using A1, A3, B1 and the Mean Value Theorem we obtain

$$\begin{aligned}\max_{1\leqq i\leqq N}|\mathbf{E}_\Delta E_{i\Delta}^2-\|\varphi\|_2^2|&\leqq \max_{1\leqq i\leqq N}\int_{-\infty}^{\infty}|\varphi^2(F^*(|x+\Delta d_i|))-\varphi^2(F^*(|x|))|\,dF(x)\\ &\leqq 2\|\varphi\|_\infty C\max_{1\leqq i\leqq N}\int_{-\infty}^{\infty}|F^*(|x+\Delta d_i|)-F^*(|x|)|\,dF(x)\\ &\leqq 2\|\varphi\|_\infty C\int_{-\infty}^{\infty}[F^*(|x|+\max_{1\leqq i\leqq N}|\Delta d_i|)\\ &\qquad -F^*(|x|-\max_{1\leqq i\leqq N}|\Delta d_i|)]\,dF(x)\\ &=O(\max_{1\leqq i\leqq N}|d_i|)=O(N^{-1/2}\sqrt{\log N})\end{aligned} \tag{3.18}$$

and similarly

$$\max_{1\leq i\leq N} |\mathbf{E}_\Delta W_{i\Delta}| = \max_{1\leq i\leq N} |\mathbf{E}_\Delta W_{i\Delta} - \mathbf{E}_0 W_{i0}|$$

$$\leq \max_{1\leq i\leq N} \int_{M_{i\Delta}}^{\infty} |\varphi(F^*(x+\Delta d_i)) - \varphi(F^*(x))| \, dF(x)$$

$$(3.19) \qquad + \max_{1\leq i\leq N} \int_{-\infty}^{m_{i\Delta}} |\varphi(F^*(-x)) - \varphi(F^*(-x-\Delta d_i))| \, dF(x)$$

$$+ \max_{1\leq i\leq N} \int_{m_{i\Delta}}^{M_{i\Delta}} |\varphi(F^*(|x+\Delta d_i|)) \operatorname{sgn}(x+\Delta d_i)$$

$$- \varphi(F^*(|x|)) \operatorname{sgn}(x)| \, dF(x),$$

where $M_{i\Delta} = \max(0, -\Delta d_i)$ and $m_{i\Delta} = \min(0, -\Delta d_i)$. The third integral in (3.19) is of order $O(\max_{1\leq i\leq N} |d_i|)$ since $\|\Phi\|_\infty < \infty$ and $F(M_{i\Delta}) - F(m_{i\Delta}) = O(\max_{1\leq i\leq N} |d_i|)$. The first and second integrals are also of order $O(\max_{1\leq i\leq N} |d_i|)$, which can be proved as in deriving (3.18). It then follows from B2, (3.17), (3.18) and (3.19) that

$$(3.20) \qquad |\mathbf{D}_\Delta T_{\Delta N} - \|\varphi\|_2^2| = O(N^{-1/2}\sqrt{\log N}).$$

But A3 implies that $\|\varphi\|_2 > 0$. Thus there exists a positive integer N''_Δ such that $\frac{1}{2}\|\varphi\|_2^2 < \mathbf{D}_\Delta T_{\Delta N} < \frac{3}{2}\|\varphi\|_2^2$ for all $N \geq N''_\Delta$. Consequently we have for all $N \geq N''_\Delta$,

$$\left| \frac{\|\varphi\|_2}{\sqrt{\mathbf{D}_\Delta T_{\Delta N}}} - 1 \right| = \frac{|(\|\varphi\|_2 - \sqrt{\mathbf{D}_\Delta T_{\Delta N}})(\|\Phi\|_2 + \sqrt{\mathbf{D}_\Delta T_{\Delta N}})|}{\sqrt{\mathbf{D}_\Delta T_{\Delta N}}(\|\varphi\|_2 + \sqrt{\mathbf{D}_\Delta T_{\Delta N}})}$$

$$(3.21) \qquad \leq \frac{3\|\varphi\|_2^{-2}}{\sqrt{2}+1} | \, \|\varphi\|_2^2 - \mathbf{D}_\Delta T_{\Delta N}| = O(N^{-1/2}\sqrt{\log N}).$$

Now let us denote the cdf of $(T_{\Delta N} - \mathbf{E}_\Delta T_{\Delta N})(\mathbf{D}_\Delta T_{\Delta N})^{-1/2}$ by $G^*_{\Delta N}$. Since $\mathbf{E}_\Delta |c_i(W_{i\Delta} - \mathbf{E}_\Delta W_{i\Delta})|^3 < \infty$, $i = 1, \cdots, N$, we obtain from Theorem 2.1 and (3.20) that there exists a constant A (independent on N) such that for all $N \geq N''_\Delta$,

$$\sup_{x\in(-\infty,\infty)} |G^*_{\Delta N}(x) - \varphi(x)| \leq A(\mathbf{D}_\Delta T_{\Delta N})^{-3/2} \sum_{i=1}^{N} |c_i|^3 \mathbf{E} W_{i\Delta} - \mathbf{E}_\Delta W_{i\Delta}|^3$$

$$(3.22) \qquad \leq 16\sqrt{2}A\|\varphi\|_2^{-3}\|\varphi\|_\infty^3 \sum_{i=1}^{N} |c_i|^3.$$

We now consider the uniform distance between $G_{\Delta N}$ and Φ. It follows from the triangle inequality and (3.22) that

$$\sup_{x\in(-\infty,\infty)} |G_{\Delta N}(x) - \Phi(x)| = \sup_{x\in(-\infty,\infty)} |G^*_{\Delta N}(x\|\varphi\|_2(\mathbf{D}_\Delta T_{\Delta N})^{-1/2}) - \Phi(x)|$$

$$\leq \sup_{x\in(-\infty,\infty)} |G^*_{\Delta N}(x\|\varphi\|_2(\mathbf{D}_\Delta T_{\Delta N})^{-1/2}) - \Phi(x\|\varphi\|_2(\mathbf{D}_\Delta T_{\Delta N})^{-1/2})|$$

$$(3.23) \qquad + \sup_{x\in(-\infty,\infty)} |\Phi(x\|\varphi\|_2(\mathbf{D}_\Delta T_{\Delta N})^{-1/2}) - \Phi(x)|$$

$$\leq O\left(\sum_{i=1}^{N} |c_i|^3\right) + \sup_{|x|\leq\sqrt{\log N}} \|\Phi(x\|\varphi\|_2(\mathbf{D}_\Delta T_{\Delta N})^{-1/2}) - \Phi(x)|$$

$$+ \sup_{|x|>\sqrt{\log N}} |\Phi(x\|\varphi\|_2(\mathbf{D}_\Delta T_{\Delta N})^{-1}) - \Phi(x)|.$$

On the right-hand side of the last inequality in (3.23), the second term, in view of (3.21) and the Mean Value Theorem, is bounded by

$$(3.24) \qquad \|\Phi'\|_\infty\sqrt{\log N} \left\| \frac{\|\varphi\|^2}{\sqrt{\mathbf{D}_\Delta T_{\Delta N}}} - 1 \right\| = O(N^{-1/2}\log N),$$

and the third term, by the triangle inequality and the monotonicity of Φ, is bounded by

$$\sup_{x>\sqrt{\log N}} |1-\Phi(x\|\varphi\|_2(\mathbf{D}_\Delta T_{\Delta N})^{-1/2})| + \sup_{x>\sqrt{\log N}} |1-\Phi(x)|$$

$$(3.25) \qquad = [1-\Phi(\sqrt{\log N}\|\varphi\|_2(\mathbf{D}_\Delta T_{\Delta N})^{-1/2})]+[1+\Phi(\sqrt{\log N})]$$

$$= [\Phi(\sqrt{\log N})-\Phi(\sqrt{\log N})\|\varphi\|_2(\mathbf{D}_\Delta T_{\Delta N})^{-1/2})]+2[1-\Phi(\sqrt{\log N})]$$

which is of order $O(N^{-1/2}\log N)$ as can be seen from (3.21) and from Lemma 2, Chapter VII of Feller [2]. The rest of the proof follows from A2, (3.23), (3.24) and (3.25).

We now prove the main result of this section.

THEOREM 3.2. *Under assumptions* A1–A3, B1 *and* B2, *it holds true for any* $\delta>0$ *that*

$$\sup_{x\in(-\infty,\infty)} |F_{\Delta N}(x)-\Phi(x)| = O(N^{-1/2+\delta}),$$

where $F_{\Delta N}$ *is the cdf of* $(S_{\Delta N}-\mathbf{E}_\Delta S_{\Delta N})\|\Phi\|_2^{-1}$.

Proof. For any $\varepsilon>0$ and any real number x, we have

$$(3.26) \qquad \begin{aligned} P_\Delta[(S_{\Delta N}-\mathbf{E}_\Delta S_{\Delta N})\|\varphi\|_2^{-1}\leqq x] &\leqq P_\Delta[(T_{\Delta N}-\mathbf{E}_\Delta T_{\Delta N})\|\varphi\|_2^{-1}\leqq x+\varepsilon] \\ &\quad + P_\Delta[|(S_{\Delta N}-\mathbf{E}_\Delta S_{\Delta N})-(T_{\Delta N}-\mathbf{E}_\Delta T_{\Delta N})|\,\|\varphi\|_2^{-1}>\varepsilon] \end{aligned}$$

and

$$(3.27) \qquad \begin{aligned} P_\Delta[(S_{\Delta N}-\mathbf{E}_\Delta S_{\Delta N})\|\varphi\|_2^{-1}\leqq x] &\geqq P_\Delta[(T_{\Delta N}-\mathbf{E}_\Delta T_{\Delta N})\|\varphi\|_2^{-1}\leqq x-\varepsilon] \\ &\quad - P_\Delta[|(S_{\Delta N}-\mathbf{E}_\Delta S_{\Delta N})-(T_{\Delta N}-\mathbf{E}_\Delta T_{\Delta N})|\,\|\varphi\|_2^{-1}>\varepsilon]. \end{aligned}$$

It follows from (3.26), (3.27), Lemma 3.4 and Theorem 3.1 that for all large N

$$(3.28) \qquad \begin{aligned} \sup_{x\in(-\infty,\infty)} |F_{\Delta N}(x)-\Phi(x)| &\leqq O(N^{-1/2}\sqrt{\log^3 N})+O(\varepsilon) \\ &\quad +(\|\varphi\|_2\varepsilon)^{-2K}O[N^{-(1-1/(2p))K}(\log N)^{(5-3/(2p))K}] \end{aligned}$$

holds for any $K, p=1,2,\cdots$, and any $\varepsilon>0$. Now, pick K large enough so that $(2K+1)^{-1}<\delta$, let $p=K(2K+1)$, and put $\varepsilon=N^{-1/2}+1/(2(2K+1))$. The theorem then follows from (3.28).

REFERENCES

[1] H. BERGSTRÖM AND M. L. PURI, *Convergence and remainder terms in linear rank statistics*, Ann. Statist., 4 (1977), pp. 671–680.

[2] W. FELLER, *An Introduction to Probability Theory and Its Applications*, Vol. 2, 2nd ed., John Wiley, New York, 1971.

[3] M. HŮSKOVÁ, *Asymptotic distribution of simple linear rank statistics for testing symmetry*, Z. Wahrsch. Verw. Gebiete, 14 (1970), pp. 308–322.

[4] ———, *The rate of convergence of simple linear rank statistics under hypothesis and alternatives*, Ann. Statist., 5 (1977), pp. 658–670.

[5] J. JURECKOVÁ AND M. L. PURI, *Order of normal approximation for rank test statistics distribution*, Ann. Probab., 3 (1975), pp. 526–533.

[6] W. C. M. KALLENBERG, *Cramér type large deviations for simple linear rank statistics*, Z. Wahrsch. Verw. Gebiete, 60 (1982), pp. 403–409.

[7] V. V. PETROV, *Sums of Independent Random Variables*, Springer-Verlag, New York, Berlin, Heidelberg, 1975.

Reprinted from
Theory of Probability and Its Applications, Volume 31
(1986), pages 145-151, by permission of the publisher,

CENTRAL LIMIT THEOREM FOR PERTURBED EMPIRICAL DISTRIBUTION FUNCTIONS EVALUATED AT A RANDOM POINT

MADAN L. PURI* and STEFAN S. RALESCU

Indiana University
Queens College, City University of New York
communicated by M. Rosenblatt

Let $\hat{F}_n$ be an estimator obtained by integrating a kernel type density estimator based on a random sample of size n from a (smooth) distribution function F. Sufficient conditions are given for the central limit theorem to hold for the target statistic $\hat{F}_n(U_n)$ where $\{U_n\}$ is a sequence of U-statistics.

1. Introduction

Given a sequence X_i, $i \geqslant 1$, of iidrv's (independent and identically distributed random variables) with cdf (cumulative distribution function) F, the natural estimator of F based on the sample $X_1, \ldots, X_n$ is the empirical df F_n defined by

$$F_n(x) = n^{-1} \sum_{i=1}^{n} u(x - X_i), \qquad x \in \mathbb{R}, \tag{1.1}$$

where $u(t) = 1$ if $t \geqslant 0$ and $= 0$ otherwise. Although F_n is in a sense already quite smooth, it does not take fully into account the smoothness of F (i.e., the existence of a density f). In fact, if F is a continuous cdf, it seems

Received November 30, 1983; revised July 25, 1984

AMS 1980 subject classifications: primary 62E20; 52G30

Key words and phrases: perturbed empirical distribution functions, U-statistics, central limit theorem.

* Research supported by the National Foundation Grant MCS8301409

reasonable to consider continuous estimators of F which are better adapted to this situation. Thus, in relatively general situations, an estimator of the form

$$\hat{F}_n(x) = n^{-1} \sum_{i=1}^{n} G_n(x - X_i), \qquad x \in \mathbb{R}, \tag{1.2}$$

where $\{G_n\}$ is a sequence of continous cdf's, suggests itself. Furthermore, if one expects this estimator to perform as satisfactory as F_n, it seems natural to require that the sequence $\{G_n\}$ be *not* "too far" from the degenerate df u, e.g., by requiring that G_n converge weakly to u (written $G_n \rightarrow^w u$). Such estimators arise quite naturally as integrals of density estimators of the kernel type. Rosenblatt [7] and Parzen [5] suggested the density kernel type estimator

$$\hat{f}_n(x) = (n\alpha_n)^{-1} \sum_{i=1}^{n} g((x - X_i)/\alpha_n), \qquad x \in \mathbb{R}, \tag{1.3}$$

where $\alpha_n > 0$ and g satisfies $g \geqslant 0$ and $\int_{-\infty}^{\infty} g(t)\, dt = 1$. The interest generated by such estimators is due primarily to their simple structure as averages over the independent elements of the nth row of a double array of rv's. We refer to Scott *et al.* [8] and Wertz [12] for a general review in this area. Using the density estimator $\hat{f}_n$, define

$$\hat{F}_n(x) = \int_{-\infty}^{x} \hat{f}_n(t)\, dt, \qquad x \in \mathbb{R}, \tag{1.4}$$

and note that this is of the form (1.2) with $G_n(x) = \int_{-\infty}^{x} g_n(t)\, dt$, where $g_n(t) = \alpha_n^{-1} g(t\alpha_n^{-1})$. Furthermore, if $\alpha_n \rightarrow 0$ as $n \rightarrow \infty$, then it is easy to check that $G_n \rightarrow^w u$.

In recent years, interest has increasingly been focused on the asymptotic properties of $\hat{F}_n$ and its use in estimation theory. The a.s. (almost sure) convergence of $\hat{F}_n$ to F was proved by Nadaraya [4], Winter [13] and Yamato [15], while Watson and Leadbetter [11] obtained the asymptotic normality of $\hat{F}_n$. Recently, Winter [14] showed that $\hat{F}_n$ has the Chung–Smirnov property, i.e., with probability one

$$\limsup_{n \rightarrow \infty} (2n/\log \log n)^{1/2} \sup_x |\hat{F}_n(x) - F(x)| \leqslant 1.$$

To understand better the utility of the integral type estimators $\hat{F}_n$, it is worthwhile to study the asymptotic behavior of the distribution of $\hat{F}_n$ (defined by (1.4)) evaluated at a random point having the structure of a U-statistic. Specifically, let $h(x_1, \ldots, x_m)$, symmetric in its m arguments, be a Borel-measurable kernel (of degree m). Based on a random sample

$X_1,\ldots, X_n$ $(n \geqslant m)$ from a population with cdf F, the corresponding U-statistic U_n for the estimation of the expected value $\xi = Eh(X_1,\ldots, X_m)$ is formed by averaging the kernel h symmetrically over the observations, namely

$$U_n = \binom{n}{m}^{-1} \sum_{C_{n,m}} h(X_{i_1},\ldots, X_{i_m}),$$

where $C_{n,m}$ denotes the set of all the $\binom{n}{m}$ combinations of m distinct elements $\{i_1,\ldots, i_m\}$ from $\{1,\ldots, n\}$.

If $\bar{X}$ denotes the sample mean, it is well known (see, e.g., Ghosh [2]) that under certain regularity conditions, $F_n(\bar{X})$, properly normalized, has normal distribution, in the limit (see also, Ralescu and Puri [6], for the corresponding problems concerning the rate of convergence of $F_n(\bar{X})$ to normality, a law of iterated logarithm, and an invariance principle for $F_n(\bar{X})$). The present note is devoted to investigating conditions for the asymptotic normality of the statistic $\hat{F}_n(U_n)$. Such a statistic is useful in estimating a functional $\theta = F(\xi)$ if F is unknown. Also, if U_n is the sample mean $\bar{X}$ (in which case $\xi = E(X_1)$), one may use $\hat{F}_n(\bar{X})$ for testing the hypothesis that a smooth F is symmetric about an unknown location ξ against certain alternatives.

2. Main Theorem

Let g be a probability density function on $\mathbb{R}$, and let $\{\alpha_n\}$ (bandwidths) be a sequence of positive real numbers such that $\alpha_n \to 0$ as $n \to \infty$. If $g_n(t) = \alpha_n^{-1} g(t\alpha_n^{-1})$ and $G_n(x) = \int_{-\infty}^{x} g_n(t)\, dt$, consider the perturbed empirical distribution function $\hat{F}_n$ defined by (1.2) corresponding to the sequence $\{G_n\}$.

Let

$$h^*(t) = E_F h(t, X_2,\ldots, X_m) \tag{2.1}$$

and

$$\zeta = \mathrm{Var}_F[h^*(X_1)]. \tag{2.2}$$

Then, our main result, dealing with the central limit theorem for a normalized $\hat{F}_n(U_n)$, is as follows:

THEOREM. *Assume that*

(i) $E_F h^2(X_1,\ldots, X_m) < \infty$ *and* $\zeta > 0$,

(ii) $\int_{-\infty}^{\infty} |t|\, g(t)\, dt < \infty$, *and*

(iii) *F is twice differentiable on $\mathbb{R}$ with a bounded second derivative F''.*

Then

$$n^{1/2}[\hat{F}_n(U_n) - \mu_n] \xrightarrow{\mathscr{D}} N(0, \sigma^2), \tag{2.3}$$

where $\mu_n = EG_n(\xi - X_1) = \int_{-\infty}^{\infty} F(\xi - t)\, g_n(t)\, dt$, *and*

$$\sigma^2 = \operatorname{Var}[u(\xi - X_1) + m(h^*(X_1) - \xi)\, F'(\xi)] > 0.$$

Proof. Letting

$$Q_n = n^{1/2}[\hat{F}_n(U_n) - \hat{F}_n(\xi) - (U_n - \xi)\, F'(\xi)],$$

we first show that

$$Q_n \xrightarrow{P} 0. \tag{2.4}$$

To this end, set

$$Q_n = A_n + B_n,$$

where

$$A_n = \int [Z_n(U_n - t) - Z_n(\xi - t)]\, dG_n(t)$$

and

$$B_n = \sqrt{n} \int \{F(U_n - t) - F(\xi - t) - F'(\xi)(U_n - \xi)\}\, dG_n(t).$$

Here $Z_n = \sqrt{n}[F_n - F]$ is the standard empirical process, and G_n is the d[illegible] associated with the kernel in constructing $\hat{F}_n$.

Since $U_n - \xi = O(a_n)$ a.s., where $a_n = (n_{-1} \log\log n)^{1/2}$ under our assumptions (see [9, p. 191]), using the fluctuation result of Stute [10], we have that

$$|A_n| \leqslant \sup |Z_n(x) - Z_n(y)| \cdot \int dG_n(t)$$

(where the sup is taken over all x and y, with $|x - y| \leqslant c a_n$)

$$= O\left(a_n \log \frac{1}{a_n}\right)^{1/2} \quad \text{a.s.}$$

for some sufficiently large $c > 0$ as $n \to \infty$. In the second term B_n, by usin[illegible] the Lagrange form of Taylor expansion with remainder of $F(U_n - t)$ w.r.t. $\xi - t$ up to the second derivative term, the integrand equals

$$[F'(\xi - t) - F'(\xi)][U_n - \xi] + F''(\alpha(t))[U_n - \xi]^2/2$$
$$= F''(\beta(t))[U_n - \xi](-t) + F''(\alpha(t))[U_n - \xi]^2/2$$

with $|\beta(t) - \xi| \leqslant |t|$ and $|\alpha(t) - \xi + t| \leqslant |U_n - \xi|$ a.s.

Hence

$$|B_n| \leqslant \sqrt{n}\, |U_n - \xi| \cdot \| F'' \|_\infty \cdot \int |t|\, dG_n(t)$$
$$+ \sqrt{n}\, |U_n - \xi|^2 \cdot \| F'' \|_\infty \cdot \int dG_n(t)/2,$$

where $\| F'' \|_\infty = \sup_x |F''(x)|$. Noting that $n^{1/2}(U_n - \xi) \to \mathcal{N}(0, m^2\zeta)$ in distribution, the first term on the right $\to 0$ in probability because of the choice of the bandwidth α_n:

$$\int |t|\, dG_n(t) = \alpha_n \int |t|\, g(t)\, dt \to 0 \qquad \text{as} \quad n \to \infty.$$

The second term on the right $\to 0$ a.s. by using the Serfling result quoted earlier. Equation (2.4) follows.

Consequently, we can write

$$\hat{F}_n(U_n) = \hat{F}_n(\xi) + (U_n - \xi)\, F'(\xi) + n^{-1/2} Q_n. \tag{2.5}$$

Now, using condition (i) and Corollary 4.2 of Geertsema [1], we obtain the decomposition

$$U_n = \xi + \frac{m}{n} \sum_{i=1}^{n} \bar{h}(X_i) + R_n, \tag{2.6}$$

where $\bar{h}(t) = h^*(t) - \xi$, and for any $\rho > \frac{1}{2}$

$$R_n = o(n^{-1}(\log n)^\rho) \qquad \text{a.s. as } n \to \infty. \tag{2.7}$$

Therefore, from (2.5) and (2.6), we get the asymptotic representation

$$\hat{F}_n(U_n) = n^{-1} \sum_{j=1}^{n} T_{n,j} + F'(\xi)\, R_n + n^{-1/2} Q_n, \tag{2.8}$$

where $T_{n,j} = G_n(\xi - X_j) + mF'(\xi)\, \bar{h}(X_j)$, $1 \leqslant j \leqslant n$, $n \geqslant 1$. Since $G_n \to^w u$ entails $\mathrm{Var}(T_{n,1}) \to \sigma^2$ as $n \to \infty$, by using (2.4), (2.7) and (2.8) we see that the theorem will follow from the Lindeberg central limit theorem if we show that the double-array $\{T_{n,j},\ 1 \leqslant j \leqslant n,\ n \geqslant 1\}$ of row-rise independent rv's satisfies the Lindeberg condition.

To this end, let $s_n^2 = \operatorname{Var}(\sum_{j=1}^n T_{n,j}) = n \operatorname{Var}(T_{n,1})$. For arbitary $\varepsilon > 0$, denote

$$L_n(\varepsilon) = s_n^{-2} \sum_{j=1}^{n} E\{(T_{n,j} - \mu_n)^2 I_{\{|T_{n,j} - \mu_n| > \varepsilon s_n\}}\}.$$

Now, using the inequality

$$(T_{n,j} - \mu_n)^2 \leqslant 2 + 2m^2[F'(\xi)]^2[\bar{h}(X_j)]^2,$$

we deduce that $L_n(\varepsilon) \leqslant L_{n,1}(\varepsilon) + L_{n,2}(\varepsilon)$, where

$$L_{n,1}(\varepsilon) = 2s_n^{-2} \sum_{j=1}^{n} P\{|T_{n,j} - \mu_n| > \varepsilon s_n\}$$

and

$$L_{n,2}(\varepsilon) = 2m^2[F'(\xi)]^2 s_n^{-2} \sum_{j=1}^{n} E\{[\bar{h}(X_j)]^2 I_{\{|T_{n,j} - \mu_n| > \varepsilon s_n\}}\}.$$

Now, since $s_n^2 \to \infty$ as $n \to \infty$, by Chebyshev's inequality it is easily seen that $L_{n,1}(\varepsilon) \to 0$ as $n \to \infty$.

To estimate $L_{n,2}(\varepsilon)$, note that for sufficiently large n,

$$\{|T_{n,j} - \mu_n| > \varepsilon s_n\} \subset \{mF'(\xi)|\bar{h}(X_j)| > \varepsilon s_n/2\}$$

which implies that

$$L_{n,2}(\varepsilon) \leqslant 2m^2[F'(\xi)]^2(\operatorname{Var}(T_{n,1}))^{-1} \int_A [\bar{h}(x)]^2\, dF(x) \tag{2.9}$$

where $A = \{2mF'(\xi)|\bar{h}(x)| > \varepsilon s_n\}$.

Then, since $\operatorname{Var}(T_{n,1}) \to \sigma^2$ as $n \to \infty$ and $\int_{-\infty}^{\infty} [\bar{h}(x)]^2\, dF(x) < \infty$, we see from (2.9) that $L_{n,2}(\varepsilon) \to 0$ as $n \to \infty$. Consequently, the conclusion $L_n(\varepsilon) \to 0$ as $n \to \infty$ obtains and the proof is completed.

EXAMPLE 1. The sample mean $\bar{X} = n^{-1} \sum_{i=1}^n X_i$, for which $h(x) = x$, clearly satisfies condition (i) of our theorem if $0 < \operatorname{Var}(X_1) < \infty$.

EXAMPLE 2. The sample variance $S^2 = (n-1)^{-1}(\sum_{i=1}^n X_i^2 - n\bar{X}^2)$ corresponding to the kernel $h(x_1, x_2) = \frac{1}{2}(x_1^2 + x_2^2 - 2x_1x_2)$ will satisfy condition (i) of our theorem provided that

$$0 < [\operatorname{Var}(X_1)]^2 < E\{(X_1 - E(X_1))^4\} < \infty.$$

EXAMPLE 3. For the Wilcoxon one-sample statistic ($h(x_1, x_2) = I_{\{x_1 + x_2 \leqslant 0\}}$), it is easily seen that if $0 < P_F\{X_1 + X_2 \leqslant 0\} < 1$, $\zeta > 0$ so that condition (i) of our theorem is automatically satisfied.

ACKNOWLEDGMENT

The authors would like to express their sincere thanks to the referee for the very careful examination of the paper and for providing a shorter and elegant proof of the assertion (2.4).

REFERENCES

1. GEERTSEMA, J. C. (1970). Sequential confidence intervals based on rank tests. *Ann. Math. Statist.* **41** 1016–1026.
2. GHOSH, J. K. (1971). A new proof of the Bahadur representation of quantiles and an application. *Ann. Math. Statist.* **42** 1957–1961.
3. HOEFFDING, W. (1948). A class of statistics with asymptotically normal distribution. *Ann. Math. Statist.* **19** 293–325.
4. NADARAYA, E. A. (1964). Some new estimates for distribution functions. *Theory Probab. Appl.* **9** 497–500.
5. PARZEN, E. (1962). On the estimation of a probability density and mode. *Ann. Math. Statist.* **33** 1065–1076.
6. RALESCU, S. S., AND PURI, M. L. (1983). On Berry–Esséen rates, a law of the iterated logarithm and an invariance principle for the proportion of the sample below the sample mean. *J. Multivariate Anal.* **14** 231–247.
7. ROSENBLATT, M. (1956). Remark on some non-parametric estimates of a density function. *Ann. Math. Statist.* **27** 832–837.
8. SCOTT, D. W., TAPIA, R. A., AND THOMPSON, J. R. (1977). Kernel density estimation revisited. *Nonlinear Anal. Theor. Math. Appl.* **1** 339–372.
9. SERFLING, R. J. (1980). *Approximation Theorems of Mathematical Statistics.* Wiley, New York.
10. STUTE, W. (1982). The oscillation behavior of empirical processes. *Ann. Probab.* **10** 86–107.
11. WATSON, G. S., AND LEADBETTER, M. R. (1964). Hazard analysis II. *Sankhyā Ser. A.* **26** 101–116.
12. WERTZ, W. (1978). *Statistical Density Estimation: A Survey.* Vandenhoeck & Ruprecht, Gottingen.
13. WINTER, B. B. (1973). Strong uniform consistency of integrals of density estimators. *Canad. J. Statist.* **1** 247–253.
14. WINTER, B. B. (1979). Convergence rate of perturbed empirical distribution functions. *J. Appl. Probab.* **16** 163–173.
15. YAMATO, H. (1973). Uniform convergence of an estimator of a distribution function. *Bull. Math. Statist.* **15** 69–78.

LIMIT THEOREMS FOR RANDOM CENTRAL ORDER STATISTICS

MADAN L. PURI* and STEFAN S. RALESCU

Department of Mathematics Indiana University, Bloomington, Indiana 47405
Department of Mathematics Queens College of the City University of New York, Flushing, NY 11367

Let $X_{n,1} \leq \ldots \leq X_{n,n}$ be the order statistics of a random sample $X_1, \ldots, X_n$ from a distribution function F. If $\{\nu_n\}$ is a sequence of integer-valued random variables such that $1 \leq \nu_n \leq n$ and $\nu_n/n \xrightarrow{P} p$, for some $p \in (0,1)$, the sequence $\{X_{n,\nu_n}\}$ is referred to as a sequence of random central order statistics (of limiting rank p) corresponding to the random central rank sequence $\{\nu_n\}$. In this paper, we establish the weak as well as strong consistency of X_{n,ν_n} in estimating the p-th quantile of F. We derive several central limit theorems for X_{n,ν_n} for regular as well as non-regular cases, and for each case, we provide remainder term estimates of the Berry-Esséen type.

---

*Research supported by the Office of Naval Research, Contract N00014-85-K-0648.

AMS 1980 subject classifications. Primary 60F05,60F15, 62G30.

Key words and phrases. Random central order statistics, weak consistency, strong consistency, central limit theorem, Bahadur representation, Berry-Esséen bounds.

1. Introduction. Let X_i, $i \geq 1$ be a sequence of i.i.d. r.v.'s (independent and identically distributed random variables) with a cdf (cumulative distribution function) F, and let $X_{n,1} \leq X_{n,2} \leq \ldots \leq X_{n,n}$ be the order statistics of $X_1,\ldots,X_n$. In classical theory, one defines the central order statistics as sequences $\{X_{n,k_n}\}$ where $k_n \in \{1,\ldots,n\}$ (deterministic integers) and k_n/n has a limit $p \in (0,1)$ as $n \to \infty$. (The ratio k_n/n is generally called the rank of X_{n,k_n} and p is called the limiting rank). A typical example of central order statistics having a limiting rank $p \in (0,1)$ is provided by the sequence of the sample p-th quantile $\{\hat{\xi}_{np}\}$, where $\hat{\xi}_{np} = X_{n,np}$ if np is an integer, and $\hat{\xi}_{np} = X_{n,[np]+1}$ if np is not an integer ($[\cdot]$ denotes the integer part). Central order statistics serve to provide consistent estimators, tolerance limits and distribution free confidence intervals for "central" parameters, e.g., quantiles. They have, in general, an asymptotically normal distribution, and they converge strongly to appropriate limits (see Smirnov (1952) for a characterization, under suitable conditions, of the class of all possible limit distributions and the corresponding domains of attraction). An important feature of central order statistics is that they can be expressed asymptotically as sums of indpendent random variables, via the Bahadur (1966) representation.

In many problems dealing with central parameters of the underlying distribution function, there are in general several candidates based on sequences of central order statistics that can be used in deriving optimal statistical procedures. A relative efficiency comparison based on these procedures may still leave us with the difficult task of having to select the "best" central rank sequence $\{k_n\}$. Thus, it would be of interest to introduce a new class of statistics for which one allows random flexibility on the sequence of ranks and to study their asymptotic properties.

Our objective in this paper is to generalize the classical asymptotic theory for the so called random central order statistics. Specifically, given a sequence of i.i.d.r.v.'s $\{X_n\}$ with a common cdf F, let $X_{n,j}$ denote the j-th order statistic from $X_1,\ldots,X_n$ and let $\{\nu_n\}$ be a sequence of integer-valued random variables such that $1 \leq \nu_n \leq n$ for each n. Then, if for some $p \in (0,1)$, $\nu_n/n \xrightarrow{P} p$, $\{X_{n,\nu_n}\}$ is called a sequence of random central order statistics, and $\{\nu_n\}$ a random central rank sequence. In section 2, we establish, under quite general conditions, the weak and strong consistency of a sequence of random central order statistics $\{X_{n,\nu_n}\}$ of the limiting rank p in estimating the (unknown) p-th quantile $\xi_p = \inf\{x: F(x) \geq p\}$. Further, confining attention to the so called regular cases (those for which $F'(\xi_p)$ exists and is positive), if $\nu_n/n \xrightarrow{P} p$ sufficiently fast, we obtain, in section 3, a central limit theorem which shows that the random flexibility of the ranks (ν_n/n) does not disturb the form of the limiting distribution of the normalized sequence $\{X_{n,\nu_n}\}$, a feature which adds greatly to the usefulness of the theory. We conclude this section with the derivation of a weak Bahadur representation for random central order statistics, a result which generalizes Ghosh (1971). In section 4, a Berry-Esséen type theorem is established for the distribution of random central order statistics in regular cases. Our results generalize as well as extend those of Reiss (1974) and Serfling (1980), Theorem C, p. 81) who derived the bound $O(n^{-1/2})$ for the departure from normality of the distribution function of the sample quantile. In section 5, we study the limit law of X_{n,ν_n} in non-regular cases. In this context, we present a central limit theorem for a suitably normalized X_{n,ν_n} which generalizes a result of Chanda (1975). We also give a remainder term estimate of the Berry-Esséen type for the corresponding normal distribution approximation.

2. Consistency of X_{n,ν_n}. Let $\{X_n\}$ be a sequence of i.i.d.r.v.'s with a common cdf $F(x) = P[X_1 \leq x]$ and let $\{\nu_n\}$ be a random central rank sequence such that $\nu_n/n \xrightarrow{P} p \in (0,1)$. We begin with the simplest result of interest.

THEOREM 2.1. (Weak Consistency). If ξ_p is the unique solution y of $F(y-) \leq p \leq F(y)$, then

$$X_{n,\nu_n} \xrightarrow{P} \xi_p, \text{ as } n \to \infty. \tag{2.1}$$

Proof. Let $F_n(x)$ be the empirical distribution function corresponding to the sample $X_1, \ldots, X_n$, i.e. $nF_n(x)$ = number of $X_i \leq x$, $1 \leq i \leq n$. For each $\alpha > 0$, we have (by the uniqueness condition of the theorem),

$$F(\xi_p - \alpha) < p < F(\xi_p + \alpha) \tag{2.2}$$

Note that

$$\begin{aligned} P\{X_{n,\nu_n} > \xi_p + \alpha\} &= 1 - P\{F_n(\xi_p + \alpha) \geq \frac{\nu_n}{n}\} \\ &= P\{[\,F_n(\xi_p + \alpha) - F(\xi_p + \alpha)] + [F(\xi_p + \alpha) - p\,] < \frac{\nu_n}{n} - p\} \end{aligned} \tag{2.3}$$

and, since $F_n(\xi_p + \alpha) \xrightarrow{P} F(\xi_p + \alpha)$ and $\frac{\nu_n}{n} \xrightarrow{P} p$, we obtain on account of (2.2) and (2.3) that $P\{X_{n,\nu_n} > \xi_p + \alpha\} \to 0$ as $n \to \infty$.

A similar argument shows that $P\,\{X_{n,\nu_n} < \xi_p - \alpha - \alpha\} \to 0$ as $n \to \infty$. The proof follows.

To achieve strong consistency of X_{n,ν_n} for the estimation of ξ_p, we assume the following:

$$\sum_{n=1}^{\infty} P\{|\frac{\nu_n}{n} - p| > \varepsilon\} < \infty \quad \text{for every } \varepsilon > 0. \tag{2.4}$$

THEOREM 2.2. (Strong Consistency) If ξ_p is the unique solution y of $F(y-) \leq p \leq F(y)$ and, in addition, (2.4) holds, then, with probability one,

$$X_{n,\nu_n} \to \xi_p, \text{ as } n \to \infty. \tag{2.5}$$

Proof: For each $\alpha > 0$, observe that

$$(2.6)\quad \begin{aligned} P\{X_{n,\nu_n} > \xi_p + \alpha\} &= P\{F_n(\xi_p + \alpha) < \frac{\nu_n}{n}\} \\ &\leq P\{|\frac{\nu_n}{n} - p| \leq \varepsilon,\ F_n(\xi_p + \alpha) < \frac{\nu_n}{n}\} + P\{|\frac{\nu_n}{n} - p| > \varepsilon\} = a_n + b_n \end{aligned}$$

say, where $\varepsilon > 0$ is chosen such that $2\varepsilon < F(\xi_p + \alpha) - p$. Then, by Markov's inequality, we get

$$(2.7)\quad a_n \leq P\{F_n(\xi_p + \alpha) - F(\xi_p + \alpha) < -\varepsilon\} \leq \frac{1}{\varepsilon^4} E\{F_n(\xi_p + \alpha) - F(\xi_p + \alpha)\}^4$$

and, since $E\{F_n(\xi_p + \alpha) - F(\xi_p + \alpha)\}^4 \leq 3/n^2$, from (2.4), (2.6) and (2.7), we obtain

$$(2.8)\quad \sum_{n=1}^{\infty} P\{X_{n,\nu_n} > \xi_p + \alpha\} \leq \frac{3}{\varepsilon^4} \sum_{n=1}^{\infty} \frac{1}{n^2} + \sum_{n=1}^{\infty} P\{|\frac{\nu_n}{n} - p| > \varepsilon\} < \infty .$$

Similarly, we have

$$(2.9)\quad \sum_{n=1}^{\infty} P\{X_{n,\nu_n} < \xi_p - \alpha\} < \infty .$$

Theorem 2.2 follows by combining (2.8) and (2.9).

Remark 2.1. Note that if $\{k_n\}$ is the numerical sequence defined by $k_n = np$ if np is an integer and $k_n = [np] + 1$ if np is not an integer, then, if $\nu_n \equiv k_n$, $n \geq 1$, X_{n,ν_n} reduces to the usual sample p-th quantile $\hat{\xi}_p$. In this case, condition (2.4) is easily checked and Theorem 2.2 yields the well-known strong consistency of the sample quantile for estimation of ξ_p (see Serfling (1980), Theorem 2.3.1).

3. The Asymptotic Distribution of X_{n,ν_n} in Regular Cases.

A well-known result on the sample quantile $\hat{\xi}_{np}$ asserts that if $F(\xi_p) = p$, F is differentiable at ξ_p and $F'(\xi_p) > 0$ (i.e. the regular cases), then

$$\frac{n^{1/2}(\hat{\xi}_{np} - \xi_p)F'(\xi_p)}{[p(1-p)]^{1/2}} \xrightarrow{\mathcal{D}} N(0,1) .$$

In the present context, it would be of interest to know whether or not in regular cases the limiting distribution of X_{n,ν_n} is the same as that of $\hat{\xi}_{np}$. In seeking a solution to this problem we would like to impose minimal assumptions on the parent distribution and allow dependence between $\{\nu_n\}$ and the original variates $\{X_n\}$ (an independence assumption will rarely be fulfilled in interesting cases). Here the main obstacle is to overcome the random factor introduced by ν_n and we have to develop new methods of proof.

The main result of this section is contained in the following theorem.

THEOREM 3.1. (Central Limit Theorem) If

(i) $F(\xi_p) = p$, F is differentiable at ξ_p and $F'(\xi_p) > 0$,

and

(ii) $n^{1/2}(\frac{\nu_n}{n} - p) \xrightarrow{P} c$, for some constant c,

then

(3.1) $$n^{1/2}(X_{n,\nu_n} - \xi_p)F'(\xi_p)\sigma_p^{-1} \xrightarrow{\mathcal{D}} N(c\sigma_p^{-1}, 1)$$

where $\sigma_p^2 = p(1-p)$, $p \in (0,1)$.

<u>Remark</u> 3.1. (3.1) shows that the constant c in (ii) above has a direct influence on the asymptotic mean of the (normalized) X_{n,ν_n} . For nonrandom central order statistics, this fact was noticed by Serfling (1980, p. 94) who emphasized its importance in the treatment of confidence intervals for quantiles.

Proof of Theorem 3.1. Our approach to proving that (3.1) holds is to show that the asymptotic distribution of $n^{1/2}(X_{n,\nu_n} - \xi_p)$ coincides with the asymptotic distribution of $[F'(\xi_p)]^{-1}\{n^{-1/2}\sum_{i=1}^{n} W_i + n^{1/2}(\frac{\nu_n}{n} - p)\}$, where $\{W_i\}_{1\leq i\leq n}$ are i.i.d.r.v.'s with mean 0 and variance σ_p^2.

To this end, set

$$G_n(x) = P\{n^{1/2}(X_{n,\nu_n} - \xi_p) \leq x\}, \quad x \in \mathbb{R}$$

and, notice that

$$\begin{aligned} G_n(x) &= P\{F_n(\xi_p + xn^{-1/2}) \geq \frac{\nu_n}{n}\} \\ &= P\{n^{1/2}[p - F_n(\xi_p)] - \rho_n(x) \leq F'(\xi_p)x - n^{1/2}(\frac{\nu_n}{n} - p)\} \end{aligned} \tag{3.2}$$

where

$$\rho_n(x) = n^{1/2}\{F_n(\xi_p + xn^{-1/2}) - F_n(\xi_p)\} - F'(\xi_p)x .$$

We now show that for each $x \in \mathbb{R}$,

$$\rho_n(x) \xrightarrow{P} 0 . \tag{3.3}$$

Let $u(t) = 1$ if $t \geq 0$ and $= 0$ if $t < 0$. In proving (3.3), we may assume $x \neq 0$. Write

$$\rho_n(x) = n^{-1/2}\sum_{i=1}^{n} Z_{ni} - F'(\xi_p)x \tag{3.4}$$

where $Z_{ni} = u(\xi_p + xn^{-1/2} - X_i) - u(\xi_p - X_i)$, $1 \leq i \leq n$, $n \geq 1$ are row-wise independent random variables with

$$p_n = E(Z_{ni}) = F(\xi_p + xn^{-1/2}) - p \text{ and } Var(Z_{ni}) = |p_n|(1 - |p_n|)$$

Then, since $p_n \to 0$ as $n \to \infty$, we obtain by Chebyshev's inequality, that for any $\varepsilon > 0$,

$$P\{|\sum_{i=1}^{n} [Z_{ni} - E(Z_{ni})]| > \varepsilon n^{1/2}\} \leq \frac{|p_n|(1-|p_n|)}{\varepsilon^2} \to 0 \text{ , as } n \to \infty$$

entailing

(3.5) $$n^{-1/2} \sum_{i=1}^{n} Z_{ni} - n^{1/2}[F(\xi_p + xn^{-1/2}) - p] \xrightarrow{P} 0 .$$

On the other hand,

$$n^{1/2}[F(\xi_p + xn^{-1/2}) - p] \to F'(\xi_p)x \text{ as } n \to \infty$$

which together with (3.5) implies (3.3).

Now, write $p - F_n(\xi_p) = \frac{1}{n} \sum_{i=1}^{n} W_i$, where $W_i = p - u(\xi_p - X_i)$, $1 \leq i \leq n$, $n \geq 1$. This is an average of i.i.d.r.v.'s to which according to the classical central limit theorem,

(3.6) $$n^{-1/2} \sum_{i=1}^{n} W_i \xrightarrow{\mathcal{D}} N(0,\sigma_p^2) .$$

Using (3.2), (3.3) and (3.6) we deduce

(3.7) $$G_n(x) \to P\{Z \leq F'(\xi_p)x\sigma_p^{-1}\} = \Phi((F'(\xi_p)x - c)\sigma_p^{-1})$$

where Z is $N(c\sigma_p^{-1}, 1)$ r.v., and Φ is the standard normal cdf. (3.1) follows from (3.7).

Another relevant question for the asymptotic theory of random central order statistics is a Bahadur-type representation. It will be interesting (but seems difficult) to investigate whether a Bahadur representation with a strong remainder term holds for the case of X_{n,ν_n} treated here. We will nevertheless prove

THEOREM 3.2. (Weak Representation Theorem) Under the assumptions (i) and (ii) of Theorem 3.1 we have

$$X_{n,\nu_n} = \xi_p + [F'(\xi_p)]^{-1}[\frac{\nu_n}{n} - F_n(\xi_p)] + R_n \tag{3.8}$$

where

$$R_n = o_p(n^{-1/2}) \text{ as } n \to \infty . \tag{3.9}$$

Proof: Defining

$$Q_n = n^{1/2}[\frac{\nu_n}{n} - F_n(\xi_p) - (X_{n,\nu_n} - \xi_p)F'(\xi_p)],$$ we need to show that

$$Q_n \xrightarrow{P} 0 . \tag{3.10}$$

To achieve this, take an arbitrary $\varepsilon > 0$. Then, using Theorem 3.1, choose $K > 0$ sufficiently large such that

$$P\{n^{1/2}|X_{n,\nu_n} - \xi_p| > K\} < \varepsilon/2, \text{ for } n > n_o . \tag{3.11}$$

Now, partition the interval $[-K,K]$ into $m - 1$ intervals $-K = \Delta_1 < \Delta_2 < \ldots < \Delta_m = K$ such that

$$\Delta_i - \Delta_{i-1} < \varepsilon_o[2F'(\xi_p)]^{-1}, \text{ for } i = 2,3,\ldots, m \text{ and } \varepsilon_o > 0 . \tag{3.12}$$

Now, set

$$\Pi_{n,i} = \{\Delta_{i-1} \le n^{1/2}(X_{n,\nu} - \xi_p) \le \Delta_i\},\ i = 2,\ldots,m;\ n \ge 1$$

and use (3.11) to get

$$(3.13)\qquad P\{Q_n > \varepsilon_o\} \le \sum_{i=2}^{m} P\{[Q_n > \varepsilon_o] \cap \Pi_{n,i}\} + \varepsilon/2 .$$

Since $X_{n,\nu_n} \le \xi_p + n^{-1/2}\Delta_i$ entails $\nu_n/n \le F_n(\xi_p + n^{-1/2}\Delta_i)$, by using the monotonicity of F_n and (3.12), we have for $i = 2,\dots,m$

$$\{[Q_n > \varepsilon_o] \cap \Pi_{n,i}\} \subset \{n^{1/2}[F_n(\xi_p + n^{-1/2}\Delta_i) - F_n(\xi_p)] - F'(\xi_p)\Delta_i > \varepsilon_o/2\}$$

which, together with (3.13), entails

(3.14)

$$P\{Q_n > \varepsilon_o\} \le \sum_{i=2}^{m} P\{n^{1/2}[F_n(\xi_p + n^{-1/2}\Delta_i) - F_n(\xi_p)] - F'(\xi_p)\,\Delta_i > \varepsilon_o/2\} + \varepsilon/2.$$

Now, by (3.3) with $x = \Delta_i$ for $i = 2,\dots,m$, the sum in the right-hand side of (3.14) is $< \varepsilon/2$ for sufficiently large n, proving that $P\{Q_n > \varepsilon_o\} < \varepsilon \;\; \forall\, n \ge n_1$. Similarly $P\{Q_n < -\varepsilon_o\} < \varepsilon \; \forall\, n \ge n_2$. The proof follows.

Remark 3.2. Note that if $c = 0$ in the condition (ii) of Theorem 3.1 (in which case the limiting law of the normalized X_{n,ν_n} is standard normal), then (3.8) and (3.9) reduce to

$$(3.15)\qquad X_{n,\nu_n} = \xi_p + [F'(\xi_p)]^{-1}[p - F_n(\xi_p)] + o_p(n^{-1/2}) ,$$

showing that asymptotically, X_{n,ν_n} may be represented as an average of i.i.d.r.v.'s. In this form (3.15) represents a generalization of a result due to Ghosh (1971) (see also Serfling (1980), p. 92) and David (1981), pp. 254-256)).

4. The Berry-Esséen Bound for X_{n,ν_n} in Regular Cases. Recently interest has

been focused on the convergence to normality for sample quantiles $\hat{\xi}_{np}$ in regular cases. For such situations, under the assumption that F has a bounded second derivative on $\mathbb{R}$, Reiss (1974) and Serfling (1980, Theorem C, p. 81) derived independently the Berry-Esséen bound $O(n^{-1/2})$. In this section we study this problem in the case of normalized random central order statistic X_{n,ν_n} under assumptions somewhat weaker than those of Reiss and Serfling (cit. op.). Our results not only include the results of these authors as a special case but also extend their results to cover general random rank sequences $\{\nu_n\}$. To be precise, we seek bounds on the quantity

$$\Delta_n = \sup_x |P\{n^{1/2}(X_{n,\nu_n} - \xi_p) F'(\xi_p) \leq x\sigma_p\} - \Phi(x)|$$

for the case when the limit law in (3.1) is standard normal (i.e., when $c = 0$). To solve this problem we would need to impose a relatively stronger version of the condition (ii) of Theorem 3.1 (with $c = 0$), namely $P\{n^{1/2}|\frac{\nu_n}{n} - p| > \varepsilon_n\} \leq \delta_n$, for some numerical sequences $\{\varepsilon_n\}$ and $\{\delta_n\}$ converging to zero. The "exact" order of approximation for Δ_n will then be obtained when $\varepsilon_n = O(n^{-1/2}) = \delta_n$.

In what follows $C_1, C_2, C_3, \ldots$ will denote positive constants.

THEOREM 4.1. (Rates of Convergence in the CLT) Let ε_n and δ_n be positive constants such that $\varepsilon_n \leq 1$ and

$$P\{n^{1/2} | \frac{\nu_n}{n} - p | > \varepsilon_n\} \leq \delta_n, \quad n \geq 1 \tag{4.1}$$

where $\{\nu_n\}$ is a random central rank sequence and $p \in (0,1)$. Assume that F"exists and is bounded in the interval $J = [\xi_p - K, \xi_p + K]$ (for some $K > 0$). Let $M = \sup_{x \in J} |F''(x)|$ and $K_o = \min\{F'(\xi_p)M^{-1}, K\}$. Then

$$\Delta_n \leq C_1 n^{-1/2} + C_2 n^{-1} + C_3 \varepsilon_n + \delta_n \tag{4.2}$$

where

$$C_1 = 4\sigma_p^{-2} + [1.19625 + 8(2\pi)^{-1/2} e^{-1}]\sigma_p^{-1} + 4\sigma_p M(2\pi)^{-1/2} e^{-1} [F'(\xi_p)]^{-2},$$

$$C_2 = 24\sigma_p^{-2} + (2\pi e)^{-1/2} \sigma_p^2 K_o^{-2} [F'(\xi_p)]^{-2}$$

and

$$C_3 = 1.5(2\pi)^{-1/2} \sigma_p^{-1} .$$

Remark 4.1. (Example) Let $\{X_n\}$ be a given sequence of i.i.d.r.v.'s with cdf F. Let $\{\delta_n\}$ be a sequence of positive constants such that $\delta_n \to 0$ as $n \to \infty$. Consider the "quantile" sequence $\{k_n\}$ defined in Remark 2.1. Define $\nu_n : \Omega \to \{1,\dots,n\}$ by $\nu_n(\omega) = k_n$ if $\omega \in \Omega_n$, where $\{\Omega_n\}$ is a sequence of events chosen such that $P\{\Omega_n\} = 1 - \delta_n$ (the definition of ν_n on Ω_n^c may be made arbitrarily provided $\nu_n \in \{1,\dots,n\}$ and is measurable).

Then, since $|\frac{k_n}{n} - p| < \frac{1}{n}$, it follows that

$$\{|\frac{\nu_n}{n} - p| \geq \frac{1}{n}\} \subset \Omega_n^c$$

which guarantees that

$$P\{n^{1/2} |\frac{\nu_n}{n} - p| \geq n^{-1/2}\} \leq P(\Omega_n^c) = \delta_n .$$

Thus, condition (4.1) is fulfilled with $\varepsilon_n = n^{-1/2}$. We remark that according to our Theorem 4.1, if $\delta_n = O(n^{-1/2})$ as $n \to \infty$, the (exact) Berry-Esséen approximation order for Δ_n is $O(n^{-1/2})$. In particular, if $\delta_n \equiv 0$, $n \geq 1$, the theorem of Reiss and Serfling will follow as a special case.

To prove Theorem 4.1 we need a sequence of lemmas.

LEMMA 4.1. If

$$(4.3) \qquad |F(x) - p| < \sigma_p^2/2$$

then

$$(4.4) \qquad |\sigma_p \sigma_{F(x)}^{-1} - 1| < \sigma_p^{-2}|F(x) - p|, \text{ where } \sigma_{F(x)}^2 = F(x)(1 - F(x)) .$$

Proof: Let $d(x) = |\sigma_p \sigma_{F(x)}^{-1} - 1|$. Note that under (4.3), $0 < F(x) < 1$, so that $d(x)$ is well defined. By setting $\lambda = \sigma_p^{-2}(F(x) - p)$, $d(x)$ simplifies, after some calculation, to

$$(4.5) \qquad |\lambda| \cdot \frac{|\lambda\sigma_p^2 - (1-2p)|}{[1+(1-2p)\lambda-\sigma_p^2 \lambda^2] + [1+(1-2p)\lambda-\sigma_p^2\lambda^2]^{1/2}} .$$

Now, since $|\lambda| < 1/2$ and $p \in (0,1)$, it is easily seen that $|\lambda\sigma_p^2 - (1 - 2p)| < 1$, while

$$1 + (1-2p)\lambda - \sigma_p^2\lambda^2 > 1 - |\lambda| - \lambda^2/4 > 7/16.$$

Thus, according to (4.5), we have

$$d(x) < |\lambda| \frac{1}{\frac{7}{16} + (\frac{7}{16})^{1/2}} < |\lambda|$$

and (4.4) obtains.

The following well-known lemma (cf. Petrov (1975), p. 16) is also needed.

LEMMA 4.2. Let $\{V_n\}$ and $\{W_n\}$ be two sequences of random variables. If $\{a_n\}$ is a sequence of positive constants, then

$$(4.6)\quad \sup_x |P\{V_n + W_n \le x\} - \Phi(x)| \le \sup_x |P\{V_n \le x\} - \Phi(x)| + P\{|W_n| \ge a_n\} + (2\pi)^{-1/2} a_n .$$

The next result gives an approximation of the true distribution of random central order statistics.

LEMMA 4.3. Assume that (4.1) holds. Then $\forall\ n > 1$,

(4.7)

$$D_n = \sup_x |P\{X_{n,\nu_n} < x\} - \Phi(n^{1/2}(F(x) - p)\sigma_p^{-1})| < C_4 n^{-1/2} + C_5 n^{-1} + C_3 \varepsilon_n + \delta_n$$

where C_3 is defined in Theorem 4.1, $C_4 = 4\sigma_p^{-2} + [1.19625 + 8(2\pi)^{-1/2} e^{-1}]\sigma_p^{-1}$ and $C_5 = 24\sigma_p^{-2}$.

Remark 4.2. As a consequence of the fact that in Lemma 4.3 absolutely no conditions are imposed on the distribution function F, the estimate (4.7) will also play a key role in the study of the asymptotic law of X_{n,ν_n} in nonregular cases; (see Section 5).

Remark 4.3. The following example shows that for a given sequence $\{\varepsilon_n\}$ such that $\varepsilon_n = 0(n^{-1/2})$, the rate of convergence in (4.7) cannot be sharpened even if the assumption (4.1) is maximally sharpened to

$$(4.1)^* \qquad P\{n^{1/2} |\frac{\nu_n}{n} - p| > \varepsilon_n\} = 0 .$$

Let $\{X_n\}$ be a sequence of i.i.d. symmetric Bernoulli r.v.'s with $P\{X_1 = 1\} = P\{X_1 = -1\} = 1/2$. By setting $\nu_n = [\frac{n}{2}]$, it is easily seen that $(4.1)^*$ is satisfied with $p = 1/2$ and $\varepsilon_n = n^{-1/2}$. Now, if n is even, we have

$$P\{X_{n,\nu_n} < 0\} = P\{F_n(0) > \frac{\nu_n}{n}\} = P\{\sum_{i=1}^{n} u(-X_i) > \frac{n}{2}\}$$

and since $\sum_{i=1}^{n} u(-X_i) \overset{\mathcal{D}}{\sim} B(n,\frac{1}{2})$ (binomial), with D_n as defined in (4.7), we have

$$D_n \geq \tfrac{1}{2} P\{B(n, \tfrac{1}{2}) = \tfrac{n}{2}\} = \frac{1}{2^{n+1}} \frac{n!}{[(\frac{n}{2})!]^2} = d_n, \text{ say.}$$

Then, using Stirling's formula, it is easily seen that $d_n \sim 1/(2\pi n)^{1/2}$ as $n \to \infty$. This shows that the assumption (4.1) in Lemma 4.3 with $\varepsilon_n = O(n^{-1/2}) = \delta_n$ is the most reasonable assumption to obtain the "correct" Berry-Esséen type bound $O(n^{-1/2})$.

Proof of Lemma 4.3. We estimate D_n by splitting it into two parts, namely for $x \in I$ and for $x \in I^c$, where $I = \{x: |F(x) - p| < \sigma_p^2/2\}$.

(i) Let $x \in I$. Set

$$S_n(x) = n^{-1/2} \sigma_{F(x)}^{-1} \sum_{i=1}^{n} [u(x - X_i] - F(x)].$$

Then,

$$P\{X_{n,\nu_n} \leq x\} = P\{F_n(x) \geq \frac{\nu_n}{n}\}$$

(4.8)

$$= P\{S_n(x) - n^{1/2}(\frac{\nu_n}{n} - p)\sigma_{F(x)}^{-1} \geq n^{1/2}(p - F(x))\sigma_{F(x)}^{-1}\}.$$

In view of Lemma 4.2 and (4.8), we have

(4.9)

$$D_{n,1}(x) = |P\{X_{n,\nu_n} \leq x\} - \Phi(n^{1/2}(F(x) - p)\sigma_{F(x)}^{-1})|$$
$$\leq \sup_{t \in \mathbb{R}} |P\{S_n(x) \leq t\} - \Phi(t)| + P\{|n^{1/2}(\frac{\nu_n}{n} - p)| \geq a_n \sigma_{F(x)}\} + a_n/(2\pi)^{1/2}.$$

for any sequence of positive constants $\{a_n\}$. Now, since $x \in I$, using Lemma 4.1, we get

(4.10) $$\sigma_p \sigma_{F(x)}^{-1} \le 1.5$$

and by setting $a_n = 1.5\varepsilon_n \sigma_p^{-1}$, we deduce from (4.1) and (4.9), that

(4.11) $$D_{n,1}(x) \le \sup_{t \in \mathbb{R}} |P\{S_n(x) \le t\} - \Phi(t)| + 1.5\varepsilon_n(2\pi)^{-1/2}\sigma_p^{-1} + \delta_n .$$

Applying the Berry-Esséen theorem (with the sharpest constant 0.7975 given by van Beeck (1972)) the first term in the right-hand side of (4.11) is bounded by $(0.7975)\sigma_{F(x)}^{-3} E|u(x - X_1) - F(x)|^3 n^{-1/2}$ and since $E|u(x - X_1) - F(x)|^3 \le \sigma_{F(x)}^2$, (4.11) implies

(4.12) $$D_{n,1}(x) \le (0.7975)\sigma_{F(x)}^{-1} n^{-1/2} + 1.5\varepsilon_n(2\pi)^{-1/2}\sigma_p^{-1} + \delta_n .$$

Consequently, using (4.10) once again, we get from (4.12) that

(4.13) $$D_{n,1}(x) \le (1.19625)\sigma_p^{-1} n^{-1/2} + 1.5\varepsilon_n(2\pi)^{-1/2}\sigma_p^{-1} + \delta_n .$$

Consider now

$$D_{n,2}(x) = |\Phi(n^{1/2}(F(x) - p)\sigma_{F(x)}^{-1}) - \Phi(n^{1/2}(F(x) - p)\sigma_p^{-1})| .$$

The mean value theorem yields

(4.14) $$D_{n,2}(x) = n^{1/2}|F(x) - p|\sigma_p^{-1}|\sigma_p\sigma_{F(x)}^{-1} - 1|(2\pi)^{-1/2} \exp\left\{ \frac{-n(F(x) - p)^2}{2\sigma_p^2}[1 + \theta(\sigma_p\sigma_F^{-1}(x) - 1)]^2 \right.$$

for some $0 < \theta < 1$.

Now Lemma 4.1 implies that

$$|1 + \theta(\sigma_p \sigma_{F(x)}^{-1} - 1)| \geq 1 - |\sigma_p \sigma_{F(x)}^{-1} - 1| \geq 1/2$$

which together with (4.14) and another application of Lemma 4.1 gives

$$D_{n,2}(x) \leq n^{1/2}(F(x) - p)^2 (2\pi)^{-1/2} \sigma_p^{-3} \exp\{-\tfrac{1}{8} n(F(x) - p)^2 \sigma_p^{-2}\}. \tag{4.15}$$

Therefore, since

$$\sup_{t>0} t \exp(-ta) = (ae)^{-1}, \quad a > 0 \tag{4.16}$$

we obtain, from (4.15) that

$$D_{n,2}(x) \leq 8(2\pi)^{-1/2} (\sigma_p e)^{-1} n^{-1/2}. \tag{4.17}$$

Combining (4.17) and (4.13), we conclude that

$$\begin{aligned} \sup_{x \in I} |P\{X_{n,\nu_n} \leq x\} - \Phi(n^{1/2}(F(x) - p)\sigma_p^{-1})| \\ \leq \{1.19625 + 8(2\pi)^{-1/2} e^{-1}\}\sigma_p^{-1} n^{-1/2} + 1/5(2\pi)^{-1/2} \sigma_p^{-1} \varepsilon_n + \delta_n. \end{aligned} \tag{4.18}$$

(ii) Let $x \in I^c$. In order to estimate $\Phi(n^{1/2}(F(x) - p)\sigma_p^{-1})$ for x in I^c, we shall use the well-known inequality:

$$1 - \Phi(t) \leq (2\pi)^{-1/2} t^{-1} \exp(-t^2/2) \leq (2\pi e)^{-1/2} t^{-2}, \quad t > 0. \tag{4.19}$$

Set $\lambda = \sigma_p^{-2}(F(x) - p)$ (as in the proof of Lemma 4.1) and assume that $\lambda \leq -1/2$. Then, according to (4.19), we have

$$\Phi(n^{1/2}(F(x)-p)\sigma_p^{-1}) \leq 4(2\pi e)^{-1/2} \sigma_p^{-2} n^{-1}. \tag{4.20}$$

On the other hand, note that

$$P\{X_{n,\nu_n} < x\} = P\{F_n(x) \geq \frac{\nu_n}{n},\ n^{1/2}\left|\frac{\nu_n}{n} - p\right| < \varepsilon_n\}$$

$$(4.21)\qquad + P\{F_n(x) \geq \frac{\nu_n}{n},\ n^{1/2}\left|\frac{\nu_n}{n} - p\right| \geq \varepsilon_n\}$$

$$\leq P\{F_n(x) > p - \varepsilon_n n^{-1/2}\} + \delta_n$$

where the last inequality follows from (4.1).

Observe that, in proving (4.7), we may without loss of generality assume that

$$(4.22)\qquad n^{1/2}\sigma_p^2 \geq 4\ ,$$

since if (4.22) does not hold, then the bound in (4.7) applies trivially. Now, by Chebyshev's inequality, we have

$$(4.23)\qquad P\{F_n(x) > p - \varepsilon_n n^{-1/2}\} \leq \sigma^2_{F(x)}(p - F(x) - \varepsilon_n n^{-1/2})^{-2} n^{-1}\ .$$

To estimate the right-hand side of (4.23), we see that, since $\lambda \leq -1/2$,

$$(4.24)\qquad \sigma^2_{F(x)}\sigma_p^{-2} = 1 + (1-2p)\lambda - \sigma_p^2\lambda^2 \leq 1 + |\lambda| \leq 6\lambda^2$$

and

$$(4.25)\qquad \sigma_p^2 + \varepsilon_n\lambda^{-1}n^{-1/2} \geq \sigma_p^2 - 2\varepsilon_n n^{-1/2} \geq \sigma_p^2/2$$

where the last inequality of (4.25) follows from (4.22).

Thus, using (4.23)-(4.25), we derive

(4.26) $$P\{F_n(x) > p - \varepsilon_n n^{-1/2}\} \leq 24\sigma_p^{-2} n^{-1} .$$

Since, for $\alpha,\beta > 0$, $|\alpha - \beta| \leq \max(\alpha,\beta)$, by combining (4.20), (4.21) and (4.26), we find that for $\lambda \leq -1/2$,

(4.27) $$|P\{X_{n,\nu_n} \leq x\} - \Phi(n^{1/2}(F(x) - p)\sigma_p^{-1})| \leq 24\sigma_p^{-2} n^{-1} + \delta_n .$$

Repeating the argument in (4.20) and (4.26) for $\lambda \geq 1/2$, we see that (4.27) continues to hold in this case also. (4.7) now follows by combining (4.27) and (4.18). The proof follows.

Proof of Theorem 4.1. Our main result (4.2) follows readily from the estimate (4.7) and the following two lemmas.

LEMMA 4.4. If $|\alpha x| \leq 1/2$, then

(4.28) $$|\Phi(x + \alpha x^2) - \Phi(x)| \leq 8|\alpha|(2\pi)^{-1/2} e^{-1} .$$

LEMMA 4.5. Under the assumptions of Theorem 4.1, we have

(4.29)
$$E_n = \sup_{x\in R} | \Phi(n^{1/2}(F(x) - p)\sigma_p^{-1}) - \Phi(n^{1/2}(x - \xi_p)F'(\xi_p)\sigma_p^{-1})| \leq C_6 n^{-1/2} + C_7 n^{-1}$$

where

$$C_6 = 4M\sigma_p(2\pi)^{-1/2} e^{-1}[F'(\xi_p)]^{-2} \text{ and } C_7 = \sigma_p^2(2\pi e)^{-1/2} K_o^{-2}[F'(\xi_p)]^{-2} .$$

Proof of Lemma 4.4. By the mean value theorem, we have with $y = x + \alpha x^2$,

$$\Phi(y) - \Phi(x) = (2\pi)^{-1/2} \alpha x^2 \exp[-1/2\, x^2[1 + \theta\alpha x]^2\} , \quad 0 < \theta < 1 ,$$

and, since $|1+\theta\alpha x| \geq 1 - |\alpha x| \geq 1/2$, we obtain

$$|\Phi(y) - \Phi(x)| \leq (2\pi)^{-1/2} |\alpha| x^2 \exp(-\frac{1}{8} x^2)$$

which, together with (4.16) yields (4.28).

Proof of Lemma 4.5. If $|x - \xi_p| \leq K_o$, then by using a second order Taylor expansion, we may express

$$E_n(x) = |\Phi(n^{1/2}(F(x) - p)\sigma_p^{-1}) - \Phi(n^{1/2}(x - \xi_p)F'(\xi_p)\sigma_p^{-1})|$$

as

$$E_n(x) = |\Phi(n^{1/2}(x-\xi_p)F'(\xi_p)\sigma_p^{-1} + 1/2\, n^{1/2}(x-\xi_p)^2 F''(\xi_p+\theta(x-\xi_p))\sigma_p^{-1}) - \Phi(n^{1/2}(x-\xi_p)F'(\xi_p)\sigma_p^{-1})| , \quad 0 < \theta < 1 , \tag{4.30}$$

and, by using Lemma 4.4, we obtain

$$\sup_{|x-\xi_p|\leq K_o} E_n(x) \leq 4M\sigma_p(2\pi)^{-1/2} e^{-1}[F'(\xi_p)]^{-2} n^{-1/2} . \tag{4.31}$$

Assume on the other hand that $|x - \xi_p| \geq K_o$. If $x \leq \xi_p - K_o$, then, according to (4.19) we have

$$\Phi(n^{1/2}(x-\xi_p)F'(\xi_p)\sigma_p^{-1}) \leq \sigma_p^2(2\pi e)^{-1/2} K_o^{-2}[F'(\xi_p)]^{-2} n^{-1} . \tag{4.32}$$

Now, since $x \leq \xi_p - K_o$, by using (4.31) (with $x = \xi_p - K_o$) together with (4.32) we deduce that

(4.33)

$$\Phi(n^{1/2}(F(x)-p)\sigma_p^{-1}) \leq 4M\sigma_p(2\pi)^{-1/2} e^{-1}[F'(\xi_p)]^{-2} n^{-1/2} + \sigma_p^2(2\pi e)^{-1/2} K_o^{-2}[F'(\xi_p)]^{-2} n^{-1}$$

Thus, from (4.32) and (4.33), we obtain for $x \leq \xi_p - K_o$, that

(4.34) $$E_n(x) \leq C_6 n^{-1/2} + C_7 n^{-1}$$

where C_6 and C_7 are defined in Lemma 4.5.

Finally, since a similar statement holds for $x \geq \xi_p + K_o$ we deduce that

(4.35) $$\sup_{|x-\xi_p| \geq K_o} E_n(x) \leq C_6 n^{-1/2} + C_7 n^{-1} .$$

Lemma 4.5 now follows immediately from (4.31) and (4.35).

Remark 4.4. Apropos the regularity conditions on F in Theorem 4.1, the requirements concerning F" may be dropped. In many situations, F is not sufficiently smooth at ξ_p and expansion (4.30) is inappropriate. Yet an estimate like (4.29) may still be valid under modified assumptions. Specifically, assume that F is differentiable at ξ_p, $(F(\xi_p) = p)$ with $F'(\xi_p) > 0$, and

(4.36) $$|F(\xi_p + h) - p - hF'(\xi_p)| = O(h^2) \quad \text{as} \quad h \to 0 .$$

Then, with E_n defined by (4.29), we may show that

(4.37) $$E_n \leq C_8 n^{-1/2} + C_9 n^{-1}$$

for some positive constants C_8 and C_9 to be specified below. To establish (4.37), we write

$$\alpha(x) = F(x) - p - (x - \xi_p)F'(\xi_p) ,$$

and use (4.36) to infer that $\lim_{x \to \xi_p} \alpha(x)\,(x-\xi_p)^{-1} = 0$.

Now pick $\delta > 0$ such that if $|x - \xi_p| \leq \delta$,

$$|\alpha(x)(x - \xi_p)^{-1}[F'(\xi_p)]^{-1}| \leq 1/2$$

and let $K_1 > 0$ be such that if $|x - \xi_p| \leq \delta$, then

$$|\alpha(x)| \leq K_1(x - \xi_p)^2 .$$

Then, by putting together the estimates found by the method of Lemma 4.5 we note that (4.37) holds with

$$C_8 = 8K_1\sigma_p(2\pi)^{-1/2} e^{-1}[F'(\xi_p)]^{-2}$$

and

$$C_9 = \sigma_p^2 \ (2\pi e)^{-1}\delta^{-2}[F'(\xi_p)]^{-2} .$$

The details are omitted. We may also note that if $F''(\xi_p)$ exists, then by using Young's form of Taylor's theorem (cf. Hardy (1952)) we have

$$F(\xi_p + h) = p + hF'(\xi_p) + \frac{h^2}{2} F''(\xi_p) + o(h^2) \quad \text{as} \quad h \to 0 ,$$

and consequently (4.36) obtains. In any case, (4.36) implies (4.37) which together with Lemma 4.3 guarantees that

$$\Delta_n \leq O(n^{-1/2}) + O(\varepsilon_n) + \delta_n \text{ , as } n \to \infty .$$

5. Limit Law and Berry-Esséen Rates for X_{n,ν_n} in Nonregular Cases.

In studying the asymptotic law of X_{n,ν_n} , of considerable interest are those distributions for which $F'(\xi_p) = 0$ and are not covered by Theorem 3.1.

Our next results are tailored to just these cases. In theorem 5.1 we shall assume that $F(\xi_p) = p$ and that

$$(5.1) \qquad \lim_{h\to 0} [F(\xi_p+h) - F(\xi_p)]h^{-\rho} = M > 0 \text{ , for some odd integer } \rho > 1 \text{ .}$$

Condition (5.1) is considerably less restrictive than its formulation makes it appear. The generality it confers is discussed by Chanda (1975) who considered nonrandom central order statistics X_{n,k_n} and showed that under (5.1), if $k_n/n = p + o(n^{-1/2})$ as $n \to \infty$, then

$$(5.2) \qquad Mn^{1/2}(X_{n,k_n} - \xi_p)^{\rho} \xrightarrow{\mathcal{D}} N(0,\sigma_p^2) \text{ .}$$

In this section we generalize (5.2) for X_{n,ν_n} and then study the accuracy of the corresponding normal approximation. To the best of our knowledge the Berry-Esséen type problem has been investigated even for the non-random case.

THEOREM 5.1. (Central Limit Theorem) If

$$(5.3) \qquad n^{1/2}(\frac{\nu_n}{n} - p) \xrightarrow{P} 0 \text{ ,}$$

and (5.1) is satisfied, then

$$(5.4) \qquad Mn^{1/2}(X_{n,\nu_n} - \xi_p)^{\rho} \xrightarrow{\mathcal{D}} N(0,\sigma_p^2) \text{ .}$$

Proof: Define

$$(5.5) \qquad \Gamma_n(x) = \Phi(n^{1/2}(F(x) - p)\sigma_p^{-1}) - \Phi(Mn^{1/2}(x - \xi_p)^{\rho}\sigma_p^{-1})$$

and

$$(5.6) \qquad \alpha(x) = (F(x) - p)(x - \xi_p)^{-\rho} \text{ .}$$

Then, since $\alpha(x) \to M > 0$ as $x \to \xi_p$, given $0 < \varepsilon < 1/2$, there exists $\eta > 0$, such that if $|x - \xi_p| \leq \eta$,

(5.7) $$|\alpha(x)M^{-1} - 1| < \varepsilon .$$

First we estimate $\sup_{x \in \mathbb{R}} |\Gamma_n(x)|$, and to this end, we distinguish the following cases.

<u>Case (i):</u> $|x - \xi_p| \leq \eta$. Using the mean value theorem, we obtain

$$\Gamma_n(x) = (2\pi)^{-1/2} \sigma_p^{-1} n^{1/2} (x - \xi_p)^{\rho}(\alpha(x)-M)\exp\{-1/2\, M^2 \sigma_p^{-2} n(x-\xi_p)^{2\rho}[1+\theta(\alpha(x)M^{-1} - 1]^2\}$$

for some $0 < \theta < 1$, and proceeding as in the derivation of (4.31), we obtain with the help of the inequality $t \exp(-t^2) \leq (2e)^{-1/2}$, $t > 0$ that, for $|x - \xi_p| \leq \eta$,

(5.8) $$|\Gamma_n(x)| \leq 2^{1/2} (\pi e)^{-1/2} \varepsilon$$

<u>Case (ii):</u> $|x - \xi_p| \geq \eta$. Assume that $x \leq \xi_p - \eta$. Then, by using (4.19), we deduce

(5.9)
$$\Phi(Mn^{1/2} (x-\xi_p)^{\rho}\sigma_p^{-1}) \leq \sigma_p^2(2\pi e)^{-1/2} M^{-2}(x-\xi_p)^{-2\rho}n^{-1} \leq \sigma_p^2(2\pi e)^{-1/2} M^{-2}\eta^{-2\rho}n^{-1} .$$

On the other hand, defining $x_1 = \xi_p - \eta$, we have

$$\Phi(n^{1/2} (F(x)-p)\sigma_p^{-1}) \leq \Phi(n^{1/2} (F(x_1) - p)\sigma_p^{-1})$$

$$\leq |\Gamma_n(x_1)| + \Phi(Mn^{1/2} (x_1 - \xi_p)^{\rho} \sigma_p^{-1})$$

which together with (5.8) (for $x = x_1$) and (5.9) implies

(5.10) $$|\Gamma_n(x)| \le 2^{1/2}(\pi e)^{-1/2}\varepsilon + \sigma_p^2(2\pi e)^{-1/2}M^{-2}\eta^{-2\rho}n^{-1} .$$

Thus, putting together the estimates obtained so far, we have

(5.11) $$\sup_{x\in\mathbb{R}} |\Gamma_n(x)| \le 2^{1/2}(\pi e)^{-1/2}\varepsilon + \sigma_p^2\,(2\pi e)^{-1/2}M^{-2}\eta^{-2\rho}n^{-1}$$

Now, condition (5.3) implies that

$$P\{n^{1/2}\,|\,\frac{\nu_n}{n} - p\,| \ge \varepsilon\} \le \varepsilon\ ,\ \forall n \ge n_o(\varepsilon).$$

Therefore, with D_n and C_i ($i = 1,2,3$) as defined in Lemma 4.3, the technique leading to (4.7) may be used to prove that for $n \ge n_o(\varepsilon)$,

(5.12) $$D_n \le C_4 n^{-1/2} + C_5 n^{-1} + (C_3 + 1)\varepsilon$$

Now, combining (5.11) and (5.12) we obtain (via the inversion $y = Mn^{1/2}(x - \xi_p)^{\rho}\sigma_p^{-1}$ which is in order since ρ is an odd integer ≥ 1) that

(5.13) $$\sup_{y\in\mathbb{R}} |P\{Mn^{1/2}(X_{n,\nu_n} - \xi_p)^{\rho} \le y\sigma_p\} - \Phi(y)| \le C_4 n^{-1/2} + C_{10}n^{-1} + C_{11}\varepsilon$$

where

$$C_{10} = C_5 + \sigma_p^2\,(2\pi e)^{-1/2}M^{-2}\eta^{-2\rho} \text{ and } C_{11} = C_3 + 1 + 2^{1/2}(\pi e)^{-1/2} .$$

(5.13) guarantees (5.4) and the theorem is proved.

Remark 5.1. Note that if (5.1) holds with $\rho = 1$, then $M = F'(\xi_p)$ and in this case, by the preceding proof, one gets an alternative proof of (3.1) when $c = 0$.

To study the speed of convergence apropos to the preceding theorem we shall need somewhat stronger assumptions than those imposed in Theorem 5.1. Precisely, we assume (4.1) and

(5.14) $$F(\xi_\rho + h) - F(\xi_p) = Mh^\rho + O(h^{2\rho}) \quad \text{as} \quad h \to 0$$

for some $M > 0$ and odd integer $\rho > 1$.

Then, we have

THEOREM 5.2. (Rates of Convergence in the CLT) Under (4.1) and (5.14)

(5.15)
$$\sup_x |P\{Mn^{1/2}(X_{n,\nu_n} - \xi_p)^\rho \le x\sigma_p\} - \Phi(x)| \le C_{12}n^{-1/2} + C_{13}n^{-1} + C_3\varepsilon_n + \delta_n$$

where

$$C_{12} = C_4 + 8\sigma_p K_o(2\pi)^{-1/2}e^{-1}M^{-2} \text{ and } C_{13} = C_5 + \sigma_p^2(2\pi e)^{-1/2}\eta^{-2\rho}M^{-2}$$

Proof: (outline) Let

$$\beta(x) = F(x) - F(\xi_p) - M(x - \xi_p)^\rho .$$

In view of (5.14), select $\eta_o > 0$ and $K_2 > 0$ such that if $|x - \xi_p| \le \eta_o$,

(5.16) $$|\beta(x)| \le K_2(x - \xi_p)^{2\rho} .$$

Then, choose $0 < \eta < \eta_o$ sufficiently small, so that

(5.17) $$\eta^\rho \le M(2K_2)^{-1} .$$

To estimate $\sup_x |\Gamma_n(x)|$ (with $\Gamma_n(x)$ defined by (5.5)), we proceed as in the proof of the preceding theorem and break $\mathbb{R}$ into two parts for which we have different arguments.

For $|x - \xi_p| < \eta$, by the mean value theorem, we get

$$\begin{aligned}\Gamma_n(x) &= \Phi(Mn^{1/2}(x-\xi_p)^\rho \sigma_p^{-1} + n^{1/2}\beta(x)\sigma_p^{-1}) - \Phi(Mn^{1/2}(x - \xi_p)^\rho \sigma_p^{-1}) \\ &= n^{1/2}\beta(x)\,(2\pi)^{-1/2}\sigma_p^{-1} \exp\{-\tfrac{1}{2}M^2 n(x-\xi_p)^{2\rho}\sigma_p^{-2}[1+\theta\beta(x)M^{-1}(x-\xi_p)^{-\rho}]^2\}\end{aligned} \tag{5.18}$$

for some $0 < \theta < 1$.

Now, from (5.16) and (5.17), we obtain

$$\begin{aligned}|1 + \theta\beta(x)M^{-1}(x-\xi_p)^{-\rho}| &\geq 1 - M^{-1}|\beta(x)(x-\xi_p)^{-\rho}| \geq 1 - K_2M^{-1}|x-\xi_p|^\rho \\ &\geq 1 - K_2M^{-1}\eta^\rho \geq 1/2,\end{aligned}$$

which together with (5.18) entails (via (4.16))

$$|\Gamma_n(x)| \leq 8K_2\sigma_p(2\pi)^{-1/2}e^{-1}M^{-2}n^{-1/2}. \tag{5.19}$$

On the other hand, if $|x - \xi_p| \geq \eta$, proceeding as in the derivation of (5.10), we obtain

$$|\Gamma_n(x)| \leq 8K_2\sigma_p(2\pi)^{-1/2}e^{-1}M^{-2}n^{-1/2} + \sigma_p^2\,(2\pi e)^{-1/2}\eta^{-2\rho}M^{-1}n^{-1}. \tag{5.20}$$

The proof of the theorem follows by combining the estimates (5.19) and (5.20) with Lemma 4.3 (again via the inversion $y = Mn^{1/2}(x-\xi_p)^\rho\sigma_p^{-1}$).

Remark 5.2. Regarding the variation of F at ξ_p in nonregular cases, we may consider a condition more general than (5.1), namely

$$\lim_{h\to 0} |F(\xi_p + h) - F(\xi_p)|\,|h|^{-\rho} = M > 0 \tag{5.1*}$$

for some $\rho > 0$ (not necessarily an integer).

Then, using arguments similar to those in the proof of Theorem 5.1, it can be shown (details omitted) that under (5.1*) and (5.3),

$$(5.21) \qquad Mn^{1/2}|X_{n,\nu_n} - \xi_p|^{\rho}\sigma_p^{-1} \xrightarrow{\mathcal{D}} |N(0,1)| \ .$$

This result again is a generalization of a result of Chanda (1975).

Finally, we note that under stronger assumptions, an error bound of the approximation (5.21) can also be obtained. To avoid repetition, we only state an analog of Theorem 5.2. Assume that (4.1) holds and

$$(5.14)^* \qquad |F(\xi_p + h) - F(\xi_p)||h|^{-\rho} = M + O(|h|^{\rho}) \quad \text{as} \quad h \to 0$$

for some $M > 0$ and $\rho > 0$. Then

$$(5.22) \qquad \sup_x |P\{Mn^{1/2}|X_{n,\nu_n} - \xi_p|^{\rho} \le x\sigma_p\} - \Phi^*(x)| = O(n^{-1/2}) + O(\varepsilon_n) + \delta_n$$

where Φ^* denotes the distribution function of a $|N(0,1)|$ random variable.

REFERENCES

Bahadur, R. R. (1966). A note on quantiles in large samples. Ann. Math. Statist. **37**, 577-580.

Chanda, K. C. (1975). Some comments on the asymptotic probability laws of sample quantiles. Calcutta Stat. Assoc. Bulletin, 123-126.

David, H. A. (1981). Order Statistics. Wiley, New York.

Ghosh, J. K. (1971). A new proof of Bahadur's representation of quantiles and an application. Ann. Math. Statist. **42**, 1957-1961.

Hardy, G. H. (1952). A Course of Pure Mathematics, 10th ed. Cambridge University Press, New York.

Petrov, V. V. (1975). Sums of Independent Random Variables. Springer-Verlag, Berlin.

Reiss, R. D. (1974). On the accuracy of the normal approximation for quantiles. Ann. Prob. 2, 741-744.

Serfling, R. J. (1980). Approximation Theorems of Mathematical Statistics. Wiley, New York.

Smirnov, N. V. (1952). Limit distributions for the terms of a variational series. American Math Society Translation No. 67.

Van Beeck, P. (1972). An application of Fourier methods to the problem of sharpening the Berry-Esséen inequality. Z. Wahrscheinlichkeitstheorie und Verw. Gebiete **23**, 187-196.

Reprinted from
IMS Lecture Notes and Monograph Series, Volume 8
(1987), pages 447-475, by permission of the publisher,

ASYMPTOTIC EXPANSIONS FOR SUMS OF NONIDENTICALLY DISTRIBUTED BERNOULLI RANDOM VARIABLES

PAUL DEHEUVELS, MADAN L. PURI* and STEFAN S. RALESCU

Université Paris VI, Paris, France
Indiana University
Queens College of the City University of New York

This paper concerns an asymptotic expansion for the distribution of the sum of independent zero–one random variables in case where this sum has variance $\sigma_n^2 \to \infty$. The expansion presented is given to the order $O(\sigma_n^{-2})$. An application to the study of the exact rate of convergence in the central limit theorem for intermediate order statistics is included.

1. Introduction and Results

Let $\{X_{nj},\ 1 \leqslant j \leqslant n\}$ be a triangular array of random variables such that

(A) for all $n \geqslant 1$, $X_{n1}, ..., X_{nn}$ are independent;

(B) for all $1 \leqslant j \leqslant n$, X_{nj} follows a Bernoulli $B(p_{nj})$ distribution, i.e.,

$$P(X_{nj}=1)=1-P(X_{nj}=0)=p_{nj} \in (0, 1). \tag{1.1}$$

Let $S_n = \sum_{j=1}^n X_{nj}$, $\mu_n = \sum_{j=1}^n p_{nj}$ and $\sigma_n^2 = \sum_{j=1}^n p_{nj}(1-p_{nj})$, $n \geqslant 1$. In this paper, we will study the rate of approximation of $F_n(x) = P(\sigma_n^{-1}(S_n - \mu_n) \leqslant x)$ by the distribution function $\phi(y)$ of a standard normal

Received August 1987; revised February 8, 1988.

AMS 1980 subject classifications: Primary 62E20, 60F05; Secondary 62G30.

Key words and phrases: normal approximation, Poisson approximation, binomial distributions, order statistics, central limit theorem, weak laws.

* Research partially supported by the Office of Naval Research Contract N00014–85–K–0648.

$N(0, 1)$ random variable, under suitable conditions on $p_n = (p_{n1}, ..., p_{nn})$ and appropriate choices of $y = y(x)$.

Before stating our results, it is worthwhile to specify the range of normal approximation for $\sigma_n^{-1}(S_n - \mu_n)$. This is achieved in the following two theorems whose proofs are simple consequences of general results on the central limit theorem (see, e.g., Zolotarev [26]).

THEOREM 1.1. *The asymptotic normality of* $\sigma_n^{-1}(S_n - \mu_n)$, *i.e.*, $\sup_x |F_n(x) - \phi(x)| \to 0$ *as* $n \to \infty$, *holds if and only if*

$$\sigma_n^2 = \sum_{j=1}^{n} p_{nj}(1 - p_{nj}) \to \infty \qquad \text{as} \quad n \to \infty. \tag{1.2}$$

THEOREM 1.2. *There exists a universal constant A such that, for all* $n \geqslant 1$,

$$\sup_x (1 + |x|^3)\, |F_n(x) - \phi(x)| \leqslant A\sigma_n^{-1}. \tag{1.3}$$

We will give details concerning these theorems in Section 2. By these theorems, in order to achieve asymptotic normality for $\sigma_n^{-1}(S_n - \mu_n)$, we need to assume that $\sigma_n^2 \to \infty$ as $n \to \infty$.

Let $\varphi(x) = \phi'(x) = (2\pi)^{-1/2} \exp(-x^2/2)$ denote the density of the $N(0, 1)$ distribution. It is natural to incorporate in (1.3) the first term of the Edgeworth expansion (see, e.g., Petrov [19, Chap. 6], Hall [12, Theorem 2.2]) of $\sigma_n^{-1}(S_n - \mu_n)$. Noting that

$$\begin{aligned} &\frac{1}{6}\sigma_n^{-3}(1 - x^2)\varphi(x) \sum_{j=1}^{n} E((X_{nj} - E(X_{nj}))^3 \\ &\qquad = \frac{1 - x^2}{6\sigma_n}\varphi(x)\left\{1 - 2\sigma_n^{-2} \sum_{j=1}^{n} p_{nj}^2(1 - p_{nj})\right\}, \end{aligned} \tag{1.4}$$

$$\sigma_n^{-2} \sum_{j=1}^{n} p_{nj}^2(1 - p_{nj}) \leqslant \max_{1 \leqslant j \leqslant n} p_{nj} \leqslant 1, \tag{1.5}$$

and

$$\sup_x (1 + |x|^3)\, |1 - x^2|\, \varphi(x)/6 \leqslant 0.70, \tag{1.6}$$

we obtain an equivalent version of (1.3) given by

$$\begin{aligned} &\sup_x (1 + |x|^3) \left| F_n(x) - \phi(x) - \frac{1 - x^2}{6\sigma_n}\varphi(x) \right. \\ &\qquad \left. \times \left\{1 - 2\sigma_n^{-2} \sum_{j=1}^{n} p_{nj}^2(1 - p_{nj})\right\} \right| \leqslant B\sigma_n^{-1}, \end{aligned} \tag{1.7}$$

where B is an absolute constant (here we can choose $B = A + 0.7 \leqslant A + 1$).

In the special case where $\max_{1 \leqslant j \leqslant n} p_{nj} \to 0$ as $n \to \infty$, (1.5) ensures that the right-hand side of (1.4) is closely approximated by the simple expression

$$\frac{1-x^2}{6\sigma_n}\varphi(x). \tag{1.8}$$

It is clear that there is no hope in obtaining a better rate than $O(\sigma_n^{-1})$ as $\sigma_n \to \infty$ for (1.7). This is due to the lattice structure of the distribution of S_n and follows from the remark that $F_n(x-\sigma_n^{-1}\mu_n)$ is constant in all intervals of the form $(\sigma_n^{-1}(k-1), \sigma_n^{-1}k)$ with k integer. Hence the variation of $\phi(x-\sigma_n^{-1}\mu_n)$ in such an interval may exceed $\{\varphi(0)-\varepsilon\}\,\sigma_n^{-1}$ for a fixed $\varepsilon>0$ and suitable choices of k as $\sigma_n \to \infty$. This implies that in (1.3) and (1.7) we must have

$$\min(A, B) \geqslant \varphi(0) = 1/\sqrt{2\pi} \geqslant 0.40. \tag{1.9}$$

In order to achieve a sharper rate in (1.7), it is necessary to reduce the "rounding errors" by replacing $F_n(x)$ by a smoothed version. For that, we introduce the lattice points $x_{nk} = \sigma_n^{-1}(k-\mu_n+\frac{1}{2})$, $k=0, \pm 1, \pm 2, ...; n \geqslant 1$. It will be convenient to denote by $\tilde{F}_n(x)$ the distribution function characterized by conditions (i) and (ii) below and providing the desired smoothed version of F_n:

(i) $\tilde{F}_n(x_{nk}) = F_n(x_{nk})$, $k=0, \pm 1, \pm 2, ...$;

(ii) $\tilde{F}_n(x)$ is continuous in x, and linear on all intervals $(x_{n(k-1)}, x_{nk})$, $k=0, \pm 1, \pm 2, ...$.

Under these notations, our main result may be stated as follows.

THEOREM 1.3. *There exists a universal constant C such that, for all* $n \geqslant 1$,

$$\sup_x \left| \tilde{F}_n(x) - \phi(x) - \frac{1-x^2}{6\sigma_n}\varphi(x)\left\{1-2\sigma_n^{-2}\sum_{j=1}^{n} p_{nj}^2(1-p_{nj})\right\}\right| \leqslant C\sigma_n^{-2}. \tag{1.10}$$

EXAMPLE 1. Assume that $X_{n1}, ..., X_{nn}$ are identically distributed. Let

$$\pi_n = p_{n1} = \cdots = p_{nn} = n^{-1}\mu_n \qquad \text{and} \qquad \sigma_n^2 = n\pi_n(1-\pi_n). \tag{C}$$

Under (C), S_n follows a binomial $B(n, \pi_n)$ distribution, and (1.10) can be restated as (see, e.g., Feller [8] and Uspenky [23] for related results):

$$\sup_x \left| \tilde{F}_n(x) - \phi(x) - \frac{1-x^2}{6\sqrt{n\pi_n(1-\pi_n)}}\{1-2\pi_n\}\varphi(x)\right| \leqslant \frac{C}{n\pi_n(1-\pi_n)}. \tag{1.11}$$

EXAMPLE 2. Assume that $p_{nj}=1/j$, so that $\mu_n=\sum_{j=1}^n j^{-1}\sim\sigma_n^2=\sum_{j=1}^n j^{-1}(1-j^{-1})\sim\log n$ as $n\to\infty$. In this case we see that there exists a constant θ such that, for all $n\geqslant 2$,

$$\sup_x \left| \tilde{F}_n(x)-\phi(x)-\frac{1-x^2}{6\sqrt{\log n}}\varphi(x)\right| \leqslant \theta/\log n. \tag{1.12}$$

The proof of Theorem 1.3 is given in Section 4.

Remark 1. There is a huge literature on the approximation of binomial distributions by normal distributions (see, e.g., Johnson and Kotz [13] and Molenaar [16]). This case corresponds to (C) and will be discussed further in Section 5, together with an application to the theory of intermediate order statistics. On the other hand, very few results are available in the non-i.i.d. case. This motivates our work.

Remark 2. In the case where $\max_{1\leqslant j\leqslant n} p_{nj}\to 0$ as $n\to\infty$, it is natural to approximate the distribution of S_n by a Poisson distribution with mean μ_n. Since the approximation of a Poisson distribution by a normal distribution is well known, we may use the corresponding results to obtain directly an evaluation such as in (1.10). We will develop this approach in Section 3 and show that it gives only suboptimal results with respect to (1.10) such as the following simple corollary of Theorem 1.3.

COROLLARY 1.3. *There exists a universal constant D such that, for all $n\geqslant 1$ and $\sigma_n\geqslant 1$, we have*

$$\sup_x \left| \tilde{F}_n(x)-\phi(x)-\frac{1-x^2}{6\sigma_n}\varphi(x)\right| \leqslant D\sigma_n^{-2}\left\{1+\sum_{j=1}^n p_{nj}^2\right\}. \tag{1.13}$$

An interesting side application of these results is the characterization of the range of values of p_{nj}, $j=1, ..., n$ for which the normal distribution gives a better approximant than a Poisson distribution for S_n. This motivates Section 3 in the sequel.

2. EXPANSIONS BY GENERAL METHODS

In this section, we prove Theorems 1.1 and 1.2. First we use the following result of Nagaev [17] (see, e.g., Hall [12, p. 7]).

LEMMA 2.1. *There exists a universal constant a such that, for any* *independent random variables $Z_1, Z_2, \ldots$ such that $E(Z_i)=0$, $i=1, 2, \ldots$, if $s_n=\sum_{j=1}^n E(Z_j^2)<\infty$ and if $T_n=\sum_{j=1}^n Z_j$, then*

$$\sup_x (1+|x|^3)\,|P(T_n \leqslant s_n x)-\phi(x)| \leqslant a s_n^{-3} \sum_{j=1}^n E(|Z_j|^3). \tag{2.1}$$

An application of (2.1) proves Theorem 1.2. Note that in the i.i.d. case (see Michel [15]) we may take $a=30.54$.

Since Theorem 1.2 implies that asymptotic normality of $\sigma_n^{-1}(S_n-\mu_n)$ holds if $\sigma_n \to \infty$, all we need for Theorem 1.1 is to prove the converse.

For this, let us assume that $\{n_l, l \geqslant 1\}$ is an increasing sequence of positive integers such that $\lim_{l\to\infty} \sigma_{n_l}^2=\sigma^2<\infty$.

Let $S_n'=\sum_{j=1}^n X_{nj} 1_{\{p_{nj}\leqslant (1/2)\}}$ and $S_n''=\sum_{j=1}^n (1-X_{nj})\, 1_{\{p_{nj}>(1/2)\}}$. Since

$$E(S_{n_l}')+E(S_{n_l}'') \leqslant 2(\operatorname{Var}(S_{n_l}')+\operatorname{Var}(S_{n_l}''))=\sigma_{n_l}^2 \to \sigma^2 \qquad \text{as} \quad l\to\infty,$$

it follows that

$$\max\{p_{nj} 1_{\{p_{nj}\leqslant(1/2)\}}\} \to 0 \qquad \text{and} \qquad \max\{(1-p_{nj})\, 1_{\{p_{nj}>(1/2)\}}\} \to 0$$
$$\text{as} \quad n=n_l \text{ and } l\to\infty.$$

LEMMA 2.2. *Let $\xi_1, \xi_2, \ldots$ be independent random variables such that $P(\xi_i=1)=1-P(\xi_i=0)=r_i$. Let $\zeta_n=\xi_1+\cdots+\xi_n$ and denote by $L(\zeta_n)$ the distribution of ζ_n, and by $\Pi(\lambda)$ the distribution of a Poisson random variable with expectation $\lambda>0$. Then*

$$d_v\left(L(\zeta_n), \Pi\left(\sum_{i=1}^n r_i\right)\right) \leqslant \left\{\sum_{i=1}^n r_i^2\right\} \Big/ \left\{\sum_{i=1}^n r_i\right\} \leqslant \max_{1\leqslant i\leqslant n} r_i, \tag{2.2}$$

where $d_v(\mu, \nu)=\sup_A |\mu(A)-\nu(A)|$ denotes the distance in variation between the probability measures μ and ν.

Proof. (2.2) is a direct corollary of Theorem 1 in Barbour and Hall [2].

By Lemma 2.2, we see that $d_v(L(S_n'), \Pi(E(S_n'))) \to 0$ and $d_v(L(S_n''), \Pi(E(S_n''))) \to 0$ as $n=n_l$ and $l\to\infty$. Since S_n' and S_n'' are independent for all $n\geqslant 1$, this suffices to prove that the limiting distribution of $\sigma_n^{-1} S_n$ is the same as that of $\sigma_n^{-1}(S_n'-E(S_n')-S_n''+E(S_n''))$ whenever it exists and it cannot be normal. The proof of Theorem 1.1 is now complete.

Remark 3. It is possible to obtain an evaluation of the constant A in Theorem 1.2. (In the i.i.d. case based on Michel [15] it is easy to see that $A\leqslant 31$). We refer to Hall [12] for further refinements.

3. Expansions by Poisson Approximations

In this section, we provide a direct proof of Corollary 1.3. Our main tool will be the following result due to Cheng [3] (see, e.g., Johnson and Kotz [13, pp. 98–101]).

Lemma 3.1. *Let* $\lambda > 0$ *and* $y_k = \lambda^{-1/2}(k - \lambda + 1/2)$. *Then, for all* $k = 0, 1, 2, \ldots$, *we have*

$$\left|\sum_{r=0}^{k} \frac{\lambda^r}{r!} e^{-\lambda} - \phi(y_k) - \frac{1 - y_k^2}{6\sqrt{\lambda}} \varphi(y_k)\right| \leqslant 0.076\lambda^{-1} + 0.043\lambda^{-3/2} + 0.13\lambda^{-2}.$$

From Lemma 3.1, the following result is straightforward.

Lemma 3.2. *Let* Z *follow a Poisson distribution with expectation* $\lambda > 0$. *There exists a universal constant* $\alpha < 0.61$ *such that, for all* $k = 0, \pm 1, \pm 2, \ldots$,

$$\left|P(Z \leqslant k) - \phi(y_k) - \frac{1 - y_k^2}{6\sqrt{\lambda}} \varphi(y_k)\right| \leqslant \alpha\lambda^{-1}, \tag{3.1}$$

where $y_k = \lambda^{-1/2}(k - \lambda + 1/2)$.

Proof. For $\lambda \leqslant \gamma$, we see that the LHS of the inequality to be proved is less than $2 + \lambda^{-1/2}(1/6\sqrt{2\pi}) \leqslant \{2\gamma + \gamma^{1/2}(1/6\sqrt{2\pi})\}\,\lambda^{-1}$. For $\lambda \geqslant \gamma$, we use the bound given in Lemma 3.1 to obtain $\{0.076 + 0.043\gamma^{-1/2} + 0.13\gamma^{-1}\}$ λ^{-1}. Take $\gamma = 0.287$ to obtain $\alpha \leqslant 0.61$.

Lemma 3.3. *We have*

$$\sup_x |\phi(\gamma x) - \phi(x)| \sim \frac{|\gamma - 1|}{\sqrt{2\pi e}} \qquad \textit{and}$$

$$\sup_x |\gamma\varphi(\gamma x) - \varphi(x)| \sim \frac{|\gamma - 1|}{\sqrt{2\pi}} \qquad \textit{as} \quad \gamma \to 1.$$

Proof. Straightforward.

Now let y_k and x_k be respectively defined for $k = 0, \pm 1, \pm 2, \ldots$, by

$$y_k = \mu_n^{-1/2}(k - \mu_n + \tfrac{1}{2}) \qquad \text{and} \qquad x_k = \sigma_n^{-1}(k - \mu_n + \tfrac{1}{2}),$$

where $\mu_n = \sum_{j=1}^n p_{nj}$ and $\sigma_n^2 = \sum_{j=1}^n p_{nj}(1-p_{nj}) = \mu_n - \sum_{j=1}^n p_{nj}^2 = \mu_n(1-\rho_n)$, where μ_n and σ_n^2 are as in Section 1 and where

$$0 < \rho_n = \left\{\sum_{j=1}^n p_{nj}^2\right\} \bigg/ \left\{\sum_{j=1}^n p_{nj}\right\} < 1.$$

LEMMA 3.4. *There exists a universal constant M such that, for all $k = 0, 1, 2, \ldots$, $\mu_n > 0$ and $\sigma_n^2 = \mu_n(1-\rho_n)$, $0 \leqslant \rho_n < 1$, we have*

$$\sup_k \left| \left\{\phi(y_k) + \frac{1-y_k^2}{6\sqrt{\mu_n}}\varphi(y_k)\right\} - \left\{\phi(x_k) + \frac{1-x_k^2}{6\sigma_n}\varphi(x_k)\right\} \right.$$
$$\leqslant M\rho_n(1+\sigma_n^{-1}). \tag{3.2}$$

Observe that $y_k = x_k(1-\rho_n)^{1/2}$. Let $\gamma = (1-\rho_n)^{1/2} \in (0, 1)$ and note that the LHS of (3.2) is less than or equal to

$$\begin{aligned} a &= \sup_x |\phi(\gamma x) - \phi(x)| + \frac{1}{6\sigma_n} \sup_x |\gamma\varphi(\gamma x) - \varphi(x)| \\ &\quad + \frac{1}{6\sigma_n} \sup_x x^2 |\gamma^3 \varphi(\gamma x) - \varphi(x)| \\ &= A_1(\gamma) + A_2(\gamma) + A_3(\gamma), \qquad \text{say.} \end{aligned}$$

Recalling that $\sigma_n \to \infty$ as $n \to \infty$, we have according to Lemma 3.3 that $A_1(\gamma) = 0(1-\gamma) = 0(\rho_n)$ and $A_2(\gamma) = 0(1-\gamma) = 0(\rho_n)$ as $\rho_n \to 0$ ($\Leftrightarrow \gamma \to 1^-$). Therefore, there exist universal constants M_1 and M_2 such that

$$A_1(\gamma) \leqslant M_1\rho_n \qquad \text{and} \qquad A_2(\gamma) \leqslant M_2\rho_n\sigma_n^{-1}.$$

The same arguments apply for $A_3(\gamma) \leqslant M_3\rho_n\sigma_n^{-1}$ and the conclusion follows by taking $M = M_1 + M_2 + M_3$.

Proof of Corollary 1.3. In the first place, we remark that (1.13) is equivalent to

$$\sup_k \left| P(S_n \leqslant k) - \phi(x_k) - \frac{1-x_k^2}{6\sigma_n}\varphi(x_k) \right| \leqslant c\mu_n^{-1} + d\rho_n(1+\sigma_n^{-1}),$$

for some appropriate constants c and d, and where x_k and ρ_n are as in Lemma 3.4.

We see by Lemma 2.2 that if Z has a Poisson distribution with expectation μ_n, then

$$\sup_k |P(S_n \leqslant k) - P(Z \leqslant k)| \leqslant d_v(L(S_n), L(Z))$$

$$\leqslant \rho_n = \left\{ \sum_{j=1}^n p_{nj}^2 \right\} \Big/ \left\{ \sum_{j=1}^n p_{nj} \right\}. \qquad (3.3)$$

By Lemma 3.2, we see that if Z has a Poisson distribution with expectation μ_n, then

$$\sup_k \left| P(Z \leqslant k) - \phi(y_k) - \frac{1 - y_k^2}{6\sqrt{\mu_n}} \varphi(y_k) \right| \leqslant 0.61 \mu_n^{-1}.$$

The above estimates jointly imply that, for some absolute constant D,

$$\sup_x \left| \tilde{F}_n(x) - \phi(x) - \frac{1 - x^2}{6\sigma_n} \varphi(x) \right| \leqslant D \left\{ (1 + \sigma_n^{-1}) \sum_{j=1}^n p_{nj}^2 \right\} \Big/ \left\{ \sum_{j=1}^n p_{nj} \right\}, \qquad (3.4)$$

which in turn proves Corollary 1.3.

Remark 4. Deheuvels and Pfeifer [7] have shown that the rate in (3.3) cannot be improved in the range where $\mu_n = \sum_{j=1}^n p_{nj} \to \infty$ and $\sigma_n^2 \sim \mu_n$ for in this case

$$\sup_k |P(S_n \leqslant k) - P(Z \leqslant k)| \sim \frac{1}{2\sqrt{2\pi e}} \left\{ \sum_{j=1}^n p_{nj}^2 \right\} \Big/ \left\{ \sum_{j=1}^n p_{nj} \right\}. \qquad (3.5)$$

By (1.10), we see that a Poisson approximation gives, with respect to the Kolmogorov distance, a better fit than a normal approximation if

$$\sum_{j=1}^n p_{nj}^2 = o\left(\left\{ \sum_{j=1}^n p_{nj} \right\}^{1/2} \right), \qquad (3.6)$$

while the converse holds if

$$\left\{ \sum_{j=1}^n p_{nj} \right\}^{1/2} = o\left(\sum_{j=1}^n p_{nj}^2 \right) \qquad (3.7)$$

Remark 5. As shown in Hall [12, Theorem 2.2], the appropriate leading term for $\tilde{F}_n(x) - \phi(x)$ is

$$L_n(x) = \sum_{j=1}^n E\left(\phi\left(x - \frac{X_{nj} - p_{nj}}{\sigma_n} \right) - \phi(x) \right) - \frac{1}{2} \varphi'(x).$$

By the Taylor expansion of ϕ, and using the fact that $\sigma_n^{-1}|X_{nj} - p_{nj}| \leqslant \sigma_n^{-1} \to 0$, we see that

$$L_n(x) = \frac{1 - x^2}{6\sigma_n} \varphi(x) \left\{ 1 - 2\sigma_n^{-2} \sum_{j=1}^n p_{nj}^2 (1 - p_{nj}) \right\} + O(\sigma_n^{-2}) \qquad \text{as} \quad \sigma_n \to \infty.$$

An application of Theorem 1.3 proves that

$$\sup_x \left| \tilde{F}_n(x) - \phi(x) - \frac{1-x^2}{6\sigma_n} \varphi(x) \right| = O(\sigma_n^{-2}) \quad (\text{resp. } o(\sigma_n^{-1})) \qquad \text{as} \quad \sigma_n \to \infty \tag{3.8}$$

whenever $\sum_{j=1}^n p_{nj}^2(1-p_{nj}) = O(\sigma_n)$ (resp. $o(\sigma_n^{-2})$). These conditions are sharp, the last one (i.e., $o(\sigma_n^{-2})$) being always satisfied if $\max_{1 \leqslant j \leqslant n} p_{nj} \to 0$.

4. Expansions Obtained by a Direct Approach

In this section, we prove Theorem 1.3. In the first place it will be convenient to introduce some notation.

Let $U_1, U_2, \ldots$ be independent random variables with a uniform distribution on (0, 1), and let $u(t) = 1_{[0,\infty)}(t) = 1$ or 0 according as $t \geqslant 0$ or $t < 0$. (Here I denotes the indicator function). In the sequel, we shall make use of the fact that, without loss of generality, it is possible to set

$$F_n(x) = P\left(\sum_{j=1}^n U_{nj} \leqslant x\sigma_n \right), \tag{4.1}$$

where $U_{nj} = u(p_{nj} - U_j) - p_{nj} = X_{nj} - p_{nj}$, $1 \leqslant j \leqslant n$.

Let $U_{(a,b)}(x)$ denote the distribution function of a random variable uniformly distributed on (a, b). It is straightforward that the approximant $\tilde{F}_n$ may be described analytically as the convolution $\tilde{F}_n = F_n * U_{(-h_n/2,\, h_n/2)}$, where $h_n = 1/\sigma_n$.

Our aim is to prove that, uniformly in x,

$$\tilde{F}_n(x) = \phi(x) + \frac{1-x^2}{6\sigma_n^3} \varphi(x) \sum_{j=1}^n E(U_{nj}^3) + O(\sigma_n^{-2}) \qquad \text{as} \quad n \to \infty. \tag{4.2}$$

To this end, let $\Gamma_n(x) = \phi(x) + ((1-x^2)/6\sigma_n^3))\, \varphi(x) \xi_n$, where $\xi_n = \sum_{j=1}^n E(U_{nj}^3)$, and let $\tilde{\Gamma}_n(x) = \Gamma_n * U_{(-h_n/2,\, h_n/2)}$.

Writing $\tilde{F}_n - \Gamma_n$ as the sum of $\tilde{F}_n - \tilde{\Gamma}_n$ and $\tilde{\Gamma}_n - \Gamma_n$, we shall treat the two terms separately. First, we prove:

Lemma 4.1. *There exists a universal constant E such that, for all $n \geqslant 1$,*

$$\sup_x |\tilde{\Gamma}_n(x) - \Gamma_n(x)| \leqslant E\sigma_n^{-2}\left(1 + \frac{1}{\sigma_n}\right).$$

Proof. First note that

$$|\sigma_n^{-3}\xi_n| = \sigma_n^{-1}\left|\sum_{j=1}^{n} p_{nj}(1-p_{nj})(1-2p_{nj})\right| \Big/ \left\{\sum_{j=1}^{n} p_{nj}(1-p_{nj})\right\} \leqslant \sigma_n^{-1}.$$

Next, observe that there exists a constant ρ such that, for all $n \geqslant 1$,

$$\sup_x |\Gamma_n''(x)| \leqslant \rho(1+\sigma_n^{-1}).$$

By Taylor's expansion, we have for each x

$$\tilde{\Gamma}_n(x) = \Gamma_n(x) + \frac{1}{2h_n}\int_{-h_n/2}^{h_n/2} t^2 \Gamma_n''(\theta_t)\,dt,$$

where θ_t denotes an intermediate point between x and $x-t$. Accordingly, we obtain the estimate

$$\sup_x |\tilde{\Gamma}_n(x) - \Gamma_n(x)| \leqslant \frac{\rho}{24} h_n^2(1+\sigma_n^{-1}) = \frac{\rho}{24}\sigma_n^{-2}(1+\sigma_n^{-1}),$$

as desired.

For (4.2), it remains only to prove that, uniformly in x,

$$\tilde{F}_n(x) - \tilde{\Gamma}_n(x) = O(\sigma_n^{-2}) \qquad \text{as} \quad n \to \infty. \tag{4.3}$$

The starting point in proving (4.3) is Esseen's smoothing lemma (Feller [9, p. 512]) according to which, if c denotes a positive constant, we have

$$\pi|\tilde{F}_n(x) - \tilde{\Gamma}_n(x)| \leqslant \int_{-c\sigma_n}^{c\sigma_n} \left|\frac{f_n(t/\sigma_n) - \gamma_n(t)}{t}\right| |\omega_n(t)|\,dt + \frac{24}{c\sigma_n}\sup_x |\Gamma_n'(x)| = I_1 + I_2,$$

where f_n is the characteristic function (ch.f.) of $\sum_{j=1}^{n} U_{nj}$, ω_n is the ch.f. of $U_{(-h_n/2,\, h_n/2)}$, and $\gamma_n(t) = e^{-t^2/2}\{1 + \xi_n/6\sigma_n^3(it)^3\}$ is the Fourier–Stieltjes transform of $\Gamma_n(x)$.

LEMMA 4.2. *There exists a universal constant K such that, for all $n \geqslant 1$,*

$$\sup_x |\Gamma_n'(x)| \leqslant K\left(1 + \frac{1}{\sigma_n}\right).$$

Proof. As in Lemma 4.1.

By (4.5), if we choose $c=\sigma_n$, we obtain

$$I_2 \leqslant \frac{24K}{c}\sigma_n^{-1}\left(1+\frac{1}{\sigma_n}\right) \leqslant 24K\sigma_n^{-2}\left(1+\frac{1}{\sigma_n}\right). \tag{4.6}$$

It follows that the proof of (4.3) will be completed if we show that, for any fixed $c>0$, we have

$$I_1 \leqslant I_{11}+I_{12}=O(\sigma_n^{-2}), \tag{4.7}$$

where

$$I_{11}=\int_{|t|\leqslant\alpha_n}\left|\frac{f_n(t/\sigma_n)-\gamma_n(t)}{t}\right| dt, \tag{4.8}$$

$$I_{12}=\int_{\alpha_n\leqslant|t|\leqslant\beta_n}\left|\frac{f_n(t/\sigma_n)-\gamma_n(t)}{t}\right| |\omega_n(t)|\, dt, \tag{4.9}$$

and where $\alpha_n=\delta\sigma_n$, $\beta_n=c\sigma_n$, and $\delta\in(0, c)$ is a constant to be specified later on.

LEMMA 4.3. *Let I_{11} be as in* (4.8). *There exists a constant $\delta\in(0, 1)$ and a constant L such that, for all $n\geqslant 1$, we have*

$$I_{11}\leqslant L\sigma_n^{-2}.$$

Proof. It is convenient to work with the function

$$g_n(t)=\log f_n(t)+\frac{1}{2}\sigma_n^2t^2$$

$$=\sum_{j=1}^n\left\{\log f_{nj}(t)+\frac{1}{2}\sigma_{nj}^2t^2\right\},$$

where $f_{nj}(t)=E(\exp(itU_{nj}))$ and $\sigma_{nj}^2=E(U_{nj}^2)$. We will make use of the following lemma.

LEMMA 4.4. *Let $0<q<1$ be fixed, and let $0<\delta\leqslant qe^{-1}$. Then, there exists a constant $M=M(q)$ such that, for all $|t|\leqslant\delta$ and $n\geqslant 1$, we have*

$$|g_n(t)-\tfrac{1}{6}\xi_n(it)^3|\leqslant M\sigma_n^2t^4. \tag{4.10}$$

Proof. Let $V_n=\sum_{j=1}^n U_{nj}$, and denote by κ_{nl}, $l=1, 2, \ldots$, the lth comulant of V_n. Noting that $\kappa_{n1}=E(V_n)=0$, $\kappa_{n2}=E(V_n^2)=\sigma_n^2$, and that $\kappa_{n3}=E(V_n^3)=\xi_n$, we get easily the formal expansion

$$g_n(t)-\frac{1}{6}\xi_n(it)^3=\sum_{l=4}^{\infty}\frac{\kappa_{nl}}{l!}(it)^l.$$

We will show that the right-hand side of the expression above is a convergent series for $|t|$ small enough. This follows from the inequalities (cf. Cramér [6, p. 27]):

$$\begin{aligned}|k_{nl}| &\leqslant l^l\sum_{j=1}^{n}E(|U_{nj}|^l)\\ &\leqslant l^l\sum_{j=1}^{n}E(U_{nj}^2)=l^l\sigma_n^2 \qquad \text{for} \quad l\geqslant 2.\end{aligned}$$

It follows that

$$\left|g_n(t)-\frac{1}{6}\xi_n(it)^3\right|\leqslant\sigma_n^2\sum_{l=4}^{\infty}\frac{l^l}{l!}|t|^l. \tag{4.11}$$

Now, Stirling's formula and $e\delta\leqslant q<1$ entail for $|t|<\delta$:

$$\begin{aligned}\sum_{l=4}^{\infty}\frac{l^l}{l!}|t|^l &\leqslant\frac{e^4|t|^4}{\sqrt{2\pi}}\sum_{i=0}^{\infty}\frac{(e\delta)^i}{\sqrt{i+4}}\\ &\leqslant\frac{e^4|t|^4}{\sqrt{2\pi}}\sum_{i=0}^{\infty}q^i=\frac{e^4|t|^4}{\sqrt{2\pi}(1-q)}.\end{aligned} \tag{4.12}$$

By (4.11) and (4.12) we get (4.10) if we choose $M=e^4(2\pi)^{-1/2}(1-q)^{-1}$.

Proof of Lemma 4.3 (continued). In the sequel, we assume that $0<q<1$ is fixed. We have

$$I_{11}=\int_{|t|\leqslant\alpha_n}\left|e^{g_n(t/\sigma_n)}-1-\frac{\xi_n}{6\sigma_n^3}(it)^3\right|e^{-t^2/2}|t|^{-1}\,dt. \tag{4.13}$$

A glance at I_{11} in this form now suggests estimating the integrand with the help of the inequality

$$|e^u-1-v|\leqslant(|u-v|+\tfrac{1}{2}|v|^2)\,e^{\max(|u|,|v|)}, \tag{4.14}$$

valid for any complex u and v. Suppose now that $0<\delta\leqslant qe^{-1}$.

Let $u=g_n(t/\sigma_n)$ and $v=\frac{1}{6}\xi_n\sigma_n^{-3}(it)^3$. By (4.10), we have for $|t|\leqslant\delta\sigma_n=\alpha_n$,

$$|u-v|\leqslant M\sigma_n^{-2}t^4. \tag{4.15}$$

Using (4.10) again, we obtain for $|t| \leqslant \delta$

$$|g_n(t)| \leqslant \tfrac{1}{6}\xi_n|t|^3 + M\sigma_n^2 t^4 \leqslant \sigma_n^2|t|^3\,(\tfrac{1}{6} + M\delta)$$
$$\leqslant \sigma_n^2 t^2 \delta(\tfrac{1}{6} + M\delta),$$

where in the last inequality we make use of the fact that $t^4 \leqslant \delta|t|^3$ and that

$$|\xi_n| \leqslant \sum_{j=1}^{n} p_{nj}(1-p_{nj})|1-2p_{nj}| \leqslant \sum_{j=1}^{n} p_{nj}(1-p_{nj}) = \sigma_n^2.$$

Let us now choose $\delta < qe^{-1}$ such that $\delta(\frac{1}{6} + M\delta) < \frac{1}{4}$. By all this, we obtain

$$|u| \leqslant \tfrac{1}{4}t^2, \qquad |t| \leqslant \delta\sigma_n = \alpha_n, \tag{4.16}$$

$$|v| = \frac{1}{6}\,|\xi_n|\,\sigma_n^{-3}|t|^3 \leqslant \frac{1}{6}\,\sigma_n^{-1}|t|^3$$
$$\leqslant \frac{\delta}{6}\,t^2 < \frac{1}{4}\,t^2, \qquad |t| \leqslant \delta\sigma_n = \alpha_n. \tag{4.17}$$

Combining (4.14)–(4.17) we find via (4.13) the estimate

$$I_{11} \leqslant \int_{|t| \leqslant \alpha_n} (M\sigma_n^{-2}t^4 + \tfrac{1}{72}\sigma_n^{-2}t^6)\,e^{-t^2/4}|t|^{-1}\,dt$$
$$\leqslant \sigma_n^{-2}\int_{-\infty}^{\infty} |t|^3\,(M + \tfrac{1}{72}t^2)\,e^{-t^2/4}\,dt = L\sigma_n^{-2}.$$

The proof of Lemma 4.3 is now completed.

In order to show that $I_2 = O(\sigma_n^{-2})$, a more delicate analysis starts. Let u and v be defined as above. We have, by (4.8), (4.9) and (4.13),

$$I_{12} = \int_{\alpha_n \leqslant |t| \leqslant \beta_n} |e^u - 1 - v|\,e^{-t^2/2}|t|^{-1}\,|\omega_n(t)|\,dt$$
$$\leqslant A_n + B_n + C_n, \tag{4.18}$$

where

$$A_n = \int_{\alpha_n \leqslant |t| \leqslant \beta_n} |t|^{-1}\,e^{-t^2/2}\,dt,$$

$$B_n = \int_{\alpha_n \leqslant |t| \leqslant \beta_n} |v|\,|t|^{-1}\,e^{-t^2/2}\,dt,$$

and

$$C_n = \int_{\alpha_n \leqslant |t| \leqslant \beta_n} |e^u \omega_n(t)|\ |t|^{-1} e^{-t^2/2}\, dt.$$

We first start with the easy terms A_n and B_n. We have

$$A_n \leqslant 2\int_{\alpha_n}^{\infty} t^{-1} e^{-t^2/2}\, dt \leqslant \frac{2}{\alpha_n}\int_{\alpha_n}^{\infty} e^{-t^2/2}\, dt$$

$$\leqslant 2\alpha_n^{-2} \exp\left(-\frac{1}{2}\alpha_n^2\right) \leqslant 2\delta^{-2}\sigma_n^{-2}, \tag{4.19}$$

and by (4.17)

$$B_n \leqslant \tfrac{1}{3}\sigma_n^{-1}\int_{\alpha_n}^{\infty} t^2 e^{-t^2/2}\, dt \leqslant R\sigma_n^{-1}\alpha_n^{-1} \leqslant R\delta^{-1}\sigma_n^{-2}, \tag{4.20}$$

where R is an appropriate constant.

We now focus our attention on C_n. In the sequel, we shall limit ourselves to $C_n^+ = \int_{\alpha_n}^{\beta_n} |e^u \omega_n(t)|\ t^{-1} e^{-t^2/2}\, dt$, since the proofs are identical for $C_n^- = C_n - C_n^+$. First, observe that

$$C_n^+ = \int_{\alpha_n}^{\beta_n} |f_n(t/\sigma_n)\, \omega_n(t)|\ t^{-1}\, dt.$$

Now, since $\omega_n(t) = 2\sigma_n t^{-1} \sin(t/2\sigma_n)$, it can be seen that $C_n^+ = \int_{\delta}^{c} |f_n(y) \sin y/2|\ y^{-2}\, dy$. Note that $|f_n(y)| = \prod_{j=1}^{n} |1 + p_{nj}(e^{iy} - 1)|$ and $|\sin y/2|$ are both periodic functions with period 2π. Let $N = 1 + [c/2\pi]$. We evidently have

$$C_n^+ \leqslant \delta^{-2}\int_0^{2\pi} \left| f_n(y) \sin\frac{y}{2}\right| dy$$

$$+ \sum_{l=1}^{N-1} (2l\pi)^{-2} \int_0^{2\pi} \left| f_n(y) \sin\frac{y}{2}\right| dy \tag{4.21}$$

$$\leqslant \left\{\delta^{-2} + \frac{1}{4\pi}\sum_{l=1}^{\infty} l^{-2}\right\} \int_0^{2\pi} \left| f_n(y) \sin\frac{y}{2}\right| dy. \tag{4.22}$$

Observe that the bound in (4.22) does not depend upon c. It remains, therefore, to obtain an upper bound for

$$E_n = \int_0^{2\pi} \left| f_n(y) \sin\frac{y}{2}\right| dy. \tag{4.23}$$

Let us now use Lemma 4.4, which ensures the existence of constants $\delta > 0$ and M such that, for all $|y| \leqslant \delta$ and $n \geqslant 1$, we have

$$\begin{aligned} f_n(y) &= \exp(g_n(y) - \tfrac{1}{2}\sigma_n^2 y^2) \\ &= \exp(-\tfrac{1}{2}\sigma_n^2 y^2 + \tfrac{1}{6}\xi_n (iy)^3 + R_n(y)), \end{aligned} \tag{4.24}$$

where $|R_n(y)| \leqslant M\sigma_n^2 y^4$.

It follows evidently from (4.24) that, for all $|y| \leqslant \delta$ and $n \geqslant 1$, we have

$$|f_n(y)| \leqslant \exp(-\tfrac{1}{2}\sigma_n^2 y^2 + M\sigma_n^2 y^4). \tag{4.25}$$

Let us now choose a constant $0 < d \leqslant \min(\delta, \pi)$ such that $Md^2 \leqslant \frac{1}{4}$. It follows from (4.25) that, for all $|y| \leqslant d$ and $n \geqslant 1$, we have

$$|f_n(y)| \leqslant \exp(-\tfrac{1}{4}\sigma_n^2 y^2). \tag{4.26}$$

We need now an upper bound for $|f_n(y)|$ when $d \leqslant y \leqslant 2\pi - d$. For this, we employ the explicit form of $f_n(y)$ to show that

$$\begin{aligned} |f_n(y)| &= \prod_{j=1}^{n} \{1 - 2p_{nj}(1 - p_{nj})(1 - \cos y)\} \\ &\leqslant \exp(-2\sigma_n^2(1 - \cos d)). \end{aligned} \tag{4.27}$$

By (4.23), (4.26), and (4.27), we have

$$\begin{aligned} E_n \leqslant\ & 2\int_0^d \exp\left(-\frac{1}{4}\sigma_n^2 y^2\right)\left(\sin\frac{y}{2}\right) dy \\ & + \exp(-2\sigma_n^2(1 - \cos d)) \int_0^{2\pi} \sin\frac{y}{2}\, dy. \end{aligned} \tag{4.28}$$

Since d is fixed, it is clear that the second term in (4.28) is less than $\Delta\sigma_n^{-2}$ for some universal constant Δ. For the first term, we change variables and obtain

$$\begin{aligned} &\int_0^d \exp\left(-\frac{1}{4}\sigma_n^2 y^2\right)\left(\sin\frac{y}{2}\right) dy \\ &\quad = \sigma_n^{-1} \int_0^{d\sigma_n} e^{-v^2/4} \sin\frac{v}{2\sigma_n}\, dv \\ &\quad \leqslant \frac{1}{2}\sigma_n^{-2} \int_0^{\infty} v e^{-v^2/4}\, dv. \end{aligned} \tag{4.29}$$

By (4.21)–(4.29), we have just proved:

LEMMA 4.5. *Let I_{12} be as in* (4.9). *Then, there exists an absolute constant Q such that, for all $c>0$ and $n>1$, we have*

$$I_{12} \leqslant Q\sigma_n^{-2}.$$

Proof of Theorem 1.3. It follows from (4.4), (4.6), and Lemmas 4.3 and 4.5.

5. THE I.I.D. CASE: APPLICATIONS TO THE THEORY OF ORDER STATISTICS

In this section, we assume that the conditions (A–B–C) in Section 1 hold, namely that $\pi_n = p_{n1} = \cdots = p_{nn} = n^{-1}\mu_n$, and $\sigma_n^2 = n\pi_n(1-\pi_n)$. The aim of this section is to present an application of Theorem 1.3 to the theory of intermediate order statistics. This field has been the object of intensive investigation recently (see, e.g., Reiss [20, 21] where related results may be found). It is noteworthy that our methods can be extended without difficulty to cover the case of order statistics from a sequence of independent but not identically distributed random variables. We limit ourselves in the sequel to the present case for the sake of concision.

We first mention a result due to Molenaar [16] and Peizer and Pratt [18], which we state in the following lemma.

LEMMA 5.1. *Let $x_k = \sigma_n^{-1}(k-\mu_n+\frac{1}{2})$, $k=0, 1, 2, ...,$ and define for an arbitrary $k \geqslant 0$ the exact normal deviate $\xi_k = \xi(n, k)$ as the unique solution of $\phi(\xi) = P(S_n \leqslant k)$. Then, for any fixed M,*

$$\xi_k = x_k + \frac{1-x_k^2}{6\sigma_n}\{1-2\pi_n\} + \frac{\{5-14\pi_n(1-\pi_n)\}\, x_k^3 + \{-2+2\pi_n(1-\pi_n)\}\, x_k}{72\sigma_n^2} + O(\sigma_n^{-3}), \qquad (5.1)$$

where the $O(\sigma_n^{-3})$ is uniform in π_n and in $|x_k| \leqslant M$.

By (5.1), and by using the expansion

$$P(S \leqslant k) - \phi(x_k) = \phi(\xi_k) - \phi(x_k) = (\xi_k - x_k)\varphi(x_k) + \tfrac{1}{2}(\xi_k - x_k)^2\, \varphi'(x_k) + O((\xi_k - x_k)^3), \qquad (5.2)$$

we can easily prove that, for any fixed M,

$$\sup_{|x| \leqslant M} \left| \tilde{F}_n(x) - \phi(x) - \frac{1-x^2}{6\sigma_n} \{1-2\pi_n\} \varphi(x) \right|$$
$$\sim D(M)\, \sigma_n^{-2} \qquad \text{as} \quad \sigma_n \to \infty, \tag{5.3}$$

where $D(M)$ is a function of M. This is all we need to show that the result of Theorem 1.3 is sharp (see also Feller [8] and Uspensky [23]).

Note that, even in the i.i.d. case, Theorem 1.3 does not follow from Lemma 5.1 because of the restriction that x_k remain bounded.

We now present some applications of our results to the theory of order statistics. We shall make use of the lemma:

LEMMA 5.2. *Let $\pi_n = n^{-1}k_n$, where $\{k_n, n \geqslant 1\}$ is a sequence of integers such that $1 \leqslant k_n \leqslant n$, $k_n \to \infty$ and $n^{-1}k_n \to 0$ as $n \to \infty$. Then*

$$\lim_{n\to\infty} \sqrt{k_n} \left| \sum_{l=k_n}^{n} \binom{n}{l} \pi_n^l (1-\pi_n)^{n-l} - \frac{1}{2} \right| = \frac{1}{3\sqrt{2\pi}}. \tag{5.4}$$

Proof. Let $S = S_n$ be a random variable following a binomial $B(n, \pi_n)$ distribution. We have $\tilde{F}_n(0) = \frac{1}{2}\{P(S \leqslant k_n) + P(S < k_n)\} = 1 - P(S \geqslant k_n) + \frac{1}{2}P(S = k_n)$. By (1.11), we know that

$$\left| \frac{1}{2} - P(S \geqslant k_n) + \frac{1}{2} P(S = k_n) - \frac{1}{6\sqrt{2\pi k_n}} \right| = o(1/k_n) \qquad \text{as} \quad n \to \infty. \tag{5.5}$$

The result follows from (5.5) and Stirling's formula by which

$$P(S = k_n) \sim \frac{1}{\sqrt{2\pi k_n}} \qquad \text{as} \quad n \to \infty.$$

Let us now consider independent random variables $Y_1, ..., Y_n$ with a common distribution G. Let $Y_{n1} \leqslant \cdots \leqslant Y_{nn}$ denote the corresponding order statistics. In recent years a great deal of work has been devoted to investigating the asymptotic distribution of Y_{nk_n}, where $1 \leqslant k_n \leqslant n$. Gnedenko [11] considered the maximal term Y_{nn} while Smirnov [22] studied the limit laws for Y_{nk_n} for the cases where (i) k_n (or $n - k_n$) is a constant and (*ii*) $k_n/n \to \lambda$, $0 < \lambda < 1$ (central terms). The behavior of the extreme terms when simultaneously $k_n \to \infty$ and $k_n/n \to 0$ (in which case k_n is called an intermediate rank sequence) was studied by Chibisov [5]. Subsequently, extending the investigation of Chibisov, Wu [25] obtained a general result according to which, subject to the mild restriction that k_n is nondecreasing, the only possible non-degenerate limit laws for the properly normalized term Y_{nk_n} are normal and lognormal (for further related details on this topic see also Cheng [4], Reiss [20, 21], Smirnov [22], Balkema and DeHaan [1], Galambos [10], Watts [24], and Leadbetter *et al.* [14]). We restrict ourselves to mentioning the result of the "finite endpoint," which corresponds to the normal limit law.

Define the left endpoint y_0 of the d.f. G by $y_0 = \inf\{y: G(y) > 0\}$. Assume that y_0 is finite, $G(y_0) = 0$, and in some interval $(y_0, y_0 + \delta)$ $(\delta > 0)$, $G'(y)$ exists and is strictly positive with $\lim_{y \downarrow y_0} G'(y) > 0$. Then, if $\{k_n\}$ is an intermediate rank sequence, we may find $a_n > 0$ and b_n such that

$$P(Y_{nk_n} - b_n \leqslant x a_n) \to \phi(x). \tag{5.6}$$

Moreover, the norming constants a_n and b_n can be chosen as follows: For large n, define y_n by $G(y_n) = k_n/n$. Then set $b_n = y_n$ and $a_n = \sigma_n/nG'(y_n)$, where $\sigma_n^2 = k_n(1 - k_n/n)$.

What relevance does Theorem 1.3 have to the speed of convergence in (5.6)? The problem of determining the rate of convergence in (5.6) is not just an academic one since there has been ample room for the suspicion that the distribution of intermediate order statistics may converge remarkably slowly to the normal distribution function.

The error bound in the approximation (5.6) heavily depends on the speed at which k_n increases to infinity and, as will be seen shortly, the precise uniform bound is given by $1/\sqrt{k_n}$.

Let us make a slightly stronger assumption on G in the right neighborhood of y_0, namely that in $(y_0, y_0 + \delta)$, $G''(y)$ exists and is bounded. We assume also that G and G' satisfy the hypothesis above, leading to (5.6).

We may now state a Berry–Esseen type theorem, the proof of which is deferred to the Appendix in the sequel.

THEOREM 5.1. *If the preceding assumptions hold, then there exists a universal constant H such that, for all $n \geqslant 1$*

$$D_n = \sup_x |P(n\{Y_{nk_n} - y_n\} G'(y_n) \leqslant x\sigma_n) - \phi(x)| \leqslant H/\sqrt{k_n}. \tag{5.7}$$

The question whether $1/\sqrt{k_n}$ is the "right" order (i.e., cannot be in general improved) can now be answered positively: Let us take G to be the uniform distribution on (0, 1). Then $y_n = k_n/n$ and clearly

$$D_n \geqslant \left| P\left(B\left(n, \frac{k_n}{n}\right) \geqslant k_n\right) - \frac{1}{2}\right|, \tag{5.8}$$

where $B(n, p)$ denotes a random variable having a binomial $B(n, p)$ distribution.

But, in view of Lemma 5.2, the right-hand side of (5.8) is $\sim (1/3\sqrt{2\pi})\cdot(1/\sqrt{k_n})$ as $n\to\infty$, thus proving the desired assertion. In addition, we have

$$H \geqslant 1/3\sqrt{2\pi} > 0.13.$$

Remark. Routine computations show that the rate in Theorem 1.3 is unchanged if we replace $\tilde{F}_n$ (a linear approximation) by a quadratic or a higher order approximation (the function $\phi(x)+((1-x^2)/6\sigma_n)\varphi(x)\{1-2\sigma_n^{-2}\sum_{j=1}^n p_{nj}^2(1-p_{nj})\}$ has bounded derivatives of all orders so that in any interval of length σ_n^{-1} the change from a linear approximation to a higher order approximation involves error terms of order not exceeding $O(\sigma_n^{-2})$).

APPENDIX: Proof of Theorem 5.1.

For each x set $v_{nx}=y_n+\sigma_n x/\{nG'(y_n)\}$ and $\sigma_{nx}^2=nG(v_{nx})(1-G(v_{nx}))$. We shall proceed in two parts by establishing that uniformly in x,

$$|P(n\{Y_{nk_n}-y_n\}G'(y_n)\leqslant x\sigma_n)-\phi(d_n(x))| = O(k_n^{-1/2}), \tag{6.1}$$

and

$$|\phi(d_n(x))-\phi(x)| = O(k_n^{-1/2}), \tag{6.2}$$

where $d_n(x)=n(G(v_{nx})-k_n/n)/\sigma_n$.

Let $C_1, C_2, \ldots$ denote positive universal constants throughout the proof. We shall find it convenient to introduce $I_n=\{x\colon |d_n(x)|\leqslant \sigma_n/2\}$. Then for $x\in I_n$ the following bound may be obtained:

$$|1-\sigma_n/\sigma_{nx}| \leqslant n\left|G(v_{nx})-\frac{k_n}{n}\right|\sigma_n^{-2}. \tag{6.3}$$

1. Let $x\in I_n$. Set $S_n(x)=\sum_{j=1}^n\{u(v_{nx}-Y_j)-G(v_{nx})\}\sigma_{nx}^{-1}$. Then, an application of Theorem 1.2 jointly with (6.3) yields

$$\begin{aligned}&\left|P(n\{Y_{nk_n}-y_n\}G'(y_n)\leqslant x\sigma_n)-\phi\left(n\left\{G(v_{nx})-\frac{k_n}{n}\right\}\Big/\sigma_{nx}\right)\right|\\ &\quad=\left|P\left(S_n(x)\leqslant n\left\{\frac{k_n}{n}-G(v_{nx})\right\}\Big/\sigma_{nx}\right)-\phi\left(n\left\{\frac{k_n}{n}-G(v_{nx})\right\}\Big/\sigma_{nx}\right)\right|\\ &\quad\leqslant C_1/\sigma_{nx}\leqslant 3C_1/2\sigma_n=O(k_n^{-1/2}).\end{aligned} \tag{6.4}$$

We now proceed to estimate

$$\delta_n(x) = \phi\left(n\left\{G(v_{nx}) - \frac{k_n}{n}\right\}\Big/\sigma_{nx}\right) - \phi(d_n(x)).$$

By the mean value theorem

$$|\delta_n(x)| = \frac{d_n(x)}{\sqrt{2\pi}}\,|1-\sigma_n/\sigma_{nx}|\,\exp\left(-\frac{d_n^2(x)}{2}\left\{1+\theta\left(\frac{\sigma_n}{\sigma_{nx}}-1\right)\right\}^2\right) \tag{6.5}$$

for some $0<\theta<1$. Now since $\sup_{t>0} te^{-at} = (ae)^{-1}$ $(a>0)$, from (6.3) and (6.5) we deduce that

$$|\delta_n(x)| \leqslant C_2/\sigma_n = O(k_n^{-1/2}), \qquad C_2 = 8/e\sqrt{2\pi}. \tag{6.6}$$

2. Let $x \notin I_n$. Set $\lambda_{nx} = d_n(x)/\sigma_n$ and assume that $\lambda_{nx} \leqslant -\frac{1}{2}$. We shall use the well-known inequality

$$1-\phi(t) \leqslant (2\pi)^{-1/2}\,t^{-1}e^{-t^2/2} \leqslant (2\pi e)^{-1/2}\,t^{-2}, \qquad t>0, \tag{6.7}$$

according to which

$$\phi(d_n(x)) \leqslant (2\pi e)^{-1/2}\,n^{-2}\sigma_n^2\left(G(v_{nx}) - \frac{k_n}{n}\right)^{-2} \leqslant C_3/\sigma_n^2 = O(k_n^{-1}), \tag{6.8}$$

where $C_3 = 4(2\pi e)^{-1/2}$.

On the other hand, note that

$$\begin{aligned} P(Y_{nk_n} \leqslant v_{nx}) &= P\left(\sum_{j=1}^{n}\{u(v_{nx}-Y_j) - G(v_{nx})\} \geqslant n\left(\frac{k_n}{n} - G(v_{nx})\right)\right) \\ &\leqslant \sigma_{nx}^2\Big/\left\{n\left(\frac{k_n}{n} - G(v_{nx})\right\}^2 \\ &\leqslant 6/\sigma_n^2 = O(k_n^{-1}). \end{aligned} \tag{6.9}$$

The first inequality in (6.9) is obtained by Chebyshev's inequality, while in the second we used the fact that $\sigma_{nx}^2/\sigma_n^2 = 1 + (1-2k_n/n)\,\lambda_{nx} - n\sigma_n^2\lambda_{nx}^2 \leqslant 1 + |\lambda_{nx}| \leqslant 6\lambda_{nx}^2$. Thus, since for α, $\beta>0$, $|\alpha-\beta| \leqslant \max(\alpha, \beta)$, by combining (6.8) and (6.9) we see that for $\lambda_{nx} \leqslant -\frac{1}{2}$,

$$\sup_{x\notin I_n} |P(Y_{nk_n} \leqslant v_{nx}) - \phi(d_n(x))| = O(1/k_n). \tag{6.10}$$

An estimate similar to (6.10) is valid for $\lambda_{nx} \geqslant \frac{1}{2}$. Hence by (6.4), (6.6), and (6.10), we deduce (6.1) and the first assertion is proved.

Now to prove (6.2), let $K = \sup_y |G''(y)|$ (where here y is restricted to $(y_0, y_0+\delta)$) and choose $a>0$ such that $G'(y_n) \geqslant a$ for all $n \geqslant 1$. Set $J_n = \{x : |x| \leqslant a^2 n/K\sigma_n\}$. Consider first $x \in J_n$. Invoking a second-order Taylor expansion, we may express $\varepsilon_n(x) = \phi(d_n(x)) - \phi(x)$ as

$$\varepsilon_n(x) = \phi\left(x + \frac{\sigma_n y^2}{2n(G'(y_n))^2}\, G''\left(y_n + \frac{\xi\sigma_n x}{nG'(y_n)}\right)\right) - \phi(x), \qquad 0<\xi<1.$$

Now the inequality (Serfling [27, p. 81]) $|\phi(t+\alpha t^2) - \phi(t)| \leqslant 8|\alpha|(2\pi)^{-1/2} e^{-1}$ valid for $|\alpha t| \leqslant \frac{1}{2}$, gives

$$\sup_{x \in J_n} |\varepsilon_n(x)| \leqslant C_4 \sigma_n/n = O(k_n^{-1/2}), \tag{6.11}$$

where $C_4 = 4Ka^{-2}(2\pi)^{-1/2} e^{-1}$.

Assume on the other hand that $x \notin J_n$, say $x \leqslant -a^2 n/K\sigma_n$. Then, according to (6.7)

$$\phi(x) \leqslant C_5 \sigma_n^2/n^2, \qquad C_5 = (2\pi e)^{-1/2} K^2/a^4. \tag{6.12}$$

Also, from (6.11) and (6.12) we have

$$\phi(d_n(x)) \leqslant \phi(d_n(-a^2 n/K\sigma_n)) \leqslant C_4 \sigma_n/n + C_5 \sigma_n^2/n^2. \tag{6.13}$$

It follows from (6.12) and (6.13) that for $x \leqslant -a^2 n/K\sigma_n$,

$$\varepsilon_n(x) = O(\sigma_n^2/n^2) + O(\sigma_n/n) = O(k_n^{-1/2}) \tag{6.14}$$

uniformly in x.

Finally, since a similar statement holds for $x \geqslant a^2 n/K\sigma_n$, we deduce

$$\sup_{x \notin J_n} |\varepsilon_n(x)| = = O(k_n^{-1/2}). \tag{6.15}$$

Combining (6.11) and (6.15) yields (6.2). The proof of Theorem 5.1 is now completed.

Acknowledgment

We thank the referee for interesting suggestions and remarks.

References

[1] Balkema, A. A., and De Haan, L. (1978). Limit distribution for order statistics. *Theory Probab. Appl.* **23** 77–92.

[2] Barbour, A. D., and Hall, P. (1984). On the rate of Poisson convergence. *Math. Proc. Cambridge Philos Soc.* **95** 475–480.

[3] CHENG, T. T. (1949). The normal approximation to the Poisson distribution and a proof of a conjecture of Ramanujan. *Bull. Amer. Math. Soc.* **55** 396–401.

[4] CHENG, B. (1964). The limiting distributions of order statistics. *Acta Math. Sinica* **14** 694–714. (Transl. in *Chinese Math.* (1965). **6** 84–104).

[5] CHIBISOV, D. M. (1964). On the limit distribution for order statistics. *Theory Probab. Appl.* **9** 142–148.

[6] CRAMÉR, H. (1962). *Random Variables and Probability Distributions*, 2nd. ed. Cambridge Univ. Press, Cambridge.

[7] DEHEUVELS, P., AND PFEIFER, D. (1986). A semigroup approach to Poisson approximation, *Ann. Probab.* **14** 663–676.

[8] FELLER, W. (1945). On the normal approximation to the binomial distribution. *Ann. Math. Statist.* **16** 319–329.

[9] FELLER, W. (1966). *An Introduction to Probability Theory and its Applications*, 2. Wiley, New York.

[10] GALAMBOS, J. (1978). *The Asymptotic Theory of Extreme Order Statistics* Wiley, New York.

[11] GNEDENKO, B. V. (1943). Sur la distribution limite du terme maximum d'une série aléatoire. *Ann. of Math.* **44** 423–453.

[12] HALL, P. (1982). *Ratès of Convergence in the Central Limit Theorem.* Pitman, Boston.

[13] JOHNSON, N. L., AND KOTZ, S. (1969). *Discrete Distributions.*HoughtonMifflin, Boston.

[14] LEADBETTER, M. R., LINDGREN, G., AND ROOTZÉN, H. (1983). *Extremes and Related Properties of Random Sequences and Processes.* Springer-Verlag, Berlin.

[15] MICHEL, R. (1981). On the constant in the nonuniform version of the Berry–Esséen theorem. *Z. Wahrsch. Verw. Gebiete* **55** 109–117.

[16] MOLENAAR, W. (1973). *Approximations to the Poisson, Binomial and Hypergeometric Distribution Functions.* Math. Centre Tracts Vol. 31, Math. Centrum, Amsterdam.

[17] NAGAEV, S. V. (1965). Some limit theorems for large deviations. *Theory Probab. Appl.* **10** 214–235.

[18] PEIZER, D. B., AND PRATT, J. W. (1968). A normal approximation for binomial, *F*, beta and other common, related tail probabilities. *J. Amer. Statist. Assoc.* **63** 1417–1456, 1457–1483.

[19] PETROV, V. V. (1975). *Sums of Independent Random Variables.* Springer-Verlag, Berlin.

[20] REISS, R. D. (1974). On the accuracy of the normal approximation for quantiles. *Ann. Probab.* **2** 741–744.

[21] REISS, R. D. (1976). Asymptotic distributions for sample quantiles. *Ann. Probab.* **4** 249–258.

[22] SMIRNOV, B. V. (1967). Some remarks on limit laws for order statistics. *Theory Probab. Appl.* **12** 337–339.

[23] USPENSKY, J. V. (1937). *Introduction to Mathematical Probability.* McGraw–Hill, New York.

[24] WATTS, V. (1980). The almost sure representation of intermediate order statistics. *Z. Wahrsch. Verw. Gebiete* **54** 281–285.

[25] WU, C.-Y. (1966). The types of limit distributions for some terms of a variational series. *Sci. Sinica* **15** 749–776.

[26] ZOLOTAREV, V. M. (1967). A generalization of the Lindeberg–Feller theorem. *Theory Probab. Appl.* **12** 608–618.

[27] SERFLING, R. J. (1980). *Approximation Theorems of Mathematical Statistics.* John Wiley, New York.

ON THE RATE OF CONVERGENCE IN NORMAL APPROXIMATION AND LARGE DEVIATION PROBABILITIES FOR A CLASS OF STATISTICS*

M. L. PURI† and M. SEOH‡

Summary. A new class of statistics is introduced to include, as special cases, unsigned linear rank statistics, signed linear rank statistics, linear combination of functions of order statistics, linear functions of concomitants of order statistics, and a rank combinatorial statistic. For this class, the rate of convergence to normality and Cramér's type large deviation probabilities are investigated. Under the assumption that underlying observations are only independent, it is shown that this rate is $O(N^{-\delta/2}\log N)$ if the first derivative of the score generating function φ satisfies Lipschitz's condition of order δ, $0<\delta\leqq 1$, and it is $O(N^{-1/2})$ if φ'' satisfies Lipschitz's condition of order $\delta\geqq\frac{1}{2}$; and that Cramér's large deviation theorem holds in the optimal range $0<x\leqq\rho_N N^{1/6}$, $\rho_N=o(1)$. The results obtained provide new results and extend as well as generalize a number of known results obtained in this direction.

1. Introduction. Let $Z_{Nj}=(Y_{Nj}^{(1)}, Y_{Nj}^{(2)},\cdots, Y_{Nj}^{(p)})$, $1\leqq j\leqq N$, be independent p-variate r.v.'s (random vectors) with joint c.d.f. (cumulative distribution function) $F_{Nj}(t)$, $t\in\mathbf{R}^p$ (the Euclidean p-dimensional space). Let g be a function on $\mathbf{R}^p$ such that $X_{Nj}^g=g(Z_{Nj})$, $1\leqq j\leqq N$, has a c.d.f. G_{Nj} which may not be continuous. Define the rank R_{Nj}^g of X_{Nj}^g among $\{X_{Nk}^g;\ 1\leqq k\leqq N\}$ by $R_{Nj}^g=\sum_{k=1}^N u(X_{Nj}^g-X_{Nk}^g)$, where $u(x)=1$ or 0 according to whether $x\geqq 0$ or $x<0$. Consider a class of statistics

$$T_N=\sum_{j=1}^N c_{Nj}a_{NR_{Nj}^g}(Z_{Nj}), \tag{1.1}$$

where c_{Nj}, $1\leqq j\leqq N$, are regression constants and $a_{Nj}(\underset{\sim}{t})$, $1\leqq j\leqq N$, are functions (called scores) generated by a known nonconstant function (called a score generating function) $\varphi(s,\underset{\sim}{t})$, $0<s<1$, $\underset{\sim}{t}\in\mathbf{R}^p$, in either of the following two ways:

$$a_{Nj}(\underset{\sim}{t})=\varphi(\mathbf{E}U_{N:j}, t),\quad 1\leqq j\leqq N,\quad \text{(approximate scores)}, \tag{1.2}$$

$$a_{Nj}(t)=\mathbf{E}\varphi(U_{N:j},\underset{\sim}{t}),\quad 1\leqq j\leqq N,\quad \text{(exact scores)}. \tag{1.3}$$

Here $U_{N:j}$ denotes the jth order statistic in a random sample of size N from the uniform distribution on $(0,1)$.

The statistic (1.1) includes as special cases the following statistics.

(a) *Unsigned linear rank statistic* $\mathcal{S}_1=\sum_{j=1}^N c_{Nj}a_{NR_{Nj}}^*$ by taking $p=1$, $g(t)=t$ and $\varphi(s,t)=\tilde{\varphi}(s)$.

(b) *Signed linear rank statistic* $\mathcal{S}_2=\sum_{j=1}^N c_{Nj}a_{NR_{Nj}^+}^*\operatorname{sgn}Y_{Nj}^{(1)}$ by taking $p=1$, $g(t)=|t|$ and $\varphi(s,t)=\tilde{\varphi}(s)\operatorname{sgn}t$ (where $\operatorname{sgn}t=2u(t)-1$).

(c) *Linear combination of functions of order statistics* $\mathcal{S}_3=(1/N)\sum_{j=1}^N a_{NR_{Nj}}^*\psi(Y_{Nj}^{(1)})(=(1/N)\sum_{j=1}^N a_{Nj}^*\psi(Y_{N:j}^{(1)}))$ by taking $p=1$, $g(t)=t$, $\varphi(s,t)=\tilde{\varphi}(s)\psi(t)$ and $c_{Nj}=N^{-1}$, $1\leqq j\leqq N$.

* Received by the editors July 2, 1984

† Department of Mathematics, Indiana University, Bloomington, Indiana 47405. The U.S.A. research was supported by Office of Naval Research contract N14-85-K-0648.

‡ Department of Mathematics and Statistics, Wright State University, Dayton, Ohio 45435, U.S.A.

(d) *Linear functions of concomitants of order statistics* $\mathcal{S}_4 = (1/N)\sum_{j=1}^N a^*_{NR_{Nj}}\psi(Y^{(2)}_{Nj})(=(1/N)\sum_{j=1}^N a^*_{Nj}\psi(Y^{(2)}_{N:j}))$ by taking $p=2$, $g(t_1, t_2)=t_1$, $\varphi(s, t_1, t_2)=\tilde{\varphi}(s)\psi(t_2)$ and $c_{Nj}=N^{-1}$, $1\leqq j\leqq N$.

Here the a^*_{Nj} are the usual scores generated by $\tilde{\varphi}(s)$; R_{Nj}, $Y^{(1)}_{N:j}$, $1\leqq j\leqq N$, are the rank of $Y^{(1)}_{Nj}$ and the jth order statistic, respectively, among $\{Y^{(1)}_{Nk}, 1\leqq k\leqq N\}$; R^+_{Nj}, $1\leqq j\leqq N$, is the rank of $|Y^{(1)}_{Nj}|$ among $\{|Y^1_{Nk}|, 1\leqq k\leqq N\}$; $Y^{(2)}_{N:j}$ is the concomitant of the jth order statistic $Y^{(1)}_{N:j}$ (jth induced order statistic, see [2]), i.e., $Y^{(2)}_{N:j}=Y^{(2)}_{N:R_{Nj}}$, if the marginal c.d.f.'s of $Y^{(1)}_{Nj}$, $1\leqq j\leqq N$, are continuous. (Note that the equalities in $\mathcal{S}_3$ and $\mathcal{S}_4$ are justified under this condition.) Furthermore, denoting $W_{jk}=c_{Nj}a_{Nk}(Z_{Nj})$, $1\leqq j$, $k\leqq N$, we can rewrite (1.1) as

(e) *Rank combinatorial statistic* $\mathcal{S}_5=\sum_{j=1}^N W_{jR^\xi_{Nj}}$ (cf. [25]).

In recent years, under different sets of assumptions, a number of authors have investigated these statistics separately, to derive asymptotic normality, rates of convergence to normality and the Cramér-type large deviation results. See, for example, Hájek [8], [9], Does [4], and Kallenberg [17] for results concerning the statistic $\mathcal{S}_1$; Huškova [13] and Müller-Funk and Witting [19] for results concerning $\mathcal{S}_2$; Stigler [22], Helmers [10], [11], and Vandermaele and Veraverbeke [24] for results concerning $\mathcal{S}_3$; Yang [26] for $\mathcal{S}_4$; and Motoo [18] and von Bahr [25] for results concerning $\mathcal{S}_5$, among others.

In this paper, we derive suitable order bounds for the remainder term in the normal approximation, and the Cramér-type large deviation probabilities for the class of statistics T_N defined in (1.1). The results obtained include, as special cases, Bergström and Puri [1], Kallenberg [17]. Vandemaele and Veraverbeke [24], and [27], [29]. They also extend and generalize Hušková [14], [15], Does [4], Müller-Funk and Witting [19], Bjerve [3] and Helmers [10], [11], among others (see Remarks 3.1–3.5 and 4.1–4.4).

2. Assumption and notation. We assume the following.

ASSUMPTION $A^{(\delta)}_l$. The score generating function $\varphi(s, t)$, $0<s<1$, $\underset{\sim}{t}\in\mathbf{R}^p$, has the lth partial derivatives $\varphi^{(l)}_1(s, t)=\theta^l(s, t)/\partial s^l$ satisfying Lipschitz's condition of order δ, $0<\delta\leqq 1$, with respect to its first argument s, i.e.,

$$|\varphi^{(l)}_1(x, \underset{\sim}{t})-\varphi^{(l)}_1(y, \underset{\sim}{t})|\leqq|x-y|^\delta\Delta_l(\underset{\sim}{t}),\qquad 0<x, y<1, \tag{2.1}$$

where Δ_l is some non-negative function. There are non-negative absolute constants K and η such that

$$\max_{1\leqq j\leqq N} \mathrm{E}|\Delta_l(Z_{Nj})|^r\leqq K^r r^{\eta r}\quad\text{for any real } r>0. \tag{2.2}$$

Furthermore, $\varphi(s, \underset{\sim}{t})$, $\varphi^{(1)}_1(s, \underset{\sim}{t}), \cdots, \varphi^l_1(s, \underset{\sim}{t})$ are jointly measurable in s and $\underset{\sim}{t}$ such that

$$|\varphi^{(k)}_1(s, \underset{\sim}{t})|\leqq\Delta_l(t),\qquad k=0, 1, \cdots, l\quad(\varphi^{(0)}_1\equiv\varphi). \tag{2.3}$$

Remark 2.1. The condition (2.1) ensures that we can choose a suitable function Δ_l satisfying (2.3) without loss of generality.

Remark 2.2. With the special form $\varphi(s, t)=\tilde{\varphi}(s)\psi(\underset{\sim}{t})$ of the score generating function, we note that the joint measurability is out of question and the condition (2.1) restricts the score generating function $\tilde{\varphi}$ to bounded ones.

The main idea (utilized essentially by everyone) is as follows: assuming $\varphi_1(\equiv\varphi^{(1)}_1)$, we first approximate the statistic T_N in (1.1) with approximate scores (1.2) by the two-term Taylor expansion, i.e.,

$$S_N=\sum_{j=1}^N c_{Nj}\{\varphi(\rho_{Njj}, Z_{Nj})+(\rho_{Nj}-\rho_{Njj})\varphi_1(\rho_{Njj}, Z_{Nj})\}, \tag{2.4}$$

where $\rho_{Nj} = R^g_{Nj}/(N+1)$ and $\rho_{Njj} = \mathbf{E}(\rho_{Nj} | Z_{Nj})$, $1 \leqq j \leqq N$, and then treat the remainder term under suitable assumptions. Next, we use Hájek's [9] projection method to approximate S_N by

$$(2.5) \qquad \hat{S}_N = \sum_{l=1}^{N} \mathbf{E}(S_N | Z_{Nl}) - (N-1)\mathbf{E}S_N.$$

Then, the desired properties of T_N can be established by showing those of $\hat{S}_N$ as applications of the well-known classical theory for sums of independent r.v.'s.

Let $\varphi(x)$ denote the standard normal c.d.f. and put

$$(2.6) \qquad \tau_N^2 = \mathbf{D}T_N, \quad \sigma_N^2 = \mathbf{D}S_N, \quad \hat{\sigma}_N^2 = \mathbf{D}\hat{S}_N.$$

Furthermore, let $\|\cdot\|$ denote the usual supremum norm and, for any r.v. S, let S^0 denote the r.v. S centered at its expectation, i.e., $S^0 = S - \mathbf{E}S$.

From now on, we omit the first subscript N in Z_{Nj}, X^g_{Nj}, c_{Nj}, ρ_{Njj}, etc., for the ease of notation whenever it causes no confusion.

3. The rate of convergence in the normal approximation. The asymptotic normality of the statistics $\mathscr{S}_1$ and $\mathscr{S}_5$ (special cases of the statistics T_N in (1.1)) is well known. However, often one needs information more precise than the asymptotic normality can provide, and one may try to find suitable order bounds for the remainder term $\|\mathbf{P}(T_N^0 \leqq \tau_N \cdot) - \varphi(\cdot)\|$ when the distribution of (1.1) is approximated by the normal one.

THEOREM 3.1. *If assumption $A_1^{(\delta)}$ is satisfied, then*

$$(3.1) \qquad \|\mathbf{P}(T_N^0 \leqq \hat{\sigma}_N \cdot) - \Phi(\cdot)\| \leqq 36CK^3 3^{3\eta}\omega_3 + 2eM_N(\log N)^{1+\eta_N - \delta/2},$$

$$(3.2) \qquad |\tau_N - \hat{\sigma}_N| \leqq M_N \hat{\sigma}_N N^{-\delta/2},$$

where C is a constant (0.7975 *is van Beek's* [23] *estimation*) *and*

$$(3.3) \qquad M_N = 64eK2^\eta \hat{\sigma}_N^{-1}\left(\sum_{j=1}^{N} c_j^2\right)^{1/2}.$$

THEOREM 3.2. *If assumption $A_2^{(\delta)}$ is satisfied with a bounded Δ_2, then*

$$(3.4) \qquad \|\mathbf{P}(T_N^0 \leqq \hat{\sigma}_N \cdot) - \varphi(\cdot)\| = O(\max\{\omega_3, \omega_3^{2\delta}\}) \quad \text{and} \quad |\tau_N - \hat{\sigma}_N| O(\hat{\sigma}_{N^{\omega_3}}),$$

where $\omega_3 = \hat{\sigma}_N^{-3} \sum_{j=1}^N |c_j|^3$.

Theorems 3.1 and 3.2 not only extend or generalize a number of existing theorems, they also provide new results as explained in the following remarks. (In what follows the general alternatives mean that the underlying distributions are only independent.)

Remark 3.1 (Unsigned linear rank statistic $\mathscr{S}_1$). Theorem 3.1 includes Theorem 1.2 in Bergström and Puri [1] that obtained the order bound $N^{-1/2} \log N$ only for approximate scores. Our assumptions are slightly weaker, and our bound is more explicit than the Bergström–Puri bound. As in our Theorem 3.2, Hušková [14], [15] and Does [4] obtained the Berry-Esséen bound of order $N^{-1/2}$ (the former dealt with bounded φ and the latter included unbounded φ) under assumptions on φ weaker than ours. However, their results are derived under the restrictions that underlying distributions are continuous and identical (Hušková, Does) or contiguous (Hušková).

Remark 3.2 (Signed linear rank statistic $\mathscr{S}_2$). Müller-Funk and Witting [19] derived a bound of order $O(N^{-1/2}(\log N)^2)$ for a restrictive case of statistic $\mathscr{S}_2$ (assuming that the underlying distributions are i.i.d. and $c_{N1} = \cdots = c_{NN}$). For the case of general alternatives, Ralescu and Puri [21] derived a bound of order $O(N^{-\alpha})$, $\alpha > \frac{1}{2}$.

Remark 3.3 (Linear combination of functions of order statistic $\mathcal{S}_3$). Bjerve [3], dealing with the trimmed linear combination of order statistics, obtained the rate $O(N^{-1/2})$ for the case of i.i.d. r.v.'s imposing rather strong smoothness conditions on the underlying distribution function. Helmers [10], [11], who improved Bjerve's results and considered the linear combination of order statistic only for the approximate scores, obtained the same rate as Bjerve under assumption on φ weaker than ours. However, both Bjerve and Helmers deal with linear combination of order statistics for the i.i.d. r.v.'s while we consider the problem under the general alternative both for approximate as well as exact scores.

Remark 3.4 (Linear functions of concomitants of order statistics $\mathcal{S}_4$). Our Theorems 3.1 and 3.2 provide rates of convergence to normality for the statistic $\mathcal{S}_4$ that to the best of our knowledge, has not been investigated so far. This would justify the application in Yang [26].

Remark 3.5 (Rank combinatorial statistic $\mathcal{S}_5$). Von Bahr [25] considered a class of statistics that includes $\mathcal{S}_5$ as a special case, and derived the order $N^{-1/2}$ under rather restrictive assumption that the vector of ranks is from the i.i.d. r.v.'s.

4. Cramér's type large deviation probabilities. In this section, we are concerned with relative error in the normal approximation of the distribution of the statistic (1.1), i.e., $P(T_N^0) > \tau_N x)[1-\varphi(x)]^{-1}$.

We assume that the variance of T_N satisfies

$$\underline{\lim}\ \tau_N^2 = \underline{\lim}\ DT_N > 0, \tag{4.1}$$

and that the regression constants satisfy

$$\max_{1\leq j\leq N} |c_j| = O(N^{-1/3}),\quad \sum_{j=1}^{N} c_j^2 = 1,\quad \sum_{j=1}^{N} |c_j|^3 = O(N^{-1/2}). \tag{4.2}$$

Let $\beta = 1/(2(3+2\eta))$. Then, we have

THEOREM 4.1. *If $Z_1, Z_2, \cdots, Z_N$ are i.i.d. r.v.'s, $c_1 = c_2 = \cdots = c_N$, and assumption $A_1^{(1)}$ is satisfied, then uniformly in the region $0 < x \leqq \rho_n N^{\beta}$, $\rho_N = o(1)$,*

$$P(T_N^0 \geqq \tau_N x)[1-\Phi(x)]^{-1} = 1 + o(1) \quad as\ N \to \infty \tag{4.3}$$

which remains true if τ_N is replaced by $\hat{\sigma}_N$.

THEOREM 4.2. *If $Z_1, Z_2, \cdots, Z_N$ are only independent and assumption $A_1^{(1)}$ is satisfied with a bounded Δ_1, then the conclusion of Theorem 4.1 holds.*

Theorems 4.1 and 4.2 provide a number of new results, as well as include known results as special cases. For reasons of clarity, we make the following remarks.

Remark 4.1 (Unsigned linear rank statistic $\mathcal{S}_1$). Kallenberg's [17] result is a special case of our Theorem 4.2. Kallenberg obtained the result under the assumption that underlying distributions are i.i.d. and continuous while ours is obtained under the general alternatives and without assuming the continuity of the underlying distribution functions.

Remark 4.2 (Signed linear rank statistic $\mathcal{S}_2$). The same result as Kallenberg's is obtained under the general alternative that to the best of our knowledge is not known.

Remark 4.3 (Linear combinations of functions of order statistics $\mathcal{S}_3$). Our Theorem 4.1 includes, as a special case, Theorem 2 in Vandemaele and Veraverbeke [24]. However, the latter applies only to the linear combinations of order statistics without continuity of underlying distributions while the former applies to the linear combinations of functions of order statistics with the continuity assumption. Our Theorem 4.2 extends these results under the general alternatives; however, our score generating function is uniformly bounded to cover only the linear combination of bounded functions of order statistics.

Remark 4.4 (Linear functions of concomitants of order statistics $\mathcal{S}_4$ and rank combinatorial statistic $\mathcal{S}_5$). Theorems 4.1 and 4.2 provide, to the best of our knowledge, the only known results concerning large deviation probabilities both for the statistics $\mathcal{S}_4$ and $\mathcal{S}_5$.

5. Preliminaries. In this section, we derive bounds on the $2r$th moments for the r.v.'s defined in the following lemmas. These bounds play important roles in the proofs of our theorems.

Let $\{Y_j\}_{j=1}^{\infty}$ be a sequence of r.v.'s and $\{d_j\}$ a sequence of real numbers.

LEMMA 5.1. *Let W_j, $j \geqq 1$, be r.v.'s of the form $W_j = g_j(Y_1, Y_2, \cdots, Y_j)$ such that $\mathrm{E}(W_j \mid Y_1, Y_2, \cdots, Y_{j-1}) = 0$ for $j \geqq 2$. If the sequence $\{d_j\}$ is nonincreasing in absolute values, then for any positive integers r and l,*

$$\mathrm{E}\left(\sum_{j=1}^{l} d_j W_j\right)^{2r} \leqq (4e)^r \left(\sum_{j=1}^{l} d_j^2\right)^r r^r \max_{1 \leqq j \leqq l} \mathrm{E} W_j^{2r}. \tag{5.1}$$

LEMMA 5.2. *Let Y_j, $j \geqq 1$, be independent and let $\tilde{V}_{jk}$ be r.v.'s of the form $\tilde{V}_{jk} = g_{jk}(Y_j, Y_k)$, $1 \leqq j$, $k < \infty$, such that for any j and k, $j \neq k$, $\mathrm{E}(\tilde{V}_{jk} \mid Y_j) = \mathrm{E}(\tilde{V}_{jk} \mid Y_k) = 0$. Then, for any positive integers l and r*

$$\mathrm{E}\left(\frac{1}{l} \sum_{j=1}^{l} \sum_{k \neq j}^{l} d_j \tilde{V}_{jk}\right)^{2r} \leqq (4e)^{2r} \left(\sum_{j=1}^{l} d_j^2\right)^r (2r)^{2r} l^{-r} \max_{\substack{1 \leqq j,k \leqq l \\ j \neq k}} \mathrm{E} \tilde{V}_{jk}^{2r}. \tag{5.2}$$

Proof of Lemma 5.1. For $r \geqq l$, the proof follows by applying Hölder's inequality to $(\sum_{j=1}^{l} d_j W_j)^{2r}$ and some routine computations. For $r < l$, the proof follows by induction on l with r fixed.

Proof of Lemma 5.2. Since the assumptions and the conclusion of this lemma are invariant under simultaneous permutation of the d_j and Y_j, we may assume that $|d_1| \geqq |d_2| \geqq \cdots \geqq |d_l|$, without loss of generality.

Define $W_1 = \tilde{W}_1 = 0$, $W_j = \sum_{k=1}^{j-1} \tilde{V}_{jk}$ and $\tilde{W}_j = \sum_{k=1}^{j-1} d_k \tilde{V}_{kj}$, $2 \leqq j \leqq l$, so that

$$\hat{V}_l = \frac{1}{l} \sum_{j=1}^{l} \sum_{k \neq j}^{l} d_j \tilde{V}_{jk} = \frac{1}{l} \left\{ \sum_{j=1}^{l} d_j W_j + \sum_{j=1}^{l} \tilde{W}_j \right\}.$$

The proof follows by noting that $\{W_j\}_{j=1}^{l}$ and $\{\tilde{W}_j\}_{j=1}^{l}$ satisfy the assumptions of Lemma 5.1, and using the following facts:

(a) $$\mathrm{E}\hat{V}_l^{2r} \leqq l^{-2r} 2^{2r-1} \left\{ \mathrm{E}\left(\sum_{j=1}^{l} d_j W_j\right)^{2r} + \mathrm{E}\left(\sum_{j=1}^{l} \tilde{W}_j\right)^{2r} \right\}$$
$$\leqq l^{-2r} 2^{2r-1} (4e)^r r^r \left\{ \left(\sum_{j=1}^{l} d_j^2\right)^r \max_{1 \leqq j \leqq l} \mathrm{E} W_j^{2r} + l^r \max_{1 \leqq j \leqq l} \mathrm{E} \tilde{W}_j^{2r} \right\},$$

(b) $$\mathrm{E} W_j^{2r} \leqq (4e)^r (j-1)^r r^r \max_{\substack{1 \leqq k \leqq l \\ k \neq j}} \mathrm{E} \tilde{V}_{jk}^{2r},$$

(c) $$\mathrm{E} \tilde{W}_l^{2r} \leqq (4e)^r \left(\sum_{k=1}^{j-1} d_k^2\right)^r r^r \max_{\substack{1 \leqq k \leqq l \\ k \neq j}} \mathrm{E} \tilde{V}_{kj}^{2r},$$

which can easily be obtained by repeated use of Lemma 5.1.

We now derive suitable bounds for remainders in approximations of S_N by $\hat{S}_N$ and of T_N by S_N. Note that

$$(5.3)\qquad \rho_j-\rho_{jj}=\frac{1}{N+1}\sum_{k\neq j}^{N}\{u(X_j^g-X_k^g)-G_k(X_j^g)\}\equiv\frac{1}{N+1}\sum_{k\neq j}^{N}H_{jk}.$$

For each j, conditionally given Z_j, $\rho_j-\rho_{jj}$ is the sum of independent r.v.'s with zero means. Thus an application of Lemma 5.1 and Hölder's inequality yield that, for any real r,

$$(5.4)\qquad \mathbf{E}\{(\rho_j-\rho_{jj})^{2r}\,|\,Z_j\}\leqq(4e)^r[r]^rN^{-r},\qquad j=1,2,\cdots,N,$$

where $[r]$ is the smallest integer $\geqq r$. It follows by elementary computations that

$$\mathbf{E}(S_N\,|\,Z_l)=c_l\varphi(\rho_{ll},Z_l)+\mathbf{E}S_N-c_l\mathbf{E}\varphi(\rho_{ll},Z_l)$$
$$+(N+1)^{-1}\sum_{j\neq l}^{N}c_j\mathbf{E}\{H_{jl}\varphi_1(\rho_{jj},Z_j)\,|\,Z_l\},\qquad 1\leqq l\leqq N.$$

Thus we can write

$$(5.5)\qquad \hat{S}_N^0=\sum_{j=1}^{N}\hat{S}_N^{(j)},\qquad S_N^0-\hat{S}_N^0=\frac{1}{N+1}\sum_{j=1}^{N}\sum_{k\neq j}^{N}c_jV_{jk},$$

where

$$(5.6)\qquad \begin{aligned}\hat{S}_N^{(j)}&=c_j\{\varphi(\rho_{jj},Z_j)-\mathbf{E}\varphi(\rho_{jj},Z_j)\}+\frac{1}{N+1}\sum_{k\neq j}^{N}c_k\mathbf{E}\{H_{kj}\varphi_1(\rho_{kk},Z_k)\,|\,Z\},\\ V_{jk}&=H_{jk}\varphi_1(\rho_{jj},Z_j)-\mathbf{E}\{H_{jk}\varphi_1(\rho_{jj},Z_j)\,|\,Z_k\},\end{aligned}$$

and H_{jk} is defined in (5.3). Since $\mathbf{E}\{V_{jk}\,|\,Z_j\}=\mathbf{E}\{V_{jk}\,|\,Z_k\}=0$ for $j\neq k$, we may apply Lemma 5.2 to obtain that, for any integer $r\geqq 1$,

$$(5.7)\qquad \mathbf{E}(S_N^0-\hat{S}_N^0)^{2r}\leqq(8e)^{2r}\left(\sum_{j=1}^{N}c_j^2\right)^r(2r)^{2r}N^{-r}\max_{1\leqq j\leqq N}\mathbf{E}\Delta_1^{2r}(Z_j).$$

We finally prove

LEMMA 5.3. *For any integer $r>0$, we have, under assumption $A_1^{(\delta)}$,*

$$(5.8)\qquad \mathbf{E}(T_N^0-S_N^0)^{2r}\leqq(16eK)^{2r}\left(\sum_{j=1}^{N}c_j^2\right)^r(2r)^{(1+\eta)2r}N^{-\delta r};$$

under assumption $A_l^{(\delta)}$, $l\geqq 1$ (for approximate scores),

$$(5.9)\qquad \mathbf{E}\left(T_N^0-\sum_{k=0}^{l}I_k^0\right)^{2r}\leqq\left(\frac{2K}{l!}\right)^{2r}\left(\sum_{j=1}^{N}c_j^2\right)^r(4er(l+1))^{r(l+\delta)}(2r)^{2r\eta}N^{-r(l+\delta-1)},$$

where the r.v.'s I_k, $0\leqq k\leqq l$, are given by

$$(5.10)\qquad I_k=\sum_{j=1}^{N}c_j\frac{(\rho_j-\rho_{jj})^k}{k!}\varphi_1^{(k)}(\rho_{jj},Z_j);$$

and, under assumption $A_l^{(\delta)}$, $l\geqq 2$,

$$(5.11)\qquad \mathbf{E}\left(\sum_{j=1}^{N}c_j\left\{\tilde{a}_{R_j}(Z_j)-\varphi\left(\frac{R_j}{N+1},Z_j\right)\right\}^0\right)^{2r}\leqq\left(\sum_{j=1}^{N}c_j^2\right)^rN^{-r}\max_{1\leqq j\leqq N}\mathbf{E}\Delta_2^{2r}(Z_j),$$

where $\tilde{a}_j(t)=\mathbf{E}\varphi(U_{N:j},\underline{t})$, $1\leqq j\leqq N$, $\underline{t}\in\mathbf{R}^p$.

Proof. Under assumption $A_l^{(\delta)}$, Taylor's expansion yields that

$$\left|\varphi(x,\underset{\sim}{t})-\sum_{k=0}^{l}\frac{(x-y)^k}{k!}\varphi_1^{(k)}(y,\underset{\sim}{t})\right|\leqq\frac{1}{l!}|x-y|^{l+\delta}\Delta_l(\underset{\sim}{t}) \tag{5.12}$$

for any $0<x,\ y<1$ and $\underset{\sim}{t}\in\mathbf{R}^p$.

Let $\tilde{T}_N$ denote the statistic (1.1) with exact scores (given by (1.3)) to distinguish it from the statistic T_N with approximate scores. Then it follows from (5.4), (5.12), and Hölder's inequality that

$$\begin{aligned}\mathbf{E}\Big(T_N^0-\sum_{k=0}^{l}I_k^0\Big)^{2r}&\leqq 2^{2r}\mathbf{E}\Big(\sum_{j=1}^{N}c_j\Big\{\varphi(\rho_j,Z_j)-\sum_{k=0}^{l}\frac{(\rho_j-\rho_{jj})^k}{k!}\varphi_1^{(k)}(\rho_{jj},Z_j)\Big\}\Big)^{2r}\\&\leqq 2^{2r}\Big(\sum_{j=1}^{N}c_j^2\Big)^r(l!)^{-2r}N^{r-1}\sum_{j=1}^{N}\mathbf{E}|\rho_j-\rho_{jj}|^{2r(l+\Delta)}\Delta_l^{2r}(Z_l)\\&\leqq\Big(\frac{2}{l!}\Big)^{2r}\Big(\sum_{j=1}^{N}c_j^2\Big)^r(4e)^{r(l+\delta)}\\&\quad\cdot(r(l+1))^{r(l+\delta)}N^{-r(l+\delta-1)}\max_{1\leqq j\leqq N}\mathbf{E}\Delta_l^{2r}(Z_j)\end{aligned} \tag{5.13}$$

which, together with (2.2), yields (5.9). Moreover,

$$\mathbf{E}(\tilde{T}_N^0-T_N^0)^{2r}\leqq 2^{2r}\Big(\sum_{i=1}^{N}c_j^2\Big)^r\mathbf{E}\Big[\sum_{j=1}^{N}\Big\{\bar{a}_{R_j}(Z_j)-\varphi\Big(\frac{R_j}{N+1},Z_j\Big)\Big\}^2\Big]^r. \tag{5.14}$$

Under assumption $A_1^{(\delta)}$, (5.12) ensures that $|\tilde{a}_j(\underset{\sim}{t})-\varphi(i/(N+1),t)|\leqq N^{-(1+\delta)/2}\Delta_1(\underset{\sim}{t})$ for any $1\leqq j\leqq N$ and $\underset{\sim}{t}$. Since $S_N=I_0+I_1$, (5.8) follows from (5.13) and (5.14). On the other hand, under assumption $A_l^{(\delta)}$, $l\geqq 2$, there are second partial derivatives such that $|\tilde{a}_j(\underset{\sim}{t})-\varphi(i/(N+1),\underset{\sim}{t})|\leqq(2N)^{-1}\Delta_2(\underset{\sim}{t})$ for any $1\leqq j\leqq N$ and $\underset{\sim}{t}$. Hence (5.11) follows from (5.14). The proof follows.

6. Proof of theorems in § 3. We now prove Theorems 3.1 and 3.2.

For convenience, we denote, for any integer r, $\omega_r=\sum_{j=1}^{N}|c_j^*|^r$, where $c_j^*=c_j\hat{\sigma}_N^{-1}$, $1\leqq j\leqq N$.

Proof of Theorem 3.1. By standard arguments, we have, for any $\varepsilon>0$,

$$\begin{aligned}&\mathbf{P}(T_N^0\leqq\hat{\sigma}_Nx)\leqq\mathbf{P}(\hat{S}_N^0\leqq\hat{\sigma}_N(x+\varepsilon))+\mathbf{P}(|T_N^0-\hat{S}_N^0|\geqq\hat{\sigma}_N\varepsilon),\\&\mathbf{P}(T_N^0\leqq\hat{\sigma}_Nx)\geqq\mathbf{P}(\hat{S}_N^0\leqq\hat{\sigma}_N(x-\varepsilon))-\mathbf{P}(|T_N^0-\hat{S}_N^0|\geqq\hat{\sigma}_N\varepsilon).\end{aligned} \tag{6.1}$$

Since $\hat{\sigma}_N^{-1}\hat{S}_N^0=\sum_{j=1}^{N}\hat{\sigma}_N^{-1}\hat{S}_N^{(j)}$ (cf. (5.5)) is a normalized sum of independent r.v.'s and since, because of (2.2),

$$\sum_{j=1}^{N}\mathbf{E}|\hat{\sigma}_N^{-1}\hat{S}_N^{(j)}|^3\leqq 4\sum_{j=1}^{N}\{8|c_j^*|^3+N^{-1}\omega_3\}\max_{1\leqq j\leqq N}\mathbf{E}\Delta_1^3(Z_j)\leqq L_N, \tag{6.2}$$

where $L_N=36K^3 3^{3\eta}\omega_3$, we may apply Lemma V.2.1 and Theorem V.2.3 of Petrov [20] to obtain that

$$|\mathbf{E}\exp(it\hat{\sigma}_N^{-1}\hat{S}_N^0)-e^{-t^2/2}|\leqq 16L_N|t|^3e^{-t^2/3},\qquad |t|\leqq(4L_N)^{-1} \tag{6.3}$$

and

$$\|\mathbf{P}(\hat{S}_N^0\leqq\hat{\sigma}_N\cdot)-\Phi(\cdot)\|\leqq CL_N \tag{6.4}$$

for some absolute constant C (van Beek's [23] estimation is 0.7975).

It follows from (2.2), (5.7), and (5.8) that

$$\mathbf{E}(T_N^0-\hat{S}_N^0)^{2r}\leqq(64eK2^{\eta})^{2r}\Big(\sum_{j=1}^{N}c_i^2\Big)^r r^{2r(1+\eta)}N^{-\delta r}. \tag{6.5}$$

Using the well-known inequality $|\Phi(x+y)-\Phi(x)|\leqq|y|$, we derive from (6.1), (6.4), and (6.5),

$$\|\mathbf{P}(T_N^0\leqq\hat{\sigma}_N\cdot)-\Phi(\cdot)\|\leqq CL_N+\varepsilon+(\hat{\sigma}_N\varepsilon)^{-2r}B_N^{(r)}, \tag{6.6}$$

where

$$B_N^{(r)}=\left[64eK2^{\eta}\left(\sum_{j=1}^{N}c_j^2\right)^{1/2}r^{1+\eta}N^{-\delta/2}\right]^{2r}.$$

Now choose $\varepsilon=(\hat{\alpha}_N\varepsilon)^{-2r}B_N^{(r)}$, i.e., $\varepsilon\leqq M_N(r^{(1+\eta)}N^{-\delta/2})^{2r/(2r+1)}$, where M_N is given by (3.3) and take $r=[-1+\log N]$ so that $\varepsilon\leqq M_Ne(\log N)^{1+\eta}N^{-\delta/2}$. This inequality and (6.6) ensure (3.1), and (6.5) yields (3.2). The proof follows.

Proof of Theorem 3.2. The theorem is trivial if $\omega_3\geqq e^{-2}$. Thus we may, without loss of generality, assume that

$$\omega_3<e^{-2},\qquad \|\Delta_2\|\leqq\tfrac{1}{6}, \tag{6.7}$$

which ensures

$$\mathbf{E}(\hat{\sigma}_N^{-1}\hat{S}_N^{(j)})^2\leqq((c_i^*)^2+N^{-1}\omega_2)/18,\qquad \mathbf{E}|\hat{\sigma}_N^{-1}S_N^{(j)}|^3\leqq(8|c_j^*|^3+N^{-1}\omega_3)/54. \tag{6.8}$$

Moreover, because $1=\sum_{j=1}^{N}\mathbf{E}(\hat{\sigma}_N^{-1}\hat{S}_N^{(j)})^2\leqq 9^{-1}\omega_2$, Hölder's inequality yields

$$N^{-1/2}\leqq 3^{-3}\omega_3,\quad \omega_1\leqq N^{2/3}\omega_3^{1/3},\quad \omega_2\leqq N^{1/3}\omega_3^{2/3}. \tag{6.9}$$

We first prove the theorem for approximate scores.

By standard arguments similar to (6.1), to prove the theorem, it suffices to show that

$$\mathbf{P}(|T_N^0-S_N^0-I_2^0-I_{21}^0|\geqq\hat{\sigma}_N\omega_3)=O(\max\{\omega_3,\omega_3^{2\delta}\}), \tag{6.10}$$

$$\|\mathbf{P}(S_N^0+I_{22}^0\leqq\hat{\sigma}_N\cdot)-\Phi(\cdot)\|=O(\omega_3), \tag{6.11}$$

where I_2 is defined by (5.10) and I_{21} and I_{22} are given by

$$I_{21}=(2(N+1)^2)^{-1}\sum\sum_{(j,k)\neq}c_jH_{jk}^2\varphi_1^{(2)}(\rho_{jj},Z_j),$$

$$I_{22}=(2(N+1)^2)^{-1}\sum\sum\sum_{(j,k,l)\neq}c_jH_{jk}H_{jl}\varphi_1^{(2)}(\rho_{jj},Z_j),$$

where $(j,k)\neq$ and $(j,k,l)\neq$ mean that summations are taken for different indices j, k, and l over the set $\{1,2,\cdots,N\}$. It follows that $I_2=I_{21}+I_{22}$.

Since $S_N=I_0+I_1$, applying (5.9) with $r=1$, $l=2$, and then using (6.9), we obtain $\mathbf{E}(\hat{\sigma}_N^{-1}(T_N^0-S_N^0-I_2^0))^2=O(\omega_3^{2+2\delta})$. Also direct computations yield $\mathbf{E}(\hat{\sigma}_N^{-1}I_{21}^0)^2=O(\omega_3^4)$. Equation (6.10) now follows easily by Chebyshev's inequality.

It follows by Esseen's smoothing lemma (see, e.g., Feller [7, p. 603] and (6.3) that, to prove (6.11), it suffices to show that, for some positive $\varepsilon\leqq(144K^33^{3\eta})^{-1}$,

$$\int_{|t|\leqq\varepsilon\omega_3^{-1}}|t|^{-1}|f(t)|\,dt=O(\omega_3), \tag{6.12}$$

where $f(t)=\mathbf{E}\exp\{it\hat{\sigma}_N^{-1}(S_N^0+I_{22}^0)\}-\mathbf{E}\exp(it\hat{\sigma}_N^{-1}\hat{S}_N^0)$.

We now need two lemmas that are extremely crucial and the proofs of which are given in the Appendix.

LEMMA 6.1. *Under* (6.7), *we have*

$$\mathbf{E}(\hat{\sigma}_N^{-1}I_{22}^0)^{2r}\leqq(2e\|\Delta_2\|/9)^{2r}(2r)^{2r}\omega_3^{2r}.$$

LEMMA 6.2. *Under (6.7), there is an absolute constant K such that*

$$|\mathbf{E}\exp(it\hat\sigma_N^{-1}\hat S_N^0)\{\hat\sigma_N^{-1}(S_N^0-\hat S_N^0+I_{22}^0)\}^r|\leqq K^r r^{3r}\omega_3(1+|t|+t^2)^r e^{-t^2/5}$$

for $|t|\leqq\omega_3^{-1}$ *and any positive integer* $r\leqq 2^{-1}\log\omega_3^{-1}$.

Now for any positive integer k, Taylor's expansion yields

$$\left|\frac{f(t)}{t}\right|\leqq\sum_{r=1}^{2k-1}\frac{|t|^{r-1}}{r!}|\mathbf{E}[\{\hat\alpha_N^{-1}(S_N^0-\hat S_N^0+I_{22}^0)\}^r\exp(it\hat\sigma_N^{-1}\hat S_N^0)]| \tag{6.13}$$
$$+O\left(\frac{|t|^{2k-1}}{(2k)!}\mathbf{E}\{\hat\sigma_N^{-1}(S_N^0-\hat S_N^0+I_{22}^0)\}^{2k}\right).$$

Taking $k=1$, and using (5.7), (6.9), Lemmas 6.1 and 6.2, we see that, for any $\varepsilon_1>0$,

$$\int_{|t|\leqq\varepsilon_1\omega_3^{-1/2}}\left|\frac{f(t)}{t}\right|dt=O(\omega_3). \tag{6.14}$$

Next, take $2k=[\log\omega_3^{-1}]$. Lemma 6.2 ensures that for any ε_1, $0<\varepsilon_1\leqq 1$,

$$\int_{\varepsilon_1\omega_3^{-1/2}\leqq|t|\leqq\varepsilon_1\omega_3^{-1}}\sum_{r=1}^{2k-1}\frac{|t|^{r-1}}{r!}|\mathbf{E}\{\hat\sigma_N^{-1}(S_N^0-\hat S_N^0+I_{22}^0)\}^r\exp(it\hat\sigma_N^{-1}\hat S_N^0)|\,dt \tag{6.15}$$
$$\leqq\omega_3\exp\left(-\frac{\varepsilon_1^2\omega_3^{-1}}{5}\right)(2k)^{6k}\omega_3^{-6k}e^{3K}=O(\omega_3).$$

It also follows from (5.7), (6.9), and Lemma 6.1 that

$$\int_{|t|\leqq\varepsilon_1\omega_3^{-1}}\frac{|t|^{2k-1}}{(2k)!}\mathbf{E}\{\hat\sigma_N^{-1}(S_N^0-\hat S_N^0+I_{22}^0)\}^{2k}dt\leqq\frac{(8e\|\Delta_2\|)^{2k}(2k)^{2k}\omega_3^{2k}}{(2k)!}(\varepsilon_1\omega_3^{-1})^{2k} \tag{6.16}$$
$$\leqq(2e\varepsilon_1)^{2k}(2k)^{2k}[(2k)!]^{-1}$$
$$\leqq(2e^2\varepsilon_1)^{2k},$$

where the last equality follows by Stirling's formula (see, e.g., Feller [6]). By taking $\varepsilon_1=1/(2e^3)$ and combining (6.13), (6.15), and (6.16), we obtain

$$\int_{\varepsilon_1\omega_3^{-1/2}\leqq|t|\leqq\varepsilon_1\omega_3^{-1}}\left|\frac{f(t)}{t}\right|dt=O(\omega_3), \tag{6.17}$$

which, together with (6.14), ensures (6.12), and consequently (6.11).

Since, because of (5.11),

$$\mathbf{P}\left(\left|\sum_{j=1}^N c_j(\tilde a_{R_j}(Z_j)-\varphi(R_j/(N+1)))^0\right|\geqq\hat\sigma_N^{-1}\omega_3\right)\leqq\omega_3,$$

the proof for exact scores follows analogously. The remaining half of (3.4) follows from (5.7), (5.8), and (6.9). The proof is complete.

7. Proofs of theorems in § 4. Before proving Theorems 4.1 and 4.2, we state some preliminary results. First note that it follows from (2.2), (4.2), (5.7), (5.8), and Hölder's inequality that, for any $r\geqq\frac{1}{2}$,

$$\mathbf{E}(T_N^0-\hat S_N^0)^{2r}\leqq(32eK)^{2r}(2r)^{2r(1+\eta)}N^{-r}. \tag{7.1}$$

Furthermore, since $|\hat\sigma_N^2-\tau_N^2|\leqq 2\tau_N(\mathbf{D}(\hat S_N-T_N))^{1/2}+\mathbf{D}(\hat S_N-T_N)$, we have

$$\hat\sigma_N^2/\tau_N^2=1+O(N^{-1/2}),\ \hat\sigma_N^2\geqq\sigma^2\quad\text{for some constant }\sigma>0. \tag{7.2}$$

Hence, to prove Theorems 4.1 and 4.2, it suffices to show that

$$\mathbf{P}(T_N^0 > \hat{\sigma}_N x)[1-\varphi(x)]^{-1} = 1 + o(1) \tag{7.3}$$

uniformly in the region $0 < x \leqq \rho_N N^\beta$ (recall that $\beta = (2(3+2\eta))^{-1}$).

We now consider the standard arguments

$$\begin{aligned} \mathbf{P}(T_N^0 > \hat{\sigma}_N x) &\leqq \mathbf{P}(\hat{S}_N^0 > \hat{\sigma}_N (x - N^{-\beta})) + \mathbf{P}(|T_N^0 - \hat{S}_N^0| > \hat{\sigma}_N N^{-\beta}), \\ \mathbf{P}(T_N^0 > \hat{\sigma}_N x) &\geqq \mathbf{P}(\hat{S}_N^0 > \hat{\sigma}_N (x + N^{-\beta})) - \mathbf{P}(|T_N^0 - \hat{S}_N^0| > \hat{\sigma}_N N^{-\beta}). \end{aligned} \tag{7.4}$$

Using Chebyshev's inequality and applying (7.1) with $2r = \delta N^{2\beta}$, $\delta = (\sigma/32e^2K)^{1/(1+\eta)}$, we have

$$\begin{aligned} \mathbf{P}(|T_N^0 - \hat{S}_N^0| > \hat{\sigma}_N N^{-\beta}) &\leqq \mathbf{E}(T_N^0 - \hat{S}_N^0)^{2r} N^{2r\beta} \hat{\sigma}_N^{-2r} \\ &\leqq \{(32eK/\sigma)(2r)^{1+\eta} N^{-(1+\eta)2\beta}\}^{2r} = \exp(-\delta N^{2\beta}). \end{aligned}$$

Hence, using Lemma VII.1.2 of Feller [6], we derive

$$\mathbf{P}(|T_N^0 - \hat{S}_N^0| > \hat{\sigma}_N N^{-\beta})[1-\varphi(x)]^{-1} \leqq \exp(-\delta N^{2\beta})[1-\varphi(\rho_N N^\beta)]^{-1} = o(1) \tag{7.5}$$

uniformly for $0 < x \leqq \rho_N N^\beta$.

Thus, in view of (7.4) and (7.5), to prove (7.3), it suffices to show that

$$\mathbf{P}(\hat{S}_N^0 > \hat{\sigma}_N x_N)[1-\varphi(x)]^{-1} = 1 + o(1), \qquad |x_N - x| = N^{-\beta}, \tag{7.6}$$

uniformly in $0 < x \leqq \rho_N N^\beta$.

Proof of Theorem 4.1. Define a function

$$h(x) = \int_{\mathbf{R}^p} \{u(g(\underset{\sim}{t}) - x) - G_j(g(\underset{\sim}{t}))\} \varphi_1(\Gamma(\underset{\sim}{t}), \underset{\sim}{t})\, dF(\underset{\sim}{t}),$$

where $\Gamma(\underset{\sim}{t}) = \mathbf{E}(\rho_j \mid Z_j = \underset{\sim}{t})$ and $F(\underset{\sim}{t})$ is the common distribution function of the Z_j. Set

$$\tilde{S}_N = \sum_{j=1}^N \tilde{S}_N^{(j)} = \sum_{j=1}^N \{\varphi(\rho_{jj}, Z_j) - \mathbf{E}\varphi(\rho_{jj}, Z_j) + h(X_j)\},$$

and denote $\tilde{\sigma}^2 = \operatorname{Var}(\tilde{S}_N^{(j)})$, $1 \leqq j \leqq N$. Since $\hat{S}_N^0 - N^{-1/2}\tilde{S}_N = -2N^{-1/2}(N+1)^{-1}\sum_{j=1}^N h(X_j)$, it suffices, using the same arguments as above, to show that, for $|x_N - x| = N^{-\beta}$,

$$\mathbf{P}(\tilde{S}_N > N^{1/2}\tilde{\sigma} x_N)[1-\varphi(x)]^{-1} = 1 + o(1) \tag{7.7}$$

uniformly in the region $0 < x \leqq \rho_N N^\beta$. Note that $\tilde{S}_N$ is the sum of i.i.d. r.v.'s with zero means and finite variance $\tilde{\sigma}^2$ such that, with $k_0 = \max\{k: 4\beta k/(2\beta+1) < 1\}$,

$$\begin{aligned} \mathbf{E}\exp|\tilde{S}_N^{(1)}|^{4\beta/(2\beta+1)} &\leqq 1 + \sum_{k=1}^\infty \frac{1}{k!} \mathbf{E}(\Delta_1(Z_1) + 2\mathbf{E}\Delta_1(z_1))^{4\beta k/(2\beta+1)} \\ &\leqq 1 + \sum_{k=1}^{k_0} \frac{(3K)^{4\beta k/(2\beta+1)}}{k!} + \sum_{k=k_0+1}^\infty \frac{16^k}{k!} \left(K \left(\frac{4\beta k}{2\beta+1} \right)^\eta \right)^{4\beta k/(2\beta+1)} \\ &\leqq 1 + e^{\tilde{C}} \sum_{k=1}^\infty \frac{\tilde{C}^k}{k!} k^{(1-2/(\eta+2))^k} < \infty, \end{aligned}$$

where $\tilde{C}$ is an absolute constant; (the last inequality follows by Sterling's formula). Hence (7.7) follows using Petrov [20, p. 251]. The proof follows.

Proof of Theorem 4.2. The method to prove (7.6) is similar to that of Kallenberg [17].

First note that $\hat{S}_N^0 = \sum_{j=1}^N \hat{S}_M^{(j)}$ is the sum of independent r.v.'s with zero means, and

$$|\hat{S}_N^{(j)}| \leqq 2|c_j| \, \|\Delta_1\| + (N+1)^{-1} \sum_{k \neq j}^{N} |c_k| \, \|\Delta_1\| \leqq 3\|\Delta_1\| \leqq 3\|\Delta_1\| A_1 N^{-1/3}, \qquad 1 \leqq j \leqq N,$$

for some constant A_1. Furthermore there is an integer N_0 such that, for all $N \geqq N_0$ and $0 < x < \rho_N N^{1/6}$, $|x_N - x| \leqq N^{-1/6}$,

$$0 < 3\|\Delta_1\| A_1 N^{-1/3} \hat{\sigma}_N^{-1} x_N \leqq \tfrac{1}{12} \tag{7.8}$$

in view of (7.2). Thus we can use Theorem 1 of Feller [5] (see also Petrov [20, p. 253]), to obtain that for all $N \geqq N_0$,

$$\mathbf{P}(\hat{S}_N^0 > x_N \hat{\sigma}_N) = \exp\{-\tfrac{1}{2} x_N^2 Q_N(x_N)\}[1 - \Phi(x_N) + \theta_N \lambda_N e^{-\lambda_N^2/2}], \tag{7.9}$$

where

$$\begin{aligned} &\lambda_N = 3\|\Delta_1\| A_1 N^{-1/3} \hat{\sigma}_N^{-1}, \quad \theta_N < 7.465; \quad Q_N(x) = \sum_{j=1}^{\infty} q_{Nj} x^j, \\ &q_{N1} = 3^{-1} \hat{\sigma}_N^{-3} \sum_{j=1}^{N} \mathbf{E}(\hat{S}_N^{(j)})^3, \qquad |q_{Nj}| < 8^{-1}(12\lambda_N)^j, \quad j \geqq 2. \end{aligned} \tag{7.10}$$

Note that $|x_N| \leqq \rho_N N^{1/6} + N^{-1/6}$. Since $\rho_N = o(1)$, we have $x_N = o(N^{1/6})$ as $N \to \infty$. Let $K > 0$ be such that

$$|x_N| \leqq K \hat{\sigma}_N N^{1/6} / (36 A_1 \|\Delta_1\|), \qquad N \geqq 1. \tag{7.11}$$

Also it follows by (7.2) and assumption (4.2) that

$$|q_{N1} x_N| \leqq 4 \hat{\sigma}_N^{-3} \|\Delta_1\| \sum_{j=1}^{N} |c_j|^3 x_N \leqq A_3 N^{-1/3}, \tag{7.12}$$

where A_3 is an absolute constant. Hence, combining (7.10), (7.11), and (7.12), we obtain

$$|Q_N(x_N)| \leqq A_3 N^{-1/3} + \sum_{j=2}^{\infty} |q_{Nj} x_N^j| \leqq A_3 N^{-1/3} + 8^{-1} \sum_{j=2}^{\infty} (K N^{-1/6})^j,$$

which implies that, as $N \to \infty$,

$$x_N^2 Q_N(x_N) = x_N^2 O(N^{-1/3}) = o(1) \quad \text{uniformly in } 0 < x \leqq \rho_N N^{1/6}. \tag{7.13}$$

Moreover, Lemma VII.1.2 of Feller [6] ensures that

$$\begin{aligned} &\theta_N \lambda_N \exp(-\lambda_N^2/2)[1 - \Phi(x_N)]^{-1} = O(1), \\ &[1 - \Phi(x_N)]^{-1}[1 - \Phi(x)] = 1 + o(1). \end{aligned} \tag{7.14}$$

The proof follows using (7.9), (7.13), and (7.14).

Appendix. We now prove Lemmas 6.1 and 6.2.

Proof of Lemma 6.1. Set $\hat{\sigma}_N^{-1} I_{22}^0 = (2(N+1))^{-1} \sum_{i=1}^N c_j^* \hat{H}_j$, where

$$\hat{H}_j = (N+1)^{-1} \sum_{k \neq j} \sum_{l \neq j,k} H_{jk} H_{jl} \varphi_1^{(2)}(\rho_{jj}, Z_j).$$

Then, an application of Lemma 5.2 yields that, for $1 \leqq j \leqq N$,

$$\mathbf{E} \hat{H}_j^{2r} \leqq (4e\|\Delta_2\|)^{2r} (2r)^{2r}$$

which, together with (6.9) and Hölder's inequality, gives the desired result.

Proof of Lemma 6.2. Since $\log \omega_3^{-1} \leqq 2(\max_{1\leqq j\leqq N} |c_j^*|)^{-1}$ when $\omega_3 \leqq 1$, it suffices to show that

$$\begin{aligned} &|\mathbf{E} \exp(it\hat{\sigma}_N^{-1}\hat{S}_N^0)\{\hat{\sigma}_N^{-1}(S_N^0 - \hat{S}_N^0 + I_{22}^0)\}^r| \\ &\quad\leqq K^r r^{3r} \omega_3 (1+|t|+t^2)^r \exp\left(\frac{-7t^2}{18} + \frac{t^2 r}{6} \max_{1\leqq j\leqq N} |c_j^*|\right). \end{aligned} \tag{A.1}$$

Let $\mu_j(t)$ be the characteristic function of $\hat{\sigma}_N^{-1}\hat{S}_N^{(j)}$. Then it follows (cf. Petrov [20, p. 110]) that

$$|\mu_j(t)| \leqq \exp\left\{-\frac{t^2}{2}\mathbf{E}(\hat{\sigma}_N^{-1}\hat{S}_N^{(j)})^2 + \frac{2}{3}|t|^3 \mathbf{E}|\hat{\sigma}_N^{-1}\hat{S}_N^{(j)}|^3\right\},$$

which, for $|t| \leqq \omega_3^{-1}$, yields

$$\prod_{j\in J} |\mu_j(t)| \leqq \exp\left\{\frac{-7t^2}{18}\frac{t^2}{18} \# J \max_{1\leqq j\leqq N} |c_j^*|\right\}, \tag{A.2}$$

where J is a subset of $\{1, 2, \cdots, N\}$ and $\#J$ is its cardinality.

Recalling that H_{jk} and V_{jk}, $1 \leqq j, k \leqq N$, are given by (5.3) and (5.6), respectively, we define r.v.'s for $1 \leqq j, k, l \leqq N$:

$$\tilde{V}_{jkl} = c_j^* \left\{\frac{N+1}{N-2} V_{jk} + \frac{1}{2} H_{jk} H_{jl} \varphi_1^{(2)}(\rho_{jj}, Z_j)\right\}. \tag{A.3}$$

Then it follows from (5.5) and the definition of I_{22} that

$$\hat{\sigma}_N^{-1}(S_N^0 - \hat{S}_N^0 + I_{22}^0) = (N+1)^{-2} \sum\sum\sum_{(j,k,l)\neq} \tilde{V}_{jkl}. \tag{A.4}$$

Note that for any distinct indices j, k, and l

$$\mathbf{E}(\tilde{V}_{jkl} | Z_j) = \mathbf{E}(\tilde{V}_{jkl} | Z_k) = \mathbf{E}(\tilde{V}_{jkl} | Z_l) = 0. \tag{A.5}$$

In view of (5.5) and (5.6), $\hat{\sigma}_N^{-1}\hat{S}_N^0$ is a sum of independent r.v.'s. This fact and (A.4) ensure that

$$\begin{aligned} &\mathbf{E}\{\exp(it\hat{\sigma}_N^{-1}\hat{S}_N^0)\hat{\sigma}_N^{-1}(S_N^0 - \hat{S}_N^0 + I_{22}^0)\} \\ &\quad= (N+1)^{-2} \sum\sum\sum_{(j,k,l)\neq} \left(\prod_{\nu\neq j,k,l} \mu_\nu(t)\right) \mathbf{E}\{\tilde{V}_{jkl} \exp(it\hat{\sigma}_N^{-1}(\hat{S}_N^{(j)} + \hat{S}_N^{(k)} + \hat{S}_N^{(l)}))\}, \end{aligned} \tag{A.6}$$

where $\mu_\nu(t)$ is the characteristic function of the r.v. $\hat{\sigma}_N^{-1}\hat{S}_N^{(\nu)}$. It follows by Lemma XV.4.1 of Feller [7] and (A.5) that

$$\mathbf{E}\{\tilde{V}_{jkl} \exp(it\hat{\sigma}_N^{-1}(\hat{S}_N^{(j)} + \hat{S}_N^{(k)} + \hat{S}_N^{(l)}))\} = O(t^2|c_j^*|\{(c_j^*)^2 + (c_k^*)^2 + (c_l^*)^2 + N^{-1}\omega_2\}). \tag{A.7}$$

Hence for $r = 1$, (A.1) follows from (A.2), (A.6), and (A.7).

We next consider the expansion of $\{\hat{\sigma}_N^{-1}(S_N^0 - \hat{S}_N^0 + I_{22}^0)\}^r$. Expanding it directly, we find that

$$\{\hat{\sigma}_N^{-1}(S_N^0 - \hat{S}_N^0 + I_{22}^0)\}^r = (N+1)^{-2r}\left(\sum\sum\sum_{(j,k,l)\neq} \tilde{V}_{jkl}\right)^r = (N+1)^{-2r} \sum_{b=3}^{3r} \sum_{\nu} q_{b\nu}^{(r)}, \tag{A.8}$$

where $q_{b\cdot}^{(r)}$ is of the form

$$q_{b\cdot}^{(r)} = \sum\sum\cdots\sum_{(j_1,j_2,\cdots,j_b)\neq} \tilde{V}_{aaa_1}\tilde{V}_{aaa_2}\cdots\tilde{V}_{aaa_r} \tag{A.9}$$

and each «a» is one of the indices $j_1, j_2, \cdots, j_b$. Thus we have

$$|\mathbf{E}\exp(it\hat{\sigma}_N^{-1}\hat{S}_N^0)\{\hat{\sigma}_N^{-1}(S_N^0-\hat{S}_N^0+I_{22}^0)\}^r|$$

$$= \left|(N+1)^{-2r}\sum_{b=3}^{3r}\left(\prod_{j\neq j_1,\cdots,j_b}\mu_j(t)\right)\sum_\nu \mathbf{E}\left\{q_{b\nu}^{(r)}\prod_{k=1}^{b}\exp(it\hat{\sigma}_N^{-1}\hat{S}_N^{(j_k)})\right\}\right| \tag{A.10}$$

$$\leqq \exp\left(-\frac{7t^2}{18}+\frac{t^2}{6}r\max_{1\leqq j\leqq N}|c_j^*|\right)\sum_{b=3}^{3r}\sum_\nu N^{-2r}\left|\mathbf{E}\left\{q_{b\nu}^{(r)}\prod_{k=1}^{b}\exp(it\hat{\sigma}_N^{-1}\hat{S}_N^{(j_k)})\right\}\right|$$

in view of (A.2).

Let $Q_b^{(r)}$, $3\leqq b\leqq 3r$, be the cardinality of the collection of all different terms $q_{b\cdot}^{(r)}$ and put $\Omega_r=\sum_{b=3}^{3r}Q_b^{(r)}$, which is the total number of terms $q_{b\cdot}^{(r)}$ in the expansion (A.8). Then clearly $\Omega_1=1$. To estimate Ω_r, we first note the recursive relation

$$\Omega_r\leqq\Omega_{r-1}\{(\alpha^2-\alpha+1)+\alpha^2+\alpha+1\}\leqq 2\alpha^3\Omega_{r-1},$$

where $\alpha=3(r-1)$. Thus there is a constant K_1 (independent of N and r) such that, for any positive integer r,

$$\Omega_r\leqq(54)^{r-1}((r-1)!)^3\leqq K_1^r r^{3r}, \tag{A.11}$$

where the second inequality follows by Stirling's formula (see, e.g., Feller [6]).

To complete the proof, it remains to show that for any positive integer r and b, $3\leqq b\leqq 3r$,

$$N^{-2r}\left|\mathbf{E}q_{b\cdot}^{(r)}\prod_{k=1}^{b}\exp(it\hat{\sigma}_N^{-1}\hat{S}_N^{(j_k)})\right|\leqq K_2^r\omega_3(1+|t|+t^2)^r, \tag{A.12}$$

where K_2 is an absolute constant. We shall prove this by induction on r. Note that (A.12) holds for $r=1$ in view of (A.7). We now suppose that (A.12) holds for $r\leqq m-1$ and let $r=m\geqq 2$.

If $b\leqq 2m=2r$, then (A.12) is trivial. Thus we consider only terms $q_{b\cdot}^{(m)}$, $b\geqq 2m+1$. Pick any s, $1\leqq s\leqq m$, and let $b=2m+s$. Let γ denote the number of indices j (in (A.9)) which appear at least twice as a subscript of one of the $\tilde{V}$-terms and then we have that $3m\geqq 2\gamma+2m+s-\gamma$, i.e.,

$$0\leqq\gamma\leqq m-s. \tag{A.13}$$

Hence at least $m+2s$ indices among $j_1, j_2, \cdots, j_{2m+s}$ appear exactly once. If one of the $\tilde{V}$-terms has three subscripts appearing only once, then the expectation in (A.12) is split into expectations of two groups to ensure (A.12) immediately by the induction hypothesis. This is the case when $s>m/2$ or $\gamma<s$.

Next we consider the remaining case when

$$1\leqq s\leqq m/2,\qquad s\leqq\gamma\leqq m-s \tag{A.14}$$

and none of the $\tilde{V}$-terms in (A.9) satisfy the condition that every one of its subscripts appears only once. Since at least $m+2s$ indices among $j_1, j_2, \cdots, j_{2m+s}$ appear exactly once, we must have at least $2s$ $\tilde{V}$-terms, of which exactly two subscripts appear only once. Thus, by rearranging the order of summations, we can write the typical form of

$q_{b}^{(r)} \prod_{k=1}^{b} \exp(it\hat{\sigma}_N^{-1}\hat{S}_N^{(j_k)})$ in the following way:

$$N^{-2m}\Big(\sum_{k_1}\sum_{l_1}\cdots\sum_{k_\alpha}\sum_{l_\alpha}\Big)\Big(\sum_{m_1}\cdots\sum_{m_\beta}\Big)\Big(\sum_{j_1}\cdots\sum_{j_\gamma}\Big)$$

$$\text{(A.15)}\qquad \times \tilde{V}_{ak_1l_1}\cdots\tilde{V}_{ak_\delta l_\delta}\tilde{V}_{k_{\delta+1}al_{\delta+1}}\cdots\tilde{V}_{k_\alpha l_\alpha a} \text{ —— (A)}$$

$$\times \tilde{V}_{aam_1}\cdots\tilde{V}_{aam_\varepsilon}\tilde{V}_{am_{\varepsilon+1}a}\cdots\tilde{V}_{m_\beta aa} \text{ —— (B)}$$

$$\times \tilde{V}_{aaa_1}\cdots\tilde{V}_{aaa\eta} \text{ —— (C)}$$

$$\times \prod_{\nu=1}^{\alpha}\exp(it\hat{\sigma}_N^{-1}(\hat{S}_N^{(k_\nu)}+\hat{S}_N^{(l_\nu)}))\prod_{\nu=1}^{\beta}\exp(it\hat{\sigma}_N^{-1}\hat{S}_N^{(m_\nu)})\prod_{\nu=1}^{\gamma}\exp(it\hat{\sigma}_N^{-1}\hat{S}_N^{(j_\nu)}),$$

where summations are taken over distinct indices; indices k, l, and m appear only once; j's appear at least twice among places marked by ⟪a⟫; $\alpha, \beta, \gamma, \delta, \varepsilon$, and η are non-negative integers such that

$$\text{(A.16)}\qquad \begin{array}{ll} 1\leqq s\leqq a/2\leqq m/2 & s\leqq\gamma\leqq m-s, \\ 0\leqq\delta\leqq\alpha\leqq m, & 0\leqq\varepsilon\leqq\beta\leqq m, \\ 2\alpha+\beta+\gamma=2m+s, & \alpha+\beta+\eta=m. \end{array}$$

We next consider the conditional expectation of (A.15) given r.v.'s $Z_{j_1}, Z_{j_2}, \cdots, Z_{j_\gamma}$. To this end, applying Lemma XV.4.1 of Feller [7] and then taking conditional expectations, we obtain that for distinct indices j, k, and l

$$\sum\sum_{(k,l)\neq} \mathbf{E}(\tilde{V}_{jkl}\tilde{e}_{kl}\,|\,Z_j)\leqq \sum\sum_{(k,l)\neq}|t|\{|c_k^*|+|c_l^*|+N^{-1}\omega_1\}\leqq 3|t|N^{5/3}\omega_3^{1/3},$$

$$\text{(A.17)}\qquad \sum\sum_{(k,l)\neq} \mathbf{E}(\tilde{V}_{kjl}\tilde{e}_{kl}\,|\,Z_j)\leqq \sum\sum_{(k,l)\neq}|t||c_k^*|\{|c_k^*|+|c_l^*|+N^{-1}\omega_1\}\leqq 3|t|N^{4/3}\omega_\beta^{2/3},$$

$$\sum\sum_{(k,l)\neq} \mathbf{E}(\tilde{V}_{klj}\tilde{e}_{kl}\,|\,Z_j)\leqq 3|t|N^{4/3}\omega_3^{2/3},$$

where $\tilde{e}_{kl}=\exp(it\hat{\sigma}_N^{-1}(\hat{S}_N^{(k)}+\hat{S}_N^{(l)}))$.

We finally split the proof into two cases when $\alpha=m$ or $\alpha<m$. We first prove (A.12) when $\alpha=m$ and then when $\alpha<m$.

When $\alpha=m$, we have only group (A) in (A.15). If $s\geqq 2$, then the expectation of (A.15) is split into two groups and the induction hypothesis ensures (A.12) immediately. If $s=1$ and $\delta=0$, then it follows by substitution of (A.17) into (A.15) that the expectation of (A.15) is bounded by

$$N^{-2m}(3|t|N^{4/3}\omega_3^{2/3})^m N\leqq(3|t|)^m\omega_3.$$

If $s=1$ and $\delta>0$ then the expectation is bounded by

$$N^{-2m}(3|t|N^{5/3}\omega_3^{1/3})^m\omega_1\leqq(3|t|)^m\omega_3.$$

This completes the proof for the case when $\alpha=m$.

We now suppose that $\alpha<m$. If some of the indices j appear only in group (A) of (A.15), then we can again split the expectation into two groups. Thus it remains to show (A.12) when all indices j in group (A) appear again in group (B) or (C). Note that $\beta+\gamma$ is the total number of indices m's and j's which appear in group (B) or (C) and that the number of indices m's and j's, which appear as one of the second or the third subscripts of the $\tilde{V}$-terms in group (B) or (C), is at most $2(\eta+\beta)$. Moreover,

$$\beta+\gamma-2(\eta+\beta)=\gamma-2(m-\alpha-\beta)-\beta=s$$

in view of (A.16). Thus at least s indices among $m_1, m_2, \cdots, m_\beta, j_1, j_2, \cdots, j_\gamma$ must appear as the first subscript of one of the $\tilde{V}$-terms in group (B) or (C). Utilizing this fact and substituting (A.17) into (A.15), we find that the expectation of (A.15) is bounded by

$$N^{-2m}(3|t|N^{5/3}\omega_3^{1/3})^{\delta}(3|t|N^{4/3}\omega_3^{2/3})^{\alpha-\delta}\omega_1^{s}N^{2m+s-(2\alpha+s)}$$
$$\leqq (3|t|)^{\alpha}N^{-2\alpha+\delta/3+4\alpha/3+2s/3}\omega_3^{2\alpha/3-\delta/3+s/3}$$
$$\leqq (3(1+|t|))^{m}N^{-(\alpha-2s)/3}\omega_3^{(\alpha+s)/3}\leqq (3(1+|t|))^{m}\omega_3$$

because $2s\leqq\alpha$ and $\delta\leqq\alpha$. This completes the proof of the case when $\alpha<m$. The proof follows.

REFERENCES

[1] H. Bergström and M. L. Puri, *Convergence and remainder terms in linear rank statistics*, Ann. Statist., 5 (1977), pp. 671–680.

[2] P. K. Bhattacharya, *Convergence of sample path of normalized sums of induced order statistics*, Ann. Statist., 2 (1974), pp. 1034–1039.

[3] S. Bjerve, *Error bounds for linear combinations of order statistics*, Ann. Statist., 5 (1977), pp. 357–369.

[4] R. J. M. M. Does, *Berry-Esseen theorems for simple linear rank statistics under null-hypothesis*, Ann. Probab., 10 (1982), pp. 982–991.

[5] W. Feller, *Generalization of a probability limit theorem of Cramér*, Trans. Amer. Math. Soc., 54 (1943), pp. 361–372.

[6] ———, *An Introduction to Probability Theory and Its Applications*, Vol. 1, 3rd ed., John Wiley, New York, 1968.

[7] ———, *An Introduction to Probability Theory and Its Applications*, Vol. II, 2nd ed., John Wiley, New York, 1971.

[8] J. Hájek, *Asymptotically most powerful rank order tests*, Ann. Math. Statist., 33 (1962), pp. 1129–1147.

[9] ———, *Asymptotic normality of simple linear rank statistics under alternatives*, Ann. Math. Statist., 39 (1968), pp. 325–346.

[10] R. Helmers, *The order of the approximation for linear combinations of order statistics with smooth weight functions*, Ann. Probab., 5 (1977), pp. 940–953.

[11] ———, *A Berry-Esseen theorem for linear combinations of order statistics*, Ann. Probab., 9 (1981), pp. 342–347.

[12] W. Hoeffding, *A combinatorial central limit theorem*, Ann. Math. Statist., 22 (1951), pp. 558–566.

[13] M. Hušková, *Asymptotic distribution of simple linear rank statistics for testing symmetry*, Z. Wahrsch. Verw. Gebiete, 14 (1970), pp. 308–322.

[14] ———, *The rate of convergence of simple linear rank statistics under hypothesis and alternatives*, Ann. Statist., 5 (1977), pp. 658–670.

[15] ———, *The Berry-Esseen theorem for rank statistics*, Comment. Math. Univ. Carolin., 20 (1979), pp. 399–415.

[16] J. Jurečková and M. L. Puri, *Order of normal approximation for rank test statistics distribution*, Ann. Probab., 3 (1975), pp. 526–533.

[17] W. C. M. Kallenberg, *Cramér type large deviations for simple linear rank statistics*, Z. Wahrsch. Verw. Gebiete, 60 (1982), pp. 403–409.

[18] M. Motoo, *On the Hoeffding's combinatorial central limit theorem*, Ann. Inst. Statist. Math., 8 (1957), pp. 145–154.

[19] U. Müller-Funk and H. Witting, *On the rate of convergence in* CLT *for signed linear rank statistics*, Nonparametric Statistical Inference, Vol. II, B. V. Gnedenko, M. L. Puri, and I. Vincze, eds., Elsevier, Amsterdam, 1982, pp. 637–652.

[20] V. V. Petrov, *Sums of Independent Random Variables*, Springer-Verlag, Berlin, New York, 1975.

[21] S. S. Ralescu and M. L. Puri, *On the rate of convergence in the central theorem for signed rank statistics*, Adv. Appl. Math., (1985), pp. 23–51.

[22] S. M. Stigler, *Linear functions of order statistics with smooth weight functions*, Ann. Statist., 2 (1974), pp. 676–693.

[23] P. van Beek, *An application of Fourier methods to the problem of sharpening the Berry-Esseen inequality*, Z. Wahrsch. Verw. Gebiete, 23 (1972), pp. 187–196

[24] M. Vandermaele and N. Veraverbeke, *Cramér-type large deviations and moderate deviations for linear combinations of order statistics*, Ann. Probab., 10 (1982), pp. 423–434.

5] V. VON BAHR, *Remainder term estimate in a combinatorial limit theorem*, Z. Wahrsch. Verw. Gebiete, 35 (1976), pp. 131–139.

6] S. YANG, *Linear functions of concomitants of order statistics with application to nonparametric estimate of a regression function*, J. Amer. Statist. Assoc., 76 (1981), pp. 658–662.

7] M. L. PURI AND M. SEOH, *On the rate of convergence to normality for generalized linear rank statistics*, Ann. Inst. Statist. Math., 37 (1985), pp. 51–67.

8] M. L. PURI AND S. S. RALESCU, *Almost sure linearity for signed rank statistics in the non-i.i.d. case*, Acta Math. Hungar., 48 (1986), pp. 273–284.

9] M. SEOH, S. S. RALESCU, AND M. L. PURI, *Cramér-type large deviations for generalized rank statistics*, Ann. Probab., 13 (1985), pp. 115–125.

30] M. SEOH AND M. L. PURI, *Asymptotic normality of a class of nonparametric statistics*, Econ. Theory, 3 (1987), pp. 313–347.

Reprinted from
Theory of Probability and Its Applications, Volume 33
(1988), pages 682-697, by permission of the publisher,

ON HILBERT-SPACE-VALUED U-STATISTICS*

M. L. PURI† and V. V. SAZONOV
(SUBMITTED BY YU. V. PROKHOROV)

In this note we show that an elementary argument similar to the one used in [1] for real-valued U-statistics can be applied to Hilbert-space-valued U-statistics to obtain the rate of convergence of their distributions to their Gaussian limits.

Let $\{X_i,\ i \geqq 1\}$ be a sequence of independent and identically distributed random variables with values in a measurable space $(S, \mathcal{S})$, where $\mathcal{S}$ is a σ-algebra. Let $h(s,t)$ be a function on $S \times S$ with values in a separable Hilbert space H; $h(s,t)$ is symmetric, i.e., $h(s,t) = h(t,s)$, $s,t \in S$, and measurable in the sense that $h^{-1}(E) \in \mathcal{S} \times \mathcal{S}$ for any Borel set $E \subset H$. We consider the U-statistic of degree 2, viz.,

$$U_n = \binom{n}{2}^{-1} \sum_{1 \leqq i < j < n} h(X_i, X_j), \qquad n = 1, 2, \cdots \tag{1}$$

(the U-statistics of degree higher that 2 can be treated similarly).

Without loss of generality, let $\mathbf{E}h(X_1,X_2) = 0$, and assume that $\sigma^2 = \mathbf{E}|h(X_1,X_2)|^2 < \infty$. Let $g(x) = \mathbf{E}(h(X_1,X_2) \mid X_1 = x)$, denote by V the covariance operator corresponding to $g(X_1)$, i.e., the operator defined by $(Vx,y) = \mathbf{E}(g(X_1),x)(g(X_1),y)$, and let $\sigma_1^2 \geqq \sigma_2^2 \geqq \cdots$ be the eigenvalues of V. Furthermore, let Y be a Gaussian random variable with zero mean and covariance operator V, and let $B_r(a) = \{x \in H\colon |x-a| < r\}$.

Then, we have the following theorem.

THEOREM. *For all* $n \geqq 1$, $a \in H$, $r \geqq 0$,

$$\begin{aligned} \Delta_n(a,r) &= \left|\mathbf{P}\big((n^{1/2}/2)U_n \in B_r(a)\big) - \mathbf{P}\big(Y \in B_r(a)\big)\right| \\ &\leqq c\Big(\Big(\sigma_1^{-1}\sigma_2^{-1}\big(\sigma + |a|^2\big)\big(\mathbf{E}\big|h(X_1,X_2)\big|^2 - 2\sigma^2\big)n^{-1}\Big)^{1/3} \\ &\quad + \sigma^3\Big(\prod_{i=1}^{6}\sigma_i^{-1}\mathbf{E}\big|g(X_1)\big|^3\big(1+|a|^3\big)n^{-1/2}\Big), \end{aligned} \tag{2}$$

where c is an absolute constant.

If, in addition, $\mathbf{E}|h(X_1,X_2)|^{2p} < \infty$ *for a positive integer* $p \geqq 2$, *then*

$$\Delta_n(a,r) = O\big(n^{-p/(2p+1)}\big). \tag{3}$$

Proof. Without loss of generality, we may assume that $\sigma_6 \neq 0$ and $\mathbf{E}|g(X_1)^3| < \infty$. Put

$$\widehat{U}_n = \binom{n}{2}^{-1} \sum_{1 \leqq i < j \leqq n} \big(g(X_i) + g(X_j)\big) = \frac{2}{n}\sum_{i=1}^{n} g(X_i), \tag{4}$$

---

*Received by the editors December 18, 1990.

†This research was supported by the Office of Naval Research under contract N00014–91–J–1020.

the projection of U_n. Letting $V_n = n^{1/2}U_n/2$ and $\widehat{V}_n = n^{1/2}\widehat{U}_n/2$, we can write

$$\Delta_n(a,r) \leqq \left|\mathbf{P}\big(V_n \in B_r(a)\big) - \mathbf{P}\big(\widehat{V}_n \in B_r(a)\big)\right| + \left|\mathbf{P}\big(\widehat{V}_n \in B_r(a)\big) - \mathbf{P}\big(Y \in B_r(a)\big)\right| = \Delta_{n1}(a,r) + \Delta_{n2}(a,r).$$

Observe now that, for any $\varepsilon > 0$, $q > 0$,

$$\begin{aligned}\Delta_{n1}(a,r) &= \left|\mathbf{P}\big(V_n \in B_r(a),\ \widehat{V}_n \notin B_r(a)\big) - \mathbf{P}\big(\widehat{V}_n \in B_r(a),\ V_n \notin B_r(a)\big)\right| \\ &\leqq \mathbf{P}\big(\widehat{V}_n \in B_{r+\varepsilon}(a)\backslash B_r(a)\big) + \mathbf{P}\big(\widehat{V}_n \in B_r(a)\backslash B_{r-\varepsilon}(a)\big) + 2\mathbf{P}\big(|V_n - \widehat{V}_n| \geqq \varepsilon\big) \\ &\leqq \Delta_{n2}(a, r-\varepsilon) + \Delta_{n2}(a, r+\varepsilon) + \mathbf{P}\big(Y \in B_{r+\varepsilon}(a)\backslash B_{r-\varepsilon}(a)\big) + 2\varepsilon^{-q}\mathbf{E}|V_n - \widehat{V}_n|^q,\end{aligned}$$

where $B_{r-\varepsilon}(a) = \varnothing$ if $\varepsilon > r$. Thus

$$\Delta_n(a,r) \leqq \sum_{i=1}^{3} \Delta_{n2}(a, r_i) + \mathbf{P}\big(Y \in B_{r+\varepsilon}(a)\backslash B_{r-\varepsilon}(a)\big) + 2\varepsilon^{-q}\mathbf{E}|V_n - \widehat{V}_n|^q, \tag{5}$$

where $r_1 = r - \varepsilon$, $r_2 = r$, $r_3 = r + \varepsilon$. By [2], we have

$$\Delta_{n2}(a, r_i) \leqq c\sigma^3\left(\prod_{i=1}^{6}\sigma_i^{-1}\right)\mathbf{E}\left|g(X_1)\right|^3\big(1 + |a|^3\big)n^{-1/2}, \qquad i = 1,2,3, \tag{6}$$

and, by [3, p. 84],

$$\mathbf{P}\big(Y \in B_{r+\varepsilon}(a)\backslash B_{r-\varepsilon}(a)\big) \leqq c\sigma_1^{-1}\sigma_2^{-1}\big(\sigma + |a|\big)\varepsilon. \tag{7}$$

Furthermore, it is easy to check that

$$\mathbf{E}\big(h(X_i, X_j), g(X_i)\big)^2 = \sigma^2 \tag{8}$$

and, for any, i, j, k, $i \neq j \neq k$,

$$\mathbf{E}\big(h(X_i, X_j), h(X_i, X_k)\big) = \mathbf{E}\big[\mathbf{E}\big(h(X_i, X_j), h(X_i, X_k)\big) \mid X_i\big] = \sigma^2. \tag{9}$$

A simple straightforward calculation using (8) and (9) gives

$$\mathbf{E}|V_n - \widehat{V}_n|^2 = 2^{-1}(n-1)^{-1}\big(\mathbf{E}\left|h(X_1, X_2)\right|^2 - 2\sigma^2\big). \tag{10}$$

Combining (5)–(7), (10) and putting

$$\varepsilon = \big(n^{-1}\big(\mathbf{E}\left|h(X_1, X_2)\right|^2 - 2\sigma^2\big)\sigma_1\sigma_2(\sigma + |a|)^{-1}\big)^{1/3},$$

we obtain (2).

For an integer $p \geqq 2$, the same reasoning as in Theorem 2.1 in [1] applied to

$$\mathbf{E}|V_n - \widehat{V}_n|^{2p} = n^p 2^{-2p}\binom{n}{2}^{-2p}\left(\sum_{1 \leqq i < j \leqq n} H(X_i, X_j),\ \sum_{1 \leqq i < j \leqq n} H(X_i, X_j)\right)^p,$$

where $H(X_i, X_j) = h(X_i, X_j) - g(X_i) - g(X_j)$, gives

$$\mathbf{E}|V_n - \widehat{V}_n|^{2p} = O(n^{-p}). \tag{11}$$

Combining (5)–(7), (11) and putting $\varepsilon = n^{-p/(2p+1)}$, we obtain (3).

An obvious consequence of the above theorem is that if $|g(X_1)|$ has moments of all orders, then

$$\Delta_n(a,r) = O\big(n^{-1/2+\delta}\big)$$

for any $\delta > 0$.

REFERENCES

[1] W. F. GRAMS AND R. J. SERFLING, *Convergence rates for U-statistics*, Ann. Statist., 1 (1973), pp. 153–160.

[2] V. V. SAZONOV, V. V. UL'YANOV, AND B. A. ZALESSKII, *A regular estimate of the convergence rate in the central limit theorem in Hilbert space*, Math. USSR Sbornik, 180 (1989), pp. 1587–1613. (In Russian.)

[3] V. I. PAULAUSKAS AND A. YU. RACHKAUSKAS, *Accuracy of Normal Approximation in the Central Limit Theorem in Banach Spaces*, Mokslas, Vilnius, 1987. (In Russian.)

Reprinted from
Theory of Probability and Its Applications, Volume 36
(1991), pages 619-621, by permission of the publisher,

ON THE CENTRAL LIMIT THEOREM IN HILBERT SPACE WITH APPLICATION TO U-STATISTICS

MADAN L. PURI and V. V. SAZONOV

Indiana University and Steklov Mathematical Institute
Bloomington, Indiana, U.S.A. and Moscow, Russia

Abstract. Estimates of the rate of convergence on balls in the Central Limit Theorem in Hilbert space are presented. An application to the study of the limit behavior of the distributions of Hilbert valued U-statistics is given.

Key words: Central Limit Theorem, Hilbert space, U-statistics.

Let $X_1, X_2, \ldots$ be independent and identically distributed (i.i.d.) random variables with values in a separable Hilbert space H. Assume that $E|X_1|^2 < \infty$ and, for simplicity, that $EX_1 = 0$. Denote by V the covariance operator corresponding to X_1, i.e. $(V x, y) = E(X_1, x)(X_1, y)$ and let $\sigma_1^2 \geq \sigma_2^2 \geq \ldots$ be the eigenvalues of V. For $a \in H$ and $r \geq 0$ denote also $B_r(a) = \{x \in H : |x - a| < r\}$, $B_r = B_r(0)$. Recently in [1] the following theorem was proved.

THEOREM 1. *For any $a \in H$, $r \geq 0$, $n = 1, 2, \ldots$*

$$\Delta_n(a,r) = |P(S_n \in B_r(a)) - P(Y \in B_r(a))| \leq c \left(\prod_{i=1}^{6} \sigma_i^{-1} \right) \sigma^3 \beta (1 + |a|^3) n^{-\frac{1}{2}} \tag{1}$$

where c (here and below, with or without indices) is a positive constant and $\sigma^2 = E|X_1|^2$, $\beta = E|X_1|^3$.

Estimate (1) is a precise estimate. Indeed, from [2] and [3] it is easy to deduce that for any $c_0 > 0$ and $1 \geq \tau_1^2 \geq \ldots \geq \tau_\sigma^2$, there exist $a \in H$, $|a| > c_0$, and a probability measure P on H such that if $X_1, X_2, \ldots$ are i.i.d. random variables distributed as P, then $EX_1 = 0$, $\sigma_i^2 = \tau_i^2$, $i = 1, \ldots, 6$, $\beta < \infty$, and

$$\liminf_{n \to \infty} n^{\frac{1}{2}} \sup_{r \geq 0} \Delta_n(a,r) \geq c \left(\prod_{1}^{6} \sigma_i^{-1} \right) \sigma^3 \beta (1 + |a|^3). \tag{2}$$

The proof of (1) basically goes along the following lines. Assume that $\sigma = 1$ (this is not a restriction since the general case is reduced to this one if instead of X_i we consider $(E|X_i|^2)^{-\frac{1}{2}} X_i$, $i = 1, 2, \ldots$). We can also assume that

$$\left(\prod_{1}^{6} \sigma_i^{-1} \right) \beta n^{-\frac{1}{2}} \leq c_1, \tag{3}$$

since otherwise the theorem is obvious.

Note that if $X_i^{(1)} = X_i 1_{\{|X_i| \le n^{\frac{1}{2}}\}}$, $i = 1, 2, ..., n$, and $S_n' = n^{-\frac{1}{2}} \sum_{i=1}^{n} X_i^{(1)}$, then, as is easy to prove, for any Borel set E

$$\Delta_{n1}(a,r) = |P(S_n \in E) - P(S_n' \in E)| \le nP(|X_1| > n^{\frac{1}{2}}) \le \beta n^{-\frac{1}{2}}.$$

Thus it is enough to prove **(1)** with S_n replaced by S_n'.

Define $R = (\sigma_6^{-8}\beta^3 n)^{\frac{1}{5}}$. We will assume that $R \le n^{\frac{1}{2}}$; the opposite case is treated similarly but slightly differently. Note that if **(3)** is satisfied, then

$$P(B_R^c) \le \beta R^{-3} \le 1/2$$

if c_1 is small enough (which we assume). In this case, denoting P (resp. P_1) the probability measure corresponding to X_1 (resp. $X_1^{(1)}$) and defining P_2 as

$$P_2(A) = P(A \cap B_R)/P(B_R),$$

we have $P_1 \ge P_2/2$ and, hence,

$$P_1 = (P_2 + P_3)/2,$$

where P_3 is a probability measure.

Now let P_4 be the probability measure corresponding to the H-valued random variable

$$Z' = \zeta(Z - EZ) + Y' + EZ, \tag{4}$$

where ζ is a bounded real random variable such that $E\zeta = 0$, $E\zeta^2 = 1/2$ and $E\zeta^3 = 1$, Z is distributed as P_2, Y' is $(0, V_2/2)$ Gaussian, V_2 is the covariance operator of P_2, and ζ, Z, Y' are independent. It is easy to check that Z' has mean EZ, covariance operator V_2 and for any $h_1, h_2, h_3 \in H$

$$E(Z', h_1), (Z', h_2)(Z', h_3) = E(Z, h_1)(Z, h_2)(Z, h_3),$$

i.e. Z and Z' have the same moments of the first three orders, and in this sense they are "close". At the same time P_4, the distribution of Z', has a "solid" Gaussian component, the distribution of Y'. Denote

$$P_5 = (P_4 + P_3)/2$$

and let $X_i^{(5)}$, $i = 1, 2, ...$ be independent random variables with distribution P_5. Define $S_n'' = n^{-\frac{1}{2}} \sum_1^n X_i^{(5)}$. The next step is to estimate

$$\Delta_{n2}(a,r) = |P(S_n' \in B_r(a)) - P(S_n'' \in B_r(a))|.$$

If, for a measure μ on H, we denote $\bar{\mu}(\cdot) = \mu(n^{\frac{1}{2}} \cdot)$, $\mu^n = \mu * ... * \mu(n \text{ times})$, then, of course, the distribution of S_n' (resp. S_n'') can be written as $((\bar{P}_2 + \bar{P}_3)/2)^n$ (resp. $((\bar{P}_4 + \bar{P}_3)/2)^n$ and

$$\begin{aligned}\Delta_{n2}(a,r) &= \left|\left(((\bar{P}_2 + \bar{P}_3)/2)^n - ((\bar{P}_4 + \bar{P}_3)/2)^n\right)(B_r(a))\right| \\ &\le 2^{-n} \left(\sum\nolimits_1 + \sum\nolimits_2\right) \binom{n}{q} (\bar{P}_2^q - \bar{P}_4^q) * \bar{P}_3^{n-q}(B_r(a)) = I_1 + I_2,\end{aligned}$$

where $\sum_1$ is the summation over all integers q such that

$$|q-(n+1)/2| < (1/2)n^{7/10}$$

and $\sum_2$ is the summation over the remaining q, $0 \le q \le n$. Well known estimates of the tails of the binomial distribution give $I_2 \le cn^{-\frac{1}{2}}$ and one has to estimate I_1 composed of terms with q "not very different" from $(n+1)/2$. Consider a term

$$\nu_q = ((\bar{P}_2^q - \bar{P}_4^q) * \bar{P}_3^{n-q}(B_r(a)).$$

One can show that under our conditions the function $\bar{P}_4^q * \bar{P}_3^{n-q}(B_{r^{\frac{1}{2}}}(a))$ as a function of r is a distribution function with density bounded by $c\sigma_1^{-1}\sigma_2^{-1}$ (this is due to the solid Gaussian component in P_4). Thus by the Esséen inequality, for any $T > 0$

$$|\nu_q| \le c\left(\int_{|t|\le T} g_n(t)|t|^{-1}dt + \sigma_1^{-1}\sigma_2^{-1}T^{-1}\right),$$

where $g_n(t) = |f_{2n}(t) - f_{4n}(t)|$,

$$f_{kn}(t) = E\exp\left\{it\left|n^{-\frac{1}{2}}\left(\sum_1^q X_j^{(k)} + \sum_1^{n-q} X_j^{(3)}\right) - a\right|^2\right\},$$

and $X_j^{(k)}$ are independent of $X_j^{(3)}$, $k = 2, 4$. Then we choose an appropriately large T and for "large" $|t|$, $T_1 \le |t| \le T$, we use the inequality $|g_n(t)| \le |f_{2n}(t)| + |f_{4n}(t)|$ and estimate the integrals

$$\int_{T_1 \le t \le T} |f_{kn}(t)|\ |t|^{-1}dt, \qquad k = 2, 4 \tag{5}$$

individually. For "small" t, we proceed as follows. Using the fact that $\bar{P}_2$ and $\bar{P}_4$ have the same moments of the first three orders, we can write

$$\begin{aligned}
&\left|\int_{|t|\le T_1} g_n(t)|t|^{-1}dt\right| \\
&\le \int_{|t|\le T_1} |t|^{-1}\left|\int \exp\left\{it|x-a|^2\right\}(\bar{P}_2^q - \bar{P}_4^q) * \bar{P}_3^{n-q}(dx)\right| dt \\
&\le \sum_{j=0}^{q-1} \int_{|t|\le T_1} |t|^{-1}\left|\iint f(1)\bar{P}_2^j * \bar{P}_4^{q-j-1} * \bar{P}_3^{n-q}(dx)(\bar{P}_2 - \bar{P}_4)(dy)\right| dt \\
&\le \frac{1}{6}\sum_{j=1}^{q-1} \int_{|t|\le T_1} |t|^{-1}\left|\int_0^1 \iint f^{(4)}(\lambda)\bar{P}_2^j * \bar{P}_4^{q-j-1} * \bar{P}_3^{n-q}(dx)(\bar{P}_2 - \bar{P}_4)(dy)(1-\lambda)^3 d\lambda\right| dt
\end{aligned} \tag{6}$$

where $f(\lambda) = \exp\{it|x+\lambda y - a|^2\}$. Estimates of the integrals in **(5)** and in the right side of **(6)** are basically similar and use versions of the Götze–Yurinskii Lemma (see [4], [5]) which gives estimates of expressions like

$$|f_{kn}(t)|, \left|\int f^{(4)}(\lambda)\bar{P}_2^j * \bar{P}_4^{q-j-1} * \bar{P}_5^{n-q}(dx)\right|.$$

Next comes the estimation of

$$\Delta_{n3}(a,r) = |P(S_n'' \in B_r(a)) - P(Y \in B_r(a))|.$$

To this aim the following estimate is proved first. Let G be the $(0,V)$ Gaussian probability measure on $H, R = \bar{P}_5 - \bar{G}$ and let Y_1 be a $(0,U)$ Gaussian H-valued random variable with $trU \le 2$. Denote $\rho_1^2 \ge \rho_2^2 \ge ...$ the eigenvalues of U and let $e_1, e_2, ...$ be the corresponding eigenvectors. Then for any $r \ge 0$, $b \in H$

$$\left|\int P(|Y_1 - x - b| \le r) R(dx)\right| \le c \left(\prod_1^6 \rho_i^{-1}\right)\left(\beta + \sum_{i=2}^{3} E|(X_1, \bar{b}_j)|^3\right) n^{-3/2} \tag{7}$$

where $\bar{b}_2 = \sum_5^6 (b, e_j)e_j$, $\bar{b}_3 = \sum_7^\infty (b, e_j)e_j$. The proof of this inequality is rather long and technical.

To estimate $\Delta_{n3}(a,r)$ we write

$$\Delta_{n3}(a,r) = |(\bar{P}_5^n - \bar{G}^n)(B_r(a))| \le \sum_{m=0}^{n-1} I_m,$$

where $I_m = |\bar{P}_5^m * \bar{G}^{n-m-1} * R(B_r(a))|$. If $m \le (n-1)/2$, we represent I_m as

$$I_m = \left|\int\int \bar{G}^{n-m-1}(B_r(a) - x - y) R(dx) \bar{P}_5^m(dy)\right|$$

and apply **(7)** with Y_1 $(0, ((n-m-1)/n)V)$ Gaussian (note that $(n-m-1)/n \ge 1/4$ if $n \ge 2$). If $m > (n-1)/2$ we write I_m as

$$I_m = \left|\int\int \bar{P}_5^m(B_r(a) - x - y) R(dx) \bar{G}^{n-m-1}(dy)\right|$$

and represent $\bar{P}_5^m$ as

$$\bar{P}_5^m = ((\bar{P}_4 + \bar{P}_3)/2)^m = \left(\sum\nolimits_1 + \sum\nolimits_2\right) 2^{-m} \binom{m}{\ell} \bar{P}_4^\ell * \bar{P}_3^{m-\ell}$$

where $\sum_1$ is the summation over all ℓ such that $|2\ell - m| < n^{7/10}$ and $\sum_2$ is the summation over all remaining $\ell \in \{0, ..., m\}$. Using known estimates of the tails of the binomial distribution we have $2^{-m}\sum_2 \binom{m}{\ell} \le cn^{-3/2}$, and hence

$$I_m \le 2^{-m} \sum\nolimits_1 \binom{m}{\ell} I_{m\ell} + cn^{-3/2},$$

where

$$\begin{aligned} I_{m\ell} &= \left|\int\int\int \bar{P}_4^\ell(B_r(a) - x - y - z) R(dx) \bar{G}^{n-m-1}(dy) \bar{P}_3^{m-\ell}(dz)\right| \\ &= \left|\int\int\int\int G_\ell(B_r(a) - x - y - z - u) R(dx) \bar{G}^{n-m-1}(dy) \bar{P}_3^{m-\ell}(dz) \bar{P}_6^\ell(du)\right|, \end{aligned}$$

since $\bar{P}_4^\ell = G_\ell * \bar{P}_6^\ell$, where G_ℓ is $(0, (\ell/2n)V_2)$ Gaussian and P_6 is the distribution of $\zeta(Z - EZ) + EZ$ in **(4)**. Observe that $\ell/n \ge c'$ if $n \ge c''$, and under condition **(3)**

$\sigma_{2j}^2 \geq \sigma_j^2$, $j = 1, ..., 6$, where $\sigma_{21}^2 \geq \sigma_{22}^2 \geq ...$ are the eigenvalues of V_2. To estimate $I_{m\ell}$ it remains now to apply **(7)** and to make some moment estimations.

The estimates of $\Delta_{nk}(a, r)$, $k = 1, 2, 3$ together prove the assertion of the theorem.

Estimate **(1)** has sense only if $E|X_1|^3 < \infty$. However one can construct similar meaningful estimates if only $E|X_1|^p < \infty$, $2 \leq p < 3$. E.g. (see [6]) in the notation of Theorem 1 for all p, $2 \leq p \leq 3$,

$$\Delta_n(a, r) \leq c(p)\left(\prod_1^6 \sigma_i^{-1}\right)^{p/3} \sigma^p E|X_1|^p(1 + |a|^p)n^{(2-p)/2}.$$

One can also construct estimates which sometimes give better bounds than **(1)**. Denote $d = ||a| - r|$, which is the distance from zero to the boundary of $B_r(a)$. Then in the notation of Theorem 1

$$\Delta_n(a, r) \leq c_1\left[\sigma^{-1}\beta(1 + d^3)^{-1} + \left(\prod_{i=1}^6 \sigma_j^{-1}\right)\sigma^3\beta(1 + \min(|a|^3, r^3))\exp\{-c_2 d\}\right]n^{-\frac{1}{2}} \quad \textbf{(8)}$$

where α is a constant $\geq 1/5$ (see [7]). A similar estimate is claimed in [8]. Estimate **(8)** implies in particular that

$$\Delta_n(a, r) \leq c'\left(\prod_1^6 \sigma_i^{-1}\right)\sigma^3\beta(1 + \min(|a|^3, r^3))(1 + d^3)^{-1}n^{-\frac{1}{2}}.$$

Consider now an application of Theorem 1 to the estimation of the rate of convergence of H-valued U-statistics. Let $T_1, T_2, ...$ be a sequence of i.i.d. random variables with values in a measurable space $(S, \mathcal{A})$, where $\mathcal{A}$ is a σ-algebra. Let $h(s, t)$ be a function on $S \times S$ with values in H which is symmetric, i.e. $h(s, t) = h(t, s)$ and measurable in the sense that $h^{-1}(E) \in \mathcal{A} \times \mathcal{A}$ for any Borel set $E \subset H$. Let U_n be the U-statistic of degree 2, viz.

$$U_n = \binom{n}{2}^{-1} \sum_{1 \leq i < j \leq n} h(T_i, T_j), \qquad n = 1, 2, ...$$

Without loss of generality let $Eh(T_1, T_2) = 0$ and assume that $E|h(T_1, T_2)|^2 < \infty$. Let $Z_j = E(h(T_j, T_{j+1})|T_j)$, $j = 1, 2, ..., \tau^2 = E|Z_1|^2$, denote by W the covariance operator corresponding to Z_1 and let $\tau_1^2 \geq \tau_2^2 \geq ...$ be its eigenvalues. Finally let Y' be a $(0, W)$ Gaussian H-valued random variable. The following result was obtained in [9].

THEOREM 2. *For all* $a \in H$, $r \geq 0$, $n = 1, 2, ...$

$$D_n(a, r) = \left|P((n^{\frac{1}{2}}/2)U_n \in B_r(a)) - P(Y' \in B_r(a))\right| \quad \textbf{(9)}$$

$$\leq c\left(\left(\tau_1^{-1}\tau_2^{-1}(r + |a|)^2(E|h(T_1, T_2)|^2 - 2\tau^2)n^{-1}\right)^{1/3} + \left(\prod_{j=1}^6 \tau_j^{-1}\right)\tau^3 E|Z_1|^3(1 + |a|^3)n^{-\frac{1}{2}}\right).$$

If, in addition, $E|h(T_1, T_2)|^{2s} < \infty$ for a positive integer $s \geq 2$, then

$$D_n(a, r) = O\left(n^{-s/(2s+1)}\right). \quad \textbf{(10)}$$

Theorem 2 implies obviously that if $|h(T_1, T_2)|$ has moments of all orders then $D_n(a,r) = O(n^{-1/2+\delta})$ for all $\delta > 0$.

Let us sketch the proof of Theorem 2. Denote

$$\hat{U}_n = \binom{n}{2} \sum_{1\le i<j\le n} (Z_i + Z_j) = \frac{2}{n}\sum_{i=1}^{n} Z_i.$$

Letting $V_n = n^{\frac{1}{2}}U_n/2$, $\hat{V}_n = n^{\frac{1}{2}}\hat{U}_n/2$, we can write

$$\begin{aligned} D_n(a,r) &\le \left|P(V_n \in B_r(a)) - P(\hat{V}_n \in B_r(a))\right| + \left|P(\hat{V}_n \in B_r(a)) - P(Y' \in B_r(a))\right| \\ &= D_{n1}(a,r) + D_{n2}(a,r). \end{aligned}$$

Now for any $\varepsilon > 0$, $q > 0$ we have

$$\begin{aligned} D_{n1}(a,r) &= \left|P(V_n \in B_r(a)), \hat{V}_n \notin B_r(a)) - P(\hat{V}_n \in B_r(a), V_n \notin B_r(a))\right| \\ &\le P(\hat{V}_n \in B_{r+\varepsilon}(a)\backslash B_r(a)) + P(\hat{V}_n \in B_r(a)\backslash B_{r-\varepsilon}(a)) + 2P(|V_n - \hat{V}_n| \ge \varepsilon) \\ &\le D_{n2}(a, r-\varepsilon) + D_{n2}(a, r+\varepsilon) \\ &\quad + P(Y \in B_{r+\varepsilon}(a)\backslash B_{r-\varepsilon}(a)) + 2\varepsilon^{-q}E|V_n - \hat{V}_n|^q \end{aligned}$$

where $B_{r-\varepsilon}(a) = \phi$ if $\varepsilon > r$. Thus

$$D_n(a,r) \le \sum_{i=1}^{3} D_{n2}(a, r_i) + P(Y \in B_{r+\varepsilon}(a)\backslash B_{r-\varepsilon}(a)) + 2\varepsilon^{-q}E|V_n - \hat{V}_n|^q,$$

where $r_1 = r - \varepsilon$, $r_2 = r$, $r_3 = r + \varepsilon$. To estimate D_{n2} we apply Theorem 1. Furthermore it is known that

$$P(Y' \in B_{r+\varepsilon}(a)\backslash B_{r-\varepsilon}(a)) \le c\tau_1^{-1}\tau_2^{-1}(\tau + |a|)\varepsilon$$

(see e.g. [10]), and a simple calculation gives

$$E|V_n - \hat{V}_n|^2 = 2^{-1}(n-1)^{-1}(E|h(T_1,T_2)|^2 - 2\tau^2).$$

Now choosing appropriately ε we obtain (9); (10) is proved similarly.

Acknowledgements.
Research of Madan L. Puri supported by the Office of Naval Research Contract N00014-91-J-1020.

REFERENCES:

1. Sazonov, V.V., Ulyanov, V.V. and Zalesskii, B.A. A precise estimate of the rate of convergence in the central limit theorem in Hilbert space. *Mat. Sbornik.*, **180**, 1587-1613 (1989); English transl. in *Math. USSR Sbornik*, **68** (1991).

2. Senatov, V.V. On the dependence of estimates of the convergence rate in the central limit theorem on the covariance operator of summands. *Teor. Verojatnost. i Primenen.*, **30**, 354-357 (1985); English transl. in *Theory Probab. Appl.*, **30**, (1985).

3. Senatov, V.V. Four examples of lower bounds in the multidimensional central limit theorem. *Teor. Verojatnost. i Primenen.*, **30**, 750-758 (1985); English transl. in *Theory Probab. Appl.*, **30** (1985).

4. Gotze, F. Asymptotic expansions for bivariate von Mises functional. *Z. Wahrscheinlichkeitstheor. verw. Gebiete* **50**, 333-355 (1979).

5. Yurinskii, V.V. On the accuracy of the normal approximation of the probability of hitting a ball. *Teor. Veroyatnost. i Primenen.* **27**, 270-278 (1982), English transl. *Theory Probab. Appl.* **27** (1982), 280-289.

6. Sazonov, V.V. and Ulyanov, V.V. Speed of convergence in the central limit theorem in Hilbert space under weakened moment conditions. *Probability Theory and Mathematical Statistics, Proceedings of the Fifth Vilnius Conference.* Editors: B. Grigelionis, Yu.V. Prohorov, V.V. Sazonov and V. Statulevicius. Mokslas, Vilnius, 394-410 (1991).

7. Sazonov, V.V. and Ulyanov, V.V. An improved estimate of the accuracy of Gaussian approximation in Hilbert space. *New Trends in Probability and Statistics*, Vol. **1**, Editors: V.V. Sazonov and T. Shervashidze, Mokslas, Vilnius, 123-136 (1991).

8. Senatov, V.V. Several remarks on the estimating the rate of convergence in the central limit theorem in Hilbert space. *Teor. Verojatnost. i Primenen.*, **36**, 382-385 (1991).

9. Puri, M.L. and Sazonov, V.V. On Hilbert space valued *U*–statistics. *Teor. Verojatnost. i Primenen.*, **36**, 604-605 (1991).

10. Paulauskas, V. and Račkauskas, A. *Approximation theory in the central limit theorem. Exact results in Banach spaces.* Mokslas, Vilnius (1987); English transl. Kluwer, Dordrecht, 1989.

Reprinted from
Statistical Sciences and Data Analysis,
edited by Kameo Matusita, Madan L. Puri
and Takesi Hayakawa,

ASYMPTOTIC EXPANSIONS IN STATISTICS: A REVIEW OF METHODS AND APPLICATIONS

RABI N. BHATTACHARYA and MADAN L. PURI

1 Introduction

Normal approximation lies at the heart of the asymptotic theory of statistics. Cramér–Edgeworth asymptotic expansions refine this approximation and, therefore, improve classical procedures for setting confidence intervals and critical regions. They also provide the principal tools for: (a) the study of higher order asymptotic optimality theory, which discriminates among (first order) asymptotically efficient estimators and test procedures; (b) the analysis of the bootstrap methodology; (c) computation of power functions in parametric models in the neighborhood of the null hypothetical values etc. In addition, they have many applications in other fields of mathematics.

Historically, Chebyshev (1890) seems to have been the first to provide a formal expansion of the probability density function of the normalized sum S_n of n i.i.d. random variables X_j in terms of the Gaussian density and Hermite polynomials. Edgeworth (1896, 1905) independently obtained the same expansion, and went much further. In the second part of his 1905 article Edgeworth provided a scheme for expanding the density of certain statistics, which are not just linear functions of S_n, assuming appropriate decays of higher order cumulants in powers of $n^{-1/2}$. This last expansion has come to be known as the *Edgeworth expansion*. It was, however, Cramér (1928) who gave the first rigorous derivation of an asymptotic expansion (of the distribution function of S_n) in powers of $n^{-1/2}$ under the condition that the magnitude of the characteristic function (c.f.) of X_j be bounded away from one at infinity. A multidimensional extension of Cramér's result was derived in Bhattacharya (1968, 1971). Bikyalis (1968) provided an L_1-

expansion for the density of S_n under conditions which were necessarily stronger. A comprehensive account of normal approximation errors and asymptotic expansions for normalized sums of independent random vectors may be found in Bhattacharya and Ranga Rao (1976).

Since very few statistics are expressible as normalized sums, Edgeworth's formal method was still not justified in general. The importance of such a justification was laid out in a survey article by Wallace (1958). Chibishov (1972, 1973) and Bhattacharya and Ghosh (1978) provided the desired justification of the formal Edgeworth expansion for large classes of classical statistics, which included M-estimators, under readily verifiable conditions on the common distribution of the underlying i.i.d. observations. In section 2 the approach of Bhattacharya and Ghosh (1978) is explained. The latter was extended to classical test statistics, such as likelihood ratios, by Chandra and Ghosh (1979, 1980).

The approach to the derivation of Edgeworth expansions in Bhattacharya and Ghosh (1978) apply to statistics which can be expressed as, or approximated adequately (for example, using a Taylor expansion) by, smooth functions of a vector of average sample characteristics, such as a finite set of sample moments. Expansions for this class of statistics may be derived by the scheme described in section 2, making use of expansions for normalized sums S_n of random vectors such as stated in sections 2 and 3. But two important classes of statistics not covered by this scheme are U-statistics and rank-based nonparametric statistics. Expansions up to order $o(n^{-1})$ for them are given in sections 4 and 5. The main result presented for U-statistics is due to Götze and van Zwet (1991), which improves earlier results of many authors, especially Callaert et al. (1980) and Bickel et al. (1986). For nonparametric statistics the results are due to Albers et al. (1976), Bickel and van Zwet (1978), Does (1983a, b), Seoh (1983), and Puri and Seoh (1984a, b).

Section 3 deals with normalized sums S_n of weakly dependent sequences of random vectors such as arising in time series models. The main result here is due to Götze and Hipp (1983), improving and extending earlier results of Nagaev (1961) and Statulevichius (1969–70). For statistics arising from such models one may then derive Edgeworth expansions following the same route as taken in Bhattacharya and Ghosh (1978) and explained in section 2.

Among other methods proposed for deriving Edgeworth-type expansions we mention Bhattacharya and Puri (1983) and a versatile approach of Götze (1985). The powerful method of Stein (1986), which has been successful in deriving Berry–Esseen bounds under weak dependence, is yet to be applied effectively for the derivation of Edgeworth expansions.

Since the important case of lattice random vectors has been left out of the main body of the article because of lack of space, some comments about it are in order in this introduction. In the second part of his 1905 paper Edgeworth considered expanding the point masses of the multinomial. But the first general results were obtained in one of the most important papers on the subject of normal approximations by Esseen (1945). In the one-dimensional lattice case Esseen derived full local expansions, as well as expansions of distribution functions which necessarily involved jump terms. The latter was based on local expansion and the Euler–McLaurin summation formula. Extensions of these expansions to multidimension were given by Ranga Rao (1961) and, in more precise form, by Bhattacharya and Ranga Rao (1976). For sums of weakly dependent random vectors, Huang has recently obtained similar local expansions following the route taken by Götze and Hipp (1983). In his article Esseen (1945) showed that the chi-square approximation to Karl Pearson's frequency chi-square statistic under a simple null hypothesis has an error $O(n^{-k/k+1})$, where k is the degrees of freedom. In general this is the best possible error bound. Note that for $k \geqslant 2$ this is of smaller order than the Berry–Esseen bound $O(n^{-1/2})$ for Borel measurable convex sets. For this reason the classical chi-square approximation turns out to be very good for most statistical purposes, even when some expected frequencies are smaller than the proverbial "five." This is confirmed in extensive numerical comparisons carried out by Yarnold (1970) and Bhattacharya and Chan (1994). In the latter article the chi-square is seen to outperform the bootstrap in most cases.

In the concluding section (section 6) we briefly review Edgeworth type expansions in economics, often derived independently of the probability-statistics literature, and indicate how the methods described in this article are of general applicability in econometrics. Among many contributions involving multivariate analysis and time-series models arising in economics only a few are mentioned. The derivations in these articles are mostly by a direct attack on the characteristic function of the statistic concerned. In many models with Gaussian errors, this approach has yielded a great deal of detailed information, especially in the so-called dynamic (i.e. autoregressive) models (see e.g. Sargan and Satchell 1986; Phillips 1978, 1986). On the other hand, such a program is difficult, if not impossible, to implement in semiparametric models where the type of the error distribution is not completely specified.

2 Asymptotic Expansions under Independence

The present section gives an overview of methods and some of the applications of asymptotic expansions for statistics based on i.i.d. observations.

Suppose $X_1, X_2, \ldots, X_n, \ldots$ is a sequence of i.i.d. observations with values in $\mathbb{R}^k$, $EX_n = 0$, $\operatorname{cov} X_n = \Sigma$, $E|X_n|^s < \infty$ for some integer $s \geqslant 3$. Here Σ is assumed nonsingular. The characteristic function (c.f.) of $S_n = (X_1 + \ldots + X_n)/n^{1/2}$ is $\hat{Q}_n(t) = \hat{Q}_1^n(t/n^{1/2})$, where Q_1 is the distribution of X_n and Q_n that of $(X_1 + \ldots + X_n)/n^{1/2}$, and the Fourier transform of a finite (signed) measure μ is denoted by $\hat{\mu}$. The cumulant generating function of the normalized sum may then be expanded in Taylor series to yield

$$\ln \hat{Q}_n(t) = -t'\Sigma t/2 + \sum_{r=1}^{s-2} n^{-r/2}\lambda_{r+2}(it'X_1)/(r+2)! + o(n^{-(s-2)/2}) \quad (2.1)$$

as $n \to \infty$. Here $\lambda_r(t'X_1)$ denotes the rth cumulant of the random variable $t'X_1$, and $\lambda_r(it'X_1) = i^r\lambda_r(t'X_1)$.

One may rewrite (2.1) as

$$\begin{aligned}\hat{Q}_n(t) &= \exp(-t'\Sigma t/2)\exp\left(\sum_{r=1}^{s-2} n^{-r/2}\lambda_{r+2}(it'X_1)/(r+2)!\right) \\ &\quad (1 + o(n^{-(s-2)/2})) \\ &= \exp(-t'\Sigma t/2)\left[1 + \sum_{j=1}^{s-2} n^{-j/2}P_j(it)\right] + o(n^{-(s-2)/2}), \quad (2.2)\end{aligned}$$

where P_j is a polynomial, obtained by collecting together coefficients of $n^{-j/2}$ in the Taylor expansion of the second exponential on the right side of the first equality in (2.2). The principal part of $\hat{Q}_n$ in (2.2) is the Fourier transform of a finite signed measure $\Gamma_{s,n}$ whose density is

$$\gamma_{s,n}(x) = \left[1 + \sum_{j=1}^{s-2} n^{-j/2}P_j(-D)\right]\varphi_{\Sigma}(x). \quad (2.3)$$

Here φ_V is the standard normal density with mean zero and covariance matrix V, and D is the gradient operator $D = (\partial/\partial x_1, \ldots, \partial/\partial x_k)$. Although such an expansion of Q_n in one dimension was formally proposed by Edgeworth (1905) and Chebyshev (1890) at the turn of the century, Cramér (1928) was the first to rigorously prove that the expansion

$$\sup_x \left| F_n(x) - \int_{-\infty}^{x} \gamma_{s,n}(y)dy \right| = o(n^{-(s-2)/2}) \tag{2.4}$$

for the distribution function (d.f.) F_n of Q_n holds in the one-dimensional case under the so-called *Cramér's condition*

$$\lim_{|t|\to\infty} \sup |E\exp(it'X_1)| < 1. \tag{2.5}$$

A more general multidimensional extension of this result may be stated as follows (see Bhattacharya and Ranga Rao, 1976, chapter 4). If (2.5) holds then

$$\sup_{B\in\mathscr{B}} \left| P(S_n \in B) - \int_B \gamma_{s,n}(y)dy \right| = o(n^{-(s-2)/2}) \tag{2.6}$$

for every class $\mathscr{B}$ of Borel sets satisfying

$$\sup_{B\in\mathscr{B}} \int_{(\partial B)^\varepsilon} \exp(-|x|^2/2)dx = o(-\ln\varepsilon)^{-r} \quad \forall r > 0 \text{ as } \varepsilon \downarrow 0. \tag{2.7}$$

Here ∂B is the boundary of B and $(\partial B)^\varepsilon$ is the ε-neighborhood of ∂B. Clearly (2.7) holds for every class $\mathscr{B}$ for which the left side is $O(\varepsilon^a)$, as $\varepsilon \downarrow 0$, for some positive a. In particular it holds for the class $\mathscr{C}$ of all Borel measurable convex subsets of $\mathbb{R}^k$ (see Bhattacharya and Ranga Rao, 1976, corollary 3.2).

Note that without the assumption (2.5) such an expansion cannot hold in general. For example, if X_j are one-dimensional lattice random variables, i.e. if there exist constants x_0 and $b > 0$ such that $P(X_j \in \{x_0 + b\mathbb{Z}\}) = 1$, then F_n has jumps of order $n^{-1/2}$.

Since few statistics are expressible as normalized sums of i.i.d. random variables or vectors, one may, at least formally, consider an analogous formal expansion for the distribution of a real-valued normalized statistic W_n whose asymptotic distributions is $\mathscr{N}(0, \sigma^2)$. Such an expansion was suggested by Edgeworth (1905). We present now an appropriate version of this scheme. Suppose there exists an approximation $\tilde{W}$ of W_n (often given by a Taylor expansion of W_n) such that

$$\sup_x |P(W_n \leqslant x) - P(\tilde{W}_n \leqslant x)| = o(n^{-(s-2)/2}). \tag{2.8}$$

Assume also that the rth cumulant of $\tilde{W}_n$ may be expressed as

$$\gamma_r(\tilde{W}_n) = \chi_{r,n} + o(n^{-(s-2)/2}), \tag{2.9}$$

where there are constants b_{rj} satisfying

$$\chi_{r,n} = \begin{cases} \sum_{j=1}^{s-2} n^{-j/2} b_{1j} & \text{if } r = 1, \\ \sigma^2 + \sum_{j=1}^{s-2} n^{-j/2} b_{2j} & \text{if } r = 2, \\ \sum_{j=r-2}^{s-2} n^{-j/2} b_{rj} & \text{if } 3 \leqslant r \leqslant s. \end{cases} \tag{2.10}$$

A large class of normalized statistics satisfy (2.8) to (2.10). Among them are M-estimators in regular parametric models (see Bhattacharya and Ghosh, 1978) and U-statistics (Bhattacharya and Puri, 1983). If (2.8) to (2.10) hold, then the c.f. of W_n may be expressed as

$$\begin{aligned} E\exp(itW_n) &= \exp\left(-\sum_{s=1}^{s-2} \chi_{r,n}(it)^r/r!\right) + o(n^{-(s-2)/2}) \\ &= \exp(-\sigma^2t^2/2)\exp\left(\sum_{r=1}^{s-2} \chi_{r,n}(it)^r/r! + \sigma^2t^2/2\right) \\ &\quad + o(n^{-(s-2)/2}) \\ &= \exp(-\sigma^2t^2/2)\exp\left(\sum_{j=1}^{s-2} n^{-j/2}p_j(it)\right) + o(n^{-(s-2)/2}) \\ &\quad \left[p_j(it) := \sum_{r=1}^{s-2} b_{rj}(it)^r/r!\right] \\ &= \exp(-\sigma^2t^2/2)\left[1 + \sum_{j=1}^{s-2} n^{-j/2}P_j(it)\right] + o(n^{-(s-2)/2}) \\ &= \eta_{s,n}(t) + o(n^{-(s-2)/2}) \end{aligned} \tag{2.11}$$

in a neighborhood of $t = 0$. For example,

$$\begin{aligned} &P_1(it) = p_1(it), \qquad P_2(it) = p_2(it) - p_1^2(it), \\ &P_3(it) = p_3(it) + p_1(it)p_2(it) - p_1^3(it)/3! \end{aligned} \tag{2.12}$$

Based on (2.11) one may attempt an expansion of the distribution of W_n by a signed measure $\Psi_{s,n}$ whose Fourier transform is $\eta_{s,n}$ and whose density is

$$\psi_{s,n}(x) = \left[1 + \sum_{j=1}^{s-2} n^{-j/2}P_j(-d/dx)\right]\varphi_{\sigma^2}(x). \tag{2.13}$$

Once again, such a *"formal" expansion* is not always valid. The methods for deriving valid expansions may be roughly divided into two groups. One may directly attempt to expand the c.f. f_n of W_n to show that, for some sequence $\{a_n\}$ satisfying $a_n/(\log n)^{1/2} \to \infty$,

$$\int_{\{|t| \leq a_n\}} \frac{1}{|t|} |f_n(t) - \eta_{s,n}(t)| dt = o(n^{-(s-2)/2}), \tag{2.14}$$

and, for some sequence $\Delta_n \to \infty$,

$$\int_{\{a_n < |t| \leq \Delta_n n^{(s-2)/2}\}} \frac{|f_n(t)|}{|t|} dt = o(n^{-(s-2)/2}), \tag{2.15}$$

The proof is completed by the use of Esseen's smoothing inequality (see Bhattacharya and Ranga Rao, 1976, pp. 101–2, lemma 12.2). For every $T > 0$ one has, with $M := \sup |\psi_{s,n}(x)|$,

$$\sup_x \left| F_n(x) - \int_{-\infty}^{x} \psi_{s,n}(y)\, dy \right| \leq (7/8\pi) \int_{-T}^{T} \frac{1}{|t|} |f_n(t) - \eta_{s,n}|\, dt + \frac{9M}{8\pi T}. \tag{2.16}$$

Asymptotic expansions for U-statistics and rank based nonparametric statistics are generally derived by this method, under appropriate assumptions on the statistic and on the common distribution of the i.i.d. observations on which the statistic is based. Such a derivation often involves ingenious estimation of the growth of the c.f. $f_n(t)$ for large values of t, and is generally provided for $s = 4$, which is adequate for many statistical purposes. In section 4 we provide some details of such estimation for rank statistics.

The second method, which is generally applicable to classical parametric as well as semiparametric statistics, expresses W_n as (or approximates it by) a polynomial $\tilde{W}_n = h_n(S_n)$ of the normalized sum $S_n = (Z_1 + \ldots + Z_n)/n^{1/2} = (S_{n1}, \ldots, S_{nk})$ with a nonvanishing leading linear term $\Sigma_i a_i S_{ni}$. Here $Z_j = (f_1(X_j), \ldots, f_k(X_j))$ is a vector function of the jth observation X_j. If the c.f. of Z_j satisfies Cramér's condition (2.5), and if the sth absolute moment of Z_j is finite, then one may apply the expansion (2.6) to get

$$\sup_x \left| P(W_n \leq x) - \int_{\{z: h_n(z) \leq x\}} \gamma_{s,n}(z) dz \right| = o(n^{-(s-2)/2}). \tag{2.17}$$

Since it is in general terribly messy and impractical to carry out the integration in (2.17), the next step is to show that this expansion is the same as the formal expansion (2.13). The details may be found in

Bhattacharya and Ghosh (1978), where it is shown, in addition, that if (a) the m-dimensional observation X_j has a nonzero absolutely continuous component with a positive density in a ball U, (b) $f_1, f_2, \ldots, f_k$ are continuously differentiable in U, (c) $1, f_1, f_2, \ldots, f_k$ are linearly independent as elements of the vector space of continuous functions on U, (d) the statistic W_n is of the form $n^{1/2}(H(\bar{Z}) - H(\mu))$ where H is $s - 1$ times continuously differentiable in a neighborhood of $\mu = EZ_j$, and (e) Z_j has a finite sth absolute moment, then

$$\sup_{B\in\mathscr{B}^1}\left|P(W_n \in B) - \int_B \psi_{s,n}(x)dx\right| = o(n^{-(s-2)/2}). \qquad (2.18)$$

where $\mathscr{B}^1$ is the class of all Borel subsets of $\mathbb{R}^1$. This result extends to vector-valued statistics W_n converging in distribution to $\mathcal{N}(0, V)$. One class of examples are maximum likelihood estimators in exponential (or curved exponential) families. More generally, under usual regularity conditions, one may show that (see Bhattacharya and Ghosh, 1978, or Bhattacharya, 1985) for normalized p-dimensional M-estimators W_n one has

$$\sup_{B\in\mathscr{B}}\left|P(W_n \in B) - \int_B \psi_{s,n}(x)dx\right| = o(n^{-(s-2)/2}). \qquad (2.19)$$

for every class $\mathscr{B}$ of Borel sets of $\mathbb{R}^p$ satisfying, for some $a > 0$,

$$\sup_{B\in\mathscr{B}} \int_{(\partial B)^\varepsilon} \exp(|x|^2/2)dx = O(\varepsilon^a) \qquad \text{as } \varepsilon \downarrow 0. \qquad (2.20)$$

This method also applies to statistics W_n of the form $n(H(\bar{Z}) - H(\mu))$, where grad H is zero at μ, and the matrix of second order derivatives does not vanish at μ (see Chandra and Ghosh, 1979, 1980). The latter case arises for likelihood ratios and chi-square statistics, for example. For some significant earlier work on asymptotic expansions for statistics we refer to Chibishov (1973).

A recent extension of the method described above has been given by Bai and Rao (1991), where for the statistic $W_n = n^{1/2}(H(\bar{Z}) - H(\mu))$ it is shown that if (a) $Z_j = (Z_{j1}, Z_{j2}, \ldots, Z_{jk})$ has a finite sth absolute moment, (b) H is $s - 1$ times continuously differentiable in a neighborhood of $\mu = EZ_j$, (c) $l_1 = (\partial/\partial x_1)H(x))_{x=\mu} \neq 0$, and (d) $\limsup_{|t|\to\infty} E|v_1(t)| < 1$, where $v_1(t) = E(\exp(itZ_{j1})|Z_{j2}, \ldots, Z_{jk})$, then the formal Edgeworth expansion for the d.f. of W_n is valid. There are statistical applications of this result where one of the components of Z_j is discrete, so that Cramér's condition is not satisfied. Of course, in general Cramér's condition for Z_j does not imply (d) either.

Since discrete observations in general, and lattice observations in particular, do not satisfy Cramér's condition, it is comforting to know that expectations of smooth functions of S_n (and of W_n, as a consequence) have asymptotic expansions under moment conditions alone. A precise statement of this result, due to Götze and Hipp (1978), is as follows. If Z_j has a finite sth absolute moment, then

$$E(f(S_n)) = \int f(x)\gamma_{s,n}(x)dx + o(n^{-(s-2)/2}) \tag{2.21}$$

for all $s - 2$ times continuously differentiable functions f such that $(1 + |x|^s)^{-1}|f(x)|$ is bounded and $(s - 2)$th order derivatives of f have at most a polynomial growth at infinity. An alternative version of this result may be found in Bhattacharya (1985).

We conclude this section with brief discussions of two types of applications of asymptotic expansions to statistics. First, consider Efron's (1979) *bootstrap distribution function* F_n^* of a normalized statistic W_n. In the case W_n is the standardized mean of n i.i.d. one-dimensional random variables, it was shown in a basic paper by Singh (1981), using asymptotic expansions, that the difference between the true distribution function F_n of W_n and F_n^* is a.s. $o(n^{-1/2})$, uniformly for all arguments of the d.f., provided the distribution of the summands is nonlattice and has a finite third moment. Thus the bootstrap provides a better approximation of the distribution of W_n than the classical normal approximation, the latter having error $o(n^{-1/2})$. The method of Singh was one-dimensional and based on Esseen's inequality (2.16), which has no proper multidimensional analogue. In a subsequent significant contribution, Babu and Singh (1984) provided an asymptotic expansion up to $o(n^{-(s-2)/2})$ for a normalized multivariate sample mean $\bar{Z}$ under Cramér's condition for Z_j. It follows in particular that the bootstrap distribution function F_n^* of the Student's statistic approximates the true d.f. F_n with an error $o(n^{-1/2})$ a.s., uniformly for all arguments. It is shown in Bhattacharya and Qumsiyeh (1989) that the bootstrap approximation for studentized statistics is superior to even a two-term empirical Edgeworth expansion. A comprehensive account of the bootstrap and its relations with the Edgeworth expansion may be found in the monograph by Hall (1992).

Turning to *higher order asymptotics*, in a parametric model satisfying the usual regularity conditions the m.l.e. (i.e. a consistent solution of the likelihood equation) is asymptotically efficient in the sense of Fisher; namely, its asymptotic variance attains the information lower bound. There are, however, many such estimators. For example, a one-step Newton–Raphson approximation to the solution of the likelihood equation, starting with any $n^{1/2}$-consistent estimator, is asymptotically efficient, or what is now called first order efficient. A notion of *second order efficiency* may be used to discriminate among first order efficient estimators. Roughly speaking, if the m.l.e. is bias-corrected up to $o(n^{-1})$, then it has the smallest expected squared error among all similarly bias-corrected estimators if terms of order $o(n^{-2})$ are neglected. Fisher (1925) himself suggested that the m.l.e. itself is second order efficient in this sense. The theory of higher order efficiency was developed much further in a series of papers by Rao (1961, 1962, 1963). More modern treatments as well as extensions were given by Ghosh and Subramanyam (1974), Efron (1975), Pfanzagl (1980), and Akahira and Takeuchi (1981). An insightful account of this and many other aspects of higher order asymptotics may be found in the recent book by Ghosh (1994).

An expository derivation of the basic expansions discussed in this section, along with a number of statistical applications, may be found in Bhattacharya and Denker (1990).

3 Expansions under Weak Dependence

For notational convenience, the symbols c, d below will denote positive constants, not always the same, which do not vary with the variable indices n, m, i, j, r, etc.

We begin with a description of the basic structure of the Fourier analytic method for the derivation of an asymptotic expansion of the nth normalized partial sum $S_n = n^{-1/2}(X_1 + \ldots + X_n)$ of a sequence of k-dimensional random vectors $X_1, X_2, \ldots$ satisfying

$$EX_j = 0 \text{ for all } j, \quad \beta_{s+1} := \sup_n E|X_n|^{s+1} < \infty. \tag{3.1}$$

For technical purposes it is necessary to consider truncated random vectors $Z_j' = X_j \cdot 1(|X_j| \leqslant n^{\beta})$, $Z_j = Z_j' - EZ_j'$, $S_n' = n^{-1/2}(Z_1' + \ldots + Z_n')$, $S_n^* = n^{-1/2}(Z_1 + \ldots + Z_n)$. Here $\beta \in (0, \frac{1}{2}]$ is an appropriate constant. Let Σ_n denote the covariance matrix of S_n^*, and let $\lambda_\nu(S_n^*)$ denote its νth cumulant for a multi-index $\nu = (\nu_1, \ldots, \nu_k)$. Let δ_n be the smallest eigenvalue of Σ_n. Assume there exist positive constants c_2, c_r such that

$$\liminf \delta_n \geqslant c_2, \ \limsup n^{(r-2)/2}|\lambda_\nu(S_n^*)| \leqslant c_r \quad \text{for } 2 \leqslant r \leqslant s+1, \tag{3.2}$$

where $r = \nu_1 + \ldots + \nu_k$. The *formal Cramér–Edgeworth expansion* of the c.f. $\hat{Q}_n(t)$ of the distribution Q_n of S_n^* is defined as follows. Write

$$\exp\left(\sum_{r=3}^{s} \frac{\lambda_r(it'S_n^*)}{r!}\right) = \exp\left(\sum_{r=1}^{s-2} n^{-r/2} \frac{(n^{r/2}\lambda_{r+2}(it'S_n^*))}{r+2!}\right)$$

$$= 1 + \sum_{r=1}^{s-2} n^{-r/2} P_{r,n}(it) + R_n, \tag{3.3}$$

obtained by a Taylor expansion. In other words,

$$P_{r,n}(it) = r!\left[(d^r/du^r)\exp\left(\sum_{r=1}^{s-2} u^r a_{r+2,n}(it)\right)\right]_{u=0},$$

$$a_{r+2,n}(it) := n^{r/2}\lambda_{r+2}(it'S_n^*)/(r+2)!. \tag{3.4}$$

Lemma 3.1 If (3.1) and (3.2) hold, then for all multi-indices $\alpha = (\alpha_1, \ldots, \alpha_k)$, $\Sigma_i\alpha_i \leqslant s + 1 + k$, there exist positive constants c, c', c'', such that for $|t| \leqslant c''n^{1/2}$ one has

$$\left|D^\alpha\left[\exp\left(\sum_{r=2}^{s} \frac{\lambda_r(it'S_n^*)}{r!}\right) - \exp\left(-\frac{t'\Sigma_n t}{2}\right)\left(1 + \sum_{r=1}^{s-2} n^{-r/2} P_{r,n}(it)\right)\right]\right|$$

$$\leqslant c'n^{-(s-1)/2}\exp(-c|t^2|). \quad \square \tag{3.5}$$

The proof of lemma 3.1 is the same as in the classical case, and only uses Taylor expansion and Cauchy's estimate for derivatives of analytic functions (see Bhattacharya and Ranga Rao, 1976, lemmas 9.7, 9.8). The formal Edgeworth expansion of the c.f. of S_n^* is now given by

$$\hat{\gamma}_{s,n}(t) = \exp(-t'\Sigma_n t/2)\left(1 + \sum_{r=1}^{s-2} n^{-r/2} P_{r,n}(it)\right), \tag{3.6}$$

and the signed measure $\Gamma_{s,n}$ having Fourier transform $\hat{\gamma}_{s,n}(it)$ has density

$$\gamma_{s,n}(x) = \left(1 + \sum_{r=1}^{s-2} n^{-r/2} P_{r,n}(-D)\right)\phi_{\Sigma_n}(x). \tag{3.7}$$

Lemma 3.2 Assume that (3.1) and (3.2) hold, and that (a continuous extension of) the logarithm of the c.f. $\hat{Q}_n(it)$ satisfies, for all α, $\Sigma\alpha_i \leqslant s + 1 + k$,

$$\left|D^\alpha\left(\log \hat{Q}_n(it) - \sum_{r=2}^{s} \lambda_r(it'S_n^*)/r!\right)\right| \leqslant cn^{-(s-2+\varepsilon_1)/2} q_1(t) \tag{3.8}$$

for $|t| \leqslant c'n^{\varepsilon_2}$. Here q_1 is a polynomial which does not vary with n, and

ε_1 and ε_2 are positive constants, $\varepsilon_2 \leqslant \frac{1}{2}$. Then there exist positive constants c_1', c_2', c_3', ε_3 ($\leqslant \varepsilon_2$) such that

$$\left| D^{\alpha}\left[\hat{Q}_n(t) - \exp\left(\sum_{r=2}^{s} \frac{\lambda_r(it'S_n^*)}{r!} \right) \right] \right| \leqslant c_1' n^{-(s-2+\varepsilon_1)} \exp(-c_2'|t|^2) \quad (3.9)$$

for $|t| \leqslant c_3' n^{\varepsilon_3}$ and for $\Sigma\alpha_i \leqslant s + 1 + k$. □

The proof of lemma 3.2 is quite analogous to that of the corresponding result in the classical case, and once again requires little beyond Taylor expansion and Cauchy's estimate (see lemma 9.3 in Bhattacharya and Ranga Rao, 1976).

Assume next that the following estimate for large values of t holds for $D^{\alpha}\hat{Q}_n(t)$:

$$\sup_{\substack{|t| \geqslant c_3' n^{\varepsilon_3} \\ \Sigma\alpha_i \leqslant s+1+k}} |D^{\alpha}\hat{Q}_n(t)| = o(n^{-r}) \quad \text{for all } r > 0. \quad (3.10)$$

Theorem 3.1 Assume (3.10) and the hypothesis of lemma 3.2. Assume also that

$$P(S_n \neq S_n') = o(n^{-(s-2)/2}), \quad |ES_n'| = o(n^{-(s-2)/2}). \quad (3.11)$$

Then

$$\sup_{B \in \mathscr{B}} \left| P(S_n \in B) - \int_B \gamma_{s,n}(x)dx \right| = o(n^{-(s-2)/2}) \quad (3.12)$$

for every class $\mathscr{B}$ of Borel sets satisfying, for some positive constant a,

$$\sup_{B \in \mathscr{B}} \int_{(\partial B)^{\varepsilon}} \exp\left(\frac{-|x|^2}{2} \right) dx = O(\varepsilon^a), \quad \text{as } \varepsilon \downarrow 0. \quad □ \quad (3.13)$$

Proof. By corollary 11.3 in Bhattacharya and Ranga Rao (1976), for every Borel set B and every $\varepsilon > 0$, one has

$$|Q_n(B) - \Gamma_{s,n}(B)| \leqslant \frac{1}{2} \| (Q_n - \Gamma_{s,n}) * K_{\varepsilon} \| + \int_{(\partial B)^{2\varepsilon}} |\gamma_{s,n}(x)| dx. \quad (3.14)$$

Here $\|\cdot\|$ denotes variation norm and K_{ε} is the distribution of εZ with Z satisfying

$$P(|Z|) \leqslant 1) \equiv K_1(\{x : |x| \leqslant\}) = 1. \quad (3.15)$$

Choose K_1 to satisfy (see Bhattacharya and Ranga Rao, 1976, corollary 10.4)

$$|D^{\alpha}\hat{K}_1(t)| \leqslant c_{\alpha}\exp(-|t|^{1/2}) \quad (t \in \mathbb{R}^k), \quad \forall\alpha. \tag{3.16}$$

By lemma 11.6 in Bhattacharya and Ranga Rao (1976), writing $|\beta| = \Sigma\beta_i$,

$$\|(Q_n - \Gamma_{s,n})K_{\varepsilon}\| \leqslant c(k) \max_{|\beta|=0,k+1} \int |D^{\beta}[(\hat{Q}_n(t) - \hat{\gamma}_{s,n}(t))\hat{K}_1(\varepsilon t)]|dt$$

$$\leqslant c'(k) \max_{\substack{(\alpha_1,\alpha_2):\alpha_1+\alpha_2=\beta,\\ |\beta|=0,k+1}} \int |D^{\alpha_1}(\hat{Q}_n(t) - \hat{\gamma}_{s,n}(t)) \cdot D^{\alpha_2}\hat{K}_1(\varepsilon t)|dt. \tag{3.17}$$

Denoting the above integral by I, one has

$$I \leqslant \int_{|t|\leqslant c_3' n^{\varepsilon_3}} |D^{\alpha_1}(\hat{Q}_n(t) - \hat{\gamma}_{s,n}(t))|dt + \int_{|t|>c_3' n^{\varepsilon_3}} |D^{\alpha_1}\hat{\gamma}_{s,n}(t)|dt$$

$$+ \int_{|t|>c_3' n^{\varepsilon_3}} |D^{\alpha_1}\hat{Q}_n(t) \cdot D^{\alpha_2}\hat{K}_1(\varepsilon t)|dt$$

$$= I_1 + I_2 + I_3, \quad \text{say}. \tag{3.18}$$

By lemmas 3.1 and 3.2,

$$I_1 = o(n^{-(s-2)/2}). \tag{3.19}$$

Since $\hat{\gamma}_{s,n}(t) \to 0$ exponentially fast as $|t| \to \infty$,

$$I_2 = o(n^{-(s-2)/2}). \tag{3.20}$$

By (3.10) and (3.16),

$$I_3 = o(n^{-(s-2)/2}). \tag{3.21}$$

if $\varepsilon = n^{-\gamma}$ for some $\gamma > 0$. The proof is now complete from (3.13), (3.14) if one takes, for example, $\gamma = (s - 2)/a$. □

We now state the main result for sums of weakly dependent random vectors due to Götze and Hipp (1983). Let $X_1, X_2, \ldots,$ be a sequence of k-dimensional random vectors defined on a probability space $(\Omega, \mathscr{F}, P)$. In addition to (3.1), we assume the following conditions:

(A1) $\liminf \delta_n > 0$, $\limsup \|\Sigma_n\| < \infty$.

(A2) There exist a sequence of σ-fields $\mathscr{D}_n$ $(n = 0, \pm 1, \pm 2, \ldots)$ and positive constants c, d such that, for each pair (r, m) of positive integers, there is a random vector $X_{r,m}$ measurable with respect to $\mathscr{D}_{r-m}^{r+m}$ satisfying

$$E|X_r - X_{r,m}| \leqslant ce^{-dm}.$$

Here $\mathscr{D}_p^q := \sigma\{\mathscr{D}_n : p \leqslant n \leqslant q\}$.

(A3) There exist positive constants c, d such that for all n, m,

$$\sup_{A \in \mathscr{D}^{n}_{-\infty}, B \in \mathscr{D}^{\infty}_{n+m}} |P(A \cap B) - P(A)P(B)| \leq ce^{-dm}.$$

(A4) There exist positive constants $d_1 < 1$ and d_2 such that for all positive integers $m < n$ and for all t satisfying $|t| \geq d_2$,

$$E|E(\exp(it'(X_{n-m} + X_{n-m+1} + \ldots + X_{n+m}))|\mathscr{D}j : j \neq n)| \leq d_1. \tag{3.22}$$

(A5 There exist positive constants d_3, d_4 such that for all m, r, p and $B \in \mathscr{D}^{r+p}_{r-p}$

$$E|P(B|\mathscr{D}_j : j \neq r) - P(B|\mathscr{D}_j : 0 < |r - j| \leq m + p)| \leq d_3 \exp(-d_4 m).$$

Under assumptions (3.1), (A2), (A3), it may be easily shown that for all β sufficiently near 1/2 the normalized sum S_n^* of truncated random vectors Z_j introduced above satisfies (3.2), (3.11). In this case the formal Edgeworth expansion (3.12) of the distribution of S_n^* is the same as that of S_n.

Theorem 3.2 (Götze and Hipp, 1983). Under the assumptions (3.1), (A1) to (A5), the formal Edgeworth expansion (3.12) stated in theorem 3.1 is valid. □

In order to prove this theorem one needs to verify the hypotheses of theorem 3.1. Assume for the sake of simplicity that X_j is measurable with respect to $\mathscr{D}^{j+r}_{j-r}$ for some fixed r. In this case it is enough that the inequality in (3.22) in assumption (A4) holds for a fixed m. Now the hypotheses of lemmas 3.1, 3.2 may be verified using assumptions (3.1), (A3), and the fact that the smallest eigenvalue of Σ_n is bounded away from zero. A crucial step in such a derivation is an algebraic identity involving cumulants and moments of arbitrary random variables (see Bulinskii and Zhurbenko, 1976; Jensen, 1989a): Define the *simple cumulant* of the random variables $V_1, V_2, \ldots, V_r$ as

$$\text{cum}(V_1, V_2, \ldots, V_r) := i^{-r}\left[\frac{\partial^r}{\partial\varepsilon_1 \ldots \partial\varepsilon_r} \log E \exp\left(i \sum_{j=1}^{r} \varepsilon_j V_j\right)\right]_{\varepsilon_1 = \ldots = \varepsilon_r = 0} \tag{3.23}$$

Write $I = \{1, 2, \ldots, r\}$. Let I', II'' be nonempty and disjoint, $I = I' \cup I''$. For a given partition $\{I_1, I_2, \ldots, I_\nu\}$ of I, define $I'_j = I_j \cap I'$ $I''_j = I_j \cap I''$,

$$M_j = E\prod_{i \in I_j} V_i, \qquad M'_j = E\prod_{i \in I'_j} V_i, \qquad M''_j = E\prod_{i \in I''_j} V_i,$$

$$L_j = M_j' M_j'', \qquad D_j = M_j - L_j. \tag{3.24}$$

Then one has the identity

$$\text{cum}(V_1, \ldots, V_r) = \sum{}^* \frac{(-1)^{\nu-1}}{\nu} \sum_{j=1}^{\nu} L_1 \ldots L_{j-1} D_j M_{j+1} \ldots M_\nu, \tag{3.25}$$

where L_0, as well as any empty product, equals one, and Σ^* denotes summation over all partitions $\{I_1, I_2, \ldots, I_\nu\}$ such that $\max\{i : i \in I_j\} < \min\{i : i \in I_{j+1}\}$ for all j.

Now take $V_j = Z_{i_j}$ where $\{i_1, i_2, \ldots, i_r\} \subset \{1, 2, \ldots, n\}$ and $i_1 < i_2 < \ldots < i_r$. Choose p and $I' = \{i_1, i_2, \ldots, i_p\}$ such that the gap $i_{p+1} - i_p$ is maximal. If this gap is large, then D_j is small. From this and (A3) one may deduce (3.2), and by a similar argument also derive (3.8).

To derive (3.10), one uses the Cramér-type condition (A4) as well as the technical looking assumption (A5). The role of (A5) becomes clear when one attempts to take a conditional expectation of $\exp(it'S_n^*)$, given all summands except those involved in the sum appearing in the exponent in (3.22). In view of (A5), the conditional expectation of the exponential in (3.22) may be replaced by that given a rather small subset. One may repeat this process for N block sums such as appearing in (3.22), but widely separated, and then use the strong mixing type condition (A3) to approximate the c.f. of S_n by that of a sum of N independent random vectors each satisfying a Cramér type condition. A similar estimation holds for evaluating derivatives of $\hat{Q}_n$. Finally an appropriate choice of N completes the verification of (3.10). For details see Götze and Hipp (1983) or Jensen (1989a).

The following are a set of important examples in which all the assumptions (3.1), (A1) to (A5) are satisfied (see Götze and Hipp, 1983, 1994). The random variables X_j below are to be centered, or S_n has to be centered.

Example 3.1 (m-dependent sequences). Let Y_j ($j = 1, 2, \ldots$) be a sequence of i.i.d. random variables having a density g with respect to the Lebesgue measure. Let

$$X_j := h(Y_j, Y_{j+1}, \ldots, Y_{j+m-1}). \tag{3.26}$$

Here h is a measurable function on $\mathbb{R}^m$ satisfying

$$\sum_{r=1}^{m} \left(\frac{\partial h(x_1, \ldots, x_m)}{\partial x_r} \right)_{(x_1,\ldots,x_m)=(y_1,\ldots,y_m)} \neq 0, \tag{3.27}$$

for all y_r ($1 \leqslant r \leqslant m$) in an interval $(-c, c)$ on which g is positive. The sequence X_j ($j = 1, 2, \ldots$) is strictly stationary. Assume that X_j has a finite moment of order $s + 1$. The assumptions (A1) to (A5) are satisfied if one takes $\mathscr{D}_j$ to be the σ-field generated by Y_j. Most m-dependent sequences, although not all, are of the form (3.26).

Example 3.2 (moving averages). Let Y_j ($j = 0, \pm 1, \pm 2, \ldots$) be a sequence of i.i.d. random variables, and

$$X_j := \sum_{r=-\infty}^{\infty} c_r Y_{j+r} \tag{3.28}$$

where c_r go to zero exponentially fast as $r \to \pm\infty$, and $\Sigma c_r \neq 0$. If Y_j has a finite moment of order $s + 1$ and satisfies Cramér's condition (2.5), then all the assumptions are satisfied. Once again take $\mathscr{D}_j$ to be $\sigma\{Y_j\}$ and m in (A4) such that $c_{-n} + \ldots + c_n \neq 0$ for all $n \geqslant m$.

Example 3.3 (autoregressive and ARMA models). Let Y_j ($j = 0, \pm 1, \pm 2, \ldots$) be a sequence of i.i.d. real-valued random variables as in the preceding example, satisfying Cramér's condition (2.5). Given, $X_1, \ldots, X_p$, let X_j be recursively defined by

$$X_j := \sum_{r=1}^{p} a_r X_{j-r} + \sum_{r=1}^{q} b_r Y_{j-r} + Y_j, \quad (j = p + 1, p + 2, \ldots) \tag{3.29}$$

where b_r are arbitrary constants, $a_p \neq 0$, and the roots of the equation

$$z^p = \sum_{r=1}^{p} a_r z^{p-r} \tag{3.30}$$

all lie in the unit circle $\{z : |z| < 1\}$. In the case b_r are all zero, the X_j's define an autoregressive process of order p, or an AR(p) process. If $b_q \neq 0$, the process is called an autoregressive moving average model of order (p, q) (ARMA(p, q)). One may assume (3.29) to hold for all $j = 0, \pm 1, \pm 2, \ldots$, and express X_j in the form (3.28). The case of arbitrary initial $X_1, \ldots, X_p$ (with finite $(s + 1)$th moments) may then be dealt with by a direct comparison. It is also possible to deal with these processes by Markovian methods (see example 3.4 below).

Example 3.4 (functions of Markov processes). Let Y_j ($j = 0, 1, \ldots$) be a Markov process with stationary transition probability $p(y, dz)$ satisfying

$$\sup |p(y, B) - p(y', B)| < 1. \tag{3.31}$$

where the supremum is over all measurable sets B and all y and y'. For example, (3.31) is satisfied when there exists a positive measure μ such that $p(y, B) \geqslant \mu(B)$ for all y and all B. Let now

$$X_j := h(Y_j) \tag{3.32}$$

where h is a measurable real valued function such that, under the (unique) stationary initial distribution, X_1 satisfies Cramér's condition (2.5). If X_1 has a finite moment of order $s + 1$, then all the assumptions are satisfied, if one takes $\mathscr{D}_j = \sigma\{Y_j\}$. As remarked in example 3.3, the stable ARMA(p, q) models may be dealt with from this Markovian point of view.

Among other important classes that may be treated by the present theorem are functionals of nonlinear autoregressive processes (Götze and Hipp, 1994) and of stationary Gaussian processes.

Remark. Writing U_j for X_j in (3.28) or (3.29), one may consider $X_j := h(U_j, \ldots, U_{j+p-1})$ for a continuously differentiable $h: \mathbb{R}^p \to \mathbb{R}^k$ such that $E\|D_i h(U_j, \ldots, U_{j+p-1})\| < \infty$ (D_i being the derivative with respect to the ith argument). If, in addition, Y_j has a continuous and positive density then (A1) to (A5) hold provided some obvious degeneracy (such as $X_j := U_j - U_{j-1}$) does not occur. Also, for expansions of distributions of statistics of the form considered in section 2, but based on weakly dependent observations such as considered in this section, the same technique as before applies (Götze and Hipp, 1994).

Theorem 3.2 is very useful in the statistical analysis of stationary time series. As an example we cite Bose (1988), who used the above expansion to analyze the bootstrap methodology in stable AR(p) models. Taniguchi (1983, 1991) derived asymptotic expansions for Gaussian ARMA processes and used them for the study of higher order efficiencies of estimators.

The moment condition (3.1) and the exponential decay in (A3) have been relaxed by Lahiri (1993). Replacing the Cramér type condition (A4) by an appropriate lattice condition, Huang (1994) has recently obtained asymptotic expansions for lattice random vectors analogous to those under independence as given, for example, in Bhattacharya and Ranga Rao (1976, chapter 5).

Among other significant contributions under dependence we mention Nagaev (1961), Statulevichius (1969–70), and Jensen (1989b).

4 Edgeworth Expansions for Nonparametric Statistics

Let X_{ni} ($1 \leqslant i \leqslant n$) be independent r.v.s. (random variables) with continuous distribution functions (df) $\{F_{ni}(x), 1 \leqslant i \leqslant n\}$. Let R_{ni} be the rank of X_{ni} in the partial sequence $X_{n1}, \ldots, X_{nn}$; R_{ni}^+, the rank of $|X_{ni}|$ in the partial sequence $|X_{n1}|, \ldots, |X_{nn}|$. Let $\operatorname{sgn} x_i = 1$ if $x_i \geqslant 0$, and -1 if $x_i < 0$. Consider the following statistics:

$$S_n = \sum_{i=1}^{n} c_{ni} a_n(R_{ni}), \quad S_n^+ = \sum_{i=1}^{n} c_{ni} a_n(R_{ni}^+) \operatorname{sgn} X_{ni}. \tag{4.1}$$

Here $\{c_{ni}, 1 \leqslant i \leqslant n\}$ are known constants; $\{a_n(i), 1 \leqslant i \leqslant n\}$ are "scores" generated by a known function $J(t)$, $0 < t < 1$ in either of the following two ways:

$$a_n(i) = E[J(U_{n:i})], \quad 1 \leqslant i \leqslant n \quad \text{(exact scores)} \tag{4.2}$$

or

$$a_n(i) = J[E(U_{n:i})] \quad \text{(approximate scores)} \tag{4.3}$$

where $U_{n:i}$ is the ith order statistic in a random sample of size n from a uniform distribution over (0, 1). Scores given by (4.2) give rise to locally most powerful rank tests whereas the scores given by (4.3) have the appeal of simplicity although in both cases they lead to asymptotically most powerful rank tests. In the nonparametric literature, S_n is called a *simple linear rank statistic*; S_n^+ is called a *signed rank statistic*. These statistics are often used in nonparametric inference problems.

Example 4.1 (regression problem). Let $F_{ni}(x) = F(x - \Delta c_{ni})$, $1 \leqslant i \leqslant n$. Then, for testing the hypothesis $H_0 : \Delta = 0$ against the one-sided alternative $H_1 : \Delta > 0$, the locally most powerful rank test is based on the statistic S_n with scores $a_n(i)$ given by

$$a_n(i) = E\{-f'(F^{-1}(U_{n:i}))/f(F^{-1}(U_{n:i}))\}, \quad 1 \leqslant i \leqslant n \quad \text{(exact scores)}, \tag{4.4}$$

and a locally asymptotically most powerful test is based on

$$a_n(i) = -f'[F^{-1}(EU_{n:i})]/f[F^{-1}(EU_{n:i})], \quad 1 \leqslant i \leqslant n \quad \text{(approximate scores)}, \tag{4.5}$$

where $f = F'$ is absolutely continuous with derivative (a.e.) f' such that $\int |f'(x)| dx < \infty$ and $F^{-1}(u) = \inf\{x : F(x) \geqslant u\}$.

Similarly, for testing the hypothesis $H_0^* : \Delta = 0$ and F is symmetric, against the alternative $H_1^* : \Delta > 0$ and F is symmetric, the locally most powerful rank test is based on the statistic S_n^+ with the scores $a_n(i)$ given by

$$a_n(i) = -E\{f'_+(F_+^{-1}(U_{n:i}))/f_+(F_+^{-1}(U_{n:i}))\}, \quad 1 \leqslant i \leqslant n, \tag{4.6}$$

and the locally asymptotically most powerful test has scores $a_n(i)$ given by

$$a_n(i) = -f'_+[F_+^{-1}(EU_{n:i})]/f_+[F_+^{-1}(EU_{n:i})], \quad 1 \leqslant i \leqslant n, \qquad (4.7)$$

where $F_+(x) = P[|X| \leqslant x] = 2F(x) - 1$ if $x > 0$, and 0 if $x \leqslant 0$, and $f_+(x) = F'_+(x)$ is absolutely continuous with derivative (a.e.) f'_+ such that $\int |f'_+(x)| dx < \infty$.

Example 4.2 (two-sample problem). Let $F_{n1} = \ldots = F_{nm} \equiv F$ and $F_{n,m+1} = \ldots = F_{n,n} \equiv G$. Then, for testing $H_0: F = G$, many rank tests are based on the statistic $S'_n = \Sigma_{i=1}^m a_n(R_{ni})$ which is a specialization of S_n when $c_{ni} = 1$ if $1 \leqslant i \leqslant m$, and $c_{ni} = 0$ if $m + 1 \leqslant i \leqslant n$.

Example 4.3 (one-sample problem). Let $F_1 = \ldots = F_n \equiv F$. Then, for testing $H_0: F(x)$ is symmetric about 0, many rank tests are based on the statistic $S_n^+ = \Sigma_{i=1}^n a_n(R_{ni}^+) \operatorname{sgn} X_{ni}$ which is a specialization of S_n^+ when $c_{ni} \equiv 1$, $1 \leqslant i \leqslant n$.

For more examples, see Hájek and Sidàk (1967).

Let us now write $S_n^* = (S_n - a_n)/b_n$ and $S_n^{+*} = (S_n^+ - a_n^+)/b_n^+$ where a_n, a_n^+, b_n and b_n^+ are normalizing constants, and let us give a generic name T_n to any of these statistics. Then it is well known that under suitable assumptions T_n has asymptotically, as $n \to \infty$, the standard normal distribution, i.e. $\lim_{n\to\infty} \sup_x |P(T_n \leqslant x) - \Phi(x)| = 0$ where $\Phi(x) = (2\pi)^{-1/2} \int_{-\infty}^x e^{-t^2/2} dt$. The results of this type justify the use of normal approximation to compute the critical values of the test procedures when n is large. They also enable us to compute the limiting powers of the test procedures and make efficiency comparisons on that basis.

However, quite often, one needs information more precise than the asymptotic normality can provide. On the one hand one may need more accurate numerical approximations; on the other hand one may wish to compare the performance of one procedure with other procedures with the same limiting powers (cf. Hodges and Lehmann, 1970). For this purpose it is enough to derive Edgeworth expansions (or higher order asymptotics) with remainder $o(n^{-1})$. They provide some qualitative insight into the regions of unreliability of the first order results. The first general results (for rank statistics) in the case of the standard one-sample problem were obtained by Albers et al. (1976). They established the Edgeworth expansion under the hypothesis of symmetry as well as under contiguous location alternatives. The same programme was carried out for the two-sample problem in

Bickel and van Zwet (1978). Under the null hypothesis, a similar result was proved by Robinson (1978). (See also Robinson, 1980, where an asymptotic expansion for rank tests for several samples is established.) Extensions of these results to the cases of simple linear rank statistics S_N and signed rank statistic S_N^+ were carried out by Does (1983a, b), and Seoh (1983) and Puri and Seoh (1984a, b) respectively.

Our attention is mainly focused on simple signed rank statistics with regression constants, the Edgeworth expansion for which was derived by Seoh (1983) and Puri and Seoh (1984a, b) with remainder $o(n^{-1})$. Parallel to this study is the work of Does (1983a, b), which deals with unsigned rank statistics with regression constants. Since the results are qualitatively the same, we shall deal with the Edgeworth expansion for signed rank statistics.

Denote $F_n(x) = P((S_n^+ - E(S_n^+))\tilde{\sigma}_n^{-1} \leq x)$, $x \in (-\infty, \infty)$ where $\tilde{\sigma}_n$ is either the exact standard deviation of S_n^+ or some normalizing constant.

We make the following assumptions on regression constants and score generating functions:

Assumption A. $X_{n1}, \ldots, X_{nn}$ are i.i.d. symmetrically distributed about zero, and the regression constants $c_{n1}, \ldots, c_{nn}$ satisfy

$$\sum_{i=1}^{n} c_{ni}^2 = 1, \quad \max_{1 \leq i \leq n} |c_{ni}| = O(n^{-1/2}). \tag{4.8a}$$

Assumption B. The score generating function J is non-constant, $\int_0^1 J^2(t)dt = 1$, and is three times differentiable and

$$|J'''(t)| \leq \Gamma\{t(1-t)\}^{-3-(1/4)+\delta}, \quad t \in (0, 1), \quad \delta < \frac{1}{14} \tag{4.8b}$$

and Γ is some positive number.

Since $(t(1-t))^{-\beta} \leq (t(1-t))^{-\gamma}$, $t \in (0, 1)$, for $0 < \beta \leq \gamma$, assumption (4.8b) is satisfied if $|J'''(t)| \leq \Gamma\{t(1-t)\}^{-\alpha}$ for some α, $0 < \alpha < 3 + \frac{1}{14}$.

By the symmetry assumption it is clear that $ES_n^+ = 0$ and

$$\sigma_n^2 = \operatorname{var} S_n^+ = \frac{1}{n}\sum_{i=1}^{n} J^2(i/(n+1)).$$

Define, for each $n \geq 1$ and real x,

$$S_n^* = \sigma_n^{-1} S_n^+, \quad F_n^*(x) = P(S_n^* \leq x), \quad -\infty < x < \infty, \tag{4.9}$$

and

$$\bar{F}_n(x) = \Phi(x) - \varphi(x)\kappa_{4n}(x^3 - 3x)/24, \tag{4.10}$$

where Φ denotes the standard normal distribution function, φ is its density function and κ_{4n} is given by

$$\kappa_{4n} = \sum_{i=1}^{n} c_{ni}^4 \left\{ \int_0^1 J^4(t)dt - 3 \right\} - 3n^{-1} \left\{ \int_0^1 J^4(t)dt - 1 \right\}. \tag{4.11}$$

Our main theorem is as follows.

Theorem 4.1 Under the assumptions A and B we have, as $n \to \infty$,

$$\sup_x |F_n^*(x) - \bar{F}_n(x)| = o(n^{-1}). \quad \square$$

Since $\bar{F}_n(x) = \Phi(x) + O(n^{-1})$ uniformly in x, we have an improvement of the Berry–Esseen bound of Puri and Seoh (1985) under stronger assumptions.

Corollary 4.1 Under assumptions A and B we have, as $n \to \infty$,

$$\sup_x |F_n^*(x) - \Phi_n(x)| = O(n^{-1}). \quad \square$$

We note that the third cumulant of S_n^* is zero and that κ_{4n} is an asymptotic expansion for the fourth cumulant of S_n^* where terms of order $o(n^{-1})$ have been neglected. Hence $\bar{F}_n$ may be said to constitute a genuine Edgeworth expansion for F_n^*. It may be noted that the assumption B allows score generating functions tending to infinity in the neighborhood 0 and 1 at the rate of $\{t(1 - t)\}^{-1/14+\varepsilon}$ for some $\varepsilon > 0$. Clearly this includes the chi-quantile function.

From now on, for simplicity of notation, we shall omit indices n in X_{ni}, R_{ni}^+ and $\operatorname{sgn} X_{ni}$, etc.

In order to prove theorem 4.1 we first derive a bound for the variance of S_n^+. Then we study the behavior of the characteristic function of S_n^* for large values of the argument.

Lemma 4.1 If the score generating function J satisfies the assumption B, then

$$\sigma_n^2 = \frac{1}{n} \sum_{i=1}^{n} J^2\left(\frac{i}{n+1}\right) = 1 + O(n^{-6/7-2\delta}). \tag{4.12}$$

Proof. Note that the assumption B ensures that for some positive number Γ_0,

$$|J^{(i)}(t)| \leq \Gamma_0 \{t(1 - t)\}^{-i-1/14+\delta}, \quad i = 0, 1, 2, 3. \tag{4.13}$$

Hence the lemma follows by mean value theorems for derivatives and integrals. □

The behavior of the characteristic function of S_n^* for large values of the argument is given by lemma 4.2 below, whose proof we omit (see Seoh, 1983).

Lemma 4.2 Suppose that the assumptions of theorem 4.1 are satisfied and let $\psi_n(t) = E\exp(itS_n^*)$. Then there are positive numbers B, β and γ such that, for $\log n \leqslant |t| \leqslant \gamma n^{3/2}$, $n \geqslant 1$,

$$|\psi_n(t)| \leqslant Bn^{-\beta \log n}. \quad \square$$

To prove theorem 4.1, we start with an application of Esseen's smoothing lemma (2.16), which implies that, for all $\gamma > 0$,

$$\sup_x |F_n^*(x) - \tilde{F}_n(x)| \leqslant \frac{1}{\pi}\int_{|t|\leqslant \gamma n^{3/2}} |\psi_n(t) - \lambda_n(t)|/|t|\,dt + O(n^{-3/2}) \tag{4.14}$$

where ψ_n denotes the characteristic function of S_n^* and λ_n denotes the Fourier–Stieltjes transform of $\tilde{F}_n$, i.e.

$$\lambda_n(t) = \int_{-\infty}^{\infty} e^{itx}\,d\tilde{F}_n(x) = e^{-t^2/2}\{1 + \kappa_{4n}t^4/24\}. \tag{4.15}$$

Since derivatives of $\psi_n(t)$ and $\lambda_n(t)$ are uniformly bounded and $\psi_n(0) = \lambda_n(0) = 1$, we have

$$\int_{|t|\leqslant n^{-3/2}} |\psi_n(t) - \lambda_n(t)|/|t|\,dt = O(n^{-3/2}). \tag{4.16}$$

Furthermore, lemma 4.2 and (4.15) ensure that

$$\int_{\log n\leqslant|t|\leqslant\gamma n^{3/2}} |\psi_n(t) - \lambda_n(t)|/|t|\,dt = O(n^{-3/2}). \tag{4.17}$$

Thus it follows from (4.14), (4.16), and (4.17) that, in order to prove theorem 4.1, it suffices to show that

$$\int_{n^{-3/2}\leqslant|t|\leqslant\log n} |\psi_n(t) - \lambda_n(t)|/|t|\,dt = O(n^{-1}). \tag{4.18}$$

To prove this one uses a conditioning argument as in Bickel and van Zwet (1978) or Does (1983a, b).

Omitting the details of computations (see Seoh, 1983), we obtain (uniformly for $|t| \leqslant \log n$)

$$\psi_n(t) = e^{-t^2/2}\left[1 + \frac{t^4}{24\sigma_n^4}\kappa_{4n}\right] + O(n^{-22/15})$$
$$+ o(n^{-1}|t|P(t)e^{-\theta t^2}) + O(n^{-1-\varepsilon}|t|P(t)) \qquad (4.19)$$

where $0 < \theta < \frac{1}{2}$, $\varepsilon > 0$, and $P(t)$ is a polynomial which does not involve n. □

Theorem 4.1 justifies the validity of the expansion with terms up to order n^{-1} used by Fellingham and Stoker (1964) who investigated the performance of the formal Edgeworth series with terms taken to order n^{-1} and n^{-2}. They compared their approximations with normal approximations and demonstrated that the approximation by an Edgeworth series is considerably more accurate in the range of upper 5 percent probability and for $n \geqslant 10$. For the performance of the formal Edgeworth expansion taken to terms of order n^{-1} in the case of absolute normal scores statistic, the reader is referred to Thompson et al. (1967).

Next, we provide the main result dealing with the two-term Edgeworth expansion for the signed rank statistic S_n^+ under near location alternatives.

Let $\{X_{ni}, 1 \leqslant i \leqslant n\}$ be independent r.v.s. such that $X_{ni} = \theta_{ni} + e_{ni}$, $1 \leqslant i \leqslant n$, where e_{ni}, $1 \leqslant i \leqslant n$, are i.i.d. r.v.s. with a common continuous df $F(x)$, and θ_{ni}, $1 \leqslant i \leqslant n$, are location parameters.

We assume that

1 The regression constants c_{ni}s satisfy the assumption (4.8), and $\sum_{i=1}^n \theta_{ni}^2 = O(1)$.
2 $f(x) = F'(x)$ is symmetric about 0, and $f'(x)$ satisfies $\int |f'(x)|dx < \infty$.
3 The function J (see assumption B) is nonconstant and twice differentiable on (0, 1), and its second derivative J'' satisfies the Lipschitz condition of order δ, $0 < \delta \leqslant 1$, i.e. $|J''(s) - J''(t)| \leqslant \Delta|s - t|^\delta$, $0 < s, t < 1$ for some constant Δ. Furthermore $\int_0^1 J^2(t)dt = 1$.

Denote

$$\tau_n^2 = \text{var}(S_n^+), \quad S_n^* = \tau_n^{-1}(S_n^* - ES_n^+) \quad \text{and let } F^+(x) = P[|e_{ni}| \leqslant x].$$

Furthermore, let

$$A_n(x) = \Phi(x) - \phi(x)\left\{\frac{k_{3n}}{6}(x^2 - 1) + \frac{k_{4n}}{24}(x^3 - 3x)\right\} \qquad (4.20)$$

where

$$k_{3n} = \sum_{i=1}^{n} c_{ni}^3 \int_{-\infty}^{\infty} \{J^3(F^+(|x|)) - 3J(F^+(|x|))\}$$

$$\operatorname{sgn} x\{f(x - \theta_{ni}) - f(x)\}\,dx$$

$$- \frac{3}{n}\sum_{i=1}^{n} c_{ni} \int_{-\infty}^{\infty} \{J^3(F^+(|x|)) - J(F^+(|x|))\}$$

$$\operatorname{sgn} x\{f(x - \theta_{ni}) - f(x)\}\,dx \tag{4.21}$$

and

$$k_{4n} = \sum_{i=1}^{n} c_{ni}^4 \left\{\int_0^1 J^4(t)\,dt\right\} - 3\left\{\int_0^1 J^4(t)\,dt - 1\right\}. \tag{4.22}$$

Then.

Theorem 4.2 Under assumptions 1 through 3, we have

$$\sup_x |P(S_n^* \leqslant x) - A_n(x)| = o(n^{-1}), \quad \text{as } n \to \infty. \quad \square$$

Since $A_n(x) = \Phi(x) + O(n^{-1})$, uniformly in x, we have

Corollary 4.2 Under assumptions 1 through 3, we have

$$\sup_x |P(S_n^* \leqslant x) - \Phi(x)| = O(n^{-1}), \quad \text{as } n \to \infty. \quad \square$$

We now define for each $n > 1$ and real x

$$\tilde{A}_n(x) = \Phi(x) - \phi(x)\left\{\frac{1}{6}\tilde{k}_{3n}(x^2 - 1) + \frac{k_{4n}}{24}(x^3 - 3x)\right\}$$

where

$$\tilde{k}_{3n} = \sum_{i=1}^{n} c_{ni}^3\theta_{ni} \int_{-\infty}^{\infty} \{J^3(F^+(|x|)) - 3J(F^+(|x|))\} \operatorname{sgn} x\, f'(x)\,dx$$

$$+ \frac{3}{n}\sum_{i=1}^{n} c_{ni}\theta_{ni} \int_{-\infty}^{\infty} \{J^3(F^+(|x|)) - J(F^+(|x|))\} \operatorname{sgn} x\, f'(x)\,dx.$$

Note that if $f(x)$ has the second integrable Radon–Nikodym derivative f'', then Taylor's expansion exsures that $\sup_x |A_n(x) - \tilde{A}_n(x)| = O(n^{-3/2})$. This proves

Corollary 4.3 Under assumptions 1 through 3, we have

$$\sup_x |P(S_n^* \leqslant x) - \tilde{A}_n(x)| = o(n^{-1}), \quad \text{as } n \to \infty$$

if $f(x)$ has the second integrable Radon–Nikodym derivative. □

For proofs of these results, see Seoh (1983) or Puri and Seoh (1984b).

Theorem 4.2 provides the Edgeworth expansion with natural parameters $E(S_n^+)$ and $\text{var}(S_n^+)$. In Seoh (1983), the Edgeworth expansion is also derived for simple normalizing constants under somewhat stronger assumptions on location parameters θ_{ni}, $1 \leqslant i \leqslant n$, namely, $\sum_{i=1}^n |\theta_{ni}|^3 = O(n^{-1/2})$, and on f (for details, see Seoh, 1983; Puri and Seoh, 1984b).

Finally, we present very briefly Edgeworth expansion up to the order $o(n^{-1})$ for the df of simple linear rank statistics S_n (defined in (2.1)) under the null hypothesis. The results are due to Does (1983a). The methods in the derivation of the results are the same as in the symmetry problem discussed earlier.

Let $\{X_{ni}, 1 \leqslant i \leqslant n\}$ be i.i.d. r.v.s. with a common continuous df $F(x)$. Let R_{ni}, $1 \leqslant i \leqslant n$ be the rank of X_{ni} among $\{X_{nk}, 1 \leqslant k \leqslant n\}$. Then, Does (1983a) considers the statistics of the form $S_n = \sum_{i=1}^n c_{ni} a_n(R_{ni})$ where c_{ni}, $1 \leqslant i \leqslant n$ are known constants, and $a_n(i) = J(i/(n+1))$ as defined in (4.3).

We assume that the cs satisfy (4.8a) and $\sum_{j=1}^n c_{jn} = 0$. Furthermore, the score generating function J satisfies (4.8b), and $\limsup_{t\to 0,1} t(1-t)|J''(t)/J'(t)| < 2$. Finally, we assume (without loss of generality) that $\int_0^1 J(t) = 0$ and $\int_0^1 J^2(t)dt = 1$. It is clear that

$$E(S_n) = 0, \quad \text{and} \quad \sigma_n^2 = \text{var}(S_n) = \frac{1}{n-1}\sum_{i=1}^n \left(J\left(\frac{i}{n+1}\right) - \bar{J}\right)^2$$

where

$$\bar{J} = \sum_{i=1}^n J\left(\frac{i}{n+1}\right)\Big/ n.$$

Define for each $n \geqslant 2$, $S_n^{**} = \sigma_n^{-1} S_n$, and $F_n^{**}(x) = P(S_n^{**} \leqslant x)$ and

$$\tilde{F}_n^{*}(x) = \Phi(x) - \phi(x)\left\{\frac{k_{3n}}{6}(x^2-1) + \frac{k_{4n}}{24}(x^3-3x) + \frac{k_{3n}^2}{72}(x^5 - 10x^3 + 15x)\right\}$$

where

$$k_{3n} = \sum_{i=1}^n c_{in}^3\left\{\int_0^1 J^3(t)dt\right\}$$

and

$$k_{4n} = \sum_{i=1}^{n} c_{in}^4 \left\{ \int_0^1 J^4(t)\,dt - 3 \right\} - \frac{3}{n} \left\{ \int_0^1 J^4(t)\,dt - 1 \right\}.$$

Then

Theorem 4.3 $\sup_x |F_n^{**}(x) - \tilde{F}_n^*(x)| = o(n^{-1})$ as $n \to \infty$. □

Note that k_{3n} and k_{4n} are asymptotic expressions for the third and fourth cumulants of S_n^{**} where terms of order $o(n^{-1})$ have been neglected. Thus $\tilde{F}_n^*$ constitutes a genuine Edgeworth expansion for $\tilde{F}_n^{**}$. Furthermore, note that the characteristic function of S_n^{**} behaves exactly as in lemma 4.2. The proof of the theorem is similar to that of theorem 4.1 and is dealt with in Does (1983a, b). We may mention that the asymptotic expansions given in theorems 4.1 and 4.2 can be used to derive deficiencies in the sense of Hodges and Lehmann (1970) for S_n^+ and S_n with respect to their parametric competitors. To the best of our knowledge this problem has not been studied thus far.

5 *U*-Statistics

Most classical statistics U_n based on i.i.d. observations $X_1, X_2, \ldots, X_n$ may be cast in the form

$$U_N := \binom{n}{r}^{-1} \sum_{1 \le i_1 < i_2 < \ldots < i_r \le n} h(X_{i_1}, X_{i_2}, \ldots, X_{i_r}), \tag{5.1}$$

where h is real-valued, measurable, and symmetric in its r arguments. The statistic U_n in (5.1) is called a *U-statistic of order r*. Often such statistics may be expressed as, or approximated by, statistics of the form $H(\bar{Z})$ as described in section 2, and in such cases they are better dealt with by the (multivariate) method described earlier. There are, however, many instances where such a representation is not possible.

Example 5.1 (Wilcoxon one-sample statistic). This is related to Example 4.3. Here

$$\begin{aligned} U_n &= \binom{n}{2}^{-1} \sum_{1 \le i < j \le n} 1_{\{X_i + X_j > 0\}} \\ &= \frac{1}{n(n-1)} \sum_{i=1}^{n} R_{ni}^+ \operatorname{sgn} X_i + \frac{1}{2}\left(\frac{n+1}{n-1}\right) - \binom{n}{2}^{-1} \sum_{i=1}^{n} 1_{\{X_i > 0\}}. \end{aligned} \tag{5.2}$$

Example 5.2 (Gini's mean difference). The statistic

$$U_n = \binom{n}{2}^{-1} \sum_{1 \leq i < j \leq n} |X_i - X_j| \tag{5.3}$$

is sometimes used as a measure of concentration.

For such examples ingenious estimations of the c.f. of the normalized U_n are necessary, followed by the use of Esseen's inequality (2.16). For simplicity we assume $r = 2$. Assume, by centering, that $Eh(X_1, X_2) = 0$. Projecting U_n onto the closure of the linear space spanned by functions of single variables X_j, $1 \leq j \leq n$, one has the *Hoeffding decomposition*

$$\begin{aligned} U_n &= \frac{2}{n}\sum_{j=1}^{n} g(X_j) + \frac{1}{\binom{n}{2}} \sum_{1 \leq i < j \leq n} (h(X_i, X_j) - g(X_i) - g(X_j)), \\ &= \frac{2}{n}\sum_{j=1}^{n} g(X_j) + \binom{n}{2}^{-1} \sum_{1 \leq i < j \leq n} \Psi(X_i, X_j), \quad \text{say}, \end{aligned} \tag{5.4}$$

where $g(x) = Eh(X_1, x)$. By orthogonality, one has

$$\theta_n^2 := \operatorname{var}(U_n) = \frac{4}{n} Eg^2(X_1) + \binom{n}{2}^{-1} E(h(X_1, X_2) - g(X_1) - g(X_2))^2. \tag{5.5}$$

Assume

$$\sigma^2 := Eg^2(X_1) > 0, \tag{5.6}$$

$$E|h(X_1, X_2)|^p < \infty \quad \text{for some } p > 4, \tag{5.7}$$

and the Cramér condition

$$\limsup_{|t| \to \infty} |E \exp(itg(X_1))| < 1. \tag{5.8}$$

Theorem 5.1 (Götze and van Zwet, 1991). Under assumptions (5.6), (5.7), (5.8), one has

$$\begin{aligned} \sup_x \Bigg| P(\theta_n^{-1} U_n \leq x) - \Phi(x) - \Bigg\{ n^{-1/2}\frac{\kappa_3}{6}(x^2 - 1) + n^{-1}\frac{\kappa_4}{24}(x^3 - 3x) \\ + n^{-1}\frac{\kappa_3^2}{72}(x^5 - 10x^3 + 15x)\Bigg\}\varphi(x)\Bigg| = o(n^{-1}) \end{aligned} \tag{5.9}$$

where $\Phi(x)$ is the standard normal distribution function and $\varphi(x)$ its density, and

$$\kappa_3 = \sigma^{-3}[Eg^3(X_1) + 3Eg(X_1)g(X_2)\psi(X_1, X_2)],$$
$$\kappa_4 = \sigma^{-4}[Eg^4(X_1) - 3\sigma^4 + 12Eg^2(X_1)g(X_2)\psi(X_1, X_2) + 12Eg(X_1)g(X_2)\psi(X_1, X_3)\psi(X_2, X_3)]. \quad \square \tag{5.10}$$

This result is an improvement on earlier results of Callaert et al. (1980) and Bickel et al. (1986).

6 Expansions in Econometrics and General Remarks

One of our aims in this survey has been to present those main results and techniques for deriving Edgeworth expansions which are widely applicable in statistics and economics.

When valid, Edgeworth expansions provide good approximations to distributions of statistics, thereby providing confidence intervals for estimates and critical values for tests corresponding more precisely to the desired levels than given by the classical procedures based on normal approximation. This aspect of Edgeworth expansions is currently often superseded by the bootstrap methodology of Efron (1979). But the most convincing demonstration of the effectiveness of the bootstrap is provided only by Edgeworth expansions (Singh, 1981; Bhattacharya and Qumsiyeh, 1989; Hall, 1992).

Expansions are also important theoretical tools for gaining insight into what the economists call the "finite sample theory" (Phillips, 1980), i.e. inference based on moderate sample sizes. It has been shown in a number of studies that the asymptotically efficient maximum likelihood estimator (m.l.e.) is not always superior to some other competing estimators (see Berkson, 1980) if the sample size is not large. An insight into this is often provided by an expansion of the distribution, or of the expected loss, which goes beyond the first approximation on which asymptotic efficiency considerations are based. For example, it may be shown using expansions that the m.l.e. in a one-parameter exponential family, when properly bias-corrected, is still at least as good as any other (similarly bias-corrected) estimate (see Ghosh, 1994) up to any order; but when not corrected appropriately for bias it may be outdone by other competitors. Similar aspects of "finite sample approximation" of much significance in economics are discussed in Bergstrom (1962) and Sawa (1969). For expansions related to the so-called k-class estimators and Nagar approximation see Kadane (1971) and Sargan (1975).

Due to special structures of models that often arise in economics econometricians have often derived expansions which are particularly suited to their problems. In most of these models, e.g. regression models, with or without lagged variables (or autoregression), the disturbances or error terms are generally taken to be Gaussian. Anderson and Sawa (1973) and Anderson (1976) obtain asymptotic expansions of distributions of the m.l.e., ordinary least squares, and two-stage least squares estimates in a regression model in which the means of the independent bivariate observations have a linear functional relationship and both variables are subject to uncorrelated normal errors with a common variance. This is applied to the distribution of estimates in a simultaneous equations model. For dynamic models with endogenous lagged variables, Sargan and Satchell (1986) and Phillips (1980, 1986) give detailed and ingenious calculations of the c.f. of the normalized statistic under consideration, in order to check its necessary integrability properties. A theorem of Chambers (1967), when properly corrected, now gives an L_1-expansion of the density of the normalized statistic. Since this theorem of Chambers has been used rather often in the literature, perhaps a word of caution is in order. The theorem in question (see Chambers, 1967, p. 369) asserts a valid expansion of the density with an error $o(n^{-(r+1)/2+\varepsilon})$ for some $\varepsilon > 0$, provided (i) the cumulants of the statistic up to order $r + 3$ have the same order of magnitude as those of the normalized sum S_n of i.i.d. random variables, and (ii) the c.f. f_n of the (normalized) statistic satisfies, for all $\alpha \in (0, \frac{1}{2})$

$$\int_{|t|>n^{\alpha}} |f_n(t)|\,dt = o(n^{-(r+1)/2+\varepsilon}). \tag{6.1}$$

As explained in section 2 (see 2.4), for a valid expansion of the distribution function, or the density, one must provide (a) an expansion of the c.f. $f_n(t)$ for $|t| \leqslant a_n$ where a_n is at least of order $(\log n)^{1/2}$, and (b) an appropriate rate of decay of $f_n(t)$ for $|t| \geqslant n^{1/2}$. While the assumption (ii) takes care of the requirement (b), one can only derive an Edgeworth expansion of $f_n(t)$ for $|t| \leqslant \delta$ for some $\delta > 0$ by using (i). This clearly is not enough for (a). Most expansions derived in the econometrics literature which allude to the above theorem of Chambers (1967) are, however, accompanied with such detailed analyses of the c.f. of the statistic that the condition (a) above would be easy to check using standard techniques similar to those in the case of S_n. Alternatively, one may apply the methods of sections 2 and 3.

Taniguchi (1983, 1991) has derived expansions for estimators in Gaussian time series models, by a direct attack.

In regression type models in economics, with nonstochastic (exogenous) regressors, with or without lagged endogenous variables, and with Gaussian disturbances, the statistic may still be expressed in the form $H(\bar{Z})$ where $Z_j = (f_i(X_j), \ldots, f_k(X_j))$; but the observations X_j are not identically distributed. It is often possible to apply to such problems multivariate expansions analogous to those discussed in sections 2 and 3. Notice, for example, that theorem 3.2 does not require stationarity. For the case of independent but not identically distributed observations see Skovgaard (1981). In the classical regression model with nonstochastic regressors and i.i.d. errors, expansions for least squares estimators may be found in Qumsiyeh (1990). Similar expansions may be derived in dynamic models where endogenous lagged variables are involved. The algorithm for computing the terms in the Edgeworth expansion as described in section 2 still works. One advantage of the general approaches considered in this article is that expansions hold for non-Gaussian errors as well.

The method of expansions considered in this chapter extends to the multisample case.

There are some specialized techniques for approximating probabilities associated with statistics which have not been discussed in this chapter. For probabilities in the tails, *large deviation methods*, originally due to Cramér (1938), are useful (see Phillips, 1986, for econometric applications). In parametric models one may also sometimes use the so-called *saddlepoint approximations* of Daniels (1954), Barndorff-Nielsen and Cox (1979) and Phillips (1978). The terms of the Edgeworth expansion, on the other hand, depend only on cumulants, or moments, of the underlying distribution, making these expansions reasonably robust in semiparametric models.

Note

The authors wish to thank Professor P. C. B. Phillips and two anonymous referees for their helpful suggestions. The research of R.N.B. was supported in part by NSF Grant DMS 9206937. The research of M.L.P. was supported by the Office of Naval Research Contract N00014-91-J-1020.

References

Akahira, M. and Takeuchi, K. (1981) *Asymptotic Efficiency of Statistical Estimators: Concepts and Higher Order Efficiency*. Lecture Notes in Statistics 7. New York: Springer-Verlag.

Albers, W., Bickel, P. J. and Van Zwet, W. R. (1976) Asymptotic expansions for the power of distribution free tests in the one-sample problem. *Annals of Statistics*, 4, 108–56.

Anderson, T. W. and Sawa, T. (1973) Distribution of estimates of coefficients of a single equation in a simultaneous system and their asymptotic expansions. *Econometrica*, 41, 683–714.

Anderson, T. W. (1976) Estimation of linear functional relationships: approximate distributions and connections with simultaneous equations in econometrics. *Journal of the Royal Statistical Society B*, 38, 1–36.

Babu, G. J. and Singh, K. (1984) On one-term Edgeworth correction by Efron's bootstrap. *Sankhya A*, 46, 219–32.

Bai, Z. D. and Rao, C. R. (1991) Edgeworth expansion of a function of sample means. *Annals of Statistics*, 19, 1295–315.

Barndorff-Nielsen, O. E. and Cox, D. R. (1979) Edgeworth and saddle-point approximations with statistical applications (with discussion). *Journal of the Royal Statistical Society B*, 41, 279–312.

Bergstrom, A. R. (1962) The exact sampling distributions at least squares and maximum likelihood estimators of the marginal propensity to consume. *Econometrica*, 30, 480–90.

Berkson, J. (1980) Minimum chi-square, not maximum likelihood (with discussion). *Annals of Statistics*, 8, 457–87.

Bhattacharya, R. N. (1968) Berry–Esseen bounds for the multi-dimensional central limit theorem. *Bulletin of the American Mathematical Society*, 75, 285–7.

Bhattacharya, R. N. (1971) Rates of weak convergence and asymptotic expansions for classical central limit theorems. *Annals of Mathematical Statistics*, 42, 241–59.

Bhattacharya, R. N. (1985) Some recent results on Cramér–Edgeworth expansions with applications. In P. R. Krishnaiah (ed.), *Multivariate Analysis VI*. Amsterdam: North-Holland, pp. 57–75.

Bhattacharya, R. N. and Chan, N. H. (1994) Comparisons of chisquare, Edgeworth expansions and bootstrap approximations to the distribution of the frequency chi square. *Sankhya*, in the press.

Bhattacharya, R. N. and Denker, M. (1990) *Asymptotic Statistics*. Boston: Birkhauser.

Bhattacharya, R. N. and Ghosh, J. K. (1978) On the validity of the formal Edgeworth expansion. *Annals of Statistics*, 6, 434–51.

Bhattacharya, R. N. and Puri, M. L. (1983) On the order of magnitude of cumulants of von Mises functionals and related statistics. *Annals of Probability*, 11, 346–54.

Bhattacharya, R. N. and Qumsiyeh, M. (1989) Second order and L_p-comparisons between the bootstrap and empirical Edgeworth expansion methodologies. *Annals of Statistics*, 17, 160–9.

Bhattacharya, R. N. and Ranga Rao, R. (1976) *Normal Approximation and Asymptotic Expansions*. New York: Wiley. Second reprint with supplement (1986), Florida: Krieger, Malabar.

Bickel, P. J., Götze, F. and van Zwet, W. R. (1986) The Edgeworth expansion for U-statistics of degree two. *Annals of Statistics*, 14, 1463–84.

Bickel, P. J. and Van Zwet, W. R. (1978) Asymptotic expansions for the power of distribution free tests in the two-sample problem. *Annals of Statistics*, 6, 937–1004.

Bikyalis, A. (1968) Asymptotic expansions of distribution functions and density functions of sums of independent and identically distributed random vectors. *Litov. Mat. Sb.*, 8, 405–22 (in Russian).

Bose, A. (1988) Edgeworth correction by bootstrap in autoregressions. *Annals of Statistics*, 16, 1709–22.

Bulinskii, A. V. and Zhurbenko, I. G. (1976) The central limit theorem for random fields. *Soviet Math. Dokl.*, 17, 14–17.

Callaert, H., Janssen, P. and Veraverbeke, N. (1980) An Edgeworth expansion for *U*-statistics. *Annals of Statistics*, 8, 299–312.

Chambers, J. M. (1967) On methods of asymptotic approximation for multivariate distributions. *Biometrika*, 54, 367–83.

Chandra, T. K. and Ghosh, J. K. (1979) Valid asymptotic expansions for the likelihood ratio statistic and other perturbed chisquare variables. *Sankhya A*, 41, 22–47.

Chandra, T. K. and Ghosh, J. K. (1980) Valid asymptotic expansions for the likelihood ratio and other statistics under contiguous alternatives. *Sankhya A*, 42, 170–84.

Chebyshev, P. L. (1890) Sur deux théorèmes relatifs aux probabilités. *Acta Mathematica*, 14, 305–15.

Chibishov, D. M. (1972) An asymptotic expansion for the distribution of a statistic admitting an asymptotic expansion. *Theor. Probab. Appl.*, 17, 620–30.

Chibishov, D. M. (1973) An asymptotic expansion for a class of estimators containing the maximum likelihood estimators. *Theor. Probab. Appl.*, 18, 295–303.

Cramér, H. (1928) On the composition of elementary errors. *Skand. Aktuarietidskr.*, 11, 13–74, 141–80.

Cramér, H. (1938) Sur un nouveau théorème-limite de théorie des probabilités. *Actualités Sci. et Ind.*, 736, 5–23.

Daniels, H. E. (1954) Saddlepoint approximations in statistics. *Annals of Mathematical Statistics*, 25, 631–50.

Does, R. J. M. M. (1983a) An Edgeworth expansion for simple linear rank statistics under the null hypothesis. *Annals of Statistics*, 11, 607–24.

Does, R. J. M. M. (1983b) Asymptotic expansions for simple linear rank statistics under contiguous alternatives. In P. Mandl and M. Huskova (eds) *Proceedings of the Third Prague Symposium on Asymptotic Statistics* Amsterdam/New York: Elsevier, pp. 221–30.

Edgeworth, F. Y. (1896) The asymmetrical probability curve. *Philosophical Magazine*, 5th series, 41, 90–9.

Edgeworth, F. Y. (1905) The law of error. *Transactions of the Cambridge*

Philosophical Society, 20, 36–65, 113–41.
Efron, B. (1975) Defining the curvature of a statistical problem (with applications to second order efficiency). *Annals of Statistics*, 3, 1189–242.
Efron, B. (1979) Bootstrap methods: another look at the jackknife. *Annals of Statistics*, 7, 1–26.
Esseen, C. G. (1945) Fourier analysis of distribution functions. A mathematical study of the Laplace–Gaussian law. *Acta Mathematica*, 77, 1–125.
Fellingham, S. A. and Stoker, D. J. (1964) An approximation for the exact distribution of Wilcoxon test for symmetry. *J. Amer. Statist.*, 19, 899–905.
Fisher, R. A. (1925) Theory of statistical estimation. *Proceedings of the Cambridge Philosophical Society*, 22, 700–25.
Ghosh, J. K. (1994) *Higher Order Asymptotics*. Hayward, CA: Institute of Mathematical Statistics.
Ghosh, J. K. and Subramanyam, K. (1974) Second order efficiency of maximum likelihood estimators. *Sankhya A*, 36, 325–58.
Götze, F. (1985) Asymptotic expansions in functional central limit theorems. *Journal of Multivariate Analysis*, 16, 1–20.
Götze, F. and Hipp, C. (1978) Asymptotic expansions in the central limit theorem under moment conditions. *Z. Wahrsch. Verw. Geb.*, 42, 67–87.
Götze, F. and Hipp, C. (1983) Asymptotic expansions for sums of weakly dependent random vectors. *Z. Wahrsch. Verw. Geb.*, 64, 211–39.
Götze, F. and Hipp, C. (1994) Asymptotic distribution of statistics in time series. *Annals of Statistics*, in the press.
Götze, F. and Van Zwet, W. (1991) Edgeworth expansion for asymptotically linear statistics. Preprint 91-034, University of Bielefeld.
Hájek, J. and Šidák, Z. (1967) *Theory of Rank Tests*. New York: Academic Press.
Hall, P. G. (1992) *Bootstrap and Edgeworth Expansion*. Second Edition, New York: Springer-Verlag.
Hodges, J. L. and Lehmann, E. L. (1970) Deficiency. *Annals of Mathematical Statistics*, 41, 783–801.
Huang, F. (1994) Berry–Esseen bounds and asymptotic expansions under dependence. Thesis in progress, Indiana University, Bloomington.
Jensen, J. L. (1989a) A note on the work of Götze and Hipp concerning asymptotic expansions for sums of weakly dependent random vectors. In P. Mandl and M. Huškovà (eds), *Proceedings of the Fourth Prague Symposium on Asymptotic Statistics*. Prague: Charles University, pp. 295–303.
Jensen, J. L. (1989b) Asymptotic expansions for strongly mixing Harris recurrent Markov chains. *Scandinavian Journal of Statistics*, 16, 47–64.
Kadane, J. B. (1971) Comparison of k-class estimators when the disturbances are small. *Econometrica*, 39, 723–37.
Lahiri, S. N. (1993) Refinements in asymptotic expansions for sums of weakly dependent random vectors. *Annals of Probability*, 21, 791–9.
Nagaev, S. V. (1961) More exact statements of limit theorems for homogeneous Markov chains. *Theor. Probab. Appl.*, 6, 62–81.

Pfanzagl, J. (1980) Asymptotic expansions in parametric statistical theory. In P. R. Krishnaiah (ed.), *Developments in Statistics, volume 3*. New York: Academic Press, pp. 1–97.

Phillips, P. C. B. (1978) Edgeworth and saddle point approximations in the first order noncircular autoregression. *Biometrika*, 65, 91–8.

Phillips, P. C. B. (1980) Finite sample theory and the distributions of alternative estimators of the marginal propensity to consume. *Rev. Econ. Studies*, 47, 183–224.

Phillips, P. C. B. (1986) Large deviation expansions in econometrics. *Advances in Econometrics*, 5, 199–226.

Puri, M. L. and Seoh, M. (1984a) Edgeworth expansions for signed linear rank statistics with regression constants. *J. Statist. Plann. Inference*, 10, 137–49.

Puri, M. L. and Seoh, M. (1984b) Edgeworth expansions for signed linear rank statistics under near location alternatives. *J. Statist.* Plann. Inference, 10, 289–309.

Puri, M. L. and Seoh, M. (1985) Berry–Esseen for signed linear rank statistics with regression constants. In P. Révész (ed.), *Colloquium on Limit Theorems in Probability Statistics*. Amsterdam: North-Holland, pp. 875–905.

Qumsiyeh, M. (1990) Edgeworth expansion in regression models. *Journal of Multivariate Analysis*, 35, 86–101.

Ranga Rao, R. (1961) On the central limit theorem in $\mathbb{R}_k$. *Bulletin of the American Mathematical Society*, 67, 359–61.

Rao, C. R. (1961) Asymptotic efficiency and limiting information. *Proceedings of the Fourth Berkeley Symposium on Mathematical Statistics and Probability*, 1, 531–46.

Rao, C. R. (1962) Efficient estimates and optimum inference procedures in large samples. *Journal of the Royal Statistical Society B*, 24, 46–63.

Rao, C. R. (1963) Criteria of estimation in large samples. *Sankhya*, 25, 189–206.

Robinson, J. (1978) An asymptotic expansion for samples from a finite population. *Annals of Statistics*, 6, 1005–11.

Robinson, J. (1980) An asymptotic expansion for permutation tests with several samples. *Annals of Statistics*, 8, 851–64.

Sargan, J. D. (1975) Gram–Charlier approximations applied to *t*-ratios of *k*-class estimators. *Econometrica*, 43, 327–46.

Sargan, J. D. and Satchell, S. E. (1986) A theorem of validity of Edgeworth expansions. *Econometrica*, 54, 189–214.

Sawa, T. (1969) The exact sampling distribution of ordinary least squares and two stage least squares estimator. *Journal of the American Statistical Association*, 64, 923–36.

Seoh, M. (1983) Rate of convergence to normality and Edgeworth expansions for signed linear rank statistics with regression constants. PhD thesis, Indiana University.

Singh, K. (1981) On the asymptotic theory of Efron's bootstrap. *Annals of Statistics*, 9, 1187–95.

Skovgaard, Ib. M. (1981) Edgeworth expansions of the distributions of maximum likelihood estimators in the general (non i.i.d.) case. *Scandinavian Journal of Statistics*, 8, 227–36.
Statulevichius, V. (1969–70) Limit theorems for sums of random variables related to a Markov chain. *Litovsk Mat. Sb.*, 9, 10 (in Russian).
Stein, C. (1986) *Approximate Computation of Expectations*. IMS Lecture Notes Monograph Series 7. Hayward, CA: Institute of Mathematical Statistics.
Taniguchi, M. (1983) On the second order asymptotic efficiency of estimators of Gaussian ARMA processes. *Annals of Statistics*, 11, 157–69.
Taniguchi, M. (1991) *Higher Order Asymptotic Theory for Time Series Analysis*. Lecture Notes in Statistics 68. New York: Springer-Verlag.
Thompson, R., Govindarajulu, Z. and Doksum, K. (1967) Distribution and power of the absolute normal scores test. *Journal of the American Statistical Association*, 62, 966–75.
Wallace, D. L. (1958) Asymptotic approximations to distributions. *Annals of Mathematical Statistics*, 29, 635–54.
Yarnold, J. K. (1970) The minimum expectation in χ^2 goodness of fit tests and the accuracy of approximation for the null distribution. *Journal of the American Statistical Association*, 65, 864–86.

Reprinted from
Advances in Econometrics and Quantitative Economics: Essays in honor of C.R. Rao (1995), pages 88-122, edited by G.S. Maddala, Peter C.B. Phillips and T.N. Srinivasan, by permission of the publisher,

NORMAL APPROXIMATION OF U-STATISTICS IN HILBERT SPACE*

YU. V. BOROVSKIKH[†], L. MADAN PURI[‡] and V. V. SAZONOV[¶]

Abstract. Let $\{U_n\}$, $n = 1, 2, \ldots$, be Hilbert space H-valued U-statistics with kernel $\Phi(\cdot\,,\cdot)$, corresponding to a sequence of observations (random variables) $X_1, X_2, \ldots$. The rate of convergence on balls in the central limit theorem for $\{U_n\}$ is investigated. The obtained estimate is of order $n^{-1/2}$ and depends explicitly on $\mathbf{E}\|\Phi(X_1, X_2)\|^3$ and on the trace and the first nine eigenvalues of the covariance operator of $\mathbf{E}(\Phi(X_1, X_2)|X_1)$.

Key words. U-statistic, Hilbert space, central limit theorem, normal (Gaussian) approximation, rate of convergence

1. Introduction. Let $X_1, \ldots, X_n$ be independent random variables with values in a measurable space $(X, \mathcal{X})$ having on it the same distribution $\mathbf{P}$. Let H be a real separable Hilbert space with inner product $(\cdot\,,\cdot)$ and norm $\|\cdot\|$. Consider the UH-statistic

$$U_n = \binom{n}{2}^{-1} \sum_{1 \leqq i < j \leqq n} \Phi(X_i, X_j)$$

with a symmetric kernel Φ: $X^2 \to H$ such that $\mathbf{E}\Phi(X_1, X_2) = 0$, $\mathbf{E}\|\Phi(X_1, X_2)\| < \infty$.

By the Hoeffding decomposition,

$$U_n = 2n^{-1} \sum_{j=1}^{n} g_1(X_j) + 2n^{-1}(n-1)^{-1} \sum_{1 \leqq i < j \leqq n} g_2(X_i, X_j), \tag{1}$$

where

$$g_1(x) = \int_X \Phi(x, y)\, \mathbf{P}(dy) = \mathbf{E}\big(\Phi(X_1, X_2) \mid X_1 = x\big),$$
$$g_2(x, y) = \Phi(x, y) - g_1(x) - g_1(y).$$

We assume that $\sigma^2 = \mathbf{E}\|g_1(X_1)\|^2 > 0$ and observe that the definitions of $g_1(x)$ and $g_2(x)$ imply

$$\int_X g_2(x, y)\, \mathbf{P}(dy) = 0 \quad \text{(a.s.)}. \tag{2}$$

---

*Received by the editors May 17, 1994. This research was supported by the Office of Naval Research Contract N 00014-91-J-1020. The results of this paper were partially obtained during the visit of the last author to the Department of Management, Hong Kong University of Science and Technology, winter 1993–1994.

http://www.siam.org/journals/tvp/41-3/97519.html

[†]St.-Petersburg Transport University, Moskovskii pr. 9, 190031 St.-Peterburg, Russia.

[‡]Indiana University, Department of Mathematics, Bloomington, Indiana 47405, U.S.A.

[¶]Steklov Mathematical Institute, Vavilov St. 42, 117966 Moscow, Russia.

Assuming also that $\sigma < \infty$ we can define the covariance operator V of $g_1(X_1)$ in the usual way:

$$(Vx, y) = \int_X (x, g_1(z))\,(y, g_1(z))\,\mathbf{P}(dz), \quad x, y \in H.$$

Let Y be a Gaussian H-valued random variable with mean zero and covariance operator $\sigma^{-2}V$. Denote by $\sigma_1^2 \geqq \sigma_2^2 \geqq \cdots$ the eigenvalues of V. Finally, for any $r \geqq 0$, $a \in H$, let $B_r(a) = \{x \in H\colon \|x - a\| < r\}$ and $B_r = B_r(0)$.

By the central limit theorem for UH-statistics, if

$$\mathbf{E}\|g_1(X_1)\|^2 < \infty, \qquad \mathbf{E}\|g_2(X_1, X_2)\|^{4/3} < \infty,$$

then (see [1], [2], [5])

$$\Delta_n(a) = \sup_r \left|\mathbf{P}\{\|2^{-1}n^{1/2}\sigma^{-1}U_n - a\| < r\} - \mathbf{P}\{\|Y - a\| < r\}\right| \longrightarrow 0, \quad n \to \infty.$$

The rate of convergence of $\Delta_n(0)$ to zero with respect to n was estimated in [1], [2], [5], [12], where under the additional moment conditions

$$\mathbf{E}\|g_1(X_1)\|^3 < \infty, \qquad \mathbf{E}\|g_2(X_1, X_2)\|^2 < \infty$$

it was shown that

$$\Delta_n(0) = \rho_n(V)\, n^{-1/2}$$

with $\rho_n(V) \to 0$ as $n \to \infty$. To prove this, it was observed that the problem of estimating the probability for a value of the UH-statistic to be in the ball B_r may be treated as the problem of estimating the rate of convergence in the central limit theorem in Hilbert space for balls with a nonzero random center which tends to zero with probability 1 as $n \to \infty$. However, it was assumed in [1], [2], [5], [12] that the number of nonzero eigenvalues of the covariance operator V is sufficiently large. The explicit dependence of $\rho_n(V)$ on V in [1], [2], [5], [12] was not determined. Estimates of order $O(n^{-1/3})$ for any $a \in H$ with explicit dependence on the second and third moments were obtained in [14].

Note that in the one-dimensional case the study of the problem considered in the present paper has a rather long history. It was initiated in 1973 by Grams and Serfling [11] and later a number of gradual improvements of their result was made (for details see [13] and [10]). In [10] it was also shown that the estimate obtained by Korolyuk and Borovskikh in [4] is essentially unimprovable.

In the present paper we study the dependence of $\Delta_n(a)$ on n, the covariance operator V and $\beta = \mathbf{E}\|\Phi(X_1, X_2)\|^3$. When $\beta < \infty$ an estimate of order $n^{-1/2}$ is obtained, the dependence of which on Φ involves only β, the first nine eigenvalues of V and $\sigma^2 = \operatorname{tr} V$. The order of dependence of our estimate on n is precise, but its dependence on σ_j^2, σ_j^2 and β apparently admits improvement. To prove the estimate we employ the methods developed in [3] and [12].

2. The theorem.

THEOREM. *There exists an absolute constant c such that, for all $n \geqq 2$ and $a \in H$,*

$$\Delta_n(a) \leqq c\,(1 + \|a\|^3)\sigma^4 \left(\prod_{j=1}^{9} \sigma_j^{-1}\right) \sigma_9^{-1} \beta^2 n^{-1/2}. \tag{3}$$

In what follows we will assume that $\sigma = 1$. The general case is reduced to this one if we replace Φ by $\sigma^{-1}\Phi$. We can assume also that $\beta < \infty$ and $\sigma_9 > 0$ since otherwise (3) would be obvious.

In the proof of the theorem the upper estimates for $\Delta_n(a)$ are calculated more precisely; except for n and V, they depend only on the moments

$$\mathbf{E}\|g_1\|^3 = \mathbf{E}\|g_1(X_1)\|^3, \qquad \mathbf{E}\|g_2\|^3 = \mathbf{E}\|g_2(X_1, X_2)\|^3,$$

which are related to β by the inequalities

$$\mathbf{E}\|g_1\|^3 \leqq \beta, \qquad \mathbf{E}\|g_2\|^3 \leqq 27\beta. \tag{4}$$

We formulate the theorem using only β for the sake of simplicity.

3. Proof of the theorem. Below $c, c_1, c_2, \dots$ denote absolute constants; the same symbol may stand for different constants. The proof is divided into several lemmas.

LEMMA 1. *We have*

$$\Delta_n(a) \leqq \Delta_{n1}(a) + n\,\mathbf{P}\{\|g_1(X_1)\| > n^{1/2}\}, \tag{5}$$

where

$$\Delta_{n1}(a) = \sup_r \Big|\mathbf{P}\{\|\overline{S}_n - a + V_n\| < r\} - \mathbf{P}\{\|Y - a\| < r\}\Big|,$$

$$\overline{S}_n = n^{-1/2}\sum_{j=1}^{n} g_1(X_j)\, I\Big(\|g_1(X_j)\| \leqq n^{1/2}\Big),$$

$I(A)$ is the indicator function of the set A and

$$V_n = (n-1)^{-1} n^{-1/2} \sum_{j=1}^{n}\sum_{i=1}^{j-1} g_2(X_i, X_j).$$

Proof. According to (1),

$$2^{-1} n^{1/2} U_n - a = S_n - a + V_n,$$

where

$$S_n = n^{-1/2}\sum_{j=1}^{n} g_1(X_j)$$

and we obviously have

$$\Delta_n(a) \leqq \Delta_{n1}(a) + \sup_r \Big|\mathbf{P}\{\|S_n - a + V_n\| < r\} - \mathbf{P}\{\|\overline{S}_n - a + V_n\| < r\}\Big|.$$

Now applying Lemma 1 of [17] with $f(x) = I(\|x\| < r)$, $u_1 = \infty$, $u_2 = n^{1/2}$ and $Z = -a + V_n$, we see that the second summand in the right-hand side of this inequality is not greater than $n\mathbf{P}\{\|g_1(X_1)\| > n^{1/2}\}$ and the lemma follows.

Let $k = n - [n^{2/3}]$, where $[n^{2/3}]$ is the integral part of $n^{2/3}$ and denote

$$\alpha_k = \sum_{1 \leqq i < j \leqq k} g_2(X_i, X_j), \qquad \beta_k = \sum_{i=1}^{k} \sum_{j=k+1}^{n} g_2(X_i, X_j),$$

$$\delta_k = \sum_{k+1 \leqq i < j \leqq n} g_2(X_i, X_j), \qquad \overline{S}_n^k = n^{-1/2} \sum_{j=1}^{k} \overline{g}_1(X_j),$$

$$\overline{g}_1(X_j) = g_1(X_j)\, I\Big(\|g_1(X_j)\| \leqq n^{1/2}\Big), \quad j = 1, \dots, n,$$

$$Z_n = 2\varepsilon_n\big(\overline{S}_n^k - a,\, \alpha_k\big), \quad \varepsilon_n = (n-1)^{-1} n^{-1/2}.$$

LEMMA 2. *For $\Delta_{n1}(a)$ we have*

(6) $$\Delta_{n1}(a) \leqq \Delta_{n2}(a) + c_1\big(1 + \|a\|^3\big)\left(\beta^2 + \sigma_1^{-1}\sigma_2^{-1}\right) n^{-1/2},$$

where

$$\Delta_{n2}(a) = \sup_r \Big| \mathbf{P}\big\{\|\overline{S}_n - a\|^2 + Z_n < r\big\} - \mathbf{P}\big\{\|Y - a\|^2 < r\big\}\Big|.$$

Proof. First we observe that

$$\begin{aligned} \big\|\overline{S}_n - a + V_n\big\|^2 &= (\overline{S}_n - a + V_n,\, \overline{S}_n - a + V_n) \\ &= \big\|\overline{S}_n - a\big\|^2 + 2(\overline{S}_n - a,\, V_n) + \|V_n\|^2. \end{aligned} \tag{7}$$

In our notation,

$$\|V_n\|^2 = (n-1)^{-2} n^{-1} \left\| \sum_{j=2}^{n} \sum_{i=1}^{j-1} g_2(X_i, X_j) \right\|^2,$$

$$2\,(\overline{S}_n - a,\, V_n) = Z_n + 2\varepsilon_n\big(\overline{S}_n^k - a,\, \beta_k + \delta_k\big) + 2\,\big(\overline{S}_n - \overline{S}_n^k,\, V_n\big),$$

and

$$\big(\overline{S}_n^k - a,\, \beta_k + \delta_k\big) = \big(\overline{S}_n^k,\, \beta_k\big) - \big(a,\, \beta_k\big) + \big(\overline{S}_n^k,\, \delta_k\big) - \big(a,\, \delta_k\big),$$

$$\big(\overline{S}_n - \overline{S}_n^k,\, V_n\big) = \varepsilon_n\big(\overline{S}_n - \overline{S}_n^k,\, \alpha_k\big) + \varepsilon_n\big(\overline{S}_n - \overline{S}_n^k,\, \beta_k\big) + \varepsilon_n\big(\overline{S}_n - \overline{S}_n^k,\, \delta_k\big).$$

In what follows we will need the following inequalities: for any $p \geqq 2$,

(8) $$\mathbf{E}\big\|\overline{S}_n^k\big\|^p \leqq c_p, \qquad \big\|\mathbf{E}\,\overline{g}_1(X_1)\big\| = \left\| \int_{\|g_1(x)\| > n^{1/2}} g_1(x)\, \mathbf{P}(dx) \right\| \leqq n^{-1/2}$$

(see [9], [16]) and

$$\begin{aligned} &\mathbf{E}\|\alpha_k\|^p \leqq c_p \mathbf{E}\|g_2\|^p k^p, \\ &\mathbf{E}\|\beta_k\|^p \leqq c_p \mathbf{E}\|g_2\|^p k^{p/2}(n-k)^{p/2}, \\ &\mathbf{E}\|\delta_k\|^p \leqq c_p \mathbf{E}\|g_2\|^p (n-k)^p \end{aligned} \tag{9}$$

(see [5]).

The Chebychev inequality and (9) imply

$$\mathbf{P}\big\{\|V_n\|^2 \geqq n^{-1/2}\big\} \leqq n^{1/2}\mathbf{E}\,\|V_n\|^2 \leqq c_1 \mathbf{E}\,\|g_2\|^2 n^{-1/2}.$$

Furthermore,

$$\mathbf{P}\Big\{\big|(\overline{S}_n^k, \varepsilon_n \beta_k)\big| > n^{-1/2}\Big\} \leqq \Pi_1 + \Pi_2, \tag{10}$$

where

$$\Pi_1 = \mathbf{P}\Big\{2k\varepsilon_n\Big|\Big(\mathbf{E}\overline{g}_1(X_1),\ \sum_{i=1}^k h(X_i)\Big)\Big| > 1\Big\},$$

$$\Pi_2 = \mathbf{P}\Big\{2\varepsilon_n\Big|\Big(\sum_{i=1}^k \overline{g}(X_i),\ \sum_{i=1}^k h(X_i)\Big)\Big| > 1\Big\},$$

$$\overline{g}(X_i) = \overline{g}_1(X_i) - \mathbf{E}\overline{g}_1(X_1), \qquad h(X_i) = \sum_{j=k+1}^n g_2(X_i, X_j), \quad i = 1, \ldots, k.$$

For Π_1 by the Chebychev inequality and (8), (9)

$$\Pi_1 \leqq (2k\varepsilon_n)^3 \big\|\mathbf{E}\, g_1(X_1)\big\|^3 \mathbf{E}\Big\|\sum_{i=1}^k h(X_i)\Big\|^3 \leqq c_1 \mathbf{E}\|g_2\|^3 n^{-1/2}.$$

To estimate Π_2 we first observe that

$$\Big(\sum_{i=1}^k \overline{g}(X_i),\ \sum_{i=1}^k h(X_i)\Big) = \sum_{i=1}^4 A_i,$$

where

$$A_1 = k\,\mathbf{E}_1\big(\overline{g}(X_1),\, h(X_1)\big),$$

$$A_2 = \sum_{i=1}^k \Big(\big(\overline{g}(X_i),\, h(X_i)\big) - \mathbf{E}_1\big(\overline{g}(X_1),\, h(X_1)\big)\Big),$$

$$A_3 = \sum_{i_1=2}^k \sum_{i_2=1}^{i_1-1} \big(\overline{g}(X_{i_1}),\, h(X_{i_2})\big), \qquad A_4 = \sum_{i_1=1}^{k-1} \sum_{i_2=i_1+1}^{k} \big(\overline{g}(X_{i_1}),\, h(X_{i_2})\big)$$

and, obviously,

$$\mathbf{E}_k|A_1|^3 \leqq k^3\Big|\mathbf{E}_1\big(\overline{g}(X_1),\, h(X_1)\big)\Big|^3$$

by [8]

$$\mathbf{E}_k|A_2|^3 \leqq c_1 k^{3/2}\mathbf{E}_1\Big|\big(\overline{g}(X_1),\, h(X_1)\big)\Big|^3$$

and by [9]

$$\mathbf{E}_k|A_3|^3 \leqq c_1 k^3 \mathbf{E}_2\Big|\big(\overline{g}(X_1),\, h(X_2)\big)\Big|^3, \qquad \mathbf{E}_k|A_4|^3 \leqq c_1 k^3 \mathbf{E}_2\Big|\big(\overline{g}(X_1),\, h(X_2)\big)\Big|^3,$$

$\mathbf{E}_k(\cdot)$ denotes the expectation with respect to $X_1, \ldots, X_k$ with fixed $X_{k+1}, \ldots, X_n$ $k = 1, \ldots, n$.

Moreover, since $\mathbf{E}\|g_1(X_1)\|^2 = 1$, we have

$$\Big|\mathbf{E}\big(\overline{g}(X_1),\, h(X_1)\big)\Big|^3 \leqq \mathbf{E}_1\big\|h(X_1)\big\|^3.$$

And, obviously,

$$\mathbf{E}_1\left|\left(\overline{g}(X_1),\, h(X_1)\right)\right|^3 \leqq 8n^{3/2}\mathbf{E}_1\left\|h(X_1)\right\|^3,$$
$$\mathbf{E}_2\left|\left(\overline{g}(X_1),\, h(X_2)\right)\right|^3 \leqq c_1\mathbf{E}_1\left\|\overline{g}(X_1)\right\|^3\mathbf{E}_1\left\|h(X_1)\right\|^3$$

and by [8]

$$\overline{\mathbf{E}}_k\left\|h(X_1)\right\|^3 \leqq c_1(n-k)^{3/2}\overline{\mathbf{E}}_k\left\|g_2(X_1, X_{k+1})\right\|^3,$$

$\overline{\mathbf{E}}_k(\cdot)$ denotes the expectation with respect to $X_{k+1},\dots,X_n$ with fixed $X_1,\dots,X_k$, $k=1,\dots,n$.

The above implies

$$\Pi_2 \leqq c_1\mathbf{E}\|g_1\|^3\mathbf{E}\|g_2\|^3 n^{-1/2}$$

and, hence we obtain an estimate for the probability in the left side of (10).

Similarly, using (8), (9) we get

$$\mathbf{P}\left\{\left|c\,(\overline{S}_n^k,\, \varepsilon_n\delta_k)\right| > n^{-1/2}\right\} \leqq \mathbf{P}\left\{c\|(\overline{S}_n^k\| > n^{1/6}\right\} + \mathbf{P}\left\{n^{1/6}\varepsilon_n\|\delta_k\| > n^{-1/2}\right\}$$
$$\leqq c^3\mathbf{E}\|\overline{S}_n^k\|^3 n^{-1/2} + n^{1/2}n^{3/2}\varepsilon_n^3\mathbf{E}\|\delta_k\|^3 \leqq c_1 n^{-1/2} + c_2\mathbf{E}\|g_2\|^3 n^{-1/2},$$
$$\mathbf{P}\left\{\left|c\,(a,\, \varepsilon_n\beta_k)\right| > n^{-1/2}\right\} \leqq c^3\|a\|^3\varepsilon_n^3 n^{3/2}\mathbf{E}\|\beta_k\|^3 \leqq c_1\mathbf{E}\|a\|^3\mathbf{E}\|g_2\|^3 n^{-1/2},$$
$$\mathbf{P}\left\{\left|c\,(a,\, \varepsilon_n\delta_k)\right| > n^{-1/2}\right\} \leqq c^3\|a\|^3\varepsilon_n^3 n^{3/2}\mathbf{E}\|\delta_k\|^3 \leqq c_1\mathbf{E}\|a\|^3\mathbf{E}\|g_2\|^3 n^{-1/2},$$
$$\mathbf{P}\left\{\left|c\,(\overline{S}_n - \overline{S}_n^k,\, \varepsilon_n\alpha_k)\right| > n^{-1/2}\right\} \leqq c^3 n^{3/2}\mathbf{E}\|\overline{S}_n - \overline{S}_n^k\|^3\varepsilon_n^3\mathbf{E}\|\alpha_k\|^3$$
$$\leqq c_1\mathbf{E}\|g_1\|^3\mathbf{E}\|g_2\|^3 n^{-1/2},$$
$$\mathbf{P}\left\{c\left|(\overline{S}_n - \overline{S}_n^k,\, \varepsilon_n\beta_k)\right| > n^{-1/2}\right\} \leqq \mathbf{P}\left\{c\|\overline{S}_n - \overline{S}_n^k\| > 1\right\} + \mathbf{P}\left\{\varepsilon_n\|\beta_k\| > n^{-1/2}\right\}$$
$$\leqq c^3\mathbf{E}\left\|\overline{S}_n - \overline{S}_n^k\right\|^3 + n^{3/2}\varepsilon_n^3\mathbf{E}\|\beta_k\|^3 \leqq c_1\mathbf{E}\|g_1\|^3 n^{-1/2} + c_2\mathbf{E}\|g_2\|^3 n^{-1/2},$$
$$\mathbf{P}\left\{\left|c\,(\overline{S}_n - \overline{S}_n^k,\, \varepsilon_n\delta_k)\right| > n^{-1/2}\right\} \leqq \mathbf{P}\left\{c\|\overline{S}_n - \overline{S}_n^k\| > 1\right\} + \mathbf{P}\left\{\varepsilon_n\|\delta_k\| > n^{-1/2}\right\}$$
$$\leqq c^3\mathbf{E}\left\|\overline{S}_n - \overline{S}_n^k\right\|^3 + n^{3/2}\varepsilon_n^3\mathbf{E}\|\delta_k\|^3 \leqq c_1\mathbf{E}\|g_1\|^3 n^{-1/2} + c_2\mathbf{E}\|g_2\|^3 n^{-1/2}.$$

Denote now

$$Y_n = 2\varepsilon_n\left(\overline{S}_n^k - a,\, \beta_k + \delta_k\right) + 2\left(\overline{S}_n - \overline{S}_n^k,\, V_n\right) + \|V_n\|^2.$$

Then, according to (7),

$$\left\|\overline{S}_n - a + V_n\right\|^2 = \left\|\overline{S}_n - a\right\|^2 + Z_n + Y_n$$

and the estimates above imply

$$\mathbf{P}\left\{|Y_n| > n^{-1/2}\right\} \leqq c_1\left(1 + \|a\|^3\right)\mathbf{E}\|g_1\|^3\mathbf{E}\|g_2\|^3 n^{-1/2}. \tag{11}$$

Since, for all $r \geqq 0$,

$$\mathbf{P}\left\{\|\overline{S}_n - a\|^2 + Z_n < r - n^{-1/2}\right\} - \mathbf{P}\left\{|Y_n| > n^{-1/2}\right\}$$
$$\leqq \mathbf{P}\left\{\|\overline{S}_n - a\|^2 + Z_n + Y_n < r\right\}$$
$$\leqq \mathbf{P}\left\{\|\overline{S}_n - a\|^2 + Z_n < r + n^{-1/2}\right\} + \mathbf{P}\left\{|Y_n| > n^{-1/2}\right\},$$

we have

(12) $\Delta_{n1}(a) \leqq \Delta_{n2}(a)+\mathbf{P}\{|Y_n| > n^{-1/2}\}+\sup_r \mathbf{P}\{r-n^{-1/2} \leqq \|Y-a\|^2 \leqq r+n^{-1/2}\}.$

In [6] it is proved that

$$\sup_r \mathbf{P}\{r - n^{-1/2} \leqq \|Y - a\|^2 \leqq r + n^{-1/2}\} \leqq c\sigma_1^{-1}\sigma_2^{-1}n^{-1/2}. \tag{13}$$

Estimates (11)–(13) imply (6).

Denote now

$$\begin{aligned} \varphi_n(t) &= \mathbf{E}\exp\left(it\,\|\overline{S}_n - a\|^2 + it\,Z_n\right),\\ \varphi(t) &= \mathbf{E}\exp\left(it\,\|Y - a\|^2\right),\\ \psi_n(t) &= \mathbf{E}\exp\left(it\,\|\overline{S}_n - a\|^2\right). \end{aligned}$$

The density of the random variable $\|Y - a\|^2$ is not greater than $c(\sigma_1\sigma_2)^{-1}$ (see [6]). Thus for $\Delta_{n2}(a)$ in (6), applying Esséen's inequality (see, e.g. [7]) for any $T > 0$ we have

$$\Delta_{n2}(a) \leqq c_1 \int_{-T}^{T} |\varphi_n(t) - \varphi(t)|\, |t|^{-1} dt + c_2(\sigma_1\sigma_2)^{-1}T^{-1}. \tag{14}$$

We choose in (14)

$$T = \frac{cn^{1/2}}{L}, \quad L = 3\sigma_9^{-2}\mathbf{E}\,\|g_1\|^3,\ \sigma_9 > 0, \tag{15}$$

where c is a small enough absolute constant and then apply Lemma 12 of [3]. Note that with L as in (15) the following conditions of Lemma 12 in [3]

$$\frac{2}{3} \leqq L^2 \leqq \frac{n}{2e}, \qquad \mathbf{P}(B_L) \geqq \frac{1}{2} \tag{16}$$

are fulfilled (for a proof see [3], [9], [16]). When (16) is satisfied the distribution of $\overline{g}_1 = g_1(X_1)\, I(\|g_1(X_1)\| \leqq n^{1/2})$ may be represented as $(Q + Q_1)/2$, where

$$Q(A) = \frac{\mathbf{P}(A \cap B_L)}{\mathbf{P}(B_L)}, \quad A \in H,$$

and Q_1 is a probability measure (see [15]).

Let V_1 be the covariance operator of Q and $\sigma_{11}^2 \geqq \sigma_{12}^2 \geqq \cdots$ be its eigenvalues. By our assumptions, $\operatorname{tr} V = 1$. Hence (see [3]) $\operatorname{tr} V_1 \leqq 2$ and

$$\frac{2\sigma_j^2}{3} \leqq \sigma_{1j}^2 \leqq 2\sigma_j^2, \quad j = 1, \ldots, 9. \tag{17}$$

To estimate the integral in (14) we will need the following generalization of Lemma 11 in [3] (see also [9, Lemma 4]).

LEMMA 3. *Denote*

$$Z = n^{-1/2}\sum_{j=1}^{l+m} \overline{g}_1(X_j), \qquad Z_1(f) = \sum_{j=1}^{l+m} f(X_j),$$

where $\overline{g}_1 = g_1(X_1)\, I(\|g_1(X_1)\| \leqq n^{1/2})$ *and* $f\colon\ X \to \mathbf{R}$ *is a measurable function,* l, m *are positive integers,* $l \leqq m$, $l + m \leqq n$. *Then, for any* $A > 0$, *even* $k \geqq 0$, *integer* $r \geqq 0$, $k_j \geqq 0$, $j = 1, \ldots, r$, *and any* $x, x_j \in H$, $j = 1, \ldots, r$, *for* $|t| \leqq c(A)\, nL^{-1}(l \log(l/L^2))^{-1/2}$ *if* $l > L^2$, *and for any* t *if* $l \leqq L^2$

$$\left|\mathbf{E}\exp\left(it\,\|Z + x\|^2 + it\, Z_1(F)\right) \|Z\|^k \prod_{j=1}^{r} (x_j, Z)^{k_j}\right|$$

$$\leqq c_1 \prod_{j=1}^{r} \|x_j\|^{k_j} \left(\exp(-c_2 l) + c_3 \left(\frac{L^2}{l}\right)^A + h^{1/2}\left(\frac{c_4 T_1^2 lm}{n^2},\, V_1\right)\right), \tag{18}$$

where c_1, c_2 *are functions of* $\overline{k} = k + k_1 + \cdots + k_r$; c_3, c_4 *are functions of* A *and* $\overline{k}$

$$T_1 = \min\left\{|t|,\ nL^{-1}\left(m \log\left(\frac{m}{L^2}\right)\right)^{-1/2}\right\}, \qquad h(s, V_1) = \prod_{j=1}^{\infty} \left(1 + 2s\sigma_{1j}^4\right)^{-1/2}.$$

Proof. This lemma reduces to Lemma 11 in [3] if $Z_1(f) = 0$. The proof of the present lemma is essentially the same as the proof of Lemma 11 in [3].

Let us estimate the integral in (14). Obviously,

$$\int_{-T}^{T} |\varphi_n(t) - \varphi(t)|\, |t|^{-1} dt \leqq I_1 + I_2, \tag{19}$$

where

$$I_1 = \int_{-T}^{T} |\psi_n(t) - \varphi(t)|\, |t|^{-1} dt, \qquad I_2 = \int_{-T}^{T} |\varphi_n(t) - \psi_n(t)|\, |t|^{-1} dt.$$

By Lemma 12 in [3]

$$I_1 \leqq c_1 U, \tag{20}$$

where

$$U = c_7(V)\left(\mathbf{E}\left(\|g_1\|^2 + (a,\, g_1)^2\right) I\left(\|g_1\| > n^{1/2}\right)\right)$$
$$+ \left(\mathbf{E}\left(\|g_1\|^3 + |(a,\, g_1)|^3\right) I\left(\|g_1\| \leqq n^{1/2}\right) + (Va,\, a)^{3/2}\right) n^{-1/2}$$
$$+ c_3(V)\, \sigma_9^{-2} \mathbf{E}\|g_1\|^3 n^{-1/2}, \quad c_s(V) = \prod_{j=1}^{s} \sigma_j^{-1},$$

and for U we have

$$U \leqq \left(1 + \|a\|^3\right) c_7(V)\, \mathbf{E}\, \|g_1\|^3 n^{-1/2} + c_3(V)\, \sigma_9^{-2} \mathbf{E}\, \|g_1\|^3 n^{-1/2}$$
$$\leqq 2\left(1 + \|a\|^3\right) \beta\, c_9(V)\, \sigma_9^{-1} n^{-1/2}.$$

Consider now I_2 in (19). Obviously,

$$\varphi_n(t) - \psi_n(t) = \mathbf{E}\exp\left\{it\, \|\overline{S}_n - a\|^2\right\} \left(e^{it\, Z_n} - 1\right).$$

Next we will use the representation

$$e^z - 1 = \sum_{s=1}^{r-1} \frac{z^s}{s!} + z^r \int_0^1 \frac{(1-u)^{r-1} e^{uz}}{(r-1)!}\, du, \tag{21}$$

valid for all complex z and positive integers r, with $r = 3$, $z = it\,Z_n$. Thus we have

$$\varphi_n(t) - \psi_n(t) = it\,J_1(t) - \frac{t^2}{2}\,J_2(t) + \frac{(it)^3}{2}\,J_3(t), \tag{22}$$

where

$$J_1(t) = \mathbf{E}\Big(Z_n \exp\big\{it\,\|\overline{S}_n - a\|^2\big\}\Big), \qquad J_2(t) = \mathbf{E}\Big(Z_n^2 \exp\big\{it\,\|\overline{S}_n - a\|^2\big\}\Big),$$
$$J_3(t) = \int_0^1 du\,(1-u)^2 \mathbf{E}\Big(Z_n^3 \exp\big\{it\,\|\overline{S}_n - a\|^2 + itu\,Z_n\big\}\Big).$$

LEMMA 4. *If* $|t| \leqq T = cn^{1/2}L^{-1}$, *then*

$$\big|J_3(t)\big| \leqq c_1\left(L^{3/2}n^{-1/2} + \prod_{j=1}^{7}\left(1 + \frac{c_2\sigma_j^4 t^2}{n^{2/3}}\right)^{-1/4}\right) \times \big(\mathbf{E}\,\|g_1\|^3 + \|a\|^3\big)\,\mathbf{E}\,\|g_2\|^3 n^{-3/2}. \tag{23}$$

Proof. The random variable Z_n does not depend on $X_{k+1}, \dots, X_n$. Hence,

$$\Big|\mathbf{E}\Big(Z_n^3 \exp\big\{it\,\|\overline{S}_n - a\|^2 + it\,uZ_n\big\}\Big)\Big| \leqq \mathbf{E}_k|Z_n|^3\Big|\overline{\mathbf{E}}_k \exp\Big\{it\,\|\overline{S}_n - a\|^2\Big\}\Big|,$$

where as above $\overline{\mathbf{E}}_k(\cdot)$ denotes expectation with respect to $X_{k+1}, \dots, X_n$ with fixed $X_1, \dots, X_k$. Representing $\overline{S}_n = (\overline{S}_n - \overline{S}_n^k) + \overline{S}_n^k$ we apply Lemma 3 with $Z = \overline{S}_n - \overline{S}_n^k$, $1 \sim n^{2/3}/4$, $m \sim 3n^{2/3}/4$, $A = \frac{3}{4}$. Noting that here $T_1 = |t|$, we obtain

$$\Big|\overline{\mathbf{E}}_k \exp\big\{it\,\|\overline{S}_n - a\|^2\big\}\Big| \leqq c_1\left(\exp(-c_2 n^{2/3}) + c_3 L^{3/2} n^{-1/2} + \prod_{j=1}^{7}\left(1 + \frac{c_4\sigma_j^4 t^2}{n^{2/3}}\right)^{-1/4}\right). \tag{24}$$

To estimate $\mathbf{E}|Z_n|^3$ we first write

$$\mathbf{E}\,|Z_n|^3 \leqq 32\,\varepsilon_n^3\Big(\mathbf{E}\big|(\overline{S}_n^k,\, \alpha_k)\big|^3 + \|a\|^3\mathbf{E}\,\|\alpha_k\|^3\Big). \tag{25}$$

Then we represent the scalar product $(\overline{S}_n^k, \alpha_k)$ in the form

$$(\overline{S}_n^k,\, \alpha_k) = n^{-1/2}\left(\sum_{j=1}^{k} \overline{g}(X_j),\ \sum_{1 \leqq i_1 < i_2 \leqq k} g_2(X_{i_1}, X_{i_2})\right) + n^{-1/2}k\left(\mathbf{E}\overline{g}_1(X_1),\ \sum_{1 \leqq i_1 < i_2 \leqq k} g_2(X_{i_1}, X_{i_2})\right). \tag{26}$$

Using (8)–(9) we have

$$\mathbf{E}\left|\left(\mathbf{E}\overline{g}_1(X_1),\ \sum_{1 \leqq i_1 < i_2 \leqq k} g_2(X_{i_1}, X_{i_2})\right)\right|^3 \leqq \big\|\mathbf{E}\overline{g}_1(X_1)\big\|^3\mathbf{E}\left\|\sum_{1 \leqq i_1 < i_2 \leqq k} g_2(X_{i_1}, X_{i_2})\right\|^3 \leqq c_1 n^{-3/2}k^3\mathbf{E}\|g_2\|^3.$$

To estimate the first term in the right side of (26) we will use the representation

$$\left(\sum_{j=1}^{k}\overline{g}_1(X_j),\ \sum_{1\leqq i_1<i_2\leqq k} g_2(X_{i_1},X_{i_2})\right)$$
$$=\sum_{j=1}^{k}\sum_{1\leqq i_1<i_2\leqq k}\left(\overline{g}_1(X_j),\ g_2(X_{i_1},X_{i_2})\right)=\sum_{i=1}^{5}G_i,$$

where

$$G_1=\sum_{j=3}^{k}\sum_{i_2=2}^{j-1}\sum_{i_1=1}^{i_2-1}\left(\overline{g}(X_j),\ g_2(X_{i_1},X_{i_2})\right),$$
$$G_2=\sum_{j=2}^{k}\sum_{i_1=1}^{j-1}\left(\overline{g}(X_j),\ g_2(X_{i_1},X_j)\right),$$
$$G_3=\sum_{j=2}^{k-1}\sum_{i_1=1}^{j-1}\sum_{i_2=j+1}^{k}\left(\overline{g}(X_j),\ g_2(X_{i_1},X_{i_2})\right),$$
$$G_4=\sum_{j=1}^{k}\sum_{i_2=j+1}^{k}\left(\overline{g}(X_j),\ g_2(X_j,X_{i_2})\right),$$
$$G_5=\sum_{j=1}^{k}\sum_{i_2=j+2}^{k}\sum_{i_1=j+1}^{i_2-1}\left(\overline{g}(X_j),\ g_2(X_{i_1},X_{i_2})\right).$$

By (9) we have

(27)
$$\begin{aligned}&\mathbf{E}|G_i|^3\leqq c_1\mathbf{E}\|g_1\|^3\mathbf{E}\|g_2\|^3n^{9/2}, && i=1,\,3,\,5,\\ &\mathbf{E}|G_i|^3\leqq c_2\mathbf{E}\|g_2\|^3n^{9/2}, && i=2,\,4.\end{aligned}$$

Combining (26) and (27) we obtain

$$\mathbf{E}|Z_n|^3\leqq c_1\left(\mathbf{E}\|g_1\|^3+\|a\|^3\right)\mathbf{E}\|g_2\|^3n^{-3/2}.$$

Together with (24) this implies (23).

For $m\sim n$ partition the interval $|t|\leqq T=cn^{1/2}L^{-1}$ into the two subsets

(28)
$$|t|\leqq T(m)=nL^{-1}\left(m\log\frac{m}{L^2}\right)^{-1/2}$$

and

(29)
$$T(m)\leqq|t|\leqq T.$$

LEMMA 5. *For t satisfying* (28)

$$\left|J_1(t)\right|\leqq c_1\left(1+\|a\|^3\right)\left(\mathbf{E}\|g_1\|^3\right)\left(\mathbf{E}\|g_2\|^3\right)^{1/3}$$
$$\times\left(L^3n^{-1/2}+\left(|t|+t^2\right)\prod_{j=1}^{7}\left(1+c_2\sigma_j^4t^2\right)^{-1/4}\right.$$

(30)
$$\left.+|t|^3\prod_{j=1}^{9}\left(1+c_2\sigma_j^4t^2\right)^{-1/4}\right)n^{-1/2},$$

and for t satisfying (29)

$$(31)\qquad \left|J_1(t)\right| \leqq c_1\left(1+\|a\|^3\right)\left(\mathbf{E}\|g_1\|^3\right)\left(\mathbf{E}\|g_2\|^3\right)^{1/3} c_9(V)\, L^2 n^{-1}.$$

Proof. Let us show first that

$$(32)\qquad J_1(t) = 4n^{-1/2}\varepsilon_n \binom{k}{2} h_1(t) + 2(k-2)\, n^{-1/2}\varepsilon_n \binom{k}{2} h_2(t) - 2\varepsilon_n \binom{k}{2} h_3(t),$$

where

$$h_1(t) = \mathbf{E}\Big(\big(\overline{g}_1(X_1),\, g_2(X_1,X_2)\big) \exp\Big\{it\,\big\|\overline{g}_{1n}(X_1)+\overline{g}_{1n}(X_2)+\overline{S}_{n2}\big\|^2\Big\}\Big),$$

$$h_2(t) = \mathbf{E}\Big(\big(\overline{g}_1(X_1),\, g_2(X_2,X_3)\big) \times \exp\Big\{it\,\big\|\overline{g}_{1n}(X_1)+\overline{g}_{1n}(X_2)+\overline{g}_{1n}(X_3)+\overline{S}_{n3}\big\|^2\Big\}\Big),$$

$$h_3(t) = \mathbf{E}\Big(\big(a,\, g_2(X_1,X_2)\big) \exp\Big\{it\,\big\|\overline{g}_{1n}(X_1)+\overline{g}_{1n}(X_2)+\overline{S}_{n2}\big\|^2\Big\}\Big),$$

$$\overline{g}_{1n}(X_j) = \overline{g}_1(X_j)\, n^{-1/2}, \qquad \overline{S}_{nr} = n^{-1/2}\sum_{j=r+1}^{n} \overline{g}_1(X_j) - a.$$

Indeed, according to the definition,

$$J_1(t) = 2n^{-1/2}\varepsilon_n \sum_{i=1}^{5} H_i(t) - 2\varepsilon_n H_6,$$

$$H_1(t) = \sum_{j=3}^{k}\sum_{i_2=2}^{j-1}\sum_{i_1=1}^{i_2-1} \mathbf{E}\Big(\big(\overline{g}_1(X_j),\, g_2(X_{i_1},X_{i_2})\big) \exp\Big\{it\,\big\|\overline{S}_n - a\big\|^2\Big\}\Big) = \binom{k}{3} h_2(t),$$

$$H_2(t) = \sum_{j=2}^{k}\sum_{i_2=1}^{j-1} \mathbf{E}\Big(\big(\overline{g}_1(X_j),\, g_2(X_{i_2},X_j)\big) \exp\Big\{it\,\big\|\overline{S}_n - a\big\|^2\Big\}\Big) = \binom{k}{2} h_1(t),$$

$$H_3(t) = \sum_{j=2}^{k-1}\sum_{i_1=1}^{j-1}\sum_{i_2=j+1}^{k} \mathbf{E}\Big(\big(\overline{g}_1(X_j),\, g_2(X_{i_1},X_{i_2})\big) \exp\Big\{it\,\big\|\overline{S}_n - a\big\|^2\Big\}\Big) = \binom{k}{3} h_2(t),$$

$$H_4(t) = \sum_{j=1}^{k-1}\sum_{i_1=j+1}^{k} \mathbf{E}\Big(\big(\overline{g}_1(X_j),\, g_2(X_{i_1},X_j)\big) \exp\Big\{it\,\big\|\overline{S}_n - a\big\|^2\Big\}\Big) = \binom{k}{2} h_1(t),$$

$$H_5(t) = \sum_{j=1}^{k-2}\sum_{i_2=j+2}^{k}\sum_{i_1=j+1}^{i_2-1} \mathbf{E}\Big(\big(\overline{g}_1(X_j),\, g_2(X_{i_1},X_{i_2})\big) \exp\Big\{it\,\big\|\overline{S}_n - a\big\|^2\Big\}\Big) = \binom{k}{3} h_2(t),$$

$$H_6(t) = \sum_{i_2=2}^{k}\sum_{i_1=1}^{i_2-1} \mathbf{E}\Big(\big(a,\, g_2(X_{i_1},X_{i_2})\big) \exp\Big\{it\,\big\|\overline{S}_n - a\big\|^2\Big\}\Big) = \binom{k}{2} h_3(t).$$

These representations imply (32). Next we estimate functions $h_i(t)$ in (32). Obviously

$$\big\|\overline{g}_{1n}(X_1)+\overline{g}_{1n}(X_2)+\overline{S}_{n2}\big\|^2 = \|\overline{S}_{n2}\|^2 + W_1 + W_2,$$

where

$$W_1 = \left\|\overline{g}_{1n}(X_1) + \overline{g}_{1n}(X_2)\right\|^2, \qquad W_2 = 2\left(\overline{g}_{1n}(X_1) + \overline{g}_{1n}(X_2),\, \overline{S}_{n2}\right).$$

By (21) with $r = 1$ and $z = \mathrm{it}(W_1 + W_2)$, taking into consideration (2) we have

$$\begin{aligned} h_1(t) = it \int_0^1 du\, \mathbf{E}\Big((\overline{g}_1(X_1),\, g_2(X_1, X_2))\, W_1 \exp\{it\, uW_1\} \\ \times \overline{\mathbf{E}}_2 \exp\left\{it \left\|\overline{S}_{n2}\right\|^2 + itu\, W_2\right\}\Big) \\ + it \int_0^1 du\, \mathbf{E}\Big((\overline{g}_1(X_1),\, g_2(X_1, X_2))\, \exp\{itu\, W_1\} \\ \times \overline{\mathbf{E}}_2\Big(\exp\left\{it \left\|\overline{S}_{n2}\right\|^2 + itu\, W_2\right\} W_2\Big)\Big). \end{aligned}$$

The expectation $\overline{\mathbf{E}}_2$ with respect to $X_3, \ldots, X_n$ we will estimate using Lemma 3. To estimate $h_1(t)$ in the interval (28) we take in Lemma 3 $l \sim n/2$, $m = n - l$, $A = 1$. Then $T_1 = |t|$ and

$$\begin{aligned} &\left|\overline{\mathbf{E}}_2 \exp\left\{it \left\|\overline{S}_{n2}\right\|^2 + itu\, W_2\right\}\right| \\ &\leqq c_1\left(\exp(-c_2 n) + c_3 L^2 n^{-1} + \prod_{j=1}^{5} \left(1 + c_2\sigma_j^4 t^2\right)^{-1/4}\right), \end{aligned} \tag{33}$$

$$\begin{aligned} &\left|\overline{\mathbf{E}}_2\Big(\exp\left\{it \left\|\overline{S}_{n2}\right\|^2 + itu\, W_2\right\} W_2\Big)\right| \leqq c_1\left(1 + \|a\|\right) \left\|\overline{g}_{1n}(X_1) + g_{1n}(X_2)\right\| \\ &\times \left(\exp(-c_2 n) + c_3 L^2 n^{-1} + \prod_{j=1}^{5} \left(1 + c_2\sigma_j^4 t^2\right)^{-1/4}\right). \end{aligned} \tag{34}$$

Moreover,

$$\begin{aligned} &\mathbf{E}\Big(\left|(\overline{g}_1(X_1),\, g_2(X_1, X_2))\right| \left\|\overline{g}_{1n}(X_1) + \overline{g}_{1n}(X_2)\right\|\Big) \\ &\leqq n^{-1/2}\Big(\mathbf{E}\Big(\left\|g_1(X_1)\right\|^2 \left\|g_2(X_1, X_2)\right\|\Big) + \mathbf{E}\Big(\left\|g_1(X_1)\right\| \left\|g_1(X_2)\right\| \left\|g_2(X_1, X_2)\right\|\Big)\Big) \\ &\leqq 2n^{-1/2}\left(\mathbf{E}\|g_1\|^3\right)^{2/3}\left(\mathbf{E}\|g_2\|^3\right)^{1/3}, \\ &\mathbf{E}\Big(\left|(\overline{g}_1(X_1),\, g_2(X_1, X_2))\right| \left\|\overline{g}_{1n}(X_1) + \overline{g}_{1n}(X_2)\right\|^2\Big) \\ &\leqq 2n^{-1/2}\left(\mathbf{E}\|g_1\|^3\right)^{2/3}\left(\mathbf{E}\|g_2\|^3\right)^{1/3}. \end{aligned}$$

Thus, by (33) and (34), in the interval (28)

$$\begin{aligned} \left|h_1(t)\right| \leqq c_1\left(1 + \|a\|\right) \left(\mathbf{E}\|g_1\|^3\right)^{2/3}\left(\mathbf{E}\|g_2\|^3\right)^{1/3} \\ \times \left(L n^{-1/2} + |t| \prod_{j=1}^{7} \left(1 + c_2\sigma_j^4 t^2\right)^{-1/4}\right) n^{-1/2}. \end{aligned} \tag{35}$$

To estimate $h_1(t)$ for t satisfying (29) we take in Lemma 3, $l \sim n/\log n$, $m = n-l$, $A = \frac{5}{4}$. Then $T_1 = T(m)$ and similar reasoning gives for t satisfying (29)

$$\begin{aligned} |h_1(t)| &\leqq c_1|t|\,(1+\|a\|)\,(\mathbf{E}\|g_1\|^3)^{2/3}(\mathbf{E}\|g_2\|^3)^{1/3} \\ &\quad\times\Bigg(\exp\Big(-\frac{c_2 n}{\log n}\Big) + c_3 L^{5/2} n^{-5/4}(\log n)^{5/4} \\ &\quad + \prod_{j=1}^{5}\Big(1+\frac{c_4\sigma_j^4 n}{L^2(\log n)\log(m/L^2)}\Big)^{-1/4}\Bigg) n^{-1/2} \\ &\leqq c_5(1+\|a\|)\,(\mathbf{E}\|g_1\|^3)^{2/3}(\mathbf{E}\|g_2\|^3)^{1/3} c_5(V)\, L^{3/2} n^{-1}. \end{aligned} \tag{36}$$

Consider now $h_2(t)$ in (32). Denote

$$\xi = \sum_{j=1}^{3}\xi_j, \qquad \bar{\xi} = \sum_{j=1}^{3}\mathbf{E}\,\overline{g}_{1n}(X_j), \quad \xi_j = \overline{g}_{1n}(X_j) - \mathbf{E}\,\overline{g}_{1n}(X_j),\ j = 1,\,2,\,3.$$

Obviously,

$$\begin{aligned} \|\overline{g}_{1n}(X_1) + \overline{g}_{1n}(X_2) + \overline{g}_{1n}(X_3) + \overline{S}_{n3}\| &= \|\xi + \overline{S}_{n3} + \bar{\xi}\|^2 \\ = \|\overline{S}_{n3} + \bar{\xi}\|^2 + W_3 + W_4 + W_5, \end{aligned}$$

where

$$W_3 = \sum_{j=1}^{3}\|\xi_j\|^2, \qquad W_4 = 2\sum_{j=1}^{3}(\xi_i,\, \overline{S}_{n3} + \bar{\xi}), \qquad W_5 = 2\sum_{1\leqq i<j\leqq 3}(\xi_i, \xi_j).$$

Applying (21) with $r = 2$ and $z = it\,W_5$ we can write

$$h_2(t) = h_{21}(t) + h_{22}(t) + h_{23}(t), \tag{37}$$

where

$$\begin{aligned} h_{21}(t) &= \mathbf{E}\Big((\overline{g}_1(X_1),\, g_2(X_2, X_3))\ \exp\Big\{it\,\|\overline{S}_{n3} + \bar{\xi}\|^2 + it\,(W_3 + W_4))\Big\}\Big), \\ h_{22}(t) &= it\,\mathbf{E}\Big(\,(\overline{g}_1(X_1),\, g_2(X_2, X_3))\, W_5\ \exp\Big\{it\,\|\overline{S}_{n3} + \bar{\xi}\|^2 + it\,(W_3 + W_4)\Big\}\Big), \\ h_{23}(t) &= -t^2\int_0^1 du\,(1-u)\,\mathbf{E}\Big(\,(\overline{g}_1(X_1),\, g_2(X_2, X_3))\, W_5^2 \\ &\quad\times\exp\Big\{it\,\|\overline{S}_{n3} + \bar{\xi}\|^2 + it\,(W_3 + W_4 + uW_5)\Big\}\Big). \end{aligned}$$

Furthermore, observe that

$$|h_{23}(t)| \leqq t^2\mathbf{E}\Big(\,|(\overline{g}_1(X_1),\, g_2(X_2, X_3))|\, W_5^2\Big|\overline{\mathbf{E}}_3\exp\Big\{it\,\|\overline{S}_{n3} + \bar{\xi}\|^2 + it\,W_4\Big\}\Big|\Big).$$

In the interval (28) by Lemma 3 with $l \sim n/2$, $m = n - l$, $A = \frac{3}{2}$

$$\begin{aligned} &\Big|\overline{\mathbf{E}}_3\exp\big\{it\,\|\overline{S}_{n3} + \bar{\xi}\|^2 + it\,W_4\big\}\Big| \\ &\leqq c_1\Big(\exp(-c_2 n) + c_3 L^3 n^{-3/2} + \prod_{j=1}^{7}(1 + c_4\sigma_j^4 t^2)^{-1/4}\Big), \end{aligned} \tag{38}$$

and for t satisfying (29) Lemma 3 with $l \sim n/\log n$, $m = n - l$, $A = \frac{7}{4}$ gives

$$\begin{aligned} &\left|\overline{\mathbf{E}}_3 \exp\left\{it\,\|\overline{S}_{n3} + \bar{\xi}\|^2 + it\,W_4\right\}\right| \\ &\qquad \leqq c_1\left(\exp\left(-\frac{c_2 n}{\log n}\right) + c_3 L^{7/2} n^{-7/4} (\log n)^{7/4}\right. \\ &\qquad\qquad \left. + \prod_{j=1}^{7}\left(1 + \frac{c_4 \sigma_j^4 n}{L^2 (\log n) \log(m/L^2)}\right)^{-1/4}\right). \end{aligned} \tag{39}$$

Moreover,

$$\mathbf{E}\Big(\big|(\bar{g}_1(X_1),\, g_2(X_2, X_3))\big|\, W_5^2\Big) \leqq c\left(\mathbf{E}\|g_1(X_1)\|^3\right)^{2/3} \left(\mathbf{E}\|g_2\|^3\right)^{1/3} n^{-3/2}. \tag{40}$$

Combining (38)–(40) we obtain an estimate for $h_{23}(t)$.

By (21) with $r = 1$, $z = it\,(W_3 + W_4)$ we have

$$h_{22}(t) = F_1(t) + F_2(t), \tag{41}$$

where

$$\begin{aligned} F_1(t) &= it\,\mathbf{E}\Big((\mathbf{E}\bar{g}_1(X_1),\, g_2(X_2, X_3))\, W_5 \exp\left\{it\,\|\overline{S}_{n3} + \bar{\xi}\|^2 + it\,(W_3 + W_4)\right\}\Big), \\ F_2(t) &= -t^2 \int_0^1 du\,\mathbf{E}\Big((\bar{g}_1(X_1) - \mathbf{E}\bar{g}_1(X_1),\, g_2(X_2, X_3))\, W_5 (W_3 + W_4) \\ &\qquad\qquad \times \exp\left\{it\,\|\overline{S}_{n3} + \bar{\xi}\|^2 + itu\,(W_3 + W_4)\right\}\Big). \end{aligned}$$

Here

$$|F_1(t)| \leqq |t|\,\mathbf{E}\Big(\big|(\mathbf{E}\bar{g}_1(X_1),\, g_2(X_2, X_3))\, W_5\big|\,\Big|\overline{\mathbf{E}}_3 \exp\left\{it\,\|\overline{S}_{n3} + \bar{\xi}\|^2 + it\,W_4\right\}\Big|\Big),$$

and

$$\mathbf{E}\Big(\big|(\mathbf{E}\bar{g}_1(X_1),\, g_2(X_2, X_3))\big|\, W_5\Big) \leqq 24\left(\mathbf{E}\|g_2\|^2\right)^{1/2} n^{-3/2}.$$

Thus for t satisfying (28) by Lemma 3 with $l \sim n/2$, $m = n - l$, $A = 1$

$$\begin{aligned} &|F_1(t)| \leqq c_1 |t|\,\left(\mathbf{E}\|g_2\|^2\right)^{1/2} \\ &\qquad \times c_1\left(\exp(-c_2 n) + c_3 L^2 n^{-1} + \prod_{j=1}^{5}\left(1 + c_4 \sigma_j^4 t^2\right)^{-1/4}\right) n^{-3/2}, \end{aligned} \tag{42}$$

and for t satisfying (29) by Lemma with $l \sim n/\log n$, $m = n - l$ and $A = \frac{5}{4}$,

$$\begin{aligned} &|F_1(t)| \leqq c_1 |t|\,\left(\mathbf{E}\|g_2\|^2\right)^{1/2} c_1\left(\exp\left(-\frac{c_2 n}{\log n}\right) + c_3 L^{5/2} n^{-5/4} (\log n)^{5/4}\right. \\ &\qquad\qquad \left. + \prod_{j=1}^{5}\left(1 + \frac{c_4 \sigma_j^4 n}{L^2 (\log n) \log(m/L^2)}\right)^{-1/4}\right) n^{-3/2}. \end{aligned} \tag{43}$$

To estimate $f_2(t)$ in (41) we first observe, that

$$|F_2(t)| \leqq t^2 \int_0^1 du\, \mathbf{E}\Big(\big|(\bar{g}_1(X_1) - \mathbf{E}\bar{g}_1(X_1),\, g_2(X_2, X_3))\, W_5\big| \times \big|\mathbf{E}_3(W_3 + W_4) \exp\big\{it \|\bar{S}_{n3} + \bar{\xi}\|^2 + itu\, W_4\big\}\big|\Big),$$

and

$$\mathbf{E}\Big(\big|(\bar{g}_1(X_1) - \mathbf{E}\bar{g}_1(X_1),\, g_2(X_2, X_3))\, W_5\big| \Big[W_3 + \sum_{j=1}^{3} \|\xi_j\| \,(1 + \|a\|)\Big]\Big) \leqq c\,\mathbf{E}\|g_1\|^3 (\mathbf{E}\|g_2\|^3)^{1/3} (1 + \|a\|)\, n^{-3/2}.$$

Applying now Lemma 3 for t satisfying (28) with $l \sim n \log n$, $m = n - l$ and $A = \frac{3}{2}$, we have

$$|F_2(t)| \leqq c_1 t^2 (1 + \|a\|)\, \mathbf{E}\|g_1\|^3 (\mathbf{E}\|g_2\|^3)^{1/3} \times \Big(\exp(-c_2 n)\, c_3 L^3 n^{-3/2} + \prod_{j=1}^{7} (1 + c_4 \sigma_j^4 t^2)^{-1/4}\Big)\, n^{-3/2}, \tag{44}$$

and for t satisfy (29), using Lemma 3 with $l \sim n \log n$, $m = n - l$ and $A = \frac{7}{4}$, we have

$$|F_2(t)| \leqq c_1 t^2 (1 + \|a\|)\, \mathbf{E}\|g_1\|^3 (\mathbf{E}\|g_2\|^3)^{1/3} \times \Big(\exp\Big(-\frac{c_2 n}{\log n}\Big) + c_3 L^{7/2} n^{-7/4} (\log n)^{7/4} + \prod_{j=1}^{7} \Big(1 + \frac{c_4 \sigma_j^4 n}{L^2 (\log n) \log(m/L^2)}\Big)^{-1/4}\Big)\, n^{-3/2}. \tag{45}$$

In (37) it remains only to estimate $h_{21}(t)$. We use the representation

$$h_{21}(t) = F_3(t) + F_4(t), \tag{46}$$

where

$$F_3(t) = \mathbf{E}\Big((\mathbf{E}\bar{g}_1(X_1),\, g_2(X_2, X_3)) \exp\big\{it \|\bar{S}_{n3} + \bar{\xi}\|^2 + it\,(W_3 + W_4)\big\}\Big),$$

$$F_4(t) = \mathbf{E}\Big((\bar{g}_1(X_1) - \mathbf{E}\bar{g}_1(X_1),\, g_2(X_2, X_3)) \times W_5 \exp\big\{it \|\bar{S}_{n3} + \bar{\xi}\|^2 + it\,(W_3 + W_4)\big\}\Big).$$

In view of (2) we can also write

$$F_3(t) = \mathbf{E}\Big((\mathbf{E}\bar{g}_1(X_1),\, g_2(X_2, X_3)) \exp\big\{it \|\bar{S}_{n3} + \bar{\xi}\|^2 + it\,\|\xi_1\|^2 + 2it\,(\xi_1, \bar{S}_{n3} + \bar{\xi})\big\} \times \prod_{j=2}^{3} \Big(\exp\big\{it\,\|\xi_j\|^2 + 2it\,(\xi_j, \bar{S}_{n3} + \bar{\xi})\big\} - 1\Big)\Big), \tag{47}$$

$$F_4(t) = \mathbf{E}\Bigg((\overline{g}_1(X_1) - \mathbf{E}\overline{g}_1(X_1),\, g_2(X_2, X_3)) \exp\Big\{it \,\|\overline{S}_{n3} + \bar{\xi}\|^2\Big\}$$

$$\times \prod_{j=1}^{3} \Big(\exp\Big\{ it\, \|\xi_j\|^2 + 2it\, (\xi_j, \overline{S}_{n3} + \bar{\xi})\Big\} - 1\Big)\Bigg). \tag{48}$$

Furthermore, applying (21) with $r = 1$ in (47) and (48), we have

$$F_3(t) = -t^2 \mathbf{E}\Bigg((\mathbf{E}\overline{g}_1(X_1),\, g_2(X_2, X_3))$$

$$\times \exp\Big\{it\, \|\overline{S}_{n3} + \bar{\xi}\|^2 + it\,\|\xi_1\|^2 + 2it\,(\xi_1, \overline{S}_{n3} + \bar{\xi})\Big\}$$

$$\times \prod_{j=2}^{3} \Big(\|\xi_j\|^2 + 2(\xi_j, \overline{S}_{n3} + \bar{\xi})\Big)$$

$$\times \int_0^1 du \exp\Big\{ itu\, \|\xi_j\|^2 + 2itu\, (\xi_j, \overline{S}_{n3} + \bar{\xi})\Big\}\Bigg), \tag{49}$$

$$F_4(t) = (it)^2 \mathbf{E}\Bigg((\overline{g}_1(X_1) - \mathbf{E}\overline{g}_1(X_1),\, g_2(X_2, X_3)) \exp\Big\{it\, \|\overline{S}_{n3} + \bar{\xi}\|^2\Big\}$$

$$\times \prod_{j=1}^{3} \Big(\|\xi_j\|^2 + 2(\xi_j, \overline{S}_{n3} + \bar{\xi})\Big)$$

$$\times \int_0^1 du \exp\Big\{ itu\, \|\xi_j\|^2 + 2itu\, (\xi_j, \overline{S}_{n3} + \bar{\xi})\Big\}\Bigg). \tag{50}$$

Denote

$$b_j = \|\xi_j\|^2 + 2(\xi_j, \bar{\xi}) - 2(\xi_j, a).$$

Then

$$\prod_{j=2}^{3} \Big(\|\xi_j\|^2 + 2(\xi_j, \overline{S}_{n3} + \bar{\xi})\Big) = \prod_{j=2}^{3} \Big(b_j + 2(\xi_j, S_{n3} + a)\Big)$$

$$= b_2 b_3 + 2b_2(\xi_3, \overline{S}_{n3} + a) + 2b_3(\xi_2, \overline{S}_{n3} + a) + 4(\xi_2, \overline{S}_{n3} + a)\,(\xi_3, \overline{S}_{n3} + a), \tag{51}$$

where according to the notation introduced above

$$\overline{S}_{n3} + a = n^{-1/2}(\overline{g}_1(X_4) + \cdots + \overline{g}_1(X_n)).$$

Note that

$$\|\xi_j\| \leqq 2, \qquad \|\xi_j\| \leqq n^{-1/2}\Big(1 + \big\|g_1(X_j)\big\|\Big),$$

$$|b_j| \leqq 4(1 + \|a\|)\,\Big(1 + \big\|g_1(X_j)\big\|\Big)\, n^{-1/2}. \tag{52}$$

Substituting now (51) in (49) we obtain four expectations. Then we apply Lemma 3 to each of them, taking into account (52) and the estimate $\|\mathbf{E}\overline{g}_1(X_1)\| \leqq n^{-1/2}$. For t satisfying (28) we take in Lemma 3 $l \sim n/2$, $m = n - l$, $A = \frac{3}{2}$ and obtain

$$\big|F_3(t)\big| \leqq c_1 t^2\, (1 + \|a\|)^2 \big(\mathbf{E}\|g_2\|^2\big)^{1/2}$$

$$\times\left(\exp(-c_2n)+c_3L^3n^{-3/2}+\prod_{j=1}^{7}\left(1+c_4\sigma_j^4t^2\right)^{-1/4}\right)n^{-3/2}, \tag{53}$$

and for t satisfying (29) we take $l\sim n/\log n$, $m=n-l$ and $A=\frac{7}{4}$ and obtain

$$\begin{aligned}|F_3(t)| &\leqq c_1t^2\,(1+\|a\|)^2\left(\mathbf{E}\|g_2\|^2\right)^{1/2}\\ &\times\left(\exp\left(-\frac{c_2n}{\log n}\right)+c_3L^{7/2}n^{-7/4}(\log n)^{7/4}\right.\\ &\left.+\prod_{j=1}^{7}\left(1+\frac{c_4\sigma_j^4n}{L^2(\log n)\,\log(n/L^2)}\right)^{-1/4}\right)n^{-3/2}.\end{aligned} \tag{54}$$

To estimate $f_4(t)$ in (50) we use the following formula which is analogous to (51):

$$\begin{aligned}\prod_{j=1}^{3}\left(\|\xi_j\|^2+2(\xi_j,\overline{S}_{n3}+\bar{\xi})\right) &= b_1b_2b_3+2b_1b_2(\xi_3,\overline{S}_{n3}+a)\\ &+2b_1b_3(\xi_2,\overline{S}_{n3}+a)+2b_2b_3(\xi_1,\overline{S}_{n3}+a)+4b_1(\xi_2,\overline{S}_{n3}+a)\,(\xi_3,\overline{S}_{n3}+a)\\ &+4b_2(\xi_1,\overline{S}_{n3}+a)\,(\xi_3,\overline{S}_{n3}+a)+4b_3(\xi_1,\overline{S}_{n3}+a)\,(\xi_2,\overline{S}_{n3}+a)\\ &+8(\xi_1,\overline{S}_{n3}+a)\,(\xi_2,\overline{S}_{n3}+a)\,(\xi_3,\overline{S}_{n3}+a).\end{aligned}$$

Substituting it in (50) we obtain eight similar expectations. To estimate these expectations we again apply Lemma 3. For t satisfying (28) we take in it $l\sim n/2$, $m=n-l$, $A=2$ and obtain,

$$\begin{aligned}|F_4(t)| &\leqq c_1|t|^3(1+\|a\|)^3\left(\mathbf{E}\|g_2\|^2\right)^{1/2}\\ &\times\left(\exp(-c_2n)+c_3L^4n^{-2}+\prod_{j=1}^{9}\left(1+c_4\sigma_j^4t^2\right)^{-1/4}\right)n^{-3/2},\end{aligned} \tag{55}$$

and for t satisfying (29) we take $l\sim n/\log n$, $m=n-l$ and $A=\frac{9}{4}$ and obtain

$$\begin{aligned}|F_4(t)| &\leqq c_1|t|^3(1+\|a\|)^3\left(\mathbf{E}\|g_2\|^2\right)^{1/2}\\ &\times\left(\exp\left(-\frac{c_2n}{\log n}\right)+c_3L^{9/2}n^{-9/4}(\log n)^{9/4}\right.\\ &\left.+\prod_{j=1}^{9}\left(1+\frac{c_4\sigma_j^4n}{L^2(\log n)\,\log(m/L^2)}\right)^{-1/4}\right)n^{-3/2}.\end{aligned} \tag{56}$$

Estimates (37)–(38), (41)–(42), (53), and (55) together for t satisfying (28) imply

$$\begin{aligned}|h_2(t)| &\leqq c_1(1+\|a\|)^3\left(\mathbf{E}\|g_1\|^3\right)\left(\mathbf{E}\|g_2\|^3\right)^{1/3}\\ &\times\left(Ln^{-1/2}+(|t|+t^2)\prod_{j=1}^{7}\left(1+c_4\sigma_j^4t^2\right)^{-1/4}\right.\\ &\left.+|t|^3\prod_{j=1}^{9}\left(1+c_4\sigma_j^4t^2\right)^{-1/4}\right)n^{-3/2},\end{aligned} \tag{57}$$

and (37), (39), (41), (43), (54), and (56) for t satisfying (29) imply

$$|h_2(t)| \leqq c_1(1+\|a\|)^3(\mathbf{E}\|g_1\|^3)(\mathbf{E}\|g_2\|^3)^{1/3}c_9(V)\,L^{3/2}n^{-2}. \tag{58}$$

It remains to estimate $h_3(t)$ in (32). Applying (21) with $r=1$ and $z = it(W_1 + W_2)$, $z = itu(W_1+W_2)$ twice, we have

$$\begin{aligned} h_3(t) = it\int_0^1 du\,\mathbf{E}\Big(\big(a,\, g_2(X_1,X_2)\big)\,W_1\exp\{ituW_1\}\,\overline{\mathbf{E}}_2\exp\big\{it\,\|\overline{S}_{n2}\|^2 + ituW_2\big\}\Big) \\ - t^2\int_0^1 dv\int_0^1 du\,u\mathbf{E}\Big(\big(a,\, g_2(X_1,X_2)\big)\,\exp\{ituv\,W_1\} \\ \times\overline{\mathbf{E}}_2\Big(\exp\big\{it\,\|\overline{S}_{n2}\|^2 + ituv\,W_2\big\}(W_1+W_2)\Big)\Big). \end{aligned}$$

Arguing in the same way as in proving (35) and (36) we obtain for t satisfying (28)

$$\begin{aligned} |h_3(t)| &\leqq c_1(1+\|a\|)^3(\mathbf{E}\|g_1\|^3)^{2/3}(\mathbf{E}\|g_2\|^3)^{1/3} \\ &\quad\times\left[Ln^{-1/2} + (|t|+t^2)\prod_{j=1}^{7}(1+c_2\sigma_j^4t^2)^{-1/4}\right]n^{-1}, \end{aligned} \tag{59}$$

and for t satisfying (29)

$$|h_3(t)| \leqq c_1(1+\|a\|^3)(\mathbf{E}\|g_1\|^3)^{2/3}(\mathbf{E}\|g_2\|^3)^{1/3}c_7(V)\,L^{3/2}n^{-3/2}. \tag{60}$$

Now (32), (35), (57), (59) imply (30) and (32), (36), (58), (60) imply (31).

LEMMA 6. *If t belongs to the interval* (28), *then*

$$\begin{aligned} |J_2(t)| &\leqq c_1(1+\|a\|^2)(\mathbf{E}\|g_1\|^3)^{2/3} \\ &\quad\times\left(L^2n^{-1/2} + n^{1/6}\prod_{j=1}^{9}\left(1+\frac{c_2\sigma_j^4t^2}{n^{1/3}}\right)^{-1/6}\right)n^{-1}, \end{aligned} \tag{61}$$

and if t satisfies (29)

$$|J_2(t)| \leqq c_1(1+\|a\|^2)(\mathbf{E}\|g_1\|^3)^{2/3}c_9(V)\,L^3n^{-3/2}. \tag{62}$$

Proof. Noting that

$$J_2(t) = \mathbf{E}\left(Z_n^2\overline{\mathbf{E}}\exp\{it\,\|\overline{S}_n - a\|^2\}\right)$$

we have

$$\begin{aligned} |J_2(t)| &\leqq 4\varepsilon_n^2\mathbf{E}\Big(\|\overline{S}_n^k\|^2\|\alpha_k\|^2\big|\overline{\mathbf{E}}_k\exp\{it\,\|\overline{S}_n - a\|^2\}\big|\Big) \\ &\quad + 4\varepsilon_n^2\|a\|^2\mathbf{E}\Big(\|\alpha_k\|^2\big|\overline{\mathbf{E}}_k\exp\{it\,\|\overline{S}_n - a\|^2\}\big|\Big). \end{aligned} \tag{63}$$

By Hölder's inequality

$$\begin{aligned} &\mathbf{E}\Big(\|\overline{S}_n^k\|^2\|\alpha_k\|^2\big|\overline{\mathbf{E}}_k\exp\{it\,\|\overline{S}_n - a\|^2\}\big|\Big) \\ &\leqq (\mathbf{E}\|\alpha_k\|^3)^{2/3}\Big(\mathbf{E}\Big(\|\overline{S}_n^k\|^6\big|\overline{\mathbf{E}}_k\exp\{it\,\|\overline{S}_n - a\|^2\}\big|^3\Big)\Big)^{1/3}. \end{aligned} \tag{64}$$

Since for any $x > 0$ and $p > 0$

$$\|\overline{S}_n^k\|^6 \leqq x^6 + x^{-p}\|\overline{S}_n^k\|^{p+6},$$

we have in (64)

$$\begin{aligned} &\mathbf{E}\Big(\|\overline{S}_n^k\|^6\Big|\overline{\mathbf{E}}_k \exp\{it\|\overline{S}_n - a\|^2\}\Big|^3\Big) \\ (65) \qquad &\leqq x^6\mathbf{E}\Big|\overline{\mathbf{E}}_k \exp\{it\|\overline{S}_n - a\|^2\}\Big|^2 + x^{-p}\mathbf{E}\|\overline{S}_n^k\|^{p+6}. \end{aligned}$$

To estimate the second expectation in (63) we also use Hölder's inequality and obtain

$$\begin{aligned} \mathbf{E}\Big(\|\alpha_k\|^2\Big|\overline{\mathbf{E}}_k \exp\{it\|\overline{S}_n - a\|^2\}\Big|\Big) &\leqq \mathbf{E}(\|\alpha_k\|^3)^{2/3}\Big(\mathbf{E}\Big|\overline{\mathbf{E}}_k \exp\{it\|\overline{S}_n - a\|^2\}\Big|^3\Big)^{1/3} \\ (66) \qquad &\leqq \mathbf{E}(\|\alpha_k\|^3)^{2/3}\Big(\mathbf{E}\Big|\overline{\mathbf{E}}_k \exp\{it\|\overline{S}_n - a\|^2\}\Big|^2\Big)^{1/3}. \end{aligned}$$

Substituting (64) in (63) and using the elementary inequality $(\alpha+\beta)^{1/3} \leqq \alpha^{1/3} + \beta^{1/3}$ we get

$$\begin{aligned} &|J_2(t)| \leqq 4\varepsilon_n^2\mathbf{E}(\|\alpha_k\|^3)^{2/3} \\ (67) \qquad &\times\Big[x^{-p/3}(\mathbf{E}\|\overline{S}_n^k\|^{p+6})^{1/3} + (x^2 + \|a\|^2)\Big(\mathbf{E}\Big|\overline{\mathbf{E}}_k \exp\{it\|\overline{S}_n - a\|^2\}\Big|^2\Big)^{1/3}\Big]. \end{aligned}$$

Applying the same argument as in the proof of Lemma 10 in [3] with $l \sim n^{2/3}$, $m = n - l$, $A = \frac{3}{2}$ for $|t| \leqq cn^{1/2}L^{-1}$, we have

$$\mathbf{E}\Big|\overline{\mathbf{E}}_k \exp\{it\|\overline{S}_n - a\|\}\Big|^2 \leqq c_1\Big(L^3n^{-1} + \prod_{j=1}^{9}\Big(1 + \frac{c_2\sigma_j^4T_1^2}{n^{1/3}}\Big)^{-1/2}\Big),$$

where T_1 is defined in Lemma 3. Furthermore, taking in (67) $p = 36$, $x = n^{1/12}$ and using (8) and (9) we obtain the estimate

$$\begin{aligned} |J_2(t)| &= c_1(1 + \|a\|^2)(\mathbf{E}\|g_2\|^3)^{2/3} \\ &\times\Big(L^2n^{-1/2} + n^{1/6}\prod_{j=1}^{9}\Big(1 + \frac{c_2\sigma_j^4T_1^2}{n^{1/3}}\Big)^{-1/6}\Big)n^{-1}, \end{aligned}$$

which implies (61) and (62).

LEMMA 7. *I_2 in (19) satisfies the inequality*

$$I_2 \leqq c_1(1 + \|a\|^3)\beta^2\sigma_9^{-1}c_9(V)n^{-1/2}. \qquad (68)$$

Proof. (22) implies

$$|\varphi_n(t) - \psi_n(t)|\,|t|^{-1} \leqq |J_1(t)| + |t|\,|J_2(t)| + t^2|J_3(t)|. \qquad (69)$$

By Lemma 4

$$\int_{-T}^{T} t^2|J_3(t)|\,dt \leqq c_1(1 + \|a\|^3)\mathbf{E}\|g_1\|^3\mathbf{E}\|g_2\|^3c_7(V)n^{-1/2}. \qquad (70)$$

By Lemma 5

$$\int_{-T}^{T} |J_1(t)|\,dt \leqq c_1\big(1+\|a\|^3\big)\big(\mathbf{E}\|g_1\|^3\big)\big(\mathbf{E}\|g_2\|^3\big)^{1/3} c_9(V)\,L^{1/2}n^{-1/2}. \tag{71}$$

By Lemma 6

$$\int_{-T}^{T} |t|\,|J_2(t)|\,dt \leqq c_1\big(1+\|a\|^2\big)\big(\mathbf{E}\|g_1\|^3\big)^{2/3} c_9(V)\,Ln^{-1/2}. \tag{72}$$

Noting now that the moments in (70)–(72) satisfy (4) and integrating both sides in (69), we obtain (68).

From (14), (19)–(20), and (68) it follows that

$$\Delta_{n2}(a) \leqq c_1\big(1+\|a\|^3\big)\,\beta^2\sigma_9^{-1}c_9(V)\,n^{-1/2}. \tag{73}$$

Estimates (5), (6), and (73) imply (3) and the theorem is proved.

REFERENCES

[1] YU. V. BOROVSKIKH, *Theory of U-statistics in Hilbert space*, Preprint No 86.78, Institute of Mathematics, Ukraine Academy of Science, Kiev, 1986 (in Russian).

[2] YU. V. BOROVSKIKH AND V. S. KOROLYUK, *UH-statistics*, Soviet Math. Dokl., 40 (1990), pp. 432–435.

[3] B. A. ZALESSKII, V. V. SAZONOV, AND V. V. UL'YANOV, *Normal approximation in Hilbert space*, I, II, Theory Probab. Appl., 33 (1988), pp. 207–227; pp. 473–487.

[4] V. S. KOROLJUK AND YU. V. BOROVSKIKH, *Approximation of nondegenerate U-statistics*, Theory Probab. Appl., 30 (1986), pp. 439–450.

[5] V. S. KOROLYUK AND YU. V. BOROVSKIKH, *Theory of U-statistics*, Kluwer, Dordrecht, 1994.

[6] V. PAULAUSKAS AND A. RAČKAUSKAS, *Approximation Theory in the Central Limit Theorem. Exact Results in Banach Spaces*, Kluwer, Dordrecht, 1989.

[7] V. V. PETROV, *Sums of Independent Random Variables*, Springer-Verlag, Berlin, 1975.

[8] I. F. PINELIS, *Estimates of moments of infinite dimensional martingales*, Math. Notes, 27 (1980), pp. 459–462.

[9] B. A. ZALESSKII, V. V. SAZONOV, AND V. V. UL'YANOV, *A precise estimate of the rate of convergence in the central limit theorem in Hilbert space*, Math. USSR Sbornik, 68 (1991), pp. 453–482.

[10] V. BENTKUS, F. GÖTZE, AND R. ZITIKIS, *Lower estimates of the convergence rate for U-statistics*, Ann. Probab., 22 (1994), No. 4.

[11] W. S. GRAMS AND R. J. SERFLING, *Convergence rate for U-statistics.*, Ann. Statist., 1 (1973), pp. 153–160.

[12] V. S. KOROLYUK AND YU. V. BOROVSKIKH, *Rate of convergence in the central limit theorem for UH-statistics*, Theory Probab. Math. Statist., 43 (1991), pp. 79–85.

[13] A. J. LEE, *U-statistics; Theory and Practice*, Marcel Dekker, New York, 1990.

[14] M. L. PURI AND V. V. SAZONOV, *On Hilbert space valued U-statistics*, Theory Probab. Appl., 36 (1991), pp. 604–605.

[15] V. V. SAZONOV, *Normal Approximation — Some Recent Advances*, Lecture Notes in Math., Vol. 879, Springer-Verlag, Berlin, 1981.

[16] V. V. SAZONOV, V. V. UL'YANOV, AND B. A. ZALESSKII, *Asymptotically precise estimate of accuracy of Gaussian approximation in Hilbert space*, J. Multivariate Anal., 28 (1989), pp. 304–330.

[17] V. V. SAZONOV AND V. V. UL'YANOV, *Speed of convergence in the central limit theorem in Hilbert space under weakened moment conditions*, in Proc. 5th Vilnius Conference on Probability Theory and Mathematical Statistics, 1989, Mokslas/VSP, Vilnius/Utrecht, 1990, pp. 394–410.

Reprinted from
Theory of Probability and Its Applications, Volume 41 (1996), pages 405-424, by permission of the publisher,

Part II

Limit Theorems (Dependent Case)

EMPIRICAL DISTRIBUTION FUNCTIONS AND FUNCTIONS OF ORDER STATISTICS FOR MIXING RANDOM VARIABLES

MADAN L. PURI* and LANH T. TRAN†

Indiana University, Bloomington, Indiana 47401

Communicated by M. Rosenblatt

Some fundamental properties of the empirical distribution functions are derived in the case of mixing random variables. These properties are then utilized to study asymptotic normality and strong laws of large numbers for functions of order statistics.

1. Introduction

Let $\{\xi_i : 1 \leqslant i < \infty\}$ be a stationary sequence of uniform $[0, 1]$ random variables defined on a probability space $(\Omega, \mathscr{F}, P)$. Let M_1^k and M_{k+n}^{∞} be respectively the σ-fields generated by $\{\xi_i : 1 \leqslant i \leqslant k\}$ and $\{\xi_i : i \geqslant k+n\}$. Let α and φ be functions of nonnegative integers satisfying $\alpha(n) \downarrow 0$ and $\varphi(n) \downarrow 0$. Then $\{\xi_i : 1 \leqslant i < \infty\}$ is said to be strong mixing if for all $A \in M_1^k$ and $B \in M_{k+n}$,

$$|P(A \cap B) - P(A)\,P(B)| \leqslant \alpha(n),$$

where k and n are arbitrary positive integers; it is φ-mixing if the above inequality holds with $\varphi(n)\,P(A)$ instead of $\alpha(n)$.

For a sample $\xi_1, \ldots, \xi_n$, define the empirical distribution Γ_n by

$$\Gamma_n(t) = n^{-1} \sum_{i=1}^{n} u(t - \xi_i), \qquad 0 \leqslant t \leqslant 1,$$

where $u(t) = 1$ or 0 according as $t \geqslant 0$ or < 0.

Revised December 14, 1979.

AMS 1970 subject classifications: Primary 60F05, 60F15; secondary 62G30.

Keywords and phrases: empirical distribution function, order statistics, mixing, strong laws of large numbers.

* Research supported by the Air Force Office of Scientific Research AFSC, USAF, under Grant AFOSR 76-2927. Reproduction in whole or part is permitted for any purpose of the U.S. Government.

† Research supported by National Science Foundation Grant MCS76–00951.

When $\xi_1, \xi_2, \ldots$, are independent, "linear in probability" and "almost sure nearly linear" bounds for Γ_n have been established by Shorack [9] and Wellner [13], respectively. In this paper, we investigate the effects of φ-mixing and strong mixing types of dependence on these bounds. For more information on mixing conditions, the reader is referred to Rosenblatt [8], Ibragimov [4], and Philipp [7].

Our main theorems under φ-mixing type of dependence are the following:

THEOREM 1.1. *Assume* $\{\xi_i : 1 \leqslant i < \infty\}$ *is* φ*-mixing and* (2.1) *holds. Given* $\varepsilon > 0$, *there exist* $0 < \beta = \beta_\varepsilon < 1$ *and a subset* $S_{n,\varepsilon}$ *of* Ω *having* $P(S_{n,\varepsilon}) > 1 - \varepsilon$ *on which*

$$1 - (1-t)/\beta \leqslant \Gamma_n(t) \leqslant t/\beta \qquad \text{for} \quad 0 \leqslant t \leqslant 1, \tag{1.1}$$

$$\beta t \leqslant \Gamma_n(t) \qquad \text{for all } t \text{ such that } 0 < \Gamma_n(t), \tag{1.2}$$

$$\Gamma_n(t) \leqslant 1 - \beta(1-t) \qquad \text{for all } t \text{ such that } \Gamma_n(t) < 1, \tag{1.3}$$

$$\beta t \leqslant \Gamma_n^{-1}(t) \leqslant 1 - \beta(1-t) \qquad \text{for} \quad 0 \leqslant t \leqslant 1, \tag{1.4}$$

$$\Gamma_n^{-1}(t) \leqslant t/\beta \qquad \text{for} \quad t \geqslant n^{-1}, \tag{1.5}$$

and

$$1 - (1-t)/\beta \leqslant \Gamma_n^{-1}(t) \qquad \text{for} \quad t \leqslant 1 - n^{-1}. \tag{1.6}$$

Remark 1.1. Theorem 1.1 generalizes Lemma A.3 of Shorack [9] where the sample elements are assumed to be independent. However, to extend the result of Shorack [9] to the φ-dependence case, new arguments are needed. Mehra and Rao [5] have extended parts of Lemma A.3 of Shorack [9] to the φ-mixing case (see Lemmas 2.2 and 2.3 of Mehra and Rao [5]) under the condition $\sum_{k=1}^{\infty} k^2 \varphi^{1/2}(k) < \infty$, which is stronger than the assumption (2.1) given below. Using an approach based on Theorem 12.2 of Billingsley [1] they proved (1.4), which is their Lemma 2.2. However, their Lemma 2.3 is weaker than our (1.1). In fact, (1.1) implies that the constants τ_1 and τ_2 in their Lemma 2.3 can be set equal to zero.

THEOREM 1.2. *Let* $\tau_1, \tau_2 > 1$ *be fixed. If* $\{\xi_i : 1 \leqslant i < \infty\}$ *is* φ*-mixing and* (2.1) *holds, then there exist* $0 < \beta = \beta(\tau_1, \tau_2) < \frac{1}{2}$ *and a set* $A \subset \Omega$ *with* $P(A) = 1$ *having the following properties: for all* $\omega \in A$ *there is an* $N \equiv N(\omega, \tau_1, \tau_2)$ *for which* $n > N$ *implies*

$$1 - \left(\frac{1-t}{\beta}\right)^{1/\tau_2} \leqslant \Gamma_n(t) \leqslant (t/\beta)^{1/\tau_1} \qquad \text{for} \quad 0 \leqslant t \leqslant 1, \tag{1.7}$$

$$\beta t^{\tau_1} \leqslant \Gamma_n(t) \qquad \text{for all } t \text{ such that } 0 < \Gamma_n(t), \tag{1.8}$$

$$\Gamma_n(t) \leqslant 1 - \beta(1-t)^{\tau_2} \qquad \textit{for all } t \textit{ such that } \Gamma_n(t) < 1, \tag{1.9}$$

$$\beta t^{\tau_1} \leqslant \Gamma_n^{-1}(t) \leqslant 1 - \beta(1-t)^{\tau_2} \qquad \textit{for} \quad 0 \leqslant t \leqslant 1, \tag{1.10}$$

$$\Gamma_n^{-1}(t) \leqslant (t/\beta)^{1/\tau_1} \qquad \textit{for} \quad t \geqslant n^{-1}, \tag{1.11}$$

and

$$1 - \left(\frac{1-t}{\beta}\right)^{1/\tau_2} \leqslant \Gamma_n^{-1}(t) \qquad \textit{for} \quad t \leqslant 1 - n^{-1}. \tag{1.12}$$

Remark 1.2. Theorem 1.2 generalizes Theorem 1 of Wellner [13], where $\xi_1, \ldots, \xi_n$ are assumed to be independent. His proof depends strongly on the independence of the sample elements and cannot be extended to the φ-mixing case considered here.

We also state here our main theorems under strong mixing types of dependence. They are as follows:

THEOREM 1.3. *Assume* $\{\xi_i : 1 \leqslant i < \infty\}$ *is strong mixing satisfying* (2.2). *Let* $\tau_1 > 1$, $\tau_2 \geqslant 1 + \delta/2$ *and* $\varepsilon > 0$. *Then there exist* $0 < \beta = \beta(\varepsilon, \tau_1, \tau_2)$ *and a subset* $S_{n,\epsilon}$ *of* Ω *having* $P(S_{n,\epsilon}) > 1 - \varepsilon$ *on which*

$$1 - \left(\frac{1-t}{\beta}\right)^{1/\tau_1} \leqslant \Gamma_n(t) \leqslant (t/\beta)^{1/\tau_1} \qquad \textit{for} \quad 0 \leqslant t \leqslant 1, \tag{1.13}$$

$$\beta t^{\tau_2} \leqslant \Gamma_n(t) \qquad \textit{for all } t \textit{ such that } 0 < \Gamma_n(t), \tag{1.14}$$

$$\Gamma_n(t) \leqslant 1 - \beta(1-t)^{\tau_2} \qquad \textit{for all } t \textit{ such that } \Gamma_n(t) < 1, \tag{1.15}$$

$$\beta t^{\tau_1} \leqslant \Gamma_n^{-1}(t) \leqslant 1 - \beta(1-t)^{\tau_1} \qquad \textit{for all } 0 \leqslant t \leqslant 1, \tag{1.16}$$

$$\Gamma_n^{-1}(t) \leqslant (t/\beta)^{1/\tau_2} \qquad \textit{for} \quad t \geqslant n^{-1}, \tag{1.17}$$

and

$$1 - \left(\frac{1-t}{\beta}\right)^{1/\tau_2} \leqslant \Gamma_n^{-1}(t) \qquad \textit{for} \quad t \leqslant 1 - n^{-1}. \tag{1.18}$$

When $\{\xi_i : 1 \leqslant i < \infty\}$ is strong mixing satisfying (2.2), we do not know whether there exist "almost sure" upper bounds and lower bounds for Γ_n and Γ_n^{-1} which account for the strength of the mixing in a spirit similar to that of (1.14), (1.15), (1.17), and (1.18). We next show that if the strong mixing coefficient decays to zero at the rate given by (2.3) below, then the "almost sure nearly linear" bounds for Γ_n and Γ_n^{-1} obtained in Theorem 1.2 continue to hold as in the φ-mixing case satisfying (2.1). Our method of argument here is based on a martingale approximation of $\sum u(t - \xi_i)$ similar to that of Philipp [7].

THEOREM 1.4. *The conclusion of Theorem* 1.2 *remains valid if* $\{\xi_i : 1 \leqslant i < \infty\}$ *is strong mixing satisfying* (2.3).

The proofs of these theorems are deferred to Sections 3 and 4. Section 2 deals with some preliminaries and auxiliary lemmas. The results obtained in Sections 3 and 4 are then used in Section 5 to obtain the asymptotic normality and strong law of large numbers for functions of order statistics.

2. PRELIMINARIES

In what follows we make the following assumptions on the mixing coefficients $\alpha(n)$ and $\varphi(n)$.

$$\sum_{n=1}^{\infty} n^2\varphi(n) < \infty, \tag{2.1}$$

$$\sum_{n=1}^{\infty} n^2\alpha^{\delta}(n) < \infty, \qquad 0 < \delta < 1, \tag{2.2}$$

and

$$\alpha(n) = O(e^{-\theta(n)\log n}), \qquad \text{where } \theta(n) \uparrow \infty \text{ arbitrarily slowly.} \tag{2.3}$$

The inequalities (2.4) and (2.5) below will be referred to occasionally. Let ξ be M_1^k measurable and η be M_{k+n}^{∞} measurable; then, if $\{\xi_i : 1 \leqslant i < \infty\}$ is φ-mixing,

$$|\text{Cov}(\xi, \eta)| \leqslant 2\,\|\xi\|_a\,\|\eta\|_b\,[\varphi(n)]^{1/a} \tag{2.4}$$

for all $1 \leqslant a,\ b \leqslant \infty$ with $a^{-1} + b^{-1} = 1$; and if $\{\xi_i : 1 \leqslant i < \infty\}$ is strong mixing, then

$$|\text{Cov}(\xi, \eta)| \leqslant 12\,\|\xi\|_a\,\|\eta\|_b\,[\alpha(n)]^{1/c} \tag{2.5}$$

for all $1 \leqslant a, b, c \leqslant \infty$, with $a^{-1} + b^{-1} + c^{-1} = 1$. The reader is referred to Ibragimov [4], Billingsley [1], and Davydov [2] for the proofs of (2.4) and (2.5).

Occasionally, c will be used to denote constants whose values are unimportant and may differ from line to line.

Then, we shall need the following results.

LEMMA 2.1. *Let* $0 < \delta < 1$ *and* $t \in [0, 1]$ *be fixed. Define* $\tau = t(1-t)$.

(i) *If* $\{\xi_i\}$ *is strong mixing and* (2.2) *holds, then*

$$E\left[\sum_{i=1}^{n} \{u(t-\xi_i) - t\}\right]^4 \leqslant c[n^2\tau^2 + n\tau]\tau^{-\delta}.$$

(ii) *If* $\{\xi_i\}$ *is* φ*-mixing and* (2.1) *holds, then*

$$E\left[\sum_{i=1}^{n}\{u(t-\xi_i)-t\}\right]^4 \leqslant c[n^2\tau^2+n\tau].$$

Here c denotes a constant independent of n and τ.

Proof. We shall prove part (i), since the proof of part (ii) is similar. Let $\eta_i = u(t-\xi_i)-t$. Clearly $E[\sum_{i=1}^{n}\{u(t-\xi_i)-t\}]^4$ equals

$$\sum E\eta_i^4 + 4\sum_{i\neq j}\sum E(\eta_i^3\eta_j) + 6\sum_{i<j}\sum E(\eta_i^2\eta_j^2)$$
$$+ 12\sum_{i<j\neq k}\sum\sum E(\eta_i\eta_j\eta_k^2) + 24\sum_{i<j<k<l}\sum\sum\sum E(\eta_i\eta_j\eta_k\eta_l).$$

Observe that $E|\eta_i|^a \leqslant 2\tau$ for $a \geqslant 1$. Using (2.2), (2.5) and following an argument similar to that of Mehra and Rao [6], each term can easily be shown to be bounded by $[n^2\tau^2+n\tau]\tau^{-\delta}$.

LEMMA 2.2. *Let* ρ *be an arbitrary positive number and* $0 \leqslant t \leqslant 1$. *If* $\{\xi_i\colon 1 \leqslant i < \infty\}$ *is* φ*-mixing and* (2.1) *holds, then*

$$E\left|\sum_{i=1}^{n}\{u(t-\xi_i)-t\}\right|^l \leqslant cn^\rho((n\tau)^{l/2}+n\tau),$$

where $l \geqslant 4$, $\tau = t(1-t)$ *and c is a constant independent of n and* τ.

Remark 2.1. From Lemma 7.4 of Doob [3, p. 225], it is easy to show that

$$E\left|\sum_{i=1}^{n}\{u(t-\xi_i)-t\}\right|^l \leqslant cn^{l/2}.$$

However, this inequality is insufficient for our purposes in regard to the proof of Theorem 1.2. Actually, Lemma 7.4 of Doob (cited above) is proved only for Markov chains satisfying Doeblin's condition and not for φ-mixing random variables. However, Doob's argument can be adapted to prove the inequality above. The proof of Lemma 2.2 due to Tran is given in the Appendix.

LEMMA 2.3. *Let* $\{\xi_i\colon 1 \leqslant i < \infty\}$ *be a stationary and ergodic sequence of uniform* $[0,1]$ *random variables defined on* $(\Omega, \mathcal{F}, P)$. *Let* $h(t)$ *be a nonnegative, nondecreasing, continuous function on* $[0,1]$ *for which* $\int_0^1 (1/h(t))\,dt < \infty$. *Then there exist an* $M > 0$ *and a set* $A \subset \Omega$ *with* $P(A) = 1$

having the following property: for all $\omega \in A$ *there is an* $N \equiv N(\omega, M)$ *for which* $n > N$ *implies*

$$\Gamma_n(t) \leqslant Mh(t) \qquad \textit{for} \quad 0 \leqslant t \leqslant 1.$$

Proof. Using arguments as in the proof of Theorem 1(A) of Wellner [12] we obtain

$$\sup_{0 \leqslant t \leqslant 1} \frac{\Gamma_n(t)}{h(t)} = \sup_{0 \leqslant t \leqslant 1} n^{-1} \sum_{i=1}^{n} \frac{u(t-\xi_i)}{h(t)}$$

$$\leqslant \sup_{0 \leqslant t \leqslant 1} n^{-1} \sum_{i=1}^{n} \frac{u(t-\xi_i)}{h(\xi_i)} \leqslant n^{-1} \sum_{i=1}^{n} \frac{1}{h(\xi_i)}.$$

Since $\{\xi_i : 1 \leqslant i < \infty\}$ is stationary and ergodic, $n^{-1} \sum_{i=1}^{n} 1/h(\xi_i)$ converges almost surely to $\int_0^1 (1/h(t))\,dt < \infty$. The proof now follows by choosing M greater than $\int_0^1 (1/h(t))\,dt$.

3. Proofs of Theorems for the φ-Mixing Case

Proof of Theorem 1.1. First note that for reasons of symmetry, it suffices to prove the upper bounds of (1.1) and (1.5). It can easily be seen that the upper bound of (1.1) implies the remaining inequalities in (1.1) and (1.4). Similarly, (1.5) implies the remaining inequalities in (1.2), (1.3), and (1.6). First we prove the upper inequality in (1.1).

Clearly

$$P[\Gamma_n(\varepsilon/2n) > 0] = P[\Gamma_n(\varepsilon/2n) \geqslant n^{-1}]$$

$$\leqslant \sum_{i=1}^{n} P[\xi_i \leqslant \varepsilon/2n] = \varepsilon/2.$$

Thus

$$P[\Gamma_n(t) \leqslant t/\beta \text{ for } 0 \leqslant t \leqslant \varepsilon/2n] \geqslant 1 - \varepsilon/2, \tag{3.1}$$

since $\Gamma_n(t)$ is nondecreasing in t.

Again, using the property that $\Gamma_n(t)$ is nondecreasing in t, we have for $0 < \varepsilon < 2$

$$P[\Gamma_n(t) > t/\beta \text{ for some } \varepsilon/2n \leqslant t \leqslant 1]$$

$$\leqslant P[\Gamma_n(1/n) > \varepsilon/2\beta n] + \sum_{i=1}^{n-1} P[\Gamma_n((i+1)/n) > i/\beta n]$$

$$= P[(\# \text{ of } \xi_i \leqslant 1/n) - 1 > (\varepsilon/2\beta) - 1]$$

$$+ \sum_{i=1}^{n-1} P[(\# \text{ of } \xi_i \leqslant (i+1)/n) - (i+1) > i\,(1/\beta - 2)]. \tag{3.2}$$

By Lemma 2.1(ii),

$$E\left[\sum_{i=1}^{n}\{u(t-\xi_i)-t\}\right]^4 \leqslant c[n^2\tau^2+n\tau], \tag{3.3}$$

where c is a constant independent of n and τ. By (3.3) and the Chebyshev inequality, the sum of the last two terms in (3.2) is bounded by

$$\begin{aligned}2c(\varepsilon/2\beta-1)^{-4}+c(1/\beta-2)^{-4}\sum_{i=1}^{n-1}\{(i+1)^2/i^4+(i+1)/i^4\}\\ \leqslant c(\varepsilon/2\beta-1)^{-4}+c(1/\beta-2)^{-4}\leqslant \varepsilon/2,\end{aligned}$$

by choosing β sufficiently small. Thus

$$P[\Gamma_n(t)\leqslant t/\beta \text{ for } \varepsilon/2n\leqslant t\leqslant 1]\geqslant 1-\varepsilon/2. \tag{3.4}$$

Now combining (3.1) and (3.4), we obtain the upper bound of (1.1).

We now prove (1.5). Let $\xi_{n1},\ldots,\xi_{nn}$ denote the order statistics of $\xi_1,\ldots,\xi_n$. Then

$$\begin{aligned}P[\xi_{n,i+1}\geqslant i/n\beta] &= P[(\#\text{ of }\xi_i\geqslant i/n\beta)>n-i-1]\\ &= P[(\#\text{ of }\xi_i\geqslant i/n\beta)-n(1-i/n\beta)\\ &\quad > i(1/\beta-1-1/i)].\end{aligned} \tag{3.5}$$

Using the Chebyshev inequality and (3.3), the last term of (3.5) is bounded by

$$c[n^2(i/n\beta)^2+n(i/n\beta)]/i^4\,(1/\beta-1-1/i)^4. \tag{3.6}$$

Pick $\beta<\frac{1}{4}$. Then $1/\beta-1-1/i>\frac{1}{2}\beta$ for all $i\geqslant 1$, and (3.6) is bounded by $c[\beta^2 i^{-2}+\beta^3 i^{-3}]$. Finally,

$$\begin{aligned}&P[\xi_{n,i+1}\geqslant i/n\beta \text{ for some } 1\leqslant i\leqslant n-1]\\ &\leqslant c\beta^2\sum_{i=1}^{n-1}\{i^{-2}+\beta i^{-3}\}\leqslant c\beta^2(1+\beta)<\varepsilon\end{aligned} \tag{3.7}$$

for β sufficiently small. (1.5) is now an immediate consequence of (3.7).

Proof of Theorem 1.2. As in Theorem 1.1, it suffices to prove the upper bounds of (1.7) and (1.11). Since the φ-mixing property implies ergodicity, the upper bound of (1.7) follows from Lemma 2.3 by setting $h(t)=t^{1/\tau_1}$ and choosing $\beta<M^{-\tau_1}$.

We now derive the upper bound of (1.11). Let us define

$$A_n = \left\{ \max_{1 \leqslant i \leqslant n-1} \frac{\xi_{n,i+1}}{(i/n\beta)^{1/\tau_1}} \geqslant 1 \right\}.$$

It suffices to show that $P(A_n \text{ i.o.}) = 0$.
Obviously,

$$P(A_n) \leqslant \sum_{i=1}^{n-1} P[\xi_{n,i+1} \geqslant (i/n\beta)^{1/\tau_1}].$$

Next,

$$\begin{aligned} P[\xi_{n,i+1} \geqslant (i/n\beta)^{1/\tau_1}] = P[\# \text{ of } \xi_i \geqslant (i/n\beta)^{1/\tau_1}) - n + n(i/n\beta)^{1/\tau_1} \\ > n((1/\beta)^{1/\tau_1}(i/n)^{1/\tau_1} - i/n - 1/n)] \end{aligned}$$

which, by Lemma 2.2 and the Chebyshev inequality, is bounded by

$$\frac{cn^{\rho}((i/n\beta)^{l/2\tau_1} n^{l/2} + (i/n\beta)^{1/\tau_1} n)}{n^l((1/\beta)^{1/\tau_1} (i/n)^{1/\tau_1} - i/n - 1/n)^l}. \tag{3.8}$$

Observe that $i/n < (i/n)^{1/\tau_1}$ for $1 \leqslant i \leqslant n$. Choose β small enough so that $1/\beta > 3^{\tau_1}$. Then $(1/\beta)^{1/\tau_1} (i/n)^{1/\tau_1} - i/n - 1/n$ is greater than $(2/3)(1/\beta)^{1/\tau_1} (i/n)^{1/\tau_1}$, and consequently, (3.8) is bounded by

$$\begin{aligned} c\beta^{-l/2\tau_1} n^{-l/2 + l/2\tau_1 + \rho} \sum_{i=1}^{n-1} i^{-l/2\tau_1} \\ + c\beta^{-1/\tau_1} n^{-l + l/\tau_1 + \rho - 1/\tau_1 + 1} \sum_{i=1}^{n-1} i^{-(l-1)/\tau_1}, \end{aligned}$$

which equals $cn^{-1-\epsilon}$ for some $\varepsilon > 0$ if we choose

$$l > \max\left(\frac{2(1+\rho)}{1 - 1/\tau_1}, \frac{2 + \rho - 1/\tau_1}{1 - 1/\tau_1}, 2\tau_1, \tau_1 + 1 \right).$$

Hence $\sum P(A_n) < \infty$, which implies $P(A_n \text{ i.o.}) = 0$ by the Borel–Cantell Lemma. The proof follows.

4. Proofs of Theorems for the Strong Mixing Case

Proof of Theorem 1.3. The upper bound of (1.13) implies the remainin inequalities in (1.13) and (1.16). Similarly, (1.17) implies the inequalities i

(1.14), (1.15), and (1.18). Since strong mixing implies ergodicity, the upper bound of (1.13) is an immediate consequence of Lemma 2.3.

We now prove (1.17). It suffices to prove (1.17) for $\tau_2 = 1 + \delta/2$. Let $1 \leqslant i \leqslant n-1$. Clearly,

$$\begin{aligned}
&P[\xi_{n,i+1} \geqslant (i/n\beta)^{1/(1+\delta/2)}] \\
&\quad = P[(\# \text{ of } \xi_i \geqslant (i/n\beta)^{1/(1+\delta/2)}) > n-i-1] \\
&\quad = P[(\# \text{ of } \xi_i \geqslant (i/n\beta)^{1/(1+\delta/2)}) - n(1-(i/n\beta)^{1/(1+\delta/2)}) \\
&\qquad > n-i-1-n(1-(i/n\beta)^{(1/(1+\delta/2))})] \\
&\quad \leqslant P[(\# \text{ of } \xi_i \geqslant (i/n\beta)^{1/(1+\delta/2)}) - n(1-(i/n\beta)^{1/(1+\delta/2)}) \\
&\qquad > (1/\beta)^{1/(1+\delta/2)} i(i/n)^{(1/(1+\delta/2)-1)} - 2i].
\end{aligned} \tag{4.1}$$

Pick β small enough so that $2 < (1/2)(1/\beta)^{1/(1+\delta/2)}$. By the Chebyshev inequality and Lemma 2.1 (i), the last term of (4.1) is bounded by

$$\begin{aligned}
&\frac{c[n^2(i/n\beta)^{2/(1+\delta/2)} + n(i/n\beta)^{1/(1+\delta/2)}]\,(i/n\beta)^{-\delta/(1+\delta/2)}}{[(1/2)(1/\beta)^{1/(1+\delta/2)} i(i/n)^{(1/(1+\delta/2)-1)}]^4} \\
&\quad \leqslant c[\beta^2 i^{-2} + \beta^{(3+\delta)/(1+\delta/2)} n^{-\delta/2+\delta} i^{-(3+\delta)/(1+\delta/2)}].
\end{aligned}$$

Finally, by summing over i, we have

$$P[\xi_{n,i+1} \geqslant (i/n\beta)^{1/(1+\delta/2)} \text{ for some } 1 \leqslant i \leqslant n-1] \tag{4.2}$$

for sufficiently small β. (1.17) is a consequence of (4.2).

Proof of Theorem 1.4. It suffices to derive the upper bounds of (1.7) and (1.11). Since strong mixing implies ergodicity, the proof of the upper bound of (1.7) is the same as in the φ-mixing case. We now turn to the proof of (1.11).

Let γ be a small positive number to be specified later. Define blocks H_j and I_j of consecutive integers as follows: H_j consists of $[j^\gamma]$, and I_j consists of $[j^\gamma]$ consecutive integers, respectively. No gaps are allowed between the blocks. The order is $H_1, I_1, H_2, I_2, \ldots$. Let

$$A_{j,t} = \sum_{i \in H_j} (u(t-\xi_i) - t),$$

$$B_{j,t} = \sum_{i \in I_j} (u(t-\xi_i) - t).$$

Let M be the index of the block H_j or I_j containing n, and let h_j be the smallest member of H_j. Let $\mathscr{F}_{j,t}$ and $\mathscr{L}_{j,t}$ be the σ-fields generated by $A_{1,t}, \ldots, A_{j,t}$ and $B_{1,t}, \ldots, B_{j,t}$, respectively.

Clearly,

$$\left|\sum_{j=1}^{n}(u(t-\xi_j)-t)\right| \leqslant \sum_{j=h_M}^{h_{M+1}-1}|u(t-\xi_j)-t| + \sum_{j=1}^{M}|E(A_{j,t}/\mathscr{F}_{j-1,t})| + \sum_{j=1}^{M}|E(B_{j,t}/\mathscr{L}_{j-1,t})| + \left|\sum_{j=1}^{M}[A_{j,t}-E(A_{j,t}/\mathscr{F}_{j-1,t})]\right| + \left|\sum_{j=1}^{M}[B_{j,t}-E(B_{j,t}/\mathscr{L}_{j-1,t})]\right|. \tag{4.3}$$

Choose β small enough that $1/\beta > 3^{\tau_1}$. Then $(1/\beta)^{1/\tau_1}(i/n)^{1/\tau_1} - i/n - 1/n$ is greater than $(2/3)(1/\beta)^{1/\tau_1}(i/n)^{1/\tau_1}$. Set $(i/n\beta)^{1/\tau_1} = t_i$, and for notational convenience we drop the subscripts in A_{j,t_i}, B_{j,t_i}, $\mathscr{F}_{j-1,t_i}$, and $\mathscr{L}_{j-1,t_i}$. Then, using (4.3), we have

$$P[\xi_{n,i+1} \geqslant t_i] = P[-(\#\text{ of }\xi_i < t_i) + nt_i > n((1/\beta)^{1/\tau_1}(i/n)^{1/\tau_1} - i/n - 1/n)]$$

$$\leqslant P\left[\left|\sum_{i=1}^{n}u(t_i-\xi_j)\right| > c(1/\beta)^{1/\tau_1}n(i/n)^{1/\tau_1}\right]$$

$$\leqslant P\left[\sum_{j=h_M}^{h_{M+1}-1}|u(t_i-\xi_j)| > (c/5)(1/\beta)^{1/\tau_1}n(i/n)^{1/\tau_1}\right] + P\left[\sum_{j=1}^{M}|E(A/\mathscr{F})| > (c/5)(1/\beta)^{1/\tau_1}n(i/n)^{1/\tau_1}\right] + P\left[\sum_{j=1}^{M}|E(B/\mathscr{L})| > (c/5)(1/\beta)^{1/\tau_1}n(i/n)^{1/\tau_1}\right] + P\left[\left|\sum_{j=1}^{M}[A-E(A/\mathscr{F})]\right| > (c/5)(1/\beta)^{1/\tau_1}n(i/n)^{1/\tau_1}\right] + P\left[\left|\sum_{j=1}^{M}[B-E(B/\mathscr{L})]\right| > (c/5)(1/\beta)^{1/\tau_1}n(i/n)^{1/\tau_1}\right]. \tag{4.4}$$

We shall show that each of the five terms on the right-hand side of (4.4) is of order $n^{-1-\epsilon}$ for some $\varepsilon > 0$.

Observe that $M \leqslant cn^{1/\gamma+1}$ since $\sum_{i=1}^{M} i^{\gamma} \leqslant cn$. Hence $h_{M+1} - h_M \leqslant M^{\gamma} \leqslant cn^{\gamma/(\gamma+1)}$. Pick γ small enough that $\gamma/(\gamma+1) < 1 - \tau_1^{-1}$. Then

$$P\left[\sum_{j=h_M}^{h_{M+1}-1} |u(t_i-\xi_j)| > (c/5)(1/\beta)^{1/\tau_1} n(i/n)^{1/\tau_1}\right]$$
$$\leqslant P\left[\sum_{j=h_M}^{h_{M+1}-1} |u(t_i-\xi_j)| > cn^{1-\tau_1^{-1}}\right] = 0, \tag{4.5}$$

if n is large enough.

Let l be a positive number to be specified later. Since $\alpha(n) = O(e^{-\theta(n)\log n}) = O(n^{-\theta(n)})$, where $\theta(n) \uparrow \infty$, there exists a constant c depending on l such that $\alpha(n) < cn^{-l}$ for all n.

Pick $1 < a < 2$. By the Markov and the Minkowski inequalities,

$$P\left[\sum_{j=1}^{M} |E(A/\mathscr{F})| > cn(i/n)^{1/\tau_1}\right] \leqslant \left\{\frac{\sum_{j=1}^{M} \|E(A/\mathscr{F})\|_{2a}}{cn(i/n)^{1/\tau_1}}\right\}^{2a}. \tag{4.6}$$

By (2.5)

$$\begin{aligned} E[|E(A/\mathscr{F})|^{2a}] &= E[A(E(A/\mathscr{F}))^{2a-1}] \\ &\leqslant c\,\|A\|_4\,[E|E(A/\mathscr{F})|^{2a}]^{(2a-1)/2a} \times j^{-\gamma l(2-a)/4a}. \end{aligned} \tag{4.7}$$

Dividing both sides of (4.7) by $[E|E(A/\mathscr{F})|^{2a}]^{(2a-1)/2a}$, we get

$$\|E(A/\mathscr{F})\|_{2a} \leqslant c\,\|A\|_4\, j^{-\gamma l(2-a)/4a}. \tag{4.8}$$

Let δ be an arbitrary positive small number. Clearly (2.3) implies that (2.2) holds for any $\delta > 0$. By Lemma 2.1(i), (4.6), and (4.8), we obtain

$$\begin{aligned} &P\left[\sum_{j=1}^{M} |E(A/\mathscr{F})| > cn(i/n)^{1/\tau_1}\right] \\ &\leqslant c\left\{\frac{\sum_{j=1}^{M} [j^{2\gamma} t_i^{(2-\delta)} + j^{\gamma} t_i^{(1-\delta)}]^{1/4} j^{-\gamma l(2-a)/4a}}{n(i/n)^{1/\tau_1}}\right\}^{2a} \\ &\leqslant c\left\{\frac{\sum_{j=1}^{M} [j^{\gamma/2} t_i^{(2-\delta)/4} + j^{\gamma/4} t_i^{(1-\delta)/4}]\, j^{-\gamma l(2-a)/4a}}{n(i/n)^{1/\tau_1}}\right\}^{2a} \end{aligned} \tag{4.9}$$

Consider the fourth term on the right-hand side of the inequality (4.4). Clearly,

$$\begin{aligned} &P\left[\left|\sum_{j=1}^{M} [A - E(A/\mathscr{F})]\right| > cn(i/n)^{1/\tau_1}\right] \\ &\leqslant P\left[\sum_{j=1}^{M} [A - E(A/\mathscr{F})] > cn(i/n)^{1/\tau_1}\right] \\ &\quad + P\left[\sum_{j=1}^{M} [-A + E(A/\mathscr{F})] > cn(i/n)^{1/\tau_1}\right]. \end{aligned} \tag{4.10}$$

It suffices to find the upper bound for the second term of (4.10) since the third term can be treated in an analogous manner. To this end, set

$$U_{k,t_i} = \sum_{j \leqslant k} [A - E(A/\mathscr{F})] \qquad \text{for} \quad k \leqslant M,$$
$$= U_M, t_i \qquad \text{for} \quad k > M,$$

$$S^2_{k,t_i} = \sum_{j \leqslant k} E([A - E(A/\mathscr{F})]^2/\mathscr{F}) \qquad \text{for} \quad k \leqslant M,$$
$$= S^2_{M,t_i} \qquad \text{for} \quad k > M.$$

Let $\lambda > 0$ with $\lambda M^\gamma \leqslant 1$, and define

$$T_{k,t_i} = \exp\left(\lambda U_{k,t_i} - \frac{\lambda^2}{2}(1 + \lambda M^\gamma) S^2_{k,t_i}\right) \qquad \text{for} \quad k \geqslant 1.$$

Clearly $EU_{1,t_i} = 0$ and $U_{k,t_i} - U_{k-1,t_i} \leqslant 2M^\gamma$ for each k.

By Lemma 5.4.1 and Corollary 5.4.1 of Stout [10] $\{T_{k,t_i}: k \geqslant 1\}$ is a nonnegative supermartingale and

$$P\{\sup_{k \geqslant 0} T_{k,t_i} > c\} < 1/c \qquad \text{for any } c > 0. \tag{4.11}$$

Set

$$\lambda = \frac{1}{cn^{(1-1/\tau_1)(1/2)} i^{1/\tau_1}}, \qquad K = c^2 i^{2/\tau_1} n^{(3/2)(1-1/\tau_1)},$$

and pick $\gamma < -(1-\tau_1)(1+\tau_1)^{-1}$ so that $\gamma(\gamma+1)^{-1} < (1/2)(1-\tau_1^{-1})$. Then $\lambda M^\gamma \leqslant 1$. By (4.11), we have

$$P\left[\sum_{j=1}^{M} [A - E(A/\mathscr{F})] > cn(i/n)^{1/\tau_1}\right]$$
$$\leqslant P[\sup_{k \geqslant 0} U_{k,t_i} > \lambda K]$$
$$= P[\sup_{k \geqslant 0} T_{k,t_i} > \exp(\lambda^2 K - \frac{\lambda^2}{2}(1 + \lambda M^\gamma) S^2_{M,t_i}]$$
$$\leqslant P[\sup_{k \geqslant 0} T_{k,t_i} > \exp(\lambda^2 K/2)] + P[S^2_{M,t_i} > K/2]$$
$$\leqslant \exp(-2^{-1} n^{(1/2)(1-1/\tau_1)}) + P[S^2_{M,t_i} > K/2]. \tag{4.12}$$

By the Markov and Minkowski inequalities,

$$P[S_{M,t_i}^2 > K/2] = P\left[\left(S_{M,t_i}^2 - \sum_{j=1}^{M} EA^2\right) > ci^{2/\tau_1}n^{(3/2)(1-1/\tau_1)} - \sum_{j=1}^{M} EA^2\right]$$

$$\leqslant \left(\frac{\sum_{j=1}^{M} \|E([A - E(A/\mathscr{F})]^2/\mathscr{F}) - EA^2\|_a}{ci^{2/\tau_1}n^{(3/2)(1-1/\tau_1)} - \sum_{j=1}^{M} EA^2}\right)^a. \tag{4.13}$$

Arguing as in (4.7) and (4.8), we obtain

$$\|E([A^2 - EA^2]/\mathscr{F})\|_a \leqslant c(EA^4)^{1/2} j^{-\nu l(2-a)/2a}. \tag{4.14}$$

Next, by (4.8) and (4.14),

$$\|E([A - E(A/\mathscr{F})]^2/\mathscr{F}) - EA^2\|_a \leqslant c\,\|E(A^2/\mathscr{F}) - EA^2\|_a + c\|(E(A/\mathscr{F}))^2\|_a$$

$$\leqslant c(EA^4)^{1/2} j^{-\nu l(2-a)/2a}. \tag{4.15}$$

By (2.5) and stationarity, for any $\delta > 0$,

$$EA^2 = E\left[\sum_{k\in H_j} (u(t_i - \xi_k) - t_i)\right]^2$$

$$= \sum_{k=1}^{[j^\gamma]} E(u(t_i - \xi_k))^2 + 2\sum_{k<l}\sum E[(u(t_i - \xi_k) - t_i)(u(t_i - \xi_l) - t_i)]$$

$$\leqslant cj^\gamma t_i(1 - t_i) + 20[E|u(t_i - \xi_1) - t_i|^{2/(1-\delta)}]^{1-\delta} \sum_{k<l}\sum \alpha^\delta(l - k)$$

$$\leqslant cj^\gamma[t_i(1 - t_i)]^{1-\delta}$$

$$\leqslant cj^\gamma i^{(1-\delta)/\tau_1} n^{-(1-\delta)/\tau_1}. \tag{4.16}$$

By (4.16),

$$ci^{2/\tau_1}n^{(3/2)(1-1/\tau_1)} - \sum_{j=1}^{M} EA^2$$

$$\geqslant ci^{2/\tau_1}n^{(3/2)(1-1/\tau_1)} - ci^{(1-\delta)/\tau_1}n^{1-(1-\delta)/\tau_1} \tag{4.17}$$

$$\geqslant ci^{2/\tau_1}n^{(3/2)(1-1/\tau_1)},$$

by choosing δ sufficiently small. By (4.10), (4.12), (4.13), (4.15), (4.17) and Lemma 2.1(i), we have

$$P\left[\left|\sum_{j=1}^{M} [A - E(A/\mathscr{F})]\right| > cn(i/n)^{1/\tau_1}\right]$$

$$\leqslant \exp(-2^{-1}n^{(1/2)(1-1/\tau_1)})$$

$$
+ c \left\{ \frac{\sum_{j=1}^{M} [j^{2\gamma} t_i^{(2-\delta)} + j^{\gamma} t_i^{(1-\delta)}]^{1/2} j^{-\gamma l(2-a)/2a}}{i^{2/\tau_1} n^{(3/2)(1-1/\tau_1)}} \right\}^{a}
$$

$$
\leqslant \exp(-2^{-1} n^{(1/2)(1-1/\tau_1)})
$$

$$
+ c \left\{ \frac{\sum_{j=1}^{M} [j^{\gamma} t_i^{(2-\delta)/2} + j^{\gamma/2} t_i^{(1-\delta)/2}] j^{-\gamma(2-a)/2a}}{i^{2/\tau_1} n^{(3/2)(1-1)/\tau_1)}} \right\}^{a}. \tag{4.18}
$$

Pick l large enough so that the summations over j in the last terms of (4.9) and (4.18) are finite. Note that (4.6) and (4.18) remain valid if A_{j,t_i} and $\mathscr{F}_{j-1,t_i}$ are replaced by B_{j,t_i} and $\mathscr{L}_{j-1,t_i}$. In view of (4.4), (4.5), (4.9), and (4.18), we obtain after summing over j and simplifying the last terms of (4.9) and (4.18) that

$$
\begin{aligned}
P[\xi_{n,i+1} \geqslant (i/n\beta)^{1/\tau_1}] \leqslant\; & c\{i^{(-\delta-2)/(4\tau_1)} n^{-1+(1/4\tau_1)(\delta+2)}\}^{2a} \\
& + c\{i^{(-\delta-3)/(4\tau_1)} n^{-1+(1/4\tau_1)(\delta+3)}\}^{2a} \\
& + \exp(-2^{-1} n^{(1/2)(1-1/\tau_1)}) \\
& + c\{i^{(-\delta-2)/(2\tau_1)} n^{(-3/2)+(1/2\tau_1)(1+\delta)}\}^{a} \\
& + c\{i^{(-\delta-3)/(2\tau_1)} n^{(-3/2)+(1/2\tau_1)(2+\delta)}\}^{a}.
\end{aligned}
$$

Without loss of generality, we assume $1 < \tau_1 < \frac{3}{2}$ since if (1.11) holds for some value of $\tau_1 < \frac{3}{2}$, then (1.11) also holds for all values of $\tau_1 \geqslant \frac{3}{2}$. Pick $\delta < 2(\tau_1 - 1)$ and then $\max[\frac{3}{2}, (\frac{3}{2} - (2+\delta)/2\tau_1)^{-1}] < a < 2$. After simplification, (4.19) yields

$$
P[\xi_{n,i+1} \geqslant (i/n\beta)^{1/\tau_1}] \leqslant c i^{-1-\epsilon_1} n^{-1-\epsilon_2} + \exp(-2^{-1} n^{(1/2)(1-1/\tau_1)}) \tag{4.20}
$$

for some $\varepsilon_1, \varepsilon_2 > 0$.

Summing up over i and n in (4.20) and using the Borel–Cantelli Lemma, we obtain $P(A_n \text{ i.o.}) = 0$, where A_n is defined as in the proof of Theorem 1.2. The proof of (1.11) is completed.

In the next section, we present some applications of Theorems 1.1, 1.2, 1.3, and 1.4.

5. Applications to Functions of Order Statistics

Consider the statistic

$$
T_n = n^{-1} \sum_{i=1}^{n} c_{ni} g(\xi_{ni}), \qquad (5.1
$$

where, for $n \geqslant 1$, $c_{n1}, \ldots, c_{nn}$ are known constants, and g is in the class $\mathscr{G}$ o left continuous functions on (0, 1) that are of bounded variation on $(\theta, 1-\theta)$ for all $\theta > 0$. When $\xi_1, \ldots, \xi_n$ are independent and identically distributed, conditions under which T_n has a limiting normal distribution have been obtained by Shorack [9]. These results have been extended to the φ-mixing case by Mehra and Rao (1975a) under the assumption $\sum_{k=1} k^2\varphi^{1/2}(k) < \infty$. Theorem 1.1 can be used for proving asymptotic normality of T_n when $\{\xi_i : 1 \leqslant i < \infty\}$ is φ-mixing under (2.1). In fact, using (1.2), (1.3) and Theorem 4.2 of Mehra and Rao [6] and following arguments as in Mehra and Rao [5], the main results of Shorack [9] can be shown to remain valid under the same conditions when $\{\xi_i : 1 \leqslant i < \infty\}$ is φ-mixing and (2.1) holds. However, the advantages of (1.2) and (1.3) over Lemmas 2.2 and 2.3 of Mehra and Rao [5] are that Shorack's results can be extended to the φ-mixing case without unnecessary complications made by Mehra and Rao in the proof of their Theorem 1.1. Our Theorem 1.2 can be used to prove strong laws of large numbers for T_n.

Define a function $J_n(t)$ on [0, 1] to equal c_{ni} for $(i-1)/n < t \leqslant i/n$ and $1 \leqslant i \leqslant n$ with $J_n(0) = c_{n1}$. Let $\mu_n = \int_0^1 gJ_n\,dI$, where I denotes the identity function and $\int^n \cdot\, dI$ denotes the integration with respect to the Lebesgue measure. For fixed b_1, b_2 and M define a "scores bounding function" B by

$$B(t) = Mt^{-b_1}(1-t)^{-b_2} \qquad \text{for} \quad 0 < t < 1.$$

For fixed $\varepsilon > 0$, define

$$D(t) = Mt^{-1+b_1+\epsilon}(1-t)^{-1+b_2+\epsilon}, \qquad 0 < t < 1,$$

and

$$h(t) = [t(1-t)]^{1-\epsilon/2}, \qquad 0 < t < 1.$$

Then, we have the following theorem.

THEOREM 5.1. *Let* $\{\xi_i : 1 \leqslant i < \infty\}$ *be* φ*-mixing satisfying* (2.1). *Under the assumptions* (a) $|g| \leqslant D$, $|J_n| \leqslant B$ *for all* n, *and* (b) $\int_0^1 Bhd|g| < \infty$, *we have*

$$\lim_{n\to\infty} (T_n - \mu_n) = 0 \qquad \text{a.s.}$$

Proof. For $t \in [0, 1]$, define $\Psi_n(t) = -\int_t^1 J_n\,dI$. Then

$$c_{ni}/n = \Psi_n(i/n) - \Psi_n((i-1)/n), \qquad 1 \leqslant i \leqslant n.$$

Set $U_n = n^{1/2}[\Gamma_n - I]$ on [0, 1], where I is the identity function. As in Shorack [9], integration by parts yields

$$T_n - \mu_n = -n^{-1/2}(S_n + \gamma_{n1} + \gamma_{n2} + \gamma_{n3}),$$

where

$$S_n = \int_{\xi_{n1}}^{\xi_{nn}} A_n U_n \, dg = \int_0^1 A_n^* U_n \, dg$$

with

$$A_n = [\Psi_n(\Gamma_n) - \Psi_n]/[\Gamma_n - I],$$
$$A_n^* = A_n \qquad \text{on}[\xi_{n1}, \xi_{nn})$$
$$\qquad 0 \qquad \text{otherwise},$$
$$\gamma_{n1} = n^{1/2} g(\xi_{n1}) \, [\Psi_n(0) - \Psi_n(\xi_{n1})],$$
$$\gamma_{n2} = n^{1/2} g(\xi_{nn}) \;\; \Psi_n(\xi_{nn}),$$

and

$$\gamma_{n3} = n^{1/2} \int_{[\xi_{n1}, \xi_{nn})^c} gJ_n \, dI.$$

Following Wellner [12], using assumptions (a), (b), and Theorem 1.2, we obtain

$$n^{-1/2} S_n \leqslant c \sup_{0 \leqslant t \leqslant 1} \frac{|\Gamma_n(t) - t|}{[t(1-t)]^{1-\epsilon/4}}. \tag{5.2}$$

We now show that the right-hand side of (5.2) goes to zero a.s. as $n \to \infty$. Observe that, for reasons of symmetry, it is enough to consider $0 \leqslant t \leqslant \frac{1}{2}$. Let $0 < \theta \leqslant \frac{1}{2}$. Then

$$\sup_{0 \leqslant t \leqslant 1/2} \frac{|\Gamma_n(t) - t|}{[t(1-t)]^{1-\epsilon/4}} \leqslant 2\theta^{\epsilon/8} \sup_{0 \leqslant t \leqslant \theta} \frac{\Gamma_n(t)}{t^{1-\epsilon/8}} + 2\theta^{\epsilon/4} + 2\theta^{-1+\epsilon/4} \sup_{\theta \leqslant t \leqslant 1/2} |\Gamma_n(t) - t|. \tag{5.3}$$

By the upper inequality of (1.7), a.s., the first term on the right-hand side of (5.3) can be made arbitrary small for large n by choosing θ small. The second term on the right-hand side can be made small by choosing small θ. The third term goes to zero a.s. as $n \to \infty$ by the Glivenko–Cantelli Theorem for mixing random variables (see Tucker [11]). Each $n^{-1/2}\gamma_{ni}$, $i = 1, 2, 3$, can be seen to be of order ξ_{n1}^{ϵ}. By Lemma 2.2, it is easy to show that ξ_{n1}^{ϵ} 0 a.s. as $n \to \infty$.

Corollary 2 and Theorem 4 of Wellner [12] can also be generalized to the φ-mixing case considered here. The proofs are easy consequences of Theorem 5.1 and are therefore omitted.

Theorem 1.3 can be used to extend the main results of Shorack [9] to the strong mixing case. Let g denote a fixed function in $\mathscr{G}$ and let J denote a fixed measurable function on (0, 1).

ASSUMPTION 1 (boundedness). Let

$$|g| \leqslant D(t)^{1/2+\delta/2(b_1+1)} \times (1-t)^{1/2+\delta/2(b_2+1)},$$

$|J| \leqslant B$, and all $|J_n| \leqslant B$ on (0, 1). $\int_0^1 \bar{B}q \, d|g| < \infty$ with

$$\bar{B}(t) = t^{-b_1(1+\delta/2)}(1-t)^{-b_2(1+\delta/2)} \text{ and } q(t) = [t(1-t)]^{(1-\delta)/2}.$$

ASSUMPTION 2 (smoothness). Except on a set of t's of $|g|$-measure 0 we have that both J is continuous at t and $J_n \to J$ uniformly in some small neighborhood of t as $n \to \infty$. Let $0 \leqslant s, t \leqslant 1$ and set

$$\sigma(s, t) = [(s \wedge t)] + \sum_{j=1}^{\infty} [F_{ij}(s, t) - st] + \sum_{j=2}^{\infty} [F_{ij}(t, s) - st],$$

where $F_{ij}(s, t) = P[\xi_1 \leqslant s, \xi_j \leqslant t]$.

Define $\sigma^2 = \int_0^1 \int_0^1 \sigma(t, s) \, J(s) \, J(t) \, dg(s) \, dg(t)$.

THEOREM 5.2. *If Assumptions* 1 *and* 2 *hold, and* $\{\xi_i : 1 \leqslant i < \infty\}$ *is strong mixing satisfying* (1.2), *then*

$$n^{1/2}(T_n - \mu_n) \to N(0, \sigma^2)$$

with μ_n *and* σ^2 *finite.*

Proof. The proof is analogous to that of Theorem 1 of Shorack (1972). Let $\chi_{n\epsilon}$ denote the indicator function of the set $S_{n,\epsilon}$ of Theorem 1.3. Define $S = \int_0^1 JU_0 \, dg$, where U_0 is a Gaussian process tied down at 0 and 1 and defined by

$$EU_0(t) = 0, \; E[U_0(s) \, U_0(t)] = \sigma(s, t), \qquad 0 \leqslant s, \; 1 \leqslant t.$$

Let $S = \int_0^1 JU_0 \, dg$. Then

$$|\chi_{n\epsilon} S_n - S| \leqslant \int_0^1 |\chi_{n\epsilon} A_n^* U_n - JU_0| \, d|g|.$$

By (1.14) and (1.15),

$$\chi_{n\epsilon} |A_n^*| \leqslant M_\epsilon B t^{-b_1\delta/2}(1-t)^{-b_2\delta/2} \qquad \text{for some constant } M_\epsilon.$$

By Corollary 4.1 of Mehra and Rao [6] for fixed ω, $\int_0^1 |\chi_{n\epsilon} A_n^* U_n - JU_0| \, d|g|$ is bounded by

$$(M_\epsilon + 1)M_\omega Bt^{-b_1\delta/2}(1-t)^{-b_2\delta/2}[t(1-t)]^{1/2-\delta/2},$$

where M_ω is a constant depending on ω.

Use Assumption 1 to find that $(M_\epsilon + 1)M_\omega Bt^{-b_1\delta/2}(1-t)^{-b_2\delta/2} \times [t(1-t)]^{1/2-\delta/2}$ is a $|g|$ integrable function. By the dominated convergence theorem, $S_n \to_p S$ as $n \to \infty$. The rest of the proof is easy and similar to Shorack's [9].

THEOREM 5.3. *Let* $\{\xi_i : 1 \leqslant i < \infty\}$ *be strong mixing satisfying* (2.3). *Then the conclusion of Theorem* 5.1 *continues to hold under the same conditions.*

With Theorem 1.4, the proof of Theorem 5.3 can be obtained by the argument used in the proof of Theorem 5.1.

6. APPENDIX

Proof of Lemma 2.2. Our method of argument is closely related to that of Lemma 7.4 of Doob [3]. By Lemma 2.1(ii), clearly Lemma 2.2 holds for $l = 4$. For $l = 4$, ρ can be set equal to zero.

Define

$$D_n = \sum_{i=1}^{n} \{u(t-\xi_i) - t\}, \quad \hat{D}_n = \sum_{i=n+k+1}^{2n+k} \{u(t-\xi_i) - t\},$$

$$E_n = \sum_{i=n+1}^{n+k} \{u(t-\xi_i) - t\},$$

where k is a positive integer to be specified later. Assume that the lemma is true for $l = m \geqslant 4$. We shall show by induction that the lemma also holds for $l = m + e$, where $0 \leqslant e \leqslant 1$.

Let $B_n = E[|D_n|^{m+e}]$. Then

$$\begin{aligned} E[|D_n + \hat{D}_n|^{m+e}] &\leqslant E[|D_n + \hat{D}_n|^m (|D_n|^e + \hat{D}_n|^e) \\ &\leqslant 2B_n + E\left[\sum_{i=0}^{m-1} \binom{m}{i} |D_n|^{i+e} |\hat{D}_n|^{m-i} \right. \\ &\quad \left. + \sum_{i=1}^{m} \binom{m}{i} |D_n|^i |\hat{D}_n|^{m-i+e}\right]. \end{aligned} \tag{6.1}$$

Let u, v be positive numbers and let $\imath = (m+e)/u$, $\jmath = (m+e)/v$. Then by (2.4),

$$E[|D_n|^u |\hat{D}_n|^v] \leqslant 2\varphi_{k+1}^{1/\imath} E^{1/\imath}[|D_n|^{u\imath}] E^{1/\jmath}[|\hat{D}_n|^{v\jmath}] + E[|D_n|^u] E[|\hat{D}_n|^v]. \tag{6.2}$$

Holder's inequality implies that the last term in (6.2) is bounded by

$$[\mathrm{E}[|D_n|^m]]^{(m+e)/m}.$$

Let η be a positive number such that $\eta(m+e)/m < \rho$. From (6.2), we obtain

$$E[|D_n|^u |\hat{D}_n|^v] \leqslant 2\varphi_{k+1}^{1/\imath} B_n + c[n^\eta((n\tau)^{m/2} + n\tau)]^{(m+e)/m}. \tag{6.3}$$

By (2.1), $\varphi_{k+1} \to 0$ as $k \to \infty$. Pick

$$0 < \varepsilon < 2^{1+\rho} - 2. \tag{6.4}$$

Substitute (6.3) into (6.1) giving u, v the appropriate values, and then choose k large enough. We obtain

$$E[|D_n + \hat{D}_n|^{m+e}] \leqslant (2 + 2^{-1}\varepsilon)B_n + c[n^\eta((n\tau)^{m/2} + n\tau)]^{(m+e)/m}. \tag{6.5}$$

We claim that there exists a constant c independent of n and τ such that

$$B_{2n} \leqslant (2+\varepsilon)B_n + c((n\tau)^{(m+e)/2} + n\tau)n^\rho \text{ for all } n \geqslant 1. \tag{6.6}$$

Observe that

$$E[|u(t-\xi_i) - t|^{m+e}] \leqslant 2\tau,$$

and

$$(n\tau)^{(m+e)/2} + (n\tau)^{(m+e)/m} \leqslant 2((n\tau)^{(m+e)/2} + n\tau).$$

By the Minkowski inequality and (6.5),

$$\begin{aligned}
B_{2n} &= E\left[\left|D_n + \hat{D}_n + E_n - \sum_{i=2n+1}^{2n+k} \{u(t-\xi_i) - t\}\right|\right]^{m+e} \\
&\leqslant \left[E^{1/(m+e)}[|D_n + \hat{D}_n|^{m+e}] \right. \\
&\quad \left. + \left(\sum_{i=n+1}^{n+k} + \sum_{i=2n+1}^{2n+k}\right) E^{1/(m+e)}[|u(t-\xi_i) - t|^{m+e}]\right]^{m+e} \qquad (6.7) \\
&\leqslant [[(2+2^{-1}\varepsilon)B_n + cn^{\eta(m+e)/m}((n\tau)^{(m+e/2} \\
&\quad + (n\tau)^{(m+e)/m}]^{1/(m+e)} + 4k\tau^{1/(m+e)}]^{m+e} \\
&\leqslant [[(2+2^{-1}\varepsilon)B_n + cn^{\eta(m+e)/m}((n\tau)^{(m+e)/2} + n\tau)]^{1/(m+e)} + 4k\tau^{1/(m+e)}]^{m+e}.
\end{aligned}$$

Now

$$\frac{(2+2^{-1}\varepsilon)B_n + cn^{\eta(m+e)/m}((n\tau)^{(m+e)/2}+n\tau)}{(4k)^{m+e}\cdot\tau} \geqslant \frac{c}{(4k)^{m+e}}.$$

Consequently, the last term of (6.7) is bounded by

$$\left[1+\frac{4k}{c^{1/(m+e)}}\right][(2+2^{-1}\varepsilon)B_n + cn^{\rho}((n\tau)^{(m+e)/2}+n\tau)]$$
$$\leqslant (2+\varepsilon)B_n + cn^{\rho}((n\tau)^{(m+e)/2}+n\tau),$$

by choosing c large enough that

$$\left[1+\frac{4k}{c^{1/(m+e)}}\right](2+2^{-1}\varepsilon) < 2+\varepsilon.$$

The proof of (6.6) is now completed.

If $n = 2^{\imath}$, where $\imath$ is a nonnegative integer, then by (6.4) and (6.6),

$$B_{2^{\imath}} \leqslant (2+\varepsilon)^{\imath} B_1^{\imath} + c\tau^{(m+e)/2}\cdot 2^{(\imath-1)(\rho+(m+e)/2)}\left(\sum_{i=0}^{\infty}\frac{2+\varepsilon}{2^{\rho+(m+e)/2}}\right)^{i}$$
$$+ c\tau\, 2^{(\imath-1)(\rho+1)}\sum_{i=0}^{\infty}\left(\frac{2+\varepsilon}{2^{\rho+1}}\right)^{i}$$
$$\leqslant (2+\varepsilon)^{\imath}B_1 + c\tau^{(m+e)/2}2^{\imath(\rho+(m+e)/2)} + c\tau\, 2^{\imath(\rho+1)}$$
$$\leqslant c\, 2^{\imath\rho}((n\tau)^{(m+e)/2}+n\tau).$$

If n is any positive integer, it can be written in the form

$$n = 2^{\imath} + v_1 2^{\imath-1} + \cdots + v_{\imath} \leqslant 2^{\imath} + 2^{\imath-1} + \cdots + 1,$$

where $2^{\imath} \leqslant n \leqslant 2^{\imath+1}$, and each v_j is either 0 or 1. Hence D_n can be written as the sum of ($+1$) groups of sums containing $2^{\imath}$, $v_1 2^{\imath-1}$,..., terms. By the Minkowski inequality and (6.8), we obtain

$$B_n \leqslant \left[\sum_{i=0}^{\imath} E^{1/(m+e)}|D_{2^{\imath-i}}|^{m+e}\right]^{m+e}$$
$$\leqslant c\left[\sum_{i=0}^{\imath}\{2^{(\imath-i)\rho}(2^{(\imath-i)(m+e)/2}\cdot\tau^{(m+e)/2}\right.$$
$$\left. + 2^{(\imath-i)}\tau)\}^{1/(m+e)}\right]^{m+e}$$

$$\leqslant c\tau^{(m+e)/2}\left[\sum_{i=0}^{\iota} 2^{(\iota-i)(1/2+(\rho/(m+e)))}\right]^{m+e}$$

$$+\, c\tau\left[\sum_{i=0}^{\iota} 2^{(\iota-i)(1+\rho/(m+e))}\right]^{m+e}$$

$$\leqslant c\tau^{(m+e)/2}\cdot 2^{\iota(\rho+(m+e)/2)}\left[\sum_{i=0}^{\iota} 2^{-i(1/2+\rho/(m+e))}\right]^{m+e}$$

$$+\, c\tau\, 2^{\iota(\rho+1)}\left[\sum_{i=0}^{\iota} 2^{-i(1+\rho)/(m+e)}\right]^{m+e}$$

$$\leqslant cn^{\rho}((n\tau)^{(m+e)/2}+n\tau).$$

Acknowledgments

The authors are indebted to the referee for helpful and valuable suggestions.

References

[1] Billingsley, P. (1968). *Convergence of Probability Measures.* Wiley, New York.
[2] Davydov, Yu. A. (1968). Convergence of distributions generated by stationary stochastic process. *Theor. Probability Appl.* **4**, 691–696.
[3] Doob, J. L. (1953). *Stochastic Processes.* Wiley, New York.
[4] Ibragimov, I. A. (1962). Some limit theorems for stationary sequences. *Theor. Probability Appl.* **4**, 347–382.
[5] Mehra, K. L., and Rao, M. S. (1975a). On functions of order statistics for mixing processes. *Ann. Statist.* **3**, 874–883,
[6] Mehra, K. L., and Rao, M. S. (1975b). Weak convergence of generalized empirical processes relative to d_q under strong mixing. *Ann. Probability* **3**, 979–991.
[7] Philipp, W. (1971). A function law of the iterated logarithm for empirical distribution functions of weakly dependent random variables. *Ann. Probability* **5**, 319–350.
[8] Rosenblatt, M. (1956). A central limit theorem and a strong mixing condition. *Proc. Nat. Acad. Sci.* U.S.A. **42**, 43–47.
[9] Shorack, G. R. (1972). Functions of order statistics. *Ann. Math. Statist.* **43**, 412–427.
[10] Stout, W. F. (1974). *Almost Sure Convergence.* Academic Press, New York.
[11] Tucker, H. G. (1959). A generalization of Glivenko–Cantelli theorem. *Ann. Math. Statist.* **30**, 828–830.
[12] Wellner, J. A. (1977a). A Glivenko-Cantelli theorem and strong laws of large numbers for functions of order statistics. *Ann. Statist.* **5**, 473–480.
[13] Wellner, J. A. (1977b). A law of the iterated logarithm for functions of order statistics. *Ann. Statist.* **5**, 481–494.

AN INVARIANCE PRINCIPLE FOR PROCESSES INDEXED BY TWO PARAMETERS AND SOME STATISTICAL APPLICATIONS

MANFRED DENKER* and MADAN L. PURI**

Göttingen, Bloomingtom

Abstract. Let $D((0, 1]^2)$ denote the space of all functions on $(0, 1]^2$ with no discontinuities of the second kind. We prove weak invariance principles in the space $D((0, 1]^2)$ for processes of the form $\int h(H_{n+m}(t))dF_n(t)$, $m, n \geqslant 1$, where F_n and G_m are two independent empirical distribution functions of independent, identically distributed sequences of random variables,

$$H_{n+m} = (n+m+1)^{-1}(nF_n+mG_m),$$

and where h belongs to a certain class of functions on the open unit interval. The appropriate topology in $D((0, 1]^2)$ is given by uniform convergence on compact sets. This type of processes is central in nonparametric statistics having applications to two-sample linear rank statistics and signed rank statistics.

1. INTRODUCTION

In this paper we prove several weak invariance principles for a certain type of stochastic integrals which arise from ranking procedures in nonparametric statistics. These integrals can be written in the form

$$\int h(H_{n+m}(t))dF_n(t) \qquad (n, m \geqslant 1) \tag{1.1}$$

(see section 2 for definitions), thus the corresponding random functions for the invariance principle are functions in two variables from $[0, 1]^2$. It turns out that the important aspect of the problem lies in the behavior of the functions h: $(0, 1) \to \mathbf{R}$. In view of the result of Dupač and Hājek [9] our final result in section 8 is optimal concerning the imposed condition on h. It is also worth noticing that Schulze–Pillot's [18] approach using the convergence in "weighted supremum norm" of the two-sample empirical

* This research was done during a visit at the Department of Mathematics at Indiana University.

** Work supported by the Office of Naval Research Contract N/00014/85/K/0648.

process cannot lead to a result of the same generality. Another difficulty arises from the behavior of the random functions near (0, 0), which is overcome by using a topology of uniform convergence on certain compact sets. In some cases of functions h, however, the usual topology of uniform convergence suffices.

Weak invariance principles, as proved in this paper, have not appeared in literature. Schulze–Pillot's work will remain unpublished, hence we shall briefly mention some of his results in sections 3 and 9. Though the motivation to the present investigation comes from statistics, most of the material presented here does not depend on it; in any case this motivation will be apparent for specialists. Sections 2 and 10 illustrate these connections.

The paper is organized as follows: Section 2 contains the definition of random functions based on the two-sample linear rank statistics. Section 3 gives a brief introduction to weak convergence in those function spaces which are necessary for the purpose of our study. The limit process which appears in subsequent sections is investigated in section 4. It is a Gaussian process, indexed by points in $\{(r, s), 0 < r, s \leqslant 1\} \cup \{(0, 0)\}$ with continuous paths. Section 5 contains a first invariance principle for (1.1) in $D([0, 1]^2)$ for certain differentiable functions h. In section 6 we derive an approximation theorem for general functions h and use this result together with the result of section 5 to derive an invariance principle for absolutely continuous h. Finally, in section 8, an invariance principle for general h is derived (like in [9]). As mentioned before, section 9 contains results of the unpublished work of Schulze–Pillot and section 10 is devoted to some applications.

2. TWO-SAMPLE LINEAR RANK STATISTICS

Let X_i, $i \geqslant 1$, and Y_i, $i \geqslant 1$, be two independent sequences of independent and identically distributed random variables with continuous distribution functions F and G, respectively. Since F and G are isomorphic to Lebesgue measure on (0, 1), without loss of generality we can assume that F and G are absolutely continuous and

$$(F(x)+G(x))/2 = x, \quad 0 \leqslant x \leqslant 1. \tag{2.1}$$

Let F_n and G_m be the empirical distribution functions based on $X_1, \ldots, X_n$ and $Y_1, \ldots, Y_m$, respectively. Let $R(i, n, m)$, $1 \leqslant i \leqslant n$, be the rank of X_i among $(X_1, \ldots, X_n, Y_1, \ldots, Y_m)$. Let $n+m = N$, and consider the scores $a(i, n, m)$ satisfying

$$a(i, n, m) = h(i/(N+1)), \quad 1 \leqslant i \leqslant N, \tag{2.2}$$

or

$$\sum_{i=1}^{N} |a(i, n, m) - h(i/(N+1))| = o(N^{1/2}), \tag{2.3}$$

where h: $(0, 1) \to \boldsymbol{R}$ denotes some Lebesgue measurable function to be specified later. Denote the linear rank statistics $T^*_{n,m}$ by

$$(2.4)\qquad T^*_{n,m} = \sum_{i=1}^{n} a\big(R(i, n, m), n, m\big),$$

where the scores $a(\cdot, n, m)$ satisfy (2.2) or (2.3). Write

$$(2.5)\quad \hat{H}_{r,s}(t) = \frac{1}{(r+s)}\Big(\sum_{i=1}^{r} I_{[X_i \leqslant t]} + \sum_{j=1}^{s} I_{[Y_j \leqslant t]}\Big) = \frac{rF_r(t)+sG_s(t)}{r+s}, \qquad r, s \in \boldsymbol{N},$$

and

$$(2.6)\qquad H_\lambda(t) = \lambda F(t) + (1-\lambda) G(t), \qquad 0 \leqslant \lambda \leqslant 1.$$

Then, if the scores are given by (2.2), we can rewrite $T^*_{n,m}$ as

$$(2.7)\qquad T^*_{n,m} = n\int_0^1 h\left(\frac{N}{N+1}\hat{H}_{n,m}(t)\right) dF_n(t).$$

Let

$$(2.8)\qquad T_{n,m} = \int_0^1 h\left(\frac{N}{N+1}\hat{H}_{n,m}(t)\right) dF_n(t) - \int_0^1 h\big(H_{n/N}(t)\big)\, dF(t).$$

For $N \in \boldsymbol{N}$ define the random functions $S_N(h)$ by

$$S_N(h, r, s) = \begin{cases} 0 & \text{if } r < 1/N, \\ N^{-1/2}[rN]\, T_{[rN],[sN]}(h) & \text{if } r \geqslant 1/N. \end{cases} \qquad (0 \leqslant r,\ s \leqslant 1)$$

By definition, $S_N(h)$ is a random function with values in $D([0, 1]^2)$, the space of all right continuous bounded functions in $[0, 1]^2$ having at most discontinuities of the first kind. We will be interested in the weak convergence of $S_N(h)$ with respect to the uniform topology in a special case, and with respect to the topology of uniform convergence on certain compact sets – in the general case.

We need to put some restrictions on the score functions h. We assume that h is right continuous and has bounded variation on every compact set contained in the open interval $(0, 1)$. We then can write $h = h_1 - h_2$, where h_1 and h_2 are non-decreasing right continuous functions having bounded variation on every compact set in $(0, 1)$. We also assume that $h_1(1/2) = h_2(1/2) = 0$ (the last assumption and the right continuity are really no restrictions).

If h_1 and h_2 are integrable with respect to $[t(1-t)]^{-1/2}\,dt$, we put

$$\|h\| = \int \big(|h_1(t)| + |h_2(t)|\big)\big(t(1-t)\big)^{-1/2}\, dt.$$

Denote by $\mathscr{H}$ the set of all functions h with $\|h\| < \infty$. It is clear that if $h \in \mathscr{H}$, then $\int h^2(t)\, dt < \infty$ and

$$(2.9)\qquad \int [t(1-t)]^{1/2}\, dh(t) \leqslant \|h\| \ll \int [t(1-t)]^{1/2}\, dh(t).$$

In the sequel, we shall use the Vinogradov symbol "$\ll$" instead of Landau's "0". Thus $a \ll b$ means that a/b is bounded.

If $h \in \mathscr{H}$ is absolutely continuous, then h can be approximated (in the norm $\| \ \|$) by functions which have bounded continuous second derivatives. In general, every h can be approximated by functions of the form $hI_{[\varepsilon, 1-\varepsilon]}$ for $\varepsilon > 0$.

Since $h \in \mathscr{H}$ has bounded variation on each compact set and satisfies $h(1/2) = 0$, we can rewrite $T_{n,m}$ (see Pyke and Shorack [15]) as

$$T_{n,m}(h) = \int_0^1 \left[F_n\left(\left(\frac{N}{N+1}\hat{H}_{n,m}\right)^{-1}(t)\right) - F\left(H^{-1}_{\lambda(n,m)}(t)\right)\right] dh(t), \tag{2.10}$$

where

$$\lambda(n, m) = n/N. \tag{2.11}$$

3. FUNCTION SPACES OF UNIFORM CONVERGENCE ON COMPACT SETS

For $0 < \varkappa \leqslant 1/2$ let

$$E_\varkappa = \left\{(r, s) \in [0, 1]^2 | \varkappa \leqslant \frac{r}{r+s} \leqslant 1-\varkappa\right\} \cup \{(0, 0)\} \tag{3.1}$$

and let

$$E = \bigcup_{0<\varkappa\leqslant 1/2} E_\varkappa = (0, 1]^2 \cup \{(0, 0)\}. \tag{3.2}$$

Note that, $E_\varkappa$ being a compact subset of $[0, 1]^2$, we can consider the space $D(E_\varkappa)$ of all functions (on $E_\varkappa$) which are "right continuous" and have at most discontinuities of the first kind (see [14]). $D(E)$ denotes the set of all "right continuous" functions having no discontinuities of the second kind. More precisely, if $f \in D(E_\varkappa)$ (resp. $D(E)$) whenever $r_k \downarrow r$ and $s_k \downarrow s$, then $\lim_{n\to\infty} f(r_k, s_k) = f(r, s)$ and otherwise $\lim_{k\to\infty} f(r_k, s_k)$ exists (if $r_k \to r$ and $s_k \to s$). Note that $f|E_\varkappa \in D(E_\varkappa)$ whenever $f \in D E)$.

Let $d_\varkappa$ denote the Skorohod metric on $D(E_\varkappa)$ and let $\delta_\varkappa$ denote the uniform metric on $D(E_\varkappa)$. These metrics extend to metrics d and δ on $D(E)$ by the following definition: for $g, f \in D(E)$ let $d_\varkappa(f, g) = d_\varkappa(f|E_\varkappa, g|E_\varkappa)$ and $\delta_\varkappa(f, g) = \delta_\varkappa(f|E_\varkappa, g|E_\varkappa)$. Then

$$d(f, g) = \sum_{k=2}^{\infty} \min(2^{-k}, d_{1/k}(f, g)) \tag{3.3}$$

and

$$\delta(f, g) = \sum_{k=2}^{\infty} \min(2^{-k}, \delta_{1/k}(f, g)). \tag{3.4}$$

Both the metrics d and δ are bounded; d metrizes the topology of Skorohod convergence on each of the $E_\varkappa$, and δ metrizes the topology on uniform convergence on every $E_\varkappa$. We note the following

LEMMA 3.1. *$D(E)$ is complete with respect to both metrics, and the space $C(E)$ of continuous functions is a closed separable subspace. Moreover, the space $(D(E), d)$ is separable.*

$D(E)$ can be described as the δ-closure of all functions which are finite linear combinations of indicator functions of the form $I_{E \cap A_1 \times A_2}$, where each A_i is either an interval of the form $[a_i, b_i)$ or $\{1\}$.

We now discuss weak convergence in $D(E)$ with respect to both metrics d and δ to the extent it is needed in the following discussion.

Let $\{Z_n, n \geqslant 1\}$ be a family of random functions with values in $D([0, 1]^2)$. Since the map $D([0, 1]^2) \to D(E)$, given by $f \to f|E$, is continuous with respect to the Skorohod topology on $D([0, 1]^2)$ and the metric d on $D(E)$, and also continuous with respect to the uniform metric on $D([0, 1]^2)$ and the metric δ on $D(E)$, weak convergence of $Z_n \to Z_0$ in $D([0, 1]^2)$ implies the weak convergence in $D(E)$ with respect to the corresponding metric. In particular, we shall show (in Section 5) the weak convergence in $D([0, 1]^2)$ of $S_N(h)$ to some Gaussian process when h satisfies certain conditions. Later, in Section 7, we shall show that the above fact implies weak convergence in $D(E)$ for a large class of score functions h. Also, as a consequence of the weak convergence of the empirical process $N^{-1/2}[rN](F_{[rN]}(t) - F(t))$ to a Kiefer process in $D([0, 1]^2)$, we infer its weak convergence also in $D(E)$.

We also note that the embedding $(D(E), \delta) \to (D(E), d)$ is continuous, so that the weak convergence in the δ-metric implies that in the d-metric. Conversely, we can argue as in [2] that a sequence Z_n converges weakly in $(D(E), \delta)$ to Z_0 if it converges weakly in $(D(E), d)$ and Z_0 has its support on $C_b(E)$, the space of bounded continuous functions. This follows from the general fact that if Z_0 has a separable range in $(D(E), d)$, then the Borel fields given by δ and d coincide when restricted to that separable range.

If $D_0 \subset D(E)$ is separable, then the Borel σ-field is generated by the projections (see [2], Section 18). In particular, a random function X with values in D_0 is measurable if $\Pi \circ X$ is measurable for every projection. The random functions $S_N(h)$ clearly have a separable range as they use only the points $(k/N, j/N)$ $(0 \leqslant k, j \leqslant N)$ in E, and hence they are measurable.

We shall always reduce the weak convergence in $D(E)$ to that in $D([0, 1]^2)$; this way we can avoid speaking of tightness. In order to do so we shall use the following well known lemma:

LEMMA 3.2. *Let Z_n and U_n, $n \geqslant 1$, be random functions with values in $D(E)$ and suppose that U_n converges weakly to U_0 with respect to δ (resp. d). If, for any $\varepsilon > 0$, $\lim \mathrm{P}(\delta(Z_n, U_n) \geqslant \varepsilon) = 0$ (resp. $\lim \mathrm{P}(d(Z_n, U_n) \geqslant \varepsilon) = 0$) then Z_n converges weakly to U_0 with respect to δ (resp. d).*

If $\{P_n, n \geqslant 0\}$ is a family of probability measures on $D(E)$ such that $P_n(C(E)) = 1$ $(n \geqslant 1)$ and $P_n \to P_0$ weakly with respect to δ or d, then $P_0(C(E)) = 1$.

The following lemma is a straightforward extension of Lemma 3.2:

LEMMA 3.3. *Let Z_n be random functions in $D(E)$. Suppose that for each $\varepsilon \geqslant 0$ there exists a sequence U_n^ε, $n \geqslant 1$, of random functions and a random function U^ε such that*

$$U_n^\varepsilon \to U^\varepsilon \tag{3.5}$$

weakly in $D(E)$ with respect to δ and such that

$$\mathrm{P}(\delta(Z_n, U_n^\varepsilon) \geqslant \eta) \leqslant \eta^{-2} a(\varepsilon) \quad \text{for any } \eta \geqslant 0. \tag{3.6}$$

If there exists a random function Z such that $U^\varepsilon \to Z$ as $\varepsilon \to 0$ in probability and if $\lim_{\varepsilon \to 0} a(\varepsilon) = 0$, then Z_n converges weakly in $D(E)$ with respect to δ.

4. THE PROCESS $Z(h)$

Let the distribution functions F and G be fixed. Let $\hat{\mathscr{H}}$ be a class of score functions $h \in \mathscr{H}$ such that, for any $0 < \lambda < 1$, the functions $F \circ H_\lambda^{-1}$ and $G \circ H_\lambda^{-1}$ are almost surely differentiable with respect to $d|h|$ ($d|v|$ denotes the total variation measure given by the signed measure v). This assumption always holds if h is absolutely continuous.

We shall write dF/dH_λ and dG/dH_λ for versions of the Radon–Nikodym derivatives satisfying

$$\frac{dF}{dH_\lambda} \circ H_\lambda^{-1} = (F \circ H_\lambda^{-1})' \text{ and } \frac{dG}{dH_\lambda} \circ H_\lambda^{-1} = (G \circ H_\lambda^{-1})' \; d|h| \text{ a.s.}$$

For $h \in \hat{\mathscr{H}}$, define a process $Z(h)$ on E by $Z(h, 0, 0) = 0$ and

$$\begin{aligned} Z(h, r, s) = &-\frac{s}{r+s} \int_0^1 \frac{dG(t)}{dH_{\lambda(r,s)}} \mathscr{K}_1(r, t)\, dh(H_{\lambda(r,s)}(t)) + \\ &+ \frac{r}{r+s} \int_0^1 \frac{dF(t)}{dH_{\lambda(r,s)}} \mathscr{K}_2(s, t)\, dh(H_{\lambda(r,s)}(t)) \quad (r, s \neq 0), \end{aligned} \tag{4.1}$$

where $\mathscr{K}_1$ and $\mathscr{K}_2$ are two independent Kiefer processes on $C([0, 1]^2)$ with expectations zero and the covariance structure given by

$$\mathrm{E}\, \mathscr{K}_1(r, t) \mathscr{K}_1(r', t') = \min(r, r') F(t)(1 - F(t')), \quad 0 \leqslant t \leqslant t' \leqslant 1, \; 0 \leqslant r, r' \leqslant 1$$

and

$$\mathrm{E}\, \mathscr{K}_2(s, t) \mathscr{K}_2(s', t') = \min(s, s') G(t)(1 - G(t')), \quad 0 \leqslant t \leqslant t' \leqslant 1, \; 0 \leqslant s, s' \leqslant 1$$

LEMMA 4.1. *The process $Z(h)$ has the same distribution as the process $-Z(h)$ and the process $Z^*(h)$, where*

$$(4.2)\quad Z^*(h, r, s) = -\frac{s}{r+s}\int_0^1 (G\circ H^{-1}_{\lambda(r,s)})'(t)\,\mathscr{K}_1(r, H^{-1}_{\lambda(r,s)}(t))\,dh(t) + \frac{r}{r+s}\int_0^1 (F\circ H^{-1}_{\lambda(r,s)})'(t)\,\mathscr{K}_2(s, H^{-1}_{\lambda(r,s)}(t))\,dh(t).$$

If h is absolutely continuous and h' is integrable, then $Z(h)$ has the same distribution as $Z_1^*(h)$ where

$$(4.3)\quad Z_1^*(h, r, s) = \int_0^1 \Big[h(H_{\lambda(r,s)}(u)) + \frac{r}{r+s}\int_t^1 h'(H_{\lambda(r,s)}(u))\,dF(u)\Big]\,\mathscr{K}_1(r, dt) + \frac{r}{r+s}\int_0^1 h'(t)\,\mathscr{K}_2(s, H^{-1}_{\lambda(r,s)}(t))\,dF(H^{-1}_{\lambda(r,s)}(t))$$

Proof. Since for any Kiefer process $\mathscr{K}$ under consideration, $-\mathscr{K}$ has the same distribution, it follows that $Z(h)$ and $-Z(h)$ have the same distribution. That $Z^*(h)$ has the same distribution as $Z(h)$ follows by a change of variables in (4.1).

Finally, if h is absolutely continuous and if h' is Lebesgue integrable, then from (4.2) we obtain

$$\begin{aligned} Z(h, r, s) = &-\frac{s}{r+s}\int_0^1 h'(t)\,\mathscr{K}_1(r, H^{-1}_{\lambda(r,s)}(t))\,dG(H^{-1}_{\lambda(r,s)}(t)) + \\ &+\frac{r}{r+s}\int_0^1 h'(t)\,\mathscr{K}_2(s, H^{-1}_{\lambda(r,s)}(t))\,dF(H^{-1}_{\lambda(r, s)}(t)) = \\ = &-\int_0^1 h'(t)\,\mathscr{K}_1(r, H^{-1}_{\lambda(r,s)}(t))\,dt + \\ &+\frac{r}{r+s}\int_0^1 h'(t)\,\mathscr{K}_1(r, H^{-1}_{\lambda(r,s)}(t)) \\ &+\frac{r}{r+s}\int_0^1 h'(t)\,\mathscr{K}_2(s, H^{-1}_{\lambda(r,s)}(t))\,dF(H^{-1}_{\lambda(r,s)}(t)) \\ = &\int_0^1 \Big[h(H_{\lambda(r,s)}(t)) + \frac{r}{r+s}\int_t^1 h'(H_{\lambda(r,s)}(u))\,dF(u)\Big]\,\mathscr{K}_1(r, dt) + \\ &+\frac{r}{r+s}\int_0^1\int_t^1 h'(H_{\lambda(r,s)}(u))\,dF(u)\,\mathscr{K}_2(s, dt), \end{aligned}$$

which proves (4.3).

The integrals with respect to the Brownian bridge component of the Kiefer process were first defined by Filippova [10] for functions h satisfying $\int h^2(x)\,dF(x) < \infty$ (resp. $\int h^2(x)\,dG(x) < \infty$). Indeed, if h' is integrable and belongs to $\mathscr{H}$, then by (2.9) it is easy show that these integrals are finite in our special case. It is also clear that the processes of the form $\int f(t)\,\mathscr{K}(r,t)\,dv(t)$, where v is some measure and f some square integrable function, have continuous paths and are in fact Wiener processes with zero expectations and covariance

$$\mathrm{E}\int f(t)\,\mathscr{K}(r,t)\,dv(t)\int f(t)\,\mathscr{K}(r',t)\,dv(t)$$

$$= 2\iint_{u<v} f(u)f(v)\min(r,r')\,u(1-v)\,dv(u)\,dv(v) +$$

$$+\iint I_{\{u=v\}} f^2(u)\min(r,r')\,u(1-u)\,dv(u)\,dv(v).$$

Since, in (4.1), $H_{\lambda(r,s)}(t)$ depends continuously on r and s, the processes $Z(h)$ must have continuous paths. They are obviously Gaussian and centered at their expectations. Their covariance structure is given by the following

LEMMA 4.2. *We have*

$$\text{(4.4)}\qquad \mathrm{E}Z(h,r,s)Z(h,r',s') = \min(r,r')\frac{ss'}{(r+s)(r'+s')}\times$$

$$\times\iint (G\circ H^{-1}_{\lambda(r,s)})'(u)(G\circ H^{-1}_{\lambda(r',s')})'(v)\min\big(F(H^{-1}_{\lambda(r,s)}(u)),\,F(H^{-1}_{\lambda(r',s')}(v)\big)\times$$

$$\times\Big(1-\max\big(F(H^{-1}_{\lambda(r,s)}(u)),\,F\big(H^{-1}_{\lambda(r',s')}(v)\big)\big)\Big)\,dh(u)\,dh(v)+$$

$$+\min(s,s')\frac{rr'}{(r+s)(r'+s')}\iint (F\circ H^{-1}_{\lambda(r,s)})'(u)(F\circ H^{-1}_{\lambda(r',s')})'(v)\times$$

$$\times\min\big(G(H^{-1}_{\lambda(r,s)}(u)),\,G\big(H^{-1}_{\lambda(r',s')}(v)\big)\times$$

$$\times\Big(1-\max\big(G(H^{-1}_{\lambda(r,s)}(u)),\,G\big(H^{-1}_{\lambda(r',s')}(v)\big)\big)\Big)\,dh(u)\,dh(v)$$

and $|\mathrm{E}Z(h,r,s)Z(h,r',s')| \ll \|h\|^2$, *where "$\ll$" depends on* r, r', s *and* s'.

Proof. Let $r<r'$ and $s<s'$ (the other cases can be proved analogously). Then

$$\mathrm{E}Z(h,r,s)Z(h,r',s') = \mathrm{E}\Bigg\{\frac{s}{r+s}\int_0^1 \frac{dG}{dH_{\lambda(r,s)}}(t)\,\mathscr{K}_1(r,t)\,dh\big(H_{\lambda(r,s)}(t)\big) -$$

$$-\frac{r}{r+s}\int_0^1 \frac{dF}{dH_{\lambda(r,s)}}(t)\,\mathscr{K}_2(s,t)\,dh\big(H_{\lambda(r,s)}(t)\big)\Bigg\}\times$$

$$\times\Bigg\{\frac{s'}{r'+s'}\int_0^1 \frac{dG}{dH_{\lambda(r,s)}}(t)\,\mathscr{K}_1(r',t)\,dh\big(H_{\lambda(r',s')}(t)\big) -$$

$$-\frac{r'}{r'+s'}\int_0^1 \frac{dF}{dH_{\lambda(r',s')}}(t)\,\mathscr{K}_2(s',t)\,dh\big(H_{\lambda(r',s')}(t)\big)\Bigg\}$$

$$= \frac{ss'}{(r+s)(r+s')} \iint \frac{dG}{dH_{\lambda(r,s)}}(u) \frac{dG}{dH_{\lambda(r',s')}}(v) \times$$

$$\times \mathrm{E}\mathscr{K}_1(r, u)\, \mathscr{K}_1(r', v)\, dh(H_{\lambda(r,s)}(u))\, dh(H_{\lambda(r',s')}(v)) +$$

$$+ \frac{rr'}{(r+s)(r'+s')} \iint \frac{dF}{dH_{\lambda(r,s)}}(u) \frac{dF}{dH_{\lambda(r',s')}}(v) \times$$

$$\times \mathrm{E}\mathscr{K}_2(s, u)\, \mathscr{K}_2(s', v)\, dh(H_{\lambda(r,s)}(u))\, dh(H_{\lambda(r',s')}(v)).$$

Since $\mathrm{E}\mathscr{K}_1(r, u)\, \mathscr{K}_1(r', v) = r \min(F(u), F(v))\,(1-\max(F(u), F(v)))$ and $\mathrm{E}\mathscr{K}_2(s, u)\, \mathscr{K}_2(s', v) = s\ \min(G(u), G(v))\ (1-\max(G(u), G(v)))$, (4.4) follows by change of variables.

If h is non-decreasing, then both the integrands in (4.4) are bounded by $C(r, r', s, s')\|h\|^2$, where $C(r, r', s, s')$ is a constant, independent of h, but dependent on r, r', s, s'. Since the norm $\|\ \|$ of h on $\mathscr{H}$ is compatible with the Jordan decomposition of h, it follows that

$$|\mathrm{E}Z(h, r, s)\, Z(h, r', s')| \ll \|h\|^2.$$

If, in (4.1), $r/(r+s)$ is fixed, so are $H_{\lambda(r,s)}$ and $s/(r+s)$. Hence, for any fixed c $(0 < c < 1)$,

$$Z\left(h, r, \frac{1-c}{c} r\right) \tag{4.5}$$

is a sum of two independent Wiener processes with covariance structure given by Lemma 4.2. In particular, the variance is given by

$$\mathrm{E}Z\left(h, r, \frac{1-c}{c} r\right)^2 = r\sigma^2,$$

where

$$\sigma^2 = 2(1-c)^2 \iint_{u<v} \{(G \circ H_c^{-1})'(u) (G \circ H_c^{-1})'(v) (F \circ H_c^{-1})(u) \times$$

$$\times (1 - F \circ H_c^{-1}(v))\}\, dh(u)\, dh(v) +$$

$$+ 2c(1-c) \iint_{u<v} \{(F \circ H_c^{-1})'(u) (F \circ H_c^{-1})'(v) (G \circ H_c^{-1})(u) \times$$

$$\times (1 - G \circ H_c^{-1}(v))\}\, dh(u)\, dh(v),$$

provided dh has no point masses.

We also need the following

PROPOSITION 4.3. *If $h_1, h_2 \in \hat{\mathscr{H}}$, then*

$$\mathrm{E}[\delta(Z(h_1), Z(h_2))]^2 \ll \|h_1 - h_2\|. \tag{4.6}$$

In particular, if $\{h_n\colon n \geq 1\}$ converges to h in norm, then $Z(h_n) \to Z(h)$ in probability with respect to δ.

Proof. We shall use the representation of $Z(h)$ given in Lemma 4.1. It is sufficient to show that $\mathrm{E}\left[\delta(Z(h), 0)\right]^2 \ll \|h\|$ for $h \in \hat{\mathscr{H}}$.

Fix $0 < \varkappa < \frac{1}{2}$ and denote by $I(z)$ the closed interval with endpoints $H_\varkappa^{-1}(z)$ and $H_{1-\varkappa}^{-1}(z)$, $0 < z < 1$. Assume first that – given a fixed $z \in (0, 1)$ $-F(H_\varkappa^{-1}(z)) \geqslant G(H_\varkappa^{-1}(z))$. In this case $I(z) = \left[H_{1-\varkappa}^{-1}(z), H_\varkappa^{-1}(z)\right]$ and it follows that

$$\max_{(r,s) \in E_\varkappa} \left|\mathscr{K}_2\left(s, H_{\lambda(r,s)}^{-1}(z)\right)\right| \leqslant \max_{0 < s \leqslant 1, y \in I(z)} |\mathscr{K}_2(s, y)|.$$

Since

$$\{\sup_{y \in I(z)} |\mathscr{K}_2(s, y)|^2 \colon 0 < s \leqslant 1\} \quad \text{and} \quad \left\{\frac{(\mathscr{K}_2(1, y))^2}{(1-G(y))^2} \colon y \in I(z)\right\}$$

are non-negative submartingales, applying Doob's inequality twice ([8], p. 314), it follows that

$$\mathrm{E} \max_{(r,s) \in E_\varkappa} \left|\mathscr{K}_2\left(s, H_{\lambda(r,s)}^{-1}(z)\right)\right|^2 \leqslant 16 \mathrm{E}\left(\mathscr{K}_2\left(1, H_\varkappa^{-1}(z)\right)\right)^2 \left(\frac{1-G\left(H_{1-\varkappa}^{-1}(z)\right)}{1-G\left(H_\varkappa^{-1}(z)\right)}\right)^2.$$

(Note that $G\left(H_\varkappa^{-1}(z)\right) \neq 1$). Now, since

$$\mathrm{E}\left(\mathscr{K}_2(1, \mathscr{H}_\varkappa^{-1}(z))\right)^2 = G\left(H_\varkappa^{-1}(z)\right)\left(1-G\left(H_\varkappa^{-1}(z)\right)\right) \leqslant (1-\varkappa)^{-1} z(1-z),$$

$$1 - z = 1 - \varkappa F\left(H_\varkappa^{-1}(z)\right) - (1-\varkappa) G\left(H_\varkappa^{-1}(z)\right) \leqslant 1 - G\left(H_\varkappa^{-1}(z)\right),$$

$$\varkappa\left[1 - G\left(H_{1-\varkappa}^{-1}(z)\right)\right] \leqslant 1 - (1-\varkappa) F\left(H_{1-\varkappa}^{-1}(z)\right) - \varkappa G\left(H_{1-\varkappa}^{-1}(z)\right) = 1 - z,$$

therefore

$$\mathrm{E} \max_{(r,s) \in E_\varkappa} \left|\mathscr{K}_2\left(s, H_{\lambda(r,s)}^{-1}(z)\right)\right|^2 \leqslant 16 \varkappa^{-2}(1-\varkappa)^{-1} z(1-z). \tag{4.7}$$

In order to prove the corresponding inequality for $\mathscr{K}_1$, let us first assume that $G\left(H_{1-\varkappa}^{-1}(t)\right) \leqslant F\left(H_{1-\varkappa}^{-1}(t)\right)$ holds as well as $F\left(H_\varkappa^{-1}(z)\right) \geqslant G\left(H_\varkappa^{-1}(z)\right)$.

Here we use the fact that $\{\mathscr{K}_1(1, y)^2/F(y)^2 \colon y \in I(z)\}$ also is a non-negative (backward) submartingale and we obtain

$$\mathrm{E} \max_{(r,s) \in E_\varkappa} \left|\mathscr{K}_1\left(r, H_{\lambda(r,s)}^{-1}(z)\right|^2 \leqslant 16 \mathrm{E}\left(\mathscr{K}_1\left(1, H_{1-\varkappa}^{-1}(z)\right)\right)^2 \left(\frac{F\left(H_\varkappa^{-1}(z)\right)}{F\left(H_{1-\varkappa}^{-1}(z)\right)}\right)^2.$$

Here $F\left(H_{1-\varkappa}^{-1}(z)\right) \neq 0$. Moreover,

$$\mathrm{E}\left(\mathscr{K}_1\left(1, H_{1-\varkappa}^{-1}(z)\right)\right)^2 = F\left(H_{1-\varkappa}^{-1}(z)\right)\left(1 - F\left(H_{1-\varkappa}^{-1}(z)\right)\right) \leqslant (1-\varkappa)^{-1} z(1-z),$$

$$z = (1-\varkappa) F\left(H_{1-\varkappa}^{-1}(z)\right) + \varkappa G\left(H_{1-\varkappa}^{-1}(z)\right) \leqslant F\left(H_{1-\varkappa}^{-1}(z)\right) \quad \text{and} \quad \varkappa F\left(H_\varkappa^{-1}(z)\right) \leqslant z.$$

Consequently,

$$\text{(4.8)} \qquad E \max_{(r,s)\in E_\varkappa} \left|\mathcal{K}_1\left(r, H^{-1}_{\lambda(r,s)}(z)\right)\right|^2 \leqslant 16\varkappa^{-2}(1-\varkappa)^{-1} z(1-z).$$

Let now $G\left(H^{-1}_{1-\varkappa}(z)\right) > F\left(H^{-1}_{1-\varkappa}(z)\right)$. Then by continuity, there exists a $t_0 \in [\varkappa, 1-\varkappa]$ with $G\left(H^{-1}_{t_0}(z)\right) = F\left(H^{-1}_{t_0}(z)\right)$. Put $I_1 = [H^{-1}_{1-\varkappa}(z), t_0]$ and $I_2 = [t_0, H^{-1}_\varkappa(z)]$. Then, on I_1, $G(t_0) \geqslant F(t_0)$ and the argument showing (4.7) (interchanging F with G and replacing $\mathcal{K}_2$ by $\mathcal{K}_1$) gives

$$E \max_{0<r\leqslant 1, y\in I_1} |\mathcal{K}_1(r, y)|^2 \leqslant 16\varkappa^{-2}(1-\varkappa)^{-1} z(1-z).$$

On I_2 we can use the argument for (4.8) to show an analogous inequality. Hence

$$\text{(4.9)} \qquad E \max_{(r,s)\in E_\varkappa} \left|\mathcal{K}_1\left(r, H^{-1}_{\lambda(r,s)}(z)\right)\right|^2 \leqslant 32\varkappa^{-2}(1-\varkappa)^{-1} z(1-z).$$

Finally, if $F\left(H^{-1}_\varkappa(z)\right) \leqslant G\left(H^{-1}_\varkappa(z)\right)$, we interchange in the preceding arguments F by G and $\mathcal{K}_1$ by $\mathcal{K}_2$ and obtain

$$E \max_{(r,s)\in E_\varkappa} \left|\mathcal{K}_1\left(r, H^{-1}_{\lambda(r,s)}(z)\right)\right|^2 \leqslant 32\varkappa^{-2}(1-\varkappa)^{-1} z(1-z),$$

$$\text{(4.10)} \qquad E \max_{(r,s)\in E_\varkappa} \left|\mathcal{K}_2\left(s, H^{-1}_{\lambda(r,s)}(z)\right)\right|^2 \leqslant 32\varkappa^{-2}(1-\varkappa)^{-1} z(1-z) \quad \text{for any } 0<z<1.$$

By our assumption, $(G \circ H^{-1}_{\lambda(r,s)})'$ and $(F \circ H^{-1}_{\lambda(r,s)})'$ are bounded functions if $(r, s) \in E_\varkappa$. From this and (4.10) we obtain finally (if h is non-decreasing) that

$$E \sup_{(r,s)\in E_\varkappa} \left(Z^*(h, r, s)\right)^2$$

$$\leqslant 2 \sup_{(r,s)\in E_\varkappa} \frac{s^2}{(r+s)^2}\left(\int (G \circ H^{-1}_{\lambda(r,s)})'(z) \times \sqrt{32\varkappa^{-2}(1-\varkappa)^{-1}} \sqrt{z(1-z)}\, dh(z) + \right.$$

$$\left. + 2 \sup_{(r,s\in E_\varkappa} \frac{r^2}{(r+s)^2}\left(\int (F \circ H^{-1}_{\lambda(r,s)})'(z) \sqrt{32\varkappa^{-2}(1-\varkappa)^{-1}} \sqrt{z(1-z)}\, dh(z) \leqslant C(\varkappa)\|h\|^2,\right.$$

where $C(\varkappa)$ denotes a constant with a polynomial dependence on $\varkappa^{-1}$.

Clearly, the same inequality holds for general h because of the definition of the norm $\|\ \|$. Choose $m \in N$ such

$$\sum_{k\leqslant m} C(1/k)^{1/2} \leqslant \|h\|^{-1/2} \leqslant 2^{m-1}.$$

Then

$$\left(\mathrm{E}\left(\delta(Z^*(h), 0)\right)^2\right)^{1/2} = \left\{\mathrm{E}\left(\sum_{k=2}^{\infty} \min\left(2^{-k}, \sup_{(r,s)\in E_{\varkappa}} Z^*(h, r, s)\right)\right)^2\right\}^{1/2}$$

$$\leqslant \sum_{k=2}^{m} C(1/k)^{1/2}\|h\| + \sum_{k>m} 2^{-k} \ll \|h\|^{1/2} + 2^{-m+1} \ll 2\|h\|^{1/2},$$

proving (4.6).

5. AN INVARIANCE PRINCIPLE FOR C^2-SCORE

THEOREM 5.1. *Let the score function h of class C^2 satisfy*

(5.1) $$|h''(t)| \leqslant K\left(t(1-t)\right)^{-1+\eta}$$

for some $\eta > 0$ and $K > 0$. Then, the processes $S_N(h)$, $N \geqslant 1$, converge weakly in $D([0, 1]^2)$ with respect to the uniform metric to the process $Z(h)$ defined in (4.1). (Here $Z(h)$ can be considered on $C([0, 1]^2)$).

The proof of this theorem is based on the following lemmas.

LEMMA 5.1. *We have*

(5.2) $$\lim_{N\to\infty} \sup_{\substack{0\leqslant r,s\leqslant 1\\ r\geqslant 1/N}} [rN]\, N^{-1/2}\left|\int_0^1 \{h(H_{\lambda(r,s)}(t)) - h(H_{\lambda_{r,s}}(t))\}\, dF(t)\right| = 0,$$

(5.3)

$$\lim_{N\to\infty} \sup_{\substack{0\leqslant r,s\leqslant 1\\ r\geqslant 1/N}} [rN]\, N^{-1/2}\left|\int_0^1 \{h(H_{\lambda(r,s)}(t)) - h(H_{\lambda_{r,s}}(t))\}\, d(F_{[rN]}(t) - F(t))\right| = 0,$$

where

(5.4) $$\lambda_{r,s} = [rN]/N(r, s).$$

Proof. By the Mean Value Theorem, for some ξ in the interval given by $H_{\lambda(r,s)}(t)$ and $H_{\lambda_{r,s}}(t)$ we have

$$\left|h(H_{\lambda(r,s)}(t)) - h(H_{\lambda_{r,s}}(t))\right| = |h'(\xi)|\,|H_{\lambda(r,s)}(t) - H_{\lambda_{r,s}}(t)|$$

$$\ll |H_{\lambda(r,s)}(t) - H_{\lambda_{r,s}}(t)| \ll \left(\left|\frac{[rN]}{N(r, s)} - \frac{r}{r+s}\right| + \left|\frac{[sN]}{N(r, s)} - \frac{s}{r+s}\right|\right) \ll N(r, s)^{-1}.$$

(Note: $h'(\xi)$ is bounded.)

LEMMA 5.2. *We have*

(5.5) $$\lim_{N\to\infty} \mathrm{E}\sup_{0\leqslant r,s\leqslant 1} \left[[rN]\, N^{-1/2} \frac{r}{r+s} \int_0^1 \{h'(H_{\lambda(r,s)}(t)) - h'(H_{\lambda_{r,s}}(t))\} \times\right.$$

$$\left.\times \{F_{[rN]}(t) - F(t)\}\, dF(t)\right]^2 = 0,$$

$$(5.6)\quad \lim_{N\to\infty} \mathrm{E} \sup_{0\leqslant r,s\leqslant 1} \Bigg[[sN]\, N^{-1/2} \frac{r}{r+s} \int_0^1 \{h'(H_{\lambda(r,s)}(t)) - h'(H_{\lambda_{r,s}}(t))\} \times \times \{G_{[sN]}(t) - G(t)\}\, dF(t) \Bigg]^2 = 0.$$

Proof. For $0 \leqslant t \leqslant 1$, $2/N \leqslant r \leqslant 1$ and $0 \leqslant s \leqslant 1$ there exists some ξ in the interval given by $H_{\lambda(r,s)}(t)$ and $H_{\lambda_{r,s}}(t)$ with

$$|h'(H_{\lambda(r,s)}(t)) - h'(H_{\lambda_{r,s}}(t))| = |h''(\xi)|\, |H_{\lambda(r,s)}(t) - H_{\lambda_{r,s}}(t)|$$
$$\leqslant \frac{2K}{[rN]} \left(\left\{ \frac{r}{r+s} (H_{\lambda(r,s)}(t)(1 - H_{\lambda(r,s)}(t)) \right\}^{-1+\eta} + \right.$$
$$\left. + \frac{[Nr]}{N(r,\, s)} \{H_{\lambda_{r,s}}(t)(1 - H_{\lambda_{r,s}}(t))\}^{-1+\eta} \right).$$

Since

$$\mathrm{E} \sup_{1\leqslant n\leqslant N} (F_n(t) - F(t))^2 \leqslant \sum_{n=1}^{N} n^{-1} F(t)(1 - F(t)) \ll (\log N)\, F(t)(1 - F(t)),$$

it follows for $q = [\log N]^2$ that

$$\mathrm{E} \sup_{\substack{q/N\leqslant r\leqslant 1\\ 0\leqslant s\leqslant 1}} \Bigg[[rN]\, N^{-1/2} \frac{r}{r+s} \int_0^1 \{h'(H_{\lambda(r,s)}(t)) - h'(H_{\lambda_{r,s}}(t))\} \times \times \{F_{[rN]}(t) - F(t)\}\, dF(t) \Bigg]^2$$

$$\leqslant \Big(\int_0^1 \sup([rN]\, N^{-1/2})^2\, |h'(H_{\lambda(r,s)}(u)) - h'(H_{\lambda_{r,s}}(u))| \times \times \{\mathrm{E} \max_{1\leqslant n\leqslant N} |F_n(u) - F(u)|^2\}^{1/2}\, dF(u) \Big)^2$$

$$\ll N^{-1} \log N \Big(\int_0^1 \sup_{r,s} \Big\{ (H_{\lambda(r,s)}(u)(1 - H_{\lambda(r,s)}(u)))^{-1+\eta} \Big(\frac{r}{r+s} F(u)(1 - F(u) \Big) + $$
$$+ (H_{\lambda_{r,s}}(u)(1 - H_{\lambda_{r,s}}(u)))^{-1+\eta} \Big(\frac{[rN]}{N(r,\, s)} F(u)(1 - F(u)) \Big)^{1/2} \Big\}\, dF(u) \Big)^2$$

$$\leqslant N^{-1} \log N \Big(\int_0^1 \sup_{r,s} \{(H_{\lambda(r,s)}(t)(1 - H_{\lambda(r,s)}(t)))^{-1/2+\eta} +$$
$$+ (H_{\lambda_{r,s}}(t)(1 - H_{\lambda_{r,s}}(t)))^{-1/2+\eta}\}\, dF(t) \Big)^2$$

$$\ll N^{-1} \log N \Big(\int_0^1 (t(1-t))^{-1/2+\eta}\, dt \Big)^2 +$$

$$+N^{-1}\log N\left\{\int_0^1\left(H_{\frac{q}{2(N+q)}}(t)\left(1-H_{\frac{q}{2(N+q)}}(t)\right)\right)^{-1/2+\eta}dF(t)\right\}^2$$

$$\ll[\log N]^{-1}\left\{\int_0^1\left(t(1-t)\right)^{-1/2+\eta}dt\right\}^2$$

by the Cauchy–Schwarz inequality.

In the derivation of the above inequalities we have used the following facts:

(i) for $0\leqslant\lambda\leqslant 1$, $H_\lambda(t)\left(1-H_\lambda(t)\right)\geqslant\lambda F(t)\left(1-F(t)\right)$;

(ii) if $q/N\leqslant r\leqslant 1$ and $0\leqslant s\leqslant 1$, then:

(a) for $G(t)\leqslant F(t)$,

$$H_{\frac{q}{(q+N)}}(t)\leqslant H_{\lambda(r,s)}(t)\leqslant F(t)\quad\text{and}\quad H_{\frac{q}{2(q+N)}}(t)\leqslant H_{\lambda_{r,s}}(t)\leqslant F(t),$$

(b) for $G(t)\geqslant F(t)$,

$$F(t)\leqslant H_{\lambda(r,s)}(t)\leqslant H_{\frac{q}{q+N}}(t)\quad\text{and}\quad F(t)\leqslant H_{\lambda_{r,s}}(t)\leqslant H_{\frac{q}{2(q+N)}}(t).$$

The proof of (5.6) is analogous to that of (5.5) and is therefore omitted.

LEMMA 5.3. *If* $F(t)\geqslant G(t)$, *then for any* $r\leqslant r'$ *and* $s'\leqslant s$, $H_{\lambda(r,s)}(t)\leqslant H_{\lambda(r',s')}(t)$, *and if* $F(t)\leqslant G(t)$, *then for any* $r\leqslant r'$ *and* $s'\leqslant s$, $H_{\lambda(r,s)}(t)\geqslant H_{\lambda(r',s')}(t)$.

Since the proof is straightforward, it has been omitted.

LEMMA 5.4 *We have*

$$(5.7)\quad \lim_{N\to\infty}\mathrm{E}\max_{1\leqslant n,m\leqslant N}\Big(N^{-1/2}(n+m)^{-1}\sum_{i=1}^{n}\sum_{j=1}^{m}\Big\{h'\Big(H_{\frac{n}{n+m}}(X_i)\Big)I_{[X_j\leqslant X_i]}-$$
$$-h'\Big(H_{\frac{n}{n+m}}(X_i)\Big)F(X_i)-\int_0^1 h'\Big(H_{\frac{n}{n+m}}(t)\Big)I_{[X_j\leqslant t]}dF(t)+\int_0^1 h'\Big(H_{\frac{n}{n+m}}(t)\Big)F(t)\,dF(t)\Big\}\Big)^2=0,$$

$$(5.8)\quad \lim_{N\to\infty}\mathrm{E}\max_{1\leqslant n,m\leqslant N}\Big(N^{-1/2}(n+m)^{-1}\sum_{i=1}^{n}\sum_{j=1}^{m}\Big\{h'\Big(H_{\frac{n}{n+m}}(X_i)\Big)I_{[Y_j\leqslant X_i]}-$$
$$-h'\Big(H_{\frac{n}{n+m}}(X_i)\Big)G(X_i)-\int_0^1 h'\Big(H_{\frac{n}{n+m}}(t)\Big)I_{[Y_j\leqslant t]}dF(t)+$$
$$+\int_0^1 h'\Big(H_{\frac{n}{n+m}}(t)\Big)G(t)\,dF(t)\Big\}\Big)^2=0.$$

Proof. Write

$$f_{n,m}(u,v)=h'\Big(H_{\frac{n}{n+m}}(u)\Big)I_{[v\leqslant u]}-h'\Big(H_{\frac{n}{n+m}}(u)\Big)F(u)-$$
$$-\int_0^1 h'\Big(H_{\frac{n}{n+m}}(t)\Big)I_{[v\leqslant t]}dF(t)+\int_0^1 h'\Big(H_{\frac{n}{n+m}}(t)\Big)F(t)\,dt$$

and

$$W_{n,m} = \sum_{i \neq j = 1}^{n} f_{n,m}(X_i, X_j).$$

We first estimate $\mathrm{E} \max\limits_{1 \leqslant n,m} N^{-1}(n+m)^{-2} W_{n,m}^2$. Define

$$\omega_{n,m} = W_{n,m} - W_{n-1,m} - W_{n,m-1} + W_{n-1,m-1} \qquad (n, m \geqslant 1),$$

where $W_{-j,k} = W_{j,-k} = 0$ for $j, k \geqslant 0$.

It is easy to check (by induction) that

$$W_{n,m} = \sum_{k=1}^{n} \sum_{j=1}^{m} \omega_{k,j}$$

and, for $1 \leqslant a < b \leqslant N$ and $1 \leqslant u < v \leqslant N$,

$$\begin{aligned}
\sum_{k=a+1}^{b} \sum_{j=u+1}^{v} \omega_{k,j} &= \sum_{k=a+1}^{b} \sum_{j=u+1}^{v} W_{k,j} - \sum_{k=a}^{b-1} \sum_{j=u+1}^{v} W_{k,j} - \\
&\qquad - \sum_{k=a+1}^{b} \sum_{j=u}^{v-1} W_{k,j} + \sum_{k=a}^{b-1} \sum_{j=u}^{v-1} W_{k,j} \\
&= \sum_{j=u+1}^{v} (W_{b,j} - W_{a,j}) + \sum_{j=u}^{v-1} (W_{a,j} - W_{b,j}) \\
&= W_{b,v} - W_{b,u} - W_{a,v} + W_{a,u} \\
&= \sum_{i \neq j=1}^{b} (f_{b,v}(X_i, X_j) - f_{b,u}(X_i, X_j)) - \\
&\qquad - \sum_{i \neq j=1}^{a} (f_{a,v}(X_i, X_j) - f_{a,u}(X_i, X_j)) \\
&= \sum_{i \neq j=1}^{a} (f_{b,v} - f_{b,u} - f_{a,v} + f_{a,u})(X_i, X_j) + \\
&\quad + \sum_{i=a+1}^{b} \sum_{j=1}^{a} (f_{b,v} - f_{b,u})(X_i, X_j) + \sum_{j=a+1}^{b} \sum_{i=1}^{a} (f_{b,v} - f_{b,u})(X_i, X_j) + \\
&\qquad + \sum_{\substack{i=a+1 \\ i \neq j}}^{b} \sum_{j=a+1}^{b} (f_{b,v} - f_{b,u})(X_i, X_j).
\end{aligned}$$

Because of independence, the square integral of the second and third summands is

$$\leqslant a(b-a) \int\!\!\int (f_{b,v} - f_{b,u})^2 (x, y)\, dF(x)\, dF(y) \tag{5.9}$$

and the square integral of the last summand is

$$(5.10) \qquad \ll (b-a)^2 \int\int (f_{b,v}-f_{b,u})^2 (x, y)\, dF(x)\, dF(y).$$

To estimate the double integrals, it suffices to estimate

$$\int\int |h'(H_{\frac{b}{b+v}}(x)) - h'(H_{\frac{b}{b+u}}(x))|^2 (I_{[y\leqslant x]} - F(x))^2\, dF(x)\, dF(y).$$

Considering different cases, it can be shown that

$$|h'(H_{\frac{b}{b+v}}(x)) - h'(H_{\frac{b}{b+u}}(x))|^2 \ll [\{H_{\frac{b}{b+v}}(x)(1-H_{\frac{b}{b+v}}(x))\}^{-2+2\eta} +$$

$$+ \{H_{\frac{b}{b+u}}(x)(1-H_{\frac{b}{b+u}}(x))\}^{-2+2\eta}] \cdot |H_{\frac{b}{b+v}}(x) - H_{\frac{b}{b+u}}(x)|^2$$

$$\leqslant (v-u)^2 [(b+u)^{-2}(b+v)^{-2} b^2 \{H_{\frac{b}{b+v}}(x)(1-H_{\frac{b}{b+v}}(x))\}^{-2+2\eta} +$$

$$+ (b+u)^{-2}(b+v)^{-2} b^2 \{H_{\frac{b}{b+u}}(x)(1-H_{\frac{b}{b+u}}(x))\}^{-2+2\eta}];$$

and since

$$b(b+v)^{-1} F(x)(1-F(x)) \leqslant H_{\frac{b}{b+v}}(x)(1-H_{\frac{b}{b+v}}(x)),$$

and

$$b(b+u)^{-1} F(x)(1-F(x)) \leqslant H_{\frac{b}{b+u}}(x)(1-H_{\frac{b}{b+u}}(x)),$$

we obtain

$$(5.11) \qquad \int\int (f_{bv}-f_{bu})^2 (x, y)\, dF(x)\, dF(y) \leqslant (v-u)^2 (b+u)^{-2} \int [t(1-t)]^{-1+2\eta}\, dt$$

$$\ll (v-u)\, v b^{-1} (b+u)^{-1} \int [t(1-t)]^{-1+2\eta}\, dt.$$

The square integral of the first summand is bounded by

$$a^2 \int\int (f_{bv} - f_{bu} - f_{av} + f_{au})^2 (x, y)\, dF(x)\, dF(y).$$

We now fix x; the it is easy to check that

$$|H_{\frac{b}{b+u}}(x) - H_{\frac{a}{a+u}}(x)| \leqslant (b-a) \frac{u}{(b+u)(a+u)},$$

$$|H_{\frac{b}{b+v}}(x) - H_{\frac{a}{a+v}}(x)| \leqslant (b-a) \frac{v}{(b+v)(a+v)},$$

$$|H_{\frac{b}{b+v}}(x) - H_{\frac{b}{b+u}}(x)| \leqslant (v-u) \frac{b}{(b+v)(b+u)},$$

and

$$|H_{\frac{a}{a+v}}(x) - H_{\frac{a}{a+u}}(x)| \leqslant (v-u) \frac{a}{(a+v)(a+u)}.$$

Thus

$$[(h'(H_{\frac{b}{b+v}}(x))-h'(H_{\frac{b}{b+u}}(x))-h'(H_{\frac{a}{a+v}}(x))+h'(H_{\frac{a}{a+u}}(x))]^2$$

$$\ll (v-u)(b-a)\left[\frac{u}{(b+u)(a+u)}+\frac{v}{(b+v)(a+v)}\right]\left[\frac{a}{(a+v)(a+u)}+\right.$$

$$\left.+\frac{b}{(b+v)(b+u)}\right]\Big(\sum_j (H_j(x)(1-H_j(x))\Big)^{-2+2},$$

where the sum extends over $j=b/(b+u)$ and $a/(a+v)$ (cf. Lemma 5.3). Consequently,

$$a^2\int\int(f_{bv}-f_{bu}-f_{av}+f_{au})^2\,dF(x)\,dF(y)$$

$$\ll (v-u)(b-a)\Big(\int(t(1-t))^{-1+2\eta}\,dt\Big)\left(\max\left(\frac{a+v}{b+u},\frac{b+u}{a+v}\right)+1\right)$$

since

$$a^2\left[\frac{u}{(b+u)(a+u)}+\frac{v}{(b+v)(a+v)}\right]\left[\frac{b}{(b+v)(b+u)}+\frac{a}{(a+v)(a+u)}\right]$$

$$\ll\left(\frac{a}{a+v}\right)^2\left(\frac{a+v}{b+u}+1\right)\text{ and }\ll\left(\frac{b}{b+u}\right)^2\left(\frac{b+u}{a+v}+1\right),$$

we have, e.g.,

$$\frac{b}{b+u}\int(I_{[y\leqslant x]}-F(x))^2\,dF(y)=\frac{b}{b+u}F(x)(1-F(x))$$

and

$$\frac{b}{b+u}F(x)(1-F(x))\leqslant H_{\frac{b}{b+u}}(x)[1-H_{\frac{b}{b+u}}(x)].$$

Together with (5.9), (5.10) and (5.11), we obtain

$$\text{(5.12)}\qquad \mathrm{E}\Big(\sum_{l=a+1}^{b}\sum_{j=u+1}^{v}\omega_{l,j}\Big)^2\ll(b-a)(v-u)\max\left(\frac{a+v}{b+u},\frac{b+u}{a+v}\right).$$

To estimate $\mathrm{E}\max\limits_{\substack{1\leqslant n\leqslant N\\0\leqslant m\leqslant N}} N^{-1}(n+m)^{-2}W_{n,m}^2$ we apply Theorem 8 of Móricz [13].

If $2^{k-1}\leqslant a\leqslant b\leqslant 2^k$ and $2^{j-1}\leqslant u\leqslant v<2^j$, then

$$\max\left(\frac{a+u}{b+u},\frac{b+u}{a+v}\right)\leqslant 4.$$

Thus by (5.12) we have for 2^k, $2^j \leqslant N$,

$$\mathrm{E} \max_{\substack{2^{k-1} \leqslant n < 2^k \\ 2^{j-1} \leqslant m < 2^j}} N^{-1}(n+m)^{-2} W_{n,m}^2 \ll 2^k \cdot 2^j \cdot (2^{k-1}+2^{j-1})^{-2} \cdot N^{-1} \log 2^k \log 2^j$$
$$\ll N^{-1}(\log N)^2.$$

Hence

$$\mathrm{E} \max_{\substack{1 \leqslant n \leqslant N \\ 0 \leqslant m \leqslant N}} N^{-1}(n+m)^{-2} W_{n,m}^2 \ll N^{-1}(\log N)^4. \tag{5.13}$$

To prove (5.7), it remains to estimate

$$\mathrm{E} \max_{\substack{1 \leqslant n \leqslant N \\ 0 \leqslant m \leqslant N}} \Big(N^{-1/2}(n+m)^{-1} \sum_{i=1}^{n} \{h'(H_{\frac{n}{n+m}}(X_i))(1-F(X_i)) -$$
$$- \int h'(H_{\frac{n}{n+m}}(t)) I_{[X_i \leqslant t]} dF(t) + \int h'(H_{\frac{n}{n+m}}(t)) F(t) dF(t)\}\Big)^2.$$

But, since h' is bounded, this maximum is clearly bounded by N^{-1}, proving (5.7). The proof of (5.8) is similar to that of (5.13) and is therefore omitted.

LEMMA 5.5. *We have*

$$\lim_{N \to \infty} \sup_{r,s \in [0,1]} [rN] N^{-1/2} \int_0^1 h''(\xi(t)) \left(\frac{N(r,s)}{N(r,s)+1} \hat{H}_{[rN],[sN]}(t) - H_{\lambda_{r,s}}(t)\right)^2 dF_{[rN]}(t)$$
$$= 0 \text{ in probability,}$$

where $\xi(t)$ is any point in the interval given by $H_{\lambda_{r,s}}(t)$ and $\dfrac{N(r,s)}{N(r,s)+1} \hat{H}_{[rN],[sN]}(t)$.

Proof. Using (5.1), we have

$$\left|\int_0^1 h''(\xi(t)) \left(\frac{N(r,s)}{N(r,s)+1} \hat{H}_{[rN],[sN]}(t) - H_{\lambda_{r,s}}(t)\right)^2 dF_{[rN]}(t)\right| \tag{5.14}$$
$$\ll \int_0^1 \Bigg[\left(H_{\lambda_{r,s}}(t)(1-H_{\lambda_{r,s}}(t))\right)^{-1+\eta} + \left(\frac{N(r,s)}{N(r,s)+1} \hat{H}_{[rN],[sN]}(t) \times\right.$$
$$\left.\times\left(1-\frac{N(r,s)}{N(r,s)+1} \hat{H}_{[rN],[sN]}(t)\right)\right)^{1+2\eta}\Bigg] \left(\frac{N(r,s)}{N(r,s)+1} \hat{H}_{[rN],[sN]}(t) - H_{\lambda_{r,s}}(t)\right)^2 dF_{[rN]}(t)$$

Next,

$$\int_0^1 \left(\frac{N(r,s)}{N(r,s)+1} \hat{H}_{[rN],[sN]}(t) \left(1-\frac{N(r,s)}{N(r,s)+1} \hat{H}_{[rN],[sN]}(t)\right)\right)^{-1+\eta} \times \tag{5.15}$$

$$\times\left(\frac{N(r,s)}{N(r,s)+1}\hat{H}_{[rN],[sN]}(t)-H_{\lambda_{r,s}}(t)\right)^2 dF_{[rN]}(t)$$

$$\leqslant \sup_t\left(\frac{N(r,s)}{N(r,s)+1}\hat{H}_{[rN],[sN]}(t)-H_{\lambda_{r,s}}(t)\right)^2\int\left(\frac{N(r,s)}{N(r,s)+1}\hat{H}_{[rN],[sN]}(t)\times\right.$$

$$\left.\times\left(1-\frac{N(r,s)}{N(r,s)+1}\hat{H}_{[rN],[sN]}(t)\right)\right)^{-1+\eta} dF_{[rN]}(t)$$

$$\leqslant \sup_t\left(\frac{N(r,s)}{N(r,s)+1}\hat{H}_{[rN],[sN]}(t)-H_{\lambda_{r,s}}(t)\right)^2\frac{N(r,s)+1}{[rN]}\int_0^1\left(t(1-t)\right)^{-1+\eta}dt.$$

Now, from Theorem 4.1 in [7], we have for each r and s

$$\mathrm{E}\left(\sqrt{N(r,s)}\int_0^1 h''(\xi(t))\left(\frac{N(r,s)}{N(r,s)+1}\hat{H}_{[rN],[sN]}(t)-H_{\lambda_{r,s}}(t)\right)^2 dF_{[rN]}(t)\right)^2\ll\frac{N(r,s)}{[rN]^2}.$$

Consequently,

$$\text{(5.16)}\quad \mathrm{E}\max_{\substack{1\leqslant[rN]\leqslant N^{1/2}/\log N\\ 0\leqslant[sN]\leqslant N^{1/2}/\log N}}\left([rN]\,N^{-1/2}\int_0^1 h''(\xi(t))\left(\frac{N(r,s)}{N(r,s)+1}\hat{H}_{[rN],[sN]}(t)-\right.\right.$$

$$\left.\left.-H_{\lambda_{r,s}}(t)\right)^2 dF_{[rN]}(t)\right)^2\ll(\log N)^{-2}.$$

Now let $\varepsilon>0$ be fixed. Then, using (5.16), we obtain

$$\text{(5.17)}\quad \mathrm{P}\left(\sup_{\substack{1\leqslant n\leqslant N\\ 0\leqslant m\leqslant N}} nN^{-1/2}\int_0^1|h''(\xi(t))|\left(\frac{n+m}{n+m+1}\hat{H}_{n,m}(t)-H_{\frac{n}{n+m}}(t)\right)^2 dF_n(t)\geqslant\varepsilon\right)$$

$$\ll\varepsilon^{-2}(\log N)^{-2}+\mathrm{P}\Big(\sup_{\substack{N^{1/2}/\log N\leqslant n\leqslant N\\ \text{or } N^{1/2}/\log N\leqslant m\leqslant N}} nN^{-1/2}\int_0^1|h''(\xi)|\times$$

$$\times\left(\frac{n+m}{n+m+1}\hat{H}_{n,m}(t)-H_{\frac{n}{n+m}}(t)\right)^2 dF_n(t)\geqslant\varepsilon\Big).$$

Now let us fix intervals $[2^{k-1},2^k)$ and $[2^{j-1},2^j)$. Then, using Lemma 5.3, we have for each $2^{k-1}\leqslant n<2^k$ and $2^{j-1}\leqslant m<2^j$,

$$\left(H_{\frac{n}{n+m}}(t)\left(1-H_{\frac{n}{n+m}}(t)\right)\right)^{-1+\eta}\leqslant\left(H_{\frac{2^k}{2^k+2^{j-1}}}(t)\left(1-H_{\frac{2^k}{2^k+2^{j-1}}}(t)\right)\right)^{-1+\eta}+$$

$$+\left(H_{\frac{2^{k-1}}{2^{k-1}+2^j}}(t)\left(1-H_{\frac{2^{k-1}}{2^{k-1}+2^j}}(t)\right)\right)^{-1+\eta}$$

and so,

$$\text{(5.18)}\quad \mathrm{E}\max_{\substack{2^{k-1}\leqslant n\leqslant 2^k\\ 2^{j-1}\leqslant m\leqslant 2^j}}\sum_{i=1}^n\left(H_{\frac{n}{n+m}}(X_i)\left(1-H_{\frac{n}{n+m}}(X_i)\right)\right)^{-1+\eta}$$

$$\leqslant \mathrm{E} \sum_{i=1}^{2^k} \left(H_{\frac{2^k}{2^k+2^j-1}}(X_i)\left(1-H_{\frac{2^k}{2^k+2^j-1}}(X_i)\right)\right)^{-1+\eta} +$$

$$+\mathrm{E} \sum_{i=1}^{2^k} \left(H_{\frac{2^{k-1}}{2^{k-1}+2^j}}(X_i)\left(1-H_{\frac{2^{k-1}}{2^{k-1}+2^j}}(X_i)\right)\right)^{-1+\eta}$$

$$\ll (2^k+2^j)\int \left(t(1-t)\right)^{-1+\eta} dt.$$

Let

$$B_{kj}(N) = \Big\{ \max_{\substack{2^{k-1}\leqslant n<2^k \\ 2^{j-1}\leqslant m<2^j}} \sum_{i=1}^{n} \left(H_{\frac{n}{n+m}}(X_i)\left(1-H_{\frac{n}{n+m}}(X_i)\right)\right)^{-1+\eta} \geqslant (2^k+2^j)(\log N)^3 \Big\}.$$

Then, using formula (5.18) and Markov's inequality, $\sum_{k,j\leqslant \log N} \mathrm{P}(B_{k,j}(N)) \ll (\log N)^{-1}$ and on $\bigcap_{k,j}(B_{kj}(N))^c$, we have for each n, m

$$n\int \left(H_{\frac{n}{n+m}}(t)\left(1-H_{\frac{n}{n+m}}(t)\right)\right)^{-1+\eta} dF_n(t) \leqslant 2(n+m)(\log N)^3. \tag{5.19}$$

Now let

$$A_N = \Big\{ \sup_{N^{1/4}\leqslant n\leqslant N} \sup_t n^{3/8}|F_n(t)-F(t)| < 1 \text{ and } \sup_{N^{1/4}\leqslant m\leqslant N} \sup_t m^{3/8}|G_m(t)-G(t)| < 1 \Big\}.$$

Then, by the law of the iterated logarithm for the empirical processes, we have $\lim_{N\to\infty} \mathrm{P}(A_N^c) = 0$ and on A_N we have for $n \geqslant N^{1/2}/\log N$ or for $m \geqslant N^{1/2}/\log N$,

$$\left|\frac{n+m}{n+m+1}\hat{H}_{n,m}(t) - H_{\frac{n}{n+m}}(t)\right| \tag{5.20}$$

$$\leqslant \frac{1}{n+m} + \left|\frac{n}{n+m}(F_n(t)-F(t)) + \frac{m}{n+m}(G_n(t)-G(t))\right|$$

$$\leqslant (n+m)^{-1}\left[1+n^{5/8}+m^{5/8}+N^{1/4}\right].$$

Using (5.14), (5.15), (5.19), and (5.20), we have on $A_N \cap [\bigcap_{k,j} B_{k,j}(N)^c]$ for $n \geqslant N^{1/2}/\log N$ or $m \geqslant N^{1/2}/\log N$,

$$\left| nN^{-1/2}\int_0^1 h''(\xi(t))\left(\frac{n+m}{n+m+1}\hat{H}_{n,m}(t) - H_{\frac{n}{n+m}}(t)\right)^2 dF_n(t)\right|$$

$$\ll \frac{n+m+1}{\sqrt{N}}(n+m)^{-2}(1+n^{5/8}+m^{5/8}+N^{1/4})^2 +$$

$$+N^{-1/2}(n+m)^{-2}(1+n^{5/8}+m^{5/8}+N^{1/4})^2(n+m)(\log N)^3 \ll N^{-1/4}(\log N)^3.$$

Thus, with N sufficiently large, we have

$$\left| \sup_{\substack{N^{1/2}/\log N \leqslant n \leqslant N \\ \text{or } N^{1/2}/\log N \leqslant m \leqslant N}} \frac{n}{\sqrt{N}} \int_0^1 h''(\xi(t)) \left(\frac{n+m}{n+m+1} \hat{H}_{n,m}(t) - H_{\frac{n}{n+m}}(t) \right)^2 dF_n(t) \right| < \varepsilon$$

on the set $A_N \cap [\bigcap_{k,j} B_{k,j}(N)^c]$. Since the measure of its complement tends to zero, (5.17) tends to zero as $N \to \infty$. This completes the proof of the Lemma.

LEMMA 5.6. *We have*

$$\lim_{N\to\infty} \mathrm{E} \max_{0 \leqslant r,s \leqslant 1} \left(\frac{[rN]\, N^{-1/2}}{[rN]+[sN]+1} \int_0^1 h'(H_{\lambda_{r,s}}(t))\, \hat{H}_{[rN],[sN]}(t)\, dF_{[rN]}(t) \right)^2 = 0.$$

Proof. Since h' is bounded, the proof follows trivially.

Proof of Theorem 5.1. Since h is of class C_2, we obtain from Taylor's theorem

$$S_N(h, r, s) = [rN]\, N^{-1/2} \left[\int_0^1 h\left(\frac{N(r, s)}{N(r, s)+1} \hat{H}_{[rN],[sN]}(t) \right) dF_{[rN]}(t) - \right.$$

$$\left. - \int_0^1 h(H_{\lambda_{r,s}}(t))\, dF(t) \right]$$

$$= [rN]\, N^{-1/2} \int_0^1 h(H_{\lambda_{r,s}}(t))\, d(F_{[rN]}(t) - F(t)) +$$

$$+ [rN]\, N^{-1/2} \int_0^1 h'(H_{\lambda_{r,s}}(t)) \left(\frac{N(r, s)}{N(r, s)+1} \hat{H}_{[rN],[sN]}(t) - H_{\lambda_{r,s}}(t) \right) \times$$

$$\times dF_{[rN]}(t) +$$

$$+ \frac{1}{2} [rN]\, N^{-1/2} \int_0^1 h''(\xi(t)) \left(\frac{N(r, s)}{N(r, s)+1} \hat{H}_{[rN],[sN]}(t) - H_{\lambda_{r,s}}(t) \right)^2 \times$$

$$\times dF_{[rN]}(t),$$

where $\xi(t)$ belongs to the interval given by $H_{\lambda_{r,s}}(t)$ and $\dfrac{N(r, s)}{N(r, s)+1} \hat{H}_{[rN],[sN]}(t)$.

Making some routine computations, we obtain

$$S_N(h, r, s) = \sum_{i=1}^{11} E_{Ni}(r, s),$$

where

$$E_{N1}(r, s) = \frac{r}{r+s} \int_0^1 h'(H_{\lambda(r,s)}(t))\, [rN]\, N^{-1/2} (F_{[rN]}(t) - F(t))\, dF(t) -$$

$$-\int_0^1 [rN]\, N^{-1/2}\left(F_{[rN]}\left(H^{-1}_{\lambda(r,s)}(t)\right)-F\left(H^{-1}_{\lambda(r,s)}(t)\right)\right)dh(t),$$

$$E_{N2}(r,s)=\frac{r}{r+s}\int_0^1 h'(H_{\lambda(r,s)}(t))\,[sN]\,N^{-1/2}(G_{[sN]}(t)-G(t))\,dF(t),$$

$$E_{N3}(r,s)=\frac{1}{2}[rN]\,N^{-1/2}\int_0^1 h''(\xi(t))\left(\frac{N(r,s)}{N(r,s)+1}\hat{H}_{[rN],[sN]}(t)-\right.$$
$$\left.-H_{\lambda_{r,s}}(t)\right)^2 dF_{[rN]}(t),$$

$$E_{N4}(r,s)=[rN]\,N^{-1/2}\int_0^1 h'(H_{\lambda_{r,s}}(t))\frac{1}{N(r,s)+1}\hat{H}_{[rN],[sN]}(t)\,dF_{[rN]}(t),$$

$$E_{N5}(r,s)=[rN]\,N^{-1/2}\int_0^1 h'(H_{\lambda_{r,s}}(t))\frac{[rN]}{N(r,s)}(F_{[rN]}(t)-F(t))\,d(F_{[rN]}(t)-F(t)),$$

$$E_{N6}(r,s)=[rN]\,N^{-1/2}\int_0^1 h'(H_{\lambda_{r,s}}(t))\frac{[sN]}{N(r,s)}(G_{[sN]}(t)-G(t))\,d(F_{[rN]}(t)-F(t)),$$

$$E_{N7}(r,s)=[rN]\,N^{-1/2}\int_0^1 \left(h(H_{\lambda_{r,s}}(t))-h(H_{\lambda(r,s)}(t))\right)d(F_{[rN]}(t)-F(t)),$$

$$E_{N8}(r,s)=[rN]\,N^{-1/2}\int_0^1 \left(h'(H_{\lambda_{r,s}}(t))-h'(H_{\lambda(r,s)}(t))\right)\frac{r}{r+s}(F_{[rN]}(t)-$$
$$-F(t))\,dF(t),$$

$$E_{N9}(r,s)=[sN]\,N^{-1/2}\int_0^1 \left(h'(H_{\lambda_{r,s}}(t))-h'(H_{\lambda(r,s)}(t))\right)\frac{r}{r+s}(G_{[sN]}(t)-$$
$$-G(t))\,dF(t),$$

$$E_{N10}(r,s)=[rN]\,N^{-1/2}\left(\frac{[rN]}{N(r,s)}-\frac{r}{r+s}\right)\int_0^1 h'(H_{\lambda_{r,s}}(t))(F_{[rN]}(t)-F(t))\,dF(t),$$

and

$$E_{N11}(r,s)=[sN]\,N^{-1/2}\left(\frac{[rN]}{N(r,s)}-\frac{r}{r+s}\right)\int_0^1 h'(H_{\lambda_{r,s}}(t))(G_{[sN]}(t)-G(t))\,dF(t).$$

Now, using Lemmas 5.5, 5.6, 5.4, 5.1, and 5.2, we notice that $E_{Ni}\to 0$ in probability with respect to the uniform metric for $i=3$ to 9. Furthermore, since

$$[rN]\,N^{-1/2}\left|\frac{[rN]}{N(r,s)}-\frac{r}{r+s}\right|=O(N^{-1/2}),$$

it follows that E_{N10} as well as $E_{N11}\to 0$ in probability with respect to the uniform metric. Thus the weak convergence of $S_N(h)$ will follow from that of $E_{N1}+E_{N2}$.

Since E_{N1} and E_{N2} are independent random functions, we have to show that E_{N1} and E_{N2} converge to the corresponding integrals replacing the empirical processes by independent Kiefer processes. We consider only $E_{N1}(r, s)$ because for $E_{N2}(r, s)$ the same argument applies.

Consider the map ψ: $D([0, 1]^2) \to D([0, 1]^2)$ defined by

$$\psi[f](r, s) = \frac{r}{r+s}\int_0^1 h'(H_{\lambda(r,s)}(t))\, f(r, t)\, dF(t) - \int_0^1 f(r, H^{-1}_{\lambda(r,s)}(t))\, dh(t).$$

If f is a continuous function, then clearly ψ is continuous at f with respect to the uniform metric on $D([0, 1]^2)$ because $d|h|$ is a finite measure.

Let $\mathscr{K}$ be a Kiefer process on $C([0, 1]^2)$ with covariance $E\mathscr{K}(u, v) = 0$ and $E\mathscr{K}(u, v)\mathscr{K}(u', v') = \min(u, u')F(v)(1-F(v'))$ for $0 \leqslant u,\ u' \leqslant 1$ and $0 \leqslant v \leqslant v' \leqslant 1$.

Since $\mathscr{K}$ has continuous paths and since $[rN]N^{-1/2}[F_{[rN]}(t)-F(t)]$ converges weakly in $D([0, 1]^2)$ to $\mathscr{K}(r, t)$ with respect to the uniform topology, Theorem 5.1 of Billingsley [2] shows that $\psi[([rN]N^{-1/2}(F_{[rN]}(t) - F(t))]$ converges weakly in $D([0, 1]^2)$ to $\psi(\mathscr{K})$ with respect to the uniform topology:

$$\begin{aligned}\psi[\mathscr{K}](r, s) &= \frac{r}{r+s}\int_0^1 h'(H_{\lambda(r,s)}(t))\, \mathscr{K}(r, t)\, dF(t) - \int_0^1 \mathscr{K}(r, H^{-1}_{\lambda(r,s)}(t))\, dh(t) \\ &= -\frac{s}{r+s}\int_0^1 \frac{dG}{dH_{\lambda(r,s)}}(t)\, \mathscr{K}(r, t)\, dh(H_{\lambda(r,s)}(t)).\end{aligned}$$

COROLLARY 5.1. *Set*

$$\bar{S}_N(h, r, s) = \begin{cases} 0 & \text{if } r < 1/N, \\ [rN]N^{-1/2}\left[\int_0^1 h\left(\dfrac{N(r, s)}{N(r, s)+1}\hat{H}_{[rN],[sN]}(t)\right)dF_{[rN]}(t) - \right. \\ \left. \qquad - \int_0^1 h(H_{\lambda(r,s)}(t))\, dF(t)\right] & \text{otherwise.}\end{cases} \tag{5.21}$$

Then $\bar{S}_N(h)$ converges weakly in $D([0, 1]^2)$ with respect to the uniform metric to the process $Z(h)$ defined in (4.1).

COROLLARY 5.2. *Let the scores $a_N(i, n, m)$ satisfy* (2.3), *where h satisfies* (5.1). *Then the statistics*

$$\hat{S}_N(h, r, s) = \begin{cases} 0 & \text{if } r < 1/N, \\ [rN]N^{-1/2}(T^*_{[rN],[sN]} - E\,T^*_{[rN],[sN]}) & \text{otherwise,}\end{cases} \tag{5.22}$$

where T^* is defined in (2.4), converge weakly in $D(E)$ with respect to the uniform metric to $Z(h)$.

Proof. Set

$$h_{N,n,m}(t) = a_N(i, n, m) \quad \text{if } t \in \left[\frac{i}{n+m+1}, \frac{i+1}{n+m+1}\right), \ i = 0, \ldots, n+m.$$

Then, by (2.3) and the definition of the Riemann integral, it follows that

$$[rN]\, N^{-1/2} \int_0^1 [h(H_{\lambda(r,s)}(t)) - h_{N,n,m}(H_{\lambda(r,s)}(t))]\, dF(t) \to 0$$

uniformly in r and s. Similarly,

$$[rN]\, N^{-1/2} \left(T^*_{[rN],[sN]} - \int_0^1 h\left(\frac{N(r, s)}{N(r, s)+1} \hat{H}_{[rN],[sN]}(t)\right) dF_{[rN]}(t)\right) \to 0$$

uniformly in r and s. The proof follows.

Replacing, in (5.22), $\mathrm{E}T^*_{[rN],[sN]}$ by $\int h_{N(r,s),[rN],[sN]}(H_{\lambda(r,s)}(t))\, dF(t)$, the convergence of $\hat{S}_N(h)$ to $Z(h)$ is again in $D([0, 1]^2)$.

6. THE CONTINUITY THEOREM FOR S_N

The random functions $S_N(h, \cdot, \cdot)$ define operators $S_N\colon \mathscr{H} \to L_2(P, D(E_\varkappa))$ for every $N \geqslant 1$ and $0 < \varkappa < 1/2$. We shall show in this section that the family $\{S_N\colon N \geqslant 1\}$ is uniformly continuous for fixed $\varkappa$. This result will provide an easy argument to extend invariance principles for $S_N(h)$ from nice score functions h to more complicated ones.

THEOREM 6.1. *Let $0 < \varkappa < 1/2$. Then there exists a constant $C(\varkappa)$ such that for every score function $h \in \mathscr{H}$ and every $N \geqslant 1$*

$$\mathrm{E} \max_{(r,s)\in E_\varkappa} (S_N(h, r, s))^2 \leqslant C(\varkappa)\|h\|^2. \tag{6.1}$$

Remark. This theorem is an extension of a theorem of Denker and Rösler [5], which says that $\mathrm{E}(S_N(h, r, s))^2 \leqslant \text{const}\cdot\|h\|^2$ for every fixed r, $s \in (0, 1)$.

The theorem will be proven by a series of lemmas. If $r < 1/N$, then by definition, $S_N(h, r, s) = 0$ for every s. Also, if $s < 1/N$, then

$$S_{N(h,r,s)} = N^{-1/2}[rN]\left(\frac{1}{[rN]}\sum_{i=1}^{[rN]} h\left(\frac{i}{[rN]+1}\right) - \int_0^1 h(t)\, dt\right) \to 0$$

uniformly in $r \in (0, 1)$ (and $0 < s < 1/N$), since $h \in \mathscr{H}$. Therefore, it suffices to show (6.1) in the case where $r \geqslant 1/N$ and $s \geqslant 1/N$. We shall assume in the following lemmas that $r, s \geqslant 1/N$; especially $(r, s) \in E_\varkappa$ means, more precisely, that $(r, s) \in E_\varkappa$ and $r, s \geqslant 1/N$.

We begin with an easy observation (cf. Lemma 5.3):

LEMMA 6.1. *If* $(r, s) \in E_\varkappa$, *then*

$$H_{\frac{\varkappa}{2+\varkappa}}(t) \leqslant H_{\lambda_{r,s}}(t) \leqslant H_{\frac{2}{2+\varkappa}}(t) \quad \text{if } G(t) \leqslant F(t) \tag{6.2}$$

and

$$H_{\frac{2}{2+\varkappa}}(t) \leqslant H_{\lambda_{r,s}}(t) \leqslant H_{\frac{\varkappa}{2+\varkappa}}(t) \quad \text{if } F(t) \leqslant G(t), \tag{6.3}$$

where, as before, $\lambda_{r,s} = [rN]/N(r,; s)$, $N(r, s) = [rN]+[sN]$.

Proof. Since $\varkappa \leqslant r/(r+s) \leqslant 1-\varkappa$,

$$\frac{[rN]+[sN]}{[rN]} = 1+\frac{[sN]}{[rN]} \leqslant 1+\frac{s}{r}\left(1+\frac{1}{[rN]}\right) \leqslant 2\varkappa^{-1}.$$

Using this estimate, where r and s are exchanged, we also have

$$\frac{[rN]}{[rN]+[sN]} = 1-\frac{[sN]}{[rN]+[sN]} \leqslant 1-\varkappa/2$$

and it follows that

$$\varkappa/2 \leqslant \frac{[rN]}{[rN]+[sN]} \leqslant 1-\varkappa/2. \tag{6.4}$$

Lemma 5.3 states that $F(t) \geqslant G(t)$, $r \leqslant r'$, $s' \leqslant s \Rightarrow H_{r/(r+s)}(t) \leqslant H_{r'/(r'+s')}(t)$ and $F(t) \leqslant G(t)$, $r \leqslant r'$, $s' \leqslant s \Rightarrow H_{r/(r+s)}(t) \geqslant H_{r'/(r'+s')}(t)$.

Assume now that $F(t) \geqslant G(t)$ (the case $F(t) \leqslant G(t)$ is similar). Choose $s' = \varkappa/2$, $r' = 1$. By (6.4), $[sN]/([rN]+[sN]) \geqslant 1-(1-\varkappa/2) = \varkappa/2$ and, consequently, by Lemma 5.3 we have

$$H_{\lambda_{r,s}}(t) \leqslant H_{\frac{r'}{r'+s'}}(t) = H_{\frac{2}{2+\varkappa}}(t).$$

On the other hand, choosing $s = 1$ and $r = \varkappa/2$, we obtain $H_{\varkappa/(2+\varkappa)}(t) \leqslant H_{\lambda_{r',s'}}(t)$.

LEMMA 6.2. *For every* $0 < z < 1$ *and every* $0 < \varkappa < {}^1/_2$, *we have*

$$\max_{(r,s)\in E_\varkappa} \left|F_{[rN]}(H^{-1}_{\lambda_{r,s}}(z)) - F(H^{-1}_{\lambda_{r,s}}(z))\right| \leqslant \max_{0<r\leqslant 1} \max_{y\in I_\varkappa(z)} |F_{[rN]}(y)-F(y)| \tag{6.5}$$

and

$$\max_{(r,s)\in E_\varkappa} \left|G_{[sN]}(H^{-1}_{\lambda_{r,s}}(z)) - G(H^{-1}_{\lambda_{r,s}}(z))\right| \leqslant \max_{0<s\leqslant 1} \max_{y\in I_\varkappa(z)} |G_{[sN]}(y)-G(y)|, \tag{6.6}$$

where

$$I_\varkappa(z) = \left\{y\colon \min\left\{H^{-1}_{\frac{2}{2+\varkappa}}(z), H^{-1}_{\frac{\varkappa}{2+\varkappa}}(z)\right\} \leqslant y \leqslant \max\left\{H^{-1}_{\frac{2}{2+\varkappa}}(z), H^{-1}_{\frac{\varkappa}{2+\varkappa}}(z)\right\}\right\}.$$

Proof. Let $F(H^{-1}_{\varkappa/(2+\varkappa)}(z)) \geqslant G(H^{-1}_{\varkappa/(2+\varkappa)}(z))$. Then, by Lemma 6.1, $H_{\lambda_{r,s}}(H^{-1}_{\varkappa/(2+\varkappa)}(z)) \geqslant z$, i.e. $H^{-1}_{\varkappa/(2+\varkappa)}(z) \geqslant H^{-1}_{\lambda_{r,s}}(z)$. Since $H_{\lambda_{r,s}}(H^{-1}_{\lambda_{r,s}}(z)) \geqslant H_{\varkappa/(2+\varkappa)}(H^{-1}_{\lambda_{r,s}}(z))$, we have

$$\left(\frac{[rN]}{[rN]+[sN]}-\frac{\varkappa}{2+\varkappa}\right)F(H^{-1}_{\lambda_{r,s}}(z)) \geqslant \left(\frac{2}{2+\varkappa}-\frac{[sN]}{[rN]+[sN]}\right)G(H^{-1}_{\lambda_{r,s}}(z));$$

equivalently, $F(H^{-1}_{\lambda_{r,s}}(z)) \geqslant G(H^{-1}_{\lambda_{r,s}}(z))$, since by (6.4),

$$\frac{[rN]}{[rN]+[sN]}-\frac{\varkappa}{2+\varkappa} > 0.$$

It follows (by Lemma 6.1) that $H_{2/(2+\varkappa)}(H^{-1}_{\lambda_{r,s}}(z)) \geqslant z$ and $H^{-1}_{\lambda_{r,s}}(z) \geqslant H^{-1}_{2/(2+\varkappa)}(z)$.

LEMMA 6.3. *For every $0 < \varkappa < 1/2$ there exists a constant $C_0(\varkappa)$ such that, for every $0 < z < 1$,*

$$\text{(6.7)} \qquad \mathrm{E} \max_{(r,s)\in E_\varkappa} [Nr]^2 \left(F_{[Nr]}(H^{-1}_{\lambda_{r,s}}(z)) - F(H^{-1}_{\lambda_{r,s}}(z))\right)^2 \leqslant C_0(\varkappa)Nz(1-z)$$

and

$$\text{(6.8)} \qquad \mathrm{E} \max_{(r,s)\in E_\varkappa} [Ns]^2 \left(G_{[Ns]}(H^{-1}_{\lambda_{r,s}}(z)) - G(H^{-1}_{\lambda_{r,s}}(z))\right)^2 \leqslant C_0(\varkappa)Nz(1-z).$$

Proof. Assume that $G(H^{-1}_{\varkappa/(2+\varkappa)}(z)) \leqslant F(H^{-1}_{\varkappa/(2+\varkappa)}(z))$. We shall show that (6.7) and (6.8) hold in this case. (The other case will follow analogously). Under this assumption, we have $I_\varkappa(z) = [H^{-1}_{2/(2+\varkappa)}(z), H^{-1}_{\varkappa/(2+\varkappa)}(z)]$ (cf. Lemma 6.2) and so

$$\text{(6.9)} \qquad \mathrm{E} \max_{(r,s)\in E_\varkappa} [Ns]^2 \left(G_{[Ns]}(H^{-1}_{\lambda_{r,s}}(z)) - G(H^{-1}_{\lambda_{r,s}}(z))\right)^2 \leqslant \mathrm{E} \max_{1\leqslant k\leqslant N} \max_{y\in I_\varkappa(z)} k^2 \left(G_k(y)-G(y)\right)^2.$$

Since $\{\max_{y\in I_\varkappa(z)} k^2(G_k(y)-G(y))^2 : 1 \leqslant k \leqslant N\}$ is a nonnegative submartingale, it follows from Doob's inequality ([8], p. 314) that

$$\text{(6.10)} \qquad \mathrm{E} \max_{1\leqslant k\leqslant N} \max_{y\in I_\varkappa(z)} k^2 (G_k(y)-G(y))^2 \leqslant 4\,\mathrm{E} \max_{y\in I_\varkappa(z)} N^2 (G_N(y)-G(y))^2.$$

Moreover, since

$$\left\{N\frac{G_N(y)-G(y)}{1-G(y)} : H^{-1}_{\frac{2}{2+\varkappa}}(z) \leqslant y \leqslant H^{-1}_{\frac{\varkappa}{2+\varkappa}}(z)\right\}$$

is a martingale, Doob's inequality again implies that

$$\mathrm{E} \max_{y\in I_\varkappa(z)} N^2(G_N(y)-G(y))^2 \leqslant \mathrm{E} \max_{y\in I_\varkappa(z)} N^2 \left(\frac{G_N(y)-G(y)}{1-G(y)}\right)^2 \left(1-G(H^{-1}_{\frac{2}{2+\varkappa}}(z))\right)^2$$

$$\leqslant 4N^2\,\mathrm{E}\left(G_N\left(H^{-1}_{\frac{\varkappa}{2+\varkappa}}(z)\right)-G\left(H^{-1}_{\frac{\varkappa}{2+\varkappa}}(z)\right)\right)^2\left(\frac{1-G\left(H^{-1}_{2/(2+\varkappa)}(z)\right)}{1-G\left(H^{-1}_{\varkappa/(2+\varkappa)}(z)\right)}\right)^2$$

$$=4NG\left(H^{-1}_{\frac{\varkappa}{2+\varkappa}}(z)\right)\left(1-G\left(H^{-1}_{\frac{\varkappa}{2+\varkappa}}(z)\right)\right)\left(\frac{1-G\left(H^{-1}_{2/(2+\varkappa)}(z)\right)}{1-G\left(H^{-1}_{\varkappa/(2+\varkappa)}(z)\right)}\right)^2.$$

Note that $1-G\left(H^{-1}_{\frac{\varkappa}{2+\varkappa}}(z)\right)\neq 0$, since otherwise

$$z=\frac{\varkappa}{2+\varkappa}F\left(H^{-1}_{\frac{\varkappa}{2+\varkappa}}(z)\right)+\frac{2}{2+\varkappa}G\left(H^{-1}_{\frac{\varkappa}{2+\varkappa}}(z)\right)\geqslant G\left(H^{-1}_{\frac{\varkappa}{2+\varkappa}}(z)\right)=1$$

because of $F\left(H^{-1}_{\varkappa/(2+\varkappa)}(z)\right)\geqslant G\left(H^{-1}_{\varkappa/(2+\varkappa)}(z)\right)$.

Using

$$H_{\frac{\varkappa}{2+\varkappa}}(y)\left(1-H_{\frac{\varkappa}{2+\varkappa}}(y)\right)\geqslant\frac{2}{2+\varkappa}G(y)(1-G(y)),$$

$$1-z=1-\frac{\varkappa}{2+\varkappa}F\left(H^{-1}_{\frac{\varkappa}{2+\varkappa}}(z)\right)-\frac{2}{2+\varkappa}G\left(H^{-1}_{\frac{\varkappa}{2+\varkappa}}(z)\right)\leqslant 1-G\left(H^{-1}_{\frac{\varkappa}{2+\varkappa}}(z)\right),$$

and

$$\left(1-G\left(H^{-1}_{\frac{2}{2+\varkappa}}(z)\right)\right)\frac{\varkappa}{2+\varkappa}\leqslant 1-z,$$

we note the left-hand side of (6.8) is bounded by

$$16N\frac{2+\varkappa}{2}z(1-z)\left(\frac{\frac{2+\varkappa}{\varkappa}(1-z)}{1-z}\right)^2=8\frac{(2+\varkappa)^3}{\varkappa^2}Nz(1-z).$$

The proof of (6.7) is more involved. Assume first that also $G\left(H^{-1}_{2/(2+\varkappa)}(z)\right)\leqslant F\left(H^{-1}_{2/(2+\varkappa)}(z)\right)$. Formula (6.10) remains valid replacing G by F. Then use the fact that

$$\left\{N\frac{1-F_N(y)-(1-F(y))}{F(y)}:\ H^{-1}_{\frac{2}{2+\varkappa}}(z)\leqslant y\leqslant H^{-1}_{\frac{\varkappa}{2+\varkappa}}(z)\right\}$$

is a martingale to obtain

$$\mathrm{E}\max_{y\in I_\varkappa(z)}N^2\left(F_N(y)-F(y)\right)^2\leqslant\mathrm{E}\max_{y\in I_\varkappa(z)}N^2\frac{\left(1-F_N(y)-(1-F(y))\right)^2}{F^2(y)}F\left(H^{-1}_{\frac{\varkappa}{2+\varkappa}}(z)\right)$$

$$\leqslant 4N^2\,\mathrm{E}\left(1-F_N\left(H^{-1}_{\frac{2}{2+\varkappa}}(z)\right)-\left(1-F\left(H^{-1}_{\frac{2}{2+\varkappa}}(z)\right)\right)\right)^2\left(\frac{F\left(H^{-1}_{\varkappa/(2+\varkappa)}(z)\right)}{F\left(H^{-1}_{2/(2+\varkappa)}(z)\right)}\right)^2$$

$$=4NF\left(H^{-1}_{\frac{2}{2+\varkappa}}(z)\right)\left(1-F\left(H^{-1}_{\frac{2}{2+\varkappa}}(z)\right)\right)\left(\frac{F\left(H^{-1}_{\varkappa/(2+\varkappa)}(z)\right)}{F\left(H^{-1}_{2/(2+\varkappa)}(z)\right)}\right)^2.$$

(Note that $F(H^{-1}_{\frac{2}{2+\varkappa}}(z)) \neq 0$, since otherwise

$$z = \frac{2}{2+\varkappa} F(H^{-1}_{\frac{2}{2+\varkappa}}(z)) + \frac{\varkappa}{2+\varkappa} G(H^{-1}_{\frac{2}{2+\varkappa}}(z)) \leqslant F(H^{-1}_{\frac{2}{2+\varkappa}}(z)) = 0).$$

We also have

$$H_{\frac{2}{2+\varkappa}}(y)(1-H_{\frac{2}{2+\varkappa}}(y)) \geqslant \frac{2}{2+\varkappa} F(y)(1-F(y)),$$

$$z = \frac{2}{2+\varkappa} F(H^{-1}_{\frac{2}{2+\varkappa}}(z)) + \frac{\varkappa}{2+\varkappa} G(H^{-1}_{\frac{2}{2+\varkappa}}(z)) \leqslant F(H^{-1}_{\frac{2}{2+\varkappa}}(z))$$

$$\text{and} \quad \frac{\varkappa}{2+\varkappa} F(H^{-1}_{\frac{\varkappa}{2+\varkappa}}(z)) \leqslant z.$$

Consequently, the left-hand side of (6.7) is bounded by

$$16N \frac{2+\varkappa}{2} z(1-z) \left(\frac{\frac{2+\varkappa}{\varkappa} z}{z} \right)^2 = 8 \frac{N(2+\varkappa)^3}{\varkappa^2} z(1-z).$$

We still have to prove formula (6.7) in the case where $G(H^{-1}_{2/(2+\varkappa)}(z)) > F(H^{-1}_{2/(2+\varkappa)}(z))$. The function $u \to (F-G)(H_u^{-1}(z))$ is continuous in u, where $\varkappa/(2+\varkappa) \leqslant u \leqslant 2/(2+\varkappa)$. By our present assumptions, there exists a u_0 such that $(F-G)(H_{u_0}^{-1}(z)) = 0$. Splitting the interval $I_\varkappa(z) = [H^{-1}_{2/(2+\varkappa)}(z), H^{-1}_{\varkappa/(2+\varkappa)}(z)]$ into two subintervals $J_1 = [H^{-1}_{2/(2+\varkappa)}(z), H_{u_0}^{-1}(z)]$ and $J_2 = [H_{u_0}^{-1}(z), H^{-1}_{\varkappa/(2+\varkappa)}(z)]$, we can argue on J_1 as previously in the proof of (6.8) for G and on J_2 as before. This completes the proof.

LEMMA 6.4. *For every* $0 < \varkappa < {}^1/_2$ *there exists a constant* $C_1(\varkappa)$ *such that for every* $h \in \mathscr{H}$ *and every* $N \geqslant 1$,

$$\mathrm{E}\left\{ \max_{(r,s)\in E_\varkappa} \frac{[rN]}{\sqrt{N}} \left[\int_0^1 h\left(\frac{N(r,s)}{N(r,s)+1} \hat{H}_{[rN],[sN]}(t) \right) dF_{[rN]}(t) - \right.\right.$$
$$\left.\left. - \int_0^1 h(H_{\lambda_{r,s}}(t))\, dF_{[rN]}(t) \right] \right\}^2 \leqslant C_1(\varkappa) \|h\|^2.$$

Proof. We may assume that h is non-decreasing. Using

$$h(z) = \begin{cases} \int I_{(1/2,z)}(y)\, dh(y) & (z > {}^1/_2) \\ -\int I_{(z,1/2)}(y)\, dh(y) & (z < {}^1/_2) \end{cases} = \int_{1/2}^{z} dh(y),$$

we have

$$\int_0^1 \left[h\left(\frac{N(r,s)}{N(r,s)+1} \hat{H}_{[Nr],[Ns]}(t) \right) - h(H_{\lambda_{r,s}}(t)) \right] dF_{[rN]}(t)$$

$$= \int_0^1 \left\{ \int_{1/2}^{\frac{N(r,s)}{N(r,s)+1}\hat{H}_{[Nr],[sN]}(t)} dh(z) - \int_{1/2}^{H_{\lambda_{r,s}}(t)} dh(z) \right\} dF_{[rN]}(t)$$

$$= \int_0^1 \Big\{ I_{\{H_{\lambda_{r,s}}(t) \leqslant \frac{N(r,s)}{N(r,s)+1}\hat{H}_{[rN],[sN]}(t)\}} \int_{H_{\lambda_{r,s}}(t)}^{\frac{N(r,s)}{N(r,s)+1}\hat{H}_{[rN],[sN]}(t)} dh(z) -$$

$$- I_{\{H_{\lambda_{r,s}}(t) \leqslant \frac{N(r,s)}{N(r,s)+1}\hat{H}_{[rN],[sN]}(t)\}} \int_{\frac{N(r,s)}{N(r,s)+1}\hat{H}_{[rN],[sN]}(t)}^{H_{\lambda_{r,s}}(t)} dh(z) \Big\} dF_{[rN]}(t)$$

$$= \int_0^1 \int_0^1 \Big\{ I_{\{H_{\lambda_{r,s}}(t) \leqslant \frac{N(r,s)}{N(r,s)+1}\hat{H}_{[rN],[sN]}(t)\}} I_{\{\hat{H}^{-1}_{[rN],[sN]}(\frac{N(r,s)+1}{N(r,s)}z) \leqslant t < H^{-1}_{\lambda_{r,s}}(z)\}} -$$

$$- I_{\{H_{\lambda_{r,s}}(t) > \frac{N(r,s)}{N(r,s)+1}\hat{H}_{[rN],[sN]}(t)\}} I_{\{H^{-1}_{\lambda_{r,s}}(z) \leqslant t < \hat{H}^{-1}_{[rN],[sN]}(\frac{N(r,s)+1}{N(r,s)}z)\}} dF_{[rN]}(t)\, dh(z)$$

$$= I_1 + I_2,$$

where I_1 denotes the above integral with integration over $0 < z \leqslant 2N/(2N+1)$ and where I_2 denotes the above integral with integration over $2N/(2N+1) < z < 1$.

Let us first estimate I_2. Since $z > 2N/(2N+1)$, we have $((N(r, s)+1)/N(r, s))z > 1$, and hence

$$I_{\{\hat{H}^{-1}_{[rN],[sN]}(\frac{N(r,s)+1}{N(r,s)}z) \leqslant t < H^{-1}_{\lambda_{r,s}}(z)\}} \equiv 0.$$

It follows that

$$|[rN]\, I_2| = \Big| - \int_{\frac{2N}{2N+1}}^{1} \int_0^1 I_{\{H^{-1}_{\lambda_{r,s}}(z) \leqslant t\}} [rN]\, dF_{[rN]}(t)\, dh(z) \Big|$$

$$\leqslant \int_{\frac{2N}{2N+1}}^{1} \int_0^1 I_{\{H^{-1}_{\lambda_{r,s}}(z) \leqslant t\}} dN(r, s)\, \hat{H}_{[rN],[sN]}(t)\, dh(z)$$

$$= N(r, s) \int_{\frac{2N}{2N+1}}^{1} \big[1 - \hat{H}_{[rN],[sN]}(H^{-1}_{\lambda_{r,s}}(z))\big]\, dh(z)$$

$$\leqslant N(r, s) \int_{\frac{2N}{2N+1}}^{1} (1-z)\, dh(z) + N(r, s) \int_{\frac{2N}{2N+1}}^{1} \big|\hat{H}_{[rN],[sN]}(H^{-1}_{\lambda_{r,s}}(z)) - z\big|\, dh(z)$$

$$\leqslant \sqrt{2}\, \frac{N(r, s)}{\sqrt{N}} \|h\| + N(r, s) \int_{\frac{2N}{2N+1}}^{1} \big|\hat{H}_{[rN],[sN]}(H^{-1}_{\lambda_{r,s}}(z)) - z\big|\, dh(z),$$

since

$$1 - z \leqslant \frac{1}{\sqrt{2N}} \sqrt{1-z} \leqslant \frac{1}{\sqrt{N}} \sqrt{z(1-z)} \quad \text{for } z \geqslant \frac{2N}{2N+1}.$$

We now estimate I_1. Let

$$\varphi_{r,s}(t, z) = \begin{cases} 1 & \text{if } H_{\lambda_{r,s}}(t) < z \leqslant \dfrac{N(r, s)}{N(r, s)+1}\hat{H}_{[rN],[sN]}(t), \\ -1 & \text{if } \dfrac{N(r, s)}{N(r, s)+1}\hat{H}_{[rN],[sN]}(t) < z \leqslant H_{\lambda_{r,s}}(t), \\ 0 & \text{otherwise.} \end{cases}$$

Then

$$|[rN]I_1| = \left|\int_0^{\frac{2N}{2N+1}}\int_0^1 \varphi_{r,s}(t, z)\, d[rN]F_{[rN]}(t)\, dh(z)\right|$$

$$\leqslant N(r, s)\int_0^{\frac{2N}{2N+1}}\int_0^1 |\varphi_{r,s}(t, z)|\, d\hat{H}_{[rN],[sN]}(t)\, dh(z)$$

$$= N(r, s)\int_0^{\frac{2N}{2N+1}} \Big|\hat{H}_{[rN],[sN]}(H^{-1}_{\lambda_{r,s}}(z)) -$$

$$-\hat{H}_{[rN],[sN]}\left(\hat{H}^{-1}_{[rN],[sN]}\left(\frac{N(r, s)+1}{N(r, s)}z\right)-\right)\Big|\, dh(z).$$

It is easy to see that

$$\left|\hat{H}_{[rN],[sN]}\left(\hat{H}^{-1}_{[rN],[sN]}\left(\frac{N(r, s)+1}{N(r, s)}z\right)-\right)-\frac{N(r, s)+1}{N(r, s)}z\right| \leqslant 2\min\left\{z, \frac{1}{N(r, s)}\right\},$$

$$2\int_0^{\frac{2N}{2N+1}} N(r, s)\min\left\{z, \frac{1}{N(r, s)}\right\} dh(z) \leqslant 2\sqrt{5}\sqrt{N}\,\|h\|,$$

$$\int_0^{\frac{2N}{2N+1}} z\, dh(z) \leqslant \sqrt{5}\sqrt{N}\,\|h\|.$$

It follows that

$$|[rN]I_1| \leqslant 3\sqrt{5}\sqrt{N}\,\|h\| + N(r, s)\int_0^{\frac{2N}{2N+1}} |\hat{H}_{[rN],[sN]}(H^{-1}_{\lambda_{r,s}}(z)) - z|\, dh(z),$$

and so

$$\left|[rN]\int_0^1 \left[h\left(\frac{N(r, s)}{N(r, s)+1}\hat{H}_{[Nr],[Ns]}(t)\right) - h(H_{\lambda_{r,s}}(t))\right] dF_{[rN]}(t)\right|$$

$$\leqslant 4\sqrt{5}\sqrt{N}\,\|h\| + \int_0^1 N(r, s)\,|\hat{H}_{[rN],[sN]}(H^{-1}_{\lambda_{r,s}}(z)) - z|\, dh(z).$$

By Lemma 6.3 it follows finally that

$$\left\| \max_{(r,s)\in E_\varkappa} \frac{[rN]}{\sqrt{N}} \int_0^1 \left[h\left(\frac{N(r,s)}{N(r,s)+1} \hat{H}_{[Nr],[Ns]}(t) \right) - h(H_{\lambda_{r,s}}(t)) \right] dF_{[rN]}(t) \right\|_{L_2(P)}$$

$$\leqslant 4\sqrt{5}\,\|h\| + N^{-1/2} \left\| \int_0^1 \max_{(r,s)\in E_\varkappa} \left| \hat{H}_{[rN],[sN]}(H^{-1}_{\lambda_{r,s}}(z)) - z \right| dh(z) \right\|_{L_2(P)}$$

$$\leqslant 4\sqrt{5}\,\|h\| + \frac{1}{N} \int_0^1 \Big(\Big\| \max_{(r,s)\in E_\varkappa} [rN] \big(F_{[rN]}(H^{-1}_{\lambda_{r,s}}(z)) - F(H^{-1}_{\lambda_{r,s}}(z)\big) \Big\|_{L_2(P)} +$$

$$+ \Big\| \max_{(r,s)\in E_\varkappa} [sN] \big(G_{[sN]}(H^{-1}_{\lambda_{r,s}}(z)) - G(H^{-1}_{\lambda_{r,s}}(z)\big) \Big\|_{L_2(P)} \Big) dh(z)$$

$$\leqslant 4\sqrt{5}\,\|h\| + 2C_0^{1/2}(\varkappa) \int_0^1 \sqrt{z(1-z)}\, dh(z) = C_1^{1/2}(\varkappa)\,\|h\|.$$

LEMMA 6.5. *For every* $0 < \varkappa < 1/2$ *there exists a constant* $C_2(\varkappa)$ *such that for every* $h \in \mathscr{H}$ *and every* $N \geqslant 1$,

$$\text{(6.12)} \qquad E\left\{ \max_{(r,s)\in E_\varkappa} \frac{[rN]}{\sqrt{N}} \int_0^1 h(H_{\lambda_{r,s}}(t))\, d(F_{[rN]} - F)(t) \right\}^2 \leqslant C_2(\varkappa)\,\|h\|^2.$$

Proof. Let h be increasing. Then, integrating by parts, we have

$$\int_0^1 h(H_{\lambda_{r,s}}(t))\, d(F_{[rN]} - F)(t) = -\int_0^1 \big(F_{[rN]}(H^{-1}_{\lambda_{r,s}}(t)) - F(H^{-1}_{\lambda_{r,s}}(t))\big)\, dh(z).$$

It follows from Lemma 6.3 that

$$\left(\mathrm{E}\left\{ \max_{(r,s)\in E_\varkappa} \frac{[rN]}{\sqrt{N}} \int_0^1 h(H_{\lambda_{r,s}}(t))\, d(F_{[rN]} - F)(t) \right\}^2 \right)^{1/2}$$

$$\leqslant \int_0^1 \left\| \max_{(r,s)\in E_\varkappa} \frac{[rN]}{\sqrt{N}} \big(F_{[rN]}(H^{-1}_{\lambda_{r,s}}(t)) - F(H^{-1}_{\lambda_{r,s}}(t))\big) \right\|_{L_2(P)} dh(z)$$

$$\leqslant C_2^{1/2}(\varkappa)\,\|h\|.$$

The proof of Theorem 6.1 follows from Lemmas 6.4 and 6.5, since

$$\big(\mathrm{E} \max_{(r,s)E_\varkappa} S_N^2(h, r, s)\big)^{1/2} = \left\| \max_{(r,s)\in E_\varkappa} \left\{ \frac{[rN]}{\sqrt{N}} \int_0^1 \left[h\left(\frac{N(r,s)}{N(r,s)+1} \hat{H}_{[rN],[sN]}(t) \right) - \right.\right.\right.$$

$$\left.\left.\left. - h(H_{\lambda_{r,s}}(t)) \right] dF_{[rN]}(t) + \frac{[rN]}{\sqrt{N}} \int_0^1 h(H_{\lambda_{r,s}}(t))\, d(F_{[rN]} - F)(t) \right\} \right\|_{L_2(P)}$$

$$\leqslant \big(C_1^{1/2}(\varkappa) + C_2^{1/2}(\varkappa)\big)\,\|h\|.$$

7. AN INVARIANCE PRINCIPLE FOR ABSOLUTELY CONTINUOUS SCORES

In this section we list some (more or less) immediate corollaries to Theorems 5.1 and 6.1 Replacing Theorem 5.1 in the arguments below leads to general invariance principles to be proved in section 8.

THEOREM 7.1. *Let* $h \in \mathscr{H}$ *be an absolutely continuous score function. Then* $S_N(h)$ $(N \geqslant 1)$ *converges weakly in* $(D(E), \delta)$ *to the process* $Z(h)$ *defined in formula* (4.1).

Proof. As remarked in section 2, the score functions of class C^2 with bounded continuous second derivatives are dense in the space of all absolutely continuous score functions. By Theorem 5.1, for such a score function h, $S_N(h) \to Z(h)$ weakly in $D([0, 1]^2)$ equipped with the uniform topology. But then, the restrictions of $S_N(h)$ to E converge weakly with respect to δ (see section 3). Hence we have established the theorem for a dense subspace of absolutely continuous score functions.

Now let $h \in \mathscr{H}$ be an arbitrary absolutely continuous score function. Given $\varepsilon > 0$, choose a C^2-function h_ε with continuous bounded second derivative such that $\|h - h_\varepsilon\| < \varepsilon$. Since $S_N(h) - S_N(h_\varepsilon) = S_N(h - h_\varepsilon)$, Theorem 6.1 implies (cf. 3.4)) that

$$\begin{aligned}\left(\mathrm{E}\left[\delta(S_N(h), S_N(h_\varepsilon))\right]^2\right)^{1/2} &= \left(\mathrm{E}\left[\sum_{k=2}^{\infty} \min(2^{-k}, \delta_{1/k}(S_N(h), S_N(h_\varepsilon)))\right]^2\right)^{1/2} \\ &\leqslant \sum_{k=2}^{\infty} \min\left(2^{-k}, \left(\mathrm{E}\left[\delta_{1/k}(S_N(h), S_N(h_\varepsilon))\right]^2\right)^{1/2}\right) \\ &\leqslant \sum_{k=2}^{\infty} \min\left(2^{-k}, \left(C(1/k)\|h - h_\varepsilon\|^2\right)^{1/2}\right) \\ &\leqslant \sum_{k=2}^{\infty} \min\left(2^{-k}, (C(1/k))^{1/2}\varepsilon\right).\end{aligned}$$

Since the constants $C(1/k)$ do not depend on the functions in $\mathscr{H}$, there exists a $k(\varepsilon)$ with

$$(C(1/k))^{1/2}\varepsilon < 2^{-k(\varepsilon)}[k(\varepsilon)]^{-1} \qquad \text{for all } k \leqslant k(\varepsilon).$$

Hence

$$\left(\mathrm{E}\left[\delta(S_N(h), S_N(h_\varepsilon))\right]^2\right)^{1/2} \leqslant 2^{-k(\varepsilon)} + \sum_{k > k(\varepsilon)} 2^{-k} = 2^{-k(\varepsilon)+1},$$

which tends to zero as $\varepsilon \to 0$.

Note that (3.6) is satisfied with $a(\varepsilon) = 2^{-2k(\varepsilon)+2}$ and (3.5) holds because of the previous discussion.

It follows from Proposition 4.3 that $Z(h_\varepsilon) \to Z(h)$ in probability. Therefore, by Lemma 3.3, $S_N(h)$ converge weakly to $Z(h)$.

The following result is an equivalent formulation of Theorem 1.

THEOREM 7.2. *For $n, m \in N$, define*

$$\tilde{S}_{n,m}(h, r, s)$$
$$= \frac{[rn]}{\sqrt{n+m}} \left(\int_0^1 h\left(\frac{[rn]+[sm]}{[rn]+[sm]+1} \hat{H}_{[rn],[sm]}(t) \right) dF_{[rn]}(t) - \int_0^1 h(H_{\frac{[rn]}{[rn]+[sm]}}(t))\, dF(t) \right).$$

If $h \in \mathscr{H}$ (cf. section 4), then the weak convergence of $S_N(h)$ to $Z(h)$ in $(D(E), \delta)$ is equivalent to the weak convergence of $\tilde{S}_{n,m}(h)$ to $\tilde{Z}(h)$ in $(D(E), \delta)$ as $n/(m+n)$ converges to some $0 < \lambda < 1$, where the distribution of $\tilde{Z}(h)$ is given by

$$(\tilde{Z}(h, r, s))_{0<r,s<1} \overset{D}{=} (Z(h, \lambda r, (1-\lambda)s))_{0<r,s<1}.$$

Proof. Given $n, m \in N$, put $N = n+m$ and $\lambda_N = n/N$. Assume that $(S_N(h), \lambda_N) \to (Z(h), \lambda)$ weakly with respect to $\delta \times |\ |$. Since the map ψ: $D(E) \times [0, 1] \to D(E)$, defined by $\psi(f, u)(r, s) = f(ur, us)$, is continuous at every pair $(f, u) \in C(E) \times [0, 1]$, $\tilde{S}_{n,m}(h) = \psi(S_N(h), \lambda_N)$ converges weakly in $(D(E), \delta)$ to $\psi(Z(h), \lambda) = \tilde{Z}(h)$ (see Theorem 5.1 in [2]).

Conversely, if $\tilde{S}_{n,m}(h) \to \tilde{Z}(h)$ as $n/(m+n) \to \lambda$, for $N \in N$ choose n, m such that $(N/n) \to \lambda_1 \in (0, 1)$ and $(N/m) \to (\lambda\lambda_1/(1-\lambda))$. Similarly as before, it follows that

$$S_N(h, r, s) = \sqrt{\frac{n+m}{N}} \tilde{S}_{nm}\left(b, \frac{Nr}{n}, \frac{Ns}{m}\right) \to \frac{1}{\sqrt{\lambda_1 \lambda}} Z(h, \lambda_1 \lambda r, \lambda_1 \lambda s)$$

and the theorem follows since $a^{-1/2} Z(h, ar, as)$ has the same distribution as $Z(h)$.

COROLLARY 7.1. *Let $h \in \mathscr{H}$ be absolutely continuous and let $\bar{S}_N(h)$ be defined as in (5.21). Then $\bar{S}_N(h)$ converges weakly in $(D(E), \delta)$ to the process $Z(h)$ defined in (4.1).*

Proof. This follows from Theorem 7.1 and Lemma 3.2 together with the estimate on p. 65 of [5] for each $E_\varkappa$:

$$\sup_{r,s \in E_\varkappa} N^{-1/2}[rN] \left| \int_0^1 (h(H_{\lambda(r,s)}(t)) - h(H_{\lambda_{r,s}}(t)))\, dF(t) \right| \leqslant \varkappa^{-2} N^{-1/2} \|h\|.$$

COROLLARY 7.2. *Let the scores $a_N(i, m, n)$ satisfy (2.3), where $h \in \mathscr{H}$ is absolutely continuous. Define the statistics $\hat{S}_N(h)$ as in (5.22). Then $\hat{S}_N(h)$ converges weakly in $(D(E), \delta)$ to $Z(h)$ defined in (4.1).*

Proof. The proof immediately follows from Theorem 7.1 together with the arguments used to prove Corollary 5.2.

8. WEAK INVARIANCE PRINCIPLES

In this section we prove similar to Theorem 7.1 invariance principles for certain score functions $h \in \hat{\mathscr{H}}$. The proofs of the theorems appearing here are independent of sections 5 and 7. We just replace Theorem 5.1 by Proposition 8.3 below, which gives the invariance principle for score functions with finite total variation norm on (0, 1). Denote by $\hat{H}_{n,m}^{-1}$ the left continuous inverse of the distribution function $\hat{H}_{n,m}$ $(n, m \geqslant 1)$.

LEMMA 8.1. *Let* $h \in \mathscr{H}$. *Then for* $(r, s) \in E_{\varkappa}$ *with* $r, s \geqslant 1/N$ *we have*

$$(8.1)\quad S_N(h, r, s) = \frac{-[rN]}{\sqrt{N}} \int_0^1 \Bigg\{(1-\lambda_{r,s}) \frac{G(\hat{H}_{[rN],[sN]}^{-1}(\hat{t})) - G(H_{\lambda_{r,s}}^{-1}(t))}{H_{\lambda_{r,s}}(\hat{H}_{[rN],[sN]}^{-1}(\hat{t})) - t} \times$$

$$\times \left(F_{[rN]}(\hat{H}_{[rN],[sN]}^{-1}(\hat{t})) - F(\hat{H}_{[rN],[sN]}^{-1}(\hat{t}))\right) -$$

$$-(1-\lambda_{r,s}) \frac{F(\hat{H}_{[rN],[sN]}^{-1}(\hat{t})) - F(H_{\lambda_{r,s}}^{-1}(t))}{H_{\lambda_{r,s}}(\hat{H}_{[rN],[sN]}^{-1}(\hat{t})) - t} \left(G_{[sN]}(\hat{H}_{[rN],[sN]}^{-1}(\hat{t})) - G(\hat{H}_{[rN],[sN]}^{-1}(\hat{t}))\right) +$$

$$+ \frac{F(\hat{H}_{[rN],[sN]}^{-1}(\hat{t})) - F(H_{\lambda_{r,s}}^{-1}(t))}{H_{\lambda_{r,s}}(\hat{H}_{[rN],[sN]}^{-1}(\hat{t})) - t} (\hat{t} - t)\Bigg\} dh(t),$$

where

$$(8.2)\quad \hat{t} = \frac{i+1}{N(r, s)} \quad \text{if } \frac{i}{N(r, s)+1} < t \leqslant \frac{i+1}{N(r, s)+1} \quad (1 \leqslant i \leqslant N),\ (r, s) \in E_{\varkappa},$$

and where the integrand is defined to be zero if

$$(8.3)\quad H_{\lambda_{r,s}}(\hat{H}_{[rN],[sN]}^{-1}(\hat{t})) - t = 0.$$

Remark. The integrand (8.1) is a measurable function as an element in $D([0, 1]^3)$ equipped with the supremum metric due to using $\hat{t}$ instead of t. This will become clear in section 9.

Proof. By the definition of $S_N(h)$, (8.2) and integrating by parts we observe that

$$\frac{\sqrt{N}}{[rN]} S_N(h, r, s) = \int_0^1 h(t) dF_{[rN]}(\hat{H}_{[rN],[sN]}^{-1}(\hat{t})) - \int_0^1 h(t) dF(H_{\lambda_{r,s}}^{-1}(t))$$

$$= -\int_0^1 \left(F_{[rN]}(\hat{H}_{[rN],[sN]}^{-1}(\hat{t})) - F(H_{\lambda_{r,s}}^{-1}(t))\right) dh(t).$$

Use the identity

$$1 = \lambda_{r,s} \frac{F(\hat{H}_{[rN],[sN]}^{-1}(\hat{t})) - F(H_{\lambda_{r,s}}^{-1}(t))}{H_{\lambda_{r,s}}(\hat{H}_{[rN],[sN]}^{-1}(\hat{t})) - t} + (1-\lambda_{r,s}) \frac{G(\hat{H}_{[rN],[sN]}^{-1}(\hat{t})) - G(H_{\lambda_{r,s}}^{-1}(t))}{H_{\lambda_{r,s}}(\hat{H}_{[rN],[sN]}^{-1}(\hat{t})) - t}$$

and the convention (8.3) to obtain

$$\frac{\sqrt{N}}{[rN]} S_N(h, r, s) = -\int_0^1 \Big\{(1-\lambda_{r,s}) \frac{G(\hat{H}^{-1}_{[rN],[sN]}(\hat{t})) - G(H^{-1}_{\lambda_{r,s}}(t))}{H_{\lambda_{r,s}}(\hat{H}^{-1}_{[rN],[sN]}(\hat{t})) - t} \times$$

$$\times [F_{[rN]}(\hat{H}^{-1}_{[rN],[sN]}(\hat{t})) - F(\hat{H}^{-1}_{[rN],[sN]}(\hat{t}))] +$$

$$+ \lambda_{r,s} \frac{F(\hat{H}^{-1}_{[rN],[sN]}(\hat{t})) - F(H^{-1}_{\lambda_{r,s}}(t))}{H_{\lambda_{r,s}}(\hat{H}^{-1}_{[rN],[sN]}(\hat{t})) - t} [F_{[rN]}(\hat{H}^{-1}_{[rN],[sN]}(\hat{t})) - F(\hat{H}^{-1}_{[rN],[sN]}(\hat{t}))] +$$

$$+ F(\hat{H}^{-1}_{[rN],[sN]}(\hat{t})) - F(H^{-1}_{\lambda_{r,s}}(t))\Big\} dh(t)$$

$$= -\int_0^1 \Big\{(1-\lambda_{r,s}) \frac{G(\hat{H}^{-1}_{[rN],[sN]}(\hat{t})) - G(H^{-1}_{\lambda_{r,s}}(t))}{H_{\lambda_{r,s}}(\hat{H}^{-1}_{[rN],[sN]}(\hat{t})) - t} [F_{[rN]}(\hat{H}^{-1}_{[rN],[sN]}(\hat{t})) -$$

$$- F(\hat{H}^{-1}_{[rN],[sN]}(\hat{t}))] +$$

$$+ \frac{F(\hat{H}^{-1}_{[rN],[sN]}(\hat{t})) - F(H^{-1}_{\lambda_{r,s}}(t))}{H_{\lambda_{r,s}}(\hat{H}^{-1}_{[rN],[sN]}(\hat{t})) - t} [\lambda_{r,s}(F_{[rN]}(\hat{H}^{-1}_{[rN],[sN]}(\hat{t})) - F(\hat{H}^{-1}_{[rN],[sN]}(\hat{t}))) +$$

$$+ \lambda_{r,s}(F(\hat{H}^{-1}_{[rN],[sN]}(\hat{t})) - F_{[rN]}(\hat{H}^{-1}_{[rN],[sN]}(\hat{t}))) +$$

$$+ (1-\lambda_{r,s})(G(\hat{H}^{-1}_{[rN],[sN]}(\hat{t})) - G_{[sN]}(\hat{H}^{-1}_{[rN],[sN]}(\hat{t}))) + \hat{t} - t]\Big\} dh(t).$$

Here we used the fact that $\hat{H}_{[rN],[sN]}(\hat{H}^{-1}_{[rN],[sN]}(\hat{t})) = \hat{t}$. So (8.1) follows.

Like the processes $Z(h)$, defined in section 4, can only be defined for special pairs (h, F) of score functions h and continuous distribution functions F, the invariance principle to be proved in this section needs a restriction on h and F which we state now as condition

(A) For every $0 < \lambda < 1$ the derivative $(F \circ H_\lambda^{-1})'$ exists, $d|h|$ a.e. (equivalently $h \in \hat{\mathscr{H}}$), and for $d|h|$ a.e. $t \in (0, 1)$ $\{(F \circ H_\lambda^{-1})' \colon 0 < \lambda < 1\}$ is uniformly (in λ) continuous at t.

This condition is satisfied if FH_λ^{-1} is differentiable on (0, 1) and has a continuous extension for some $\lambda_0 \in (0, 1)$. Indeed, we have

$$(FH_\lambda^{-1})' = \frac{(1-\lambda_0)(FH_{\lambda_0}^{-1})' \circ H_\lambda \circ H_\lambda^{-1}}{1 - \lambda - (\lambda_0 - \lambda)(FH_{\lambda_0}^{-1})' \circ H_{\lambda_0} \circ H_\lambda^{-1}}.$$

Thus $(F \circ H_\lambda^{-1})'$ is continuous, has a continuous extension to [0, 1] and for a.e. t it is uniformly continuous. It follows now that the condition for (h, F) is always satisfied if $F = G$ or if the following holds: F and G have continuous densities f and g and there exist intervals (α_i, β_i), $1 \leq i \leq n$ satisfying:

(i) $\{f > 0\} \cup \{g > 0\} = \bigcup_{i=1}^{n} (\alpha_i, \beta_i)$,

(ii) $\lim_{x \uparrow \beta_i} \frac{g(x)}{f(x)} = \lim_{x \downarrow \alpha_i} \frac{g(x)}{f(x)} \quad (1 \leqslant i < n-1)$,

(iii) $\lim_{x \uparrow \beta_n} \frac{g(x)}{f(x)}$ and $\lim_{x \downarrow \alpha_n} \frac{g(x)}{f(x)}$ exist.

We leave it as an exercise to show the implication. (These remarks are contained in [18]; cf. also section 9).

LEMMA 8.2. *Let* $0 < \varkappa < 1/2$. *If* h *and* F *satisfy condition* (A), *then*

$$\lim_{N \to \infty} \max_{\substack{(r,s) \in E_\varkappa \\ N(r,s) \geqslant \sqrt[4]{N}}} \left| \frac{F(\hat{H}^{-1}_{[rN],[sN]}(\hat{t})) - F(H^{-1}_{\lambda_{r,s}}(t))}{H_{\lambda_{r,s}}(\hat{H}^{-1}_{[rN],[sN]}(\hat{t})) - t} - (F \circ H^{-1}_{\lambda(r,s)})'(t) \right| = 0 \tag{8.4}$$

for $d|h|$ *almost all* t, P *a.e. and*

$$\lim_{N \to \infty} \max_{\substack{(r,s) \in E_\varkappa \\ N(r,s) \geqslant \sqrt[4]{N}}} \left| \frac{G(\hat{H}^{-1}_{[rN],[sN]}(\hat{t})) - G(H^{-1}_{\lambda_{r,s}}(t))}{H_{\lambda_{r,s}}(\hat{H}^{-1}_{[rN],[sN]}(\hat{t})) - t} - (G \circ H^{-1}_{\lambda(r,s)})'(t) \right| = 0 \tag{8.5}$$

for $d|h|$ *almost all* t, P *a.e.*

Moreover

$$\max_{(r,s) \in E_\varkappa} \left| \frac{F(\hat{H}^{-1}_{[rN],[sN]}(\hat{t})) - F(H^{-1}_{\lambda_{r,s}}(t))}{H_{\lambda_{r,s}}(\hat{H}^{-1}_{[rN],[sN]}(\hat{t})) - t} - (F \circ H^{-1}_{\lambda(r,s)})'(t) \right| \leqslant 3\varkappa^{-1} \tag{8.6}$$

and

$$\max_{(r,s) \in E_\varkappa} \left| \frac{G(\hat{H}^{-1}_{[rN],[sN]}(\hat{t})) - G(H^{-1}_{\lambda_{r,s}}(t))}{H_{\lambda_{r,s}}(\hat{H}^{-1}_{[rN],[sN]}(\hat{t})) - t} - (G \circ H^{-1}_{\lambda(r,s)})'(t) \right| \leqslant 3\varkappa^{-1} \tag{8.7}$$

for $d|h|$ *almost all* t.

Proof. We first show (8.6). The proof of (8.7) follows replacing F by G. By (2.1), $F(x)+G(x) = 2x$. Hence $F'(x) \leqslant 2$ whenever F is differentiable at x. Since $h \in \hat{\mathscr{H}}$, $F \circ H^{-1}_{\lambda(r,s)}$ is $d|h|$ a.e. differentiable. Consequently,

$$(F \circ H^{-1}_{\lambda(r,s)})'(t) = \lim_{\varepsilon \downarrow 0} \frac{F(H^{-1}_{\lambda(r,s)}(t+\varepsilon)) - F(H^{-1}_{\lambda(r,s)}(t))}{\varepsilon}$$

$$\leqslant \lambda(r,s)^{-1} \lim_{\varepsilon \downarrow 0} \frac{\lambda(r,s)\left[F(H^{-1}_{\lambda(r,s)}(t+\varepsilon)) - F(H^{-1}_{\lambda(r,s)}(t))\right]}{\varepsilon} +$$

$$+ \frac{(1-\lambda(r,s))\left[G(H^{-1}_{\lambda(r,s)}(t+\varepsilon)) - G(H^{-1}_{\lambda(r,s)}(t))\right]}{\varepsilon} \leqslant \lambda(r,s)^{-1} \leqslant \varkappa^{-1}.$$

Similarly, if $u \geqslant v$, then $0 \leqslant \lambda_{r,s}(F(u)-F(v)) \leqslant H_{\lambda_{r,s}}(u) - H_{\lambda_{r,s}}(v)$ implies

$$\frac{F(u)-F(v)}{H_{\lambda_{r,s}}(u) - H_{\lambda_{r,s}}(v)} \leqslant \lambda_{r,s}^{-1} \leqslant 2\varkappa^{-1}$$

by (6.4). Formula (8.6) follows from both estimates.

We now prove (8.4) ((8.5) is similar). By (6.4) we have $[rN]^{-1} \leqslant 2\varkappa^{-1} N(r,s)^{-1} \leqslant 2\varkappa^{-1} N^{-1/4}$ and $[sN]^{-1} \leqslant 2\varkappa^{-1} N^{-1/4}$, and therefore

$$\left|1-\frac{r}{r+s}\frac{N(r,s)}{[rN]}\right| \leqslant \frac{1}{[rN]} \leqslant 2\varkappa^{-1} N^{-1/4},$$

$$\left|1-\frac{s}{r+s}\frac{N(r,s)}{[sN]}\right| \leqslant \frac{1}{[sN]} \leqslant 2\varkappa^{-1} N^{-1/4}.$$

It follows that

$$\begin{aligned}&|H_{\lambda(r,s)}(\hat{H}^{-1}_{[rN],[sN]}(\hat{t}))-t-H_{\lambda(r,s)}(\hat{H}^{-1}_{[rN],[sN]}(\hat{t}))+H_{\lambda(r,s)}(H^{-1}_{\lambda_{r,s}}(t))|\\ &\quad=\left|\left(1-\frac{r}{r+s}\frac{N(r,s)}{[rN]}\right)\frac{[rN]}{N(r,s)}(F(\hat{H}^{-1}_{[rN],[sN]}(\hat{t}))-F(H^{-1}_{\lambda_{r,s}}(t)))+\right.\\ &\qquad\left.+\left(1-\frac{s}{r+s}\frac{N(r,s)}{[sN]}\right)\frac{[sN]}{N(r,s)}(G(H^{-1}_{[rN],[sN]}(\hat{t}))-G(H^{-1}_{\lambda_{r,s}}(t)))\right|\\ &\quad\leqslant 2\varkappa^{-1} N^{-1/4}|H_{\lambda_{r,s}}(\hat{H}^{-1}_{[rN],[sN]}(\hat{t}))-t|\end{aligned}$$

and

$$\left|1-\frac{H_{\lambda(r,s)}(\hat{H}^{-1}_{[rN],[sN]}(\hat{t}))-H_{\lambda(r,s)}(H^{-1}_{\lambda_{r,s}}(t))}{H_{\lambda_{r,s}}(\hat{H}^{-1}_{[rN],[sN]}(\hat{t}))-t}\right| \leqslant 2\varkappa^{-1} N^{-1/4}.$$

Therefore, it suffices to show that

$$\left|\frac{F(\hat{H}^{-1}_{[rN],[sN]}(\hat{t}))-F(H^{-1}_{\lambda_{r,s}}(t))}{H_{\lambda(r,s)}(\hat{H}^{-1}_{[rN],[sN]}(\hat{t}))-H_{\lambda(r,s)}(H^{-1}_{\lambda_{r,s}}(t))}-(F\circ H^{-1}_{\lambda(r,s)})'(t)\right| \to 0 \tag{8.9}$$

uniformly in $(r,s)\in E_\varkappa$ such that $N(r,s) \geqslant N^{1/4}$

Since h and F satisfy condition (A), we have for $d|h|$ a.e. t: Given $\varepsilon>0$ $\exists \eta>0$ such that for all $0<\lambda<1$ and all $|\delta|<\eta$ for which $(F\circ H_\lambda^{-1})'(t+\delta)$ is defined,

$$|(F\circ H_\lambda^{-1})'(t+\delta)-(F\circ H_\lambda^{-1})'(t)|<\varepsilon.$$

Since $F\circ H_\lambda^{-1}$ is absolutely continuous, it follows that

$$\begin{aligned}&\left|\frac{F(H_\lambda^{-1}(t+\delta))-F(H_\lambda^{-1}(t))}{\delta}-(F\circ H_\lambda^{-1})'(t)\right|\\ &\qquad=\left|\delta^{-1}\int_t^{t+\delta}[(F\circ H_\lambda^{-1})'(u)-(F\circ H^{-1})'(t)]\,du\right|<\varepsilon,\end{aligned}$$

and this holds uniformly in $0<\lambda<1$.

Since

$$\left|\frac{r}{r+s}-\frac{[rN]}{N(r,s)}\right| \leqslant N^{-1/4}$$

uniformly in $(r, s) \in E_\varkappa$ and $N(r, s) \geqslant N^{1/4}$, we have $H_{\lambda(r,s)}(H_{\lambda_{r,s}}^{-1}(t)) \to t$ as $N \to \infty$ (uniformly in $(r, s) \in E_\varkappa$) with $N(r, s) \geqslant N^{1/4}$ and hence, for $d|h|$ a.e. t,

$$\frac{F(H_{\lambda_{r,s}}^{-1}(t)) - F(H_{\lambda(r,s)}^{-1}(t))}{H_{\lambda(r,s)}(H_{\lambda_{r,s}}^{-1}(t)) - t} - (F \circ H_{\lambda(r,s)}^{-1})'(t) \to 0$$

uniformly in (r, s) as $N \to \infty$ $dJh|$ a.e.

Also, since $H_{\lambda(r,s)}(\hat{H}_{[rN],[sN]}^{-1}(\hat{t})) - t \to 0$ uniformly in $(r, s) \in E_\varkappa$ with $N(r, s) \geqslant N^{1/4}$,

$$\frac{F(\hat{H}_{[rN],[sN]}^{-1}(\hat{t})) - F(H_{\lambda(r,s)}^{-1}(t))}{H_{\lambda(r,s)}(\hat{H}_{[rN],[sN]}^{-1}(\hat{t})) - t} - (F \circ H_{\lambda(r,s)}^{-1})'(t) \to 0$$

uniformly in (r, s) as $N \to \infty$ $d|h|$ a.e.

It follows that (8.9) is

$$\left| \left(\frac{F(\hat{H}_{[rN],[sN]}^{-1}(\hat{t})) - F(H_{\lambda(r,s)}^{-1}(t))}{H_{\lambda(r,s)}(\hat{H}_{[rN],[sN]}^{-1}(\hat{t})) - t} - (F \circ H_{\lambda(r,s)}^{-1})'(t) \right) \times \right.$$

$$\times \frac{H_{\lambda(r,s)}(\hat{H}_{[rN],[sN]}^{-1}(\hat{t})) - t}{H_{\lambda(r,s)}(\hat{H}_{[rN],[sN]}^{-1}(\hat{t})) - H_{\lambda(r,s)}(H_{\lambda_{r,s}}^{-1}(t))} +$$

$$+ \left(\frac{F(H_{\lambda(r,s)}^{-1}(t)) - F(H_{\lambda_{r,s}}^{-1}(t))}{t - H_{\lambda(r,s)}(H_{\lambda_{r,s}}^{-1}(t))} - (F \circ H_{\lambda(r,s)}^{-1})'(t) \right) \times$$

$$\left. \times \frac{t - H_{\lambda(r,s)}(H_{\lambda_{r,s}}^{-1}(t))}{H_{\lambda(r,s)}(H_{[rN],[sN]}^{-1}(\hat{t})) - H_{\lambda(r,s)}(H_{\lambda_{r,s}}^{-1}(t))} \right|$$

and tends to zero as $N \to \infty$ uniformly in $(r, s) \in E_\varkappa$ with $N(r, s) \geqslant N^{1/4}$, $d|h|$ a.e. This proves (8.4).

PROPOSITION 8.3. *Let h and F satisfy condition* (A) *and let h have bounded total variation norm on* (0, 1). *Assume that the underlying probability space is rich enough such that there exist two independent Kiefer processes $\mathscr{K}_1$ and $\mathscr{K}_2$ such that uniformly in (r, t), resp. (s, t),*

$$\frac{[rN]}{\sqrt{N}}[F_{[rN]}(t) - F(t)] \to \mathscr{K}_1(r, t) \quad \textit{and} \quad \frac{[sN]}{\sqrt{N}}[G_{[sN]}(t) - G(t)] \to \mathscr{K}_2(s, t)$$

in probability.

Then, for any $0 < \varkappa < 1/2$,

$$\max_{(r,s) \in E_\varkappa} \left| S_N(h, s, r) + \frac{s}{r+s} \int_0^1 (G \circ H_{\lambda(r,s)}^{-1})'(t)\, \mathscr{K}_1(r, H_{\lambda(r,s)}^{-1}(t))\, dh(t) - \right.$$

$$\left. - \frac{r}{r+s} \int_0^1 (F \circ H_{\lambda(r,s)}^{-1})'(t)\, \mathscr{K}_2(s, H_{\lambda(r,s)}^{-1}(t))\, dh(t) \right|$$

converges to zero in probability as $N \to \infty$.

Proof. We shall use the representation (8.1) of $S_N(h)$ from Lemma 8.1. Observe first that, by (8.2) and the proof of (8.6),

$$\left|\frac{F(\hat{H}^{-1}_{[rN],[sN]}(\hat{t}))-F(H^{-1}_{\lambda_{r,s}}(t))}{H_{\lambda_{r,s}}(\hat{H}^{-1}_{[rN],[sN]}(\hat{t}))-t}(t-\hat{t})\right| \leqslant 2\varkappa^{-1} N(r, s)^{-1}. \tag{8.10}$$

Next write

$$A_N(r, s, t) = \left|\frac{[rN]}{\sqrt{N}}[F_{[r,N]}(\hat{H}^{-1}_{[rN],[sN]}(\hat{t}))-F(\hat{H}^{-1}_{[rN],[sN]}(\hat{t}))] - \mathscr{K}_1(r, H^{-1}_{\lambda(r,s)}(t))\right|, \tag{8.11}$$

$$B_N(r, s, t) = \left|\frac{[sN]}{\sqrt{N}}[G_{[sN]}(\hat{H}^{-1}_{[rN],[sN]}(\hat{t}))-G(\hat{H}^{-1}_{[rN],[sN]}(\hat{t}))] - \mathscr{K}_2(s, H^{-1}_{\lambda(r,s)}(t))\right|. \tag{8.12}$$

Since all paths of $\mathscr{K}_1$ and $\mathscr{K}_2$ are uniformly continuous, $A_N(r, s, t)$, $B_N(r, s, t) \to 0$ as $N \to \infty$ uniformly in t and $(r, s) \in E_\varkappa$ with $N(r, s) \geqslant N^{1/4}$. In view of (8.6) and (8.7) it follows that

$$\max_{\substack{(r,s)\in E_\varkappa \\ N(r,s)\geqslant N^{1/4}}} \left|\frac{[rN]}{\sqrt{N}}\int_0^1 (1-\lambda_{r,s})\left\{\frac{G(\hat{H}^{-1}_{[rN],[sN]}(\hat{t}))-G(H^{-1}_{\lambda_{r,s}}(t))}{H_{\lambda_{r,s}}(\hat{H}^{-1}_{[rN],[sN]}(\hat{t}))-t}\times\right.\right. \tag{8.13}$$

$$\times[F_{[rN]}(\hat{H}^{-1}_{[rN],[sN]}(\hat{t}))-F(\hat{H}^{-1}_{[rN],[sN]}(\hat{t}))]-$$

$$\left.-\frac{F(\hat{H}^{-1}_{[rN],[sN]}(\hat{t}))-F(H^{-1}_{\lambda_{r,s}}(t))}{H_{\lambda_{r,s}}(\hat{H}^{-1}_{[rN],[sN]}(\hat{t}))-t}[G_{[sN]}(\hat{H}^{-1}_{[rN],[sN]}(\hat{t}))-G(\hat{H}^{-1}_{[rN],[sN]}(\hat{t}))]\right\}dh(t)-$$

$$-\int_0^1\left\{(1-\lambda(r, s))\frac{G(\hat{H}^{-1}_{[rN],[sN]}(\hat{t}))-G(H^{-1}_{\lambda_{r,s}}(t))}{H_{\lambda_{r,s}}(\hat{H}^{-1}_{[rN],[sN]}(\hat{t}))-t}\mathscr{K}_1(r, H^{-1}_{\lambda(r,s)}(t))-\right.$$

$$\left.\left.-\lambda(r, s)\frac{F(\hat{H}^{-1}_{[rN],[sN]}(\hat{t}))-F(H^{-1}_{\lambda_{r,s}}(t))}{H_{\lambda_{r,s}}(\hat{H}^{-1}_{[rN],[sN]}(\hat{t}))-t}\mathscr{K}_2(s, H^{-1}_{\lambda(r,s)}(t))\right\}dh(t)\right|$$

$$\leqslant C(\varkappa)[\max A_N(r, s, t)+\max B_N(r, s, t)+N^{-1/4}] \to 0 \quad \text{as } N\to\infty,$$

where $C(\varkappa)$ denotes some constant depending on $\varkappa$ and the total variation norm $\|dh\|$ of dh.

By Lemma 8.2, (8.4), and (8.5), and the dominated convergence theorem,

$$\lim_{N\to\infty}\int_0^1 \max_{\substack{(r,s)\in E_\varkappa \\ N(r,s)\geqslant N^{1/4}}}\left|\frac{F(\hat{H}^{-1}_{[rN],[sN]}(\hat{t}))-F(H^{-1}_{\lambda_{r,s}}(t))}{H_{\lambda_{r,s}}(\hat{H}^{-1}_{[rN],[sN]}(\hat{t}))-t}-(F\circ H^{-1}_{\lambda(r,s)})'(t)\right| d|h|(t) = 0 \text{ a.s.}$$

and similarly for G.

Now, we set

$$R_N = \Big\{ C(\varkappa)\,[\max A_N(r, s, t) + \max B_N(r, s, t) + N^{-1/4}] < \varepsilon/3,$$

$$\max_{(u,v)\in[0,1]^2} |\mathscr{K}_i(u, v)| \leqslant M \qquad (i = 1, 2),$$

$$\int \max_{\substack{N(r,s)\geqslant N^{1/4}\\ (r,s)\in E_\varkappa}} \left| \frac{F(\hat{H}^{-1}_{[rN],[sN]}(\hat{t})) - F(H^{-1}_{\lambda_{r,s}}(t))}{H_{\lambda_{r,s}}(\hat{H}^{-1}_{[rN],[sN]}(\hat{t})) - t} - (F \circ H^{-1}_{\lambda(r,s)})'(t) \right| d|h|(t) < \frac{\varepsilon}{3M}$$

and

$$\int \max_{\substack{N(r,s)\geqslant N^{1/4}\\ (r,s)\in E_\varkappa}} \left| \frac{G(\hat{H}^{-1}_{[rN],[sN]}(\hat{t})) - G(H^{-1}_{\lambda_{r,s}}(t))}{H_{\lambda_{r,s}}(\hat{H}^{-1}_{[rN],[sN]}(\hat{t})) - t} - (G \circ H^{-1}_{\lambda(r,s)})'(t) \right| d|h|(t) < \frac{\varepsilon}{3M} \Big\}.$$

Here we may choose M, depending on N, in such a way that $M \to \infty$ and $P(R_N^c) \to 0$ as $N \to \infty$.

By (8.10) and (8.13) it follows that on R_N

$$\begin{aligned} \text{(8.14)} \qquad \max_{\substack{(r,s)E_\varkappa\\ N(r,s)>N^{1/4}}} \Big| S_N(h, r, s) + (1-\lambda(r, s)) \int_0^1 (G \circ H^{-1}_{\lambda(r,s)})'(t) \times \\ \times \mathscr{K}_1(r, H^{-1}_{\lambda(r,s)}(t))\, dh(t) - \\ - \lambda(r, s) \int_0^1 (F \circ H^{-1}_{\lambda(r,s)})'(t)\, \mathscr{K}_2(s, H^{-1}_{\lambda(r,s)}(t))\, dh(t) \Big| \leqslant \varepsilon. \end{aligned}$$

It is easy to see that we can extend the maximum in (8.14) over all $(r, s) \in E_\varkappa$ and this proves the proposition.

THEOREM 8.1. *Let h and F satisfy condition* (A) *and let h have bounded total variation on* $[0, 1]$. *Then $S_N(h)$ converges weakly in $(D(E), \delta)$ to the process $Z(h)$ defined in* (4.1).

Proof. We may assume that the probability space is rich enough. By Skorohod's theorem and the weak convergence of empirical processes to Kiefer processes, there exist independent Kiefer processes $\mathscr{K}_1$ and $\mathscr{K}_2$ such that the assumption of Proposition 8.3 is satisfied. Denote by $Z(h)$ the process defined in (4.1) with the two specified Kiefer processes. Then by Proposition 8.3 we have, for any $0 < \varkappa < 1/2$ and any $\varepsilon > 0$,

$$\lim_N \mathrm{P}\big(\max_{(r,s)\in E_\varkappa} |S_N(h, r, s) - Z(h, r, s)| \geqslant \varepsilon \big) = 0.$$

Let $\varepsilon > 0$ and $\eta > 0$ be given. Choose k_0 such that $\varepsilon - 2^{k_0} \geqslant \varepsilon/2$ and then N_0 such that, for $N \geqslant N_0$,

$$\mathrm{P}\left(\max_{(r,s)\in E_\varkappa} |S_N(h, r, s) - Z(h, r, s)| \geqslant \frac{\varepsilon}{2k_0} \right) < \eta/k_0.$$

It follows that

$$P\big(\delta(S_N(h), Z(h)) \geqslant \varepsilon\big) = P\big(\sum_{k=2}^{\infty} \min\left[2^{-k}, \delta_{1/k}(S_N(h), Z(h))\right] \geqslant \varepsilon\big)$$

$$\leqslant P\big(\sum_{k=2}^{k_0} \min\left[2^{-k}, \delta_{1/k}(S_N(h), Z(h))\right] \geqslant \frac{\varepsilon}{2}\big) \leqslant \sum_{k=2}^{k_0} P\big(\delta_{1/k}(S_N(h), Z(h)) \geqslant \frac{\varepsilon}{2k_0}\big) < \eta.$$

The weak convergence follows now from Lemma 3.2.

THEOREM 8.2. *Let h and F satisfy condition* (A). *Then $S_N(h)$ converges weakly in $(D(E), \delta)$ to $Z(h)$ defined in* (4.1).

Proof. If $h \in \mathscr{H}$, then, for any $\theta > 0$,

$$h_\theta(t) = \begin{cases} h(t) & \text{if } \theta \leqslant t \leqslant 1-\theta, \\ 0 & \text{if } 0 < t < \theta \text{ or } 1-\theta < t < 1 \end{cases}$$

has bounded total variation. Moreover, if (h, F) satisfies condition (A), so does (h_θ, F). We can proceed as in the proof of Theorem 7.1 to prove our claim.

COROLLARY 8.1. *Let h and F satisfy condition* (A) *and let $\bar{S}_N(h)$ be defined as in* (5.21). *Then $\bar{S}_N(h)$ converges weakly in $(D(E), \delta)$ to $Z(h)$ defined in* (4.1).

Proof. This follows from Theorem 8.2 as in Corollary 7.1.

COROLLARY 8.2. *Let the scores $a_N(i, m, n)$ be defined by* (2.3), *where (h, F) is as before. Define the random function $\hat{S}_N(h)$ by* (5.22). *Then $\hat{S}_N(h)$ converges weakly in $(D(E), \delta)$ to $Z(h)$ defined in* (4.1).

Proof. This follows as Corollaries 5.3 or 7.2, using Theorem 8.2 instead of 5.1 or 7.1.

9. THE INVARIANCE PRINCIPLE OF T. SCHULZE-PILLOT

To the best of our knowledge, no invariance principle of the form given in sections 5, 7 and 8 has been proved except some of the results of Schulze–Pillot [18] in his Ph. D. thesis and, maybe, some very special cases. Since the results of Schulze–Pillot have never been published elsewhere, we shortly sketch his results.

Schulze–Pillot uses the approach of Pyke and Shorack [15] to prove limit theorems for the two-sample linear rank statistics. The linear rank statistic can be represented as an integral over the two-sample empirical process. We used this representation in (8.1), however in the present proof only in the case of score functions with bounded total variation on (0, 1). Contrary to this easy case, a general score function needs a more careful handling of the two-sample empirical process, that is, an invariance principle with respect to some dominating function q has to be proven.

This requires to enlarge E to $E \times [0, 1]$, to replace $D(E)$ by $D_q(E \times [0, 1])$ and to define a suitable metric on $D_q(E \times [0, 1])$. We will start with this description. Denote by $\mathscr{A}$ the set of all functions q: $[0, 1] \to \boldsymbol{R}_+$ which are continuous, non-decreasing on the interval $[0, {}^1/_2]$ and non-increasing on the interval $[{}^1/_2, 1]$, and (see [15]) which satisfy

$$\int_0^1 [q(t)]^{-2} dt < \infty . \tag{9.1}$$

As in section 3 let us define a metric d_q as

$$d_q(f, g) = \sum_{k=2}^{\infty} \min \left\{ 2^{-k}; \sup_{\substack{(r,s) \in E_{1/k} \\ 0 \leqslant t \leqslant 1}} \frac{|f(r, s, t) - g(r, s, t)|}{q(t)} \right\} \tag{9.2}$$

for two functions f and g defined on $E \times [0, 1]$. In order to make d_q well defined, use the convention $0/0 = 0$ and $a/0 = \infty$ for $0 < a < \infty$. Clearly, d_q defines the topology of uniform convergence with respect to q on each of the sets $E_\lambda \times [0, 1]$, $0 < \lambda \leqslant {}^1/_2$. This means that $d_q(f_n, f) \to 0$ if and only if, for every $0 < \lambda \leqslant {}^1/_2$, $[q(t)]^{-1} (f_n(r, s, t) - f(r, s, t))$ converges to zero uniformly in $(r, s) \in E_\lambda$ and $0 \leqslant t \leqslant 1$.

In analogy to the definition of $D(E)$ denote by $D_q(E \times [0, 1])$, the d_q-closure of all simple functions of the form

$$f = \sum_{k=1}^{n} \alpha_k I_{[E \cap (A_1^k \times A_2^k)] \times A_3^k},$$

where A_i^k for $i = 1, 2$ is again an interval $[a_i^k, b_i^k)$ with $0 \leqslant a_i^k, b_i^k \leqslant 1$ or $\{1\}$, and where $A_3^k = [a_3^k, b_3^k]$ with $0 < a_3^k, b_3^k \leqslant 1$. It is not hard to see that functions in $D_q(E \times [0, 1])$ have only discontinuities of the first kind and that $C_q(E \times [0, 1])$, the space of continuous functions on $E \times [0, 1]$, is a closed separable subset.

Let $\mathscr{D}_q$ denote the Borel-σ-field generated by d_q. If D^1 is a separable subspace, then $\mathscr{D}_q \cap D'$ is generated by the projections (cf. [2], § 18)). Thus, the random functions X taking value in $D^1 \subset \mathscr{D}_q(E \times [0, 1])$ are measurable if, for every projection π: $D_q(E \times [0, 1]) \to \boldsymbol{R}$, $\pi f = f(r, s, t)$, where $(r, s) \in E$ and $0 \leqslant t \leqslant 1$, the function $\pi \circ X$ is measurable.

In particular, the two-sample empirical process appearing in (8.1) is measurable.

In the situation described in section 2, Pyke and Shorack [15] defined the two-sample empirical process as

$$N^{1/2} [F_n(\hat{H}_{n,m}^{-1}(t)) - F(H_{n/N}^{-1}(t))]. \tag{9.3}$$

More generally, Schulze–Pillot defined the two sample sequential empirical process $\Lambda_N(r, s, t)$ by

$$(9.4)\qquad \Lambda_N(r, s, t) = \frac{[rN]}{\sqrt{N}}\,[F_{[rN]}\,\hat{H}^{-1}_{[rN],[sN]}(\hat{t}) - FH^{-1}_{\lambda_{r,s}}(t)]$$

for $N \in \mathbf{N}$, $(r, s, t) \in E \times [0, 1]$. Here, similarly to (8.2),

$$\hat{t} - \frac{i}{N} \text{ if } t \in \left[\frac{i}{N+1}, \frac{i+1}{N+1}\right),\ 0 \leqslant i < n+m, \quad \text{and} \quad \hat{t} = 1 \text{ if } t = 1.$$

Since $\hat{t}$ can assume only the values $i/(N+1)$, $F_{[rN]}\big(\hat{H}^{-1}_{[rN],[sN]}(\hat{t})\big)$ has a separable range and so has $F\big(H^{-1}_{\lambda_{r,s}}(t)\big)$. It follows that $\Lambda_N(r, s, t)$ *is* $D_q(E \times [0, 1])$*-measurable*. We may, therefore, speak of weak convergence of Λ_N with respect to d_q.

Let $\mathscr{K}_1$ and $\mathscr{K}_2$ be two independent, standard Kiefer processes on $C[0, 1]^2$, i.e. $\mathrm{E}\mathscr{K}_i(s, t) = 0$ $(0 \leqslant s, t \leqslant 1)$ and $\mathrm{E}\mathscr{K}_i(s, t)\,\mathscr{K}_i(s', t') = t(1-t') \times \min(s, s')$ $(0 \leqslant t, t' \leqslant 1,\ 0 \leqslant s, s' \leqslant 1)$. For each $0 < \lambda \leqslant 1$, FH_λ^{-1} and GH_λ^{-1} are monotone and hence $a_\lambda(t) = (FH_\lambda^{-1})'(t)$ and $b_\lambda(t) = (GH_\lambda^{-1})'(t)$ exist a.e. with respect to t and satisfy $a_\lambda \geqslant 0$, $b_\lambda \geqslant 0$ and $\lambda a_\lambda + (1-\lambda) b_\lambda = 1$. We now put

$$(9.5)\qquad \Lambda(r, s, t)$$
$$= \big(1-\lambda(r, s)\big)\, b_{\lambda(r,s)}(t)\, \mathscr{K}_1\big(r, FH^{-1}_{\lambda(r,s)}(t)\big) - \lambda(r, s)\, a_{\lambda(r,s)}\, \mathscr{K}_2\big(s, GH^{-1}_{\lambda(r,s)}(t)\big)$$

for all $(r, s, t) \in E \times [0, 1]$, where a_λ and b_λ are defined. (Note that integrating over Λ with respect to dh yields a random variable with the same distribution as in (4.2) provided $h \in \hat{\mathscr{H}}$.) For the hypothesis of the next theorem compare section 8.

THEOREM 9.1 [18]. *Suppose that, for every* $\lambda \in [0, 1]$, FH_λ^{-1} *is differentiable in the open interval* $(0, 1)$ *and that, for some* $0 < \lambda_0 < 1$, a_{λ_0} *has a continuous extension on* $[0, 1]$. *Then*:

(a) *for any* $q \in Q$, Λ *is* $D_q(E \times [0, 1])$*-measurable and has continuous paths*;

(b) Λ_N *converges weakly in* $\big(D_q(E \times [0, 1]), d_q\big)$ *to* Λ, *provided* $q \in Q$.

It is clear that Λ is a Gaussian process and its covariance structure is easily deduced from its definition.

Let $h \in \mathscr{H}$ and denote by $d|h|$ the total variation measure of h. Let $q \in Q$ such that $\int q d|h| < \infty$. Then every $f \in D_q(E \times [0, 1])$ is integrable with respect to $d|h|$ in its last coordinate, i.e., for every $(r, s) \in E$, $\int |f(r, s, t)|\, d|h|(t) < \infty$.

The map I_h: $D_q(E \times [0, 1]) \to D(E)$, given by $I_h(f)(r, s) = \int f(r, s, t)\, dh(t)$ is continuous and, by (4.2) and (9.5),

$$(9.6)\qquad I_h(\Lambda) = Z(h)$$

whenever this is well defined (cf. section 4, i.e. $h \in \hat{\mathscr{H}}$ in particular if h is absolutely continuous).

THEOREM 9.2 [18]. *Let $h \in \mathscr{H}$ and assume that there exists a $q \in Q$ with $\int q d|h| < \infty$. Then under the assumptions of Theorem 9.1, $S_N(h)$ converges weakly in $(D(E), \delta)$ to $Z(h)$ given in* (4.1).

If h can be written as a difference $h_1 - h_2$ of two monotone functions satisfying

(9.7) $$|h_i| \leqslant K\big(t(1-t)\big)^{-1/2+\eta}$$

for some $K > 0$ and $\eta > 0$, then there exists a $q \in Q$ with $\int q(t)\, d|h|(t) < \infty$.

Formula (9.7) is not necessary to guarantee the existence of such a function $q \in Q$ as remarked in [15]. But not all functions $h = h_1 - h_2$ with

$$\int \frac{|h_i(t)|}{\sqrt{t(1-t)}}\, dt < \infty \qquad (i = 1, 2)$$

satisfy the above condition on q. This is one extension of Theorem 9.2 by Theorem 8.2. On the other hand, since we do not depend on the weak convergence of the two-sample sequential empirical process (9.4), we do not need the stronger assumptions on the differentiability in Theorem 9.1 for proving Theorems 8.1 and 8.2. Finally, we would like to remark that Theorem 5.1 seems to be completely new.

10. APPLICATIONS

Pyke and Shorack [16] have shown that $S_N(h, m_N, n_N)$ converges to a normal $N(0, \sigma^2)$ distribution for some specified σ^2 if m_N and n_N are random variables satisfying certain conditions. We first give a proof their result, using Theorems 8.2 and 7.1. This also extends a result of Schulze–Pillot [18].

THEOREM 10.1. *Let m_N and n_N be integer valued random variables $(N \in \mathbf{N})$ satisfying*

(10.1) $$0 < m_0 \leqslant m_1 \leqslant \ldots \leqslant m_N \leqslant N,$$
$$0 < n_0 \leqslant n_1 \leqslant \ldots \leqslant n_N \leqslant N \qquad (N \geqslant 1)$$

and

(10.2) $$\lim N^{-1} m_N = \lambda_0 \quad \text{and} \quad \lim N^{-1} n_N \to 1 - \lambda_0$$

in probability for some $0 < \lambda_0 < 1$.

If (h, F) satisfies the assumption of Theorem 8.2 or of Theorem 7.1, then $S_N(h, m_N/N, n_N/N)$ converges weakly to a normal distribution $N(0, \sigma^2)$, where σ^2 is given by

(10.3) $$\sigma^2 = \mathrm{E}\,[Z(h, \lambda_0, 1-\lambda_0)]^2,$$

i.e. the limiting distribution is given by $Z(h, \lambda_0, 1-\lambda_0)$ which is the sum of two independent normally distributed random variables.

Proof. We have $N^{-1}(m_N, n_N) \in E$ unless $m_N = 0$ or $n_N = 0$, and so the limit distribution for $S_N(h, m_N/N, n_N/N)$ can be obtained from $S_N(h)$ as follows. Let $\psi\colon D(E) \times E \to \boldsymbol{R}$ denote the map $\psi(f, r, s) = f(r, s)$. Since, by Theorem 8.2, $S_N(h) \to Z(h)$ weakly in $(D(E), \delta)$, we obtain (using Theorem 4.4 of [2]) that

$$\left(S_N(h), \frac{m_N}{N}, \frac{n_N}{N}\right) \to (Z(h), \lambda_0, 1-\lambda_0)$$

in $(D(E) \times E, \delta \times |\ | \times |\ |)$, where $|\ |$ denotes the usual metric in $\boldsymbol{R}$.

We observe that ψ is continuous (with respect to this metric) at all points $f \in D(E)$ which are themselves continuous and all points $(r, s,) \in (0, 1]^2$. Hence

$$S_N\left(h, \frac{m_N}{N}, \frac{n_N}{N}\right) = \psi\left(S_N(h), \frac{m_N}{N}, \frac{n_N}{N}\right) \to \psi(Z(h), \lambda_0, 1-\lambda_0) = Z(h, \lambda_0, 1-\lambda_0)$$

weakly (in $\boldsymbol{R}$) by Theorem 5.1 of [2]. Letting $c = \lambda_0 = r$ in (4.5) we see that $Z(h, \lambda_0, 1-\lambda_0)$ is normally distributed. The variance is explicitly given there.

In order to prove an invariance principle in the situation of the last theorem we need a bit stronger assumptions.

THEOREM 10.2. *Under the assumptions of Theorem* 10.1, *replacing* (10.2) *by*

$$\frac{m_N}{N} \to \lambda_0 \ \text{and}\ \frac{n_N}{N} \to 1-\lambda_0 \ \text{a.s.,} \tag{10.4}$$

it follows that $S_N(h, m_{[uN]}/N, n_{[uN]}/N)$ converges weakly to a Wiener process W with variance given by (10.3). *This convergence is in $D([0, 1])$ with respect to the uniform metric.*

Proof. We proceed as in [18]. Define a random function with values in E by

$$Y_N(u) = \left(\frac{m_{[uN]}}{N}, \frac{n_{[uN]}}{N}\right)$$

(i.e. Y_N has values in $D([0, 1], E)$, the space of functions $v\colon [0, 1] \to E$ having at most discontinuity of the first kind). By (10.4), $Y_N \to \varphi$ a.s. with respect to the sup-metric $\|\ \|_\infty$ in $D([0, 1], E)$, where $\varphi(u) = (u\lambda_0, u(1-\lambda_0))$.

Again by Theorem 4.4 in [2] we have

$$(S_N(h), Y_n) \to (Z(h), \varphi) \tag{10.5}$$

weakly with respect to the metric $\delta \times \|\ \|_\infty$.

Define now ψ: $D(E)\times D([0,1],E)\to D([0,1])$ by $\psi(f,\xi)(u) - f(\xi_1(u),\xi_2(u))$, where $f\in D(E)$, $\xi=(\xi_1,\xi_2)\in D([0,1],E)$, $0\leqslant u\leqslant 1$.

Unfortunately ψ is not continuous at points (f,φ), $f\in C(E)$, with respect to the uniform metric on $D([0,1])$ and the metric $\delta\times\|\ \|_\infty$ on $D(E)\times D([0,1],E)$. However, we can argue as follows. Let $k\in N$ be so large that $\lambda_0\in[k^{-1},1-k^{-1}]$. Define ψ_k: $D(E)\times D([0,1],E)\to D([0,1])$ by

$$\psi_k(f,\xi)=\begin{cases}\psi(f,\xi) & \text{if } \xi([0,1])\subset E_{1/k},\\ \psi(f,\varphi) & \text{otherwise.}\end{cases}$$

We claim that ψ_k is continuous on $C(E)\times\{\varphi\}$. Let $f_0\in C(E)$. Then, given $\varepsilon>0$, there is an $\eta>0$ such that

$$\sup_{\substack{|r-r'|<\eta\\ |s-s'|<\eta\\ (r,s)\in E_{1/2k}\\ (r',s')\in E_{1/2k}}} |f_0(r,s)-f_0(r',s')|<\varepsilon/2 \quad\text{and}\quad 1/k-\eta\geqslant\frac{1}{2k}.$$

Now let $f\in D(E)$ satisfy

$$\sup_{(r,s)\in E_{1/k}} |f(r,s)-f_0(r,s)|<\varepsilon/2$$

and let $\xi\in D([0,1],E)$ satisfy $\|\xi-\varphi\|_\infty<\eta$. Then, if $\xi([0,1])\subset E_{1/k}$, we have

$$\begin{aligned}
&\sup_{0\leqslant u\leqslant 1}|\psi_k(f,\xi)(u)-\psi_k(f_0,\varphi)(u)|\\
&\quad\leqslant \sup_{0\leqslant u\leqslant 1}|\psi(f,\xi)(u)-\psi(f_0,\xi)(u)|+\sup_{0\leqslant u\leqslant 1}|\psi(f_0,\xi)(u)-\psi(f_0,\varphi)(u)|\\
&\quad= \sup_{0\leqslant u\leqslant 1}\big|f(\xi(u))-f_0(\xi(u))\big|+\sup_{0\leqslant u\leqslant 1}\big|f_0(\xi(u))-f_0(\varphi(u))\big|\\
&\quad\leqslant \sup_{(r,s)\in E_{1/k}}|f(r,s)-f_0(r,s)|+\sup_{\substack{(r,s),(r',s')\in E_{1/2k}\\ |r-s|,|r'-s'|<\eta}}|f_0(r,s)-f_0(r',s')|<\varepsilon.
\end{aligned}$$

On the other hand, if $\xi([0,1])\not\subset E_{1/k}$, we have

$$\begin{aligned}
\sup_{0\leqslant u\leqslant 1}|\psi_k(f,\xi)(u)-\psi_k(f_0,\varphi)(u)| &= \sup_{0\leqslant u\leqslant 1}|\psi(f,\varphi)(u)-\psi(f_0,\varphi)(u)|\\
&= \sup_{0\leqslant u\leqslant 1}\big|f(\varphi(u))-f_0(\varphi(u))\big|<\varepsilon/2
\end{aligned}$$

since $\varphi(u)=\big(u\lambda_0,u(1-\lambda_0)\big)\in E_{1/k}$ (because $u\lambda_0/(u\lambda_0+u(1-\lambda_0))=\lambda_0\in[k^{-1},1-k^{-1}]$). Together with (10.5), we conclude from Theorem 5.1 in [2] that $\psi_k(S_N(h),Y_N)\to\psi_k(Z(h),\varphi)=\psi(Z(h),\varphi)$ for every $k\in N$ with $\lambda_0\in[k^{-1},1-k^{-1}]$ weakly in $(D([0,1],\|\ \|_\infty)$, where $\|\ \|_\infty$ also is used for the supnorm on $D([0,1])$. Note that $\psi(Z(h),\varphi)(u)=Z(h,u\lambda_0,u(1-\lambda_0))$ will have the desired properties because of (4.5).

It, therefore, remains to show that for every $\varepsilon > 0$,

$$\lim_{k\to\infty}\overline{\lim_{N\to\infty}}\ P\big(\|\psi_k(S_N(h), Y_N)-\psi(S_N(h), Y_N)\|_\infty \geqslant \varepsilon\big) = 0. \tag{10.6}$$

This follows from the estimate

$$\lim_{k\to\infty}\overline{\lim_{N\to\infty}}\ P\big(\|\psi_k(S_N(h), Y_N)-\psi(S_N(h), Y_N)\|_\infty \geqslant \varepsilon\big)$$

$$\leqslant \lim_{k\to\infty}\overline{\lim_{N\to\infty}}\ P\big(Y_N([0, 1]) \not\subset E_{1/k}\big)$$

$$= \lim_{k\to\infty}\overline{\lim_{N\to\infty}}\ P\left(\exists u\in[0, 1] \text{ such that } \frac{m_{[uN]}}{m_{[uN]}+n_{[uN]}}\notin[k^{-1}, 1-k^{-1}]\right)$$

$$\leqslant \lim_{k\to\infty} P\left(\exists l\in N \text{ such that } \frac{m_l}{m_l+n_l}\notin[k^{-1}, 1-k^{-1}]\right)$$

$$= P\left(\bigcap_k \{\exists l\in N \text{ such that } \frac{m_l}{m_l+n_l}\notin[k^{-1}, 1-k^{-1}]\right) = 0$$

since $m_l/l \to \lambda_0$ and $n_l/l \to 1-\lambda_0$ a.s.

THEOREM 10.3. *Let Y_N be random variables with values in* (0, 1) *such that $Y_N \to \lambda_0$ a.s. Then $S_N(h, Y_N r, Y_N s)$ converges weakly in $(D(E), \delta)$ to the process $Z(h, \lambda_0 r, \lambda_0 s)$, $(r, s)\in E$, provided (h, F) satisfies the assumption of Theorem* 8.2 *or Theorem* 7.1.

Proof. Again by Theorem 4.4 of [2], we get that $(S_N(h), Y_N) \to (Z(h), \lambda_0)$ weakly with respect to the metric $\delta\times|\ |$.

Define the map ψ: $D(E)\times[0, 1]\to D(E)$ by $\psi(f, u)(r, s) = f(ur, us)$. Clearly, ψ is continuous at every point $(f, u)\in C(E)\times[0, 1]$. Our claim follows again from Theorem 5.1 in [2], since $\psi(S_N(h), Y_N)(r, s) = S_N(h, Y_N r, Y_N s)$.

As in [16], tests for symmetry may be considered as a variation of the two-sample problem. Let $\xi_1, \xi_2, \ldots$ be independent, identically distributed random variables with common continuous distribution function π satisfying $0 < \pi(0) < 1$. Define, for $N\in N$,

$$Z_{N,i} = \begin{cases} 1 & \text{if the } i\text{-th smallest of } |\xi_1|, \ldots, |\xi_N| \text{ is from a positive } \xi, \\ 0 & \text{otherwise} \end{cases}$$

With $a_N(i)$, $1 \leqslant i \leqslant N$, as scores, the statistic

$$T_N = \frac{1}{N}\sum_{i=1}^{N} a_N(i) Z_{Ni} \tag{10.7}$$

serves to test the symmetry of π.

In order to write this statistic as a two-sample linear rank statistic, define $X_k(\omega) = \xi_{i_k(\omega)}(\omega)$, $1 \leqslant k \leqslant m_N(\omega)$, where $i_k(\omega) = \min\{j > i_{k-1}(\omega) \mid \xi_j(\omega) \geqslant 0\}$ and $m_N(\omega) = |\{j \mid \xi_j(\omega) \geqslant 0\}|$. Similarly, Y_k is defined using the $\xi_j < 0$. Let $n_N(\omega) = N - m_N(\omega)$. The sequences $\{X_k\}_{k\geqslant 1}$, $\{Y_k\}_{k\geqslant 1}$ and $\{P_k\}_{k\geqslant 1}$ are independent, where $P_k = 1_{[0,\infty)} \circ \xi_k$. The common distribution of the X_k is given by

$$F(x) = \frac{\pi(x) - \pi(0)}{1 - \pi(0)}, \qquad x \geqslant 0,$$

and that of the Y_k's is given by

$$G(x) = \frac{\pi(0) - \pi(-x)}{\pi(0)}, \qquad x > 0,$$

i.e. the distributions are conditional distributions given $\xi_1 \geqslant 0$, resp. $\xi_1 < 0$. If $\pi_0 = \mathrm{P}(\xi_1 \geqslant 0)$, then $P_k (k \geqslant 1)$ are independent and identically distributed Bernoulli random variables with parameter π_0.

Clearly, $Z_{N,i} = 1$ if the i-th smallest among $X_1, \ldots, X_{m_N}, Y_1, \ldots, Y_{n_N}$ is an X-observation and then i is the rank of that observation. Therefore

$$T_N = \frac{m_N}{N} \int_0^1 h\left(\frac{N}{N+1} \hat{H}_{m_N, n_N}(t)\right) dF_{m_N}(t)$$

if the scores are given by (2.2) (cf. 2.4), and hence we may study the statistic

(10.8) $\quad \Sigma_N(u)$

$$= \frac{[uN]}{\sqrt{N}} \left(\frac{m_{[uN]}}{[uN]} \int_0^1 h\left(\frac{[uN]}{[uN]+1} \hat{H}_{m_{[uN]}, n_{[uN]}}(t)\right) dF_{m_{[uN]}}(t) - \pi_0 \int_0^1 h(H_{\pi_0}(t))\, dF(t)\right).$$

We first prove the following

Lemma 10.1. *If (h, F) satisfies the assumptions of section 8, then*

$$\frac{d}{d\lambda}\left[\int_0^1 F(H_\lambda^{-1}(t))\, dh(t)\right](\lambda_0) = \int_0^1 a_{\lambda_0}(t)\left(G(H_{\lambda_0}^{-1}(t)) - F(H_{\lambda_0}^{-1}(t))\right) dh(t),$$

where a_λ is defined as in (9.5), i.e. $a_\lambda = (F \circ H_\lambda^{-1})'$ whenever it exists.

Proof. We have

$$\frac{d}{d\lambda}\left[\int_0^1 F \circ H_\lambda^{-1}(t)\, dh(t)\right](\lambda_0) = \lim_{\lambda \to \lambda_0} \int_0^1 \frac{F \circ H_\lambda^{-1}(t) - F \circ H_{\lambda_0}^{-1}(t)}{\lambda - \lambda_0}\, dh(t)$$

as in the proof of Lemma 8.2. Hence it suffices to show that, for $d|h|$ almost all t,

$$\frac{F \circ H_\lambda^{-1}(t) - F \circ H_{\lambda_0}^{-1}(t)}{\lambda - \lambda_0} \to a_{\lambda_0}(t)\left(G \circ H_{\lambda_0}^{-1}(t) - F \circ H_{\lambda_0}^{-1}(t)\right).$$

Write

$$\frac{F\circ H_\lambda^{-1}(t)-F\circ H_{\lambda_0}^{-1}(t)}{\lambda-\lambda_0}$$

$$=\frac{F\circ H_{\lambda_0}^{-1}\circ H_{\lambda_0}\circ H_\lambda^{-1}(t)-F\circ H_{\lambda_0}^{-1}(t)}{H_{\lambda_0}\circ H_\lambda^{-1}(t)-t}\,\frac{H_{\lambda_0}\circ H_\lambda^{-1}(t)-t}{\lambda-\lambda_0}.$$

The first factor converges $d|h|$ a.s. to $(F\circ H_{\lambda_0}^{-1})'(t)=a_{\lambda_0}(t)$ and the second factor can be written in the form

$$\frac{H_{\lambda_0}\circ H_\lambda^{-1}(t)-t}{\lambda-\lambda_0}=\frac{(\lambda_0-\lambda)F(H_\lambda^{-1}(t))+(\lambda-\lambda_0)G(H_\lambda^{-1}(t))}{\lambda-\lambda_0}$$

$$=G(H_\lambda^{-1}(t))-F(H_\lambda^{-1}(t))$$

and converges to $G(H_{\lambda_0}^{-1}(t))-F(H_{\lambda_0}^{-1}(t))$ as $\lambda\to\lambda_0$ by continuity.

The following theorem extends a result of Schulze–Pillot [18].

THEOREM 10.4. *In the situation described above, let (h, F) satisfy the assumptions of section* 8. *Then, the statistics $\Sigma_N(u)$, defined by* (10.8), *converges weakly in $(D([0,1]), \|\ \|_\infty)$ to the process*

$$\Sigma(u)=Z(h, u\pi_0, u(1-\pi_0))-\sqrt{\pi_0(1-\pi_0)}\Big[\int_0^1 F(H_{\pi_0}^{-1}(t))\,dh(t)+$$

$$+\pi_0\int_0^1 a_{\pi_0}(t)\big(G(H_{\pi_0}^{-1}(t))-F(H_{\pi_0}^{-1}(t))\big)\,dh(t)\Big]\,W(u),$$

where W denotes a standard Brownian motion independent of $Z(h)$, which is defined in (4.1).

Proof. We will proceed as in the proof of Theorem 10.2. Define the random functions $Y_N(u)=(m_{[uN]}/N,\ n_{[uN]}/N)$ as before and write

$$\lambda_N(u)=\frac{m_{[uN]}}{[uN]},\qquad V_N(u)=\frac{[uN]}{\sqrt{N}}(\lambda_N(u)-\pi_0)\qquad(0\leqslant u\leqslant 1)\tag{10.9}$$

and

$$l_N(u)=-\int F(H_{\lambda_N(u)}^{-1}(t))\,dh(t)-\tag{10.10}$$

$$-\pi_0\frac{\int_0^1\big(F(H_{\lambda_N(u)}^{-1}(t))-F(H_{\pi_0}^{-1}(t))\big)\,dh(t)}{\lambda_N(u)-\pi_0}\quad\text{if } \lambda_N(u)-\pi_0\neq 0,$$

$$l_N(u)=0\ \text{ if } \lambda_N(u)=\pi_0\qquad(0\leqslant u\leqslant 1).$$

Integration by parts gives

$$\Sigma_N(u)=S_N\left(h,\frac{m_{[uN]}}{N},\frac{n_{[uN]}}{N}\right)+V_N(u)\,l_N(u).\tag{10.11}$$

Let us define the map ψ: $D(E)\times D([0,1])\times D([0,1],E)\times D([0,1]) \to D([0,1])$ by $\psi(f,g,\varphi,\eta)(u)=f(\varphi_1(u),\varphi_2(u))+g(u)\eta(u)$ for $0\leqslant u\leqslant 1$, $f\in D(E)$, g, $\eta\in D([0,1])$ and $\varphi=(\varphi_1,\varphi_2)\in D([0,1],E)$. Here the difficulty of the non-continuity of ψ occurs again, but we can proceed as before.

Let k be so large that $\pi_0\in[k^{-1},1-k^{-1}]$ and define

$$\psi_k(f,g,\varphi,\eta)=\begin{cases}\psi(f,g,\varphi,\eta) & \text{if } \varphi([0,1])\subset E_{1/k},\\ \psi(f,g,\varphi_0,\eta) & \text{otherwise,}\end{cases}$$

where $\varphi_0(u)=(u\pi_0, u(1-\pi_0))$, and

$$v_k(t)=\begin{cases}0, & 0\leqslant t\leqslant 1/2k,\\ 2kt-1, & 1/2k\leqslant t\leqslant 1/k,\\ 1, & 1/k\leqslant t\leqslant 1.\end{cases}$$

By Theorem 8.2 we have $S_N(h)\to Z(h)$ weakly in $(D(E),\delta)$. The strong law of large numbers implies that $\lambda_N(1)\to\pi_0$ a.s., consequently $Y_N\to\varphi_0$ uniformly in u a.s. The application of Donsker's theorem yields $V_N \to\sqrt{\pi_0(1-\pi_0)}\,W$ weakly in $(D([0,1]),\|\ \|_\infty)$, and since $\{P_k\}_{k\geqslant 1}$ is independent of $(X_m, Y_n\colon n,m\geqslant 1)$, W is a standard Brownian motion, independent of $Z(h)$. Setting

$$\gamma=-\int_0^1 F(H_{\pi_0}^{-1}(t))\,dh(t)-\pi_0\int_0^1 a_{\pi_0}(t)\big(G(H_{\pi_0}^{-1}(t))-F(H_{\pi_0}^{-1}(t))\big)\,dh(t),$$

the strong law of large numbers and Lemma 10.1 imply that $l_N v_k\to\gamma v_k$ a.s. From Theorems 4.4 and 5.2. of [2] it follows that

$$(S_N(h), V_N, Y_N, l_N v_k)\to\big(Z(h),\sqrt{\pi_0(1-\pi_0)}\,W,\varphi_0,\gamma v_k\big)$$

weakly with respect to the metrics δ on $D(E)$, $\|\ \|_\infty$ on $D([0,1])$ and on $D([0,1],E)$. ψ_k is continuous at all points (f,g,φ_0,η) with $f\in C(E)$ and g, $\eta\in C([0,1])$ (cf. the proof of Theorem 10.2). Using Theorem 5.1 of [2], we obtain, for $0\leqslant u\leqslant 1$,

$$\big(\psi_k(S_N(h), V_N, Y_N, l_N v_k)(u)\big)\to\big(Z(h,u\pi_0,u(1-\pi_0))+\sqrt{\pi_0(1-\pi_0)}\,\gamma v_k(u)\,W(u)\big)$$

weakly in $(D([0,1]),\|\ \|_\infty)$.

Since

$$\mathrm{P}\big(\|\Sigma-Z(h,\cdot\,\pi_0,\cdot(1-\pi_0))-\sqrt{\pi_0(1-\pi_0)}\,\gamma v_k(\cdot)\,W(\cdot)\|_\infty\geqslant\varepsilon\big)$$
$$=\mathrm{P}\big(\sup_{0\leqslant u\leqslant k^{-1}}\big|\sqrt{\pi_0(1-\pi_0)}\,\gamma(1-v_k(u))\,W(u)\big|\geqslant\varepsilon\big)\to 0 \quad \text{as } k\to\infty$$

(since $\lim_{u\to 0}W(u)=0$ a.s.), we obtain

$$\psi_k\big(Z(h,\cdot\,\pi_0,\cdot(1-\pi_0)),\sqrt{\pi_0(1-\pi_0)}\,W,\varphi_0,\gamma v_k\big)\to\Sigma$$

weakly in $(D([0,1], \|\ \|_{\infty})$ and hence it is left to show that $\psi_k(S_N(h), V_N, Y_N, l_N v_k) - \psi(S_N(h), V_N, Y_N, l_N)$ converges to zero in probability with respect to the metric $\|\ \|_{\infty}$ on $D([0,1])$ (cf. the proof of Theorem 10.2).

Let $\varepsilon > 0$. Then

$$P\left(\|\psi_k(S_N(h), V_N, Y_N, l_N v_k) - \psi(S_N(h), V_N, Y_N, l_N)\|_{\infty} \geqslant \varepsilon\right)$$

$$\leqslant P\left(Y_N([0,1]) \not\subset E_{k-1}\right) + P\left(\sup_{0 \leqslant u \leqslant 1} |V_N(u) l_N(u)(1 - v_k(u))| \geqslant \varepsilon\right).$$

The first term tends to zero using the same arguments as in the proof of Theorem 10.2. Since $1 - v_k(u) \to 0$ as $k \to \infty$ uniformly on compact sets in (0, 1), by Lemma 10.1 and by the weak convergence of V_N to $\sqrt{\pi_0(1-\pi_0)}\, W$, we infer that $\max V_N(u) l_N(u)$ is bounded and

$$\max_{u \leqslant u_0} V_N(u) l_N(u) \to 0 \qquad \text{as } u_0 \to 0.$$

Consequently, the second term tends to zero as well. Therefore

$$\lim_{k \to \infty} \overline{\lim_{N \to \infty}} P\left(\|\psi_k(S_N(h), V_N, Y_N, l_N v_k) - \psi(S_N(h), V_N, Y_N, l_N)\|_{\infty} \geqslant \varepsilon\right) = 0.$$

This completes the proof of the theorem.

REFERENCES

[1] P. Bickel and M. F. Wichura, *Convergence criteria for multiparameter stochastic processes and some applications*, Ann. Math. Statist. 42 (1971), p. 1656-1670.

[2] P. Billingsley, *Convergence of Probability Measures*, Wiley, New York 1968.

[3] H. I. Braun, *Weak convergence of sequential linear rank statistics*, Ann. Statist. 4 (1976), p. 554-576.

[4] H. Chernoff and I. R. Savage, *Asymptotic normality and efficiency of certain nonparametric test statistic*, Ann. Math. Statist. 29 (1958), p. 972-994.

[5] M. Denker and U. Rösler, *Some contributions to Chernoff-Savage theorems*, Statistics and Decision 3 (1985), p. 49-75.

[6] — *A note on asymptotic normality for rank statistics*, Proc. Conf. on Limit Theorems in Prob. and Stat., Veszprém 1982, p. 405-425.

[7] M. Denker, M. L. Puri and U. Rösler, *A sharpening of the remainder term in the higher dimensional CLT for multivariate rank statistics*, J. Multiv. Anal. 17 (1985), p. 148–167.

[8] J. L. Doob, *Stochastic processes*, Wiley, New York 1953.

[9] V. Dupač and J. Hájek, *Asymptotic normality of simple linear rank statistics under alternatives II*, Ann. Math. Statist. 40 (1969), p. 1992-2017.

[10] A. A. Filippova, *Mises' theorem on the asymptotic behavior of functionals of empirical distribution functions and its statistical applications*, Teoria Veroyat. i ee Prim. 7 (1961), p. 24-57.

[11] J. Hájek, *Asymptotic normality of simple linear rank statistics under alternatives*, Ann. Math. Statist. 39 (1968), p. 325-346.

[12] T. L. Lai, *On Chernoff-Savage statistics and sequential rank tests*, Ann. Statist. 3 (1975), p 825-845.

[13] F. Moricz, *Moment inequalities for the maximum of partial sums of random fields*, Acta Sci. Math. Hungar. 39 (1977), p. 353-366.

[14] G. Neuhaus, *On weak convergence of stochastic processes with multidimensional time parameter*, Ann. Math. Statist. 42 (1971), p. 1285-1295.

[15] R. Pyke and G. R. Shorack, *Weak convergence of a two sample empirical process and a new approach to Chernoff-Savage theorems*, ibidem 39 (1968), p. 755-771.

[16] – *Weak convergence and a Chernoff-Savage theorem for random sample sizes*, ibidem 39 (1968), p. 1675-1685.

[17] – *A note on Chernoff-Savage theorems*, ibidem 40 (1969), p. 1116-1119.

[18] T. Schulze-Pillot, *Über Invarianzprinzipien für lineare Rang-Statistiken*, Dissertation, Göttingen 1978.

Institut für Mathematische Stochastik
Universität Göttingen
Lotzestr. 13
3400 Göttingen, FRG

Dept. of Mathematics
Indiana University
Bloomington, Indiana 47405
USA

LIMITING BEHAVIOR OF U-STATISTICS, V-STATISTICS, AND ONE SAMPLE RANK ORDER STATISTICS FOR NONSTATIONARY ABSOLUTELY REGULAR PROCESSES

MICHEL HAREL and MADAN L. PURI*

Institut Universitaire de Technologie de Limoges et U.A. C.N.R.S., Rouen, France and Indiana University

The weak convergences of U- and V-statistics were established by Yoshihara (1976, *Z. Warsch. Verw. Gebiete* **35** 237–252) for stationary absolutely regular processes. Later Yoshihara (1978, *Z. Warsch. Verw. Gebiete* **43** 101–127) also proved the weak convergence of one sample rank order statistics under similar conditions. In this paper, we extend some of Yoshihara's results to the nonstationary cases.

1. Introduction

Let X_{ni}, $1 \leqslant i \leqslant n$, $n \geqslant 1$ be real-valued r.v.'s (random variables) with continuous d.f.'s (distribution functions) $F_{ni}(x)$, $x \in \mathbb{R}$. The d.f. of the $\mathbb{R}^n$-valued r.v. $(X_{n1}, ..., X_{nn})$ is denoted by $\mathbf{F}_n$.

In the first part of the paper, we will study the asymptotic behavior of the U-statistic

$$U(\mathbf{F}_n) = \binom{n}{k}^{-1} \sum_{(i)}^{(n)} g(X_{ni_1}, ..., X_{ni_k}), \qquad n \geqslant k \geqslant 1, \tag{1.1}$$

where the summation $\sum_{(i)}^{(n)}$ extends over all possible $1 \leqslant i_1 < \cdots < i_k \leqslant n$ and g: $\mathbb{R}^k \to \mathbb{R}$ is a Borel measurable function which is symmetric in its k $(\geqslant 1)$ arguments.

Received February 1988; accepted September 2, 1988.

AMS 1980 subject classifications: Primary 60B10, 60F05.

Key words and phrases: weak convergence, U-statistics, V-statistics, one sample rank order statistics, absolute regularity, strong mixing, Brownian process.

* Research supported by the Office of Naval Research Contract N00014-85-K-0648.

From the study of $U(\mathbf{F}_n)$, we will deduce the asymptotic behavior of the V-static

$$V(\mathbf{F}_n)=n^{-k}\sum_{i_1=1}^{n}\cdots\sum_{i_k=1}^{n}g(X_{ni_1},\ldots,X_{ni_k}) \tag{1.2}$$

called the Von Mises' statistic.

We assume that the underlying r.v.'s are absolutely regular with rates

$$\beta(m)=O(m^{-(2+\delta)/\delta}) \qquad \text{for some} \quad \delta>0. \tag{1.3}$$

Recall that the sequence is absolutely regular if

$$\sup_{m\leqslant n}\max_{1\leqslant j\leqslant n-m}E\{\sup_{A\in\sigma(X_{ni},i\geqslant j+m)}|P(A\mid\sigma(X_{ni},1\leqslant i\leqslant j))-P(A)|\}=\beta(m)\downarrow 0.$$

Here $\sigma(X_{ni},\ 1\leqslant i\leqslant j)$ and $\sigma(X_{ni},\ i\geqslant j+m)$ are the σ-fields generated by $(X_{n1},\ldots,X_{nj})$ and $(X_{n,j+m},\ X_{n,j+m+1},\ldots,X_{nn})$, respectively. Also recall that $\{X_{ni}\}$ satisfies the strong mixing condition if $\sup_{m\leqslant n}\sup_{1\leqslant j\leqslant n-m}\{|P(A\cap B)-P(A)\,P(B)|;\ A\in\sigma(X_{ni},\ 1\leqslant i\leqslant j),\ B\in\sigma(X_{ni},\ i\geqslant j+m)\}=\alpha(m)\downarrow 0$. Since $\alpha(m)\leqslant\beta(m)$, it follows that if $\{X_{ni}\}$ is absolutely regular, then it is also strong mixing.

In the second part of the paper, we will study the asymptotic behavior of the one sample rank order statistic $\mathscr{S}_{n,m}$

$$\mathscr{S}_{n,m}=\sum_{i=1}^{m}c_{ni}s(X_{ni})\,J\left(\frac{R_{n,m,i}}{m+1}\right), \qquad n\geqslant m\geqslant 1, \tag{1.4}$$

where J is a score function, $s(x)=\operatorname{sgn}(x)$, and the c_{ni} are regression constants defined by a continuous function $h(x)$ on $[0,1]$ as

$$c_{ni}=h\left(\frac{i}{n}\right), \qquad 1\leqslant i\leqslant n,\ n\geqslant 1 \tag{1.5}$$

and $R_{n,m,i}=\sum_{j=1}^{m}I_{[|X_{nj}|\leqslant|X_{ni}|]}$, $1\leqslant i\leqslant m\leqslant n$, where $I_{[\ ]}$ denotes the indicator function. We assume also that the r.v.'s are absolutely regular but with rates

$$\beta(m)=O(m^{-4}). \tag{1.6}$$

The convergence of the U-statistic as well as the V-statistic was established by Yoshihara [8] and afterwards, for the one sample rank order statistic also, by Yoshihara [9] but only for stationary r.v.'s. Later, Denker and Keller [3] proved some limit theorems of the above-mentioned statistics for the processes which are uniformly mixing in both directions of time. In this paper we extend the results of Yoshihara [8, 9] from the stationary cases to the nonstationary cases.

Our results are the adaptions of some of the ideas of Yoshihara (loc. cit.) and a new central limit theorem (Lemma 2.3) for nonstationary unbounded strongly mixing random variables which is an extension of a central limit theorem due to Withers [6] for nonstationary uniformly bounded strongly mixing random variables.

To prove the weak convergence of the U-statistic $U(\mathbf{F}_n)$ in Theorem 3.1, we first decompose it as $\theta(\mathbf{F}_n)+\sum_{p=1}^{k}\binom{k}{p}U_n^{(p)}$, where $\theta(\mathbf{F}_n)$ is a centering constant and the $U_n^{(p)}$'s are some random variables (defined in (2.2)). Using Lemma 2.3, we prove that for $p=1$, $n^{1/2}U_n^{(p)}$ has a limiting normal distribution, and for $p\geqslant 2$, $n^{1/2}U_n^{(p)}\to 0$ in probability, as $n\to\infty$ (Lemma 2.2). The limiting behavior of the Von Mises V-statistic $V(\mathbf{F}_n)$ is the same as that of the U-statistic $U(\mathbf{F}_n)$, since $n^{1/2}\,|U(\mathbf{F}_n)-V(\mathbf{F}_n)|\to 0$ in probability (Corollary 3.1). To prove the weak convergence of rank statistics in Theorem 4.1, we have to verify that the finite-dimensional distributions of the process $L_n(t)$ defined in (4.7) satisfy the conditions of our Lemma 2.3, and the probability measure defined by $L_n(t)$ is tight.

For reasons of brevity and to avoid repetitious arguments, we have either omitted or given brief outlines of the proofs that are similar to those of Yoshihara.

2. Preliminaries

2.1. *Definitions*

For every p $(1\leqslant p\leqslant k)$ and $n\geqslant k$, let $1\leqslant i_1\neq\cdots\neq i_p\leqslant n$ be arbitrary integers. Put

$$g_{p,n}^{(i_1,\ldots,i_p)}(x_1,\ldots,x_p)=\sum_{(i_{p+1},\ldots,i_k)\in I_{p,n}(i_1,\ldots,i_p)}\lambda(x_1,\ldots,x_p), \tag{2.1}$$

where

$$\lambda(x_1,\ldots,x_p)=\int_{\mathbb{R}^{k-p}}g(x_1,\ldots,x_k)\,dF_{ni_{p+1}}(x_{p+1})\cdots dF_{ni_k}(x_k),$$

and $I_{p,n}(i_1,\ldots,i_p)=\{(i_{p+1},\ldots,i_k);\;\; 1\leqslant i_{p+1}\neq\cdots\neq i_k\leqslant n,\;\; i_l\notin\{i_1,\ldots,i_p\},\; p+1\leqslant l\leqslant k\}$ and $g_{0,n}=\sum_{(i_1,\ldots,i_k)\in I_{0,n}}\int_{\mathbb{R}^k}g(x_1,\ldots,x_k)\,dF_{ni_1}\cdots dF_{ni_k}$, where $I_{0,n}=\{(i_1,\ldots,i_k);\,1\leqslant i_1\neq\cdots\neq i_k\leqslant n\}$.

For every p $(1\leqslant p\leqslant k)$, set

$$U_n^{(p)}=n^{-[k]}\sum_{1\leqslant i_1\neq\cdots\neq i_p\leqslant n}\int_{\mathbb{R}^p}g_{p,n}^{(i_1,\ldots,i_p)}(x_1,\ldots,x_p)\times\prod_{j=1}^{p}d(I_{[X_{ni_j}\leqslant x_j]}-F_{ni_j}(x_j)) \tag{2.2}$$

where $n^{-[k]}=n(n-1)\cdots(n-k+1)$.

Also for every p, $(0 \leqslant p \leqslant k)$, we write

$$g^*_{p,n}(x_1, ..., x_p) = \sum_{(i_{p+1}, ..., i_k) \in \{1, ..., n\}^{k-p}} \times \int_{\mathbb{R}^{k-p}} g(x_1, ..., x_k)\, dF_{ni_{p+1}} \cdots dF_{ni_k} \tag{2.3}$$

and

$$V_n^{(p)} = n^{-k} \sum_{i_1=1}^{n} \cdots \sum_{i_p=1}^{n} g^*_{p,n}(x_1, ..., x_p) \times \prod_{j=1}^{p} d(I_{[X_{ni_j} \leqslant x_j]} - F_{ni_j}(x_j)). \tag{2.4}$$

Let $p \geqslant 1$ and $1 \leqslant i_1 < i_2 < \cdots < i_p \leqslant n$ be arbitrary integers. For any j $(1 \leqslant j \leqslant p-1)$, $P_{j,n}^{(i_1, ..., i_p)}$ is the probability measure defined by

$$P_{j,n}^{(i_1, ..., i_p)}(A^{(j)} \times B^{(p-j)}) = P[(X_{ni_1}, ..., X_{ni_j}) \in A^{(j)}] \times P[(X_{ni_{j+1}}, ..., X_{ni_p}) \in B^{(p-j)}]$$

and $P_{0,n}^{(i_1, ..., i_p)}$ is the probability measure defined by

$$P_{0,n}^{(i_1, ..., i_p)}(A^{(p)}) = P[(X_{ni_1}, ..., X_{ni_p}) \in A^{(p)}]$$

for any $A^{(j)} \in \sigma(X_{ni_1}, ..., X_{ni_j})$ $(1 \leqslant j \leqslant p)$ and any $B^{(p-j)} \in \sigma(X_{ni_{j+1}}, ..., X_{ni_p})$ $(1 \leqslant j \leqslant p-1)$.

For any function h: $\mathbb{R}^k \to \mathbb{R}$, we denote by $h \otimes h$ the function $h \otimes h$: $\mathbb{R}^{2k} \to \mathbb{R}$ defined by $h \otimes h(x_1, ..., x_{2k}) = h(x_1, ..., x_k)\, h(x_{k+1}, ..., x_{2k})$.

2.2. *Basic Lemmas*

LEMMA 2.1. *For every $p \geqslant 1$ and $(i_1, ..., i_p)$ such that $i_1 < i_2 < \cdots < i_p$ and any j $(0 \leqslant j \leqslant p-1)$, let $h(x_1, ..., x_p)$ be a Borel function such that*

$$\int_{\mathbb{R}^p} |h(x_1, ..., x_p)|^{1+\delta}\, dP_{j,n}^{(i_1, ..., i_p)} \leqslant M$$

for some $\delta > 0$, then

$$\left| \int_{\mathbb{R}^p} h(x_1, ..., x_p)\, dP_{0,n}^{(i_1, ..., i_p)} - \int_{\mathbb{R}^p} h(x_1, ..., x_p)\, dP_{j,n}^{(i_1, ..., i_p)} \right| \leqslant 4M^{1/(2+\delta)} \beta^{\delta/(1+\delta)}(i_{j+1} - i_j). \tag{2.5}$$

As a special case, if $h(x_1, \ldots, x_p)$ is bounded, say, $|h(x_1, \ldots, x_p| \leqslant M$, then we can replace the right side of (2.5) by $2M\beta(i_{j+1} - i_j)$.

Proof. Follows from Lemma 1 of Yoshihara [8].

LEMMA 2.2. *If there is a positive number δ' such that for $r = 2 + \delta'$,*

$$\sup_n \max_{(i_1, \ldots, i_k) \in \{1, \ldots, n\}^k} \int_{\mathbb{R}^k} |g(x_1, \ldots, x_k)|^r \, dF_{ni_1} \cdots dF_{ni_k} \leqslant M_0 < \infty \tag{2.6}$$

$$\sup_n \max_{(i_1, \ldots, i_k) \in \{1, \ldots, n\}^k} E(|g(X_{ni_1}, \ldots, X_{ni_k})|^r) \leqslant M_0 < \infty \tag{2.7}$$

and for some δ $(0 < \delta < \delta')$, $\beta(n) = O(n^{-(2+\delta)/\delta})$, then we have

$$E(U_n^{(p)})^2 = O(n^{-1-\gamma}), \qquad 2 \leqslant p \leqslant k, \tag{2.8}$$

$$E(V_n^{(p)})^2 = O(n^{-1-\gamma}), \qquad 1 \leqslant p \leqslant k, \tag{2.9}$$

where $\gamma = 2(\delta - \delta')/\delta'(2 + \delta) > 0$.

Proof. It suffices to prove (2.8) because the proof of (2.9) is similar. Also it suffices to prove (2.8) for $p = 2$; the proofs for $p = 3, \ldots, k$ are analogous and are therefore omitted. We first note that

$$U_n^{(2)} = n^{-[k]} \sum_{1 \leqslant i_1 \neq i_2 \leqslant n} \int_{\mathbb{R}^2} g_{2,n}^{(i_1, i_2)}(x_1, x_2) \, d(I_{[X_{ni_1} \leqslant x_1]} - F_{ni_1}(x_1)) \, d(I_{[X_{ni_2} \leqslant x_2]} - F_{ni_2}(x_2)), \tag{2.10}$$

so we have

$$E(U_n^{(2)})^2 = (n^{-[k]})^2 \sum_{1 \leqslant i_1 \neq i_2 \leqslant n} \sum_{1 \leqslant j_1 \neq j_2 \leqslant n} J((i_1, i_2), (j_1, j_2)), \tag{2.11}$$

where

$$\begin{aligned} J((i_1, i_2), (j_1, j_2)) = E \Big\{ & \int_{\mathbb{R}^2} g_{2,n}^{(i_1, i_2)}(x_1, x_2) \, d(I_{[X_{ni_1} \leqslant x_1]} - F_{ni_1}(x_1)) \\ & \times d(I_{[X_{ni_2} \leqslant x_2]} - F_{ni_2}(x_2)) \Big\} \\ & \times \Big\{ \int_{\mathbb{R}^2} g_{2,n}^{(j_1, j_2)}(y_1, y_2) \, d(I_{[X_{nj_1} \leqslant y_1]} - F_{nj_1}(y_1)) \\ & \times d(I_{[X_{nj_2} \leqslant y_2]} - F_{nj_2}(y_2)) \Big\} \end{aligned} \qquad (2.12$$

and

$$g_{2,n}^{(i_1,i_2)}(x_1, x_2) = \sum_{(i_3,\dots,i_k)\in I_{2,n}(i_1,i_2)} \int_{\mathbb{R}^{k-2}} g(x_1, {}_2, \dots, x_k)\, dF_{ni_3}(x_3)\cdots dF_{ni_k}(x_k).$$

We deduce

$$\begin{aligned} J((i_1, i_2), (j_1, j_2)) = & \sum_{(i_3,\dots,i_k)\in I_{2,n}(i_1,i_2)} \sum_{(j_3,\dots,j_k)\in I_{2,n}(j_1,j_2)} \\ & \times E\Bigg(\Bigg\{\int_{\mathbb{R}^2}\int_{\mathbb{R}^{k-2}} g(x_1, x_2, \dots, x_k) \\ & \times dF_{ni_3}(x_3)\cdots dF_{ni_k}(x_k)(d(I_{[X_{ni_1}\leqslant x_2]} - F_{ni_1}(x_1)) \\ & \times d(I_{[X_{ni_2}\leqslant x_2]} - F_{ni_2}(x_2))\Bigg\} \\ & \times\Bigg\{\int_{\mathbb{R}^2}\int_{\mathbb{R}^{k-2}} g(y_1, y_2, \dots, y_k)\, dF_{nj_3}(y_3)\cdots dF_{nj_k}(y_k) \\ & \times (d(I_{[X_{nj_1}\leqslant y_1]} - F_{nj_1}(y_1)) \\ & \times d(I_{[X_{nj_2}\leqslant y_2]} - F_{nj_2}(y_2))\Bigg\}\Bigg), \end{aligned} \tag{2.13}$$

since for any $(i_1, \dots, i_k)\in I_{0,n}$,

$$\begin{aligned} &\int_{\mathbb{R}}\Bigg(\int_{\mathbb{R}^{k-2}} g(x_1, x_2, \dots, x_k)\, dF_{ni_3}(x_3)\cdots dF_{ni_k}(x_k) \\ &\quad -\int_{\mathbb{R}^{k-1}} g(x_1, x_2, \dots, x_k)\, dF_{ni_2}(x_2)\cdots dF_{ni_k}(x_k) \\ &\quad -\int_{\mathbb{R}^{k-1}} g(x_1, x_2, \dots, x_k)\, dF_{ni_1}(x_1)\, dF_{ni_3}(x_3)\cdots dF_{ni_k}(x_k) \\ &\quad +\int_{\mathbb{R}^{k}} g(x_1, x_2, \dots, x_k)\, dF_{ni_1}(x_1)\cdots dF_{ni_k}(x_k)\Bigg)\, dF_{ni_j}(x_j) = 0, \qquad j = 1, 2. \end{aligned}$$

So from Lemma 2.1 we have the inequalities

(i) If $1 \leqslant i_1 < i_2 \leqslant j_1 < j_2 \leqslant n$ and $j_2 - j_1 \geqslant i_2 - i_1$, then

$$\begin{aligned} &J((i_1, i_2), (j_1, j_2)) \\ &\quad\leqslant \sum_{(i_3,\dots,i_k)\in I_{2,n}(i_1,i_2)} \sum_{(j_3,\dots,j_k)\in I_{2,n}(j_1,j_2)} M_0^{1/(1+\delta')}\beta^{\delta'/(2+\delta')}(j_2 - j_1) \\ &\quad= \left(\frac{n^{[k]}}{n(n-1)}\right)^2 M_0^{1/(1+\delta')}\beta^{\delta'/(2+\delta')}(j_2 - j_1) \end{aligned} \tag{2.14}$$

and similarly if $1 \leqslant i_1 < i_2 \leqslant j_1 < j_2 \leqslant n$ and $i_2 - i_1 \geqslant j_2 - j_1$, then

$$J((i_1, i_2), (j_1, j_2)) \leqslant \left(\frac{n^{[k]}}{n(n-1)}\right)^2 M_0^{1/(1+\delta')} \beta^{\delta'/(2+\delta')}(i_2 - i_1). \quad (2.15)$$

Thus, using (2.14), (2.15), and (1.3), we obtain

$$\left| \sum_{1 \leqslant i_1 < i_2 \leqslant j_1 < j_2 \leqslant n} J((i_1, i_2), (j_1, j_2)) \right| = O(n^{2k-1-\gamma}). \quad (2.16)$$

Similarly,

$$\left| \sum_{1 \leqslant i_1 < j_1 \leqslant i_2 < j_2 \leqslant n} J((i_1, i_2), (j_1, j_2)) \right| = O(n^{2k-1-\gamma}) \quad (2.17)$$

$$\left| \sum_{1 \leqslant i_1 < j_1 < j_2 < i_2 \leqslant n} J((i_1, i_2), (j_1, j_2)) \right| = O(n^{2k-1-\gamma}) \quad (2.18)$$

$$\left| \sum_{1 \leqslant i_1, j_1 \leqslant n} \sum_{i_2 = 1}^{n} J((i_1, i_2), (j_1, i_2)) \right| = O(n^{2k-2}). \quad (2.19)$$

From (2.16)–(2.19) and (2.11) we have (2.8) for $p = 2$.

LEMMA 2.3. *Let $\{X_{ni}^*\}$ be r.v.'s (with mean 0) which are strong mixing with rates satisfying*

$$\sum_{m=1}^{\infty} (\alpha(m))^{\delta''/(2+\delta'')} < \infty \qquad \textit{for some} \quad \delta'' > 0. \quad (2.20)$$

Suppose that for any $K > 0$ there exists a sequence $\{Y_{ni}^K\}$ of r.v.'s satisfying (2.20) such that

$$\sup_{n \in \mathbb{N}^*} \max_{1 \leqslant i \leqslant n} |Y_{ni}^K| \leqslant B_K < \infty, \forall K > 0, \quad (2.21)$$

where B_K is some constant > 0;

$$\sup_{n \in \mathbb{N}^*} \max_{1 \leqslant i \leqslant n} E|X_{ni}^* - Y_{ni}^K|^{2+\delta''} \to 0 \qquad \textit{as} \quad K \to \infty \quad (2.22)$$

$$E\left(\sum_{i=1}^{n} X_{ni}^*\right)^2 \Big/ n \to C < \infty \textit{ as } n \to \infty, \textit{ where } C \textit{ is some constant} > 0; \quad (2.23)$$

$$E \sum_{i=1}^{n} (Y_{ni}^K - E(Y_{ni}^K))^2/n \to C_K < \infty \qquad \forall K > 0 \textit{ as } n \to \infty, \quad (2.24)$$

where C_K is some constant > 0;

$$C_K \to C \qquad \textit{as} \qquad K \to \infty; \quad (2.25)$$

then $n^{-1/2}\sum_{i=1}^{n} X_{ni}^{*}$ *converges in law to the normal distribution with mean* 0 *and variance* C^2.

Proof. From Corollary 1 of Withers [6], Lemma 2.3 holds if $\{X_{ni}^{*}\}$ is uniformly bounded. Thus, $\forall K>0$, $n^{-1/2}\sum_{i=1}^{n}(Y_{ni}^{K}-E(Y_{ni}^{K}))$ converges in law to the normal distribution with mean 0 and variance C_K^2.

Now denoting $Z_{ni}^{K}=X_{ni}^{*}-Y_{ni}^{K}$ and using Ibragimov [5], we represent the sum $S_n=n^{-1/2}C^{-1}(\sum_{i=1}^{n}X_{ni}^{*})$ as $S_n=S'_n+S''_n$, where $S'_n=n^{-1/2}C_K^{-1}\sum_{i=1}^{n}(Y_{ni}^{K}-E(Y_{ni}^{K}))\,C_K C^{-1}$ and $S''_n=n^{-1/2}C^{-1}\sum_{i=1}^{n}(Z_{ni}^{K}-E(Z_{ni}^{K}))$. Then, using the well-known inequality on moments of strong mixing sequences of r.v.'s [4, Proposition 8] and omitting the details of computations, we obtain

$$E\,|S''_n|^2 \leqslant 2n^{-1}C^{-2}\sum_{i=0}^{n-1}\alpha(i))^{\delta''/(2+\delta'')}(n-i)2^{2+2\delta''}$$
$$\times \sup_{n\in\mathbb{N}^*}\max_{1\leqslant i\leqslant n}(E\,|Z_{ni}^{K}|^{2+\delta''})^{2/(2+\delta'')}.$$

Let $D_K=\sup_{n\in\mathbb{N}^*}\max_{1\leqslant i\leqslant n}(E\,|Z_{ni}^{K}|^{2+\delta''})^{2/(2+\delta'')}$. Then, from (2.24), it follows that

$$E\,|S''_n|^2 \leqslant 2^{3+\delta''}C^{-2}\left(\sum_{i=0}^{\infty}(\alpha(i))^{\delta''/(2+\delta'')}\right)D_K\to 0 \qquad \text{as} \quad K\to\infty.$$

For any $\varepsilon>0$, we can always choose K so large that $E\,|S''_n|^2\leqslant\varepsilon$ and $|1-C_K^2/C^2|\leqslant\varepsilon$. If now $f_n(t)=E(e^{itS_n})$ is the characteristic function of S_n, then

$$\begin{aligned}|f_n(t)-e^{-t^2/2}| &\leqslant |e^{-(t^2/2)(C_K^2/C^2)}-e^{-t^2/2}|\\ &\quad + |e^{-(t^2/2)(C_K^2/C^2)}-Ee^{itS'_n}| + |t|\;E\,|S''_n|\\ &\leqslant |t|\sqrt{\varepsilon}+\varepsilon t^2+|e^{-(t^2/2)(C_K^2/C^2)}-Ee^{itS'_n}|\end{aligned}$$

from the convergence in law of S'_n to the normal law, the last expression converges to zero as $n\to\infty$.

3. Convergence of the U-Statistic and the V-Statistic

Let $\mathbf{F}_{n,i,j}$ be the d.f. of (X_{ni}, X_{nj}), $1\leqslant i<j\leqslant n$, $n\geqslant 1$.

THEOREM 3.1. *If there are two positive numbers* δ, δ' $(0<\delta<\delta')$, *such that* (1.3), (2.6), *and* (2.7) *hold, and if, further, for any* $i, j\in\mathbb{N}^*$, *with* $i<j$,

there exists a continuous d.f. $\mathbf{F}_{ij}$ *on* $\mathbb{R}^2$ *with continuous marginals* F_i^* *and* F_j^* *such that*

$$\lim_{n\to\infty} \max_{1\leq i<j\leq n} |\mathbf{F}_{n,i,j}(x_1, x_2) - \mathbf{F}_{ij}(x_1, x_2)| = 0 \qquad \textit{for all} \quad (x_1, x_2) \in \mathbb{R}^2 \tag{3.1}$$

and

$$\mathbf{F}_{ij} = \mathbf{F}_{1,j-i+1} \qquad \textit{for all} \quad i<j. \tag{3.2}$$

Furthermore, if

$$\int_{\mathbb{R}^k} |g(x_1, ..., x_k)|^r \prod_{j=1}^{k} dF(x_j) \leq M_0 < \infty, \tag{3.3}$$

where M_0 *is some constant* >0 *and* $F = F_i^*$ *for all* $i \in \mathbb{N}^*$ *and* g *is right continuous and has left-hand limits (rcll) or left continuous and has right-hand limits (lcrl), then* $n^{1/2}(U(F_n) - \theta(F_n))$ *converges in lax to the normal distribution with mean* 0 *and variance* $k^2\sigma^2$ *if* $\sigma^2 > 0$, *where* $\theta(F_n) = n^{-[k]} g_{0,n}$ *and*

$$\begin{aligned}\sigma^2 = &\left[\int_{\mathbb{R}^k} g^2(x_1, ..., x_k) \prod_{l=1}^{k} dF(x_l) - \theta^2(F)\right] \\ &+ 2 \sum_{i=2}^{\infty} \left[\int_{\mathbb{R}^{2k}} g(x_1, ..., x_k)\, g(x_{k+1}, ..., x_{2k}) \right.\\ &\left.\times dF_{1i}(x_1, x_{k+1}) \prod_{l=2}^{k} dF(x_l) \prod_{p=k+2}^{2k} dF(x_p) - \theta^2(F)\right] < \infty, \end{aligned} \tag{3.4}$$

where $\theta(F) = \int_{\mathbb{R}^k} g(x_1, ..., x_k) \prod_{l=1}^{k} dF(x_l)$.

Note. The condition (3.2) implies that the limiting distribution $F_{ij}(x, y)$ of some random variable, say (X_i, X_j) is such that the marginal d.f. F_i^* of X_i are identical to F_1^* $(=F)$ for all $i \in \mathbb{N}^*$.

Proof. We rewrite (1.1) as

$$\begin{aligned} U(F_n) &= n^{-[k]} \sum_{(i_1, ..., i_k) \in I_{0,n}} \int_{\mathbb{R}^k} g(x_1, ..., x_k) \prod_{j=1}^{k} d(I_{[X_{ni_j} \leq x_j]}) \\ &= \theta(F_n) + \sum_{p=1}^{k} \binom{k}{p} U_n^{(p)}. \end{aligned}$$

From (2.8) of Lemma 2.2, it is sufficient to prove that $n^{1/2} U_n^{(1)}$ converges in law to the normal distribution with mean 0 and variance σ^2.

From (2.2), we note that $U_n^{(1)} = n^{-1} \sum_{i=1}^{n} X_{ni}^*$, where

$$X_{ni}^* = n^{-[k-1]} \left(g_{1,n}^{(i)}(X_{ni}) - \int_{\mathbb{R}} g_{1,n}^{(i)}(x_1)\, dF_{ni}(x_1) \right).$$

To prove the assertion, we have to show that $\{X_{ni}^*\}$ satisfies the conditions of Lemma 2.3. Note that (2.20) comes from (1.3) for $\delta < \delta'' < \delta'$. Now define Y_{ni}^K for any $K > 0$ by

$$\begin{aligned} Y_{ni}^K = n^{-[k-1]} & \sum_{(i_2, \ldots, i_k) \in I_{1,n}(i)} \\ & \times \left(\int_{\mathbb{R}^{k-1}} g^K(X_{ni}, x_2, \ldots, x_k) \prod_{j=2}^{k} dF_{ni_j}(x_j) \right. \\ & \left. - \int_{\mathbb{R}^k} g^K(x_1, \ldots, x_k) \prod_{j=2}^{k} dF_{ni_j}(x_j)\, dF_{ni}(x_1) \right), \end{aligned} \tag{3.5}$$

where g^K is defined by

$$g^K = \begin{cases} g & \text{if } \ |g| \leqslant K \\ 0 & \text{if } \ |g| > K. \end{cases} \tag{3.6}$$

Then $\sup_{n \in \mathbb{N}^*} \max_{1 \leqslant i \leqslant n} |Y_{ni}^K| \leqslant 2K$ which proves (2.21).

We now prove (2.22). Set $g^{-K} = g - g^K$. Then

$$\begin{aligned} & (E|X_{ni}^* - Y_{ni}^K|^{2+\delta''})^{1/(2+\delta'')} \\ & \leqslant n^{-[k-1]} \sum_{(i_2, \ldots, i_k) \in I_{1,n}(i)} \left(\int_{\mathbb{R}} 2^{1+\delta''} \left(\int_{\mathbb{R}^{k-1}} |\bar{g}^K(x_1, \ldots, x_k)|^{2+\delta''} \right. \right. \\ & \quad \left. + \int_{\mathbb{R}} |\bar{g}^K(y_1, x_2, \ldots, x_k)|^{2+\delta''}\, dF_{ni}(y_1) \right) \\ & \quad \left. \times \prod_{j=2}^{k} dF_{ni_j}(x_j))\, dF_{ni}(x_1) \right)^{1/(2+\delta'')} \end{aligned}$$

From (2.6) we deduce

$$(E\,|X_{ni}^* - Y_{ni}^K|^{2+\delta''})^{1/(2+\delta'')} \leqslant 2^{2+\delta''} \frac{M_0}{(K^{2+\delta''})^{\varepsilon}} \to 0 \qquad \text{as} \quad K \to \infty,$$

where $(2+\delta'')(1+\varepsilon) = 2 + \delta'$ and (2.22) is proved.

To prove (2.23) and (2.24) we first prove the following lemma.

LEMMA 3.1. *Let $h: \mathbb{R}^k \to \mathbb{R}$ be a function such that h is rcll (or lcrl) and suppose that*

$$\sup_n \max_{(i_1, ..., i_k) \in \{1, ..., n\}^k} \int_{\mathbb{R}^k} |h(x_1, ..., x_k)|^r \prod_{j=1}^k dF_{ni_j}(x_j) < M < \infty \tag{3.7}$$

$$\int_{\mathbb{R}^k} |h(x_1, ..., x_k)|^r \prod_{j=1}^k dF(x_j) < M < \infty, \tag{3.8}$$

where M is some constant > 0, then

$$\lim_{n \to \infty} \max_{1 \leqslant i_1 < i_{k+1} \leqslant n} \Bigg|(n^{-[k-1]})^2 \Bigg(\sum_{(i_2, ..., i_k) \in I_{1,n}(i_1)} \sum_{(i_{k+2}, ..., i_{2k}) \in I_{1,n}(i_{k+1})} \times \int_{\mathbb{R}^{2k}} (h \otimes h)(x_1, ..., x_{2k}) \times \prod_{\substack{j \neq 1 \\ j \neq k+1}} dF_{ni_j}(x_j)\, d\mathbf{F}_{n, i_1, i_{k+1}}(x_1, x_{k+1}) \Bigg) - \int_{\mathbb{R}^{2k}} (h \otimes h)(x_1, ..., x_{2k}) \times \prod_{\substack{j \neq 1 \\ j \neq k+1}} dF(x_j)\, d\mathbf{F}_{i_1 i_{k+1}}(x_1, x_{k+1}) \Bigg| = 0. \tag{3.9}$$

Proof. From (3.1) and (3.2) it follows that (3.9) is true if h is a step function. For any $K > 0$, we define the function h^K by

$$h^K = \begin{cases} h & \text{if } |h| \leqslant K \\ 0 & \text{if } |h| > K. \end{cases}$$

As h^K is rcll (or lcrl) and bounded, it is well known that h^K can be uniformly approached by a step function. Let $\varepsilon > 0$. Choose K sufficiently large so that

$$\max \left\{ \frac{M^2}{K^{r-1}}, \frac{M}{K^{r-2}} \right\} < \frac{\varepsilon}{6}. \tag{3.10}$$

Then, there exists a step function g_ε^K such that

$$\sup_{(x_1, ..., x_k)} |h^K(x_1, ..., x_k) - g_\varepsilon^K(x_1, ..., x_k)| < \frac{\varepsilon}{6K} \tag{3.11}$$

and there exists $N_0 \in \mathbb{N}$ such that $\forall n \geqslant N_0$,

$$\max_{1 \leqslant i_1 < i_{k+1} \leqslant n} \Bigg| (n^{-[k-1]})^2 \Bigg(\sum_{(i_2, \ldots, i_k) \in I_{1,n}(i_1)} \sum_{(i_{k+2}, \ldots, i_{2k}) \in I_{1,n}(i_{k+1})}$$

$$\times \int_{\mathbb{R}^{2k}} (g_\varepsilon^K \otimes g_\varepsilon^K)(x_1, \ldots, x_{2k})$$

$$\times \prod_{\substack{j \neq 1 \\ j \neq k+1}} dF_{ni_j}(x_j)\, d\mathbf{F}_{n, i_1, i_{k+1}}(x_1, x_{k+1}) \Bigg)$$

$$- \int_{\mathbb{R}^{2k}} (g_\varepsilon^K \otimes g_\varepsilon^K)(x_1, \ldots, x_{2k})$$

$$\times \prod_{\substack{j \neq 1 \\ j \neq k+1}} dF_{i_1 i_{k+1}}(x_1, x_{k+1}) \Bigg| < \varepsilon/3. \tag{3.12}$$

From (3.7), (3.8), (3.10)–(3.12) and the decomposition $h \otimes h = h \otimes h - h \otimes h^K + h \otimes h^K - h^K \otimes h^K + h^K \otimes h^K - h^K \otimes g_\varepsilon^K + h^K \otimes g_\varepsilon^K - g_\varepsilon^K \otimes g_\varepsilon^K + g_\varepsilon^K \otimes g_\varepsilon^K$, we deduce that $\forall n \geqslant N_0$,

$$\max_{1 \leqslant i_1 < i_{k+1} \leqslant n} \Bigg| (n^{-[k-1]})^2 \Bigg(\sum_{(i_2, \ldots, i_k) \in I_{1,n}(i_1)} \sum_{(i_{k+2}, \ldots, i_{2k}) \in I_{1,n}(i_{k+1})}$$

$$\times \int (h \otimes h)(x_1, \ldots, x_{2k}) \prod_{\substack{j \neq 1 \\ j \neq k+1}} dF_{ni_j}(x_j)$$

$$\times dF_{n, i_1, i_{k+1}}(x_1, x_{k+1}) \Bigg) - \int_{\mathbb{R}^{2k}} (h \otimes h)(x_1, \ldots, x_{2k})$$

$$\times \prod_{\substack{j \neq 1 \\ j \neq k+1}} dF(x_j)\, dF_{i_1 i_{k+1}}(x_1, x_{k+1}) \Bigg|$$

$$< \frac{M^2}{K^{r-1}} + \frac{M}{K^{r-2}} + K\frac{\varepsilon}{6K} + K\frac{\varepsilon}{6K} + \varepsilon/3 < \varepsilon.$$

As ε is chosen arbitrarily, (3.9) is proved.

We now prove (2.23). Denote

$$\rho(1) = \int_{\mathbb{R}^k} g^2(x_1, \ldots, x_k) \prod_{l=1}^{k} dF(x_l) - \theta^2(F)$$

$$\rho(i) = 2\Bigg[\int_{\mathbb{R}^{2k}} g(x_1, \ldots, x_k)\, g(x_{k+1}, \ldots, x_{2k})\, d\mathbf{F}_{1i}(x_1, x_{k+1})$$

$$\times \prod_{l=2}^{k} dF(x_l) \prod_{p=k+2}^{2k} dF(x_p) - \theta^2(F) \Bigg], \qquad \forall i \geqslant 2,$$

$$\psi(i, i) = \int_{\mathbb{R}} (g_{1,n}^{(i)}(x_1))^2 \, dF_{ni}(x_1)$$

$$- \left(\int_{\mathbb{R}} (g_{1,n}^{(i)}(x_1) \, dF_{ni}(x_1) \right)^2, \qquad \forall i \geqslant 1,$$

$$\psi(i, j) = 2 \int_{\mathbb{R}^2} (g_{1,n}^{(j)}(x_1) \, g_{1,n}^{(i+j)}(y_1) \, dF_{n,j,i+j}(x_1, y_1)$$

$$- \left(\int_{\mathbb{R}} (g_{1,n}^{(j)}(x_1 \, dF_{nj}(x_1)) \right.$$

$$\left. \times \left(\int_{\mathbb{R}} g_{1,n}^{(i+j)}(y_1) \, dF_{n,i+j}(y_1) \right), \forall (i, j) \in \mathbb{N}^* \right)^2, \qquad i < j.$$

Then,

$$\left| E\left(\left(\sum_{i=1}^{n} X_{ni}^* \right)^2 \Big/ n \right) - \sigma^2 \right|$$

$$= \left| \frac{1}{n} (n^{-[k-1]})^2 \sum_{i=0}^{n-1} \sum_{j=1}^{n-i} \psi(i, j) - \sum_{i=1}^{\infty} \rho(i) \right|$$

$$\leqslant \left| \frac{1}{n} (n^{-[k-1]})^2 \sum_{i=0}^{n-1} \sum_{j=1}^{n-i} \psi(i, j) - \frac{1}{n} \sum_{i=1}^{n} (n-i) \, \rho(i) \right|$$

$$+ \sum_{i=n+1}^{\infty} |\rho(i)| + \sum_{i=1}^{n} \sum_{k=i}^{\infty} |\rho^k|$$

$$= |A_n| + B_n + C_n.$$

From property (3.9) of Lemma 3.1, it follows that $|A_n| \to 0$ as $n \to \infty$.

From (3.3) and the well-known inequality on moments of strong mixing sequences of r.v.'s, we deduce $|\rho(i)| \leqslant (\alpha(i))^{\delta'/(2+\delta')}(M_0)^{2/(2+\delta')}$ which implies $B_n \to 0$ and $C_n \to 0$ as $n \to \infty$ by using (1.3). (2.23) is proved.

The proof of (2.24) follows analogously, and is therefore omitted.

COROLLARY 3.1. *Under the conditions of Theorem* 3.1,

$$n^{1/2} \, |U(\mathbf{F}_n) - V(\mathbf{F}_n)| \to 0 \qquad \text{in probability as} \quad n \to \infty.$$

Hence $n^{1/2}(V(\mathbf{F}_n) - \theta(\mathbf{F}_n))$ *converges in law to the normal distribution with mean* 0 *and variance* $k^2\sigma^2$, *where* σ^2 *is defined in* (3.4).

Proof. From (2.2) and (2.4), we have $n^{1/2}(U_n^{(1)} - V_n^{(1)}) = O_P(n^{-1/2})$, $E(V_n^1) = n^{-[k]} g_{0,n}^*$ and, from (2.1) and (2.3),

$$n^{1/2}(\theta(\mathbf{F}_n) - n^{-[k]} g_{0,n}^*) = O(n^{-1/2}),$$

which implies that $n^{1/2}(V_n^{(1)}-\theta(\mathbf{F}_n))$ converges in law to the normal distribution with mean 0 and variance σ^2.

On the other hand, since

$$U(\mathbf{F}_n)=\theta(\mathbf{F}_n)+\sum_{p=1}^{k}\binom{k}{p}U_n^{(p)} \quad \text{and} \quad V(\mathbf{F}_n)=n^{-[k]}g_{0,n}^{*}+\sum_{p=1}^{k}\binom{k}{p}V_n^{(p)},$$

we conclude (by using properties (2.8) and (2.9) of Lemma 2.2) that $n^{1/2}\,|U(\mathbf{F}_n)-V(\mathbf{F}_n)|\to 0$ in probability as $n\to\infty$.

4. Convergence of the Rank Statistic $\mathscr{S}_{n,m}$

From now on we assume that $F_{ni}=F_n$ for any i $(1\leqslant i\leqslant n)$. For any real number x, define $H_n(|x|)$ and $Y_{ni}(x)$ as

$$H_n(|x|)=F_n(|x|)-F_n(-|x|), \quad Y_{ni}(x)=I_{[|X_{ni}|\leqslant|x|]}-H_n(|x|).$$

Let

$$X_{n,m,i}^{*}=\frac{1}{m+1}\{(m-1)\,H_n(|X_{ni}|)+1\},$$

$$Z_{n,m,i}=\frac{R_{n,m,i}}{m+1}-X_{n,m,i}^{*}=\frac{1}{m+1}\sum_{\substack{1\leqslant j\leqslant m\\ j\neq i}}Y_{nj}(X_{ni}).$$

Let F be a d.f. on $\mathbb{R}$ and define the d.f. H on $\mathbb{R}^+$ by $H(|x|)=F(|x|)-F(-|x|)$.

For a score function $J(u)$ which is square integrable put

$$\mu_n=\mu_J(F_n)=\int_{\mathbb{R}}s(x)\,J(H_n(|x|)\,dF_n(x). \tag{4.1}$$

For h defined in (1.4) put

$$\mu_{n,m}=\mu_n\sum_{i=1}^{m}c_{ni}=\mu_n\sum_{i=1}^{m}h\left(\frac{i}{n}\right), \qquad m\leqslant n. \tag{4.2}$$

For any sequence of d.f.'s $\{\mathbf{F}_{1i};\, i\geqslant 2\}$ on $\mathbb{R}^2$ with marginals F, we denote

$$\sigma^2(\{\mathbf{F}_{1i}\})=\lim_{n\to\infty}\left\{\int_{\mathbb{R}}g^2(x)\,dF(x)+2\sum_{l=2}^{n}\int_{\mathbb{R}^2}g(x)\,g(y)\,d\mathbf{F}_{1l}(x,y)\right\} \tag{4.3}$$

if the limit exists, where

$$g(x)=\int_{\mathbb{R}} s(y)\{I_{[|x|\leqslant|y|]}-H(|y|)\}\, J'(H(|y|)\, dF(y)$$

$$+s(x)\, J(H(|x|))-\int_{\mathbb{R}} s(y)\, J(H(|y|)\, dF(y).$$

For every $n\geqslant 1$, let

$$V_n(t)=\begin{cases} 0 & \text{for} \quad t=0 \\ (\mathscr{S}_{n,k}-\mu_{n,k})/\sigma n^{1/2} & \text{for} \quad t=k/n,\ k=1, ..., n \\ \text{linearly interpolated} & \text{for} \quad t\in[k-1/n, k/n],\ k=1, ..., n, \end{cases} \tag{4.4}$$

where σ is the positive constant defined in (4.3).

The process $V_n(t)=\{V_n(t),\ 0\leqslant t\leqslant 1\}$ belongs to the space C_1 of all continuous functions on $[0, 1]$ with which we associate the usual uniform metric.

Now we give the following theorem which is a generalization from the stationary case to the nonstationary case of Theorem 3.1 in Yoshihara [9].

THEOREM 4.1. *Suppose the sequence $\{X_{ni}\}$ is absolutely regular with rate (1.6) and the sequence $\{F_n\}$ satisfies the conditions (3.1) and (3.2) of Theorem 3.1. Let J be a score function having a bounded second derivative. If $\sigma^2(\{\mathbf{F}_{1i}\})$ defined by (4.3), (below) is strictly positive, then V_n defined in (4.3) converges weakly in the uniform topology on C_1 to the process $V_0=\{V_0(t), 0\leqslant t\leqslant 1\}$, where*

$$V_0(t)=\int_0^t h(u)\, dW(u), \qquad 0\leqslant t\leqslant 1 \tag{4.5}$$

and $W=\{W(t),\ 0\leqslant t\leqslant 1\}$ is a standard Brownian motion process, and $\sigma^2(\{\mathbf{F}_{1i}\}<\infty$.

Proof. We need at first some lemmas. For any n $(n\geqslant 1)$ and for any i $(1\leqslant i\leqslant n)$ let

$$A_{ni}=\int_{\mathbb{R}} s(x)\{I_{[|X_{ni}|\leqslant|x|\}}-H_n(|x|)\}\, J'(H_n(|x|)\, dF_n(x)$$

$$+\{s(X_{ni})\, J(H_n(|X_{ni}|))-E(s(X_{ni})\, J(H_n(|X_{ni}|)))\}. \tag{4.6}$$

It is obvious that $E(A_{ni})=0$.

Now we consider the process $L_n(t)$ defined on C_1 by

$$L_n(t) = n^{-1/2}\left(\sum_{i=1}^{[nt]} A_{ni} - (nt-[nt])\, A_{n,[nt]+1}\right), \tag{4.7}$$

where $[nt]$ denotes the integer part of the real number nt.

LEMMA 4.1. *Suppose that $\{X_{ni}\}$ satisfies the conditions of Theorem 4.1. Then, the process $L_n(t)$ converges weakly in uniform topology to a Gaussian process $L_0(t)$ with trajectories a.s. in C_1 with mean 0 and variance $t\sigma^2(\{\mathbf{F}_{1i}\})$, where $\sigma^2(\{F_{1i}\})$ is defined in (4.3).*

Proof. The process L_n defines a probability measure P_n on C_1. From Theorem 8.1 of Billingsley [2], we have to prove that (i) the finite-dimensional distributions of P_n converge in law to normal distributions and (ii) P_n is tight.

First we prove (i) which is equivalent to proving that $\sum_{l=1}^{p} \lambda_l L_n(t_l)$ converges in law to a normal distribution for any $p \in \mathbb{N}^*$, any $t_l \in [0, 1]$ and any $\lambda_l \in \mathbb{R}$ $(1 \leqslant l \leqslant p)$. Without loss of generality we can take $p=2$ and suppose that $t_1 < t_2$.

As J and J' are bounded, the sequence of r.v.'s $\{A_{ni}\}$ are uniformly bounded and we have

$$\begin{aligned}\sum_{l=1}^{2} \lambda_l L_n(t_l) = n^{-1/2}\Bigg[&\sum_{i=1}^{[nt_1]} (\lambda_1+\lambda_2) A_{ni} \\ &+ \sum_{i=[nt_1]+1}^{[nt_2]} \lambda_2 A_{ni} + (nt_1 - [nt_1]) A_{n,[nt_1]+1} \\ &+ (nt_2 - [nt_2]) A_{n,[nt_2]+1}\Bigg].\end{aligned} \tag{4.8}$$

We define the sequence of r.v.'s $\{B_{ni}\}$ by

$$B_{ni} = \begin{cases} (\lambda_1+\lambda_2) A_{ni} & \text{if } \ i \leqslant [nt_1] \\ \lambda_2 A_{ni} & \text{if } \ [nt_1] < i \leqslant [nt_2] \\ 0 & \text{if } \ i > [nt_2] \end{cases} \tag{4.9}$$

and then we have

$$\sum_{l=1}^{2} \lambda_l L_n(t_l) = n^{-1/2} \sum_{i=1}^{n} B_{ni} + O_p(n^{-1/2}). \tag{4.10}$$

From Lemma (2.3) we have only to verify (2.20) and (2.23). (2.20) is immediate from (1.6) for $\delta'' > \frac{2}{3}$.

Now we prove (2.23). We have

$$E\left(\sum_{i=1}^{n} B_{ni}\right)^2\Big/ n=\frac{1}{n}\left[(\lambda_1+\lambda_2)^2\left(\sum_{i=1}^{[nt_1]}\sum_{j=1}^{[nt_1]} E(A_{ni}A_{nj})\right)\right.$$
$$+(\lambda_1+\lambda_2)\lambda_2\left(\sum_{i=1}^{[nt_1]}\sum_{j=[nt_1]+1}^{[nt_2]} E(A_{ni}A_{nj})\right)$$
$$+\lambda_2^2\sum_{i=[nt_1]+1}^{[nt_2]}\sum_{j=[nt_1]+1}^{[nt_2]} E(A_{ni}A_{nj}). \tag{4.11}$$

As the $\{A_{ni}\}$ are uniformly bounded, for any $\delta''>0$, we can find a constant $M>0$ such that

$$\frac{1}{n}\left|(\lambda_1+\lambda_2)\lambda_2\sum_{i=1}^{[nt_1]}\sum_{j=[nt_1]+1}^{[nt_2]} E(A_{ni}A_{nj})\right.$$
$$\left.+\lambda_2^2\sum_{i=[nt_1]+1}^{[nt_2]}\sum_{j=[nt_1]+1}^{[nt_2]} E(A_{ni}A_{nj})\right|$$
$$\leqslant\left(\sum_{i=1}^{[nt_1]}\sum_{j=[nt_1]+1}^{[nt_2]}\alpha^{\delta''/(2+\delta'')}(j-i)\right.$$
$$\left.+\sum_{i=[nt_1]+1}^{[nt_2]}\sum_{j=[nt_1]+1}^{[nt_2]}\alpha^{\delta''/(2+\delta'')}(j-i)M\right)\Big/ n. \tag{4.12}$$

From condition (1.6), the last expression converges to 0 as $n\to\infty$ for δ'' sufficiently large.

It remains to prove that $(1/n)\sum_{i=1}^{[nt_1]}\sum_{j=1}^{[nt_1]} E(A_{ni}A_{nj})$ converges to some finite constant.

We first prove the convergence by replacing J' and J in the expression of A_{ni} in (4.6) by indicator functions such as $J(t)=I_{[a\leqslant t\leqslant b]}$ and $J'(t)=I_{[a'\leqslant t\leqslant b']}$ and we then can write

$$A_{ni}=\int_{(a'\vee H_n(|X_{ni}|))\wedge b'}^{b'} d(F_n\circ H_n^{-1}(t))-\int_{a'}^{b'} t\,d(F_n\circ H_n^{-1}(t))$$
$$-\int_{(a'\vee H_n(|X_{ni}|))\wedge b'}^{b'} d(F_n\circ H_n^{*}(t))$$
$$+\int_{a'}^{b'} t\,d(F_n\circ H_n^{*}(t))+s(X_{ni})I_{[a\leqslant H_n(|X_{ni}|)\leqslant b]}$$
$$-\int_a^b d(F_n\circ H_n^{-1}(t))+\int_a^b d(F_n\circ H_n^{*}(t)),$$

where $H_n^*(t)=-H_n^{-1}(t)$.

Denote

$$D_{ni}(X_{ni}) = s(X_{ni}) I_{[a \leqslant H_n(|X_{ni}|) \leqslant b]} - \int_a^b d(F_n \circ H_n^{-1}(t)) + \int_a^b d(F_n \circ H_n^*(t))$$

and $C_{ni}(|X_{ni}|) = A_{ni} - D_{ni}(X_{ni})$.

Let $\bar{\mathbf{F}}_{n,i,j}$ be the d.f. of $(|X_{ni}|, |X_{nj}|)$ (resp. $\mathbf{F}^1_{n,i,j}$ of $(|X_{ni}|, X_{nj})$ and $\mathbf{F}^2_{n,i,j}$ of $(X_{ni}, |X_{nj}|)$) and $\bar{\mathbf{F}}_{ij}$ (resp. $\mathbf{F}^1_{ij}$ and $\mathbf{F}^2_{ij}$) be the d.f. on $\mathbb{R}^+ \times \mathbb{R}^+$ (resp. $\mathbb{R}^+ \times \mathbb{R}$ and $\mathbb{R} \times \mathbb{R}^+$) defined by $\mathbf{F}_{ij}(x, y) = \mathbf{F}_{ij}(x, y) - \mathbf{F}_{ij}(-x, y) - \mathbf{F}_{ij}(x, -y) + \mathbf{F}_{ij}(-x, -y)$ for any $x \geqslant 0$, $y \geqslant 0$ (resp. $\mathbf{F}^1_{ij}(x, y) = \mathbf{F}_{ij}(x, y) - \mathbf{F}_{ij}(-x, y)$ for any $x \geqslant 0$, $y \in \mathbb{R}$, and $\mathbf{F}^2_{ij}(x, y) = \mathbf{F}_{ij}(x, y) - \mathbf{F}_{ij}(x, -y)$ for any $x \in \mathbb{R}$, $y \geqslant 0$) and $(i, j) \in \mathbb{N}^* \times \mathbb{N}^*$, where $i < j$.

We then have

$$\begin{aligned} E(A_{ni}A_{nj}) = & \int_0^{+\infty} \int_0^{+\infty} C_{ni}(x)\, C_{nj}(y)\, d\bar{F}_{n,i,j}(x, y) \\ & + \int_{-\infty}^{+\infty} \int_{-\infty}^{+\infty} D_{ni}(x)\, D_{nj}(y)\, dF_{n,i,j}(x, y) \\ & + \int_0^{+\infty} \int_{-\infty}^{+\infty} C_{ni}(x)\, D_{nj}(y)\, dF^1_{n,i,j}(x, y) \\ & + \int_{-\infty}^{+\infty} \int_0^{+\infty} D_{ni}(x)\, C_{nj}(y)\, dF^2_{n,i,j}(x, y). \end{aligned}$$

By using properties (3.1) and (3.2) and the following inequality

$$|F_n \circ H_n^{-1}(t) - F \circ H^{-1}(t)| \leqslant |F_n \circ H_n^{-1}(t) - F \circ H_n^{-1}(t)| + |F \circ H_n^{-1}(t) - F \circ H^{-1}(t)|$$

which implies

$$\sup_n |F_n \circ H_n^{-1}(t) - F \circ H^{-1}(t)| \xrightarrow[n \to \infty]{} 0 \qquad \text{for all} \quad t \in [0, 1],$$

we deduce that

$$\begin{aligned} \lim_{n \to \infty} \max_{1 \leqslant i < j \leqslant n} \Big| E(A_{ni}A_{nj}) & - \int_0^{+\infty} \int_0^{+\infty} C(x)\, C(y)\, d\bar{\mathbf{F}}_{ij}(x, y) \\ & - \int_{-\infty}^{+\infty} \int_{-\infty}^{+\infty} D(x)\, D(y)\, d\mathbf{F}_{ij}(x, y) \\ & - \int_0^{+\infty} \int_{-\infty}^{+\infty} C(x)\, D(y)\, d\mathbf{F}^1_{ij}(x, y) \\ & - \int_{-\infty}^{+\infty} \int_0^{+\infty} D(x)\, C(y)\, d\mathbf{F}^2_{ij}(x, y) \Big| = 0, \end{aligned}$$

where

$$C(x)=\int_{(a' \vee H(x)) \wedge b'}^{b'} d(F\circ H^{-1}(t))-\int_{a'}^{b'} t\, d(F\circ H^{-1}(t))$$

$$-\int_{(a' \vee H(x)) \wedge b'}^{b'} d(F\circ H^{*}(t))+\int_{a'}^{b'} t\, d(F\circ H^{*}(t))$$

$$D(x)=s(x)I_{[a\leqslant H(|x|)\leqslant b]}-\int_{a}^{b} d(F\circ H^{-1}(t))+\int_{a}^{b} d(F\circ H^{*}(t)),$$

where $H^*(t)=-H^{-1}(t)$.

We obtain a similar property if J and J' are replaced by step functions. As J and J' are continuous, we can uniformly approach them by a step function and we deduce that

$$\lim_{n\to\infty}\max_{1\leqslant i<j\leqslant n}\Bigg|E(A_{ni}A_{nj})-\Bigg(\int_0^{+\infty}\int_0^{+\infty} C^*(x)\,C^*(y)\,d\bar{\mathbf{F}}_{ij}(x,y)$$

$$+\int_0^{+\infty}\int_{-\infty}^{+\infty} C^*(x)\,D^*(y)\,d\mathbf{F}^1_{ij}(x,y)$$

$$+\int_{-\infty}^{+\infty}\int_0^{+\infty} D^*(x)\,C^*(y)\,d\mathbf{F}^2_{ij}(x,y)$$

$$+\int_{-\infty}^{+\infty}\int_{-\infty}^{+\infty} D^*(x)\,D^*(y)\,d\mathbf{F}_{ij}(x,y)\Bigg)\Bigg|=0, \tag{4.13}$$

where

$$C^*(x)=\int_{-\infty}^{+\infty} s(y)\{I_{[x\leqslant|y|]}-H(|y|)\}\,J'(H(|y|))\,dF(y),$$

and

$$D^*(x)=\Bigg\{s(x)\,J(H(|x|))-\int_0^{+\infty} J(H(x))\,dF(x)$$

$$+\int_{-\infty}^{0} J(H(-x))\,dF(x)\Bigg\}.$$

If we denote

$$\rho(1)=\int_0^{+\infty}(C^*(x))^2\,dH(x)+\int_{-\infty}^{+\infty}(D^*(x))^2\,dF(x)$$

$$+2\int_{-\infty}^{+\infty} C(|x|)\,D(x)\,dF(x)$$

$$\rho(i)=2\left(\int_0^{+\infty}\int_0^{+\infty} C^*(x)\,C^*(y)\,d\bar{\mathbf{F}}_{1i}(x,y)\right.$$
$$+\int_{-\infty}^{+\infty}\int_{-\infty}^{+\infty} D^*(x)\,D^*(y)\,d\mathbf{F}_{1i}(x,y)$$
$$+\int_0^{+\infty}\int_{-\infty}^{+\infty} C^*(x)\,D^*(y)\,d\mathbf{F}^1_{1i}(x,y)$$
$$\left.+\int_{-\infty}^{+\infty}\int_0^{+\infty} D^*(x)\,C^*(y)\,d\mathbf{F}^2_{1i}(x,y)\right),$$

we can write

$$\left|\frac{1}{n}\sum_{i=1}^{[nt_1]}\sum_{j=1}^{[nt_1]} E(A_{ni}A_{nj})-\frac{[nt_1]}{n}\sum_{i=1}^{+\infty}\rho(i)\right|$$
$$\leqslant\left|\frac{[nt_1]}{n}\frac{1}{[nt_1]}\sum_{i=0}^{[nt_1]-1}\sum_{j=1}^{[nt_1]-i}\varphi(i)\,E(A_{nj}A_{n,j+i})\right.$$
$$\left.-\frac{[nt_1]}{n}\frac{1}{[nt_1]}\sum_{i=1}^{[nt_1]}([nt_1]-i)\,\rho(i)\right|$$
$$+\frac{[nt_1]}{n}\sum_{i=[nt_1]+1}^{\infty}|\rho(i)|+\frac{[nt_1]}{n}\sum_{i=1}^{[nt_1]}\sum_{k=i}^{\infty}|\rho^k|$$
$$=|A_n|+B_n+C_n,$$

where

$$\varphi(i)=\begin{cases}1 & \text{if } i=0\\ 2 & \text{if } i\neq 0.\end{cases}$$

From (4.13) we deduce that $|A_n|\to 0$ and from the inequality on moments of strong mixing sequence of r.v.'s we deduce $|\rho(i)|\leqslant (\alpha(i))^{\delta''/(2+\delta'')}M$ for some constant $M>0$ which implies that $B_n\to 0$ and $C_n\to 0$ (as $n\to\infty$) by (1.6) for $\delta''>\frac{3}{2}$. It is also immediate that

$$\left|\frac{[nt_1]}{n}\sum_{i=1}^{+\infty}\rho(i)-t_1\left(\sum_{i=1}^{+\infty}\rho(i)\right)\right|\to 0 \quad \text{as} \quad n\to\infty. \tag{4.14}$$

We conclude that $(1/n)\sum_{i=1}^{[nt_1]}\sum_{j=1}^{[nt_1]} E(A_{ni}A_{nj})$ converges to $t_1(\sum_{i=1}^{+\infty}\rho(i))$ as $n\to\infty$, where $\sum_{i=1}^{+\infty}\rho(i)$ is equal to $\sigma^2(\{\mathbf{F}_{1i}\})$ defined in (4.3).

From (4.8)–(4.12) and (4.14), we deduce that $E(\sum_{l=1}^{2}\lambda_l L_n(t_l))^2$ converges to $(\lambda_1+\lambda_2)^2\,t_1\sigma^2(\{\mathbf{F}_{1i}\})$ which implies that $\sum_{l=1}^{2}\lambda_l L_n(t_l)$ converges in law to the normal distribution with mean 0 and variance $(\lambda_1+\lambda_2)^2\,t_1\sigma^2(\{\mathbf{F}_{1i}\})$ and (i) is proved.

To prove (ii), we have to verify (cf. Billingsley [2]) that $\forall\varepsilon>0$, $\exists\eta>0$ $(0<\eta<1)$, and an integer N_0 such that $\forall n\geqslant N_0$,

$$P[\sup_{|s-t|<\eta} |L_n(t)-L_n(s)|\geqslant\varepsilon]\leqslant\varepsilon \tag{4.15}$$

which follows by using Theorem 10 of Doukhan and Portal [4], Lemma 2 of Balacheff and Dupont [1], and routine analysis.

We now consider the process $\hat{L}_n(t)$ defined in C_1 by

$$\hat{L}_n(t)=n^{-1/2}\left(\sum_{i=1}^{[nt]} c_{ni}A_{ni}-c_{n,[nt]+1}(nt-[nt])\right.$$
$$\left.\times A_{n,[nt]+1})(\sigma(\{\mathbf{F}_{1i}\})\right)^{-1}. \tag{4.16}$$

LEMMA 4.2. *Suppose that $\{X_{ni}\}$ satisfies the conditions of Theorem 4.1. Then, the process $\hat{L}_n(t)$ converges weakly in uniform topology to a Gaussian process $\hat{L}_0(t)=\int_0^t h(u)\,dW(u)$ with trajectories a.s. in C_1, where W is a standard Brownian motion process.*

Proof. From Theorem 1 of Yoshihara and Negishi [7] we have to verify that $\forall\varepsilon>0$

$$\lim_{\eta\to 0}\limsup_{n\to\infty} P\left[\max_{1\leqslant k\leqslant n}\left|\sum_{i=1}^{k}\eta_i A_{ni}(\sigma(\{\mathbf{F}_{1i}\})^{-1}\right|\geqslant\varepsilon M\sqrt{n}\right]=0, \tag{4.20}$$

where η_i is any set of real numbers, $\eta=\sup_i\eta_i$ and $M=\sup_n\max_{1\leqslant i\leqslant n} E(A_{ni}^2(\sigma^2(\{\mathbf{F}_{1i}\}))^{-1})$. By using Theorem 10 of Doukhan and Portal [4] for $q=2$, we have

$$E\left(\sum_{i=1}^{k}\eta_i A_{ni}\right)^4\leqslant C_4(\alpha)(Mk\eta+M^2k^2\eta^2)$$

which implies from Theorem 12.2 of Billingsley for $\eta\geqslant n^{-2}$ that

$$P\left[\max_{1\leqslant k\leqslant n}\left|\sum_{i=1}^{k}\eta_i A_{ni}\right|\geqslant\varepsilon M\sqrt{n}\right]\leqslant K\varepsilon^{-4}(M^{-3}+M^{-2})\eta^{3/2}. \tag{4.21}$$

Since the right side of (4.21) converges to zero as $\eta\to 0$, (4.20) follows.

Now we proceed as in Yoshihara [9]. As the score function J possesses a bounded derivative, we have by the Taylor expansion

$$J\left(\frac{R_{n,m,i}}{m+1}\right)=J(X^*_{n,m,i})+Z_{n,m,i}J'(X^*_{n,m,i})+\frac{1}{2}Z^2_{n,m,i}J''(\mu_{n,m,i}), \tag{4.22}$$

where

$$\mu_{n,m,i} \in \left[\frac{R_{n,m,i}}{m+1} \wedge X^*_{n,m,i}, \frac{R_{n,m,i}}{m+1} \vee X^*_{n,m,i}\right].$$

We note that

$$|X^*_{n,m,i} - H_n(|X_{ni}|)| \leqslant \frac{1}{m+1}.$$

So from (1.4), (4.1), (4.6), and (4.22) it follows (omitting the routine computations) that

$$\left|\mathscr{S}_{n,m} - \mu_{n,m} - \sum_{i=1}^{m} c_{ni} A_{ni}\right| \leqslant |V_m| + M_1 \sum_{i=1}^{m} Z^2_{n,m,i} + M_2,$$

where M_1 and M_2 are some constants >0, $M_0 = \sup_{t \in [0,1]} |J''(t)|$ and

$$V_m = \frac{1}{m+1} \sum_{i=1}^{m} \sum_{\substack{j=1\\ j\neq i}}^{m} c_{ni} \Big[s(X_{ni}) \{ I_{[|X_{ni}| \leqslant |X_{nj}|]}$$
$$- H_n(|X_{ni}|)\} J'(H_n(|X_{ni}|))$$
$$- \int_{\mathbb{R}} s(x) \{ I_{[|X_{ni}| \leqslant |x|]} - H_n(|x|)\} J'(H_n(|x|))\, dF_n(x) \Big].$$

LEMMA 4.3. *Under the conditions of Theorem* 4.1, *we have*

$$E(|V_m|^2) = O(1), \qquad E(|V_m|^4) = O(1). \tag{4.23}$$

Proof. Follows using the same technique as in Lemma 2.2.

LEMMA 4.4. *Under the conditions of Theorem* 4.1, *we have*

$$E\left(\left|\sum_{i=1}^{m} Z^2_{n,m,i}\right|^2\right) = O(m^{-1}) \tag{4.24}$$

$$E\left(\sum_{i=1}^{m} Z^2_{n,m,i}\right) = O(m^{-1/2}). \tag{4.25}$$

Proof. Put $g_n(x, y) = I_{[|y| \leqslant |x|]} - H_n(|x|)$. Let i_{rs} $(\leqslant m)$ $(r = 1, ..., 3;$ $s = 1, 2)$ be mutually different positive integers. Reorder $\{i_{rs}\}$ as $1 \leqslant k_1 < k_2 < \cdots < k_6 \leqslant m$ and put $E[\prod_{j=1}^{2} g_n(X_{ni_{j1}}, X_{ni_{j2}})\, g_n(X_{ni_{j3}}, X_{ni_{j2}})] = E[L(X_{nk_1}, ..., X_{nk_6})] = M(k_1, ..., k_6)$.

Let $d^{(c)}$ be the cth largest difference among $k_{j+1} - k_j$ $(j = 1, ..., 5)$. Since $\int_{\mathbb{R}} g_n(x, y)\, dF_n(x) = 0$, we deduce from Lemma 2.1 that

$$M(k_1, ..., k_6) \leqslant K \sum_{\alpha=1}^{3} \beta(k_{j_\alpha+1} - k_{j_\alpha}),$$

where K is some constant >0 if for some j_α $(1 \leqslant j_\alpha \leqslant 5)$, $k_{j_\alpha+1} - k_{j_\alpha} = d^{(\alpha)}$ $(1 \leqslant \alpha \leqslant 3)$. Consequently,

$$\sum_{1 \leqslant k_1 < \cdots < k_6 \leqslant m} M(k_1, ..., k_6) \leqslant K' m^3 \sum_{j=1}^{m} (j+1)^2 \beta(j),$$

where K' is some constant >0. From (1.6) we deduce $\sum_{j=1}^{\infty} (j+1)^2 \beta(j) < +\infty$.

Using similar arguments we estimate the sums in the other cases and obtain (4.24). (4.25) follows immediately from (4.24).

LEMMA 4.5. *Under the conditions of Theorem* 4.1, *we have*

$$\forall \varepsilon > 0, \qquad P[\sup_{t \in [0,1]} |V_n(t) - \hat{L}_n(t)| > \varepsilon] \to 0 \qquad as \quad n \to \infty. \quad (4.26)$$

Proof. To prove (4.26), it is sufficient to prove

$$P\left[\max_{1 \leqslant m \leqslant n} \left| \mathscr{S}_{n,m} - \mu_{n,m} - \sum_{i=1}^{m} c_{ni} A_{ni} \right| \geqslant 3\varepsilon n^{1/2}\right] \to 0 \qquad \text{as} \quad n \to \infty,$$

for every $\varepsilon > 0$, which follows by using (4.23) and (4.25). (Since the method is the same as in the proof of the Lemma 3.5 of Yoshihara [9], the details are omitted.)

Theorem 4.1 now follows from Lemmas 4.2 and 4.5.

Define the score function $J_n(u)$ by

$$J_n(u) = \begin{cases} J(u) & \text{if} \quad 0 \leqslant u \leqslant n/n+1 \\ J(n/n+1) & \text{otherwise.} \end{cases}$$

COROLLARY 4.1. *If the conditions of the Theorem* 4.1 *are satisfied except that the score function J is replaced by J_n in the expression of $\mathscr{S}_{n,m}$ defined in* (1.4) *and if J is twice differentiable and satisfies the condition*

$$|J''(u)| \leqslant M n^{1/2-\delta} \qquad if \quad 0 \leqslant u \leqslant \frac{n}{n+1} \text{ for some } \delta (0 < \delta < \tfrac{1}{2}), \quad (4.27)$$

where M is some constant >0, then the conclusion of Theorem 4.1 *remains true.*

Proof. Same as that of Theorem 3.2 of Yoshihara [9].

Acknowledgment

The authors are very much grateful to Professor G. Kallianpur for his very helpful suggestions.

References

[1] Balacheff, S., and Dupont, G. (1980). Normalité asymptotique des processus empiriques tronqués et des processus de rang. In *Lecture Notes in Mathematics* Vol. 82, Springer-Verlag, New York/Berlin.

[2] Billingsley, P. (1968). *Convergence of Probability Measures.* Wiley, New York.

[3] Denker, M., and Keller, G. (1983). On U-statistics and V. Mises' statistics for weakly dependent processes. *Z. Wahrsch. Verw. Gebiete* **64** 505–522.

[4] Doukhan, P., and Portal, F. (1983). Moments de variables aléatoires mélangeantes. *C.R. Acad. Sci. Paris Sér. I.* **297** 129–132.

[5] Ibragimov, I. A. (1962). Some limit theorems for stationary processes. *Theory Probab. Appl.* **7**, No. 4 349–382.

[6] Withers, C. S. (1975). Convergence of empirical processes of mixing rv's on [0, 1]. *Ann. Statist.* **3**, No. 5 1101–1108.

[7] Yoshihara, K., and Negishi, H. (1975). Weak convergence theorems for a type of weighted sums of random variables. *Sci. Rep. Yokohama Nat. Univ. Sect.* 1 **22** 1–9.

[8] Yoshihara, K. (1976). Limiting behavior of U-statistics for stationary absolutely regular processes. *Z. Wahrsch. Verw. Gebiete* **35** 237–252.

[9] Yoshihara, K. (1978). Limiting behavior of one sample rank order statistics for absolutely regular processes. *Z. Wahrsch. Verw. Gebiete* **43** 101–127.

WEAK INVARIANCE OF GENERALIZED U-STATISTICS FOR NONSTATIONARY ABSOLUTELY REGULAR PROCESSES

MICHEL HAREL and MADAN L. PURI*

Département Technique de Commercialisation, Institut Universitaire de Technologie de Limoges, 87065 Limoges, France, and U.A. ORSAY, Rouen, France

Department of Mathematics, Indiana University, Bloomington, IN 47405, USA

Yoshihara (1976) established the weak invariance theorem of the generalized U-statistic for absolutely regular variables but only for the stationary case. In this paper we extend the results from the stationary case to the nonstationary case.

weak convergence * generalised U-statistic * absolute regularity * strong mixing * Brownian process * Skorohod topology

1. Introduction

Let c $(\geqslant 2)$ be an integer and for each j $(1 \leqslant j \leqslant c)$ consider an integer $m_j \geqslant 1$; let g be a Borel measurable function $g:(\mathbb{R}^k)^{m_0} \to \mathbb{R}$ where $m_0 = m_1 + \cdots + m_c$ and assume that g is symmetric in m_j $(\geqslant 1)$ arguments of the jth set for $j = 1, \ldots, c$, i.e. g is symmetric in each of the c sets of arguments. We consider c independent triangular arrays of r.v.'s (random variables) with values in $\mathbb{R}^k$ and let the random variables corresponding to the jth triangular array be denoted by $\underset{\sim}{X}_{n,i,j}$ $(1 \leqslant i \leqslant n,\ n \geqslant 1)$. Let $\underset{\sim}{X}_{n,i,j}$ have the continuous d.f. (distribution function) $F_{n,i,j}$, $1 \leqslant i \leqslant n$, $1 \leqslant j \leqslant c$, $n \geqslant 1$.

Denote $\boldsymbol{n} = (n_1, \ldots, n_c)$ and assume that for any j, $n_j \geqslant m_j$, let $\mathcal{S}_{\boldsymbol{n}} = \prod_{j=1}^{c} \mathcal{S}_{n_j,j}$ where $\mathcal{S}_{n_j,j}$ is the set of all the strictly increasing sequences of m_j elements in $\{1, \ldots, n_j\}$ $(1 \leqslant i_{j1} < i_{j2} < \cdots < i_{jm_j} \leqslant n_j)$.

The generalized U-statistic of degree $\boldsymbol{m}$ $(=(m_1, \ldots, m_0))$ is defined as

$$U(\boldsymbol{n}) = \left[\prod_{j=1}^{c} \binom{n_j}{m_j}^{-1} \right] \sum_{\mathcal{S}_{\boldsymbol{n}}} g(\underset{\sim}{X}_{n_j,i_{jl},j},\ 1 \leqslant l \leqslant m_j,\ 1 \leqslant j \leqslant c). \tag{1.1}$$

* Research supported by the Office of Naval Research Contract N00014-85-0648.

Generalizing the definition given by Yoshihara (1976) in the stationary case, we say that a triangular array of r.v.'s $\eta_{n,i}$ $(1 \leq n, 1 \leq i \leq n)$ is absolutely regular with rates β if

$$\beta(m) = \sup_{m \leq n} \sup_{1 \leq l \leq n-m} E\left\{ \sup_{A \in \mathcal{M}_{n,l+m}^{n}} |P(A|\mathcal{M}_{n,1}^{l}) - P(A)| \right\} \downarrow 0 \quad \text{as } m \to \infty,$$

where $\mathcal{M}_{n,a}^{b}$ is the σ-field generated by $(\eta_{n,a}, \ldots, \eta_{n,b})$.

Also recall that $\{\eta_{n,i}\}$ is strong mixing if

$$\alpha(m) = \sup_{m \leq n} \sup_{1 \leq l \leq n-m} \{\sup \{|P(A \cap B) - P(A)P(B)|;$$

$$A \in \mathcal{M}_{n,1}^{l}, B \in \mathcal{M}_{n,l+m}^{n}\}\} \downarrow 0 \quad \text{as } m \to \infty,$$

since $\alpha(m) \leq \beta(m)$, it follows that if $\{\eta_{n,i}\}$ is absolutely regular, then it is also strong mixing.

We now suppose that for each j, the triangular array of r.v.'s $\xi_{n,i,j}$ is absolutely regular with rate β_j (consequently, it is strong mixing with rate $\alpha_j \leq \beta_j$).

Our aim is to study the asymptotic behavior of $U(\boldsymbol{n})$ when the $n_j \to \infty$ such that for any j, $n_j/(n_1 + \cdots + n_c) \to \lambda_j$ $(0 < \lambda_j < 1)$, $1 \leq j \leq c$.

The convergence of the generalized U-statistic was established by Yoshihara (1976) for absolutely regular r.v.'s but only for the stationary case. In this paper, we generalize the results of Yoshihara (1976) from the stationary case to the nonstationary case. Our methods are the adaptations of some of the ideas of Yoshihara (loc. cit.) and a new central limit theorem (Lemma 2.3) for nonstationary unbounded strongly-mixing random variables which is an extension of a central limit theorem due to Withers (1975) for nonstationary uniformly bounded strongly-mixing random variables. For a special case of the generalized U-statistic studied in this paper (i.e. for the case $c = 1$) and its applications to some Markov processes and ARMA models, the reader is referred to Harel and Puri (1989a, b).

2. Preliminaries

2.1. Notations

For each $\boldsymbol{d} = (d_1, \ldots, d_c)$ with $0 \leq d_j \leq m_j$, $1 \leq j \leq c$, we set

$$J_{\boldsymbol{n}}(\boldsymbol{d}) = \{\boldsymbol{i};\ \boldsymbol{i} = (i_1, \ldots, i_c),\ i_j = (i_{j1}, \ldots, i_{jd_j}),$$

$$1 \leq i_{j1} \neq \cdots \neq i_{jd_j} \leq n_j \text{ if } d_j \neq 0 \text{ and } i_{jd_j} = 0 \text{ if } d_j = 0, 1 \leq j \leq c\}.$$

Let $\boldsymbol{i} \in J_{\boldsymbol{n}}(\boldsymbol{d})$ exist, and let

$$g_{\boldsymbol{d},\boldsymbol{n}}^{(\boldsymbol{i})}(\underset{\sim}{x}_{i,j}, i = 1, \ldots, d_j, j = 1, \ldots, c)$$

$$= \sum_{\boldsymbol{i}^* \in I_{\boldsymbol{d},\boldsymbol{n}}^{(\boldsymbol{i})}} \int_{\mathbb{R}_c^*} g(\underset{\sim}{x}_{i,j}, i = 1, \ldots, m_j, j = 1, \ldots, c) \prod_{j=1}^{c} \prod_{\nu = d_j + 1}^{m_j} \mathrm{d}F_{n_j, i_{j\nu}, j}(\underset{\sim}{x}_{\nu,j}) \quad (2.1)$$

where

$$I_{d,n}^{(i)} = \{i^* = (i_1^*, \ldots, i_c^*), i_j^* = (i_{j,d_j+1}, \ldots, i_{jm_j}), 1 \leq i_{j,d_j+1} \neq \cdots \neq i_{jm_j} \leq n_j,$$
$$i_{jl} \notin \{i_{j1}, \ldots, i_{jd_j}\}, \text{ if } d_j+1 \leq l \leq m_j, 1 \leq j \leq c\}, \tag{2.2}$$
$$\mathbb{R}_c^* = (\mathbb{R}^k)^{\sum_{j=1}^c (m_j - d_j)}$$

and we set $g_{0,n} = g_{d,n}^{i}$ for $d = (0, \ldots, 0)$. By convention if $d = (d_1, \ldots, d_c)$ where $d_j = 1$ and $d_l = 0$, $l \neq j$, then d will be denoted by $d^{(j)}$.

For every d, we put

$$U_n^{(d)} = \prod_{j=1}^{c} n_j^{-[m_j]} \sum_{i \in J_n(d)} \int_{\tilde{\mathbb{R}}_c} g_{d,n}^{(i)}(x_{l,j}, l = 1, \ldots, d_j, j = 1, \ldots, c)$$
$$\times \prod_{j=1}^{c} \prod_{l=1}^{d_j} \mathrm{d}(I_{[X_{n_j, i_{jl}, j} \leq x_{l,j}]} - F_{n_j, i_{jl}, j}(x_{l,j})) \tag{2.3}$$

where $I_{[\ ]}$ denotes the indicator function on $\mathbb{R}^k$, $a \leq b$ means $a_l \leq b_l$, $l = 1, \ldots, k$, $\tilde{\mathbb{R}}_c = (\mathbb{R}^k)^{\sum_{j=1}^c d_j}$ and $m^{-[p]} = m(m-1) \cdots (m-p+1)$.

Let $a = (a_1, \ldots, a_c)$ with $a_j \geq 1$, $1 \leq j \leq c$, and denote $L_n(a) = \{i; i = (i_1, \ldots, i_c),$ $i_j = (i_{j1}, \ldots, i_{ja_j}), 1 \leq i_{j1} < \cdots < i_{ja_j} \leq n_j, j = 1, \ldots, c\}$.

Let $i \in L_n(a)$ and let for any j $(1 \leq j \leq c)$ and any l_j $(1 \leq l_j \leq a_j)$, $P_{j,l_j,n}^{i}$ be the probability measure induced by the random vectors

$$Y_{n,j,l_j}^{i} = (X_{n_1, i_{11}, 1}, \ldots, X_{n_1, i_{1a_1}, 1}, X_{n_2, i_{21}, 2}, \ldots, X_{n_{j-1}, i_{j-1, a_{j-1}}, j-1},$$
$$X_{n_j, i_{j1}, j}, \ldots, X_{n_j, i_{jl_j}, j})$$

and

$$Y_{n,j,l_j}^{*i} = (X_{n_j, i_{j,l_j+1}, j}, \ldots, X_{n_j, i_{ja_j}, j}, X_{n_{j+1}, i_{j+1,1}, j+1}, \ldots, X_{n_{j+1}, i_{ca_c}, c})$$

respectively and defined by

$$P_{j,l_j,n}^{i}(A_{l_j}^{(j)} \times A_{l_j}^{*(j)}) = P[Y_{n,j,l_j}^{i} \in A_{l_j}^{(j)}] \times P[Y_{n,j,l_j}^{*i} \in A_{l_j}^{*(j)}]$$

and

$$P_{0,n}^{i}(A^{(c)}) = P[(X_{n_1, i_{11}, 1}, \ldots, X_{n_c, i_{ca_c}, c}) \in A^{(c)}]$$

where

$$A_{l_j}^{(j)} \in \sigma(Y_{n,j,l_j}^{i}), \qquad A_{l_j}^{*(j)} \in \sigma(Y_{n,j,l_j}^{*i}), \qquad A^{(c)} \in \sigma(X_{n_1, i_{11}, 1}, \ldots, X_{n_c, i_{ca_c}, c}),$$

and where $\sigma(X)$ is the σ-field generated by X.

For any function $h: (\mathbb{R}^k)^{m_0} \to \mathbb{R}$, we denote $h \otimes h$ the function $h \otimes h: (\mathbb{R}^k)^{2m_0} \to \mathbb{R}$ defined by

$$h \otimes h(x_1, \ldots, x_{2km_0}) = h(x_1, \ldots, x_{km_0}) h(x_{km_0+1}, \ldots, x_{2km_0}).$$

For any $p \geq 1$, let $f: [0, 1]^p \to \mathbb{R}$. For $\rho = (\rho(1), \ldots, \rho(p)) \in \{0, 1\}^p$, define

$$f_\rho(t) = \lim_{\substack{s_i \uparrow t_i\, \rho(i)=1 \\ s_i \downarrow t_i\, \rho(i)=0}} f(s), \quad s \in [0, 1]^p, \ t \in [0, 1]^p,$$

if it exists where $s=(s_1,\ldots,s_p)$, $t=(t_1,\ldots,t_p)$, and denote by D_p (D_p^* resp.) the space of all maps $f:[0,1]^p\to\mathbb{R}$ such that for all $\rho\in\{0,1\}^p$, f_ρ exists and $f_\rho=f$ for $\rho=(0,\ldots,0)$ ($\rho=(1,\ldots,1)$ resp.). We associate on D_p the Skorohod topology (for a definition of Skorohod topology see Neuhaus, 1971, and Balacheff and Dupont, 1980).

2.2. Basic lemmas

We need the following lemmas.

Lemma 2.1. *For every $\boldsymbol{a}=(a_1,\ldots,a_c)$ with $a_j=b_jm_j$, $b_j\geq 1$, $1\leq j\leq c$, let $\boldsymbol{i}\in L_{\boldsymbol{n}}(\boldsymbol{a})$ exist, and for every j $(1\leq j\leq c)$ and any l_j $(1\leq l_j\leq a_j)$, let*

$$h(\underset{\sim}{x}_1,\ldots,\underset{\sim}{x}_{b_1m_1},\underset{\sim}{x}_{b_1m_1+1},\ldots,\underset{\sim}{x}_{b_1m_1+\cdots+b_cm_c})$$

be a Borel function such that

$$\int_{\mathbb{R}_{a,c}}|h(\underset{\sim}{x}_1,\ldots,\underset{\sim}{x}_{b_1m_1+\cdots+b_cm_c})|^{1+\delta}\,\mathrm{d}P^{\boldsymbol{i}}_{j,l_j,\boldsymbol{n}}\leq M, \tag{2.4}$$

$$\int_{\mathbb{R}_{a,c}}|h(\underset{\sim}{x}_1,\ldots,\underset{\sim}{x}_{b_1m_1+\cdots+b_cm_c})|^{1+\delta}\,\mathrm{d}P^{\boldsymbol{i}}_{0,\boldsymbol{n}}\leq M,\quad \textit{for some } \delta>0, \tag{2.5}$$

where $\mathbb{R}_{a,c}=(\mathbb{R}^k)^{\sum_{i=1}^c b_im_i}$. Then

$$\left|\int_{\mathbb{R}_{a,c}}h(\underset{\sim}{x}_1,\ldots,\underset{\sim}{x}_{b_1m_1+\cdots+b_cm_c})\,\mathrm{d}P^{\boldsymbol{i}}_{j,l_j,\boldsymbol{n}}-\int_{\mathbb{R}_{a,c}}h(\underset{\sim}{x}_1,\ldots,\underset{\sim}{x}_{b_1m_1+\cdots+b_cm_c})\,\mathrm{d}P^{\boldsymbol{i}}_{0,\boldsymbol{n}}\right|$$
$$\leq 4M^{1/(2+\delta)}\beta_j^{\delta/(1+\delta)}(i_{j,l_j+1}-i_{j,l_j}). \tag{2.6}$$

As the special case, if h is bounded, say $|h(\underset{\sim}{x}_1,\ldots,\underset{\sim}{x}_{b_1m_1+\cdots+b_cm_c})|\leq M$, then we can replace the right side of (2.6) *by $2M\beta_j(i_{j,l_j+1}-i_{j,l_j})$.*

Proof. Follows from Lemma 1 of Yoshihara (1976). □

Let $\mathcal{S}_j^*$ be the set of all nondecreasing sequences of m_j elements in $\{1,\ldots,n_j\}$ $(1\leq i_{j,1}\leq i_{j,2}\leq\cdots\leq i_{j,m_j}\leq n_j)$ and let $\mathcal{S}_{\boldsymbol{n}}^*=\prod_{j=1}^c\mathcal{S}_j^*$ exist.

For any $\boldsymbol{n}$ and any family $s=(i_{j,l},1\leq l\leq m_j,1\leq j\leq c)$ which belongs to $\mathcal{S}_{\boldsymbol{n}}^*$, we denote

$$a_{\boldsymbol{n},s}(h,r)=\int|h(\underset{\sim}{x}_{l,j},1\leq l\leq m_j,1\leq j\leq c)|^r\prod_{j=1}^c\prod_{l=1}^{m_j}\mathrm{d}F_{n_j,i_{j,l},j}(\underset{\sim}{x}_{l,j}) \tag{2.7}$$

and

$$b_{\boldsymbol{n},s}(h,r)=E(|h(\underset{\sim}{X}_{n_j,i_{jl},j})|)^r \tag{2.8}$$

for any $r>0$ and any Borel function $h:(\mathbb{R}^k)^{m_0}\to\mathbb{R}$.

Lemma 2.2. *If there exists a positive number* δ' *such that* (i) *for* $r = 4 + \delta'$, *the family* $(a_{\boldsymbol{n},s}(h, r), s \in \mathcal{S}_{\boldsymbol{n}}^*, \boldsymbol{n} \in (\mathbb{N}^*)^c)$ *and* $(b_{\boldsymbol{n},s}(h, r), s \in \mathcal{S}_{\boldsymbol{n}}^*, \boldsymbol{n} \in \mathbb{N}^*)^c)$ *are bounded by a finite constant M,* (ii) *for some* δ, $\delta > 0$, $\delta' > 12\delta^{-1} - 2$, *and* (iii) *the rate of absolute regularity satisfies* $\max_{1 \le j \le c} \beta_j(m) = O(m^{-6-\delta})$, *then*

$$E(U_{\boldsymbol{n}}^{(\boldsymbol{d})})^4 = O(n^{-6}) \quad \textit{for any } \boldsymbol{d} \textit{ such that } \sum_{j=1}^{c} d_j = 2 \textit{ and } \max_{1 \le j \le c} d_j = 1, \tag{2.9}$$

$$E(U_{\boldsymbol{n}}^{(\boldsymbol{d})})^4 = O(n^{-4}) \quad \textit{for any } \boldsymbol{d} \textit{ such that } \sum_{j=1}^{c} d_j = 2 \textit{ and } \max_{1 \le j \le c} d_j = 2, \tag{2.10}$$

$$E(U_{\boldsymbol{n}}^{(\boldsymbol{d})})^2 = O(n^{-2-c(\boldsymbol{d})}) \quad \textit{for any } \boldsymbol{d} \textit{ such that } \sum_{j=1}^{c} d_j > 2, \tag{2.11}$$

where $c(\boldsymbol{d}) = \operatorname{card}\{j \in \{1, \ldots, c\}, d_j \ne 0\}$.

Proof. It is sufficient to prove (2.9) for $\boldsymbol{d} = (1, 1, 0, \ldots, 0)$, the proof for $\boldsymbol{d} \ne (1, 1, 0, \ldots, 0)$ is analogous and is therefore omitted.

Suppose $\boldsymbol{d} = (1, 1, 0, \ldots, 0)$. Then, we first note that

$$U_{\boldsymbol{n}}^{(\boldsymbol{d})} = \prod_{j=1}^{c} n_j^{-[m_j]} \sum_{i_1=1}^{n_1} \sum_{i_2=1}^{n_1} \int_{\mathbb{R}^{2k}} g_{\boldsymbol{d},\boldsymbol{n}}^{(\boldsymbol{i})}(\boldsymbol{x}_{1,1}, \boldsymbol{x}_{1,2})\, \mathrm{d}(I_{[\boldsymbol{X}_{n_1,i_1,1} \le \boldsymbol{x}_{1,1}]} - F_{n_1,i_1,1}(\boldsymbol{x}_{1,1}))$$
$$\mathrm{d}(I_{[\boldsymbol{X}_{n_2,i_2,2} \le \boldsymbol{x}_{1,2}]} - F_{n_2,i_2,2}(\boldsymbol{x}_{1,2})) \tag{2.12}$$

where $\boldsymbol{i} = (i_1, i_2, 0, \ldots, 0)$, so that

$$E(U_{\boldsymbol{n}}^{(\boldsymbol{d})})^4 = \left(\prod_{j=1}^{c} n_j^{-[m_j]} \right)^4 \sum{}^{*} J(i_{jl}, j = 1, 2, l = 1, \ldots, 4) \tag{2.13}$$

where $\sum^*$ is the sum over all $1 \le i_{jl} \le n_j$, $j = 1, 2$, $l = 1, \ldots, 4$.

$$J(i_{jl}, j = 1, 2, l = 1, \ldots, 4)$$
$$= E\left\{ \prod_{l=1}^{4} \int_{\mathbb{R}^{2k}} g_{\boldsymbol{d},\boldsymbol{n}}^{(\boldsymbol{l})}(\boldsymbol{x}_{l,1}, \boldsymbol{x}_{l,2})\, \mathrm{d}(I_{[\boldsymbol{X}_{n_1,i_{1l},1} \le \boldsymbol{x}_{l,1}]} - F_{n_1,i_{1l},1}(\boldsymbol{x}_{l,1})) \right.$$
$$\left. \mathrm{d}(I_{[\boldsymbol{X}_{n_2,i_{2l},2} \le \boldsymbol{x}_{l,2}]} - F_{n_2,i_{2l},2}(\boldsymbol{x}_{l,2})) \right\}, \tag{2.14}$$

$\boldsymbol{l} = (l_1, \ldots, l_c)$, $l_m = i_{ml}$, $m = 1, 2$, $l_m = 0$, $m \ne 1, 2$, $l = 1, \ldots, 4$, and

$$g_{\boldsymbol{d},\boldsymbol{n}}^{(\boldsymbol{l})} = \sum_{\boldsymbol{l}^* \in I_{\boldsymbol{d},\boldsymbol{n}}^{(\boldsymbol{l})}} \int_{(\mathbb{R}^k)^{(\sum_{j=1}^{c} m_j) - 2}} g(\boldsymbol{x}_{i,j}, i = 1, \ldots, m_j, j = 1, \ldots, c)$$
$$\times \prod_{j=1}^{c} \prod_{\nu=\varepsilon(j)}^{m_j} \mathrm{d}F_{n_j, i_{j\nu}, j}(\boldsymbol{x}_{\nu,j}),$$

where $\varepsilon(j) = 1$ if $j \ne 1, 2$ and $\varepsilon(j) = 2$ if $j = 1, 2$.

Suppose $n_1 \le n_2$. If i_{jl} $(j = 1, 2, l = 1, \ldots, 4)$ are mutually different, reorder $\{i_{jl}\}$ as $1 \le k_1 < k_2 < \cdots < k_8 \le n$ and put

$$J(i_{jl}, j = 1, 2, l = 1, \ldots, 4) = H(k_1, \ldots, k_8). \tag{2.15}$$

Let $m^{(p)}$ be the pth largest difference among $k_{j+1}-k_j$ $(j=1,\dots,7)$ if for some j_α $(1\le j_\alpha\le 7, 1\le\alpha\le 2)$, $k_{j_\alpha+1}-k_\alpha=m^{(\alpha)}$. Then, from Lemma 2.1, (2.14) and proceeding as in the proof of Lemma 2.2, we obtain

$$H(k_1,\dots,k_8)\le M^{1/(1+\delta')}\left(\prod_{j=1}^{c} n_j^{[m_j]}\right)^4 (n_1n_2)^4 \sum_{\alpha=1}^{2}\max_{1\le j\le c}\beta_j^{(2+\delta')/(4+\delta')}(k_{j_\alpha+1}-k_\alpha)$$

and then

$$\sum_{1\le k_1<\dots<k_8\le n_2} H(k_1,\dots,k_8)$$

$$\le 2M^{1/(1+\delta')}\left(\prod_{j=1}^{c} n_j^{[m_j]}\right)^4 (n_1n_2)^{-4}n_2^2\sum_{l=1}^{n_2}(l+1)^5\max_{1\le j\le c}\beta_j^{(2+\delta')/(4+\delta')}(l)$$

$$=\mathrm{O}(n^{4\sum_{j=1}^{c}m_j-6}). \tag{2.16}$$

We can use a similar method to estimate the sums in the other cases and so from (2.13) and (2.16), we have (2.9) for $\boldsymbol{d}=(1,1,0,\dots,0)$.

Now suppose $\boldsymbol{d}=(2,0,\dots,0)$ and then we have

$$E(U_{\boldsymbol{n}}^{(\boldsymbol{d})})^4=\left(\prod_{j=1}^{c} n_j^{-[m_j]}\right)^4 \sum{}^{*} J(i_{jl}, j=1,\dots,4, l=1,2)$$

where $\sum^*$ is the sum over all $1\le i_{j1}\le i_{j2}\le n_1$, $j=1,\dots,4$.

$$J(i_{jl}, j=1,\dots,r, l=1,2)$$

$$=E\left\{\prod_{j=1}^{4}\int_{\mathbb{R}^{2k}} g_{\boldsymbol{d},\boldsymbol{n}}^{(\boldsymbol{j})}(\underset{\sim}{x}_{1,j}\underset{\sim}{x}_{2,j})\,\mathrm{d}(I_{[\underset{\sim}{X}_{n_1,i_{j1},1}\le\underset{\sim}{x}_{1,j}]}-F_{n_1,i_{j1},1}(\underset{\sim}{x}_{1,j}))\right.$$

$$\left.\mathrm{d}(I_{[\underset{\sim}{X}_{n_1,i_{j2},1}\le\underset{\sim}{x}_{2,j}]}-F_{n_1,i_{j2},1}(\underset{\sim}{x}_{2,j}))\right\},$$

$\boldsymbol{j}=(j_1,\dots,j_c)$, $j_1=(i_{j1},i_{j2})$, $1\le i_{j1}\le i_{j2}\le n_1$, $j_l=0$, $l\ne 1$. If i_{jl} $(j=1,\dots,4, l=1,2)$ are mutually different, reorder $\{i_{jl}\}$ as $1\le k_1<k_2<\dots<k_8\le n_1$ and put

$$J(i_{jl}, j=1,\dots,4, l=1,2)=H(k_1,\dots,k_8).$$

Then, proceeding as in the preceding case, we obtain

$$\sum_{1\le k_1<\dots<k_8\le n_1} H(k_1,\dots,k_8)$$

$$\le 4M^{1/(1+\delta')}\left(\prod_{j=1}^{c} n_j^{[m_j]}\right)^4 (n_1(n_1-1)]^{-4}n_1^4\sum_{l=1}^{n_1}(l+1)^3\max_{1\le j\le c}\beta_j^{(2+\delta')/(4+\delta')}(l)$$

$$=\mathrm{O}(n^{4\sum_{j=1}^{c}m_j-4}) \tag{2.17}$$

and similar results for the other cases. Thus (2.10) is proved for $\boldsymbol{d}=(2,0,\dots,0)$.

The proof of (2.11) is similar to that of (2.9) and (2.10) and is therefore omitted. □

Lemma 2.3. *Let $\{X^*_{ni}\}$, $1 \leq i \leq n$, $n \geq 1$, be r.v.'s (with means 0) which are strong mixing with rate $\alpha(m)$ satisfying*

$$\sum_{m=1}^{\infty} (\alpha(m))^{\delta''/(2+\delta'')} < \infty \quad \textit{for some } \delta''. \tag{2.18}$$

Suppose that for any $K > 0$, there exists a sequence $\{Y^K_{ni}\}$ of real r.v.'s satisfying (2.18) such that

$$\sup_{n \in \mathbb{N}^*} \max_{1 \leq i \leq n} |Y^K_{ni}| \leq B_K < \infty \quad \forall K > 0, \tag{2.19}$$

where B_K is some constant >0;

$$\sup_{n \in \mathbb{N}^*} \max_{1 \leq i \leq n} E|X^*_{ni} - Y^K_{ni}|^{2+\delta''} \to 0, \quad K \to \infty, \tag{2.20}$$

$$E\left(\sum_{i=1}^{n} X^*_{ni}\right)^2 \Big/ n \to C^2 < \infty, \quad n \to \infty, \tag{2.21}$$

where C is some constant >0;

$$E\left(\sum_{i=1}^{n} Y^K_{ni} - E(Y^K_{ni})\right)^2 \Big/ n \to C^2_K < \infty, \quad n \to \infty, \ \forall K > 0, \tag{2.22}$$

where C_K is some constant >0;

$$C_K \to C, \quad K \to \infty; \tag{2.23}$$

then, $n^{-1/2} \sum_{i=1}^{n} X^*_{ni}$ converges in law to the normal distribution with mean 0, and variance C^2.

Remark. Actually we are considering the sequence $\{X^*_{n_N i}\}$, $1 \leq i \leq n_N$, where $n_N \to \infty$ as $N \to \infty$, but for convenience we are writing $\{X^*_{ni}\}$.

Proof of Lemma 2.3. From Corollary 1 of Withers (1975), since $\{Y^K_{ni}\}$ is uniformly bounded, we deduce that $\forall K > 0$, $n^{-1/2} \sum_{i=1}^{n} (Y^K_{ni} - E(Y^K_{ni}))$ converges in law to the normal distribution with mean 0 and variance C^2_K.

Now denote $Z^K_{ni} = X^*_{ni} - Y^K_{ni}$ and following Ibragimov (1962), we represent the sum

$$S_n = n^{-1/2} C^{-1} \left(\sum_{i=1}^{n} X^*_{ni} \right)$$

as $S_n = S'_n + S''_n$ where

$$S'_n = n^{-1/2} C_K^{-1} \left(\sum_{i=1}^{n} Y^K_{ni} - E(Y^K_{ni}) \right) C_K C^{-1}$$

and

$$S''_n = n^{-1/2} C^{-1} \left(\sum_{i=1}^{n} Z^K_{ni} - E(Z^K_{ni}) \right).$$

Evaluating $E|S''_n|^2$, we have

$$E|S''_n|^2 \leq 2n^{-1}C^{-2} \sum_{i=0}^{n-1} \sum_{j=1}^{n-i} |E(Z^K_{nj} - E(Z^K_{nj}))(Z^K_{n,j+i} - E(Z^K_{n,j+i}))|$$

$$\leq 2n^{-1}C^{-2} \sum_{i=0}^{n-1} (\alpha(i))^{\delta''/(2+\delta'')} \sum_{j=1}^{n-i} (2^{2+2\delta''} E|Z^K_{nj}|^{2+\delta''} E|Z^K_{n,j+i}|^{2+\delta''})^{1/(2+\delta'')}$$

(from the well known inequality on moments of strong mixing sequences of r.v.'s; see Doukhan and Portal, 1987, Proposition 2.8)

$$\leq 2n^{-1}C^{-2} \sum_{i=0}^{n-1} (\alpha(i))^{\delta''/(2+\delta')}(n-i)2^{2+2\delta''} \sup_{n\in\mathbb{N}^*} \max_{1\leq i\leq n} E|Z^K_{ni}|^{2+\delta''}.$$

Denoting

$$D_K = \sup_{n\in\mathbb{N}^*} \max_{1\leq i\leq n} E|Z^K_{ni}|^{2+\delta''},$$

we deduce from (2.18) and (2.20) that

$$E|S''_n|^2 \leq 2^{3+\delta''}C^{-2}\left(\sum_{i=0}^{\infty} (\alpha(i))^{\delta''/(2+\delta'')}\right) D_K \to 0 \quad \text{as } K\to\infty.$$

For any $\varepsilon > 0$, we can always choose K so large that

$$E|S''_n|^2 \leq \varepsilon \quad \text{and} \quad |1 - C^2_K/C^2| \leq \varepsilon.$$

If now $f_n(t) = E(\mathrm{e}^{\mathrm{i}tS_n})$ is the characteristic function of the quantity S_n, then

$$|f_n(t) - \mathrm{e}^{-t^2/2}|$$

$$\leq |\mathrm{e}^{-(t^2/2)(C^2_K/C^2)} - \mathrm{e}^{-t^2/2}| + |\mathrm{e}^{-(t^2/2)(C^2_K/C^2)} - E\ \mathrm{e}^{\mathrm{i}tS'_n}| + |t|E|S''_n|$$

$$\leq |t|\sqrt{\varepsilon} + \varepsilon t^2 + |\mathrm{e}^{-(t^2/2)(C^2_K/C^2)} - E\ \mathrm{e}^{\mathrm{i}tS'_n}|.$$

From the convergence in law of S'_n to the normal law, the last expression converges to zero as $n\to\infty$, and the lemma is proved. □

Let $G_{n,i,l,j}$ be the d.f. of $(\underset{\sim}{X}_{n,i,j}, \underset{\sim}{X}_{n,l,j})$, $1\leq i<l\leq n_j$, $1\leq j\leq c$. In what follows $\lim_n$ means $n_j\to\infty$ $\forall j=1,\ldots,c$, and $n_j/\sum_{i=1}^{c} n_i \to \lambda_j$ as $n_j\to\infty$, $j=1,\ldots,c$.

Lemma 2.4. *Let $h:(\mathbb{R}^k)^{m_0}\to\mathbb{R}$ be a function which belongs to D_{km_0} or $D^*_{km_0}$ and suppose that for any $l\in\mathbb{N}^*$ with $1<l$ and any $j\in\{1,\ldots,c\}$ there exists a d.f. $G_{l,j}$ on $\mathbb{R}^{2k}$ with marginals F_j. Furthermore assume that*

$$\lim_n \max_{1\leq j\leq c} \max_{1\leq i<l\leq n} |G_{n_j,i,l,j}(\underset{\sim}{x}_1, \underset{\sim}{x}_2) - G_{l-i,j}(\underset{\sim}{x}_1, \underset{\sim}{x}_2)| = 0 \tag{2.24}$$

for all $(x_1, x_{2k}) \in \mathbb{R}^{2k}$, *the family* $(a_{\mathbf{n},s}(h, r),\ s \in \mathcal{S}_n^*,\ \mathbf{n} \in (\mathbb{N}^*)^c)$ *is bounded by a finite constant* M_0, *and*

$$c(h, r) = \int_{(\mathbb{R}^k)^{m_0}} |h(x_{1,1}, \ldots, x_{1,2}, \ldots, x_{c,m_c})|^r \prod_{j=1}^{c} \prod_{l=1}^{m_j} \mathrm{d}F_j(x_{l,j})$$
$$< M_0 < \infty \quad \text{for } r > 2. \tag{2.25}$$

Then for any $j \in \{1, \ldots, c\}$,

$$\lim_{\mathbf{n}} \max_{(\mathbf{i},\mathbf{l}) \in J_{\mathbf{n}}(\mathbf{d}^{(j)})} \Bigg| \left(n_j \prod_{j=1}^{c} n_j^{-[m_j]}\right)^2 \Bigg(\sum_{\mathbf{i}^* \in I^{(\mathbf{i})}_{\mathbf{d}^{(j)},\mathbf{n}}} \sum_{\mathbf{l}^* \in I^{(\mathbf{l})}_{\mathbf{d}^{(j)},\mathbf{n}}} \int_{\mathbb{R}^{2km_0}} (h \otimes h)(x_1, \ldots, x_{2m_0})$$
$$\mathrm{d}G_{n_j,i,l,j}(x_1, x_{m_0+1})\, \mathrm{d}F^*_{\mathbf{n},\mathbf{i}^*}(x_2, \ldots, x_{m_0})\, \mathrm{d}F^*_{\mathbf{n},\mathbf{l}^*}(x_{m_0+2}, \ldots, x_{2m_0}) \Bigg)$$
$$- \int_{\mathbb{R}^{2km_0}} (h \otimes h)(x_1, \ldots, x_{2m_0})\, \mathrm{d}G_{l_j - i_j, j}(x_1, x_{m_0+1})$$
$$\mathrm{d}F^*(x_2, \ldots, x_{m_0})\, \mathrm{d}F^*(x_{m_0+2}, \ldots, x_{2m_0})) \Bigg| = 0 \tag{2.26}$$

where

$$F^*_{\mathbf{n},\mathbf{i}^*} = \left(\prod_{r \neq j} \prod_{p=1}^{m_r} F_{n_p, i_{rp}, r} \right) \prod_{p=2}^{m_j} F_{n_j, i_{jp}, j}$$

if

$$\mathbf{i}^* = (i_1^*, \ldots, i_c^*), \quad i_r^* = (i_{r1}, \ldots, i_{rm_r}) r \neq j, \quad i_j^* = (i_{j2}, \ldots, i_{jm_j}),$$
$$F^* = \left(\prod_{r \neq j} F_r^{m_r} \right) \times F_j^{m_j - 1}.$$

Proof. From (2.24) we easily deduce that (2.26) is true if h is a step function.

Without loss of generality, we only prove (2.26) if h is a nondecreasing function. For any $K > 0$, we denote by h^K the function defined by

$$h^K = \begin{cases} h, & \text{if } |h| \leq K, \\ 0, & \text{if } |h| > K. \end{cases}$$

As h belongs to D_{km_0} or $D^*_{km_0}$ and is bounded, it is well known that h^K can be uniformly approached by a step function. Let $\varepsilon > 0$ exist.

Choose K sufficiently large so that

$$\max\left\{ \frac{M_0^2}{K^{r-1}}, \frac{M_0}{K^{r-2}} \right\} < \tfrac{1}{6}\varepsilon. \tag{2.27}$$

There exists a step function g_ε^K such that

$$\sup_{(x_1, \ldots, x_{m_0})} |h^K(x_1, \ldots, x_{m_0}) - g_\varepsilon^K(x_1, \ldots, x_{m_0})| < \varepsilon / 6K$$

and there exists $N_0 \in \mathbb{N}$ such that $\forall n \geq N_0$,

$$\max_{(\boldsymbol{i},\boldsymbol{l}) \in J_n(\boldsymbol{d}^{(j)})} \Bigg| \left(n_j \prod_{j=1}^{c} n_j^{-[m_j]} \right)^2 \Bigg(\sum_{\boldsymbol{i}^* \in I^{(\boldsymbol{i})}_{\boldsymbol{d}^{(j)},n}} \sum_{\boldsymbol{l}^* \in I^{(\boldsymbol{l})}_{\boldsymbol{d}^{(j)},n}} \int_{\mathbb{R}^{2km_0}} (g_\varepsilon^K \otimes g_\varepsilon^K)(\underset{\sim}{x}_1, \ldots, \underset{\sim}{x}_{2m_0})$$

$$\mathrm{d}G_{n_j,i,l,j}(\underset{\sim}{x}_1, \underset{\sim}{x}_{m_0+1})\, \mathrm{d}F^*_{\boldsymbol{n},\boldsymbol{i}^*}(\underset{\sim}{x}_2, \ldots, \underset{\sim}{x}_{m_0})\, \mathrm{d}F^*_{\boldsymbol{n},\boldsymbol{l}^*}(\underset{\sim}{x}_{m_0+2}, \ldots, \underset{\sim}{x}_{2m_0}) \Bigg)$$

$$- \int (g_\varepsilon^K \otimes g_\varepsilon^K)(\underset{\sim}{x}_1, \ldots, \underset{\sim}{x}_{2m_0})\, \mathrm{d}G_{l_j-i_j,j}(\underset{\sim}{x}_1, \underset{\sim}{x}_{m_0+1})$$

$$\mathrm{d}F^*(\underset{\sim}{x}_2, \ldots, \underset{\sim}{x}_{m_0})\, \mathrm{d}F^*(\underset{\sim}{x}_{m_0+2}, \ldots, \underset{\sim}{x}_{2m_0}) \Bigg| < \tfrac{1}{3}\varepsilon. \tag{2.28}$$

From (2.27), (2.28) and the decomposition

$$h \otimes h = h \otimes h - h \otimes h^K + h \otimes h^K - h^K \otimes h^K + h^K \otimes h^K$$
$$- h^K \otimes g_\varepsilon^K + h^K \otimes g_\varepsilon^K - g_\varepsilon^K \otimes g_\varepsilon^K + g_\varepsilon^K \otimes g_\varepsilon^K,$$

we deduce that $\forall n \geq N_0$ the left side of (2.28) in which we replace $g_\varepsilon^K \otimes g_\varepsilon^K$ by $h \otimes h$ is less than

$$\frac{M_0^2}{K^{r-1}} + \frac{M_0}{K^{r-2}} + K\frac{\varepsilon}{6K} + K\frac{\varepsilon}{6K} + \tfrac{1}{3}\varepsilon < \varepsilon.$$

As ε is chosen arbitrarily, (2.26) is proved. □

3. Convergence of the U-statistic

3.1. *Definitions*

For any double sequence of distribution functions $G_{l,j}$ on $\mathbb{R}^{2k}$ with marginals F_j, $1 < l$, $j \in \{1, \ldots, c\}$, let $\boldsymbol{F} = (F_1, \ldots, F_c)$ and $\boldsymbol{G}_{\boldsymbol{l}} = (G_{l_1,1}, \ldots, G_{l_c,c})$, where $\boldsymbol{l} = (l_1, \ldots, l_c)$, we define

$$\theta(\boldsymbol{F}) = \int_{(\mathbb{R}^k)^{m_0}} g(\underset{\sim}{x}_{1,1}, \ldots, \underset{\sim}{x}_{m_1,1}, \underset{\sim}{x}_{2,1}, \ldots, \underset{\sim}{x}_{m_c,c}) \prod_{j=1}^{c} \prod_{i=1}^{m_j} \mathrm{d}F_j(\underset{\sim}{x}_{i,j}) \tag{3.1}$$

and

$$\sigma_j^2(\{\boldsymbol{G}_{\boldsymbol{l}}\}) = \lim_{n_j \to \infty} \Bigg\{ \Bigg[\Bigg[\int_{\mathbb{R}^k} (e_j(\underset{\sim}{x}))^2\, \mathrm{d}F_j(\underset{\sim}{x}) - \theta^2(\boldsymbol{F}) \Bigg]$$

$$+ 2 \sum_{l_j=2}^{n_j} \Bigg[\Bigg[\int_{\mathbb{R}^{2k}} (e_j(\underset{\sim}{x}) e_j(\underset{\sim}{z}))\, \mathrm{d}G_{l_j,j}(\underset{\sim}{x}, \underset{\sim}{z}) - \theta^2(\boldsymbol{F}) \Bigg] \Bigg] \Bigg\} \tag{3.2}$$

if it exists, $j = 1, \ldots, c$ where $e_j = g^{\boldsymbol{F}}_{\delta_{j1} \ldots \delta_{jc}}$, $\delta_{ab} = 1$ or 0 according $a = b$ or not and

$$g^{\boldsymbol{F}}_{d_1,\ldots,d_c}(\underset{\sim}{x}_{i,j}, i = 1, \ldots, d_j, j = 1, \ldots, c)$$
$$= \int_{\mathbb{R}^*_c} g(\underset{\sim}{x}_{i,j}, i = 1, \ldots, m_j, j = 1, \ldots, c) \prod_{j=1}^{c} \prod_{i=d_j+1}^{m_j} \mathrm{d}F_j(\underset{\sim}{x}_{i,j}).$$

For convenience we shall write σ_j for $\sigma_j(\{\boldsymbol{G}_l\})$, if no confusion is possible. For every $\boldsymbol{t}=(t_1,\ldots,t_c)\in[0,1]^c$ and $\boldsymbol{n}=(n_1,\ldots,n_c)$, denote $[\boldsymbol{nt}]=([n_1t_1],\ldots,[n_ct_c])$ where $[s]$ is the largest integer $\leqslant s$.

Let $Z(\boldsymbol{n})=\{Z(t;\boldsymbol{n}),t\in[0,1]^c\}$ be the process defined by

$$Z(\boldsymbol{t};\boldsymbol{n})=\begin{cases}\psi([\boldsymbol{nt}];n)[U([\boldsymbol{nt}])-\theta(\boldsymbol{F}_{\boldsymbol{n}})] & \text{for all } [\boldsymbol{nt}]\geqslant \boldsymbol{m},\\ 0 & \text{otherwise,}\end{cases} \tag{3.3}$$

where for $\boldsymbol{p}=(p_1,\ldots,p_c)$ $(p_j>0, j=1,\ldots,c)$ and $n=n_1+\cdots+n_c$,

$$\psi(\boldsymbol{p};n)=n^{-1/2}\left(\sum_{j=1}^{c}\sigma_j\lambda_j^{1/2}\right)\left(\sum_{j=1}^{c}\sigma_j\lambda_j^{1/2}p_j^{-1}\right)^{-1},$$

σ_j is the constant defined in (3.2), $\theta(\boldsymbol{F}_{\boldsymbol{n}})=(\prod_{j=1}^{c}n_j^{-[m_j]})g_{0,\boldsymbol{n}}$ and $\boldsymbol{a}\leqslant\boldsymbol{b}$ means $a_j\leqslant b_j$ for all $j\in\{1,\ldots,c\}$.

Denote by $\{W_j;j=1,\ldots,c\}=\{W_j(t_j);0\leqslant t_j\leqslant 1,j=1,\ldots,c\}$, c independent copies of a standard Brownian motion on $[0,1]$.

3.2. Convergence of the U-statistic

Theorem 3.1. *Assume that there are two positive numbers δ,δ', where $\delta'>\max(12\delta^{-1}-2, 2(5+\delta)^{-1})$ such that the families $(a_{\boldsymbol{n},s}(g,r), s\in\mathcal{S}^*_{\boldsymbol{n}}, \boldsymbol{n}\in(\mathbb{N}^*)^c)$ and $(b_{\boldsymbol{n},s}(g,r), s\in\mathcal{S}^*_{\boldsymbol{n}}, \boldsymbol{n}\in(\mathbb{N}^*)^c)$ are bounded, $c(g,r)$ is finite where $r=4+\delta'$ and the rates of absolute regularity satisfy*

$$\max_{1\leqslant j\leqslant c}\beta_j(m)=O(m^{-6-\delta}). \tag{3.4}$$

Further assume the condition (2.24) is satisfied and g belongs to D_{km_0} or $D^*_{km_0}$. Then, the limits in (3.2) exist.

Furthermore if

$$\max_{1\leqslant j\leqslant c}\sigma_j^2(\{\boldsymbol{G}_l\})>0,$$

then $Z(\boldsymbol{n})$ converges in law with respect to the Skorohod topology on D_c to a Gaussian process $W=\{W(\boldsymbol{t});\boldsymbol{t}\in[0,1]^c\}$ where

$$W(\boldsymbol{t})=\begin{cases}\left(\sum_{j=1}^{c}m_j\sigma_j\lambda_j^{1/2}\right)\left(\sum_{j=1}^{c}m_j\sigma_j\lambda_j^{-1/2}\right)\left[\sum_{j=1}^{c}m_j\sigma_j\lambda_j^{-1/2}W_j(t_j)\right], & \boldsymbol{t}>\boldsymbol{0},\\ 0, \quad \textit{with probability one if } t_j=0 \textit{ for } j\ (1\leqslant j\leqslant c).\end{cases} \tag{3.5}$$

We write (1.1) as

$$U(\boldsymbol{n})=\prod_{j=1}^{n}\binom{n_j}{m_j}^{-1}\sum_{i^*\in I^{(i)}_{d^{(0)},\boldsymbol{n}}}\int_{\mathbb{R}^{2km_0}}g(\underset{\sim}{x}_{i,j}, i=1,\ldots,m_j, j=1,\ldots,c)$$

$$\times\prod_{j=1}^{c}\prod_{\nu=1}^{m_j}\mathrm{d}(I_{[\underset{\sim}{X}_{n_j,i_{j\nu}j}\leqslant\underset{\sim}{x}_{\nu,j}]})$$

$$=\prod_{j=1}^{c}\binom{n_j}{m_j}^{-1}\sum_{i^*\in I^{(i)}_{d^{(0)},\boldsymbol{n}}}\int_{\mathbb{R}^{2km_0}}g(\underset{\sim}{x}_{i,j}, i=1,\ldots,m_j, j=1,\ldots,c)$$

$$\times \prod_{j=1}^{c} \prod_{\nu=1}^{m_j} \mathrm{d}(I_{[\underset{\sim}{X}_{n_j,i_{j\nu},j} \leqslant \underset{\sim}{x}_{\nu j}]} - F_{n_j,i_{j\nu}j}(\underset{\sim}{x}_{\nu,j}) + F_{n_j,i_{j\nu},j}(\underset{\sim}{x}_{\nu,j}))$$

$$= \theta(\boldsymbol{F}_{\boldsymbol{n}}) + \sum_{\boldsymbol{d} \in I_0(\boldsymbol{m})} \prod_{j=1}^{c} \binom{m_j}{d_j} U_{\boldsymbol{n}}^{(\boldsymbol{d})}$$

where $I_0(\boldsymbol{m}) = \{\boldsymbol{d} = (d_1, \ldots, d_c);\ \ 0 \leqslant d_j \leqslant m_j,\ \ j = 1, \ldots, c,\ \ \boldsymbol{d} \neq (0, \ldots, 0)\}$, $\boldsymbol{d}^0 = (0, \ldots, 0)$ and $\boldsymbol{i} \in J_{\boldsymbol{n}}(\boldsymbol{d}^{(0)})$.

For any $j \in \{1, \ldots, c\}$ we define the process $V_{\boldsymbol{n},j}(t_j)$ by

$$V_{\boldsymbol{n},j}(t_j) = n_j^{-1/2} \left(\sum_{i=1}^{[n_j t_j]} A_{\boldsymbol{n},i,j} \right) \tag{3.6}$$

where

$$A_{\boldsymbol{n},i,j} = n_j \left(\prod_{j=1}^{c} n_j^{-[m_j]} \right) \int_{\mathbb{R}^k} g_{\boldsymbol{d}^{(j)},\boldsymbol{n}}^{(\boldsymbol{i})}(\underset{\sim}{x})\, \mathrm{d}(I_{[\underset{\sim}{X}_{n_j,i,j} \leqslant \underset{\sim}{x}]} - F_{n_j,i,j}(\underset{\sim}{x})),$$

$\boldsymbol{i} = (i_1, \ldots, i_c)$, $i_1 = i$, $i_j = 0$, $j \neq 1$. We remark that $V_{\boldsymbol{n},j}(1) = n_j^{1/2} U_{\boldsymbol{n}}^{(\boldsymbol{d}^{(j)})}$.

First we prove the following lemma.

Lemma 3.1. *Under the conditions of Theorem* 3.1 *for any* $j \in \{1, \ldots, c\}$, $V_{\boldsymbol{n},j}(t_j)$ *converges in law to the Skorohod topology on* D_1 *to a Gaussian process* $V_{0,j}(t_j)$ *with trajectories a.s. continuous with mean* 0 *and variance* $t_j \sigma_j^2(\{\boldsymbol{G}_l\})$ *where* $\sigma_j^2(\{\boldsymbol{G}_l\})$ *is defined in* (3.2).

Proof. It suffices to prove for one value of j and by convenience, we take $j = 1$. The process $V_{\boldsymbol{n},1}$ defines a probability measure $P_{\boldsymbol{n}}$ on D_1. By Theorem 15.1 of Billingsley (1968), we have to prove that (i) the finite dimensional distributions of $P_{\boldsymbol{n}}$ converge in law to normal distributions and (ii) $P_{\boldsymbol{n}}$ is tight.

To prove (i), we have to show that $\sum_{l=1}^{p} \lambda_l V_{\boldsymbol{n},1}(t_1^{(l)})$ converges in law to a normal distribution for any $p \in \mathbb{N}^*$, any $t_1^{(l)} \in [0, 1]$ and any $\lambda_l \in \mathbb{R}$ $(1 \leqslant l \leqslant p)$. Without loss of generality we can take $p = 2$ and suppose $t_1^{(1)} < t_1^{(2)}$.

Then, we have

$$\sum_{l=1}^{2} \lambda_l V_{\boldsymbol{n},1}(t_1^{(l)}) = n_1^{-1/2} \left[\sum_{i=1}^{[n_1 t_1^{(1)}]} (\lambda_1 + \lambda_2) A_{\boldsymbol{n},i,1} + \sum_{i=[n_1 t^{(1)}]+1}^{[n_1 t^{(2)}]} \lambda_2 A_{\boldsymbol{n},i,1} \right]. \tag{3.7}$$

We define the sequence of r.v.'s $\{B_{\boldsymbol{n},i,1}\}$ by

$$B_{\boldsymbol{n},i,1} = \begin{cases} (\lambda_1 + \lambda_2) A_{\boldsymbol{n},i,1}, & \text{if } i \leqslant [n_1 t_1^{(1)}], \\ \lambda_2 A_{\boldsymbol{n},i,1}, & \text{if } [n_1 t_1^{(1)}] < i \leqslant [n_1 t_1^{(2)}], \\ 0, & \text{if } i > [n_1 t_1^{(2)}], \end{cases} \tag{3.8}$$

and we have

$$\sum_{l=1}^{2} \lambda_l V_{\boldsymbol{n},1}(t_1^{(l)}) = n_1^{-1/2} \sum_{i=1}^{n_1} B_{\boldsymbol{n},i,1}. \tag{3.9}$$

From Lemma 2.3 we have to verify (2.18)-(2.23). (2.18) is immediate from (3.4) for $\delta''>2(5+\delta)^{-1}\vee(12\delta^{-1}-2)$. Now we choose such a δ'' for which $\delta'>\delta''$.

We define $A^K_{\boldsymbol{n},i,1}$ for any $K>0$ by

$$A^K_{\boldsymbol{n},i,1}=n_1\left(\prod_{j=1}^{c} n_j^{-[m_j]}\right)\int_{\mathbb{R}^k}(g^K)^{(i)}_{\boldsymbol{d}^{(1)},\boldsymbol{n}}(\underset{\sim}{x})\,\mathrm{d}(I_{[\underset{\sim}{X}_{n_j,i,1}\leq \underset{\sim}{x}]}-F_{n_1,i,1}(\underset{\sim}{x}))$$

where g^K is defined by

$$g^K=\begin{cases} g, & \text{if } |g|\leq K,\\ 0, & \text{if } |g|>K,\end{cases} \tag{3.10}$$

$(g^K)^{(i)}_{\boldsymbol{d}^{(1)},\boldsymbol{n}}$ is defined as in (2.1) in which we replace g by g^K and $B_{\boldsymbol{n},i,1}$ by

$$B^K_{\boldsymbol{n},i,1}=\begin{cases}(\lambda_1+\lambda_2)A^K_{\boldsymbol{n},i,1}, & \text{if } i\leq[n_1t_1^{(1)}],\\ \lambda_2A^K_{\boldsymbol{n},i,1}, & \text{if } [n_1t_1^{(1)}]<i\leq[n_1t_1^{(2)}],\\ 0, & \text{if } i>[n_1t_1^{(2)}].\end{cases} \tag{3.11}$$

We have

$$\sup_{\boldsymbol{n}}\max_{1\leq i\leq n_1}|B^K_{\boldsymbol{n},i,1}|\leq 2(|\lambda_1|+|\lambda_2|)K$$

which proves (2.19).

We now prove (2.20). If $\bar{g}^K=g-g^K$, then we have

$$\begin{aligned}
&(E|B_{\boldsymbol{n},i,1}-B^K_{\boldsymbol{n},i,1}|^{2+\delta''})^{1/(2+\delta'')}\\
&\leq(|\lambda_1|+|\lambda_2|)n_1\left(\prod_{j=1}^{c} n_j^{-[m_j]}\right)\\
&\quad\times\sum_{i^*\in I^{(i)}_{d^{(1)},\boldsymbol{n}}}\left(\int_{\mathbb{R}^k}\left|\int_{\mathbb{R}^{km_0-k}}(\bar{g}^K(\underset{\sim}{x}_{l,j},l=1,\ldots,m_j,j=1,\ldots,c)\right.\right.\\
&\qquad-\int_{\mathbb{R}^{km_0}}\left(\bar{g}^K(\underset{\sim}{z}_{l,j},l=1,\ldots,m_j,j=1,\ldots,c)\prod_{j=1}^{c}\prod_{l=1}^{m_j}\mathrm{d}F_{n_j,i_{jl},j}(\underset{\sim}{z}_{l,j})\right)\\
&\qquad\left.\left.\times\prod_{j=2}^{c}\prod_{l=1}^{m_j}\mathrm{d}F_{n_j,i_{jl},j}(\underset{\sim}{x}_{l,j})\prod_{l\neq1}\mathrm{d}F_{n_1,i_{1l},1}(\underset{\sim}{x}_{l,1})\right|^{2+\delta''}\mathrm{d}F_{n_1,i,1}(\underset{\sim}{x}_{i,1})\right)^{1/(2+\delta'')}\\
&\leq(|\lambda_1|+|\lambda_2|)n_1\left(\prod_{j=1}^{c} n_j^{-[m_j]}\right)\\
&\quad\times\sum_{i^*\in I^{(i)}_{d^{(1)},\boldsymbol{n}}}\left(\int_{\mathbb{R}^{km_0}}2^{2+\delta''}|\bar{g}^K(\underset{\sim}{x}_{l,j},l=1,\ldots,m_j,j=1,\ldots,c)|^{2+\delta''}\right.\\
&\qquad\left.\times\prod_{j=1}^{c}\prod_{l=1}^{m_j}\mathrm{d}F_{n_j,i_{jl},j}(\underset{\sim}{x}_{l,j})\right)^{1/(2+\delta'')}.
\end{aligned}$$

If the family defined in (2.7) is bounded by the constant M_0 ($<\infty$), we deduce

$$(E|B_{n,i,1}-B^K_{n,i,1}|^{2+\delta''})^{1/(2+\delta'')} \leqslant 2^{2+\delta''}\frac{M_0}{(K^{2+\delta''})^{\varepsilon}} \to 0 \quad \text{as } K\to\infty,$$

where ε is given by $(2+\delta'')(1+\varepsilon)=2+\delta'$. (2.20) is proved.

We now prove (2.21). Note that

$$E\left(\sum_{i=1}^{n_1} B_{n,i,1}\right)^2 \Big/ n_1 = \frac{1}{n_1}\Bigg[(\lambda_1+\lambda_2)^2 \sum_{i=1}^{[n_1t_1^{(1)}]}\sum_{j=1}^{[n_1t_1^{(1)}]} E(A_{n,i,1}A_{n,j,1})$$
$$+(\lambda_1+\lambda_2)\lambda_2 \sum_{i=1}^{[n_1t_1^{(1)}]}\sum_{j=[n_1t_1^{(1)}]+1}^{[n_1t_1^{(2)}]} E(A_{n,i,1}A_{n,j,1})$$
$$+\lambda_2^2 \sum_{i=[n_1t_1^{(1)}]+1}^{[n_1t_1^{(2)}]}\sum_{j=[n_1t_1^{(1)}]+1}^{[n_1t_1^{(2)}]} E(A_{n,i,1}A_{n,j,1})\Bigg]. \tag{3.12}$$

As the family defined in (2.7) is bounded by M_0, we deduce from Proposition 2.8 of Doukhan and Portal (1987) that

$$\frac{1}{n_1}|(\lambda_1+\lambda_2)\lambda_2| \sum_{i=1}^{[n_1t_1^{(1)}]}\sum_{j=[n_1t_1^{(1)}]+1}^{[n_1t_1^{(2)}]} E(A_{n,i,1})A_{n,j,1}$$
$$\leqslant (|\lambda_1|+|\lambda_2|)|\lambda_2| M_0^{2/(2+\delta')} n_1^{-1} \sum_{i=1}^{[n_1t_1^{(1)}]}\sum_{j=[n_1t_1^{(1)}]+1}^{[n_1t_1^{(2)}]} \alpha_1^{\delta'/(2+\delta')}(j-i). \tag{3.13}$$

From condition (3.4) the last expression converges to 0 as $n_1\to\infty$.

It remains to prove that

$$\frac{1}{n_1}\sum_{i=1}^{[n_1t_1^{(1)}]}\sum_{j=1}^{[n_1t_1^{(1)}]} E(A_{n,i,1}A_{n,j,1})$$

and

$$\frac{1}{n_1}\sum_{i=[n_1t_1^{(1)}]+1}^{[n_1t_1^{(2)}]}\sum_{j=[n_1t_1^{(1)}]+1}^{[n_1t_1^{(2)}]} E(A_{n,i,1}A_{n,j,1})$$

converge respectively to some finite constants as $n_1\to\infty$. Now denote

$$\rho(1)=\int_{\mathbb{R}^k}(e_1(\underset{\sim}{x}))^2\,\mathrm{d}F_1(\underset{\sim}{x})-\theta^2(\boldsymbol{F})$$

and

$$\rho(i)=2\left[\iint_{\mathbb{R}^{2k}}(e_j(\underset{\sim}{x})e_j(\underset{\sim}{z}))\,\mathrm{d}G_{i,1}(\underset{\sim}{x},\underset{\sim}{z})-\theta^2(\boldsymbol{F})\right], \quad i\geqslant 2.$$

Then

$$\left|\frac{1}{n_1}\left(\sum_{i=1}^{[n_1t_1^{(1)}]}\sum_{j=1}^{[n_1t_1^{(1)}]} E(A_{n,i,1}A_{n,j,1}) - \frac{[n_1t_1^{(1)}]}{n_1}\sum_{i=1}^{+\infty}\rho(i)\right|$$

$$\leqslant \left|\frac{[n_1t^{(1)}]}{n_1}\frac{1}{[n_1t_1^{(1)}]}\sum_{i=0}^{[n_1t_1^{(1)}]-1}\sum_{j=1}^{[n_1t_1^{(1)}]-i}(\varphi(i)E(A_{n,j,1}A_{n,j+i,1})\right.$$

$$\left. -\frac{[n_1t_1^{(1)}]}{n_1}\frac{1}{[n_1t_1^{(1)}]}\sum_{i=1}^{[n_1t_1^{(1)}]}([n_1t_1^{(1)}]-i)\rho(i)\right|$$

$$+\frac{[n_1t_1^{(1)}]}{n_1}\sum_{i=[n_1t_1^{(1)}]+1}^{\infty}|\rho(i)| + \frac{[n_1t_1^{(1)}]}{n_1}\sum_{i=1}^{[n_1t_1^{(1)}]}\sum_{k=i}^{\infty}|\rho(k)|$$

$$= |A_{n,1}| + B_{n_1} + C_{n_1} \tag{3.14}$$

where

$$\varphi(i) = \begin{cases} 1, & \text{if } i = 0, \\ 2, & \text{if } i \neq 0. \end{cases}$$

From Lemma 2.4 we deduce that $|A_{n,1}| \to 0$ as $n_1 \to \infty$ and from the well known inequality on moments for strong mixing sequences of r.v.'s (Doukhan and Portal, 1987, Proposition 2.8), we deduce

$$|\rho(i)| \leqslant (\alpha_1(i))^{\delta'/(2+\delta')} M_0^{2/(2+\delta')} \tag{3.15}$$

which entails $B_{n_1} \to 0$ and $C_{n_1} \to 0$ as $n_1 \to \infty$ by (3.4).

It is also immediate that

$$\left|\frac{[n_1t_1^{(1)}]}{n_1}\sum_{i=1}^{+\infty}\rho(i) - t_1^{(1)}\sum_{i=1}^{+\infty}\rho(i)\right| \to 0 \quad \text{as } n_1 \to \infty.$$

We conclude that

$$\frac{1}{n_1}\sum_{i=1}^{[n_1t_1^{(1)}]}\sum_{j=1}^{[n_1t_1^{(1)}]} E(A_{n,i,1}A_{n,j,1}) \to t_1^{(1)}\left(\sum_{i=1}^{+\infty}\rho(i)\right) \quad \text{as } n_1 \to \infty, \tag{3.16}$$

where $\sum_{i=1}^{+\infty}\rho(i) = \sigma_1^2(\{G_l\})$ is defined in (3.2).

Now from the decomposition

$$\sum_{i=1}^{[n_1t_1^{(2)}]}\sum_{j=1}^{[n_1t_1^{(2)}]} = \sum_{i=1}^{[n_1t_1^{(1)}]}\sum_{j=1}^{[n_1t_1^{(1)}]} + \sum_{i=1}^{[n_1t_1^{(1)}]}\sum_{j=[n_1t_1^{(1)}]+1}^{[n_1t_1^{(2)}]}$$

$$+ \sum_{i=[n_1t_1^{(1)}]+1}^{[n_1t_1^{(2)}]}\sum_{j=1}^{[n_1t_1^{(1)}]} + \sum_{i=[n_1t_1^{(1)}]+1}^{[n_1t_1^{(2)}]}\sum_{j=[n_1t_1^{(1)}]+1}^{[n_1t_1^{(2)}]}$$

and using (3.13) and (3.16), we conclude that

$$\frac{1}{n_1}\sum_{i=[n_1t_1^{(1)}]+1}^{[n_1t_1^{(2)}]}\sum_{j=[n_1t_1^{(1)}]+1}^{[n_1t_1^{(2)}]} E(A_{n,i,1}A_{n,j,1}) \to (t_1^{(2)} - t_1^{(1)})\sigma_1^2(\{G_l\}) \quad \text{as } n_1 \to \infty. \tag{3.17}$$

From (3.8) and (3.12)-(3.16) we deduce that

$$E\left(\sum_{i=1}^{n_1}(B_{n,i,1}-E(B_{n,i,1}))\right)^2\Big/n_1\to\{(\lambda_1+\lambda_2)^2t_1^{(1)}+\lambda_2^2(t_1^{(2)}-t_1^{(1)})\}\sigma_1^2(\{G_l\})$$

where $\sigma_1^2(\{G_l\})$ is defined in (3.2), and (2.21) is proved.

To prove (2.22), we replace g by g^K defined in (3.10) and proceeding as above we deduce that

$$E\left(\sum_{i=1}^{n_1}B_{n,i,1}^K-E(B_{n,i,1}^K)\right)^2\Big/n_1$$
$$\to\{(\lambda_1+\lambda_2)^2t_1^{(1)}+\lambda_2^2(t_1^{(2)}-t_1^{(1)})\}\sigma_{1,K}^2(\{G_l\})\quad\forall K>0$$

as $n_1\to\infty$, where $B_{n,i,1}^K$ is defined in (3.11), and

$$\sigma_{1,K}^2(\{G_l\})=\lim_{n_1\to\infty}\left\{\left[\int_{\mathbb{R}^k}(e_1^K(x))^2\,\mathrm{d}F_1(x)-\theta_K^2(F)\right]\right.$$
$$\left.+2\sum_{l_1=1}^{n_1}\left[\int(e_1^K(x)e_1^K(z))\,\mathrm{d}G_{l_1,1}(x,z)-\theta_K^2(F)\right]\right\},$$

$$e_1^K(x_{1,1})=\int_{\mathbb{R}^{km_0-k}}g^K(x_{i,j},i=1,\ldots,m_j,j=1,\ldots,c)$$
$$\times\prod_{j=2}^{c}\prod_{i=1}^{m_j}\mathrm{d}F_j(x_{i,j})\prod_{i=2}^{m_1}\mathrm{d}F_1(x_{i,1}),$$

$$\theta_K^2(F)=\int_{\mathbb{R}^{km_0}}g^K(x_{i,j},i=1,\ldots,m_j,j=1,\ldots,c)\prod_{j=1}^{c}\prod_{i=1}^{m_j}\mathrm{d}F_j(x_{i,j}).$$

From the definition of $\sigma_{1,K}^2(\{G_l\})$ and $\sigma_1^2(\{G_l\})$, the properties (3.4) and (2.25), it easily follows that $\sigma_{1,K}^2(\{G_l\})$ converges to $\sigma_1^2(\{G_l\})$ as $K\to\infty$, that is (2.23) and the fact that $\sigma_1^2(\{G_l\})<\infty$ is also an easy consequence of (3.4) and (2.25). Conditions (2.18)-(2.23) of Lemma 2.2 are verified and we conclude that $\sum_{l=1}^{2}\lambda_lV_{n,1}(t_1^{(l)})$ converges in law to the normal distribution with mean 0 and variance $\{(\lambda_1+\lambda_2)^2t_1^{(1)}+\lambda_2^2(t_1^{(2)}-t_1^{(1)})\}\sigma_1^2(\{G_l\})$. (i) is proved.

We now prove (ii). From Theorem 8.2 of Billingsley (1986) we have to verify that $\forall\varepsilon>0$, $\exists\eta>0$ $(0<\eta<1)$ and an integer N_0 such that $\forall n_1\geqslant N_0$,

$$P\left[\sup_{|s_1-t_1|<\eta}|V_{n,1}(t_1)-V_{n,1}(s_1)|\geqslant\varepsilon\right]<\varepsilon. \tag{3.18}$$

If n_1t_1 and n_1s_1 are integers and $t_1\geqslant s_1$, we obtain the following inequality by using Theorem 10 for $q=4$ of Doukhan and Portal (1987),

$$E(V_{n,1}(t_1)-V_{n,1}(s_1))^4\leqslant\left((t_1-s_1)^2+\frac{t_1-s_1}{n_1}\right)M\sum_{m=0}^{+\infty}m^2\alpha_1^{\delta''/(2+\delta'')}(m)$$

where M is some constant >0.

From (3.4), we have

$$\sum_{m=0}^{\infty} m^2 \alpha_1^{\delta''/(2+\delta'')}(m) < M_1 < +\infty$$

where M_1 is some constant >0. If $t_1 > s_1$, and $n_1 t_1$ and $n_1 s_1$ are integers, we have $t_1 - s_1 \geq 1/n_1$. We deduce

$$E(V_{n,1}(t_1) - V_{n,1}(s_1))^4 \leq 2MM_1(t_1 - s_1)^2. \tag{3.19}$$

From (3.18) and Lemma 2 of Balacheff and Dupont (1980), we obtain for n_1 sufficiently large

$$P\left[\sup_{|s_1-t_1|<\eta} |V_{n,1}(t_1) - V_{n,1}(s_1)| \geq \varepsilon\right]$$

$$\leq P\left[\sup_{|[n_1 s_1]/n_1 - [n_1 t_1]/n_1| < 2\eta} \left|V_{n,1}\left(\frac{[n_1 t_1]}{n_1}\right) - V_{n,1}\left(\frac{[n_1 s_1]}{n_1}\right)\right| > \frac{1}{2}\varepsilon\right] \leq \frac{16}{\varepsilon^4} K\eta$$

where K is some constant >0, which entails (3.18) for η sufficiently small and N_c sufficiently large. This proves (ii) and also the lemma. □

Proof of Theorem 3.1. For any $j \in \{1, \ldots, c\}$, let $W_{n,j}(t_j)$ be defined by

$$W_{n,j}(t_j) = \begin{cases} ([n_j t_j]/m_j \sigma_j n_j^{1/2}) m_j U^{d(j)}_{[nt^{(j)}]}, & m_j \leq [n_j t_j] \leq n_j, \\ 0, & t_j \leq (m_j - 1) n_j^{-1}, \end{cases}$$

where $t^{(j)} = (1, \ldots, 1, t_j, 1, \ldots, 1)$ and put

$$w_j(t, n) = \psi([nt]; n)\{m_j \sigma_j n_j^{1/2}/[n_j t_j]\}.$$

Then, since

$$\lim_{n} w_j(t, n) = \left(\sum_{j=1}^{c} m_j \sigma_j \lambda_j^{1/2}\right)\left(\sum_{j=1}^{c} m_j \sigma_j \lambda_j^{-1/2} t_j^{-1}\right)^{-1} (m_j \sigma_j \lambda_j^{-1/2} t_j^{-1}),$$

we deduce (from Lemma 3.1) that $\sum_{j=1}^{c} w_j(t, n) W_{n,j}(t_j)$ converges in law to $W(t$ defined in (3.5).

As

$$U(n) - \theta(F_n) - \sum_{j=1}^{c} m_j U_n^{d^{(j)}} = \sum_{d \in I_1(m)} \prod_{j=1}^{c} \binom{m_j}{d_j} U_n^{(d)}$$

where

$$I_1(m) = \left\{d = (d_1, \ldots, d_c); \sum_{j=1}^{c} d_j > 1\right\},$$

we have only to prove that

$$\lim_{n} \max_{m \leq p \leq n} \psi(p; n) \left| \sum_{d \in I_1(m)} \prod_{j=1}^{c} \binom{m_j}{d_j} U_n^{(d)} \right| = O_P. \tag{3.20}$$

Let $C(1)$ and $C(2)$ be defined by

$$C(1) = \left\{ d = (d_1, \ldots, d_c);\ \sum_{l=1}^{c} d_l = 2 \right\},$$

$$C(2) = \left\{ d = (d_1, \ldots, d_c);\ \sum_{l=1}^{c} d_l > 2 \right\}.$$

Then, we have the following decomposition:

$$\sum_{d \in I_1(m)} \prod_{j=1}^{c} \binom{m_j}{d_j} U_n^{(d)} = \sum_{j=1}^{2} \sum_{d \in C(j)} \prod_{l=1}^{c} \binom{m_l}{d_l} U_n^{(d)}. \tag{3.21}$$

We also have

$$\psi(p; n)(\psi(n; n))^{-1} = O(1) \quad \text{for } p \leq n \tag{3.22}$$

and

$$\psi(n; n) = n^{1/2} \left(\sum_{j=1}^{c} m_j \sigma_j \lambda_j^{1/2} \right) \left(\sum_{j=1}^{c} m_j \sigma_j \lambda_j^{-1/2} \right)^{-1} (1 + O(1)). \tag{3.23}$$

Consequently, using (3.22) and (3.23) we deduce that $\forall \varepsilon > 0$ and for n sufficiently large, we can find a constant K such that

$$\begin{aligned} P\left[\max_{m \leq p \leq n} \psi(p; n) \left| \sum_{d \in I_1(m)} \prod_{j=1}^{c} \binom{m_j}{d_j} U_p^{(d)} \right| > \varepsilon \right] &\leq P\left[n^{1/2} \left(\sum_{j=1}^{c} m_j \sigma_j \lambda_j^{1/2} \right) \left(\sum_{j=1}^{c} m_j \sigma_j \lambda_j^{-1/2} \right)^{-1} \right. \\ &\quad \left. \times \max_{m \leq p \leq n} \left| \sum_{d \in I_1(m)} \prod_{j=1}^{c} \binom{m_j}{d_j} U_p^{(d)} \right| > K\varepsilon \right]. \end{aligned} \tag{3.24}$$

For convenience, let us write

$$K_0 = \left(\sum_{j=1}^{c} m_j \sigma_j \lambda_j^{1/2} \right) \left(\sum_{j=1}^{c} m_j \sigma_j \lambda_j^{-1/2} \right)^{-1}.$$

Then, from (3.21), we can write

$$\begin{aligned} P\left[n^{1/2} K_0 \max_{m \leq p \leq n} \left| \sum_{d \in I_1(m)} \prod_{j=1}^{c} \binom{m_j}{d_j} U_p^{(d)} \right| > K\varepsilon \right] \\ \leq \sum_{j=1}^{2} P\left[n^{1/2} K_0 \max_{m \leq p \leq n} \left| \sum_{d \in C(j)} \prod_{l=1}^{c} \binom{m_l}{d_l} U_p^{(d)} \right| > K\varepsilon 2^{-1} \right]. \end{aligned} \tag{3.25}$$

From (2.9) and (2.10) of Lemma 2.2, we deduce that

$$P\left[n^{1/2}K_0 \max_{m\leq p\leq n}\left|\sum_{d\in C(1)}\prod_{l=1}^{c}\binom{m_l}{d_l}U_p^{(d)}\right|>2^{-1}K\varepsilon\right]$$

$$\leq P\left[n^{1/2}K_0 \max_{m\leq p\leq n}\left|\sum_{d\in D(1)}\prod_{l=1}^{c}\binom{m_l}{d_l}U_p^{(d)}\right|>4^{-1}K\varepsilon\right]$$

$$+P\left[n^{1/2}K_0 \max_{m\leq p\leq n}\left|\sum_{d\in C(1)-D(1)}\prod_{l=1}^{c}\binom{m_l}{d_l}U_p^{(d)}\right|>4^{-1}K\varepsilon\right]$$

$$\leq\left(4\sum_{\substack{i,j=1\\ i\neq j}}^{c}\binom{m_i}{1}\binom{m_j}{1}\right)^4(K_0\varepsilon)^{-4}n^2\times n^2\times \mathrm{O}(n^{-6})$$

$$+\left(4\sum_{i=1}^{c}\binom{m_i}{2}\right)^4(K_0\varepsilon)^{-4}n^2n+\mathrm{O}(n^{-4})=\mathrm{O}(n^{-1}) \tag{3.26}$$

where

$$D(1)=\left\{d\in C(1);\ \max_{1\leq j\leq c} d_j=1\right\}.$$

From (2.11) of Lemma 2.2 we obtain proceeding as above that

$$P\left[n^{1/2}K_0 \max_{m\leq p\leq n}\left|\sum_{d\in C(2)}\binom{m_l}{d_l}U_p^{(d)}\right|>2^{-1}K\varepsilon\right]=\mathrm{O}(n^{-1}). \tag{3.27}$$

From (3.24)–(3.27) we deduce (3.20). The theorem follows. □

Let $Z^*(\boldsymbol{n})=\{Z(\boldsymbol{t};\boldsymbol{n}), \boldsymbol{t}\in[0,1]^c\}$ be the process defined by

$$Z^*(\boldsymbol{t},\boldsymbol{n})=[\mathrm{Var}(U(\boldsymbol{n}))]^{-1/2}[U([\boldsymbol{n}/\boldsymbol{t}])-\theta(\boldsymbol{F}_n)]$$

where $[\boldsymbol{n}/\boldsymbol{t}]=([n_1/t_1]),\ldots,[n_c/t_c])$.

Further let

$$W^*=\{W^*(\boldsymbol{t}), \boldsymbol{t}\in[0,1]^c\}\quad (W^*(\boldsymbol{t})=\boldsymbol{w}'\hat{W}(\boldsymbol{t}), \boldsymbol{t}\in[0,1]^c) \tag{3.28}$$

be the process defined by

$$\boldsymbol{w}=(w_1,\ldots,w_c)',\quad w_j=m_j\sigma_j\lambda_j^{-1/2}\left(\sum_{j=1}^{c}m_j^2\sigma_j^2\lambda_j^{-1}\right)^{-1/2},\ j\in\{1,\ldots,c\},$$

and

$$\hat{W}(\boldsymbol{t})=(W_1(t_1),\ldots,W_c(t_c)),\quad \boldsymbol{t}\in[0,1]^c.$$

Corollary 3.1. *Under the conditions of Theorem 3.1, $Z^*(\boldsymbol{n})$ converges in law with respect to the Skorohod topology on D_c to the Gaussian process W^*.*

Proof. From Lemmas 2.2, 3.1 and the decomposition

$$U(\boldsymbol{n})=\theta(\boldsymbol{F}_n)+\sum_{d\in I_0(m)}\prod_{j=1}^{c}\binom{m_j}{d_j}U_n^{(d)}$$

we easily deduce that

$$\mathrm{Var}(U(\boldsymbol{n})) = \left(\sum_{j=1}^{c} n_j^{-1} m_j^2 \sigma_j^2\right)(1+\mathrm{O}_n(1)) + \mathrm{O}(n^{-2}) \tag{3.29}$$

by showing $E(U_{\boldsymbol{n}}^{(\boldsymbol{d})})^2 = \mathrm{O}(n^{-2})$ for any $\boldsymbol{d}$ such that $\sum_{j=1}^{c} d_j = 2$.

Let us now define

$$W_{\boldsymbol{n},j}^*(t_j) = \begin{cases} (m_j\sigma_j)^{-1} n_j \quad U_{[\boldsymbol{n}/\boldsymbol{t}^{(j)}]}, & 0 < t_j \leqslant 1, \\ 0, & t_j = 0, \end{cases}$$

and

$$w_j^*(\boldsymbol{t}, \boldsymbol{n}) = [\mathrm{Var}(U(\boldsymbol{n}))]^{-1/2} m_j \sigma_j n_j^{-1/2}.$$

From Lemma 3.1 and (3.30) we deduce $\sum_{j=1}^{c} w_j^*(\boldsymbol{t}, \boldsymbol{n}) W_{\boldsymbol{n},j}^*(t_j)$ converges in law to $W^*(\boldsymbol{t})$ defined in (3.28). We finish the proof of the corollary by showing that

$$\lim_{n} \max_{m \leqslant p \leqslant n} [\mathrm{Var}(U(\boldsymbol{n}))]^{-1/2} \left| \sum_{\boldsymbol{d} \in I_1(\boldsymbol{m})} \prod_{j=1}^{c} \binom{m_j}{d_j} U_{\boldsymbol{n}}^{(\boldsymbol{d})} \right| = \mathrm{O}_P,$$

which follows proceeding as in the proof of Theorem 3.1. □

Acknowledgment

The authors would like to express their sincere thanks to Professor C.C. Heyde and the referee for their critical examination of the first draft of the paper. Their constructive comments and suggestions for improvements are gratefully acknowledged.

References

S. Balacheff and G. Dupont, Normalité asymptotique des processus empiriques tronqués et des processes de rang, Lecture Notes in Math., Vol. 821 (Springer, Berlin, 1980).

P. Billingsley, Convergence of Probability Measures (Wiley, New York, 1968).

P. Doukhan and F. Portal, Principe d'invariance faible pour la fonction de répartition empirique dans un cadre multidimensionel et mélangeant, Probab. Math. Statist. 8 (1987) 117–132.

M. Harel and M.L. Puri, Limiting behavior of U-statistics, V-statistics and one-sample rank order statistics for non-stationary absolutely regular processes, J. Multivariate Anal. 30 (1989a) 181–204.

M. Harel and M.L. Puri, Weak convergence of the U-statistic and weak invariance of the one-sample rank order statistic for Markov processes and ARMA models, J. Multivariate Anal. 31 (1989b) 259–265.

I.A. Ibragimov, Some limit theorems for stationary processes, Theory Probab. Appl. VII 4 (1962) 349–382.

G. Neuhaus, On weak convergence of stochastic processes with multidimensional time parameter, Ann. Math. Statist. 42 (1971) 1285–1295.

C.S. Withers, Convergence of empirical processes of mixing r.v.'s on [0, 1], Ann. Statist. 3 (1975) 1101–1108.

K. Yoshihara, Limiting behavior of U statistics for stationary absolutely regular processes, Z. Wahrsch. Verw. Gebiete 35 (1976) 237–252.

THE SPACE $\tilde{D}_k$ AND WEAK CONVERGENCE FOR THE RECTANGLE-INDEXED PROCESSES UNDER MIXING

MICHEL HAREL and MADAN L. PURI*

Institut Universitaire de Technologie de Limoges et U.A. C.N.R.S. Orsay, and Indiana University, Bloomington, Indiana 47401

In this paper we study the weak convergence of the weighted empirical processes indexed by rectangles of $[0,1]^k$ under both weak and strong mixing conditions. This is accomplished by generalizing the Skorohod topology on a space of functions defined on a set of rectangles.

1. Introduction

Let $X_{ni} = (X_{ni}^{(1)}, \ldots, X_{ni}^{(k)})$, $1 \le i \le n$, $n \ge 1$ be $\mathbb{R}^k$-valued random variables with continuous d.f.s (distribution functions) F_{ni} and continuous marginal d.f.s $F_{ni}^{(j)}$ of $X_{ni}^{(j)}$, $1 \le j \le k$, $1 \le i \le n$. Let $F_n^{(j)} = n^{-1}\sum_{i=1}^n F_{ni}^{(j)}$, and let $\{H_{ni}\}$ be a sequence of measures on $[0,1]^k$ defined by

$$H_{ni}(t_1, \ldots, t_k) = F_{ni}\left(F_n^{(1)^{-1}}(t_1), \ldots, F_n^{(k)^{-1}}(t_k)\right), \qquad 1 \le i \le n. \quad (1.1)$$

Let $\tilde{W}_n$ be an empirical process defined by

$$\tilde{W}_n(B) = n^{-1/2} \sum_{i=1}^{n} \left(\prod_{j=1}^{k} I_{[a_j < F_n^{(j)}(X_{ni}^{(j)}) \le b_j]} - H_{ni}(B) \right), \qquad (1.2)$$

where $B = \prod_{j=1}^k (a_j, b_j] \subset [0,1]^k$ when $I_{[\ ]}$ denotes the indicator of [ ].

Our aim is to study the asymptotic behavior of $\tilde{W}_n$ for a certain Skorohod topology (to be defined in Section 3) when the sequence $\{X_{ni}\}$ is

$$\varphi\text{-mixing with rates } \varphi(m) = O(m^{-1-\varepsilon}) \qquad (1.3)$$

*Research supported by the Office of Naval Research Contract N00014-85-K-0648.

or

$$\text{strong mixing with rates } \sum_{m=1}^{\infty} m^{2(k+1)}\alpha^{\varepsilon}(m) < \infty \quad \text{for some } \varepsilon \in \left(0, \frac{1}{2(k+2)}\right). \tag{1.4}$$

Recall that $\{X_{ni}\}$ is φ-mixing if $\sup\{|P(B|A) - P(B)|; A \in \sigma(X_{ni}, 1 \le i \le j), B \in \sigma(X_{ni}, i \ge j+m)\} = \varphi(m)\downarrow 0$ for positive j and m; and it is strong mixing if $\sup\{|P(A \cap B) - P(A)\cdot P(B)|; A \in \sigma(X_{ni}, 1 \le i \le j), B \in \sigma(X_{ni}, i \ge j+m)\} = \alpha(m)\downarrow 0$ for positive integers j and m. Here $\sigma(X_{ni}, 1 \le i \le j)$ and $\sigma(X_{ni}, i \ge j+m)$ are the σ-fields generated by $(X_{n1},\dots,X_{nj})$ and $(X_{n,j+m}, X_{n,j+m+1},\dots)$, respectively.

Bass and Pyke [4] extended the M_2 Skorohod [18] topology for functions on $[0,1]$ to set functions that are outer continuous with inner limits, but the empirical processes indexed by rectangles that we are considering have trajectories concentrated on the set of functions for which an extension of the Skorohod [18] J_1 topology is more appropriate. Most of the work dealing with the empirical processes indexed by points is concentrated on the Skorohod's J_1 topology. In their basic paper, Bass and Pyke [4] posed an open problem of determining a general class of sets on which an extension of Skorohod's J_1 topology is possible. In this paper we provide an extension of the Skorohod J_1 topology to functions indexed by rectangles and thus answer a part of the problem posed by Bass and Pyke [4]. We may also mention an interesting paper of Straf [19] which deals with the extension of J_1 topology to very general index sets. This requires the existence of a group Λ of homeomorphisms on an arbitrary space and does not seem to give real answers perhaps of its very theoretical nature. Also Straf's methods seem to be applicable on the space of functions defined only on $[0,1)^k$ and not on $[0,1]^k$, and they do not seem to be applicable to empirical processes indexed by rectangles.

The weak convergence of the weighted univariate empirical process indexed by points was established for the independent case by Pyke and Shorack [13], and for the φ-mixing case by Fears and Mehra [8] and later by Ahmad and Lin [1]. The generalization to the φ-mixing multivariate nonstationary case was carried on by Harel [9]. Shorack and Wellner [17] established the weak convergence of the weighted univariate empirical process indexed by intervals when the underlying random variables are independent. Their results were later generalized by Einmahl, Ruymgaart, and Wellner [7] to the multivariate case by using directly the well known Skorohod construction when the underlying random variables are independent, identical and uniformly distributed over $[0,1]^k$. Later Ruymgaart and Wellner [16] considered the case when the random variables are not uniformly distributed, but left open the problem of the convergence of the weighted empirical processes (see Ruymgaart and Wellner [16, remark, p. 221]. Our method of constructing Skorohod topology on the space of functions indexed by rectangles leads also to the extension of the results of Ruymgaart and Wellner [16] for more general classes of distributions (not necessarily uniform distributions).

We may also mention for reference the work of Alexander [2] on weighted empirical processes indexed by Vapnik–Cervonenkis classes of sets for the independent case. For the weak convergence of (nonweighted) empirical processes the reader is referred to the interesting papers of Neuhaus [10, 11] for the independent case, Rüschendorf [14] and Balacheff and Dupont [3] for the φ-mixing cases.

2. The $\tilde{D}_k$ and $\tilde{C}_k$ Spaces and Preliminaries

We write $t = (t_1, \ldots, t_k)$, and half-open rectangles $R(t, t') = \prod_{j=1}^k (t_j, t'_j]$. By convention, any point $t_j \in [0, 1]$ will be called a half-open interval and will be written as $(t_j, t_j]$. Note that $R(t, t) = t$. For $(t, t') \in ([0, 1]^k)^2$, $t \leq t'$ will mean $t_j \leq t'_j$ $\forall j = 1, \ldots, k$, and $t < t'$ will mean $t_j < t_j$ $\forall j = 1, \ldots, k$. For $(t, t') \in ([0, 1]^k)^2$ with $t < t'$ or $t \leq t'$ we can associate a rectangle $R(t, t')$ defined as before.

Let $\mathscr{I}(k) = \{R(t, t'); R(t, t') \subset [0, 1]^k\}$, and associate with the space $\mathscr{I}(k)$ the Hausdorff metric d_H, where $d_H(R(t, t'), R(s, s')) = \max_{1 \leq j \leq k} \max\{|s_j - t_j|, |s'_j - t'_j|\}$.

Consider a family of k strictly increasing finite sequences of elements of $[0, 1]$ such that 0 is the first element of each sequence. For example, let $B = \{t_{ji}\}$, $1 \leq i \leq n_j$, $1 \leq j \leq k$ be k sequences such that $t_{j1} = 0$ and $t_{ji} < t_{jt+1}$ $\forall i \in \{1, \ldots, n_j - 1\}$ and $j \in \{1, \ldots, k\}$. Now associate with B a set $I^{(j)}(i, l)$ defined as

$$I^{(j)}(i, l) = \begin{cases} \left\{(t_j, t'_j] \subset [0, 1], t_{ji} \leq t_j < t_{ji+1}, t_{jl} \leq t'_j < t_{jl+1}\right\} \\ \qquad \text{if } 1 \leq i, l \leq n_j - 1 \\ \left\{(t_j, t'_j] \subset [0, 1], t_{ji} \leq t_j < t_{ji+1}, t_{jl} \leq t'_j \leq 1\right\} \\ \qquad \text{if } 1 \leq i \leq n_j - 1, l = n_j \end{cases} \tag{2.1}$$

and

$$I^{(j)}(n_j, n_j) = \left\{(t_j, t'_j] \subset [0, 1]; t_{jn_j} \leq t_j \leq t'_j \leq 1\right\}.$$

(Note that $I^{(j)}(i, l) = \phi$ if $t_{jl} \leq t_{ji}$.)

Then a partition G of $\mathscr{I}(k)$ defined as

$$G = \left\{ \prod_{j=1}^{k} I^{(j)}(i_j, l_j),\ 1 \le i_j \le n_j,\ 1 \le l_j \le n_j, \text{ and } \prod_{j=1}^{k} I^{(j)}(i_j, l_j) \ne \phi \right\} \tag{2.2}$$

will be called a grid of $\mathscr{I}(k)$ with base B.

Let S^* be a finite subset of $\mathscr{I}(k)$, where $S^* = \{R(a^{(1)}, b^{(1)}), \dots, R(a^{(p)}, b^{(p)})\}$. Then the base generated by S^* is the smallest base of the grid B such that $\{a_j^{(1)}, \dots, a_j^{(p)}\} \cup \{b_j^{(1)}, \dots, b_j^{(p)}\} \subset \{t_{j1}, \dots, t_{jn_j}\}$ $\forall j = 1, \dots, k$.

Let $R(t, t') \in \mathscr{I}(k)$ and $(\rho, \varepsilon) \in \{0,1\}^k \times \{0,1\}^k$. Then the (ρ, ε) quadrant of $\mathscr{I}(k)$ with top $R(t, t')$ is a subset $Q(R(t, t'), \rho, \varepsilon)$ of $\mathscr{I}(k)$ defined as

$$Q(R(t, t'), \rho, \varepsilon) = \prod_{j=1}^{k} Q_j(t_j, t_j', \rho_j, \varepsilon_j),$$

where

$$Q_j(t_j, t_j', \rho_j, \varepsilon_j) = \begin{cases} \{(s_j, s_j'] \subset [0,1],\ s_j < t_j,\ s_j' < t_j'\} & \text{if } \rho_j = \varepsilon_j = 0 \\ \{(s_j, s_j'] \subset [0,1],\ s_j \ge t_j,\ s_j' < t_j'\} & \text{if } \rho_j = 1,\ \varepsilon_j = 0 \\ \{(s_j, s_j'] \subset [0,1],\ s_j < t_j,\ s_j' \ge t_j'\} & \text{if } \rho_j = 0,\ \varepsilon_j = 1 \\ \{(s_j, s_j'] \subset [0,1],\ s_j \ge t_j,\ s_j' \ge t_j'\} & \text{if } \rho_j = \varepsilon_j = 1. \end{cases}$$

Let $R(t, t') \in \mathscr{I}(k)$ and $(\rho, \varepsilon) \in \{0,1\}^k \times \{0,1\}^k$ such that $Q(R(t, t'), \rho, \varepsilon) \ne \phi$. Then, it is easy to see that if G is a grid, there exists an $S \in G$ such that $S \cap Q(R(t, t'), \rho, \varepsilon)$ is a neighborhood of $R(t, t')$ in $Q(R(t, t'), \rho, \varepsilon)$ in the topology induced on $Q(R(t, t'), \rho, \varepsilon)$ by the Hausdorff metric.

Let S be a nonempty subset of $\mathscr{I}(k)$. S is called a *pavement* of $\mathscr{I}(k)$ if S is of the form $S = \prod_{j=1}^{k} S_j$, where for any $1 \le j \le k$, $\exists (a_j^{(1)}, b_j^{(1)}, a_j^{(2)}, b_j^{(2)}) \in [0,1]^4$ such that $S_j = \{(t_j, t_j'],\ a_j^{(1)} \le t_j \le a_j^{(2)},\ b_j^{(1)} \le t_j' \le b_j^{(2)}\}$.

For any $R(s, s') \in S$, we call the pair (J, L) the *indicator* of $R(s, s')$ into S if $(J, L) \subset \{1, \dots, k\} \times \{1, \dots, k\}$ is such that

$$s_j \in \{a_j^{(1)}, a_j^{(2)}\}\ \forall j \in J \quad \text{and} \quad s_j \in (a_j^{(1)}, a_j^{(2)})\ \forall j \notin J$$

$$s_l' \in \{b_l^{(1)}, b_l^{(2)}\}\ \forall l \in L \quad \text{and} \quad s_l' \in (b_l^{(1)}, b_l^{(2)})\ \forall l \notin L.$$

We say F, a subset of S, is the *face* of $R(s, s')$ in S with indicator (J, L) if F is of the form

$$F = \{R(u,u') \in S;\ u_j = s_j\ \forall j \in J \text{ and } u'_l = s'_l\ \forall l \in L\}.$$

Note that if $(J, L) = (\phi, \phi)$, then $F = S$, and if $(J, L) = \{1, \dots, k\} \times \{1, \dots, k\}$, then $F = R(s, s')$.

We say that $f\colon \mathscr{I}(k) \to \mathbb{R}$ admits a (ρ, ε) limit in $R(t, t')$ if and only if the restriction $f|Q(R(t, t'), \rho, \varepsilon)$ admits a limit in $R(t, t')$ with respect to the metric d_H and the usual metric on $\mathbb{R}$. We shall denote the (ρ, ε) limit of f in $R(t, t')$ by $f(R(t, t') + O(\rho, \varepsilon))$. If $(\rho, \varepsilon) = ((1, \dots, 1), (1, \dots, 1))$, then the (ρ, ε) limit of f is $f(R(t, t'))$.

Denote by $\tilde{D}_k$, the set of maps $f\colon \mathscr{I}(k) \to \mathbb{R}$ such that for any $R(t, t') \in \mathscr{I}(k)$ and any $(\rho, \varepsilon) \in \{0, 1\}^k \times \{0, 1\}^k$ for which $Q(R(t, t'), \rho, \varepsilon) \neq \phi$, f admits a (ρ, ε) limit in $R(t, t')$.

Finally denote by $\tilde{C}_k$, the set of maps $f\colon \mathscr{I}(k) \to \mathbb{R}$ which are continuous in d_H and the usual metric in $\mathbb{R}$.

2.1. *Properties of the Spaces $\tilde{D}_k$ and $\tilde{C}_k$*

Let $f \in \tilde{D}_k$ and $R(t, t') \in \mathscr{I}(k)$, and for any $(J, L) \subset \{1, \dots, k\} \times \{1, \dots, k\}$, set

$$H(f, R(t,t'), J, L) = \max\left\{\begin{array}{c} |f(R(t,t') + O(\rho,\varepsilon)) - f(R(t,t') + O(\rho',\varepsilon'))|, \\ Q(R(t,t'),\rho,\varepsilon) \neq \phi, Q(R(t,t'),\rho',\varepsilon') \neq \phi, \\ \text{where } \forall j \in J,\ \rho_j = \rho'_j \text{ and } \forall l \in L,\ \varepsilon_l = \varepsilon'_l \end{array}\right\}. \tag{2.3}$$

Note that if $(J', L') \subset (J, L)$, then $H(f, R(t, t'), J, L) \leq H(f, R(t, t'), J', L')$ and if $(J, L) = \{1, \dots, k\} \times \{1, \dots, k\}$, then $H(f, R(t, t'), J, L) = 0$. If $(J, L) = (\phi, \phi)$, then we shall denote $H(f, R(t, t'), J, L)$ by $H(f, R(t, t'))$.

Lemma 2.1. *Let $f \in \tilde{D}_k$, $R(t, t') \in \mathscr{I}(k)$, $(J, L) \subset \{1, \dots, k\} \times \{1, \dots, k\}$ and $\eta > 0$. Let $B(R(t, t'), \alpha)$ be an open ball with center $R(t, t')$ and radius α such that $\forall(\rho, \varepsilon) \in \{0, 1\}^k \times \{0, 1\}^k$, $\forall(s, s') \in B(R(t, t'), \alpha) \cap Q(R(t, t'), \rho, \varepsilon)$,*

$$|f(R(s,s')) - f(R(t,t') + O(\rho,\varepsilon))| < \eta. \tag{2.4}$$

Then, for any $R(u, u') \in B(R(t, t'), \alpha)$ for which $t_j = u_j$ $\forall j \in J$, $t_j \neq u_j$ $\forall j \notin J$; $t'_l = u'_l$ $\forall l \in L$ and $t'_l \neq u'_l$ $\forall l \notin L$, we have

$$H(f, R(u,u'), J, L) \leq 2\eta. \tag{2.5}$$

Proof. For any $(\rho, \varepsilon) \in \{0, 1\}^k \times \{0, 1\}^k$ such that $Q(R(u, u'), \rho, \varepsilon) \neq \phi$, we denote by $(\bar{\rho}, \bar{\varepsilon})$, an element of $\{0, 1\}^k \times \{0, 1\}^k$ defined as

$$\text{if } j \in J,\ \bar{\rho}_j = \rho_j \qquad\qquad \text{if } l \in L,\ \bar{\varepsilon}_l = \varepsilon_l$$

$$\text{if } j \notin J,\ \bar{\rho}_j = \begin{cases} 0 & \text{if } u_j < t_j \\ 1 & \text{if } u_j > t_j \end{cases} \qquad \text{if } l \notin L,\ \bar{\varepsilon}_l \begin{cases} 0 & \text{if } u'_l > t_l \\ 1 & \text{if } u'_j < t_l \end{cases}$$

then $Q(R(u, u'), \rho, \varepsilon) \cap Q(R(t, t'), \bar{\rho}, \bar{\varepsilon}) \neq \phi$.

Let $\eta_1 > 0$ and $\alpha' > 0$ be such that

$$|f(R(s, s')) - f(R(u, u') + O(\rho, \varepsilon))| < \eta_1$$

$\forall(\rho, \varepsilon) \in \{0, 1\}^k \times \{0, 1\}^k$ and $\forall R(s, s') \in B(R(u, u'), \alpha') \cap Q(R(u, u'), \rho, \varepsilon)$.

As $B(R(u, u'), \alpha') \cap Q((R(u, u'), \rho, \varepsilon) \cap B(R(t, t'), \alpha) \cap Q(R(t, t'), \bar{\rho}, \bar{\varepsilon}) \neq \phi$ we can find $R(s, s')$ such that (using (2.4))

$$|f(R(s, s')) - f(R(u, u') + O(\rho, \varepsilon))| < \eta_1$$

and

$$|f(R(s, s')) - f(R(t, t') + O(\bar{\rho}, \bar{\varepsilon}))| < \eta.$$

From the above inequalities, we deduce

$$|f(R(t, t') + O(\bar{\rho}, \bar{\varepsilon})) - f(R(u, u') + O(\rho, \varepsilon))| < \eta + \eta_1.$$

Since η_1 is arbitrary, the left side of the above inequality $\leq \eta$.

Let (ρ', ε') be another element of $\{0, 1\}^k \times \{0, 1\}^k$ such that $Q(R(u, u'), \rho', \varepsilon') \neq \phi$. Since $(\bar{\rho}', \bar{\varepsilon}') = (\bar{\rho}, \bar{\varepsilon})$, we deduce, proceeding analogously,

$$|f(R(u, u') + O(\rho', \varepsilon')) - f(R(t, t') + O(\bar{\rho}, \bar{\varepsilon}))| \leq \eta$$

and

$$|f(R(u, u') + O(\rho', \varepsilon')) - f(R(u, u') + O(\rho, \varepsilon))| \leq 2\eta.$$

Now using (2.3), we conclude (2.5).

LEMMA 2.2. *Let $f \in \tilde{D}_k$ and $(J, L) \subset \{1, \ldots, k\} \times \{1, \ldots, k\}$. Set $T \subset \mathscr{I}(k)$ be such that $\forall(R(t, t'), R(s, s')) \in T \times T$ with $R(t, t') \neq R(s, s')$, we have $t_j = s_j\ \ \forall j \in J, t_j \neq s_j\ \ \forall j \notin J, t'_l = s'_l\ \ \forall l \in L,$ and $t'_l \neq s'_l\ \ \forall l \notin L$. Then, for any $\eta > 0$, the set of elements of T such that $H(f, R(t, t'), J, L) \geq \eta$ is finite.*

Proof. Let $\eta > 0$ be fixed, and let T_η be the set of elements of T such that $H(R(t, t'), J, L) \geq \eta$. If T_η is infinite, then it has a limiting point

$R(t,t')$, and we can find a sequence $R(t^{(n)}, t'^{(n)})$ of points of T_η which admits $R(t, t')$ as limit, and is such that $\forall n \in \mathbb{N}, t_j^{(n)} \neq t_j$ for $\forall j \notin J$ and $t_l'^{(n)} \neq t_l'$ $\forall l \notin L$.

Let $\eta' < \eta/2$ and $\alpha > 0$ be such that $B(R(t, t'), \alpha)$ satisfies $|f(R(s, s')) - f(R(t, t'), \rho, \varepsilon)| < \eta'$ $\forall(\rho, \varepsilon) \in \{0, 1\}^k \times \{0, 1\}^k$ and $\forall R(s, s') \in B(R(t,t'), \alpha) \cap Q(R(t, t'), \rho, \varepsilon)$. Also let $n_0 \in \mathbb{N}$ be such that $R(t^{(n)}, t'^{(n)}) \in B(R(t, t'), \alpha)$ $\forall n \geq n_0$. Then, using Lemma 2.1, it follows that $H(f, R(t^{(n)}, t'^{(n)}), J, L) \leq 2\eta' < \eta$ which contradicts the hypothesis.

THEOREM 2.1. *Let R be the set of grids of* $\mathscr{I}(k)$ *and let* $f : \mathscr{I}(k) \to \mathbb{R}$. *Then* $f \in \tilde{D}_k$ *if and only if* $\forall \eta > 0, \exists \delta > 0, \exists G \in R$ *such that* $\forall S \in G$ *and* $\forall(R(t, t'), R(s, s')) \in S \times S$ *we have*

$$d_H(R(t,t'), R(s,s')) \leq \delta \Rightarrow |f(R(t,t') - f(R(s,s'))| < \eta. \quad (2.6)$$

Proof. (Sufficiency). Let $f : \mathscr{I}(k) \to \mathbb{R}$, and let $f \notin \tilde{D}_k$. Then we show that (2.4) is not true, that is we show that $\forall \delta > 0$, $\forall G \in R$, $\exists S \in G$, $\exists (R(t, t'), R(s, s') \in S \times S$ such that $d_H(R(t, t'), R(s, s')) \leq \delta$ and $|f(R(t, t') - f(R(s, s'))| \geq \eta$. Since $f \notin \tilde{D}_k$, it follows that $\exists R(u, u') \in \mathscr{I}(k)$ and $\exists(\rho, \varepsilon) \in \{0, 1\}^k \times \{0, 1\}^k$ such that $Q(R(u, u'), \rho, \varepsilon) \neq \phi$ and f has no (ρ, ε) limit in $R(u, u')$. This means that $\forall \eta_1 > 0$, $\exists(R(t, t'), R(s, s')) \in Q^2(R(u, u'), \rho, \varepsilon)$ such that $d_H(R(t, t'), R(s, s')) < \eta_1$ and $|f(R(t, t')) - f(R(s, s'))| \geq \eta$.

Let $\delta > 0$ and $G \in R$ be fixed. Then we can find $S \in G$ such that $Q(R(u, u'), \rho, \varepsilon) \cap S$ is a neighborhood of $R(u, u')$ in $Q(R(u, u'), \rho, \varepsilon)$. Then, there exists an $\alpha > 0$ such that

$$Q(R(u,u'), \rho, \varepsilon) \cap B(R(u,u'), \alpha) \subset Q(R(u,u'), \rho, \varepsilon) \cap S.$$

Now choose $\eta_1 = \min(\delta, \alpha/2)$. Then

$$\forall(R(t,t'), R(s,s')) \in (Q(R(u,u'), \rho, \varepsilon) \cap B(R(u,u'), \alpha))^2$$

such that $d_H(R(t, t'), R(s, s')) < \eta_1$. We see that $(R(t, t'), R(s, s')) \in S \times S$ and $d_H(R(t, t'), R(s, s') \leq \delta$.

We can now choose $(R(t, t'), R(s, s'))$, such that $|f(R(t, t') - f(R(s, s'))| \geq \eta$ and this leads to contradiction. Sufficiently is established.

(*Necessary part*). Let $f \in \tilde{D}_k$. Choose an $\eta > 0$, and let $\mathscr{H}$ be a class of subsets T of $\mathscr{I}(k)$ such that

(a) $\forall R(t, t') \in T$, $H(f, R(t, t')) \geq \eta$

(b) $\forall(R(t, t'), R(s, s')) \in T \times T$, $R(t, t') \neq R(s, s') \Rightarrow t_j \neq s_j$ $\forall j \in \{1, \ldots, k\}$ and $t_l' \neq s_l'$ $\forall l \in \{1, \ldots, k\}$.

Note that T is finite $\forall T \in \mathscr{H}$. Let S_1^* $(\in \mathscr{H})$ be a maximal element (in $\mathscr{H}$).

By iteration, we construct a sequence $S_1^*, \ldots, S_{2k}^*$ of sets in $\mathscr{I}(k)$ as follows: For any $R(s^{(1)}, s'^{(1)}) \in S_1^*$ and any $(J, L) \in \{1, \ldots, k\} \times \{1, \ldots, k\}$ such that Card J + Card $L = 1$, let $\mathscr{H}(R(s^{(1)}, s'^{(1)}), J, L)$ be a class of subsets T^* of $\mathscr{I}(k)$ such that

(c) $\forall R(t, t') \in T^*$, $H(f, R(t, t'), J, L) \geq \eta$

(d) $\forall R(t, t') \in T^*$, $t_j = s_j^{(1)}$ $\forall j \in J$ and $t_j' = s_j'^{(1)}$ $\forall l \in L$ and

(e) $\forall (R(t, t'), R(s, s')) \in T \times T$, $R(t, t') \neq R(s, s') \Rightarrow t_j \neq s_j$ $\forall j \notin J$ and $t_l' \neq s_l'$ $\forall l \notin L$.

Now let $S^*(R(s^{(1)}, s'^{(1)}), J, L)$ be a maximal element in $\mathscr{H}(R(s^{(1)}, s'^{(1)}), J, L)$. We set $S_2^* = S_1^* \cup_{(1)} \cup_{(2)} S^*(R(s^{(1)}, s'^{(1)}), J, L)$, where $\cup_{(1)}$ in the union over $R(s^{(1)}, s'^{(1)}) \in S_1^*$ and $\cup_{(2)}$ is the union over $(J, L) \in \mathscr{P}_1$, where $\mathscr{P}_1$ is the class of subsets of $J \subset \{1, \ldots, k\}$ and $L \subset \{1, \ldots, k\}$ such that Card J + Card $L = 1$.

Proceeding this way we get a sequence $S_1^* \subset \cdots \subset S_{2k}^*$ of sets in $\mathscr{I}(k)$. Let G be a grid generated by S_{2k}^*. Now denote

$$J(R(t,t')) = \{j \in \{1, \ldots, k\} \colon \exists R(s, s') \in S_{2k}^* \text{ and } s_j = t_j\}$$

and

$$L(R(t,t')) = \{l \in \{1, \ldots, k\} \colon \exists R(s, s') \in S_{2k}^* \text{ and } s_l' = t_l'\}.$$

Then, we first prove the following lemma.

LEMMA 2.3. *Let $R(t, t') \in \mathscr{I}(k)$. Then for every (J, L) such that $(J(R(t, t'), L(R(t, t')) \subset (J, L)$, we have*

$$H(f, R(t, t'), J, L) < \eta. \tag{2.7}$$

Proof. Let $\mathscr{F}$ be a set of sequences $(R(s^{(1)}, s'^{(1)}), j_1), \ldots, (R(s^{(h)}, s'^{(h)}), j_h)$, where $1 \leq h \leq 2k$, $j_l \in \{1, \ldots, 2k\}$, and $1 \leq l \leq h$, such that,

$$s_{j_l}^{(l)} = t_{j_l} \quad \text{if } j_l \leq k,\ s_{j_l-k}'^{(l)} = t_{j_l-k}' \text{ if } j_l > k\ (1 \leq l \leq h)$$

$$R(s^{(1)}, s'^{(1)}) \in S_1^*, \qquad R(s^{(l)}, s'^{(l)}) \in S^*\big(R(s^{(l-1)}, s'^{(l-1)}), J_1, L_1\big) \tag{2.8}$$

where $J_1 = \cup_{p<l-1, j_p \leq k}\{j_p\}$, $L_1 = \cup_{p<l-1, j_p>k}\{j_p - k\}$, and $j_l \notin \{j_1, \ldots, j_{l-1}\}$. Note that $\mathscr{F}$ has at least one maximal sequence and it is easy to check that

$$\forall l \in \{1, \ldots, h\}, \qquad s_{j_l}^{(h)} = t_{j_l} \quad \text{if } j_l \leq k \text{ and } s_{j_l-k}'^{(h)} = t_{j_l-k}' \text{ if } j_l > k \tag{2.9}$$

Since $J_1 \subset J(R(t,t'))$ and $L_1 \subset L(R(t,t'))$, it suffices to prove that any maximal sequence satisfies

$$H(f, R(t,t'), J_2, L_2) < \eta,$$
$$\text{where } (J_2, L_2) = (J_1, L_1) \text{ with } h = l - 1. \quad (2.10)$$

There are three cases to be disposed of.

Case 1. Let $h = 2k$. Then from (2.9), $R(t,t') = R(s^{(2k)}, s'^{(2k)})$, $J(R(t,t')) = L(R(t,t')) = \{1, \ldots, k\}$ and so

$$H(f, R(t,t'), J(R(t,t')), L(R(t,t'))) = 0\,(<\eta).$$

Case 2. Let $\mathcal{F} = \phi$. Then $H(f, R(t,t')) < \eta$, and also for any $(J, L) \subset \{1, \ldots, k\} \times \{1, \ldots, k\}$, $H(f, R(t,t'), J, L) < \eta$.

Case 3. Let $1 \le h < 2k$. Then, for any $R(s,s') \in S^*(R(s^{(h)}, s'^{(h)}), J_2, L_2)$ and $j \notin \{j_1, \ldots, j_h\}$, we have $s_j \ne t_j$ if $j \le k$, $s'_{j-k} \ne t'_{j-k}$ if $j > k$. Now if $H(f, R(t,t'), J_2, L_2) \ge \eta$, then the sequence would not be maximal, and so we have the contraction. This proves the lemma.

Now let G be a grid generated by S^*_{2k}, and let $S \in G$.
Let $\bar{S}$ denote the closure of S under d_H. (Note that $\bar{S}$ is a pavement.) If $R(t,t') \in \bar{S}$, and (J, L) is the indicator of $R(t,t')$ in $\bar{S}$, then we have $H(f, R(t,t'), J, L) < \eta$.
Now we prove that for any $S \in G$ $\exists \delta > 0$ such that $\forall (R(t,t'), R(s,s')) \in S \times S$,

$$d_H(R(t,t'), R(s,s')) \le \eta \Rightarrow |f(R(t,t')) - f(R(s,s'))| < \eta. \quad (2.11)$$

Suppose (2.11) is not true. Then, we can find an $S \in G$ such that $\forall \delta > 0$, $\exists (R(t,t'), R(s,s') \in S \times S$, such that

$$d_H(R(t,t'), R(s,s')) \le \delta \quad \text{and} \quad |f(R(t,t')) - f(R(s,s'))| > \eta. \quad (2.12)$$

In this case, we can extract two sequences $R(t^{(n)}, t'^{(n)})$ and $R(s^{(n)}, s'^{(n)})$ in S such that they converge to the same limit $R(t,t')$ in $\bar{S}$; furthermore $\exists (\rho, \varepsilon) \in \{0,1\}^k \times \{0,1\}^k$ and $(\rho', \varepsilon') \in \{0,1\}^k \times \{0,1\}^k$ with $R(t^{(n)}, t'^{(n)}) \in Q(R(t,t'), \rho, \varepsilon)$ and $R(s^{(n)}, s'^{(n)})$ in $Q(R(s,s'), \rho', \varepsilon')$ for any $n \ge 1$ such that $|f(R(t^{(n)}, t'^{(n)})) - f(R(s^{(n)}, s'^{(n)})| > \eta$.
As $f \in \tilde{D}_k$, $\lim_{n\to\infty} f(R(t^{(n)}, t'^{(n)})) = f(R(t,t') + O(\rho, \varepsilon))$ and $\lim_{n\to\infty} f(R(s^{(n)}, s'^{(n)})) = f(R(t,t') + O(\rho', \varepsilon'))$.
Consequently,

$$|f(R(t,t') + O(\rho, \varepsilon)) - f(R(t,t') + O(\rho', \varepsilon'))| \ge \eta. \quad (2.13)$$

Let (J, L) be the indicator of $R(t, t')$ in $\bar{S}$. If $(J, L) = (\phi, \phi)$, $J(R(t, t')) = \phi$, and $L(R(t, t')) = \phi$, then $\mathscr{F}$ is ϕ and so $H(f, R(t, t')) < \eta$, which is not compatible with (2.13).

If $(J, L) \neq (\phi, \phi)$, the $\rho_j = \rho_j'$ $\forall j \in J$, and $\varepsilon_l = \varepsilon_l'$ $\forall l \in L$, and so

$$\begin{aligned}&|f(R(t,t') + O(\rho, \varepsilon)) - f(R(t,t') + O(\rho', \epsilon'))|\\ &\quad \leq H(f(R(t,t'), J, L)\\ &\quad \leq H(f(R(t,t'), J(R(t,t')), L(R(t,t')) < \eta\end{aligned}$$

(from Lemma 2.3) and this is not compatible with (2.13). This proves the theorem.

Definition (Balacheff and Dupont [3]). A grid G' is *finer* than a grid G if and only if $\forall S' \in G'$, $\exists S \in G$ such that $S' \subset S$.

Property of a Finer Grid. Given any $\delta > 0$ and any grid G, we can find another grid G' finer than G such that the diameter of each element S' of G' is less than or equal to δ.

For any grid G with base $B = \{\{t_{ji}\}, 1 \leq i \leq n_j, 1 \leq j \leq k\}$, we associate the number $m(G)$, called permeability $m(G)$ of G, defined as

$$m(G) = \inf_{1 \leq j \leq k} \inf_{1 \leq i \leq n_j} \left\{|t_{ji} - t_{ji+1}|, t_{ji} \neq t_{ji+1}\right\}, \qquad t_{jn_j+1} = 1$$

by convention. (Note that this concept of permeability is an extension of the similar concept given by Neuhaus [10] for a rather specialized situation connected with the space D_k.) Now denote

$$R_\eta = \{G; m(G) > \eta \text{ for any } \eta > 0\}. \tag{2.14}$$

Corollary 2.1. *Let f: $\mathscr{I}(k) \to \mathbb{R}$. Then*

(a) *$f \in \tilde{D}_k$ if and only if $\forall \eta > 0$, $\exists$ a grid $G \in R$ such that $\forall S \in G$ and $\forall (R(t, t'), R(s, s')) \in S \times S$, we have $|f(R(t, t')) - f(R(s, s'))| < \eta$.*

(b) *$f \in \tilde{D}_k$ if and only if $\forall \eta > 0$, $\exists \delta > 0$ and $G \in R_\delta$ such that $\forall S \in G$ and $\forall (R(t, t'), R(s, s')) \in S \times S$, we have $|f(R(t, t')) - f(R(s, s'))| < \eta$.*

Proof. (a) is a consequence of Theorem 2.1 and the property of the finer grid, and (b) is a consequence of (a).

For any function f: $\mathscr{I}(k) \to \mathbb{R}$, and any $\delta > 0$, we define the "modulus of continuity" $\omega(f, \delta)$ in $\tilde{D}_k$ and $\tilde{\omega}(f, \delta)$ in $\tilde{C}_k$ as

$$\omega'(f, \delta) = \inf_{G \in R_\delta} \max_{S \in G} \sup_{(R(t,t'), R(s,s')) \in S \times S} |f(R(t,t')) - f(R(s,s'))| \tag{2.15}$$

and

$$\tilde{\omega}(f,\delta) = \sup\{|f(R(t,t')) - f(R(s,s'))|;\\ (R(t,t'),R(s,s')) \in \mathscr{I}(k) \times \mathscr{I}(k),\\ d_H(R(t,t'),R(s,s')) \leq \delta\}. \tag{2.16}$$

Note that a function $g: \delta \to \omega'(f,\delta)$, where $\delta \in (0,1]$ is nondecreasing.

COROLLARY 2.2. *Let $f: \mathscr{I}(k) \to \mathbb{R}$. Then*

(a) $f \in \tilde{D}_k$ *if and only if* $\lim_{\delta \to 0} \omega'(f,\delta) = 0$

(b) $f \in \tilde{C}_k$ *if and only if* $\lim_{\delta \to 0} \tilde{\omega}(f,\delta) = 0$.

Proof. (a) is a consequence of Corollary 2.1(b), and (b) follows by definition.

Note that for any bounded function $f: \mathscr{I}(k) \to \mathbb{R}$, and any $\delta \in (0, \frac{1}{2})$, $\omega'(f,\delta) \leq \tilde{\omega}(f, 2\delta)$.

3. SKOROHOD TOPOLOGY ON $\tilde{D}_k$

3.1. *Preliminaries*

In what follows, Λ denotes the space of maps $h: [0,1] \to [0,1]$ which are nondecreasing, continuous, and bijective. $\lambda^{(k)}$ denotes the space of maps $\lambda: [0,1]^k \to [0,1]^k$, where $\lambda(t_1, \ldots, t_k) = (\lambda_1(t_1), \ldots, \lambda_k(t_k))$, $\lambda_j \in \Lambda$, $1 \leq j \leq k$.

$I_{(k)}$ denotes the identity map on $[0,1]^k$,

$$|||\,\lambda\,||| = \max_{1 \leq j \leq k} \sup_{0 \leq t_j < s_j \leq 1} \left| \log \frac{\lambda_j(t_j) - \lambda_j(s_j)}{t_j - s_j} \right|$$

for any $R(t,t') \in \mathscr{I}(f)$, $\lambda(R(t,t'))$ denotes an element $R(s,s')$ of $\mathscr{I}(k)$ defined by $\lambda_j(t_j) = s_j$ and $\lambda_j(t_j') = s_j'$, $1 \leq j \leq k$. For any bounded maps $f,g: \mathscr{I}(k) \to \mathbb{R}$, we denote

$$d(f,g) = \inf_{\lambda \in \Lambda^{(k)}} \max\{\|f - g \circ \lambda\|, \|\lambda - I_{(k)}\|\};$$

$$d_0(f,g) = \inf_{\lambda \in \Lambda^{(k)}} \max\{\|f - g \circ \lambda\|, |||\,\lambda\,|||\},$$

where $\|f - g \circ \lambda\| = \sup_{R(t,t') \in \mathscr{I}(k)}\{|f(R(t,t')) - g \circ \lambda(R(t,t'))|\}$ and $\|\lambda - I_{(k)}\| = \sup_{t \in [0,1]^k}\{|\lambda(t) - I_{(k)}(t)|\}$. We shall call the topologies associated with d and d_0 as Skorohod and modified Skorohod topologies respectively.

LEMMA 3.1. *Let f_n: $\tilde{D}_k \to \mathbb{R}$ be a sequence of maps such that $f_n \to f \in \tilde{D}_k$ in Skorohod topology. Let $R(t, t') \in \mathscr{I}(k)$ be such that the restriction of f to the face of $R(t, t')$ in $\mathscr{I}(k)$ is continuous in $R(t, t')$. Then* $\lim_{n \to \infty} f_n(R(t, t')) = f(R(t, t'))$.

Proof. Proof follows by using the inequality

$$\begin{aligned} &|f_n(R(t,t')) - f(R(t,t'))| \\ &\quad \le |f_n(R(t,t')) - f(\lambda^{-1}(R(t,t')))| \\ &\qquad + |f(\lambda^{-1}(R(t,t'))) - f(R(t,t'))|, \end{aligned}$$

where λ^{-1} is the inverse function of λ.

Remark 3.1. Following Billingsley [5] and Neuhaus [10], the following facts can easily be established:

(i) Skorohod topology as well as modified Skorohod topology implies uniform topology.

(ii) Uniform topolopgy is finer than the Skorohod topology.

(iii) The modified Skorohod topology is finer than the Skorohod topology.

(iv) The Skorohod topology and the modified Skorohod topology are equivalent in $\tilde{D}_k$.

(v) The space $(\tilde{D}_k, d)$ is separable.

(vi) The space $(\tilde{D}_k, d_0)$ is complete.

(vii) $\forall \delta > 0$, $\omega'(f, \delta)$ is upper semicontinuous in $f \in \tilde{D}_k$ with respect to the Skorohod topology.

LEMMA 3.2. *For any $R(t, t') \in \mathscr{I}(k)$, let $\varphi_{R(t,t')}$: $\tilde{D}_k \to \mathbb{R}$ be a map defined by $\varphi_{R(t,t')} = f(R(t, t'))$. If the restriction of the face of $R(t, t')$ into $\mathscr{I}(k)$ is continuous, then $\varphi_{R(t,t')}$ is continuous with respect to the Skorohod topology.*

Proof. Consequence of Lemma 3.1.

THEOREM 3.1. *Let $K \subset \tilde{D}_k$. Then the closure of $\overline{K}$ of K (with respect to the Skorohod topology) is compact if and only if*

$$\sup_{f \in K} \|f\| < \infty \tag{3.1}$$

and

$$\lim_{\delta \to \infty} \sup_{f \in K} \omega'(f, \delta) = 0. \tag{3.2}$$

Proof. Let $\overline{K}$ be compact. Then (3.1) and (3.2) follow from Remark 3.1(vii). We now prove the sufficiency part.

Let $K \subset \tilde{D}_k$, and let (3.1) and (3.2) hold. Since $(\tilde{D}_k, d_0)$ is complete, we have to show that $\forall \eta > 0$, $\exists$ a finite subset $K(\eta)$ of $\tilde{D}_k$ which is η-net in K with respect to d_0.

Choose an $\eta > 0$, and an integer m such that $\eta > 1/m$, and $\omega'(f, 1/m) < \eta \ \forall f \in K$.

Let $\mathscr{L}_m = \{j/m;\ 0 \le j \le m\}$ and let $R(\mathscr{L}_m)$ be the set of grids G of $\mathscr{I}(k)$, with each grid having the base of the form $\{t_{ji},\ 1 \le i \le n_j,\ 1 \le j \le k\}$, where $t_{ji} \in \mathscr{L}_m \ \ \forall i$ and $\forall j$.

Let H be an η-net set of $[-\sup\|f\|, \sup\|f\|]$ and let $K(\eta)$ be the space of step functions of the form $\Sigma_{S \in G} \alpha_S I_S$, where $G \in R(\mathscr{L}_m)$ and for any $S \in G$, α_S belongs to H. Then it is easy to check that $K(\eta)$ is 2η-net in K with respect to d, and hence $K(\eta)$ is η-net in K with respect to d_0.

4. Weak Convergence of Probability Measures in $\tilde{D}_k$

4.1. *Measurability on* $\tilde{D}_k$

For any $T \subset \mathscr{I}(k)$, let φ_T denote the projection of $\tilde{D}_k$ in $\mathbb{R}^{\mathrm{T}}$, and let $\tilde{\mathscr{D}}_k$ be the Borel σ-field generated by the Skorohod topology in $\tilde{D}_k$. For any $(J, L) \subset \{1, \dots, k\} \times \{1, \dots, k\}$, let $F_{J,L} = \{R(t, t');\ t_j = 1\ \forall j \in J$ and $t'_l = 1\ \forall l \in L\}$. Then, we have

Theorem 4.1. *$\tilde{\mathscr{D}}_k$ is the restriction to $\tilde{D}_k$ of the σ-field $\mathscr{B}^{\mathscr{I}(k)}$ on $\mathbb{R}^{\mathscr{I}(k)}$, where $\mathscr{B}$ is the usual Borel σ-field on $\mathbb{R}$.*

Proof. First we show that $\mathscr{B}^{\mathscr{I}(k)} \subset \tilde{\mathscr{D}}_k$. To prove this we have to show that $\forall R(t, t') \in \mathscr{I}(k)$, the map $\varphi_{R(t,t')}$: $\hat{D}_k \to \mathbb{R}$ is measurable. If $t = t' = (1, \dots, 1)$, then the measurability of $\varphi_{R(t,t')}$ follows as a consequence of Lemma 3.1. If t or $t' \neq (1, \dots, 1)$, then setting $J = \{j;\ j \in \{1, \dots, k\}, t_j \neq 1\}$ and $L = \{l;\ l \in \{1, \dots, k\},\ t'_l \neq 1\}$, we notice that $L \subset J$. Then, from the continuity of the right with respect to t_j and t'_j in $R(t, t')$, it follows that $\varphi_{R(t,t')} = \lim_{\varepsilon \to 0} h_\varepsilon$, where

$$h_\varepsilon(f) = \frac{1}{\varepsilon^{(\mathrm{Card}\, J + \mathrm{Card}\, L)}}$$

$$\times \int_{\prod_{j \in J} [t_j, t_j + \varepsilon) \prod_{l \in L} [t'_l, t'_l + \varepsilon)} \left(f(R(u, u'))(du_j)_{j \in J} (du'_l)_{l \in L} \right)$$

and

$$\varepsilon < \inf\Big(\inf_{j \in J} (1 - t_j),\ \inf_{l \in L} (1 - t_l'),\ \inf_{l \in L} (t_l' - t_l) \Big),$$

where $u_j = 1$ if $j \notin J$, and $u_l' = 1$ if $l \notin L$.

We prove that h_ε is continuous with respect to the Skorohod topology. For any $f \in \tilde{D}_k$, denote

$$C(f) = \{R(s, s');\ R(s, s') \in \mathscr{I}(k) \text{ and the restriction of } f \text{ to the face of } R(s, s') \text{ into } \mathscr{I}(k) \text{ is continuous in } R(s, s')\}, \tag{4.1}$$

and the map

$$\Pi_{J, L}: \mathscr{I}(k) \to [0, 1]^J \times [0, 1]^L \tag{4.2}$$

defined by

$$\Pi_{J, L}(R(s, s')) = \big((s_j)_{j \in J}, (s_j')_{l \in L}\big).$$

If $f_n \to f_0$ in the Skorohod topology, then $f_n(R(s, s')) \to f_0(R(S, s'))$ $\forall R(s, s') \in C(f_0)$, and from Lemma 2.2 we deduce that $\Pi_{J, L}(C^*(f_0))$, where C^* denotes the complement of C, has the Lebesque measure 0 in $[0, 1]^J \times [0, 1]^L$. It follows (from the Lebesgue theorem) that $h_\varepsilon(f_n) \to h_\varepsilon(f_0)$, and so h_ε is continuous in the Skorohod topology. This implies that h_ε is measurable and hence $\varphi_{R(t, t')}$ is measurable. Thus $\mathscr{B}^{\mathscr{I}(k)} \subset \tilde{\mathscr{D}}_k$. Now we prove that $\tilde{\mathscr{D}}_k \subset \mathscr{B}^{\mathscr{I}(k)}$.

Let $T \subset \mathscr{I}(k)$, and suppose T is dense in $\mathscr{I}(k)$ and also for $(J, L) \subset \{1, \ldots, k\} \times \{1, \ldots, k\}$, $F_{J, L} \cap T$ is dense in $F_{J, L}$. Then, we prove that $\tilde{\mathscr{D}}_k$ is generated by $\{\varphi_{R(t, t')};\ R(t, t') \in T\}$.

Following Billingsley [5, p. 121], because d_0 is separable, it suffices to show that for any $f \in \tilde{\mathscr{D}}_k$ and $r > 0$, the open ball $B(f, r) = \{g \in \tilde{D}_k;\ d_0(f, g) < r\}$ belongs to the σ-field generated by $\{\varphi_{R(t, t')};\ R(t, t') \in T\}$. Take a sequence $\{R(t^{(n)}, t'^{(n)}\}$ of T, and suppose that it is dense in $\mathscr{I}(k)$, and the intersection of this sequence with $F_{J, L}$ is also dense in $F_{J, L}$. For any $\eta < r$ and any $N^* \in \mathbb{N}$, denote

$$\begin{aligned} A_{N^*}(\eta) = \big\{g \in \tilde{D}_k,\ &|||\lambda||| < r - \eta,\\ &|g\big(R(t^{(i)}, t'^{(i)})\big) - f\big(\lambda\big(R(t^{(i)}, t'^{(i)})\big)| < r - \eta,\\ &0 \le i \le N^*, \text{ for some } \lambda \in \Lambda^{(k)}\big\}. \end{aligned} \tag{4.3}$$

Then, it follows that

$$A_{N^*}(\eta) = \varphi^{-1}_{\{R(t^{(0)}, t'^{(0)}), \dots, R(t^{(N^*)}, t'^{(N^*)})\}}(H_{N^*}(\eta)), \tag{4.4}$$

where

$$\begin{aligned} H_{N^*}(\eta) = \big\{(a_0, \dots, a_{N^*}) \in \mathbb{R}^{N^*+1};\ |||\lambda||| < r - \eta \text{ and} \\ |a_i - f(\lambda(R(t^{(i)}, t'^{(i)})))| < r - \eta, \\ 0 \le i \le N^* \text{ for some } \lambda \in \Lambda^{(k)}\big\} \end{aligned} \tag{4.5}$$

which implies that $A_{N^*}(\eta)$ belongs to the σ-field generated by $\{\varphi_{R(t,t')}; R(t,t') \in T\}$.

Next, we show that $B(f, r) = \bigcup_{\eta \in Q \cap (0,r)} (\bigcap_{N^* \in \mathbb{N}} A_{N^*}(\eta))$, where Q is the set of rational numbers.

It is clear that $B(f, r) \subset \bigcup_{\eta \in Q \cap (0,r)} (\bigcap_{N^* \in \mathbb{N}} A_{N^*}(\eta))$. It remains to show that for each $\eta \in Q \cap (0, r)$,

$$\bigcap_{N^* \in \mathbb{N}} A_{N^*}(\eta) \subset B(f, r). \tag{4.6}$$

Let $g \in \bigcap_{N^* \in \mathbb{N}} A_{N^*}(\eta)$. Choose N^* and $\lambda(N^*) \in \Lambda^{(k)}$ such that

$$|||\lambda^{(N^*)}||| < r - \eta$$

and

$$|g(R(t^{(i)}, t'^{(i)})) - f(\lambda^{(N^*)}(R(t^{(i)}, t'^{(i)})))| < r - \eta \quad \forall 0 \le i \le N^*.$$

Following Billingsley [5, p. 122], we can find a subsequence $\{\lambda^{(i_{N^*})}; N^* \in \mathbb{N}\}$ of $\{\lambda^{(N^*)}, N^* \in \mathbb{N}\}$ such that $\lambda^{(i_{N^*})} \to \lambda \in \Lambda^{(k)}$, and $|||\lambda||| < r - \eta$.

For any $R(t, t') \in T$, we can find an i such that $t_j \le t_j^{(i)}$ and $t'_j \le t_j'^{(i)}$ and then we have

$$|g(R(t^{(i)}, t'^{(i)})) - f(\lambda^{(i_{N^*})}(R(t^{(i)}, t'^{(i)})))| < r - \eta$$

$\forall N^* \ge i,\ i \in \{0, \dots, i_{N^*}\}$. It follows that there exists a $(\rho(R(t, t')), \varepsilon(R(t, t'))) = ((\rho_j)_{1 \le j \le k}, (\varepsilon_l)_{1 \le l \le k}) \in \{0, 1\}^k \times \{0, 1\}^k$ such that

$$t_j = 1 \Rightarrow \rho_j = 1\ \forall 1 \le j \le k \qquad \text{and} \qquad t'_l = 1 \Rightarrow \varepsilon_l = 1\ \forall 1 \le l \le k \tag{4.7}$$

$$Q(R(t, t'), \rho, \varepsilon) \neq \phi \tag{4.8}$$

and

$$|g(R(t, t')) - f(\lambda(R(t, t')) + O(\rho, \varepsilon))| \le r - \eta. \tag{4.9}$$

We now prove that $\|g - f \circ \lambda\| \le r - \eta$. To prove this, suppose there exists $R(u, u') \in \mathscr{I}(k)$ such that $|g(R(u, u')) - f(\lambda(R(u, u'))| > r - \eta$. Then we can find an $R(t, t') \in T$ such that for any $(\rho, \varepsilon) \in \{0, 1\}^k \times \{0, 1\}^k$ such that (4.9) is not satisfied. This leads to contradiction. Thus $\|g - f \circ \lambda\| \le r - \eta$. Now since $|||\lambda||| < r - \eta$, it follows that $g \in B(f, r)$. (4.6) holds.

COROLLARY 4.1. *Let T be denumerable and dense in $\mathscr{I}(k)$. Also let $F_{J,L} \cap T$ be dense in $F_{J,L}$ $\forall (J, L) \subset \{1, \dots, k\} \times \{1, \dots, k\}$. Let $\mathscr{D}_T \subset \tilde{\mathscr{D}}_k$, where $\mathscr{D}_T = \{\varphi_U^{-1}(H_U)$, U is a finite subset of T, and H_U a Borel-subset of $\mathbb{R}^U\}$, and φ_U is the projection of D_{k+1} on $\mathbb{R}^U$. If P and Q are probability measures on $\tilde{\mathscr{D}}_k$, and if $P = Q$ on $\mathscr{D}_T$, then $P = Q$ on $\tilde{\mathscr{D}}_k$.*

Proof. Consequence of Theorem 4.1.

For any probability measures on $(\tilde{D}_k, \tilde{\mathscr{D}}_k)$, denote

$$T_P = \Big\{R(t, t');\ R(t, t') \in \mathscr{I}(k);\ P\big(\{f;\ \varphi_{R(t,t')} \text{ is discontinuous on } f\}\big) = 0\Big\}$$

$$T_P' = \Big\{R(t, t');\ R(t, t') \in \mathscr{I}(k);\ P(\{f; \text{ the restriction of } f \text{ to the face of } R(t, t') \text{ into } \mathscr{I}(k) \text{ is discontinuous in } R(t, t')\}) = 0\Big\}.$$

Then, it is clear that $T_P' \subset T_P$.

THEOREM 4.2. *$T_P \cap F_{J,L}$ is dense in $F_{J,L}$ $\forall (J, L) \subset \{1, \dots, k\} \times \{1, \dots, k\}$.*

Proof. It suffices to prove that $T_P' \cap F_{J,L}$ is dense in $F_{J,L}$. To prove this it suffices to show that $T_P' \cap F_{\phi,\phi}$ is dense in $F_{\phi,\phi}$. For any $R(t, t') \in F_{\phi,\phi}$, let

$$J_{R(t,t')} = \{f;\ f \text{ is discontinuous in } R(t, t')\},$$

$$C = \Big\{R(t, t');\ R(t, t') \in F_{\phi,\phi};\ P\big(J_{R(t,t')}\big) \neq 0\Big\},$$

and

$$J_{\{R(t,t'),\eta\}} = \{f;\ H(f, R(t, t')) \ge \eta\}.$$

Obviously,

$$C = \bigcup_{l \ge 1} \bigcup_{m \ge 1} \Big\{R(t, t') \in F_{\phi,\phi};\ P\big(J_{\{R(t,t'),1/m\}}\big) \ge 1/l\Big\}.$$

Now using Lemma 2.2 and proceeding as in Billingsley [5, p. 124], we find that $\{R(t, t');\ R(t, t') \in F_{\phi,\phi};\ P(J_{\{R(t,t')\}}) \ge \eta_1\}$ has a Lebesque measure zero. This implies that C has a Lebesgue measure zero. Consequently $T_P' \cap F_{\phi,\phi}$ is dense in $F_{\phi,\phi}$.

THEOREM 4.3. *The sequence* $\{P_n\}_{n\geq 1}$ *of probability measures on* $(\tilde{D}_k, \tilde{\mathscr{D}}_k)$ *converges weakly to a probability measure* P *on* $(\tilde{D}_k, \tilde{\mathscr{D}}_k)$ *if and only if*

$$\varphi_U(P_n) \text{ converges weakly to } \varphi_U(P) \text{ for all finite subsets } U \text{ of } T_P; \tag{4.10}$$

$$\text{the sequence is tight.} \tag{4.11}$$

Proof. The necessary part is obvious. To prove the sufficiency part, note that since the sequence $\{P_n\}$ is tight, it is weakly relatively compact, we have to prove (cf. Billingsley [5, Theorem 2.3]) that any subsequence of $\{P_n\}$ which converges admits P as limit. For this we use the same line of argument as in Billingsley [5, Theorem 15.1], the fact that for any probability measure Q and any finite subset U of T_Q, φ_U is a.s. Q-continuous, and the Corollary 4.1.

COROLLARY 4.2. *Let* $\{P_n\}_{n\geq 1}$ *be a sequence of probability measures on* $(\tilde{D}_k, \tilde{\mathscr{D}}_k)$ *such that*

$$\lim_{\delta\to 0} \limsup_{n\to\infty} P_n(f;\, \omega'(f,\delta) \geq \varepsilon) = 0 \qquad \forall \varepsilon > 0 \tag{4.12}$$

and there exists a probability measure P *on* $(\tilde{D}_k, \tilde{\mathscr{D}}_k)$ *such that*

$$\varphi_U(P_n) \to \varphi_U(P) \text{ weakly for any finite subset } U \text{ of } T_P. \tag{4.13}$$

Then,

$$P_n \to P \text{ weakly with respect to the Skorohod topology.} \tag{4.14}$$

Proof. Follows from Theorems 3.1 and 4.3.

COROLLARY 4.3. *Let* $\{P_n\}_{n\geq 1}$ *be a sequence of probability measures on* $(\tilde{D}_k, \tilde{\mathscr{D}}_k)$ *such that*

$$\lim_{\delta\to 0} \limsup_{n\to\infty} P_n(f;\, \tilde{\omega}(f,\delta) \geq \varepsilon) = 0 \qquad \forall \varepsilon > 0 \tag{4.15}$$

and

$$(P_n) \text{ converges weakly to some probability measure } P_U \text{ on } R^U \text{ for every finite subset } U \text{ of } \mathscr{I}(k). \tag{4.16}$$

Then,

$$P_n \text{ converges weakly to some probability measure } P \text{ in Skorohod topology with } P(\tilde{C}_k) = 1. \tag{4.17}$$

Proof. Follows from Corollary 4.2 and the inequality $\omega'(f,\delta) \leq \tilde{\omega}(f,\delta)$

5. Convergence of the Process $\tilde{W}_n$

5.1. *Preliminaries and Some Basic Tools*

In this section we study the asymptotic behavior of the empirical process $\tilde{W}_n$ defined in (1.2). Before we do that we introduce the space D_k. Let f: $[0,1]^k \to \mathbb{R}$. For $\rho \in \{0,1\}^k$, define

$$f_\rho(t) = \lim_{\substack{s_i \uparrow t_i,\ \rho(i)=1 \\ s_i \downarrow t_i,\ \rho(i)=0}} f(s)\big((s,t) \in [0,1]^k\big),$$

if it exists; in which case, call $f_\rho(t)$ the ρ-limit of f at t. Denote by D_k, the space of all maps f: $[0,1]^k \to \mathbb{R}$ such that for all $\rho \in \{0,1\}^k$, f_ρ exists and $f_\rho = f$ for $\rho = (0,\dots,0)$.

For any map f: $[0,1]^k \to \mathbb{R}$, and any rectangle $B = \prod_{j=1}^k (a_j, b_j]$, we denote a difference operator $\Delta_B f$ by

$$\Delta_B f = \Sigma(-1)^{\operatorname{card} I} f((b_i)_{i\in I}, (a_i)_{i\notin I}), \tag{5.1}$$

where card I is the cardinal of I, and Σ is over all the 2^k subsets $I \subset \{1,\dots,k\}$.

We will study the asymptotic behavior of $\tilde{W}_n$ via the empirical process W_n defined as

$$W_n(t) = n^{-1/2} \sum_{i=1}^{n} \left(\prod_{j=1}^{k} I_{[F_n^{(j)}(X_{ni}^{(j)}) \le t_j]} - H_{ni}(t) \right). \tag{5.2}$$

The process W_n has been studied by Balacheff and Dupont [3], among others. (See the references in Balacheff and Dupont [3].)

It is well known that the process W_n has a.s. (almost sure) trajectories in D_k. Set $B = \prod_{j=1}^k (a_j, b_j] \in \mathscr{I}(k)$, and by convention we put $\tilde{W}_n(B) = 0$ if there exists at least one j for which $b_j = a_j$. Then the process $\tilde{W}_n$ has a.s. trajectories in the space $\tilde{D}_k$.

Our results of Section 5 as well as Section 6 are based on the following two lemmas.

Lemma 5.1. *Let the sequence $\{X_{ni}\}$ of real-valued random variables centered at their expectations be φ-mixing with rates $\Sigma_{m\ge 1} m^{-1}\varphi^{1/2q}(m) < \infty$. Let N_n be the number of indexes i $(1 \le i \le n)$ for which X_{ni} is not identically zero. Set $S_n = \Sigma_{i=1}^n X_{ni}$, and $\|X_{ni}\|_l = (\int |X_{ni}|^{2l}\, dP_n)^{1/2l}$. Then, there exists a constant $C_q(\varphi)$ depending only on $q \in N^* = \{1,2,\dots\}$ and φ such that*

$$E(S_n^{2q}) \le C_q(\varphi) \sum_{l=1}^{q} N_n^{q/l} \Big(\sup_{1\le i\le n} \|X_{ni}\|_l \Big)^{2q}. \tag{5.3}$$

Proof. The proof is a slight modification of Theorem 2.1 of Neumann [12] and is therefore omitted. See also Harel and Puri ([21], Lemma 4.1).

LEMMA 5.2. *Let the sequence $\{X_{ni}\}$ of real-valued random variables centered at their expectations be strong mixing with rates $\sum_{m>1} m^{2q-2}\alpha^{\varepsilon}(m) < \infty$, $q \geq 1$, $\varepsilon \in (0, 1/2q)$, and $|X_{ni}| \leq 1$, $1 \leq i \leq n$, $n \geq 1$. Let N_n be the number of indexes i $(1 \leq i \leq n)$ for which X_{ni} is not identically zero. Set $S_n = \sum_{i=1}^{n} X_{ni}$ and $\|X_{ni}\|_{\varepsilon} = (\int |X_{ni}|^{2/(1-\varepsilon)})^{1-\varepsilon}$. Then, there exists a constant $C_q(\alpha)$ depending only on q and α such that*

$$E(S_n^{2q}) \leq C_q(\alpha) \sum_{l=1}^{q} N^l \Big(\sup_{1 \leq i \leq n} \|X_{ni}\|_{\varepsilon}^l \Big). \tag{5.4}$$

Proof. The proof is essentially the same as in Doukhan and Portal [6] and is therefore omitted.

Now for any grid G with base $B = \{t_{ji}, 1 \leq i \leq n_j, 1 \leq j \leq k\}$, the number $\tau = \max_{1 \leq j \leq k} \max_{1 \leq j \leq n_j - 1} \{|t_{ji} - t_{ji+1}|\}$ is called the *pace* of the grid G.

Let $\{G_n\}_{n \geq 1}$ be a sequence of grids with paces $\{\tau_n\}_{n \geq 1}$. $\{G_n\}_{n \geq 1}$ is called asymptotically dense in $[0, 1]^k$ if $\tau_n \to 0$ as $n \to \infty$. We denote the *lower boundary* of a subset C of $[0, 1]^k$ by $\mathbf{C}$, where $\mathbf{C} = \{t; t \in C, t_j = 0$ for at least one j, $1 \leq j \leq k\}$. For any base $B = \{t_{ji}, 1 \leq i \leq n_j, 1 \leq j \leq k\}$, we say that a subset $\tilde{B}$ is tied to B if $\tilde{B} = \{t;\ t_j \in \{t_{j1}, \dots, t_{jn_j}\}, 1 \leq j \leq k\}$ and $R(t, t')$ is *tied* to B if $t_j \in \{t_{j1}, \dots, t_{jn_j}\}$ and $t_j' \in \{t_{j1}, \dots, t_{jn_j}\}$ $\forall 1 \leq j \leq k$ and $|R(t, t')| \neq 0$, where $|R(t, t')|$ is the Lebesgue measure of $R(t, t')$, i.e., $|R(t, t')| = \prod_{j=1}^{k}(t_j' - t_j)$. Note that $\tilde{B}$ is unique whereas $R(t, t')$ is not unique.

For any base B of a grid G and the corresponding subset $\tilde{B}$ tied to B, denote

$$\omega_B(f, \delta) = \sup\{|f(t) - f(t')|; (t, t') \in \tilde{B} \times \tilde{B}, \|t - t'\| \leq \delta\}$$
$$\text{for any } \delta > 0, \text{ where } \|t\| = \sup\{|t_j|, 1 \leq j \leq k\}. \tag{5.5}$$

Now for any bounded function f: $[0, 1]^k \to \mathbb{R}$, denote

$$\omega(f, \delta) = \sup\{|f(t) - f(t')|, (t, t') \in [0, 1]^k \times [0, 1]^k, \|t - t'\| \leq \delta\}. \tag{5.6}$$

Let P_n, $n \geq 1$ be a sequence of probability measures on $(D_k, \mathscr{D}_k)$, where $\mathscr{D}_k$ is the σ-field generated by the Skorohod topology (on D_k). We say that the sequence $\{G_n\}$ of grids with bases $\{B_n\}$ *accompanies* the measure P_n if and only if $\forall \varepsilon > 0$, $\exists \varepsilon' > 0$ and $\forall \delta \in [0, 1/2)$, $\exists N_0 \geq 1$ such that $P_n[f \in D_k; \omega(f, \delta) \geq \varepsilon$ and $\omega_{B_n}(f, 2\delta) < \varepsilon'\}] = 0$ $\forall n \geq N_0$.

For the ease of convenience we state the following lemma due to Balacheff and Dupont [3] which will be used in the sequel.

LEMMA 5.3. *Let ν be a positive finite measure on $[0,1]^k$ with continuous marginals. Let $\{P_n\}_{n\geq 1}$ be a sequence of probability measures on $(D_k, \mathscr{D}_k)$ such that $P_n[f \in D_k;\ f|[0,1]^k = 0] = 1\ \forall n \geq 1$ where $f|[0,1]^k$ means the restriction of f on the lower boundary of $[0,1]^k$. Suppose $\{G_n\}_{n\geq 1}$, a sequence of grids with bases $\{B_n\}_{n\geq 1}$, is asymptotically dense in $[0,1]^k$, and accompanies P_n. Furthermore, suppose that for any $R(t^{(n)}, t'^{(n)}) \in \mathscr{I}(k)$ tied to B_n,*

$$P_n\left[f \in D_k;\ |\Delta_{R(t^{(n)}, t'^{(n)})} f| > \lambda\right] \leq \lambda^{-\gamma}\left[\nu\left(R\left(t^{(n)}, t'^{(n)}\right)\right)\right]^{\beta},$$
$$\beta > 1, \gamma > 0. \quad (5.7)$$

Then, $\forall \varepsilon > 0$, $\exists \delta \in (0,1)$ and $N_0 \geq 1$ such that

$$P_n\left[f \in D_k;\ \omega(f, \delta) \geq \varepsilon\right] \leq \varepsilon\ \forall n \geq N_0. \quad (5.8)$$

Finally, we need the notion of μ-boundedness.

We shall say that the sequence $\{H_{ni}\}$ is μ-bounded if there exists a finite and positive measure μ on $[0,1]^k$ with continuous marginal distributions such that for every $1 \leq i \leq n$, $n \geq 1$, $H_{ni}(R(t,t')) \leq \mu(R(t,t'))$ for any rectangle $R(t,t')$ in $\mathscr{I}(k)$.

5.2. *Convergence of $\tilde{W}_n$*

THEOREM 5.1. *Assume that the sequence $\{X_{ni}\}$ is* (a) *φ-mixing with rates* (1.3) *or* (b) *strong mixing with rates* (1.4); *the sequence $\{H_{ni}\}$ is* (c) *μ-bounded, where μ is absolutely continuous with bounded density f_μ or* (d) *$\{H_{ni}\}$ has uniform marginals for all $n \geq 1$ and $1 \leq i \leq n$. Furthermore, assume that* (e) *the covariance function C_n of the empirical process W_n defined in* (5.2) *converges to a function C. Then, $\tilde{W}_n$ converges weakly in the Skorohod topology to a Gaussian process $\tilde{W}_0$ with trajectories a.s. in $\tilde{C}_k$.*

Proof. Let the probability measure $\tilde{Q}_n$ (resp. Q_n) on $(\tilde{D}_k, \tilde{\mathscr{D}}_k)$ (resp. $D_k, \mathscr{D}_k$)) be associated with $\tilde{W}_n$ (resp. W_n). To prove this theorem, we have to verify (4.12) and (4.13). Then following Withers [20] $\varphi_U(\tilde{Q}_n) \to$ weakly to a Gaussian measure $\tilde{Q}_U$ if (i) $C_n \to$ some function C, (ii) $\sum_{m\geq 1}\alpha(m) < \infty$, and (iii) $m^{1-a}\alpha([m^b]) \to 0$ (as $m \to \infty$), where $0 < 2b < a < 1 - b$. Now in our situation (i) holds by assumption (e); (ii) follows from (1.3) and (1.4); and (iii) from (1.3) and (1.4) by taking $a = 3/4 - \varepsilon/8$, $b = 1/4$ and ε sufficiently small (since taking $\alpha(m) = m^{-1-\varepsilon}$, $m^{1-a}\alpha([m^b]) \leq Am^{-\varepsilon/8}$, where $A > 0$ is some constant). Thus (4.13) is proved.

We now prove (4.12). Since $|\tilde{W}_n(B)| = |\Delta_B W_n|$, where Δ_B is defined in (5.1), we have

$$\sup\{|\tilde{W}_n(R(t,t')) - \tilde{W}_n(R(s,s'))|;\ d_H(R(t,t'),R(s,s')) \le \delta\}$$
$$\le \sup\{2k|W_n(t) - W_n(s)|;\ \|t-s\| \le \delta\}. \tag{5.9}$$

To prove (4.12), it suffices to prove (5.8) for W_n. (5.8) will follow if we prove (5.7).

Let $B_n = \{i/n;\ 0 \le i \le n\}^k$, $n \ge 1$, be a sequence of bases of grids G_n, $n \ge 1$. Note that G_n is asymptotically dense in $[0,1]^k$, and we prove that G_n accompanies Q_n. Now for every $t \in [0,1]^k$, let $(\underline{t}, \bar{t})$ be the points of $\tilde{B}_n$, where $\tilde{B}_n$ is tied to B_n such that $\underline{t} < t \le \bar{t}$ and $\|\bar{t} - \underline{t}\| \le 1/n$. Then, with the conditions (c) or (d) we obtain, after some computations, that $|W_n(t) - W_n(t')| \le 2kK(\mu)/\sqrt{n} + |W_n(\bar{t}) - W_n(\underline{t}')|$ $\forall t \in [0,1]^k$ and $\forall t' \in [0,1]^k$, where $K(\mu) = \sup_{t \in [0,1]^k} f_\mu(t)$ if we have (c), and $K(\mu) = 1$ if we have (d). Consequently, for every $\delta \in (0, 1/2]$, we have $\omega(W_n, \delta) \le 2kK(\mu)/\sqrt{n} + \omega_B(W_n, 2\delta)$. It follows that G_n accompanies Q_n. It remains to show that Q_n satisfies (5.7).

Suppose we have condition (c). Let $\sum_{m=1}^{\infty} m^{-1}\varphi^{1/4}(m) < \infty$ (implied by (1.3)) and let $R(t^{(n)}, t'^{(n)})$ be tied to B_n. Using Lemma 5.1, with $q = 2$, we obtain

$$E\left[\tilde{W}_n(R(t^{(n)}, t'^{(n)}))\right]^4$$
$$\le C_2(\varphi)\left[\left(K(\mu)\prod_{j=1}^{k}(t_j' - t_j)\right)^2 + n^{-1}\left(K(\mu)\prod_{j=1}^{k}(t_j' - t_j)\right)\right]. \tag{5.10}$$

Let $\nu = (C_2(\varphi)(K(\mu) + K^2(\mu)))^{\beta^{-1}} U^k$, where U^k is the uniform probability measure on $[0,1]^k$ and $\beta = 1 + k^{-1}$. Then, by the Markov inequality, we obtain

$$Q_n\left[f \in D_k;\ |\Delta_{R(t^{(n)}, t'^{(n)})} f| > \lambda\right] \le \lambda^{-4}\left[\nu(R(t^{(n)}, t'^{(n)}))\right]^{\beta}$$

which implies (5.7) for the φ-mixing case with rates (1.3). Equation (5.8) follows.

For the strong mixing case with rates (1.4), we use Lemma 5.2 for $q = 2$ and $\varepsilon < (2k+4)^{-1}$ and obtain

$$E\left[\tilde{W}_n(R(t^{(n)}, t'^{(n)}))\right]^4$$
$$\le C_2(\alpha)\left[K(\mu)\prod_{j=1}^{k}(t_j - t_j')\right)^{2(1-\varepsilon)} + n^{-1}\left(K(\mu)\prod_{j=1}^{k}(t_j - t_j')\right)^{1-\varepsilon}\right]$$

which (with $\beta = (1 - \varepsilon) + k^{-1}$) implies (5.7) and hence (5.8) by proceeding as above.

Now let us suppose we have condition (d). Then, for the φ-mixing case with rates $\sum_{m=1}^{\infty} m^{-1}\varphi^{1/2(k+1)}(m) < \infty$ (implied by (1.3)), we use Lemma 5.1 with $q = k + 1$, and obtain

$$E\left[\tilde{W}_n\left(R(t^{(n)}, t'^{(n)})\right)\right]^{2(k+1)} \le C_{k+1}(\varphi) \sum_{l=1}^{k+1} n^{-(k+1)}\left[n^{(k+1)/l} \prod_{j=1}^{k} (t_j' - t_j)^{(k+1)/kl}\right]$$

and proceeding as in the φ-mixing case dealt with above, we get the desired result.

For the strong mixing case with rates (1.4), we use Lemma 5.2 with $q = k + 1$ and $\varepsilon < 1/2(k + 2)$ and obtain

$$E\left[\tilde{W}_n\left(R(t^{(n)}, t'^{(n)})\right)\right]^{2(k+1)} \le C_{k+1}(\alpha) \sum_{l=1}^{k+1} n^{-(k+1-l)} \prod_{j=1}^{k} (t_j' - t_j)^{l(1-\varepsilon)/k}$$

and then proceed as above for the first φ-mixing case. The proof follows.

6. Convergence of the Weighted Empirical Process

We start with the definition of the weight function.

Definition 6.1. A function $r: [0, 1] \to \mathbb{R}^+$ is called a weight function if it satisfies the following conditions:

(i) r is continuous.

(ii) $r(u) = 0$ if $u = 0$ or $u = 1$.

We will consider a *modified empirical process* $\hat{W}_n$ defined as

$$\hat{W}_n(R(t,t)) = \begin{cases} \tilde{W}_n(R(t,t')) & \text{if } |R(t,t')| \ge n^{-1} \text{ and } |R(t,t')| \le 1 - n^{-1} \\ 0 & \text{otherwise,} \end{cases} \tag{6.1}$$

where $|R(t, t')|$ is defined in Section 5.

For any weight function r, we introduce a *weighted modified empirical process* $\hat{W}_n/r$ defined as

$$\frac{\hat{W}_n}{r}(R(t,t')) = \begin{cases} \hat{W}_n(R(t,t'))/r(|R(t,t')|) & \text{if } |R(t,t')| \neq 0 \\ 0 & \text{otherwise.} \end{cases} \tag{6.2}$$

Then, the following lemma is a slight variation of a proposition in Harel [9].

LEMMA 6.1. *For any $n \geq 1$, let Y_n be a process with values in $\tilde{D}_k$, and measurable with respect to $\tilde{\mathscr{D}}_k$. Suppose $Y_n \to Y_0$ in law, where Y_0 is a Gaussian process with trajectories a.s. continuous. Let P_n be the probability measure associated with Y_n. Let r be a weight function such that*

$$Y_n \cdot 1/r \text{ has trajectories a.s. in } \tilde{D}_k, \ (n \geq 1) \tag{6.3}$$

$$\begin{aligned} &\forall \varepsilon > 0, \exists \theta > 0, \\ &\exists N_0 \geq 1 \text{ such that } P_n[\sup\{|Y_n(R(t,t'))(1/r)(|R(t,t')|)|\} \geq \varepsilon] \\ &\quad \leq \varepsilon \ \forall n \geq N_0, \end{aligned} \tag{6.4}$$

where sup is over $R(t,t')$ with the condition that $|R(t,t')| \leq \theta$ or $1 - |R(t,t')| \leq \theta$. Then $Y_n \cdot 1/r$ converges weakly in Skorohod topology to the Gaussian process $Y_0 \cdot 1/r$ with trajectories a.s. in $\tilde{C}_k$.

THEOREM 6.1. *If the sequence $\{X_{ni}\}$ satisfies the assumptions of Theorem 5.2, then for any weight function r satisfying*

$$r(u) \geq A[u(1-u)]^{1/2-\delta}, \qquad A > 0, \tag{6.5}$$

where

$$0 < \tfrac{1}{2} - \delta < \tfrac{1}{8}k \qquad \textit{if the condition } (\mathrm{c}) \textit{ of Theorem 5.2 is satisfied} \tag{6.6}$$

or

$$0 < \frac{1}{2} - \delta < \frac{1}{2k(k+1)}$$

$$\textit{if the condition } (\mathrm{d}) \textit{ of Theorem 5.2 is satisfied.} \tag{6.7}$$

$\hat{W}_n/r$ converges weakly in the Skorohod topology to the Gaussian process $\tilde{W}_0/r$ with trajectories a.s. in $\tilde{C}_k$ (where $\tilde{W}_0$ is the same as in Theorem 5.2).

Proof. Convergence of $\hat{W}_n$ to $\tilde{W}_0$ follows from the definition of $\hat{W}_n$ and Theorem 5.2. The theorem will follow if (6.3) and (6.4) are satisfied. Equation (6.3) follows from the definition of $\hat{W}_n$. We now prove (6.4).

For any $\theta \in [0,1]$, let $C_\theta^{(1)}$ and $C_\theta^{(2)}$ be two subsets of $\mathscr{I}(k)$, where

$$C_\theta^{(1)} = \{R(t,t');\ |R(t,t')| \le \theta\}$$
$$C_\theta^{(2)} = \{R(t,t');\ |R(t,t')| \ge 1-\theta\}.$$

Equation (6.4) will follow if we show that $\forall \eta > 0$, $\exists \theta > 0$, $\exists N_0 \ge 1$ such that

$$P_n\left[\sup_{R(t,t')\in C_\theta^{(i)}} \left| \hat{W}_n(R(t,t')) \cdot \frac{1}{r(|R(t,t')|)} \right| > \eta \right] \le \eta$$
$$\forall n \ge N_0,\ i = 1,2. \quad (6.8)$$

First let us take $i = 1$. Without any loss of generality, let $r(u) = u^{1/2-\delta}$. Set (for any $n \ge 1$)

$$m_n = \max\left\{m;\ \frac{1}{n} \le \frac{1}{2^m},\ m \ge 0\right\} \tag{6.9}$$

and

$$m(\theta) = \max\{m;\ \theta \le 1/2^m,\ m \ge 0\}. \tag{6.10}$$

For any $p \ge 0$, consider a base B_p of some grid G_p, where

$$B_p = \left\{t_{ji};\ 1 \le i \le n_j^{(p)},\ 1 \le j \le k,\ t_{ji} = i/2^p,\ n_j^{(p)} = 2^p\right\}. \tag{6.11}$$

We need a few lemmas.

LEMMA 6.1. *Let $\theta \in [0,1]$ and suppose a function f: $[0,1] \to \mathbb{R}$ is given. Then, for any two points*

$$(u,v) \in \left\{0, \frac{1}{2^m}, \frac{2}{2^m}, \dots, \frac{2^m-1}{2^m}, 1\right\} \times \left\{0, \frac{1}{2^m}, \frac{2}{2^m}, \dots, \frac{2^m-1}{2^m}, 1\right\}$$

with $|u-v| \le 2^{-m(\theta)}$, where m is an integer $> m(\theta)$, the inequality

$$|f(u) - f(v)| \le 4 \sum_{r=m(\theta)}^{m} \sup |f(u_1 + 2^{-4}) - f(u_1)|,$$

holds, where the sup *is taken for all*

$$u_1 \in \left\{0, \frac{1}{2^r}, \frac{2}{2^r}, \dots, \frac{2^r-1}{2^r}, 1\right\} \quad \text{and} \quad u_1 + 2^{-r} \in [0,1].$$

Proof. Follows from Neuhaus [10, Lemma 5.1].

LEMMA 6.2. *Let $\theta \in [0,1]$ and suppose a function $f\colon [0,1]^k \to \mathbb{R}$ be given. Then, for any $(t,t') \in B_m \times B_m$ with $t < t'$ and $|R(t,t')| \le 2^{-m(\theta)}$, the inequality*

$$\frac{|\Delta_{R(t,t')}f|}{r(|R(t,t')|)} \le 4^k \sum_{\substack{r_1 \le m \\ r_1 + \cdots + r_k \ge m(\theta)}} \cdots \sum_{r_k \le m} \sup\left\{ \frac{|\Delta_{\prod_{j=1}(u_j, u_j + 2^{-r_j}]}f|}{\left(\prod_{j=1}^{k} 2^{-r_j}\right)^{1/2-\delta}} \right\} \tag{6.13}$$

holds, where the sup *is taken over all*

$$u_j \in \left\{0, \frac{1}{2^{r_j}}, \ldots, \frac{2^{r_j}-1}{2^{r_j}}\right\} \qquad \textit{and} \qquad u_j + 2^{-r_j} \in [0,1],\ 1 \le j \le k,$$

and where m is an integer $> m(\theta)$.

Proof. From Lemma 6.1, we deduce (by iteration)

$$|\Delta_{R(t,t')}f| \le 4^k \sum_{\substack{r_1 \le m \\ r_1 + \cdots + r_k \ge m(\theta)}} \cdots \sum_{r_k \le m} \sup\{|\Delta_{\prod_{j=1}^{k}(u_j, u_j+2^{-r_j}]}f|. \tag{6.14}$$

Since each term on the left side of (6.14) is less than or equal to a finite number of terms on the right side of (6.14) for which the Lebesgue measure $|R(t,t')| \ge \prod_{j=1}^{k} 2^{-r_j}$ and, since r is a nondecreasing function, we obtain (6.13).

LEMMA 6.3. $\forall \eta > 0, \exists \theta > 0, \exists N \ge 1$ *such that*

$$P_n\left[\sup_{\substack{R(t,t') \in \tilde{B}_{m_n} \cap C_\theta^{(1)} \\ |R(t,t')| \ne 0}} \left\{ \left| \tilde{W}_n(R(t,t')) \cdot \frac{1}{r(|R(t,t')|)} \right| \ge \eta \right\}\right] \le \eta\ \forall n \ge N. \tag{6.15}$$

Proof. See Appendix.

Now we prove (6.8) for $i = 1$. We assume that condition (c) of Theorem 5.2 holds. Let $\theta \in [0,1]$ and n be fixed. For any $R(t,t') \in \mathcal{I}(k)$ for which $|R(t,t')| \ge n^{-1}$, let $\underline{t}, \bar{t}, t'$, and $\bar{t}'$ be points of $\tilde{B}_{m_n}$ such that

$$\underline{t}_j \le t_j \le \bar{t}_j,\ \underline{t}_j' \le t_j' \le \bar{t}_j',\ \bar{t}_j - \underline{t}_j \le 2^{-m_n} \quad \text{and} \quad \bar{t}_j' - \underline{t}_j' \le 2^{-m_n},$$
$$1 \le j \le k$$

For any $(J, L) \subset \{1, \ldots, k\} \times \{1, \ldots, k\}$ for which $J \cap L = \phi$, we define an element $R(t, t')(J, L)$ of $\mathscr{I}(k)$ by

$$R(t,t')(J,L) = \prod_{j\in J}(\underline{t}_j, t_j] \prod_{j\in L}(t_j', \bar{t}_j'] \prod_{j\notin J\cup L}(\underline{t}_j, \bar{t}_j'].$$

Then, we have the inequality:

$$\left|\hat{W}_n(R(t,t')) \cdot \frac{1}{r(|R(t,t')|)}\right| \leq \sum_{\substack{(J,L)\subset\{1,\ldots,k\}\times\{1,\ldots,k\}\\ J\cap L=\phi}} \left|\tilde{W}_n(R(t,t')(J,L))\frac{1}{r(|R(t,t')|)}\right|. \quad (6.16)$$

If $J = L = \phi$, then we have the inequality

$$\left|\tilde{W}_n(R(t,t')(\phi,\phi))\frac{1}{r(|R(t,t')|)}\right| \leq 3^k\left|\tilde{W}_n(R(\underline{t},\bar{t}'))\frac{1}{r(|R(\underline{t},\bar{t}')|)}\right|. \quad (6.17)$$

If $J \cup L \neq \phi$, then we have the inequality

$$\left|\tilde{W}_n(R(t,t')(J,L))\frac{1}{r(|R(t,t')|)}\right| \leq \tilde{W}_n(G_k) \times \left.\frac{1}{r(|R(t,t')|)}\right| + n^{1/2}K(\mu)U^k(G_k)\cdot\frac{1}{r(|R(t,t')|)}, \quad (6.18)$$

where

$$G_k = \prod_{j\in J}(\underline{t}_j, \bar{t}_j] \prod_{j\in L}(\underline{t}_j', \bar{t}_j'] \prod_{j\notin J\cup L}(\underline{t}_j, \bar{t}_j'].$$

Equation (6.18) follows due to the fact that for any $(B_1, B_2) \in \mathscr{I}(k) \times \mathscr{I}(k)$, where $B_2 \subset B_1$,

$$|\tilde{W}_n(B_2)| \leq |\tilde{W}_n(B_1)| + n^{1/2}\mu(B_1). \quad (6.19)$$

As r is non-decreasing, we deduce

$$\left|\tilde{W}_n(R(t,t')(J,L))\frac{1}{r(|R(t,t')|)}\right|$$
$$\le 3^k|\tilde{W}_n(G_k)|\times(r(|G_k|))^{-1}$$
$$+n^{1/2}\Big(K(\mu)\prod_{j\in J}(\bar{t}_j-\underline{t}_j)\prod_{j\in L}(\bar{t}_j-\underline{t}'_j)\Big)^{(1/2+\delta)}\}. \quad (6.20)$$

Now

$$n^{1/2}\Big(\prod_{j\in J}(\bar{t}_j-\underline{t}_j)\prod_{j\in L}(\bar{t}'_j-\underline{t}'_j)\Big)^{(\delta+1/2)}=n^{1/2}(2^{-m_n})^{(\delta+1/2)l}\to 0 \quad (6.21)$$

as $n\to\infty$, where $l=\text{Card } J\cup L$.

Using (6.16), (6.17), (6.20), and (6.21), we obtain

$$\sup_{R(t,t')\in C_\theta^{(1)}}\left|\hat{W}_n(R(t,t'))\cdot\frac{1}{r(|R(t,t')|)}\right|$$
$$\le K\sup_{\substack{R(t,t')\in\bar{B}_m\cap C_\theta^{(1)}\\ |R(t,t')|\neq 0}}\left|\tilde{W}_n(R(t,t'))\cdot\frac{1}{r(|R(t,t')|)}\right|$$
$$+O\Big(n^{1/2}(2^{-m_n})^{(\delta+1/2)l}\Big), \quad (6.22)$$

where $K>0$ is some constant.

Using (6.22) along with Lemma 6.3, we obtain (6.8) for $i=1$ when condition (c) of Theorem 5.2 is satisfied. The proof when condition (d) of Theorem 5.2 is satisfied is essentially similar and is therefore omitted.

Now we prove (6.8) for $i=2$. Without loss of generality, we take $r(u)=(1-u)^{1/2-\delta}$. Following the ideas of Einmahl, Ruymgaart, and Wellner [7], we start with the equality $\tilde{W}_n(B)=-\tilde{W}_n(B^*)$, where B^* is the complement of B and $\tilde{W}_n(B^*)=n^{-1/2}\sum_{i=1}^n[I_{[F_n(X_{ni})\in B^*]}-H_{ni}(B^*)]$, where $F_n=(F_n^{(1)},\dots,F_n^{(k)})$. For any $R(t,t')\in\mathscr{I}(k)$ with $|R(t,t')|\le 1-n^{-}$ and using the union–intersection principle (see Einmahl, Ruymgaart, and Wellner [7]), we obtain

$$|\tilde{W}_n(R(t,t'))|\le\sum_{l\in L}|\tilde{W}_n(R_l(t,t'))|, \quad (6.23)$$

where L is a finite index set, $R_l(t,t')$ is an element of $\mathscr{I}(k)$ with the condition that there exists $J\subset\{1,\dots,k\}$ and $M\subset\{1,\dots,k\}$ with

$J \cap M = \phi$ and $J \cup M \neq \phi$, such that,

$$R_l(t,t') = \prod_{j \in J} (0, t_j] \cdot \prod_{j \in M} (t'_j, 1] \cdot \prod_{j \notin J \cup M} (0,1].$$

From (6.23), we obtain

$$\sup_{R(t,t') \in C_\theta^{(2)}} \left| \hat{W}_n(R(t,t')) \frac{1}{r(|R(t,t')|)} \right|$$

$$\leq \sum_{l \in L} \left| \tilde{W}_n(R_l(t,t')) \frac{1}{\left(|R_l(t,t')| V n^{-1}\right)^{1/2-\delta}} \right. \tag{6.24}$$

where $a \vee b = \max(a,b)$.

If $|R_l(t,t')| \geq n^{-1}$, then $R_l(t,t') \in C_\theta^{(1)}$ and from (6.8) for $i = 1$, we obtain the desired result for $\tilde{W}_n(R_l(t,t')) \cdot (r(|R(t,t')|))^{-1}$.

If $R_l(t,t')| < n^{-1}$, then using the inequality (6.19), we obtain

$$\left| \tilde{W}_n(R_l(t,t')) \cdot \frac{1}{(n^{-1})^{1/2-\delta}} \right| \leq \left| \tilde{W}_n\left(R_l^{(n)}(t^{(n)}, t'^{(n)})\right) \frac{1}{(n^{-1})^{1/2-\delta}} \right|$$
$$+ n^{1/2} \mu\left(R_l^{(n)}(t^{(n)}, t'^{(n)})\right) / (n^{-1})^{1/2-\delta}, \tag{6.25}$$

where $R_l^{(n)}(t^{(n)}, t'^{(n)}) \in \mathscr{I}(k)$, $|R_l^{(n)}(t^{(n)}, t'^{(n)})| = n^{-1}$ and $R_l(t,t') \subset R_l^{(n)}(t^{(n)}, t'^{(n)})$.

Since $R_l^{(n)}(t^{(n)}, t'^{(n)}) \in C_\theta^{(1)}$, we obtain the desired result from (6.8) for $i = 1$ for $\tilde{W}_n(R_l^{(n)}(t^{(n)}, t'^{(n)})) \cdot 1/r(|R_l^{(n)}(t^{(n)}, t'^{(n)})|)$. Since $|R_l^{(n)}(t,t')| = n^{-1}$,

$$n^{1/2} \frac{\mu\left(R_l^{(n)}(t^{(n)}, t'^{(n)}\right)}{(n^{-1})^{1/2-\delta}} \leq K(\mu) \frac{n^{1/2} \cdot n^{-1}}{(n^{-1})^{1/2-\delta}} = K(\mu) n^{-\delta} \to 0$$

as $n \to \infty$.

Using (6.24), and the properties of $\tilde{W}_n(R_l(t,t')) \cdot 1/(|R_l(t,t')| v n^{-1})^{1/2-\delta}$ for each $l \in L$ obtained in the discussion following (6.24), we obtain the desired result, viz. (6.8) for $i = 2$. The proof follows.

Appendix

Proof of Lemma 6.3. (a) Let us suppose that $\{X_{ni}\}$ is φ-mixing with rates (1.3) and $\{H_{ni}\}$ is μ-bounded (viz. conditions (a) and (c) of Theorem 5.1).

From Lemma 6.2, we obtain if $m(\theta) \le m_n$ that

$$\sup_{\substack{R(t,t')\in \bar{B}_{m_n}\cap C_\theta^{(1)} \\ |R(t,t')|\neq 0}} \left| \tilde{W}_n(R(t,t')) \cdot \frac{1}{r(|R(t,t')|)} \right|$$

$$\le 4^k \sum_{\substack{r_1 \le m_n \\ r_1 + \cdots + r_k \ge m(\theta)}} \cdots \sum_{r_k \le m_n} \sup \left| \tilde{W}_n\left(\prod_{j=1}^{k} \left(u_j, u_j + 2^{-r_j}\right] \right) \right.$$

$$\left. \times \frac{1}{\left(\prod_{j=1}^{k} 2^{-r_j} \right)^{1/2-\delta}} \right|, \quad \text{(A.1)}$$

where the sup is taken as in (6.13). Let a be a real number such that

$$0 < a < 1 \qquad \text{and} \qquad a^4 \cdot 2^{1/k - 4(1/2-\delta)} > 1. \qquad \text{(A.2)}$$

If the left side of (A.1) exceeds η, then $\exists (r_1, \dots, r_k) \in \{1, \dots, m_n\} \times \cdots \times \{1, \dots, m_n\}$ with $r_1 + \cdots + r_k \ge m(\theta)$ such that

$$\sup \left\{ \left| \tilde{W}_n(A_k) \cdot \frac{1}{\left(\prod_{j=1}^{k} 2^{-r_j} \right)^{1/2-\delta}} \right| \right\} > B_k, \qquad \text{(A.3)}$$

where

$$A_k = \prod_{j=1}^{k} \left(u_j, u_j + 2^{-r_j}\right], \qquad B_k = \frac{a^{\sum_{i=1}^{k} r_i - m(\theta)}}{4^k B_k^{(1)}} \eta$$

and

$$B_k^{(1)} = \sum_{r_1 + \cdots + r_k \ge m(\theta)} a^{\sum_{i=1}^{k} r_i - m(\theta)} = O\left(m^{k-1}(\theta)\right).$$

From (A.3), we deduce

$$P_n\left[\sup_{\substack{|R(t,t')|\neq 0\\ R(t,t')\in B_{m_n}\cap C_\theta^{(1)}}}\left\{\left|\tilde{W}_n(R(t,t'))\frac{1}{r(|R(t,t')|)}\right|\geq\eta\right\}\right]$$

$$\leq \sum_{\substack{r_1\leq m_n\\ r_1+\cdots+r_k\geq m(\theta)}}\cdots\sum_{r_k\leq m_n}\sum_{u_1=0}^{2^{r_1}-1}\cdots\sum_{u_k=0}^{2^{r_k}-1}$$

$$\times P_n\left[\left|\tilde{W}_n(A_k)\frac{1}{\left(\prod_{j=1}^{k}2^{-r_j}\right)^{1/2-\delta}}\right|>B_k\right]$$

$$\leq \sum_{\substack{r_1\leq m_n\\ r_1+\cdots+r_k\geq m(\theta)}}\cdots\sum_{r_k\leq m_n}\sum_{u_1=0}^{2^{r_1}-1}\cdots\sum_{u_k=0}^{2^{r_k}-1}(B_k^{-4})$$

$$\times E\left[\tilde{W}_n(A_k)\left(\prod_{j=1}^{k}2^{-r_j}\right)^{\delta-1/2}\right]^4 \qquad (A.4)$$

by the Markov inequality.

Using Lemma 5.1 for $q = 2$, the right side of (A.4) is less than or equal to

$$\sum_{\substack{r_1\leq m_n\\ r_1+\cdots+r_k\geq m(\theta)}}\cdots\sum_{r_k\leq m_n}K_1(B_k)^{-4}\left(2^{-\sum_{j=1}^k r_j}\right)^{1/k-4(1/2-\delta)},$$

$K_1>0$ a constant,

$$\leq\left(\frac{4^k}{\eta}B_k^{(1)}\right)^4\sum_{\substack{r_1\leq m_n\\ r_1+\cdots+r_k\geq m(\theta)}}\cdots\sum_{r_k\leq m_n}b^{-m(\theta)}\cdot(ba^4)^{m(\theta)-\sum_{j=1}^k r_j},$$

$b = 2^{1/k-4(1/2-\delta)} > 1,$

$$\leq K_1\left(\frac{4^k}{\eta}B_k^{(1)}\right)^4 B_k^{(2)}b^{-m(\theta)},$$

$$\text{where } B_k^{(2)} = \sum_{r_1+\cdots+r_k\geq m(\theta)}(ba^4)^{-\sum_{i=1}^k r_i+m(\theta)}. \qquad (A.5)$$

Jsing (A.2), we note that $B_k^{(2)} = O(m^{k-1}(\theta))$ and, since $(m^{k-1}(\theta))^5\cdot$ $\,^{-m}(\theta) \to \infty$ as $\theta \to 0$ $(m(\theta) \to \infty$ as $\theta \to 0)$, it follows that the extreme side of (A.5) $\to 0$ as $\theta \to 0$, which implies that $n \to \infty$. This proves (6.15) with conditions (a) and (c) of Theorem 5.2.1.

(b) Let $\{X_{ni}\}$ be strong mixing with rates (1.4), and let $\{H_{ni}\}$ be μ-bounded (viz. conditions (b) and (c) of Theorem 5.1). Choose a real number a such that

$$0 < a < 1 \qquad \text{and} \qquad a^4 \cdot 2^{1/k - \varepsilon - 4(1/2 - \delta)} > 1. \tag{A.6}$$

Now using Lemma 5.2 for $q = 2$, and proceeding as above, we obtain

$$P_n\left[\sup_{\substack{R(t,t') \in \tilde{B}_{m_n} \cap C_\theta^{(1)} \\ |R(t,t')| \neq 0}} \left| \tilde{W}_n\left(R(t,t') \cdot \frac{1}{r(|R(t,t')|)}\right| \geq \eta\right]$$

$$\leq K_2\left(\frac{4^k}{\eta} B_k^{(1)}\right)^4 B_k^{(2)} b^{-m(\theta)}, \tag{A.7}$$

where $b = 2^{1/k - \varepsilon - 4(1/2 - \delta)}$ and $K_2 > 0$ is a constant. Using (A.7) and arguing as in (A.5), we find that the right side of (A.7) $\to 0$ as $\theta \to 0$ (and therefore $n \to \infty$). This proves (6.15) with conditions (b) and (c) of Theorem 5.1.

(c) Let $\{X_{ni}\}$ be φ-mixing with rates (1.3) and let $\{H_{ni}\}$ have uniform marginals (viz. conditions (a) and (d) of Theorem 5.1).

Choose a real number a such that

$$1 < a < 1 \qquad \text{and} \qquad a^{2(k+1)} 2^{1/k - 2(k+1)(1/2 - \delta)} > 1. \tag{A.8}$$

(This is always possible since from (6.7), $1/k - 2(k+1)(\frac{1}{2} - \delta) > 0$, and so we choose $a \in (0, 1)$ such that (A.8) is satisfied.) Using Lemma 5.1 for $q = k + 1$ and proceeding as above, we note that the left side of (A.7) $\leq K_3((4^k/\eta) B_k^{(1)})^{2k+2} B_k^{(3)} b^{-m(\theta)}$, where

$$B_k^{(3)} = \sum_{r_1 + \cdots + r_k \geq m(\theta)} \left(ba^{2k+2}\right)^{-\sum_{i=1}^k r_i + m(\theta)} = O\left(m^{k-1}(\theta)\right),$$

$$b = 2^{1/k - 2(k+1)(1/2 - \delta)}$$

and $K_3 > 0$ is some constant. Using (A.8) and arguing as in (A.5), we prove (6.15) with conditions (a) and (d) of Theorem 5.1.

(d) Finally, let $\{X_{ni}\}$ be strong mixing with rates (1.4) and let $\{H_{ni}\}$ have uniform marginals (viz. conditions (b) and (d) of Theorem 5.1). Here choosing $0 < a < 1$ and $a^{2(k+2)} 2^{(2 - \varepsilon(k+1))k^{-1} - 2(k+2)(1/2 - \delta)} > 1$ using Lemma 5.2 for $q = k + 2$, and proceeding as in (b), we get (6.15). This proves Lemma 6.3.

1. I. Ahmad and P. E. Lin, On the Chernoff–Savage theorem for dependent random sequences. *Ann. Inst. Statist. Math* **32** (1980), 211–222.
2. K. S. Alexander, "Some Limit Theorems for Weighted and Non-identically Distributed Empirical Process," Ph.D. thesis, MIT, 1982.
3. S. Balacheff and G. Dupont, Normalité asymptotique des processus empiriques tronqués et des processus de range, in "Lecture Notes in Mathematics," Vol. 821, pp. 19–45, Springer-Verlag, New York/Berlin, 1980.
4. R. F. Bass and R. Pyke, The space $D(A)$ and weak convergence for set-indexed processes, *Ann. Probab.* **13** (1985), 860–884.
5. P. Billingsley, "Convergence of Probability Measures," Wiley, New York, 1968.
6. P. Doukhan and F. Portal, Principle d'invariance faible pour la fonction de répartition empirique dans un cadre multidimensionnel et mélangeant, *Probab. Math. Statist.* **8** Fasc. 2 (1987).
7. J. H. J. Einmahl, R. H. Ruymgaart, and J. A. Wellner, Criteria for weak convergence of weighted multivariate empirical processes indexed by points or rectangles, *Acta. Sci. Math. (Szeged)* (1984).
8. T. R. Fears and K. Mehra, Weak convergence of a two sample empirical process and a Chernoff–Savage theorem for ϕ mixing sequence, *Ann. Statist.* **2**, *No.* 3 (1974), 586–596.
9. M. Harel, Convergence en loi pour la topologie de Skorohod du processus empirique multidimensionnel normalisé tronqué et semi-corrigé, *in* "Lecture Notes in Mathematics," Vol. 821, pp. 46–85, Springer-Verlag, New York/Berlin, 1980.
10. G. Neuhaus, On weak convergence of stochastic processes with multidimensional time parameters, *Ann. Math. Statist.* **42** (1971), 1285–1295.
11. G. Neuhaus, Convergence of the reduced empirical process for non i.i.d. random vectors, *Ann. Statist.* **3** (1975), 528–531.
12. N. Neumann, "Ein Schwaches Invarianzprinzip für den gewichteten empirischen Prozess von gleichmässig mischenden Zufallsvariablen," Ph.D. thesis, Göttingen, West Germany, 1982.
13. R. Pyke and G. R. Shorack, Weak convergence of a two sample empirical process and a new approach to Chernoff-Savage theorems. *Ann. Math. Statist.* **39** (1968), 755–771.
14. L. Rüschendorf, On the empirical process of multivariate, dependent random variables. *J. Multivariate Anal.* **4** (1974), 469–478.
15. L. Rüschendorf, Asymptotic distributions of multivariate rank order statistics, *Ann. Statist.* **4** (1976), 912–923.
16. F. M. Ruymgaart and J. A. Wellner, Some properties of weighted multivariate empirical processes, *Statist. Decisions* **2** (1984), 199–223.
17. G. R. Shorack and J. A. Wellner, Limit theorems and inequalities for the uniform empirical process indexed by intervals, *Ann. Probab.* **10** (1982), 639–652.
18. A. V. Skorohod, Limit theorems for stochastic processes, *Theory Probab. Appl.* **1** (1956), 289–319.
19. M. L. Straf, Weak convergence of stochastic processes with several parameters, *in* "Proceedings, Sixth Berkeley Symp. Math. Statist. Probab. *2*, pp. 187–221, Univ. of California Press, Berkeley, 1972.
20. C. S. Withers, Convergence of empirical processes of mixing rv's on [0, 1], *Ann. Statist.* **3** (1975), 1101–1108.
21. M. Harel and Madan L. Puri, Weak convergence of serial rank statistics under dependence with applications in time series and Markov processes, *Ann. Probab.* **18** (1990), 1361–1387.

WEAK INVARIANCE OF THE MULTIDIMENSIONAL RANK STATISTIC FOR NONSTATIONARY ABSOLUTELY REGULAR PROCESSES

MICHEL HAREL and MADAN L. PURI*

Institut Universitaire de Technologie de Limoges, U.A. C.N.R.S. Orsay, France and Indiana University

In this paper, we study the weak invariance of the multidimensional rank statistic when the underlying random variables are nonstationary absolutely regular.

1. Introduction

Let $\mathbf{X}_{ni} = (X_{ni}^{(1)},..., X_{ni}^{(k)})$, $1 \leqslant i \leqslant n$, $n \geqslant 1$ be $\mathbb{R}^k$-valued rv's (random variables) with continuous df's (distribution functions) $F_{ni}(\mathbf{x})$, $\mathbf{x} \in \mathbb{R}^k$ and continuous marginal distribution functions $F_{ni}^{(j)}(x_j)$, $x_j \in \mathbb{R}$, $1 \leqslant i \leqslant n$, $n \geqslant 1$, $1 \leqslant j \leqslant k$. Our aim is to establish a weak invariance theorem of the rank statistic

$$\mathcal{S}_{n,m} = \sum_{i=1}^{m} c_{ni} a_{n,m}(R_{n,m,i}^{(1)},..., R_{n,m,i}^{(k)}), \tag{1.1}$$

where the c_{ni} are the regression constants defined by a continuous function $h(x)$ on $[0, 1]$ as

$$c_{ni} = h\left(\frac{i}{n}\right), \qquad 1 \leqslant i \leqslant n, n \geqslant 1 \tag{1.2}$$

the $a_n(\cdots)$ are the scores defined by a continuous function J on $[0, 1]^k$ as

$$a_{n,m}(i_1,..., i_k) = J\left(\frac{i_1}{m+1},..., \frac{i_k}{m+1}\right), (i_1,..., i_k) \in \{1,..., m\}^k$$

Received June 12, 1990; revised August 16, 1990.

AMS 1980 subject classification: 60F17.

Key words and phrases. Weak invariance, rank statistics, absolutely regular processes nonstationary random variables.

* Research supported by the Office of Naval Research Contracts N00014-85-K-0648 and N00014-91-J-1020.

and $R^{(j)}_{n,m,i}$ is the rank of $X^{(j)}_{ni}$ among $\{X^{(j)}_{nl}, 1 \leqslant l \leqslant m\}$. We assume that the underlying rv's are absolutely regular with rates

$$\beta(m) = O(n^{-4-\varepsilon}), \qquad \varepsilon > 0. \tag{1.3}$$

Recall that the sequence $\{\mathbf{X}_{ni}\}$ is absolutely regular if

$$\sup_{n \geqslant 1} \max_{1 \leqslant j \leqslant n-m} E\{ \sup_{A \in \sigma(\mathbf{X}_{ni},\, i \geqslant j+m)} |P(A \mid \sigma(\mathbf{X}_{ni}, 1 \leqslant i \leqslant j)) - P(A)| \} = \beta(m) \downarrow 0.$$

Here $\sigma(\mathbf{X}_{ni}, 1 \leqslant i \leqslant j)$ and $\sigma(\mathbf{X}_{ni}, i \geqslant j+m)$ are the σ-fields generated by $(\mathbf{X}_{n1}, ..., \mathbf{X}_{nj})$ and $(\mathbf{X}_{n,j+m}, \mathbf{X}_{n,j+m+1}, ..., \mathbf{X}_{nn})$, respectively.

Also recall that $\{\mathbf{X}_{ni}\}$ satisfies the strong mixing condition if

$$\sup_{n \geqslant 1} \max_{1 \leqslant j \leqslant n-m} [\sup\{|P(A \cap B) - P(A)\, P(B)|; A \in \sigma(\mathbf{X}_{ni}, 1 \leqslant i \leqslant j),$$
$$B \in \sigma(\mathbf{X}_{ni}, i \geqslant j+m)\}] = \alpha(m) \downarrow 0.$$

Since $\alpha(m) \leqslant \beta(m)$, it follows that if $\{\mathbf{X}_{ni}\}$ is absolutely regular, then it is also strong mixing.

For the special case $m = n$, the asymptotic distribution of the statistic $\mathscr{S}_{n,m}$ was studied by Ruymgaart and van Zuijlen [8] when the random variables $\mathbf{X}_{ni}$ are independent, and by Harel [5] when the random variables $\mathbf{X}_{ni}$ are φ-mixing. Here, in this paper, by adapting some of the ideas of Yoshihara [11], we prove a weak invariance theorem for the statistic $\mathscr{S}_{n,m}$ when the random variables $\mathbf{X}_{ni}$ are nonstationary absolutely regular.

2. Weak Invariance of the Rank Statistic $\mathscr{S}_{n,m}$

For any $\mathbf{x} = (x_1, ..., x_k) \in \mathbb{R}^k$, define $F_n(\mathbf{x})$ as $F_n(\mathbf{x}) = n^{-1} \sum_{i=1}^n F_{ni}(\mathbf{x})$ and for any j $(1 \leqslant j \leqslant k)$, $F^{(j)}_n(x_j) = n^{-1} \sum_{i=1}^n F^{(j)}_{ni}(x_j)$. For a score function $J(t_1, ..., t_k)$ which is square integrable, put

$$\mu^*_{ni} = \mu_J(F_{ni}) = \int_{\mathbb{R}^k} J(F^{(1)}_n(x_1), ..., F^{(k)}_n(x_k))\, dF_{ni}(x_1, ..., x_k). \tag{2.1}$$

For h defined in (1.2), put

$$\mu_{n,m} = \sum_{i=1}^{m} c_{ni} \mu^*_{n,i} = \sum_{i=1}^{m} h\left(\frac{i}{n}\right) \mu^*_{n,i}, \qquad m \leqslant n. \tag{2.2}$$

For any sequence of df's $\{\mathbf{F}_l\}$ on $(\mathbb{R}^k)^2$ with marginals F on $\mathbb{R}^k$ and assuming that J admits a derivative for each coordinate, we denote

$$\sigma^2(\{\mathbf{F}_l\}) = \lim_{n\to\infty} \left\{ \int_{\mathbb{R}^k} g^2(\mathbf{x})\, dF(\mathbf{x}) + 2 \sum_{l=1}^{n} \int_{(\mathbb{R}^k)^2} g(\mathbf{x})\, g(\mathbf{y})\, d\mathbf{F}_l(\mathbf{x}, \mathbf{y}) \right\} \tag{2.3}$$

if the limit exists where

$$g(\mathbf{x}) = \sum_{j=1}^{k} \int_{\mathbb{R}^k} (I_{[x_j \leqslant y_j]} - F^{(j)}(y_j)) \frac{\partial J}{\partial u_j} (F^{(1)}(y_1), ..., F^{(k)}(y_k))\, dF(\mathbf{y})$$

$$+ J(F^{(1)}(x_1), ..., F^{(k)}(x_k)) - \int_{\mathbb{R}^k} J(F^{(1)}(y_1), ..., F^{(k)}(y_k))\, dF(\mathbf{y}), \tag{2.4}$$

where $\mathbf{x} = (x_1, ..., x_k)$, $\mathbf{y} = (y_1, ..., y_k)$, and $F^{(j)}$ is the marginal df of F, $1 \leqslant j \leqslant k$.

For every $n \geqslant 1$, let

$$V_n(s) = (\sigma n)^{-1/2} [A_{n,[ns]} + (ns - [ns])\, A_{n,[ns]+1}], \tag{2.5}$$

where σ is the positive constant defined in (2.3), $A_{ni} = \mathscr{S}_{n,i} - \mu_{n,i}$, and $[ns]$ is the integral part of ns. The process $V_n(s) = \{V_n(s), 0 \leqslant s \leqslant 1\}$ belongs to the space C_1 of all continuous functions on $[0, 1]$ on which we associate the usual uniform metric.

Let $F_{n,i,j}$ be the df of $(\mathbf{X}_{ni}, \mathbf{X}_{nj})$. Then we have

THEOREM 2.1. *Suppose the sequence $\{\mathbf{X}_{ni}\}$ is absolutely regular with rate* (1.3). *Furthermore assume that for any $l \in \mathbb{N}^*$ with $l > 1$, there exists a continuous df $\mathbf{F}_l$ on $(\mathbb{R}^k)^2$ with continuous marginals F on $\mathbb{R}^k$ and continuous marginals $F^{(j)}$ on $\mathbb{R}$, $1 \leqslant j \leqslant k$ such that*

$$\lim_{n\to\infty} \max_{1 \leqslant i < j \leqslant n} |F_{n,i,j}(F_n^{(1)^{-1}}(t_1), ..., F_n^{(k)^{-1}}(t_k), F_n^{(1)^{-1}}(t'_1), ..., F_n^{(k)^{-1}}(t'_k))$$

$$- \mathbf{F}_{j-i}(F^{(1)^{-1}}(t_1), ..., F^{(k)^{-1}}(t_k), F^{(1)^{-1}}(t'_1), ..., F^{(k)^{-1}}(t'_k))| = 0 \tag{2.6}$$

and

$$\max_{1 \leqslant i \leqslant n} \sup_{\mathbf{x} \in \mathbb{R}^k} |F_{ni}(\mathbf{x}) - F(\mathbf{x})| = O(n^{-\gamma}), \text{ where } \gamma > 1. \tag{2.7}$$

Let J be a score function having bounded second derivatives. If $\sigma^2(\{\mathbf{F}_l\})$ defined in (2.3) *is strictly positive, then V_n defined in* (2.5) *converges weakly in the uniform topology on C_1 to the process $V_0 = \{V_0(s), 0 \leqslant s \leqslant 1\}$, where*

$$V_0(s) = \int_0^s h(u)\, dW(u), \qquad 0 \leqslant s \leqslant 1$$

and $W = \{W(s), 0 \leqslant s \leqslant 1\}$ is a standard Brownian motion process and $\sigma^2(\{\mathbf{F}_l\}) < \infty$.

Remark. Let the sequence $\{\mathbf{X}_{ni}\}$ be absolutely regular, and let the sequence of distribution functions $\{F_{n,i,j}\}$ satisfy the following conditions:

There exists a sequence $\{\mathbf{F}_l\}$ of distribution functions defined on $(\mathbb{R}^k)^2$ such that

$$\lim_{n\to\infty} \max_{1\leqslant i<j\leqslant n} |F_{n,i,j}(\mathbf{x}_1, \mathbf{x}_2) - \mathbf{F}_{j-i}(\mathbf{x}_1, \mathbf{x}_2)| = 0 \qquad \text{for all } (\mathbf{x}_1, \mathbf{x}_2) \in (\mathbb{R}^k)^2 \tag{2.8}$$

and

$$F_{ni} = F_n \qquad \text{for all } 1 \leqslant i \leqslant n,\ n \geqslant 1 \tag{2.9}$$

then the condition (2.6) is satisfied, and moreover, the condition (2.7) is not necessary for the proof of Theorem 2.1.

3. Preliminary Lemmas

At first we give a lemma from Yoshihara [11] which we will only use in the proof of Theorem 2.1 (Section 4). Let $p \geqslant 1$ and $1 \leqslant i_1 < i_2 < \cdots < i_p \leqslant n$ be arbitrary integers. For any j $(1 \leqslant j \leqslant p-1)$, let $P_{j,n}^{(i_1,\dots,i_p)}$ be the probability measure defined by $P_{j,n}^{(i_1,\dots,i_p)}(A^{(j)} \times B^{(p-j)}) = P[(\mathbf{X}_{ni_1}, \dots, \mathbf{X}_{ni_j}) \in A^{(j)}]$ $P[(\mathbf{X}_{ni_{j+1}}, \dots, \mathbf{X}_{ni_p}) \in B^{(p-j)}]$ and $P_{0,n}^{(i_1,\dots,i_p)}$ be the probability beasure defined by

$$P_{0,n}^{(i_1,\dots,i_p)}(A^{(p)}) = P[(\mathbf{X}_{ni_1}, \dots, \mathbf{X}_{ni_p}) \in A^{(p)}]$$

for any $A^{(j)} \in \sigma(\mathbf{X}_{ni_1}, \dots, \mathbf{X}_{ni_j})$, $1 \leqslant j \leqslant p$, and $B^{(p-j)} \in \sigma(\mathbf{X}_{ni_{j+1}}, \dots, \mathbf{X}_{ni_p})$, $1 \leqslant j \leqslant p-1$.

Lemma 3.1 (Yoshihara, [11]). *For every $p \geqslant 1$ and $(i_1, \dots, i_p)$ such that $i_1 < i_2 < \cdots < i_p$ and any j $(0 \leqslant j \leqslant p-1)$, let $h(\mathbf{x}_1, \dots, \mathbf{x}_p)$ be a Borel function such that*

$$\int_{\mathbb{R}^p} |h(\mathbf{x}_1, \dots, \mathbf{x}_p)|^{1+\delta}\, dP_{j,n}^{(i_1,\dots,i_p)} \leqslant M$$

for some $\delta > 0$, then

$$\left| \int_{\mathbb{R}^p} h(\mathbf{x}_1, \dots, \mathbf{x}_p)\, dP_{0,n}^{(i_1,\dots,i_p)} - \int_{\mathbb{R}^p} h(\mathbf{x}_1, \dots, \mathbf{x}_p)\, dP_{j,n}^{(i_1,\dots,i_p)} \right| \leqslant 4M^{1/(2+\delta)} \beta^{\delta/(1+\delta)}(i_{j+1} - i_j). \tag{3.1}$$

As a special case, if $h(\mathbf{x}_1, \dots, \mathbf{x}_p)$ is bounded, say $|h(\mathbf{x}_1, \dots, \mathbf{x}_p)| \leqslant M$, then we can replace the right side of (3.1) *by* $2M\beta(i_{j+1} - i_j)$.

Now we give preliminary results to the proof of Theorem 2.1.

For any n $(n \geqslant 1)$ and for any i $(1 \leqslant i \leqslant n)$ let

$$A^*_{ni} = \sum_{j=1}^{k} \int_{\mathbb{R}^k} (I_{[X^{(j)}_{ni} \leqslant x_j]} - F^{(j)}_{ni}(x_j)) \frac{\partial J}{\partial u_j} (F^{(1)}_n(x_1), ..., F^{(k)}_n(x_k))\, dF_{ni}(x_1, ..., x_k)$$
$$+ J(F^{(1)}_n(X^{(1)}_{ni}), ..., F^{(k)}_n(X^{(k)}_{ni})) - E(J(F^{(1)}_n(X^{(1)}_{ni}), ..., F^{(k)}_n(X^{(k)}_{ni}))). \tag{3.2}$$

It is obvious that $E(A^*_{ni}) = 0$.

Consider the process $L_n(t)$ defined on C_1 by

$$L_n(t) = n^{-1/2} \left(\sum_{i=1}^{[nt]} A^*_{ni} + (nt - [nt])\, A^*_{n,[nt+1]} \right). \tag{3.3}$$

LEMMA 3.2. *Suppose that* $\{\mathbf{X}_{ni}\}$ *satisfies the conditions of Theorem* 2.1. *Then the process* $L_n(t)$ *converges weakly in uniform topology to a Gaussian process* $L_0(t)$ *with trajectories a.s. in* C_1 *with mean* 0 *and variance* $t\sigma^2(\{\mathbf{F}_l\})$, *where* $\sigma^2(\{\mathbf{F}_l\})$ *is defined in* (2.3).

Proof. The process L_n defines a probability measure P_n on C_1. From Theorem 8.1 of Billingsley [2], we have to prove that (i) the finite dimensional distributions of P_n converge in law to normal distributions and (ii) P_n is tight.

First we prove (i) which is equivalent to proving that $\sum_{l=1}^{p} \lambda_l L_n(t_l)$ converges in law to a normal distribution for any $p \in \mathbb{N}^*$, any $t_l \in [0, 1]$ and any $\lambda_l \in \mathbb{R}$ $(1 \leqslant l \leqslant p)$. Without loss of generality we can take $p = 2$ and suppose that $t_1 < t_2$.

As J and $\partial J/\partial u_j$ $(1 \leqslant j \leqslant k)$ are bounded, the sequence of rv's $\{A_{ni}\}$ are uniformly bounded and we have

$$\sum_{l=1}^{2} \lambda_l L_n(t_l)$$
$$= n^{-1/2} \left[\sum_{i=1}^{[nt_1]} (\lambda_1 + \lambda_2)\, A^*_{ni} + \sum_{i=[nt_1]+1}^{[nt_2]} \lambda_2 A^*_{ni} + (nt_1 - [nt_1])\, A^*_{n,[nt_1]+1} \right.$$
$$\left. + (nt_2 - [nt_2])\, A^*_{n,[nt_2]+1} \right]. \tag{3.4}$$

We now define the sequence of rv's $\{B_{ni}\}$ by

$$B_{ni} = \begin{cases} (\lambda_1 + \lambda_2)\, A^*_{ni} & \text{if } i \leqslant [nt_1] \\ \lambda_2 A^*_{ni} & \text{if } [nt_1] < i \leqslant [nt_2] \\ 0 & \text{if } i > [nt_2]. \end{cases} \tag{3.5}$$

Then, we have

$$\sum_{l=1}^{2} \lambda_l L_n(t_l) = n^{-1/2} \sum_{i=1}^{n} B_{ni} + O(n^{-1/2}). \tag{3.6}$$

From Corollary 1 of Withers [9], we only have to prove that $E(\sum_{i=1}^{n} B_{ni})^2/n$ converges to some constant as $n \to \infty$. We have

$$\begin{aligned} E\left(\sum_{i=1}^{n} B_{ni}\right)^2 = \frac{1}{n} [(\lambda_1+\lambda_2)^2 &\left(\sum_{i=1}^{[nt_1]} \sum_{j=1}^{[nt_1]} E(A_{ni}^* A_{nj}^*)\right) \\ &+ 2(\lambda_1+\lambda_2)\,\lambda_2 \sum_{i=1}^{[nt_1]} \sum_{j=[nt_1]+1}^{[nt_2]} E(A_{ni}^* A_{nj}^*) \\ &+ \lambda_2^2 \sum_{i=[nt_1]+1}^{[nt_2]} \sum_{j=[nt_1]+1}^{[nt_2]} E(A_{ni}^* A_{nj}^*). \end{aligned} \tag{3.7}$$

As the $\{A_{ni}^*\}$ are uniformly bounded and using the well-known inequality on the moment of strong mixing rv's (see [4, Proposition 2.8]), we obtain for any $\delta > 0$,

$$\begin{aligned} &\frac{1}{n} \left| (\lambda_1+\lambda_2)\,\lambda_2 \sum_{i=1}^{[nt_1]} \sum_{j=[nt_1]+1}^{[nt_2]} E(A_{ni}^* A_{nj}^*) \right| \\ &\leqslant \left(\sum_{i=1}^{[nt_1]} \sum_{j=[nt_1]+1}^{[nt_2]} \alpha^{\delta/(2+\delta)}(j-i) \right) M/n, \end{aligned}$$

$$\text{where } M \text{ is a constant} > 0. \tag{3.8}$$

From condition (1.3), the last expression converges to 0 as $n \to \infty$ for δ sufficiently large ($\delta > 2/(3+\varepsilon)$). It remains to prove that $(1/n) \sum_{i=1}^{[nt_1]} \sum_{j=1}^{[nt_1]} E(A_{ni}^* A_{nj}^*)$ converges to some finite constant.

We first prove the convergence by replacing J and $\partial J/\partial u_j$ $(1 \leqslant j \leqslant k)$ in the expression of A_{ni}^* in (3.2) by indicator functions such as $J(u_1, ..., u_k) = \prod_{j=1}^{k} I_{[a_j^0 \leqslant u_j \leqslant b_j^0]}$ and $(\partial J/\partial u_j)(u_1, ..., u_k) = \prod_{l=1}^{k} I_{[a_l^j \leqslant u_l \leqslant b_l^j]}$. We can then write

$$\begin{aligned} A_{ni}^* = \sum_{j=1}^{k} \int_{a_1^j}^{b_1^j} \cdots \int_{a_{j-1}^j}^{b_{j-1}^j} \int_{(a_j^j \wedge F_n^{(j)}(X_{ni}^{(j)})) \wedge b_j^j}^{b_j^j} \int_{a_{j+1}^j}^{b_{j+1}^j} \cdots \\ \times \int_{a_k^j}^{b_k^j} dF_{ni}(F_n^{(1)^{-1}}(u_1), ..., F_n^{(k)^{-1}}(u_k)) \end{aligned}$$

$$-\sum_{j=1}^{k}\int_{a_1^j}^{b_1^j}\cdots\int_{a_k^j}^{b_k^j} F_{ni}^{j}(F_n^{(j)^{-1}}(u_j)\,dF_{ni}(F_n^{(1)^{-1}}(u_1),\ldots,F_n^{(k)^{-1}}(u_k))$$

$$+\prod_{l=1}^{k} I_{[a_l^0\leqslant F_n^{(l)}(X_{ni}^{(l)})\leqslant b_l^0]}-\int_{a_1^0}^{b_1^0}\cdots\int_{a_k^0}^{b_k^0} dF_{ni}(F_n^{(1)^{-1}}(u_1),\ldots,F_n^{(k)^{-1}}(u_k))$$

$$=C_{ni}-E(C_{ni}),$$

where C_{ni} is equal to the sum of the first and the third terms in the decomposition of A_{ni}^* (above). We then have

$$E(C_{ni}C_{nj})=\sum_{l=1}^{k}\sum_{p=1}^{k}\int_{\tilde a_1^l}^{\tilde b_1^l}\cdots\int_{\tilde a_k^l}^{\tilde b_k^l}\int_{\tilde a_1^p}^{\tilde b_1^p}\cdots\int_{\tilde a_k^p}^{\tilde b_k^p} F_{ni}^{(l)}(x_l)\,F_{nj}^{(p)}(x_p)$$
$$\times dF_{n,i,j}(x_1,\ldots,x_k,y_1,\ldots,y_k)$$
$$-\sum_{l=1}^{k}\int_{\tilde a_1^l}^{\tilde b_1^l}\cdots\int_{\tilde a_k^l}^{\tilde b_k^l}\int_{\tilde a_1^0}^{\tilde b_1^0}\cdots\int_{\tilde a_k^0}^{\tilde b_k^0}(F_{ni}^{(l)}(x_l)+F_{nj}^{(l)}(y_l))$$
$$\times dF_{n,i,j}(x_1,\ldots,x_k,y_1,\ldots,y_k)$$
$$+\int_{\tilde a_1^0}^{\tilde b_1^0}\cdots\int_{\tilde a_k^0}^{\tilde b_k^0}\int_{\tilde a_1^0}^{\tilde b_1^0}\cdots\int_{\tilde a_k^0}^{\tilde b_k^0} dF_{n,i,j}(x_1,\ldots,x_k,y_1,\ldots,y_k),$$

where

$$\tilde b_p^l=F_n^{(l)^{-1}}(b_p^l),\qquad 1\leqslant p\leqslant k,\ 0\leqslant l\leqslant k$$
$$\tilde a_p^l=F_n^{(l)^{-1}}(a_p^l),\qquad 1\leqslant p\leqslant k,\ 0\leqslant l\leqslant k.$$

We can write

$$E(C_{ni}C_{nj})$$
$$=\sum_{l=1}^{k}\sum_{p=1}^{k}\int_{a_1^l}^{b_1^l}\cdots\int_{a_k^l}^{b_k^l}\int_{a_1^p}^{b_1^p}\cdots\int_{a_k^p}^{b_k^p}$$
$$\times F_{ni}^{(l)}(F_n^{(l)^{-1}}(u_l)\,F_{nj}^{(p)}(F_n^{(p)^{-1}}(v_p))\,dF_{n,i,j}(F_n^{(1)^{-1}}(u_1),\ldots,F_n^{(k)^{-1}}(v_k))$$
$$-\sum_{l=1}^{k}\int_{a_1^l}^{b_1^l}\cdots\int_{a_k^l}^{b_k^l}\int_{a_1^0}^{b_1^0}\cdots\int_{a_k^0}^{b_k^0}[F_{ni}^{(l)}(F_n^{(l)^{-1}}(u_l))+F_{nj}^{(l)}(F_n^{(l)^{-1}}(v_l))]$$
$$\times dF_{n,i,j}(F_n^{(1)^{-1}}(u_1),\ldots,F_n^{(k)^{-1}}(u_k),F_n^{(1)^{-1}}(v_1),\ldots,F_n^{(k)^{-1}}(v_k))$$
$$+\int_{a_1^0}^{b_1^0}\cdots\int_{a_k^0}^{b_k^0}\int_{a_1^0}^{b_1^0}\cdots\int_{a_k^0}^{b_k^0}$$
$$\times dF_{n,i,j}(F_n^{(1)^{-1}}(u_1),\ldots,F_n^{(k)^{-1}}(u_k),F_n^{(1)^{-1}}(v_1),\ldots,F_n^{(k)^{-1}}(v_k)).$$

By using (2.6), (2.7), and the df's F_{ni} are continuous and nondecreasing, we deduce

$$\lim_{n\to\infty} \max_{1\leqslant i<j\leqslant n} \left| E(C_{ni}C_{nj}) - \sum_{l=1}^{k} \sum_{p=1}^{k} \int_{a_1^l}^{b_1^l} \cdots \int_{a_k^l}^{b_k^l} \int_{a_1^p}^{b_1^p} \cdots \int_{a_k^p}^{b_k^p} \right.$$

$$\times u_l v_l d\mathbf{F}_{j-i}(F^{(1)^{-1}}(u_1), \ldots, F^{(k)^{-1}}(u_k), F^{(1)^{-1}}(v_1), \ldots, F^{(k)^{-1}}(v_k))$$

$$+ \sum_{l=1}^{k} \int_{a_1^l}^{b_1^l} \cdots \int_{a_k^l}^{b_k^l} \int_{a_1^0}^{b_1^0} \cdots \int_{a_k^0}^{b_k^0} (u_l + v_l)$$

$$\times d\mathbf{F}_{j-i}(F^{(1)^{-1}}(u_1), \ldots, F^{(k)^{-1}}(u_k), F^{(1)^{-1}}(v_1), \ldots, F^{(k)^{-1}}(v_k))$$

$$- \int_{a_1^0}^{b_1^0} \cdots \int_{a_k^0}^{b_k^0} \int_{a_1^0}^{b_1^0} \cdots \int_{a_k^0}^{b_k^0}$$

$$\left. \times d\mathbf{F}_{j-i}(F^{(1)^{-1}}(u_1), \ldots, F^{(k)^{-1}}(u_k), F^{(1)^{-1}}(v_1), \ldots, F^{(k)^{-1}}(v_k)) \right|$$

$$\left(= \lim_{n\to\infty} \max_{1\leqslant i<j\leqslant n} |E(C_{ni}C_{nj}) - C_{ij})| = 0\right).$$

Continuing in the same way, we obtain

$$\lim_{n\to\infty} \max_{1\leqslant i<j\leqslant n} |E(A_{ni}^* A_{nj}^*) - C_{ij} + C^2| = 0, \tag{3.9}$$

where

$$C = \sum_{l=1}^{k} \int_{a_1^l}^{b_1^l} \cdots \int_{a_k^l}^{b_k^l} u_l dF(F^{(1)^{-1}}(u_1), \ldots, F^{(k)^{-1}}(u_k))$$

$$+ \int_{a_1^0}^{b_1^0} \cdots \int_{a_k^0}^{b_k^0} dF(F^{(1)^{-1}}(u_1), \ldots, F^{(k)^{-1}}(u_k)).$$

We obtain a similar property if J and $\partial J/\partial u_l$ $(1 \leqslant l \leqslant k)$ are replaced by step functions.

As J and $\partial J/\partial u_l$ $(1 \leqslant l \leqslant k)$ are continuous, we can uniformly approach them by a step function and we deduce that

$$\lim_{n\to\infty} \max_{1\leqslant i<j\leqslant n} \left| E(A_{ni}^* A_{nj}^*) - \int_{\mathbb{R}^{2k}} g(\mathbf{x})\, g(\mathbf{y})\, d\mathbf{F}_{j-i}(\mathbf{x}, \mathbf{y}) \right| = 0,$$

where $g(\mathbf{x})$ is defined in (2.4).

Let us denote

$$\rho(0) = \int_{\mathbb{R}^k} g^2(\mathbf{x})\, dF(\mathbf{x})$$

$$\rho(i) = 2 \int_{\mathbb{R}^{2k}} g(\mathbf{x})\, g(\mathbf{y})\, dF_i(\mathbf{x}, \mathbf{y}), \qquad i \geqslant 1,$$

then

$$\left|\frac{1}{n}\sum_{i=1}^{[nt_1]}\sum_{j=1}^{[nt_1]} E(A^*_{ni}A^*_{nj}) - \frac{[nt_1]}{n}\sum_{i=0}^{+\infty}\rho(i)\right|$$

$$\leqslant \left|\frac{[nt_1]}{n}\frac{1}{[nt_1]}\sum_{i=0}^{[nt_1]-1}\sum_{j=1}^{[nt_1]-i}\varphi(i)\,E(A^*_{nj}A^*_{n,j+i})\right.$$

$$\left. -\frac{[nt_1]}{n}\frac{1}{[nt_1]}\sum_{i=0}^{[nt_1]}([nt_1]-i)\,\rho(i)\right| + \frac{[nt_1]}{n}\sum_{i=[nt_1]+1}^{\infty}|\rho(i)|$$

$$+\frac{1}{n}\sum_{i=0}^{[nt_1]} i\,|\rho(i)| = |A_n| + B_n + C_n, \quad \text{where } \varphi(i) = \begin{cases} 1 & \text{if } i=0 \\ 2 & \text{if } i \neq 0. \end{cases}$$

From (3.9) we deduce that $|A_n| \to 0$ and, from the well-known inequality (see Doukhan and Portal [4, Proposition 2.8]) on moments of strong mixing sequences of rv's, we deduce $|\rho(i)| \leqslant \alpha(i)^{\delta/(2+\delta)} M$ for some constant $M > 0$ which implies that $B_n \to 0$ and $C_n \to 0$ (as $n \to \infty$) by (1.3) for $\delta > \frac{3}{2}$. It is also immediate that

$$\left|\frac{[nt_1]}{n}\sum_{i=0}^{+\infty}\rho(i) - t_1\left(\sum_{i=0}^{+\infty}\rho(i)\right)\right| \to 0 \qquad \text{as } n \to \infty. \tag{3.10}$$

We conclude that $(1/n)\sum_{i=1}^{[nt_1]}\sum_{j=1}^{[nt_1]} E(A^*_{ni}A^*_{nj})$ converges to $t_1(\sum_{i=0}^{+\infty}\rho(i))$ as $n \to \infty$, where $\sum_{i=0}^{+\infty}\rho(i)$ is equal to $\sigma^2(\{\mathbf{F}_l\})$ defined in (2.3).

From (3.4)–(3.8) and (3.10), we deduce that $E(\sum_{l=1}^{2}\lambda_l L_n(t_l))^2$ converges to $\{(\lambda_1+\lambda_2)^2 t_1 + \lambda_2^2(t_2-t_1)\}\,\sigma^2(\{\mathbf{F}_l\})$ which implies that $\sum_{l=1}^{2}\lambda_l L_n(t_l)$ converges in law to the normal distribution with mean 0 and variance $\{(\lambda_1+\lambda_2)^2 t_1 + \lambda_2^2(t_2-t_1)\}\,\sigma^2(\{\mathbf{F}_l\})$ and (i) is proved.

To prove (ii), we have to verify (cf. Billingsley [2]) that $\forall \varepsilon > 0$, $\exists \eta > 0$ $(0 < \eta < 1)$, and an integer N_0 such that $\forall n \geqslant N_0$,

$$P[\sup_{|s-t|<\eta} |L_n(t) - L_n(s)| \geqslant \varepsilon] \leqslant \varepsilon. \tag{3.11}$$

If ns and nt are integers, we obtain the following inequality by using Theorem 2.10 for $q = 4$ of Doukhan and Portal [4] for $s < t$,

$$E(L_n(s) - L_n(t))^4 \leqslant \left((t-s)^2 + \frac{(t-s)}{n}\right) M \sum_{m=0}^{+\infty} m^2 \alpha^{\delta/(4+\delta)}(m).$$

Equation (1.3) entails

$$\sum_{m=0}^{+\infty} m^2 \alpha^{\delta/(4+\delta)}(m) = M_0 < +\infty. \tag{3.12}$$

If ns and nt are integers, $s<t$, and $t-s\geqslant n^{-1}$, we have

$$E(L_n(s)-L_n(t))^4\leqslant 2M(t-s)^2 M_0. \tag{3.13}$$

From Lemma 2 of Balacheff and Dupont [1], we obtain that $\forall\varepsilon>0$, $\exists\eta>0$, and there exists an integer N_0 sufficiently large such that $\forall n\geqslant N_0$,

$$P\left[\sup_{|[ns]/n-[nt]/n|<2\eta}\left|L_n\left(\frac{[ns]}{n}\right)-L_n\left(\frac{[nt]}{n}\right)\right|>\varepsilon/2\right]\leqslant 2MM_0\eta^2K\,16\varepsilon^{-4},$$

where K is some constant >0.

From the definition of $L_n(s)$ in (3.3) we obtain

$$\sup_{|s-t|<\eta}|L_n(s)-L_n(t)|\leqslant 2\max_{|[ns]/n-[nt]/n|<2\eta}\left|L_n\left(\frac{[ns]}{n}\right)-L_n\left(\frac{[nt]}{n}\right)\right|. \tag{3.14}$$

By using (3.13) and (3.14), we deduce

$$P[\sup_{|s-t|<\eta}|L_n(s)-L_n(t)|>\varepsilon]\leqslant 64MM_0K\varepsilon^{-4}\eta^2 \tag{3.15}$$

and (3.11) is proved. Lemma 3.2 follows.

We now consider the process $\hat{L}_n(t)$ defined on C_1 by

$$\hat{L}_n(t)=n^{-1/2}\left(\sum_{i=1}^{[nt]}c_{ni}A^*_{ni}+c_{n,[nt]+1}(nt-[nt])\,A^*_{n,[nt]+1})\right)(\sigma(\{\mathbf{F}_l\})^{-1}. \tag{3.16}$$

LEMMA 3.3. *Suppose that $\{\mathbf{X}_{ni}\}$ satisfies the conditions of Theorem 2.1. Then $\hat{L}_n(t)$ converges weakly in uniform topology to a Gaussian process $\hat{L}_0(t)=\int_0^t h(u)\,dW(u)$ with trajectories a.s. in C_1, where W is a standard Brownian motion process.*

Proof. From Theorem 1 of Yoshihara and Negishi [10] we have to verify that

$$\lim_{\eta\to 0}\limsup_{n\to\infty}P\left[\max_{1\leqslant k\leqslant n}\left|\sum_{i=1}^{k}\eta_iA^*_{ni}(\sigma(\{\mathbf{F}_l\})^{-1}\right|\geqslant\varepsilon M\sqrt{n}\right]=0, \tag{3.17}$$

where η_i is any set of real numbers, $\eta=\sup_{1\leqslant i\leqslant n}\eta_i$, and

$$M=\sup_n\max_{1\leqslant i\leqslant n}E(A^{*2}_{ni}(\sigma^2(\{\mathbf{F}_l\}))^{-1}).$$

By using Theorem 2.10 of Doukhan and Portal [4] for $q=4$, we have

$$E\left(\sum_{i=1}^{k}\eta_iA^*_{ni}\right)^4\leqslant C_4(\alpha)(Mk\eta+M^2k^2\eta^2)$$

which implies from Theorem 12.2 of Billingsley that for $\eta \geqslant n^{-2}$,

$$P\left\{\max_{1\leqslant k\leqslant n}\left|\sum_{i=1}^{k}\eta_i A_{ni}^*\right| \geqslant \varepsilon M\sqrt{n}\right\} \leqslant K\varepsilon^{-4}(M^{-3}+M^{-2})\,\eta^{3/2}. \quad (3.18)$$

Since the right side of (3.18) converges to zero as $\eta \to 0$, (3.17) follows.

4. Proof of Theorem 2.1

For any $j \in \{1, ..., k\}$ and any $x_j \in \mathbb{R}$ define $Y_{ni}^{(j)}(x_j)$ as

$$Y_{ni}^{(j)}(x_j) = I_{[X_{ni}^{(j)} \leqslant x_j]} - F_{ni}^{(j)}(x_j).$$

Let

$$\tilde{X}_{n,m,i}^{(l)} = \frac{1}{m+1}\left(\sum_{\substack{j\leqslant m\\ j\neq i}} F_{nj}^{(l)}(X_{ni}) + 1\right), \qquad l \in \{1, ..., k\},\ 1 \leqslant m \leqslant n.$$

$$Z_{n,m,i}^{(l)} = \frac{R_{n,m,i}^{(l)}}{m+1} - \tilde{X}_{n,m,i}^{(l)} = \frac{1}{m+1}\sum_{\substack{j\leqslant m\\ j\neq i}} Y_{nj}^{(l)}(X_{ni}^{(l)}), \quad l \in \{1, ..., k\},\ 1 \leqslant m \leqslant n.$$

From condition (2.7) we easily deduce that for n sufficiently large

$$|\tilde{X}_{n,m,i}^{(l)} - F_n^{(l)}(X_{ni}^{(l)})| \leqslant \frac{2}{m+1}. \quad (4.1)$$

From the Taylor expansion, it follows that

$$\begin{aligned} J\left(\frac{R_{n,m,i}^{(1)}}{m+1}, ..., \frac{R_{n,m,i}^{(k)}}{m+1}\right) &= J(\tilde{X}_{n,m,i}^{(1)}, ..., \tilde{X}_{n,m,i}^{(k)}) \\ &+ \sum_{l=1}^{k} Z_{n,m,i}^{(l)} \frac{\partial J}{\partial u_l}(\tilde{X}_{n,m,i}^{(1)}, ..., \tilde{X}_{n,m,i}^{(k)}) \\ &+ \frac{1}{2}\sum_{l=1}^{k}\sum_{p=1}^{k} Z_{n,m,i}^{(l)} Z_{n,m,i}^{(p)} \frac{\partial^2 J}{\partial u_l\,\partial u_p}(\mu_{n,m,i}^{(1)}, ..., \mu_{n,m,i}^{(k)}), \end{aligned} \quad (4.2)$$

where

$$\mu_{n,m,i}^{(l)} \in \left[\frac{R_{n,m,i}^{(l)}}{m+1} \wedge \tilde{X}_{n,m,i}^{(l)}, \frac{R_{n,m,i}^{(l)}}{m+1} \vee \tilde{X}_{n,m,i}^{(l)}\right], l \in \{1, ..., k\}.$$

So from (1.1), (2.2), (3.2), and (4.2) it follows that

$$\left|\mathscr{S}_{n,m}-\mu_{m,n}-\sum_{i=1}^{m} c_{ni}A_{ni}^{*}\right|$$

$$=\left|\sum_{i=1}^{m} c_{ni}\left[J(\tilde{X}_{n,m,i}^{(1)},\ldots,\tilde{X}_{n,m,i}^{(k)})+\sum_{l=1}^{k} Z_{n,m,i}^{(l)}\frac{\partial J}{\partial u_l}(\tilde{X}_{n,m,i}^{(1)},\ldots,\tilde{X}_{n,m,i}^{(k)})\right.\right.$$

$$+\frac{1}{2}\sum_{l=1}^{k}\sum_{p=1}^{k} Z_{n,m,i}^{(l)}Z_{n,m,i}^{(p)}\frac{\partial^2 J}{\partial u_l\,\partial u_p}(\mu_{n,m,i}^{(1)},\ldots,\mu_{n,m,i}^{(k)})$$

$$-\sum_{l=1}^{k}\int_{\mathbb{R}^k}(I_{[X_{ni}^{(l)}\leqslant x_l]}-F_{ni}^{(l)}(x_l))\frac{\partial J}{\partial u_l}(F_n^{(1)}(x_1),\ldots,F_n^{(k)}(x_k))\,dF_{ni}(x_1,\ldots,x_k)$$

$$\left.\left.-J(F_n^{(1)}(X_{ni}^{(1)}),\ldots,F_n^{(k)}(X_{ni}^{(k)}))\right]\right|$$

$$\leqslant\left|\sum_{i=1}^{m} c_{ni}\{J(\tilde{X}_{n,m,i}^{(1)},\ldots,\tilde{X}_{n,m,i}^{(k)})-J(F_n^{(1)}(X_{ni}^{(1)}),\ldots,F_n^{(k)}(X_{ni}^{(k)}))\}\right|$$

$$+\left|\sum_{i=1}^{m} c_{ni}\left[\sum_{l=1}^{k} Z_{n,m,i}^{(l)}\frac{\partial J}{\partial u_l}(\tilde{X}_{n,m,i}^{(1)},\ldots,\tilde{X}_{n,m,i}^{(k)})\right.\right.$$

$$-\sum_{l=1}^{k}\int_{\mathbb{R}^k}(I_{[X_{ni}^{(l)}\leqslant x_l]}-F_{ni}^{(l)}(x_l))$$

$$\left.\left.\times\frac{\partial J}{\partial u_l}(F_n^{(1)}(x_1),\ldots,F_n^{(k)}(x_k))\,dF_{ni}(x_1,\ldots,x_k)\right]\right|$$

$$+M_0\sum_{i=1}^{m}|c_{ni}|\sum_{l=1}^{k}\sum_{p=1}^{k}|Z_{n,m,i}^{(l)}Z_{n,m,i}^{(p)}|$$

$$\left(\text{where } 2M_0=\max_{\substack{1\leqslant l\leqslant k\\1\leqslant p\leqslant k}}\sup_{\mathbf{u}\in[0,1]^k}\frac{\partial^2 J}{\partial u_p\,\partial u_l}(\mathbf{u})\right)$$

$$\leqslant\sum_{i=1}^{m}|c_{ni}|\left\{\sum_{l=1}^{k}|\tilde{X}_{n,m,i}^{(l)}-F_n^{(l)}(X_{ni}^{(l)})|\left|\frac{\partial J}{\partial u_l}(F_n^{(1)}(X_{ni}^{(1)}),\ldots,F_n^{(k)}(X_{ni}^{(k)}))\right|\right.$$

$$\left.+M_0\sum_{l=1}^{k}\sum_{p=1}^{k}|\tilde{X}_{n,m,i}^{(l)}-F_n^{(l)}(X_{ni}^{(l)})|\,|\tilde{X}_{n,m,i}^{(p)}-F_n^{(p)}(X_{ni}^{(p)})|\right\}$$

$$+\left|\sum_{i=1}^{m} c_{ni}\left[\sum_{l=1}^{k} Z_{n,m,i}^{(l)}\frac{\partial J}{\partial u_l}(F_n^{(1)}(X_{ni}^{(1)}),\ldots,F_n^{(k)}(X_{ni}^{(k)}))\right.\right.$$

$$-\sum_{l=1}^{k}\int_{\mathbb{R}^k}(I_{[X_{ni}^{(l)}\leqslant x_l]}-F_{ni}^{(l)}(x_l))$$

$$\left.\left.\times\frac{\partial J}{\partial u_l}(F_n^{(1)}(x_1),\ldots,F_n^{(k)}(x_k))\,dF_{ni}(x_1,\ldots,x_k)\right]\right|$$

$$+ M_0 \sum_{i=1}^{m} |c_{ni}| \sum_{l=1}^{k} |\tilde{X}^{(l)}_{n,m,i} - F^{(l)}_n(X^{(l)}_{ni})| \cdot |Z^{(l)}_{n,m,i}|$$

$$+ M_0 \sum_{i=1}^{m} |c_{ni}| \sum_{l=1}^{k} \sum_{p=1}^{k} |Z^{(l)}_{n,m,i} Z^{(p)}_{n,m,i}|$$

and, by using (4.1),

$$\leqslant \frac{2}{m+1} \sum_{i=1}^{m} |c_{ni}| \sum_{l=1}^{k} \left| \frac{\partial J}{\partial u_l} (F^{(1)}_n(X^{(1)}_{ni}), \ldots, F^{(k)}_n(X^{(k)}_{ni})) \right|$$

$$+ M_0 \frac{4k^2}{(m+1)^2} \sum_{i=1}^{m} |c_{ni}| + \sum_{l=1}^{k} |V^{(l)}_m|$$

$$+ \frac{2}{m+1} \sum_{i=1}^{m} |c_{ni}| \sum_{l=1}^{k} \left| \int_{\mathbb{R}^k} (I_{[X^{(l)}_{ni} \leqslant x_l]} - F^{(l)}_{ni}(x_l)) \right.$$

$$\left. \times \frac{\partial J}{\partial u_l} (F^{(1)}_n(x_1), \ldots, F^{(k)}_n(x_k))\, dF_{ni}(x_1, \ldots, x_k) \right|$$

$$+ \frac{2M_0}{m+1} \sum_{i=1}^{m} |c_{ni}| \sum_{l=1}^{k} |Z^{(l)}_{n,m,i}|$$

$$+ M_0 \sum_{i=1}^{m} |c_{ni}| \sum_{l=1}^{k} \sum_{p=1}^{k} |Z^{(l)}_{n,m,i}|\, |Z^{(p)}_{n,m,i}|,$$

where

$$V^{(l)}_m = \frac{1}{m+1} \sum_{i=1}^{m} \sum_{\substack{j=1 \\ j \neq i}}^{m} c_{ni} \left[(I_{[X^{(l)}_{nj} \leqslant X^{(l)}_{ni}]} \right.$$

$$- F^{(l)}_{nj}(X^{(l)}_{ni})) \frac{\partial J}{\partial u_l} (F^{(1)}_n(X^{(1)}_{ni}), \ldots, F^{(k)}_n(X^{(k)}_{ni}))$$

$$- \int_{\mathbb{R}^k} (I_{[X_{nj} \leqslant x_l]} - F^{(l)}_{nj}(x_l))$$

$$\left. \times \frac{\partial J}{\partial u_l} (F^{(1)}_n(x_1), \ldots, F^{(k)}_n(x_k))\, dF_{nj}(x_1, \ldots, x_k) \right].$$

As the coefficients c_{ni}, the functions $\partial J/\partial u_l$ and the rv $Z^{(l)}_{n,m,i}$ are bounded there exist constants M_1 and M_2 such that

$$\left| \mathscr{S}_{n,m} - \mu_{n,m} - \sum_{i=1}^{m} c_{ni} A^*_{ni} \right|$$

$$\leqslant \sum_{l=1}^{k} |V^{(l)}_m + M_1 \sum_{i=1}^{m} \sum_{l=1}^{k} \sum_{p=1}^{k} |Z^{(l)}_{n,m,i}|\, |Z^{(p)}_{n,m,i}| + M_2. \qquad (4.3)$$

To prove the theorem, we need at first three lemmas.

LEMMA 4.1. *Under the conditions of Theorem* 2.1, *the following relations hold*

$$E((V_m^{(l)})^4) = O(1), \qquad 1 \leqslant l \leqslant k. \tag{4.4}$$

Proof. Put

$$g_j^{(l)}(\mathbf{x}, \mathbf{y}) = (I_{[y_l \leqslant x_l]} - F_{nj}^{(l)}(x_l)) \frac{\partial J}{\partial u_l} (F_n^{(1)}(x_1), ..., F_n^{(k)}(x_k))$$

$$- \int_{\mathbb{R}^k} (I_{[y_l \leqslant z_l]} - F_{nj}^{(l)}(z_l))$$

$$\times \frac{\partial J}{\partial u_l} (F_n^{(1)}(z_1), ..., F_n^{(k)}(z_k))\, dF_{nj}(z_1, ..., z_k),$$

where $\mathbf{x} = (x_1, ..., x_k)$ and $\mathbf{y} = (y_1, ..., y_k)$. We have:

$$\int_{\mathbb{R}^k} g_j^{(l)}(\mathbf{x}, \mathbf{y})\, dF_{nj}(\mathbf{y}) = 0, \qquad \forall \mathbf{x} \in \mathbb{R}^k. \tag{4.5}$$

We note

$$J^{(l)}(i_{jp}, j = 1, 2, p = 1, ..., 4) = E\left[\prod_{p=1}^{4} g_{i_{2p}}^{(l)}(\mathbf{X}_{ni_{1p}}, \mathbf{X}_{ni_{2p}})\right],$$
$$1 \leqslant i_{jp} \leqslant n,\ 1 \leqslant j \leqslant 2,\ 1 \leqslant p \leqslant 4.$$

If i_{jp} $(j = 1, 2, p = 1, ..., 4)$ are mutually different, reorder $\{i_{jp}\}$ as $1 \leqslant k_1 < k_2 < \cdots < k_8 \leqslant n$ and put

$$J^{(l)}(i_{jp}, j = 1, 2, p = 1, ..., 4) = H(k_1, ..., k_8).$$

Let $m^{(p)}$ be the pth largest difference among $k_{j+1} - k_j$ $(j = 1, ..., 7)$ if for some j_α $(1 \leqslant j_\alpha \leqslant 7, 1 \leqslant \alpha \leqslant 2)$, $k_{j_\alpha + 1} - k_j = m^{(\alpha)}$. Then, from Lemma 3.1, (4.5), and (4.6), we obtain

$$H(k_1, ..., k_8) = M_0 \sum_{\alpha=1}^{4} \beta(k_{j_\alpha+1} - k_{j_\alpha}), \quad \text{where } M_0 \text{ is some constant} > 0$$

and then

$$\sum_{1 \leqslant k_1 < \cdots < k_8 \leqslant n} H(k_1, ..., k_8) \leqslant M_0 n^4 \sum_{l=1}^{n} (l+1)^3 \beta(l) = O(n^4). \tag{4.7}$$

We obtain similar results in the other cases, and so from (4.6), (4.7), and the definition of $V_m^{(l)}$, we have (4.4).

Lemma 4.2. *Under the conditions of Theorem 2.1, the following condition holds*

$$E\left[\left(\sum_{i=1}^{m}(Z_{n,m,i}^{(l)})^2\right)^2\right]=O(m^{-1}), \qquad 1\leqslant l\leqslant k. \tag{4.8}$$

Proof. The proof is similar to the proof of Lemma 4.1 and is therefore omitted.

Lemma 4.3. *Under the conditions of Theorem 2.1, we have* $\forall \varepsilon>0$,

$$P[\sup_{s\in[0,1]}|V_n(s)-\hat{L}_n(s)|>\varepsilon]\to 0 \qquad \text{as } n\to\infty. \tag{4.9}$$

Proof. To prove (4.9), it is sufficient to prove

$$P\left[\max_{1\leqslant m\leqslant n}\left|\mathcal{S}_{n,m}-\mu_{n,m}-\sum_{i=1}^{m}c_{ni}A_{ni}^*\right|\geqslant\varepsilon n^{1/2}\right]\to 0 \qquad \text{as } n\to\infty \tag{4.10}$$

for every $\varepsilon>0$.

In order to prove (4.10), from (4.3), it suffices to show that for every $\varepsilon>0$

$$P[\max_{1\leqslant m\leqslant n}|V_m^{(l)}|>\varepsilon n^{1/2}]\to 0 \qquad \text{as } n\to\infty,\ l=1,\ldots,k, \tag{4.11}$$

$$P\left[\max_{1\leqslant m\leqslant n}\sum_{i=1}^{m}|Z_{n,m,i}^{(l)}|\,|Z_{n,m,i}^{(p)}|>\varepsilon n^{1/2}\right]\to 0$$
$$\text{as } n\to\infty,\ l=1,\ldots,k,\ p=1,\ldots,k. \tag{4.12}$$

From (4.4),

$$P[\max_{1\leqslant m\leqslant n}|V_m^{(l)}|>\varepsilon n^{1/2}]\leqslant(\varepsilon^4n^2)^{-1}\sum_{m=1}^{n}E((V_m^{(l)})^4)=O(n^{-1})$$

and, from (4.8),

$$\begin{aligned}P&\left[\max_{1\leqslant m\leqslant n}\sum_{i=1}^{m}|Z_{n,m,i}^{(l)}|\,|Z_{n,m,i}^{(p)}|>\varepsilon n^{1/2}\right]\\&\leqslant\sum_{m=1}^{n}P\left[\sum_{i=1}^{m}|Z_{n,m,i}^{(l)}|\,|Z_{n,m,i}^{(p)}|>\varepsilon n^{1/2}\right]\\&\leqslant(\varepsilon^2n)^{-1}\sum_{m=1}^{n}E\left[\left(\sum_{i=1}^{m}|Z_{n,m,i}^{(l)}|\,|Z_{n,m,i}^{(p)}|\right)^2\right]\end{aligned}$$

$$\leqslant (\varepsilon^2 n)^{-1} \sum_{m=1}^{n} E\left[\left(\sum_{i=1}^{m} (Z_{n,m,i}^{(l)})^2\right)\left(\sum_{j=1}^{m} (Z_{n,m,i}^{(p)})^2\right)\right]$$

$$\leqslant (\varepsilon^2 n)^{-1} \sum_{m=1}^{n} \left\{E\left[\sum_{i=1}^{m} (Z_{n,m,i}^{(l)})^2\right]^2\right\}^{1/2} \left\{E\left[\sum_{j=1}^{m} (Z_{n,m,i}^{(p)})^2\right]^2\right\}^{1/2}$$

$$\leqslant (\varepsilon^2 n)^{-1} \sum_{m=1}^{n} m^{-1/2} m^{-1/2} = O(n^{-1} \log n).$$

So (4.11) and (4.20) hold. Thus we have the lemma.

Theorem 2.1 now follows from Lemmas 3.3 and 4.3.

5. Applications

5.1. Consider a sequence $\{\mathbf{X}_{ni}, i \in \mathbb{Z}\}$ of $\mathbb{R}^k$-valued processes such that for all $n \in \mathbb{N}^*$, $\{\mathbf{X}_{ni}\}$ is a Markov process with stationary transition probability $P_n(\mathbf{x}; A)$, where $A \in \mathscr{B}^k$, $\mathscr{B}$ is the σ-field on $\mathbb{R}$, and $\mathbf{x} \in \mathbb{R}^k$ and with invariant measure μ_n.

Theorem 5.1. *Let $\{\mathbf{X}_{ni}, i \in \mathbb{Z}\}$ be a Markov process such that for every $n \in \mathbb{N}^*$, $\{\mathbf{X}_{ni}\}$ is either* (a) *aperiodic, Harris recurrent and geometrically ergodic with rates $0 < \rho_n < \rho_0$, $\rho_0 \in (0, 1)$ or* (b) *aperiodic and Doeblin recurrent.*

Suppose there exists a probability measure μ_0 on $\mathbb{R}^k$ and a transition probability P_0 such that

$$\mu_n((-\infty; \mathbf{x}]) \to \mu_0((-\infty; \mathbf{x}]) \qquad \textit{as } n \to \infty \tag{5.1}$$

for all $\mathbf{x} \in \mathbb{R}^k$,

$$P_n(\mathbf{x}; (-\infty, \mathbf{y}]) \to P_0(\mathbf{x}; (-\infty, \mathbf{y}]) \qquad \textit{as } n \to \infty \tag{5.2}$$

for all $(\mathbf{x}, \mathbf{y}) \in \mathbb{R}^{2k}$, where $(-\infty, \mathbf{y}] = \prod_{j=1}^{k} (-\infty, y_i]$. Then for the process V_n defined in (2.5) *associated with the sequence $\{\mathbf{X}_{n1}, ..., \mathbf{X}_{nn}\}$, with regression constants and score functions satisfying the conditions of Theorem* 2.1, *the conclusions of this theorem* (loc. cit.) *hold.*

Proof. (i) Suppose (a) holds. We have to show that the conditions (1.3), (2.8), and (2.9) are verified. From Nummelin and Tuominen [12], a Markov process which is aperiodic, Harris recurrent, and geometrically ergodic with rates ρ_n satisfies

$$\int_{\mathbb{R}^k} \|P_n^m(\mathbf{x}; \cdot) - \mu_n(\cdot)\| \, \mu_n(d\mathbf{x}) = O(\rho_n^m)$$

where $\| \ \|$ denotes the norm of total variation and P_n^m is the m-step transition probability.

From Proposition 1 of Davydov [3] we have

$$\beta(m) = \sup_{n \in \mathbb{N}^*} \int_{\mathbb{R}^k} \|P_n^m(\mathbf{x}; \cdot) - \mu_n(\cdot)\| \, \mu_n(d\mathbf{x})$$

and we deduce that the sequence $\{X_{ni}\}$ is absolutely regular with a geometrical rate ρ_0 which implies (1.3). The conditions (2.8) and (2.9) are now a consequence of condition (1.3) after some computations left to the reader.

(ii) Suppose (b) holds. From Davydov [3], a Markov process which is Doeblin recurrent and aperiodic is geometrically φ-mixing. This implies that $\{\mathbf{X}_{ni}\}$ is absolutely regular with a geometrical rate.

5.2. EXAMPLE. Consider the sequence of multivariate ARMA processes $\{\mathbf{X}_{ni}, i \in \mathbb{Z}\}$, $n \in \mathbb{N}^*$,

$$\mathbf{X}_{ni} + \sum_{j=1}^{p_1} \mathbf{A}_{n,j} \mathbf{X}_{n,i-j} = \boldsymbol{\varepsilon}_i + \sum_{l=1}^{p_2} \mathbf{B}_{n,l} \boldsymbol{\varepsilon}_{i-l}, \tag{5.3}$$

where $\{\boldsymbol{\varepsilon}_i, i \in \mathbb{N}\}$ is a k-variate white noise process, i.e., a sequence of i.i.d random variables with common density function $f(\mathbf{x})$, $\mathbf{A}_{n,1}, ..., \mathbf{A}_{n,p_1}$, $\mathbf{B}_{n,1}, ..., \mathbf{B}_{n,p_2}$ are real $k \times k$ matrices, $\mathbf{A}_{n,p_1}$ and $\mathbf{B}_{n,p_2}$ are of full rank for a n sufficiently large.

From Pham and Tran [7], $\mathbf{X}_{n,i}$ admits a Markovian representation

$$\begin{aligned} &\mathbf{X}_{n,i} = \mathbf{H}_n \mathbf{Z}_{n,i}, \qquad \mathbf{Z}_{n,i} = \mathbf{F}_n \mathbf{Z}_{n,i-1} + \mathbf{G}_n \boldsymbol{\varepsilon}_i \\ &\mathbf{H}_n = (h_{i,j}^{(n)}),\ 1 \leqslant i, j \leqslant k, \qquad \mathbf{F}_n = (f_{i,j}^{(n)}),\ 1 \leqslant i, j \leqslant k \\ &\mathbf{G}_n = (g_{i,j}^{(n)}),\ 1 \leqslant i, j \leqslant k \text{ are appropriate matrices.} \end{aligned} \tag{5.4}$$

We suppose that there exist matrices $\mathbf{H} = (h_{i,j})_{1 \leqslant i,j \leqslant k}$, $\mathbf{F} = (f_{i,j})_{1 \leqslant i,j \leqslant k}$, $\mathbf{G} = (g_{i,j})_{1 \leqslant i,j \leqslant k}$, such that

$$\lim_{n \to \infty} h_{i,j}^{(n)} = h_{i,j}, \ \lim_{n \to \infty} f_{i,j}^{(n)} = f_{i,j}, \qquad \lim_{n \to \infty} g_{i,j}^{(n)} = g_{i,j} \tag{5.5}$$

for every $1 \leqslant i, j \leqslant k$.

From Pham and Tran [7] and Mokkadem [6], if the eigenvalues of **F** are modulus strictly less than 1 and if $E(\|\varepsilon_i\|^s) < \infty$ for some $s > 0$, then the process $\mathbf{X}_{n,i}$ is absolutely regular at an exponential rate.

References

[1] Balacheff, S., and Dupont, G. (1980). Normalité asymptotique des processus empiriques tronqués et des processus de rang. In *Lecture in Mathemativcs*, Vol. 821, pp. 19–45, Springer-Verlag, New York/Berlin.

[2] Billingsley, P. (1968). *Convergence of Probability Measures.* Wiley, New York.

[3] Davydov, Yu. A. (1973). Mixing conditions for Markov chains. *Theory Probab. Appl.* **18**, No. 2, 312–328.

[4] Doukhan, P., and Portal, F. (1987). Principe d'invariance faible pour la fonction de répartition empirique dans un cadre multidimensionnel et mélangeant. *Probab. Math. Statist.* **8**, No. 2, 117–132.

[5] Harel, M. (1988). Weak convergence of multidimensional rank statistics under φ-mixing conditions. *J. Statist. Plann. Inference* **20** 41–63.

[5′] Harel, M., and Puri, M. L. (1990). Weak convergence of serial rank statistics under dependence with applications in time series and Markov processes. *Ann. Probab.* **18** 1361–1387.

[6] Mokkadem, A. (1987). Sur un modèle autorégressif non linéaire. Ergodicité et ergodicité géométrique. *J. Time Ser. Anal.* **8** 195–204.

[7] Pham, T. D., and Tran, L. T. (1985). Some mixing properties of time series models. *Stochastic Process. Appl.* **19** 197–303.

[8] Ruymgaart, F. H., and Van Zuijlen, M. C. (1978). Asymptotic normality of multivariate linear rank statistics in the non-i.i.d. case. *Ann. Statist.* **6** 588–602.

[9] Withers, C. S. (1975). Convergence of empirical processes of mixing rv's on [0, 1]. *Ann. Statist.* **3** 1101–1108.

[10] Yoshihara, K., and Negishi, H. (1975). Weak convergence theorems for a type of weighted sums of random variables. *Sci. Rep. Yokohama Nat. Univ. Sect.* 1 **22** 1–9.

[11] Yoshihara, K. (1976). Limiting behavior of U-statistics for stationary absolutely regular processes. *Z. Wahrsch. Verw. Gebiete* **35** 237–252.

[12] Nummelin E., and Tuominen, P. (1982). Geometric ergodicity of Harris recurrent Markov chains with applications to renewal theory. *Stochastic Process. Appl.* **12** 187–202.

WEAK CONVERGENCE OF THE SIMPLE LINEAR RANK STATISTIC UNDER MIXING CONDITIONS IN THE NONSTATIONARY CASE*

M. HAREL† and M. L. PURI†‡

Abstract. The asymptotic distribution theory of simple linear rank statistics for the case when the underlying random variables are nonstationary is studied both for the φ-mixing and strong mixing cases.

Key words. weak convergence; linear rank statistics, empirical processes, φ-mixing, strong mixing

1. Introduction. Let X_{ni}, $1 \leqq i \leqq n$, $n \geqq 1$, be real-valued r.v.'s (random variables) with continuous d.f.'s (distribution functions) $F_{ni}(x)$, $x \in \mathbf{R}$, and let c_{ni} $(1 \leqq i \leqq n,\ n \geqq 1)$ be an array of regression constants defined by a function g on $[0, 1]$ as

$$c_{ni} = g(i/n), \qquad 1 \leqq i \leqq n, \quad n \geqq 1. \tag{1.1}$$

Denote by $\widehat{H}_n(x) = n^{-1} \sum_{i=1}^{n} c_{ni} \mathbf{1}_{[X_{ni} \leqq x]}$ the weighted empirical process where $\mathbf{1}_{[]}$ denotes the indicator function and by $\widehat{F}_n(x) = n^{-1} \sum_{i=1}^{n} \mathbf{1}_{[X_{ni} \leqq x]}$ the usual empirical process.

The corresponding expectations are denoted by

$$H_n = \mathbf{E}\left(\widehat{H}_n\right) \quad \text{and} \quad F_n = \mathbf{E}\left(\widehat{F}_n\right).$$

We will study the asymptotic behavior of the simple linear rank statistic of the form

$$S_n(J) = n^{1/2} \int_{-\infty}^{+\infty} J\left(\frac{n}{n+1}\, \widehat{F}_n(x)\right) d\widehat{H}_n(x) \int_{-\infty}^{+\infty} J\big(F_n(x)\big)\, dH_n(x), \tag{1.2}$$

where J is a score function defined on the open unit interval.

The problem of finding a sufficiently large class of score functions for which the linear rank statistic is asymptotically normal was first considered by Chernoff and Savage [3]. Their results were later on strengthened considerably by several authors, mainly by Govindarajulu et al. [8], Pyke and Shorack [10], Hájek [9] and Dupač and Hájek [6] for the independent case, by Fears and Mehra [7] for the φ-mixing case with stationary random variables, and by Denker and Rösler [4] for the φ-mixing as well as strong mixing case but under a *stationary* set-up. In this paper we investigate the asymptotic distribution theory of the simple linear rank statistics (1.2) for the case when the underlying random variables are *nonstationary*.

2. Preliminaries. In this section, we give some propositions that are minor variations of Denker and Rösler [4] and so their proofs will be either omitted or briefly outlined.

---

*Received by the editors July 5, 1990.

†Department of Mathematics, Indiana University, Bloomington, Indiana 47405.

‡This research was supported by Office of Naval Research contract N00014-91-J-1020.

For $\delta \geqq 0$, set $\eta = \delta(4+2\delta)^{-1}$.

Let μ_δ denote the measure on [0, 1] given by its density $(z(1-z))^{-1/2-\eta}$ relative to the Lebesgue measure. For a monotone function J, let $\|J\|_\delta$ be the $\mathbf{L}_1$-norm of J in $\mathbf{L}_1(\mu_\delta)$. By the Jordan decomposition of any right continuous function, J has a unique decomposition: $J = J_1 - J_2$ where J_1 and J_2 are monotone functions and $J_1(\frac{1}{2}) = 0$. For such a function, we set

$$\|J\|_\delta = \|J_1\|_\delta + \|J_2\|_\delta,$$

where J_1 and J_2 belong to $\mathbf{L}_1(\mu_\delta)$.

Denote by $\mathcal{H}_\delta$ the space of all right continuous functions J with $\|J\|_\delta < \infty$ and $J(\frac{1}{2}) = 0$ and let $\mathcal{G}_\delta$ be the set of all $J \in \mathcal{H}_\delta$ for which the measure ν defined by $J = \int d\nu$ is absolutely continuous with respect to Lebesgue measure. It is well known that $\mathcal{G}_\delta$ is the $\| \ \|_\delta$-norm closure of $\mathbf{C}_{2,b}$: the space of functions with bounded second derivatives. The a priori assumption of having the space $\mathcal{H}_\delta$ of right continuous functions J with $J(\frac{1}{2}) = 0$ is no restriction because $\bar{J}$ defined by $\bar{J}(x) = J(x+) = \lim_{y\downarrow x} J(y)$ is a well-defined right continuous function and if $\bar{J}$ belongs to $\mathcal{H}_\delta$, $\mathcal{S}_n(J)$ and $\mathcal{S}_n(\bar{J})$ are asymptotically equivalent (see [4]).

We consider the array $G_{ni} (1 \leqq i \leqq n,\ n \geqq 1)$ of d.f.'s on [0, 1] defined by

$$G_{ni} = F_{ni} \circ F_n^{-1}. \tag{2.1}$$

Denote by $\widehat{G}_n$ the empirical process in [0, 1] derived from $\widehat{H}_n$ and defined by

$$\widehat{G}_n(t) = n^{-1} \sum_{i=1}^{n} c_{ni} \mathbf{1}_{[F_n(X_{ni}) \leqq t]}, \qquad t \in [0,\ 1], \tag{2.2}$$

and $\hat{I}_n$ the empirical process on [0, 1] derived from $\widehat{F}_n$ and defined by

$$\hat{I}_n(t) = n^{-1} \sum_{i=1}^{n} \mathbf{1}_{[F_n(X_{ni}) \leqq t]}, \qquad t \in [0,\ 1]. \tag{2.3}$$

We also denote $G_n = \mathbf{E}(\widehat{G}_n)$ and $I_n = \mathbf{E}(\hat{I}_n)$. The linear rank statistic defined in (1.2) can then be written as

$$\mathcal{S}_n(J) = n^{1/2} \int_0^1 J\Big(\frac{n}{n+1}\, \hat{I}_n(t)\Big)\, d\widehat{G}_n(t) - n^{1/2} \int_0^1 J(I_n(t))\, dG_n(t). \tag{2.4}$$

The connection between the dependence structure of the processes and the class of functions for which asymptotic normality holds is expressed by the following condition:

$$\begin{cases} n\,\mathbf{E}\,(\widehat{G}_n(t) - G_n(t))^2 \leqq C\Lambda_n^2 (t(1-t))^{1-2\eta}, \\ n\,\mathbf{E}\,(\hat{I}_n(t) - I_n(t))^2 \leqq C(t(1-t))^{1-2\eta}, \end{cases} \tag{2.5}$$

for all $t \in (0, 1)$ and $n \geqq 1$ where $\eta = \delta(4+2\delta)^{-1}$, $\Lambda_n = \sup_{1 \leqq i \leqq n} |g(i/n)|$ for g defined in (1.1) and C is some positive constant.

PROPOSITION 2.1. *Let $K > 0$ and $\delta \geqq 0$ be given. Then there exists a constant C_1 such that the following holds: if $\{X_{ni}\}$ is an array of r.v.'s satisfying* (2.5) *and if $\{c_{ni}\}$ are regression constants defined by* (1.1), *we have*

$$n\,\mathbf{E}\Big\{ \int_{-\infty}^{+\infty} J\Big(\frac{n}{n+1}\, \widehat{F}_n(x)\Big)\, d\widehat{H}_n(x)$$

$$(2.6)\qquad - \int_{-\infty}^{+\infty} J\Big(\frac{n}{n+1} F_n(x)\Big)\, d\widehat{H}_n(x)\Big\}^2 \leqq C_1 \Lambda_n^2 \|J\|_\delta^2,$$

$$(2.7)\qquad n\,\mathbf{E}\Big\{\int_{-\infty}^{+\infty} J\Big(\frac{n}{n+1} F_n(x)\Big)\, d(\widehat{H}_n - H_n)(x)\Big\}^2 \leqq C_1 \Lambda_n^2 \|J\|_\delta^2,$$

$$(2.8)\qquad n\Big\{\int_{-\infty}^{+\infty} \Big(J(F_n(x)) - J\Big(\frac{n}{n+1} F_n(x)\Big)\Big)\, dH_n(x)\Big\}^2 \leqq C_1 \Lambda_n^2 \|J\|_\delta^2,$$

$$(2.9)\qquad n\,\mathbf{E}\,(S_n(J))^2 \leqq C_1 \Lambda_n^2 \|J\|_\delta^2.$$

Proof. We only prove (2.6) because our method of proof is similar to that of Proposition 2 in [4]. It suffices to consider an increasing function $J \in \mathcal{H}_\delta$. Define

$$\varphi(x,\ t) = \begin{cases} 1 & \text{if } \dfrac{n}{n+1} F_n(x) \leqq t < \dfrac{n}{n+1} \widehat{F}_n(x), \\ -1 & \text{if } \dfrac{n}{n+1} \widehat{F}_n(x) \leqq t < \dfrac{n}{n+1} F_n(x), \\ 0 & \text{otherwise} \end{cases}$$

and denote by $\widehat{F}_n^{-1}(t) = \inf\{x \in \mathbf{R}\colon \widehat{F}_n(x) \geqq t\}$ the left continuous inverse of $\widehat{F}_n$. Since for $t \leqq n/(n+1)$,

$$\varphi(x,\ t) = \begin{cases} 1 & \text{if } \widehat{F}_n^{-1}\Big(\dfrac{n+1}{n} t\Big) \leqq x < F_n^{-1}\Big(\dfrac{n+1}{n} t\Big), \\ -1 & \text{if } F_n^{-1}\Big(\dfrac{n+1}{n} t\Big) \leqq x < \widehat{F}_n^{-1}\Big(\dfrac{n+1}{n} t\Big), \\ 0 & \text{otherwise} \end{cases}$$

it follows that for fixed $t \leqq n/(n+1)$,

$$\begin{aligned} \Big|\int_{-\infty}^{\infty} \varphi(x,\ t)\, d\widehat{H}_n(x)\Big| &\leqq \Lambda_n \int_{-\infty}^{\infty} |\varphi(x,\ t)|\, d\widehat{F}_n(x) \\ &= \Lambda_n \Big|\Big\langle \frac{n+1}{n} t\Big\rangle - \widehat{F}_n\Big(F_n^{-1}\Big(\frac{n+1}{n} t\Big) -\Big)\Big| \\ &\leqq \Lambda_n \Big(\Big|\widehat{F}_n\Big(F_n^{-1}\Big(\frac{n+1}{n} t\Big) -\Big) - \frac{n+1}{n} t\Big| + 2\inf\Big\{1/n,\ t\Big\}\Big) \end{aligned}$$

where $\langle t\rangle = (k-1)/n$ if $(k-1)/n < t \leqq k/n$.

From the assumption (2.5) on the sequence $\{X_{ni}\}$, it follows for $t \leqq n/(n+1)$ that

$$\begin{aligned} \Big(\int_{-\infty}^{+\infty} \varphi(x,\ t)\, d\widehat{H}_n(x)\Big)^2 &\leqq 8\Lambda_n^2 \Big(\frac{1}{n}(1 \wedge nt)\Big)^2 + 2\,\mathbf{E}\,\Big(\hat{I}_n\Big(\frac{n+1}{n} t\Big) - I_n\Big(\frac{n+1}{n} t\Big)\Big)^2 \\ &\leqq 8\Lambda_n^2 \Big(\frac{1}{n}(1 \wedge nt)\Big)^2 + 2\frac{C}{n}\Lambda_n^2 (t(1-t))^{1-2\eta}. \end{aligned}$$

Finally, interchanging the order of integration, and using the Cauchy–Schwarz inequal ity, we obtain

$$n\,\mathbf{E}\left(\int_{-\infty}^{+\infty} J\left(\frac{n}{n+1}\,\widehat{F}_n(x)\right) d\widehat{H}_n(x) - \int_{-\infty}^{+\infty} J\left(\frac{n}{n+1}\,F_n(x)\right) d\widehat{H}_n(x)\right)^2$$

$$= n\,\mathbf{E}\left(\int_0^{n/(n+1)} \int_{-\infty}^{+\infty} \varphi(x,\ t)\, d\widehat{H}_n(x) dJ(t)\right)^2$$

$$\leqq n\left(\int_0^{n/(n+1)} \left(\mathbf{E}\left(\int_{-\infty}^{+\infty} \varphi(x,\ t)\, d\widehat{H}(x)\right)^2\right)^{1/2} dJ(t)\right)^2$$

$$\leqq nC_1\Lambda_n^2\left(\int_0^{n/(n+1)} \frac{1}{n}(1 \wedge nt)\, dJ(t)\right.$$

$$\left. + \int_0^{n/(n+1)} n^{-1/2}\big(t(1-t)\big)^{1/2-\eta} dJ(t)\right)^2 \leqq C_1\Lambda_n^2 \|J\|_\delta^2$$

where C_1 is some positive constant, since

$$n^{1/2} \int_0^{n/(n+1)} (1/n)\,(1 \wedge nt) dJ(t) \leqq C_1 \int_0^1 \big(t(1-t)\big)^{1/2} dJ(t).$$

The inequality (2.6) is proved.

PROPOSITION 2.2. *Let $\{X_{ni}\}$ satisfy condition (2.5) for some $\delta > 0$, and let the regression constants c_{ni} satisfy (1.1) and $\sup_{n\in\mathbf{N}} \Lambda_n < +\infty$. Assume that $K \subset \mathcal{H}_\delta$ is a subset possessing the following property: for every $J \in K$, there exists a normal distribution $\mathcal{N}(0,\ \sigma^2)$ where $0 < \sigma < +\infty$ such that $\mathcal{S}_n(J)$ converges in law to $\mathcal{N}(0,\ \sigma^2)$, then the $\|\ \|_\delta$-norm closure of K has the same property.*

Proof. Let $J_1 \in \overline{K}$ where $\overline{K}$ is the closure of K, and $J \in K$. By Proposition 2.1 and the fact that $\mathcal{S}_n(J_1) = \mathcal{S}_n(J) + \mathcal{S}_n(J_1 - J)$, the distributions of $\mathcal{S}_n(J_1)$ and $\mathcal{S}_n(J)$ are closed, uniformly in n, in the weak topology for sufficiently small $\|J - J_1\|_\delta$. This proves the proposition.

For two probability measures $\mathbf{P}$ and $\mathbf{Q}$ on $\mathbf{R}$, denote by $D_2(\mathbf{P},\ \mathbf{Q}) = \inf(\mathbf{E}\,(X - Y)^2)^{1/2}$ where the infinitum extends over all r.v.'s X and Y defined on the same probability space and having distributions $\mathbf{P}$ and $\mathbf{Q}$, respectively.

Let $\mathcal{L}(Z)$ denote the distribution of the r.v. Z.

PROPOSITION 2.3. *Let $\{X_{ni}\}$ satisfy condition (2.5) for some $\delta > 0$ and let c_{ni} satisfy (1.1) and $\sup_{n\in\mathbf{N}^*} \Lambda_n < +\infty$. Assume that there exists an operator σ: $\mathcal{H}_\delta \to \mathbf{R}$ which is uniformly bounded and satisfies the Lipschitz condition for the $\|\ \|_\delta$ norm.*

If for every $J \in K \subset \mathcal{H}_\delta$, $\mathcal{S}_n(J)$ converges in law to a normal distribution $\mathcal{N}(0,\ \sigma^2(J))$, then the $\|\|_\delta$-norm closure of K has the same property.

Proof. Let $J_1 \in \overline{K}$ and $J \in K$. Then, we have

$$D_2\Big(\mathcal{L}\big(\mathcal{S}_n(J_1)\big),\ \mathcal{N}\big(0,\ \sigma^2(J_1)\big)\Big) \leqq D_2\Big(\mathcal{L}\big(\mathcal{S}_n(J_1)\big),\ \mathcal{L}\big(\mathcal{S}_n(J)\big)\Big)$$

$$+ D_2\Big(\mathcal{L}\big(\mathcal{S}_n(J)\big),\ \mathcal{N}\big(0,\ \sigma^2(J)\big)\Big) + D_2\Big(\mathcal{N}\big(0,\ \sigma^2(J)\big),\ \mathcal{N}\big(0,\ \sigma^2(J_1)\big)\Big)$$

$$\leqq D_2\Big(\mathcal{L}\big(\mathcal{S}_n(J)\big),\ \mathcal{N}\big(0,\ \sigma^2(J)\big)\Big) + C'\|J - J_1\|_\delta$$

for some $C' > 0$ using Proposition 2.1 and the Lipschitz condition on σ.

Since the convergence in law from $\mathcal{S}_n(J)$ to $\mathcal{N}(0,\ \sigma^2(J))$ implies (see [4])

$$\lim_{n\to\infty} D_2\Big(\mathcal{L}\big(\mathcal{S}_n(J)\big),\ \mathcal{N}\big(0,\ \sigma^2(J_1)\big)\Big) = 0,$$

the theorem follows.

3. Convergence of the linear rank statistic. Recall that the sequence $\{X_{ni}\}$ is φ-mixing if

$$\sup_{n\geqq 1}\sup_{1\leqq j\leqq n-m}\Big\{\sup\Big\{\Big|\mathbf{P}(B|\,A)-\mathbf{P}(B)\Big|;\ A\in\sigma\big(X_{ni},\ 1\leqq i\leqq j\big),$$
$$B\in\sigma\big(X_{ni},\ i\geqq j+m\big)\Big\}\Big\}=\varphi(m)\downarrow 0$$

for positive integers j and m.

Here $\sigma(X_{n1},\ldots,X_{nj})$ and $\sigma(X_{n,j+m},\ldots,X_{nn})$ are the σ-fields generated by $(X_{n1},\ldots,X_{nj})$ and $(X_{n,j+m},\ldots,X_{nn})$, respectively.

Also recall that $\{X_{ni}\}$ satisfies the strong mixing condition if

$$\sup_{n\geqq 1}\sup_{1\leqq j\leqq n-m}\Big\{\sup\Big\{\Big|\mathbf{P}(A\cap B)-\mathbf{P}(A)\,\mathbf{P}(B)\Big|;\ A\in\sigma\big(X_{ni},\ 1\leqq i\leqq j\big),$$
$$B\in\sigma\big(X_{ni},\ i\geqq j+m\big)\Big\}\Big\}=\alpha(m)\downarrow 0.$$

Since $\alpha(m)\leqq\varphi(m)$, it follows that if $\{X_{ni}\}$ is φ-mixing, then it is also strong mixing.

We will study the asymptotic behavior of $\mathcal{S}_n(J)$ when the r.v.'s $\{X_{ni}\}$ are φ-mixing with rates

$$\sum_{m=1}^{+\infty} m\big(\varphi(m)\big)^{(2+3\delta)/(4+2\delta)}<+\infty\qquad\text{for some}\quad 0\leqq\delta<2 \tag{3.1}$$

or strong mixing with rates

$$\sum_{m=1}^{+\infty} m^2\alpha(m)^{\delta/(2+\delta)}<+\infty\qquad\text{for some}\quad \delta>0. \tag{3.2}$$

Let $F_{n,i,j}$ be the d.f. of (X_{ni},X_{nj}), $1\leqq i\leqq j\leqq n$, $n\geqq 1$. For any sequence of d.f.'s $\{G_l^*,\ l\geqq 1\}$ on $[0,1]^2$ with uniform marginals, we denote

$$\sigma_J^2(\{G_l^*\})=\lim_{n\to\infty}\left\{\int_0^1 f^2(u)\,du+2\sum_{l=1}^{n}\int_0^1\int_0^1 f(u)f(v)\,dG_l^*(u,\ v)\right\} \tag{3.3}$$

if the limit exists, where

$$f(u)=\int_0^1\big(\mathbf{1}_{[u\leqq v]}-v\big)\,dJ(v)+J(u)-\int_0^1 J(v)\,dv \tag{3.4}$$

and $J\in\mathcal{H}_\delta$ for some $\delta>0$.

THEOREM 3.1. *Suppose the sequence $\{X_{ni}\}$ is φ-mixing with rate* (3.1) *or strong mixing with rate* (3.2), *the function g which defines the regression constants in* (1.1) *belongs to $\mathbf{C}^*_{1,b}$ the space of functions which admit a derivative of bounded variation, and suppose that for any $l>1$, there exists a continuous d.f. G_l^* on $[0,1]^2$ with uniform marginals such that*

$$\lim_{n\to\infty}\max_{1\leqq i<j\leqq n}\Big|F_{n,i,j}\big(F_n^{-1}(u),\ F_n^{-1}(v)\big)-G_{j-i}^*(u,\ v)\Big|=0 \tag{3.5}$$

for all (u, v) $[0, 1]^2$.

Then, for every $J \in \mathcal{G}_\delta$ *with* $2 > \delta \geqq 0$ *if we have* (3.1), *and* $\delta > 0$ *if we have* (3.2)

$$\lim_{n\to\infty} D_2\Big(\mathcal{L}(\mathcal{S}_n(J)),\ \mathcal{N}(0,\ \tilde{\sigma}_J^2(\{G_l^*\}))\Big) = 0 \tag{3.6}$$

where

$$\tilde{\sigma}_J^2(\{G_l^*\}) = \sigma_J^2(\{G_l^*\}) \times \Big(\int_0^1\int_0^1 (u \wedge v)\, g'(u)\, g'(v)\, dudv - 2\int_0^1 ug'(u)\, du + g^2(1)\Big) \tag{3.7}$$

and $\tilde{\sigma}_J^2(\{G_l^*\}) < \infty$.

Remark 3.1. Let the sequence of distribution functions $\{F_{n,i,j}\}$ satisfy the following conditions:

(i) There exists a sequence of d.f.'s F_l^* and $\mathbf{R}^2$ such that

$$\lim_{n\to\infty} \max_{1\leqq i<j\leqq n} \Big|F_{n,i,j}(x_1,\ x_2) - F_{j-i}^*(x_1,\ x_2)\Big| = 0 \qquad \text{for all} \quad (x_1,\ x_2) \in \mathbf{R}^2,$$

(ii) $F_{ni} = F_n$ for all $1 \leqq i \leqq n$, $n \geqq 1$,

then the condition (3.5) is satisfied whenever the sequence $\{X_{ni}\}$ is strong mixing.

Proof. To prove Theorem 3.1, we first need a few lemmas.

For any n $(n \geqq 1)$, and i $(1 \leqq i \leqq n)$ and any $J \in \mathbf{C}_{2,b}$, let

$$A_{ni}(J) = \int_0^1 \Big(\mathbf{1}_{[F_n(X_{ni})\leqq t]} - G_{ni}(t)\Big) J'(t)\, dG_{ni}(t) + J\big(F_n(X_{ni})\big) - \int_0^1 J(t)\, dG_{ni}(t). \tag{3.8}$$

It is obvious that $\mathbf{E}\,(A_{ni}) = 0$.

Now consider for any $J \in \mathbf{C}_{2,b}$, the process $L_n(J)(s)$ defined on $\mathbf{C}_1$ the space of continuous functions on $[0, 1]$, by

$$L_n(J)(s) = n^{-1/2}\left(\sum_{i=1}^{[ns]} A_{ni} + \big(ns - [ns]\big)\, A_{n,[ns]+1}\right), \tag{3.9}$$

where $[ns]$ denotes the integer part of the real number ns.

LEMMA 3.1. *Suppose that* $\{X_{ni}\}$ *satisfies the conditions of Theorem* 3.1 *and* J *belongs to* $\mathbf{C}_{2,b}$, *then the process* $L_n(J)(s)$ *converges weakly in uniform topology to a Gaussian process* $L_0(J)(s)$ *with trajectories a.s. in* $\mathbf{C}_1$ *with mean* 0 *and variance* $s\sigma_J^2(\{G_l^*\})$ *where* $\sigma_J^2(\{G_l^*\})$ *is defined in* (3.3), *and* $\sigma_J^2(\{G_l^*\}) < \infty$.

Proof. The process $L_n(J)$ defines a probability measure $\mathbf{P}_n$ on $\mathbf{C}_1$. From Theorem 8.1 of [2] we have to prove that (i) the finite-dimensional distribution of $\mathbf{P}_n$ converges in law to a normal distribution and (ii) $\mathbf{P}_n$ is tight.

First we prove (i), which is equivalent to proving that $\sum_{l=1}^{p} \lambda_l L_n(J)(s_l)$ converges in law to a normal distribution for any $p \in \mathbf{N}^*$, any $l \in [0, 1]$ and any $\lambda_l \in \mathbf{R}$ $(1 \leqq l \leqq p)$. Without loss of generality, we can take $p = 2$ and suppose that $s_1 < s_2$.

We have

$$\sum_{l=1}^{2} \lambda_l L_n(J)(s_l) = n^{-1/2}\Bigg[\sum_{i=1}^{[ns_1]} (\lambda_1+\lambda_2)\, A_{ni}(J) + \sum_{i=[ns_1]+1}^{[ns_2]} \lambda_2 A_{ni}(J)$$

$$+ \lambda_1\big(ns_1 - [ns_1]\big)\, A_{n,[ns_1]+1}(J)$$

$$+ \lambda_2\big(ns_2 - [ns_2]\big)\, A_{n,[ns_2]+1}(J)\Bigg]. \tag{3.10}$$

We define the sequence of r.v.'s $\{B_{ni}(J)\}$ by

$$B_{ni}(J) = \begin{cases} (\lambda_2+\lambda_2)\, A_{ni}(J) & \text{if } \ i \leqq [ns_1], \\ \lambda_2 A_{ni}(J) & \text{if } \ [ns_1] < i \leqq [ns_2], \\ 0 & \text{if } \ i > [ns_2]. \end{cases} \tag{3.11}$$

As J and J' are bounded, we deduce

$$\sum_{l=1}^{2} \lambda_l L_n(J)(s_l) = n^{-1/2} \sum_{i=1}^{n} B_{ni}(J) + O(n^{-1/2}). \tag{3.12}$$

From [11, Corollary 1] we have to verify that

$$\mathbf{E}\,\frac{1}{n}\left(\sum_{i=1}^{n} B_{ni}(J)\right)^2 \longrightarrow \Big\{(\lambda_1+\lambda_2)^2 s_1 + \lambda_2^2(s_2-s_1)\Big\}\,\sigma_J^2(\{G_l^*\})$$

$$\text{as} \quad n \to \infty. \tag{3.13}$$

We have

$$\mathbf{E}\,\frac{1}{n}\left(\sum_{i=1}^{n} B_{ni}(J)\right)^2 = \frac{1}{n}\Bigg[(\lambda_1+\lambda_2)^2 \sum_{i=1}^{[ns_1]}\sum_{j=1}^{[ns_1]} \mathbf{E}\left(A_{ni}(J)\, A_{nj}(J)\right)$$

$$+ (\lambda_1+\lambda_2)\lambda_2 \sum_{i=1}^{[ns_1]} \sum_{j=[ns_1]+1}^{[ns_2]} \mathbf{E}\left(A_{ni}(J)\, A_{nj}(J)\right)$$

$$+ \lambda_2^2 \sum_{i=[ns_1]+1}^{[ns_2]} \sum_{j=[ns_1]+1}^{[ns_2]} \mathbf{E}\left(A_{ni}(J)\, A_{nj}(J)\right)\Bigg]. \tag{3.14}$$

Suppose the sequence $\{X_{ni}\}$ is φ-mixing, then from the boundedness of J and J' and from the well-known inequality on the moment of φ-mixing r.v.'s (see [5, Prop. 2.2]), we obtain

$$\frac{1}{n}\left|(\lambda_1+\lambda_2)\lambda_2 \sum_{i=1}^{[ns_1]} \sum_{j=[ns_1]+1}^{[ns_2]} \mathbf{E}\left(A_{ni}(J)\, A_{nj}(J)\right)\right|$$

$$\leqq \frac{M}{n} \sum_{i=1}^{[ns_1]} \sum_{j=[ns_1]+1}^{[ns_2]} \varphi^{1/p}(j-i) \tag{3.15}$$

where M is some positive constant and $p = (4+2\delta)/(2+3\delta)$. From (3.1), the last expression goes to zero as $n \to \infty$.

If the sequence $\{X_{ni}\}$ is strong mixing, the left-hand side of (3.15) is majorized by

$$\frac{M'}{n} \sum_{i=1}^{[ns_1]} \sum_{j=[ns_1]+1}^{[ns_2]} \alpha^{\delta/(2+\delta)}(j-i),$$

where $M' > 0$ is some constant and from (3.2), this converges to 0 as $n \to \infty$.

It remains to prove that

$$\tag{3.16} \begin{aligned} &\frac{1}{n} \sum_{i=1}^{[ns_1]} \sum_{j=1}^{[ns_1]} \mathbf{E}\left(A_{ni}(J)\, A_{nj}(J)\right) \longrightarrow s_1 \sigma_J^2(\{G_l^*\}) \quad \text{as} \quad n \to \infty, \\ &\frac{1}{n} \sum_{i=[ns_1]+1}^{[ns_2]} \sum_{j=[ns_1]+1}^{[ns_2]} \mathbf{E}\left(A_{ni}(J)\, A_{nj}(J)\right) \longrightarrow (s_2 - s_1) \sigma_J^2(\{G_l^*\}) \quad \text{as} \quad n \to \infty. \end{aligned}$$

We first prove the convergence of (3.16) when J and J' are replaced by indicator functions. Suppose

$$J(t) = \mathbf{1}_{[a \leqq t \leqq b]} \qquad \text{and} \qquad J' = \mathbf{1}_{[a' \leqq t \leqq b']}.$$

Then we can write

$$\begin{aligned} A_{ni}(J) = {} & G_{ni}(b') - G_{ni}\left(\{a' \vee F_n(X_{ni})\} \wedge b'\right) \\ & - \int_a^b t\, dG_{ni}(t) + \mathbf{1}_{[a \leqq F_n(X_{ni}) \leqq b]} - \left(G_{ni}(b) - G_{ni}(a)\right) = D_{ni}(X_{ni}^*) \end{aligned}$$

where $X_{ni}^* = F_n(X_{ni})$. Let $G_{n,i,j}$ be the d.f. of $(F_n(X_{ni}), F_n(X_{nj}))$. Then, we have

$$\mathbf{E}\left(A_{ni}(J) A_{nj}(J)\right) = \int_0^1 \int_0^1 D_{ni}(u)\, D_{nj}(v)\, dG_{n,i,j}(u,\ v).$$

From condition (3.5) we easily deduce that

$$\tag{3.17} \lim_{n\to\infty} \max_{1 \leqq i < j \leqq n} \left| \mathbf{E}\left(A_{ni}(J) A_{nj}(J)\right) - \int_0^1 \int_0^1 D(u) D(v)\, dG_{j-i}^*(u,\ v) \right| = 0$$

where

$$D(u) = \mathbf{1}_{[a \leqq u \leqq b]} - (b-a) + b' - \{(a' \vee u) \wedge b'\} - \tfrac{1}{2}(b'-a')^2.$$

We obtain the same result if J and J' are replaced by step functions.

As J and J' are continuous and bounded, we can uniformly approach them by step functions and we deduce that

$$\tag{3.18} \lim_{n\to\infty} \max_{1 \leqq i < j \leqq n} \left| \mathbf{E}\left(A_{ni}(J) A_{nj}(J)\right) - \int_0^1 \int_0^1 f(u) f(v)\, dG_{j-i}^*(u,\ v) \right| = 0$$

where $f(u)$ is defined in (3.4).

Now denote $\rho(0) = \int_0^1 f^2(u)du$ and $\rho(i) = 2\int_0^1 \int_0^1 f(u)f(v)\,dG_i^*(u,v)$, $i \geqq 1$. Then,

$$\left| n^{-1} \sum_{i=1}^{[ns_1]} \sum_{j=1}^{[ns_1]} \mathbf{E}\left(A_{ni}(J)\, A_{nj}(J)\right) - [ns_1]\, n^{-1} \sum_{i=0}^{+\infty} \rho(i) \right|$$

$$\leqq \left| n^{-1}[ns_1]\left([ns_1]\right)^{-1} \sum_{i=0}^{[ns_1]-1} \sum_{j=1}^{[ns_1]-i} \mathbf{E}\left(A_{nj}(J)\, A_{n,j+i}(J)\right) \right.$$

$$\left. - [ns_1]\left(n[ns_1]\right)^{-1} \sum_{i=0}^{[ns_1]} \left([ns_1]-i\right)\rho(i) \right| + [ns_1]n^{-1} \sum_{i=[ns_1]+1}^{\infty} |\rho(i)|$$

$$+ [ns_1]n^{-1} \sum_{i=0}^{[ns_1]} \sum_{k=i}^{\infty} |\rho(k)| = |A_n| + B_n + C_n.$$

From (3.18) we deduce that $|A_n| \to 0$ as $n \to \infty$ and from the well-known inequalities on the moment of mixing r.v.'s (see [5, Props. 2.2 and 2.8]) and (3.1) or (3.2) we deduce that $B_n \to 0$ and $C_n \to 0$ and $n \to \infty$.

It is also immediate that

$$\left| [ns_1]n^{-1} \sum_{i=0}^{+\infty} \rho(i) - s_1 \sum_{i=0}^{+\infty} \rho(i) \right| \longrightarrow 0 \qquad \text{as} \quad n \to \infty.$$

We conclude that $n^{-1} \sum_{i=1}^{[ns_1]} \sum_{j=1}^{[ns_1]} \mathbf{E}\left(A_{ni}(J)A_{nj}(J)\right)$ converges to $s_1(\sum_{i=0}^{+\infty} \rho(i))$ as $n \to \infty$ where $\sum_{i=0}^{+\infty} \rho(i)$ is equal to $\sigma_J^2(\{G_l^*\})$. Similarly

$$\frac{1}{n} \sum_{i=[ns_1]+1}^{[ns_2]} \sum_{j=[ns_1]+1}^{[ns_2]} \mathbf{E}\left(A_{ni}(J)\, A_{nj}(J)\right) \longrightarrow (s_2 - s_1)\sigma_J^2(\{G_l^*\})$$

as $n \to \infty$, so (3.16) is proved.

From (3.14)–(3.16), we deduce (3.13) and we conclude that $\mathbf{E}\,(\sum_{l=1}^{2} \lambda_l L_n(J)(s_l))^2$ converges to $\{(\lambda_1+\lambda_2)^2 s_1 + \lambda_2^2(s_2-s_1)\}\sigma_J^2(\{G_l^*\})$ which implies that $\sum_{l=1}^{2} \lambda_l L_n(J)(s_l)$ converges in law to the normal distribution with mean 0 and variance $\{(\lambda_1+\lambda_2)^2 s_1 + \lambda_2^2(s_2 - s_1)\}\sigma_J^2(\{G_l^*\})$ and (i) is proved.

We now prove (ii).

From [2, Thm. 8.2] we have to verify that $\forall \varepsilon > 0\ \exists \eta > 0\ (0 < \eta < 1)$ and an integer N_0 such that $\forall n \geqq N_0$

$$\mathbf{P}\left[\sup_{|s-s'|<\eta} |L_n(J)(s) - L_n(J)(s')| \geqq \varepsilon \right] \leqq \varepsilon. \tag{3.19}$$

If ns and ns' are integers, by using Theorem 11 from [5] for $q = 4$ for the strong mixing case and Lemma 5.1 in Harel (1988) for $q = 2$ for the φ-mixing case, we obtain for $s \geqq s'$

$$\mathbf{E}\left(L_n(J)(s) - L_n(J)(s')\right)^4 \leqq \left((s-s')^2 + n^{-1}(s-s')\right) MC(\beta) \tag{3.20}$$

where

$$C(\beta) = \sum_{m=1}^{+\infty} m^{-1}\varphi^{1/4}(m) \tag{3.21}$$

if the sequence $\{X_{ni}\}$ is φ-mixing, and

$$C(\beta) = \sum_{m=1}^{+\infty} m^{2}\alpha^{\delta/(2+\delta)}(m) \tag{3.22}$$

if the sequence $\{X_{ni}\}$ is strong mixing and M is some positive constant.

If $s > s'$ and ns and ns' are integers, we have $s - s' \geqq n^{-1}$ and

$$\mathbf{E}\left(L_n(J)(s) - L_n(J)(s')\right)^4 \leqq 2M(s-s')^2 C(\beta).$$

From [1, Lemma 2] we obtain that for any $\varepsilon > 0$ there exist $\eta > 0$, and an integer N_0 sufficiently large such that $\forall n \geqq N_0$,

$$\begin{aligned}\mathbf{P}&\left[\sup_{|[ns]/n-[ns']/n|<2\eta} \left|L_n(J)\left([ns]/n\right) - L_n(J)\left([ns']/n\right)\right| > \varepsilon/2\right] \\ &\leqq 2MC(\beta)K\eta^2\varepsilon^{-4}\end{aligned} \tag{3.23}$$

where K is some positive constant.

From the definition of $L_n(J)(s)$ in (3.9), we obtain

$$\begin{aligned}&\sup_{|s-s'|<\eta} \left|L_n(J)(s) - L_n(J)(s')\right| \\ &\leqq 2 \max_{|[ns]/n-[ns']/n|<2\eta} \left|L_n(J)\left([ns]/n\right) - L_n(J)\left([ns']/n\right)\right|.\end{aligned} \tag{3.24}$$

By using (3.23) and (3.24), we deduce

$$\mathbf{P}\left[\sup_{|s-s'|<\eta} \left|L_n(J)(s) - L_n(J)(s')\right| \geqq \varepsilon\right] \leqq MC(\beta)K'\eta^2\varepsilon^{-4} \tag{3.25}$$

where K' is some positive constant and (3.19) is proved. The fact that $\sigma_J^2(\{G_l^*\}) < \infty$ is a simple consequence of $J \in \mathbf{C}_{2,b}$ and (3.1) or (3.2).

Now we consider for any $J \in \mathbf{C}_{2,b}$ the r.v. $V_n(J)$ defined by

$$V_n(J) = n^{-1/2}\sum_{i=1}^{n} c_{ni}A_{ni}. \tag{3.26}$$

LEMMA 3.2 *Suppose that $\{X_{ni}\}$ satisfies the conditions of Theorem 3.1, J belongs to $\mathbf{C}_{2,b}$ and g admits a derivative g'. Then $V_n(J)$ converges in law to the normal distribution with mean 0 and variance $\widetilde{\sigma}_J^2(\{G_l^*\})$ where $\widetilde{\sigma}_J^2(\{G_l^*\})$ is defined in (3.7) and $\widetilde{\sigma}_J^2(\{G_l^*\}) < \infty$.*

Proof. For any n define a measure λ_n on [0, 1] by setting

$$\lambda_n\left(\{i/n\}\right) = c_{ni} - c_{n,i+1}, \qquad 1 \leqq i \leqq n-1, \quad \text{and} \quad \lambda_n\left(\{1\}\right) = c_{nn}.$$

By definition, we have

$$V_n(J) = \int_0^1 L_n(J)(u)\lambda_n(du).$$

We now prove that

$$\int_0^1 L_n(J)(u)\lambda_n(du) \qquad \text{converges in law to}$$

$$(3.27) \qquad -\int_0^1 L_0(J)(u)g'(u)du + L_0(J)(1)g(1) \qquad \text{as} \quad n \to \infty.$$

Let h_n: $\mathbf{C}_1 \to \mathbf{R}$, $n \geqq 1$, be defined as $h_n(f) = \int_0^1 f(u)\lambda_n(du)$ and h_0: $\mathbf{C}_1 \to \mathbf{R}$ be defined as $h_0(f) = -\int_0^1 fg'(u)du + f(1)g(1)$. Let $\{f_n,\ n \geqq 1\}$ be a sequence of functions in $\mathbf{C}_1$ and suppose that $f_n \to f_0$ in uniform topology as $n \to \infty$ where $f_0 \in \mathbf{C}_1$. We show that $h_n(f_n) \to h_0(f_0)$ as $n \to \infty$.

$$\begin{aligned}
&\left|\int_0^1 f_n(u)\lambda_n(du) - \left(-\int_0^1 f_0(u)g'(u) + f_0(1)g(1)\right)\right| \\
&\leqq \left|\int_0^1 (f_n(u) - f_0(u))\lambda_n(du)\right| \\
&\quad + \left|\int_0^1 f_0(u)\lambda_n(du) + \int_0^1 f_0(u)g'(u) - f_0(1)g(1)\right| \\
&\leqq \sup_{u\in[0,1]} \left|f_n(u) - f_0(u)\right| \left|\int_0^1 \lambda_n(du)\right| \\
&\quad + \left|\sum_{i=1}^{n-1} f_0(i/n)\Big(g(i/n) - g((i+1)/n)\Big) + f_0(1)g(1)\right. \\
&\qquad \left. + \int_0^1 f_0(u)g'(u)\,du - f_0(1)g(1)\right| \longrightarrow 0 \qquad \text{as} \quad n \to \infty
\end{aligned}$$

from the hypothesis $f_n \to f_0$ in uniform topology, g' is an integrable function and $\sup_{n\in\mathbf{N}^*} \Lambda_n < \infty$.

Consequently, $h_n(f_n) \to h_0(f_0)$ as $n \to \infty$ and by [2, Thm. 5.5] (3.27) follows. It remains to show that

$$\mathbf{E}\left(-\int_0^1 L_0(J)(u)g'(u)du + L_0(J)(1)g(1)\right)^2 = \widetilde{\sigma}_J(\{G_l^*\}) < \infty.$$

We have

$$\begin{aligned}
&\mathbf{E}\left(-\int_0^1 L_0(J)(u)g'(u)\,du + L_0(J)(1)g(1)\right)^2 \\
&= \int_0^1\int_0^1 \mathbf{E}\left[L_0(J)(u)L_0(J)(v)\right] g'(u)g'(v)\,du\,dv \\
&\quad - 2\int_0^1 \mathbf{E}\left[L_0(J)(1)L_0(J)(u)\right] g'(u)\,du + \mathbf{E}\left[L_0(J)(1)L_0(J)(1)\right] g^2(1).
\end{aligned}$$

As $\mathbf{E}\left[L_0(J)(u)L_0(J)(v)\right] = (u \wedge v)\sigma_J^2(\{G_l^*\}) < \infty$, the property follows and Lemma 3.2 is proved.

Proof of the theorem. We first prove that the theorem is true for $J \in \mathbf{C}_{2,b}$. We have the following decomposition:

$$S_n(J) = V_n(J) + A_n(J) + B_n(J) + C_n(J),$$

where

$$A_n(J) = n^{1/2} \int_0^1 J'\big(I_n(t)\big)\big(\hat{I}_n(t) - I_n(t)\big)\, d\big(\widehat{G}_n - G_n\big)(t),$$

$$B_n(J) = -(n+1)^{-1} n^{1/2} \int_0^1 J'\big(I_n(t)\big)\hat{I}_n(t)\, d\widehat{G}_n(t),$$

$$C_n(J) = 2^{-1} n^{1/2} \int_0^1 J''\Big(\theta_n\big(I_n(t)\big)\Big)\Big(n(n+1)^{-1}\big(\hat{I}_n(t) - I_n(t)\big)\Big)^2 d\widehat{G}_n(t),$$

where $\theta_n(I_n(t)) \in [I_n(t) \wedge \hat{I}_n(t),\ I_n(t) \vee \hat{I}_n(t)]$.

Suppose $J \in \mathbf{C}_{2,b}$. Then the weak convergence of $V_n(J)$ is established in Lemma 3.2. The random variables $A(J)$, $B(J)$, $C(J)$ converge to zero in probability and in $\mathbf{L}_2$, since

$$\mathbf{E}\left(A_n^2(J)\right) \leqq Kn^{-1} \sup_{t\in[0,1]} |J'(t)|^2 \sup_{n\in\mathbf{N}^*} \Lambda_n^2 C(\beta), \tag{3.28}$$

$$|B_n(J)| \leqq Kn^{-1/2} \sup_{t\in[0,1]} |J'(t)| \sup_{n\in\mathbf{N}^*} \Lambda_n, \tag{3.29}$$

$$\mathbf{E}\left(C_n^2(J)\right) \leqq Kn^{-1} \sup_{t\in[0,1]} |J'(t)|^2 \sup_{n\in\mathbf{N}^*} \Lambda_n^2 C(\beta) \tag{3.30}$$

where $C(\beta) = \sum_{m=1}^{+\infty} m(\varphi(m))^{(2+3\delta)/(4+2\delta)}$ if we have (3.1), and $C(\beta) = m^2 \times \sum_{m=1}^{+\infty}(\alpha(m))^{\delta/(2+\delta)}$ if we have (3.2), and K is some positive constant.

We only prove the inequality (3.28) for the φ-mixing case, because the method is similar to the proof of the three inequalities in [4, p. 66]. For any $(i,\ j,\ l,\ q) \in \mathbf{N}^4$, we put

$$\beta(i,\ j,\ l,\ q) = \left(\int_0^1 J'\big(I_n(t)\big)c_{ni}\big(\mathbf{1}_{[X_{ni}\leqq t]} - F_{ni}(t)\big)\, d\big(\mathbf{1}_{[X_{nj}\leqq t]} - F_{nj}(t)\big)\right)$$
$$\times \left(\int_0^1 J'\big(I_n(u)\big)c_{nl}\big(\mathbf{1}_{[X_{nl}\leqq u]} - F_{nl}(u)\big)\, d\big(\mathbf{1}_{[X_{nq}\leqq u]} - F_{nq}(u)\big)\right).$$

Suppose $i \leqq j \leqq l \leqq q$ and let $p = (2+3\delta)/(4+2\delta)$, then from the condition of φ-mixing, we have the following three inequalities:

$$\beta(i,\ j,\ l,\ q) \leqq \sup_{t\in[0,1]} |J'(t)|^2 \Lambda_n^2 \varphi(j-i),$$

$$\beta(i,\ j,\ l,\ q) \leqq \Big(2 \sup_{t\in[0,1]} |J'(t)|^2 \Lambda_n^2\Big)\Big(\varphi(l-j) + 4\varphi^{1/p}(j-i)\varphi^{1/p}(q-l)\Big),$$

$$\beta(i,\ j,\ l,\ q) \leqq \sup_{t\in[0,1]} |J'(t)|^2 \Lambda_n^2 \varphi(q-l).$$

If i, j, l, q are differently ordered, we obtain similar inequalities.

From this, we deduce

$$\mathbf{E}\left(A_n^2(J)\right) \leqq 4!\, n^{-3} \sup_{t\in[0,1]} |J'(t)|^2 \Lambda_n^2 \sum_{i,j,l,q} \beta(i,\ j,\ l,\ q).$$

Put $j' = j - i$, $l' = l - j$ and $q' = q - l$. We have

$$\mathbf{E}\left(A_n^2(J)\right) \leqq 4!\,n^{-3} \sup_{t\in[0,1]} \left|J'(t)\right|^2 \Lambda_n^2$$
$$\times \sum_{i=1}^{n} \left(\sum_{\substack{0\leqq l'\leqq j' \\ 0\leqq q'\leqq j'}} 2\varphi(j') + \sum_{\substack{0\leqq j'\leqq q' \\ 0\leqq l'\leqq q'}} 2\varphi(q') + \sum_{\substack{0\leqq j'\leqq l' \\ 0\leqq q'\leqq l'}} 2\varphi(l') + 4\varphi^{1/p}(j')\varphi^{1/p}(q') \right)$$

and after some computations, we obtain

$$\mathbf{E}\left(A_n^2(J)\right) \leqq 288 n^{-3} \sup_{t\in[0,1]} \left|J'(t)\right|^2 \Lambda_n^2 \Big(\sum_{m\geqq 1} \varphi^{1/p}(m) \Big) \Big(\sum_{m\geqq 1} m\varphi^{1/p}(m)/m \Big) (n+n^2)$$
$$\leqq K n^{-1} \sup_{t\in[0,1]} \left|J'(t)\right|^2 \sup_{n\in\mathbf{N}^*} \Lambda_n^2 C(\beta)$$

where K is some positive constant and $C(\beta) = \sum_{m\geqq 1} m(\varphi(m))^{(2+3\delta)/(4+2\delta)}$. The inequality (3.28) is proved for the φ-mixing case. Hence the theorem is true for $J \in \mathbf{C}_{2,b}$.

Following Proposition 2.3 it remains to prove that the operator σ: $\mathcal{H}_\delta \to \mathbf{R}$ defined by $\sigma(J) = \widetilde{\sigma}_J(\{G_l^*\})$ satisfies the Lipschitz condition for the $\|\ \|_\delta$ norm and the condition (2.5) is satisfied. The first property follows easily from the definition of $\widetilde{\sigma}_J(\{G_l^*\})$ in (3.7) and the definition of $\sigma_J(\{G_l^*\})$ in (3.3) and (3.4).

We now prove (2.5) if we have (3.1). For $p = (4+2\delta)/(2+3\delta)$ and $q = (1-p^{-1})^{-1}$ we have

$$n\,\mathbf{E}\left(\widehat{G}_n(t) - G_n(t)\right)^2 = n\,\mathbf{E}\left[n^{-1} \sum_{i=1}^{n} c_{ni}\left(\mathbf{1}_{[F_n(X_{ni})\leqq t]} - G_{ni}(t)\right)\right]^2$$
$$= n^{-1} \sum_{1\leqq i,j\leqq n} \mathbf{E}\left[c_{ni}c_{nj}\left(\mathbf{1}_{[F_n(X_{ni})\leqq t]} - G_{ni}(t)\right)\left(\mathbf{1}_{[F_n(X_{nj})\leqq t]} - G_{nj}(t)\right)\right]$$
$$\leqq 2n^{-1}\Lambda_n^2 \sum_{i=0}^{n-1}\sum_{j=1}^{n-i} \mathbf{E}\left|\left(\mathbf{1}_{[F_n(X_{ni})\leqq t]} - G_{ni}(t)\right)\left(\mathbf{1}_{[F_n(X_{n,j+i})\leqq t]} - G_{n,j+1}(t)\right)\right|$$
$$\leqq 2n^{-1}\Lambda_n^2 \sum_{i=0}^{n-1} \varphi^{1/p}(i) \sum_{j=1}^{n-i} \left(G_{n,j}(t)(1-G_{n,j}(t))\right)^{1/p}$$
$$\times \left(G_{n,j+i}(t)(1-G_{n,j+i}(t))\right)^{1/q}$$
$$\leqq 2n^{-1}\Lambda_n^2 \sum_{i=1}^{n-1} \varphi^{1/p}(i) \left(\sum_{j=1}^{n} G_{n,j}(t)(1-G_{n,j}(t)) \right)^{1/p}$$
$$\times \left(\sum_{j=1}^{n} G_{n,j}(t)(1-G_{n,j}(t)) \right)^{1/q}$$
$$\leqq 2n^{-1}\Lambda_n^2 \sum_{i=1}^{n-1} \varphi^{1/p}(i) \left(\sum_{j=1}^{n} G_{n,j}(t)(1-G_{n,j}(t)) \right)$$
$$\leqq 2n^{-1}\Lambda_n^2 \sum_{i=1}^{n-1} \varphi^{1/p}(i) n \left(\frac{1}{n} \sum_{j=1}^{n} G_{nj}(t) \right) \left(1 - n^{-1} \sum_{j=1}^{n} G_{nj}(t) \right)$$

$$= 2\Lambda_n^2 \sum_{i=1}^{n-1} \varphi^{1/p}(i)t(1-t) \leqq 2\left(\sum_{i=1}^{+\infty} \varphi^{1/p}(i)\right)\Lambda_n^2 t(1-t),$$

which implies (2.5) if we have (3.1).

Finally, we prove (2.5) if we have (3.2). We have in analogy with preceding arguments

$$n\,\mathbf{E}\left(\widehat{G}(t) - G_n(t)\right)^2 \leqq 2n^{-1}\Lambda_n^2 \sum_{i=0}^{n-1} (\alpha(i))^{\delta/(2+\delta)} \sum_{j=1}^{n-i} \Big(G_{nj}(t)(1 - G_{nj}(t))\Big)^{1/(2+\delta)} \Big(G_{n,j+i}(t)(1 - G_{n,j+i}(t))\Big)^{1/(2+\delta)}$$

$$\leqq 2n^{-1}\Lambda_n^2 \sum_{i=0}^{n-1} (\alpha(i))^{\delta/(2+\delta)} \left(\sum_{j=1}^{n} G_{nj}(1 - G_{nj}(t))\right)^{1/(2+\delta)} \times \left(\sum_{j=1}^{n} G_{n,j}(t)(1 - G_{n,j}(t))\right)^{1/(2+\delta)}$$

$$= 2n^{-1}\Lambda_n^2 \sum_{i=0}^{n-1} (\alpha(i))^{2/(2+\delta)} \left(\sum_{j=1}^{n} G_{nj}(t)(1 - G_{nj}(t))\right)^{2/(2+\delta)}$$

$$\leqq 2n^{-1}\Lambda_n^2 \sum_{i=0}^{n-1} (\alpha(i))^{\delta/(2+\delta)} n\left(n^{-1}\sum_{j=1}^{n} G_{nj}(t)\left(1 - n^{-1}\sum_{j=1}^{n} G_{nj}(t)\right)\right)^{2/(2+\delta)}$$

$$= 2\Lambda_n^2 \sum_{i=0}^{n-1} (\alpha(i))^{\delta/(2+\delta)} (t(1-t))^{2/(2+\delta)}$$

$$\leqq 2\Lambda_n^2 \sum_{i=0}^{+\infty} (\alpha(i))^{2/(2+\delta)} (t(1-t))^{1-2\eta} \qquad \text{for} \quad \eta = \delta(4+2\delta)^{-1}.$$

Thus (2.5) is proved if we have (3.2).

COROLLARY 3.1. *If among conditions in Theorem 3.1 the function g is replaced by a function for which there exists a decomposition $g = g_c + g_d$ where $g_c \in \mathbf{C}^*_{1,b}$ and g_d is a step function with p jumps, say at $a_1, \dots, a_p$ such that $a_i \in (0, 1)$ $(1 \leqq i \leqq p)$, then the conclusion of Theorem 3.1 remains true but $\widetilde{\sigma}_J(\{G_l^*\})$ defined in (3.7) is replaced by $\widehat{\sigma}^2_J(\{G_l^*\})$ where*

$$\widehat{\sigma}^2_J(\{G_l^*\}) = \sigma^2_J(\{G_l^*\})\Bigg(\int_0^1\int_0^1 (u \wedge v) g_c'(u) g_c'(v)\,du\,dv - 2\sum_{i=1}^{p} (g_d(a_i-) - g_d(a_i+)) \int_0^1 (u \wedge a_i) g_c'(u)\,du - 2\int_0^1 u g_c'(u)\,du + \sum_{1<i,j\leqq p} (a_i \wedge a_j)\big(g_d(a_i-) - g_d(a_i+)\big)\big(g_d(a_j-) - g_d(a_j+)\big) + 2\sum_{i=1}^{p} a_i\big(g_d(a_i-) - g_d(a_i+)\big) + g^2(1)\Bigg), \tag{3.31}$$

where $\sigma^2_J(\{G_l^*\})$ is defined in (3.3).

4. Convergence of the two-sample linear rank statistic. Let $\{Y_{n_1 i}\}$, $1 \leqq i \leqq n_1$, and $\{Z_{n_2 j}\}$, $1 \leqq j \leqq n_2$, be two independent sequences of weakly dependent random variables with continuous d.f.'s $F^{(1)}_{n_1 i}(x)$ and $F^{(2)}_{n_2 j}(x)$, respectively, $x \in \mathbf{R}$. Given $n = n_1 + n_2$ we set $X_{ni} = Y_{n_1 i}$ if $i \leqq n_1$ and $X_{ni} = Y_{n_2 i - n_1}$ if $i > n_1$. Denote by $\widehat{F}^{(1)}_{n_1}(x) = n_1^{-1} \sum_{i=1}^{n_1} c_{n_1 i} \mathbf{1}_{[X_{ni} \leqq x]}$ the empirical process based on the first sequence of r.v.'s $\{Y_{n_1 i}\}$ and weighted by the regression constants $c_{n_1 i}$. We put $F^{(1)}_{n_1} = \mathbf{E}(\widehat{F}^{(1)}_{n_1})$.

Then $\mathcal{S}^*_n(J)$ defined by

$$\mathcal{S}^*_n(J) = n^{1/2}\Big(\int_{-\infty}^{+\infty} J\Big(\frac{n}{n+1}\widehat{F}_n(x)\Big)\, d\widehat{F}^{(1)}_{n_1}(x) - \int_{-\infty}^{+\infty} J\left(F_n(x)\right) dF^{(1)}_{n_1}(x)\Big) \tag{4.1}$$

is the two-sample linear rank statistic. We suppose that the regression constants $c_{n_1 i}$ $(1 \leqq i \leqq n_1)$ are defined by a function h on $[0,\ 1]$ as

$$c_{n_1 i} = h(i/n_1), \qquad 1 \leqq i \leqq n_1, \quad n_1 \geqq 1.$$

We assume that $n_1 n^{-1} \to \lambda_0 \in (0,\ 1)$.

We have $F_n = n^{-1}(\sum_{i=1}^{n_1} F^{(1)}_{n_1 i} + \sum_{j=1}^{n_2} F^{(2)}_{n_2 j})$.

Let $F^{(1)}_{n_1, i, l}$ be the d.f. of $(Y_{n_1 i}, Y_{n_1 l})$ and $F^{(2)}_{n_2, j, k}$ be the d.f. of $(Z_{n_2 j}, Z_{n_2 k})$.

THEOREM 4.1. *Suppose the sequences* $\{Y_{n_1 i}\}$ *and* $\{Z_{n_2 j}\}$ *are* φ*-mixing with rate* (3.1) *or strong mixing with rate* (3.2), *the function* h *satisfies* $h = h_c + h_d$ *with* $h_c \in \mathbf{C}^*_{1,b}$ *and* h_d *is a step function and if for each* $p > 1$, *there exist two continuous d.f.'s* $\widehat{G}^{(1)}_p$ *and* $\widehat{G}^{(2)}_p$ *on* $\mathbf{R}^2$ *with marginals* $F^{(1)}$ *and* $F^{(2)}$ *such that*

$$\lim_{n\to\infty} \max_{1 \leqq i < j \leqq n} \left| F^{(l)}_{n_l, i, j}(F_n^{-1}(t_1),\ F_n^{-1}(t_2)) - \widehat{G}^{(l)}_{j-i}(H^{-1}(t_1),\ H^{-1}(t_2)) \right| = 0 \tag{4.2}$$

for all $(t_1,\ t_2) \in [0,\ 1]^2$, $l = 1,\ 2$, *where*

$$H = \lambda_0 F^{(1)} + (1 - \lambda_0)\, F^{(2)}. \tag{4.3}$$

Then, for every $J \in \mathcal{G}_\delta$ *with* $2 > \delta \geqq 0$ *if we have* (3.1) *and* $\delta > 0$ *if we have* (3.2)

$$\lim D_2\Big(\mathcal{L}(\mathcal{S}^*_n(J)),\ \mathcal{N}\big(0,\ \tilde{\sigma}_J(\{G^{(l)}_p\})\big)\Big) = 0 \tag{4.4}$$

where

$$\tilde{\sigma}^2_J(\{\widehat{G}^{(l)}_p\}) = \bar{\sigma}^2_J\Big(\{\widehat{G}^{(l)}_p\}\Big)\, L(h) \tag{4.5}$$

where

$$\begin{aligned} \bar{\sigma}^2_J\Big(\{G^{(l)}_p\}\Big) = \lambda_0^{-1}\Big\{ &\int_0^1 f_1^2(u)\, d(F^{(1)} \circ H^{-1})(u) \\ &+ 2\sum_{p \geqq 2} \int_0^1 \int_0^1 f_1(u) f_1(v)\, d\Big(\widehat{G}^{(1)}_p(H^{-1}(u),\ H^{-1}(v))\Big)\Big\} \\ + (1-\lambda_0)\Big\{ &\int_0^1 f_2^2(u)\, d(F^{(2)} \circ H^{-1})(u) \\ &+ 2\sum_{p \geqq 2} \int_0^1 \int_0^1 f_2(u) f_2(v)\, d\Big(\widehat{G}^{(2)}_p(H^{-1}(u),\ H^{-1}(v))\Big)\Big\} \end{aligned} \tag{4.6}$$

Here

$$f_1(u) = f_1^*(u) - \int_0^1 f_1^*(v)\, d\big(F^{(1)} \circ H^{-1}(v)\big),$$

$$f_2(u) = f_2^*(u) - \int_0^1 f_2^*(v)\, d\big(F^{(2)} \circ H^{-1}(v)\big),$$

with

$$f_1^*(u) = J(u) + \lambda_0 \int_u^1 J'(v)\, d\big(F^{(1)} \circ H^{-1}(v)\big),$$

$$f_2^*(u) = \int_u^1 J'(v)\, d\big(F^{(2)} \circ H^{-1}(v)\big),$$

and

$$\begin{aligned} L(h) = {} & \int_0^1 \int_0^1 h_c'(u) h_c'(v)\, du\, dv - 2 \sum_{i=1}^{p} \big(h_d(a_i-) - h_d(a_i+)\big) \int_0^1 h_c(u)\, du \\ & - 2 \int_0^1 h_c'(u)\, du + \sum_{1 \leqq i,j \leqq p} \big(h_d(a_i-) - h_d(a_i+)\big)\big(h_d(a_j-) - h_d(a_j+)\big) \\ & + 2 \sum_{i=1}^{p} \big(h_d(a_i-) - h_d(a_i+)\big) + h_c^2(1), \end{aligned} \tag{4.7}$$

where a_i, $1 \leqq i \leqq p$, are the discontinuous points of h_d.

Remark. Let the sequences $\{F_{n_1,i,j}^{(1)}\}$ and $\{F_{n_2,i,j}^{(2)}\}$ satisfy the following conditions:

(i) there exist two sequences of d.f.'s $G_p^{(l)}$ on $\mathbf{R}^2$, $l = 1, 2$, such that

$$\lim_{n \to \infty} \max_{1 \leqq i < j \leqq n} \big| F_{n_l,i,j}^{(l)}(x_1, x_2) - G_{j-1}^{(l)}(x_1, x_2) \big| = 0$$

for all $(x_1, x_2) \in \mathbf{R}^2$, $l = 1, 2$,

(ii) $F_{n_1 i}^{(1)} = F_{n_1}^{(1)}$ for all $1 \leqq i \leqq n_1$,

(iii) $F_{n_2 j}^{(2)} = F_{n_2}^{(2)}$ for all $1 \leqq j \leqq n_2$.

Then condition (4.2) is satisfied when the sequences $\{Y_{n_1 i}\}$ and $\{Z_{n_2 j}\}$ are strong mixing.

Proof of Theorem 4.2. For any $n_1 \geqq 1$, for any i $(1 \leqq i \leqq n_1)$ and for any $J \in \mathbf{C}_{2,b}$, let

$$\begin{aligned} B_{n_1 i}(J) = {} & n_1 n^{-1} \int_0^1 \Big(\mathbf{1}_{[F_n(Y_{n_1 i}) \leqq t]} - G_{n_1 i}^{(1)}(t) \Big) J'(t)\, dG_{n_1 i}^{(1)}(t) \\ & + J\Big(F_n\big(Y_{n_1 i}\big)\Big) - \int_0^1 J(t)\, dG_{n_1 i}^{(1)}(t) \end{aligned} \tag{4.8}$$

where $G_{n_1 i}^{(1)} = F_{n_1 i}^{(1)} \circ F_n^{-1}$, and for any $n_2 \geqq 1$, any j $(1 \leqq j \leqq n_2)$, any $J \in \mathbf{C}_{2,b}$ and any $u \in [0, 1]$, let

$$\begin{aligned} C_{n_2 j}(J)(u) = {} & n_2^{-1} \sum_{l=1}^{[n_2 u]} \int_0^1 \Big(\mathbf{1}_{[F_n(Z_{n_2 j}) \leqq t]} - G_{n_2 j}^{(2)}(t) \Big) J'(t)\, dG_{n_2 l}^{(2)}(t) \\ & + n_2^{-1}\big(n_2 u - [n_2 u]\big) \int_0^1 \Big(\mathbf{1}_{[F_n(Z_{n_2 j}) \leqq t]} - G_{n_2 j}^{(2)}(t) \Big)\, dG_{n_2,[n_2 u]+1}^{(2)}(t), \end{aligned} \tag{4.9}$$

where $G_{n_2j}^{(2)} = F_{n_2j}^{(2)} \circ F_n^{-1}$.

Now consider for any $J \in \mathbf{C}_{2,b}$ the processes $W_{n_1}(J)(s)$ and $W_{n_2}^*(J)(s,u)$ defined, respectively, on $\mathbf{C}_1$ and $\mathbf{C}_1^*$ (= the space of continuous functions on $[0,\,1]^2$) by

$$W_{n_1}(J)(s) = n_1^{1/2}\left(\sum_{i=1}^{[n_1 s]} B_{n_1 i}(J) + \left(n_1 s - [n_1 s]\right) B_{n_1,[n_1 s]+1}(J)\right), \tag{4.10}$$

$$W_{n_2}^*(J)(s,\,u) = n_2^{1/2}\left(\sum_{i=1}^{[n_2 s]} C_{n_2 j}(J)(u) + \left(n_2 s - [n_2 s]\right) C_{n_2,[n_2 s]+1}(J)(u)\right). \tag{4.11}$$

By similar techniques as in Lemma 3.1 one can prove that the process $W_{n_1}(J)(s)$ converges weakly in uniform topology to a Gaussian process $W_0(J)(s)$ with trajectories a.s. in $\mathbf{C}_1$ with mean 0 and variance

$$\begin{aligned} s\Bigg[&\int_0^1\int_0^1 f_1^2(u)\,d(F^{(1)}\circ H^{-1})(u) \\ &+2\sum_{p\geqq 1}\int_0^1\int_0^1 f_1(u)f_1(v)\,d\Big(G_p^{(1)}\big(H^{-1}(u),\,H^{-1}(v)\big)\Big)\Bigg] \end{aligned} \tag{4.12}$$

and $W_{n_2}^*(J)(s,u)$ converges weakly in uniform topology to a Gaussian process $W_0^*(J)(s,\,u)$ with trajectories a.s. in $\mathbf{C}_1^*$ with mean 0 and variance

$$\begin{aligned} su\Bigg[&\int_0^1 f_2^2(u)\,d(F^{(2)}\circ H^{-1})(u) \\ &+2\sum_{p\geqq 1}\int_0^1\int_0^1 f_2(u)f_2(v)\,d\Big(G_p^{(2)}\big(H^{-1}(u),\,H^{-1}(v)\big)\Big)\Bigg]. \end{aligned} \tag{4.13}$$

From this and following Lemma 3.2, it is easy to prove that $V_n^*(J)$ converges in law to the normal distribution with mean 0 and variance $\widetilde{\sigma}_J(\{\widehat{G}_p^{(l)}\})$ where $\widetilde{\sigma}_J(\{\widehat{G}_p^{(l)}\})$ is defined in (4.5) and $V_n^*(J)$ is a random variable defined by

$$\begin{aligned} V_n^*(J) = n^{1/2}\Bigg(&\int_{-\infty}^{+\infty} J\big(\widehat{F}_n(x)\big)\,d\big(\widehat{F}_n^{(1)}(x) - F_n^{(1)}(x)\big) \\ &+\int_{-\infty}^{+\infty} J'\big(\widehat{F}_n(x)\big)\,\big(\widehat{F}_n(x) - F_n(x)\big)\,dF_n^{(1)}(x)\Bigg). \end{aligned} \tag{4.14}$$

Since $\mathcal{S}_n^*(J) = V_n^*(J) + U_n(J)$ where $\mathbf{E}\,(U_n(J))^2 = O(n^{-1/2})$ we prove Theorem 4.1 following the line of argument in the proof of Theorem 3.1.

REFERENCES

[1] S. Balacheff and G. Dupont, *Normalité asymptotique des processus empiriques tronqués et des processus de rang*, Lect. Notes Math., 821 (1980), pp. 19–46.

[2] P. Billingsley, *Convergence of Probability Measures*, Wiley, New York, 1968.

[3] H. Chernoff and I. R. Savage, *Asymptotic normality and efficiency of certain nonparametric test statistics*, Ann. Math. Statist., 29 (1958), pp. 972–994.

[4] M. Denker and U. Rösler, *Some contributions to Chernoff–Savage theorems*, Statist. and Decision, 3 (1985), pp. 49–75.

[5] P. Doukhan and F. Portal, *Principe d'invariance faible pour la fonction de répartition empirique dans un cadre multidimensionnel et mélangeant*, Probab. Math. Statist., 8 (1987), pp. 117–132.

[6] V. Dupač and J. Hájek, *Asymptotic normality of simple linear rank statistics under alternative* II, Ann. Math. Statist., 40 (1969), pp. 1992–2017.

[7] T. R. Fears and K. L. Mehra, *Weak convergence of a two-sample empirical process and a Chernoff–Savage theorem for φ-mixing sequences*, Ann. Statist., 2 (1974), pp. 586–596.

[8] Z. Govindarajulu, L. Le Cam, and M. Raghavachari, *Generalizations of theorems of Chernoff and Savage on the asymptotic normality of test statistics*– In: Proceedings of the Fifth Berkeley Symposium on Mathematical Statistics and Probability. V. I. Berkeley–Los Angeles: Univ. California Press, 1967, pp. 609–638.

[9] J. Hájek, *Asymptotic normality of simple linear rank statistics under alternatives*, Ann. Math. Statist., 39 (1968), pp. 325–346.

[10] R. Pyke and G. R. Shorack, *Weak convergence of a two-sample empirical process and a new approach to Chernoff–Savage theorems*, Ann. Math. Statist., 39 (1968), N 3, pp. 755–771.

[11] C. S. Withers, *Convergence of empirical processes of mixing r.v.'s on* [0, 1], Ann. Statist., 3 (1975), pp. 1101–1108.

Reprinted from
Theory of Probability and Its Applications, Volume 38
(1993), pages 405-422, by permission of the publisher,

WEAK CONVERGENCE OF WEIGHTED EMPIRICAL U-STATISTICS PROCESSES FOR DEPENDENT RANDOM VARIABLES

MICHEL HAREL, COLM ART O'CINNEIDE and MADAN L. PURI*

UA CNRS 743, Orsay, France, Purdue University, Indiana University

A weak convergence of the weighted empirical U-statistic with respect to the Skorohod topology is obtained for absolutely regular random variables. It is a generalization of the results of F. H. Ruymgaart and M. C. A. van Zuijlen (1991, *J. Statist. Plann. Inference*) for i.i.d. random variables.

1. Introduction

Let $X_1, X_2, \ldots$ be real-valued random variables (r.v.'s) with F as their common distribution function (d.f.) and let $h: \mathbb{R}^k \to \mathbb{R}$ be a Borel measurable function which is symmetric in its $k \geqslant 1$ arguments. Furthermore, let $J_n(k)$ be the set of all subsets of k different elements from $\{1, 2, \ldots, n\}$ so that card $J_n(k) = \binom{n}{k}$, where card A denotes the cardinal of the set A.

Let us write $X_I = h(X_{i_1}, \ldots, X_{i_k})$ for $I = \{i_1, \ldots, i_k\} \in J_n(k)$, and define

$$H_n(t) = \frac{1}{\binom{n}{k}} \sum_{I \in J_n(k)} I_{[X_I \leqslant t]}, \tag{1.1}$$

$$H(t) = \int_{\mathbb{R}^k} I_{[h(x_1, \ldots, x_k) \leqslant t]} \prod_{j=1}^{k} F(dx_j), \qquad t \in \mathbb{R}. \tag{1.2}$$

We assume that $H(t)$ is continuous.

Received March 18, 1992; revised February 10, 1993.

AMS 1980 subject classification: 60F05.

Key words and phrases: weighted empirical U-statistics, weak convergence, Skorohod topology, absolute regularity.

* Research supported by the Office of Naval Research under Contract N00014-91-J-1020.

Our aim is to study the weak convergence of the empirical U-statistic process defined by

$$U_n(u)=\begin{cases}\hat{U}_n(H^{-1}(u)) & \text{if} \quad 0<u<1\\ 0 & \text{if} \quad u=0 \text{ or } u=1, \qquad n\in\mathbb{N}^*=\{1,2,\ldots\},\end{cases} \tag{1.3}$$

where

$$\hat{U}_n(t)=n^{1/2}(H_n(t)-H(t)), \qquad t\in\mathbb{R}$$

when this process is weighted by a continuous and positive function $r(u)$.

We also assume that the underlying r.v.'s are absolutely regular with rates

$$\beta(m)=O(m^{-6-\epsilon}). \tag{1.4}$$

Recall that the sequence is absolutely regular if

$$\sup_{j\geqslant 1} E\{\sup_{A\in\sigma(X_i;i\geqslant j+m)} |P(A \mid \sigma(X_i,1\leqslant i\leqslant j))-P(A)|\}=\beta(m)\downarrow 0.$$

Here $\sigma(X_i; 1\leqslant i\leqslant j)$ and $\sigma(X_i, i\geqslant j+m)$ are the σ-fields generated by $(X_1, \ldots, X_j)$ and $(X_{j+m}, X_{j+m+1}, \ldots)$, respectively.

Also recall that the sequence is strong mixing if

$$\sup_{j>1}\{|P(A\cap B)-P(A)\,P(B)|; A\in\sigma(X_i, 1\leqslant i\leqslant j),$$
$$B\in\sigma(X_i, i\geqslant j+m)\}=\alpha(m)\downarrow 0.$$

Since $\alpha(m)\leqslant\beta(m)$, it follows that if $\{X_i\}$ is absolutely regular, it is also strong mixing.

Recently, the asymptotic behavior of the weighted empirical U-statistic was studied by Ruymgaart and Van Zuijlen (1991) for i.i.d. random variables. Our paper extends their results to the case when the random variables are dependent. We mention that the asymptotic behavior of U-statistics was studied for absolutely regular and stationary random variables by Yoshihara (1976) and for nonstationary random variables by Harel and Puri (1989a, 1989b, 1990). However, these authors did not consider the weighted empirical U-statistic.

2. Weak Convergence of the Weighted Empirical U-Statistic

A function $r: [0,1]\to\mathbb{R}^+$ is called a weight function if it satisfies the following conditions:

(i) r is continuous;

(ii) $r(0)=0$ and $r(1)=0$.

For any bounded function $f: [0, 1] \to \mathbb{R}$, and $\delta > 0$, define $\omega(f, \delta) = \sup\{|f(u) - f(v)|, |u - v| \leqslant \delta\}$. Denote by $D[0, 1]$ the space of functions on $[0, 1]$ that are right-continuous and have left-hand limits.

Now let α be fixed such that

$$0 < \alpha < \left\{\left(\frac{1}{2} + \delta\right) \Big/ \left(\frac{3}{2} - \delta\right)\right\} \wedge \left\{\frac{\delta}{(\frac{1}{2} - \delta)}\right\}, \qquad 0 < \delta < \frac{1}{2}. \tag{2.1}$$

We consider a *modified U-statistic process* defined by

$$\tilde{U}_n(u) = \begin{cases} U_n(u) & \text{if } n^{-1-\alpha} \leqslant u \leqslant 1 - n^{-1-\alpha} \\ 0 & \text{otherwise.} \end{cases} \tag{2.2}$$

For any weight function, we introduce a *weighted modified U-statistic process* $\tilde{U}_n \cdot (1/r)$ defined as

$$\tilde{U}_n(u) \cdot \frac{1}{r(u)} = \begin{cases} \tilde{U}_n(u)/r(u) & \text{if } u \neq 0 \text{ and } u \neq 1 \\ 0 & \text{otherwise.} \end{cases} \tag{2.3}$$

THEOREM 2.1. *Suppose that the sequence $\{X_i\}$ is absolutely regular with rates* (1.4). *For each weight function $r(u)$ satisfying*

$$r(u) \leqslant A(u(1-u))^{1/2-\delta}, \qquad 0 < \delta < \tfrac{1}{2}, \tag{2.4}$$

where A is some positive constant, there exists a Gaussian process U_r with trajectories a.s. continuous such that $\tilde{U}_n \cdot (1/r)$ converges in law to U_r in $D[0, 1]$ with respect to the Skorohod topology.

The proof of the theorem follows from the following three propositions.

PROPOSITION 2.1. *Suppose that the sequence $\{X_i\}$ satisfies the conditions of Theorem* 2.1. *Then there exists a Gaussian process U with trajectories a.s. continuous such that U_n converges in law to U in $D[0, 1]$ with respect to the Skorohod topology. Furthermore we have $\forall \epsilon > 0$, $\exists \delta > 0$ and N_0 such that $\forall n \geqslant N_0$,*

$$Q_n[f \in D[0, 1], \omega(f, \delta) \geqslant \epsilon] \leqslant \epsilon, \tag{2.5}$$

where Q_n is the law of U_n.

PROPOSITION 2.2. *Let $\{X_i\}$ satisfy the conditions of Theorem* 2.1. *Then, for each weight function $r(u)$ verifying* (2.1), *we have $\forall \epsilon > 0$, $\exists \theta > 0$, $\exists N_0$, $\forall n \geqslant N_0$ such that*

$$P\left[\sup_{u \in C_0} \left|\tilde{U}_n(n) \cdot \frac{1}{r(u)}\right| \geqslant \epsilon\right] \leqslant \epsilon, \tag{2.6}$$

where

$$C_\theta = \{u \in [0, 1], u \leqslant \theta \text{ or } u \geqslant 1 - \theta\}.$$

PROPOSITION 2.3. *Let Y_n, $n \in \mathbb{N}^*$, be a process with values in $D[0, 1]$, suppose that Y_n converges in law (in Skorohod topology) to a Gaussian process Y with trajectories a.s. continuous, and suppose also that $\forall \epsilon > 0$, $\exists \delta > 0$ and N_0 such that $\forall n \geqslant N_0$*

$$Q_n^*[f \in D[0, 1], \omega(f, \delta) \geqslant \epsilon] \leqslant \epsilon, \tag{2.7}$$

where Q_n^* is the law of Y_n. Let the weight function r be such that $Y_n \cdot (1/r)$, $n \in \mathbb{N}$, has a.s. trajectories in $D[0, 1]$. Furthermore assume that $\forall \epsilon > 0$, $\exists \theta > 0$ and N_0 such that $\forall n \geqslant N_0$

$$P\left[\sup_{u \in C_\theta} \left| Y_n(u) \cdot \frac{1}{r(u)} \right| \geqslant \epsilon \right] \leqslant \epsilon. \tag{2.8}$$

Then $Y_n \cdot (1/r)$ converges weakly in Skorohod topology to the Gaussian process $Y \cdot (1/r)$ with trajectories a.s. continuous.

3. PROOF OF PROPOSITION 2.1

We define

$$g(x_1, ..., x_k; u) = I_{[h(x_1, ..., x_k) \leqslant H^{-1}(u)]}, \qquad u \in [0, 1], x_i \in \mathbb{R}, i = 1, ..., k, \tag{3.1}$$

$$g_l(x_1, ..., x_l; u) = \int_{\mathbb{R}^{k-l}} g(x_1, ..., x_l, x_{l+1}, ..., x_k; u) \prod_{j=l+1}^{k} F(dx_j), \qquad 1 \leqslant l \leqslant k. \tag{3.2}$$

Note that $g_k(u) = u$.

Put

$$V_{l,n}(u) = n^{-[l]} \int \sum_{\{i_1, i_2, ..., i_l\} \subset J_n^*(l)} g_l(x_{i_1}, ..., x_{i_l}; u) \times \prod_{j=1}^{l} d(I_{[X_{i_j} \leqslant x_{i_j}]} - F(x_j)), \qquad 1 \leqslant l \leqslant k, \tag{3.3}$$

where $n^{-|l|} = (n(n-1)(n-2) \cdots (n-l+1))^{-1}$. $J_n^*(l)$ is the set of all subsets $\{i_1, i_2, ..., i_l\}$ of l elements from $\{1, 2, ..., n\}$ such that $i_1 < i_2 < \cdots < i_l$. Then we have

$$U_n(u) = n^{1/2} \left\{ \sum_{l=1}^{k} \binom{k}{l} V_{l,n}(u) \right\}. \tag{3.4}$$

LEMMA 3.1. *Under the conditions of Theorem 2.1, the process $n^{1/2}V_{1,n}$ converges in law to a Gaussian process U in $D[0, 1]$ with respect to the Skorohod topology and $\forall \varepsilon > 0$, $\exists \delta > 0$ and N_0 such that $\forall n \geqslant N_0$*

$$P_n[f \in D[0, 1], \omega(f, \delta) \geqslant \varepsilon] \leqslant \varepsilon, \tag{3.5}$$

where P_n is the law of $n^{1/2}V_{1,n}$.

Proof. From Theorem 15.1 in Billingsley (1968), we must prove that (i) the finite dimensional distributions of P_n converge in law to a normal distribution and (ii) P_n is tight.

First we prove (i). This is equivalent to showing that $\sum_{l=1}^{p} n^{1/2}\lambda_l V_{1,n}(u_l)$ converges in law to a normal distribution for any $p \in \mathbb{N}^*$, any $u_l \in [0, 1]$, and any $\lambda_l \in \mathbb{R}$ $(1 \leqslant l \leqslant p)$, where $\mathbb{N}^*$ is defined in (1.3). Without loss of generality we can take $p = 2$ and suppose $u_1 < u_2$. We have $E(\sum_{l=1}^{p} \lambda_l n^{1/2} V_{1,n}(u_l)) = 0$ and

$$\begin{aligned}
\sum_{l=1}^{p} \lambda_l n^{1/2} V_{1,n}(u_l) &= n^{-1/2}\left[\lambda_1\left(\sum_{i=1}^{n} g_1(X_i; u_1) - \int g_1(x; u_1)\, F(dx)\right)\right. \\
&\quad \left. + \lambda_2 \sum_{i=1}^{n}\left(g_1(X_i; u_2) - \int g_1(x; u_2)\, F(dx)\right)\right] \\
&= n^{-1/2} \sum_{i=1}^{n}\left(\lambda_1 g_1(X_i; u_1) + \lambda_2 g_1(X_i; u_2)\right. \\
&\quad \left. - \int (\lambda_1 g_1(x; u_1) + \lambda_2 g_1(x; u_2))\, F(dx)\right).
\end{aligned}$$

As

$$|\lambda_1 g_1(X_i; u_1) + \lambda_2 g_1(X_i; u_2)| < |\lambda_1| + |\lambda_2| < \infty$$

and the sequence of r.v.'s $\{\lambda_1 g_1(X_i; u_1) + \lambda_2 g_1(X_i; u_2), i \geqslant 1\}$ is strong mixing with rates (1.4), we deduce from Theorem 18.5.4, page 347, of Ibragimov and Linnik (1971) (see also Theorem 1 of Oodaira and Yoshihara, 1972) that $\sum_{l=1}^{p} \lambda_l n^{1/2} V_{1,n}(u_l)$ converges in law to a normal distribution with mean zero and variance $\sigma^2 = E(A_1)^2 + 2\sum_{j=2}^{\infty} E(A_1 A_j) < \infty$, where $A_j = \lambda_1 g_1(X_j; u_1) + \lambda_2 g_1(X_j; u_2)$.

Now we prove (ii). From Lemma 5.1 in the Appendix, we have

$$\begin{aligned}
&E(n^{1/2}V_{1,n}(v) - n^{1/2}V_{1,n}(u))^4 \\
&\quad \leqslant C((v-u)^{2-2\epsilon} + n^{-1}(v-u)^{1-\epsilon}), \qquad 0 < \epsilon < 1, C > 0.
\end{aligned} \tag{3.6}$$

Define for any $n \geqslant 1$, a sequence u_i, $i = 0, \ldots, n$, by

$$u_0 = 0 < u_1 < u_2 < \cdots < u_i < u_{i+1} < \cdots < u_n = 1$$

and

$$u_{i+1} - u_i = n^{-1}.$$

For any $v \in [0, 1]$, there exists an $l \in \{0, ..., n-1\}$ such that $v \in (u_l, u_{l+1})$ and we can write

$$\begin{aligned} n^{1/2}(V_{1,n}(v) - V_{1,n}(u_l)) &= n^{-1/2} \sum_{i=1}^{n} \left(\int_{\mathbb{R}^{k-1}} I_{[H^{-1}(u_l) \leqslant h(x_1, ..., x_{k-1}, X_i) \leqslant H^{-1}(v)]} \right. \\ &\quad \left. \times \prod_{j=1}^{k-1} F(dx_j) - (v - u_l) \right) \\ &\leqslant n^{-1/2} \sum_{i=1}^{n} \left(\int_{\mathbb{R}^{k-1}} I_{[H^{-1}(u_l) \leqslant h(x_1, ..., x_{k-1}, X_i) \leqslant H^{-1}(u_{l+1})]} \right. \\ &\quad \left. \times \prod_{j=1}^{k-1} F(dx_j) - (v - u_l) \right) \\ &\leqslant n^{1/2}(V_{1,n}(u_{l+1}) - V_{1,n}(u_l)) + n^{1/2}(u_{l+1} - u_l) \\ &\leqslant n^{1/2}(V_{1,n}(u_{l+1}) - V_{1,n}(u_l)) + n^{-1/2}. \end{aligned}$$

We have also

$$\begin{aligned} n^{1/2}(V_{1,n}(u_l) - V_{1,n}(v)) &\leqslant n^{1/2}(v - u_l) \\ &\leqslant n^{1/2} \, |V_{1,n}(u_{l+1}) - V_{1,n}(u_l)| + n^{-1/2}, \end{aligned}$$

which implies

$$n^{1/2} \, |V_{1,n}(v) - V_{1,n}(u_l)| \leqslant n^{1/2} \, |V_{1,n}(u_{l+1}) - V_{1,n}(u_l)| + n^{-1/2}. \tag{3.7}$$

From (3.6) and (3.7) we easily deduce (3.5), which implies that P_n is tight.

LEMMA 3.2. *Under the conditions of Theorem 2.1, for any $u \in [0, 1]$*

$$E(V_{l,n}(u))^6 = O(n^{-5}), \qquad 2 \leqslant l \leqslant k. \tag{3.8}$$

Proof. We only consider the case $l = 2$. The proofs for the cases $l = 3, ..., k$ are analogous and so are omitted.

We first note that

$$V_{2,n}(u) = [n(n-1)]^{-1} \sum_{1 \leqslant i < j \leqslant n} g_2(X_i, X_j; u) - g_1(X_i; u) + u.$$

Let i_{rs} $(\leqslant n)$ $(r=1, ..., 6; s=1, 2)$ be mutually different positive integers. Reorder $\{i_{rs}\}$ as $1 \leqslant k_1 < k_2 < \cdots < k_{12} \leqslant n$ and put

$$E\left[\prod_{j=1}^{6} \{g_2(X_{i_{j1}}, X_{i_{j2}}; u) - g_1(X_{i_{j1}}; u) - g_1(X_{i_{j2}}; u) + u\}\right]$$

$$= E\left[\prod_{j=1}^{6} \tilde{g}_2(X_{i_{j1}}, X_{i_{j2}})\right] = J(k_1, ..., k_{12}) = E[G(X_{k_1}, ..., {}_{k_{12}})]. \quad (3.9)$$

Then, we have $|G(x_1, ..., x_{12})| \leqslant 1$. Let $d^{(c)}$ be the cth largest difference among $(k_{j+1} - k_j)$, $j=1, ..., 11$. Since

$$\int_{\mathbb{R}^k} G(x_1, ..., x_{12})\, dP_j(x_1, ..., x_{12}) = 0, \qquad j=1, ..., 11,$$

where the probability measure P_j is defined in (5.4), we have from Lemma 5.2

$$J(k_1, ..., k_{12}) \leqslant 2M\beta(k_{12} - k_{11}) \qquad \text{if} \quad k_{12} - k_{11} = d^{(1)}$$

and

$$J(k_1, ..., k_{12}) \leqslant 2M\beta(k_2 - k_1) \qquad \text{if} \quad k_2 - k_1 = d^{(1)},$$

where $M=1$, and where β is the rate of absolute regularity.

Hence from (1.4)

$$\sum_{\substack{1 \leqslant k_1 < \cdots < k_{12} \leqslant n \\ k_{12} - k_{11} = d^{(1)} \text{ or } k_2 - k_1 = d^{(1)}}} J(k_1, ..., k_{12}) \leqslant n^6 \sum_{j=1}^{n} (j+1)^5 \beta(j) = O(n^6).$$

If for some j_α $(2 \leqslant j_\alpha \leqslant 10, 1 \leqslant \alpha \leqslant 6)$, $k_{j_\alpha+1} - k_{j_\alpha} = d^{(\alpha)}$ $(1 \leqslant \alpha \leqslant 6)$, then from Lemma 5.2

$$J(k_1, ..., k_{12}) \leqslant 2 \sum_{\alpha=1}^{6} \beta(k_{j_{\alpha+1}} - k_{j_\alpha})$$

and hence

$$\sum_{\substack{1 \leqslant k_1 < \cdots < k_{12} \leqslant n \\ k_{j_\alpha+1} - k_{j_\alpha} = d^{(\alpha)}\, (1 \leqslant \alpha \leqslant 6)}} J(k_1, ..., k_{12}) \leqslant 12n^6 \sum_{j=1}^{n} (j+1)^5 \beta(j) = O(n^6).$$

Consequently,

$$\sum_{1 \leqslant k_1 < \cdots < k_{12} \leqslant n} J(k_1, ..., k_{12}) = O(n^6). \quad (3.10$$

Suppose now that there exists for any $r \in \{1, ..., 6\}$ one $s_l \in \{1, 2\}$ such that

$$i_{1s_1} = i_{2s_2} = \cdots = i_{6s_6}$$

and if furthermore $s_1 = s_2 = \cdots = s_6 = 1$, then we have

$$\left| \sum_{j=1}^{n} \sum_{i_{11}=1}^{n} \cdots \sum_{i_{61}=1}^{n} E(\tilde{g}_2(X_j, i_{12})\, \tilde{g}_2(X_j, i_{22}) \cdots \tilde{g}_2(X_j, i_{62})) \right| \leqslant n^7. \quad (3.11)$$

As we have 2^6 possibilities to choose $(s_1, ..., s_6)$, we note that the sums in the case (3.10) are $O(n^7)$. By using similar methods to estimate the sums in the other cases, we find that they are also $O(n^7)$.

We deduce that

$$E\left(\sum_{1 \leqslant i < j \leqslant n} \tilde{g}_2(X_i, X_j) \right)^6 = O(n^7), \quad (3.12)$$

which implies

$$E(V_{2,n}(u))^6 = O(n^{-5})$$

and Lemma 3.2 is proved for $l = 2$.

LEMMA 3.3. *Under the conditions of Theorem* 2.1, *we have* $\forall \epsilon > 0$

$$P[\sup_{u \in [0,1]} (n^{1/2} V_{l,n}(u)) \geqslant \epsilon] = O(n^{-1/2}), \qquad 2 \leqslant l \leqslant k. \quad (3.13)$$

Proof. Define for any $n \geqslant 1$, a sequence u_i, $i = 0, ..., n$, by

$$u_0 = 0 < u_1 < u_2 < \cdots < u_i < u_{i+1} < \cdots < u_n = 1$$

and

$$u_{i+1} - u_i = n^{-1}.$$

First, consider the case $l = 2$. For any $v \in [0, 1]$, there exists $l \in \{0, ..., n-1\}$ such that $v \in (u_l, u_{l+1})$ and we can write

$$n^{1/2} V_{2,n}(v) = n^{1/2}(V_{2,n}(v) - V_{2,n}(u_{l+1}) + V_{2,n}(u_{l+1}))$$

or

$$|n^{1/2} V_{2,n}(v)| \leqslant |n^{1/2} V_{2,n}(u_{l+1})| + |n^{1/2} V_{2,n}(u_{l+1}) - n^{1/2} V_{2,n}(v)|. \quad (3.14)$$

We have by definition

$$n^{1/2}V_{2,n}(u_{l+1})-n^{1/2}V_{2,n}(v)$$

$$=n^{1/2}\binom{n}{2}^{-1}\sum_{1\leqslant i<j\leqslant n}(g_2(X_i,X_j;u_{l+1})-g_2(X_i,X_j;v)-g_1(X_i;u_{l+1})$$

$$+g_1(X_i;v)-g_1(X_j;u_{l+1})+g_1(X_j;v)+u_{l+1}-v)$$

$$\leqslant n^{1/2}\binom{n}{2}^{-1}\left\{\sum_{1\leqslant i<j\leqslant n}(g_2(X_i,X_j;u_{l+1})-g_2(X_i,X_j;u_l)\right.$$

$$\left.-g_1(X_i;u_{l+1})+g_1(X_i;u_l)-g_1(X_j;u_{l+1})+g_1(X_j;u_l)+u_{l+1}-u_l)\right\}$$

$$+n^{1/2}n^{-1}\left\{\sum_{j=1}^{n}(g_1(X_j;u_{l+1})-g_1(X_j;u_l)-u_{l+1}+u_l)\right\}$$

$$+n^{1/2}n^{-1}\left\{\sum_{i=1}^{n}(g_1(X_i;u_{l+1})-g_1(X_i;u_l)-u_{l+1}+u_l)\right\}$$

$$+2n^{1/2}(u_{l+1}-u_l)$$

$$\leqslant n^{1/2}(V_{2,n}(u_{l+1})-V_{2,n}(u_l))$$

$$+2n^{1/2}(V_{1,n}(u_{l+1})-V_{1,n}(u_l))+2n^{1/2}(u_{l+1}-u_l).$$

Similarly we have

$$n^{1/2}V_{2,n}(v)-n^{1/2}V_{2,n}(u_{l+1})$$

$$\leqslant 2n^{1/2}n^{-1}\sum_{i=1}^{n}(g_1(X_i;u_{l+1})-g_1(X_i;v))$$

$$\leqslant 2n^{-1/2}\sum_{i=1}^{n}(g_1(X_i;u_{l+1})-g_1(X_i;u_l))$$

$$\leqslant 2n^{-1/2}\left\{\sum_{i=1}^{n}(g_1(X_i;u_{l+1})-g_1(X_i;u_l)-u_{l+1}+u_l)\right\}$$

$$+2n^{1/2}(u_{l+1}-u_l)+|n^{1/2}V_{2,n}(u_{l+1})-n^{1/2}V_{2,n}(u_l)|.$$

It follows from the preceding two inequalities that

$$n^{1/2}\,|V_{2,n}(u_{l+1})-V_{2,n}(v)|$$

$$\leqslant n^{1/2}\,|V_{2,n}(u_{l+1})-V_{2,n}(u_l)|$$

$$+2n^{1/2}\,|V_{1,n}(u_{l+1})-V_{1,n}(u_l)|+2n^{1/2}(u_{l+1}-u_l). \qquad (3.15$$

From (3.14) and (3.15), we deduce

$$\sup_{v \in [0,1]} |n^{1/2} V_{2,n}(v)| \leqslant 3n^{1/2} \sup_{0 \leqslant l \leqslant n} |V_{2,n}(u_l)| + 2n^{1/2} \sup_{0 \leqslant l \leqslant n-1} |V_{1,n}(u_{l+1}) - V_{1,n}(u_l)| + 2n^{-1/2}. \tag{3.16}$$

From Lemmas 3.2 and 5.1, it follows that $\forall \epsilon > 0$

$$P[\sup_{u \in [0,1]} |n^{1/2} V_{2,n}(u)| \geqslant \epsilon] = O(n^{-1/2}) \tag{3.17}$$

and (3.13) is proved for $l = 2$.

Now consider the case $l = 3$. For any $v \in [0, 1]$, there exists $l \in \{0, ..., n-1\}$ such that $v \in (u_l, u_{l+1})$ and we can write

$$|n^{1/2} V_{3,n}(n)| \leqslant |n^{1/2} V_{3,n}(u_{l+1})| + |n^{1/2} V_{3,n}(u_{l+1}) - n^{1/2} V_{3,n}(v)|. \tag{3.18}$$

We have (by definition)

$$\begin{aligned}
&n^{1/2} V_{3,n}(u_{l+1}) - n^{1/2} V_{3,n}(v) \\
&= n^{1/2} \binom{n}{3}^{-1} \sum_{1 \leqslant i < j < l \leqslant n} (g_3(X_i, X_j, X_l; u_{l+1}) \\
&\quad - g_3(X_i, X_j, X_l; v) - g_2(X_i, X_j; u_{l+1}) + g_2(X_i, X_j; v) \\
&\quad - g_2(X_i, X_l; u_{l+1}) + g_2(X_i, X_l; v) - g_2(X_j, X_l; u_{l+1}) \\
&\quad + g_2(X_j, X_l; v) + g_1(X_i; u_{l+1}) - g_1(X_i; v) + g_1(X_j; u_{l+1}) \\
&\quad - g_1(X_j; v) + g_1(X_l; u_{l+1}) - g_1(X_l; v) - u_{l+1} + v) \\
&\leqslant n^{1/2}(V_{3,n}(u_{l+1}) - V_{3,n}(u_l)) + 3n^{1/2}(V_{2,n}(u_{l+1}) - V_{2,n}(u_l)) \\
&\quad + 3n^{1/2}(V_{1,n}(u_{l+1}) - V_{1,n}(u_l)) + 6n^{1/2}(u_{l+1} - u_l).
\end{aligned}$$

Similarly we have

$$\begin{aligned}
&n^{1/2}(V_{3,n}(v) - V_{3,n}(u_{l+1})) \\
&\leqslant 3n^{1/2}(V_{2,n}(u_{l+1}) - V_{2,n}(u_l)) + 4n^{1/2}(u_{l+1} - u_l) \\
&\leqslant 3n^{1/2}(V_{2,n}(u_{l+1}) - V_{2,n}(u_l)) + 4n^{1/2}(u_{l+1} - u_l) \\
&\quad + n^{1/2} |V_{3,n}(u_{l+1}) - V_{3,n}(u_l)| + 3n^{1/2} |V_{1,n}(u_{l+1}) - V_{1,n}(u_l)|.
\end{aligned}$$

It follows from the preceding two inequalities that

$$\sup_{v \in [0,1]} |n^{1/2} V_{3,n}(v)| \leqslant 3n^{1/2} \sup_{0 \leqslant l \leqslant n} |V_{3,n}(u_l)| + 6n^{1/2} \sup_{0 \leqslant l \leqslant n} |V_{2,n}(u_l)| + 3n^{1/2} \sup_{0 \leqslant l \leqslant n-1} |V_{1,n}(u_{l+1}) - V_{1,n}(u_l)| + 6n^{-1/2} \tag{3.19}$$

From Lemmas 3.2 and 5.1 and from (3.17), it follows that $\forall\epsilon>0$

$$P[\sup_{u\in[0,1]} |n^{1/2}V_{3,n}(u)|\geqslant\epsilon]=O(n^{-1/2})$$

and (3.13) is proved for $l=3$.

The proofs in the cases $l=4, ..., k$ are analogous and so are omitted.

Now Proposition 2.1 is an easy consequence of Lemmas 3.1 and 3.3 and the equality (3.4).

4. Proof of Proposition 2.2

Proposition 2.2 follows from the following two lemmas.

Define the modified process $\tilde{V}_{1,n}$ by

$$\tilde{V}_{1,n}(u)=\begin{cases}V_{1,n}(u) & \text{if } n^{-1-\alpha}\leqslant u\leqslant 1-n^{-1-\alpha}\\ 0 & \text{otherwise.}\end{cases}$$

Lemma 4.1. *Under the conditions of Theorem* 2.1, $\forall\epsilon>0$, $\exists\theta>0$, $\exists N_0$, *such that* $\forall n\geqslant N_0$

$$P\left[\sup_{u\in C_\theta}\left|n^{1/2}\tilde{V}_{1,n}(u)\frac{1}{r(u)}\right|\geqslant\epsilon\right]\leqslant\epsilon. \tag{4.1}$$

Proof. From now on, we agree to denote by the letter C some quantity bounded in absolute value. Define for any $n\geqslant 1$, a sequence $\{u_i^{(n)}=u_i,$ $i=0, ..., i_n\}$ of equidistant points such that

$$\left(\frac{1}{n}\right)^{1+\alpha'}\leqslant u_{i+1}^{(n)}-u_i^{(n)}\leqslant\left(\frac{1}{n}\right)^{1+\alpha}, \qquad i=1, ..., i_n, \tag{4.2}$$

$0<\alpha<\alpha'$, $\alpha>\alpha'/2$, $u_0^{(n)}=0$, $u_{i_n-1}^{(n)}<\theta\leqslant u_{i_n}^{(n)}$. For any $n\geqslant 1$, for any i, j, $1\leqslant i<j\leqslant i_n$, we have (using the inequality $|\sum_{e\in E}x_e|^4\leqslant(2^3)^{\text{card } E}\sum_{e\in E}x_e^4$)

$$\begin{aligned}
&E\left(n^{1/2}\frac{V_{1,n}(u_j)}{r(u_j)}-n^{1/2}\frac{V_{1,n}(u_i)}{r(u_i)}\right)^4\\
&\leqslant 64\left\{E\left(n^{1/2}\frac{(V_{1,n}(u_j)-V_{1,n}(u_i))}{r(u_j)}\right)^4\right.\\
&\quad\left.+E\left((n^{1/2}V_{1,n}(u_i)\left(\frac{1}{r(u_i)}-\frac{1}{r(u_j)}\right)\right)^4\right\}\\
&\leqslant C[(u_j-u_i)^{2-2\epsilon}+n^{-1}(u_j-u_i)](u_j)^{-2+4\delta}\\
&\quad+[(u_i)^{2-2\epsilon}+n^{-1}(u_i)^{1-\epsilon}][(u_i)^{-1/2+\delta}-(u_j)^{-1/2+\delta}]^4\\
&\leqslant 2C\{(u_j-u_i)^{2-2\epsilon}(u_j)^{-2+4\delta}+(u_i)^{2-2\varepsilon}[(u_i)^{-1/2+\delta}-(u_j)^{-1/2+\delta}]^4\},
\end{aligned} \tag{4.3}$$

where the second inequality comes from Lemma 5.1, the third from (4.2).

Put $u_i = u$ and $u_j = v$. The left-hand side of (4.3) is less than or equal to

$$C\{(v-u)^{2-2\epsilon}\, v^{-2+4\delta} + u^{2-2\epsilon}(v^{-1/2+\delta} - u^{-1/2+\delta})^4\}$$

$$= C\left\{\left[(v-u)\left(\frac{1}{v^{1/2-\delta}}\right)^{1+\beta}\right]^{2-2\epsilon} + \left[u\left(\frac{1}{u^{1/2-\delta}} - \frac{1}{v^{1/2}-\delta}\right)^{1+\beta}\right]^{2-2\epsilon}\right\}$$

(where $(1+\beta)(2-2\epsilon)=4$).

From the mean theorem, there exists w_1 $(u \leqslant w_1 \leqslant v)$ such that

$$(v-u)\frac{1}{w_1^{(1/2-\delta)(1+\beta)}} = \int_u^v \frac{dt}{t^{(1/2-\delta)(1+\beta)}}$$

and w_2 such that

$$w_2^{1-(1/2-\delta)\beta}\left(\frac{1}{u^{1/2-\delta}} - \frac{1}{v^{1/2-\delta}}\right) = \int_v^u \frac{t}{t^{(1/2-\delta)\beta}}\left(\frac{\partial}{\partial t}\frac{1}{t^{1/2-\delta}}\right) dt$$

$$= \frac{1}{\frac{1}{2}-\delta}\int_u^v \frac{dt}{t^{(1/2-\delta)(1+\beta)}}.$$

Then, the left-hand side of (4.3) is less than or equal to

$$C\left\{\left[(v-u)\frac{1}{w_1^{(1/2-\delta)(1+\beta)}}\right]^{2-2\epsilon} + \left[u\frac{1}{u^{(1/2-\delta)\beta}}\left(\frac{1}{u^{1/2-\delta}} - \frac{1}{v^{1/2-\delta}}\right)\right]^{2-2\epsilon}\right\}$$

$$\leqslant C\left\{\left(\int_u^v \frac{dt}{t^{(1/2-\delta)(1+\beta)}}\right)^{2-2\epsilon} + \left[w_2^{1-(1/2-\delta)\beta}\left(\frac{1}{u^{1/2-\delta}} - \frac{1}{v^{1/2-\delta}}\right)\right]^{2-2\epsilon}\right\}$$

$$\leqslant C\left\{\left(\int_u^v \frac{dt}{t^{(1/2-\delta)(1+\beta)}}\right)^{2-2\epsilon} + \left(\int_v^u \frac{t}{t^{(1/2-\delta)}}\left(\frac{\partial}{\partial t}\frac{1}{t^{1/2-\delta}}\right) dt\right)^{2-2\epsilon}\right\}$$

$$\leqslant C\left\{\left(\int_u^v \frac{dt}{t^{(1/2-\delta)(1+\beta)}}\right)^{2-2\epsilon} + \left(\frac{1}{\frac{1}{2}-\delta}\int_u^v \frac{dt}{t^{(1/2-\delta)(1+\beta)}}\right)^{2-2\epsilon}\right\}.$$

As $\delta \in (0, \frac{1}{2})$, we have $(\frac{1}{2}-\delta)(1+\beta) < 1$ if $\epsilon < 2\delta$. Thus $\int_u^v (dt/t^{(1/2-\delta)(1+\beta)})$ define a diffuse measure ν on $(0, 1)$ by $\nu([u, v]) = \int_u^v (dt/t^{(1/2-\delta)(1+\beta)})$.

Hence

$$E\left(n^{1/2}\frac{V_{1,n}(u_j)}{r(u_j)} - n^{1/2}\frac{V_{1,n}(u_i)}{r(u_i)}\right)^4 \leqslant C\nu^{2-2\epsilon}(u_j - u_i). \tag{4.4}$$

If $u_i = 0$, we have only

$$E\left(n^{1/2}\frac{V_{1,n}(u_j)}{r(u_j)}\right)^4 \leqslant (u_j)^{2-2\epsilon}\,(u_j)^{-2+4\delta} \leqslant C\nu^{2-2\epsilon}(u_j).$$

From the Markov inequality we obtain $\forall\eta>0$

$$P[n^{1/2}\,|V_{1,n}(u_j)-V_{1,n}(u_i)|>\eta]>c\eta^{-4}\nu^{2-2\epsilon}([u_i,u_j]),\qquad 0\leqslant i<j\leqslant i_n.$$

From Lemma 5.3 we deduce that

$$P\left[\sup_{\substack{1\leqslant i\leqslant i_n\\ u_i\leqslant\theta}}\left|n^{1/2}V_{1,n}(u_i)\frac{1}{r(u_i)}\right|>\eta\right]<C\eta^{-4}\nu^{2-2\epsilon}(2\theta).\tag{4.5}$$

Let $v\in[0,1]$ be such that $v\geqslant u_1$. Let i be such that $u_i\leqslant v<u_{i+1}$. Then, we have

$$\begin{aligned}
&\left|n^{1/2}V_{1,n}(v)\frac{1}{r(v)}\right|\\
&\leqslant\left|n^{1/2}V_{1,n}(v)\frac{1}{r(u_i)}\right|\\
&\leqslant\left|n^{1/2}V_{1,n}(u_i)\frac{1}{r(u_i)}\right|+n^{1/2}\,|V_{1,n}(u_{i+1})-V_{1,n}(u_i)|\,\frac{1}{r(u_i)}\\
&\quad+n^{1/2}\left((u_{i+1}-u_i)\frac{1}{r(u_i)}\right)\\
&\quad\text{(by using methods similar to those for obtaining (3.7))}\\
&\leqslant 2n^{1/2}\,|V_{1,n}(u_i)|\,\frac{1}{r(u_i)}+n^{1/2}\frac{r(u_{i+1})}{r(u_i)}\,|V_{i,n}(u_{i+1})|\,\frac{1}{r(u_{i+1})}\\
&\quad+n^{1/2}\frac{u_{i+1}-u_i}{r(u_i)}.
\end{aligned}\tag{4.6}$$

From (4.2), we deduce

$$n^{1/2}\frac{u_{i+1}-u_i}{r(u_i)}=o(n^{-\alpha}).\tag{4.7}$$

From (4.6) and (4.7), we deduce

$$\sup_{\substack{v\leqslant\theta\\ v\geqslant u_1^{(n)}}}\left|n^{1/2}V_{1,n}(v)\frac{1}{r(v)}\right|\leqslant 3\sup_{\substack{1\leqslant i\leqslant i_n\\ u_i\leqslant\theta}}\left|n^{1/2}V_{1,n}(u_i)\frac{1}{r(u_i)}\right|+o(n^{-\alpha}),\tag{4.8}$$

which implies

$$P\left[\sup_{\substack{v\leqslant\theta\\ v\geqslant u_1^{(n)}}}\left|n^{1/2}V_{1,n}(v)\frac{1}{r(v)}\right|>\eta\right]\leqslant C\eta^{-4}\nu^{2-2\epsilon}(2\theta).\tag{4.9}$$

Finally, we have

$$P\left[\sup_{v\leqslant u_1^{(n)}}\left|n^{1/2}\tilde{V}_{1,n}(v)\frac{1}{r(v)}\right|=0\right]=1. \tag{4.10}$$

From (4.9) and (4.10) we easily deduce (4.1) for $u\leqslant\theta$, and by symmetry (4.1) follows also for $u\geqslant 1-\theta$.

Now define $\tilde{V}_{l,n}$ by

$$\tilde{V}_{l,n}(u)=\begin{cases}V_{l,n}(u) & \text{if } n^{-1-\alpha}\leqslant u\leqslant 1-n^{-1-\alpha}\\ 0 & \text{otherwise.}\end{cases} \tag{4.11}$$

LEMMA 4.2. *Under the conditions of Theorem* 2.1, $\forall\epsilon>0$, $\exists\theta>0$, $\exists N_0$, *such that* $\forall n\geqslant N_0$

$$P\left[\sup_{v\in C_\theta}\left|n^{1/2}\tilde{V}_{l,n}(v)\frac{1}{r(v)}\right|\geqslant\epsilon\right]\leqslant\epsilon,\qquad l=2,...,k. \tag{4.12}$$

Proof. Again consider the sequence $\{u_i^{(n)}=u_i,\ i=0,...,i_n\}$ defined in (4.2). Consider the case $l=2$. From Lemma 3.2, and α satisfying (2.1), we deduce

$$P\left[\sup_{1\leqslant l\leqslant i_n}\left|n^{1/2}V_{2,n}(u_l)\frac{1}{r(u_l)}\right|>\epsilon\right]=O(n^{-\eta'}),\qquad \eta'>0. \tag{4.13}$$

By applying methods similar to those used in deriving the inequality (3.15), we obtain

$$\begin{aligned}\sup_{v\leqslant\theta}|n^{1/2}V_{2,n}(v)|\leqslant{}&3n^{1/2}\sup_{1\leqslant l\leqslant i_n}|V_{2,n}(u_l)|\\ &+2n^{1/2}\sup_{1\leqslant l\leqslant i_{n-1}}|V_{1,n}(u_{l+1})-V_{1,n}(u_l)|\\ &+2n^{-\delta+\alpha(1/2-\delta)}.\end{aligned} \tag{4.14}$$

From (4.2), we deduce

$$\begin{aligned}\sup_{v\leqslant\theta}\left|n^{1/2}\tilde{V}_{2,n}(v)\frac{1}{r(v)}\right|\leqslant{}&6n^{1/2}\sup_{1\leqslant l\leqslant i_n}\left|\tilde{V}_{2,n}(u_l)\frac{1}{r(u_l)}\right|\\ &+8n^{1/2}\sup_{1\leqslant l\leqslant i_n}\left|\tilde{V}_{1,n}(u_l)\frac{1}{r(u_l)}\right|\\ &+2n^{-\alpha-\delta+(1/2-\delta)\alpha'}.\end{aligned} \tag{4.15}$$

From (4.13), (4.16), and (4.5), we easily deduce (4.12) for $v \leqslant \theta$ and $l=2$. By symmetry, (4.12) follows also for $v \geqslant 1-\theta$ and $l=2$. The case $l \geqslant 3$ is similar and is therefore omitted.

Proof of Proposition 2.3. From Theorems 15.1 and 15.5 of Billingsley (1968), we must prove that the finite dimensional distributions of Q_n^* converge in law to a normal distribution and $\forall \varepsilon > 0$, $\exists \delta > 0$ and N_0 such that $\forall n \geqslant N_0$

$$Q_n^*\left[f \in D[0, 1], \omega\left(f \cdot \frac{1}{r}, \delta \right) \geqslant \epsilon \right] \leqslant \epsilon. \tag{4.16}$$

It is sufficient to prove (4.16) because the first condition is an easy consequence of the hypothesis.

We can write

$$\begin{aligned}
\omega\left(f \cdot \frac{1}{r}, \delta \right) &= \sup_{|u-v| \leqslant \delta} \left| f(u) \frac{1}{r}(u) - f(v) \frac{1}{r}(v) \right| \\
&\leqslant \sup_{u \in C_{\theta-\delta}} \sup_{|u-v| < \delta} \left| f(u) \frac{1}{r}(u) - f(v) \frac{1}{r}(v) \right| \\
&\quad + \sup_{u \notin C_{\theta-\delta}} \sup_{|u-v| < \delta} \left| (f(u) - f(v)) \frac{1}{r}(u) + f(v) \left(\frac{1}{r}(u) - \frac{1}{r}(v) \right) \right| \\
&\leqslant 2 \sup_{u \in C_\theta} \left| f(u) \frac{1}{r}(u) \right| + \frac{\omega(f, \delta)}{m} + \frac{\omega(r, \delta)}{m^2} \sup_{u \in [0, 1]} |f(t)|,
\end{aligned}$$

where $m = \min_{u \notin C_{\theta - 2\delta}} r(u)$ and we obtain (4.16) easily by using (2.7).

The proof of Theorem 2.1 follows immediately from Proposition 2.1 through Proposition 2.3.

5. Appendix

Lemma 5.1. *If the sequence* $\{X_i, i \geqslant 1\}$ *is absolutely regular with rates* (1.4), *then for any* $0 < \epsilon < 1$, *for any* (s, t), $s \leqslant t$, *we have*

$$\begin{aligned}
&E(n^{1/2} \hat{V}_{1,n}(t) - n^{1/2} \hat{V}_{1,n}(s))^4 \\
&\quad \leqslant C[(H(t) - H(s))^{2-2\epsilon} + n^{-1}(H(t) - H(s))^{1-\epsilon}],
\end{aligned} \tag{5.1}$$

where C is some constant >0 *and* $\hat{V}_{1,n}(t) = V_{1,n}(H(t))$.

Proof. From Theorem 2.10 for $q=4$ of Doukhan and Portal (1987) we have the inequality

$$E(n^{1/2}\hat{V}_{1,n}(t)-n^{1/2}\hat{V}_{1,n}(s))^4 \leqslant C_1(\|Y(s,t)\|_{\epsilon}^{2-2\epsilon}+n^{-1}\|Y(s,t)\|_{\epsilon}^{1-\epsilon}), \tag{5.2}$$

where $\|Y(s,t)\|_{\epsilon} = (E|Z(t)-Z(s)|^{2/(1-\epsilon)})$, $Z(t)=\hat{g}_1(X_1;t)-H(t)$, and $\hat{g}_1(X_i;t)=g_1(X_1;H(t))$.

As $0\leqslant g_1(X_1;t)-g_1(X_1;s)-H(t)+H(s)\leqslant 1$ we have

$$\begin{aligned}
E|Z(t)-Z(s)|^{2/(1-\epsilon)} &\leqslant E(g_1(X_1;t)-g_1(X_1;s)-H(t)+H(s))^2\\
&=E(g_1(X_1;t)-g_1(X_1;s))^2\\
&\quad -2(H(t)-H(s))\,E(g_1(X_1;t)-g_1(X_1;s))\\
&\quad +(H(t)-H(s))^2\\
&\leqslant E(g_1(X_1;t)-g_1(X_1;s))\\
&\quad -2(H(t)-H(s))\,E(g_1(X_1;t)-g_1(X_1;s))\\
&\quad +(H(t)-H(s))^2\\
&\leqslant H(t)-H(s)-2(H(t)-H(s))^2+(H(t)-H(s))^2\\
&=(H(t)-H(s))(1-(H(t)-H(s)))\\
&\leqslant H(t)-H(s),
\end{aligned}$$

which implies

$$\|Y(s,t)\|_{\epsilon}\leqslant H(t)-H(s).$$

From (5.2) and (5.3) we deduce (5.1) and the lemma is proved.

Let $1\leqslant i_1<i_2<\cdots<i_k\leqslant n$ arbitrary integers. For any j $(1\leqslant j\leqslant k-1)$, P_j is the probability measure defined by

$$\begin{aligned}
P_j^{(i_1,\dots,i_k)}(A^{(j)}\times B^{(k-j)}) &= P[(X_{i_1},\dots,X_{i_j})\in A^{(j)}]\\
&\quad\times P[(X_{i_{j+1}},\dots,X_{i_k})\in B^{(k)}]
\end{aligned}\tag{5.4}$$

and $P_0^{(i_1,\dots,i_k)}$ is the probability measure defined by

$$P_0^{(i_1,\dots,i_k)}(A^{(k)})=P[(X_{i_1},\dots,X_{i_k})\in A^{(k)}]$$

for any $A^{(j)}\in\sigma(X_{i_1},\dots,X_{i_j})$ $(1\leqslant j\leqslant k)$ and any $B^{(k-j)}\in\sigma(X_{i_{j+1}},\dots,X_{i_k})$ $(1\leqslant j\leqslant k-1)$.

The following lemma is proved by Yoshihara (1976).

LEMMA 5.2. *For every* $(i_1, ..., i_k)$ *such that* $i_1 < i_2 < \cdots < i_k$ *and any* j $(0 \leqslant j \leqslant k-1)$, *let* $h(x_1, ..., x_k)$ *be a bounded Borel function such that*

$$\sup_{(x_1, ..., x_k) \in \mathbb{R}^k} |h(x_1, ..., x_k)| \leqslant M;$$

then

$$\left| \int_{\mathbb{R}^k} h(x_1, ..., x_k)\, dP_0^{(i_1, ..., i_k)} - \int_{\mathbb{R}^k} h(x_1, ..., x_k)\, dP_j^{(i_1, ..., i_k)} \right| \leqslant 2M\beta(i_{j+1} - i_j) \tag{5.5}$$

where β *is the rate of absolute regularity.*

The following lemma is a variant of Lemma 1 of Balacheff and Dupont (1980).

LEMMA 5.3. *Let* ν *be a continuous positive finite measure on* $[0, 1]$. *Let* P *be a probability measure on* $D[0, 1]$ *such that* $P(f \in D[0, 1], f(0) = 0) = 1$. *Let* T *be a finite subset of* $[0, 1]$ *which includes* 0. *Furthermore suppose that for any interval* $[u, v]$ *of* $[0, 1]$ *such that* $u \in T$, $v \in T$, *and* $u < v$,

$$P[f \in D[0, 1], |f(v) - f(u)| \geqslant \lambda] \leqslant \lambda^{-\gamma}(\nu[u, v]))^{\beta}, \qquad \beta > 1 \qquad \text{and} \qquad \gamma > 0\ \forall \lambda > 0. \tag{5.6}$$

Then there exists a constant C *such that* $\forall \lambda > 0$

$$P[M_T(f) \geqslant \lambda] \leqslant C\lambda^{-\gamma}(\nu([0, u_T]))^{\beta}, \tag{5.7}$$

where

$$M_T(f) = \sup\{|f(u)|, u \in T\}$$

and

$$u_T = \max\{u, u \in T\}.$$

ACKNOWLEDGMENT

The authors express their sincere thanks to the referees for very helpful suggestions which contributed considerably to improvement of the paper.

REFERENCES

1. BALACHEFF, S., AND DUPONT, G. (1980). *Normalité asymptotique des processus empiriques tronqués et des processes de rang.* Lecture Notes in Mathematics, Vol. 821. Springer-Verlag, Berlin.
2. BILLINGSLEY, P. (1968). *Convergence of Probability Measures.* Wiley, New York.
3. DOUKHAN, P., AND PORTAL, F. (1987). Principe d'invariance faible pour la fonction de répartition empirique dans un cadre multidimensionel et mélangeant. *Probab. Math. Statist.* **8** 117–132.
4. HAREL, M. AND PURI, M. L. (1989a). Limiting behavior of *U*-statistics, *V*-statistics and one-sample rank order statistics for non-stationary absolutely regular processes. *J. Multivariate Anal.* **30** 181–204.
5. HAREL, M., AND PURI, M. L. (1989b). Weak convergence of the *U*-statistic and weak invariance of the one-sample rank order statitic for Markov processes and ARMA models. *J. Multivariate Anal.* **31** 259–265.
6. HAREL, M. AND PURI, M. L. (1990). Weak invariance of generalized *U*-statistics for non-stationary absolutely regular processes. *Stochastic Process. Appl.* **34** 341–360.
7. IBRAGIMOV, J. A., AND LINNIK, YU. V. (1971). *Independent and Stationary Sequences of Random Variables.* Wolters–Noordhoff, Groningen.
8. OODAIRA, H., AND HOSHIHARA, K. (1972). Functional central limit theorems for strictly stationary processes satisfying the strong mixing condition. *Kodai Math. Sem. Rep.* **24**, No. 3 259–269.
9. RUYMGAART, F. H., AND VAN ZUIJLEN, M. C. A. (1991). Empirical *U*-statistics processes. *J. Statist. Plann. Inference* **32** 259–269.
10. YOSHIHARA, K. (1976). Limiting behavior of *U*-statistics for stationary absolutely regular processes. *Z. Wahrsch. Verw. Gebiete* **35** 237–252.

LAW OF THE ITERATED LOGARITHM FOR PERTURBED EMPIRICAL DISTRIBUTION FUNCTIONS EVALUATED AT A RANDOM POINT FOR NONSTATIONARY RANDOM VARIABLES[1]

MICHEL HAREL[2] and MADAN L. PURI[3]

Received April 1, 1993; revised October 7, 1993

We consider perturbed empirical distribution functions $\hat{F}_n(x)=1/n\sum_{i=1}^n G_n(x-X_i)$, where $\{G_n, n\geqslant 1\}$ is a sequence of continuous distribution functions converging weakly to the distribution function of unit mass at 0, and $\{X_i, i\geqslant 1\}$ is a nonstationary sequence of absolutely regular random variables. We derive the almost sure representation and the law of the iterated logarithm for the statistic $\hat{F}_n(U_n)$ where U_n is a U-statistic based on $X_1,..., X_n$. The results obtained extend or generalize the results of Nadaraya,[7] Winter,[16] Puri and Ralescu,[9,10] Oodaira and Yoshihara,[8] and Yoshihara,[19] among others.

KEY WORDS: Perturbed empirical distribution functions; absolutely regular processes; strong mixing; almost sure representation; U-statistic; law of the iterated logarithm.

1. INTRODUCTION

Let $\{X_i, i\geqslant 1\}$ be a sequence of absolutely regular r.v.s (random variables) with continuous c.d.f.s (cumulative distribution functions) $\{F_i, i\geqslant 1\}$ defined on the real line $\mathbb{R}$ and let $\tilde{F}_n$ be the corresponding empirical distribution function based on $X_1,..., X_n$, i.e., $\tilde{F}_n(x)=n^{-1}\sum_{i=1}^n u(x-X_i)$ where $u(t)=1$ or 0 according as $t\geqslant 0$ or $t<0$. Suppose that F_i converges to a fixed d.f. F as $i\to\infty$. If $\{X_i, i\geqslant 1\}$ is a stationary sequence (in which case $F_i\equiv F$ for all i), $\tilde{F}_n$ is a natural estimator of F based on the sample $X_1,..., X_n$. However if F is a smooth distribution function, then it seems

---

[1] Research supported by the Office of Naval Research Contract N00014-91-J-1020.

[2] U.R.A. C.N.R.S. 743, Orsay, France and Indiana University, Bloomington, Indiana, 47405.

[3] Indiana University, Bloomington, Indiana, 47405.

reasonable to consider smooth estimators which are better adapted to this situation. For the case $\{X_i, i \geqslant 1\}$ is a sequence of iid (independent and identically distributed) r.v.s, Watson and Leadbetter,[14] Nadaraya,[7] Yamato,[18] and Winter,[16,17] have considered an estimator of the form

$$\hat{F}_n(x) = n^{-1} \sum_{i=1}^{n} G_n(x - X_i), \qquad x \in \mathbb{R}, \quad n \geqslant 1 \tag{1.1}$$

where $G_n(x) = \int_{-\infty}^{x} g_n(t)\,dt$, $g_n(t) = \alpha_n^{-1} g(t\alpha_n^{-1})$, $\{\alpha_n; n \geqslant 1\}$ is a sequence of positive real numbers such that $\alpha_n \to 0$ as $n \to \infty$, and g is a probability density function. Watson and Leadbetter[14] established the asymptotic normality of $\hat{F}_n$, while Nadaraya,[7] Winter,[16] and Yamato,[18] proved the a.s. (almost sure) convergence of $\hat{F}_n$ to F. Later Winter[17] showed that $\hat{F}_n$ has the Chung-Smirnov property, i.e. with probability one,

$$\limsup_{n \to \infty} (2n/\log n)^{1/2} \sup_x |\hat{F}_n(x) - F(x)| \leqslant 1$$

For some related results dealing with the estimation of probability density function, see Schuster,[11] Scott *et al.*,[12] and Wertz,[15] for a general review in this area.

Let $h(x_1, ..., x_m)$, symmetric in its arguments be a measurable kernel (of degree m), and let U_n be the corresponding U-statistics given by

$$U_n = \binom{n}{m}^{-1} \sum_{C_{n,m}} h(X_{i_1}, ..., X_{i_m}) \tag{1.2}$$

where $C_{n,m}$ denotes the set of all $\binom{n}{m}$ combination of m distinct elements $\{i_1, ..., i_m\}$ from $\{1, ..., n\}$. Let

$$\theta(F) = \int_{\mathbb{R}^m} h(x_1, ..., x_m) \prod_{i=1}^{m} dF(x_i) \tag{1.3}$$

When the r.v.s are i.i.d., Puri and Ralescu[10] derived the asymptotic normality of $\hat{F}_n(U_n)$ which is useful in estimating $\xi = F(\theta(F))$ when F is unknown. Later Sun[13] extended this result when the r.v.s are stationary and absolutely regular. Lea and Puri[6] established for the iid case, (i) the almost sure (a.s.) representation, (ii) the law of the iterated logarithm and (iii) an invariance principle for $\hat{F}_n(U_n)$. In this paper, we extend the results (i) and (ii) when the r.v.s are nonstationary and absolutely regular.

With our general framework (nonstationarity and absolute regularity), the results have applications for a larger class of processes such as ARMA processes and Markov processes for which the initial measure is not necessary to be the invariant measure. For example, suppose that the sequence $\{X_i, i \geq 1\}$ of r.v.'s is an ARMA process defined by

$$X_i = aX_{i-1} + b\varepsilon_i + \varepsilon_{i-1}, \qquad i \geq 1 \tag{1.4}$$

where X_0 admits a strictly positive density, $\{\varepsilon_i, i \geq 1\}$ is a sequence of i.i.d. r.v.'s with strictly positive density such that $E(\varepsilon_i) = 0$ and a and b are real numbers such that $|a| < 1$. Then, from Harel and Puri,(4) this process is absolutely regular and for any initial measure, the conclusions of the theorems and propositions here hold if the sequence of r.v.'s is an ARMA process defined by Eq. (1.4). The results of our paper can also be applied for nonstationary Markov processes (see, e.g. Harel and Puri(5)).

2. SOME PRELIMINARY RESULTS

In the following, we suppose that the sequence $\{X_i, i \geq 1\}$ is absolutely regular with rates

$$\beta(m) = O(\rho^m), \qquad 0 < \rho < 1 \tag{2.1}$$

Recall that a nonstationary sequence $\{X_{ni}, 1 \leq i \leq n, n \geq 1\}$ is absolutely regular if

$$\sup_{m \leq n} \max_{1 \leq j \leq n-m} E\{\sup_{A \in \sigma(X_{ni}, i \geq j+m)} |P(A \mid \sigma(X_{ni}, 1 \leq i \leq j)) - P(A)|\}$$
$$= \beta(m) \downarrow 0 \text{ as } m \to \infty$$

Here $\sigma(X_{ni}, 1 \leq i \leq j)$ and $\sigma(X_{ni}, i \geq j+m)$ are the σ-fields generated by $(X_{n1},..., X_{nj})$ and $(X_{n,j+m}, X_{n,j+m+1},..., X_{nn})$ respectively. Also recall that $\{X_{ni}\}$ satisfies the strong mixing condition if $\sup_{m \leq n} \sup_{1 \leq j \leq n-m}\{|P(A \cap B) - P(A)\,P(B)|;\ A \in \sigma(X_{ni}, 1 \leq i \leq j),\ B \in \sigma(X_{ni}, i \geq j+m)\} = \alpha(m) \downarrow 0$ as $m \to \infty$. Since $\alpha(m) \leq \beta(m)$, it follows that if $\{X_{ni}\}$ is absolutely regular, then it is also strong mixing.

Let $\{d_n\}$ be a sequence of positive real numbers such that $d_n \sim kn^{-1/2}(\log\log n)^{1/2}$ as $n \to \infty$, for some $k > 0$. Define

$$K_n(t) = \sup_{|x| \leq d_n} |\tilde{F}_n(t+x) - \tilde{F}_n(t) - \bar{F}_n(t+x) - \bar{F}_n(t)| \tag{2.1'}$$

where $\bar{F}_n = n^{-1}\sum_{i=1}^n F_i$, and let F_i' denote the density of F_i.

Lemma 2.1. For $\delta > 0$, if $\{F_i', i \geq 1\}$ is uniformly bounded in $(\theta(F) - \delta, \theta(F) + \delta)$ and

$$K_n^* = \sup_{|t - \theta(F)| < \delta/2} K_n(t) \tag{2.2}$$

then, for any v $(0 < v < \frac{1}{8})$ and for any $s > 1$ there exist two positive constants $C_v^{(1)}$ and $C_v^{(2)}$ and an integer n_s such that

$$P[K_n^* \geqslant C_v^{(1)} n^{-3/4+v}] \leqslant C_v^{(2)} n^{-s} \tag{2.3}$$

for all $n \geqslant n_s$. Hence

$$K_n^* = O(n^{-3/4+v}) \text{ a.s. as } n \to \infty$$

Proof. We start in a way similar way to Lemma 2.1 of Lea and Puri.[6] For $\eta \in [\theta(F) - \delta/2, \theta(F) + \delta/2]$, let

$$H_n(x) = \tilde{F}_n(x) - \tilde{F}_n(\eta) - \bar{F}_n(x) + \bar{F}_n(\eta) \tag{2.4}$$

Let $\{b_n, n \geqslant 1\}$ be a sequence of positive integers such that $b_n \sim n^{1/4}$ as $n \to \infty$. Consider a particular n. For any integer $r \in [-b_n, b_n]$, let $\eta_{r,n} = \eta + d_n b_n^{-1} r$ and $\theta_{r,n} = \bar{F}_n(\eta_{r+1,n}) - \bar{F}_n(\eta_{r,n})$. Since $\tilde{F}_n$ and $\bar{F}_n$ are non-decreasing, it follows from Eq. (2.4) that for $x \in [\eta_{r,n}, \eta_{r+1,n}]$,

$$H_n(x) \leqslant \tilde{F}_n(\eta_{r+1,n}) - \tilde{F}_n(\eta) - \bar{F}_n(\eta_{r,n}) + \bar{F}_n(\eta) = H_n(\eta_{r+1,n}) + \theta_{r,n} \tag{2.5}$$

Similarly

$$H_n(x) \geqslant H_n(\eta_{r,n}) - \theta_{r,n} \text{ for } x \in [\eta_{r,n}, \eta_{r+1,n}] \tag{2.6}$$

It follows from Eqs. (2.1), (2.5), and (2.6) that

$$K_n(\eta) \leqslant J_n + \beta_n \tag{2.7}$$

where

$$J_n = \max_{-b_n \leqslant r \leqslant b_n} |H_n(\eta_{r,n})|, \qquad \beta_n = \max_{-b_n \leqslant r \leqslant b_n} \theta_{r,n} \tag{2.8}$$

Now, since $\{F_i, i \geqslant 1\}$ is uniformly bounded in a neighborhood of η, $\bar{F}'_n$ is also uniformly bounded in a neighborhood of η, it follows that for large n

$$\beta_n = O(n^{-3/4}(\log \log n)^{1/2}) \tag{2.9}$$

Let C_2 be the upper bound of $\sup_{n \geqslant 1} \bar{F}'_n(x)$ in $(\xi - \delta, \xi + \delta)$. Since $d_n \to 0$ as $n \to \infty$, there exists a positive integer N_1 such that $\sup_{n \geqslant 1} \bar{F}_n(\eta) - \bar{F}_n(\eta - d_n) \leqslant C_2 d_n$ for all $n \geqslant N_1$. Applying Lemma 5.2 (see Appendix), with $m = n^{2v - \varepsilon}$, $0 < \varepsilon < 2v$, and $a = \frac{7}{8}$, we have

$$P[|H_n(\eta_{r,n})| \geqslant \delta_n^{(v)}] \leqslant 2n^{2v-\varepsilon} \exp\{-r_n^{(v)}\} + 2n\beta([n^{2v-\varepsilon}]) \tag{2.10}$$

where $\delta_n^{(v)} = C_1 n^{-3/4+v}$, [a] is the integer part of a, C_1 is some constant > 0 and

$$r_n^{(v)} = n^{7/8(1-2v+\varepsilon)} \delta_n^{(v)} / 5M_0 \tag{2.11}$$

After some routine computations, we obtain

$$r_n^{(v)} = O(n^{1/4 - v + \varepsilon})$$

which implies

$$n^{2v - \varepsilon} \exp\{-r_n^{(v)}\} = O(n^{-(s+3/4)}) \tag{2.12}$$

We also have

$$\beta([n^{2v-\varepsilon}]) = O(e^{(\log \rho) n^{2v-\varepsilon}}) \tag{2.13}$$

Thus, from Eq. (2.10), we deduce

$$P[|H(\eta_{r,n})| \geqslant \delta_n^{(v)}] = O(n^{-(s+3/4)}) + O(ne^{n^{(2v-\varepsilon)} \log \rho}) \tag{2.14}$$

It follows from Eqs. (2.8) and (2.14) that

$$P[J_n \geqslant \delta_n^{(v)}] = O(b_n n^{-(s+3/4)}) + O(b_n ne^{n^{(2v-\varepsilon)} \log \rho}) \tag{2.15}$$

It is easy to see that if C_1 is chosen sufficiently large, there exists a positive constant C_2 and a positive integer N_1 such that for $n \geqslant N_1$

$$P[J_n \geqslant \delta_n^{(v)}] \leqslant C_2 n^{-(s+1/2)} \tag{2.16}$$

From Eqs. (2.7), (2.9), and (2.16), it follows that there exists a positive integer N_2 such that

$$P[K_n(\eta) \geqslant 2\delta_n^{(v)}] \leqslant C_2 n^{-(s+1/2)} \qquad \text{for} \quad n \geqslant N_2 \tag{2.17}$$

Now, let

$$r_{j,n} = \theta(F) + \frac{j\delta}{2\sqrt{n}}, \qquad j = -[\sqrt{n}], \ldots, 0, \ldots, [\sqrt{n}]$$

Then by Eq. (2.17), we have

$$\begin{aligned} P[\max_{-\sqrt{n} \leqslant j \leqslant \sqrt{n}} K_n(r_{j,n}) \geqslant 2\delta_n^{(v)}] &\leqslant 2C_2 n^{1/2} n^{-(s+1/2)} \\ &= 2C_2 n^{-s} \qquad \text{for} \quad n \geqslant N_2 \end{aligned} \tag{2.18}$$

For any $t \in [\theta(F)-\delta/2, \theta(F)+\delta/2]$ and for any $x \in [-d_n, d_n]$ where $x \neq 0$, there exist two integers j_1 and j_2 such that

$$[-\sqrt{n}] \leqslant j_1 < j_2 \leqslant [\sqrt{n}]$$

and

$$\theta(F)+\frac{(j_1-1)\,\delta}{2\sqrt{n}} < \underline{t} \leqslant \theta(F)+\frac{j_1\delta}{2\sqrt{n}} \leqslant \theta(F)+\frac{(j_2-1)\,\delta}{2\sqrt{n}} < \bar{t} \leqslant \theta(F)+\frac{j_2\delta}{2\sqrt{n}}$$

where $\underline{t}=\min\{t, t+x\}$ and $\bar{t}=\max\{t, t+x\}$.

From the triangular inequality, we obtain

$$|V_n(t)-V_n(t+x)| \leqslant \left| V_n(\underline{t}) - V_n\left(\theta(F)+\frac{j_1\delta}{2\sqrt{n}}\right)\right| + \left| V_n\left(\theta(F)+\frac{j_1\delta}{2\sqrt{n}}\right) - V_n\left(\theta(F)+\frac{(j_2-1)\,\delta}{2\sqrt{n}}\right)\right| + \left| V_n\left(\theta(F)+\frac{(j_2-1)\,\delta}{2\sqrt{n}}\right) - V_n(t+x)\right|$$

where $V_n(t)=\tilde{F}_n(t)-\bar{F}_n(t)$.

As $d_n \geqslant \delta/2\sqrt{n}$ for n sufficiently large, we deduce

$$K_n^* \leqslant 3 \max_{-\sqrt{n} \leqslant j \leqslant \sqrt{n}} K_n(r_{j,n}) \tag{2.19}$$

Choosing $C_v^{(1)}=6C_1$, $C_v^{(2)}=2C_2$ and $n \geqslant N_2$, it follows, from Eqs. (2.18) and (2.19), that

$$P[K_n^* \geqslant C_v^{(1)} n^{-3/4+v}] \leqslant C_v^{(2)} n^{-s}$$

Thus Eq. (2.3) follows. Also

$$\sum_{n=1}^{\infty} P[K_n^* \geqslant C_v^{(1)} n^{-3/4+v}] \leqslant C_v^{(2)} \sum_{n=1}^{\infty} n^{-s} < \infty$$

Consequently, using the Borel-Cantelli Lemma, we obtain

$$K_n^* = O(n^{-3/4+v}) \text{ a.s.} \qquad \text{as} \quad n \to \infty$$

This completes the proof.

Lemma 2.2. If (i) $\{F'_i, i \geq 1\}$ is uniformly bounded in a neighborhood of $\theta(F)$ and (ii) $\int x^2 g(x)\,dx < \infty$, then

$$|\hat{F}_n(\theta(F)) - \tilde{F}_n(\theta(F))| = O(n^{-3/4+v}) + O(n\alpha_n^2(\log\log n)^{-1}) \text{ a.s. } (0 < v < \tfrac{1}{8})$$

Proof. We proceed as in Lemma 2.2 of Lea and Puri.[6] First note that

$$\hat{F}_n(\theta(F)) = n^{-1} \sum_{i=1}^{n} G_n(\theta(F) - X_i) = \int_{-\infty}^{+\infty} G_n(\theta(F) - t)\,d\tilde{F}_n(t)$$

$$= \int_{-\infty}^{+\infty} \tilde{F}_n(\theta(F) - t)\,g_n(t)\,dt \tag{2.20}$$

and

$$\int_{-\infty}^{+\infty} g_n(t)\,dt = 1 \tag{2.21}$$

Condition (i) ensures that there exists $\delta > 0$ and $M > 0$ such that $\sup_{i \in \mathbb{N}} |F_i(\theta(F) - t) - F_i(\theta(F))| \leq M\,|t|$ for all $|t| < \delta$.

Using Lemma 2.1 and Eqs. (2.20) and (2.21), we have

(a) $|\hat{F}_n(\theta(F)) - \tilde{F}_n(\theta(F))|$

$$\leq \int_{-\infty}^{+\infty} |\tilde{F}_n(\theta(F) - t) - \tilde{F}_n(\theta(F))|\,g_n(t)\,dt$$

$$\leq \int_{|t| \leq d_n} |\tilde{F}_n(\theta(F) - t) - \tilde{F}_n(\theta(F))|\,g_n(t)\,dt + \int_{|t| > d_n} |\tilde{F}_n(\theta(F) - t) - \tilde{F}_n(\theta(F))|\,g_n(t)\,dt$$

$$\leq \int_{|t| \leq d_n} |\bar{F}_n(\theta(F) - t) - \bar{F}_n(\theta(F))|\,g_n(t)\,dt + \int_{|t| > d_n} g_n(t)\,dt + O(n^{-3/4+v}) \text{ a.s.}$$

For large n, the first term on the right-hand side of the last inequality in (a) is bounded by

(b) $$M \int_{-\infty}^{+\infty} |t|\,g_n(t)\,dt = M\alpha_n \int_{-\infty}^{+\infty} |y|\,g(y)\,dy = O(\alpha_n)$$

and the second term is bounded by

$$\text{(c)}\quad n\alpha_n^2(\log\log n)^{-1}\int_{|y|>d_n/\alpha_n} y^2 g(y)\,dy = O(n\alpha_n^2(\log\log n)^{-1})$$

since $\int_{-\infty}^{+\infty} y^2 g(y)\,dy < \infty$. Thus, from (a)–(c) we have

$$\begin{aligned}|\hat{F}_n(\theta(F)) - \tilde{F}_n(\theta(F))| &= O(\alpha_n) + O(n\alpha_n^2(\log\log n)^{-1}) + O(n^{-3/4+v})\\ &= O(n\alpha_n^2(\log\log n)^{-1}) + O(n^{-3/4+v})\end{aligned}$$

a.s. as $n\to\infty$ and Lemma 2.2 is proved.

Because of the hypothesis of nonstationarity, we have to define some new notations for the U-statistics.

For every $p(1\leqslant p\leqslant m)$ and $n\geqslant m$, let $1\leqslant i_1\neq\cdots\neq i_p\leqslant n$ be arbitrary integers. Put

$$h_{p,n}^{(i_1,\dots,i_p)}(x_1,\dots,x_p) = \sum_{(i_{p+1},\dots,i_m)\in I_{p,n}(i_1,\dots,i_p)} \lambda(x_1,\dots,x_p;i_{p+1},\dots,i_m) \tag{2.22}$$

where

$$\lambda(x_1,\dots,x_p;i_{p+1},\dots,i_m) = \int_{\mathbb{R}^{m-p}} h(x_1,\dots,x_m)\,dF_{i_{p+1}}(x_{p+1})\dots dF_{i_m}(x_m)$$

and $I_{p,n}(i_1,\dots,i_p) = \{(i_1,\dots,i_p);\quad 1\leqslant i_{p+1}\neq\cdots\neq i_m\leqslant n,\quad i_l\notin(i_1,\dots,i_p),$ $p+1\leqslant l\leqslant m\}$ and

$$h_{0,n} = \sum_{(i_1,\dots,i_m)\in I_{0,n}} \int_{\mathbb{R}^m} h(x_1,\dots,x_m)\,dF_{i_1}(x_1),\dots,dF_{i_m}(x_m)$$

where $I_{0,n} = \{(i_1,\dots,i_m),\ 1\leqslant i_1\neq\cdots\neq i_m\leqslant n\}$.

For every $p(1\leqslant p\leqslant m)$, set

$$U_n^{(p)} = n^{-[m]}\sum_{1\leqslant i_1\neq\cdots\neq i_p\leqslant n}\int_{\mathbb{R}^p} h_{p,n}^{(i_1,\dots,i_p)}(x_1,\dots,x_p)$$

$$\times\prod_{j=1}^{p} d(I_{[X_{i_j}\leqslant x_j]} - F_{i_j}(x_j)) \tag{2.23}$$

where $n^{-[m]} = (n(n-1)\cdots(n-m+1))^{-1}$.

Lemma 2.3. If there exists a positive number δ such that for $r=4+\delta$

$$\sup_n \max_{(i_1,\dots,i_m)\in\{1,\dots,n\}^m} \int_{\mathbb{R}^m} |h(x_1,\dots,x_m)|^r \, dF_{i_1}(x_1) \times \cdots \times dF_{i_m}(x_m) \leqslant M_0 < \infty \tag{2.24}$$

$$\sup_n \max_{(i_1,\dots,i_m)\in\{1,\dots,n\}^m} E(|h(X_{i_1},\dots,X_{i_m})|^r) \leqslant M_0 < \infty \tag{2.25}$$

and for some $\delta'(0<\delta'<\delta)$, $\beta(n)=O(n^{-3(4+\delta')/(2+\delta')})$, then we have

$$E(U_n^{(2)})^4 = O(n^{-3-\gamma}) \tag{2.26}$$

where $\gamma=6(\delta-\delta')/(4+\delta)(2+\delta')>0$ and

$$E(U_n^{(p)})^2 = O(n^{-3}), \qquad 3\leqslant p\leqslant m \tag{2.27}$$

Proof. We first note that

$$U_n^{(2)} = n^{-[m]} \sum_{1\leqslant i_1\neq i_2\leqslant n} \int_{\mathbb{R}^2} h_{2,n}^{(i_1,i_2)}(x_1,x_2)\, d(I_{[X_{i_1}\leqslant x_1]} - F_{i_1}(x_1))\, d(I_{[X_{i_2}\leqslant x_2]} - F_{i_2}(x_2))$$

So we have

$$E(U_n^{(2)})^4 = (n^{[m]})^4 \sum_{1\leqslant i_{11},i_{12}\leqslant n} \sum_{1\leqslant i_{21},i_{22}\leqslant n} \sum_{1\leqslant i_{31},i_{32}\leqslant n} \times \sum_{1\leqslant i_{41},i_{42}\leqslant n} J((i_{11},i_{12}),(i_{21},i_{22}),(i_{31},i_{32}),(i_{41},i_{42})) \tag{2.28}$$

where

$$J((i_{11},i_{12}),(i_{21},i_{22}),(i_{31},i_{32}),(i_{41},i_{42})) = E\left\{\prod_{j=1}^{4} \int_{\mathbb{R}^2} h_{2,n}^{(i_{j1},i_{j2})}(x_{j1},x_{j2})\, d(I_{[X_{i_{j1}}\leqslant x_{j1}]} - F_{i_{j1}}(x_{j1}))\, d(I_{[X_{i_{j2}}\leqslant x_{j2}]} - F_{i_{j2}}(x_{j2}))\right\} \tag{2.29}$$

and

$$h_{2,n}^{(i_{j1},i_{j2})}(x_{i1},x_{i2}) = \sum_{(i_3,\dots,i_m)\in I_{2,n}(i_{j1},i_{j2})} g(x_{i1},x_{i2},x_3,\dots,x_m)\, dF_{i_3}(x_3)\cdots dF_{i_m}(x_m)$$

Let $i_{rs}(\leqslant n)(r=1,\ldots,4, s=1,2)$ be mutually different. Reorder i_{rs} as $1 \leqslant k_1 < k_2 < \cdots < k_8 \leqslant n$ and put

$$J((i_{11}, i_{12}),\ldots, (i_{41}, i_{42})) = H(k_1,\ldots, k_8) = E[g(X_{k_1},\ldots, X_{k_8})] \tag{2.30}$$

Let $d^{(c)}$ be the cth largest difference among $(k_{j+1} - k_j)$, $j = 1,\ldots, 7$

Since

$$\int_{\mathbb{R}^8} g(x_1,\ldots, x_8)\, P_j^{(k_1,\ldots,k_8)}(x_1,\ldots, x_8) = 0 \ (j = 1, 7) \text{ (see Appendix)}$$

It follows from Lemma 5.1 (see Appendix) that

$$H(k_1,\ldots, k_8) \leqslant \left(\frac{n^{[m]}}{n(n-1)}\right)^4 M_0[\beta(k_8 - k_7)]^{(2+\delta)/(4+\delta)}$$
$$\text{if} \quad k_8 - k_7 = d^{(1)} \tag{2.31}$$

and

$$H(k_1,\ldots, k_8) \leqslant \left(\frac{n^{[m]}}{n(n-1)}\right)^4 M_0[\beta(k_2 - k_1)]^{(2+\delta)/(4+\delta)}$$
$$\text{if} \quad k_2 - k_1 = d^{(1)} \tag{2.32}$$

Hence

$$\sum_{\substack{1 \leqslant k_1 < \cdots < k_8 \leqslant n \\ k_8 - k_7 = d^{(1)} \text{ or } k_2 - k_1 = d^{(1)}}} H(k_1,\ldots, k_8)$$
$$\leqslant \left(\frac{n^{[m]}}{n(n-1)}\right)^4 M_0 n^4 \sum_{j=1}^{n} (j+1)^3\, [\beta(j)]^{(2+\delta)/(4+\delta)} \tag{2.33}$$

If for some $j_\alpha (2 \leqslant j_\alpha \leqslant 6, 1 \leqslant \alpha \leqslant 4)$, $k_{j_\alpha + 1} - k_{j_\alpha} = d^{(\alpha)} (1 \leqslant \alpha \leqslant 4)$, then from Lemma 5.1 (see Appendix)

$$H(k_1,\ldots, k_8) \leqslant \left(\frac{n^{[m]}}{n(n-1)}\right)^4 M \sum_{\alpha=1}^{4} [\beta(k_{j_\alpha + 1} - k_{j_\alpha})]^{(2+\delta)/(4+\delta)} \tag{2.34}$$

and hence

$$\sum_{\substack{1 \leqslant k_1 < \cdots < k_8 \leqslant n \\ k_{j_\alpha + 1} - k_{j_\alpha} = d^{(\alpha)} (1 \leqslant \alpha \leqslant 4)}} H(k_1,\ldots, k_8)$$
$$\leqslant 4 \left(\frac{n^{[m]}}{n(n-1)}\right)^4 M_0 n^4 \sum_{j=1}^{n} (j+1)^3\, [\beta(j)]^{(2+\delta)/(4+\delta)} \tag{2.35}$$

Consequently

$$\sum_{1 \leqslant k_1 < \cdots < k_8 \leqslant n} H(k_1, \ldots, k_8)$$

$$\leqslant M\left(\frac{n^{[m]}}{n(n-1)}\right)^4 n^4 \sum_{j=1}^{n} (j+1)[\beta(j)]^{(2+\delta)/(4+\delta)} = O(n^{4m-3-\gamma}) \quad (2.36)$$

where $M > 0$ is some constant.

We can use a similar method to estimate the sums in the other cases and so we have Eq. (2.26).

The proof of Eq. (2.27) is analogous and so it is omitted.

Lemma 2.4. Under the conditions of Lemma 2.3, we have

$$U_n - \theta(F) = mU_n^{(1)} + R_n \quad (2.37)$$

where $R_n = O((n^{-1} \log \log n)^{1/2})$ a.s. as $n \to \infty$.

Proof. We rewrite Eq. (1.2) as

$$U_n = \theta(F) + \sum_{p=1}^{m} \binom{m}{p} U_n^{(p)} = \theta(F) + mU_n^{(1)} + \sum_{p=2}^{m} \binom{m}{p} U_n^{(p)}$$

then we have to prove that for every $\varepsilon > 0$

$$P[|W_n| > \varepsilon(n \log \log n)^{1/2} \text{ i.o.}] = 0 \quad (2.38)$$

where $W_n = n(U_n - \theta(F) - mU_n^{(1)})$. Let $n_k = [k^{(2+\delta')(4+\delta)/3(\delta-\delta')}]$ and $n_{k_0} \geqslant m$. Then, from Lemma 2.3 and the Bonferroni inequality

$$\sum_{k=k_0}^{\infty} P[\max_{m \leqslant n \leqslant n_k} |W_n| \geqslant \varepsilon(n_k \log \log n_k)^{1/2}]$$

$$\leqslant \sum_{k=k_0}^{\infty} \left(\sum_{n=m}^{n_k} P[|W_n| \geqslant \varepsilon(n_k \log \log n_k)^{1/2}] \right)$$

$$\leqslant M \sum_{k=k_0}^{\infty} k^{-2} < \infty \quad (2.39$$

where $M > 0$ is some constant. Consequently, from the Borel-Cantell Lemma, we have

$$P[|W_n| \geqslant \varepsilon(n \log \log n)^{1/2} \text{ i.o.}]$$

$$\leqslant P[\max_{n_k \leqslant n \leqslant n_{k+1}} |W_n| \geqslant \varepsilon(n_k \log \log n_k) \text{ i.o.}]$$

$$\leqslant P\left[\max_{m \leqslant n \leqslant n_{k+1}} |W_n| > \frac{\varepsilon}{4}(n_{k+1} \log \log n_{k+1}) \text{ i.o.}\right] = 0$$

3. A LAW OF THE ITERATED LOGARITHM FOR THE NONSTATIONARY *U*-STATISTIC

$$\sigma^2(\{F_l^*\}) = \sigma^2 = \left[\int_{\mathbb{R}^m} h^2(x_1, \ldots, x_m) \prod_{l=1}^{k} dF(x_l) - \theta^2(F)\right]$$
$$+ 2\sum_{i=1}^{\infty} \left[\int_{\mathbb{R}^{2m}} h(x_1, \ldots, x_m)\, h(x_{m+1}, \ldots, x_{2m})\, dF_i^*(x_1, x_{m+1})\right.$$
$$\left. \times \prod_{l=2}^{m} dF(x_l) \prod_{p=m+2}^{2m} dF(x_p) - \theta^2(F)\right] \tag{3.1}$$

if the limit exists and is finite.

Let $F_{i,j}$ be the distribution function of (X_i, X_j), $1 \leqslant i < j$.

Theorem 3.1. Suppose the sequence $\{X_i,\ i \geqslant 1\}$ is absolutely regular with rates satisfying Eq. (2.1). Furthermore, assume that for any $l > 1$, there exists a continuous d.f. F_l^* on $\mathbb{R}^2$ with marginals F such that

$$\|F_{i,j} - F_{j-i}^*\| = O(\rho_0^{\max(i,\, j-i)}), \qquad 1 \leqslant i < j \tag{3.2}$$

for some $0 < \rho_0 < 1$ where $\|\ \|$ denotes the norm of total variation.

Suppose also, that there exists a δ' such that Eqs. (2.24) and (2.25) are satisfied. Then we have

$$\limsup_{n \to \infty} \frac{n^{1/2}\,|U_n - \theta(F)|}{2^{1/2} m\sigma (\log\log n)^{1/2}} = 1 \text{ a.s.} \tag{3.3}$$

Proof. From Lemma 2.4, we have only to prove that

$$\limsup_{n \to \infty} \frac{|U_n^{(1)}|}{2^{1/2}\sigma(\log\log n)^{1/2}} = 1 \text{ a.s.} \tag{3.4}$$

where $U_n^{(1)} = n^{-1}\sum_{i=1}^{n} X_{ni}^*$ and

$$X_{ni}^* = n^{-[m-1]}\left(h_{1,n}^{(i)}(X_i) - \int_{\mathbb{R}} h_{1,n}^{(i)}(x)\, dF_i(x)\right) \tag{3.5}$$

From Lemma 5.6 (in the Appendix), we have to verify that the sequence X_{ni}^* satisfies the conditions in Eqs. (5.12)–(5.14). Condition in Eq. (5.13) follows from condition in Eq. (2.24). Condition in Eq. (5.14) follows from condition in Eq. (2.1).

We have to prove condition in Eq. (5.12), that is, for any n, m such that $n \geqslant m$ and any $J \subset \{1,\dots, n\}$ such that *card* $J = m$

$$E\Big(\sum_{i \in J} X_{ni}^*\Big)^2 = m\sigma^2(1 + o(1)) \tag{3.6}$$

We prove Eq. (3.6) for *card* $J = n$, because for *card* $J < n$, the proof is similar. Denote

$$\rho(1) = \int_{\mathbb{R}^m} h^2(x_1,\dots, x_m) \prod_{l=1}^{m} dF(x_l) - \theta^2(F)$$

$$\rho(i) = 2\Bigg[\int_{\mathbb{R}^m} h(x_1,\dots, x_m)\, h(x_{m+1},\dots, x_{2m})\, dF_i^*(x_1, x_{m+1})$$

$$\times \prod_{l=2}^{m} dF(x_l) \prod_{p=m+2}^{2m} dF(x_p) - \theta^2(F)\Bigg], \qquad \forall i \geqslant 2$$

$$\psi(i, i) = \int_{\mathbb{R}} (h_{1,n}^{(i)}(x))^2\, dF_i(x) - \Big(\int_{\mathbb{R}} h_{1,n}^{(i)}(x)\, dF_i(x)\Big)^2, \qquad \forall i \geqslant 1$$

$$\psi(i, j) = 2\int_{\mathbb{R}^2} (h_{1,n}^{(j)}(x)\, h_{1,n}^{(j+i)}(y))\, dF_{i,i+j}(x, y)$$

$$-\Big(\int_{\mathbb{R}} h_{1,n}^{(j)}(x)\, dF_j(x)\Big)\Big(\int_{\mathbb{R}} h_{1,n}^{(i+j)}(y)\, dF_{i+j}(y)\Big), \qquad i < j$$

Then

$$\left|E\left(\Big(\sum_{i=1}^{n} X_{ni}^*\Big)^2\Big/ n\right) - \sigma^2\right|$$

$$= \left|\frac{1}{n}(n^{-[m-1]})^2 \sum_{i=0}^{n-1}\sum_{j=1}^{n-i} \psi(i, j) - \sum_{i=1}^{\infty} \rho(i)\right|$$

$$\leqslant \left|\frac{1}{n}(n^{-[m-1]})^2 \sum_{i=0}^{n-1}\sum_{j=1}^{n-i} \psi(i, j) - \frac{1}{n}\sum_{i=1}^{\infty} (n-i)\,\rho(i)\right|$$

$$+ \sum_{i=n+1}^{\infty} |\rho(i)| + \sum_{i=1}^{n}\sum_{k=i}^{\infty} |\rho(k)|$$

$$= |A_n| + B_n + C_n.$$

From condition in Eq. (3.2), it follows that $|A_n| = o(1)$.

From condition in Eq. (2.24), we deduce that $|\rho(i)| \leqslant$

$(\alpha(i))^{\delta'/(2+\delta')}M_0^{2/(2+\delta')}$ which implies, by using condition in Eq. (2.1), that $B_n \to 0$ and $C_n \to 0$ as $n \to \infty$. Hence we have $|E(\sum_{i=1}^n X_{ni}^*)^2/n - \sigma^2| = o(1)$ and Theorem 3.1 is proved.

4. A LAW OF THE ITERATED LOGARITHM AND ALMOST SURE REPRESENTATION FOR $\hat{F}_n(U_n)$

In what follows we make the following assumptions on F and g.

$$\{F'_i, 1 \geqslant 1\} \text{ is uniformly bounded in a neighborhood of } \theta(F) \tag{4.1}$$

$$\int_{-\infty}^{\infty} |t|\, g(t)\, dt < \infty \tag{4.2}$$

Proposition 4.1. Under assumptions in Eqs. (4.1) and (4.2), we have for any $v (0 < v < \frac{1}{8})$

$$\sup_{|x-\theta(F)| \leqslant d_n} |\hat{F}_n(x) - \hat{F}_n(\theta(F)) - \bar{F}_n(x) + \bar{F}_n(\theta(F))| = O(n^{-3/4+v}) + O(\alpha_n) \tag{4.3}$$

a.s. as $n \to \infty$.

Proof. We proceed as in Theorem 3.2 of Lea and Puri.[6] Suppose that $\{F'_i, i \geqslant 1\}$ is uniformly bounded in $(\theta(F) - \delta, \theta(F) + \delta)$ for some $\delta > 0$. Note that $|x - \theta(F)| \leqslant d_n \Rightarrow |x - \theta(F)| \leqslant \frac{1}{2}\delta$ for large n. Now,

$$\begin{aligned}|E(\hat{F}_n(x)) - \bar{F}_n(x)| &\leqslant \int_{-\infty}^{+\infty} |\bar{F}_n(x-t) - \bar{F}_n(x)|\, g_n(t)\, dt \\ &\leqslant \int_{|t| < \delta/2} |\bar{F}_n(x-t) - \bar{F}_n(x)|\, g_n(t)\, dt + \int_{|t| \geqslant \delta/2} g_n(t)\, dt\end{aligned}$$

But

$$\begin{aligned}\int_{|t| \geqslant \delta/2} g_n(t)\, dt &= \int_{|y| \geqslant \delta\alpha_n^{-1}/2} g(y)\, dy \\ &\leqslant \frac{2\alpha_n}{\delta} \int_{|y| \geqslant \delta\alpha_n^{-1}/2} |y|\, g(y)\, dy = O(\alpha_n) \qquad \text{as} \quad n \to \infty \end{aligned} \tag{4.4}$$

and by Taylor's theorem

$$\int_{|t|<\delta/2} |\bar{F}_n(x-t)-\bar{F}_n(x)|\; g_n(t)\,dt$$

$$= \int_{|t|<\delta/2} |\bar{F}'_n(x_t)|\; |t|\; g_n(t)\,dt$$

$$\leqslant \text{const. } \alpha_n \int_{-\infty}^{+\infty} |y|\; g(y)\,dy = O(\alpha_n) \qquad \text{as} \quad n\to\infty \tag{4.5}$$

where x_t is between x and t. Combining Eqs. (4.4) and (4.5), we have

$$|E(\hat{F}_n(x))-\bar{F}_n(x)| = O(\alpha_n) \qquad \text{as} \quad n\to\infty \tag{4.6}$$

Moreover, for $|x|\leqslant d_n$

$$|\hat{F}_n(x)-\hat{F}_n(\theta(F))-E(\hat{F}_n(x))+E(\hat{F}_n(\theta(F)))|$$

$$= \left| \int_{-\infty}^{+\infty} (\tilde{F}_n(x-t)-\tilde{F}_n(\theta(F)-t)-\bar{F}_n(x-t)+\bar{F}_n(\theta(F)-t))\; g_n(t)\,dt \right|$$

$$\leqslant \int_{|t|\geqslant\delta/2} g_n(t)\,dt$$

$$+\int_{|t|<\delta/2} |\tilde{F}_n(x-t)-\tilde{F}_n(\theta(F)-t)-\bar{F}_n(x-t)+\bar{F}_n(\theta(F)-t)|\; g_n(t)\,dt \tag{4.7}$$

But

$$\int_{|t|\geqslant\delta/2} g_n(t)\,dt = \int_{|y|\geqslant\delta/2} g(y)\,dy$$

$$\leqslant \frac{2\alpha_n}{\delta}\int_{|y|\geqslant\delta/2\alpha_n} |y|\; g(y)\,dy = O(\alpha_n) \qquad \text{as} \quad n\to\infty \tag{4.8}$$

and the second term of the last inequality in Eq. (4.7) is bounded by

$$\sup_{|t-\theta(F)|\leqslant\delta/2} K_n(t) = O(n^{-3/4+v}) \text{ a.s.} \qquad \text{as} \quad n\to\infty, \tag{4.9}$$

by Lemma 2.1.

From Eqs. (4.7)–(4.9), it follows that

$$\sup_{|x-\theta(F)|\leqslant d_n} |\hat{F}_n(x)-\hat{F}_n(\theta(F))-E(\hat{F}_n(x))+E(\hat{F}_n(\theta(F))|$$

$$= O(n^{-3/4+v})+O(\alpha_n) \text{ a.s.} \qquad \text{as} \quad n\to\infty \tag{4.10}$$

The proof now follows from Eqs. (4.6), (4.10), and the following inequality

$$
\begin{aligned}
&|\hat{F}_n(x) - \hat{F}_n(\theta(F)) - \bar{F}_n(x) + \bar{F}_n(\theta(F))| \\
&\quad \leqslant |\hat{F}_n(x) - \hat{F}_n(\theta(F)) - E(\hat{F}_n(x)) + E(\hat{F}_n(\theta(F)))| \\
&\qquad + |E(\hat{F}_n(x)) - \bar{F}_n(x) + E(\hat{F}_n(\theta(F)))| - \hat{F}_n(\theta(F))|
\end{aligned}
$$

Theorem 4.1. Suppose that $F_i''(\theta(F))\ \forall i \geqslant 1$ exists and is finite. Then, under the conditions of Theorem 3.1 and Eq. (4.2), we have

$$\hat{F}_n(U_n) = \hat{F}_n(\theta(F)) + m\,\frac{\bar{F}_n'(\theta(F))}{n} \sum_{i=1}^{n} X_{ni}^* + R_n \tag{4.11}$$

where X_{ni}^* is defined in Eq. (3.5) and

$$R_n = O(n^{-3/4+v}) + O(\alpha_n) \text{ a.s.}$$

as $n \to \infty$ and $0 < v < \frac{1}{8}$.

Proof. First, using Theorem 3.1, we obtain

$$|U_n - \theta(F)| \leqslant (1+\varepsilon)\,2^{1/2} m\sigma n^{-1/2} (\log\log n)^{1/2} \text{ a.s.}$$
$$\text{as} \quad n \to \infty, \quad \text{for some} \quad \varepsilon > 0 \tag{4.12}$$

Next, by Proposition 4.1, we get

$$\hat{F}_n(U_n) - \hat{F}_n(\theta(F)) = \bar{F}_n(U_n) - \bar{F}_n(\theta(F)) + O(n^{-3/4+v}) + O(\alpha_n) \tag{4.13}$$

a.s. as $n \to \infty$. Now using Young's form of Taylor's theorem, and Eq. (4.12), we obtain,

$$\bar{F}_n(U_n) - \bar{F}_n(\theta(F)) = (U_n - \theta(F))\,\bar{F}_n'(\theta(F)) + O(n^{-1}\log\log n). \tag{4.14}$$

The proof follows by using Eqs. (4.13), (4.14), and Lemma 2.4.

Theorem 4.2. Suppose that $F_i''(\theta(F))$ exists for $\forall i \geqslant 1$ and is finite, the conditions of Theorem 3.1 and condition Eq. (4.2) are satisfied, $\theta(F) > 0$, and $\int_{-\infty}^{\infty} y^2 g(y)\,dy < \infty$. Then

$$\hat{F}_n(U_n) = n^{-1} \sum_{i=1}^{n} Y_{ni} + R_n \tag{4.15}$$

where $Y_{ni} = u(\theta(F) - X_i) + m\bar{F}'_n(\theta(F))\, X^*_{ni}$ and

$$R_n = O(n^{-3/4+v}) + O(n\alpha_n^2 \log\log n)^{-1} \text{ a.s.}$$

as $n \to \infty$ and $0 < v < \frac{1}{8}$.

Proof. Applying Theorem 4.1, we obtain

$$\hat{F}_n(U_n) = \hat{F}_n(\theta(F)) + m\,\frac{\bar{F}_n(\theta(F))}{n} \sum_{i=1}^{n} X^*_{ni} + R_n \tag{4.16}$$

where

$$R_n = O(n^{-3/4+v}) + O(\alpha_n) \text{ a.s. as } n \to \infty$$

Next using Lemma 2.2, we obtain

$$\hat{F}_n(\theta(F)) = \tilde{F}_n(\theta(F)) + O(n^{-3/4+v}) + O(n\alpha_n^2(\log\log n)^{-1}) \text{ a.s.} \tag{4.17}$$

The proof follows using Eqs. (4.16) and (4.17).

Theorem 4.3. Under the conditions of Theorem 4.2, if $\alpha_n = o(n^{-3/4+v}(\log\log n)^{1/2})$ for $0 < v < \frac{1}{8}$ and $F'(\theta(F))$ exists and is finite, then

$$\overline{\lim}\, n^{1/2} \frac{\hat{F}_n(U_n) - F(\theta(F))}{\sqrt{2\sigma_*^2 \log\log n}} = 1 \text{ a.s.} \tag{4.18}$$

where $\sigma_*^2 = \int_{\mathbb{R}} A^2(x)\, F(dx) + 2\sum_{k=1}^{\infty} \int_{\mathbb{R}^2} A(x)\, A(y)\, F_k^* dx, dy)$ and $A(x) = u(\theta(F) - x) + mF'(\theta(F))[h_1(x) - \theta(F)]$, and $h_1(x) = \int h(x, x_2, \ldots, x_m) \prod_{i=2}^{m} dF(x_i)$.

Proof. The proof follows from Theorem 4.2 and Lemma 5.6 (in the Appendix).

5. APPENDIX

Let $p \geq 2$ and $1 \leq i_1 < i_2 < \cdots < i_p$ be arbitrary integers. For any $j(1 \leq j \leq p-1)$, $P_j^{(i_1,\ldots,i_p)}$ is the probability measure defined by

$$P_j^{(i_1,\ldots,i_p)}(A^{(j)} \times B^{(p-j)}) = P[(X_{i_1}, \ldots, X_{i_j}) \in A^{(j)}]\, P[(X_{i_{j+1}}, \ldots, X_{i_p}) \in B^{(p-j)}]$$

and $P_0^{(i_1,\ldots,i_p)}$ is the probability measure defined by

$$P_0^{(i_1,\ldots,i_p)}(A^{(p)}) = P[(X_{i_1}, \ldots, X_{i_p}) \in A^{(p)}]$$

for any $A^{(j)} \in \sigma(X_{i_1},..., X_{i_j})$ $(1 \leqslant j \leqslant p)$ and any $B^{(p-j)} \in \sigma(X_{i_{j+1}},..., X_{i_p})$, $(1 \leqslant j \leqslant p-1)$.

Lemma 5.1. For every $p \geqslant 1$ and $(i_1,..., i_p)$ such that $1 \leqslant i_1 < i_2 < \cdots < i_p$ and any $j (1 \leqslant j \leqslant p-1)$, let $h(x_1,..., x_p)$ be a Borel function such that

$$\int_{\mathbb{R}^p} |h(x_1,..., x_p)|^{1+\delta} \, dP_j^{(i_1,..., i_p)} \leqslant M$$

for some $\delta > 0$, then

$$\left| \int_{\mathbb{R}^p} h(x_1,..., x_p) \, dP_0^{(i_1,..., i_p)} - \int_{\mathbb{R}^p} h(x_1,..., x_p) \, dP_j^{(i_1,..., i_p)} \right|$$
$$\leqslant 4M^{1/(2+\delta)} \beta^{\delta/(1+\delta)} (i_{j+1} - i_j) \tag{5.1}$$

As a special case, if $h(x_1,..., x_p)$ is bounded, say, $|h(x_1,..., x_p)| \leqslant M^*$, then we can replace the right side of Eq. (5.1) by $2M^* \beta(i_{j+1} - i_j)$.

Proof. Follows from Lemma 1 of Yoshihara.[19]

Lemma 5.2. Let $\{Y_i, i \geqslant 1\}$ be a sequence of absolutely regular random variables with mean 0. Assume that $\sup_{i \geqslant 1} |Y_i| \leqslant M_0$ and $\sup_{i \geqslant 1} [\mathrm{Var}\ Y_i]^{1/2} \leqslant M$. Then, for any a $(0 < a < 1)$ and $m \leqslant n$ sufficiently large,

$$P\left[n^{-1} \left| \sum_{i=1}^{n} Y_i \right| \geqslant u \right] \leqslant 2m \exp\left\{ -\frac{ak}{5M_0} \right\} + 2n\beta(m) \tag{5.2}$$

where $k = [n/m]$.

Proof. Let $\{\tilde{Y}_i, i \geqslant 1\}$ be a sequence of independent random variables such that for any i, $\tilde{Y}_i$ has the same distribution as Y_i. Then, from Collomb's[2b] extension of the Bernstein inequality (see Bennett[1]), we have for any $I \subset \{1,..., n\}$ such that *card* $I = m$

$$P\left[\left| \sum_{i \in I} \tilde{Y}_i \right| \geqslant v \right] \leqslant 2 \exp\left\{ -\frac{vk^{-a}}{5M_0} \right\} \tag{5.3}$$

Let now $S^{(j)} = Y_j + Y_{j+m} + \cdots + Y_{j+k_j m}$ $(j = 1,..., m)$ where for each j $(1 \leqslant j \leqslant m)$ $k_j = k_{j,n}$ is the largest integer for which $j + k_j m \leqslant n$. Then

$$P\left[\left| \sum_{i=1}^{n} Y_i \right| \geqslant v \right] \leqslant P\left[\sum_{j=1}^{m} |S^{(j)}| \geqslant v \right] \leqslant \sum_{j=1}^{m} P[|S^{(j)}| \geqslant m^{-1} v] \tag{5.4}$$

For each j $(1 \leqslant j \leqslant m-1)$, let A_j be the Borel subset of $\mathbb{R}^{k_j+1}$ defined by

$$A_j = \{(y_0, \ldots, y_{k_j}) \colon |y_0 + \cdots + y_{k_j}| \geqslant m^{-1} v\}$$

and put

$$g_j(y_0, \ldots, y_{k_j}) = \begin{cases} 1 & \text{if } (y_0, \ldots, y_{k_j}) \in A_j \\ 0 & \text{otherwise} \end{cases}$$

Then $|g_j(y_0, \ldots, y_{k_j})| \leqslant 1$ and after iterating $k_j + 1$ times the inequality of Eq. (5.1) for the particular case $M^* = 1$, we get

$$\begin{aligned} P[|S^{(j)}| \geqslant m^{-1} v] &= E(g_j(Y_j, \ldots, Y_{j+k_j m})) \\ &\leqslant \int_{\mathbb{R}^{k_j+1}} g_j(y_0, \ldots, y_{k_j})\, dF_j(y_0) \cdots dF_{j+k_j m}(y_{k_j}) \\ &\quad + 2(k_j + 1)\, \beta(m) \\ &= P\left[\left|\sum_{i \in I} \tilde{Y}_i\right| \geqslant m^{-1} v\right] + 2(k_j + 1)\, \beta(m) \\ &\leqslant 2 \exp\left\{-\frac{m^{-1} v k_j^{-a}}{5 M_0}\right\} \\ &\quad + 2(k_j + 1)\, \beta(m) \text{ (from Eq. (5.3)).} \end{aligned} \tag{5.5}$$

Now from Eqs. (5.4) and (5.5), we get

$$P\left[\left|\sum_{i=1}^{n} Y_i\right| \geqslant v\right] \leqslant 2m \exp\left\{-\frac{u k_j^{1-u}}{5 M_0}\right\} + 2n\beta(m) \tag{5.6}$$

If in Eq. (5.6), we replace v by un, we obtain Eq. (5.2).

The following lemma is the Lemma 3 of Chow and Teicher[(2a)] [p. 298].

Lemma 5.3. Let $\varphi_n^*(t)$ be the characteristic function of $S_n = \sum_{i=1}^{n} Z_j$ where $\{Z_j, j \geqslant 1\}$ are independent (not necessarily stationary) r.v.'s with zero means, and variances σ_j^2, $j \geqslant 1$. If $\Gamma_n^{2+\delta} = \sum_{j=1}^{n} \gamma_j^{2+\delta}$ and $s_n^2 = \sum_{j=1}^{n} \sigma_j^2$ where $\gamma_j^{2+\delta} = E\,|Z_j - E(Z_j)|^{2+\delta}$, then for $0 \leqslant \delta \leqslant 1$

$$\left|\varphi_n^*\left(\frac{t}{s_n}\right) - e^{-t^2/2}\right| \leqslant 3 \left|\frac{t\Gamma_n}{s_n}\right|^{2+\delta} e^{-t^2/2} \quad \text{for } |t| < \frac{s_n}{2\Gamma_n} \tag{5.7}$$

Lemma 5.4. Let $\{Y_{ni}, 1 \leqslant i \leqslant n, n \geqslant 1\}$ be a sequence of absolutely regular random variables (with means 0) satisfying the strong mixing condition. Suppose that $\sum_{m\geq 1} \alpha(m) < \infty$ and there exists $\sigma^2 > 0$ such that

$$\operatorname{Var}\left(\sum_{i=1}^{n} Y_{ni}\right) = n\sigma^2(1+o(1)) \tag{5.8}$$

Then the process $\{Y_{ni}, 1 \leq i \leq n, n \geq 1\}$ obeys the law of the iterated logarithm, that is $\limsup_{n\to\infty} |\sum_{i=1}^{n} (Y_{ni}/\chi(n))| = 1$ a.s., if the following requirements are fulfilled for some $\rho > 0$ and for all sufficiently large n:

$$\sup_{x\in\mathbb{R}} \left| P\left(\sum_{i=1}^{n} Y_{ni} < x\sigma\sqrt{n}\right) - \Phi(x)\right| = O\left(\frac{1}{(\log n)^{1+\rho}}\right) \tag{5.9}$$

where

$$\Phi(x) = \frac{1}{\sqrt{2\pi}} \int_{-\infty}^{x} e^{-t^2/2}\, dt$$

and

$$P\left[\max_{1\leq j\leq n} \left|\sum_{i=1}^{j} Y_{ni}\right| \geq b\chi(n)\right] = O\left(\frac{1}{(\log n)^{1+\rho}}\right) \tag{5.10}$$

where $b > 1$ is an arbitrary number and $\chi(n) = (2\sigma^2 n \log\log n)^{1/2}$.

This Lemma was proved by Oodaira and Yoshihara[8] [Th. 1] for the stationary case. The proof is similar for the nonstationary case and is therefore omitted.

Lemma 5.5. For a sequence of random variables satisfying the strong mixing condition, the condition in Eq. (5.10) holds if Eq. (5.9) holds and there exists a function $r = r(n)$ such that $r(n) \leq n (n \geq 1)$, $r(n) \to \infty$ as $n \to \infty$ and

$$\max\left\{\frac{n}{r} P\left[\sum_{i=1}^{r} |Y_{ni}| \geq \varepsilon\chi(n)\right],\ \frac{n}{r}\alpha(r)\right\} = O\left(\frac{1}{(\log n)^{1+\rho}}\right) \tag{5.11}$$

for any ε $(0 < \varepsilon < (b-1)/b$ where $b > 1$ is an arbitrarily fixed number).

Proof. From Oodaira and Yoshihara[8] [Remark 1, Th. 1].

Lemma 5.6. Let $\{Y_{ni}, 1 \leq i \leq n, n \geq 1\}$ be a sequence of absolutely regular random variables satisfying the strong mixing condition with means 0. Suppose for any n, m such that $n \geqslant m$ and any $J \subset \{1, \ldots, n\}$ such that $card\ J = m$

$$E\left(\sum_{j \in J} Y_{nj}\right)^2 = m\sigma^2(1 + o(1)) \tag{5.12}$$

Then the process $\{Y_{ni},\ 1 \leqslant i \leqslant n,\ n \geqslant 1\}$ obeys the law of the iterated logarithm if the following requirements are fulfilled for some δ and δ' such that

$$\sup_{n \geqslant 1} \max_{1 \leqslant i \leqslant n} E\,|Y_{ni}|^{2+\delta} = M < \infty \tag{5.13}$$

$$\sum_{m=1}^{\infty} \{\alpha(m)\}^{\delta'/(2+\delta')} < \infty \tag{5.14}$$

Proof. Define

$$f_N(x) = \begin{cases} x & \text{if } |x| \leqslant N \\ 0 & \text{otherwise} \end{cases}$$

We define $\bar{f}_N(x) = x - f_N(x)$. Let $N = n^{1/2(1+\delta')} (\log n)^{-3}$ and

$$r(n) = [n^{\delta'/2(1+\delta')}(\log n)^3]$$

For convenience and without loss of generality, we denote Y_i for Y_{ni}. Then for any $b > 0$

$$\frac{n}{r} P[|Y_1| + \cdots + |Y_r| \geqslant b_\chi(n)]$$

$$= \frac{n}{r} P\left[\sum_{i=1}^{r} |\bar{f}_N(Y_i)| \geqslant b\chi(n)\right]$$

$$\leqslant \frac{n}{b^2\{\chi(n)\}^2 r} E\left(\sum_{i=1}^{r} |\bar{f}_N(Y_i)|\right)^2$$

$$\leqslant \frac{n}{b^2\{\chi(n)\}^2 r}\left\{\sum_{1 \leqslant i,j \leqslant r} E[|\bar{f}_N(Y_i)| \cdot |\bar{f}_N(Y_j)|]\right\}$$

$$\leqslant \frac{n}{b^2\{\chi(n)\}^2 r}\left\{2\sum_{i=0}^{r-1}\sum_{j=1}^{r-i} E[|\bar{f}_N(Y_i)| \cdot |\bar{f}_N(Y_{j+i})|]\right\}$$

$$\leqslant \frac{n}{b^2\{\chi(n)\}^2 r}\left\{2\sum_{i=1}^{r} E[|\bar{f}_N(Y_i)|^2] + 2\sum_{i=1}^{r-1}\sum_{j=1}^{r-i} E[|\bar{f}_N(Y_i)| \cdot |\bar{f}_N(Y_{j+i})|]\right\}$$

$$\leqslant \frac{2n}{b^2\{\chi(n)\}^2 r}\left\{\frac{1}{N^\delta}\sum_{i=1}^{r} E[|\bar{f}_N(Y_i)|^{2+\delta}]\right.$$

$$+12\sum_{i=1}^{r-1}(\alpha(i))^{\delta'/(2+\delta')}\sum_{j=1}^{r-i} E[|\bar{f}_N(Y_i)|^{2+\delta'}]^{1/(2+\delta')}$$

$$\left.\times E[|\bar{f}_N(Y_{j+i})|^{2+\delta'}]^{1/(2+\delta')}\right\} \text{(from Lemma 5.7)}$$

$$\leqslant \frac{2n}{b^2\{\chi(n)\}^2 r}\left\{\frac{r}{N^\delta}M+12\sum_{i=1}^{r-1}(\alpha(i))^{\delta'/(2+\delta')}\sum_{j=1}^{r-i}\frac{1}{N^{2(\delta-\delta')}}\right.$$

$$\left.\times E[|\bar{f}_N(Y_i)|^{2+\delta'}]^{1/(2+\delta')} E[|\bar{f}_N(Y_{j+i})|^{2+\delta'}]^{1/(2+\delta')}\right\}$$

$$\leqslant \frac{2n}{b^2\{\chi(n)\}^2 r}\left\{\frac{rM}{N^\delta}+\frac{12r\sum_{i=1}^{r-1}(\alpha(i))^{\delta'/(2+\delta')}}{N^{2(\delta-\delta')}}M^{2/(2+\delta')}\right\}$$

$$= O(n^{-\gamma}) \tag{5.15}$$

holds for some $\gamma > 0$ and

$$\frac{n}{r}\alpha(r) = \frac{n}{r}O(r^{-(2+\delta')/\delta'}) = O\left(\frac{1}{(\log n)^3}\right)$$

From Lemmas 5.4 and 5.5 it suffices to show

$$\sup_{x\in\mathbb{R}}\left|P\left(\sum_{i=1}^{n} Y_i < \sqrt{n}\,\sigma x\right)-\Phi(x)\right| = O\left(\frac{1}{(\log n)^3}\right) \tag{5.16}$$

Define p, q and k by $p(n)=[n^{1/2+\alpha}]$, $q(n)=[n^{1/2-\alpha}]$ and $k(n)=[n/(p+q)]$ where α is a small number. Let $N'=n^{\delta'/16(1+\delta')}$ if $0\leqslant\delta\leqslant 2$ and $N'=n^{1/16(1+\delta')}$ if $\delta>2$. Put

$$S'_n = \frac{1}{\sigma\sqrt{n}}\sum_{j=1}^{n}(f_{N'}(Y_j)-E(f_{N'}(Y_j)))$$

$$S'' = \frac{1}{\sigma\sqrt{n}}\sum_{j=1}^{n}(\bar{f}_{N'}(Y_j)-E(\bar{f}_{N'}(Y_j)))$$

$$Z_i = \sum_{j=1}^{p}(f_{N'}(Y_{(i-1)(p+q)+j}))-E(f_{N'}(Y_{(i-1)(p+q)+j})), \qquad i=1,\dots,k$$

$$T'_n = \frac{1}{\sqrt{n}\,\sigma}\sum_{i=1}^{k} Z_i$$

$$T''_n = S'_n - T'_n$$

Then, it can easily be proved that for some $\gamma > 0$

$$E|S''_n|^2 \leqslant \frac{1}{\sigma^2}\left(\frac{1}{N'^{\delta}} M + \frac{K}{N'^{2(\delta-\delta')(2+\delta')}} M^{2/(2+\delta')}\right) = O(n^{-\gamma}) \qquad (5.17)$$

(where M is defined in Eq. (5.13)), K is some constant > 0. $E|T''_n|^2 = O(n^{-\gamma})$, $|E(T'^2_n) - 1| = O(n^{-\gamma})$.

Now, let $\varphi_n(t)$ be the characteristic function of $S_n/\sigma\sqrt{n}$ where $S_n = \sum_{i=1}^{n} Y_i$. Then

$$\begin{aligned}|\varphi_n(t) - e^{-t^2/2}| &\leqslant |\varphi_n(t) - E(e^{itS'_n})| + |E(e^{itS'_n}) - E(e^{itT'_n})| \\ &\quad + \left|E(e^{itT'_n}) - \prod_{j=1}^{k} E(e^{itZ_j/\sqrt{kE(Z_j)^2}})\right| \\ &\quad + \left|(e^{-t^2/2}) - \prod_{j=1}^{k} E(e^{itZ_j/\sqrt{kE(Z_j)^2}})\right| \\ &= I_1 + I_2 + I_3 + I_4, \text{ say.}\end{aligned} \qquad (5.18)$$

Then, we have

$$I_1 \leqslant |t|\, E(|S''_n|) = O(n^{-\gamma/2}), \qquad I_2 \leqslant |t|\, E(|T''_n|) = O(n^{-\gamma/2}) \qquad (5.19)$$

From Lemma 5.3, we have

$$\begin{aligned}&\left|e^{-t^2/2} - \prod_{j=1}^{k} E(e^{itZ_j/\sqrt{kE(Z_j)^2}})\right| \\ &\leqslant 3\frac{|t|^{2+\delta}}{k^{\delta/2}} e^{-t^2/2} \sup_{1\leqslant j\leqslant k} E|Z_j|^{2+\delta} \min_{1\leqslant j\leqslant k} \frac{1}{(E|Z_j|^2)^{(2+\delta)/2}} \\ &\quad \text{for } |t| \leqslant \frac{(\sum_{j=1}^{k} E(Z_j)^2)^{1/2}}{2\sum_{j=1}^{k} E|Z_j|^{2+\delta}}\end{aligned} \qquad (5.20)$$

Since

$$\begin{aligned}E(Z_j^4) &\leqslant K(N')^4 p^2 \sum_{j=1}^{p} j\alpha(j) \leqslant K(N')^4 p^2 \sum_{j=1}^{p} j^{-2/\delta'} \\ &\leqslant K(N')^4 p^2 \max(1, p^{-2/\delta})\end{aligned}$$

and $E(Z_j^2) = p\sigma^2(1 + o(1))$ (from condition Eq. (5.12)), for all sufficiently large n, we have

$$\begin{aligned}&\sup_{1\leqslant j\leqslant k} E|Z_j|^{2+\delta} \frac{1}{k^{\rho/2}\min_{1\leqslant j\leqslant k}(E(Z_j^2))^{(2+\rho)/2}} \\ &\leqslant \sup_{1\leqslant j\leqslant k} (E(Z_j^4))^{(2+\delta)/4} \frac{2^{(2+\rho)/2}}{k^{\rho/2}(p\sigma^2)^{(2+\rho)/2}} = O(n^{-\gamma})\end{aligned} \qquad (5.21)$$

olds for all sufficiently large n where $\rho = \min(2, \delta)$ and γ is a positive umber.

We deduce that

$$\left| e^{-t^2/2} - \prod_{j=1}^{k} E(e^{itZ_j/\sqrt{kE(Z_j)^2}}) \right| \leqslant K' n^{-\gamma} |t|^{2+\delta} e^{-t^2/2} \tag{5.22}$$

olds for all sufficiently large n and for all t such that

$$|t| \leqslant \frac{1}{3} n^{1/2 - 1/(1+2\delta)} \sigma \times \frac{1}{\min_{1 \leqslant i \leqslant n} E\,|Y_i|^{2+\delta}}$$

Now from Eq. (5.14) and using Lemma 5.7, we obtain

$$E(e^{itT'_n}) - \prod_{j=1}^{k} E(e^{itZ_j/\sqrt{kE(Z_j^2)}}) \Bigg| \leqslant k\alpha(q) = n^{1/2-\alpha} o((n^{1/2-\alpha})^{-(2+\delta')/\delta'}) \tag{5.23}$$

From Eqs. (5.17)–(5.23) it follows that

$$\begin{aligned}
&\left| P\left(\sum_{i=1}^{n} Y_i < \sqrt{n}\,\sigma x \right) - \Phi(x) \right| \\
&\leqslant \int_{-(\log n)^3}^{(\log n)^3} \left| \frac{\varphi_n(t) - e^{-t^2/2}}{t} \right| dt + \frac{K_1}{(\log n)^3} \\
&\leqslant \int_{-(\log n)^3}^{(\log n)^3} K_2 n^{-\gamma} |t|^{1+\delta}\, dt + \int_{-(\log n)^3}^{(\log n)^3} \{E(|S''_n|) + E(|T''_n|)\}\, dt \\
&\quad + K_3 \left\{ \int_{0 \leqslant |t| \leqslant n^{-1/4}} dt + \int_{n^{-1/4} \leqslant |t| \leqslant (\log n)^3} \frac{k\alpha(q)}{|t|}\, dt + \frac{K_4}{(\log n)^3} \right\} \\
&= O\left(\frac{1}{(\log n)^3} \right)
\end{aligned} \tag{5.24}$$

where K_i, $i = 1, \ldots, 4$ are constants > 0, and Lemma 5.6 is proved.

Lemma 5.7. Let $\{Y_{ni},\ 1 \leqslant i \leqslant n, n \geqslant 1\}$ be a sequence of random variables satisfying the strong mixing condition with coefficient $\alpha(m)$. Let U be $\sigma(Y_{ni}, 1 \leqslant i \leqslant j \leqslant n)$-measurable and V be $\sigma(Y_{ni}, j+m \leqslant i \leqslant n)$-measurable. If $E\,|U|^2 < \infty$ and $E\,|V|^2 < \infty$ where $r^{-1} + p^{-1} + s^{-1} = 1$ $(1 \leqslant r, s, p < \infty)$ then,

$$|E(UV) - E(U)\,E(V)| \leqslant 12(\alpha(m))^{p^{-1}} \{E\,|U|^r\}^{r^{-1}} \{E\,|V|^s\}^{s^{-1}} \tag{5.25}$$

Proof. This is Proposition 2.8 of Doukhan and Portal.[3]

REFERENCES

1. Bennett, G. (1962). Probability inequalities for the sum of independent random variables. *J. Amer. Statist. Assoc.* **57**, 33–45.

2a. Chow, Y. S. and Teicher, H. (1978). *Probability Theory*. Springer Verlag.

2b. Collomb, G. (1984). Propriétés de convergence presque complète du prédicteur à noyau. *Z. Wahrsch. Verw. Gebiete* **66**, 441–460.

3. Doukhan, P. and Portal, F. (1987). Principe d'invariance faible pour la fonction de répartition empirique dans un cadre multidimensionnel et mélangeant. *Prob. and Math. Stat.* **8**, (2), 117–132.

4. Harel, M. and Puri, M. L. (1991). Théorème central limite de l'estimateur de densité et de l'erreur quadratique intégrée pour des variables aléatoires nonstationnaires. *C.R.Academy of Sciences* Série 1, **312**, 145–148.

5. Harel, M. and Puri, M. L. (1989). Weak convergence of the *U*-statistic and weak invariance of the one sample rank order statistic for Markov processes and ARMA models. *J. of Mult. Anal.* **31**, 258–265.

6. Lea, C. and Puri, M. L. (1988). Asymptotic properties of perturbed empirical distribution functions evaluated at a random point. *J. Statistical Planning and Inference* **19**, 201–215.

7. Nadaraya, E. A. (1964). Some new estimates for distribution function. *Theory Probab. Appl.* **9**, 497–500.

8. Oodaira, H. and Yoshihara, K. (1971). The law of the iterated logarithm for stationary processes satisfying mixing conditions. *Kōdai Math. Sem. Rep.* **23**, 311–334.

9. Puri, M. L. and Ralescu, S. (1983). On Berry-Esséen rates, a law of the iterated logarithm and an invariance principle for the proportion of the sample below the sample mean. *J. Multivariate Anal.* **14**, 231–247.

10. Puri, M. L. and Ralescu, S. (1986). Central limit theorem for perturbed empirical distribution functions evaluated at a random point. *J. Multivariate Anal.* **16**, 273–279.

11. Schuster, E. F. (1969). Estimation of probability density function and its derivatives. *Ann. Math. Statist.* **40**, 1187–1195.

12. Scott, D. W., Tapia, R. A. and Thompson, J. R. (1977). Kernel density estimation revisited. *J. Nonlinear Analysis Theory Meth. Applic.* **1**, 339–372.

13. Sun, S. (1993). Asymptotic behavior of the perturbed empirical distribution function evaluated at a random point for absolutely regular sequences *J. Multivariate Anal.* (to appear).

14. Watson, G. S. and Leadbetter, M. R. (1964). Hazard analysis II. Ser. A, *Sankhyā*, **26**, 101–116.

15. Wertz, W. (1978). *Statistical Density Estimation: A Survey*. Vandenhoeck and Ruprecht, Göttingen.

16. Winter, B. B. (1973). Strong uniform consistency of integrals of density estimators. *Canad. J. Statist.* **1**, 247–253.

17. Winter, B. B. (1979). Convergence rate of perturbed empirical distribution functions. *J. Appl. Probab.* **16**, 163–173.

18. Yamato, H. (1973). Uniform convergence of an estimator of a distribution function. *Bull. Math. Statist.* **15**, 69–78.

19. Yoshihara, K. (1976). Limiting behavior of *U*-statistics for stationary absolutely regular processes. *Z. Wahrsch. Verw. Gebiete* **35, 237–252.**

VALID EDGEWORTH EXPANSIONS OF M-ESTIMATORS IN REGRESSION MODELS WITH WEAKLY DEPENDENT RESIDUALS

MASANOBU TANIGUCHI and MADAN L. PURI

Osaka University, Indiana University

Consider a linear regression model $y_t = x_t\beta + u_t$, where the u_t's are weakly dependent random variables, the x_t's are known design nonrandom variables, and β is an unknown parameter. We define an M-estimator $\hat{\beta}_n$ of β corresponding to a smooth score function. Then, the second-order Edgeworth expansion for $\hat{\beta}_n$ is derived. Here we do not assume the normality of $\{u_t\}$, and $\{u_t\}$ includes the usual ARMA processes. Second, we give the second-order Edgeworth expansion for a transformation $T(\hat{\beta}_n)$ of $\hat{\beta}_n$. Then, a sufficient condition for T to extinguish the second-order terms is given. The results are applicable to many statistical problems.

1. INTRODUCTION

The higher order asymptotic performance of estimators in the usual linear regression model with i.i.d. residuals has been studied by many authors. Navidi (1989) gave the second-order Edgeworth expansion for the least-squares estimator of the regression coefficient without assuming normal errors. Qumsiyeh (1990) gave the general sth-order Edgeworth expansion for the least-squares estimator and also the second-order Edgeworth expansion for the Studentized least-squares estimator without assuming normal errors. Ringland (1983) derived the second-order Edgeworth expansion for Studentized M-estimators in a one-way layout model. Here he restricts the design matrix elements to 0's and 1's only. For a significantly larger class of design vector, Lahiri (1992) gave the second-order Edgeworth expansions for an M-estimator corresponding to a smooth score and its studentized estimator. All this literature, however, deals with the i.i.d. error cases; in many applications, the assumption of independence seems restrictive.

Research by Puri was supported by the Office of Naval Research contract N00014-91-J-1020. Address correspondence to: Masanobu Taniguchi, Department of Mathematical Science, Faculty of Engineering Science, Osaka University, Toyonaka 560, Japan.

In this paper, we consider a linear regression model with weakly dependent residuals. Of course, this model includes a linear regression model with the usual ARMA process errors. For this model, we define an M-estimator of the regression coefficient and give its second-order Edgeworth expansion. We use the techniques of Götze and Hipp (1983), where the general Edgeworth expansion for sums of weakly dependent random vectors with zero mean is given. Throughout this paper, we do not assume the normality of error processes. Here it may be noted that, for the usual AR process without regression part, Bose (1988) gave a valid Edgeworth expansion for the least-squares estimator of the AR coefficients.

We also give the second-order Edgeworth expansion for the transformations of the M-estimator. If a transformation of the M-estimator extinguishes the second-order terms of the Edgeworth expansion, then we call it the second-order normalizing transformation. A sufficient condition for a transformation to be the second-order normalizing transformation is given. Our results are applicable to many statistical problems. Finally, we give the second-order Edgeworth expansion for the Studentized statistic. To avoid unnecessarily complex formulas, we restrict ourselves to the case where the regression parameter is scalar.

2. MODEL AND ASSUMPTIONS

We consider the following linear regression model:

$$y_t = x_t\beta + u_t, \qquad t = 1, \dots, n, \tag{2.1}$$

where $\{u_t\}$ is a weakly dependent stochastic process, $\{x_t\}$ is a known design sequence, and β is an unknown parameter that belongs to a compact set $\mathbf{C}$ of $\mathbf{R}$. The process $\{u_t\}$ satisfies the following.

(A.1) $\{u_t\}$ is a linear process generated as

$$u_t = \sum_{j=0}^{\infty} c_j \varepsilon_{t-j}, \tag{2.2}$$

where $\{\varepsilon_t\}$ is a sequence of i.i.d. random variables with $E\{\varepsilon_t\} = 0$, $E\{\varepsilon_t^4\} < \infty$, and probability density function $g(x) > 0$ on $\mathbf{R}$.

(A.2) The coefficients c_j, $j = 0, 1, \dots$, satisfy

$$|c_j| \le \delta^{-1} \exp\{-\delta |j|\}, \quad \text{for some } \delta > 0,$$

and $\sum_{j=0}^{\infty} c_j z^j \neq 0$, for all $|z| \le 1$.

It is easily checked that the usual ARMA processes satisfy conditions (A.1) and (A.2). We next impose some assumptions on $\{x_t\}$.

(B.1) $\{x_t\}$ is uniformly bounded; that is, there exists $\gamma_1 > 0$ such that $|x_t| \le \gamma_1$ for all t.

(B.2) There exists $\gamma_2 > 0$ such that

$$\sum_{t=1}^{n} x_t^2 \geq \gamma_2 n.$$

(B.3) For $k = 2,\ldots,6$, there exist regression spectral measures $M_k(\lambda_1,\ldots,\lambda_{k-1})$ such that

$$\frac{1}{n}\sum_{t=1}^{n} x_t x_{t+\ell_1} \cdots x_{t+\ell_{k-1}}$$

$$= \int_{-\pi}^{\pi} \cdots \int_{-\pi}^{\pi} e^{i(\lambda_1\ell_1+\cdots+\lambda_{k-1}\ell_{k-1})} dM_k(\lambda_1,\ldots,\lambda_{k-1}) + O(n^{-1}).$$

Condition (B.3) is a "higher order version" for one of Grenander's conditions (for Grenander's conditions, see Hannan, 1970). For examples, constant function and linear combination of harmonic functions satisfy conditions (B.2) and (B.3).

Now we are interested in the estimation of β. Let ψ be a real-valued function defined on the real line. An M-estimator $\hat{\beta}_n$ of β corresponding to ψ is defined as a solution of the equation

$$\sum_{t=1}^{n} x_t \psi(y_t - x_t\beta) = 0. \tag{2.3}$$

Here the score function ψ satisfies the following assumptions.

(C.1) $\psi(x)$ is twice differentiable and the second derivative $\psi''(x)$ satisfies the Lipschitz condition.

(C.2) $E\{\psi(u_t)\} = 0$.

(C.3) For all $t = 1,\ldots,n$, and $m = O(\log n)$, there exists $c > 0$ such that

$$\left| E\left[\sum_{j=0}^{m} x_{t+j} c_j \psi'(c_j \varepsilon_t)\right]\right| \geq c.$$

(C.4) For each $j = 1,2,\ldots,k-1$, and any k-tuple $a_1,a_2,\ldots,a_k$ (which index the set $\{\psi,\psi',\psi''\}$ with $\psi = \psi(u_t)$, $\psi' = \psi'(u_t)$, and $\psi'' = \psi''(u_t)$), we have

$$\sum_{t_1,\ldots,t_{k-1}=-\infty}^{\infty} \{1 + |t_j|\}|c_{a_1\ldots a_k}(t_1,\ldots,t_{k-1})| < \infty, \qquad k = 1,\ldots,4,$$

where $c_{a_1\ldots a_k}(t_1,\ldots,t_{k-1})$ is the kth-order cumulant of $a_1,\ldots,a_k$ with time differences $t_1,\ldots,t_{k-1}$. For example,

$$c_{\psi\psi'\psi''}(t_1,t_2) = \mathrm{cum}\{\psi(u_t),\psi'(u_{t+t_1}),\psi''(u_{t+t_2})\}.$$

From condition (C.4), we can define the kth-order cumulant spectral density

$$f_{a_1\ldots a_k}(\lambda_1,\ldots,\lambda_{k-1}) = \frac{1}{(2\pi)^{k-1}} \sum_{t_1,\ldots,t_{k-1}=-\infty}^{\infty} c_{a_1\ldots a_k}(t_1,\ldots,t_{k-1}) \times \exp\{-i(\lambda_1 t_1 + \cdots + \lambda_{k-1}t_{k-1})\}, \tag{2.4}$$

which is the kth-order cumulant of the Fourier transform of the process and shows an effect of non-Gaussianity of the process (for details, see Brillinger, 1975).

Condition (C.2) is assumed for the consistency of $\hat{\beta}_n$, and condition (C.3) is a kind of Cramer's condition for the validity of Edgeworth expansion (cf. Götze and Hipp, 1983).

It follows from conditions (B.3) and (C.4) that

$$\begin{aligned}
\mathrm{Var}\left[n^{-1/2}\sum_{t=1}^{n} x_t\psi(u_t)\right] &= n^{-1}\sum_{t=1}^{n}\sum_{s=1}^{n} x_t x_s \,\mathrm{Cov}\{\psi(u_t),\psi(u_s)\}\\
&= \sum_{\ell=-n+1}^{n-1}\left\{\frac{1}{n}\sum_{s=1}^{n-|\ell|} x_s x_{s+\ell}\right\} c_{\psi\psi}(\ell)\\
&= \int_{-\pi}^{\pi}\sum_{\ell=-n+1}^{n-1} c_{\psi\psi}(\ell)e^{i\ell\lambda}dM_2(\lambda) + O(n^{-1})\\
&= 2\pi\int_{-\pi}^{\pi} f_{\psi\psi}(-\lambda)dM_2(\lambda) + O(n^{-1})\\
&= I + O(n^{-1}) \text{ (say).} \qquad \textbf{(2.5)}
\end{aligned}$$

I is the information for β by $\hat{\beta}_n$ based on ψ. For this we assume the following.

(D.1) I is bounded away from 0.

Putting $\mathbf{U}_t = [x_t\psi(u_t), -x_t^2\{\psi'(u_t) - E\psi'(u_t)\}]'$, we first derive the Edgeworth expansion for the distribution of $\mathbf{Z} = n^{-1/2}\sum_{t=1}^{n}\mathbf{U}_t$ by using some of the ideas of Götze and Hipp (1983). Henceforth, we denote the euclidean norm by $\|\cdot\|$.

PROPOSITION 1. *Suppose that conditions* (A.1), (A.2), (B.1)–(B.3), (C.1)–(C.4), *and* (D.1) *hold. Let* $\mathfrak{D}_j$ *be the σ-field generated by* ε_j, *and let* $\mathfrak{D}_p^q$ *be the σ-field generated by* $\mathfrak{D}_j$, $p \le j \le q$. *Then, we have the following.*

(GH.1) $E\{\mathbf{U}_t\} = 0$, $t = 1,2,\dots$.

(GH.2) $E\|\mathbf{U}_t\|^4 \le \beta < \infty$, $t = 1,2,\dots$.

(GH.3) *There exists* $d > 0$ *such that for* $t, m = 1,2,\dots$ *with* $m > d^{-1}$ *there exists a* $\mathfrak{D}_{t-m}^{t+m}$*-measurable bivariate random vector* $\bar{\mathbf{U}}_{t,m}$ *for which*

$$E\|\mathbf{U}_t - \bar{\mathbf{U}}_{t,m}\| \le d^{-1}\exp(-dm).$$

(GH.4) *There exists* $d > 0$ *such that for all* $t, m = 1,2,\dots, A \in \mathfrak{D}_{-\infty}^{t}$, $B \in \mathfrak{D}_{t+m}^{\infty}$

$$|P(A\cap B) - P(A)P(B)| \le d^{-1}\exp(-dm).$$

(GH.5) *There exist* $d, \delta > 0$ *such that for all* $t = 1,2,\dots,n$, *and* $m = O(\log n)$, *and all* $\mathbf{t} \in \mathbf{R}^2$ *with* $\|\mathbf{t}\| \ge \delta$,

$$E\left|E\left\{e^{i\mathbf{t}'(\mathbf{U}_{t-m}+\cdots+\mathbf{U}_{t+m})}\,\middle|\,\mathfrak{D}_j : j \ne t\right\}\right| \le e^{-d}.$$

(GH.6) *There exists $d > 0$ such that for all $t, m, \ell = 1, 2, \ldots$ and $A \in \mathfrak{D}_{t-\ell}^{t+\ell}$,*

$$E|P(A|\mathfrak{D}_j : j \neq t) - P(A|\mathfrak{D}_j : 0 < |t-j| \leq m + \ell)| \leq d^{-1} \exp(-dm).$$

We give the proofs of propositions and theorems and so forth in the Appendix. Next we evaluate the asymptotic moments (cumulants) of $\mathbf{Z} = (Z_1, Z_2)'$. By using an argument similar to (2.5), it is not difficult to show

$$\begin{aligned} E\{Z_1 Z_2\} &= -(2\pi) \int \int_{-\pi}^{\pi} f_{\psi\psi'}(-\lambda) dM_3(\lambda, \mu) + O(n^{-1}) \\ &= J + O(n^{-1}) \qquad \text{(say)}, \end{aligned} \tag{2.6}$$

$$\begin{aligned} \text{cum}\{Z_1, Z_1, Z_1\} &= \frac{(2\pi)^2}{\sqrt{n}} \int \int_{-\pi}^{\pi} f_{\psi\psi\psi}(-\lambda, -\mu) dM_3(\lambda, \mu) + O(n^{-1}) \\ &= \frac{1}{\sqrt{n}} K + O(n^{-1}) \quad \text{(say)}, \end{aligned} \tag{2.7}$$

$$\begin{aligned} E\{Z_2^2\} &= (2\pi) \int\int \int_{-\pi}^{\pi} f_{\psi'\psi'}(-\mu) dM_4(\lambda, \mu, \gamma) + O(n^{-1}) \\ &= M + O(n^{-1}) \qquad \text{(say)}. \end{aligned} \tag{2.8}$$

As we saw in Proposition 1, Götze and Hipp's (1983) conditions for the Edgeworth expansion of $\mathbf{Z}$ are satisfied. From their results, we obtain the following lemma.

LEMMA 1. *Suppose that conditions* (A.1), (A.2), (B.1)–(B.3), (C.1)–(C.4), *and* (D.1) *hold. Let $\mathbf{B}^p$ be the class of all Borel subsets of $\mathbf{R}^p$, and let ∂A be the boundary of A. Then, we have the following*:

(1) *There exists a positive constant c such that*

$$\sup_{B \in \mathbf{B}^2} |P\{\mathbf{Z} \in B\} - Q_{\mathbf{Z}}^{(2)}(B)| = o(n^{-1/2}) + c \sup_{B \in \mathbf{B}^2} Q_{\mathbf{Z}}^{(2)}\{(\partial B)^{2\varepsilon}\}, \tag{2.9}$$

where $\varepsilon = n^{-1/2-\delta}$ with $\delta > 0$. Here

$$Q_{\mathbf{Z}}^{(2)}\{B\} = \int_B N(\mathbf{z} : \Omega) \left[1 + \frac{1}{6\sqrt{n}} \sum_{j=1}^{2} \sum_{k=1}^{2} \sum_{\ell=1}^{2} A_{jk\ell} H_{jk\ell}(\mathbf{z}) \right] d\mathbf{z}, \tag{2.10}$$

where

$$\begin{aligned} \mathbf{z} &= (z_1, z_2)' \\ N(\mathbf{z} : \Omega) &= (2\pi)^{-1} |\Omega|^{-1/2} \exp - \tfrac{1}{2} \mathbf{z}' \Omega^{-1} \mathbf{z}, \\ H_{jk\ell}(\mathbf{z}) &= \frac{-1}{N(\mathbf{z} : \Omega)} \frac{\partial^3}{\partial z_j \partial z_k \partial z_\ell} N(\mathbf{z} : \Omega), \\ \Omega &= \begin{pmatrix} I & J \\ J & M \end{pmatrix}. \end{aligned}$$

(2)

$$E\{I + \|\mathbf{Z}\|^2\}1_{\{\|\mathbf{Z}\| > \sqrt{\log n}\}} = o(n^{-1/2}), \tag{2.11}$$

where 1_A *is the indicator function of* A.

In formula (2.10), it may be noted that we can explicitly evaluate the coefficient $A_{jk\ell}$ in terms of $M_k(\cdot)$ and $f_{a_1 \dots a_k}(\cdot)$. For example, $A_{111} = K$ (see (2.7)). However, in the following discussion, we do not need the explicit forms of $A_{jk\ell}$, $(j,k,\ell) \neq (1,1,1)$, hence we omit them.

3. THE RESULTS

In this section we will give the second-order Edgeworth expansion for the M-estimator $\hat{\beta}_n$ by using the fundamental results in Lemma 1. Henceforth, we denote the true value of β in model (2.1) by β_0.

THEOREM 1. *Suppose that conditions* (A.1), (A.2), (B.1)–(B.3), (C.1)–(C.4), *and* (D.1) *hold. Then, we have the following.*

(1) *There exists a statistic* $\hat{\beta}_n$ *that solves* (2.3) *such that for some* $d_1 > 0$

$$P^n_{\beta_0}[\,|\hat{\beta}_n - \beta_0| < d_1 n^{-1/2}\sqrt{\log n}\,] = 1 - o(n^{-1/2}), \tag{3.1}$$

uniformly for $\beta_0 \in \mathbf{C}$.

(2) *For* $\{\hat{\beta}_n\}$ *satisfying* (3.1),

$$\sup_{B \in \mathbf{B}_0} \left| P^n_{\beta_0}[\sqrt{nI}(\hat{\beta}_n - \beta_0) \in B] - \int_B \varphi(x) p^n_2(x)\,dx \right| = o(n^{-1/2}), \tag{3.2}$$

uniformly for $\beta_0 \in \mathbf{C}$, *where* $\mathbf{B}_0$ *is a subclass of* $\mathbf{B}^1$ *satisfying*

$$\sup_{B \in \mathbf{B}_0} \int_{(\partial B)^\varepsilon} \varphi(x) p^n_2(x)\,dx = O(\varepsilon). \tag{3.3}$$

Here $\varphi(x) = (1/\sqrt{2\pi}) \exp\{-\frac{1}{2}x^2\}$, *and* $p^n_2(x) = 1 + (\alpha_1/\sqrt{n})x + (\gamma_1/(6\sqrt{n})) \times (x^3 - 3x)$, *where*

$$\alpha_1 = -\frac{J+K}{2I^{3/2}}, \qquad \gamma_1 = -\frac{3J+2K}{I^{3/2}}.$$

Remark 1. In the case when $\hat{\beta}_n$ is the least-squares estimator (i.e., $\psi(x) \equiv x$), the asymptotic variance matrix Ω defined in Lemma 1 is singular. In such a case, reducing the discussion to the case when $\mathbf{Z} = Z_1$ (one-dimensional vector), we can similarly prove Theorem 1.

Remark 2. In the case when $\hat{\beta}_n$ is the least-squares estimator and the process $\{u_t\}$ is Gaussian, we can see that $J = K = 0$, whence $\alpha_1 = \gamma_1 = 0$.

Therefore, using expansion (3.2), we can discuss the second-order non-Gaussian robustness of $\hat{\beta}_n$.

From the preceding theorem, in general, we can see that

$$\sup_{B\in\mathbf{B}_0} |P^n_{\beta_0}[\sqrt{nI}(\hat{\beta}_n - \beta_0) \in B] - \Phi(B)| = O(n^{-1/2}), \tag{3.4}$$

where $\Phi(B) = \int_B \varphi(x)\, dx$. So we seek a function $T(\cdot)$ so that a standardized version $\tilde{T}$ of $T(\hat{\beta}_n)$ satisfies

$$\sup_{B\in\mathbf{B}_0} |P^n_{\beta_0}[\tilde{T} \in B] - \Phi(B)| = o(n^{-1/2}). \tag{3.5}$$

Then, $T(\cdot)$ is said to be the second-order normalizing transformation of $\hat{\beta}_n$. For $T(\beta)$, we assume the following:

(E.1) $T(\beta)$ is three times continuously differentiable with respect to β.

Henceforth, we denote the jth-order derivative of $T(\beta)$ by $T^{(j)}(\beta)$ and write $T^{(j)} = T^{(j)}(\beta_0)$.

THEOREM 2. *Suppose that conditions* (A.1), (A.2), (B.1)–(B.3), (C.1)–(C.4), (D.1), *and* (E.1) *hold. Then, we have the following.*

(1) *For* $\{\hat{\beta}_n\}$ *satisfying* (3.1),

$$\sup_{B\in\mathbf{B}_0} \left| P^n_{\beta_0}\left[\sqrt{nI}\left\{T(\hat{\beta}_n) - T(\beta_0) - \frac{c}{n}\right\}\Big/ T^{(1)} \in B\right] - \int_B \varphi(x) q^n_2(x)\, dx\right|$$
$$= o(n^{-1/2}), \tag{3.6}$$

where c is a constant and

$$q^n_2(x) = 1 + \frac{c_1}{\sqrt{n}}\, x + \frac{c_2}{6\sqrt{n}}\,(x^3 - 3x),$$

with

$$c_1 = \frac{-J-K}{2I^{3/2}} + \frac{1}{2I^{1/2}}\frac{T^{(2)}}{T^{(1)}} - \frac{cI^{1/2}}{T^{(1)}},$$

$$c_2 = \frac{-3J-2K}{I^{3/2}} + \frac{3T^{(2)}}{T^{(1)}I^{1/2}}.$$

(2) *If* $T(\beta) = T_0(\beta)$ *and* $c = c_0$ *satisfy*

$$\frac{T_0^{(2)}(\beta_0)}{T_0^{(1)}(\beta_0)} = \frac{3J+2K}{3I}, \tag{3.7}$$

$$c_0 = -\frac{KT_0^{(1)}(\beta_0)}{6I^2}, \tag{3.8}$$

then

$$\sup_{B \in \mathbf{B}_0} \left| P^n_{\beta_0}\left[\sqrt{nI}\left\{ T_0(\hat{\beta}_n) - T_0(\beta_0) - \frac{c_0}{n} \right\} \Big/ T_0^{(1)} \in B \right] - \Phi(B) \right|$$
$$= o(n^{-1/2}); \tag{3.9}$$

that is, T_0 is the second-order normalizing transformation of $\hat{\beta}_n$.

We give an example of the results.

Example

Let $x_t = 1 + \cos \alpha t$ ($\alpha \neq 0 \bmod 2\pi$) in model (2.1). We deal with the M-estimator defined by (2.3). In this case, it is easily checked that $M_2(\lambda)$ has jumps 1, $\frac{1}{4}$, $\frac{1}{4}$ at $\lambda = 0$, α, and $-\alpha$, respectively, and that $M_3(\lambda, \mu)$ has jumps 1, $\frac{1}{4}$, $\frac{1}{4}$, $\frac{1}{4}$, $\frac{1}{4}$, $\frac{1}{4}$, $\frac{1}{4}$ at $(\lambda, \mu) = (0,0)$, $(\alpha,0)$, $(-\alpha,0)$, $(0,\alpha)$, $(0,-\alpha)$, $(-\alpha,\alpha)$, and $(\alpha,-\alpha)$, respectively. Thus, we can evaluate the fundamental quantities as follows:

$$I = 2\pi[f_{\psi\psi}(0) + \tfrac{1}{2}f_{\psi\psi}(\alpha)], \tag{3.10}$$

$$J = -(2\pi)[\tfrac{3}{2}f_{\psi\psi'}(0) + \tfrac{1}{2}f_{\psi\psi'}(-\alpha) + \tfrac{1}{2}f_{\psi\psi'}(\alpha)],$$

$$K = (2\pi)^2[f_{\psi\psi\psi}(0,0) + \tfrac{1}{2}f_{\psi\psi\psi}(\alpha,0) + \tfrac{1}{2}f_{\psi\psi\psi}(-\alpha,0) + \tfrac{1}{2}f_{\psi\psi\psi}(\alpha,-\alpha)], \tag{3.11}$$

where $f_{\psi\psi}(\lambda)$ and $f_{\psi\psi\psi}(\lambda, \mu)$ are the second- and third-order cumulant spectral densities of the process $\{\psi(u_t)\}$, respectively, and $f_{\psi\psi'}(\lambda)$ is the cross-spectral density between $\{\psi(u_t)\}$ and $\{\psi'(u_t)\}$. Put

$$a = \frac{3J + 2K}{3I}.$$

The differential equation $(\partial/\partial\beta)\log T_0^{(1)}(\beta) = a$ in (3.7) leads to $T_0^{(1)}(\beta) = \exp(a\beta + k_1)$, where k_1 is a constant. Thus, the second-order normalizing transformation of the M-estimator $\hat{\beta}_n$ is given by

$$T_0(\beta) = \begin{cases} a^{-1}e^{a\beta} & \text{if } a \neq 0. \\ \beta & \text{if } a = 0. \end{cases}$$

The correction constant is given by $c_0 = -(a/(4I))e^{a\beta_0}$.

Next we consider the Edgeworth expansion for the regression t-statistic in the model

$$\begin{cases} y_t = \beta_0 + u_t, & t = 1, \ldots, n, \\ u_t = \theta u_{t-1} + \epsilon_t, & |\theta| < 1, \end{cases} \tag{3.12}$$

where $\{\epsilon_t\}$ is a sequence of i.i.d. random variables with $E\{\epsilon_t\}=0$, $\mathrm{Var}\{\epsilon_t\}=1$, $E\{\epsilon_t^8\}<\infty$, and probability density function $g(x)>0$ on $\mathbf{R}$. Let $\hat{\beta}_n = n^{-1}\sum_{t=1}^n y_t$; then, $\sqrt{n}(\hat{\beta}_n-\beta) = n^{-1/2}\sum_{t=1}^n u_t$. It is known that

$$\lim_{n\to\infty} \mathrm{Var}[\sqrt{n}(\hat{\beta}_n-\beta_0)] = 2\pi f_\theta(0), \tag{3.13}$$

where $f_\theta(\lambda) = (1/(2\pi))|1-\theta e^{i\lambda}|^{-2}$ is the spectral density of $\{u_t\}$. Thus, we consider the following Studentized statistic,

$$\tilde{t} = \sqrt{n}(1-\hat{\theta}_n)(\hat{\beta}_n-\beta_0), \tag{3.14}$$

where $\hat{\theta}_n = \{\sum_{t=1}^{n-1}\hat{u}_t\hat{u}_{t+1}\}/\{\sum_{t=1}^n \hat{u}_t^2\}$, with $\hat{u}_t = y_t - \hat{\beta}_n$.

THEOREM 3. *In model* (3.12),

$$\sup_{B\in\mathbf{B}_0}\left|P_{\beta_0}^n[\tilde{t}\in B] - \int_B \varphi(x) r_2^n(x)\,dx\right| = o(n^{-1/2}), \tag{3.15}$$

where $r_2^n(x) = 1 + [(\kappa_3)/(6\sqrt{n})](x^3-3x)$ *and* κ_3 *is the third-order cumulant of* ϵ_t.

Remark 3. Implication of this theorem seems interesting. The second-order Edgeworth approximation of $\tilde{t}$ is independent of θ (i.e., the dependent structure of the residual process). Also, if $\kappa_3=0$, $\tilde{t}$ becomes the second-order normalizing transformation. If a consistent estimator of κ_3 is given, we can estimate $r_2^n(x)$ with an error $o_p(n^{-1/2})$. Thus, more accurate specifications of the level of significance of $\tilde{t}$ or of the confidence coefficient of a confidence interval can be provided.

REFERENCES

Bhattacharya, R.N. & J.K. Ghosh (1978) On the validity of the formal Edgeworth expansion. *Annals of Statistics* 6, 434–451.

Bhattacharya, R.N. & R.R. Rao (1976) *Normal Approximation and Asymptotic Expansions.* New York: Wiley.

Bose, A. (1988) Edgeworth correction by bootstrap in autoregressions. *Annals of Statistics* 4, 1709–1722.

Brillinger, D.R. (1975) *Time Series: Data Analysis and Theory.* New York: Holden-Day.

Chibisov, D.M. (1972) An asymptotic expansion for the distribution of a statistic admitting an asymptotic expansion. *Theory of Probability and Its Applications* 17, 620–630.

Götze, F. & C. Hipp (1983) Asymptotic expansions for sums of weakly dependent random vectors. *Zeitschrift für Wahrscheinlickheits Theorie und Verwandte Gebiete* 64, 211–239.

Hannan, E.J. (1970) *Multiple Time Series.* New York: Wiley.

Lahiri, S.N. (1992) Bootstrapping M-estimators of a multiple linear regression parameter. *Annals of Statistics* 20, 1548–1570.

Navidi, W. (1989) Edgeworth expansions for bootstrapping regression models. *Annals of Statistics* 17, 1472–1478.

Qumsiyeh, M.B. (1990) Edgeworth expansion in regression models. *Journal of Multivariate Analysis* 35, 86–101.

Ringland, J.T. (1983) Robust multiple comparisons. *Journal of the American Statistical Association* 78, 145–151.
Taniguchi, M. (1987) Validity of Edgeworth expansions of minimum contrast estimators for Gaussian ARMA processes. *Journal of Multivariate Analysis* 21, 1–28.
Taniguchi, M. (1991) Higher Order Asymptotic Theory for Time Series Analysis. *Lecture Notes in Statistics*, vol. 68. Heidelberg: Springer-Verlag.

APPENDIX

Proof of Proposition 1. Conditions (GH.1), (GH.4), and (GH.6) are trivial. Condition (GH.2) follows from the uniform boundedness of x_t and condition (C.4). To prove condition (GH.3), we define

$$\bar{\mathbf{U}}_{t,m} = \begin{bmatrix} x_t\psi(\bar{u}_{t,m}) \\ -x_t^2\{\psi'(\bar{u}_{t,m}) - \mu\} \end{bmatrix},$$

where $\bar{u}_{t,m} = \sum_{j=0}^{m} c_j\varepsilon_{t-j}$ and $\mu = E\psi'(u_t)$. Expanding $\bar{\mathbf{U}}_{t,m} - \mathbf{U}_t$ around u_t, we obtain

$$\bar{\mathbf{U}}_{t,m} - \mathbf{U}_t = \begin{bmatrix} x_t\{(\bar{u}_{t,m} - u_t)\psi'(u_t) + \frac{1}{2}(\bar{u}_{t,m} - u_t)^2\psi''(\tilde{u}_1)\} \\ -x_t^2\{(\bar{u}_{t,m} - u_t)\psi''(\tilde{u}_2)\} \end{bmatrix}.$$

Uniform boundedness of $\{x_t\}$ and Lipschitz continuity of ψ'' lead to

$$\begin{aligned} \|\bar{\mathbf{U}}_{t,m} - \mathbf{U}_t\| \le{}& d_1|\bar{u}_{t,m} - u_t||\psi'(u_t)| \\ &+ (\bar{u}_{t,m} - u_t)^2\{d_2 + d_3|\bar{u}_{t,m} - u_t|\} \\ &+ |\bar{u}_{t,m} - u_t|\{d_4 + d_5|\bar{u}_{t,m} - u_t|\}, \end{aligned} \tag{A.1a}$$

for some positive constants $d_1, \ldots, d_5$. Because $u_t - \bar{u}_{t,m} = \sum_{j=m+1}^{\infty} c_j\varepsilon_{t-j}$ with $E\varepsilon_t^4 < \infty$ and $|c_j| \le \delta^{-1}e^{-\delta|j|}$, the expectation of the right-hand side of (A.1a) is dominated by $ae^{-\delta|m|}(a > 0)$. If $a \le \delta^{-1}$, then $ae^{-\delta|m|} \le \delta^{-1}e^{-\delta|m|}$. If $a > d^{-1}$, $ae^{-\delta|m|} = ae^{-a^{-1}a\delta|m|} < ae^{-a^{-1}|m|}$. Thus, relation (GH.3) is proved. Define

$$U_1^m = \sum_{j=-m}^{m} x_{t+j}\psi(u_{t+j}),$$

$$U_2^m = -\sum_{j=-m}^{m} x_{t+j}^2\{\psi'(u_{t+j}) - E\psi'(u_{t+j})\}.$$

To check relation (GH.5), we evaluate

$$E\{e^{i(t_1U_1^m + t_2U_2^m)}|\mathfrak{D}_j, j \ne t\}. \tag{A.1b}$$

Setting $z = (t_1U_1^m + t_2U_2^m)/\|\mathbf{t}\|$, we rewrite (A.1b) as

$$E\{e^{i\|\mathbf{t}\|z}|\mathfrak{D}_j, j \ne t\} = E_{\varepsilon_t}\{e^{i\|\mathbf{t}\|z}\}, \tag{A.1c}$$

where E_{ϵ_t} is the expectation with respect to ϵ_t. From condition (C.3), there exists $\delta > 0$ and some interval $L = (a, b)$ such that

$$\sum_{j=0}^{m} x_{t+j} c_j \psi'(c_j x) \geq \delta, \quad \text{a.e. } x \in L.$$

Because

$$\frac{\partial z}{\partial \epsilon_t} = \left\{ t_1 \sum_{j=0}^{m} x_{t+j} c_j \psi'(u_{t+j}) - t_2 \sum_{j=0}^{m} x_{t+j}^2 c_j \psi''(u_{t+j}) \right\} \Big/ \|\mathbf{t}\|,$$

we can see that

$$\left| \frac{\partial z}{\partial \epsilon_t} \right| = 0, \quad \text{a.e. } x \in L$$

is equivalent to

$$t_1/t_2 = \frac{d}{dx} \left[\log \sum_{j=0}^{m} x_{t+j} c_j \psi'(c_j x) \right], \quad \text{a.e. } x \in L. \tag{A.1d}$$

Thus, for a.e. $\mathbf{t} \in \mathbf{R}^2$,

$$\left| \frac{\partial z}{\partial \epsilon_t} \right| \neq 0, \quad \text{a.e. } x \in L. \tag{A.1e}$$

We return to evaluate

$$E_{\epsilon_t}\{e^{i\|\mathbf{t}\|z}\} = \int_L e^{i\|\mathbf{t}\|z} g(\epsilon_t) d\epsilon_t + \int_{\mathbf{R}-L} e^{i\|\mathbf{t}\|z} g(\epsilon_t) d\epsilon_t. \tag{A.1f}$$

In view of (A.1e) and the Riemann–Lebesgue theorem, for every $\epsilon > 0$, there exists $\hat{d}_1 > 0$ such that for all $\|\mathbf{t}\| \geq \hat{d}_1$, the absolute value of the first integral of (A.1f) is dominated by ϵ. Because the absolute value of the second integral of (A.1f) is dominated by

$$\int_{\mathbf{R}-L} g(\epsilon_t) d\epsilon_t \quad (<1),$$

there exists $d_2 > 0$ such that

$$|E_{\epsilon_t}\{e^{i\|\mathbf{t}\|z}\}| \leq e^{-d_2}$$

for all $\|\mathbf{t}\| \geq \hat{d}_1$. Finally,

$$E|E_{\epsilon_t}\{e^{i\|\mathbf{t}\|z}\}| \leq e^{-d_2} P\{\|\mathbf{t}\| \geq \hat{d}_1\} + 1 - P\{\|\mathbf{t}\| \geq \hat{d}_1\},$$

which implies relation (GH.5). ∎

Proof of Part (1) in Theorem 1. This part can be proved along the line of Bhattacharya and Ghosh (1978). Write $\ell_n(\beta) = \sum_{t=1}^{n} \psi(y_t - x_t \beta)$. We consider the equation

$$0 = n^{-1} \ell_n(\beta_0) + n^{-1}(\beta - \beta_0) \frac{\partial}{\partial \beta} \ell_n(\beta_0) + R_n(\beta), \tag{A.2}$$

where $R_n(\beta)$ is the usual remainder in the Taylor expansion, for which it holds that

$$|R_n(\beta)| \leq \frac{1}{2n}|\beta - \beta_0|^2 \sup_{|\beta' - \beta_0| \leq |\beta - \beta_0|} \left| \frac{\partial^2}{\partial \beta^2} \ell_n(\beta') \right|. \tag{A.3}$$

In view of Lemma 1 and condition (C.1), we can see that there exist positive constants d_1 and d_2 such that

$$P^n_{\beta_0}[|Z_1| > d_1\sqrt{\log n}] = o(n^{-1/2}), \tag{A.4}$$

$$P^n_{\beta_0}[|Z_2| > d_1\sqrt{\log n}] = o(n^{-1/2}), \tag{A.5}$$

$$P^n_{\beta_0}[|R_n(\beta)| > |\beta - \beta_0|^2 d_2\sqrt{\log n}] = o(n^{-1/2}). \tag{A.6}$$

Therefore, on a set having $P^n_{\beta_0}$-probability at least $1 - o(n^{-1/2})$, for some constants $d_3 > 0$ and $d_4 > 0$, we can rewrite (A.2) as

$$\beta - \beta_0 = [I + \eta_n]^{-1}[\delta_n + d_4|\beta - \beta_0|^2 \xi_n], \tag{A.7}$$

where η_n and δ_n are random variables whose absolute values are less than $d_3 n^{-1/2} \times \sqrt{\log n}$ and ξ_n is a random variable whose absolute value is less than $d_2\sqrt{\log n}$. Then, there exist a sufficiently large $d_5 > 0$ and an integer n_0 such that if $n > n_0$ and $|\beta - \beta_0| \leq d_5 n^{-1/2}\sqrt{\log n}$, the right-hand side of (A.7) is less than $d_5 n^{-1/2}\sqrt{\log n}$. Applying the Brouwer fixed point theorem to the right-hand side of (A.7), we have proved (1) of Theorem 1. ■

Proof of Part (2) in Theorem 1. First, for $\hat{\beta}_n$ satisfying (3.1) we give the stochastic expansion. From the equation $\ell_n(\hat{\beta}_n) = 0$, we have

$$0 = Z_1 + \frac{1}{\sqrt{n}} Z_2 V_n - I V_n + \frac{1}{2\sqrt{n}}\left\{\frac{1}{n}\frac{\partial^2}{\partial \beta^2}\ell_n(\tilde{\beta})\right\} V_n, \tag{A.8}$$

where $|\tilde{\beta} - \beta_0| \leq |\hat{\beta}_n - \beta_0|$ and $V_n = \sqrt{n}(\hat{\beta}_n - \beta_0)$. Because $\psi''(\cdot)$ satisfies the Lipschitz condition, (A.8) can be written as

$$V_n = \frac{Z_1}{I} + \frac{1}{I\sqrt{n}} Z_2 V_n + \frac{1}{2I\sqrt{n}}\left\{\frac{1}{n}\frac{\partial^2}{\partial \beta^2}\ell_n(\beta_0) + \frac{L}{\sqrt{n}} V_n\right\} V_n^2, \tag{A.9}$$

where $E|L| < \infty$. From (1) of Theorem 1 and (A.9), it follows that

$$V_n = I^{-1}Z_1 + \frac{1}{\sqrt{n}} a_n(1), \tag{A.10}$$

where $P^n_{\beta_0}[|a_n(1)| > d_6\{\log n\}^2] = o(n^{-1/2})$, for some $d_6 > 0$. Substituting (A.10) into the right-hand side of (A.9), and noting that

$$E\left\{\frac{1}{n}\frac{\partial^2}{\partial \beta^2}\ell_n(\beta_0)\right\} = -3J - K + O(n^{-1}), \tag{A.11}$$

we have

$$V_n = I^{-1}Z_1 + \frac{1}{I^2\sqrt{n}} Z_1 Z_2 - \frac{3J + K}{2I^3\sqrt{n}} Z_1^2 + \frac{1}{n} a_n(2) = H(\mathbf{Z}) + \frac{1}{n} a_n(2), \quad \text{(say)}, \tag{A.12}$$

for some $d_7 > 0$ and $\frac{1}{4} < \alpha < \frac{1}{2}$, whence we obtain the following.

LEMMA 2.

$$\sqrt{n}(\hat{\beta}_n - \beta_0) = H(\mathbf{Z}) + n^{-1}a_n(2), \tag{A.14}$$

where $P^n_{\beta_0}[\,|a_n(2)| > \rho_n\sqrt{n}\,] = o(n^{-1/2})$, *for some sequence* $\rho_n \to 0$, $\rho_n\sqrt{n} \to \infty$ *as* $n \to \infty$.

From Chibisov (1972), we can see that the Edgeworth expansion for $\sqrt{n}(\hat{\beta}_n - \beta_0)$ (up to order $n^{-1/2}$) is equal to that for $H(\mathbf{Z})$ on $B \in \mathbf{B}_0$ defined by (3.3). Thus, we seek the Edgeworth expansion for $H(\mathbf{Z})$.

Consider the transformation

$$W_1 = Z_1,$$

$$W_2 = Z_2 - JI^{-1}Z_1. \tag{A.15}$$

Evidently, (A.15) is a continuous bijective transformation. We denote (A.15) by $\mathbf{W} = \chi(\mathbf{Z})$, where $\mathbf{W} = (W_1, W_2)'$. It follows from Lemma 1 that

$$\sup_{B\in\mathbf{B}^2} |P^n_{\beta_0}\{\mathbf{Z} \in \chi^{-1}(B)\} - Q^{(2)}_{\mathbf{Z}}\{\chi^{-1}(B)\}|$$
$$= c \sup_{B\in\mathbf{B}^2} Q^{(2)}_{\mathbf{Z}}[\{\partial\chi^{-1}(B)\}^{2\varepsilon}] + o(n^{-1/2}). \tag{A.16}$$

Putting $Q^{(2)}_{\mathbf{W}}(B) = Q^{(2)}_{\mathbf{Z}}\{\chi^{-1}(B)\}$, we can show that

$$Q^{(2)}_{\mathbf{W}}(B) = \int_B N(w_1 : I)N(w_2 : \Omega)\left[1 + \frac{1}{\sqrt{n}}\sum_{j=1}^{2}\sum_{k=1}^{2}\sum_{\ell=1}^{2} c_{jk\ell}H_{jk\ell}(\mathbf{w})\right] d\mathbf{w} \tag{A.17}$$
$$= \int_B q_n(\mathbf{w})\, d\mathbf{w} \quad \text{(say)},$$

where

$$\mathbf{w} = (w_1, w_2)',$$

$$N(w : I) = (2\pi)^{-1/2}I^{-1/2}\exp - \frac{w^2}{2I},$$

$$\Omega = M - J^2I^{-1}.$$

Because χ is continuous, we have

$$\partial\chi^{-1}(B) \subset \chi^{-1}(\partial B),$$

$$\{\partial\chi^{-1}(B)\}^{2\varepsilon} \subset \{\chi^{-1}(\partial B)\}^{2\varepsilon},$$

and there exists $a > 0$ such that

$$\{\chi^{-1}(\partial B)\}^{2\varepsilon} \subset \chi^{-1}\{(\partial B)^{a\varepsilon}\},$$

which yields the following.

LEMMA 3.

$$\sup_{B\in\mathbf{B}^2} |P^n_{\beta_0}\{\mathbf{W} \in \beta\} - Q^{(2)}_{\mathbf{W}}(B)| = c \sup_{B\in\mathbf{B}^2} Q^{(2)}_{\mathbf{W}}\{(\partial B)^{a\varepsilon}\} + o(n^{-1/2}), \tag{A.18}$$

where a is a positive constant and $\varepsilon = n^{-1/2-\delta}$ *with* $\delta > 0$.

Recall Lemma 2. By transformation (A.15), we rewrite $H(\mathbf{Z})$ as

$$H(\mathbf{W}) = \frac{1}{I} W_1 + \frac{W_1 W_2}{I^2\sqrt{n}} - \frac{J+K}{2I^3\sqrt{n}} W_1^2. \tag{A.19}$$

We define the transformation

$$S_1 = H(\mathbf{W}),$$
$$S_2 = W_2, \tag{A.20}$$

and denote this by $\mathbf{S} = \boldsymbol{\Psi}(\mathbf{W})$, where $\mathbf{S} = (S_1, S_2)'$. For sufficiently large n, we can take a set

$$M_n = \{\mathbf{W} : |W_i| \le c_i\sqrt{\log n},\ i = 1,2\}$$

such that $\boldsymbol{\Psi}$ is a bijective C^∞-mapping on M_n. From Lemma 3,

$$\sup_{B\in\mathbf{B}^1} |P^n_{\beta_0}[\mathbf{W} \in \boldsymbol{\Psi}^{-1}(B \times \mathbf{R})] - Q^{(2)}_{\mathbf{W}}[\boldsymbol{\Psi}^{-1}(B \times \mathbf{R})]|$$
$$= c \sup_{B\in\mathbf{B}^1} Q^{(2)}_{\mathbf{W}}[(\partial\boldsymbol{\Psi}(B \times \mathbf{R}))^{a\varepsilon}] + o(n^{-1/2}). \tag{A.21}$$

Then it is shown that

$$\begin{aligned} Q^{(2)}_{\mathbf{W}}[\boldsymbol{\Psi}^{-1}(B \times \mathbf{R})] &= \int_{\boldsymbol{\Psi}^{-1}(B\times\mathbf{R})} q_n(\mathbf{w})\, d\mathbf{w} \\ &= \int_{M_n\cap\boldsymbol{\Psi}^{-1}(B\times\mathbf{R})} q_n(\mathbf{w}) d\mathbf{w} + o(n^{-1/2}) \\ &= \int_{N_n\cap(B\times\mathbf{R})} q_n\{\boldsymbol{\Psi}^{-1}(\mathbf{S})\}|\tilde{J}|\, d\mathbf{S} + o(n^{-1/2}), \end{aligned}$$

where $N_n = \boldsymbol{\Psi}(M_n)$ and $|\tilde{J}|$ is the Jacobian.

Because we can solve so that

$$W_1 = IS_1 - \frac{1}{\sqrt{n}} S_1 S_2 + \frac{J+K}{2\sqrt{n}} S_1^2 + o(n^{-1/2}),$$

uniformly on M_n, it is not difficult to show that

$$q_n\{\boldsymbol{\Psi}^{-1}(\mathbf{S})\}|\tilde{J}| = N(IS_1 : I)N(S_2 : \Omega_2)\left\{I + \frac{1}{\sqrt{n}} p_1(\mathbf{S}) + o(n^{-1/2})\right\},$$

uniformly on N_n, where $p_1(\mathbf{S})$ is a polynomial of $\mathbf{S}$. Thus, we obtain

$$\begin{aligned} &Q^{(2)}_{\mathbf{W}}[\boldsymbol{\Psi}^{-1}(B \times \mathbf{R})] \\ &\quad = \int_{(B\times\mathbf{R})\cap N_n} N(IS_1, I)N(S_2 : \Omega_2)\left\{I + \frac{1}{\sqrt{n}} p_1(\mathbf{S}) + o(n^{-1/2})\right\} d\mathbf{S} + o(n^{-1/2}) \\ &\quad = \int_B N(IS_1 : I)\left[\int_{\mathbf{R}} \left\{I + \frac{1}{\sqrt{n}} p_1(\mathbf{S})\right\} N(S_2 : \Omega_2) dS_2\right] dS_1 + o(n^{-1/2}). \end{aligned} \tag{A.22}$$

Calculating the square bracket of (A.22), and noting that

$Q_{\mathbf{W}}^{(2)}[\{\partial\mathbf{\Psi}^{-1}(B \times \mathbf{R})\}^{a\varepsilon}] \le Q_{\mathbf{W}}^{(2)}[\mathbf{\Psi}^{-1}\{(\partial B)^{b\varepsilon} \times \mathbf{R}\}]$,

for some $b > 0$, we have

$$\sup_{B\in\mathbf{B}} \left| P_{\beta_0}^n\{\sqrt{nI}(\hat{\beta}_n - \beta_0) \in B\} - \int_B \varphi(x)p_2^n(x)\,dx \right|$$

$$= c \sup_{B\in\mathbf{B}} \int_{(\partial B)^{b\varepsilon}} \varphi(x)p_2^n(x)dx + o(n^{-1/2}),$$

which completes the proof of part (2). ■

Proof of Theorem 2.

(1) Expanding $\tilde{T} = \sqrt{nI}\{T(\hat{\beta}_n) - T(\beta_0) - (c/n)\}/T^{(1)}$ in a Taylor series, we obtain

$$\tilde{T} = \sqrt{nI}(\hat{\beta}_n - \beta_0) - \frac{c\sqrt{I}}{\sqrt{n}T^{(1)}} + \frac{1}{2\sqrt{n}}\frac{T^{(2)}}{T^{(1)}\sqrt{I}}\{\sqrt{nI}(\hat{\beta}_n - \beta_0)\}^2$$
$$+ \frac{1}{6n}\frac{T^{(3)}(\tilde{\beta})}{IT^{(1)}}\{\sqrt{nI}(\hat{\beta}_n - \beta_0)\}^3, \quad \mathbf{(A.23)}$$

where $|\tilde{\beta} - \beta_0| \le |\hat{\beta}_n - \beta_0|$. It follows from part (1) of Theorem 1 and relation (E.1) that

$$P_{\beta_0}^n[|T^{(3)}(\tilde{\beta})\{\sqrt{nI}(\hat{\beta}_n - \beta_0)\}^3| > d(\log n)^{3/2}] = o(n^{-1/2}) \quad \mathbf{(A.24)}$$

for some $d > 0$. Substitution of (A.14) into (A.23) yields

$$\tilde{T} = \frac{1}{\sqrt{I}}Z_1 + \frac{1}{\sqrt{n}}\frac{Z_1Z_2}{I^{3/2}} - \frac{3J+K}{2I^{5/2}\sqrt{n}}Z_1^2 - \frac{c\sqrt{I}}{\sqrt{n}T^{(1)}}$$
$$+ \frac{1}{2\sqrt{n}}\frac{T^{(2)}}{T^{(1)}\sqrt{I}}\left\{\frac{Z_1^2}{I}\right\} + \frac{1}{n}b_n(2), \quad \mathbf{(A.25)}$$

where

$$P_{\beta_0}^n[|b_n(2)| > dn^{\alpha}] = o(n^{-1/2}),$$

for some $d > 0$ and $\frac{1}{4} < \alpha < \frac{1}{2}$. Thus, we can apply all the discussion for $\sqrt{n}(\hat{\beta}_n - \beta_0)$ to $\tilde{T}$ given by (A.25) similarly, which leads to (3.6).

(2) If we set $c_1 = c_2 = 0$ in the Edgeworth expansion of (3.6), we get equations (3.7) and (3.8).

Proof of Theorem 3. Because the line of the proof is similar to that of Theorem 1, we give only a sketch of it. First, we show

$$\hat{\theta}_n = \left\{\sum_{t=1}^{n-1} u_t u_{t+1}\right\} \Big/ \left\{\sum_{t=1}^{n} u_t^2\right\} + O_p(n^{-1}). \quad \mathbf{(A.26)}$$

From (A.26), we obtain

$$\tilde{T} = (1-\theta)Z_1 - \frac{1}{\sqrt{n}}(1-\theta^2)Z_1Z_3 + \frac{1}{\sqrt{n}}\theta(1-\theta^2)Z_1Z_2 + o_p(n^{-1/2}), \quad \mathbf{(A.27)}$$

where

$$Z_1 = \frac{1}{\sqrt{n}} \sum_{t=1}^{n} u_t,$$

$$Z_2 = \frac{1}{\sqrt{n}} \sum_{t=1}^{n} \{u_t^2 - 1/(1-\theta^2)\},$$

$$Z_3 = \frac{1}{\sqrt{n}} \sum_{t=1}^{n} \{u_t u_{t+1} - \theta/(1-\theta^2)\}.$$

It is not difficult to show

$$E\{Z_1 Z_2\} = \frac{\kappa_3}{(1-\theta^2)(1-\theta)} + O(n^{-1}),$$

$$E\{Z_1 Z_3\} = \frac{\kappa_3 \theta}{(1-\theta^2)(1-\theta)} + O(n^{-1});$$

hence, we can show

$$E\{\tilde{t}\} = o(n^{-1/2}), \qquad \mathrm{Var}(\tilde{t}) = 1 + o(n^{-1/2}),$$

$$\mathrm{cum}\{\tilde{t}, \tilde{t}, \tilde{t}\} = \frac{1}{\sqrt{n}} \kappa_3 + o(n^{-1/2}).$$

For $\mathbf{U}_t$ in Proposition 1, we set

$$\mathbf{U}_t = \left[u_t, u_t^2 - \frac{1}{1-\theta^2}, u_t u_{t+1} - \frac{\theta}{1-\theta^2} \right]'.$$

Then we can similarly follow the proofs of Proposition 1 and Theorem 1 for $\tilde{t}$ given by (A.27). ■

CONDITIONAL U-STATISTICS FOR DEPENDENT RANDOM VARIABLES

MICHEL HAREL and MADAN L. PURI*

Université de Limoges, U.R.A. C.N.R.S. 745, Toulouse, France, Indiana University

W. Stute (*Ann. Probab.* **19**, No. 2 (1991), 812–825) introduced a class of so-called U-statistics, which may be viewed as a generalization of the Nadaraya–Watson estimates of a regression function. In this paper, we extend the results from the independent case to the dependent case.

1. INTRODUCTION

Stute [10] introduced a class of so-called conditional U-statistics, which may be viewed as a generalization of the Nadaraya–Watson estimates of a regression function. This extension is similar to Hoeffding's [3] generalization of sample means to what we now call U-statistics.

Assume that (X_i, Y_i) are random vectors in the space $\mathbb{R}^p \times \mathbb{R}^m$, where $X_i = (X_{i1}, ..., X_{ip})$ and $Y_i = (Y_{i1}, ..., Y_{im})$, $i = 1, ..., n$. Let h be any function of k-variates (the U-kernel), $k \leqslant n$ such that $h(Y_1, ..., Y_k)$ is integrable. We are interested in the estimation of

$$m(x_1, ..., x_k) = E[h(Y_1, ..., Y_k) \mid X_1 = x_1, ..., X_k = x_k]. \tag{1.1}$$

When $p = m = k = 1$ and $h = I_d$, then

$$m(x_1) = E[Y_1 \mid X_1 = x_1]$$

is the regression of Y_1, given $X_1 = x_1$.

Received January 25, 1994; revised March 1995.

AMS subject classifications: 60F05, 60F15, 62J99.

Key words and phrases: Conditional U-statistics, asymptotic normality, strong convergence, absolute regularity.

* Research supported by the Office of Naval Research Contract N00014-91-J-1020.

For estimation of $m(x_1)$, Nadaraya [4] and Watson [11] independently proposed:

$$m_n(x_1) = \frac{\sum_{i=1}^n Y_i K[(x_1 - X_i)/a_n]}{\sum_{i=1}^n K[(x_1 - X_i)/a_n)]}.$$

Here K is the so-called smoothing kernel satisfying $\int K(u)\,du = 1$ and (a_n) is a sequence of bandwidths tending to zero at appropriate rates.

Schuster [7] under conditions requiring the existence of the density f of X_i and finiteness of $E(|Y_1|^3)$, proved the central limit theorem for $m_n(x)$. See also Rosenblatt [6]. Then, Stute [8] proved the asymptotic normality of $m_n(x)$ only under the condition of the finiteness of $E(Y_1^2)$, while X_1 need not have a density at all. Later Yoshihara [13] proved the central limit theorem for $m_n(x)$ when the r.v.'s are φ-mixing under finiteness of $E(|Y_1|^{2+\delta})$ $(\delta > 0)$. But the φ-mixing condition has applications which are too limited; for example, an ARMA process is never φ-mixing but generally geometrically absolutely regular.

For an arbitrary k, we now consider statistics of the form

$$\begin{aligned} u_n(\mathbf{x}) &= u_n(x_1, ..., x_k) \\ &= \frac{\sum_\beta h(Y_{\beta_1}, ..., Y_{\beta_1}) \prod_{j=1}^k K[(x_j - X_{\beta_j})/a_n]}{\sum_\beta \prod_{j=1}^k K[(x_j - X_{\beta_j})/a_n]} \\ &= \frac{\left[\begin{array}{l}\sum_\beta h((Y_{\beta_1 1}, ..., Y_{\beta_1 m}), ..., (Y_{\beta_k 1}, ..., Y_{\beta_k m})) \\ \qquad \times \prod_{j=1}^k K[(x_{j1} - X_{\beta_j 1})/a_n, ..., (x_{jp} - X_{\beta_j p})/a_n]\end{array}\right]}{\sum_\beta \prod_{j=1}^k K[(x_{j1} - X_{\beta_j 1})/a_n, ..., (x_{jp} - X_{\beta_j p})/a_n]}. \end{aligned}$$

Here the summation extends over all permutations $\beta = (\beta_1, ..., \beta_k)$ of length k. Stute [10] derived the limit distribution of $u_n(x)$ when the r.v.'s are independent under finiteness of $E(|h(Y_1, ..., Y_k)|^{2+\delta})$ $(\delta > 0)$. In this paper, we extend the result of Stute [10] for absolutely regular r.v.'s. When $p = m = k = 1$ and $h = I_k$ our result extends the result of Yoshihara [13] from the φ-mixing condition to the absolutely regular condition which permits a broad range of applications.

In what follows, we assume that the function h is symmetric and the sequence of r.v.'s $(X_1, Y_1), ..., (X_n, Y_n)$ is absolutely regular with rates

$$\beta(m) = O(\rho^m) \qquad \text{for some} \quad 0 < \rho < 1. \tag{1.2}$$

Recall that a sequence of random vectors $\{X_{ni,\, 1 \leqslant i \leqslant n,\, n \geqslant 1}\}$ is absolutely regular if

$$\sum_{m \leqslant n} \max_{1 \leqslant j \leqslant n-m} E\{\sup_{A \in \sigma(X_{ni},\, i \geqslant j+m)} |P(A \mid \sigma(X_{ni}, A \leqslant i \leqslant j)) - P(A)|\} = \beta(m) \downarrow 0$$

Here $\sigma(X_{ni}, 1 \leqslant i \leqslant j)$ and $\sigma(X_{ni}, i \geqslant j+m)$ are the σ-fields generated by $(X_{n1}, ..., X_{nj})$ and $(X_{n,j+m}, X_{n,j+m+1}, ..., X_{nn})$, respectively. Also recall that $\{X_{ni}\}$ satisfies the strong mixing condition if $\sup_{m \leqslant n} \sup_{1 \leqslant j \leqslant n-m} \{|P(A \cap B) - P(A)P(B)|; A \in \sigma(X_{ni}, 1 \leqslant i \leqslant j), B \in \sigma(X_{ni}, i \geqslant j+m)\} = \alpha(m) \downarrow 0$. Since $\alpha(m) \leqslant \beta(m)$, it follows that if $\{X_{ni}\}$ is absolutely regular, then it is also strong mixing.

In Section 4, we will show how our results can be applied to some Markov processes and particularly to some ARMA processes.

2. Asymptotic Normality

Let $\mathbf{x} = (x_1, ..., x_k)$ be fixed throughout. In this section, h will be assumed to be square-integrable. Set

$$U_n(h, \mathbf{x}) \equiv U(\mathbf{x}) \equiv U_n = \frac{(n-k)!}{n!} \sum_{\beta} h(Y_{\beta_1}, ..., Y_{\beta_k}) \times \prod_{j=1}^{k} K\left(\frac{x_j - X_{\beta_j}}{a_n}\right) \Big/ \prod_{j=1}^{k} EK\left(\frac{x_j - X_1}{a_n}\right). \tag{2.1}$$

Then

$$u_n(\mathbf{x}) = U_n(h, \mathbf{x}) / U_n(1, \mathbf{x}).$$

Note that $U_n(h, \mathbf{x})$ for each $k \geqslant 1$ is a classical U-statistic with a kernel depending on n.

Next, we denote the distribution function (d.f.) of (X_i, Y_i) by H and the marginals by F and G. Consider a sequence of functionals

$$\begin{aligned} \theta_n(h, \mathbf{x}) &\equiv \theta_n \\ &= \int m(z_1, ..., z_k) \prod_{j=1}^{k} K\left(\frac{x_j - z_j}{a_n}\right) \\ &\quad \times F(dz_1)...F(dz_k) \Big/ \prod_{j=1}^{k} \int K\left(\frac{x_j - x}{a_n}\right) F(dx) \\ &= \frac{E[h(\tilde{Y}_1, ..., \tilde{Y}_k) \prod_{j=1}^{k} K((x_j - \tilde{X}_j)/a_n)]}{E[\prod_{j=1}^{k} K((x_j - \tilde{X}_j)/a_n)]}, \end{aligned} \tag{2.2}$$

where $(\tilde{X}_i, \tilde{Y}_i)$, $i = 1, ..., k$ are i.i.d. random vectors with d.f. H.

We also suppose that the r.v.'s $(Y_1 \mid X_1), (X_2 \mid Y_2), ..., (Y_n \mid X_n)$ are independent and we denote by $\tilde{G}(x; \cdot)$ the conditional d.f. of $(Y_1 \mid X_1 = x)$.

We also assume that

$$E\,|h(Y_1, ..., Y_k)|^{2+\delta} < +\infty \qquad \text{for some} \quad \delta > 0. \tag{2.3}$$

For every c $(0 \leqslant c \leqslant k)$, put

$$\begin{aligned} g_{c,n}&((z_1, y_1), ..., (z_c, y_c)) \\ &\equiv g_c((z_1, y_1), ..., (z_c, y_c)) \\ &= \int h(y_1, ..., y_k) \prod_{j=1}^{k} K\left(\frac{x_j - z_j}{a_n}\right) \Big/ \prod_{j=1}^{k} EK\left(\frac{x_j - X_1}{a_n}\right) \\ &\quad \times \prod_{j=c+1}^{k} \tilde{G}(z_j; dy_j)\, F(dz_j). \end{aligned} \tag{2.4}$$

We have $g_0 = \theta_n$ and

$$\begin{aligned} g_k&((z_1, y_1), ..., (z_k, y_k)) \\ &= h(y_1, ..., y_k) \prod_{j=1}^{k} K\left(\frac{x_j - z_j}{a_n}\right) \Big/ \prod_{j=1}^{k} EK\left(\frac{x_j - X_1}{a_n}\right) \end{aligned} \tag{2.5}$$

Let $n^{-[r]} = \{n(n-1)\cdots(n-r+1)\}^{-1}$. Set

$$\begin{aligned} U_n^{(c)} = n^{-[c]} \sum_{\beta^{(c)}} \int g_c((z_1, y_1), ..., (z_c, y_c)) \\ \times \prod_{j=1}^{c} d(I_{[(X_{\beta_j}, Y_{\beta_j}) \leqslant (z_j, y_j)]} - H(z_j, y_j)), \end{aligned} \tag{2.6}$$

where $\beta^{(c)}$ is the summation over all the permutations $\beta^{(c)} = (\beta_1, ..., \beta_c)$ of length c. Then

$$U_n = \theta_n + \sum_{c=1}^{k} \binom{k}{c} U_n^{(c)}. \tag{2.7}$$

Let

$$\sigma^2 = \sigma^2(h, \mathbf{x}) = \lim_{n \to \infty} a_n^p \{E(g_1^2(X_1, Y_1)) - \theta_n^2\} \tag{2.8}$$

if the limit exists. We note also that $\lim_{n \to \infty} a_n^p \theta_n^2 = 0$.

LEMMA 2.1. *Assume that*

(i) $a_n \to 0$ *and* $na_n \to \infty$.

(ii) *K is bounded and has compact support.*

(iii) *F admits a density f which is continuous at each x_j, $1 \leqslant j \leqslant k$, with $f(x_j) > 0$.*

Then, we have $\sigma^2 < \infty$ and $(na_n^p)^{1/2} U_n^{(1)} \to \mathcal{N}(0, \sigma^2)$ in distribution, where σ^2 is defined in (2.8).

Proof. First, we show that

$$\lim_{n \to \infty} E((na_n^p)^{1/2} (U_n^{(1)}))^2 = \sigma^2. \tag{2.9}$$

From (2.6), we have

$$\begin{aligned}
U_n^{(1)} &= n^{-1} \sum_{i=1}^{n} g_1(X_i, Y_i)\, d(I_{[(X_i, Y_i) \leqslant (x_i, y_i)]} - H(x_i, y_i)) \\
&= n^{-1} \sum_{i=1}^{n} g_1(X_i, Y_i) \\
&\quad - n^{-1} \sum_{i=1}^{n} \int h(y_1, ..., y_k) \prod_{j=1}^{k} K\left(\frac{x_j - z_j}{a_n}\right) \Big/ \prod_{j=1}^{k} EK\left(\frac{x_j - X_1}{a_n}\right) \\
&\quad \times \prod_{j=1}^{k} \tilde{G}(z_j; dy_j)\, F(dz_j) \\
&= n^{-1} \sum_{i=1}^{n} g_1(X_i, Y_i) \\
&\quad - n^{-1} \sum_{i=1}^{n} \int m(z_1, ..., z_k) \Big/ \prod_{j=1}^{k} EK\left(\frac{x_j - X_1}{a_n}\right) \prod_{j=1}^{k} F(dz_j) \\
&= n^{-1} \sum_{i=1}^{n} (g_1(X_i, Y_i) - \theta_n).
\end{aligned}$$

We can write

$$\begin{aligned}
&na_n^p E(U_n^{(1)})^2 \\
&\quad = n^{-1} a_n^p E\left(\sum_{i=1}^{n} (g_1(X_i, Y_i) - \theta_n)\right)^2 \\
&\quad = n^{-1} a_n^p \sum_{i=1}^{n} E(g_1(X_i, Y_i) - \theta_n)^2 \\
&\qquad + n^{-1} a_n^p \sum_{1 \leqslant i \neq j \leqslant n} E\{(g_1(X_j, Y_j) - \theta_n)\}
\end{aligned}$$

$$= n^{-1} a_n^p \sum_{i=1}^{n} E(g_1(X_i, Y_i) - \theta_n)^2$$

$$+ 2n^{-1} a_n^p \sum_{\substack{1 \leqslant i < j \leqslant n \\ j-i \leqslant r}} E\{(g_1(X_i, Y_i) - \theta_n)(g_1(X_j, Y_j) - \theta_n)\}$$

$$+ 2n^{-1} a_n^p \sum_{\substack{1 \leqslant i < j \leqslant n \\ j-i > r}} E\{(g_1(X_i, Y_i) - \theta_n)(g_1(X_j, Y_j) - \theta_n)\}.$$

First, note that

$$\lim n^{-1} a_n^p \sum_{i=1}^{n} E(g_1(X_i, Y_i) - \theta_n)^2 = \lim_{n \to \infty} a_n^p E(g_1(X_1, Y_1) - \theta_n)^2 = \sigma^2. \tag{2.10}$$

For the sake of brevity we set

$$M = \prod_{j=1}^{k} E\left(K\left(\frac{x_j - X_1}{a_n}\right)\right).$$

From condition (ii), we easily deduce that

$$M = O(a_n^{kp}).$$

Then, we have

$$E\{(g_1(X_i, Y_i) - \theta_n)(g_1(X_j, Y_j) - \theta_n)\}$$

$$= M^{-2} \int (m(z_1, \ldots, z_k) - \theta_n)(m(z_{k+1}, \ldots, z_{2k}) - \theta_n)$$

$$\times \prod_{m=1}^{k} K\left(\frac{x_m - z_m}{a_n}\right) \prod_{l=1}^{k} K\left(\frac{x_{k+l} - z_{k+l}}{a_n}\right)$$

$$\times \prod_{m=2}^{k} F(dz_m) \prod_{\substack{l=k+2 \\ l \neq j}}^{k} F(dz_{k+l})\, F_{i,j}(dz_1, dz_{k+1})$$

$$= M^{-2} a_n^{2kp} \int (m(x_1 - a_n u_1, \ldots, x_k - a_n u_k) - \theta_n)$$

$$\times (m(x_{k+1} - a_n u_{k+1}, \ldots, x_{2k} - a_n u_{2k}) - \theta_n)$$

$$\times \prod_{m=1}^{k} K(u_m) \prod_{l=k+1}^{2k} K(u_l)$$

$$\times \frac{f_{i,j}(F^{-1}(x_1 - a_n u_1), F^{-1}(x_{k+1} - a_n u_{k+1}))}{f \circ F^{-1}(x_1 - a_n u_1) f \circ F^{-1}(x_{k+1} - a_n u_{k+1})} du_1 \cdots d_2 k$$

$\leqslant C a_n^{2kp} M^{-2}$ (from conditions (2.4) and (iii)),

$$\leqslant C, \tag{2.11}$$

where $F_{i,j}$ and $f_{i,j}$ are respectively the distribution function and the density function of (X_i, X_j) and, by convention, $x_i - a_n u_i = (x_{i1} - a_n u_{i1}, ..., x_{ip} - a_n u_{ip})$.

For any $\delta > 0$

$$E\,|g_1(X_i, Y_i) - \theta_n|^{2+\delta}$$

$$\leqslant M^{-(2+\delta)} \int |m(z_1, ..., z_k) - \theta_n|^{2+\delta} \left| \prod_{m=1}^{k} K\left(\frac{x_m - z_m}{a_n}\right)\right|^{2+\delta} \prod_{m=1}^{k} F(dz_m)$$

$$\leqslant M^{-(2+\delta)} a_n^{kp} \int |m(x_1 - a_n u_1, ..., x_k - a_n u_k) - \theta_n|^{2+\delta}$$

$$\times \left| \prod_{m=1}^{k} K(u)\right|^{2+\delta} \prod_{m=1}^{k} f(x_m - a_n u_m)\, du_1 \cdots du_k$$

$$\leqslant c a_n^{-k(1+\delta)p}. \tag{2.12}$$

From (2.11), (2.12), and Lemma 5.1 in the Appendix it follows that

$$n^{-1} a_n^p \sum_{1 \leqslant i \neq j \leqslant n} E\{(g_1(X_i, Y_i) - \theta_n)(g_1(X_j, Y_j) - \theta_n)\}$$

$$\leqslant 2Cra_n^p + 2(Ca_n^{-k(1+\delta)p})^{2/(2+\delta)} \sum_{i=r+1}^{n} [\beta(i)]^{\delta/(2+\delta)}.$$

If we take $r = [a_n^{-(1/2)p}]$, we get (2.9) from (1.2) and (2.10).

From Lemma 5.2 in the Appendix, we obtain the following inequality:

$$E\left|\sum_{i=1}^{m} g_1(X_i, Y_i) - \theta_n\right|^{2+\delta}$$

$$\leqslant Cm^{(2+\delta)/2} \sup_{1 \leqslant i \leqslant m} E\,|g_1(X_i, Y_i) - \theta_n|^{2+\delta}, \qquad m \leqslant n. \tag{2.13}$$

Let now $r = [n^{2/3}]$, $q = [n^{1/3}]$, and $l = [n/(r+1)]$. Put

$$\eta_j = \sum_{i=(j-1)(r+q)+1}^{(j-1)(r+q)+r} (g_1(X_i, Y_i) - \theta_n), \qquad j = 1, ..., l,$$

$$\theta_j = \sum_{i=(j-1)(r+q)+r+1}^{(j-1)(r+q)+r} (g_1(X_i, Y_i) - \theta_n), \qquad j = 1, ..., l,$$

$$\theta_{k+1} = \sum_{i=l(r+q)+1}^{n} (g_1(X_i, Y_i) - \theta).$$

Then, we have

$$n^{-1/2}a_n^{p/2}\sum_{i=1}^{n}(g_1(X_i, Y_i)-\theta_n)=n^{-1/2}a_n^{p/2}\sum_{j=1}^{l}\eta_j+n^{-1/2}a_n^{p/2}\sum_{j=1}^{l+1}\theta_j.$$

From (2.13), we deduce

$$n^{-1/2}a_n^{p/2}\sum_{j=1}^{l+1}\theta_j\xrightarrow{P}0.$$

To prove Lemma 2.1, it remains to show that $n^{-1/2}a_n^{p/2}\sum_{j=1}^{l}\eta_j$ converges in law to $\mathcal{N}(0, \sigma^2)$ random variable.

From Lemma 5.1 in the Appendix, we obtain

$$\left|E\left\{\exp\left(itn^{-1/2}a_n^{p/2}\sum_{j=1}^{l}\eta_j\right)\right\}-\prod_{j=1}^{k}\left[E\{\exp(itn^{-1/2}a_n^{p/2}\eta_j)\}\right]\right|\leqslant Cl\beta(q).$$

Hence it suffices to show that

$$\prod_{j=1}^{k}\left[E\{\exp(itn^{-1/2}a_n^{p/2}\eta_j)\}\right]\text{ converges to }e^{-t^2\sigma^2/2}. \tag{2.14}$$

Using (2.13), we obtain

$$\begin{aligned}&E\{\exp(itn^{-1/2}a_n^{p/2}\eta_j)\}\\&=1-\frac{t^2a_n^p}{2n}E(\eta_j)^2+O\left(\frac{|t|^{2+\delta}a_n^{(2+\delta)p/2}}{n^{(2+\delta)/2}}E(\eta_j)^{2+\delta}\right)\\&=1-\frac{a_n^pt^2}{2n}E(\eta_n)^2+o(|t|^{2+\delta}n^{-(1/6)(2+\delta)}a_n^{-(k-1)p+((1-2k)/2)\delta p}).\end{aligned}$$

From condition (ii) and

$$\lim\frac{la_n^p}{n}E(\eta_j)^2=\sigma^2\qquad\text{(from (2.10))}$$

we get (2.14). The proof follows.

LEMMA 2.2. *Under the conditions of Lemma* 2.1

$$(na_n^p)^{1/2}(U_n-\theta_n)\to\mathcal{N}(0, k\sigma^2)\qquad\textit{in distribution.}$$

Proof. From the decomposition (2.7) and Lemma 2.1, it is sufficient to prove that

$$E(U_n^{(c)})^2=O(n^{-2}),\qquad 2\leqslant c\leqslant k. \tag{2.15}$$

We shall only consider the case $c=2$. The proofs for the cases $c=3, \ldots, k$ are analogous and are therefore omitted.

We first note that

$$U_n^{(2)} = n^{-[2]} \sum_{1 \leqslant i_1 < i_2 \leqslant n} \{g_2((X_{i_1}, Y_{i_1}), (X_{i_2}, Y_{i_2})) - g_1(X_{i_1}, Y_{i_1}) - g_1(X_{i_2}, Y_{i_2}) + \theta_n\}.$$

So we have

$$E(U_n^{(2)})^2 = \sum_{1 \leqslant i_1 < i_2 \leqslant n} \sum_{1 \leqslant j_1 < j_2 \leqslant n} J((i_1, i_2), (j_1, j_2)), \tag{2.16}$$

where

$$\begin{aligned} J((i_1, i_2), (j_1, j_2)) &= E\{g_2((X_{i_1}, Y_{i_1}), (X_{i_2}, Y_{i_2})) - g_1(X_{i_1}, Y_{i_1}) - g_1(X_{i_2}, Y_{i_2}) + \theta_n\} \\ &\quad \times \{g_2((X_{j_1}, Y_{j_1}), (X_{j_2}, Y_{j_2})) - g_1(X_{j_1}, Y_{j_1}) - g_1(X_{j_2}, Y_{j_2}) + \theta_n\}. \end{aligned}$$

Since

$$\int \{g_2((z_1, y_1), (z_2, y_2)) - g_1(z_1, y_2) - g_1(z_2, y_2) + \theta_n\}\, H(dz_1, dy_1) = 0,$$

therefore, from Lemma 5.1 we have the following inequalities: If $1 \leqslant i_1 < i_2 \leqslant j_1 < j_2 \leqslant n$ and $j_2 - j_1 \geqslant i_2 - i_1$, then

$$J((i_1, i_2), (j_1, j_2)) \leqslant M\beta^{\delta/(2+\delta)}(j_2 - j_1) \tag{2.17}$$

and, similarly, if $1 \leqslant i_1 < i_2 \leqslant j_1 < j_2 \leqslant n$ and $i_2 - i_1 \geqslant j_2 - j_1$, then

$$J((i_1, i_2), (j_1, j_2)) \leqslant M\beta^{\delta/(2+\delta)}(i_2 - i_1). \tag{2.18}$$

Thus, from (2.17), (2.18), and assumption (1.2)

$$\begin{aligned} &\left| \sum_{1 \leqslant i_1 < i_2 \leqslant j_1 < j_2 \leqslant n} j((i_1, i_2), (j_1, j_2)) \right| \\ &\quad \leqslant \Bigg\{ \sum_{\substack{1 \leqslant i_1 < i_2 \leqslant j_1 < j_2 \leqslant n \\ i_2 - i_1 \geqslant j_2 - j_1}} + \sum_{\substack{1 \leqslant i_1 < i_2 \leqslant j_1 < j_2 \leqslant n \\ i_1 - i_1 \leqslant j_2 - j_1}} \Bigg\} |J((i_1, i_2), (j_1, j_2))| \\ &\quad \leqslant Cn^2 \sum_{r=1}^{n} (r+1)\, \beta^{\delta/(2+\delta)}(r) = O(n^2). \end{aligned} \tag{2.19}$$

Similarly, we have

$$\left|\sum_{1\leq i_1<j_1\leq i_2<j_2\leq n} J((i_1,i_2),(j_1,j_2))\right| = O(n^2), \tag{2.20}$$

$$\left|\sum_{1\leq i_1<j_1<j_2<i_2\leq n} J((i_1,i_2),(j_1,j_2))\right| = O(n^2), \tag{2.21}$$

and

$$\left|\sum_{1\leq i_1,j_1\leq n}\sum_{i_2=1}^{n} J((i_1,i_2),(j_1,j_2))\right| \leq Cn^2\left(1+\sum_{r=1}^{n}\beta^{\delta/(2+\delta)}(r)\right) = O(n^2). \tag{2.22}$$

Hence from (2.19)–(2.22) and (2.16), we have (2.15) for $c=2$. ▮

In the following, we shall investigate the asymptotic behavior of the two-dimensional random vector

$$(U_n(h_1,\mathbf{x})-\theta_n(h_1),\ U_n(h_2,\mathbf{x})-\theta_n(h_2)),$$

where h_1 and h_2 are two kernels satisfying the smoothness assumptions of Lemma 2.2. We would like to apply the Cramér–Wald device. So, let c_1, c_2 denote any two real numbers. Clearly,

$$c_1U_n(h_1,\mathbf{x})+c_2U_n(h_2,\mathbf{x}) = U_n(c_1h_1+c_2h_2,\mathbf{x}) \equiv U_n(h,\mathbf{x}),$$

where $h=c_1h_1+c_2h_2$ and Lemma 2.2 applies. Specification of $\sigma^2(h)$ immediately leads to the following.

Lemma 2.3. *Under the stated assumptions*

$$(na_n^p)^{1/2}\,[U_n(h_1,\mathbf{x})-\theta_n(h_1),\ U_n(h_2,\mathbf{x})-\theta_n(h_2)] \to \mathcal{N}(0,\Sigma)$$

in distribution, with

$$\Sigma = \begin{bmatrix} \sigma^2(h_1,h_1) & \sigma^2(h_1,h_2) \\ \sigma^2(h_1,h_2) & \sigma^2(h_2,h_2) \end{bmatrix}$$

and where for two functions h_1 *and* h_2

$$\sigma^2(h_1,h_2) = \lim_{n\to\infty} a_n^p E\{(g_1((X_1,Y_1);h_1)-\theta_n(h_1))(g_1((X_1,Y_1);h_2)-\theta_n(h_2))\}.$$

From this lemma, we will deduce the limit distribution of $U_n(\mathbf{x})$.

THEOREM 2.1. *Under the assumptions of Lemma 2.1, we have*

$$(na_n^p)^{1/2}\,(u_n(\mathbf{x})-\theta_n)\to\mathcal{N}(0,\rho^2)\qquad \text{in distribution,}$$

where

$$\rho^2=k\{\sigma^2(h,h)-2m(\mathbf{x})\,\sigma^2(h,1)+m^2(\mathbf{x})\,\sigma^2(1,1)\}.$$

Proof. We have

$$u_n(\mathbf{x})=U_n(h,\mathbf{x})/U(1,\mathbf{x}).$$

Define $g(x_1,x_2)=x_1/x_2$ for $x_2\neq 0$. Then

$$D=\left(\frac{\partial g}{\partial x_1},\frac{\partial g}{\partial x_2}\right)=(x_2^{-1},\ -x_1x_2^{-2}).$$

Since $\theta_n(h,\mathbf{x})\to m(\mathbf{x})$ and $\theta_n(1,\mathbf{x})=1$, we may infer from Lemma 2.3 that

$$(na_n^p)^{1/2}\,(u_n(\mathbf{x})-\theta_n(h,\mathbf{x}))\to\mathcal{N}(0,\rho^2)\qquad\text{in distribution,}$$

where

$$\rho^2=(1,\ -m(\mathbf{x}))\,\Sigma\begin{pmatrix}1\\ -m(\mathbf{x})\end{pmatrix}$$

and

$$\Sigma=\begin{pmatrix}\sigma^2(h,h) & \sigma^2(h,1)\\ \sigma^2(h,1) & \sigma^2(1,1)\end{pmatrix}. \qquad \blacksquare$$

Under appropriate smoothness assumptions on the marginal density f, Theorem 2.1 immediately yields asymptotic normality of $u_n(\mathbf{x})-m(\mathbf{x})$. Now assume that

$$f \text{ is twice differentiable in neighborhoods of } x_j,\qquad 1\leqslant j\leqslant k, \tag{2.23}$$

and

$$K \text{ is symmetric at zero;} \tag{2.24}$$

m admits an expansion

$$m(\mathbf{y}+\boldsymbol{\Delta})=m(\mathbf{y})+\{m'(\mathbf{y})^t\,\boldsymbol{\Delta}\}+\tfrac{1}{2}\boldsymbol{\Delta}^t\{m''(\mathbf{y})\,\boldsymbol{\Delta}+o(\boldsymbol{\Delta}^t\boldsymbol{\Delta})\} \tag{2.25}$$

as $\boldsymbol{\Delta}\to 0$, for all $\mathbf{y}$ in the neighborhood of $\mathbf{x}$. Then we have the following.

COROLLARY 2.1. *If, in addition to conditions of Lemma 2.1, (2.23)–(2.25) hold, then*

$$(na_n^p)^{1/2}\,(u_n(\mathbf{x})-m(\mathbf{x}))\to \mathcal{N}(0,\rho^2) \qquad \text{in distribution,}$$

provided that $na_n^{5p}\to 0$.

Proof. See Corollary 2.4 of Stute [10].

3. CONSISTENCY

As for the regression estimators, we need to develop consistency results. We provide them here for the dependent case similar to the results established by Stute [10] for the independent case. Our conditions on the U-kernel h are not so restrictive as the conditions of Stute [10] on his Theorems 2 and 3.

THEOREM 3.1. *Under the conditions of Lemma 2.1, we have for* $\mu_1\otimes\cdots\otimes\mu_k$, *for almost all* $\mathbf{x}$

$$u_n(\mathbf{x})\to m(\mathbf{x}) \qquad \text{in probability,}$$

where μ *is the probability measure defined by the d.f.* F.

Proof. We know that almost surely

$$\theta_n(\mathbf{x})\to m(\mathbf{x}). \tag{3.1}$$

We also know that

$$u_n(\mathbf{x})=U_n(h,\mathbf{x})/U_n(1,\mathbf{x}).$$

So, we have to show that

$$U_n(h,\mathbf{x})\to m(\mathbf{x}), \qquad U_n(1,\mathbf{x})\to 1 \quad \text{in probability.}$$

Since $U_n(1,\mathbf{x})$ is a special case of $U_n(h,\mathbf{x})$, we have only to deal with $U_n(h,\mathbf{x})$.

From the decomposition (2.7) and (3.1), we have only to prove that

$$\sum_{c=1}^{k}\binom{k}{c}U_n^{(c)}\to 0 \qquad \text{in probability.}$$

But this is a consequence of (2.15). Theorem 3.1 follows.

THEOREM 3.2. *In addition to the conditions of Theorem* 3.1, *assume that*

$$\sum_{n=1}^{\infty} n^{1-\gamma} \exp(-na_n^p) < \infty \qquad \textit{for some} \quad 0 < \gamma < 1 \tag{3.2}$$

and suppose that h is bounded. Then, for almost all $\mathbf{x}$

$$u_n(\mathbf{x}) \to u(\mathbf{x}) \qquad \textit{with probability } 1.$$

Proof. From (2.15), we have

$$E(U_n - kU_n^{(1)})^2 = O(n^{-2}).$$

Then, from the Borel–Cantelli lemma, it suffices to show that

$$U_n^{(1)} \to 0 \qquad \text{with probability 1.} \tag{3.3}$$

Clearly,

$$U_n^{(1)} = n^{-1} \sum_{i=1}^{n} \{T_{i,n} - E(T_{i,n})\},$$

where

$$T_{i,n} = \int h(Y_i, y_2, ..., y_k) \prod_{j=2}^{k} K\left(\frac{x_j - z_j}{a_n}\right)$$
$$\times K\left(\frac{x_1 - X_i}{a_n}\right) \Big/ \prod_{j=1}^{k} EK\left(\frac{x_j - X_1}{a_n}\right) \prod_{j=2}^{k} \tilde{G}(z_j; dy_j)\, F(dz_j).$$

We note that there exist two positive constants b and c such that

$$|T_{i,n}| \leqslant b/a_n^p$$
$$E(T_{i,n}^2) \leqslant c/a_n^p.$$

If $U_1, U_2, ..., U_n$ are independent random variables with $|U_i| \leqslant m$, $E(U_i) = 0$, and $E(U_i^2) \leqslant \sigma_i^2$, then an inequality due to Bennett [1, p. 39] states that

$$P\left[\left|n^{-1} \sum_{i=1}^{n} U_i\right| \geqslant \varepsilon\right] \leqslant 2\exp\{-n\varepsilon^2/2(\sigma^2 + m\varepsilon)\},$$

where $\sigma^2 = n^{-1} \sum_{i=1}^{n} \sigma_i^2$. Put $q = q_n = [n^{1-\gamma}] + 1$ and write

$$U_n^{(1)} = \sum_{j=1}^{q} V_{n,j},$$

where

$$V_{n,j} = \sum_{p=0}^{l_j} \{T_{j+pq,n} - E(T_{j+pq,n})\}$$

and l_j is the largest integer such that $j + l_j q \leqslant n$. Then

$$P[|U_n^{(1)}| \geqslant \varepsilon] \leqslant P\left[n^{-1} \sum_{j=1}^{q} |V_{n,j}| \geqslant \varepsilon\right] \leqslant \sum_{j=1}^{q} P[|V_{n,j}| \geqslant \varepsilon n]. \tag{3.4}$$

For any j, $1 \leqslant j \leqslant q$, define

$$B_j = \left\{(y_1, ..., y_{l_j}) \,\middle|\, \left|\sum_{p=1}^{l_j} y_p\right| \geqslant \varepsilon n q^{-1}\right\}$$

and put

$$g(y_1, ..., y_{l_j}) = \begin{cases} 1 & \text{if } (y_1, ..., y_{l_j}) \in B_j \\ 0 & \text{otherwise.} \end{cases}$$

By Lemma 5.1,

$$\begin{aligned} P[|V_{n,j}| \geqslant \varepsilon n] &= Eg(T_{j,n} - E(T_{j,n}), ..., T_{j+l_j q,n} - E(T_{j+l_j q,n})) \\ &\leqslant P\left[\left|\sum_{p=0}^{l_j} \tilde{T}_{j+pq,n} - E(\tilde{T}_{j+pq,n})\right| \geqslant n\varepsilon q^{-1}\right] + 2l_j\beta(q) \\ &\leqslant 2\exp\left\{\frac{-n\varepsilon^2 a_n^p}{2c + 4b\varepsilon}\right\} + 2l_j\beta(q) \\ &\leqslant 2\exp\{-\alpha n a_n^p\} + 2l_j\beta(q) \qquad \left(\alpha = \frac{\varepsilon^2}{(2c + 4b\varepsilon)}\right), \end{aligned} \tag{3.5}$$

where $\tilde{T}_{1,n}, ..., \tilde{T}_{n,n}$ are independent and $\tilde{T}_{i,n}$ is distributed as $T_{i,n}$.

From (3.4) and (3.5), we deduce

$$P[|U_n^{(1)}| \geqslant \varepsilon] \leqslant 2n^{1-\gamma}\{\exp\{-\alpha n a_n^p\} + n^\gamma \beta([n^{1-\gamma}] + 1)\}.$$

From the Borel–Cantelli lemma and conditions (1.2) and (3.2), we deduce (3.3) and Theorem 3.2 is proved.

4. Applications

Consider the model

$$Y_n = \psi(X_n) + \epsilon_n, \qquad n \geqslant 1, \tag{4.1}$$

where X_n denotes a $\mathbb{R}^p$-vector of observed values, ψ is measurable known function, ϵ_n is a multivariate white noise corresponding to the measurement errors (that is, $\{\epsilon_n, n \in \mathbb{N}\}$ is a sequence of i.i.d. random $\mathbb{R}^m$-vectors with strictly positive density) and Y_n is an $\mathbb{R}^m$ predictor vector. If the sequence $(X_n)_{n \geqslant 1}$ of random vectors is absolutely regular with a geometrically rate, $E(|\psi(X_n)|^{2+\delta}) < +\infty$ and condition (ii) of Lemma 2.1 is satisfied. Thus we can apply Theorems 2.1, 3.1, and 3.2 for appropriate functions h and K and appropriate sequence a_n.

It is well known that any Markov process which is Harris recurrent, aperiodic, and geometrically aperiodic is absolutely regular with a geometrical rate.

For example, consider the model

$$X_n + \sum_{j=1}^{p_1} A_j X_{n-j} = e_n + \sum_{j=1}^{p_2} B_l e_{n-l}, \qquad n \in \mathbb{Z}, \tag{4.2}$$

where $A_1, ..., A_{p_1}$ and $B_1, ..., B_{p_2}$ are $p \times p$ real matrices, A_{p_1} and B_{p_2} are invertible and $e_n = (e_{n1}, ..., e_{np})$ is a multivariate white noise, where each e_{nj}, $n \geqslant 1$, $1 \leqslant j \leqslant p$, admits the same density g such that $\int |x|^\delta g(x)\,dx < \infty$ and $\int |g(x) - g(x-\theta)|\,dx = O(|\theta|^\gamma)$ for some $\delta > 0$ and $\gamma > 0$.

From Pham and Tran [5], X_n admits a Markovian representation

$$X_n = HZ_n, \qquad Z_n = FZ_{n-1} + Ge_n,$$

where Z_n is a sequence of random vectors and H, F, G are appropriate matrices. If the eigenvalues of the matrices H have a modulus less than 1, then X_n is absolutely regular with a geometrical rate.

If $p = 1$, $m = 1$, and $k = 2$, the example of Stute [10] can be applied to the particular model

$$Y_n = aX_n + \epsilon_n, \qquad a \in \mathbb{R}, \tag{4.3}$$

where X_n is an ARMA process defined by

$$X_n = bX_{n-1} + e_n, \qquad \text{where} \quad |b| < 1. \tag{4.4}$$

Example 4.1. Put $h(y_1, y_2) = y_1 y_2$. Then

$$\begin{aligned} m(x_1, x_2) &= E(Y_1 \mid X_1 = x_1)\, E(Y_2 \mid X_2 = x_2) \\ &= a^2 x_1 x_2. \end{aligned}$$

When $x_1 = x_2$, the variance ρ^2 defined in Theorem 2.1 yields

$$\rho^2 = 4 \operatorname{Var}(Y_1 \mid X_1 = x_1)\, a^2 x_1^2 \int K^2(u)\, du/f(x_1)$$

$$= 4\sigma^2 a^2 x_1^2 \int K^2(u)\, du/f(x_1)$$

while for $x_1 \neq x_2$, we get

$$\rho^2 = [\operatorname{Var}(Y_1 \mid X_1 = x_1)\, a^2 x_1^2/f(x_1) + \operatorname{Var}(Y_1 \mid X_1 = x_2)\, a^2 x_2^2/f(x_2)]$$

$$\times \int K^2(u)\, du$$

$$= \sigma^2 \int K^2(u)\, du [a^2 x_1^2/f(x_1) + a^2 x_2^2/f(x_2)].$$

EXAMPLE 4.2. Suppose $E(\varepsilon_1^4) < +\infty$. For $h(y_1, y_2) = \frac{1}{2}(y_1 - y_2)^2$, we obtain $m(x_1, x_1) = \operatorname{Var}(Y_1 \mid X_1 = x_1) = \sigma^2$. In this case

$$\rho^2 = \{E[(Y - ax_1)^4 \mid X = x_1] - \operatorname{Var}(Y \mid X = x_1)\} \int K^2(u)\, du/f(x_1)$$

$$= (\tau_4 - \sigma^4) \int K^2(u)\, du/f(x_1),$$

where $\tau_4 = E(\varepsilon_1^4)$.

We have seen how the examples of Stute [10] can be applied now to more general models, but it is obvious from (4.1) and (4.2) that we have the possibility of using our results for a much larger set of models and applications when $p > 1$ and $m > 1$.

APPENDIX

LEMMA 5.1 (Davydov [2]). *Let $\{X_{ni}, 1 \leqslant i \leqslant n, n \geqslant 1\}$ be a nonstationary sequence of r.v.'s which is strong mixing. Let Z be $\sigma(X_{ni}, 1 \leqslant i \leqslant j)$-measurable $(1 \leqslant j \leqslant n)$ and let V be $\sigma(X_{ni}, i \geqslant j + m)$-measurable. If $E(|Z|^p) < \infty$, $E(|V|^q) < \infty$, and $r^{-1} + p^{-1} + q^{-1} = 1$ $(r, p, q > 0)$ then*

$$|E(ZV) - E(Z)\, E(V)| \leqslant (\alpha(m))^{1/r} \{E\, |Z|^p\}^{1/p} \{E\, |V|^q\}^{1/q},$$

where C is some constant > 0, and of course if the sequence is absolutely regular, the inequality (5.1) holds if we replace $(\alpha(m))^{1/r}$ by $(\beta(m))^{1/r}$.

LEMMA 5.2 (Theorem 1 of Yokoyama [12]). *Let* $\{X_{ni}, 1 \leqslant i \leqslant n, n \geqslant 1\}$ *be a nonstationary sequence of r.v.'s which is strong mixing with* $E(X_{ni}) = 0$, $1 \leqslant i \leqslant n$, $n \geqslant 1$, *and* $\sup_{1 \leqslant i \leqslant n} E\,|X_{ni}|^{r+\delta} < C_n$ *for some* $r > 2$ *and* $\delta > 0$. *If*

$$\sum_{i=0}^{\infty} (i+1)^{r/2}\,[\alpha(i)]^{\alpha/(r+\delta)} < \infty$$

then

$$E\left|\sum_{i=1}^{n} X_{ni}\right|^{r} \leqslant C_n n^{r/2}, \qquad n \geqslant 1.$$

REFERENCES

[1] BENNETT, G. (1962). Probability inequalities for the sum of independent random variables. *J. Amer. Statist. Assoc.* **19** 33–45.

[2] DAVYDOV, YU. A. (1970). Invariance principle for empirical stationary processes. *Theory Probab. Appl.* **14** 487–498.

[3] HOEFFDING, W. (1948). A class of statistics with asymptotically normal distribution. *Ann. Math. Statist.* **19** 293–325.

[4] NADARAYA, E. A. (1964). On estimating regression. *Theory Probab. Appl.* **9** 141–142.

[5] PHAM, T. D., AND TRAN, L. T. (1985). Some mixing properties of time series models. *Stochastic Process Appl.* **19** 297–303.

[6] ROSENBLATT, M. (1969). Conditional probability density and regression estimates. In *Multivariate Analysis II* (Krishnaiah, Ed.), pp. 25–31. Academic Press, New York.

[7] SCHUSTER, E. F. (1972). Joint asymptotic distribution of the estimate regression function at a finite number of distinct points. *Ann. Math. Statist.* **43** 84–88.

[8] STUTE W. (1984). Asymptotic normality on nearest neighbor regression function estimates. *Ann. Statist.* **12** 917–918.

[9] STUTE W. (1986). On almost sure convergence of conditional empirical distribution functions, *An. Probab.* **14** 891–901.

[10] STUTE W. (1991). Conditional *U*-statistics. *Ann. Probab.* **19**, No. 2, 812–825.

[11] WATSON, G. S. (1964). Smooth regression analysis. *Snakhyā Ser. A* **26** 359–372.

[12] YOKOYAMA, R. (1980). Moment bounds for stationary mixing sequences. *Z. Wahrsch. Verw. Geb.* **52** 45–57.

[13] YOSHIHARA, K. (1988). Asymptotic normality of nearest neighbor regression function estimates based on some dependent observations. *Yokohama Math. J.* **36** 55–68.

WEAK CONVERGENCE OF SEQUENCES OF FIRST PASSAGE PROCESSES AND APPLICATIONS

STEFAN S. RALESCU[a,*,1] and MADAN L. PURI[b,2]

[a] *Department of Mathematics, Queens College, City University of New York, 65-30 Kissena Boulevard, Flushing, NY 11367, USA*

[b] *Department of Mathematics, Swain Hall East, Indiana University, Bloomington, Indiana 47405, USA.*

Abstract

Suppose $\{X_n\}_{n\geqslant 1}$ are stochastic processes all of whose paths are nonnegative and lie in the space of right continuous functions with finite left limits. Moreover, assume that X_n (properly normalized) converges weakly to a process X, i.e., for some deterministic function μ and $\theta_n \to 0$,

$$\theta_n^{-1}(X_n - \mu) \xrightarrow{d} X.$$

This paper considers the description of the weak limiting behavior of the sequence of first passage processes $\tilde{X}_n^{-1}(t) = \inf\{x : \tilde{X}_n(x) \geqslant t\}$ where $\tilde{X}_n(x) = \rho(nx)X_n(x)$ and $\rho(\cdot)$ is such that $\tilde{X}_n(x)$ has nondecreasing paths. We present a number of important motivating examples including empirical processes associated with U-statistics, empirical excursions above a given barrier, stopping rules in renewal theory and weak convergence in extreme value theory and point out the wide applicability of our result. Weak functional limit theorems for general quantile-type processes are derived. In addition, we investigate the asymptotic behavior of integrated kernel quantiles and establish: (i) an invariance principle; (ii) a strong law of large numbers; and (iii) a Bahadur-type representation which has many consequences, among which is a law of the iterated logarithm.

Keywords: Weak convergence; First passage processes; Empirical and quantile processes; U-staistics structure; Integrated kernel quantiles; Bahadur representation

AMS classification: Primary 62G05; Secondary 60F05, 62G30

1. Introduction

In many areas of probability and statistics, there has been a long-standing need to prove weak convergence of inverse processes. For example, classical quantile processes have been investigated (from probabilistic as well as statistical points of view) by many

* Corresponding author.

[1] Research supported in part by Grant PSC–CUNY 665358.

[2] Research supported by the Office of Naval Research, Contract N00014-91-J-1020.

authors and a powerful literature is now available. For surveying theoretical results and applications we refer to Csörgö and Révész (1981), Csörgö (1983), and Shorack and Wellner (1986).

The present investigation deals for the most part with the problem of weak convergence of a sequence of first passage processes to a continuous limiting process but, in order to include applications of weak convergence of maximal processes to extremal processes in Section 3, $\xrightarrow{d}$ in Skorohod's M_1 topology is used. (For an excellent discussion of the M_1 topology the reader is referred to Whitt (1980).) Let $\{X_n(x)\}_{n\geqslant 1}$ be a sequence of stochastic processes all of whose paths are nonnegative and lie in the space $D_r(-\infty,\infty)$ of right continuous functions with finite left limits on $(-\infty,\infty)$. Important in many applications is the behavior of the sequence of first passage processes $\tilde{X}_n^{-1}$ associated with X_n which is formed by inversion as follows. Define $\tilde{X}_n(x)=\rho(nx)X_n(x)$ for some nonnegative function $\rho(\cdot)\in D_r(-\infty,\infty)$ such that for $n\geqslant 1$, $\tilde{X}_n(x)$ has nondecreasing paths. Set $\tilde{X}_n^{-1}(t)=\inf\{x:\tilde{X}_n(x)\geqslant t\}$ and assume for the moment that $\tilde{X}_n^{-1}$ is well defined in a closed subinterval of $[0,\infty)$. In many applications the process X_n is naturally generated by a sample $\{\xi_1,\ldots,\xi_n\}$ corresponding to a sequence $\xi_1,\xi_2,\ldots$ of i.i.d. random variables. In many instances of large sample theory, one is interested in showing that $\tilde{X}_n^{-1}$ (properly normalized), regarded as a random element with values in $D_\ell(I)$ (= the set of all left continuous functions on a finite subinterval I of $[0,+\infty)$ with right limits), converges weakly to a limiting process on I. Here, it should be remarked that various specialized results dealing with weak convergence of first passage processes have been obtained by several authors and in this respect the contributions of Vervaat (1972) and Whitt (1980) are particularly helpful in understanding the relationship between increasing processes and their inverses (see also Shorack and Wellner (1986) in connection with the weak convergence of the quantile process corresponding to the product–limit estimator). We should also note that when $\rho(\cdot)$ is nondecreasing, the monotonicity assumption on X_n is not critical since we can first consider the supremum process $X_n^{\uparrow}(x)=\sup_{0\leqslant s\leqslant x}X_n(s)$ (Vervaat, 1972; Whitt, 1980) and then proceed to invert $\tilde{X}_n(x)=\rho(nx)X_n^{\uparrow}(x)$. It should be pointed out that the result embodied by Theorem 2.1 has a history and precursors and particular forms of it were more or less available to those "in the know" (see e.g., Shorack and Wellner, 1986, pp. 657–659 or Massey and Whitt, 1994, where Theorem 5.1 there has some of the flavor of part of Theorem 2.1 here). However, in our opinion, the full potential of the basic weak convergence results was never achieved and, in view of recent work on quantile and related processes, another look at the questions of weak convergence seems justified. In this paper we broaden the work of the above mentioned authors with the intention to provide a general basic framework for treatment of weak convergence of sequences of first passage processes with special emphasis on general quantile processes.

To provide a few explanatory comments on the method used, it should be noted that a direct approach to the weak convergence of $\tilde{X}_n^{-1}$ may cause serious technical difficulties. To bypass this aspect, an indirect route is adopted by focussing to reduce the problem to a search for weak convergence of the X_n processes. Indeed, one motivation for this paper was to stress the usefulness and the simplicity of such a method; see Vervaat (1972), Whitt (1980) (and also Shorack and Wellner, 1986) for a more detailed account. In order to illustrate this aspect, it is worth pointing out that in their monograph, Shorack and Wellner (1986) have made a far reaching investigation of various kinds of empirical processes and in many instances they proved, for special constructions of the random variables $\xi_1, \ldots, \xi_n$ inducing X_n, a result of the form

$$\sup_{a \leqslant x \leqslant b} \left| \sqrt{n}(X_n(x) - \gamma(x)) - X(x) \right| \xrightarrow{P} 0 \tag{1.1}$$

for an appropriate centering function $\gamma(\cdot)$ and a continuous process $X(\cdot)$.

This paper starts with an attempt to derive the limiting distribution of a sequence of first passage processes by showing that such a result is entailed by the specialized statement (1.1). In Section 2, we present a unified, self-contained treatment of weak convergence of sequences of first passage processes in a more general framework.

The main message of our paper is that if the weak asymptotic behavior of X_n is well tractable (using intrinsic properties of X_n), then the weak asymptotic behavior of the first passage process $\tilde{X}_n^{-1}$ can be inferred from the corresponding result for X_n. It should be noted that there are many contexts in statistics where a substantial proportion of statistical theory is devoted to establish a weak invariance principle for an X_n-like process but where the corresponding results for the inverse process are not available. In all such cases the result serves to demonstrate the close links which exist between weak convergence of X_n and weak convergence of $\tilde{X}_n^{-1}$.

In more detail, the probabilistic and statistical significance of the present paper derives from the following contributions.

(i) We obtain a general invariance principle which has a very wide range of statistical applications and which unifies the theory of weak convergence of sequences of first passage processes.

(ii) We illustrate the versatility of our results (i) by giving versions of weak invariance principles for U-statistic quantile processes, perturbed quantiles, conditional quantile processes, empirical excursions above a given barrier, stopping rules in renewal theory and maximal processes and we stress that these problems have not been previously considered and that these results are new. Other cases, such as the length of the shorths (handled previously by Grübel (1988) by a different approach) may be treated in an identical manner (Section 3).

(iii) We apply the result described in (i) to a general class of integrated kernel quantiles. In the past decade there has been considerable interest in the study of smooth sample quantiles as competitors to classical quantiles in smooth statistical models (Falk, 1984, 1985; Yang, 1985; Veraverbeke, 1987; Gijbels and Veraverbeke, 1988; etc.). The rationale for considering such competitors is motivated in part by the possible lack of second order efficiency of classical sample quantiles when the random model follows smooth probability laws and also by the success achieved in bootstrap approximations where smooth quantiles yield improved speed of convergence. For a full account of these facts we refer to the monograph of Reiss (1989). We prove a general invariance principle which broadly extends all previously known results and provides a basic framework for treatment of the wide class of integrated kernel quantiles. In addition, for the nonclassical empirical d.f. which is of U-statistic structure we obtain strong consistency and an almost sure asymptotic representation for integrated kernel quantiles.

The paper is organized as follows. In Section 2 we present our main result on the weak convergence of sequences of first passage processes in a general framework. In Section 3 we discuss selected novel statistical applications together with previously known results. These examples are intended only as a guide and are not exhaustive. Finally, in Section 4 we concentrate on a further general application of the theory to the study of integrated kernel quantiles. In addition, we show that under some (reasonable) conditions, a U-statistics integrated kernel quantile (a) is strongly consistent; and (b) enjoys a strong Bahadur representation which allows us, for example, to obtain a law of the iterated logarithm, thus broadly extending results well known for classical sample quantiles.

2. Weak convergence

Let $C[\alpha,\beta]$ denote the space of continuous functions on $[\alpha,\beta]$. As a word of notational convention, if $\{U_n\}_{n\geqslant 1}$ is a sequence of random functions in $D_r[\alpha,\beta]$ (= the set of all right continuous functions on $[\alpha,\beta]$ with left limits) and U is a random element in $C[\alpha,\beta]$, we write $U_n \overset{P}{\Longrightarrow} U$ on $[\alpha,\beta]$ to express weak convergence in probability (uniform on $[\alpha,\beta]$) i.e., $\sup_{\alpha\leqslant x\leqslant\beta}|U_n(x)-U(x)|\overset{P}{\to}0$. A similar interpretation is adopted when U_n is in $D_\ell[\alpha,\beta]$.

Let $\{X_n(\cdot)\}_{n\geqslant 1}$ be a sequence of nonnegative stochastic processes on a common probability space with paths in $D_r(-\infty,\infty)$ and suppose that there are positive norming constants $\theta_n\to 0$ as $n\to\infty$, a centering nonnegative function $\mu(\cdot)\in C[a,b]$ and a continuous process $X(\cdot)$ such that

$$\frac{X_n-\mu}{\theta_n}\overset{P}{\Longrightarrow}X \quad \text{on } [a,b], \tag{2.1}$$

where $-\infty<a<b<\infty$ are given real numbers. The intermediate statement (2.1) was inspired by (1.1) and is considered in order to avoid inessential technical complications related to weak convergence considerations. In practice, if (2.1) is not directly available, one establishes the classical variant of (2.1)

$$\frac{X_n'-\mu}{\theta_n}\overset{d}{\longrightarrow}X' \tag{2.1*}$$

in the sense of weak convergence in distribution on $D_r[a,b]$ and then one uses the Skorokhod–Dudley construction to replace (2.1)* by (2.1) for the appropriate equivalent processes $X_n\overset{d}{=}X_n'$ and $X\overset{d}{=}X'$.

In order to state and prove our main result of this section we shall need the following assumptions:

(A) there exists a nonnegative function $\rho(\cdot)\in D_r(-\infty,\infty)$ such that

(i) $\tilde{X}_n(x)\equiv\rho(nx)X_n(x)$ has nondecreasing paths;

(ii) $\sup_{a\leqslant x\leqslant b}\left|\frac{\rho(nx)}{\rho(n)}-\ell(x)\right|=\mathrm{o}(\theta_n)$ as $n\to\infty$ for some nonnegative function $\ell(\cdot)\in C[a,b]$; and

(iii) $\psi(x)\equiv\ell(x)\mu(x)$ is strictly increasing, differentiable on (a,b) such that $g\equiv\psi'\circ\psi^{-1}$ is positive and continuous on $[c,d]=\psi[a,b]$.

The stage is now set to formulate the statement of our main result.

Theorem 2.1. *Suppose that* (2.1) *holds and* (*A*) *is satisfied. Then for any* $c < d' < d$

$$\frac{g\left[\tilde{X}_n^{-1}(\rho(n)\cdot) - \psi^{-1}\right]}{\theta_n} \xrightarrow{P} -\ell(\psi^{-1})X(\psi^{-1}) \quad \text{on} \quad [c, d']. \tag{2.2}$$

Remark 1. It should be observed that from (2.4) it follows $\sup_{c \leqslant t \leqslant d} |Y_n(t) - t| \to 0$ a.s. where $Y_n = \tilde{X}_n(\psi^{-1})/\rho(x)$. This in turn implies that for each $c < d' < d$, Y_n^{-1} is well defined on $[c, d']$ for all sufficiently large n (in this connection, we point out a small error in the statement of Exercise 1 as given by Shorack and Wellner, 1986, p. 659). Therefore, $\tilde{X}_n^{-1}(\rho(n)\cdot)$ is well defined as a random element of $D_\ell[c, d']$, the space of all left continuous functions on $[c, d']$ with right limits.

2. If $X_n(\cdot)$ is monotone for each $n \geqslant 1$, we may specialize to the case $\rho(x) \equiv \ell(x) \equiv 1$. This is the situation in Applications 1–5 of Section 3 drawn mainly from the theory of quantile processes. Observe that conditions (A)(i) and (ii) are fulfilled in this case. It should be emphasized however that the norming function ρ is not limited to the choice of a constant and this is illustrated in Applications 6–8 of Section 3 which provide examples in support of the full force of Theorem 2.1.

Proof. We begin by first noting that in (2.1) and (2.2) we can replace convergence in probability by almost sure convergence (uniformly on $[a, b]$) because convergence in probability of a sequence of random variables $\{Y_n\}$ to Y takes place if and only if every subsequence contains a subsequence which converges to Y a.s.

Observe that

$$\sup_{a \leqslant x \leqslant b} \left| \frac{\frac{\rho(nx)}{\rho(n)} X_n(x) - \ell(x)\mu(x)}{\theta_n} - \ell(x)X(x) \right|$$
$$\leqslant \left(\sup_{a \leqslant x \leqslant b} |\ell(x)| \right) \left(\sup_{a \leqslant x \leqslant b} \left| \frac{X_n(x) - \mu(x)}{\theta_n} - X(x) \right| \right)$$
$$+ \left(\sup_{a \leqslant x \leqslant b} |X_n(x)| \right) \left(\sup_{a \leqslant x \leqslant b} \left| \frac{\rho(nx)}{\rho(n)} - \ell(x) \right| \Big/ \theta_n \right). \tag{2.3}$$

We have

$$\sup_{a \leqslant x \leqslant b} |X_n(x)| \leqslant \theta_n \sup_{a \leqslant x \leqslant b} \left| \frac{X_n(x) - \mu(x)}{\theta_n} - X(x) \right| + \theta_n \sup_{a \leqslant x \leqslant b} |X(x)| + \sup_{a \leqslant x \leqslant b} |\mu(x)|$$

from which it follows (in view of (2.1) and (A)(ii)) that the second term on the right side of (2.3) converges to zero a.s. Then, on using a change of variable according to $x = \psi^{-1}(t)$, (2.1) and (2.3) together yield

$$\sup_{c \leqslant t \leqslant d} \left| \frac{\tilde{X}_n(\psi^{-1}(t))/\rho(n) - t}{\theta_n} - \ell(\psi^{-1}(t))X(\psi^{-1}(t)) \right| \longrightarrow 0 \text{ a.s.} \tag{2.4}$$

By (2.4) and Lemma 1 of Vervaat (1972) with obvious modifications to cover the present case, it follows that

$$\sup_{c\leqslant t\leqslant d'}\left|\frac{\left(\frac{\tilde{X}_n\circ\psi^{-1}}{\rho(n)}\right)^{-1}(t)-t}{\theta_n}+\ell(\psi^{-1}(t))X(\psi^{-1}(t))\right|\longrightarrow 0 \text{ a.s.} \tag{2.5}$$

Note that

$$\inf\left\{u:\frac{\tilde{X}_n(\psi^{-1}(u))}{\rho(n)}\geqslant t\right\}=\psi(\tilde{X}_n^{-1}(\rho(n)t))$$

and so, by (2.5)

$$\sup_{c\leqslant t\leqslant d'}\left|\frac{\psi(\tilde{X}_n^{-1}(\rho(n)t))-t}{\theta_n}+\ell(\psi^{-1}(t))X(\psi^{-1}(t))\right|\longrightarrow 0 \text{ a.s.} \tag{2.6}$$

Put

$$Z_n(t)=\frac{g(t)\left[\tilde{X}_n^{-1}(\rho(n)t)-\psi^{-1}(t)\right]}{\theta_n} \quad\text{and}\quad Z_n^*(t)=\frac{\psi(\tilde{X}_n^{-1}(\rho(n)t))-t}{\theta_n}.$$

Write

$$Z_n(t)=Z_n^*(t)W_n(t)g(t). \tag{2.7}$$

From (2.7) we get

$$\tilde{X}_n^{-1}(\rho(n)t)-\psi^{-1}(t)=[\psi(\tilde{X}_n^{-1}(\rho(n)t))-t]W_n(t)$$

and if we replace $\tilde{X}_n^{-1}(\rho(n)t)$ by $\psi^{-1}(\psi(\tilde{X}_n^{-1}(\rho(n)t)))$, then we can apply the one-term Taylor expansion to obtain

$$W_n(t)=\frac{\mathrm{d}}{\mathrm{d}t}\psi^{-1}(t)|_{t=t_n}, \tag{2.8}$$

where $t_n=t_n(t,\omega)$ is such that

$$\psi(\tilde{X}_n^{-1}(\rho(n)t))\wedge t<t_n<\psi(\tilde{X}_n^{-1}(\rho(n)t))\vee t.$$

Plugging (2.8) into (2.7) we find with $h=\psi'$

$$Z_n(t)-Z_n^*(t)=Z_n^*(t)\left\{\frac{h(\psi^{-1}(t))-h(\psi^{-1}(t_n))}{h(\psi^{-1}(t_n))}\right\}, \tag{2.9}$$

where

$$|t_n-t|\leqslant|\psi(\tilde{X}_n^{-1}(\rho(n)t))-t|. \tag{2.10}$$

Hence, by (2.9)

$$\sup_{c\leqslant t\leqslant d'}|Z_n(t)-Z_n^*(t)|\leqslant\left(\sup_{c\leqslant t\leqslant d'}|Z_n^*(t)|\right)\left(\sup_{c\leqslant t\leqslant d'}\left|\frac{h(\psi^{-1}(t))-h(\psi^{-1}(t_n))}{h(\psi^{-1}(t_n))}\right|\right). \tag{2.11}$$

We have

$$\sup_{c\leqslant t\leqslant d'} |Z_n^*(t)| \leqslant \sup_{c\leqslant t\leqslant d'} \left|Z_n^*(t) + \ell(\psi^{-1}(t))X(\psi^{-1}(t))\right| + \sup_{c\leqslant t\leqslant d'} \left|\ell(\psi^{-1}(t))X(\psi^{-1}(t))\right| \tag{2.12}$$

and on account of (2.6), the first term on the RHS of (2.12) converges to zero a.s. Furthermore, by writing $\psi(\tilde{X}_n^{-1}(\rho(n)t)) - t = \theta_n Z_n^*(t)$, we see from (2.10) that

$$\sup_{c\leqslant t\leqslant d'} |t_n(t) - t| \leqslant \theta_n \sup_{c\leqslant t\leqslant d'} |Z_n^*(t)| \longrightarrow 0 \text{ a.s.,} \tag{2.13}$$

where the a.s. convergence follows from (2.6), (2.12) and the fact that $\theta_n \to 0$ as $n \to \infty$.

Combining (2.13) with (A)(iii) which implies uniform continuity over $[c, d']$, we obtain

$$\sup_{c\leqslant t\leqslant d'} \left|\frac{g(t) - g(t_n(t))}{g(t_n(t))}\right| \longrightarrow 0 \text{ a.s.} \tag{2.14}$$

Then, (2.9), (2.12) and (2.14) together imply

$$\sup_{c\leqslant t\leqslant d'} |Z_n(t) - Z_n^*(t)| \longrightarrow 0 \text{ a.s.} \tag{2.15}$$

and the result follows from this last step (2.15) and (2.6).

3. Applications

In this section we show the usefulness of our approach by presenting selected applications for which the result developed in Theorem 2.1 may help to provide a probabilistic underpinning for inferential statistical procedures which rely on invariance principles.

Application 1 (*Weak convergence of U-statistic quantile processes*). Let $\xi_1, \xi_2, \ldots$ be i.i.d. random variables with d.f. F and let $m \geqslant 1$ be a fixed integer. Consider the empirical d.f. of U-statistic structure

$$H_n^{(m)}(x) = \binom{n}{m}^{-1} \#\{(i_1, \ldots, i_m) : 1 \leqslant i_1 < \ldots < i_m \leqslant n,\ H_m(\xi_{i_1}, \ldots, \xi_{i_m}) \leqslant x\},$$

where H_m is an m-variate symmetric real valued function. Serfling (1980, 1984) discovered several interesting applications of the corresponding empirical process

$$Z_n^{(m)}(x) = \sqrt{n}\left[H_n^{(m)}(x) - H^{(m)}(x)\right],$$

where $H^{(m)}(x) = P\{H_m(\xi_1, \ldots, \xi_m) \leqslant x\}$. The case $m = 1$ corresponds to the standard empirical process for which weak convergence results in weighted sup-norm metrics have been obtained by Chibisov (1964), O'Reilly (1974) and Shorack (1979) (see also Shorack and Wellner (1986) and the references therein). For $m \geqslant 2$, Silverman (1983) has applied the Komlós–Major–Tusnády construction for proving weak convergence of $Z_n^{(m)}$ in weighted supremum metrics on the space $D_r(-\infty,\infty)$ endowed with the Skorokhod topology. Accordingly, if

$$EH_m^2(\xi_1,\dots,\xi_m) < \infty \quad \text{and} \quad Var\{E\,H_m(\xi_1,\dots,\xi_m|\xi_1)\} > 0,$$

there exists a special construction of the random variables $H_m(\xi_{i_1},\dots,\xi_{i_m})$ and a normal process Z with mean 0 and

$$cov[Z(x),Z(y)] = cov\left[P(H_m(\xi_1,\dots,\xi_m) \leqslant x|\xi_1),\, P(H_m(\xi_1,\dots,\xi_m) \leqslant y|\xi_1)\right]$$

such that

$$\sup_{-\infty<x<\infty} |Z_n^{(m)}(x) - Z(x)| \longrightarrow 0 \text{ a.s.} \tag{3.1}$$

(Silverman, 1983; Shorack and Wellner, 1986; Akritas, 1986).

So, by (3.1) we see that the a.s. version of (2.1) holds with $X_n = H_n^{(m)}$, $\mu(x) = H^{(m)}(x)$ and $\theta_n = n^{-1/2}$. Thus, if $H^{(m)}$ has a density $h^{(m)}$ such that $g \equiv h^{(m)} \circ (H^{(m)})^{-1}$ is positive and continuous on $[c,d] \subset [0,1]$, by Theorem 2.1 a limit law ensues for the U-statistic quantile process

$$V_n(t) = \sqrt{n}\; g(t)\left[(H_n^{(m)})^{-1}(t) - (H^{(m)})^{-1}(t)\right]$$

and we have

$$\sup_{c\leqslant t\leqslant d'} |V_n(t) - V(t)| \longrightarrow 0 \text{ a.s.} \tag{3.2}$$

with $c < d' < d$, where $V(t) = -Z((H^{(m)})^{-1}(t))$.

This example is interesting, in part, because it complements previous work by Janssen, Serfling and Veraverbeke (Janssen et al., 1984). To illustrate this aspect, suppose that one is interested in estimating a functional $T(H^{(m)})$ by $\Gamma_n = T(H_n^{(m)})$ and the goal is to establish the asymptotic normality (a.n.) of Γ_n. Then Janssen et al. have shown that the verification of the conditions ensuring the a.n. result can be reduced to establishing that

$$\sup_{c\leqslant t\leqslant d} |R_n(t)| = o_P(n^{-\frac{1}{2}}) \tag{3.2*}$$

for appropriate $0 < c < d < 1$, where

$$R_n(t) = \left(H_n^{(m)}\right)^{-1}(t) - \left(H^{(m)}\right)^{-1}(t) - \left(t - H_n^{(m)}\left((H^{(m)})^{-1}(t)\right)\right)/g(t).$$

Now, from the above, we see that (3.2)* can be obtained in a natural way from (3.1) with a change of variable $x = (H^{(m)})^{-1}(t)$ and (3.2). The result (3.2) will be applied in Section 4 to obtain weak functional limit theorems for the integrated quantile estimator of U-statistic structure.

Application 2 (*Perturbed empirical quantiles*). A fundamental problem in nonparametric statistics is to directly estimate the d.f. F of a random sample $\xi_1,\dots,\xi_n$ by a random distribution F_n^*. Traditionally, F_n^* has been taken to be the empirical d.f. F_n

(cf. Application 1 with $m=1$). Very often, however, when data points reflect a possible smooth d.f., it is more natural and desirable to use a smooth perturbation of the e.d.f. as an estimator of F rather than the discontinuous random function F_n. A perturbed e.d.f. estimator $\hat{F}_n$ based on $\xi_1,\ldots,\xi_n$ is defined as $\hat{F}_n(x)=n^{-1}\sum_{i=1}^n G_n(x-\xi_i)$ where $\{G_n\}_{n\geqslant 1}$ is a sequence of smooth d.f.'s converging weakly to the unit mass at zero. Typical examples of interest include kernel type estimators (Nadaraya, 1964; Winter 1979). These estimators are known to have a better asymptotic performance than that of F_n in the sense of asymptotic relative deficiency (Reiss, 1981; Falk, 1983).

If F satisfies a Lipschitz condition and the sequence $\{G_n\}_{n\geqslant 1}$ fulfills

$$\sqrt{n}\int |x|\,\mathrm{d}G_n(x) \longrightarrow 0 \qquad \text{as } n\to\infty,$$

it is known (Nixdorf, 1985) that

$$\sqrt{n}(\hat{F}_n - F) \xrightarrow{d} X, \tag{3.3}$$

where $X(x)=W^{\circ}(F(x))$ with $W^{\circ}=$ a Brownian bridge in $C[0,1]$.

Thus, if F has a density f such that $g\equiv f\circ F^{-1}$ is positive and continuous on $[c,d]\subset[0,1]$, in view of Theorem 2.1 and the Skorohkod–Dudley theorem we get (for a special construction) that

$$\sqrt{n}\,g\left[\hat{F}_n^{-1} - F^{-1}\right] \overset{P}{\Longrightarrow} W^{\circ} \quad \text{on} \quad [c,d'],\ c<d'<d. \tag{3.4}$$

In a similar way we can approach the weak convergence of quantile processes corresponding to recursive perturbations of the e.d.f. as considered by Nixdorf.

Application 3 (*The length of the shorths*). In the present example we consider the length of the shortest interval which contains at least a fraction t of the sample $(0<t<1)$. More precisely, suppose $\xi_1,\xi_2,\ldots$ are i.i.d. with d.f. F and density f and consider the shortest t-part of the sample $\xi_1,\xi_2,\ldots$ i.e.

$$\hat{H}_n(t)=\inf\left\{\xi_{n,i+j}-\xi_{n,i} : 1\leqslant i\leqslant i+j\leqslant n,\ (j+1)/n\geqslant t\right\},$$

where $\xi_{n,1}\leqslant\xi_{n,2}\leqslant\ldots\leqslant\xi_{n,n}$ are the ordered statistics.

It is notable that $\hat{H}_n(t)$ may be regarded as a functional of the e.d.f. F_n via $\hat{H}_n(t)=\inf\{\lambda>0:\hat{G}_n(\lambda)\geqslant t\}$ where $\hat{G}_n(\lambda)=\sup\{F_n(x+\lambda)-F_n(x):-\infty<x<\infty\}$ denotes the empirical concentration function. In order to study the asymptotic behavior of $\hat{H}_n(t)$ as $n\to\infty$, let $0<t_0<t_1<1$ be given and assume the following regularity on the density f

$$\begin{aligned}&f>0,\ f \text{ is increasing on } (-\infty,0) \text{ and symmetric about } 0;\ f' \text{ exists and } f'\\ &\geqslant c_0 \text{ on } (a,b) \text{ where } c_0>0 \text{ and } -\infty<a<b<0; F(a)<\tfrac{1}{2}(1-t_1)\\ &\text{and } F(b)>\tfrac{1}{2}(1-t_0).\end{aligned} \tag{3.5}$$

By choosing $0<\lambda_0<\lambda_1$ such that $2F(a)<2F(-\frac{\lambda_1}{2})<1-t_1$, and $1-t_0<2F(-\frac{\lambda_0}{2})$ $<2F(b)$, if $G(\lambda)=\sup\{F(x+\lambda)-F(x);\ -\infty<x<\infty\}$ denotes the concentration function corresponding to F, it follows from Grübel (1988, Theorem 2) that

$$\sqrt{n}(\hat{G}_n - G) \xrightarrow{d} X \quad \text{on } D_{\ell}[\lambda_0,\lambda_1], \tag{3.6}$$

where X is a normal process with mean 0 and covariance $cov\{X(\lambda),X(\mu)\} = 3F(\frac{\lambda}{2}) + 3F(\frac{\mu}{2}) - 4F(\frac{\lambda}{2})F(\frac{\mu}{2}) + F(\frac{\lambda\wedge\mu}{2}) - F(\frac{\lambda\vee\mu}{2}) - 2$. Now, since $\hat{H}_n(t) = \hat{G}_n^{-1}(t)$, by (3.5), (3.6) and the Skorokhod–Dudley theorem we have on account of Theorem 2.1 that

$$\sqrt{n}(\hat{H}_n - H) \xrightarrow{d} Y \quad \text{on } D_\ell[t_0, t_1], \tag{3.7}$$

where Y is a normal process with mean zero and covariance

$$cov\{Y(t), Y(s)\} = (t + s - 2ts + t \wedge s - t \vee s) \Big/ \left[2g\left(\frac{1+t}{2}\right) g\left(\frac{1+s}{2}\right)\right],$$

where $g \equiv f \circ F^{-1}$. This result agrees with the main theorem of Grübel (1988, Theorem 3) who obtained it by using a different proof based on the concept of compact differentiation of statistical functionals.

Application 4 (*Conditional quantile processes*). Suppose that we have i.i.d. observations $(X_1, Y_1), \ldots, (X_n, Y_n)$ in $\mathbb{R}^2$ with d.f. $H(x,y) = P\{X_1 \leqslant x,\ Y_1 \leqslant y\}$. Recent developments in nonparametric regression provide a strong support to investigate the conditional d.f. $m(y|x_0) = P\{Y_1 \leqslant y|X_1 = x_0\}$ at a fixed point x_0. Letting $F_n(x)$ denote the empirical d.f. of $X_1, \ldots, X_n$, Stute (1986) proposed the nearest-neighbour type estimator

$$m_n(y|x_0) = \frac{\sum_{i=1}^n I\{Y_i \leqslant y\} K\left(\frac{F_n(x_0) - F_n(X_i)}{a_n}\right)}{\sum_{i=1}^n K\left(\frac{F_n(x_0) - F_n(X_i)}{a_n}\right)},$$

where K is a smooth probability kernel and $a_n > 0$ represents a sequence of bandwidths such that $a_n \to 0$ as $n \to \infty$. Under appropriate regularity conditions, he proved that the process $\sqrt{na_n}\,[m_n(\cdot|x_0) - \overline{m}_n(\cdot|x_0)]$ converges weakly to a time-transformed Brownian bridge, where

$$\overline{m}_n(y|x_0) = \frac{1}{a_n} \int I(u \leqslant y) K\left(\frac{F(x_0) - F(u)}{a_n}\right) \mathrm{d}_u H(x,u)$$

and $F(x) = P\{X_1 \leqslant x\}$. In addition, he considered conditional sample quantiles and obtained their asymptotic normality (Stute, 1986, Theorem 3). The problem of estimating conditional quantile functions has been addressed by several authors. Chaudhuri (1991) gives relevant references and comments on them in detail.

The main objective here is to gain theoretical insight into the asymptotic behavior of the conditional quantile process $q_n(u) = \sqrt{na_n}[m_n^{-1}(u|x_0) - m^{-1}(u|x_0)]$ where $m_n^{-1}(u|x_0) = \inf\{y : m_n(y|x_0) \geqslant u\}$, $0 < u < 1$, by proving that $q_n(\cdot)$ converges weakly to a Gaussian process. This result, which generalizes Theorem 3 of Stute (1986), is a start towards developing asymptotic methods for $m^{-1}(u|x_0)$ (u = fixed), $\sup_{u\in I} m^{-1}(u|x_0)$ and $\inf_{u\in I} m^{-1}(u|x_0)$ where $I = [\varepsilon, 1-\varepsilon]$, $\varepsilon > 0$. The following regularity conditions are assumed.

1. H has uniform marginals.

2. (i) $\sup_{|t-s|\leqslant\delta}|m(t|x)-m(s|x)| = o((\log\delta^{-1})^{-1})$ as $\delta\to 0$ uniformly in a neighborhood of x_0.
   (ii) for each y, $m(y|\cdot)$ is twice continuously differentiable in a neighborhood J of x_0 such that $\sup_{x\in J}\sup_y |m''(y|x)| < \infty$.
3. $K\geqslant 0$, $\int K(u)\mathrm{d}u = 1$, $\int uK(u)\mathrm{d}u = 0$ and K is twice continuously differentiable vanishing outside a finite interval.
4. $na_n^5 \longrightarrow 0$ as $n\to\infty$.
5. $m(\cdot|x_0)$ is differentiable such that $g\equiv\varphi\circ m^{-1}(\cdot|x_0)$ is positive and continuous on $[c,d]\subset[0,1]$ where $\varphi(u) = \frac{\partial}{\partial y}m(y|x_0)|_{y=u}$.

Under assumptions 1–5, we can apply Corollary 2 of Stute (1986) together with Theorem 2.1 to infer that

$$g(\cdot)q_n(\cdot) \xrightarrow{d} Y \quad \text{on } D_\ell[c,d'],\ c<d'<d, \tag{3.8}$$

where Y is a Gaussian process on $[0,1]$ with $cov\{Y(u),Y(v)\} = (u\wedge v - uv)\int K^2(t)\,\mathrm{d}t$.

Application 5 (*Residual empirical process*). Our setting will be that of an explosive autoregressive process given by

$$\xi_0 = 0, \qquad \xi_i = \rho\xi_{i-1}+\varepsilon_i, \quad |\rho|>1,\ i\geqslant 1, \tag{3.9}$$

where ε is a r.v. with d.f. F and the ε_i's are independent copies of ε. This model arises in time series analysis and has been discussed previously by many authors including Basawa and Scott (1983) and more recently by Koul and Levental (1989). The object is to estimate a p-quantile $F^{-1}(p)$ of the d.f. F on the basis of the observed $\xi_1,\ldots,\xi_n$.

First, one starts by selecting an estimator $\hat\rho$ of ρ based on $\xi_1,\ldots,\xi_n$ which satisfies $\rho^n(\hat\rho-\rho) = o_P(n^{1/2})$. (Clearly, the classical least square estimator of ρ is a popular candidate which meets this requirement.) Next, with the help of $\hat\rho$ one constructs the residual empirical process associated with the explosive autoregressive model (3.9)

$$X_n(x) = n^{-1}\#\{1\leqslant i\leqslant n, \xi_i - \hat\rho\,\xi_{i-1}\leqslant x\}, \quad -\infty<x<\infty. \tag{3.10}$$

In order to estimate $F^{-1}(p)$, one can schedule to use the inverse process $X_n^{-1}(t)$ evaluated at $t=p$. From Theorem 1 of Koul and Levental (1989), if $E\log^+(|\varepsilon|)<\infty$ and F has a uniformly bounded derivative $f>0$ a.e., it follows that $\sqrt{n}(X_n-F)$ converges in distribution to $B(F(\cdot))$ where B is the Brownian bridge on $[0,1]$ – and in this case our Theorem 2.1 applies ensuring B as the limiting process of $\sqrt{n}f(F^{-1})(X_n^{-1}-F^{-1})$. This invariance principle embodies the weak asymptotic behavior of the residual empirical quantile process and the asymptotic normality of $X_n^{-1}(p)$ follows immediately.

In the following examples, we discuss the implications of Theorem 2.1 in a variety of different settings by focussing on cases with nonconstant norming function ρ.

Application 6 (*An empirical excursion above a given boundary*). For a variety of reasons, it is sometimes of interest to describe the behavior of the first passage time $T_n(t) = \inf\{s>0 : F_n(s)\geqslant tg(s)\}$, $t>0$, at which an e.d.f. F_n passes (crosses) a certain boundary (here $g(\cdot)$ is nonincreasing with $\lim_{s\to\infty}g(s)<1$). Now, apart from exceptional cases, explicit formulae are not available and asymptotic approximations become important. Since the class of admissible barriers is rather large and in order to simplify the presentation, we consider first the case in which $g(s)=s^{-1/2}$ and $\xi_1,\ldots,\xi_n$ are uniform $(0,1)$ r.v.'s whose empirical d.f. is F_n:

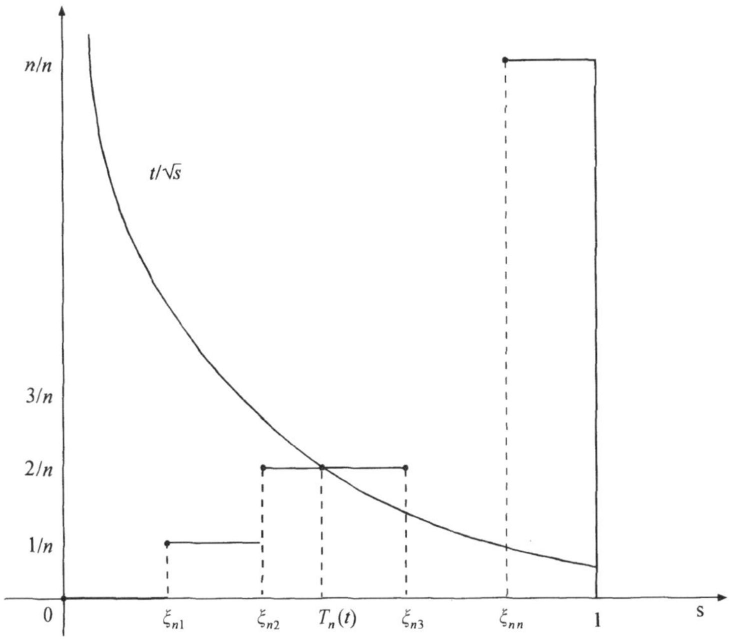

Fig. 1.

Here $\xi_{n1} \leqslant \xi_{n2} \leqslant \ldots \leqslant \xi_{nn}$ denote the uniform order statistics. In view of the well known fact

$$\sqrt{n}(F_n - I) \xrightarrow{d} W^0,$$

where I is the identity function on $[0,1]$ and W^0 is the Brownian bridge in $C[0,1]$, on choosing $\rho(x)=\ell(x)=\sqrt{x}$, it is immediately obvious that assumption (A) of Theorem 2.1 holds. We get from Theorem 2.1 that for all $0<c<d<1$

$$\sqrt{n}(T_n - \lambda) \overset{P}{\Longrightarrow} Z \quad \text{on } [c,d], \tag{3.11}$$

where $\lambda(t)=\sqrt[3]{t^2}$ and $Z=-\frac{2}{3}(W^0\circ\lambda)$.

In a more general setting, we can work with a smooth d.f. F and a positive norming function on $(0,\infty)$ that varies regularly at infinity with exponent $v>0$, i.e., $\rho(x)=x^v L(x)$ with $L(\cdot)$ increasing, slowly varying at infinity and such that for $x>0$

$$\sqrt{n}\left|\frac{L(nx)}{L(n)}-1\right| \longrightarrow 0 \quad \text{as } n\to\infty. \tag{3.12}$$

As a nonconstant example of L satisfying (3.12) consider $L(x)=x/(x+1)$. To verify (ii) of (A) we should recall the well-known fact that statement (3.12) for each $x>0$ entails uniform convergence on any $[a,b]\subset(0,\infty)$.

Proceeding as above we obtain from (2.2) a limiting result for $V_n(t) = \inf\{x : F_n(x) \geqslant \frac{L(n)}{L(nx)} tx^{-v}\}$.

Application 7 (*A stopping rule in renewal theory*). Let $\xi_1, \xi_2, \ldots$ be a sequence of non-negative r.v.'s, $S(t) = \sum_{k=1}^{[t]} \xi_k$ ($[\cdot]$= integer part) and assume that for some $-\infty < \theta < +\infty$ and $\sigma > 0$

$$\frac{S(n\cdot) - n\theta t}{\sigma\sqrt{n}} \xrightarrow{d} W, \tag{3.13}$$

where W is the Wiener process on $[0, \infty)$. If $E(\xi_1) = \theta > 0$ and $0 < \sigma^2 = Var(\xi_1) < \infty$, Donsker's theorem asserts that (3.13) holds if $\xi_1, \xi_2, \ldots$ are i.i.d. Suppose that our goal is to investigate the stopping rule defined for $c > 0$ by

$$\tau(c) = \inf\{t > 0 : S(t) \geqslant c/t\}.$$

This stopping variable may arise in certain problems in probability and statistics, e.g. sequential statistical procedures (see Gut (1974), p. 293), where the finiteness of moments of a similar stopping rule was analyzed). This is a setting where $\rho(x) = \ell(x) = 1$ and on account of Theorem 2.1 the following functional central limit theorem obtains:

$$\frac{\tau(n^2\cdot) - n\omega}{(2\theta)^{-1}\sigma\sqrt{n}} \xrightarrow{d} W', \tag{3.14}$$

where $\omega(t) = \sqrt{t/\theta}$ and $W' = -W \circ \omega$.

Application 8 (*Extreme value theory*). This example is interesting, in part, because it illustrates that a variant of our functional central limit theorem can hold with non-continuous limiting process. In the present setting, let $\{X_n(t), t \geqslant 0\}$ be a sequence of stochastic processes with paths lying in the subset of nondecreasing, nonnegative unbounded functions of $D[0, \infty)$. In view of the anticipated interest on examples from extreme value theory and the fact that in this case the limit processes are no longer continuous, the appropriate topology to use is Skorohod's (1956) M_1 topology on $\mathcal{D}[0, \infty)$ (Vervaat, 1972; Whitt, 1980).

Assume that there exists $\mu(\cdot)$ on $[0, \infty)$ strictly increasing, differentiable with $\mu(0) \geqslant 0$ such that for positive norming constants $\theta_n \to 0$ as $n \to \infty$

$$(X_n - \mu)\theta_n^{-1} \xrightarrow{d} X \, (M_1) \tag{3.15}$$

for some random element X of $\mathcal{D}[0, \infty)$. Further, suppose that $\rho(\cdot)$ is a nonnegative nondecreasing function, regularly varying at infinity with index $v > 0$ such that

$$\lim_{n\to\infty} \theta_n^{-1} \left(\frac{\rho(nx)}{\rho(x)} - x^v \right) = 0 \quad \text{for all } x > 0. \tag{3.16}$$

Then, since M_1 convergence in $\mathcal{D}[0, \infty)$ can be reduced to M_1 convergence in $\mathcal{D}[a, b]$ for all $0 \leqslant a < b < \infty$ which are continuity points of the limit, an inspection of the proof of Theorem 2.1 reveals that under (3.15) and (3.16) a similar technique is applicable with Vervaat's Lemma 1 replaced by Theorem 7.5 of Whitt (1980) and yields

$$\frac{g[\tilde{X}_n^{-1}(\rho(n)\cdot) - \psi^{-1}(\cdot)]}{\theta_n} \xrightarrow{d} Y(M_1), \tag{3.17}$$

where $\tilde{X}_n^{-1}$ is defined here by $\tilde{X}_n^{-1}(t) = \inf\{x : \tilde{X}_n(x) > t\}$, $g = \psi' \circ \psi^{-1}$, $\psi(x) = x^{\nu}\mu(x)$ and $Y(t) = -(\psi^{-1}(t))^{\nu} X(\psi^{-1}(t))$.

As an illustration of the potential applications of (3.17), consider a sequence $\{\xi_n\}_{n\geqslant 1}$ of i.i.d. r.v.'s with d.f. F for which, if $M_n = \max(\xi_1,\ldots,\xi_n)$, there exists $a_n > 0$ and b_n such that as $n \to \infty$

$$P\{a_n^{-1}(M_n - b_n) \leqslant x\} \longrightarrow G_\alpha(x) \quad \text{for } -\infty < x < \infty, \tag{3.18}$$

where

$$G_\alpha(x) = \begin{cases} \exp(-(-x)^\alpha), & x < 0 \\ 1, & x \geqslant 0 \end{cases} \qquad \text{for some } \alpha > 0.$$

Interesting examples of d.f.'s for which (3.18) holds include (i) polynomial growth at a finite endpoint d.f. $F(x) = 1 - C(x_0 - x)^\alpha$ if $x_0 - C^{1/\alpha} \leqslant x \leqslant x_0$ and (ii) truncated exponential d.f. $F(x) = C(1 - \mathrm{e}^{-x})$ if $0 \leqslant x \leqslant x_0$ (with $C = (1 - \mathrm{e}^{-x_0})^{-1}$).

It is well known (Resnick, 1987, p. 59) that if one sets $x_F = \sup\{x : F(x) < 1\}$, then $x_F < \infty$ and one can take $b_n = x_F$ and $a_n = x_F - \gamma_n$ with $\gamma_n = F^{-1}(1 - \frac{1}{n})$. Furthermore, from the theory of extremal processes (Lamperti, 1964; Resnick, 1987, Chapter 4) if $X_n(t) = M_{[nt]}$, $t > 0$ with $M_{[s]} = \xi_1$, if $0 < s < 1$, the appropriate limiting process to consider is the extremal process X generated by G_α and we infer

$$\frac{X_n - x_F}{x_F - \gamma_n} \xrightarrow{d} X(M_1). \tag{3.19}$$

Therefore, if ρ satisfies (3.16) with $\theta_n = x_F - \gamma_n$, a weak limit emerges for

$$W_n(t) = \inf\left\{x > 0 : M_{[nx]} > \frac{\rho(n)}{\rho(nx)}\, t\right\}.$$

In closing, we should remark that alternative examples from renewal theory with limiting stable processes could also be built upon (3.17).

Remark. Our treatment could include further applications of interest based on many highly interesting examples which are now available in the literature (see e.g., Shorack and Wellner, 1986). However, we leave these implicit since their treatment would follow a pattern similar to the examples already developed.

4. Integrated kernel quantiles

Consider the basic problem of estimating a functional $\mu(x) = \mu(x, F)$ of a d.f. F by an estimator $\mu_n(x)$ and assume that $\mu(\cdot)$ and $\mu_n(\cdot)$ are nondecreasing, nonnegative elements of $D_r(-\infty,\infty)$ such that $\lim_{x\to\infty} \mu_n(x) = \alpha > 0$. Without loss of generality we shall assume $\alpha = 1$. In many statistical models it is plausible that μ_n is consistent in the sense that $\sqrt{n}[\mu_n(x) - \mu(x)]$ converges weakly to a normal process over a finite interval $[a, b]$ and in this case, the equally important weak asymptotic behavior of the corresponding quantile process $\sqrt{n}[\mu_n^{-1}(t) - \mu^{-1}(t)]$ can be inferred via Theorem 2.1.

An interesting competitor to $\mu_n^{-1}(t)$ which may provide increased second order efficiency is the generalized integrated (empiric) kernel quantile estimator

$$Q_n(t) = \int_0^1 \mu_n^{-1}(s)\frac{1}{\lambda_n}k\left(\frac{t-s}{\lambda_n}\right) \mathrm{d}s, \tag{4.1}$$

where $\{\lambda_n\}$ is a sequence of positive constants (windows), $\lambda_n \to 0$ as $n \to \infty$ and k is a known kernel function ($k \geqslant 0$, $\int k = 1$). In the case when $\mu(x, F) = F(x)$ and $\mu_n(x) = F_n(x)=$ the e.d.f., this estimator has been proposed by Parzen (1979) and its asymptotic properties have been studied by Falk (1984, 1985) and Yang (1985). In the censored case, this kernel-type estimator has been studied by Padgett (1986), Lio, Padgett and Yu (1986) and Gijbels and Veraverbeke (1988). Moreover, Veraverbeke (1987) extended the study of $Q_n(t)$ to the case when $\mu_n =$ the empirical d.f. of U-statistic structure $H_n^{(m)}$ of Application 1 in Section 3.

The aim of this section is twofold. First, the purpose is to apply Theorem 2.1 to establish an invariance principle for the integrated kernel quantile estimator, thus broadly generalizing and extending the work of the authors mentioned above. We then take up the asymptotic analysis of $Q_n(t)$ a few steps further by proving: (a) a strong law of large numbers; and (b) a strong Bahadur representation theorem which has many asymptotic applications beyond asymptotic normality such as laws of the iterated logarithm and functional central limit results.

We begin by concentrating on developing general conditions that will ensure weak convergence of $Q_n(t)$, $0 < t < 1$. In the next theorem, we establish a form of the invariance principle which generalizes all previously known results.

Theorem 4.1. *Assume that $\mu(\cdot)$ has a derivative f which is positive and continuous on $[a, b]$. Let $\mu[a, b]=[c, d]$. Suppose that the kernel k has a finite support $\subset [-c_0, c_0]$, $\int xk(x)\,\mathrm{d}x = 0$ and $\int x^2k(x)\,\mathrm{d}x < \infty$. In addition, let the sequence of windows satisfy $n\lambda_n^4 \to 0$ as $n \to \infty$.*

If

$$\sup_{a \leqslant x \leqslant b} \left|\sqrt{n}\,[\mu_n(x) - \mu(x)] - X(x)\right| \xrightarrow{P} 0 \quad (or\ a.e.) \tag{4.2}$$

for some normal process X, then for each $\varepsilon > 0$

$$\sup_{c+\varepsilon \leqslant t \leqslant d-\varepsilon} \left|\sqrt{n}\,[Q_n(t) - \mu^{-1}(t)] - Y(t)\right| \xrightarrow{P} 0 \quad (or\ a.e.), \tag{4.3}$$

where $Y(t) \equiv X(\mu^{-1}(t))/f(\mu^{-1}(t))$.

Proof. By using a change of variable, since k has a finite support $\subset [-c_0, c_0]$, it follows that for all n sufficiently large we can write uniformly over $[c, d]$:

$$Q_n(t) = \int_{-c_0}^{c_0} \mu_n^{-1}(t - \lambda_n z)k(z)\mathrm{d}z. \tag{4.4}$$

Consider now the decomposition

$$\sqrt{n}\{Q_n(t)-\mu^{-1}(t)\}-Y(t)$$
$$=\int_{-c_0}^{c_0}\left\{\sqrt{n}[\mu_n^{-1}(t-\lambda_n z)-\mu^{-1}(t-\lambda_n z)]-Y(t-\lambda_n z)\right\}k(z)\,\mathrm{d}z$$
$$+\int_{-c_0}^{c_0}\sqrt{n}[\mu^{-1}(t-\lambda_n z)-\mu^{-1}(t)]k(z)\,\mathrm{d}z+\int_{-c_0}^{c_0}[Y(t-\lambda_n z)-Y(t)]k(z)\,\mathrm{d}z$$
$$\equiv \mathrm{I}+\mathrm{II}+\mathrm{III}. \tag{4.5}$$

Concerning $\mathrm{I}=\mathrm{I}(t)$, it is easily seen that

$$\sup_{c+\varepsilon\leqslant t\leqslant d-\varepsilon}|\mathrm{I}|\leqslant\sup_{c\leqslant u\leqslant d}\left|\sqrt{n}[\mu_n^{-1}(u)-\mu^{-1}(u)]-Y(u)\right|\xrightarrow{P}0\quad(\text{or a.s.}) \tag{4.6}$$

by (4.2) and Theorem 2.1.

As to the contribution of $\mathrm{III}=\mathrm{III}(t)$, if ω_Y denotes the modulus of continuity of Y, we have

$$\sup_{c+\varepsilon\leqslant t\leqslant d-\varepsilon}|\mathrm{III}|\leqslant\omega_Y(c_0\lambda_n)\longrightarrow 0\quad\text{a.s. as}\quad n\to\infty. \tag{4.7}$$

Finally, under the conditions of the theorem, it can be shown that

$$\sup_{c+\varepsilon\leqslant t\leqslant d-\varepsilon}|\mathrm{II}|\leqslant(\text{const.})\sqrt{n}\,\lambda_n^2\int_{-c_0}^{c_0}z^2k(z)\,\mathrm{d}z \tag{4.8}$$

and the RHS of (4.8) $\longrightarrow 0$ as $n\to\infty$ since $n\lambda_n^4\to 0$. We omit the routine details which are similar to those used by Veraverbeke (1987, bottom of p. 178 – top of p. 179). Combining (4.5)–(4.8) gives the desired result.

Theorem 4.1 immediately gives asymptotic normality results for $Q_n(t)$ for a fixed t. We now focus our attention on the kernel-type estimator $Q_n(t)$ (for a fixed $0<t<1$) corresponding to a U-statistic structure as previously defined in Application 1 of Section 3. In closing this section, our purpose is to establish strong consistency together with a strong Bahadur (a.s.) representation from which many useful applications may be considered. These results are of independent interest and are not dependent on the rest of the paper. We stress that these results are new even in the case that $H_n^{(m)}$ reduces to the usual e.d.f. F_n. In our asymptotic approach we build upon the insightful treatment of Choudhury and Serfling (1988) who introduced and studied the generalized order statistics $W_{n,1}\leqslant W_{n,2}\leqslant\ldots\leqslant W_{n,n_{(m)}}$ which are the ordered values of $H^{(m)}(\xi_{i_1},\ldots,\xi_{i_m})$ taken over the $n_{(m)}=n(n-1)\ldots(n-m+1)$ m-tuples $(i_1,\ldots,i_m)$ of distinct elements from $\{1,\ldots,n\}$.

Let $(H^{(m)})^{-1}(t)=\xi_t$ and assume that

$$\xi_t \text{ is the unique solution of } H^{(m)}(x-)\leqslant t\leqslant H^{(m)}(x). \tag{4.9}$$

To obtain that $Q_n(t) \to \xi_t$ a.s., we start by noting the identity of events $\{H_n^{(m)}(y) \geqslant t\} = \{W_{n,k_n} \leqslant y\}$ where

$$k_n = k_n(t) = \begin{cases} n_{(m)}t & \text{if } n_{(m)}t = \text{integer}, \\ [n_{(m)}t + 1] & \text{otherwise} \end{cases} \tag{4.10}$$

(here $[\cdot]$ = integer part).

Now, by arguments similar to those which led to the proof of Lemma 3.1 of Choudhury and Serfling (1988), it can be shown that under (4.9), if $\{\ell_n\}$ is a sequence of integers ($1 \leqslant \ell_n \leqslant n_{(m)}$) such that $\ell_n/n_{(m)} \to t$ as $n \to \infty$, then

$$W_{n,\ell_n} \longrightarrow \xi_t \quad \text{a.s.} \tag{4.11}$$

Next, observe that for all n sufficiently large

$$W_{n,k_n'} = \left(H_n^{(m)}\right)^{-1}(t - c_0\lambda_n) \leqslant Q_n(t) \leqslant W_{n,k_n''} = \left(H_n^{(m)}\right)^{-1}(t + c_0\lambda_n), \tag{4.12}$$

where $k_n' = k_n(t - c_0\lambda_n)$ and $k_n'' = k_n(t + c_0\lambda_n)$ defined by (4.10) clearly satisfy $k_n'/n_{(m)} \to t$, $k_n''/n_{(m)} \to t$ as $n \to \infty$. Consequently, (4.11) and (4.12) yield the desired strong consistency.

We summarize the previous development in the following result.

Theorem 4.2. *Let $Q_n(t)$ be given by (4.1) with $\mu_n = H_n^{(m)}$ as defined previously and assume that k is a (density) kernel with support $\subset [-c_0, c_0]$. Then, if (4.9) is satisfied*

$$Q_n(t) \longrightarrow \xi_t \quad a.s. \tag{4.13}$$

Remark. It is noteworthy that for the validity of (4.13), no speed of convergence to zero for the sequence of windows $\{\lambda_n\}$ was required.

Turning now to the Bahadur representation of $Q_n(t)$, we shall need to assume that $H^{(m)}$ is twice differentiable at ξ_t with $(H^{(m)})'(\xi_t) = h^{(m)}(\xi_t) > 0$. Let $\{k_n\}$ be a sequence of integers ($1 \leqslant k_n \leqslant n_{(m)}$) satisfying $k_n/n_{(m)} - t = o(\varepsilon_n)$ as $n \to \infty$, where $\varepsilon_n \to 0$ ($\varepsilon_n > 0$) such that for all n sufficiently large, $n\varepsilon_n^2(\log n)^{-1}(h^{(m)})^2(\xi_t)/m \geqslant c_1 > 1$. Then by Theorem 3.1 of Choudhury and Serfling (1988) we have the a.s. representation

$$W_{n,k_n} = \xi_t + \frac{1}{h^{(m)}(\xi_t)}\left[\frac{k_n}{n_{(m)}} - H_n^{(m)}(\xi_t)\right] + O(\varepsilon_n^2 \vee \varepsilon_n^{1/2} n^{-1/2}). \tag{4.14}$$

We are now able to prove the following result.

Theorem 4.3. *Suppose that the condition of Theorem 4.2 on k is satisfied and $H^{(m)}$ is twice differentiable with $h^{(m)}(\xi_t) > 0$. Then, if $\lambda_n = O\left(\sqrt{\frac{\varepsilon_n}{n}}\right)$ as $n \to \infty$ and $n\varepsilon_n^3 = O(1)$ as $n \to \infty$, we have a.s.*

$$Q_n(t) = \xi_t + \frac{1}{h^{(m)}(\xi_t)}\left[t - H_n^{(m)}(\xi_t)\right] + O\left(\sqrt{\frac{\varepsilon_n}{n}}\right), \quad n \to \infty. \tag{4.15}$$

Proof. With k_n' as defined above, it is clear that $k_n'/n_{(m)} - t = O(\lambda_n + n_{(m)}^{-1})$ as $n \to \infty$ and a similar statement holds for k_n''. Therefore, in view of (4.14), a "squeezing" argument applied to (4.12) immediately yields the validity of (4.15).

It should be noted that if $\varepsilon_n \sim (\text{const.})(n^{-1}\log n)^{1/2}$, we obtain a Bahadur representation with the remainder term $O(n^{-3/4}(\log n)^{1/4})$ a.s.. As observed earlier, the advantage of such a strong representation result is that it allows us to get many interesting asymptotic conclusions for $Q_n(t)$. We illustrate this point by stating a law of the iterated logarithm which can be derived by standard results on U-statistics (Serfling, 1980, Chapter 5) in combination with (4.15).

Theorem 4.4. *Suppose that the conditions of Theorem* 4.3 *are satisfied. Let* $\varphi_y(x_1,\ldots,x_m) = 1\{h^{(m)}(x_1,\ldots,x_m) \leqslant y\}$ *and*

$$\Phi_t(s) = Var\left\{E\left\{\sum\nolimits_A \varphi_{\xi_t}(\xi_{i_1},\ldots,\xi_{i_m})/m!\,|\xi_1 = s\right\}\right\},$$

where $\sum_A$ *denotes summation over all permutations of* $(1,\ldots,m)$*. If*

$$\sigma_t^2 = Var\{\Phi_t(\xi_1)\} > 0$$

we have a.s.

$$\limsup_{n\to\infty} \frac{\sqrt{n}(Q_n(t)-\xi_t)h^{(m)}(\xi_t)}{(2m^2\sigma_t^2\,\log\,\log n)^{\frac{1}{2}}} = 1. \tag{4.16}$$

Acknowledgements

Our thanks go to a referee and an associate editor for careful reading of the manuscript and a number of useful suggestions.

References

M.G. Akritas, Empirical processes associated with V-statistics and a class of estimators under random censoring, Ann. Statist. 14 (1986) 619–637.

I.V. Basawa and D.J. Scott, Asymptotic optimal inference for nonergodic models, Lecture Notes in Statistics, Vol. 17 (Springer, Verlag, New York, 1983).

P. Chaudhuri, Nonparametric estimates of regression quantiles and their local Bahadur representation, Ann. Statist. 19 (1991) 760–777.

D.M. Chibisov, Some theorems on the limiting behaviour of empirical distribution functions, Selected Transl. Math. Statist. Probab. 6 (1964) 147–156.

J. Choudhury and R.J. Serfling, Generalized order statistics, Bahadur representations, and sequential nonparametric fixed-width confidence intervals, J. Statist. Plann. Inference 19 (1988) 269–282.

M. Csörgö, Quantile processes with statistical applications, Regional Conf. Series on Applied Mathematics, Vol. 42, (SIAM, Philadelphia, PA, 1983).

M. Csörgö and P. Révész, Strong Approximations in Probability and Statistics (Academic Press, New York, 1981).

M. Falk, Relative efficiency and deficiency of kernel type estimators of smooth distribution functions, Statistica Neerlandica 37 (1983) 73–83.

M. Falk, Relative deficiency of kernel type estimators of quantiles, Ann. Statist. 12 (1984) 261–268.

M. Falk, (1985) Asymptotic normality of kernel quantile estimators, Ann. Statist. 13 (1985) 428–433.

I. Gijbels and N. Veraverbeke, Weak asymptotic representations for quantiles of the product–limit estimator, J. Statist. Plann. Inference 18 (1988) 161–176.

R. Grübel, The length of the shorth, Ann. Statist. 16 (1988) 619–628.

A. Gut, On the moments and limit distributions of some first passage times, Ann. Probab. 2 (1974) 277–308.

P. Janssen, R. Serfling and N. Veraverbeke, Asymptotic normality for a general class of statistical functions and applications to measures of spread, Ann. Statist. 12 (1984) 1369–1379.

H.L. Koul and S. Levental, Weak convergence of the residual empirical process in explosive autoregression, Ann. Statist. 17 (1989) 1784–1794.

J. Lamperti, On extreme order statistics, Ann. Math. Statist. 35 (1964) 1726–1737.

Y.L. Lio, W.J. Padgett and K.F. Yu, On the asymptotic properties of a kernel–type quantile estimator from censored samples, J. Statist. Plann. Inference 14 (1986) 169–177.

W.A. Massey, and W. Whitt, Unstable asymptotics for nonstationary queues, Math. Oper. Res. 19 (1994) 267–291.

E.A. Nadaraya, Some new estimates for distribution functions, Theory Probab. Appl. 9 (1964) 497–500.

R. Nixdorf, Central limit theorems in $C[0,1]$ for a class of estimators of a distribution function, Statistica Neerlandica 39 (1985) 251–260.

N.E. O'Reilly, On the weak convergence of empirical processes in sup-norm metrics, Ann. Probab. 2 (1974) 642–651.

W.J. Padgett, A kernel-type estimator of a quantile function from right-censored data, J. Amer. Statist. Assoc. 81 (1986) 215–222.

E. Parzen, Nonparametric statistical data modeling, J. Amer. Statist. Assoc. 74 (1979) 105–131.

R.D. Reiss, Nonparametric estimation of smooth distribution functions, Scand. J. Statist. 8 (1981) 116–119.

R.D. Reiss, Approximate Distributions of Order Statistics (Wiley, New York, 1989).

S.I. Resnick, Extreme Values, Regular Variation and Point Processes, Applied Probability, Vol. 4, (Springer, New York. 1987).

R.J. Serfling, Approximation Theorems of Mathematical Statistics (Wiley, New York, 1980).

R.J. Serfling, Generalized L-, M- and R-statistics. Ann. Statist. 12 (1984) 78–86.

G.R. Shorack, Weak convergence of empirical and quantile processes in sup-norm metrics via KMT constructions, Stochastic Process. Appl. 9 (1979) 95–98.

G.R. Shorack and J.A. Wellner, Empirical Processes with Applications to Statistics (Wiley, New York, 1986).

B.W. Silverman, Convergence of a class of empirical distribution functions of dependent random variables, Ann. Probab. 11 (1983) 745–751.

W. Stute, Conditional empirical processes, Ann. Statist. 14 (1986) 638–647.

N. Veraverbeke, A kernel-type estimator for generalized quantiles, Statist. Probab. Lett. 5 (1987) 175–180.

W. Vervaat, Functional central limit theorems for processes with positive drift and their inverses, Z. Wahrsch. verw. Geb. 23 (1972) 245–253.

W. Whitt, Some useful functions for functional limit theorems, Math. Oper. Res. 5 (1980) 67–85.

B.B. Winter, Convergence rate of perturbed empirical distribution functions, J. Appl. Probab. 16 (1979) 163–173.

S.S. Yang, A smooth nonparametric estimator of a quantile function, J. Amer. Statist. Assoc. 80 (1985) 1004–1011.

CONDITIONAL EMPIRICAL PROCESSES DEFINED BY NONSTATIONARY ABSOLUTELY REGULAR SEQUENCES

MICHEL HAREL and MADAN L. PURI*

IUFM du Limousin, U.M.R.L. 55830 C.N.R.S., Toulouse, France, and Indiana University, U.S.A. E-mail: harel@unilim.fr

K. I. Yoshihara (1990, *Comput. Math. Appl.* **19**, No. 1, 149–158) proved the weak invariance of the conditional nearest neighbor regression function estimator called the conditional empirical process based on φ-mixing observations. In this paper, we extend the result for nonstationary and absolutely regular random variables which have applications for Markov processes, for which the initial measure is not necessary, the invariant measure.

AMS 1991 subject classification: 62G05.

Key words and phrases: empirical distribution function; conditional empirical process; Skorohod topology; Gaussian process; absolute regularity.

1. INTRODUCTION

Let $\{\mathbf{Z}_i=(X_i, Y_i); i\geqslant 1\}$ be a sequence of random variables with continuous d.f.'s (distribution functions) $H_i(\mathbf{z})$, $i\geqslant 1$, $\mathbf{z}\in\mathbb{R}^2$, and we assume that $H_i(\mathbf{z})$ admits a strictly positive density $h_i(\mathbf{z})$, $i\geqslant 1$, and H_i has the two marginals F_i and G_i where the respective densities are noted by f_i and g_i.

Suppose that there exists a random vector (X, Y) in $\mathbb{R}^2$ with finite expectation $E\,|Y|$ and which admits H as distribution function, the regression function $m(x)=E(Y\mid X=x)$, $x\in\mathbb{R}$ of Y on X is a.s. in x uniquely defined in view of the equation

$$m(X)=E(Y\mid X). \tag{1.1}$$

Suppose also that H_i converges to the distribution function H which admits a density h and H has the two marginals F and G where the respective densities are noted by f and g. Consider the nearest neighborhood estimate

$$\tilde{m}_n(x_0)=(na_n)^{-1}\sum_{i=1}^{n} Y_i K\left(\frac{\tilde{F}_n(x_0)-\tilde{F}_n(X_i)}{a_n}\right) \tag{1.2}$$

* Research supported by the Office of Naval Research Contract N00014-91-J-1020.

of $m(x_0)$ where K is an appropriate kernel function, a_n is a sequence of bandwidths such that

$$a_n \to 0 \quad \text{as} \quad n \to \infty, \tag{1.3}$$

where $\tilde{F}_n$ is the empirical distribution function of $X_1, \ldots, X_n$.

Yoshihara (1988) under some conditions showed when the r.v.s. are φ mixing and stationary that for some σ_0^2

$$(na_n)^{1/2}\,[\tilde{m}_n(x_0) - m(x_0)]$$

converges to the normal law with mean 0 and variance σ_0^2 for almost all $x_0 \in \mathbb{R}$ such that $f(x_0) > 0$.

Let

$$m(y \mid x) = P(Y \leqslant y \mid X = x), \qquad (x, y) \in \mathbb{R}^2 \tag{1.4}$$

be the conditional distribution function and as an estimator of $m(\cdot \mid x)$ we consider the conditional empirical process

$$m_n(y \mid x_0) = (na_n)^{-1} \sum_{i=1}^{n} I_{[Y_i \leqslant y]} K\left(\frac{\tilde{F}_n(x_0) - \tilde{F}_n(X_i)}{a_n}\right), \qquad y \in \mathbb{R}, \tag{1.5}$$

where $\tilde{F}_n$ is the empirical distribution function of $X_1, X_2, \ldots, X_n$.

The main result of this paper is to show that under some conditions

$$(na_n)^{1/2}\,[m_n(G^{-1}(\cdot) \mid x_0) - m(G^{-1}(\cdot) \mid x_0)]$$

converges weakly in Skorohod topology to B_0 where B_0 is a certain Gaussian process (depending on x_0). Stute (1986) showed this result when the r.v.'s (random variables) are independent and identically distributed, then Yoshihara (1990) generalized it for φ-mixing and identically distributed r.v.'s. The generalization for absolutely regular and nonstationary r.v.'s give applications for the important case of Markov processes, for which the initial measure is not necessary, the invariant measure. Examples of models are given in Section 2.

For the particular case where $Y_i \equiv X_{i+1}$, the conditional empirical process $m_n(\ \mid\ )$ has the expression

$$m_n(y \mid x_0) = (na_n)^{-1} \sum_{i=1}^{n} I_{[X_{i+1} \leqslant y]} K\left(\frac{\tilde{F}_n(x_0) - \tilde{F}_n(X_i)}{a_n}\right). \tag{1.6}$$

Recall that the process satisfies the absolute regularity condition if

$$\max_{j \geqslant 1} E\{\sup_{A \in \sigma(\mathbf{Z}_i,\, i \geqslant j+m)} |P(A \mid \sigma(\mathbf{Z}_i, 1 \leqslant i \leqslant j)) - P(A)|\}$$
$$= \beta(m) \downarrow 0 \quad \text{as} \quad m \to \infty. \tag{1.7}$$

Here $\sigma(\mathbf{Z}_i, 1 \leqslant i \leqslant j)$ and $\sigma(\mathbf{Z}_i, i \geqslant j+m)$ are the σ-fields generated by $(\mathbf{Z}_1, ..., \mathbf{Z}_j)$ and $(\mathbf{Z}_{j+m}, \mathbf{Z}_{j+m+1}, ...)$, respectively. Also recall that $\{\mathbf{Z}_i\}$ satisfies the strong mixing condition if

$$\max_{j \geqslant 1}[\sup \{|P(A \cap B) - P(A)\, P(B)|; A \in \sigma(\mathbf{Z}_i, 1 \leqslant i \leqslant j),\ ]$$

$$B \in \sigma(\mathbf{Z}_i, i \geqslant j+m)\} = \alpha(m) \downarrow 0 \qquad \text{as} \quad m \to \infty$$

and $\{\mathbf{Z}_i\}$ satisfies the φ-mixing condition if

$$\max_{j \geqslant 1}[\sup \{|P(A \mid B) - P(A)|; B \in \sigma(\mathbf{Z}_i, 1 \leqslant i \leqslant j), A \in \sigma(\mathbf{Z}_i, i \geqslant j+m)\}]$$

$$= \varphi(m) \downarrow 0 \qquad \text{as} \quad m \to \infty.$$

Since $\alpha(m) \leqslant \beta(m) \leqslant \varphi(m)$, it follows that if $\{\mathbf{Z}_i\}$ is absolutely regular, then it is also strong mixing and if $\{\mathbf{Z}_i\}$ is φ-mixing, it is also absolutely regular.

2. CONVERGENCE OF THE CONDITIONAL EMPIRICAL PROCESS

Denote $\bar{F}_n$ and $\bar{G}_n$ by

$$\bar{F}_n = n^{-1} \sum_{i=1}^{n} F_i, \qquad \bar{G}_n = n^{-1} \sum_{i=1}^{n} G_i.$$

For any $i \in \mathbb{N}$, put

$$V_i(u, v) = H_i(\bar{F}_n^{-1}(u), \bar{G}_n^{-1}(v)).$$

Let us denote

$$\tilde{H}_n(x, y) = n^{-1} \sum_{i=1}^{n} I_{[X_i \leqslant x,\, Y_i \leqslant y]}$$

the empirical distribution function of $(X_1, Y_1), (X_2, Y_2), ..., (X_n, Y_n)$ and

$$\tilde{V}_n(u, v) = n^{-1} \sum_{i=1}^{n} I_{[\bar{F}_n(X_i) \leqslant u,\, \bar{G}_n(Y_i) \leqslant v]}$$

the empirical distribution of $(\bar{F}_n(X_1), \bar{G}_n(Y_1)), ..., (\bar{F}_n(X_n), \bar{G}_n(Y_n))$. We have

$$E(\tilde{V}_n(u, v)) = \bar{V}_n(u, v) = n^{-1} \sum_{i=1}^{n} V_i(u, v).$$

Denote also $\tilde{M}_n(u) = n^{-1} \sum_{i=1}^n I_{[\bar{F}_n(X_i) < u]}$ and $\bar{M}_n(u) = E(\tilde{M}_n(u))$, then we may write

$$m_n(y \mid x_0) = a_n \int I_{(-\infty, y]}(z)\, K\left(\frac{\tilde{F}_n(x_0) - \tilde{F}_n(x)}{a_n}\right) \tilde{H}_n(dx, dz)$$

$$= a_n^{-1} \int I_{[0, \bar{G}_n(y)]}(v)\, K\left(\frac{\tilde{M}_n(\bar{F}_n(x_0)) - \tilde{M}_n(u)}{a_n}\right) \tilde{V}_n(du, dv)$$

$$= \tilde{m}_n(\bar{G}_n(y) \mid \bar{F}_n(x_0)). \tag{2.1}$$

Let us denote

$$\tilde{m}(v \mid u_0) = m(G^{-1}(v) \mid F^{-1}(u_0)). \tag{2.2}$$

In the following, we suppose that the sequence $\{\mathbf{Z}_i = (X_i, Y_i), i \geqslant 1\}$ satisfies

$$\sup_{i \geqslant 1} E(|Y_i|^{5+\delta}) < \infty \qquad \text{for some} \quad \delta > 0 \tag{2.3}$$

and further we suppose also that

$$\sup_{|v-w| \leqslant \delta} |\tilde{m}(v \mid u) - \tilde{m}(w \mid u)| = o((\log \delta^{-1})^{-1}) \qquad \text{as} \quad \delta \to 0 \tag{2.4}$$

uniformly in a neighborhood of u_0.

For each v, $\tilde{m}(v \mid \cdot)$ (defined in (2.2)) is twice continuously differentiable in a neighborhood V of u_0 such that

$$\sup_{u \in V} \sup_{v} |\tilde{m}''(v \mid u)| < \infty. \tag{2.5}$$

K is a twice continuously differentiable kernel function on $\mathbb{R}$ such that

$$K(u) = 0 \quad \text{for } |u| \geqslant 1, \qquad \int K(u)\, du = 1 \quad \text{and} \quad \int uK(u)\, du = 0 \tag{2.6}$$

and

$$na_n^4 \to \infty, \qquad na_n^5 \to 0. \tag{2.7}$$

As usual, let $\mathscr{D}(0, 1)$ be the space of all right-continuous functions on $[0, 1]$ with left hand limits.

Consider a sequence $\{\mathbf{Z}_i\}_{i \geqslant 1}$ of real random vectors which is a Markov process with transition probability $P(\mathbf{z}; A)$ where $\mathbf{z} = (x, y) \in \mathbb{R}^2$ and $A \in \mathscr{B}$, $\mathscr{B}$ is the Borel σ-field of $\mathbb{R}^2$.

Recall that the Markov process is *geometrically ergodic* if it is ergodic and if there exists $0<\rho<1$ such that

$$\|P^n(\mathbf{z};\ \cdot\ )-\mu(\cdot)\|=O(\rho^n) \qquad \text{for all a.s.} \quad \mathbf{z}\in\mathbb{R}^2, \tag{2.8}$$

where μ is the invariant measure and P^n the n-step transition probability.

We say that the process $\{\mathbf{Z}_i\}_{i\geqslant 1}$ has ν for the initial measure if the law of probability of $\mathbf{Z}_1$ is defined by ν and for any $i>1$, the law of probability P_i of $\mathbf{Z}_i$ is defined by νP^{i-1}. For any probability measure ν and any transition probability Q we denote by $Q\otimes\nu$ the probability measure defined on $\mathbb{R}^4$ by

$$Q\otimes\nu(A\times B)=\int_B Q(\mathbf{z};A)\,\nu(d\mathbf{z}) \qquad \text{for any} \quad A\times B\in\mathscr{B}\times\mathscr{B}.$$

The Markov process is called *strongly aperiodic* if for any $\mathbf{z}\in\mathbb{R}$, the transition probability $P(\mathbf{z};\ \cdot\ )$ is equivalent to the Lebesgue measure. The Markov process is called *Harris recurrent* if there exists a σ-finite measure ν on $\mathbb{R}^2$ with $\nu(\mathbb{R}^2)>0$ such that $\nu(A)>0$ implies $P(\mathbf{z};\mathbf{Z}_i\in A i.o.)=1$ for all $\mathbf{z}\in\mathbb{R}^2$.

THEOREM 2.1. *Let $\{\mathbf{Z}_i\}_{i\geqslant 1}$ be a Markov process which is strongly aperiodic, Harris recurrent and geometrically ergodic. Suppose that the invariant measure μ has a continuous density h and the transition probability P has a continuous density $p(\cdot\,;\,\cdot)$ and furthermore suppose that the conditions* (2.3)–(2.5) *are satisfied. Then for any initial measure, and for almost all $0<u_0<1$, $(na_n)^{1/2}\{m_n(G^{-1}(v)\mid F^{-1}(u_0))-\tilde{m}(v\mid u_0)\}$ converges in law to a Gaussian process $B_0(u_0)$ as $n\to\infty$ with respect to the Skorohod topology on $\mathscr{D}(0,1)$ where B_0 is centered, has continuous sample paths vanishing at the lower boundary of* $[0,1]$ *and covariance*

$$\{\tilde{m}(v_1\wedge v_2\mid u_0)-\tilde{m}(v_1\mid u_0)\,m(v_2\mid u_0)\}\int K^2(u)\,du. \tag{2.9}$$

EXAMPLE 2.1. We consider an ARMA process

$$X_i=aX_{i-1}+b\varepsilon_i+\varepsilon_{i-1}, \qquad i\in\mathbb{N}^*, \tag{2.10}$$

where X_0 admits a strictly positive density, $\{\varepsilon_i\}$ is a sequence of i.i.d. real valued random variables with strictly density such that $E(\varepsilon_i)=0$, and a and b are real numbers such that $|a|<1$. Then the conditions of Theorem 2.1 are satisfied for the process defined in (2.10), because we have the particular case where $Y_i\equiv X_{i+1}$. The law of the process on which observations are taken is defined by the initial measure (i.e. the measure which defines the lax of X_0) and the transition probability. We can estimate the conditional distribution function by the conditional distribution process $m_n(|)$ for any initial measure of X_0 which admits a strictly positive density.

THEOREM 2.2. *Let $\{\mathbf{Z}_i\}_{i\geqslant 1}$ be a Markov process which is aperiodic and Doeblin recurrent. Suppose that the invariant measure μ has a continuous density function h and the transition probability P has a condition density $p(\ ;\ )$ and furthermore the conditions* (2.3)–(2.5) *are satisfied. Then for any initial measure, the conditions of Theorem* 2.1 *also hold for the conditional distribution process m_n.*

EXAMPLE 2.2. We consider the process

$$X_i = f(X_{i-1}) + \varepsilon_{i-1}, \qquad i \in \mathbb{N}^*, \tag{2.11}$$

where X_0 admits a strictly positive density and $\{\varepsilon_i\}_{i\geqslant 1}$ is a sequence of i.i.d. real random variables with strictly positive density such that $E(\varepsilon_i)=0$ and f is a bounded function. That is a particular case of Markov process satisfying the conditions of Theorem 2.2 where $Y_i \equiv X_{i+1}$.

Now more generally, consider the model

$$Y_i = \psi(X_i) + e_i, \qquad i \geqslant 1,$$

where X_i denotes a random variable of observed values, ψ is a continuous known function such that (2.3) is satisfied, e_i is a white noise of identically absolutely regular random variables with a geometrical rate and strictly positive density and Y_i is a predictor variable. For any initial measure of $\{(X_i, Y_i)\}_{i\geqslant 1}$, the conclusion of Theorem 2.1 holds for the conditional empirical process defined in (2.1) when the process $\{X_i\}$ is a process of the Example 2.1 or the Example 2.2; defined in (2.10) or (2.11).

In fact, the Theorems 2.1 and 2.2 will be deduced from the following theorem proved under a broader framework.

We suppose that the sequence $\{\mathbf{Z}_i = (X_i, Y_i), i \geqslant 1\}$ is absolutely regular with rates

$$\beta(m) = O(\rho^m), \qquad 0 < \rho < 1. \tag{2.12}$$

We say that a sequence of d.f.'s $\{H_\ell^*\}$ on $\mathbb{R}^4$ admits F_ℓ^*, G_ℓ^* and H as marginals on $\mathbb{R}^2$ and F and G on $\mathbb{R}$ if we have

$$\lim_{\substack{y\to+\infty\\ v\to+\infty}} H_\ell^*(x, y, u, v) = F_\ell^*(x, u)$$

$$\lim_{\substack{x\to+\infty\\ u\to+\infty}} H_\ell^*(x, y, u, v) = G_\ell^*(y, v)$$

$$\lim_{\substack{u \to +\infty \\ v \to +\infty}} H_\ell^*(x, y, u, v) = H(x, y)$$

$$\lim_{\substack{x \to +\infty \\ y \to +\infty}} H_\ell^*(x, y, u, v) = H(u, v)$$

and the marginals of H are F and G. Put $H_{i,j}$, the distribution function of $(\mathbf{Z}_i, \mathbf{Z}_j)$.

THEOREM 2.3. *Suppose that the sequence* $\{\mathbf{Z}_i\}$ *is absolutely regular with rates* (2.12) *and satisfying* (2.3). *Furthermore, assume that for any* $\ell > 1$, *there exists a continuous* d.f. H_ℓ^* *on* $\mathbb{R}^4$ *admitting a density* h_ℓ^* *with marginals* F_ℓ^*, G_ℓ^* *and* H *on* $\mathbb{R}^2$ *and* F *and* G *on* $\mathbb{R}$ *such that the density of* H *is bounded in a neighborhood of* $F^{-1}(u_0) \times \mathbb{R}$ *and*

$$\|H_{i,j} - H_{j-i}^*\| = O(\rho_0^i), \qquad 1 \leqslant i < j \leqslant n, \quad n \geqslant 1, \quad 0 < \rho_0 < 1 \tag{2.13}$$

for which there exists a sequence $\{\tilde{\mathbf{Z}}_i = (\tilde{X}_i, \tilde{Y}_i)\}$ *of stationary random variables such that* $\{\tilde{\mathbf{Z}}_i\}$ *is absolutely regular with rates* (2.12) *and* $(\tilde{\mathbf{Z}}_i, \tilde{\mathbf{Z}}_j)$ *has* H_{j-i}^* *as d.f.* $(i < j+1)$. *Suppose also that the conditions* (2.3)–(2.5) *are satisfied. Then for almost all* $0 < u_0 < 1$, *the conclusions of Theorem* 2.1 *hold for the conditional distribution process* m_n.

Remark 2.1. In Yoshihara (1988), the condition (2.7) was replaced by the weaker condition $na_n^3 \to \infty$ and $na_n^5 \to 0$, but we have to strengthen his condition to obtain the generalization to the absolutely regular and non-stationary case (see the Proof of Lemma 4.3). We denote

$$\sigma_0^2 = Var(Y \mid X = x_0) \int K^2(u)\, du, \tag{2.14}$$

where (X, Y) is a random vector which admits H as a distribution function.

The Proof of Theorem 2.1 needs the following propositions.

PROPOSITION 2.1. *Under the conditions of Theorem* 2.1, *then* $(na_n)^{1/2}$ $\{\tilde{m}_n(x_0) - m(x_0)\}$ *converges to the normal law with mean* 0 *and variance* σ_0^2 *for almost all* $x_0 \in \mathbb{R}$ *such that* $f(x_0) > 0$.

PROPOSITION 2.2. *Under the conditions of Theorem* 2.1, *for almost all* $0 < u_0 < 1$,

$$(na_n)^{1/2} \{\tilde{m}_n(v \mid u_0) - \tilde{m}(v \mid u_0)\}$$

converges in law to the Gaussian process $B_0(u_0)$ *as* $n \to \infty$ *with respect to the Skorohod topology where* $\tilde{m}_n$ *is defined in* (2.1).

3. PROOFS OF THEOREMS 2.1, 2.2, AND 2.3

3.1. *Proof of Theorem* 2.1

From Theorem 2.3, we have only to prove (2.12) and (2.13). First, we prove (2.12). From Davydov (1973), and the condition of strong aperiodicity, we have

$$\beta(m) = \sup_n \int P_n(d\mathbf{z}) \, \|P^m(\mathbf{z};\ \cdot) - P_{m+n}(\cdot)\|$$

$$\leqslant \sup_n \int P_n(d\mathbf{z}) \, \|P^m(\mathbf{z};\ \cdot) - \mu(\cdot)\| + \|P_{m+n}(\cdot) - \mu(\cdot)\|.$$

As the process is geometrically ergodic, we can find $0 < \rho < 1$ such that

$$\|P^m(\mathbf{z};\ \cdot) - \mu(\cdot)\| = O(\rho^m) \qquad \text{for all a.s.} \quad \mathbf{z} \in \mathbb{R}^2. \tag{3.1}$$

From Theorem 2.1 of Nummelin and Tuominen (1982), we deduce

$$\beta(m) = O(\rho^m) \tag{3.2}$$

which is (2.12).

Now, we prove (2.13). We have from (3.1)

$$\begin{aligned} &\|P^m \otimes P_n - P^m \otimes \mu\| \\ &= 2 \sup_{A \times B \in \mathscr{B} \times \mathscr{B}} \left| \int_B P^m(\mathbf{z}; A)\, P_n(d\mathbf{z}) - \int_B P^m(\mathbf{z}; A)\, \mu(d\mathbf{z}) \right| \\ &\leqslant 2 \sup_{A \times B \in \mathscr{B} \times \mathscr{B}} P^m(\mathbf{z}; A) \left| \int_B P_n(d\mathbf{z}) - \mu(d\mathbf{z}) \right| \\ &\leqslant 2\, \|P_n - \mu\| = O(\rho^n), \end{aligned}$$

that is,

$$\|P^m \otimes P_n - P^m \otimes \mu\| = O(\rho^n). \tag{3.3}$$

Thus the conclusions of Theorem 2.1 holds.

3.2. *Proof of Theorem* 2.2

From Theorem (4.1) of Davydov (1973) the process $\{X_i\}$ is geometrically φ mixing which implies geometrically absolute regularity. The proof is now similar to Theorem 2.1.

3.3. *Proof of Theorem* 2.3

By definition, we have

$$\begin{aligned} &m_n(G^{-1}(v) \mid F^{-1}(u_0)) - \tilde{m}(v \mid u_0) \\ &\quad = \tilde{m}_n(\bar{G}_n \circ G^{-1}(v) \mid \bar{F}_n \circ F^{-1}(u_0)) - \tilde{m}(v \mid u_0) \\ &\quad = \tilde{m}_n(\bar{G}_n \circ G^{-1}(v) \mid \bar{F}_n \circ F^{-1}(u_0)) - \tilde{m}_n(\bar{G}_n \circ G^{-1}(v) \mid u_0) \\ &\qquad + \tilde{m}_n(\bar{G}_n \circ G^{-1}(v) \mid u_0) - \tilde{m}(v \mid u_0). \end{aligned}$$

By Proposition 2.2, we know that $(na_n)^{1/2} \{\tilde{m}_n(v \mid u_0) - \tilde{m}(v \mid u_0)\}$ converges in law to $B_0(u_0)$. We deduce from Theorem 5.5 of Billingsley (1968) that

$$(na_n)^{1/2} \{\tilde{m}_n(\bar{G}_n \circ G^{-1}(v) \mid u_0) - \tilde{m}(v \mid u_0)\}$$

converges also in law to $B_0(u_0)$.

We can write

$$\begin{aligned} &\tilde{m}_n(\bar{G}_n \circ G^{-1}(v) \mid \bar{F}_n \circ F^{-1}(u_0)) - \tilde{m}(\bar{G}_n \circ G^{-1}(v) \mid u_0) \\ &\quad = a_n^{-1} \int I_{[O,\, \bar{G}_n \circ G^{-1}(v)]}(w) \\ &\qquad \times \left\{ K\left(\frac{\tilde{M}_n(\bar{F}_n \circ F^{-1}(u_0)) - \tilde{M}_n(\bar{F}_n \circ F^{-1}(u))}{a_n} \right) \right. \\ &\qquad \left. - K\left(\frac{\tilde{M}_n(u_0) - \tilde{M}_n(u)}{a_n} \right) \right\} \tilde{G}_n(du, dw) = I_n. \end{aligned}$$

From (2.13) and using the fact that K is twice continuously differentiable, we deduce that $I_n = o_p(1)$.

4. PROOF OF PROPOSITION 2.1

From now on, the letter C, with or without subscript, will denote some positive quantity.

Writing $K(\tilde{F}_n(x_0) - \tilde{F}_n(X_i)/a_n)$ as

$$K\{(\bar{F}_n(x_0) - \bar{F}_n(X_i) + \tilde{F}_n(x_0) - \tilde{F}_n(X_i) - \bar{F}_n(x_0) + \bar{F}_n(X_i))/a_n\}$$

and using Taylor's expansion, we get

$$\begin{aligned}\tilde{m}_n(x_0) &= (na_n)^{-1} \sum_{i=1}^{n} Y_i K\left(\frac{\bar{F}_n(x_0)-\bar{F}_n(X_i)}{a_n}\right)\\ &\quad + a_n^{-2}n^{-1} \sum_{i=1}^{n} Y_i\{\tilde{F}_n(x_0)-\tilde{F}_n(X_i)-\bar{F}_n(x_0)+\bar{F}_n(X_i)\}\\ &\quad \times K'((\bar{F}_n(x_0)-\bar{F}_n(X_i))/a_n)\\ &\quad + a_n^{-3}n^{-1} \sum_{i=1}^{n} Y_i\{\tilde{F}_n(x_0)-\tilde{F}_n(X_i)-\bar{F}_n(x_0)+\bar{F}_n(X_i)\}\ K''(\Delta_i)\\ &= I_1+I_2+I_3, \qquad (4.1)\end{aligned}$$

where Δ_i is some random number between $a_n^{-1}\{\tilde{F}_n(x_0)-\tilde{F}_n(X_i)\}$ and $a_n^{-1}\{\bar{F}_n(x_0)-\bar{F}_n(X_i)\}$.

We need the following lemmas.

LEMMA 4.1. *Under the conditions of Theorem* 2.3,

$$(na_n)^{1/2}\, I_3 \xrightarrow{P} 0 \qquad as \quad n\to\infty. \qquad (4.2)$$

Proof. Since K vanishes outside $(-1, 1)$, the above expansion of $m_n(x_0)$ holds true with integration restricted to those x for which $|\tilde{F}_n(x_0)-\tilde{F}_n(x)| < a_n$. Let $b_n=(na_n^3)^{1/3}$. Then, by Lemmas 6.7 and 6.8

$$\begin{aligned}&P[\sup_{x:\,|\bar{F}_n(x_0)-\bar{F}_n(x)|\leqslant C_1a_n} (na_n^{-1})^{1/2}\,|\tilde{F}_n(x_0)-\tilde{F}_n(x)\\ &\qquad -\bar{F}_n(x_0)+\bar{F}_n(x)|\geqslant b_n]\\ &\quad = P[\sup_{x:\,|\bar{F}_n(x_0)-\bar{F}_n(x)|\leqslant C_1a_n} |M_n^*(\bar{F}_n(x_0))-M_n^*(\bar{F}_n(x))|\geqslant b_na_n^{1/2}]\\ &\quad \leqslant Cb_n^{-4}=C(na_n^3)^{-2/3}, \qquad (4.3)\end{aligned}$$

where C is some constant >0.

On the other hand, as $\{Y_i\}$ is strong mixing with rates (2.12) and condition (2.3) is satisfied, we obtain

$$E\left\{n^{-1}\left(\sum_{i=1}^{n} Y_i-E(Y_i)\right)\right\}^2 \leqslant C_2n^{-1}, \qquad (4.4)$$

where C_2 is some constant >0.

From the Markov inequality, we deduce that for any $\varepsilon>0$

$$P\left[\left|n^{-1}\sum_{i=1}^{n} Y_i-E\left(n^{-1}\sum_{i=1}^{n} Y_i\right)\right|>\varepsilon\right]\leqslant \varepsilon^2C_2n^{-1} \qquad (4.5)$$

which implies that $n^{-1}\sum_{i=1}^{n} Y_i - E(n^{-1}\sum_{i=1}^{n} Y_i) \xrightarrow{p} 0$ and we have $E(n^{-1}\sum_{i=1}^{n} Y_i) \to E(Y) < \infty$. Thus

$$n^{-1}\sum_{i=1}^{n} Y_i \xrightarrow{p} E(Y) < \infty. \tag{4.6}$$

Thus, for an arbitrary number $\varepsilon > 0$, and any positive number η, we can find n sufficiently large such that

$$P(B_n(\eta)) \geqslant 1 - \varepsilon,$$

where

$$B_n(\eta) = \left\{\omega,\ E(Y) - \eta \leqslant n^{-1}\sum_{i=1}^{n} Y_i \leqslant E(Y) + \eta\right\}.$$

From (4.3) and the fact that K'' is bounded, upon observing that $na_n^3 \to \infty$ we can find an integer n sufficiently large such that on the set $B_n(\eta)$ for any $\delta > 0$

$$(na_n)^{1/2} I_3 < \delta. \tag{4.7}$$

As δ is arbitrary, (4.2) follows.

Denote $m_i(x) = E(Y_i \mid X_i = x)$ and $\bar{m}_n(x) = n^{-1}\sum_{i=1}^{n} m_i(x)$, $x \in \mathbb{R}$.

LEMMA 4.2. *Under the conditions of Theorem 2.3, $(na_n)^{1/2} I_2$ is asymptotically equivalent to*

$$-(a_n^{-1/2}\bar{m}_n(x_0)\, n^{-1/2})\left\{\left(\sum_{i=1}^{n} K\left(\frac{\bar{F}_n(x_0) - \bar{F}_n(X_i)}{a_n}\right)\right)\right.$$
$$\left. -\left(\sum_{i=1}^{n} E\left(K\left(\frac{\bar{F}_n(x_0) - \bar{F}_n(X_i)}{a_n}\right)\right)\right)\right\}. \tag{4.8}$$

Proof. Define $\{Z_n, n \geqslant 1\}$ by

$$Z_n = n^{-1} a^{-3/2} \sum_{i=1}^{n} (Y_i - m_i(X_i))(\tilde{W}_n(x_0)$$
$$- \tilde{W}_n(X_i))\, K'\left(\frac{\bar{F}_n(x_0) - \bar{F}_n(X_i)}{a_n}\right),$$

where

$$\tilde{W}_n(x) = n^{-1/2} \sum_{i=1}^{n} (I_{[X_i \leqslant x]} - F_i(x)), \qquad x \in \mathbb{R}. \tag{4.9}$$

Proceeding as in Lemma 4.2 of Yoshihara (1988), we prove that

$$Z_n \xrightarrow{P} 0 \qquad \text{as} \quad n \to \infty.$$

Now consider the function

$$k_n(x, y) = \bar{m}_n(x)\, K'\left(\frac{\bar{F}_n(x_0) - \bar{F}_n(x)}{a_n}\right)(I_{[y \leqslant x_0]} - I_{[y \leqslant x]})$$

and the random variable

$$T_n = \int k_n(x, y)\, \tilde{W}_n(dx)\, \tilde{W}_n(dy).$$

Then, using the method of the Proof of Lemma 2.2 in Harel and Puri (1989) and using (2.12), we obtain

$$E(T_n^2) = O(1) \qquad \text{as} \quad n \to \infty$$

which implies

$$a^{-3/2} \int k_n(x, y)\, \tilde{W}_n(dy) \{\tilde{F}_n(dx) - \bar{F}_n(dx)\} \xrightarrow{P} 0$$

since $na_n^3 \to \infty$.

Thus, we obtain that $(na_n)^{1/2} I_2$ is asymptotically equivalent to

$$a_n^{-3/2} \int \bar{m}_n(x)(\tilde{W}_n(x_0) - \tilde{W}_n(x))\, K'\left(\frac{\bar{F}_n(x_0) - \bar{F}_n(x)}{a_n}\right) \bar{F}_n(dx).$$

Now we prove that

$$I_4 = a_n^{-3/2} \int |\bar{m}_n(x) - \bar{m}_n(x_0)|\, |\tilde{W}_n(x_0) - \tilde{W}_n(x)|$$

$$\times \left| K'\left(\frac{\bar{F}_n(x_0) - \bar{F}_n(x)}{a_n}\right)\right| \bar{F}_n(dx) \xrightarrow{P} 0 \tag{4.10}$$

for a.s. $x_0 \in \mathbb{R}$.

From $\bar{F}_n^{-1}(\bar{F}_n(x_0)) = x_0$ for a.s. $x_0 \in \mathbb{R}$ and any $n \in \mathbb{N}^*$, we obtain

$$\lim_{n \to \infty} \int_{-1}^{1} |\bar{m}_n(\bar{F}_n^{-1}(s) - ua_n) - \bar{m}_n(\bar{F}_n^{-1}(s))| \, |K'(u)| \, du = 0$$

for a.s. $s \in \mathbb{R}$. Put

$$e_n(x_0) = \left(a_n^{-1} \int_0^1 |\bar{m}_n(\bar{F}_n^{-1}(u)) - \bar{m}_n(x_0)| \left| K'\left(\frac{\bar{F}_n(x_0) - u}{a_n}\right)\right| du \right)^{-1/6}.$$

Then, by Lemma 6.8, we have that for any $\varepsilon > 0$,

$$\begin{aligned} P(I_4 > \varepsilon) &\leqslant P\left(a_n^{-3/2} \sup_{|\bar{F}_n(x_0) - u| \leqslant a_n} |\tilde{W}_n(\bar{F}_n(x_0)) - \tilde{W}_n(u)| \right. \\ &\qquad \left. \times \int_0^1 |\bar{m}_n(\bar{F}_n^{-1}(u)) - \bar{m}_n(x_0)| \left| K'\left(\frac{\bar{F}_n(x_0) - u}{a_n}\right)\right| du > \varepsilon \right) \\ &\leqslant P(\sup_{|\bar{F}_n(x_0) - u| \leqslant a_n} |\tilde{W}_n(F(x_0)) - \tilde{W}_n(u)| \geqslant e_n(x_0)\, a_n^{1/2}) \\ &\leqslant C e_n^{-4}(x_0) = o(1). \end{aligned}$$

Hence (4.10) is proved and it follows that $(na_n)^{1/2} I_2$ is asymptotically equivalent to

$$\begin{aligned} &a_n^{-3/2} m(x_0) \int [\tilde{W}_n(x_0) - \tilde{W}_n(x)] \, K'\left(\frac{\bar{F}_n(x_0) - \bar{F}_n(x)}{a_n}\right) \bar{F}_n(dx) \\ &\quad = -a_n^{-3/2} m(x_0) \int [\tilde{W}_n(x)] \, K'\left(\frac{\bar{F}_n(x_0) - \bar{F}_n(x)}{a_n}\right) \bar{F}_n(dx) \\ &\quad = -a_n^{-1/2} m(x_0) \int K\left(\frac{\bar{F}_n(x_0) - \bar{F}_n(x)}{a_n}\right) \tilde{W}_n(dx) \end{aligned}$$

which leads to (4.8).

LEMMA 4.3. *Under the conditions of Theorem 2.3*

$$\begin{aligned} I_4 &= (a_n n)^{-1/2} \sum_{i=1}^{n} (Y_i - m_i(x_0)) \, K\left(\frac{\bar{F}_n(x_0) - \bar{F}_n(X_i)}{a_n}\right) \\ &\quad - (a_n n)^{-1/2} \sum_{i=1}^{n} E\left\{ (Y_i - m_i(x_0)) \, K\left(\frac{\bar{F}_n(x_0) - \bar{F}_n(X_i)}{a_n}\right) \right\} \end{aligned} \tag{4.11}$$

has asymptotically (as $n \to \infty$) the normal distribution with mean 0 and variance σ_0^2 (denoted as $\mathcal{N}(0, \sigma_0^2)$).

Proof. Let

$$\tilde{B}_{ni} = a_n^{-1/2}(\tilde{Y}_i - m(x_0))\, K\left(\frac{F(x_0) - F(\tilde{X}_i)}{a_n}\right) - a_n^{-1/2} E\left\{(\tilde{Y}_i - m(x_0))\, K\left(\frac{F(x_0) - F(\tilde{X}_i)}{a_n}\right)\right\}, \qquad i = 1, ..., n,$$

$$B_{ni}^* = a_n^{-1/2}(Y_i - m_i(x_0))\, K\left(\frac{\bar{F}_n(x_0) - \bar{F}_n(X_i)}{a_n}\right) - a_n^{-1/2} E\left\{(Y_i - m_i(x_0))\, K\left(\frac{\bar{F}_n(x_0) - \bar{F}_n(X_i)}{a_n}\right)\right\}, \qquad i = 1, ..., n,$$

and

$$B_{ni} = a_n^{-1/2}(Y_i - m(x_0))\, K\left(\frac{F(x_0) - F(X_i)}{a_n}\right) - a_n^{-1/2} E\left\{(Y_i - m(x_0))\, K\left(\frac{F(x_0) - F(X_i)}{a_n}\right)\right\}, \qquad i = 1, ..., n.$$

Then, we have

$$I_4 = n^{-1/2} \sum_{i=1}^{n} B_{ni}^* = n^{-1/2} \sum_{i=1}^{n} B_{ni} + n^{-1/2} \sum_{i=1}^{n} (B_{ni}^* - B_{ni}).$$

From (2.13), it follows that

$$\sup_{x \in \mathbb{R}} |F(x) - \bar{F}_n(x)| = O(n^{-1})$$

which along with (2.3) implies that

$$E\,|B_{ni}^* - B_{ni}| = O(a_n^{-3/2} n^{-1})$$

or

$$E\left| n^{-1/2} \sum_{i=1}^{n} (B_{ni} - B_{ni}^*) \right| = O(a_n^{-3/2} n^{-1/2}).$$

We deduce that

$$n^{-1/2} \sum_{i=1}^{n} (B_{ni} - B_{ni}^*) \xrightarrow{p} 0.$$

Now, we have to show that

$$n^{-1/2}\sum_{i=1}^{n} B_{ni} \qquad \text{converges in law to } \mathcal{N}(0, \sigma_0^2).$$

Let us now denote

$$\varphi(x, y) = (y - m(x_0))\, K\left(\frac{F(x_0) - F(x)}{a_n}\right) - \int (z - m(x_0))\, K\left(\frac{F(x_0) - F(u)}{a_n}\right) H(du, dz).$$

For any $M > 0$, denote also by

$$B_{ni}^M = a_n^{-1/2}(Y_i I_{[|Y_i| \leqslant M]} - m(x_0))\, K\left(\frac{F(x_0) - F(X_i)}{a_n}\right) - a_n^{-1/2} E\left\{(Y_i I_{[|Y_i| \leqslant M]} - m(x_0))\, K\left(\frac{F(x_0) - F(X_i)}{a_n}\right)\right\}$$

$$\bar{B}_{ni}^M = B_{ni} - B_{ni}^M$$

$$\varphi^M(x, y) = (I_{[|y| \leqslant M]}\, y - m(x_0))\, K\left(\frac{F(x_0) - F(x)}{a_n}\right) - \int (I_{[|z| \leqslant M]}\, z - m(x_0))\, K\left(\frac{F(x_0) - F(u)}{a_n}\right) H(du, dz)$$

$$\bar{\varphi}^M(x, y) = \varphi(x, y) - \varphi^M(x, y)$$

$$D_n^{M, \ell, p}(i) = 2a_n^{-1} \int \varphi_\ell(x, y)\, \varphi_p(u, z)\, H_i^*(dx, dy, du, dz), \qquad 0 \leqslant \ell, p \leqslant 1, \quad i \geqslant 1$$

$$D_n^{M, \ell, p}(0) = \int \varphi_\ell(x, y)\, \varphi_p(x, y)\, H(dx, dy), \qquad 0 \leqslant \ell, p \leqslant 1,$$

where $\varphi_0(x, y) = \varphi^M(x, y)$, $\varphi_1(x, y) = \bar{\varphi}^M(x, y)$

$$D_n(i) = \sum_{0 \leqslant \ell, p \leqslant 1} D_n^{M, \ell, p}(i), \qquad i \geqslant 0.$$

We then have

$$\left| n^{-1} \sum_{1 \leqslant i, j \leqslant n} E(B_{ni} B_{in}) - n^{-1} \sum_{i=0}^{n} (n-i)\, D_n(i) \right|$$

$$= \left| n^{-1} \sum_{1 \leqslant i, j \leqslant n} \{ E(B_{ni}^M) + E(\bar{B}_{ni}^M) \} \{ E(B_{nj}^M) + E(\bar{M}_{nj}^M) \} \right.$$

$$\left. - n^{-1} \sum_{0 \leqslant \ell, p \leqslant 1} \sum_{i=1}^{n} (n-i)\, D_n^{M, \ell, p}(i) \right| .$$

From condition (2.3), we have

$$\sup_{n \geqslant 1} \max_{1 \leqslant i \leqslant n} E(\bar{B}_{ni}^M)^2 = O(M^{-2\delta}) \qquad \text{and}$$

$$\int (\bar{\varphi}^M(x, y))^2\, H(dx, dy) = O(M^{-2\delta}). \tag{4.12}$$

We can write

$$\left| n^{-1} \sum_{1 \leqslant i < j \leqslant n} E(B_{ni}^M B_{nj}^M) - n^{-1} \sum_{i=1}^{n} (n-i)_n^{M, 0, 0}(i) \right|$$

$$\leqslant n^{-1} \left| \sum_{1 \leqslant j - i \leqslant k} E(B_{ni}^M B_{nj}^M) - \sum_{i=1}^{k} (n-i)\, D_n^{M, 0, 0}(i) \right|$$

$$+ n^{-1} \left| \sum_{k < j - i \leqslant n} E(B_{ni}^M B_{nj}^M) - \sum_{i=k+1}^{n} (n-i)\, D_n^{M, 0, 0}(i) \right|$$

$$= I_n^{(1)} + J_n^{(1)}.$$

From condition (2.13) and the fact that K' is bounded, if we take $M = a_n^{-1}$,

$$I_n^{(1)} = kO(a_n^{-3} n^{-1})$$

and from condition (2.12)

$$J_n^{(1)} = O(\rho_1^k), \qquad \text{where} \quad \rho_1 = \rho^{\delta/(2+\delta)}.$$

If we take $k = a_n^{-1}$, then

$$I_n^{(1)} + J_n^{(1)} = O(a_n^{-4} n^{-1}) + O(a_n^{-1} \rho_1^{a_n^{-1}}) = O(a_n^{-4} n^{-1}). \tag{4.13}$$

We deduce similarly that

$$\left| n^{-1} \sum_{i=1}^{n} E(B_{ni}^M)^2 - D_n^{M, 0, 0}(0) \right| = O(n^{-1} a_n^{-4}). \tag{4.14}$$

We now write

$$\left|n^{-1} \sum_{1 \leqslant i<j \leqslant n} E(\bar{B}_{ni}^M B_{nj}^M) - n^{-1} \sum_{i=1}^{n} (n-i)\, D_n^{M,1,0}(i)\right|$$
$$\leqslant \left|n^{-1} \sum_{1 \leqslant j-i \leqslant k} E(\bar{B}_{ni}^M B_{nj}^M) - \sum_{i=1}^{k} (n-i)\, D_n^{M,1,0}(i)\right|$$
$$+\left|n^{-1} \sum_{k<j-i \leqslant n} E(\bar{B}_{ni}^M B_{nj}^M) - \sum_{i=k+1}^{n} (n-i)\, D_n^{M,1,0}(i)\right|$$
$$= I_n^{(2)} + J_n^{(2)}. \tag{4.15}$$

From (4.12), we have

$$I_n^{(2)} = O(k a_n^{\delta-2}),$$

and from condition (2.12)

$$J_n^{(2)} = O(\rho_1^k).$$

If we take $k = a_n^{-1}$ and noting that $M = a_n^{-1}$, we have

$$I_n^{(2)} + J_n^{(2)} = O(a_n^{2\delta-4}) + O(a_n^{-1} \rho_1^{a_n^{-1}}) = O(a_n^{\delta-3}). \tag{4.16}$$

Similarly, we can show that

$$\left|n^{-1} \sum_{i=1}^{n} E(\bar{B}_{ni}^M B_{ni}^M) - D_n^{M,1,0}(0)\right| = O(a_n^{\delta-3}), \tag{4.17}$$

$$\left|n^{-1} \sum_{1 \leqslant i,\, j \leqslant n} E(B_{ni}^M \bar{B}_{nj}^M) - n^{-1} \sum_{i=0}^{n} (n-i)\, D_n^{M,0,1}(i)\right| = O(a_n^{\delta-3}), \tag{4.18}$$

and

$$\left|n^{-1} \sum_{1 \leqslant i,\, j \leqslant n} E(\bar{B}_{ni}^M \bar{B}_{nj}^M) - n^{-1} \sum_{i=0}^{n} (n-i)\, D_n^{M,1,1}(i)\right| = O(a_n^{2\delta-2}). \tag{4.19}$$

We deduce from (4.13), (4.14), (4.16)–(4.19) that

$$\left|n^{-1} \sum_{1 \leqslant i,\, j \leqslant n} E(B_{ni} B_{nj}) - n^{-1} \sum_{i=0}^{n} (n-i)\, D_n(i)\right| = O(a_n^{\delta-3}).$$

Hence

$$\lim_{n \to \infty} \left|(n)^{-1} \sum_{1 \leqslant i,\, j \leqslant n} E(B_{ni} B_{nj}) - (n)^{-1} \sum_{i=0}^{n-1} (n-i)\, D_n(i)\right| = 0 \tag{4.20}$$

which implies that

$$\lim_{n\to\infty}\left|E\left(n^{-1/2}\sum_{i=1}^{n}B_{ni}\right)^2-E\left(n^{-1/2}\sum_{i=1}^{n}\tilde{B}_{ni}\right)^2\right|=0.$$

Now, we prove that

$$\lim_{n\to\infty}E\left(n^{-1/2}\sum_{i=1}^{n}\tilde{B}_{ni}\right)^2=\sigma_0^2, \tag{4.21}$$

where σ_0^2 is defined by (2.14).

We have

$$\begin{aligned}E(\tilde{B}_{ni})^2 &= a_n^{-1}\int|y-m(x_0)|^2\,K^2\left(\frac{F(x_0)-F(x)}{a_n}\right)H(dx,dy)\\ &\quad -a_n^{-1}\left\{\int(y-m(x_0))\,K\left(\frac{F(x_0)-F(x)}{a_n}\right)H(dx,dy)\right\}^2\\ &= a_n^{-1}\int\psi(x)\,K^2\left(\frac{F(x_0)-F(x)}{a_n}\right)F(dx)\\ &\quad -a_n^{-1}\left\{\int_0^1\varphi(x)\,K\left(\frac{F(x_0)-F(x)}{a_n}\right)F(dx)\right\}^2\\ &= \int_{-1}^{+1}\psi(F^{-1}(F(x_0)-a_nu)\,K^2(u)\,du)\\ &\quad -a_n\left\{\int\varphi(F^{-1}(F(x_0)-a_nu)\,K(u)\,du)\right\}^2,\end{aligned} \tag{4.22}$$

where $\psi(x)=E\{|X_2-m(x_0)|^2\mid X_1=x\}$ and $\varphi(x)=E\{X_2-m(x_0)\mid X_1=x\}$.

From (4.22), we easily deduce that

$$\lim_{n\to\infty}n^{-1}\sum_{i=1}^{n}E(\tilde{B}_{ni})^2=\sigma_0^2. \tag{4.23}$$

Now, we can write

$$\begin{aligned}E\left(n^{-1/2}\sum_{i=1}^{n}\tilde{B}_{ni}\right)^2 &= n^{-1}\sum_{i=1}^{n}E(\tilde{B}_{ni})^2+n^{-1}\sum_{1\leqslant i\neq j\leqslant n}E(\tilde{B}_{ni}\tilde{B}_{nj})\\ &= n^{-1}\sum_{i=1}^{n}E(\tilde{B}_{ni})^2+2n^{-1}\sum_{\substack{1\leqslant i<j\leqslant n\\ j-i\leqslant k}}E(\tilde{B}_{ni}\tilde{B}_{nj})\\ &\quad +2n^{-1}\sum_{\substack{1\leqslant i<j\leqslant n\\ j-i>k}}E(\tilde{B}_{ni}\tilde{B}_{nj}).\end{aligned} \tag{4.24}$$

Then, we have

$$
\begin{aligned}
E(\tilde{B}_{ni}\tilde{B}_{nj}) &= a_n^{-1} \int \varphi(x, y)\, g(u, z)\, H_{j-i}^*(dx, dy, du, dz) \\
&\leqslant C a_n^{-1} \left\{ \int \varphi_{i,j}^*(x, u)\, K\left(\frac{F(x_0) - F(x)}{a_n}\right) \right. \\
&\qquad \left. \times K\left(\frac{F(x_0) - F(u)}{a_n}\right) F_{j-i}^*(dx, du) \right\} \\
&= C a_n \left\{ \int \varphi_{i,j}^*(F^{-1}(F(x_0) - a_n v), F^{-1}(F(x_0) - a_n w)) \right. \\
&\qquad \times \frac{f_{j-i}^*(F^{-1}(F(x_0) - a_n v), F^{-1}(F(x_0)) - a_n w)}{f \circ F^{-1}(F(x_0) - a_n v)\, f \circ F^{-1}(F(x_0) - a_n w)}\, K(v) \\
&\qquad \left. \times K(w)\, dv\, dw \right\} \\
&\leqslant C a_n. \qquad (4.25)
\end{aligned}
$$

This follows from (2.3) and the fact that f is continuous in a neighborhood of x_0 with $f(x_0) > 0$, where

$$
\varphi_{i,j}^*(x, u) = E\{(\tilde{Y}_j - m(x_0))(\tilde{Y}_i - m(x_0)) \mid \tilde{X}_i = x, \tilde{X}_j = u\}
$$

and f_{j-i}^* is the density of F_{j-i}^*.

For any $\delta \geqslant 0$,

$$
\begin{aligned}
E\, |\tilde{B}_{ni}|^{2+\delta} &\leqslant C a_n^{-(2+\delta)/2} \int |y - m(x_0)|^{2+\delta} \left| K\left(\frac{F(x_0) - F(x)}{a_n}\right) \right|^{2+\delta} \\
&\qquad \times H(dx, dy) \\
&= C a_n^{-(2+\delta)/2} \int h^*(x) \left| K\left(\frac{F(x_0) - F(x)}{a_n}\right) \right|^{2+\delta} F(dx) \\
&\leqslant C a_n^{-\delta/2} \int_{-1}^{+1} h^*(F^{-1}(F(x_0) - a_n u))\, |K(u)|^{2+\delta}\, du \\
&\leqslant C a_n^{-\delta/2}, \qquad (4.26)
\end{aligned}
$$

where $h^*(x) = E\{|\tilde{Y}_1 - m(x_0)|^{2+\delta} \mid \tilde{X}_1 = x\}$.

From (4.25) and (4.26), it follows that

$$E\left(n^{-1}\sum_{1\leqslant i\neq j\leqslant n}\tilde{B}_{ni}\tilde{B}_{nj}\right)$$
$$\leqslant 2Cka_n+2(Ca_n^{-\delta/2})^{2/(2+\delta)}\sum_{i=k+1}^{n}[\beta(i)]^{\delta/(2+\delta)}. \tag{4.27}$$

If we take $k=[a_n^{-1/2}]$, we get (4.21) from (2.12) and (4.23).

By using (4.26) and Lemma 6.9 which generalizes Lemma 3.4 of Yoshihara (1988) from the φ mixing case to the strong mixing case, we obtain the following inequality which generalizes Lemma 5.1 of Yoshihara (1988) to the strong mixing case.

$$E\left|\sum_{i=1}^{m}B_{ni}\right|^{2+\delta}\leqslant Cm^{(2+\delta)/2}\sup_{1\leqslant i\leqslant m}E|B_{ni}|^{2+\delta},\qquad m\leqslant n. \tag{4.28}$$

Now we can proceed as in Yoshihara (1988) (Proof of Theorem 2). Let $p=[n^{2/3}]$, $q=[n^{1/3}]$ and $k=[n/(p+q)]$. Put

$$\eta_j=\sum_{i=(j-1)(p+1)+1}^{(j-1)(p+q)+p}B_{ni},\qquad j=1,\ldots,k.$$
$$\theta_j=\sum_{i=(j-1)(p+q)+p+1}^{(j-1)(p+q)+p}B_{ni},\qquad j=1,\ldots,k.$$
$$\theta_{k+1}=\sum_{i=k(p+q)+1}^{n}B_{ni}.$$

Then, we have

$$n^{-1/2}\sum_{i=1}^{n}B_{ni}=n^{-1/2}\sum_{j=1}^{k}\eta_j+n^{-1/2}\sum_{j=1}^{k+1}\theta_j.$$

From (4.28), we deduce that

$$n^{-1/2}\sum_{j=1}^{k+1}\theta_j\xrightarrow{P}0.$$

To prove Lemma 4.3, it remains now to show that $n^{-1/2}\sum_{j=1}^{k}\eta_j$ converges in law to $\mathcal{N}(0,\sigma_0^2)$ random variable.

From Lemma 6.1 we obtain

$$E\left\{\exp\left(itn^{-1/2}\sum_{j=1}^{k}\eta_j\right)\right\}-\prod_{j=1}^{k}[E\{\exp(itn^{-1/2}\eta_j)\}]\leqslant Ck\beta(q),$$

where C is some constant >0. Hence it is enough to prove that

$$\prod_{j=1}^{k} [E\{\exp(itn^{-1/2}\eta_j)\}] \qquad \text{converges to } e^{-(t^2/2)\sigma_0^2}. \tag{4.29}$$

Using (4.28), we obtain

$$\begin{aligned} E\{\exp(itn^{-1/2}\eta_j)\} &= 1 - \frac{t^2}{2n} E(\eta_j)^2 + O\left(\frac{|t|^{2+\delta}}{n^{(2+\delta)/2}} E\,|\eta_j|^{2+\delta}\right) \\ &= 1 - \frac{t^2}{2n} E(\eta_j)^2 + o(|t|^{2+\delta} n^{-1/6(2+\delta)} a_n^{-\delta/2}). \end{aligned}$$

From thc fact that $na_n^4 \to \infty$ and

$$\lim_{n\to\infty} \frac{kE(\eta_j)^2}{n} = \sigma_0^2 \qquad \text{(from (4.21))}$$

we get (4.29). The proof follows.

Lemma 4.4. *Under the conditions of Theorem* 2.3.

$$(na_n^{-1})^{1/2}\left[\left\{n^{-1}\sum_{i=1}^{n}\int yK\left(\frac{\bar{F}_n(x_0)-\bar{F}_n(x)}{a_n}\right)F_{i,i+1}(dx,dy)\right\}-m(x_0)\right]$$
$$\to 0 \qquad as \quad n\to 0. \tag{4.30}$$

Proof. We have

$$\begin{aligned} D_n = {} & (na_n^{-1})^{1/2}\left[n^{-1}\sum_{i=1}^{n}\int yK\left(\frac{\bar{F}_n(x_0)-\bar{F}_n(x)}{a_n}\right)H_i(dx,dy)\right. \\ & \left. -n^{-1}\sum_{i=1}^{n}\int yK\left(\frac{F(x_0)-F(x)}{a_n}\right)H_i(dx,dy)\right] \\ & +(na_n^{-1})^{1/2}\left[n^{-1}\sum_{i=1}^{n}\int yK\left(\frac{F(x_0)-F(x)}{a_n}\right)H_i(dx,dy)\right. \\ & \left. -n^{-1}\sum_{i=1}^{n}\int yK\left(\frac{F(x_0)-F(x)}{a_n}\right)H(dx,dy)\right] \\ & +(na_n^{-1})^{1/2}\left[n^{-1}\sum_{i=1}^{n}\int yK\left(\frac{F(x_0)-F(x)}{a_n}\right)H(dx,dy)-m(x_0)\right] \\ = {} & H_n^{(1)}+H_n^{(2)}+H_n^{(3)}. \end{aligned} \tag{4.31}$$

From (2.3) and (2.13), it follows that

$$H_n^{(1)} = O(a_n^{-3/2} n^{-1/2}) \tag{4.32}$$

Denote

$$\ell(x, y) = yK\left(\frac{F(x_0) - F(x)}{a_n}\right),$$

$$\ell^{(M)}(x, y) = I_{[|y| \leqslant M]}\, yK\left(\frac{F(x_0) - F(x)}{a_n}\right),$$

and

$$\overline{\ell}^{(M)}(x, y) = \ell(x, y) - \ell^{(M)}(x, y) \qquad \text{for any} \quad M > 0.$$

We then have

$$H_n^{(2)} = (na_n)^{1/2}\Bigg[n^{-1}\sum_{i=1}^{n}\int \ell^{(M)}(x, y)(H_i(dx, dy) - H(dx, dy))$$

$$+ n^{-1}\sum_{i=1}^{n}\left(\int \overline{\ell}^{(M)}(x, y)\, H_i(dx, dy) - \int \overline{\ell}^{(M)}(x, y)\, H(dx, dy)\right)\Bigg].$$

From (2.3) and (2.13), if we take $M = n^{1/4}$, we get

$$H_n^{(2)} = O(n^{-1/4} a_n^{-1/2}) + O(n^{-(1/8)-(\delta/4)} a_n^{-1/2}). \tag{4.33}$$

From Stute (1984, Corollary, p. 919), it follows that

$$H_n^{(3)} \to 0 \qquad \text{as} \quad n \to \infty. \tag{4.34}$$

Equations (4.32)–(4.34) yield (4.30). The Proof of the Proposition 2.1 now follows from Lemmas 4.1 to 4.4.

5. PROOF OF PROPOSITION 2.2

Put

$$\beta_n(v) = \beta_n(v \mid u_0) = (na_n)^{1/2}\,\{\tilde{m}_n(v \mid u_0) - \tilde{m}(v \mid u_0)\}. \tag{5.1}$$

We shall prove Proposition 2.2 by showing the following facts

(i) the finite-dimensional distributions of β_n converge to those of B_0

(ii) $\{\beta_n : n \geqslant 1\}$ is uniformly C-tight, i.e., for each $\varepsilon > 0$ and every $\rho > 0$, there exists some $\delta > 0$ and n_0 such that for all $n \geqslant n_0$

$$P(\sup_{|v_1 - v_2| \leqslant \delta} |\beta_n(v_1) - \beta_n(v_2)| \geqslant \rho) \leqslant \varepsilon. \tag{5.2}$$

The assertion (i) easily follows from Proposition 2.1 and the Cramér-Wold device.

It remains to show (ii). Put

$$K_n(u) = K\left(\frac{\tilde{M}_n(u_0) - \tilde{M}_n(u)}{a_n}\right)$$

$$V(u, v) = H(F^{-1}(u), G^{-1}(v))$$

and

$$\tilde{m}_n^*(w \mid u_0) = a_n^{-1} \int I_{[0, w]}(v)\, K_n(u)\, G(du, dv).$$

Write

$$\begin{aligned} \beta_n(v \mid u_0) &= (na_n)^{1/2} \{\tilde{m}_n(v \mid u_0) - \tilde{m}_n^*(v \mid u_0)\} \\ &\quad + (na_n)^{1/2} \{\tilde{m}_n^*(v \mid u_0) - \tilde{m}(v \mid u_0)\} \\ &= \beta_{n1}(v \mid u_0) + \beta_{n2}(v \mid u_0). \end{aligned} \tag{5.3}$$

We show that both $\{\beta_{n1}\}$ and $\{\beta_{n2}\}$ are uniformly C-tight.

First, we note that upon integrating by parts, we have

$$\begin{aligned} \tilde{m}_n(v \mid u_0) - \tilde{m}_n^*(v \mid u_0) &= a_n^{-1}\{\tilde{V}_n(1, v) - V(1, v)\}\, K\left(\frac{\tilde{M}_n(u_0) - 1}{a_n}\right) \\ &\quad - a_n^{-1} \int \{\tilde{V}_n(u, v) - V(u, v)\}\, K_n(du). \end{aligned} \tag{5.4}$$

Since by Lemma 6.6 in the Appendix $\tilde{M}_n(u_0) \to u_0$ $(0 < u_0 < 1)$ with probability one, $a_n \to 0$ and K has finite support, the first summand is zero with probability one for all $n \geqslant n_0$, say, not depending on v. Similarly, for $n \geqslant n_1$

$$\begin{aligned} &\int \{\tilde{V}_n(u, v) - V(u, v)\}\, K_n(du) \\ &\quad = \int \{\tilde{V}_n(u, v) - \bar{V}_n(u, v) - \tilde{V}_n(u_0, v) + \bar{V}_n(u_0, v)\}\, K_n(du) \\ &\quad\quad + \int \{\bar{V}_n(u, v) - V(u, v) - \bar{V}_n(u_0, v) + V(u_0, v)\}\, K_n(du). \end{aligned} \tag{5.5}$$

Since by assumption $K=0$ outside $(-1, 1)$, the last integral remains unchanged when restricting the domain of integration to those u's for which

$$|\tilde{M}_n(u) - \tilde{M}_n(u_0)| \leqslant a_n.$$

Let $0 < \varepsilon < 1$ be given arbitrarily. Then by Lemma 6.7, we have, up to an event of probability less than or equal to ε, that

$$|u_0 - u| \leqslant c_1 a_n,$$

whenever $|\tilde{M}_n(u_0) - \tilde{M}(u)| \leqslant a_n$. Hence from (2.13) and the boundaries of the density of H, neglecting an event of probability $\leqslant \varepsilon$, we obtain for all large n that

$$\sup_{|v_1 - v_2| \leqslant \delta} |\beta_{n1}(v_1 \mid u_0) - \beta_{n1}(v_2 \mid u_0)| \leqslant a_n^{-1/2} \|K\| \, w_n^*(\delta) + o_p(1), \quad (5.6)$$

where $\|K\|$ denotes the total variation of K and w_n^* is defined in Lemma 6.5. Consequently, the uniform C-tightness of $\{\beta_{n1}, n \geqslant 1\}$ follows from (6.11)(b) of Lemma 6.5 and (5.6).

As to β_{2n}, using Taylor's expansion, we write

$$\begin{aligned}
\tilde{m}_n^*(w \mid u_0) &= a_n^{-1} \int I_{[0,w]}(v)\, K\left(\frac{u_0 - u}{a_n}\right) V(du, dv) \\
&\quad + a_n^{-2} \int I_{[0,w]}(v) \{\tilde{M}_n(u_0) - \tilde{M}_n(u) - u_0 + u\} \\
&\quad \times K'\left(\frac{u_0 - u}{a_n}\right) V(du, dv) \\
&\quad + \frac{1}{2} a_n^{-3} \int I_{[0,w]}(v) \{\tilde{M}_n(u_0) - \tilde{M}_n(u) - u_0 + u\} \\
&\quad \times K''(\Delta)\, V(du, dv) \\
&= a_n^{-1} \tilde{m}(w \mid u_0) + I_2(w, n) + I_3(w, n), \qquad (5.7)
\end{aligned}$$

where Δ is some constant between $a_n^{-1}\{\tilde{M}_n(u_0) - \tilde{M}_n(u)\}$ and $a_n^{-1}\{u_0 - u\}$.

By Lemma 4.1, we get

$$(na_n)^{1/2} I_3(w, n) \xrightarrow{P} 0. \quad (5.8)$$

Hence, to prove the assertion (ii) it remains to show that $\{(na_n)^{1/2} I_2(, n)\, n \geqslant 1\}$ is uniformly C-tight. We have

$$
\begin{aligned}
&(na_n)^{1/2} I_2(w, n)\\
&\quad = a_n^{-3/2} \int \tilde{m}(w \mid u)\{U_n(u_0) - U_n(u)\}\, K'\left(\frac{u_0 - u}{a_n}\right) du\\
&\quad = a_n^{-3/2} \int \{\tilde{m}(w \mid u) - \tilde{m}(w \mid u_0)\}\{U_n(u_0) - U_n(u)\}\, K'\left(\frac{u_0 - u}{a_n}\right) du\\
&\qquad + a_n^{-3/2} \tilde{m}(w \mid u_0) \int \{U_n(u_0) - U_n(u)\}\, K'\left(\frac{u_0 - u}{a_n}\right) du, \qquad (5.9)
\end{aligned}
$$

where $U_n(u) = n^{1/2} \tilde{M}_n(u)$.

It follows from (4.10) that the first summand converges to zero in probability uniformly in y whenever $\tilde{m}(\cdot \mid u)$ is equicontinuous in a neighborhood of u_0.

Finally for large n

$$
\begin{aligned}
&a_n^{-3/2} \tilde{m}(w \mid u_0) \int \{U_n(u_0) - U_n(u)\}\, K'\left(\frac{u_0 - u}{a_n}\right) du\\
&\quad = -a_n^{-1/2} \tilde{m}(w \mid u_0) \int K'\left(\frac{u_0 - u}{a_n}\right) dU_n(u)\\
&\quad = -\tilde{m}(w \mid u_0)(a_n n)^{-1/2}\\
&\qquad \times \left\{\sum_{i=1}^{n} K\left(\frac{u_0 - \bar{F}_n(x_i)}{a_n}\right) - E\left(K\left(\frac{u_0 - \bar{F}_n(X_i)}{a_n}\right)\right)\right\}. \qquad (5.10)
\end{aligned}
$$

From Proposition 2.1 we deduce that

$$
(a_n n)^{-1/2} \left\{\sum_{i=1}^{n} K\left(\frac{u_0 - \bar{F}_n(X_i)}{a_n}\right) - E\left(K\left(\frac{u_0 - \bar{F}_n(X_i)}{a_n}\right)\right)\right\}
$$

converges in law to the normal distribution with mean 0 and covariance

$$
u_0 \int K^2(u)\, du
$$

and so is stochastically bounded. Now C-tightness of β_{n2} follows from this fact and uniform continuity of $\tilde{m}(\cdot \mid u_0)$.

Proposition 2.2 is proved.

6. APPENDIX

Let $\{X_{ni}, 1 \leqslant i \leqslant n, n \geqslant 1\}$ be a nonstationary sequence of r.v.'s. Let $p \geqslant 1$ and $1 \leqslant i_1 < i_2 < \cdots < i_p \leqslant n$ be arbitrary integers. For any $j (1 \leqslant j \leqslant p-1)$, $P_{j,n}^{i_1, \ldots, i_p}$ is the probability measure defined by

$$P_{j,n}^{i_1, \ldots, i_p}(A^{(j)} \times B^{(p-j)})$$
$$= P[(X_{ni_1}, \ldots, X_{ni_j}) \in A^{(j)}] \times P[(X_{ni_{j+1}}, \ldots, X_{ni_p}) \in B^{(p-j)}]$$

and $P_{o,n}^{i_1, \ldots, i_p}$ is the probability measure defined by

$$P_{o,n}^{i_1, \ldots, i_p}(A^{(p)}) = P[X_{ni_1}, \ldots, x_{ni_p} \in A^{(p)}]$$

for any $A^{(j)} \in \sigma(X_{ni_1}, \ldots, X_{ni_j})$ $(1 \leqslant j \leqslant p)$ and any $B^{(p-j)} \in \sigma(X_{ni_{j+1}}, \ldots, X_{ni_p})$ $(1 \leqslant j \leqslant p-1)$.

We state the following two lemmas, the proof of the first one follows essentially by proceeding as in Lemma 1 of Yoshihara (1976) and the second lemma is from Doukhan and Portal (1987, Proposition 2.8).

LEMMA 6.1. *For every $p \geqslant 1$ and $(i_1, \ldots, i_p)$ such that $i_1 < i_2 < \cdots < i_p$ and any j $(0 \leqslant j \leqslant p-1)$, let $h(x_1, \ldots, x_p)$ be a Borel function such that*

$$\int_{\mathbb{R}^p} |h(x_1, \ldots, x_p)|^{1+\delta} \, dP_{j,n}^{(i_1, \ldots, i_p)} \leqslant M$$

for some $\delta > 0$. Then

$$\left| \int_{\mathbb{R}^p} h(x_1, \ldots, x_p) \, dP_{o,n}^{(i_1, \ldots, i_p)} - \int_{\mathbb{R}^p} h(x_1, \ldots, x_p) \, dP_{j,n}^{(i_1, \ldots, i_p)} \right|$$
$$\leqslant 4M^{1/(2+\delta)} \beta^{\delta/(1+\delta)}(i_{j+1} - i_j). \tag{6.1}$$

As a special case, if $h(x_1, \ldots, x_p)$ is bounded, say, $|h(x_1, \ldots, x_p)| \leqslant M$, then we can replace the right side of (6.1) *by $2M\beta(i_{j+1} - i_j)$.*

LEMMA 6.2. *Let $\{X_{ni}\}$ be strong mixing. Let Z_1 be $\sigma(X_{ni}, 1 \leqslant i \leqslant j)$-measurable $(1 \leqslant j \leqslant n)$ and Z_2 be $\sigma(X_{ni}, i \geqslant j+m)$-measurable. If $E(|Z_1|^p) < \infty$, $E(|Z_2|^q) < \infty$ and $r^{-1} + p^{-1} + q^{-1} = 1 (r, p, q > 0)$ then*

$$|E(Z_1 Z_2) - E(Z_1) \, E(Z_2)| \leqslant 12(\alpha(m))^{1/r} \{E|Z_1|^p\}^{1/p} \{E|Z_2|^q\}^{1/q}. \tag{6.2}$$

Let $W_n = n^{1/2}(\tilde{V}_n - \bar{V}_n)$. For a rectangle I, put

$$W_n(I) = n^{1/2}(\tilde{\mu}_n(I) - \mu_n(I)),$$

where $\tilde{\mu}_n$ and μ_n are the probability measures pertaining to $\tilde{V}_n$ and $\bar{V}_n$.

Under the conditions of Theorem 2.3, we have

LEMMA 6.3. *Under the conditions of Theorem 2.3, let* $B=[a_1, b_1]\times[a_2, b_2]$ *and* $B'=[c_1, d_1]\times[c_2, d_2]$ *be two rectangles included in* $[0,1]^2$ *such that* $B'\subset B$, *then*

$$|W_n(B')|\leqslant|W_n(B)|+n^{1/2}\mu_n(B). \tag{6.3}$$

Proof. We have

$$\begin{aligned}
W_n(B')=n^{-1/2}\Bigg\{\sum_{i=1}^{n} & I_{[c_1\leqslant \bar F_n(X_i)\leqslant d_1]}I_{[c_2\leqslant \bar G_n(Y_n)\leqslant d_2]}\\
&-(H_i(\bar F_n^{-1}(d_1), \bar G_n^{-1}(d_2))+H_i(\bar F_n^{-1}(d_1), \bar G_n^{-1}(c_2))\\
&+H_i(\bar F_n^{-1}(c_1), \bar G_n^{-1}(d_2))-H_i(\bar F_n^{-1}(c_1), \bar G_n^{-1}(_2)))\Bigg\}\\
\leqslant n^{-1/2}\Bigg\{\sum_{i=1}^{n} & I_{[a_1\leqslant \bar F_n(X_i)\leqslant b_1]}I_{[a_2\leqslant \bar G_n(Y_n)\leqslant b_2]}\\
&-(H_i(\bar F_n^{-1}(b_1), \bar F_n^{-1}(b_2))+H_i(\bar F_n^{-1}(b_1), \bar F_n^{-1}(a_2))\\
&+H_i(\bar F_n^{-1}(a_1), \bar F_n^{-1}(a_2))-H_i(\bar F_n^{-1}(a_1), \bar F_n^{-1}(a_2)))\Bigg\}\\
+n^{-1/2}\Bigg\{\sum_{i=1}^{n} & H_i(\bar F_n^{-1}(b_1), \bar F_n^{-1}(b_2))-H_i(\bar F_n^{-1}(b_1), \bar F_n^{-1}(a_2))\\
&-H_i(\bar F_n^{-1}(a_1), \bar F_n^{-1}(a_2))+H_i(\bar F_n^{-1}(a_1), \bar F_n^{-1}(a_2))\\
&-(H_i(\bar F_n^{-1}(d_1), \bar G_n^{-1}(d_2))+H_i(\bar F_n^{-1}(d_1), \bar G_n^{-1}(c_2))\\
&+H_i(\bar F_n^{-1}(c_1), \bar G_n^{-1}(d_2))-H_i(\bar F_n^{-1}(c_1), \bar G_n^{-1}(c_2)))\Bigg\}\\
\leqslant |W_n(B)| & +n^{1/2}\mu_n(B).
\end{aligned} \tag{6.4}$$

On the other hand

$$\begin{aligned}
-W_n(B')=-n^{-1/2}\Bigg\{\sum_{i=1}^{n} & I_{[c_1\leqslant \bar F_n(X_i)\leqslant d_1]}I_{[c_2\leqslant \bar G_n(Y_i)\leqslant d_2]}\\
&-(H_i(\bar F_n^{-1}(d_1), \bar G_n^{-1}(d_2))+H_i(\bar F_n^{-1}(d_1), \bar G_n^{-1}(c_2))\\
&+H_i(\bar F_n^{-1}(c_1), \bar G_n^{-1}(d_2))-H_i(\bar F_n^{-1}(c_1), \bar G_n^{-1}(c_2)))\Bigg\}\\
&\leqslant n^{-1/2}\mu_n(B')\leqslant\mu_n(B)\leqslant n^{1/2}\mu_n(B)+|W_n(B)|.
\end{aligned} \tag{6.5}$$

From (6.4) and (6.5), we deduce (6.3).

LEMMA 6.4. *Under the conditions of Theorem 2.3 we have*

$$E(W_n(I))^{2q} \leq C\left(\sum_{\ell=1}^{q} n^{-q-\ell}(\mu_n(I))^{2\ell/(2+\delta)}\right) \tag{6.6}$$

for any rectangle I of $[0,1]^2$ and $\delta>0$.

Proof. We will prove it for only $q=2$, the proof for $q>2$ is similar and is therefore omitted. Denote

$$\begin{aligned} A_{ni} &= I_{[u_1 \leq \bar{F}_n(X_i) \leq u_2]} \cdot I_{[v_1 \leq \bar{G}_n(Y_i) \leq v_2]} \\ &\quad - (H_i(\bar{F}_n^{-1}(u_2), \bar{G}_n^{-1}(v_2)) - H_i(\bar{F}_n^{-1}(u_2), \bar{G}_n^{-1}(v_1)) \\ &\quad - H_i(\bar{F}_n^{-1}(u_1), \bar{G}_n^{-1}(v_2)) + H_i(\bar{F}_n^{-1}(u_1), \bar{G}_n^{-1}(v_1))) \\ &= B_{ni} - C_{ni}, \end{aligned}$$

where $I=[u_1, u_2]\times[v_1 v_2]$.

We have

$$\begin{aligned} &E(W_n(I))^4 \\ &\leq n^{-2}\left(\sum_{1\leq i,j,k,\ell\leq n} |E(A_{n,i}A_{n,j}A_{n,k}A_{n,\ell})|\right) \\ &\leq n^{-2}4!\left(\sum_{i=0}^{n-1}\sum_{j=1}^{n-i}\sum_{k\leq i}\sum_{\ell\leq i} |E(A_{n,j}A_{n,j+i}A_{n,j+i+k}A_{n,j+i+k+m})|\right. \\ &\quad + n^{-2}\sum_{i=0}^{n-1}\sum_{j=1}^{n-i}\sum_{k\leq i}\sum_{\ell\leq i} |E(A_{n,j-i-k-\ell}A_{n,j-i-k}A_{n,j-i}A_{n,j})| \\ &\quad \left. + n^{-2}\sum_{i=0}^{n-1}\sum_{j=1}^{n-i}\sum_{k\leq i}\sum_{\ell\leq i} |E(A_{n,j-\ell}A_{n,j}A_{n,j+i}A_{n,j+i+k})|\right) \\ &= 4!(K_n^{(1)} + K_n^{(2)} + K_n^{(3)}). \end{aligned} \tag{6.7}$$

Then from Lemma 6.2, it follows that

$$\begin{aligned} K_n^{(1)} &\leq n^{-2}\sum_{i=0}^{n-1} i^2(\beta(i))^{\delta/(2+\delta)} \sum_{j=1}^{n-i} (C_{nj})^{1/(2+\delta)}(C_{n,j+i})^{1/(2+\delta)} \\ &\leq n^{-2}\sum_{i=0}^{n-1} i^2(\beta(i))^{\delta/(2+\delta)} \left(\sum_{j=1}^{n-i} C_{nj}\right)^{1/(2+\delta)} \left(\sum_{j=1}^{n-i} C_{n,j+i}\right)^{1/(2+\delta)} \\ &\leq n^{-2}\sum_{i=0}^{n-1} i^2(\beta(i))^{\delta/(2+\delta)} \left(\sum_{j=1}^{n} C_{nj}\right)^{1/(2+\delta)} \left(\sum_{j=1}^{n} C_{nj}\right)^{1/(2+\delta)} \end{aligned}$$

$$\leqslant n^{-2} \sum_{i=0}^{n-1} (\beta(i))^{\delta/(2+\delta)} \left(\sum_{j=1}^{n} C_{nj} \right)^{2/(2+\delta)}$$

$$= n^{-2} \sum_{i=0}^{n-1} i^2 (\beta(i))^{\delta/(2+\delta)} \left(n \left(n^{-1} \sum_{j=1}^{n} C_{nj} \right) \right)^{2/(2+\delta)}$$

$$\leqslant \left(\sum_{i=0}^{n-1} i^2 (\beta(i))^{\delta/(2+\delta)} \right) n^{-1} (\mu_n([u_1, u_2] \times [v_1, v_2]))^{2/(2+\delta)}$$

$$\leqslant Cn^{-1}(\mu_n(I))^{2/(2+\delta)}. \tag{6.8}$$

Similarly

$$K_n^{(2)} \leqslant C(\mu_n(I))^{2/(2+\delta)}. \tag{6.9}$$

From Lemma 6.2, we also obtain

$$K_n^{(3)} \leqslant n^{-2} \left(\sum_{i=0}^{n-1} \sum_{j=1}^{n-i} \sum_{k \leqslant i} \sum_{\ell \leqslant i} |E(C_{n,j-\ell} C_{nj}) \, E(C_{n,j+i} C_{n,j+i+k})| \right)$$

$$+ n^{-2} \left(\sum_{i=0}^{n-1} i^2 (\beta(i))^{\delta/(2+\delta)} \sum_{j=1}^{n-i} (C_{nj})^{1/(2+\delta)} (C_{n,j+i})^{1/(2+\delta)} \right)$$

$$\leqslant n^{-1} \left(\sum_{i=0}^{n-1} \sum_{j=1}^{n-i} |E(C_{nj} C_{n,j+i})| \sum_{k=0}^{n-1} \sum_{\ell=1}^{n-k} |C_{n\ell} C_{n,\ell+k}| \right)$$

$$+ Cn^{-1}(\mu_n(I))^{2/(2+\delta)}$$

$$\leqslant n^{-2} \left(\sum_{i=0}^{n-1} (\beta(i))^{\delta/(2+\delta)} \sum_{j=1}^{n-i} (C_{nj})^{1/(2+\delta)} (C_{n,j+i})^{1/(2+\delta)} \right)^2$$

$$+ Cn^{-1}(\mu_n(I))^{2/(2+\delta)}$$

$$\leqslant C((\mu_n(I))^{4/(2+\delta)} + n^{-1}(\mu_n(I))^{2/(2+\delta)}). \tag{6.10}$$

From (6.7) through (6.10), we obtain (6.6) and Lemma 6.4 is proved.

For any $\eta > 0$, and any $n \in \mathbb{N}$, let $c_0 > 0$ be given, put

$$\omega_n(\eta) = \sup \{ |W_n(u, v) - W_n(u', v')|, |v - v'| < \nu,$$
$$|u - F(x_0)| < c_0 a_n, |u' - F(x_0)| < c_0 a_n \}.$$

and

$$w_n^*(\eta) = \sup \{ |W_n(u, v) - W_n(u, v') - W_n(u', v) + W_n(u', v')|,$$
$$|v - v'| < \eta, |u - F(x_0)| < c_0 a_n, |u' - F(x_0)| < c_0 a_n \}.$$

LEMMA 6.5. *Under the conditions of Theorem 2.3, we have*

$$\lim_{\eta \to 0} \limsup_{n \to \infty} P(\omega_n(\eta) \geqslant \varepsilon) = 0, \qquad \forall \varepsilon > 0. \tag{6.11a}$$

$\forall \xi > 0$, $\forall \varepsilon > 0$, $\exists \eta > 0$, $\xi n_0 \in \mathbb{N}$, *such that* $\forall n \geqslant n_0$

$$P(w_n^*(\eta) \geqslant \varepsilon) \leqslant \xi (a_n)^{(2-\delta)/(2+\delta)}. \tag{6.11b}$$

Proof. Let n be fixed. We consider a subdivision of $[0, 1]^2$, $(u_i, v_j) \in [0, 1]^2$, $1 \leqslant i \leqslant n$, $1 \leqslant j \leqslant n$ such that

$$u_0 = 0 < u_1 < \cdots < u_n = 1$$

$$v_0 = 0 < v_1 < \cdots < v_n = 1$$

$$u_i - u_{i-1} = n^{-1}, \qquad i = 1, ..., n$$

$$v_j - v_{j-1} = n^{-1}, \qquad j = 1, ..., n$$

and we consider also a subdivision of $[0, 1]$, $v'_0 = 0 < v'_1 < \cdots < v'_k = 1$ such that

$$v'_\ell - v'_{\ell-1} < \eta/2$$

$$v'_\ell + n^{-1} - v'_{\ell-1} \geqslant \eta/2, \qquad 1 \leqslant \ell \leqslant k-1$$

$$v'_\ell \in \{v_0, v_1, ..., v_n\}.$$

We define $(a, b) \in [0, 1]^2$ by

$$(a, b) \in \{u_0, u_1, ..., u_n\}$$

$$F(x_0) - a < c_0 a_n, \qquad F(x_0) - a - n^{-1} \geqslant c_0 a_n$$

$$b - F(x_0) < c_0 a_n, \qquad b + n^{-1} - F(x_0) \geqslant c_0 a_n.$$

For any $\ell \in \{1, ..., k\}$ and any $i \in \{0, ..., n\}$ we define a sequence of r.v.'s

$$Z^{(\ell)}(i, j) = W_n(u_i, v'_{\ell-1} + jn^{-1}) - W_n(u_i, v'_{\ell-1}),$$
$$0 \leqslant i \leqslant n, \quad 1 \leqslant j \leqslant p,$$

where $v'_{\ell-1} + pn^{-1} = v'_\ell$.

For any $j \in |[0, ..., n\}$ we define another sequence of r.v.'s

$$Z^{(0)}(i, j) = W_n(a + (i-1)\, n^{-1}, v_j) - W_n(a - n^{-1}, v_j),$$
$$0 \leqslant i \leqslant m, \quad 1 \leqslant j \leqslant n,$$

where $a + (m-1)\, n^{-1} = n + n^{-1}$.

From Lemma 6.4 for $q=4$, we deduce

$$E(Z^{(\ell)}(i,j)-Z^{(\ell)}(i',j'))^8 \leqslant (\mu_n[i'n^{-1}, in^{-1}]\times[0,j])^{8/(2+\delta)} \leqslant ((i-i')\,n^{-1})^{4/(2+\delta)}, \quad \text{where} \quad 0\leqslant i'\leqslant i\leqslant n, \quad 1\leqslant \ell\leqslant k \tag{6.12}$$

and

$$E(Z^{(0)}(i,j)-Z^{(0)}(i,j'))^8 \leqslant \mu_n([0,i]\times[j'n^{-1}, jn^{-1}])^{8/(2+\delta)} \leqslant ((j-j')\,n^{-1})^{4/(2+\delta)}.$$

Then by Theorem 12.2 of Billingsley (1968), we have $\forall\varepsilon>0$, $\forall \ell\in\{1,...,k\}$, $\forall i\in\{0,...,m\}$

$$P[\sup_{0\leqslant j\leqslant p} |Z^{(\ell)}(i,j)|>\varepsilon]\leqslant C\left(\frac{\eta}{2}\right)^{4/(2+\delta)} \tag{6.13}$$

and $\forall\varepsilon>0$, $\forall j\in\{0,...,n\}$

$$P[\sup_{0\leqslant i\leqslant m} |Z^{(0)}(i,j)|>\varepsilon]\leqslant C(2c_0a_n)^{4/(2+\delta)}. \tag{6.14}$$

Define now for any $\ell\in\{1,...,k\}$

$$Z_*^{(\ell)}(i,j)=W_n(a+(i-1)\,n^{-1}, v'_{\ell-1}+jn^{-1})-W_n(a-n^{-1}, v'_{\ell-1}+jn^{-1}) - W_n(a+(i-1)\,n^{-1}, v'_{\ell-1})+W_n(a-n^{-1}, v'_{\ell-1}).$$

From Lemma 6.4 for $q=4$, we deduce

$$E(Z_*^{(\ell)}(i,j)-Z_*^{(\ell)}(i',j)-Z_*^{(\ell)}(i,j')+Z_*^{(\ell)}(i',j')) \leqslant (\mu_n([i'n^{-1}, in^{-1}]\times[j'n^{-1}, jn^{-1}]))^{8/(2+\delta)} \leqslant ((i-i')\,n^{-1}(j-j'))^{4/(2+\delta)}, \quad \text{where} \quad 0\leqslant i'<i\leqslant m, \quad 0\leqslant j'<j\leqslant p. \tag{6.15}$$

Then by Lemma 1 of Balacheff and Dupont (1980) we have $\forall\varepsilon>0$,

$$P[\sup_{\substack{0\leqslant i\leqslant m\\ 0\leqslant j\leqslant p}} |Z_*^{(\ell)}(i,j)|>\varepsilon]\leqslant C(c_0a_n\eta)^{4/(2+\delta)}. \tag{6.16}$$

For any (u_i, v_j) and any $(u_{i'}, v_{j'})$ such that $|v_j-v_{j'}|<\eta$ and $u_i, u_{i'}\in\{a, a+n^{-1}, ..., b-n^{-1}, b\}$; Suppose $u_i<u_{i'}$ and $v'_{\ell-1}\leqslant v_{j'}<v_j\leqslant v'_\ell$

$$
\begin{aligned}
W_n(u_i, v_j) - W_n(u'_i, v'_j)
&= W_n(u_i, v_j) - W_n(a, v_j) \\
&\quad - W_n(u_i, v'_{\ell-1}) + W_n(a, v'_{\ell-1}) - W_n(u_{i'}, v_{j'}) + W_n(a, v_{j'}) \\
&\quad + W_n(u_{i'}, v'_{\ell-1}) - W_n(a, v'_{\ell-1}) + W_n(a, v_j) + W_n(u_i, v'_{\ell-1}) \\
&\quad - W_n(a, v_{j'}) - W_n(u_{i'}, v'_{\ell-1}).
\end{aligned} \tag{6.17}
$$

If $v'_{\ell-1} < v_{j'} < v'_\ell \leqslant v_j$, we write

$$
\begin{aligned}
W_n(u_i, v_j) - W_n(u_{i'}, v_{j'})
&= W_n(u_i, v'_\ell) + W_n(u_i, v_j) + W_n(u_i, v'_\ell) \\
&\quad - W_n(u_i, v'_{\ell-1}) + W_n(u_i, v_j) - W_n(u_{i'}, v_{j'}).
\end{aligned} \tag{6.18}
$$

From (6.17) and (6.18), we deduce

$$
\begin{aligned}
&\sup_{\substack{|v_j - v_{j'}| < \eta \\ |F(x_0) - u_i| < c_0 a_n \\ |F(x_0) - u_{i'}| < c_0 a_n}} |W_n(u_i, v_j) - W_n(u_{i'}, v_{j'})| \\
&\leqslant \sup_{1 \leqslant \ell \leqslant k} \sup_{\substack{0 \leqslant i \leqslant m \\ 0 \leqslant j \leqslant p}} |Z_*^{(\ell)}(i, j)| + 3 \sup_{1 \leqslant \ell \leqslant k} \sup_{0 \leqslant j \leqslant p} |Z^{(\ell)}(i_a, j)| \\
&\quad + \sup_{1 \leqslant \ell \leqslant k} \sup_{0 \leqslant i \leqslant m} |Z^{(0)}(i, j_\ell)|,
\end{aligned} \tag{6.19}
$$

where $i_a = a$ and $v_{j_\ell} = v'_{\ell-1}$.

By using (6.13), (6.14), and (6.16), we obtain $\forall \varepsilon > 0$

$$
\begin{aligned}
&P\Big[\sup_{\substack{|v_j - v_{j'}| < \eta/2 \\ |F(x_0) - u_i| < c_0 a_n \\ |F(x_0) - u_{i'}| < c_0 a_n}} |W_n(u_i, v_j) - W_n(u_{i'}, v_{j'})| > \varepsilon\Big] \\
&\leqslant C\left\{k(c_0 a_n \eta)^{4/(2+\delta)} + 3k\left(\frac{\eta}{2}\right)^{4/(2+\delta)} + k(2c_0 a_n)^{4/(2+\delta)}\right\} \\
&\leqslant C\left\{(2c_0 a_n)^{4/(2+\delta)}\left(\frac{k\eta}{2}\right)\left(\frac{\eta}{2}\right)^{(2-\delta)/(2+\delta)}\right. \\
&\quad \left. + 3\left(\frac{k\eta}{2}\right)\left(\frac{\eta}{2}\right)^{(2-\delta)/(2+\delta)} + k(2c_0 a_n)^{4/(2+\delta)}\right\} \\
&\leqslant C\left\{(2c_0 a_n)^{4/(2+\delta)}\left(\frac{\eta}{2}\right)^{(2-\delta)/(2+\delta)}\right. \\
&\quad \left. + 3\left(\frac{\eta}{2}\right)^{(2-\delta)/(2+\delta)} + k(2c_0 a_n)^{4/(2+\delta)}\right\}
\end{aligned} \tag{6.20}
$$

which implies $\forall\varepsilon>0$, $\forall\lambda>0$, $\exists\eta>0$, $\exists n_0\in\mathbb{N}$, $\forall n\geqslant n_0$,

$$P[\sup_{\substack{|v_j-v_{j'}|<\eta/2\\|F(x_0)-u_i|<c_0a_n\\|F(x_0)-u_{i'}|<c_0a_n}}|W_n(u_i,v_j)-W_n(u_{i'},v_{j'})|\geqslant\varepsilon]<\lambda. \tag{6.21}$$

Let (u,v) and $(u',v')\in[0,1]$ be such that $|v-v'|<\eta$ and $|F(x_0)-u|<c_0a_n$, $|F(x_0)-u'|<c_0a_n$. There exist $(i,i')\in\{1,2,...,n\}^2$ and $(j,j')\in\{1,2,...,n\}^2$ such that

$$u_{i-1}\leqslant u\leqslant u_i,\quad u_{i'-1}\leqslant u'\leqslant u_{i'},\quad v_{j-1}\leqslant v\leqslant v_j,\quad v_{j'-1}\leqslant v'\leqslant v_{j'}.$$

We have

$$\begin{aligned}&|W_n(u,v)-W_n(u',v')|\\&\quad\leqslant|W_n(u_i,v_j)-W_n(u_{i'},v_{j'})|+|W_n(u_i,v_j)-W_n(u_i,v)|\\&\qquad+|W_n(u_i,v)-W_n(u,v)|+|W_n(u_{i'},v_{j'}-W_n(u_{i'},v'))|\\&\qquad+|W_n(u_{i'},v)-W_n(u',v')|.\end{aligned}$$

By using Lemma 6.3, we get

$$\begin{aligned}&|W_n(u,v)-W_n(u',v')|\\&\quad\leqslant|W_n(u_i,v_j)-W_n(u_{i'},v_{j'})|+|W_n(u_i,v_j)-W_n(u_i,v_{j-1})|\\&\qquad+|W_n(u_i,v_j)-W_n(u_{i-1},v_j)|+|W_n(u_{i'},v_{j'})-W_n(u_{i'},v_{j'-1})|\\&\qquad+|W_n(u_{i'},v_{j'})-W_n(u_{i'-1},v_{j'})|+4n^{-1/2}\end{aligned} \tag{6.22}$$

From (6.21), we deduce $\forall\varepsilon>0$, $\forall\lambda>0$, $\exists\eta>0$, $\exists n_0\in\mathbb{N}$, $\forall n\geqslant n_0$,

$$P[\sup_{\substack{|v_j-v_{j'}|<\eta/2\\|F(x_0)-u_i|<c_0a_n\\|F(x_0)-u_{i'}|<c_0a_n}}|W_n(u,v)-W_n(u',v')|\geqslant\varepsilon]<\lambda \tag{6.23}$$

and (6.11a) is proved.

The proof of (6.11b) is similar.

LEMMA 6.6. *We have*

$$P[\lim_{n\to\infty}\tilde{M}_n(x_0)=x_0]=1,\qquad 0<x_0<1. \tag{6.24}$$

Proof. From Lemma 6.4 we easily deduce that

$$E(\tilde{M}_n(x_0)-x_0)^4 \leqslant C(n^{-3}(x_0)^{2/(2+\delta)}+n^{-2}(x_0)^{4/(2+\delta)})$$
$$\leqslant C(n^{-3}+n^{-2}). \tag{6.25}$$

By the Markov inequality

$$P[\,|\tilde{M}_n(x_0)-x_0|>\varepsilon] \leqslant Cn^{-2}\varepsilon^{-4} \tag{6.26}$$

and so by the Borel–Cantelli lemma

$$P[\,|\tilde{M}_n(x_0)-x_0| \geqslant \varepsilon \text{ i.o.}]=0 \tag{6.27}$$

for each positive ε, which implies (6.24). So we have the lemma.

LEMMA 6.7. *Let $\varepsilon>0$ be arbitrarily. Then there is a constant C, such that, up to an even of probability less than or equal to ε, we have*

$$|t-s| \leqslant Ca_n \tag{6.28}$$

whenever

$$|\tilde{M}_n(t)-\tilde{M}_n(s)| \leqslant a_n. \tag{6.29}$$

Proof. From Theorems 6.1 and 14.3 of Billingsley (1968) and the fact that the sequence of probability measures P_n defined by W_n is tight, we deduce that there exists some finite number M such for all n sufficiently large,

$$P(A_n(M)) \geqslant 1-\varepsilon,$$

where $A_n(M)=\{\sup_{0\leqslant t\leqslant 1}|n^{-1}\tilde{M}_n(t)| \leqslant Mn^{-1/2}\}$.
From (6.29), on the set $A_n(M)$ we have

$$|t-s| \leqslant |t-\tilde{M}_n(t)|+|\tilde{M}_n(t)-\tilde{M}_n(s)|+|\tilde{M}_n(s)-s|$$
$$=|n^{-1}\tilde{M}_n(t)|+|\tilde{M}_n(t)-\tilde{M}_n(s)|+|n^{-1}\tilde{M}_n(s)|$$
$$\leqslant a_n+2Mn^{-1/2} \leqslant Ca_n$$

and the proof is completed.

LEMMA 6.8. *If the sequence* $\{X_{ni}\}$ *is strong mixing with rates* (2.12), *then for any* $\lambda > 0$,

$$P[\sup_{s \leqslant t \leqslant s + C_1 a_n} |M_n^*(t) - M_n^*(s)| \geqslant \lambda a_n^{1/2}] \leqslant C\lambda^{-4}, \tag{6.30}$$

where $M_n^* = n^{1/2}(\tilde{M}_n - \bar{M}_n)$.

Proof. From Lemma 6.4, if $t - s \geqslant \varepsilon n^{-1}$, then $E|M_n^*(t) - M_n^*(s)|^4 \leqslant C\varepsilon^{-1}(t-s)^{4/(2+\delta)}$ for all n. Hence following the method of the proof of (22.20) in Billingsley (1968), we get

$$P[\sup_{s \leqslant t \leqslant s + mp} |M_n^*(t) - M_n^*(s)| \geqslant \lambda a_n^{1/2}] \leqslant C\varepsilon^{-1}\lambda^{-4}(mp)^{4/(2+\delta)}. \tag{6.31}$$

Now putting $mp = C_1 a_n$, we have (6.30) from (6.31).

The following lemma is Theorem 1 of Yokoyama (1980).

LEMMA 6.9. *Let* $\{Y_j\}$ *be a strong mixing sequence with* $E(Y_j) = 0, j \geqslant 1$, *and* $\sup_{j \geqslant 1} E|Y_j|^{r+\delta} < \infty$ *for some* $r > 2$ *and* $\delta > 0$. *If*

$$\sum_{i=0}^{\infty} (i+1)^{(r/2)-1} [\alpha(i)]^{\delta/(r+\delta)} < \infty \tag{6.32}$$

then

$$E\left|\sum_{i=1}^{n} Y_i\right|^r \leqslant Cn^{r/2}, \qquad n \geqslant 1. \tag{6.33}$$

REFERENCES

1. S. Balacheff and G. Dupont, Normalité asymptotique des processus empiriques tronqués et des processus et rang, *in* "Lecture Notes in Mathematics," Vol. 821, pp. 19–45, Springer-Verlag, New York/Berlin, 1980.
2. G. Bennett, Probability inequalities for the sum of independent random variables, *J. Amer. Statist. Assoc.* **57** (1962), 33–45.
3. P. Billingsley, "Convergence of Probability Measures," Wiley, New York, 1968.
4. Yu. A. Davydov, Mixing conditions for Markov chains, *Theory Probab. Appl.* **18** (1973), 312–328.
5. L. P. Devroye and T. J. Wagner, On the L_1 convergence of kernel estimators of regression functions with applications in discrimination, *Z. Wahrsch. Verw. Gebiete* **51** (1980), 15–25.
6. P. Doukhan and F. Portal, Principe d'invariance faible pour la fonction de répartition empirique dans un cadre multidimensionnel et mélangeant, *Probab. Math. Statist.* **8**, Fa. 2, (1987), 117–132.

7. M. Harel and M. L. Puri, Weak convergence of weighted multivariate empirical processes under mixing conditions, *in* "Mathematical Statistics and Probability Theory, Vol. A. Theoretical Aspects" (M. L. Puri, P. Révész, and W. Wertz, Eds.), pp. 121–141, Reidel, Dordrecht, 1987.
8. M. Harel and M. L. Puri, Limiting behavior of U-statistics, V-statistics and one sample rank order statistics for nonstationary absolutely regular processes, *J. Multivariate Anal.* **30** (1989), 181–204.
9. M. Harel and M. L. Puri, Théorème central limite de l'estimateur de densité et de l'erreur quadratique intégrée pour des variables aléatoires non stationaires, *C. R. Acad. Sci. Paris. Sér. I Math.* **312** (1991), 145–148.
10. G. Neuhaus, On weak convergence of stochastic processes with multidimensional time parameter, *Ann. Math. Stat.* **42**, No. 4 (1971), 1285–1295.
11. E. Nummelin and P. Tuominen, Geometric ergodicity of Harris recurrent Markov chains with applications to renewal theory, *Stochastic Processes Appl.* **12** (1982), 187–202.
12. E. M. Stein, "Singular Integrals and Differentiability Properties of Functions," Princeton Univ. Press, Princeton, NJ, 1970.
13. W. Stute, Asymptotic normality on nearest neighbor regression function estimates, *Ann. Statist.* **12** (1984), 917–926.
14. W. Stute, On almost sure convergence of conditional empirical distribution functions, *Ann. Probab.* **14** (1986), 891–901.
15. C. S. Withers, Convergence of empirical processes of mixing rv's on [0, 1], *Ann. Statist.* **3** (1975), 1101–1108.
16. R. Yokoyama, Moment bounds for stationary mixing sequences, *Z. Wahrsch. Verw. Gebiete* **52** (1980), 45–57.
17. K. Yoshihara, Limiting behavior of U-statistics for stationary absolutely regular processes, *Z. Wahrsch. Verw. Gebiete* **35** (1976), 237–252.
18. K. Yoshihara, Asymptotic normality of nearest neighbor regression function estimates based on some dependent observations, *Yokohama Math. J.* **36** (1988), 55–68.
19. K. Yoshihara, Conditional empirical processes defined by φ-mixing sequences, *Comput. Math. Appl.* **19**, No. 1 (1990), 149–158.

Part III

Extreme Value Theory

A STRONG INVARIANCE PRINCIPLE CONCERNING THE J-UPPER ORDER STATISTICS FOR STATIONARY m-DEPENDENT SEQUENCES

GEORGE HAIMAN and MADAN L. PURI

CNRS, U.R.A. 1321 and U.F.R. de Mathématiques, Université de Lille 1, Villeneuve d'Ascq 59655, France

Department of Mathematics, Indiana University, Bloomington, IN 47405. U.S.A.

Received 21 December 1988; revised manuscript received 27 February 1989

Recommended by R. Bartoszynski

Abstract: It is shown that in the case of stationary m-dependent processes, the J-th ($J \geq 1$) record times $\{T_n, n \geq 1\}$ and the corresponding J-upper statistics $\{X_{T_n - J+1, T_n}, \dots, X_{T_n, T_n}\}$ can almost surely be identified via a translation of the time index n to the corresponding elements defined on a sequence of independent and identically distributed random variables. A construction method of approximating sequences of record times and the corresponding upper order statistics introduced by Haiman (1987a) for the case $J = 1$ is extended and applied under a weaker condition concerning the joint distribution of (X_1, X_i), $i = 2, \dots, m$.

AMS 1980 *Subject Classification:* Primary 60F15, 62G30.

Key words: Strong invariance; m-dependent processes; record times; order statistics.

1. Introduction

Let $\{X_n, n \geq 1\}$ be a stationary m-dependent ($m \geq 1$) sequence with continuous distribution function $F(x) = P(X_1 < x)$. Let $X_{1,n} < X_{2,n} < \cdots < X_{n-1,n} < X_{n,n}$ denote the order statistics of $X_1, \dots, X_n$. Let J be a positive integer, and $\theta_0 > 0$ be a fixed real number. Denote

$$T_1 = \inf\{n; n \geq J, X_{n-J+1,n} > \theta_0\}, \qquad \boldsymbol{\theta}_1 = \{X_{T_1 - J+1, T_1}, \dots, X_{T_1, T_1}\} \quad (1.1)$$

and for any $k \geq 2$, set

$$\begin{aligned} T_k &= \inf\{n; n > T_{k-1}, X_{n-J+1,n} > X_{n-J,n-1}\}, \\ \boldsymbol{\theta}_k &= \{X_{T_k - J+1, T_k}, \dots, X_{T_k, T_k}\} =: \{\theta_k^J, \theta_k^{J-1}, \dots, \theta_k^1\}. \end{aligned} \quad (1.2)$$

* Research supported by the Office of Naval Research Contract N00014-85-K-0648.

θ_k^J is called the k-th J-th-upper record. The T_k's are the J-th record times and $\{T_k, \boldsymbol{\theta}_k\}_{k\geq 1}$ describe the history of the J largest values of the sample.

Let us recall the definition of J-ordered sequences (Haiman and Puri, 1988).

Definition. A sequence $\{\boldsymbol{r}_k = (r_k^J, \ldots, r_k^1) \in \mathbb{R}^J, k \geq 1\}$ is called J-ordered if the following conditions hold:

(i) For any $k \geq 1$, $r_k^J < \cdots < r_k^1$, and

(ii) either $r_k^1 < r_{k+1}^1$ and then $r_{k+1}^J = r_k^{J-1}$, $r_{k+1}^{J-1} = r_k^{J-2}, \ldots, r_{k+1}^2 = r_k^1$; or there exists j, $2 \leq j \leq J$ such that $r_k^j < r_{k+1}^j < r_k^{j-1}$, and then $r_{k+1}^i = r_k^{i-1}$ if $i > j$ and $r_{k+1}^i = r_k^i$ if $i \leq j$.

It is well known that when $\{X_n, n \geq 1\}$ is a sequence of i.i.d. (independent and identically distributed) random variables, then for any nondecreasing sequence of positive integers $\{t_k, k \geq 1\}$ and any J-ordered sequence $\{\boldsymbol{r}_k = (r_k^J, \ldots, r_k^1), k \geq 1\}$, we have the following Markov behavior of the sequence $\{T_k, \boldsymbol{\theta}_k\}_{k \geq 1}$:

$$P\{T_{n+1} = t_{n+1}, \theta_{n+1}^j < r_{n+1}^j \mid T_1 = t_1, \boldsymbol{\theta}_1 = \boldsymbol{r}_1, \ldots, T_n = t_n, \boldsymbol{\theta}_n = \boldsymbol{r}_n\}$$
$$= [P\{X_1 < r_n^J\}]^{t_{n+1} - t_n - 1} \cdot P\{r_n^j < X_1 < r_{n+1}^j\}, \quad n \geq 1, j = 1, \ldots, J. \quad (1.3)$$

Our main result is the following:

Theorem 1.1. *Let* $\{X_n, n \geq 1\}$ *satisfy the following hypothesis* (H):

(H) *There exists* $\beta > 0$ *such that for any* $2 \leq i \leq m$,

$$\limsup_{z \uparrow b} \left\{ \sup_{\substack{z < v < w < b \\ z < u < b}} \frac{P\{X_1 > u, v < X_i < w\}}{(\log(G(u)^{-1}))^{-2-\beta} P\{v < X_1 < w\}} \right\} < \infty, \quad (1.4)$$

with $b = \sup\{x; F(x) < 1\}$, $G(x) = 1 - F(x)$. *Then, for any* $J \geq 1$, *there exists a* $\theta_0 > 0$ *and a sequence* $\{S_n, \boldsymbol{R}_n\}_{n \geq 1}$ *defined on the same probability space as* $\{X_n, n \geq 1\}$ *(possibly enlarged) where* S_n *is a strictly increasing sequence of integers, and* $\{\boldsymbol{R}_n = (R_n^J, \ldots, R_n^1), \theta_0 < R_n^J < \cdots < R_n^1, n \geq 1\}$ *a J-ordered sequence, such that*:

(a) $\{S_k, \boldsymbol{R}_k\}_{k \geq 1}$ *has the same probability law as* $\{T_k, \boldsymbol{\theta}_k\}_{k \geq 1}$ *in the i.i.d. case, and in particular* $\{S_k, \boldsymbol{R}_k\}$ *satisfies* (1.3) *with* T_k *and* $\boldsymbol{\theta}_k$ *replaced by* S_k *and* $\boldsymbol{R}_k$ *respectively.*

(b) *Almost surely, there exists an* n_0 *and* q *such that for any* $n \geq n_0$, *we have*

$$S_n = T_{n-q} \quad \text{and} \quad \boldsymbol{R}_n = \boldsymbol{\theta}_{n-q}. \quad (1.5)$$

Remarks. (1) Let

$$\{\boldsymbol{M}_n\}_{n \geq J} := \{(X_{n-J+1,n}, \ldots, X_{n,n})\}_{n \geq J}. \quad (1.6)$$

Since $\{T_n, \boldsymbol{\theta}_n\}_{n \geq 1}$ completely determines $\{\boldsymbol{M}_n\}_{n \geq \nu}$, $\nu = \inf\{k \geq J, X_{k-J+1,k} > \theta_0\}$, a straightforward consequence of Theorem 1.1 is the following:

Theorem 1.2. *Let* (H) *hold. Then, for any fixed* $J \geq 1$, *there exists a probability* *space which carries, in addition to the sequence* $\{X_n,\ n\geq 1\}$, *an i.i.d. sequence* $\{X_n^*,\ n\geq 1\}$ *with same margins, such that:*

(a) *There exists a* Θ_0 (>0) *such that, if* $\{T_k, \boldsymbol{\Theta}_k\}_{k\geq 1}$ *is the J-th record sequence based on* $\{X_n,\ n\geq 1\}$ *and* Θ_0, *and* $\{S_k, \boldsymbol{R}_k\}_{k\geq 1}$ *is the J-th record sequence based on* $\{X_n^*,\ n\geq 1\}$ *and* Θ_0, *then there exists almost surely an* n_0 *and* q *such that for all* $n\geq n_0$, (1.5) *is satisfied.*

(b) *If* $\{\boldsymbol{M}_n^*\}_{n\geq J} := \{(X^*_{n-J+1,n}, \ldots, X^*_{n,n})\}_{n\geq J}$ *then, there exists almost surely an* n_1 *and* q' *such that for any* $n\geq n_1$, *we have* $\boldsymbol{M}_n^* = \boldsymbol{M}_{n+q'}$.

(2) For $J=1$, Haiman (1987a) proved Theorem 1.1 under the more restrictive Hypothesis (H′), viz.

(H′) There exists $0<\beta<1$ such that for any $2\leq i\leq m$,

$$\limsup_{z\uparrow b}\left\{\sup_{\substack{z<v<w<b\\ z<u<b}}\frac{P\{X_1>u,v<X_i<w\}}{(G(u))^{\beta}P\{v<X_1<w\}}\right\}<\infty. \tag{1.7}$$

(3) For $J=1$, Haiman (1987b) and for $J\geq 1$, Haiman and Puri (1988), proved Theorem 1.1 for Gaussian sequences under several sets of hypotheses, one of which being that the strong mixing coefficient $\alpha(n)$ decreases to zero as $n\to\infty$ faster than $1/n^{6+\varepsilon}$ for some $\varepsilon>0$. Of course this result implies Theorem 1.1 for stationary m-dependent Gaussian sequences, since in this case $\alpha(n)=0$ for any $n>m$.

(4) As it will be seen in the next section, the proof of Theorem 1.1 is based on the following comparison theorem. The following result (needed in the sequel) is a slightly more general version of Theorem 1 of Haiman (1987a) which incidentally contains an error.

Theorem A (Haiman 1987a). *There exists a universal constant* $0<p_0<1$, *and a function* $\mu(u)$, $0<\mu(u)<1$, *defined for* $u\geq u_0$, *where* u_0 *is the solution of the equation* $mP\{X_1>u_0\}=p_0$, *such that:*

(i)
$$\max_{n\geq 1}\left|\frac{P\{\max(X_1,\ldots,X_n)<u\}}{\mu^n(u)}-1\right|\leq Km^2\sup_{2\leq i\leq m}P\{X_1>u,X_i>u\} \tag{1.8}$$

and

$$\mu(u)=1-P\{\max(X_1,\ldots,X_{m-1})<u,X_m>u\}$$
$$+\frac{1}{m}\mathrm{O}((mP\{X_1>u\})^2),\quad |\mathrm{O}(X)|\leq K'|x|, \tag{1.9}$$

where K *in* (1.8) *and* K' *in* (1.9) *are universal constants.*

(ii) *m being fixed, suppose the following hypothesis* (H_1) *holds:*

(H_1) *there exists a function f defined on* $[0,1]$ *with* $v/f(v)\to 0$ *as* $v\to 0$, *such that for any* $2\leq i\leq m$,

$$\limsup_{u\uparrow b}\frac{P\{X_1>u,X_i>u\}}{G(u)f(G(u))}<\infty; \tag{1.10}$$

then

$$\mu(u) = P\{X_1 < u\} + O(G(u)F(G(u))), \quad u \uparrow b. \tag{1.11}$$

(iii) *m being fixed, suppose the following hypothesis* (H_2) (*more restrictive than* (H_1)) *holds*:

(H_2) *there exists a function f defined on* [0, 1] *with* $v/f(v) \to 0$ *as* $v \to 0$, *such that for any* $2 \leq i \leq m$,

$$\limsup_{z \uparrow b} \left\{ \sup_{\substack{z<v<w<b \\ z<u<b}} \frac{P\{X_1 > u, v < X_i < w\}}{f(G(u))P\{v < X_1 < w\}} \right\} < \infty; \tag{1.12}$$

then, there exists $u_1 > 0$ *such that for any* $u_1 < u < b$,

$$\max_{\substack{u<v<w<b \\ n \geq 1}} \left| \frac{P\{\max(X_1, \ldots, X_n) < u, v < X_{n+1} < w\}}{\mu^n(u)(1-\mu(u))(G(u))^{-1}P\{v < X_1 < w\}} - 1 \right|$$

$$= O(f(G(u))), \quad u \to \infty. \tag{1.13}$$

Note that hypothesis (H) of Theorem 1.1 is a particular case of (H_2) with $f(v) = (\log(1/v))^{-2-\beta}$ (whereas in (H′)–(1.7), $F(v) = v^{\beta}$).

As it will be seen in the next section, this choice of $f(v)$ is optimal with respect to the method of construction of the sequence $\{S_n, R_n\}_{n \geq 1}$ used in the proof of Theorem 1.1.

2. Proof of Theorem 1.1

We shall first prove (a) by using the construction mechanism of the sequence $\{S_n, R_n\}$, as developed in Haiman (1987a). This mechanism is based on the following lemma:

Lemma 2.1 (Haiman, 1987a). *Let Y be a random variable taking values in a measurable space* $(\mathcal{Y}, \mathcal{F})$. *Let* $\varphi \in \mathcal{F}$ *and let* $\hat{P}$ *be a probability measure on* $\mathcal{Y}$ *such that* $0 < \hat{P}(\varphi) < 1$. *Assume that on* φ *the probability law of Y, denoted by* P_Y, *has a Radon–Nikodym derivative* $dP_Y/d\hat{P}$ *with respect to* $\hat{P}$ *such that*

$$\max_{y \in \varphi} \left| \frac{dP_Y}{d\hat{P}}(y) - 1 \right| (1 - \hat{P}(\varphi))^{-1} \leq q < 1. \tag{2.1}$$

Let Q be a Bernoulli random variable independent of Y such that $P(Q=0) = q$. *Then, there exist two random elements* Y' *and* $\bar{Y}$, *taking values respectively in* $\mathcal{Y}$ *and* $\bar{\varphi}$, *independent of Y and Q, and such that if we put*

$$\hat{Y} = \begin{cases} Y & \text{if } Q = 1 \text{ and } Y \in \varphi, \\ \bar{Y} & \text{if } Q = 1 \text{ and } Y \in \bar{\varphi}, \\ Y' & \text{if } Q = 0, \end{cases} \tag{2.2}$$

then the probability law of $\hat{Y}$ *is* $\hat{P}$ ($\bar{\varphi}$ *is the complement of* φ).

For any integer $s \geq 0$ and real $r \geq 0$ let $Y_{s,r}$ be defined by

$$Y_{s,r} = (t, \varrho) \text{ if and only if } \{X_i < r, i = s+1, \ldots, s+t-1, X_{s+t} = \varrho\} \quad (2.3)$$

with $(t, \varrho) \in \mathscr{Y} = \mathscr{Y}(r) := \mathbb{N} \times (r, \infty)$. Let

$$\varphi = \varphi(r) = \left\{(t, \varrho) \in \mathscr{Y};\ 1 \leq t \leq \frac{\tau}{G(r)} \log_2\left(\frac{1}{G(r)}\right)\right\} \quad (2.4)$$

where τ is a given constant >0, $\log_2(x) = \log\log(x)$ and let $\hat{P}$ be the probability law endowed by $Y_{s,r}$ when $\{X_n\}_n$ is i.i.d.

Proposition 2.1. *If* $2+\beta-\tau>0$ *then there exists a* $\theta_0>0$ *such that for any* $r>\theta_0$ *and* $s \geq 1$ *there exists an algorithm*

$$\mathscr{A}_r : \{X_{s+1}, X_{s+2}, \ldots\} \to (\Delta S, R) \in \mathscr{Y} \quad (2.5)$$

such that

$$P\{\Delta S = t, R > \varrho\} = F(r)^{t-1}(1 - F(\varrho)), \quad t \geq 1, \varrho > r, \quad (2.6)$$

and

$$P\{\Delta S = \min(t \geq 1, X_{s+t} > r), R = X_{s+\Delta s}\} = (1-q)P\{Y_{s,r} \in \varphi\} \quad (2.7)$$

with

$$q = (\text{const.})\left(\log_2 \frac{1}{G(r)}\right)\left(\log \frac{1}{G(r)}\right)^{-(2+\beta-\tau)} \uparrow 0 \quad \textit{as } r \uparrow \infty. \quad (2.8)$$

Proof. We apply Lemma 2.1 with $Y := Y_{s,r}$, φ and $\hat{P}$ defined in (2.3) and (2.4).

By Theorem A, (1.13), for any sufficiently large r we have

$$\max_{\substack{t \geq 1 \\ \varrho > r}} \left| \frac{P_y(t, \varrho)}{\mu^{t-1}(r)(1-\mu(r))\,(\mathrm{d}F/\mathrm{d}\varrho)(\varrho)(G(r))^{-1}} - 1 \right| \leq (\text{const.})(\log(G(r)^{-1}))^{-2-\beta} \quad (2.9)$$

with

$$\mu(r) = F(r) + \mathrm{O}[G(r)(\log(G(r)^{-1}))^{-2-\beta}]. \quad (2.10)$$

On the other hand,

$$\hat{P}(\bar{\varphi}) = (F(r))^{(\tau/G(r))\log_2(1/G(r))} \sim \left(\log \frac{1}{G(r)}\right)^{-\tau} \quad \text{as } r \uparrow \infty. \quad (2.11)$$

By combining (2.9), (2.10) and (2.11) we get (2.1) with q satisfying (2.8). We then take $(\Delta S, R) := \hat{Y}$ in (2.2) which satisfies (2.6) and (2.7).

In view of the construction of $\{S_n R_n\}_{n \geq 1}$ we need the following modification of Proposition 2.1.

Proposition 2.2. *If* $2+\beta-\tau>0$ *then there exists a* $\theta_0>0$ *such that for any* $r>\theta_0$ *and* $s \geq 1$ *there exists an algorithm*

$$\mathscr{B}_r : \{X_{s+1}, X_{s+2}, \ldots\} \to (\Delta\mathscr{S}, \mathscr{R}) \in \mathscr{Y} \quad (2.12)$$

such that:

(i) *if* $1 \leq \Delta\mathscr{S} \leq m-1$ *then* $(\Delta\mathscr{S}, \mathscr{R})$ *is independent of* $\sigma(X_{s+1}, \ldots)$,

(ii) *for any* $t \geq 1$ *and* $\varrho > r$,

$$P(\Delta\mathscr{S} = t, \mathscr{R} > \varrho) = F(r)^{t-1}(1 - F(\varrho)), \quad (2.13)$$

(iii) $\mathscr{B}_r$ *is defined by means of two independent Bernoulli random variables,* Q *and* L, *independent of* $\sigma\{X_1, \ldots\}$ *and such that*

$$\{\Delta\mathscr{S} = m - 1 + \min\{t \geq 0, X_{s+m+t} > r\}, \mathscr{R} = X_{s+\Delta\mathscr{S}}\}$$

$$= \left\{L = 1, Q = 1, \inf\{k \geq 0;\ X_{s+m+k} > r\} \leq \frac{\tau}{G(r)} \log_2 \frac{1}{G(r)}\right\} \tag{2.14}$$

with

$$P\{L = 1\} = F(r)^{m-1} \tag{2.15}$$

and

$$P\{Q = 1\} = 1 - q,$$

where q *satisfies* (2.8).

Proof. If $L=1$ then $\mathscr{B}_r = \mathscr{A}_r$ and if $L=0$ we define $(\Delta\mathscr{S}, \mathscr{R})$ independently of $\sigma\{X_s, \ldots\}$, such that (2.13) is satisfied. Let us construct $\{S_n, \boldsymbol{R}_n\}_{n\geq 1}$ as follows: We take $(S_1, \boldsymbol{R}_1)$ independent of $\{X_n, n \geq 1\}$. Suppose $(S_1, \boldsymbol{R}_1), \ldots, (S_n, \boldsymbol{R}_n)$ are constructed and take

$$S_{n+1} = S_n + \Delta\mathscr{S}_n \quad \text{and} \quad \boldsymbol{R}_{n+1} = \Omega(R_n^{J-1}, \ldots, R_n^1, \mathscr{R}_n) \tag{2.16}$$

with $\Delta\mathscr{S}_n := \Delta\mathscr{S}$ and $\mathscr{R}_n := \mathscr{R}$, $\Delta\mathscr{S}$ and $\mathscr{R}$ being obtained by applying $\mathscr{B}_{R_n^J}$ to $\{X_{S_n+1}, X_{S_n+2}, \ldots\}$ and where

$$\Omega(x_1, \ldots, x_j) = (x_{1j}, \ldots, x_{jj}) = \text{the ordered sample } (x_1, \ldots, x_j). \tag{2.17}$$

Clearly $\{(S_n, \boldsymbol{R}_n)\}_{n\geq 1}$ satisfies Theorem 1.1(a). Let us now prove that it also satisfies statement (b). We need the following lemmas.

Lemma 2.2. *For any* $J \geq 1$ *let* $\{\theta_n^J\}_{n\geq 1}$ *be the sequence of the J-th records when* $\{X_n\}$ *is i.i.d. For any* $0 < \alpha < 1$ *we have*

$$P\{G(\theta_n^J) > \mathrm{e}^{-\alpha n/J} \text{ i.o.}\} = 0 \tag{2.18}$$

and for any $\beta > 1$ *we have*

$$P(G(\theta_n^J) < \mathrm{e}^{-\beta n/J} \text{ i.o.}\} = 0. \tag{2.19}$$

Proof. See Haiman and Puri (1988).

Lemma 2.3. *Let* $\{(\tilde{T}_n, \tilde{\boldsymbol{\theta}}_n), n \geq 1\}$ *be a sequence where* $\tilde{T}_n$ *is a strictly increasing sequence of integers and* $\{\tilde{\boldsymbol{\theta}}_n = (\tilde{\theta}_n^J, \ldots, \tilde{\theta}_n^1)\}_{n\geq 1}$ *a J-ordered sequence, such that there exists a.s.* $p \geq 1$ *such that for any* $n \geq p$ *we have*

$$\tilde{T}_{n+1} = \tilde{T}_n + \min\{k \geq 1;\ X_{\tilde{T}_n+k} > \tilde{\theta}_n^J\} \tag{2.20}$$

and

$$\tilde{\boldsymbol{\theta}}_{n+1} = \Omega(\tilde{\theta}_n^{J-1}, \ldots, \tilde{\theta}_n^1, X_{\tilde{T}_n+1}). \tag{2.21}$$

Then a.s. there exist two integers n_0 and q such that for any $n \geq n_0$ we have

$$T_n = \tilde{T}_{n-q} \quad \text{and} \quad \boldsymbol{\theta}_n = \tilde{\boldsymbol{\theta}}_{n-q}. \tag{2.22}$$

Proof. See Haiman (1987a), Lemma 4 of §2.

Let L_n, Q_n and $\Delta\mathcal{S}_n$ be the random variables used in the construction of $S_{n+1}, \boldsymbol{R}_{n+1}$.

Lemma 2.4. *If* $1 < \tau < 1 + \beta$ *then*

$$P\{L_n = 0 \text{ i.o.}\} = P\{Q_n = 0 \text{ i.o.}\} = P\left\{\Delta\mathcal{S}_n \geq \frac{\tau}{G(R_n^J)} \log_2\left(\frac{1}{G(R_n^J)}\right) \text{ i.o.}\right\} = 0. \tag{2.23}$$

Proof. By (2.18),

$$P\{L_n = 0 \text{ i.o.}\} = P\{(L_n = 0, G(R_n^J) < e^{-\alpha n/J}) \text{ i.o.}\}, \quad 0 < \alpha < 1, \tag{2.24}$$

and by (2.15),

$$P\{L_n = 0, G(R_n^J) < e^{-\alpha n/J}\} = \int_{G(r) < e^{-\alpha n/J}} (1 - F(r)^{m-1}) \, dP_{R_n^J}(r) \leq \int_{G(r) < e^{-\alpha n/J}} (1 - (1 - G(r))^{m-1}) \, dP_{R_n^J}(r) \leq (\text{const.}) e^{-\alpha n/J}. \tag{2.25}$$

This implies $P\{L_n = 0 \text{ i.o.}\} = 0$ by the Borel-Cantelli lemma. Similarly we obtain, by combining (2.8) and (2.12)–(2.16),

$$P\{Q_n = 0, G(r_n^J) < e^{-\alpha n/J}\} \leq (\text{const.}) \frac{1}{n^{2+\beta-\tau}} \tag{2.26}$$

and

$$P\left\{\Delta\mathcal{S}_n \geq \frac{\tau}{G(R_n^J)} \log_2\left(\frac{1}{G(R_n^J)}\right), G(R_n^J) < e^{-\alpha n/J}\right\} \leq (\text{const.}) n^{-\tau} \tag{2.27}$$

from which we deduce the other equalities in (2.23), since $\sum_{n=1}^{\infty} 1/n^{2+\beta-\tau} < \infty$ and $\sum_{n=1}^{\infty} 1/n^{\tau} < \infty$.

We note here the optimality of hypothesis (H)–(1.4) which leads, by means of Theorem A(iii), to the right hand terms in (2.26) and (2.27) which are general terms of Riemann sums, instead of geometrically decreasing, as we obtained in Haiman (1987a), under (H')–(1.7).

By combining Lemma 2.4, (2.14) and (2.16) we deduce that

$$P\{((\Delta\mathcal{S}_n, \mathcal{R}_n) \neq (m - 1 + \min\{t \geq m, X_{S_n+t} > R_n^J\}, X_{S_n+\Delta\mathcal{S}_n})) \text{ i.o.}\} = 0. \tag{2.28}$$

Thus the last missing piece is to show that

$$P\{\max(X_{S_n+1}, \ldots, X_{S_n+m-1}) > R_n^J \text{ i.o.}\} = 0 \tag{2.29}$$

which combined with (2.28) and Lemma 2.3 implies Theorem 1.1(b).

Let us now prove (2.29). Put

$$A_n = \{\max(X_{S_n+1}, \dots, X_{S_n+m-1}) > R_n^J\} \tag{2.30}$$

and

$$B_n = \left\{L_n = 1, Q_n = 1, \Delta\mathscr{S}_n < \frac{\tau}{G(R_n^J)} \log_2\left(\frac{1}{G(R_n^J)}\right), G(R_n^J) < e^{-\alpha n/J}\right\} \cap A_n. \tag{2.31}$$

By (2.18) and (2.23) we have

$$P\{A_n \text{ i.o.}\} = P\{B_n \text{ i.o.}\} \tag{2.32}$$

and it remains to prove

$$P\{B_n \text{ i.o.}\} = 0. \tag{2.33}$$

By stationarity and m-dependence we have

$$P\{B_n\} = \iint_{\substack{G(r^J) < e^{-\alpha n/J} \\ r^J < r^{J-1}}} dP_{R_n^J, R_n^{J-1}}(r^J, r^{J-1})(\sum\sum(r^J, r^{J-1})) \tag{2.34}$$

where

$$\begin{aligned}\sum\sum = \sum_{t=1}^{\varphi(r^J)} \sum_{t'=1}^{m} \int_{r^J}^{r^{J-1}} & dP(X_1 < r^J, \dots, X_{t-1} < r^J, X_t = r', \\ & X_{t+1} < r', \dots, X_{t+t'-1} < r', X_{t+t'} > r') \\ + \int_{r^{J-1}}^{\infty} & dP(X_1 < r^J, \dots, X_{t-1} < r^J, X_t = r', \\ & X_{t+1} < r^{J-1}, \dots, X_{t+t'-1} < r^{J-1}, X_{t+t'} > r^{J-1})\end{aligned} \tag{2.35}$$

and

$$\varphi(r^J) = \frac{\tau}{G(r^J)} \log_2\left(\frac{1}{G(r^J)}\right).$$

If we write in (2.35),

$$\sum_{t=1}^{\varphi(r^J)} = \sum_{t=1}^{m-1} + \sum_{t=m}^{\varphi(r^J)}, \tag{2.36}$$

we deduce

$$\sum_{t=1}^{m-1} < (\text{const.})\left(\int_{r^J}^{r^{J-1}} \max_{i=2,\dots,m} dP(X_1 = r', X_i > r') + \int_{r^{J-1}}^{\infty} \max_{i=2,\dots,m} dP(X_1 = r', X_i > r^{J-1})\right) \tag{2.37}$$

and

$$\sum_{t=m}^{\varphi(r^J)} < (\text{const.})\left(\sum_{t=1}^{\varphi(r^J)-m} (F(r^J))^t\right)M, \tag{2.38}$$

where

$$M = \left(\int_{r^J}^{r^{J-1}} \max_{i=2,\ldots,m} \mathrm{d}P(X_1 = r', X_i > r') + \int_{r^{J-1}}^{\infty} \max_{i=2,\ldots,m} \mathrm{d}P(X_1 = r', X_i > r^{J-1})\right).$$

Thus a routine calculus exercise show that

$$\sum_{t=1}^{\varphi(r^J)} < (\text{const.})(\log(G(r^J)^{-1}))^{-2-\beta} \tag{2.39}$$

from which we obtain, by (2.34),

$$P(B_n) < (\text{const.})n^{-2-\beta} \tag{2.40}$$

and finally (2.29), by the Borel-Cantelli lemma.

Acknowledgement

We would like to thank Professor Jean Bretagnolle for his valuable comments and suggestions.

References

Haiman, G. (1987a). Etude des extrêmes d'une suite stationnaire m-dépendante avec une application relative aux accroissements du processus de Wiener. *Ann. Inst. Henri Pioncaré* **23**(3), 425–458.

Haiman, G. (1987b). Almost sure asymptotic behavior of the record and record time sequences of a stationary Gaussian process. In: M.L. Puri et al., Eds., *Mathematical Statistics and Probability Theory*, Vol. A, 105–120. Reidel, Dordrecht-Boston.

Haiman, G. and M.L. Puri (1988). A strong invariance principle concerning the J-upper order statistics for a stationary Gaussian sequence. Preprint, Department of Mathematics, Indiana University, Bloomington, IN.

A STRONG INVARIANCE PRINCIPLE CONCERNING THE J-UPPER ORDER STATISTICS FOR STATIONARY GAUSSIAN SEQUENCES

GEORGE HAIMAN and MADAN L. PURI[1]

Université de Lille and Indiana University

It is shown that in the case of stationary Gaussian processes, the Jth ($J \geq 1$) record times $\{T_n, n \geq 1\}$ and the corresponding J-upper order statistics $\{X_{T_n - J+1, T_n}, \dots, X_{T_n, T_n}\}$ can almost surely be identified via a translation of the time index n to the corresponding elements defined on a sequence of independent and identically distributed random variables. A construction method for approximating sequences of record times and the corresponding upper order statistics introduced by Haiman (1987a, b) for the case $J = 1$ is extended and applied under weaker conditions concerning the covariance function, and also under different sets of new hypotheses.

1. Introduction. Let $\{X_n, -\infty < n < \infty\}$ be a stationary Gaussian sequence centered at 0 with covariance function

$$\Gamma(n) = E(X_i X_{i+n}), \qquad \Gamma(0) = 1, \qquad n = 0, \pm 1, \pm 2, \dots. \tag{1.1}$$

Let $X_{1,n} < X_{2,n} < \cdots < X_{n-1,n} < X_{n,n}$ denote the order statistics of $X_1, \dots, X_n$. Let $J \geq 1$ be a fixed integer and $\Theta_0 > 0$ a fixed real number. Denote

$$\begin{aligned} T_1 &= \inf\{n; n \geq J, X_{n-J+1,n} > \Theta_0\}, \\ \underline{\Theta}_1 &= \{X_{T_1-J+1,T_1}, \dots, X_{T_1,T_1}\} \end{aligned} \tag{1.2}$$

and, for any $k \geq 2$, set

$$\begin{aligned} T_k &= \inf\{n; n > T_{k-1}, X_{n-J+1,n} > X_{n-J,n-1}\}, \\ \underline{\Theta}_k &= \{X_{T_k-J+1,T_k}, \dots, X_{T_k,T_k}\} =: \{\Theta_k^J, \Theta_k^{J-1}, \dots, \Theta_k^1\}. \end{aligned} \tag{1.3}$$

We shall call Θ_k^J the kth Jth-*upper record*. Note that our definition is a slight variant of the usual definition of the Jth-upper record [see, e.g., Resnick (1987), page 243, and Deheuvels (1988)]. Note also that when $J > 1$, Θ_k^1 is not the kth-upper record. T_k's are the Jth record times, and $\{T_k, \underline{\Theta}_k\}_{k \geq 1}$ describe the history of the J largest values of the sample $X_1, \dots, X_n$ for $n \geq T_1$. In the sequel, we will call $\{T_k, \underline{\Theta}_k\}$ the Jth record sequence based on $\{X_n, n \geq 1\}$ and Θ_0.

DEFINITION. A sequence $\{\underline{r}_k = (r_k^J, \dots, r_k^1) \in \mathbb{R}^J, k \geq 1\}$ is called J-ordered if the following conditions hold: For any $k \geq 1$, $r_k^J < \cdots < r_k^1$ and, either

Received November 1988; revised July 1991.

[1]Research supported by the Office of Naval Research under Contract N00014-91-J-1020.

AMS 1991 *subject classifications*. Primary 60F15, 60G15; secondary 62G30.

Key words and phrases. Strong invariance, Gaussian processes, record times, order statistics.

$r_k^1 < r_{k+1}^1$ and then $r_{k+1}^J = r_k^{J-1}, r_{k+1}^{J-1} = r_k^{J-2}, \ldots, r_{k+1}^2 = r_k^1$; or there exists j, $2 \le j \le J$, such that $r_k^j < r_{k+1}^j < r_k^{j-1}$, and then $r_{k+1}^i = r_k^{i-1}$ if $i > j$ and $r_{k+1}^i = r_k^i$ if $i \le j$.

Note that when $\{X_n, n \ge 1\}$ is a sequence of iid (independent and identically distributed) random variables with a continuous distribution, then for any increasing sequence of positive integers $\{t_k, k \ge 1\}$ and any J-ordered sequence $\{\underline{r}_k = (r_k^J, \ldots, r_k^1), \ k \ge 1\}$, we have the following Markov behavior [see Deheuvels (1974)] of the sequence $\{T_k, \underline{\Theta}_k\}_{k \ge 1}$:

$$P\left[T_{n+1} = t_{n+1}, \Theta_{n+1}^j < r_{n+1}^j \middle| T_1 = t_1, \underline{\Theta}_1 = \underline{r}_1, \ldots, T_n = t_n, \underline{\Theta}_n = \underline{r}_n\right]$$

$$(1.4) \qquad = \left[P\left(X_1 < r_n^J\right)\right]^{t_{n+1} - t_n - 1} \cdot P\left[r_n^j < X_1 < r_{n+1}^j\right],$$

$$n \ge 1, j = 1, \ldots, J.$$

Consider now Hypotheses H_1, H_2 and H_3.

HYPOTHESIS H_1.

$$(1.5) \qquad \sum_{n=1}^{\infty} |\Gamma(n)| < 1/2.$$

(1.6) *There exists an* $\varepsilon > 0$ *such that* $\limsup_{n \to \infty} |\Gamma(n)| n^{4+\varepsilon} = 0$.

HYPOTHESIS H_2. *$\{X_n, n \ge 1\}$ has a spectral density $f(\lambda) \ge 0$, $-\pi < \lambda < \pi$, such that*

$$(1.7) \qquad \int_{-\pi}^{\pi} [f(\lambda)]^{-1} \, d\lambda < \infty.$$

(1.8) *There exists an* $\varepsilon > 0$ *such that* $\limsup_{n \to \infty} |\Gamma(n)| n^{6+\varepsilon} = 0$.

To formulate Hypothesis H_3, we need the following preliminaries:

Let $\rho(\tau)$ be the coefficient of regularity of the sequence $\{X_n, -\infty < n < \infty\}$, defined by

$$(1.9) \qquad \rho(\tau) = \max |E(\eta_1 \eta_2)|,$$

where the maximum is taken over all $\eta_1 \in H(-\infty, 0)$ and $\eta_2 \in H(\tau, \infty)$, $\tau \ge 1$, with $E(\eta_1^2) = E(\eta_2^2) = 1$, and where $H(r, s)$ denotes the Hilbert space of random variables generated by $\{X_n; r \le n \le s\}$.

HYPOTHESIS H_3. *There exists an* $\varepsilon > 0$ *such that*

$$(1.10) \qquad \limsup_{n \to \infty} \rho(n) n^{6+\varepsilon} = 0.$$

Our main result is the following:

THEOREM 1.1. *Let Hypothesis* H_1, H_2 *or* H_3 *hold. Then, for any fixed* $J \geq 1$, *there exists a probability space which carries, in addition to the sequence* $\{X_n, n \in (-\infty, \infty)\}$, *an iid sequence* $\{X_n^*, n \geq 1\}$ *of random variables each having a* $\mathscr{N}(0,1)$ *distribution. Moreover, there exists a* Θ_0 (> 0) *such that if* $\{T_k, \underline{\Theta}_k\}_{k \geq 1}$ *is the Jth record sequence based on* $\{X_n, n \in (-\infty, \infty)\}$ *and* Θ_0 *and* $\{S_k, \underline{R}_k\}_{k \geq 1}$ *is the Jth record sequence based on* $\{X_n^*, n \geq 1\}$ *and* Θ_0, *then there exist almost surely an* n_0 *and* q *such that for all* $n \geq n_0$, *we have*

$$S_n = T_{n-q} \quad \text{and} \quad \underline{R}_n = \underline{\Theta}_{n-q}. \tag{1.11}$$

Moreover, if $\{\underline{M}_n\}_{n \geq J} := \{(X_{n-J+1,n}, \ldots, X_{n,n})\}_{n \geq J}$ *and* $\{\underline{M}_n^*\}_{n \geq J} := \{(X_{n-J+1,n}^*, \ldots, X_{n,n}^*)\}_{n \geq J}$, *then there exists almost surely an* n_1 *and* q' *such that for any* $n \geq n_1$, *we have* $\underline{M}_n^* = \underline{M}_{n-q'}$.

The proof of this theorem is given in Section 2.

REMARK 1. For $J = 1$, Haiman (1987a) proved Theorem 1.1 under:

HYPOTHESIS H_1'. (i) $\sum_{n=1}^{\infty} |\Gamma(n)| < 1/2$ *and* (ii) *there exists an* $\varepsilon > 0$ *such that* $\limsup_{n \to \infty} |\Gamma(n)| n^{4((1+\delta)/(1-\delta))+\varepsilon} = 0$ *with* $\delta = \max_{n \geq 1} |\Gamma(n)| < 1$.

It may be observed that H_1'(ii) is more restrictive than (1.6) [since $((1 + \delta)/(1 - \delta)) > 1$].

REMARK 2. Note that under Hypotheses H_1, H_2 or H_3, f is continuous and $f(\lambda) = 1 + 2\sum_{n=1}^{\infty} \Gamma(n) \cos n\lambda > 0$. Consequently, (1.5) $\Rightarrow \inf_\lambda f(\lambda) > 0 \Rightarrow$ (1.7) and (1.8) $\Rightarrow$ (1.6).

REMARK 3. A weakly stationary sequence $\{Y_n, n \geq 1\}$ is called strong mixing (or α-mixing) if

$$\alpha(\tau) = \max |P(A \cap B) - P(A) \cdot P(B)| \to 0 \quad \text{as } \tau \to \infty, \tag{1.12}$$

where the maximum is over $A \in \sigma\{Y_j;\ j \leq t\}$ and $B \in \sigma\{Y_l;\ l \geq t + \tau\}$.

When $\{Y_n, n \geq 1\}$ is Gaussian, then for any $\tau > 0$ [see Ibragimov and Rozanov (1974), page 133],

$$\alpha(\tau) \leq \rho(\tau) \leq 2\pi\alpha(\tau). \tag{1.13}$$

Thus in (1.10), $\rho(n)$ may be replaced by the strong mixing coefficient $\alpha(n)$.

REMARK 4. It is easy to show [cf. Ibragimov and Rozanov (1974), page 174], that for any $n \geq 1$, $\Gamma(n) \leq \rho(n) \int_{-\pi}^{\pi} f(\lambda)\, d\lambda$. Thus (1.10) $\Rightarrow$ (1.8).

Moreover, for Gaussian sequences, complete regularity [i.e., $\lim_{n \to \infty} \rho(n) = 0$] implies regularity [see Ibragimov and Rozanov (1974), pages 130–136] which is equivalent to the well-known condition

$$(1.14) \qquad \int_{-\pi}^{\pi} \log f(\lambda)\, d\lambda > -\infty.$$

Thus (1.10) $\Rightarrow$ (1.14).

It may be shown [see Grenander and Rosenblatt (1957), pages 65–69 and 82–83] that (1.7) $\Rightarrow$ (1.14), which gives a compensation phenomenon similar to that observed in Remark 2.

To summarize, we have the following relationship between the hypotheses:

$$(1.5) \Rightarrow (1.7) \Rightarrow \int_{-\pi}^{\pi} \log f(\lambda)\, d\lambda > -\infty$$

$$\Uparrow$$

$$(1.6) \Leftarrow (1.8) \Leftarrow \text{Hypothesis } H_3.$$

The following theorem gives the full characterization in terms of spectral density of the weakly stationary sequences (not necessarily Gaussian) satisfying a condition similar to (1.10).

Theorem 1.2 [Ibragimov and Rozanov (1974), page 212]. *A necessary and sufficient condition for*

$$\rho(\tau) = O(\tau^{-r-\beta}), \qquad \tau \to \infty, r \in \mathbb{N}, 0 < \beta < 1,$$

is that

$$f(\lambda) = |P(e^{i\lambda})|^2 \omega(\lambda),$$

where $P(Z)$ is a polynomial with zeros on $|Z| = 1$, $\omega(\lambda) \geq m > 0$, and the rth derivative of ω satisfies the Lipschitz condition of order β (for every λ_1, λ_2, $|\omega(\lambda_1) - \omega(\lambda_2)| < c|\lambda_1 - \lambda_2|^{\beta}$).

Theorem 1.1 has direct applications in extreme value theory by enabling one to reduce limiting theorems for dependent sequences to the iid case by means of a strong invariance principle. For the general classical results dealing with the extreme value theory for stationary processes, the reader is referred to Leadbetter and Rootzén (1988) and the references cited therein.

We shall prove Theorem 1.1 by first constructing on the probability space (eventually enlarged by products) on which $\{X_n, n \geq 1\}$ is defined, a sequence $\{S_n, \underline{R}_n\}_{n \geq 1}$ such that: (a) $\{S_k, \underline{R}_k\}_{k \geq 1}$ has the same distribution as the Jth record sequence based on an iid sequence of $\mathcal{N}(0,1)$ random variables and Θ_0 [i.e., satisfying (1.4) with T_k and $\underline{\Theta}_k$ replaced by S_k and $\underline{R}_k$, respectively]; (b) there exists almost surely an n_0 and a q such that for any $n \geq n_0$, we have $S_n = T_{n-q}$ and $\underline{R}_n = \underline{\Theta}_{n-q}$. The construction of the iid sequence $\{X_n^*, n \geq 1\}$ such that $\{S_k, \underline{R}_k\}_{k \geq 1}$ is the jth record sequence based on $\{X_n^*, n \geq 1\}$ and Θ_0 is then straightforward.

2. Proof of Theorem 1.1 for $J = 1$.

2.1. *Some preliminaries.* In this section, we shall first prove some basic lemmas which are needed to prove Theorem 1.1. To make the results self-contained and to provide proper motivation, we shall first prove Theorem 1.1 for $J = 1$.

The following lemma is a specialization of the classical comparison theorem of Berman (1964).

LEMMA 2.1. *Suppose $\xi_1, \dots, \xi_n$ are standard normal random variables with positive definite covariance matrix $\Lambda^1 = (\Lambda^1_{ij})$ and $\eta_1, \dots, \eta_n$ are standard normal random variables with positive definite covariance matrix $\Lambda^0 = (\Lambda^0_{ij})$. Further, let $a_1 < b_1, \dots, a_n < b_n$ be real numbers. Then*

$$|P\{a_1 < \xi_1 < b_1, \dots, a_n < \xi_n < b_n\} - P\{a_1 < \eta_1 < b_1, \dots, a_n < \eta_n < b_n\}|$$

$$\leq \frac{2}{\pi} \sum_{1 \leq i < j \leq n} \left|\Lambda^1_{ij} - \Lambda^0_{ij}\right| \left(1 - \delta^2_{ij}\right)^{-1/2} \exp\left(-\frac{c_i^2 + c_j^2}{2(1 + \delta_{ij})}\right), \tag{2.1}$$

where $c_i = \min(|a_i|, |b_i|)$, $i = 1, \dots, n$, and $\delta_{ij} = \max(|\Lambda^1_{ij}|, |\Lambda^0_{ij}|)$.

PROOF. Let f_1 be the joint density of $(\xi_1, \dots, \xi_n)$ and f_0 the joint density of $(\eta_1, \dots, \eta_n)$ (which exist because Λ^1 and Λ^0 are positive definite). Writing $\underline{a} = (a_1, \dots, a_n)$ and $\underline{b} = (b_1, \dots, b_n)$, we have

$$P(a_1 < \xi_1 < b_1, \dots, a_n < \xi_n < b_n) = \int_{\underline{a}}^{\underline{b}} \cdots \int f_1(y_1, \dots, y_n)\, d\underline{y}$$

and

$$P(a_1 < \eta_1 < b_1, \dots, a_n < \eta_n < b_n) = \int_{\underline{a}}^{\underline{b}} \cdots \int f_0(y_1, \dots, y_n)\, d\underline{y}.$$

For any $0 < h < 1$, put

$$\Lambda^h = (\Lambda_{ij}: 1 \leq i, j \leq n) = h\Lambda^1 + (1 - h)\Lambda^0, \tag{2.2}$$

which is positive definite.

Let f_h be the density of $\mathcal{N}(0, \Lambda^h)$ random variable and put

$$F(h) = \int_{\underline{a}}^{\underline{b}} \cdots \int f_h(y_1, \dots, y_n)\, d\underline{y}.$$

The difference in the first term in (2.1) equals $F(1) - F(0)$ and

$$F(1) - F(0) = \int_0^1 F'(h)\, dh = \int_0^1 \left[\int_{\underline{a}}^{\underline{b}} \cdots \int \frac{\partial f_h(y_1, \dots, y_n)}{\partial h}\, d\underline{y}\right] dh. \tag{2.3}$$

From (2.2), it follows that

$$\frac{\partial \Lambda_{ij}^h}{\partial h} = \Lambda_{ij}^1 - \Lambda_{ij}^0$$

and hence that

$$F'(h) = \sum_{i<j} \int \int_{\underline{a}}^{\underline{b}} \cdots \int \frac{\partial f_h}{\partial \Lambda_{ij}^h} \frac{\partial \Lambda_{ij}^h}{\partial h} \, d\underline{y} = \sum_{i<j} \left(\Lambda_{ij}^1 - \Lambda_{ij}^0\right) \int_{\underline{a}}^{\underline{b}} \cdots \int \frac{\partial f_h}{\partial \Lambda_{ij}^h} \, d\underline{y}. \tag{2.4}$$

But since f_h is Gaussian, we have

$$\frac{\partial f_h}{\partial \Lambda_{ij}^h} = \frac{\partial^2 f_h}{\partial y_i \, \partial y_j},$$

which combined with (2.4) implies

$$F'(h) = \sum_{i<j} \left(\Lambda_{ij}^1 - \Lambda_{ij}^0\right) \int_{\underline{a}}^{\underline{b}} \cdots \int \frac{\partial^2 f_h}{\partial y_i \, \partial y_j} \, d\underline{y}. \tag{2.5}$$

By integrating first with respect to y_i and y_j in (2.4), we obtain

$$\begin{aligned} F'(h) = \sum_{i<j} &\left(\Lambda_{ij}^1 - \Lambda_{ij}^0\right) \\ &\times \int_{\underline{a}'}^{\underline{b}'} \cdots \int \big[f_h(y_i = b_i, y_j = b_j) - f_h(y_i = b_i, y_j = a_j) \\ &\qquad - f_h(y_i = a_i, y_j = b_j) + f_h(y_i = a_i, y_j = a_j) \big] \, d\underline{y}', \end{aligned} \tag{2.6}$$

where $f_h(y_i = u_i, y_j = u_j)$ is the function of $n - 2$ variables obtained by fixing $y_i = u_i$ and $y_j = u_j$, and where $d\underline{y}' = \prod_{k \neq i, k \neq j} dy_k$.

If $u_i = a_i$ or b_i and $u_j = a_j$ or b_j, $i < j$, then

$$\begin{aligned} &\int_{\underline{a}}^{\underline{b}} \cdots \int f_h(y_i = u_i, y_j = u_j) \, d\underline{y}' \\ &\le \int_{-\infty}^{+\infty} \cdots \int_{-\infty}^{+\infty} \cdots \int f_h(y_i = u_i, y_j = u_j) \, d\underline{y}' \\ &= \frac{1}{2\pi\left(1 - \left(\Lambda_{ij}^h\right)^2\right)^{1/2}} \exp\left\{ - \frac{\left(u_i^2 - 2\Lambda_{ij}^h u_i u_j + u_j^2\right)}{2\left(1 - \left(\Lambda_{ij}^h\right)^2\right)} \right\}. \end{aligned} \tag{2.7}$$

On the other hand,

$$u_i^2 - 2\Lambda_{ij}^h u_i u_j + u_j^2 \ge \left(u_i^2 + u_j^2\right)\left(1 - \left|\Lambda_{ij}^h\right|\right) \tag{2.8}$$

and

$$|\Lambda_{ij}^h| \le \max\left(|\Lambda_{ij}^1|, |\Lambda_{ij}^0|\right) = \delta_{ij}. \tag{2.9}$$

By combining (2.3), (2.6), (2.7), (2.8) and (2.9), we deduce (2.1). □

Let $0 < K < 1$ be a constant and for any $r > 0$, define $\mu_K(r)$ by $P\{X_1 > \mu_K(r)\} = P\{X_1 > r\}^{1+K} = [1 - \Phi(r)]^{1+K}$, where $\Phi(\cdot)$ is the standard normal distribution function. Denote by $g(x_1, \dots, x_t)$ the joint density of $X_1, \dots, X_t$. For any integer $t \ge 1$ and any $0 < r < s$, put

$$\begin{aligned} Q(t,r,s) &= \int_D g(x_1,\dots,x_t)\, d\underline{x}_t, \\ D &= \{-\mu_K(r) < x_i < s,\ i = 1,\dots,t\}, \end{aligned} \tag{2.10}$$

and denote by $Q_I(t, r, s)$ the value of $Q(t, r, s)$ when the X_n are iid. [Though $Q(t, r, s)$ and $Q_I(t, r, s)$ depend on K, we have suppressed this fact for notational convenience.]

LEMMA 2.2. *If $\sum_{n \ge 1} |\Gamma(n)| < \infty$, then there exist an $r_1 > 0$, two positive constants c_1 and c_2 and a $\tau_0 > 0$ such that for all $r \ge r_1$, $0 < \tau < \tau_0$ and $0 < K < 1$,*

$$\sup_{\substack{1 \le t \le (\tau/G(r)) \log(1/G(r)) \\ 0 < r \le s < \mu_K(r)}} \left| \frac{Q(t,r,s)}{Q_I(t,r,s)} - 1 \right| < c_1 [G(r)]^{c_2}, \tag{2.11}$$

where $G(r) = 1 - \Phi(r)$.

PROOF. We first apply Lemma 2.1 with $\Lambda_{ij}^0 = 0$ and $\Lambda_{ij}^1 = \Gamma(i-j)$, $i \ne j$, for which clearly Λ^0 and Λ^1 are positive definite and $a_i = -\mu_K(r)$, $b_i = s$, $1 \le i \le n$. Then, for any $t \ge 1$ and $0 < r < s \le \mu_K(r)$, we have

$$\begin{aligned} &|Q(t,r,s) - Q_I(t,r,s)| \\ &\quad\le \frac{2}{\pi} \sum_{1 \le i < j \le t} |\Gamma(i-j)| \left(1 - (\Gamma(i-j))^2\right)^{-1/2} \\ &\qquad \times \exp\left\{ -\frac{s^2}{1 + |\Gamma(i-j)|} \right\} \\ &\quad\le (\text{const.})\, t \sum_{k=1}^{t} |\Gamma(k)| \exp\left\{ \frac{-s^2}{1 + |\Gamma(k)|} \right\} \\ &\quad\le (\text{const.})\, t \exp\left\{ \frac{-s^2}{1+\delta} \right\} \\ &\quad\le (\text{const.})\, t \exp\left\{ \frac{-r^2}{1+\delta} \right\}, \end{aligned} \tag{2.12}$$

where $\delta = \max_{i \geq 1}|\Gamma(i)| < 1$ and (const.) is a generic constant (independent of K and τ) which may change from step to step. Note that $|Q(t,r,s) - Q_1(t,r,s)| \leq (\text{const.})t \exp(-r^2/(1+\delta))$ when we have $0 < r < s$ only.

Now, if $0 < r < s < \mu_K(r)$ and $1 \leq t \leq (\tau/G(r))\log(1/G(r))$, we have, for $G(r) < 1/2$,

$$\begin{aligned} Q_I(t,r,s) &= \left(P\{-\mu_K(r) < X_1 < s\}\right)^t = \left(1 - [G(r)]^{1+K} - G(s)\right)^t \\ &\geq \left(1 - 2G(r)\right)^{(\tau/G(r))\log(1/G(r))} \\ &\geq [G(r)]^{2\tau} \exp\left\{\frac{2\tau G(r)}{1 - 2G(r)} \log G(r)\right\}, \end{aligned}$$

by using $-\log(1-u) \leq u + u^2/(2(1-u))$ for $0 \leq u < 1$.

Thus, since $|G(r)\log G(r)| \leq 1/e$, there exists $r_1 > 0$ and $\tau_1 > 0$ such that for any $r \geq r_1$, $0 < \tau < \tau_1$ and $0 < r \leq s < \mu_K(r)$, we have

$$Q_I(t,r,s) > \frac{[G(r)]^{2\tau}}{2} \quad \text{for } 1 \leq t \leq \frac{\tau}{G(r)} \log \frac{1}{G(r)}. \tag{2.13}$$

Combining (2.12) and (2.13) and using the fact that $G(r) \sim (2\pi)^{-1/2}(1/r) e^{-r^2/2}$ as $r \to \infty$, we obtain for $r \geq r_1$, $0 < \tau < \tau_1$, $0 < r \leq s < \mu_K(r)$ and $1 \leq t \leq (\tau/G(r))\log(1/G(r))$,

$$\begin{aligned} \left|\frac{Q(t,r,s)}{Q_I(t,r,s)} - 1\right| &\leq (\text{const.})\left(\frac{\tau}{G(r)} \log \frac{1}{G(r)}\right)[G(r)]^{-2\tau} \exp\left\{\frac{-r^2}{1+\delta}\right\} \\ &\leq (\text{const.}) r^{3+2\tau} \exp\left\{r^2\left(\frac{-1}{1+\delta} + \frac{1+2\tau}{2}\right)\right\}. \end{aligned}$$

Now, noting that $-1/(1+\delta) + (1+2\tau)/2 < 0$ is equivalent to $2\tau < (1-\delta)/(1+\delta)$ and using the fact that for any $0 < c'' < c'$,

$$r^{3+2\tau} \exp\{-(c'/2)r^2\}/[G(r)]^{c''} \to 0 \quad \text{as } r \to \infty,$$

we readily obtain (2.11) by choosing τ and c_2 in such a way that

$$0 < \tau < \min\left(\tau_1, \frac{1-\delta}{2(1+\delta)}\right) \quad \text{and} \quad 0 < c_2 < \frac{1}{2}\left(\frac{1}{1+\delta} - \frac{1+2\tau}{2}\right). \qquad \square$$

For any integer t and any $0 < r < s$, put

$$P(t,r,s) = \int_{D'} g(x_1, \ldots, x_{t-1}, x_t = s) \prod_{i=1}^{t-1} dx_i \tag{2.14}$$

with

$$D' = \{-\mu_K(r) < x_i < r, i = 1, \ldots, t-1\},$$

and denote by $P_I(t,r,s)$ the probability in (2.14) when the X_n are iid. [Though

$P(t, r, s)$ and $P_I(t, r, s)$ depend on K, we have suppressed this fact for notational convenience.]

LEMMA 2.3. *If $\sum_{n \geq 1} |\Gamma(n)| < \infty$ and $\int_{-\pi}^{\pi} \log f(\lambda)\, d\lambda > -\infty$, then there exists an $r_2 > 0$ and two positive constants c_1 and c_2, a $K_0 > 0$ and a $\tau_0 > 0$ such that for all $r \geq r_2$, $0 < \tau \leq \tau_0$ and $0 < K \leq K_0$,*

$$\sup_{\substack{1 \leq t \leq (\tau/G(r))\log(1/G(r)) \\ 0 < r \leq s < \mu_K(r)}} \left| \frac{P(t, r, s)}{P_I(t, r, s)} - 1 \right| < c_1 [G(r)]^{c_2}. \tag{2.15}$$

[Note that the conditions of the lemma are satisfied if either one of Hypotheses H_1, H_2 or H_3 are satisfied (see Section 1, Remarks 3 and 4).]

PROOF. With the notation in Lemmas 2.1 and 2.2, let

$$\Lambda_{ij}^0 = \Gamma(i - j), \qquad 1 \leq i, j \leq t - 1, \qquad \Lambda_{it}^0 = 0 \quad \text{if } i \neq t \text{ and } \Lambda_{tt}^0 = 1.$$

Set

$$\Lambda_{ij}^1 = \Gamma(i - j), \qquad 1 \leq i, j \leq t, \tag{2.16}$$

and consider first with $Q(t, r, s)$ as defined in (2.10),

$$\begin{aligned} &P(t, r, s) - Q(t - 1, r, r)\varphi(s) \\ &\quad =: F(1) - F(0) = \int_0^1 F'(h)\, dh, \qquad \varphi(s) = \frac{1}{\sqrt{2\pi}} e^{-s^2/2}. \end{aligned} \tag{2.17}$$

[Note that F in (2.17) is not the same as in (2.3).]

Let f_h be the density of an $\mathcal{N}(0, \Lambda^h)$ variable associated with $\Lambda^h = h\Lambda^1 + (1 - h)\Lambda^0$. Then, by proceeding as in (2.4) to (2.6), we have

$$\begin{aligned} F'(h) &= \sum_{i<j} \left(\Lambda_{ij}^1 - \Lambda_{ij}^0\right) \\ &\quad \times \underbrace{\int_{-\mu_K(r)}^{r} \cdots \int_{-\mu_K(r)}^{r}}_{t-1 \text{ times}} \frac{\partial f_h}{\partial \Lambda_{ij}^h}(y_1, \ldots, y_{t-1}, y_t = s)\, dy_1 \cdots dy_{t-1} \\ &= \sum_{i=1}^{t-1} \Gamma(t - i) \int_{-\mu_K(r)}^{r} \cdots \int_{-\mu_K(r)}^{r} \frac{\partial^2 f_h}{\partial y_i\, \partial y_t}(y_1, \ldots, y_{t-1}, y_s)\, dy_1 \cdots dy_{t-1} \\ &= \sum_{i=1}^{t-1} \Gamma(t - i) \underbrace{\int_{-\mu_K(r)}^{r} \cdots \int_{-\mu_K(r)}^{r}}_{t-2 \text{ times}} \left[\frac{\partial f_h}{\partial y_t}(y_1, \ldots, y_i = r, y_{i+1}, \ldots, y_t \geqq s) \right. \\ &\qquad \left. - \frac{\partial f_h}{\partial y_t}(y_1, \ldots, y_i = -\mu_K(r), y_{i+1}, \ldots, y_t = s) \right] \prod_{\substack{k=1 \\ k \neq i}}^{t-1} dy_k. \end{aligned} \tag{2.18}$$

Now, let for any $0 \le h \le 1$,

$$X_j^h = X_j, \qquad j = 1, \dots, t-1, \quad \text{and} \quad X_t^h = hX_t + Y^h, \tag{2.19}$$

where Y^h is independent of $X_1, \dots, X_t$, and has $\mathscr{N}(0, 1-h^2)$ distribution. Let S^h be the lower triangular matrix such that the random variables

$$Z_i^h = \sum_{j=1}^{i} S_{ij}^h X_j^h, \qquad i = 1, \dots, t, \tag{2.20}$$

form an orthonormal system [i.e., $E(Z_i^h Z_j^h) = 0$ if $i \ne j$ and $E(Z_i^h Z_j^h) = 1$ if $i = j$]. Then

$$f_h(y_1, \dots, y_t) = \frac{|S^h|}{(2\pi)^{t/2}} \exp\left\{ -\frac{1}{2} \sum_{i=1}^{t} \left(\sum_{j=1}^{i} S_{ij}^h y_j \right)^2 \right\}. \tag{2.21}$$

From (2.19) and (2.20) it follows that for any $0 \le h \le 1$,

$$S_{ij}^h = S_{ij}^1 \quad \text{for } 1 \le i, j \le t-1.$$

Note that we can choose Z_t^h in (2.20) of the form

$$Z_t^h = a\left(\sum_{j=1}^{t-1} S_{tj}^1 X_j + \frac{S_{tt}^1}{h} X_t^h \right),$$

since for any $1 \le k \le t-1$,

$$E\left(Z_t^h Z_k^h\right) = 0,$$

and the condition $\|Z_t^h\|^2 := E((Z_t^h)^2) = 1$ is satisfied for

$$a^2 = \frac{h^2}{h^2 + \left(S_{tt}^1\right)^2 (1-h)^2} < 1.$$

Denote by $X_t^{X_r, \dots, X_s}$ the orthogonal projection of X_t on $H(r, s)$. Then we have

$$Z_t^1 = \left(X_t - X_t^{X_1, \dots, X_{t-1}}\right) \cdot \left\| X_t - X_t^{X_1, \dots, X_{t-1}} \right\|^{-1}$$

from which it follows that $S_{tt}^1 = \|X_t - X_t^{X_1, \dots, X_{t-1}}\|^{-1} > 1$ and [see Grenander and Rosenblatt (1957), page 69]

$$\begin{aligned} \left|S_{tt}^1\right|^2 &\le \frac{1}{\begin{pmatrix} \text{mean square error} \\ \text{of prediction} \\ \text{one step ahead} \end{pmatrix}} \\ &= \frac{1}{2\pi} \exp\left\{ -\frac{1}{2\pi} \int_{-\pi}^{\pi} \log f(\lambda)\, d\lambda \right\} =: L^2. \end{aligned} \tag{2.22}$$

Thus, for $0 \le h \le 1$,

$$\left|S_{ij}^h\right| \le \left|S_{ij}^1\right|, \qquad 1 \le i, j \le t. \tag{2.23}$$

Going back to (2.18), and using (2.21), (2.22) and (2.23) and the fact that for any $0 < \alpha < 1$, there exists $c(\alpha)$ such that

$$xe^{-x^2/2} \le c(\alpha)e^{-\alpha x^2/2}, \qquad x \ge 0,$$

we have, with $z_i^h = \sum_{j=1}^{i} S_{ij}^h y_j^h$, $i = 1, \ldots, t$, $0 < \alpha < 1$,

$$\left|\frac{\partial f_h}{\partial y_t}(\underline{y}_t)\right| = \left|\sum_{j=1}^{t} S_{tj}^h y_j\right| \left|S_{tt}^h\right| f_h(\underline{y}_t) \le \frac{|S^h|}{(2\pi)^{t/2}} L \left|z_t^h\right| \exp\left\{-\frac{1}{2}\sum_{i=1}^{t}\left(z_i^h\right)^2\right\}$$

$$(2.24) \qquad \le \frac{|S^h|}{(2\pi)^{t/2}} Lc(\alpha) \exp\left\{-\frac{1}{2}\alpha\sum_{i=1}^{t}\left(z_i^h\right)^2\right\}$$

$$= \frac{|S^h|}{(2\pi)^{t/2}} Lc(\alpha) \exp\left\{-\frac{1}{2}\alpha\sum_{i=1}^{t}\left(\sum_{j=1}^{i} S_{ij}^h y_j\right)^2\right\}$$

$$= Lc(\alpha) f_h\left(\sqrt{\alpha}\, y_1, \ldots, \sqrt{\alpha}\, y_t\right).$$

Next, if in (2.18) we write

$$\underbrace{\int_{-\mu_K(r)}^{r} \cdots \int_{-\mu_K(r)}^{r}}_{t-2 \text{ times}} \frac{\partial f_h}{\partial y_t}(y_1, \ldots, y_i = r, y_{i+1}, \ldots, y_t = s)\, d\underline{y}'$$

$$= \int_{-\mu_K(r)}^{r} \cdots \int_{-\mu_K(r)}^{r} \frac{\partial f_h}{\partial y_t}(y_i = r, y_t = s)\, d\underline{y}',$$

then by (2.24), for any $1 \le i \le t-1$, we have

$$\left|\int_{-\mu_K(r)}^{r} \cdots \int_{-\mu_K(r)}^{r} \frac{\partial f_h}{\partial y_t}(y_i = r, y_t = s)\, d\underline{y}'\right|$$

$$\le Lc(\alpha) \underbrace{\int_{-\infty}^{+\infty} \cdots \int_{-\infty}^{+\infty}}_{t-2 \text{ times}} f_h\left(\sqrt{\alpha}\, y_i = \sqrt{\alpha}\, r, \sqrt{\alpha}\, y_t = \sqrt{\alpha}\, s\right) d\underline{y}'$$

$$(2.25a) \qquad = \frac{Lc(\alpha)}{\alpha^{(t-2)/2} 2\pi\left(1 - h^2\Gamma^2(t-i)\right)}$$

$$\times \exp\left\{-\frac{1}{2}\alpha\left(\frac{r^2 - 2h\Gamma(t-i)rs + s^2}{1 - h^2\Gamma^2(t-i)}\right)\right\}$$

$$=: A_i(h),$$

and similarly for the term containing $\mu_K(r)$ in (2.18), for which we obtain

$$\left|\int_{-\mu_K(r)}^{r}\cdots\int_{-\mu_K(r)}^{r}\frac{\partial f_h}{\partial y_t}(y_i=\mu_K(r),y_t=s)\,d\underline{y}'\right|$$

$$\leq \frac{L\cdot c(\alpha)}{\alpha^{(t-2)/2}2\pi(1-h^2\Gamma^2(t-i))} \tag{2.25b}$$
$$\times \exp\left\{-\frac{1}{2}\alpha\left(\frac{(\mu_K(r))^2-2h\Gamma(t-i)\mu_K(r)s+s^2}{1-h^2\Gamma^2(t-i)}\right)\right\}$$
$$=: A_i'(h).$$

By (2.17) and (2.18),

$$\left|\frac{P(t,r,s)-Q(t-1,r,r)\varphi(s)}{Q(t-1,r,r)\varphi(s)}\right|$$
$$\leq \int_0^1\frac{|F'(h)|}{Q(t-1,r,r)\varphi(s)}\,dh \tag{2.26}$$
$$\leq \frac{\int_0^1\sum_{i=1}^{t-1}|\Gamma(t-i)|(A_i(h)+A_i'(h))\,dh}{Q(t-1,r,r)\varphi(s)}$$
$$\leq \sum_{n\geq 1}|\Gamma(n)|\times\max_{1\leq i\leq t-1}\left[\frac{\sup_{0<h<1}(A_i(h)+A_i'(h))}{Q(t-1,r,r)\varphi(s)}\right].$$

By Lemma 2.2 and (2.13), there exists an $r_0>0$ such that for any $0<\tau\leq\tau_0$, $0<K<1$ and $r_0\leq r$, we have

$$Q(t-1,r,r)>\frac{1}{2}Q_I(t-1,r,r)>\frac{1}{4}[G(r)]^{2\tau}, \tag{2.27}$$
$$2\leq t\leq\frac{\tau}{G(r)}\log\frac{1}{G(r)}+1.$$

Next, if we put, for any $1\leq j$,

$$B_j=B_j(b,r,s,\alpha):=\exp\left\{-\frac{1}{2}\alpha\left(\frac{r^2-2h\Gamma(j)rs+s^2}{1-h^2\Gamma^2(j)}\right)+\frac{s^2}{2}\right\},$$
$$r\leq s\leq\mu_K(r),$$

we have, by (2.25) and (2.27),

$$\frac{A_i(h)}{Q(t-1,r,r)\varphi(s)}\leq\frac{Lc(\alpha)}{\alpha^{(t-2)/2}(2\pi)^{1/2}(1-h^2\Gamma^2(t-i))}B_{t-i}4[G(r)]^{-2\tau}, \tag{2.28}$$
$$1\leq i\leq t,\ 2\leq t\leq\frac{\tau}{G(r)}\log\frac{1}{G(r)}+1,$$
$$0<\tau\leq\tau_0,\ 0<K<1,\ 0<r_0\leq r.$$

By using the fact that $\mu_K \sim \sqrt{1+K}\,r$ as $r \uparrow \infty$ and writing B_j in the form

$$B_j = \exp\left\{-\frac{1}{2}\frac{r^2 b_j}{(1-h^2\Gamma^2(j))}\right\}, \qquad j \geq 1,$$

it follows by routine computations that there exists $r_1 \geq r_0$ such that for any $r \geq r_1$,

$$(2.29) \quad b_j \geq \begin{cases} (1-\varepsilon)\left[1-2\sqrt{\varepsilon}(1+k)\right]-(1+k)^2\varepsilon, & \text{if } h^2\Gamma^2(j) \leq \varepsilon, \\ (1-|\Gamma(j)|)^2 - 2(|\Gamma(j)|k + \varepsilon(1-|\Gamma(j)|(1+k))), & \\ & \text{if } h^2\Gamma^2(j) > \varepsilon, \end{cases}$$

where $k := \sqrt{1+K} - 1$ and $\alpha = 1-\varepsilon$.

Notice that for k_0, $k_0 < (1-\delta)^2/4\delta$, we have for all $k \leq k_0$,

$$\max_{j\geq 1}\left[(1-|\Gamma(j)|)^2 - 2k|\Gamma(j)|\right] \geq \frac{(1-\delta)^2}{2}, \qquad \delta := \max_{n\geq 1}|\Gamma(n)| < 1.$$

By (2.29), it is possible to choose $\varepsilon_0 > 0$ such that for any $0 < \varepsilon \leq \varepsilon_0$ and $0 < k \leq k_0$, we have $\inf_{j\geq 1} b_j \geq (1-\delta)^2/4$.

From there it follows from (2.28) that there exists an $r_1 > 0$ such that for any $2 \leq t \leq (\tau/G(r))\log(1/G(r)) + 1$, $0 < \varepsilon \leq \varepsilon_0$, $0 < \tau \leq \tau_0$, $0 < K \leq K_0$ [where $K_0 = (1+k_0)^2 - 1$] and $r_1 \leq r$, we have

$$(2.30) \quad \begin{aligned} &\frac{\max_{1\leq i\leq t-1}\sup_{0<h<1} A_i(h)}{Q(t-1,r,r)\varphi(s)} \\ &\leq \frac{2L}{(1-\delta^2)}\cdot\frac{c(1-\varepsilon)}{(1-\varepsilon)^{t/2}}\exp\left\{-\frac{1}{8}r^2(1-\delta^2)\right\}[G(r)]^{-2\tau}. \end{aligned}$$

Next, if we take $\varepsilon = \varepsilon(r) = G(r)/2$, then by the same arguments as in (2.13), there exists $0 < \tau_1 \leq \tau_0$ such that for any $0 < \tau \leq \tau_1$ and $r > 0$, we have

$$(1-\varepsilon(r))^{t/2} > \frac{[G(r)]^\tau}{2}, \qquad 2 \leq t \leq \frac{\tau}{G(r)}\log\frac{1}{G(r)} + 1.$$

If r_2 is such that $(G(r_2)/2) \leq \varepsilon_0$, then by combining this inequality and (2.30), we obtain, for any $2 \leq t \leq (\tau/G(r))\log(1/G(r)) + 1$, $r \geq r_2 \geq r_1$, $0 < \tau \leq \tau_1$ and $0 < K \leq K_0$,

$$\begin{aligned} &\frac{\max_{1\leq i\leq t-1}\sup_{0<h<1} A_i(h)}{Q(t-1,r,r)\varphi(s)} \\ &\leq \frac{4L}{(1-\delta^2)}c\left(1-\frac{G(r)}{2}\right)\exp\left\{-\frac{1}{8}r^2(1-\delta^2)\right\}[G(r)]^{-3\tau}. \end{aligned}$$

Thus, since

$$\lim_{\alpha\to 1} c(\alpha) = 1 \quad \text{and} \quad G(r) \sim \frac{1}{\sqrt{2\pi}}\frac{1}{r}e^{-r^2/2} \quad \text{as } r \to \infty,$$

we deduce that there exist $\tau_2 < \tau_1$ and two constants $u_1 > 0$ and $u_2 > 0$ such that for any $r > 0$, $0 < \tau < \tau_2$ and $0 < K < K_0$, we have

$$\max_{1 \le t \le (\tau/G(r))\log(1/G(r))} \left(\frac{\max_{1\le i\le t-1} \sup_{0<h<1} A_i(h)}{Q(t-1,r,r)\varphi(s)} \right) \le u_1[G(r)]^{u_2}.$$

From (2.25b), we obtain in the same way and under the same conditions the analogous inequality for the term corresponding to $A'_i(h)$ in (2.26):

$$\max_{\substack{1 \le t \le (\tau/G(r))\log(1/G(r)) \\ 0<r<s<\mu_K(r)}} \left(\frac{\max_{1\le i\le t-1} \sup_{0<h<1} A'_i(h)}{Q(t-1,r,r)\varphi(s)} \right) \le u_1[G(r)]^{u_2}.$$

Thus, we deduce that there exist $\tau_2 > 0$ and two constants $v_1 > 0$ and $v_2 > 0$ such that for any $0 < \tau \le \tau_2$ and $0 < K \le K_0$, we have

$$\max_{\substack{1 \le t \le (\tau/G(r))\log(1/G(r)) \\ 0<r\le s<\mu_K(r)}} \left| \frac{P(t,r,s)}{Q(t-1,r,r)\varphi(s)} - 1 \right| < v_1[G(r)]^{v_2}. \tag{2.31}$$

(2.15) follows by combining (2.31) and Lemma 2.2, since $P_I(t,r,s) = Q(t-1,r,r)\varphi(s)$. □

LEMMA 2.4. *Let Y be a random variable taking values in a measurable space $(\mathbb{E}, \mathscr{E})$. Let $\psi \in \mathscr{E}$ and let $\hat{P}$ be a probability measure on $(\mathbb{E}, \mathscr{E})$ such that $0 < \hat{P}(\psi) < 1$. Assume that on ψ, the probability measure endowed by Y has a Radon–Nikodym derivative $(dP_Y/d\hat{P})(y)$ with respect to $\hat{P}$ such that*

$$\sup_{y\in\psi} \left| \frac{dP_Y}{d\hat{P}}(y) - 1 \right| \left(1 - \hat{P}(\psi)\right)^{-1} =: q < 1. \tag{2.32}$$

Let Q be a Bernoulli random variable independent of Y such that $P(Q = 0) = q$. Then there exist two $\mathscr{E}$-measurable random variables Y' and $\overline{Y}$, taking values, respectively, in $\mathbb{E}$ and $\overline{\psi}$ (the complement of ψ), independent of Y and Q, and such that if we put

$$\hat{Y} = \begin{cases} Y, & \text{if } Q = 1 \text{ and } Y \in \psi, \\ \overline{Y}, & \text{if } Q = 1 \text{ and } Y \in \overline{\psi}, \\ Y', & \text{if } Q = 0, \end{cases} \tag{2.33}$$

then the probability measure endowed by $\hat{Y}$ is $\hat{P}$.

PROOF. See Haiman [(1987a), lemma, page 448]. □

In the construction of $\{(S_n, R_n),\ n \ge 1\}$ we shall utilize the recursive method used in Haiman (1987a). Let us recall this method by first constructing (S_1, R_1).

2.2. *Construction of* (S_1, R_1). We apply Lemma 2.4 with

$$\mathbb{E} := \mathbb{N} \times \{(-\infty, -\mu_K(\Theta_0)) \cup (\Theta_0, +\infty)\}, \qquad \Theta_0 > 0, \tag{2.34}$$

and Y defined by

$$\begin{aligned} &\{Y = (t, \rho), (t, \rho) \in \mathbb{E}\} \\ &\quad \Leftrightarrow \{-\mu_K(\Theta_0) < X_i < \Theta_0, i = 1, \dots, t-1, X_t = \rho\}, \end{aligned} \tag{2.35}$$

$$\psi := \left\{(t, \rho) \in \mathbb{E}, 1 \le t \le \frac{\tau}{G(\Theta_0)} \log \frac{1}{G(\Theta_0)}, \Theta_0 < \rho \le \mu_K(\Theta_0)\right\}, \tag{2.36}$$

$\hat{P}$ being the probability distribution of Y when the X_n are iid $\mathcal{N}(0,1)$. Thus

$$dP_Y(t, \rho) = P(t, \rho, \Theta)\delta\rho, \qquad d\hat{P}_Y(t, \rho) = P_I(t, \rho, \Theta)\,\delta\rho,$$

with $P(\cdot, \cdot, \cdot)$ and $P_I(\cdot, \cdot, \cdot)$ as in (2.14). By Lemma 2.3, there exist $\tau_0 > 0$, $K_0 > 0$, $c_1 > 0$ and $c_2 > 0$ such that for any $0 < \tau \le \tau_0$ and $0 < K \le K_0$,

$$\sup_{y \in \psi} \left| \frac{dP_Y}{d\hat{P}}(y) - 1 \right| \le c_1 [G(\Theta_0)]^{c_2}.$$

Next, by the same arguments as in (2.13) we see that, with $\bar{\psi}$ denoting the complement of ψ,

$$\begin{aligned} \hat{P}(\bar{\psi}) &\ge \left\{1 - G(\Theta_0) - [G(\Theta_0)]^{1+K}\right\}^{(\tau/G(\Theta)) \log(1/G(\Theta_0))} \\ &> \frac{[G(\Theta_0)]^{2\tau}}{2} \quad \text{for any } 0 < \tau \le \tau_1. \end{aligned}$$

Combining these inequalities we deduce that for any $0 < \tau \le \inf(\tau_0, \tau_1, c_2/4)$ and $0 < K \le K_0$, we have

$$\sup_{y \in \psi} \left| \frac{dP_Y}{d\hat{P}}(y) - 1 \right| \left[\hat{P}(\bar{\psi})\right]^{-1} < \alpha [G(\Theta_0)]^{\beta} =: q \to 0 \quad \text{as } \Theta_0 \to \infty, \tag{2.37}$$

with $\alpha = 2c_1$ and $\beta = c_2/2$.

Thus, for Θ_0 large enough, we have $q < 1$ and we can apply Lemma 2.4. Let $\hat{Y}$ be the corresponding random vector given by (2.33) and put

$$(\Delta\hat{S}, \hat{\mathcal{R}}) := \hat{Y}. \tag{2.38}$$

Let $Y^+ = (\Delta S^+, \mathcal{R}^+)$ be a random vector independent of $\sigma\{X_n, n \ge 1\} \times \sigma\{Y, Q, Y', \bar{Y}\}$ and such that

$$f_{Y^+}(t, \rho) := \frac{d}{d\rho} P\{\Delta S^+ = t, \mathcal{R}^+ < \rho\} = \Phi(\Theta_0)^{t-1} \varphi(\rho), \qquad t \ge 1, 1 < \Theta_0 < \rho. \tag{2.39}$$

Put

$$S_1 = \Delta\hat{S} 1_{\{\hat{\mathcal{R}} > 0\}} + (\Delta\hat{S} + \Delta S^+) 1_{\{\hat{\mathcal{R}} < 0\}}$$

and

$$R_1 = \hat{\mathscr{R}} 1_{\{\hat{\mathscr{R}} > 0\}} + \mathscr{R}^+ 1_{\{\hat{\mathscr{R}} < 0\}}. \tag{2.40}$$

Then, it is obvious that for any $s \geq 1$ and $r > \Theta_0$,

$$P\{S_1 = s, R_1 > r\} = [\Phi(\Theta_0)]^{s-1}(1 - \Phi(r)). \tag{2.41}$$

We shall further make use of the preceding construction of (S_1, R_1) under the following more general form.

Let $r > \Theta_0$. Let $n \geq 1$ be fixed and define $Y_r(n)$ by the following relation:

$$\begin{gathered} Y_r(n) = (t, \rho) \text{ if and only if} \\ \{-\mu_K(r) < X_i < r, i = n+1, \dots, n+t-1;\, X_{n+t} = \rho\}, \\ (t, \rho) \in \mathbb{E}(r) := \mathbb{N} \times \{(-\infty, -\mu_K(r)) \cup (r, +\infty)\}. \end{gathered} \tag{2.42}$$

Let $\tau > 0$ be given and let

$$\begin{aligned} \psi &= \psi(r) \\ &:= \{(t, \rho) \in \mathbb{E}(r); 1 \leq t \leq \tau[G(r)]^{-1}(-\log(G(r))), \\ &\qquad r < \rho < \mu_K(r)\}. \end{aligned} \tag{2.43}$$

Let η_{d-1} be a random vector defined on $(\Omega, \sigma\{X_n, n \geq 1\}, P)$ and taking values in $\mathscr{E}_d := (\mathbb{N}^+ \times \mathbb{R}^+)^d$, $d \geq 2$, and of the form $\eta_{d-1} = ((S_1, R_1), \dots, (S_{d-1}, R_{d-1}), (N_{d-1}, M_{d-1}))$ (these notations are motivated by the iterative construction of Proposition 2.1). For $d = 2$, S_1 and R_1 are the random variables constructed above. Moreover, if we denote by Y_1, Q_1, $\overline{Y}_1$, Y_1' and Y_1^+ the corresponding random variables and vectors used in the construction of S_1 and R_1, and if we put

$$\begin{gathered} Y_1 := ((Y_1)_1, (Y_1)_2), \qquad (Y_1)_1 \in \mathbb{N}, \\ (Y_1)_2 \in (-\infty, -\mu_K(\Theta_0)) \cup (\Theta_0, +\infty), \end{gathered} \tag{2.44}$$

then

$$M_1 := \max\{R_1, Q_1 1_{\{|(Y_1)_2| > \mu_K(\Theta_0)\}} |(Y_1)_2|\} \tag{2.45}$$

and

$$N_1 := Q_1 1_{\{Y_1 \in \bar{\psi}(\Theta_0)\}} 1_{\{(Y_1)_1 - S_1 > 0\}} ((Y_1)_1 - S_1). \tag{2.46}$$

PROPOSITION 2.1. *Assume that η_{d-1}, $d \geq 2$, defined above is such that for any integers $s_1 < \cdots < s_{d-1}$, $\nu_{d-1} \geq 0$ and any $0 < \rho_1 < \cdots < \rho_{d-1} \leq m_{d-1}$, the events*

$$\begin{aligned} \{\eta_{d-1}; (S_1 = s_1, R_1 < \rho_1), \dots, (S_{d-1} = s_{d-1}, R_{d-1} < \rho_{d-1}), \\ (N_{d-1} = \nu_{d-1}, M_{d-1} \leq m_{d-1})\} \end{aligned}$$

are $\sigma\{X_n, n \leq s_{d-1} + z_{d-1};\, z_{d-1} = [(G(m_{d-1}))^{A-1}] + \nu_{d-1}\} \times \sigma'$-measurable, where σ' is a σ-field independent of $\sigma\{X_n, n \geq 1\}$. Furthermore, put

$Z_{d-1} = [(G(M_{d-1}))^{A-1}\} + N_{d-1}$ *and assume that the conditional distribution of* $Y_d := Y_{R_{d-1}}(S_{d-1} + Z_{d-1})$ *given* $\eta_{d-1} = l_{d-1}$ *exists for a.a. (almost all with respect to the probability distribution of* η_{d-1}) l_{d-1} *and has a density function with respect to* $\hat{P}$ *denoted by*

$$\frac{dP_{Y_d}^{\eta_{d-1}=l_{d-1}}}{d\hat{P}}(y)$$

such that for a.a. $l_{d-1} \in \mathscr{E}_d$,

$$\text{(2.47)} \qquad \sup_{y \in \psi(r_{d-1})} \left| \frac{dP_{Y_d}^{\eta_{d-1}=l_{d-1}}}{d\hat{P}}(y) - 1 \right| \hat{P}^{-1}(\bar{\psi}(r_{d-1})) =: q_d < 1.$$

Then, there exists a $\sigma\{X_n,\ n > s_{d-1} + z_{d-1}\} \times \sigma'$*-measurable random vector* $(\Delta S_d, \mathscr{R}_d)$, $d \geq 2$, *taking values in* $\mathbb{N} \times (\mathbb{R}^+, \mathscr{B}^+)$, *such that*

$$\text{(2.48)} \qquad P\{\Delta S_d = t,\ \mathscr{R}_d > \rho | \eta_{d-1} = l_{d-1}\} = \left[\Phi(r_{d-1})\right]^{t-1}(1 - \Phi(\rho)), \qquad t \geq 1,\ \rho > r_{d-1},$$

and

$$\text{(2.49)} \qquad \begin{aligned} P\Big\{\Delta S_d &= \min\{t \geq 1,\ X_{s_{d-1}+z_{d-1}+t} > r_{d-1}\}, \\ \mathscr{R}_d &= X_{s_{d-1}+z_{d-1}+\Delta S_d} | \eta_{d-1} = l_{d-1}\Big\} \\ &= 1 - \left(1 - \left[\Phi(r_{d-1})\right]^{z_d}\right)(1 - q_d) \\ &\quad \times P\left\{Y_{r_{d-1}}(s_{d-1} + z_{d-1}) \in \psi(r_{d-1}) | \eta_{d-1} = l_{d-1}\right\}. \end{aligned}$$

Proof. We apply Lemma 2.4 by taking, for a.a. l_{d-1}, $d \geq 2$, the conditional probability distribution of Y_d, given $\eta_{d-1} = l_{d-1}$. Note that the probability distribution of η_{d-1} is concentrated on the set of l_{d-1} such that $l_{d-1} = ((s_1, r_1), \ldots, (s_{d-1}, r_{d-1}), (\nu_{d-1}, m_{d-1}))$ with $1 \leq s_1 < \cdots < s_{d-1}$, $\nu_{d-1} \geq 0$, and $0 < \Theta_0 < r_1 < \cdots < r_{d-1} \leq m_{d-1}$. Thus, in the following, l_{d-1} will be assumed to belong to this set. Let $\mathbb{E} := \mathbb{E}(r_{d-1})$ as in (2.42). Writing $\psi := \psi(r_{d-1})$ as in (2.43) we note that (2.47) implies (2.32). Put

$$\text{(2.50)} \qquad \left(\Delta \hat{S}_d, \hat{\mathscr{R}}_d\right) := \hat{Y}_d$$

with $\hat{Y}_d := \hat{Y}$ as in (2.33) (in which we also put, in order to keep homogeneity, $Q := Q_d$, $Y' := Y'_d$ and $\bar{Y} := \bar{Y}_d$), these random variables and vectors being chosen independent of $\sigma\{X_n,\ n \leq s_{d-1} + z_{d-1}\} \times \sigma'$.

Let $(\Delta S_d^+, \mathscr{R}_d^+)$ be a random vector independent of

$$\sigma\{X_n;\ n < s_{d-1} + z_{d-1}\} \times \sigma'$$

and such that

$$\text{(2.51)} \qquad \begin{aligned} &\frac{d}{d\rho} P\{\Delta S_d^+ = t,\ \mathscr{R}_d^+ < \rho | \eta_{d-1} = l_{d-1}\} \\ &= \left[\Phi(r_{d-1})\right]^{t-1} \varphi(\rho) =: f_{Y_{d^+}}(t, \rho), \qquad t \geq 1,\ r_{d-1} < \rho. \end{aligned}$$

Let

$$\Delta\tilde{S}_d = \Delta\hat{S}_d \cdot 1_{\{\hat{\mathscr{R}}_d > 0\}} + \left(\Delta\hat{S}_d + \Delta S_d^+\right)1_{\{\hat{\mathscr{R}}_d < 0\}} \tag{2.52}$$

and

$$\tilde{R}_d = \hat{\mathscr{R}}_d \cdot 1_{\{\hat{\mathscr{R}}_d > 0\}} + \mathscr{R}_d^+ 1_{\{\hat{\mathscr{R}}_d < 0\}}. \tag{2.53}$$

Let L_d be a Bernoulli random variable such that

$$P\{L_d = 1 | \eta_{d-1} = l_{d-1}\} = 1 - \left[\Phi(r_{d-1})\right]^{z_{d-1}} \tag{2.54}$$

and let $(\Delta\underline{S}_d, \underline{\mathscr{R}}_d)$ be a random vector such that

$$\begin{aligned} &\frac{d}{d\rho}P\{\Delta\underline{S}_d = t, \underline{\mathscr{R}}_d < \rho | \eta_{d-1} = l_{d-1}\} \\ &\quad = \left[\Phi(r_{d-1})\right]^{t-1}\varphi(\rho)\left[P(L_d = 1|\eta_{d-1} = l_{d-1})\right]^{-1}, \\ &\qquad\qquad t \in \{1, \dots, z_{d-1}\}, r_{d-1} < \rho. \end{aligned} \tag{2.55}$$

L_d, $\Delta\underline{S}_d$ and $\underline{\mathscr{R}}_d$ are mutually independent, independent of $\sigma\{X_n, n \le s_{d-1} + z_{d-1}\} \times \sigma'$ and depend on the other previously defined random variables only through η_{d-1}. Put

$$\Delta S_d := L_d \Delta\underline{S}_d + (1 - L_d)\left(Z_{d-1} + \Delta\tilde{S}_d\right)$$

and

$$\mathscr{R}_d := L_d \underline{\mathscr{R}}_d + (1 - L_d)\tilde{\mathscr{R}}_d. \tag{2.56}$$

It may easily be checked that these random variables satisfy (2.49). □

2.3. *Construction of* (S_d, R_d) *for* $d \ge 2$. We apply Proposition 2.1 recursively to define

$$S_d := S_{d-1} + \Delta S_d, \qquad R_d := \mathscr{R}_d, \qquad d \ge 2, \tag{2.57}$$

$$M_d := \max\left\{R_d, M_{d-1}, (1 - L_d)Q_d 1_{\{|(Y_d)_2| > \mu_K(R_{d-1})\}}|(Y_d)_2|\right\} \tag{2.58}$$

and

$$\begin{aligned} N_d := {}& (1 - L_d)Q_d 1_{\{Y_d \in \bar{\psi}(R_{d-1})\}} 1_{\{\Delta S_d < Z_{d-1} + (Y_d)_1\}} \\ &\times\left(Z_{d-1} + (Y_d)_1 - \Delta S_d\right), \end{aligned} \tag{2.59}$$

where

$$\begin{aligned} Y_d &:= ((Y_d)_1, (Y_d)_2), \qquad (Y_d)_1 \in \mathbb{N}, \\ (Y_d)_2 &\in \left(-\infty, -\mu_K(R_{d-1})\right) \cup \left(R_{d-1}, \infty\right), \end{aligned} \tag{2.60}$$

M_1 and N_1 are defined, respectively, in (2.45) and (2.46), and the other random variables are those defined in Proposition 2.1 and in its proof.

It is assumed that at any step of the construction, $d = 1, 2, \dots,$ the hypotheses of Proposition 2.1 are satisfied. Before proving this fact, it is an easy exercise to check that the process $\{S_k, R_k\}_{k\geq 1}$ constructed in this way has the same probability structure as the record sequence based on an iid $\mathscr{N}(0,1)$ sequence and Θ_0, that is, we have:

$$P\{S_{n+1}=s_{n+1}, R_{n+1}<\rho_{n+1}|S_1=s_1, R_1=r_1,\dots,S_n=s_n, R_n=r_n\}$$
$$=[\Phi(r_n)]^{s_{n+1}-s_n-1}\cdot P(r_n<X_1<\rho_{n+1}),$$
$$n\geq 1, 1\leq s_1<\cdots<s_n<s_{n+1}, 0<\Theta_0<r_1<\cdots<r_n<\rho_n.$$

We now prove that $\{\eta_d=(S_1,R_1,\dots,S_d,R_d,N_d,M_d), d\geq 1\}$ satisfies the hypotheses of Proposition 2.1.

Let $1<s_1<s_2<\cdots<s_n$ and $\Theta_0<r_1<\cdots<r_n$ be given. Let $\mathscr{S}$ be the set of vectors ζ of the form

$$\zeta=\{(s_{i_1},r_{i_1}),\dots,(s_{i_l},r_{i_l})\}, \tag{2.61}$$

with

$$1\leq i_1<i_2<\cdots<i_l\leq n,$$

or, if (s,m_n) is such that $s_{i_{k-1}}<s<s_{i_k}$ for some $1\leq k<l$ and $m_n>r_n$, then

$$\zeta=\{(s_{i_1},r_{i_1}),\dots,(s_{i_{k-1}},r_{i_{k-1}}),(s,m_n),(s_{i_k},r_{i_k}),\dots,(s_{i_l},r_{i_l})\}. \tag{2.61$'$}$$

Let E be an event of the form

$$E=\{a_i<X_{t_i}<b_i, i=1,\dots,n\} \tag{2.62}$$

and $\zeta\in\mathscr{S}$ such that

$$\{t_1,\dots,t_n\}\cap\{s_{i_1},\dots,s_{i_l}\}=\varnothing \quad \text{if } \zeta \text{ is of the form (2.61)} \tag{2.63}$$

or

$$\{t_1,\dots,t_n\}\cap\{s_{i_1},\dots,s_{i_{k-1}},s,s_{i_k},\dots,s_{i_l}\}=\varnothing \quad \text{if } \zeta \text{ is of the form (2.61$'$)}. \tag{2.63$'$}$$

Let

$$P(E|\zeta):=\begin{cases} P\{E|X_{s_{i_1}}=r_1,\dots,X_{s_{i_l}}=r_l\}, & \text{if } \zeta \text{ is of the form (2.61)},\\ P\{E|X_{s_{i_1}}=r_1,\dots,X_{s_{i_{k-1}}}=r_{i_{k-1}}, \\ \qquad X_s=m_n, X_{s_{i_k}}=r_{i_k},\dots,X_{s_{i_l}}=r_{i_l}\}, & \text{if } \zeta \text{ is of the form (2.61$'$)}, \end{cases} \tag{2.64}$$

which exists for any $\zeta\in\mathscr{S}$, since the covariance function Γ of the process is positive definite. Put

$$P_\zeta(E)=P(E|\zeta)g(\zeta), \tag{2.65}$$

where $g(\zeta)$ is the Gaussian joint density of $X_{s_{i_1}}, \dots, X_{s_{i_l}}$ in which we take $X_{s_{i_j}} = r_{i_j}$, $j = 1, \dots, l$ ($X_s = m_n$).

LEMMA 2.5. (i) *For any* $d \geq 1$, *the random vector* η_d *has a density function with respect to the Lebesgue measure given by*

$$
\begin{aligned}
f_{\eta_d}(s_1, r_1, \dots, s_d, r_d, \nu_d, m_d) &= f_{\eta_d}(l_d) \\
&= \sum_{\zeta \in \mathscr{S}(l_d)} \sum_{1 \leq j \leq J(\zeta, l_d)} A_j(\zeta, l_d) P_\zeta\big(E_j(l_d)\big) \\
&\quad + \sum_{1 \leq k \leq J'(l_d)} B_k(l_d) P(E'_k(l_d)) \\
&\quad + C(l_d) 1_{\{\nu_d = 0,\, m_d = r_d\}},
\end{aligned}
\tag{2.66}
$$

where $\mathscr{S}(l_d)$ *denotes the set of* ζ *defined in* (2.61) *and* (2.61′) *associated with* $(s_1, r_1), \dots, (s_d, r_d)$ *and* m_d, $J(l_d)$ *and* $J'(l_d)$ *are finite integers and the* E_j*'s and the* E'_k*'s are of the form* (2.62) *with*

$$
\Big\{a_i < X_{t_i} < b_i,\ i = 1, \dots, n(j);\ n(j) < \tau d\big[G(r_d)\big]^{-1}(-\log G(r_d)),
$$

$$
t_{n(j)} < s_d + \nu_d, \max\big\{(|a_i|, |b_i|), 1 \leq i \leq n(j)\big\} \leq m_d\Big\}
$$

and the A_j*'s, the* B_k*'s and* C *are* > 0.

(ii) *For all* $l_d \in \mathscr{E}_d$, *the conditional probability distribution of* Y_{d+1} *given* $\eta_d = l_d$ *has a density function with respect to the Lebesgue measure, denoted by* $f_{Y_{d+1}}(t, \rho | l_d)$, *such that for any* $t \geq 1$ *and* $r_d < \rho$,

$$
\begin{aligned}
& f_{Y_{d+1}}(t, \rho | l_d) f_{\eta_d}(l_d) \\
&= \Bigg\{ \sum_{\zeta \in \mathscr{S}(l_d)} \sum_{1 \leq j \leq J(\zeta, l_d)} A_j(\zeta, l_d) \frac{\partial}{\partial \rho} P_\zeta\big(E_j(l_d) \cap F_{r_d, s_d + z_d}(t, \rho)\big) \\
&\quad + \sum_{1 \leq k \leq J'(l_d)} B_k(l_d) \frac{\partial}{\partial \rho} P\big(E'_k(l_d) \cap F_{r_d, s_d + z_d}(t, \rho)\big) \\
&\quad + C(l_d) 1_{\{\nu_d = 0,\, m_d = r_d\}} \frac{\partial}{\partial \rho} P\big(F_{r_d, 1}(t, \rho)\big) \Bigg\},
\end{aligned}
\tag{2.67}
$$

with

$$
\begin{aligned}
F_{r_d, s}(t, \rho) = \{&-\mu_K(r_d) < X_i < r_d, \\
&i = s + 1, \dots, s + t - 1,\ r_d < X_{s+t} < \rho\}
\end{aligned}
\tag{2.68}
$$

and

$$z_d = \left[(G(m))^{A-1}\right] + \nu_d.$$

PROOF. Let us recall the construction formulas of $(S_d, \mathscr{R}_d)$ for any $d \geq 1$. We have:

$$\left(\Delta \hat{S}_d, \hat{\mathscr{R}}_d\right) =: \hat{Y}_d = Q_d\left(Y_d 1_{\{Y_d \in \psi(R_{d-1})\}} + \overline{Y}_d 1_{\{Y_d \in \bar{\psi}(R_{d-1})\}}\right) + (1 - Q_d)Y'_d$$

$$\text{[see (2.33)]},$$

$$\Delta \tilde{S}_d = \Delta \hat{S}_d \cdot 1_{\{\hat{\mathscr{R}}_d > 0\}} + \left(\Delta \hat{S}_d + \Delta S_d^+\right)1_{\{\hat{\mathscr{R}}_d < 0\}} \qquad \text{[see (2.52)]},$$

$$\tilde{\mathscr{R}}_d = \hat{\mathscr{R}}_d \cdot 1_{\{\hat{\mathscr{R}}_d > 0\}} + \mathscr{R}_d^+ 1_{\{\hat{\mathscr{R}}_d < 0\}} \qquad \text{[see (2.53)]}.$$

If $d \geq 2$,

$$\Delta S_d = L_d \Delta \underline{S}_d + (1 - L_d)\left(Z_{d-1} + \Delta \tilde{S}_d\right) \qquad \text{[see (2.56)]},$$

$$\mathscr{R}_d = L_d \underline{\mathscr{R}}_d + (1 - L_d)\tilde{\mathscr{R}}_d \qquad \text{[see (2.56)]}$$

and

$$S_d := S_{d-1} + \Delta S_d, \qquad R_d := \mathscr{R}_d \qquad \text{[see (2.57)]}.$$

If $d = 1$,

$$S_1 = \Delta \hat{S} 1_{\{\hat{\mathscr{R}} > 0\}} + (\Delta \hat{S} + \Delta S^+) 1_{\{\hat{\mathscr{R}} < 0\}}$$

$$R_1 = \tilde{\mathscr{R}} 1_{\{\hat{\mathscr{R}} > 0\}} + \mathscr{R}^+ 1_{\{\mathscr{R} < 0\}} \qquad \text{[see (2.40)]}.$$

Let us first consider the terms of (2.66) when $d = 1$.

By the proof of Lemma 2.4 [see Haiman (1987a)], if the conditional probability distribution of $\hat{Y}_d$ in (2.33) given $R_{d-1} = r_{d-1}$ is absolutely continuous with respect to the Lebesgue measure of E_d and has a density function $f_{\hat{Y}_d}$, where

$$f_{\hat{Y}_1}(s_1, r_1) = [\Phi(\Theta_0)]^{s_1 - 1}\varphi(r_1), \qquad s_1 \geq 1, \Theta_0 < r_1,$$

then Y_d, $\overline{Y}_d$ and Y'_d are also absolutely continuous with respect to the Lebesgue measure, with density functions denoted by f_{Y_d}, $f_{\overline{Y}_d}$ and $f_{Y'_d}$, respectively.

In particular, for any $s_1 \geq 1$ and $r_1 \in (-\infty, -\mu_K(\theta_0)) \cup (\Theta_0, +\infty)$, we have, with the notation in (2.60)–(2.65), in which we take $\zeta := (s_1, r_1)$,

$$f_{Y_1}(s_1, r_1) = \frac{d}{dr_1} P\{-\mu_K < X_i < \Theta_0, i = 1, \dots, s_1 - 1, X_{s_1} < r_1\}. \tag{2.69}$$

Let $t_0 = \tau(G(\Theta_0))(-\log(G(\Theta_0)))$ and put

$$a_X(t, r_1) := \sum_{k=1}^{t} P\{-\mu_K(\Theta_0) < X_i < \Theta_0, i = 1, \dots, k-1, \quad \mu_K(\Theta_0) < |X_k| < r_1\}, \qquad t \le t_0, \tag{2.70}$$

$$\alpha'_X(k, r_1, m) := \frac{\partial}{\partial m} P\{-\mu_K(\Theta_0) < X_i < \Theta_0, i = 1, \dots, k-1, \quad \mu_K(\Theta_0) < r_1 < |X_k| < m\}, \quad 1 \le k \le t_0, \mu_K(\Theta_0) < r_1 < m, \tag{2.71}$$

$$a'_X(t, r_1, m) := \sum_{k=1}^{t} \alpha'(k, r_1, m), \qquad 1 \le t \le t_0, \tag{2.72}$$

$$b_X := P\{-\mu_K(\Theta_0) < X_i < \Theta_0, i = 1, \dots, t_0\}, \tag{2.73}$$

$$a_{\bar{Y}}(t, r_1) := \sum_{j=1}^{\min(t, t_0)} P\{(\bar{Y}_1)_1 = j, (\bar{Y}_1)_2 < -\mu_K(\Theta_0)\} \times f_{Y_1^+}(t-j, r_1) + f_{\bar{Y}_1}(t, r_1) \tag{2.74}$$

with $f_{Y_1^+} = f_{Y^+}$ as in (2.39),

$$a_{Y'}(t, r_1) = \sum_{j=1}^{t} P\{(Y')_1 = j, (Y_1')_2 < -\mu_K(\Theta_0)\} \times f_{Y_1^+}(t-j, r_1) + f_{Y'}(t, r_1). \tag{2.74'}$$

With this notation, it is easy to check that if $q_1 := q$ defined in (2.37) $= P(Q_1 = 0)$, we have: If $\nu_1 = 0$ and $m_1 = r_1$,

$$f_{\eta_1}(l_1) = q_1 a_{Y'}(s_1, r_1) + (1 - q_1)$$

$$\times \begin{cases} (a_X(t_0, r_1) + b_X(t_0)) a_{\bar{Y}}(s_1, r_1), & \text{if } r_1 > \mu_K(\Theta_0),\ s_1 > t_0, \quad (1) \\ a_X(s_1, r_1) \times a_{\bar{Y}}(s_1, r_1), & \text{if } r_1 > \mu_K(\Theta_0),\ s_1 \le t_0, \quad (2) \\ b_X a_{\bar{Y}}(s_1, r_1), & \text{if } r_1 < \mu_K(\Theta_0),\ s_1 > t_0, \quad (3) \\ f_{Y_1}(s_1, r_1), & \text{if } r_1 < \mu_K(\Theta_0),\ s_1 \le t_0. \quad (4) \end{cases} \tag{2.75}$$

If $\nu_1 > 0$ or $m_1 > r_1$, we have the following situations:

If $\nu_1 > 0$ and $m_1 = r_1$, then

$$f_{\eta_1}(l_1) = (1 - q_1)$$

$$(2.75)\qquad \times \left| \begin{array}{lll} a_X(s_1 + n_1, r_1) \cdot a_{\bar{Y}}(s_1, r_1), & \text{if } r_1 > \mu_K(\Theta_0),\ s_1 + \nu_1 < t_0, & (5) \\ b_X \cdot a_{\bar{Y}}(s_1, r_1), & \text{if } s_1 + \nu_1 = t_0. & (6) \end{array} \right.$$

If $\nu_1 = 0$ and $r_1 < m_1$, then

$$f_{\eta_1}(l_1) = (1 - q_1)$$

$$(2.75)\qquad \times \left| \begin{array}{lll} a'_X(t_0, r_1, m_1) a_{\bar{Y}}(s_1, r_1), & \text{if } r_1 > \mu_K(\Theta_0),\ s_1 > t_0, & (7) \\ a'_X(s_1, r_1, m_1) a_{\bar{Y}}(s_1, r_1), & \text{if } r_1 > \mu_K(\Theta_0),\ s_1 \le t_0, & (8) \\ a'_X(t_0, \mu_K(\Theta_0), m_1) a_{\bar{Y}}(s_1, r_1), & \text{if } r_1 < \mu_K(\Theta_0),\ s_1 > t_0, & (9) \\ a'_X(s_1, \mu_K(\Theta_0), m_1) a_{\bar{Y}}(s_1, r_1), & \text{if } r_1 < \mu_K(\Theta_0),\ s_1 \le t_0. & (10) \end{array} \right.$$

If $\nu_1 > 0$ and $r_1 < m_1$, then

$$f_{\eta_1}(l_1) = (1 - q_1)$$

$$(2.75)\qquad \times \left| \begin{array}{lll} a'_X(s_1 + n_1, r_1, m_1) a_{\bar{Y}}(s_1, r_1), & \text{if } r_1 > \mu_K(\Theta_0) \text{ and } s_1 + n_1 < t_0, & (11) \\ a'_X(s_1 + n_1, \mu_K(\Theta_0), m_1) a_{\bar{Y}}(s_1, r_1), & \text{if } r_1 < \mu_K(\Theta_0) \text{ and } s_1 + n_1 < t_0. & (12) \end{array} \right.$$

Thus, the decomposition (2.66) corresponding to (2.75)-(1) is

$$C(l_1) = q_1 a_{Y'}(s_1, r_1),$$

$$E'_k = \{-\mu_K(\Theta_0) < X_i < \Theta, i = 1, \dots, k - 1, \mu_K(\Theta_0) < |X_k| < r_1\},$$

$$B_k = (1 - q_1) a_{\bar{Y}}(s_1, r_1), \qquad k = 1, \dots, t_0 =: J'(l_1) - 1;$$

$$E'_{J'} = \{-\mu_K(\theta_0) < X_i < \Theta, i = 1, \dots, t_0\};$$

$$B_{J'} = (1 - q_1) a_{\bar{Y}}(s_1, r_1).$$

There are no terms corresponding to $\Sigma\Sigma$ in (2.66). For the other terms in (2.75)-(2)–(12), one obtains similar identifications.

Thus, it is not difficult to see that the above decomposition may be extended to any $d \geq 1$. The term C corresponds to the event $\{[L_i = 1) \cup ((L_i = 0) \cap$

$(Q_i = 0))]$, $i = 1, \ldots, d\}$ whereas the other terms (Σ and $\Sigma\Sigma$) correspond to the complement of this event.

The factors $P_\zeta(E_j(l_d))$ correspond to events $\{Y_t \in \psi(R_{t-1})\}$ or $\{Y_t \in \bar{\psi}(R_{t-1}), M_t > R_t\}$, $1 \le t \le d$, and the factors $P(E'_k(l_d))$ to events $\{Y_t \in \bar{\psi}(R_{t-1}), M_t = R_t\}$, $1 \le t \le d$. The factors A_j and B_k are related to probabilities of events generated by $\{Q_t, L_t, \bar{Y}_t, Y'_t, Y^+_t, t = 1, \ldots, d\}$, which are independent of $\sigma\{X_i, i \ge 1\}$.

The decomposition (2.67) is a straightforward consequence of (2.66). □

Let us observe that the fact that for any $s_1 < \cdots < s_d$, $\nu_d \ge 0$ and $0 < \rho_1 < \cdots < \rho_d \le m_d$, the events

$$\{\eta_d; (S_1 = s_1, R_1 < \rho_1), \ldots, (S_d = s_d, R_d < \rho_d), (N_d = \nu_d, M_d < m_d)\}, \qquad d \ge 1,$$

are $\sigma\{X_n, n \le s_d + z_d\} \times \sigma'$-measurable is a straightforward consequence of the method of construction of (S_n, R_n) [see (2.33)–(2.57)].

In order to prove (2.47), we shall utilize Lemma 2.5 together with the following:

LEMMA 2.6. *With the notations in* (2.68) *and* (2.69), *if one among Hypotheses* H_1, H_2 *or* H_3 *is satisfied, then there exist an* $A_0 > 0$ *and two constants* C_1 *and* $C_2 > 0$ *independent of* l_d, *such that for any* $0 < A < A_0$, *we have*

$$\sup_{\zeta \in \mathscr{S}(l_d)} \max_{1 \le j \le J(\zeta, l_d)} \left| \left\{ \frac{\partial}{\partial \rho} P_\zeta\left(E_j(l_d) \cap F_{r_d, s_d + z_d}(t, \rho)\right)\right\} \times \left\{P_\zeta(E_d(l_d)) \cdot P(t, r_d, \rho)\right\}^{-1} - 1 \right| < C_1 [G(r_d)]^{C_2} \tag{2.76}$$

and

$$\max_{1 \le k \le J'(l_d)} \left| \left\{ \frac{\partial}{\partial \rho} P\left(E'_k(l_d) \cap F_{r_d, s_d + z_d}(t, \rho)\right)\right\} \times \left\{P(E'_k(l_d)) \cdot P(t, r_d, \rho)\right\}^{-1} - 1 \right| < C_1 [G(r_d)]^{C_2}, \qquad t \ge 1, r_d < \rho, \tag{2.77}$$

with $P(\cdot, \cdot, \cdot)$ *as in* (2.14).

Before proving Lemma 2.6, let show that this lemma combined with Lemma 2.5 implies that at any step $d \ge 2$ of the construction, (2.47) is satisfied. By

(2.43) and the same arguments as in (2.13), we have

$$(2.78)\qquad \begin{aligned}\hat{P}(\bar{\psi}(r_{d-1})) &\geq \left\{1 - G(r_{d-1}) - [G(r_{d-1})]^{1+K}\right\}^{(\tau/G(r_{d-1}))\log(1/G(r_{d-1}))}\\ &\geq \frac{[G(r_{d-1})]^{2\tau}}{2}.\end{aligned}$$

Next, with the notation in (2.47) and $P_I(\cdot, \cdot, \cdot)$ defined in (2.14), we have

$$(2.79)\qquad \frac{dP_{Y_d}^{\eta_{d-1}=l_{d-1}}}{d\hat{P}}(t, \rho) = \frac{f_Y(t, \rho | l_{d-1})}{P_I(t, r_{d-1}, \rho)}, \qquad (t, \rho) \in \psi(r_{d-1}).$$

Observe that P and P_I satisfy (2.15) with $r := r_{d-1}$ and $s := \rho$. Thus $P(t, r_d, \rho)$ in (2.76) and (2.77) may be replaced by $P_I(t, r_d, \rho)$.

Notice that $(\partial/\partial\rho)P(F_{r_d,1}(t, \rho)) = P(t, r_d, \rho)$. Further, observe that if $p_i > 0$ and $\hat{p}_i > 0$, $i = 1, \dots, n$, are such that $\max_{1 \leq i \leq n} |(p_i/\hat{p}_i) - 1| \leq q$, then for any $\alpha_i > 0$, $i = 1, \dots, n$, we have

$$\left|\frac{\Sigma\alpha_i p_i}{\Sigma\alpha_i \hat{p}_i} - 1\right| \leq q.$$

By using these remarks in (2.68) together with (2.76)–(2.79), we obtain (2.47) with

$$q_d \leq 2C_1[G(r_d)]^{C_2 - 2\tau}.$$

Thus, if we take $\tau < \inf(\tau_0, C_2/2)$ with τ_0 as in Lemma 2.3, we have

$$(2.80)\qquad q_d \leq v_1[G(r_d)]^{v_2}, \qquad v_1 = 2C_1, v_2 = C_2 - 2\tau, d = 1, 2, \dots$$

(so that for a sufficiently large Θ_0 we have $q_d < 1$ for any $d \geq 1$).

Let us now prove Lemma 2.6.

PROOF OF LEMMA 2.6. Let

$$(2.81)\qquad \begin{aligned}\{j_1 < \cdots < j_k;\ k \leq \tau d[G(r_d)]^{-1}(-\log(G(r_d)))\\ + d + 1,\ j_k \leq s_d + \nu_d\}\end{aligned}$$

be the sequence formed by the t_i of (2.67) and the s_{i_j} (or s) of ζ in (2.60) and (2.61). Put

$$(2.82)\qquad \begin{aligned}j_{k+1} = s_d + z_d + \nu_d, \dots, j_{k+t+1} = s_d + z_d + \nu_d + t,\\ 1 \leq t \leq \tau[G(r_d)]^{-1}(-\log(G(r_d))).\end{aligned}$$

Let $[R(u, v)]$ be the $(k + t + 1) \times (k + t + 1)$-matrix spanned by

$$(2.83)\qquad R(u, v) = \Gamma(j_u - j_v) = E(X_0 X_{j_u - j_v}), \qquad u, v = 1, \dots, k + t + 1,$$

and let

$$R_I(u,v) = \begin{cases} R(u,v), & \text{if } 1 \le u \le k,\ 1 \le v \le k \\ & \text{or } k < u \le k+t+1, \\ & k < v \le k+t+1, \\ 0, & \text{otherwise}. \end{cases} \tag{2.84}$$

[Though k, $j_1, \dots, j_k$ and t depend on $l_d = ((s_1, r_1), \dots, (s_d, r_d), (\nu_d, m_d))$, we have suppressed this fact for notational convenience.] Since Γ is positive definite, R and R_I are positive definite. Thus, the inverses of these matrices exist; we denote them, respectively, by R^{-1} and R_I^{-1}.

We shall deduce Lemma 2.6 as a consequence of the following lemma.

LEMMA 2.7. *There exists a universal constant* [*denoted by* (const.)], *such that*: (i) *If Hypothesis* H_1 *is satisfied, then*

$$\begin{aligned} &\max_{1 \le i,\, j \le k+t+1} \left| R^{-1}(i,j) - R_I^{-1}(i,j) \right| \\ &\qquad \le (\text{const.}) k(t+1) \max_{j \ge [G(m_d)]^{A-1}} |\Gamma(j)| \end{aligned} \tag{2.85}$$

and

$$\left|\text{Det}\left(RR_I^{-1}\right) - 1\right| \le (\text{const.}) \left(k^2 \max_{j \ge [G(m_d)]^{A-1}} |\Gamma(j)|\right)^2. \tag{2.86}$$

(ii) *If Hypotheses* H_2 *or* H_3 *are satisfied, then*

$$\begin{aligned} &\max_{1 \le i,\, j \le k+t+1} \left| R^{-1}(i,j) - R_I^{-1}(i,j) \right| \\ &\qquad \le (\text{const.}) k^2(t+1)^2 \max_{j \ge [G(m_d)]^{A-1}} |\Gamma(j)| \end{aligned} \tag{2.87}$$

and

$$\left|\text{Det}\left(RR_I^{-1}\right) - 1\right| \le (\text{const.}) \left((k(t+1))^3 \max_{j \ge (G(m_d))^{A-1}} |\Gamma(j)|\right)^2. \tag{2.88}$$

PROOF. See Appendix A. □

Put

$$\begin{aligned} F_{r_d, s}(t, (\rho, \rho + \delta\rho)) = \{ &-\mu_K(r_d) < X_i < r_d, \\ &i = s+1, \dots, s+t-1,\ \rho < X_{s+t} < \rho + \delta\rho \}, \\ &t \ge 1,\ r_d < \rho < \rho + \delta\rho. \end{aligned}$$

We shall prove Lemma 2.6 with (2.76) and (2.77) replaced, respectively, by

(2.89)
$$\sup_{\zeta \in \mathscr{S}(l_d)} \max_{1 \le j \le J(\zeta, l_d)} \Big| P_\zeta\big(E_j(l_d) \cap F_{r_d, s_d+z_d}(t, \rho, \rho + \delta\rho)\big) \times \big\{P_\zeta(E_d(l_d)) \cdot P\big(F_{r_d,1}(t, (\rho, \rho + \delta\rho))\big)\big\}^{-1} - 1\Big| < C_1[G(r_d)]^{C_2}$$

and

(2.90)
$$\max_{1 \le k \le J'(l_d)} \Big| P\big(E'_k(l_d) \cap F_{r_d, s_d+z_d}(t, (\rho, \rho + \delta\rho))\big) \times \big\{P(E'_k(l_d)) P\big(F_{r_d,1}(t, (\rho, \rho + \delta\rho))\big)\big\}^{-1} - 1\Big| < C_1[G(r_d)]^{C_2}.$$

Clearly (2.89) and (2.90) are equivalent to (2.76) and (2.77), since

$$\frac{\partial}{\partial\rho} P_\zeta\big(E_j(l_d) \cap F_{r_d, s_d+z_d}(t, \rho)\big) = \lim_{\delta\rho \downarrow 0} \frac{1}{\delta\rho} P_\zeta\big(E_j(l_d) \cap F_{r_d, s_d+z_d}(t, (\rho, \rho + \delta\rho))\big),$$

$$\frac{\partial}{\partial\rho} P\big(E'_k(l_d) \cap F_{r_d, s_d+z_d}(t, \rho)\big) = \lim_{\delta\rho \downarrow 0} \frac{1}{\delta\rho} P\big(E'_k(l_d) \cap F_{r_d, s_d+z_d}(t, (\rho, \rho + \delta\rho))\big)$$

and

$$\frac{\partial}{\partial\rho} P\big(F_{r_d,1}(t, \rho)\big) = P(t, r_d, \rho) = \lim_{\delta\rho \downarrow 0} \frac{1}{\delta\rho} P\big(F_{r_d,1}(t, (\rho, \rho + \delta\rho))\big).$$

Put

(2.91)
$$\underline{x}' = (x_{j_1}, \ldots, x_{j_{k+t+1}}).$$

Let $g(\underline{x}')$ be the Gaussian joint density of the random variables $X_{j_1}, \ldots, X_{j_{k+t+1}}$ and let $g_\zeta(\underline{y}')$ be the function deduced from $g(\underline{x})$ by fixing $x_{s_{i_j}} = r_{i_j}$, $j = 1, \ldots, l$ $(x_s = m)$ (thus depending on the vector $\underline{y}'$ formed by the $k + t + 1 - l$

remaining variables). If we put $V = R^{-1} - R_I^{-1}$, then we have

$$(2.92)\quad \begin{aligned} g(\underline{x}') &= (2\pi)^{-(k+t+1)/2}(\text{Det } R)^{-1/2} \exp\left\{-\frac{1}{2}\underline{x}'R^{-1}\underline{x}\right\} \\ &= \left(g^I(\underline{x}) \times \left(\frac{\text{Det } R}{\text{Det } R_I}\right)^{-1/2} \exp\left\{-\frac{1}{2}\underline{x}'V\underline{x}\right\}\right), \end{aligned}$$

where

$$g^I(\underline{x}) = (2\pi)^{-(k+t+1)/2}(\text{Det } R_I)^{-1/2} \exp\left\{-\frac{1}{2}\underline{x}'R_I^{-1}\underline{x}\right\}.$$

Next

$$(2.93)\quad \begin{aligned} &P\big(E_k'(l_d) \cap F_{r_d, s_d+z_d}(t, (\rho, \rho + \delta\rho))\big) \\ &= \int_{E_k'(l_d)\cap F_{r_d, s_d+z_d}(t,(\rho,\rho+\delta\rho))} g(\underline{x}')\, d\underline{x}' \end{aligned}$$

[where we use the same notation for the domain of integration of $g(\underline{x}')$ and the corresponding event in (2.90)].

Observe that

$$(2.94)\quad \begin{aligned} &\int_{E_k'(l_d)\cap F_{r_d, s_d+z_d}(t,(\rho,\rho+\delta\rho))} g^I(\underline{x}')\, d\underline{x}' \\ &= P\big(E_d'(l_d)\big)P\big(F_{r_d,1}(t, (\rho, \rho + \delta\rho))\big) \end{aligned}$$

and that for $\underline{x}'$ in the domain of integration in the left side above, we have

$$(2.95)\quad \begin{aligned} \max|\underline{x}'V\underline{x}| \le &\max_{1\le i,\, j\le k+t+1} \left|R^{-1}(i,j) - R_I^{-1}(i,j)\right| \\ &\times (k+t+1)^2(m_d^2). \end{aligned}$$

Furthermore, by (2.81), (2.82) (and again using the fact that $G(u) \sim (1/\sqrt{2\pi})(1/u)e^{-u^2/2}$ and $u^\alpha e^{-u} \to 0$ for any $\alpha > 0$ as $u \uparrow \infty$), for any $\eta > 0$ arbitrarily small, there exists a $C(\eta) > 0$ such that

$$(2.96)\quad \begin{aligned} (k+t+1)^2 &< C(\eta)[G(r_d)]^{-2-\eta}, \\ k(t+1) &\le C(\eta)[G(r_d)]^{-2-\eta}. \end{aligned}$$

Next,

$$(2.97)\quad \max_{j\ge[G(m_d)]^{A-1}} |\Gamma(j)| \le \begin{cases} [G(m_d)]^{(1-A)(4+\varepsilon)}, & \text{by } (1.6), \\ [G(m_d)]^{(1-A)(6+\varepsilon)}, & \text{by } (1.8) \text{ or } (1.10). \end{cases}$$

Combining (2.92), (2.93) and (2.94) we have for

$$D = E'_k(l_d) \cap F_{r_d, s_d + z_d}(t, (\rho, \rho + \delta\rho)),$$

$$\begin{aligned}
&\Big| P\big(E'_k(l_d) \cap F_{r_d, s_d+z_d}(t,(\rho,\rho+\delta\rho))\big) \\
&\qquad \times \Big\{P(E'_k(l_d)) \times P\big(F_{r_d,1}(t,(\rho,\rho+\delta\rho))\big)\Big\}^{-1} - 1\Big| \\
&= \left| \left(\int_D g(\underline{x}')\, d\underline{x}'\right)\left(\int_D g^I(\underline{x}')\, d\underline{x}'\right)^{-1} - 1 \right|, \qquad (2.98)\\
&\le \max_{\underline{x}' \in D} \left| \frac{g(\underline{x}')}{g^I(\underline{x}')} - 1 \right| = \max_{\underline{x}' \in D} \left| \left(\frac{\text{Det } R}{\text{Det } R_I}\right)^{-1/2} \exp\left\{-\frac{1}{2}\underline{x}'V\underline{x}\right\} - 1 \right| \\
&= O_1\left(\left(\frac{\text{Det } R}{\text{Det } R_I}\right) - 1\right) + O_2 \max_{\underline{x}' \in D}((\underline{x}'V\underline{x})),
\end{aligned}$$

where $|O_i(u)| < \sigma|u|$, $i = 1, 2$, for some $\sigma > 0$ independent of l_d [we use the fact that $(1 + u)^{1/2} \sim 1 + (u/2)$ and $\exp v \sim 1 + v$ as u and $v \to 0$].

Now, combining (2.98), (2.85)–(2.88) and (2.93)–(2.97), it is easy to check that $O_1((|R|/|R_I|) - 1) = o(\max_{\underline{x}' \in D}(\underline{x}'V\underline{x}))$ as $G(r_d) \to 0$, so that the last term in (2.98) is dominated by

$$2\sigma(m_d)^2 C(\eta)[G(r_d)]^{-2-\eta} \times \begin{cases} (\text{const.})C(\eta)[G(r_d)]^{-2-\eta}(G(m_d))^{(1-A)(4+\varepsilon)} \\ \quad \text{under } (1.6), \\ (\text{const.})C^2(\eta)[G(r_d)]^{-4-2\eta}(G(m_d))^{(1-A)(6+\varepsilon)} \\ \quad \text{under } (1.8) \text{ or } (1.10). \end{cases} \qquad (2.99)$$

Finally (2.99) $\Rightarrow$ (2.90) since η is arbitrary, $(m_d)^2 \sim -2 \log G(m_d)$ as $m_d \to \infty$ and $G(m_d) \le G(r_d)$ for $r_d \le m_d$.

The above proof of (2.90) may readily be adapted to prove (2.89) by replacing (2.93) by

$$\begin{aligned} &P_\zeta\big(E_j(l_d) \cap F_{r_d, s_d+z_d}(t,(\rho,\rho+\delta\rho))\big) \\ &\qquad = \int_{E_j(l_d)\cap F_{r_d,s_d+z_d}(t,(\rho,\rho+\delta\rho))} g_\zeta(\underline{y}')\, d\underline{y}'. \end{aligned} \qquad (2.100)$$

□

The above lemma completes the proof of the first part of Theorem 1.1 for $J = 1$.

We now prove (1.11), that is, that there exists almost surely an n_0 and q such that for all $n \ge n_0$ we have $S_n = T_{n-q}$ and $R_n = \Theta_{n-q}$. We need the following lemma.

LEMMA 2.8. *For any fixed $J \geq 1$, let $\{\Theta_n^J,\ n \geq 1\}$ be the sequence of Jth-upper records based on the sequence $\{X_n,\ n \geq 1\}$ of iid random variables with common continuous distribution function $F(x) = P(X_1 < x)$, $x \in \mathbb{R}$ and Θ_0. Then for any $0 < \alpha < 1$ we have*

$$(2.101) \qquad P\{G(\Theta_n^J) > e^{-\alpha n/J}\ i.o.\} = 0$$

and for any $\beta > 1$ we have

$$(2.102) \qquad P\{G(\Theta_n^J) < e^{-\beta n/J}\ i.o.\} = 0.$$

PROOF. Observe first that $\{F(\Theta_n^J),\ n \geq 1\}$ is the sequence of Jth-upper records based on the sequence of independent random variables $V_n := F(X_n)$ uniformly distributed on $[0, 1]$ and $F(\Theta_0)$.

Consequently, for any $0 < \varepsilon < 1$, (2.101) and (2.102) are, respectively, equivalent to

$$(2.103) \qquad P\{F(\Theta_n^J) < 1 - e^{-(n(1-\varepsilon)/J)}\ \text{i.o.}\} = 0$$

and

$$(2.104) \qquad P\{F(\Theta_n^J) > 1 - e^{-(n(1+\varepsilon)/J)}\ \text{i.o.}\} = 0.$$

Consider $Y_n = \log(1/(1 - V_n))$ which is exponentially distributed, and denote by Z_n^J $[= \log(1/(1 - F(\Theta_n^J)))]$ the nth Jth upper record based on this sequence. An application of Theorem 7 of Deheuvels (1984a) shows that $\{Z_n^J,\ n \geq 1\}$ forms a Markov chain with stationary transition probabilities such that for any $n \geq 1$, $Z_{n+1}^J - Z_n^J$ is independent of Z_n^J and

$$(2.105) \qquad P(Z_{n+1}^J - Z_n^J > t - s | Z_n^J = s) = \exp\{-J(t-s)\}, \qquad 0 < s < t.$$

Thus Z_n^J is the sequence of arrival points of a Poisson process of intensity J and by the strong law of large numbers, we have

$$(2.106) \qquad \lim_{n\to\infty} \frac{Z_n^J}{n} = E(Z_{n+1}^J - Z_n^J) = \frac{1}{J} \quad \text{a.s.}$$

Next,

$$P\{F(\Theta_n^J) > 1 - e^{-(n(1+\varepsilon)/J)}\} = P\left\{\frac{Z_n^J}{n} > \frac{1+\varepsilon}{J}\right\}$$

and

$$P\{F(\Theta_n^J) < 1 - e^{-(n(1-\varepsilon)/J)}\} = P\left\{\frac{Z_n^J}{n} < \frac{1-\varepsilon}{J}\right\},$$

which by (2.106) implies, respectively, (2.103) and (2.104). □

LEMMA 2.9. *With the notation in* (2.33)–(2.40) *we have*

$$P\{(\Delta S_d, \mathscr{R}_d) \in \bar{\psi}(R_{d-1})\ i.o.\} = P\{Q_d = 0\ i.o.\} = P\{L_d = 1\ i.o.\} = 0. \tag{2.107}$$

PROOF. Put

$$A_d = \{((\Delta S_d, \mathscr{R}_d) \in \bar{\psi}(R_{d-1})) \cap (G(R_d) < e^{-\alpha d})\}. \tag{2.108}$$

By (2.101), we have

$$P\{(\Delta S_d, \mathscr{R}_d) \in \bar{\psi}(R_{d-1})\ \text{i.o.}\} = 0 \Leftrightarrow P\{A_d\ \text{i.o.}\} = 0. \tag{2.109}$$

Now, with the notations in Proposition 2.1,

$$\begin{aligned} P\{(\Delta S, \mathscr{R}_d) &\in \bar{\psi}(r_{d-1}) | \eta_{d-1} = l_{d-1}\} \\ &= \hat{P}\{\bar{\psi}(r_{d-1})\} \\ &= 2[G(r_{d-1})]^K \\ &\quad + \left(1 - G(r_{d-1}) - [G(r_{d-1})]^{1+K}\right)^{\tau(1/G(r_{d-1}))(-\log G(r_{d-1}))} \\ &\le 2[G(r_{d-1})]^K + [G(r_{d-1})]^{\tau}. \end{aligned} \tag{2.110}$$

Thus, there exists a $d_0 > 0$ such that for any $d \ge d_0$, we have

$$\begin{aligned} P(A_d) &= \int P(A_d | \eta_{d-1} = l_{d-1})\, dP_{\eta_{d-1}}(l_{d-1}) \\ &\le \int_{G(r_{d-1}) < e^{-\alpha(d-1)}} \left(2[G(r_{d-1})]^K + [G(r_{d-1})]^{\tau}\right) dP_{R_{d-1}}(r_{d-1}) \\ &\le 2e^{-\alpha K(d-1)} + e^{-\alpha(d-1)\tau}, \qquad d \ge 1. \end{aligned} \tag{2.111}$$

Since the RHS of (2.111) is the general term of a convergent sum, the Borel–Cantelli lemma implies that $P\{A_d$ i.o.$\} = 0$. By (2.80),

$$P\{Q_d = 0 | \eta_{d-1} = l_{d-1}\} = q_d \le v_1[G(r_{d-1})]^{v_2}, \tag{2.112}$$

which by the same arguments as above implies

$$P\{Q_d = 0\ \text{i.o.}\} = 0. \tag{2.113}$$

Since $P\{A_d$ i.o.$\} = 0$ and $P\{Q_d = 0$ i.o.$\} = 0$, we have

$$P\{Z_d \ne [G(R_d)]^{A-1}\ \text{i.o.}\} = 0. \tag{2.114}$$

Thus, by (2.54) and using the same arguments as in the proof of (2.113),

$$P\Big\{(L_d = 1) \cap \Big(Z_d = [G(r_{d-1})]^{A-1}\Big) | \eta_{d-1} = l_{d-1}\Big\}$$
$$= 1 - [\Phi(r_{d-1})]^{[G(r_{d-1})]^{A-1}}$$
$$= 1 - [1 - G(r_{d-1})]^{[G(r_{d-1})]^{A-1}}$$
$$\leq 1 - \exp\left\{-G(r_{d-1})^{A}\left(1 + \frac{G(r_{d-1})}{2(1 - G(r_{d-1}))}\right)\right\}$$
$$\leq G(r_{d-1})^{A}\left(1 + \frac{G(r_{d-1})}{2(1 - G(r_{d-1}))}\right),$$

which by the preceding arguments implies

$$P\{L_d = 1 \text{ i.o.}\} = 0. \tag{2.115}$$ □

From (2.33)–(2.40) and (2.107) it follows that there exists a.s. $n_0 > 0$ such that for any $n \geq n_0$,

$$S_{n+1} = S_n + \min\Big\{t > [G(R_n)]^{A-1}; X_{S_n+t} > R_n\Big\} \tag{2.116}$$

and

$$R_{n+1} = X_{S_{n+1}}.$$

LEMMA 2.10. *For any fixed $J \geq 1$, let $\{(T_k, \underline{\Theta}_k),\ k \geq 1\}$ be the Jth record sequence based on $\{X_n,\ n \geq 1\}$ and Θ_0. Let $\{(T'_n, \underline{\Theta}'_n),\ n \geq 1\}$ be a sequence of random vectors such that there exists a.s. $p \geq 1$ and such that for any $n \geq p$, we have*

$$T'_{n+1} = \inf\{k;\ k \geq T'_n,\ X_{k-J+1,k} > X_{k-J,k-1}\}$$

and

$$\underline{\Theta}'_{n+1} = \big\{X_{T'_{n+1}-J+1,T'_{n+1}}, \dots, X_{T'_{n+1},T'_{n+1}}\big\}. \tag{2.117}$$

Then, there exist a.s. two integers n_0 and q such that for any $n \geq n_0$, we have

$$T_n = T'_{n-q} \quad \text{and} \quad \underline{\Theta}_n = \underline{\Theta}'_{n-q}. \tag{2.118}$$

PROOF. See Section 2, Lemma 4 in Haiman (1987b). □

Put, for any $n \geq 1$,

$$C_n = \Big\{\max\Big(X_{S_n+t}\colon 1 \leq t \leq [G(R_{n-1})]^{(A-1)(1+K)}\Big) > R_n\Big\}. \tag{2.119}$$

LEMMA 2.11. *If one among Hypotheses H_1, H_2 or H_3 is satisfied, then there exists a $0 < A_0 < 1$ such that for any $0 < A \leq A_0$, there exists a $0 <$*

$K_0(A)$ *such that for all* $0 < K \le K_0(A)$,

$$P\{C_n \ i.o.\} = 0. \tag{2.120}$$

PROOF. See Appendix B. □

3. Proof of Theorem 1.1 for the general case ($J > 1$).

3.1. *Construction of* $\{S_n, \underline{R}_n\}_{n \ge 1}$. $(S_1, \underline{R}_1)$ is constructed independently of $\{X_n, n \ge 1\}$ (i.e., by using a sequence $\{X_n^*, n \ge 1\}$ independent of $\{X_n, n \ge 1\}$, of iid $\mathcal{N}(0,1)$ random variables, and by taking

$$\begin{aligned} S_1 &= \inf\{n;\, n \ge J,\, X^*_{n-J+1,n} > \Theta_0\}, \\ \underline{R}_1 &= \{X^*_{S_1-J+1,S_1}, \dots, X^*_{S_1,S_1}\}, \end{aligned} \tag{3.1}$$

where Θ_0 is a fixed real number). In order to construct $(S_n, \underline{R}_n)$ for $n \ge 2$ we shall adapt step by step the method of construction in the $J = 1$ case.

Observe [see (1.2) and (2.3)] that T_k, $k \ge 1$, are the times m at which $X_{m-j+1,m}$ changes. Moreover, for any $n \ge 2$, the J-ordered random sequence $\underline{\Theta}_2, \dots, \underline{\Theta}_n$ is uniquely determined by $\Theta_1^J, \dots, \Theta_1^1$ and the sequence of $n-1$ (not necessarily increasing) values corresponding to the consecutive changes, $\Theta_2, \Theta_3, \dots, \Theta_n, \Theta_1^J < \Theta_2, \Theta_n \le \Theta_n^1$, such that if $\underline{\Theta}_1 = \{\Theta_1^J, \Theta_1^{J-1}, \dots, \Theta_1^1\}$, then

$$\{\Theta_1^J, \Theta_1^{J-1}, \dots, \Theta_1^1\} \cup \{\Theta_2, \dots, \Theta_n\} = \{X_{T_n-(J+n-2),T_n}, \dots, X_{T_n,T_n}\}. \tag{3.2}$$

For any $r > \Theta_0$ and $n \ge 1$, let $Y_r(n)$ be given by (2.42) and $\psi(r)$ by (2.43).

Let $\boldsymbol{\eta}_{d-1}$, $d \ge 2$, be a random vector defined by

$$\boldsymbol{\eta}_{d-1} := ((S_1, \underline{R}_1), \dots, (S_{d-1}, \underline{R}_{d-1}), (N_{d-1}, M_{d-1})). \tag{3.3}$$

$\boldsymbol{\eta}_{d-1}$ takes values in the set of $((s_1, \underline{r}_1), \dots, (s_{d-1}, \underline{r}_{d-1}), (\nu_{d-1}, m_{d-1}))$ such that $J < s_1 < \cdots < s_{d-1}$, $\nu_{d-1} > 0$ and $\{r_k, 1 \le k \le d-1\}$ is a J-ordered sequence. For $d = 2$, S_1 and $\underline{R}_1$ are the random vectors given above, $M_1 = R_1^1$ (we have $\underline{R}_k := \{R_k^J, \dots, R_k^1\}$, $k \ge 1$) and $N_1 = 0$.

PROPOSITION 3.1. *Assume that* $\boldsymbol{\eta}_{d-1}$, $d \ge 2$, *is such that the events of probability* > 0 *of the form* $\{\boldsymbol{\eta}_{d-1};\, (S_1 = s_1, \underline{R}_1 \in b_1), \dots, (S_{d-1} = s_{d-1}, \underline{R}_{d-1} \in b_{d-1}), N_{d-1} = \nu_{d-1}, M_{d-1} \le m_{d-1}\}$, *where* $s_1 < \cdots < s_{d-1}$, $\nu_{d-1} \ge 0$, $b_1, \dots, b_{d-1}$ *are Borel subsets of* $(\mathbb{R}^+)^J$ *and* $m_{d-1} > 0$ *are* $\sigma\{X_n \le s_{d-1} + z_{d-1};\, z_{d-1} := [(G(m_{d-1}))^{A-1}] + \nu_{d-1}\} \times \sigma'$*-measurable (where* σ' *is independent of* $\sigma\{X_n, n \ge 1\}$*). Furthermore, set* $Z_{d-1} = [(G(M_{d-1}))^{A-1}] + N_{d-1}$ *and assume that the conditional distribution of* $Y_d := Y_{R_{d-1}^J}(S_{d-1} + Z_{d-1})$ *given* $\boldsymbol{\eta}_{d-1} = \boldsymbol{l}_{d-1}$ *exists for all* $\boldsymbol{l}_{d-1} \in \mathscr{E}_d$ *and has a density function with respect to* $\hat{P}$ *denoted by*

$$\frac{dP_{Y_d}^{\boldsymbol{\eta}_{d-1} = \boldsymbol{l}_{d-1}}}{d\hat{P}}(y)$$

such that for all $\boldsymbol{l}_{d-1} \in \mathscr{E}_d$,

$$(3.4)\qquad \max_{y \in \psi(r^J_{d-1})} \left| \frac{dP^{\boldsymbol{\eta}_{d-1}=\boldsymbol{l}_{d-1}}_{Y_d}}{d\hat{P}}(y) - 1 \right| \hat{P}^{-1}\big(\bar{\psi}(r^J_{d-1})\big) =: q_d < 1.$$

Then, there exists a $\sigma\{X_n, n > s_{d-1} + z_{d-1}\} \times \sigma'$*-measurable random variable* $(\Delta S_d, \mathscr{R}_d)$ *taking values in* $\mathbb{N} \times \mathbb{R}^+$ *such that* (2.49) *is satisfied with* η_{d-1} *and* l_{d-1} *replaced by* $\boldsymbol{\eta}_{d-1}$ *and* $\boldsymbol{l}_{d-1}$ *and* r_{d-1} *by* r^J_{d-1}.

PROOF. See the proof of Proposition 2.1 in which η_{d-1} and l_{d-1} are replaced by $\boldsymbol{\eta}_{d-1}$ and $\boldsymbol{l}_{d-1}$ and r_{d-1} by r^J_{d-1}. □

Construction of $(S_n, \underline{R}_n)$ *for* $n \geq 2$. We apply Proposition 3.1 and define

$$(3.5)\qquad S_d = S_{d-1} + \Delta S_d, \qquad \underline{R}_d = \Omega\big(R^{J-1}_{d-1}, \dots, R^1_1, \mathscr{R}_d\big), \qquad d \geq 2,$$

where $\Omega(x_1, \dots, x_j) = (x_{1j}, \dots, x_{jj})$ denotes the ordered sample $(x_1, \dots, x_j)$,

$$(3.6)\qquad M_d = \max\Big[R^1_d, M_{d-1}, (1 - L_d)Q_d 1_{\{|(Y_d)_2| > \mu_K(R^J_{d-1})\}} \times |(Y_d)_2|\Big]$$

and

$$(3.7)\qquad \begin{aligned} N_d &= (1 - L_d)Q_d \times 1_{\{Y_d \in \bar{\psi}(R^J_{d-1})\}} \\ &\quad \times 1_{\{\Delta S_d < Z_{d-1} + (Y_d)_1\}}\big(Z_{d-1} + (Y_d)_1 - \Delta S_d\big), \end{aligned}$$

where L_d, Q_d and Y_d are the corresponding random vectors used in the construction of $(\Delta S_d, \mathscr{R}_d)$.

Let $r_2, r_3, \dots, r_n$, $r^J_1 < r_2$, $r_n \leq r^1_n$, $n \geq 2$, be the sequence which together with $\underline{r}_1$ uniquely determines the J-ordered sequence $\underline{r}_1, \dots, \underline{r}_n$ [see remark on (3.2)].

Let $\mathscr{S}$ be the set of ζ defined in (2.60) with the above (not necessarily increasing) r_i, $i = 2, \dots, n$, $2 \leq i_1 < i_2 < \cdots < i_l \leq n$ (instead of $1 \leq i_1 < \cdots < i_l \leq n$) and $m_n > r^1_n$ [instead of $m_n > r_n$ in (2.61)].

The fact that at any step of the construction $d = 2, \dots$ the hypotheses of Proposition 3.1 are satisfied may be deduced by the same arguments as in the case $J = 1$ by the following substitutions in Lemmas 2.5 and 2.6.

$$\eta_d := \boldsymbol{\eta}_d, \qquad l_d := \boldsymbol{l}_d \quad \text{in (2.66)–(2.68)},$$

$$(3.8)\qquad \big((s_1, r_1), \dots, (s_d, r_d)\big) := \big((s_2, r_2), \dots, (s_d, r_d)\big) \quad \text{in (2.66)},$$

$$r_d := r^J_d \quad \text{in (2.67)–(2.69) and (2.76), (2.77)}.$$

Let us now prove that the above construction of $\{(S_n, \underline{R}_n)\}_{n \geq 1}$ satisfies (1.11).

3.2. *Proof of Theorem* 1.1 ($J > 1$). With the notation in Proposition 3.1 (and its proof) we have the following version of Lemma 2.9:

LEMMA 3.1.

$$P\{(\Delta S_n, \mathscr{R}_d) \in \bar{\psi}(R_{d-1}^J)\ i.o.\} = P\{Q_d = 0\ i.o.\} = 0, \tag{3.9}$$

$$P\{L_d = 1\ i.o.\} = 0. \tag{3.10}$$

PROOF. The proof of (3.9) is similar to the proof of Lemma 2.9 for $J = 1$. To prove (3.10) we need the following:

LEMMA 3.2. *There exists a sequence* $\{\eta_n\}_{n \geq 1}$, $\lim_{n \to \infty} \eta_n = 0$, *such that for any* $\varepsilon > 0$,

$$P\left\{G(R_n^1) \leq \left(\frac{n}{J}\right)^{-(1+\varepsilon)} \exp\left(-\frac{n}{J}(1 + \eta_n)\right)\ i.o.\right\} = 0. \tag{3.11}$$

PROOF. Observe that when $J > 1$, R_n^1 has the same probability distribution as $\max(X_1, \ldots, X_{S_n})$, where S_n is the nth Jth record time based on an independent identically $\mathcal{N}(0, 1)$ distributed sequence of r.v.'s. Thus, the probability distribution of $G(R_n^1)$ is the same as that of $\min(U_1, \ldots, U_{\nu_n}) = U_{1, \nu_n}$, where $\{U_n\}_n$ is a sequence of independent random variables uniformly distributed on $[0, 1]$ and ν_n is the nth Jth record time based on this sequence.

By Deheuvels [(1986), page 134], for any $\varepsilon > 0$ there exists a.s. an $n_0 > 0$ such that for any $n \geq n_0$,

$$U_{1,n} > n^{-1}(\log n)^{-(1+\varepsilon)} \tag{3.12}$$

and by Deheuvels [(1984b), Theorem 7], there exist a.s. two sequences, $\eta_1(n) > 0$ and $\eta_2(n) > 0$, $\eta_i(n) \to 0$ as $n \to \infty$, $i = 1, 2$, such that

$$P\left\{\log \nu_n \geq \frac{n}{J}(1 + \eta_1(n))\ \text{i.o.}\right\} = 0 \tag{3.13}$$

and

$$P\left\{\log \nu_n \leq \frac{n}{J}(1 - \eta_2(n))\ \text{i.o.}\right\} = 0. \tag{3.14}$$

By combining (3.12), (3.13) and (3.14), we get (3.11). □

By (3.9), it is equivalent to prove that

$$P\{L_d\ \text{i.o.}\} = 0 \quad \text{and} \quad P\{B_d\ \text{i.o.}\} = 0, \tag{3.15}$$

where

$$\begin{aligned} B_d = \Bigg(&\{L_d = 1\} \cap \{M_d = R_d^1\} \cap \{N_d = 0\} \\ &\cap \{G(R_{d-1}^J) < e^{-(\alpha(d-1)/J)}\} \\ &\cap \left\{G(R_{d-1}^1) > \left(\frac{d-1}{J}\right)^{-(1+\varepsilon)} \exp\left(-\frac{d-1}{J}(1 + \eta_{d-1})\right)\right\}\Bigg), \end{aligned} \tag{3.16}$$

$$0 < \alpha < 1.$$

By (2.54),

$$P\{B_d\} = \int P(B_d | \boldsymbol{\eta}_{d-1} = \boldsymbol{l}_{d-1})\, dP_{\boldsymbol{\eta}_{d-1}}(\boldsymbol{l}_{d-1})$$

(3.17)

$$= \int_D \int \left(1 - \left(\Phi\left(r_{d-1}^J\right)\right)^{[G(r)]^{A-1}}\right) dP_{(R_{d-1}^1, R_{d-1}^J)}\left(r, r_{d-1}^J\right),$$

with

$$(3.18)\qquad D = \begin{cases} G\left(r_{d-1}^J\right) \le e^{-(\alpha(d-1)/J)}, \\ G(r) \ge \left(\dfrac{d-1}{J}\right)^{-(1+\varepsilon)} \exp\left\{-\dfrac{d-1}{J}(1+\eta_{d-1})\right\}. \end{cases}$$

Next

$$(3.19)\qquad 1 - \left(\Phi\left(r_{d-1}^J\right)\right)^{[G(r)]^{A-1}} = 1 - \exp\left\{[G(r)]^{A-1} \log\left(1 - G\left(r_{d-1}^J\right)\right)\right\}$$

and $\log(1 - G(r_{d-1}^J)) \sim -G(r_{d-1}^J)$ as $d \to \infty$.

Hence, there exists a positive constant (independent of d) such that, on D,

$$-\left[\{G(r)\}^{A-1}\right] \log\left(1 - G\left(r_{d-1}^J\right)\right)$$

$$\le (\text{const.}) e^{-(\alpha(d-1)/J)} \times \exp\left\{-\frac{(d-1)}{J}(1+\eta_{d-1})(A-1)\right\}$$

$$(3.20)\qquad \times \left(\frac{d-1}{J}\right)^{(1-A)(1+\varepsilon)}$$

$$= (\text{const.}) \left(\frac{d-1}{J}\right)^{(1-A)(1+\varepsilon)} \exp\left\{-\frac{d-1}{J}\left(\alpha + (1+\eta_{d-1})(A-1)\right)\right\}$$

$$\le \exp\left\{-\frac{d-1}{J}\frac{A}{2}\right\}, \qquad d \ge d_0, \text{ sufficiently large, } 0 < 1 - \alpha < \frac{A}{2}.$$

Thus, since $e^x - 1 \sim x$ as $x \to 0$, by combining (3.17)–(3.20), for $d \ge d_0$, we have $P(B_d) \le \exp\{-((d-1)/J)(A/2)\}$, which implies (3.15) by Borel–Cantelli. □

From (2.33)–(2.40) for $J > 1$ [see (3.8), (3.9) and (3.10)] it follows that there exists a.s. $n_0 > 0$ such that for any $n \ge n_0$,

$$S_{n+1} = S_n + \min\left\{t > \left[\left(G\left(R_n^1\right)\right)^{A-1}\right], X_{S_n+t} > R_n^J\right\}$$

and

$$(3.21)\qquad \underline{R}_{n+1} = \Omega\left(R_n^{J-1}, \dots, R_n^1, X_{S_{n+1}}\right).$$

Thus, if we put $\mathscr{C}_n' = \{\max(X_{S_n+t};\ 1 \le t \le [(G(R_n^1))^{A-1}]) > R_n^J\}$ and if we prove that $P(\mathscr{C}_n'$ i.o.$\} = 0$, then by (3.21) and Lemma 2.10, the result will be established.

Next, the fact that $P(\mathscr{C}'_n \text{ i.o.}\} = 0$ results as a straightforward consequence of the transcription of Lemma 2.11 with $R_{n-1} := R^J_{n-1}$ and $R_n := R^J_n$ in (2.119), the first equality in (3.9) and the following lemma.

LEMMA 3.3. *For any* $K > 0$,

$$P\left\{G(R^1_n) < \left(G(R^J_{n-1})\right)^{1+K} i.o.\right\} = 0. \tag{3.22}$$

PROOF. By (3.11), a.s. there exists an n_0 such that for $n > n_0$,

$$G(R^1_n) > \left(\frac{n}{J}\right)^{-(1+\varepsilon)} \exp\left(-\frac{n}{J}(1+\eta_n)\right), \qquad \eta_n \to 0,\, n \to \infty, \tag{3.23}$$

and by (2.99), for any $0 < \alpha < 1$ a.s. there exists an n_1 such that for $n > n_1$,

$$G(R^J_{n-1}) < \exp\left\{-\frac{(n-1)}{J}\alpha\right\}. \tag{3.24}$$

Thus, if $\alpha(1+K) > 1$,

$$\exp\left\{-\frac{n-1}{J}(1+K)\alpha\right\} \times \left[\left(\frac{n}{J}\right)^{-(1+\varepsilon)} \exp\left(-\frac{n}{J}(1+\eta_n)\right)\right]^{-1} \to 0$$

as $n \to \infty$, which implies (3.22). □

APPENDIX A

PROOF OF LEMMA 2.7

Proof under Hypothesis H_1. Put

$$U = R_I^{-1}(R - R_I). \tag{A1}$$

Then we have

$$R^{-1} - R_I^{-1} = (I+U)^{-1}R_I^{-1} - R_I^{-1} = (-U + U^2 - \cdots)R_I^{-1}. \tag{A2}$$

It may easily be checked [by using recursively the inequality $|AB(ij)| \le n \max_{1\le k\le n}|A(ik)| \max_{1\le l\le n}|B(lj)|$, $1 \le i, j \le n$] that $V^{2q+1} := U^{2q+1} \times R_I^{-1}$, $q \ge 0$, are of the form

$$V^{2q+1} = \begin{array}{|c|c|} \hline 0 & V_h^{2q+1} \\ \hline V_l^{2q+1} & 0 \\ \hline \end{array} \begin{array}{l} \}k \quad \text{with } \max_{i,j}|V_h^{2q+1}(i,j)| \le \varrho, \\ \}t+1 \quad \text{with } \max_{i,j}|V_l^{2q+1}(i,j)| \le \varrho, \end{array} \tag{A3}$$

where $\varrho = \rho(\rho\gamma k(t+1))^{2q+1}$

and where the maximum is taken over $1 \le i \le k$, $1 \le j \le t+1$ and

$$\text{(A4)}\qquad \begin{aligned} \rho &:= \max_{1 \le i,\, j \le k+t+1} \left| R_I^{-1}(i,j) \right|, \\ \gamma &:= \max_{1 \le i,\, j \le k+t+1} \left| (R - R_I^{-1})(i,j) \right|, \end{aligned}$$

and that $V^{2q} := U^{2q} \times R_I^{-1}$, $q \ge 1$, are of the form

$$\text{(A5)}\qquad V^{2q} = \begin{array}{|c|c|} \hline V_h^{2q} & 0 \\ \hline 0 & U_l^{2q} \\ \hline \end{array} \begin{array}{l} \}k \quad \text{with} \quad \max_{i,j} \left| V_h^{2q}(i,j) \right| \le \varrho', \\ \}t+1 \quad \text{with} \quad \max_{i,j} \left| V_l^{2q}(i,j) \right| \le \varrho', \end{array}$$

where $\varrho' = \rho(\rho\gamma k(t+1))^{2q}$.

Thus, for any $n \ge 1$,

$$\text{(A6)}\qquad \max_{1 \le i,\, j \le k+t+1} \left| U^n R_I^{-1}(i,j) \right| \le \rho(\rho\gamma k(t+1))^n.$$

Let us now give bounds for ρ and γ defined in (A.4). We have

$$\text{(A7)}\qquad R_I^{-1} = (I + (R_I - I))^{-1} \sim I - S + S^2 \cdots$$

with $S := R_I - I$.

By (1.5) we have

$$\text{(A8)}\qquad \max_{1 \le v \le k+t+1} \sum_{u=1}^{k+t+1} |S(u,v)| < 2 \sum_{n=1}^{\infty} |\Gamma(n)| := \lambda < 1$$

and

$$\text{(A9)}\qquad \max_{1 \le u \le k+t+1} \sum_{v=1}^{k+t+1} |S(u,v)| < \lambda,$$

from which it may be easily seen that for any $n \ge 1$,

$$\text{(A10)}\qquad \max_{1 \le i,\, j \le k+t+1} |S^n(i,j)| < \lambda^n$$

and hence

$$\text{(A11)}\qquad \rho \le \sum_{n=1}^{\infty} \lambda^n = \frac{1}{1-\lambda}.$$

Next,

$$\text{(A12)}\qquad \gamma \le \max |\Gamma(j)|, \qquad j \ge (G(m_d))^{A-1}.$$

Thus, by combining (A2), (A6), (A11) and (A12) and since by (2.96) and (2.97),

$$k(t+1) \max_{j \ge (G(m_d))^{A-1}} |\Gamma(j)| \to 0 \quad \text{as } d \to \infty,$$

we deduce that

$$\text{(A13)}\qquad \begin{aligned} &\max_{1\le i,\, j\le k+t+1} \left|\left((R^{-1}-R_I^{-1})+UR_I^{-1}\right)(i,j)\right| \\ &\quad\times\left[k(t+1)\max_{j\ge (G(m_d))^{A-1}} |\Gamma(j)|\right]^{-1} \to 0 \end{aligned}$$

as $d \to \infty$, from which (2.85) follows.

Let us now prove (2.86). We have

$$\text{(A14)}\qquad \left|\operatorname{Det}(RR_I^{-1})-1\right| = \left|\operatorname{Det}(I+U)-1\right|$$

and

$$\operatorname{Det}(I+U) = \operatorname{Det}(e_1+U_{\cdot,1},\dots,e_{k+t+1}+U_{\cdot,k+t+1}),$$

where e_n denotes the nth column of I and $U_{\cdot,n}$ is the nth column of U. By the multilinearity of the determinant we have

$$\text{(A15)}\qquad \operatorname{Det}(I+U) = 1 + \sum_{n=1}^{k+t+1} \begin{pmatrix} \text{determinants of all submatrices} \\ \text{of } U \text{ obtained by removing } n \\ \text{rows and columns of the same rank.} \end{pmatrix}.$$

Next, since U is of the form

$U =$ 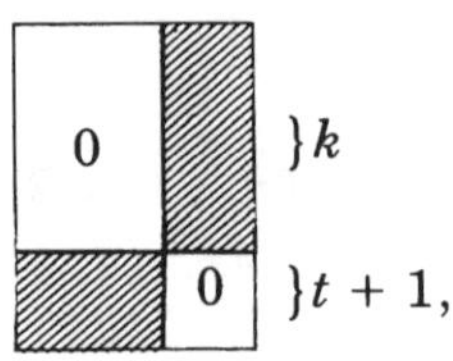

among these determinants, only those belonging to the submatrices having the two zero blocks of equal size are not equal to 0.

Each such determinant of size $2p$, $1 \le p \le \inf(k, t+1)$ (there are $C_k^p C_{t+1}^p$ of them) is bounded by $(p!) \times u^p$, where

$$\text{(A16)}\qquad u := \max_{1\le i,\, j\le k+t+1} |U(i,j)| \le \rho\gamma \max(k, t+1) = \rho\gamma k.$$

Thus

$$\text{(A17)}\qquad \begin{aligned} \left|\operatorname{Det}(I+U)-1\right| &\le \sum_{p=1}^{\inf(k,t+1)} C_k^p C_{t+1}^p (p!)^2 u^{2p} \le \sum_{p=1}^{\infty} \left(k(t+1)u^2\right)^p \\ &\le \sum_{p=1}^{\infty} \left(k^3(t+1)\rho^2\gamma^2\right)^p \le \sum_{p=1}^{\infty} \left(\rho^2 k^4 \gamma^2\right)^p, \end{aligned}$$

from which (2.86) follows since by (2.96) and (2.97),

$$k^2\gamma \le k^2 \max_{j\ge (G(m_d))^{A-1}} |\Gamma(j)| \to 0 \quad \text{as } d\to\infty.$$

Proof under Hypotheses H_2 *or* H_3. Let $R_1^{-1/2}$ be the lower triangular matrix of order k such that $R_1^{-1/2}(X_{j_1}, \ldots, X_{j_k})'$ forms a complete orthonormal system of $H\{j_1, j_k\}$ and $R_2^{-1/2}$ the lower triangular matrix order $t+1$ such that $(R_2^{-1/2}(X_{j_{k+1}}, \ldots, X_{j_{k+t+1}}))'$ forms a complete orthonormal system of $H\{j_{k+1}, j_{k+t+1}\}$.

Consider

$$R_I^{-1/2} = \begin{pmatrix} R_1^{-1/2} & 0 \\ 0 & R_2^{-1/2} \end{pmatrix} \begin{matrix} \}k \\ \}t+1 \end{matrix}$$

and observe that

(A18) $$(R_I^{-1/2})' R_I^{-1/2} = R_I^{-1}.$$

Put

(A19) $$\hat{U} = R_I^{-1/2}(R - R_I)(R_I^{-1/2})'.$$

Then we have

(A20) $$\begin{aligned} R^{-1} - R_I^{-1} &= \left(R_I(I + R_I^{-1}(R - R_I))\right)^{-1} - R_I^{-1} \\ &= (R_I^{-1/2})'(-\hat{U} + \hat{U}^2 - \cdots)R_I^{-1/2}. \end{aligned}$$

$\hat{U}^{2q+1}$, $q \geq 0$, are of the form

$\hat{U}^{2q+1} =$ 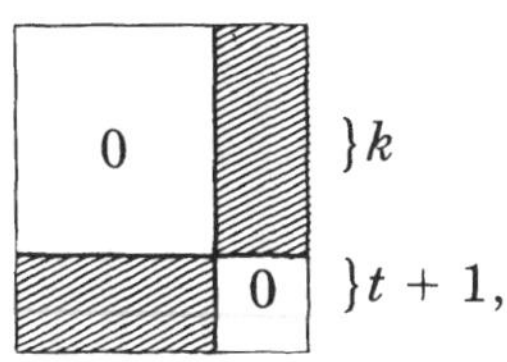

$\hat{U}^{2q}$, $q \geq 1$, are of the form

$\hat{U}^{2q} =$ 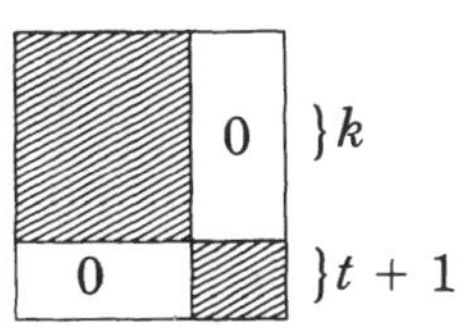

and, as before, it may be easily checked that

(A21) $$\max_{1 \leq i,\, j \leq k+t+1} \left|\hat{U}^{2q+1}(i, j)\right| \leq \hat{u}^{2q+1} \times (k(t+1))^q, \qquad q \geq 0,$$

and

(A22) $$\max_{1 \leq i,\, j \leq k+t+1} \left|\hat{U}^{2q}(i, j)\right| \leq \hat{u}^{2q}(k(t+1))^q, \qquad q \geq 1,$$

with

(A23) $$\hat{u} := \max_{1 \leq i,\, j \leq k+t+1} \left|\hat{U}(i, j)\right| \leq \hat{\rho}^2 \gamma k \times (t+1),$$

where

$$\text{(A24)}\qquad \begin{aligned}\hat{\rho} &:= \max_{1\le i,\, j\le k+t+1} \left|R_I^{-1/2}(i,j)\right|,\\ \gamma &:= \max_{1\le i,\, j\le k+t+1} \left|(R-R_I)(i,j)\right|.\end{aligned}$$

Let us now give a bound for $\hat{\rho}$ in (A24).

Bound for $\hat{\rho}$ under Hypothesis H_2. For any $2 \le n \le k$, let $X_{j_n}^*$ be the projection of X_{j_n} on $H(j_1, j_{n-1})$ and put

$$\text{(A25)}\qquad X_{j_n}^* = \sum_{p=1}^{n-1} a_{n,p} X_{j_p}.$$

If we denote by σ_p the one-step prediction error [of the projection of X_0 on $H(-\infty,-1)$] and by σ_i the interpolation error of the projection of X_0 on $H(-\infty,-1) \cup H(1,+\infty)$, then, if (1.10) or (1.7) are satisfied (see Remark 4 in Section 1) we have

$$\text{(A26)}\qquad \sigma_p = \sqrt{2\pi}\,\exp\left\{\frac{1}{4\pi}\int_{-\pi}^{\pi}\log f(\lambda)\,d\lambda\right\} > 0.$$

If (1.7) is satisfied, then we have

$$\text{(A27)}\qquad \sigma_i = 2\pi\left(\int_{-\pi}^{\pi}\frac{d\lambda}{f(\lambda)}\right)^{-1/2} > 0$$

and

$$\text{(A28)}\qquad \|X_{j_n} - X_{j_n}^*\| \ge \sigma_p \ge \sigma_i > 0.$$

Put

$$\begin{pmatrix} Y_1\\ \vdots\\ Y_k\end{pmatrix} = R_1^{-1/2}\begin{pmatrix} X_{j_1}\\ \vdots\\ X_{j_k}\end{pmatrix},$$

which, by the definition of $R_1^{-1/2}$, is an orthonormal complete system of $H(X_{j_1},\dots,X_{j_k})$ such that $Y_1 = X_{j_1}$ and for any $2 \le n \le k$,

$$\text{(A29)}\qquad Y_n = \frac{X_{j_n} - X_{j_n}^*}{\|X_{j_n} - X_{j_n}^*\|}.$$

Thus, for any $2 \le n \le k$,

$$\text{(A30)}\qquad R_1^{-1/2}(n,n) = \frac{1}{\|X_{j_n} - X_{j_n}^*\|} \le \frac{1}{\sigma_p} \le \frac{1}{\sigma_i}$$

and for any $1 \le p \le n-1$,

$$\text{(A31)}\qquad R_1^{-1/2}(n,p) = \frac{-a_{n,p}}{\|X_{j_n} - X_{j_n}^*\|}.$$

Let $1 \le p_0 \le n-1$ and $\hat{X}_{j_{p_0}}$ be the projection of $X_{j_{p_0}}$ on $H(X_{j_p},\ p = 1,\dots,n-1;\ p \ne p_0)$ (the Hilbert space generated by X_{j_p}, $p = 1,\dots,n-1$; $p \ne p_0$).

Then, we have

(A32) $$X^*_{j_n} = \sum_{\substack{p=1\\ p\ne p_0}}^{n-1} a_{n,p} X_{j_p} + a_{n,p_0}\left(\eta_{j_{p_0}} + \hat{X}_{j_{p_0}}\right),$$

where $\eta_{j_{p_0}}$ is orthogonal to $H(X_{j_p},\ p = 1,\dots,n-1;\ p \ne p_0)$. By the orthogonal projection theorem, we have

(A33) $$|a_{n,p_0}| \times \|\eta_{j_{p_0}}\| \le \|X_{j_n}\| = 1$$

and since $\|\eta_{j_{p_0}}\| \ge \sigma_i$, we deduce that for any $1 \le p \le n-1$,

(A34) $$|a_{n,p}| \le \frac{1}{\sigma_i} = \frac{\left(\int_{-\pi}^{\pi} \frac{d\lambda}{f(\lambda)}\right)^{1/2}}{2\pi} < \infty.$$

By combining (A30), (A31) and (A34), which are also valid for $R_2^{-1/2}$, we obtain that

(A35) $$\hat{\rho} \le \frac{1}{2\pi}\int_{-\pi}^{\pi} \frac{d\lambda}{f(\lambda)}.$$

Bound for $\hat{\rho}$ *under Hypothesis* H_3. Let $-i_{n+1} < -i_n < -i_{n-1} < \cdots < -i_1 < 0$ and put (with the notation defined in the proof of Lemma 2.3)

(A36) $$X_0^{X_{-i_n},\dots,X_{-i_1}} := \hat{X}_{0/n} := a_{i_1}^{(n)} X_{-i_1} + \cdots + a_{i_n}^{(n)} X_{-i_n}.$$

Let

(A37) $$\hat{X}_{0/n+1} := a_{i_1}^{(n+1)} X_{-i_1} + \cdots + a_{i_n}^{(n+1)} X_{-i_n} + a_{i_{n+1}}^{(n+1)} X_{-i_{n+1}}$$

and observe that

(A38) $$\begin{aligned}\hat{X}_{0/n} &= a_{i_1}^{(n+1)} X_{-i_1} + \cdots + a_{i_n}^{(n+1)} X_{-i_n}\\ &\quad + a_{i_{n+1}}^{(n+1)}\left(b_{i_1}^{(n)} X_{-i_n} + \cdots + b_{i_n}^{(n)} X_{-i_n}\right),\end{aligned}$$

where

(A39) $$X_{-i_{n+1}}^{X_{-i_n},\dots,X_{-i_1}} := \sum_{k=1}^{n} b_{i_k}^{(n)} X_{-i_k}.$$

Thus, by combining (A36) and (A38) we obtain

(A40) $$a_{i_k}^{(n+1)} = a_{i_k}^{(n)} - a_{i_{n+1}}^{(n+1)} b_{i_k}^{(n)}, \qquad k = 1,\dots,n.$$

Let us now suppose that there exists a sequence $u_n > 0$, $\sum_{n=1}^{\infty} u_n < \infty$ such that for any $n \ge 1$ and $-i_n < \cdots < -i_1 < 0$, we have

(A41) $$\left|a_{i_n}^{(n)}\right| \le u_n.$$

Thus, if we put $M_n := \sup_{1 \le i \le n-1} |a^{(n)}_{i_{n-1}}|$ and since $|b^{(n)}_{i_1}| \le u_n$, we have by (A40),

(A42) $$M_{n+1} \le M_n(1 + u_{n+1})$$

and hence

(A43) $$\sup_{n \ge 1} |M_n| < M_1 \prod_{p=1}^{\infty} (1 + u_{1+p}) := M < \infty.$$

Next, with the notation in (A25) and (A36), by making the identifications

(A44) $$a_{n,p} = a^{(n-1)}_{j_{n-p}}, \qquad p = 1, \ldots, n-1,$$

we deduce, by (A31) and (A43), that

(A45) $$\hat{\rho} \le \frac{M}{\sigma_p} = \frac{M}{\sqrt{2\pi}} \exp\left\{ \frac{-1}{4\pi} \int_{-\pi}^{\pi} \log f(\lambda)\, d\lambda \right\} < \infty.$$

Let us now show that if Hypothesis H_3 is satisfied, then there exists a sequence $u_n > 0$, $\sum_{n=1}^{\infty} u_n < \infty$ such that (A41) holds.

Let

(A46) $$\eta_1 = X_{-i_1}, \qquad \eta_2 = \left(X_{-i_2} - X^{X_{-i_1}}_{-i_2}\right) \Big/ \left\| X_{-i_2} - X^{X_{-i_1}}_{-i_2} \right\|, \ldots,$$
$$\eta_n = \left(X_{-i_n} - X^{X_{-i_1}, \ldots, X_{-i_{n-1}}}_{-i_n}\right) \Big/ \left\| X_{-i_n} - X^{X_{-i_1}, \ldots, X_{-i_{n-1}}}_{-i_n} \right\|$$

be an orthonormal complete system of the Hilbert space generated by $X_{-i_1}, \ldots, X_{-i_n}$.

With the notations in (A36) we have

(A47) $$\hat{X}_{0/n} = \sum_{i=1}^{n} E(X_0 \cdot \eta_i) \eta_i,$$

where, by expressing the η_i's as linear combinations of the X_{-i_k}, $k = 1, \ldots, n$, only η_n contains X_{-i_n}.

Thus, by (A46),

(A48) $$a^{(n)}_{i_n} = E(X_0 \cdot \eta_n) \left\| X_{-i_n} - X^{X_{-i_1}, \ldots, X_{-i_{n-1}}}_{-i_1} \right\|^{-1}.$$

Next, by observing that $\eta_n \in H(-\infty, -i_n)$,

(A49) $$|E(X_0 \cdot \eta_n)| \le \rho(i_n) \le \rho(n) \le Cn^{-(6+\varepsilon)}$$

[where $\rho(n)$ is as in (1.9)], we obtain finally

(A49′) $$\begin{aligned} \left| a^{(n)}_{i_n} \right| &\le \frac{c}{\sigma_p} n^{-(6+\varepsilon)} \\ &= \frac{C}{\sqrt{2\pi}} \exp\left\{ \frac{-1}{2\pi} \int_{-\pi}^{\pi} \log f(\lambda)\, d\lambda \right\} n^{-(6+\varepsilon)} \\ &=: u_n, \end{aligned}$$

where C is a positive constant.

Let us now return to (A18)–(A24). We have by (A23) and (A12),

$$\hat{u} \le \hat{\rho}^2 k(t+1) \max_{j \ge (G(m_d))^{A-1}} |\Gamma(j)| \tag{A50}$$

and if we put

$$\hat{V} = \left(R_I^{-1/2}\right)' \hat{U} R_I^{-1/2}, \tag{A51}$$

it is easy to check that

$$\max_{1 \le i,\, j \le k+t+1} \left|\hat{V}(i,j)\right| \le \hat{\rho}^4 k^2 (t+1)^2 \max_{j \ge (G(m_d))^{A-1}} |\Gamma(j)|. \tag{A52}$$

Thus, from (A20) and (A50) we deduce, by routine calculations, that

$$\begin{aligned} &\max_{1 \le i,\, j \le k+t+1} \left|\left(R^{-1} - R_I^{-1} + \hat{V}\right)(i,j)\right| \\ &\times \left[k^2(t+1)^2 \max_{j \ge (G(m_d))^{A-1}} |\Gamma(j)|\right]^{-1} \to 0 \quad \text{as } d \to \infty, \end{aligned} \tag{A53}$$

which implies (2.87).

In order to prove (2.88) we can use (A14)–(A17).

Thus we have

$$\left|\operatorname{Det}\left(RR_I^{-1}\right) - 1\right| \le \sum_{p=1}^{\infty} \left(k(t+1)u^2\right)^p \tag{A54}$$

with u as in (A16), from which (2.88) follows. □

APPENDIX B

PROOF OF LEMMA 2.11. Put

$$\begin{aligned} \Lambda(R_n) = &\left[\tau(G(R_n))^{-1}(-\log G(R_n))\right] \\ &+ \left[G(R_n)^{-(1+K)(1-A)}\right], \qquad n \ge 1, \end{aligned} \tag{B1}$$

and

$$\mathscr{G}_n = \left\{\max\left(\left|X_{S_n+Z_n+i}\right|; i = 1, \dots, \Lambda(R_n)\right) > \mu_K(R_n)\right\}, \qquad n \ge 1. \tag{B2}$$

We shall make use of the following propositions.

PROPOSITION B.1. *If one among the Hypotheses* H_1, H_2 *or* H_3 *is satisfied, then there exist a* $\tau_0 > 0$ *and* $A_0 > 0$ *such that for any* $0 < \tau < \tau_0$ *and* $0 < A < A_0$, *there exists* $0 < K_0(A)$ *such that if* $0 < K < K_0(A)$, *then*

$$P\{\mathscr{G}_n \ i.o.\} = 0. \tag{B3}$$

PROOF. For any $r > 0$, $n \ge 1$ and $t \ge 1$, put

$$\mathscr{F}_{r,n}(t) = \{\max(|X_{n+k}|, k = 1, \dots, t) < \mu_K(r)\} \tag{B4}$$

and observe that

(B5)
$$\overline{\mathscr{G}_n} = \mathscr{F}_{R_n, S_n + Z_n}(\Lambda(R_n)),$$

where, if $(1 + K)(1 - A) < 1$ [and since by (2.101) and Lemma 2.8, $G(R_n) \downarrow 0$ a.s. as $n \to \infty$], we have

(B6)
$$\Lambda(R_n) = \tau(G(R_n))^{-1}(-\log G(R_n))(1 + o(1)) \quad \text{as } n \to \infty.$$

The proof of Lemma 2.5 implies the following analogues of (2.67):

(B7)
$$\begin{aligned} P\{\mathscr{F}_{R_d, S_d+Z_d}(t) | \eta_d = l_d\} f_{\eta_d}(l_d) &= \sum_{\zeta \in \mathscr{S}(r_d)} \sum_{1 \le j \le J(\zeta, l_d)} A_j(\zeta, l_d) P_\zeta(E_j(l_d) \cap \mathscr{F}_{r_d, s_d+z_d}(t)) \\ &+ \sum_{1 \le k \le J'(l_d)} B_k(l_d) P(E'_k(l_d) \cap \mathscr{F}_{r_d, s_d+z_d}(t)) \\ &+ C(l_d) 1_{\{\nu_d = 0, m_d = r_d\}} P(\mathscr{F}_{r_d, 1}(t)) \\ &+ C(l_d) 1_{\{\nu_d = 0, m_d = r_d\}} P(\mathscr{F}_{r_d, 1}(t)). \end{aligned}$$

Similarly, by using the proof of Lemma 2.6, we obtain:

LEMMA 2.6′. *If one among Hypotheses* H_1, H_2 *or* H_3 *is satisfied, then there exist an* $0 < A_0 < 1$ *and two positive constants* C_1 *and* C_2 *independent of* l_d, *such that for any* $0 < A \le A_0$, *we have*

(B8)
$$\begin{aligned} \max_{\zeta \in \mathscr{S}(l_d)} \max_{1 \le j \le J(\zeta, l_d)} \Big| & P_\zeta(E_j(l_d) \cap \mathscr{F}_{r_d, s_d+z_d}(t)) \\ & \cdot \{P_\zeta(E_j(l_d)) P(\mathscr{F}_{r_d, 1}(t))\}^{-1} - 1 \Big| \\ & \le C_1 [G(r_d)]^{C_2} \end{aligned}$$

and

(B9)
$$\begin{aligned} \max_{1 \le k \le J'(l_d)} \Big| P(E'_k(l_d) \cap \mathscr{F}_{r_d, s_d+z_d}(t)) \cdot \{P(E'_k(l_d)) P(\mathscr{F}_{r_d, 1}(t))\}^{-1} - 1 \Big| \\ \le C_1 [G(r_d)]^{C_2}. \end{aligned}$$

By combining (B7), (B8) and (B9) and by again using the fact that

(B10)
$$\begin{aligned} & \left(\max_{\substack{1 \le i \le n \\ p_i > 0,\, \hat{p}_i > 0,\, 1 \le i \le n,}} \left| \frac{p_i}{\hat{p}_i} - 1 \right| \le q \right) \\ & \Rightarrow \left| \frac{\Sigma \alpha_i p_i}{\Sigma \alpha_i \hat{p}_i} - 1 \right| \le q \quad \text{for any } \alpha_i > 0,\ 1 \le i \le n, \end{aligned}$$

we obtain

$$\text{(B11)} \qquad \max_{t \geq 1} \left| \frac{P\{\mathscr{F}_{R_d, S_d + Z_d}(t) | \eta_d = l_d\}}{P\{\mathscr{F}_{r_d, 1}(t)\}} - 1 \right| < C_1 [G(r_d)]^{C_2}.$$

Next, if we apply Lemma 2.2 with $\mathscr{F}_{r_d, 1}(t) := Q(t, r_d, r_d)$ and $(1 + K)(1 - A) < 1$ [in order to have (B6)], it follows that there exist two positive constants c_1 and c_2 and $\tau_0 > 0$ such that for any $0 < \tau \leq \tau_0$,

$$\text{(B12)} \qquad \max_{1 \leq t \leq \Lambda(r_d)} \left| \frac{P\{\mathscr{F}_{r_d, 1}(t)\}}{(1 - 2G(\mu_K(r_d)))^t} - 1 \right| < c_1 [G(r_d)]^{c_2},$$

since $Q_I(t, r_d, r_d) = (1 - 2G(\mu_K(r_d)))^t$. Furthermore,

$$\text{(B13)} \qquad \max_{1 \leq t \leq \Lambda(r_d)} \left| (1 - 2G(\mu_K(r_d)))^t - 1 \right| \leq 3\tau [G(r_d)]^K (-\log G(r_d))$$

[by the classical inequality $(1 - at) < (1 - a)^t < 1$, $0 < a < 1$, $t > 0$.] Thus, by combining (B5), (B11), (B12) and (B13), we obtain that if $0 < K \leq K_0(A) = A/(1 - A)$, $A < A_0$, then

$$\text{(B14)} \qquad \begin{aligned} &|P\{\mathscr{G}_d | \eta_d = l_d\}| \\ &\leq C_1 [G(r_d)]^{C_2} + c_1 [G(r_d)]^{c_2} + 3\tau [G(r_d)]^K (-\log G(r_d)) \\ &\leq a [G(r_d)]^b \quad \text{for some positive constants } a \text{ and } b. \end{aligned}$$

Finally, for any $0 < \alpha < 1$,

$$\text{(B15)} \qquad \begin{aligned} &P\{\mathscr{G}_d \cap (G(R_d) < e^{-\alpha d})\} \\ &\leq \int_{\{G(r_d) < e^{-\alpha d}\}} a [G(r_d)]^b \, dP_{\eta_{d-1}}(l_{d-1}) \leq a e^{-\alpha d b}, \end{aligned}$$

from which, by using the Borel–Cantelli lemma and (2.101), we deduce (B3). □

Now, for any $0 < r < \rho < \rho + \delta\rho \leq \mu_K(r)$, $n \geq 1$, $t \geq 1$ and $t' \geq 1$, put

$$\text{(B16)} \qquad \begin{aligned} H_r^{n,t,t'}(\rho, \delta\rho) = &\{-\mu_K(r) < X_{n+s} < r; s = 0, \ldots, t - 1\} \\ &\cap \{\rho < X_t < \rho + \delta\rho\} \\ &\cap \{-\mu_K(r) < X_{n+t+s'} < \rho + \delta\rho; s' = 1, \ldots, t' - 1\} \\ &\cap \{\rho + \delta\rho < X_{n+t+t'} < \mu_K(r)\}. \end{aligned}$$

PROPOSITION B.2. *If one among Hypotheses* H_1, H_2 *or* H_3 *is satisfied, then there exists a* $0 < A_0 < 1$ *such that for any* $0 < A \leq A_0$ *and* $K \leq A/(1 - A)$

we have

$$\text{(B17)}\quad \begin{aligned}&\max_{\substack{1\le t\le \tau(G(r_d))^{-1}(-\log G(r_d))\\ 1\le t'\le (G(r_d))^{-(1-A)(1+K)}}} dP\left\{H_{r_d}^{s_d+z_d,t,t'}(\rho)|\eta_d = l_d\right\}\\ &\qquad\le (\text{const.})\, dP\left\{H_{r_d}^{1,t,t'}(\rho)\right\},\end{aligned}$$

where $dP\{H_{r_d}^{n,t,t'}(\rho)\} := P\{H_{r_d}^{n,t,t'}(\rho,\delta\rho)\}$ *and* (const.) *is a positive constant independent of* ρ *and* l_d.

Proof. As in the proof of Proposition B.1, we may replace in (B7), (B8) and (B9), $\mathscr{F}_{r_d, s_d+z_d}(t)$ and $\mathscr{F}_{r_d,1}(t)$, respectively, by $H_{r_d}^{s_d+z_d,t,t'}(\rho)$ and $H_{r_d}^{1,t,t'}(\rho)$ in order to obtain, by (B10), that there exist c_1 and $c_2 > 0$ such that the inequality corresponding to (B11) [which clearly implies (B17)] holds. □

Proposition B.3. *If one among Hypotheses* H_1, H_2 *or* H_3 *is satisfied, then there exist* $c > 0$ *and* $\varepsilon > 0$ *such that for any* $r > 0$, *we have*

$$\text{(B18)}\quad \sum_{(r)}\sum := \sum_{\substack{1\le t\le \tau(G(r))^{-1}(-\log G(r))\\ 1\le t'<(G(r))^{-(1-A)(1+K)}}}\sum \int_r^{\mu_K r} dP\left\{H_r^{1,t,t'}(\rho)\right\} \le c[G(r)]^{\varepsilon}.$$

Proof. For any $0 < r < \rho < \rho + \delta\rho < \mu_K(r)$, $t \ge 1$ and $t' \ge 1$, we have

$$\begin{aligned} dP\left\{H_r^{1,t,t'}(\rho)\right\} &\le P\{\rho < X_0 < \rho + \delta\rho,\, X_{t'} > \rho + \delta\rho\}\\ &\sim \frac{\delta\rho}{2\pi(1-\Gamma^2(t'))^{1/2}}\int_\rho^\infty \exp\left\{-\frac{\rho^2 2\Gamma(t')\rho x + x^2}{2(1-\Gamma(t'))}\right\}dx \qquad (\delta\rho\to\infty)\\ \text{(B19)}\qquad &\le \frac{\delta\rho}{2\pi(1-\Gamma^2(t'))^{1/2}}\exp\left\{-\frac{\rho^2}{2(1+|\Gamma(t')|)}\right\}\\ &\quad\times\int_\rho^\infty \exp\left\{-\frac{x^2}{2(1+|\Gamma(t')|)}\right\}dx \qquad [\text{by (2.8)}]\\ &\le \frac{\delta\rho}{(1-\Gamma^2(t'))^{1/2}}\times\frac{d}{d\rho}\left\{(-G^2)'\left(\frac{\rho}{(1+|\Gamma(t')|)^{1/2}}\right)\right\}\\ &= \frac{1}{(1-|\Gamma(t')|)^{1/2}}\, d\left(-G^2\left(\frac{\rho}{(1+|\Gamma(t')|)^{1/2}}\right)\right).\end{aligned}$$

Thus, by splitting the summation with respect to t' in (B18) into

$$\text{(B20)}\qquad \sum_{(r)}\sum = \sum_{t=1}^{\tau(G(r))^{-1}(-\log G(r))}\left(\sum_{t'=1}^{[T]} + \sum_{t'=T+1}^{[(G(r))]^{-(1-A)(1+K)}}\right),$$

where $1 \le T \le G(r)^{-(1-A)(1+K)}$, and by using (B19), we obtain

$$\sum_{(r)}\sum \le \frac{\tau}{(1-\delta)^{1/2}} \times \left(G^2\left(\frac{r}{(1+\delta)^{1/2}}\right) \times (G(r))^{-1}(-\log G(r))T \right.$$

$$\text{(B21)} \qquad + G^2\left(\frac{r}{(1+\delta(T))^{1/2}}\right)$$

$$\left. \times (G(r))^{-1-(1-A)(1+K)}(-\log G(r)) \right)$$

$$=: G_{1,n} + G_{2,n},$$

where $\delta = \max_{1\le t}|\Gamma(t)|$ and $\delta(T) = \max_{T\le t}|\Gamma(t)|$. Next,

$$G^2\left(\frac{r}{(1+\delta)^{1/2}}\right) \times (G(r))^{-1}$$

$$\sim \frac{(1+\delta)}{\sqrt{2\pi}} \cdot \frac{1}{r}\exp\left\{-r^2\left(\frac{1}{1+\delta} - \frac{1}{2}\right)\right\}, \qquad r \to \infty.$$

Observe that $(1/(1+\delta)) - (1/2) > 0$ since $0 < \delta < 1$ and take $T = T(r) = e^{\eta r^2}$ with

$$0 < \eta < \frac{1}{2}\inf\left(\left(\frac{1}{1+\delta} - \frac{1}{2}\right), (1-A)(1+K)\right).$$

Thus [since $-\log(G(r)) \sim (r^2/2)$ and for any $a, b > 0$, $r^a e^{-r^{2b}} \to 0$ as $r \to \infty$], there exist a $c > 0$ and $\varepsilon > 0$ such that $G_{1,n} \le (c/2)(G(r))^{\varepsilon}$. Likewise, if we consider $G_{2,n}$, we have

$$G^2\left(\frac{r}{(1+\delta(T))^{1/2}}\right)(G(r))^{-1-(1-A)(1+K)}$$

$$\sim \frac{(1+\delta(T))}{\sqrt{2\pi}} r^{(1-A)(1+K)-1}$$

$$\text{(B22)} \qquad \times \exp\left\{-r^2\left[\frac{A + A\delta(T) - 2\delta(T) + K(1-A)(1+\delta(T))}{2(1+\delta(T))}\right]\right\},$$

$$r \to \infty.$$

Thus, since by Hypotheses H_1, H_2 or H_3, $\lim_{T\to\infty} \delta(T) = 0$, for $r \ge r_0$ sufficiently large, the factor of $-r^2$ in (B22) is larger than

$$\frac{A}{2} + \frac{K(1-A)}{2} > 0.$$

This, by the same arguments as above, completes the proof of (B18). □

We can now prove Lemma 2.11. Put

$$\text{(B23)}\quad \begin{aligned} D_n = \Big\{ &((\Delta S_{n+1}, \mathscr{R}_{n+1}) \in \psi(R_n)) \\ &\cap \big(Z_n = [G(R_n)^{A-1}]\big) \cap (G(R_n) < e^{-\alpha n})\Big\}, \qquad n \geq 1. \end{aligned}$$

By Lemmas 2.8, 2.9 and 2.10 and Proposition B.1, it is equivalent to prove (2.120) and

$$\text{(B24)}\qquad P\{\mathscr{E}_n \text{ i.o.}\} = 0,$$

where

$$\text{(B25)}\qquad \mathscr{E}_n = C_{n+1} \cap D_n \cap \overline{\mathscr{G}}_n, \qquad n \geq 2.$$

We have, with the notation in (B18),

$$\text{(B26)}\quad P(\mathscr{E}_n) = \int_{G(r_n) < e^{-\alpha n}} \left(\sum_t \sum_{t'} \int_{r_n}^{\mu_K(r_n)} dP\{H_{r_n}^{s_n + z_n, t, t'}(\rho) | \eta_n = e_n\} \right) dP_{\eta_n}(e_n),$$

where $\Sigma\Sigma$ is over $1 < t < \tau(G(r_n))^{-1}(-\log G(r_n))$ and $1 \leq t' \leq (G(r_n))^{1+A(1+K)}$. Using (B17) and (B18) in (B26), we obtain

$$\text{(B27)}\qquad P(\mathscr{E}_n) \leq (\text{const.}) \cdot c \cdot e^{-\alpha \varepsilon n}$$

from which (B24) follows by the Borel–Cantelli lemma. □

Acknowledgment. We would like to express our sincere thanks to the referee for his valuable comments and suggestions which have resulted in many improvements in the presentation of our results.

REFERENCES

BERMAN, S. M. (1964). Limit theorems for the maximum term in stationary sequences. *Ann. Math. Statist.* **35** 502–516.

DEHEUVELS, P. (1974). Majoration et minoration presque sûre optimale des éléments de la statistique ordonnée d'un echantillon croissant de variables aléatoires indépendantes. *Rend. Cont. Acad. Naz. Lincei Ser. VIII* **5** 707–719.

DEHEUVELS, P. (1984a). The characterisation of distributions by order statistics and record values, a unified approach. *J. Appl. Probab.* **21** 326–334.

DEHEUVELS, P. (1984b). Strong approximations of records and record times. In *Statistical Extremes and Applications* (J. Tiago de Oliveira, ed.) 491–496. Reidel, Dordrecht.

DEHEUVELS, P. (1986). Strong laws for the k-th order statistics when $k \leq c \log_2 n$. *Probab. Theory Related Fields* **72** 133–154.

DEHEUVELS, P. (1988). Strong approximations of k-th records and k-th record times by Wiener processes. *Probab. Theory Related Fields* **77** 195–209.

GRENANDER, U. and ROSENBLATT, M. (1957). *Statistical Analysis of Stationary Time Series.* Wiley, New York.

HAIMAN, G. (1987a). Almost sure asymptotic behavior of the record and record time sequences of a stationary Gaussian process. In *Mathematical Statistics and Probability Theory* (M. L. Puri, P. Révész and W. Wertz, eds.) **A** 105–120. Reidel, Dordrecht.

HAIMAN, G. (1987b). Étude de extrèmes d'une suite stationnaire m-dépendante avec une application relative aux accroissements du processus de Wiener. *Ann. Inst. H. Poincaré Sect. B (N.S.)* **23** 425–458.

IBRAGIMOV, I. and ROZANOV, Y. (1974). *Processus Aléatoires Gaussiens*. Editions Mir, Paris.
LEADBETTER, M. R. and ROOTZÉN, H. (1988). Extremal theory for stochastic processes. *Ann. Probab*. **16** 431–478.
RESNICK, S. I. (1987). *Extreme Values, Regular Variation and Point Processes*. Springer, New York.

UFR DE MATHÉMATIQUES
UNIVERSITÉ DE LILLE
VILLENEUVE D'ASCQ, 59655
FRANCE

DEPARTMENT OF MATHEMATICS
INDIANA UNIVERSITY
BLOOMINGTON, INDIANA 47405

EXTREMES OF MARKOV SEQUENCES

GEORGE HAIMAN[a,*], MAXIME KIKI[a] and MADAN L. PURI[b,1]

[a] *Université des Sciences et Technologies de Lille, Laboratoire de Statistique et Probabilitiés, Bâtiment M2 59665, Villeneuve D'ascq, Cedex, France*

[b] *Department of Mathematics, Indiana University, Bloomington, Indiana 47405, USA*

Abstract

This paper establishes bounds for

$$\Delta_n = \max_{u_n \leqslant x \leqslant \omega} |P\{\text{Max}(X_0, \ldots, X_n) < x) - P(X_0 < x)^{n+1}|,$$

where $\{X_n\}_{n \geqslant 0}$ is a stationary Markov sequence with transition density $f(x, y)$ satisfying some regularity conditions, $\omega = \sup\{u;\ P(X_0 < u) < 1\}$, and u_n is defined for any fixed $\tau > 0$ by $P\{X_0 > u_n\} = \tau/n$, $n > \tau$. In particular, if there exists a probability density $g(y)$ such that

$$\eta g(y) \leqslant f(x, y) \leqslant M g(y), \quad x, y \in \mathbb{R},$$

then for any $0 < \beta < 1$ there exists $n_0(\tau, \beta)$ such that for $n \geqslant n_0$

$$\Delta_n \leqslant \beta n^{-1/2} (\log n)^{3/2}.$$

Key words: Extremes; Regenerative and Markov processes

1. Introduction and results

The starting point of this paper is a result of Rootzén (1988), recalled below, concerning the maxima of stationary Markov chains. Earlier classical references more or less closely connected with this topic are Serfozo (1980) and O'Brien (1987) (for a complete bibliography see Rootzén's paper). More recent papers are Smith (1992) and Jakubowski (1993). In order to present out results, we need some preliminaries.

* Corresponding author. Research partially supported by URA 1321, Université Paris 6.

[1] Research was supported by the Office of Naval Research Contract N00014-91-J-1020.

1.1. Preliminaries

Let $\{X_n;\ n=0,1,\ldots\}$ be a $(\mathbb{R},\mathscr{R})$ valued Markov chain (where $\mathscr{R}$ is the σ field generated by the Borel sets of $\mathbb{R}$) with stationary transition probability $P(\cdot,\cdot)$.

The distribution of $\{X_t\}_{t\geqslant 0}$ is determined by the transition probability and the distribution of X_0 (initial distribution). A probability π is stationary for $P(\cdot,\cdot)$ if $\pi(A)=\int P(x,A)\,\mathrm{d}\pi(x)$ for all $A\in\mathscr{R}$.

Let $R\in\mathscr{R}$ and set

$$\tau(R)=\inf\{t\geqslant 1,\ X_t\in R\}. \tag{1.1}$$

R is called *recurrent* if $P_x(\tau(R)<\infty)=1$ for all $x\in\mathbb{R}$. (We denote P_x the distribution of the chain with initial probability which gives probability 1 to the set $\{x\}$, $x\in\mathbb{R}$.)

R is called a *regeneration set* if it is recurrent and if there exist an integer $r\geqslant 1$, $\varepsilon\in(0,1)$ and a probability λ on $\mathscr{R}$ such that the r step transition probability satisfy

$$P^r(x,A)\geqslant\varepsilon\lambda(A)\quad\forall x\in R,\ A\in\mathscr{R}. \tag{1.2}$$

This terminology is justified by the fact (Asmussen, 1987, p. 151) that if a regeneration set R exists, then it is possible to construct $\{X_t\}$ simultaneously with a renewal process $S_0,S_1,\ldots$ with respect to which $\{X_t\}$ becomes regenerative if $r=1$ and 1-dependent regenerative if $r>1$.

$\{X_t\}$ (not necessarily Markov) is called *regenerative* if there exists a positive integer valued random sequence $\{Y_n\}_{n\geqslant 0}$ defined on the same probability space such that:

(i) For any positive integers $n\geqslant 1$ and $y_0,y_1,\ldots,y_n$ and any $A_1,\ldots,A_{s_n}\in\mathscr{R}$, the events

$$B_k=\{Y_k=y_k,\ X_{s_{k-1}}\in A_{s_{k-1}},\ldots,X_{s_k-1}\in A_{s_k-1}\}, \tag{1.3}$$

$$s_k=\sum_{i=0}^{k}y_i,\quad k=0,\ldots,n,\quad s_{-1}=0$$

are independent.

(ii) For any integers $1\leqslant k<l$ and $u>0$ and any $A_1,\ldots,A_u\in\mathscr{R}$,

$$P\{Y_k=u,\ X_{s_{k-1}}\in A_1,\ldots,X_{s_k-1}\in A_u\}$$
$$=P\{Y_l=u,\ X_{s_{l-1}}\in A_1,\ldots,X_{s_l-1}\in A_u\}. \tag{1.4}$$

Remark 1.1. This definition is described in Asmussen (1987) as 'there exists a renewal process $S_0,S_1,\ldots$ such that the cycles $c_k=\{X_t;\ S_{k-1}\leqslant t<S_k\}$, $k\geqslant 1$, are i.i.d.'. Clearly (i) and (ii) imply that Y_n, $n\geqslant 1$, are i.i.d. and $\{S_k:=Y_0+\cdots+Y_k\}_{k\geqslant 1}$ is a renewal process.

The sequence $\{X_t\}$ is called *1-dependent regenerative* if there exists a renewal process $\{S_k=Y_0+\cdots+Y_k\}_{k\geqslant 0}$ such that the events B_k in (1.3) are 1-dependent (i.e. for any $k\geqslant 0$ and $n\geqslant 2$, $\sigma(B_1,\dots,B_k)$ and $\sigma(B_{k+2},\dots,B_{k+n})$ are independent) and condition (ii) holds. Asmussen's construction of $\{X_t\}_{t\geqslant 0}$ simultaneously with $\{S_k\}_{k\geqslant 0}$ is then briefly the following.

Realise $X_0(\omega)=x$ and take the usual version of $\{X_t\}$ up to time $\tau(R)$ where R is hit. Then with probability ε let a renewal occur at $\tau(R)+r$ (i.e. $S_0=\tau(R)+r$) and then let $X_{\tau(R)+r}$ be independent of $X_0,X_1,\dots,X_{\tau(R)}$ and have distribution λ. Let, with probability $1-\varepsilon$, $X_{\tau(R)+r}$ have distribution (no renewal at $\tau(R)+r$ in this case)

$$[P^r(X_{\tau(r)},\cdot)-\varepsilon\lambda(\cdot)]/1-\varepsilon. \tag{1.5}$$

After that define $X_{\tau(R)+s}$, $0<s<r$, in order it has the right distribution.

Then repeat the procedure with the new initial value $X_{\tau(R)+r}:=x$ and so on. It is clear that at any renewal epoch S_n, $n\geqslant 0$, the distribution of X_{S_n} is λ and X_{S_n} is independent of $X_0,\dots,X_{S_{n-1}}$.

Let $\{X_t\}_{t\geqslant 0}$ be a Markov process with a regeneration set R ($\{X_t\}$ is then also called Harris recurrent) and let $\{S_n\}_{n\geqslant 0}$ be the renewal process obtained by the previous construction. Let

$$\xi_0=\max(X_0,\dots,X_{S_0-1}) \quad\text{and}\quad \text{for } n\geqslant 1,\ \xi_n=\max(X_{S_{n-1}},\dots,X_{S_n-1}). \tag{1.6}$$

It is not difficult to check that if $r=1$ then by (1.3) and (1.4) $\{\xi_n\}_{n\geqslant 1}$ is an i.i.d. sequence and if $r>1$, then $\{\xi_n\}_{n\geqslant 1}$ is a 1-dependent stationary sequence.

Since the distribution of X_0 is not necessarily λ, the distribution of ξ_0 is in general different from the common distribution of $\xi_1,\xi_2,\dots$. Let

$$v_t=\min\{k\geqslant 0;\ S_k\geqslant t\} \tag{1.7}$$

and

$$M_n=\mathrm{Max}(X_0,\dots,X_n).$$

Theorem 1.1 (Rootzén, 1988 p. 375). *Suppose* $r=1$ *in* (1.2), $\mu:=E(Y_1)<\infty$ $(Y_1=S_1-S_0)$ *and put* $G(x)=P(\xi\leqslant x)^{1/\mu}$. *Then for any* $\delta>0$ *we have*

$$|P(M_n\leqslant x)-G(x)^n|\leqslant\mu\left(\delta+\frac{1}{n}\right)+P(|v_n/n-1/\mu|\geqslant\delta)$$
$$+P(\xi_0>\max(\xi_1,\dots,\xi_{[n(1/\mu+\delta)]})) \tag{1.8}$$

with $[x]=$ *integer part of* x. *Hence, if* $\lim_{k\to\infty}P(\xi_0>\max(\xi_1,\dots,\xi_k))=0$ *(this is the case for example if the distribution of* X_0 *is* λ*), then, since by the law of large numbers* $\lim_{n\to\infty}P(|v_n/n-1/\mu|>\delta)=0$, *we have*

$$\sup_x|P(M_n\leqslant x)-G(x)^{n+1}|\to 0 \quad \text{as } n\to\infty. \tag{1.9}$$

1.2. The results

In the following we consider a Markov process $\{X_t\}_{t\geqslant 0}$ with transition density $f(x, y)$ with respect to the Lebesgue measure (i.e. $P(x, \mathrm{d}y) = f(x, y)\,\mathrm{d}y$). We suppose that the k step transition density $f^k(x, y)$ ($f^1(x, y) = f(x, y)$), given recursively by

$$f^k(x, y) = \int f^{k-1}(x, z) f(z, y)\,\mathrm{d}z, \quad k \geqslant 2, \tag{1.10}$$

satisfies the following hypothesis (H):

(H) There exists an integer $s \geqslant 1$, two constants $0 < \eta < 1$ and $M > 1$ and a probability density with respect to the Lebesgue measure $g(y)$ such that

$$\eta g(y) \leqslant f^s(x, y) \leqslant M g(y), \quad x, y \in \mathbb{R}. \tag{1.11}$$

We first prove the following result.

Theorem 1.2. *Let* $\{X_n\}_{n\geqslant 0}$ *satisfy hypothesis* (H). *Then, there exists a unique stationary distribution which has a density* $f(y)$ *with respect to the Lebesgue measure such that, for* $n \geqslant s$,

$$|f^n(x, y) - f(y)| \leqslant A\rho^n f(y), \quad -\infty < x, y < \infty, \tag{1.12}$$

where $A = 2(M/(1-\eta)\eta)^2$ *and* $\rho = (1-\eta)^{1/s}$.

Proof. See Section 3.

Let $\{X_t\}_{t\geqslant 0}$ satisfy hypothesis (H) and X_0 have the stationary density $f(y)$. If $s > 1$, suppose further that there exists $y_0 < \omega := \sup\{x;\, P(X_0 < x) < 1\}$ and a function $h(y)$, defined for $y \geqslant y_0$, such that

$$f(x, y) \leqslant h(y), \quad y \geqslant y_0 \tag{1.13}$$

and $1_{\{y\geqslant y_0\}} h(y)$ is Lebesgue integrable.

Let $\tau > 0$ be fixed and define $u_n = u_n(\tau)$ by $P(X_0 > u_n) = \tau/n$, $n > \tau$. Put

$$\Delta_n = \max_{u_n \leqslant x \leqslant \omega} |P(M_n \leqslant x) - P(X_0 < x)^{n+1}|. \tag{1.14}$$

Our main result is the following theorem.

Theorem 1.3. *Under the above hypotheses, if* $s > 1$ *and*

$$\limsup_{u\to\omega} \left[\left(\int_u^\omega h(y)\,\mathrm{d}y \right) \left(-\log \int_u^\omega g(y)\,\mathrm{d}y \right)^{3/2} \left(\int_u^\omega g(y)\,\mathrm{d}y \right)^{-1/2} \right] > 0, \tag{1.15}$$

then there exists a constant $C>0$ *such that*

$$\Delta_n \leqslant C \int_{u_n}^{\omega} h(y)\,\mathrm{d}y, \quad n \geqslant 1. \tag{1.16a}$$

If the limit above is zero or $s=1$, *then, for any* $0<\beta<1$, *there exists* $n_0(\tau, \beta)$ *such that, for* $n \geqslant n_0$,

$$\Delta_n \leqslant \beta \frac{(\log n)^{3/2}}{n^{1/2}}. \tag{1.16b}$$

Proof. See Section 2.

We conclude this section by some remarks and examples.

Remark 1.2. (a) Let, with A and ρ as in (1.14),

$$r_0 := \min\{n;\ A\rho^n < 1\}.$$

By Theorem 1.2, for any $r \geqslant r_0$ and $\varepsilon \leqslant 1 - A\rho^r$,

$$f^r(x, y) \geqslant \varepsilon f(y), \quad x, y \in \mathbb{R}. \tag{1.17}$$

Thus, $\{X_t\}_{t\geqslant 0}$ satisfy (1.2) where λ is the stationary distribution and R is $\mathbb{R}$. In this case Asmussen's construction may be simplified.

Realise $X_0(\omega)=x$ and then let with probability ε a renewal occur at r (i.e. $S_0=r$) and then let X_r be independent of $X_0, \ldots, X_{r-1}$ and have probability density $f(y)$. Let, with probability $1-\varepsilon$, X_r have density (no renewal at r in this case)

$$\frac{f^r(x, y) - \varepsilon f(y)}{1-\varepsilon}.$$

After that define X_s, $0<s\leqslant r-1$ in order it has the right distribution. Then repeat the procedure with the new initial value $X_r=x$ and so on. In this case the probability distribution of the renewal process is such that

$$P\{Y_0=S_0=kr\}=P\{Y_1=S_1-S_0=kr\}=(1-\varepsilon)^{k-1}\varepsilon, \quad k=1,2,\ldots \tag{1.18}$$

and hence

$$E(Y_1)=\mu=r/\varepsilon \quad \text{and} \quad V(Y_1)=\mu^2(1-\varepsilon). \tag{1.19}$$

(b) Observe that using (1.17), ξ_0 and ξ_n, $n\geqslant 1$ in (1.6) have the same distribution so that $\{\xi_n\}_{n\geqslant 0}$ is stationary 1-dependent. (This is not the case in general if the construction is based on the first inequality in (1.11), with $\mathrm{d}\lambda(y)=g(y)\,\mathrm{d}y$, $\varepsilon=\eta$ and $r=s$.)

These facts play an important role in the proof of Theorem 1.3.

(c) If we compare Theorems 1.1 and 1.3, on the one hand some of the hypotheses of Theorem 1.3 are stronger (existence of transition density, right-hand inequality in

(1.11) and the fact that $R=\mathbb{R}$). On the other hand Theorem 1.3 includes the case $r>1$ and (1.16a), (1.16b) are much more explicit than (1.8).

(d) Note that (1.16a) and (1.16b) imply that if the stationary probability is in the domain of attraction of an extreme value distribution, then, for some constants $a_n>0$ and b_n (which are the same as in the independent case), $a_n(M_n-b_n)$ converges weakly to a limit belonging to the same domain of attraction. If the speed of convergence in the independent case is known (see Galambos, 1987), then (1.16a) and (1.16b) provide estimates of the speed of convergence in the Markov case.

1.3. Examples

It may be checked that the following examples of processes satisfy the hypotheses of Theorem 1.3.

Example 1. $\{X_n\}_{n\geq 0}$ is defined recursively by

$$X_{n+1}=T(X_n)+\sigma(X_n)\varepsilon_{n+1}, \quad n\geq 0, \tag{1.20}$$

where T and σ are such that there exist $C\geq 0$ and $0<A<B$ such that $\sup_x |T(x)|\leq C$ and $A<1/|\sigma(x)|<B$, $x\in\mathbb{R}$.

$\{\varepsilon_n\}_{n\geq 1}$ is an i.i.d. sequence with common density $h(y)$ with respect to the Lebesgue measure such that

$$h(y)\sim C|y|^{-\alpha}, \quad |y|\to\infty, \ \alpha>0, \ C>0.$$

Then the transition density is

$$f(x,y)=\frac{1}{|\sigma(x)|}h\left(\frac{y-T(x)}{\sigma(x)}\right), \quad x,y\in\mathbb{R}$$

and one can find η, M and $g(y)$ such that (1.11) holds with $s=1$.

Example 2 (*Variant of Example* 1). $\{X_n\}_{n\geq 0}$ is defined recursively by (1.20) with $\sigma(x)$ a constant and T bounded. Then one can take $h(y)\sim C|y|^p\exp(-\alpha|y|)$, $|y|\to\infty$, $\alpha>0$, $p\geq 0$, $c>0$.

Example 3. $\{X_n\}_{n\geq 0}$ takes values in (a,b), $-\infty<a<b<\infty$, and has a bounded (or continuous on $(a,b)^2$) transition density $f(x,y)$ such that, for some integer $s\geq 1$ and $0<\varepsilon<1$,

$$f^s(x,y)\geq\varepsilon. \tag{1.21}$$

This is in particular the case if $a=0$, $b=1$ and

$$f(x,y)=xh(xy)\left(\int_0^x h(z)\,\mathrm{d}z\right)^{-1} \quad x,y\in(0,1), \tag{1.22}$$

where $h(x)$ is a bounded (or continuous) probability density on $(0,1)$ such that

$$h(x) \geqslant \varepsilon, \quad x \in (0,1),$$

for some $\varepsilon > 0$.

Then it may be seen that there exist $0 < \eta < M$ such that

$$\eta \leqslant f(x,y) \leqslant M, \quad x, y \in (0,1), \tag{1.23}$$

which implies (1.11) with $s=1$ and $g(y)=1$, $y \in (0,1)$.

If $f(x,y)$ satisfy (1.22) with $h(x)=2(1-x)$, $0<x<1$, then $f(x,y)=2(1-xy)/(2-x)$ and it may be seen that (1.11) does not hold for $s=1$ but (1.21) holds for $s=2$.

Condition (1.13) is also satisfied since $f(x,y) \leqslant 2$, $x, y \in (0,1)$.

2. Proof of Theorem 1.3

The proof consists in showing first that an inequality similar to (1.8) (without the last term) may be established also in the case under consideration, and then, in giving explicit expressions for $G(x)$, μ and $P\{|\nu_n/n - 1/\mu| \geqslant \delta\}$.

Roughly, $P\{M_n < x\} \sim P\{\max_{i \leqslant \nu_n} \xi_i < x\}$, $\nu_n \sim n/\mu$ with $\mu = E(Y_0) = r/\varepsilon$ and by Theorem 2.1 (the key of the proof), if $\{\xi_i\}$ is a stationary 1-dependent sequence, then

$$P\left\{\max_{i \leqslant t} \xi_i < x\right\} \sim G^t(x),$$

where an explicit expression of $G(x)$ may be given.

In order to get the result we then evaluate the differences between 'equivalent' terms.

We need the following preliminary results.

Lemma 2.1. *With the definitions in* (1.8), (1.9) *and* (1.19), *for any* $0 < \delta < 1/\mu$ *and* $n > (1/\mu - \delta)^{-1}$, *we have*

$$P(M_n < x) \leqslant (1-\varepsilon)^{n/r - 1} + P\left\{\operatorname*{Max}_{0 \leqslant k \leqslant [n(1/\mu - \delta)] - 1} \xi_k < x\right\} + P\left\{\frac{\nu_n}{n} < \frac{1}{\mu} - \delta\right\} \tag{2.1}$$

and, for any $\delta > 0$,

$$P(M_n < x) \geqslant P\left\{\operatorname*{Max}_{0 \leqslant k \leqslant [n(1/\mu + \delta)] - 1} \xi_k < x\right\} - P\left\{\frac{\nu_n}{n} > \frac{1}{\mu} + \delta\right\}, \tag{2.2}$$

where $[u]$ = *integer part of* u.

Proof. We have

$$P\{M_n<x\}=P(M_n<x,\ v_n=0)+P(M_n<x,\ v_n\geqslant 1)$$

$$\leqslant P\{v_n=0\}+P\left\{\operatorname*{Max}_{0\leqslant k\leqslant v_n-1}\xi_k<x,\ v_n\geqslant 1\right\}$$

and

$$P\left(\operatorname*{Max}_{0\leqslant k\leqslant v_n-1}\xi_k<x,\ v_n\geqslant 1\right)=P\left\{\operatorname*{Max}_{0\leqslant k\leqslant v_n-1}\xi_k<x,\ v_n\geqslant 1,\ \frac{v_n}{n}<\frac{1}{\mu}-\delta\right\}$$

$$+P\left\{\operatorname*{Max}_{0\leqslant k\leqslant v_n-1}\xi_k<x,\ v_n\geqslant 1,\ \frac{v_n}{n}\geqslant\frac{1}{\mu}-\delta\right\}.$$

Thus we obtain (2.1) by observing that $P\{v_n=0\}=(1-\varepsilon)^{\lceil n/r\rceil^{-1}}$ where $\lceil x\rceil-1<x\leqslant\lceil x\rceil$.

The proof of (2.2) is similar and simpler since

$$P(M_n<x)\geqslant P\left(\operatorname*{Max}_{0\leqslant k\leqslant v_n}\xi_k<x\right).\qquad\square$$

In the following lemmas we develop expressions for the last two terms in (2.1) and (2.2). We need the following theorem.

Theorem 2.1 (Haiman, 1981). *Let $\{\xi_k\}_{k\geqslant 0}$ be a 1-dependent stationary sequence with continuous marginal distribution function.*

Then there exists a function $\mathscr{M}(x)$, $0<\mathscr{M}(x)<1$, defined for $x\geqslant x_0$, where x_0 is given by $P\{\xi_0>x_0\}=p_0$, with p_0 universal constant, $0<p_0<1$, such that

$$\mathscr{M}(x)=1-P\{\xi_0>x\}+P\{\xi_0>x,\xi_1>x\}+\mathrm{O}(P^2(\xi_0>x)),\quad x\geqslant x_0 \tag{2.3}$$

and

$$\operatorname*{Max}_{n\geqslant 0}\left|P\left(\operatorname*{Max}_{0\leqslant k\leqslant n}\xi_k<x\right)\times\mathscr{M}^{-(n+1)}(x)-1\right|\leqslant C_2P(\xi_0>x,\ \xi_1>x),\quad x\geqslant x_0, \tag{2.4}$$

with $|\mathrm{O}(x)|\leqslant C_1|x|$ and C_1, C_2 universal constants.

Lemma 2.2. *For $1/n<\delta<1/\mu$, where $\mu=r/\varepsilon$, we have*

$$P\left\{\frac{v_n}{n}<\frac{1}{\mu}-\delta\right\}\leqslant C\exp\left\{-\frac{1}{\varepsilon}\left(\frac{nr}{\varepsilon}\right)^{1/2}\frac{\delta-1/n}{(1-\delta\mu)^{1/2}}\right\}, \tag{2.5}$$

with

$$C=\exp\left(\frac{1}{2}\frac{(1-\varepsilon)e}{(1-e(1-\varepsilon))^2}\right)$$

and

$$P\left\{\frac{\nu_n}{n}>\frac{1}{\mu}+\delta\right\}\leqslant C'\exp\left\{-n^{1/2}\delta\frac{\mu^{3/2}}{r}(1+\mu\delta)^{-1/2}\right\}, \tag{2.6}$$

with $C'=\exp((1-\varepsilon)/2\varepsilon^2)$.

Proof. We have, since ν_n is an integer,

$$P\left\{\frac{\nu_n}{n}<\frac{1}{\mu}-\delta\right\}\leqslant P\left\{\nu_n\leqslant\left[n\left(\frac{1}{\mu}-\delta\right)\right]\right\}.$$

Put $q:=[n(1/\mu-\delta)]$. Then

$$P\{\nu_n\leqslant q\}=P\{S_q\geqslant n\}=P\{S_q-(q+1)\mu\geqslant n-(q+1)\mu\}$$

$$\leqslant P\left\{S_q-(q+1)\mu\geqslant n\mu\left(\delta-\frac{1}{n}\right)\right\},$$

since $q\leqslant n(1/\mu-\delta)$, which implies $n-(q+1)\mu\geqslant n\mu(\delta-1/n)$.

We now apply Bernstein's inequality

$$P\left\{\xi>\frac{t+\log M_\xi(\eta)}{\eta}\right\}<\mathrm{e}^{-t},\quad t>0,\ \eta>0, \tag{2.7}$$

with $\xi=S_q-(q+1)\mu$ and

$$M_\xi(\eta)=E(\mathrm{e}^{\eta\xi})=\left(\mathrm{e}^{\eta r(1-1/\varepsilon)}\frac{\varepsilon}{1-\mathrm{e}^{\eta r}(1-\varepsilon)}\right)^{q+1}.$$

Since we have $1-\varepsilon<\mathrm{e}^{-1}$, $M_\xi(\eta)$ is defined for $\eta<1/r$.

Next, using Taylor's formula, we get

$$\frac{\log M_\xi(\eta)}{\eta}\leqslant(q+1)\frac{\eta r^2}{2}\frac{(1-\varepsilon)\mathrm{e}}{(1-\mathrm{e}(1-\varepsilon))^2},\quad \eta<\frac{1}{r}.$$

Let $\eta=(1/r)(q+1)^{-1/2}$. Then (2.7) implies

$$P\{\xi>r(q+1)^{1/2}(t+C)\}<\mathrm{e}^{-t},\quad t>0,$$

from which (2.5) follows.

The proof of (2.6) is completely similar. □

In the proof of Theorem 1.3, for any integer n we will consider the construction of the renewal process (see Section 1) made with $r=r(n)=[\alpha\log n]$, where α is a fixed positive constant to be chosen later.

Then, for $n\geqslant n_0$ large enough, $A\rho^r<\frac{1}{2}$. We then take a fixed $\varepsilon=\frac{1}{2}$.

Lemma 2.3. *For $n \geqslant n_0$ and $x \geqslant \max(x_0, y_0)$, with y_0 in (1.15), $\mathcal{M}(x)$ in (2.3) has the form*

$$\mathcal{M}(x)=\mathcal{M}(n,x)=1-2rP(X_0>x)\left[1+2\mathrm{O}_1(A\rho^r)+2s\mathrm{O}_2\left(\int_x^{\omega} h(y)\,\mathrm{d}y\right)\right]$$
$$+\mathrm{O}_3[16r^2P^2(X_0>x)(5M+C_1)],\quad |\mathrm{O}_i(x)|\leqslant|x|,\ i=1,2,3. \tag{2.8}$$

Proof. Since $P\{\xi_0>x\}=P\{\xi_1>x\}$, if we put

$$A=\{\mathrm{Max}(X_0,\ldots,X_{S_0-1})<x\}\quad\text{and } B=\{\mathrm{Max}(X_0,\ldots,X_{S_1-1})<x\},$$

we have

$$1-P\{\xi_0>x\}+P\{\xi_0>x,\xi_1>x\}=1-P\{A\}+P\{B\}.$$

Next, since $P\{A|Y_0=kr\}=P\{A|Y_0=kr,\ Y_1=tr\}$, we have

$$P(A)-P(B)=\sum_{t=1}^{\infty}P\{Y_1=tr\}\sum_{k=1}^{\infty}[P\{A|Y_0=kr,\ Y_1=tr\}$$
$$-P\{B|Y_0=kr,\ Y_1=tr\}]P(Y_0=kr) \tag{2.9}$$

with $Y_0=S_0$, $Y_1=S_1-S_0$.

Put

$$\mathcal{N}=\mathcal{N}(x,k,t):=P\{A|Y_0=kr,\ Y_1=tr\}-P\{B|Y_0=kr,\ Y_1=tr\}$$

and let $X'_0,\ldots,X'_{(k+t)r}$ be random variables with joint distribution equal to the conditional distribution of $X_0,\ldots,X_{(k+t)r}$ given $Y_0=kr$ and $Y_1=tr$.

We then have, by Bonferroni's inequalities,

$$\mathcal{N}=P\{\mathrm{Max}(X'_0,\ldots,X'_{kr-1})<x\}-P\{\max(X'_0,\ldots,X'_{(k+t)r-1})<x\}$$
$$=P(X'_{kr}>x)+\cdots+P(X'_{(k+t)r-1}>x)$$
$$+\mathrm{O}\left(\sum\sum_{0\leqslant i<j\leqslant(k+t)r-1}P(X'_i>x,\ X'_j>x)\right) \tag{2.10}$$

with $|\mathrm{O}(x)|\leqslant|x|$.

Remark 2.1. Observe that by Theorem 2.1 and since $r=r(n)=[\alpha\log n]$, x_0 in Lemma 2.3 also depends on n.

Next,

$$P\{\xi_0>x\}=\sum_{k=1}^{\infty}P(\xi_0>x\,|\,S_1=kr)P(S_1=kr)$$
$$\leqslant\sum_{k=1}^{\infty}krP\{X_0>x\}P(S_1=kr)=P(X_0>x)\times\frac{r}{\varepsilon}.$$

Let $x_r = x_{r(n)}$ satisfy $(r/\varepsilon)P(X_0 > x_r) = p_0$. Then the above inequality implies that for n large enough Lemma 2.3 holds for $x \geqslant x_r$.

In order to evaluate the terms of (2.10), we need the following lemma.

Lemma 2.4. *Put* $i = lr + m$, $0 \leqslant l < k$, $0 \leqslant m \leqslant r-1$. *Then the probability density of* X'_i, $g_i(x)$, *is*

$$g_i(x) = \begin{cases} f(x) & \text{if } i = lr,\ 0 \leqslant l < k, \\ f(x)(1 + \mathrm{O}((\frac{A\rho^r}{1 - A\rho^r}))) & \text{if } i = lr + m,\ 0 \leqslant l \leqslant k-1,\ 0 < m \leqslant r-1, \end{cases} \tag{2.11}$$

where $f(x)$ *is the density of* X_0 *and* $|\mathrm{O}(x)| \leqslant |x|$. *The joint probability density of* (X'_i, X'_j), $0 \leqslant i < j \leqslant (k+t)r - 1$, *denoted by* $g_{ij}(x, y)$, *is such that*

$$g_{ij}(x, y) = g_i(x) g_j(y) \quad \text{if } i \leqslant (k-1)r \quad \text{and } j \geqslant kr,$$

and, in the complementary case for i and j,

$$g_{ij}(x, y) \leqslant 8 f(x) f^{j-i}(x, y) \leqslant \begin{cases} 8M f(x) f(y) & \text{if } j - i \geqslant s, \\ 8 f(x) h(y) & \text{if } 1 \leqslant j - i < s,\ s > 1. \end{cases} \tag{2.12}$$

Proof. See Section 3.

Proof of Lemma 2.3 (*Conclusion*). Combining (2.10)–(2.12) (and observing that the probability distribution of $(X'_{kr}, \ldots, X'_{(k+t)r-1})$ is the same as the distribution of $X_0, \ldots, X_{tr-1}$ given $Y_0 = tr$) we get

$$\mathcal{N} = trP(X_0 > x)(1 + 2\mathrm{O}_1(A\rho^r)) + \mathrm{O}_2\left[8M\frac{(k+t)^2}{2} r^2 P^2(X_0 > x)\right]$$
$$+ \mathrm{O}_3\left[(k+t)rsP(X_0 > x)\int_x^\omega h(y)\,\mathrm{d}y\right] \tag{2.13}$$

with $|\mathrm{O}_i(x)| \leqslant |x|$, $i = 1, 2, 3$.

From (2.9) and since $\varepsilon = \frac{1}{2}$ we obtain

$$P(A) - P(B) = \sum_{t=1}^{\infty}\sum_{k=1}^{\infty} \mathcal{N}(k, t) P\{Y_0 = kr\} P\{Y_1 = tr\}$$
$$= 2rP(X_0 > x)\left[1 + 2\mathrm{O}_1(A\rho^r) + \mathrm{O}_3\left(2s\int_x^\omega h(y)\,\mathrm{d}y\right)\right]$$
$$+ \mathrm{O}_2(80Mr^2 P^2(X_0 > x)) \tag{2.14}$$

and, using (2.11),

$$P(\xi_0>x)=P(A^c)\leqslant\sum_{k=1}^{\infty}\left(\sum_{i=0}^{kr-1}P(X_i'>x)\right)P(Y_0=kr)\leqslant 2rP(X_0>x)(1+2\mathrm{O}(A\rho^r)$$

$$\leqslant 4rP(X_0>x). \tag{2.15}$$

Combining (2.3), (2.14) and (2.15) we get (2.8). □

Lemma 2.5. *Put* $H(x)=H(n,x)=\mathcal{M}(x)^{1/\mu}=\mathcal{M}(x)^{1/2r}$. *Then, for* $n>\mu$,

$$\max_x|\mathcal{M}(x)^{n(1/\mu-\delta)-1}-H^n(x)|\leqslant\delta\mu+\frac{\mu}{n} \tag{2.16}$$

and

$$\max_x|\mathcal{M}(x)^{n(1/\mu+\delta)+1}-H^n(x)|\leqslant\delta\mu+\frac{\mu}{n}. \tag{2.17}$$

Proof. Put $Z=H(x)^n$. Then, for $0<\delta<1/\mu-1/n$, we have

$$|\mathcal{M}(x)^{n(1/\mu-\delta)-1}-H(x)^n|=|Z^{1-\gamma}-Z|=Z^{1-\gamma}-Z,\quad \gamma=\mu\left(\delta+\frac{1}{n}\right)<1$$

and (2.16) follows by maximising $Z^{1-\gamma}-Z$ subject to $0\leqslant Z\leqslant 1$. The proof of (2.17) is completely similar. □

The following lemmas are easy applications of Taylor's formula.

Lemma 2.6. *For* $0\leqslant v\leqslant\frac{1}{2}$ *and* $0<a<1$ *we have*

$$1-av+2a(a-1)v^2\leqslant(1-v)^a\leqslant 1-av. \tag{2.18}$$

Lemma 2.7. *For any fixed* $\tau>0$ *and* u, $|u|<\frac{1}{2}$ *we have, for* $n\geqslant 4\tau$,

$$\left(1-\frac{\tau}{n}(1+u)\right)^n=\left(1-\frac{\tau}{n}\right)^n+\mathrm{O}(2\tau e^{-\tau/2})u,\quad |\mathrm{O}(x)|\leqslant|x|. \tag{2.19}$$

Proof of Theorem 1.3. Let $\tau>0$ be fixed and $u_n=u_n(\tau)$ be defined by $P(X_0>u_n)=\tau/n$, $n>\tau$.

Then, by Remark 2.1, for $n\geqslant n_0(\tau)$ large enough $u_n>x_r$ and since $rP(X_0>u_n)\leqslant\alpha\log n\frac{\tau}{n}\to 0$ as $n\to\infty$ we have $1-\mathcal{M}(x)<\frac{1}{2}$, $x\geqslant u_n$. Thus, using (2.18) with $v=1-M(x)$ and $a=1/2r$ we get

$$H(x)=1-P(X_0>x)\left[1+2\mathrm{O}_1(A\rho^r)+\mathrm{O}_2\left(rP(X_0>x)+\int_x^{\infty}h(y)\,\mathrm{d}y\right)\right],\quad x\geqslant u_n$$

with $|\mathrm{O}_1(x)|\leqslant|x|$ and $|\mathrm{O}_2(x)|\leqslant K|x|$, K universal constant.

Next, applying (2.19) with $u=(1-H(x))/P(X_0>x)-1$, we obtain that for $n\geqslant n_1(\tau)$ large enough

$$H^n(x)=P(X_0<x)^n+\mathrm{O}(2\tau\,\mathrm{e}^{-\tau/2})\left(2\mathrm{O}_1(A\rho^r)+\mathrm{O}_2\left(rP(X_0>x)+\int_x^\omega h(y)\,\mathrm{d}y\right)\right),$$
$$x\geqslant u_n. \tag{2.20}$$

Combining (2.4), (2.15) and (2.16) we have

$$\begin{aligned}P\left\{\operatorname*{Max}_{0\leqslant k\leqslant[n(1/\mu-\delta)]-1}\xi_k<x\right\}&=\mathscr{M}(x)^{[n(1/\mu-\delta)]}(1+\mathrm{O}(C_2P(\xi_0>x)))\\&\leqslant\mathscr{M}(x)^{n(1/\mu-\delta)-1}(1+C_2P(\xi_0>x))\\&\leqslant\left(H(x)^n+\mu\delta+\frac{\mu}{n}\right)(1+4C_2rP(X_0>x))\end{aligned}$$

and similarly, using (2.17),

$$P\left\{\operatorname*{Max}_{0\leqslant k\leqslant[n(1/\mu+\delta)]}\xi_k<x\right\}\geqslant\left(H(x)^n-\mu\delta-\frac{\mu}{n}\right)(1-4C_2rP(X_0>x)). \tag{2.22}$$

Observe that, for $x\geqslant u_n$,

$$P(X_0<x)^n\geqslant\left(1-\frac{\tau}{n}\right)^n\simeq\mathrm{e}^{-\tau},\quad n\to\infty \tag{2.23}$$

and, since $1+u\leqslant1/(1-u)\leqslant1+2u$, $0<u<\frac{1}{2}$, for n large enough,

$$\begin{aligned}P(X_0<x)^n&=P(X_0<x)^{n+1}(1-P(X_0>x))^{-1}\\&=P(X_0<x)^{n+1}\left(1+2\mathrm{O}\left(\frac{\tau}{n}\right)\right),\quad |\mathrm{O}(x)|\leqslant|x|.\end{aligned} \tag{2.24}$$

Let $r(n)=\alpha\log n$ with $\alpha=\frac{1}{2}1/|\log\rho|$. Then, in (2.20) we have $\rho^r=n^{-1/2}$. Let

$$\delta=\frac{1}{2}\left(\frac{\varepsilon^3}{\alpha}\frac{\log n}{n}\right)^{1/2}.$$

Then the right-hand terms in (2.5) and (2.6) are $\mathrm{O}(n^{-1/2})$. We now evaluate Δ_n of (1.14), by combining (2.1), (2.2), (2.4)–(2.6) and (2.20)–(2.24).

In (2.21) and (2.22) we have

$$\mu\delta=\left(\frac{\log n}{2}\right)^{3/2}\left(\frac{\varepsilon}{|\log\rho|n}\right)^{1/2}\quad\text{and}\quad\frac{\mu}{n}=\mathrm{o}(\mu\delta),\quad n\to\infty.$$

Next, the contribution of the term $\int_x^\omega h(y)\,\mathrm{d}y$ in (2.20) is $\int_{u_n}^\omega h(y)\,\mathrm{d}y$. If the limit in (1.15) is strictly positive, then, since by (1.11) $\eta\int_x^\omega g(y)\,\mathrm{d}y\leqslant\int_x^\omega f(y)\,\mathrm{d}y\leqslant M\int_x^\omega g(y)\,\mathrm{d}y$, we have $0<\limsup_{n\to\infty}(\int_{u_n}^\omega h(y)\,\mathrm{d}y)(\mu\delta)^{-1}\leqslant\infty$.

If the limit is zero, then $\int_{u_n}^\omega h(y)\,\mathrm{d}y=\mathrm{o}(\mu\delta)$, $n\to\infty$.

Since the other terms are all $o(\mu\delta)$, $n\to\infty$, either $\delta_n = O(\int_{u_n}^{\omega} h(y)\,dy)$ or $\lim_{n\to\infty} \delta_n/\mu\delta = 1$. In the second case, since ε is arbitrary, we get (1.16b). The fact that n_0 depends on β (through ε) may be seen in particular from (2.1). □

3. Proofs of Theorem 1.2 and Lemma 2.4

3.1. Proof of Theorem 1.2

For any $r \geqslant 1$ and y put

$$m^{(r)}(y) = \inf_x f^r(x, y) \quad \text{and} \quad M^{(r)}(y) = \max_x f^r(x, y). \tag{3.1}$$

Formula (1.10) then implies that $\{m^r(y)\}$ and $\{M^r(y)\}$ are, respectively, increasing and decreasing. Then, since for any y $m^r(y) \leqslant M^r(y)$ we deduce that

$$m^i(y) \leqslant M^j(y), \quad i, j \geqslant 1, \quad y \in \mathbb{R}.$$

For any fixed x and y, let $\psi_{x,y}(u) := f^s(x, u) - f^s(y, u)$, $u \in \mathbb{R}$, and put

$$S_{xy}^+ = \{u;\ \psi_{xy}(u) \geqslant 0\} \quad \text{and} \quad S_{xy}^- = \{u;\ \psi_{xy}(u) < 0\}.$$

We then have

$$0 \leqslant \int_{S_{xy}^+} \psi_{xy}(u)\,du = 1 - \int_{S_{xy}^-} f^s(x, u)\,du - \int_{S_{xy}^+} f^s(y, u)\,du,$$

which combined with the first inequality in (1.13) provides

$$0 \leqslant \int_{S_{xy}^+} \psi_{xy}(u)\,du \leqslant 1 - \eta \int g(u)\,du = 1 - \eta. \tag{3.2}$$

Next, $0 \leqslant M^s(u) - m^s(u) \leqslant M^s(u) \leqslant Mg(u)$ and for any $k \geqslant 1$,

$$\begin{aligned} M^{(k+1)s}(u) - m^{(k+1)s}(u) &= \sup_{x,y} \left[f^{(k+1)s}(x, u) - f^{(k+1)s}(y, u) \right] \\ &= \sup_{x,y} \left[\int f^{ks}(z, u)(f^s(x, z) - f^s(y, z))\,dz \right] \\ &= \sup_{x,y} \left[\int_{S_{xy}^+} f^{ks}(z, u)\psi_{x,y}(z)\,dz + \int_{S_{xy}^-} f^{ks}(z, u)\psi_{x,y}(z)\,dz \right] \\ &\leqslant \sup_{x,y} \left(M^{ks}(u) \int_{S_{xy}^+} \psi_{x,y}(z)\,dz + m^{ks}(u) \int_{S_{xy}^-} \psi_{x,y}(z)\,dz \right) \end{aligned} \tag{3.3}$$

since $\psi_{xy} \geqslant 0$ on S_{xy}^+ and $\psi_{xy} < 0$ on S_{xy}^-.

Observe that $\int_{S^+_{xy}} \psi(z)\,dz = -\int_{S^-_{xy}} \psi(z)\,dz$ which combined with (3.2) provides

$$M^{(k+1)s}(u) - m^{(k+1)s}(u) \leqslant (M^{ks}(u) - m^{ks}(u)) \sup_{x,y} \int_{S^+_{xy}} \psi_{xy}(z)\,dz. \tag{3.4}$$

Thus, using (3.2) with $\lambda = 1-\eta$, we obtain

$$M^{ks}(u) - m^{ks}(u) \leqslant \lambda^{k-1} \times Mg(u), \quad k \geqslant 1, \ u \in \mathbb{R}.$$

If $1 \leqslant i \leqslant s-1$, we have

$$\begin{aligned}
M^{ks+i}(u) - m^{ks+i}(u) &= \sup_{x,y} (f^{ks+i}(x,u) - f^{ks+i}(y,u)) \\
&\leqslant \int f^i(z,u)(M^{ks}(z) - m^{ks}(z))\,dz \\
&\leqslant M\lambda^{k-1} \int f^i(z,u) g(z)\,dz \leqslant M\lambda^{k-1} \int f^i(z,u) \frac{f^s(x,z)}{\eta}\,dz \\
&= \frac{M}{\eta} \lambda^{k-1} f^{(i+s)}(x,u) \leqslant \frac{M^2}{\eta} \lambda^{k-1} g(u).
\end{aligned}$$

Thus, for any $n \geqslant s$, we have

$$M^n(u) - m^n(u) \leqslant C\lambda^{[n/s]} g(u), \quad C = \frac{M^2}{\eta\lambda}. \tag{3.5}$$

Next

$$m^n(u) \leqslant f^n(x,u) \leqslant M^n(u), \quad n \geqslant 1, \ x, u \in \mathbb{R}, \tag{3.6}$$

and since $m^n(u)$ is increasing and $M^n(u)$ decreasing, these sequences have a common limit, $f(u)$, which coincides, for any x, with $\lim_{n\to\infty} f^n(x,u)$. It is clear that $f(u) \geqslant 0$, $\int f(u)\,du = 1$ and $\eta g(u) \leqslant f(u) \leqslant Mg(u)$. Moreover, from (3.6) and (3.5) we have, for $n \geqslant s$,

$$\begin{aligned}
|f^n(x,y) - f(y)| &= |f^n(x,y) - m^n(y) + m^n(y) - f(y)| \\
&\leqslant f^n(x,y) - m^n(y) + f(y) - m^n(y) \\
&\leqslant 2(M^n(y) - m^n(y)) \leqslant 2C\lambda^{n/s-1} g(y), \quad x, y \in \mathbb{R}.
\end{aligned}$$

Since $g(u) \leqslant f(u)/\eta$, we obtain (1.14). □

Proof of Lemma 2.4. Let $i = lr$, $0 \leqslant l < k$ and denote by $g(z_0, z_r, \ldots, z_{lr})$ the joint density of $X'_0, X'_r, \ldots, X'_{lr}$. We have

$$g(z_0, z_r, \ldots, z_{lr}) = f(z_0) \frac{f^r(z_0, z_r) - \varepsilon f(z_r)}{1-\varepsilon} \cdots \frac{f(z_{(l-1)r}, z_{lr}) - \varepsilon f(z_{lr})}{1-\varepsilon}$$

and thus, the density of X'_{lr} evaluated at x is

$$\int \cdots \int g(z_0, \dots, z_{(l-1)r}, x)\, dz_0 \cdots dz_{l-1)r} = f(x)$$

by a recurrence over l.

Let $i=lr+m$, $0 \leqslant l < k-1$, $0 < m \leqslant r-1$. Then, by the classical conditional density formula, the density $g_i(x)$ of X'_i, evaluated at x, is

$$g_i(x) = \iint f(z_{lr}) f^m(z_{lr}, x) f^{r-m}(x, z_{(l+1)r}) \times (f(z_{lr}) f^r(z_{lr}, z_{(l+1)r}))^{-1}$$

$$\times f(z_{lr}) \frac{f^r(z_{lr}, z_{(l+1)r}) - \varepsilon f(z_{(l+1)r})}{1-\varepsilon} dz_l\, dz_{(l+1)r}$$

$$= \iint f(z_{lr}) f^m(z_{lr}, x) f^{r-m}(x, z_{(l+1)r})$$

$$\times \left(1 + \frac{\varepsilon}{1-\varepsilon} \frac{f^r(z_{lr}, z_{(l+1)r}) - f(z_{lr}),}{f^r(z_{lr}, z_{(l+1)r})}\right) dz_{lr}\, dz_{(l+1)r}$$

$$= f(x)\left(1 + \frac{\varepsilon}{1-\varepsilon} \mathrm{O}\left(\frac{A\rho^r}{(1-A\rho^r)}\right)\right), \quad |\mathrm{O}(x)| \leqslant |x|,$$

since, by (1.14),

$$\left|\frac{f^r(x,y) - f(y)}{f^r(x,y)}\right| \leqslant \frac{A\rho^r}{1-A\rho^r}.$$

The case $i=(k-1)r+m$, $0<m\leqslant r-1$ is similar and since $\varepsilon=\frac{1}{2}$ we get (2.11).

The first equality in (2.12) is straightforward by the construction of $\{X_n\}$.

Next, if $i=lr+m$, $j=l'r+m'$, $0 \leqslant i < j \leqslant (k-1)r$, we obtain in the same way

$$g_{ij}(x,y) = f(x) f^{j-i}(x,y) \left(1 + \mathrm{O}\left(\frac{\varepsilon}{1-\varepsilon} \frac{A\rho^r}{1-A\rho^r}\right)\right)^2$$

$$\times \left(1 + \mathrm{O}'\left(\frac{\eta}{1-\eta} \frac{A\rho^{(l'-l)r}}{1-A\rho(l-l')r}\right)\right),$$

where $|\mathrm{O}(x)| \leqslant |x|$, $|\mathrm{O}'(x)| \leqslant x$ and $\eta = 1-(1-\varepsilon)^{l'-1}$.

Thus we deduce the first inequality in (2.12). The second inequality follows from the fact that (1.15) implies

$$f^k(x,y) \leqslant h(y), \quad x \in \mathbb{R}, \ y \geqslant y_0, \ k \geqslant 1.$$

For more details see Kiki (1994). □

Acknowledgement

We wish to thank Professors T. Hsing and H. Rootzén for the careful examination of the manuscript, and for providing many valuable comments and suggestions. In particular, we are indebted to Professor Rootzén one of whose remarks lead to a sharper bound in (1.16b).

References

Asmussen, S. (1987). *Applied Probability and Queues*. Wiley, New York.

Galambos, J. (1987). *The Asymptotic Theory of extreme order statistics*, 2nd ed. Krieger, New York.

Haiman, G. (1981). Valeurs extrêmales de suites de variables aléatoires *m*-dépendantes. *Ann. Inst. H Poincaré* **XVII**, 309–330.

Jakubowski, A. (1993). An asymptotic independent representation in limit theorems for maxima of nonstationary random sequences. *Ann. Probab.* **21**, 819–830.

Kiki, M. (1994). Thése de Doctorat.

O'Brien, G.L. (1987). Extreme values for stationary and Markov sequences. *Ann. Probab.* **15**, 281–291.

Rootzén, H. (1988). Maxima and exceedances of stationary Markov chains. *Adv. in Appl. Probab.* **20**, 371–390.

Serfozo, R. (1980). High level exceedances of regenerative and semi stationary processes. *J. Appl. Probab.* **17**, 423–431.

Smith, R.L. (1992). The extremal index for a Markov chain. *J. Appl. Probab.* **29**, 37–45.

RECORDS AND 2-BLOCK RECORDS OF 1-DEPENDENT STATIONARY SEQUENCES UNDER LOCAL DEPENDENCE

GEORGE HAIMAN, NELLY MAYEUR, VALÉRY NEVZOROV and MADAN L. PURI

Université de Lille 1 and URA CNRS 1321, Université de Lille 1, St. Petersburg University, Indiana University

ABSTRACT. – We first study the records of several examples of 1-dependent stationary sequences $\{X_n\}$, some of them which satisfy a local independence condition introduced in Haiman (1987a), whereas the other do not satisfy this condition.

For the second category (called with local dependence), we then extend a result of the above cited paper by proving, (under some regularity conditions on the joint distribution of (X_1, X_2, X_3)), that the 2-block record times of $\{X_n\}$ a.s. coincide, (via a translation of the time index and from some index on) with the record times of an appropriately constructed i.i.d. sequence $\{\hat{X}_n\}$, whereas the 2-block record values of $\{X_n\}$ and the record values of $\{\hat{X}_n\}$ are imbricate. © Elsevier, Paris

Key words and phrases: 1-dependent stationary sequences, records, 2-block records, clusters, local dependence, strong approximation, extremes.

AMS Classifications: 60 F 15, 60 G 15, 62 G 30.

RÉSUMÉ. – On étudie d'abord les records pour plusieurs exemples de suites stationnaires 1-dépendantes $\{X_n\}$, certaines satisfaisant une condition d'indépendance locale introduite dans Haiman (1987a) et d'autres ne la satisfaisant pas. Pour la deuxième catégorie nous étendons un résultat de l'article précité en démontrant (sous des conditions de régularité portant sur la loi de (X_1, X_2, X_3)) que les instantes de records par blocs de deux de $\{X_n\}$ coïncident presque sûrement (modulo une translation d'indice et à partir d'un certain rang) avec les instants de records d'une suite i.i.d. $\{\hat{X}_n\}$ convenablement construite, alors que les valeurs de records par blocs de deux de $\{X_n\}$ et les records de $\{\hat{X}_n\}$ sont imbriqués.

1. INTRODUCTION

Let $\{X_n\}_{n\geq 1}$, be a strictly stationary sequence of random variables (r.v.'s) with continuous marginal d.f. $F(x) = P\{X_1 \leq x\}$. Many papers have been devoted to the study of the records of $\{X_n\}_{n\geq 1}$, especially when $\{X_n\}_{n\geq 1}$ is i.i.d.. For an overview of these results, the reader is referred to Nevzorov (1987) and the references cited therein.

The classical formal definition of record times T_n and record values θ_n is:

$$T_1 = 1,\ \theta_1 = X_1,$$

and for $n \geq 1$,

$$T_{n+1} = \inf\{k > T_n,\ X_k > \theta_n\} \text{ and } \theta_{n+1} = X_{T_{n+1}}.$$

However, in the sequel we will use the following more general definition of records with respect to an initial threshold θ_0, $\alpha \leq \theta_0 < \omega$, where $\alpha = \inf\{x; F(x) > 0\}$ and $\omega = \sup\{x; F(x) < 1\}$, (the left and right end points of F).

DEFINITION 1.1. – (Haiman (1987a))

$$T_1 = \inf\{k \geq 1; X_k > \theta_0\},\ \ \theta_1 = X_{T_1}$$

and for $n \geq 1$ (1.1)

$$T_{n+1} = \inf\{k > T_n,\ X_k > \theta_n\} \text{ and } \theta_{n+1} = X_{T_{n+1}}.$$

This definition is justified by the following proposition:

PROPOSITION 1.1. – (Haiman (1987a))
If $\{(T'_n, \theta'_n)\}$ *is any other sequence such that for* $n \geq n_0$

$$T'_{n+1} = \inf\{k > T'_n,\ X_k > \theta'_n\} \text{ and } \theta'_{n+1} = X_{T'_{n+1}},$$

then there exist n_1 *and* q *such that for* $n \geq n_1$, *we have*

$$T'_n = T_{n-q} \text{ and } \theta'_n = \theta_{n-q}.$$

For m-dependent sequences (i.e. for any $t \geq 1$, $\sigma(\ldots, X_t)$ and $\sigma(X_{t+m-1}, \ldots)$ are independent), the following theorem was obtained in (Haiman (1987a)), under the additional "local independence" hypothesis: There exists a $\beta > 0$ such that for $2 \leq k \leq m+1$

$$\limsup_{z \nearrow \omega} \left\{ \sup_{\substack{z<u \\ z<v}} P\{X_1 > u | X_k = v\} (P\{X_1 > u\})^{-\beta} \right\} < \infty. \tag{1.2}$$

THEOREM 1.1. – *Let* $\{X_n\}_{n\geq 1}$ *satisfy hypothesis (1.2). Then there exists a probability space which carries, in addition to* $\{X_n\}_{n\geq 1}$, *an i.i.d. sequence* $\{\hat{X}_n\}_{n\geq 1}$ *having the same marginal distribution and such that, if* $\{(\hat{T}_n, \hat{\theta}_n)\}_{n\geq 1}$ *are the records of* $\{\hat{X}_n\}_{n\geq 1}$, *then a.s. there exist two r.v.'s* N_0 *and* Q *such that for* $n \geq N_0$,

$$T_n = \hat{T}_{n-Q} \text{ and } \theta_n = \hat{\theta}_{n-Q}. \tag{1.3}$$

The method of proof of theorem 1.1 was adapted in order to obtain the same statement (1.3) for some classes of Gaussian stationary sequences (Haiman (1987 b) and Haiman and Puri (1993)) and also for some Markov sequences (Haiman, Kiki and Puri (1994)). A multivariate version of theorem 1.1 was obtained in Haiman (1992).

Let now $\{X_n\}_{n\geq 1}$ be any sequence of r.v.'s for which the record times T_n defined in (1.1) are a.s. finite for any $n \geq 1$. Let θ_n be the corresponding record values. Consider another sequence $\{\hat{X}_n\}_{n\geq 1}$ whose records $(\hat{T}_n, \hat{\theta}_n)$ are similarly defined a.s. for any $n \geq 1$. Let

$$M_n = \max(X_1, \ldots, X_n)$$

and

$$\hat{M}_n = \max(\hat{X}_1, \ldots, \hat{X}_n),\ n \geq 1.$$

It is clear that the sequences $\{(T_n, \theta_n)\}_{n\geq 1}$ and $\{(\hat{T}_n, \hat{\theta}_n)\}_{n\geq 1}$ completely determine the sequences $\{M_n\}_{n\geq 1}$ and $\{\hat{M}_n\}_{n\geq 1}$. Then, elementary arguments based on the monotonicity of the sequences $\{\theta_n\}_{n\geq 1}$ and $\{\hat{\theta}_n\}_{n\geq 1}$ lead to the following theorem 1.2 (we improve here the statements of Haiman (1992), th. 1.1, p.147 and Haiman and Puri (1993), th.1.1, p.88):

THEOREM 1.2. – *If there exist r.v.'s N_0 and Q such that a.s. for $n \geq N_0$, (1.3) is satisfied, then there exists a random variable N_1 such that for $n \geq N_1$, we have*

$$M_n = \hat{M}_n \ a.s. \tag{1.4}$$

It may be seen (Haiman (1987a), p. 437) that the proof of theorem 1.1 in the general case ($m \geq 1$) easily follows from the proof in the case $m = 1$, where (1.2) becomes

$$\limsup_{z\nearrow\omega}\left\{\sup_{\substack{z<u\\ z<v}} P\{X_1 > u|X_2 = v\}(P\{X_1 > u\})^{-\beta}\right\} < \infty. \tag{1.5}$$

It also may be seen that for $m = 1$ condition (1.5) may be replaced by the weaker condition: there exists $\beta > 0$ and $k > 0$ such that

$$\limsup_{u\nearrow\omega}\left\{\sup_{u<v<\varphi(u)} P\{X_1 > u|X_2 = v\}(P\{X_1 > u\})^{-\beta}\right\} < \infty \tag{1.6}$$

where $\varphi(u)$ is the solution of the equation

$$1 - F(\varphi(u)) = (1 - F(u))^{1+k}.$$

This is a consequence of the construction method used in Haiman (1987a) and the fact (see Haiman and Puri (1993), lemma 3.3, p. 122) that when $\{X_n\}_{n\geq 1}$ is i.i.d., we have, for any $k > 0$,

$$P\{1 - F(\theta_{n+1}) < (1 - F(\theta_n))^k \ i.o.\} = 0.$$

The paper is constructed as follows: in section 2, we study the records of several examples of 1-dependent sequences $\{X_n\}_{n\geq 1}$, some of them which satisfy condition (1.6) whereas the other do not satisfy this condition. We call sequences belonging to the second category "with local dependence" and show (example 7) that the records of such sequences may behave very different from the records of i.i.d. sequences.

In section 3, we consider sequences which do not necessarly satisfy condition (1.6). We show that under some regularity conditions (satisfied by all our examples) a general version of theorem 1.1 (theorem 3.1), may be obtained.

2. STUDY OF EXAMPLES

Let $(\mathrm{E}, \mathcal{E})$ be some measurable space and $f : \mathrm{E}^2 \to \mathbb{R}$ some measurable function. The most classical examples of 1-dependent stationary sequences are of the form

$$X_n = f(\xi_n, \xi_{n+1}), \quad n \geq 1, \tag{2.1}$$

where $\{\xi_n\}_{n\geq 1}$ is a sequence of i.i.d. $(\mathrm{E}, \mathcal{E})$-valued random elements.

Our examples are all of this form. However, "exotic" examples which do not have a representation like (2.1), (for a recent discussion on this subject, see for example, Matus (1993)) may be given.

The following examples 1-4 satisfy condition (1.6).

2.1. Example 1

Let $\{Y_n\}_{n\geq 1}$ be a sequence of independent $\mathcal{N}(0,1)$ normally distributed r.v.'s and let

$$X_n = aY_n + bY_{n+1}, \quad n \geq 1, \tag{2.2}$$

where a and b are two constants such that $0 < |a| < 1$ and $a^2 + b^2 = 1$. It may be shown (see Mayeur (1996)) that if $\rho = \mathrm{Cov}(X_n, X_{n+1}) = ab < 0$, then $\{X_n\}_{n\geq 1}$ satisfies condition (1.5) whereas if $\rho > 0$, it does not satisfy this condition but satisfies condition (1.6) for any $k > 0$.

2.2. Examples 2, 3 and 4

Let $\{U_n\}_{n\geq 1}$ be a sequence of i.i.d. uniformly on $[0,1]$ distributed r.v.'s. We now consider examples of the form $X_n = \varphi(U_n, U_{n+1})$ and show that these examples satisfy the following condition, which in turn implies (1.6): There exists $\beta > 0$ such that

$$\limsup_{u \nearrow \omega} \left\{ \sup_{u<v<\omega} P\{X_1 > u | X_2 = v\} (P\{X_1 > u\})^{-\beta} \right\} < \infty. \tag{2.3}$$

2.2.1. Example 2. – $X_n = U_n + U_{n+1}, \quad n \geq 1.$

By elementary calculations, we obtain

$$P\{X_1 \leq x\} = \begin{cases} 0 & \text{if } x \leq 0 \\ x^2/2 & \text{if } 0 \leq x \leq 1 \\ -x^2/2 + 2x - 1 & \text{if } 1 \leq x \leq 2 \\ 1 & \text{if } x \geq 2 \end{cases} \tag{2.4}$$

and, for $1 \leq u \leq v \leq 2$,

$$\begin{aligned} P\{X_1 \leq u, X_2 \leq v\} &= (u-1) + u(v-u) - \frac{1}{2}(v-1)^2 + \frac{1}{2}(u-1)^2 \\ &\quad + uv(2-v) - \frac{1}{2}(u+v)[1-(v-1)^2] \\ &\quad + \frac{1}{3}[1-(v-1)^3]. \end{aligned} \tag{2.5}$$

Note that since all our examples are of the form $X_n = \varphi(U_n, U_{n+1})$, we have

$$\begin{aligned} P\{X_1 \leq u, X_2 \leq v\} &= \int_0^1 P\{\varphi(U_1, U_2) \leq u, \varphi(U_2, U_3) \leq v | U_2 = x\}\, dx \\ &= \int_0^1 P\{\varphi(U_1, x) \leq u\} \times P\{\varphi(x, U_3) \leq v\}\, dx. \end{aligned} \tag{2.6}$$

Combining (2.4) and (2.5), for $1 \leq u \leq v \leq 2$, we obtain

$$\begin{aligned} P\{X_1 > u | X_2 = v\} &= \frac{d}{dv} P\{X_1 > u, X_2 < v\} \Big/ \frac{d}{dv} P\{X_2 < v\} \\ &= \frac{1}{2}(2+v) - u. \end{aligned}$$

Next, if we put $u = 2 - \varepsilon$, $v = 2 - \eta$, $0 \leq \eta \leq \varepsilon \leq 1$, (here $\omega = 2$), we get

$$\frac{P\{X_1 > u | X_2 = v\}}{(P\{X_1 > u\})^\beta} = 2^\beta \frac{\varepsilon - \frac{1}{2}\eta}{\varepsilon^{2\beta}} \tag{2.7}$$

which is bounded by $2^{1/2}$ for $\beta = \frac{1}{2}$. Thus, (2.3) is satisfied.

2.2.2. Example 3. – $X_n = U_n \times U_{n+1}, \quad n \geq 1$.

We obtain

$$P\{X_1 \leq x\} = \begin{cases} 0 & \text{if } x \leq 0 \\ x(1 - \log x) & \text{if } 0 \leq x \leq 1 \\ 1 & \text{if } x \geq 1 \end{cases}$$

and for $0 \leq u \leq v \leq 1$,

$$P\{X_1 \leq u, X_2 \leq v\} = u + u(\log v - \log u) + u - uv.$$

Thus, if we put $u = 1 - \varepsilon$, $v = 1 - \eta$, $0 < \eta < \varepsilon < 1$, (here $\omega = 1$), we get

$$\frac{P\{X_1 > u | X_2 = v\}}{(P\{X_1 > u\})^\beta} = \frac{-\log v + u - \frac{u}{v}}{(1 - u + u \log u)^\beta (-\log v)}$$
$$\sim 2^\beta \frac{\varepsilon - \frac{1}{2}\eta}{\varepsilon^{2\beta}}, \quad \varepsilon, \quad \eta \to 0,$$

which coincides with the last term in (2.7).

2.2.3. Example 4. – $X_n = \inf(U_n, U_{n+1}), \quad n \geq 1.$

We obtain

$$P\{X_1 \leq x\} = \begin{cases} 0 & \text{if } x \leq 0 \\ 2x - x^2 & \text{if } 0 \leq x \leq 1 \\ 1 & \text{if } x \geq 1 \end{cases}$$

and for $0 \leq u \leq v \leq 1$,

$$P\{X_1 \leq u, X_2 \leq v\} = u(1 - u) + uv(2 - v).$$

Thus, if we put $u = 1 - \varepsilon$, $v = 1 - \eta$, $0 < \eta < \varepsilon < 1$, we get

$$\frac{P\{X_1 > u | X_2 = v\}}{(P\{X_1 > u\})^\beta} = \frac{1 - v - u + uv}{(1 - 2u + u^2)^\beta (1 - v)} = \frac{1}{\varepsilon^{2\beta - 1}}$$

which equals 1 if $\beta = \frac{1}{2}$. Thus (2.3) is satisfied.

The following examples 5, 6 and 7 do not satisfy condition (1.6).

2.3. Example 5

Let again $\{U_n\}_{n \geq 1}$ be a sequence of i.i.d. uniformly on $[0, 1]$ distributed r.v.'s. Let

$$X_n = \max(U_n, U_{n+1}), \quad n \geq 1. \tag{2.8}$$

We obtain

$$P\{X_1 \leq x\} = \begin{cases} 0 & \text{if } x \leq 0 \\ x^2 & \text{if } 0 \leq x \leq 1 \\ 1 & \text{if } x \geq 1, \end{cases}$$

and

$$P\{X_1 \leq u, X_2 \leq v\} = u^2 v, \quad 0 \leq u \leq v \leq 1.$$

Thus

$$\frac{P\{X_1 > u, X_2 > u\}}{P\{X_1 > u\}} = \frac{1 - 2u^2 + u^3}{1 - u^2} = 1 - \frac{u^2}{1+u} \to \frac{1}{2}, \quad u \to 1.$$

Condition (1.6) is not satisfied since it implies

$$\lim_{u \to \omega} \frac{P\{X_1 > u, X_2 > u\}}{P\{X_1 > u\}} = 0. \tag{2.9}$$

Remark 2.1. – In this example, the records of $\{X_n\}_{n\geq 1}$ coincide with the records of $\{U_{n+1}\}_n$. Thus, the sequence $\{U_{n+1}\}_n$ trivially satisfies the requirements of $\{\hat{X}_n\}_{n\geq 1}$ in theorem 1.1. This shows that condition (1.6) in this theorem is sufficient but not necessary. (See also remark 3.5).

2.4. Example 6

Let $\{Z_n\}_{n\geq 0}$ be a sequence of i.i.d. exponentially with parameter 1 distributed r.v.'s and $\{J_n\}_{n\geq 1}$ a sequence of i.i.d. Bernoulli $\mathcal{B}(1,p)$ r.v.'s. We suppose $\{Z_n\}_{n\geq 0}$ and $\{J_n\}_{n\geq 1}$ independent. Let

$$X_n = \begin{cases} Z_n & \text{if } J_n = 0 \\ Z_{n-1} & \text{if } J_n = 1. \end{cases} \tag{2.10}$$

The sequence $\{X_n\}_{n\geq 1}$ is of the form (2.1), hence it is stationary 1-dependent.

We obtain

$$F(x) = P\{X_1 \leq x\} = 1 - e^{-x}$$

and, for $x_1,\ x_2 \geq 0$,

$$P\{X_1 \leq x_1, X_2 \leq x_2\} = (1-q)F(x_1)F(x_2) + q\psi(x_1, x_2),$$

where $q = p(1-p)$ and $\psi(x_1, x_2) = 1 - e^{-\inf(x_1, x_2)}$.

Thus,

$$\lim_{u\to\infty} \frac{P\{X_1 > u, X_2 > u\}}{P\{X_1 > u\}} = \lim_{u\to\infty} \frac{qe^{-u} + (1-q)e^{-2u}}{e^{-u}} = q > 0.$$

2.5. Example 7

Let $\{Z_n\}_{n\geq 0}$ and $\{J_n\}_{n\geq 1}$ be as in example 6 and let $\{V_n\}_{n\geq 1}$ be a sequence of i.i.d. uniformly on $[0,1]$ distributed r.v.'s, independent of $\{Z_n\}_{n\geq 0}$ and $\{J_n\}_{n\geq 1}$. Let

$$X_n = \begin{cases} Z_n & \text{if } J_n = 0 \\ Z_{n-1} + V_n & \text{if } J_n = 1. \end{cases} \tag{2.11}$$

We then obtain (for further details, see Mayeur (1996), p. 47), for $x \geq 1$,

$$P\{X_1 > x\} = e^{-x}(1 + p(e - 2)),$$

and

$$P\{X_1 > x, X_2 > x\} = e^{-x}p(1-p)+e^{-2x}[(1+p(e-2))^2-p(1-p)(e-1)].$$

Thus

$$\lim_{x \to \infty} \frac{P\{X_1 > x, X_2 > x\}}{P\{X_1 > x\}} = \frac{p(1-p)}{1+p(e-2)} > 0.$$

Remark 2.2. – The record values $\{\theta_n\}$ of $\{X_n\}$ in example 6 behave as the record values of an i.i.d. sequence of exponential $E(1)$ r.v.'s (i.e. $\theta_{n+1} - \theta_n$, $n \geq 1$ form a sequence of i.i.d. exponential $E(1)$ r.v.'s, (see Nevzorov (1987))). As $n \to \infty$, the number $N(n)$ of records in $X_1, \ldots, X_n$ is approximately the same as in a sample of $np(1-p)$ i.i.d. r.v.'s (i.e. $\mathbb{E}(N) \sim \log np(1-p)$, $n \to \infty$).

Remark 2.3. – The record times $\{T_n\}$ of $\{X_n\}$ in example 7 have the particularity that

$$P\{T_{n+1} - T_n = 1 \; i.o.\} = 1, \tag{2.12}$$

whereas the record times of an i.i.d. sequence are such that

$$P\{T_{n+1} - T_n \nearrow \infty\} = 1. \tag{2.13}$$

However, an easy consequence of theorem 3.1 (section 3) is that, for any $k \geq 2$, we have

$$P\{T_{n+1} - T_n = k \; i.o.\} = 0. \tag{2.14}$$

The proof of (2.12) follows from the fact (see also remark 2.2) that

$$\begin{aligned} P\{(J_{T_n^Z} = 0, J_{T_n^Z+1} = 1, J_{T_{n+1}^Z} = 0, J_{T_{n+1}^Z+1} = 1) \\ \cap (\theta_{n+1}^Z - \theta_n^Z > 1) \; i.o.\} = 1, \end{aligned} \tag{2.15}$$

where $\{(T_n^Z, \theta_n^Z)\}$ is the sequence of records of $\{Z_n\}_{n \geq 1}$ in (2.11). Note that when an event inside the brackets of (2.15) occurs, there is a $k \geq 1$ such that $T_k = T_{n+1}^Z$ and $T_{k+1} - T_k = 1$. For further details, see Mayeur (1996), p. 49.

Remark 2.4. – Observe that in examples 5, 6 and 7, we have

$$\lim_{x\to\omega} \frac{P\{X_1 > u, X_2 > u\}}{P\{X_1 > u\}} = \alpha \tag{2.16}$$

with $0 < \alpha \leq \frac{1}{2}$. We close this section by prooving the following proposition 2.1.

PROPOSITION 2.1. – *If $\{X_n\}$ is a 1-dependent stationary sequence such that (2.16) is satisfied for some $\alpha > 0$, then $\alpha \leq \frac{1}{2}$.*

Proof. – The limit in (2.16) is α, $0 \leq \alpha \leq 1$, if and only if

$$\lim_{u\to\omega} \frac{P\{X_1 > u, X_2 \leq u\}}{P\{X_1 > u\}} = 1 - \alpha.$$

Next, using 1-dependence and stationarity,

$$\begin{aligned} &\frac{P\{X_1 > u, X_2 > u, X_3 > u\}}{P\{X_1 > u\}} \\ &\quad = \frac{P\{X_1 > u, X_2 > u\}}{P\{X_1 > u\}} - \frac{P\{X_1 > u, X_2 > u, X_3 \leq u\}}{P\{X_1 > u\}} \\ &\quad \leq \frac{P\{X_1 > u, X_3 > u\}}{P\{X_1 > u\}} = P\{X_1 > u\} \underset{u\to\omega}{\longrightarrow} 0, \end{aligned} \tag{2.17}$$

where

$$\frac{P\{X_1 > u, X_2 > u, X_3 \leq u\}}{P\{X_1 > u\}} \leq \frac{P\{X_1 > u, X_2 \leq u\}}{P\{X_1 > u\}} \underset{u\to\omega}{\longrightarrow} 1 - \alpha. \tag{2.18}$$

But, by (2.17), the limit as $u \to \omega$ of the left hand term in (2.18) must be α. Thus $\alpha \leq \frac{1}{2}$. ■

3. EXTENSION OF THEOREM 1.1.

We now consider a 1-dependent sequence $\{X_n\}_{n\geq 1}$ with continuous marginal d.f. $F(x) = P\{X_1 \leq x\}$, which we suppose strictly increasing near the right end point ω of X_1, i.e. for $x \geq x_0$, $x_0 < \omega$. Furthermore, we suppose that $\{X_n\}_{n\geq 1}$ satisfies the following hypotheses **H1** and **H2**:

H1.) The function $\tilde{H}(u) = P\{F(X_1) > u, F(X_2) > u\}$ is continuously differentiable near $u = 1$ and

$$\tilde{H}'(1) = \lim_{x\to\omega} \frac{P\{X_1 > x,\ X_2 > x\}}{P\{X_1 > x\}} = \alpha,\ 0 \leq \alpha \leq 1/2. \tag{3.1}$$

H2.) There exists a constant $K > 0$ such that for $u \leq v \leq \omega$, we have

$$\begin{aligned} &P\{X_2 > u, \max(X_2, X_3) \in (v, v + dv)\} \\ &\quad \leq K\ P\{X_1 \leq u, X_2 > u, \max(X_2, X_3) \in (v, v + dv)\}. \end{aligned} \tag{3.2}$$

Remark 3.1. – It is rather easy to check, (even if for example 7 some calculations are quite lenghly) that all examples satisfy hypotheses **H1** and **H2**. If we compare condition (1.6) of theorem 1.1 and **H1**-**H2**, **H1** is weaker than (1.6) and then **H2** appears as a compensation of this fact.

Remark 3.2. – Consider the function

$$\begin{aligned} H(u) :&= P\{F(X_1) \leq u, F(X_2) > u\} \\ &= P\{F(X_1) > u\} - \tilde{H}(u) \\ &= 1 - u - \tilde{H}(u), \quad F(x_0) < u \leq 1. \end{aligned} \tag{3.3}$$

Hypothesis **H1** implies that there exists u_0, $F(x_0) \leq u_0 < 1$, such that $H(u)$ is strictly decreasing to zero for $u_0 \leq u \leq 1$. However, examples of 1-dependent stationary sequences for which such an u_0 does not exist (and thus **H1** is not satisfied) may be given. The following example has the form $X_n = f(\xi_n, \xi_{n+1})$, with $\{\xi_n\}_{n\geq 1}$ a sequence of i.i.d. r.v.'s. Observe that in this case, it is sufficient to construct $X_1 = f(U, V)$ and $X_2 = f(V, Z)$, U, V, Z i.i.d., in such a way that there exists an infinity of $u_n < v_n < \omega$, $u_n \nearrow \omega$ for which

$$P\{X_1 \leq u_n, X_2 > u_n\} < P\{X_1 \leq v_n, X_2 > v_n\}. \tag{3.4}$$

Let U have strictly positive density on $\mathbb{R}$ and let

$$X_1 = \begin{cases} U & \text{if } [V] = 2l + 1,\ l \in \mathbb{Z} \\ V + 1 & \text{if } [V] = 2l \end{cases}$$

and

$$X_2 = \begin{cases} V & \text{if } [Z] = 2l + 1,\ l \in \mathbb{Z} \\ Z + 1 & \text{if } [Z] = 2l \end{cases}$$

Observe that if $u_n < v_n$ satisfy (3.4), then

$$\begin{aligned} P\{X_1 \leq u_n, X_2 > u_n\} = &P\{X_1 \leq u_n, X_2 > v_n\} \\ &+ P\{X_1 \leq u_n, u_n < X_2 \leq v_n\} \end{aligned}$$

and

$$\begin{aligned} P\{X_1 \leq v_n, X_2 > v_n\} = &P\{X_1 \leq u_n, X_2 > v_n\} \\ &+ P\{u_n \leq X_1 \leq v_n, X_2 > v_n\}, \end{aligned}$$

whence (3.4) is equivalent to

$$P\{X_1 \leq u_n, u_n < X_2 \leq v_n\} < P\{u_n \leq X_1 \leq v_n, X_2 > v_n\}. \tag{3.5}$$

If we take $u_n = 2n$ and $v_n = 2n + 1$, then

$$P\{X_1 \leq 2n, 2n < X_2 \leq 2n + 1\} = 0$$

whereas

$$\begin{aligned} &P\{2n \leq X_1 \leq 2n + 1, X_2 > 2n + 1\} \\ &\quad \geq P\{[U] = 2n, [V] = 2n + 1, [Z] = 2n\} \\ &\quad = P\{[U] = 2n\} \times P\{[V] = 2n + 1\} \times P\{[Z] = 2n\} > 0. \end{aligned}$$

Thus, we have (3.5).

Let

$$h(u) := 1 - H(u). \tag{3.6}$$

The function $h(u)$ is strictly increasing to 1 for $u_0 \leq u \leq 1$ (see remark 3.2). We first state our results in terms of the 1-dependent stationary sequence

$$Z_n = h(F(X_n)), \ \ n \geq 1, \tag{3.7}$$

whose right end point is 1.

DEFINITION 3.1. – *For any fixed ρ_0, $h(u_0) \leq \rho_0 < 1$, define the "γ-2-block records" of $\{Z_n\}$, $\{(\tau_n, \rho_n)\}_{n \geq 1}$, as follows:*

$$\begin{aligned} \tau_1 &= \inf\{k \geq 1, Z_k > \rho_0\}, \\ \rho_1 &= \gamma_{\rho_0}(\max(Z_{\tau_1}, Z_{\tau_1 + 1})), \end{aligned} \tag{3.8}$$

where for any $\rho_0 \leq \rho < z$,

$$\gamma_\rho(z) = z + P\{\xi_1 > \rho, \rho < \xi_2 \leq z, \xi_3 > z\} \tag{3.9}$$

with $(\xi_1, \xi_2, \xi_3) \stackrel{\mathcal{L}}{=} (Z_1, Z_2, Z_3)$.

For $n \geq 1$

$$\tau_{n+1} = \inf\{k > \tau_n, \ Z_k > \rho_n\}$$

and (3.10)

$$\rho_{n+1} = \gamma_{\rho_n}(\max(Z_{\tau_n}, Z_{\tau_{n+1}})).$$

Remark 3.3. – The sequence $\{(\tau_n, \rho_n)\}$ depends on the choice of the initial threshold ρ_0. However, we shall prove the following consistency result.

PROPOSITION 3.1. – *Let $\{(\tau'_n, \rho'_n)\}$ be a sequence such that for some n_0, $\rho'_{n_0} \geq h(u_0)$ and for $n \geq n_0$, (τ'_n, ρ'_n) are defined by the recurrence formulas (3.10), in which the τ and ρ are replaced by τ' and ρ'.*

Then there exists r.v.'s M and K such that for $n \geq M$,

$$\tau'_n = \tau_{n-K} \text{ and } \rho'_n = \rho_{n-K}. \tag{3.11}$$

Our main result is the following theorem.

THEOREM 3.1. – *Let $\{X_n\}$ satisfy hypotheses* **H1** *and* **H2**. *Then, there exists a probability space which carries, in addition to $\{X_n\}$, an i.i.d. uniformly on $[0,1]$ distributed sequence $\{\hat{Z}_n\}$. Moreover, there exist two random variables N and Q such that a.s. for $n \geq N$, we have*

$$\tau_n = \hat{T}_{n-Q} \text{ and } \rho_n = \hat{\theta}_{n-Q}, \tag{3.12}$$

with $\{(\hat{T}_n, \hat{\theta}_n)\}$ the record sequence of $\{\hat{Z}_n\}$.

In terms of the initial sequence $\{X_n\}$, we then deduce the following corollary.

Consider the family of functions $\{G_u; u \in [F^{-1}(u_0), \omega]\}$, where $G_u : [u, \omega] \to [u, \omega]$ is defined as

$$G_u(v) = F^{-1} \circ h^{-1} \circ \gamma_{h \circ F(u)}(h \circ F(v)), \quad u < v < \omega, \tag{3.13}$$

with $\gamma_\rho(z)$ defined in (3.9) and u_0 of remark 3.2. We then have, for

$$F^{-1}(u_0) \leq u < \omega, \; G_u(u) = u \text{ and } G_u(v) \geq v, \; u \leq v < \omega.$$

For any fixed ρ_0^X, $F^{-1}(u_0) \leq \rho_0^X < \omega$, define the "G-2-block records" of $\{X_n\}$, $\{(\tau_n^X, \rho_n^X)\}_{n \geq 1}$, as follows:

$$\tau_1^X = \inf\{k \geq 1, X_k > \rho_0^X\},$$

$$\rho_1^X = G_{\rho_0^X}(\max(X_{\tau_1^X}, X_{\tau_1^X+1})),$$

and for $n \geq 1$ (3.14)

$$\tau_{n+1}^X = \inf\{k > \tau_n^X; X_k > \rho_n^X\},$$

and

$$\rho_{n+1}^X = G_{\rho_n^X}(\max(X_{\tau_n^X}, X_{\tau_n^X+1})).$$

COROLLARY 3.1. – *Let $\{X_n\}$ satisfy hypotheses* **H1** *and* **H2**. *Then, there exists a probability space which carries, in addition to $\{X_n\}$ an i.i.d. sequence $\{\hat{X}_n\}_{n\geq 1}$ with marginal distribution*

$$P\{\hat{X}_1 \leq x\} = 1 - P\{X_1 \leq x, X_2 > x\}, \quad x \geq F^{-1}(u_0).$$

Moreover, there exist random variables N and Q such that a.s. for $n \geq N$, we have

$$\tau_n^X = \hat{T}_{n-Q}^X \text{ and } \rho_n^X = \hat{\theta}_{n-Q}^X, \tag{3.15}$$

with $\{(\hat{T}_n^X, \hat{\theta}_n^X)\}$ the record sequence of $\{\hat{X}_n\}$.

DEFINITION 3.2. – *For any fixed $\mathcal{T}_0$, $h(u_0) \leq \mathcal{T}_0 < 1$, define the 2-block record sequence of $\{Z_n\}$ with respect to the initial threshold $\mathcal{T}_0$, $\{(\mathcal{T}_n, \mathcal{R}_n)\}_{n\geq 1}$, as*

$$\mathcal{T}_1 = \inf\{k \geq 1; Z_k > \mathcal{T}_0\}, \quad \mathcal{R}_1 = \max(Z_{\mathcal{T}_1}, Z_{\mathcal{T}_1+1})$$

and for $n \geq 1$

$$\mathcal{T}_{n+1} = \inf\{k > \mathcal{T}_n; Z_k > \mathcal{R}_n\}$$

and (3.16)

$$\mathcal{R}_{n+1} = \max(Z_{\mathcal{T}_{n+1}}, Z_{\mathcal{T}_{n+1}+1}).$$

$\mathcal{T}_n$ are the 2-block record times and $\mathcal{R}_n$ the 2-block record values.

Remark 3.4. – Let $\{(\mathcal{T}_n', \mathcal{R}_n')\}$ be a sequence such that for $n \geq n_0$, $\mathcal{T}_n'$ and $\mathcal{R}_n'$ are defined by the recurrence formulas (3.16), in which $\mathcal{T}$ and $\mathcal{R}$ are replaced by $\mathcal{T}'$ and $\mathcal{R}'$. Then, it can be easily seen that there exist m and k such that for $n \geq m$,

$$\mathcal{T}_n' = \mathcal{T}_{n-k} \text{ and } \mathcal{R}_n' = \mathcal{R}_{n-k}. \tag{3.17}$$

The next result is the following theorem 3.2.

THEOREM 3.2. – *If $\{X_n\}$ satisfies hypotheses* **H1** *and* **H2**, *then, there exist r.v.'s M and L such that a.s. for $n \geq M$, we have*

$$\tau_n = \mathcal{T}_{n-L} \text{ and } \mathcal{R}_{n-L} \leq \rho_n \leq \mathcal{R}_{n-L+1} \leq \rho_{n+1}. \tag{3.18}$$

The analogous result holds for the sequence of 2-block records of $\{X_n\}$, $\{(\mathcal{T}_n^X, \mathcal{R}_n^X)\}$ and $\{(\tau_n^X, \rho_n^X)\}$, i.e.

$$\tau_n^X = \mathcal{T}_{n-L}^X \text{ and } \mathcal{R}_{n-L}^X \leq \rho_n^X \leq \mathcal{R}_{n-L+1}^X.$$

Remark 3.5. – Combining (3.12) and (3.18), we also may say that there exist r.v.'s M' and L' such that a.s. for $n \geq M'$, we have

$$\hat{T}_n = \mathcal{T}_{n-L'} \text{ and } \mathcal{R}_{n-L'} \leq \hat{\theta}_n \leq \mathcal{R}_{n-L'+1} \leq \hat{\theta}_{n+1}.$$

This in particular means that the 2-block record times of $\{X_n\}$ (resp. $\{Z_n\}$) a.s. assymptotically coincide via an index translation to the record times of an i.i.d. sequence $\{\hat{X}_n\}$ (resp. $\{\hat{Z}_n\}$) and that the 2-block record values of $\{X_n\}$ (resp. $\{Z_n\}$) and the record values of $\{\hat{X}_n\}$ (resp. $\{\hat{Z}_n\}$) are imbricate.

In example 5 of section 2, we have $Z_n = 1 - X_n^2 + X_n^3 = X_n + O((1 - X_n)^2)$ and the 2-block records of $\{X_n\}$ a.s. coincide, via an index translation, to the records of the i.i.d. sequence $\{U_{n+1}\}$. This behaviour does not contradict the statements of Theorems 3.1 and 3.2.

We now prove proposition 3.1 and theorems 3.1 and 3.2 by means of the following intermediary results.

LEMMA 3.1. – *There exists a probability space which carries, in addition to* $\{X_n\}$, *a sequence* $\{(S_n, R_n)\}_{n\geq 1}$ *having the same Markov structure as the records of an i.i.d. sequence* $\{\bar{Z}_n\}$ *with* $U[0,1]$ *distributed margins, i.e. for any* $n \geq 0$,
$1 \leq t_1 < t_2 < \ldots < t_n$, $s \geq 1$ *and* $s_0 = u_0 < u_1 < \ldots < u_{n+1} < 1$,

$$\begin{aligned} P\{S_{n+1} - S_n = s; R_{n+1} < u_{n+1} | S_1 = t_1, R_1 = u_1; \ldots; S_n = t_n, R_n = u_n\} \\ = P\{S_{n+1} - S_n = s; R_{n+1} < u_{n+1} | R_n = u_n\} \\ = u_n^{s-1}(u_{n+1} - u_n). \qquad (3.19) \end{aligned}$$

Furthermore, there exists a r.v. N *such that a.s. for* $n \geq 1$, S_n *and* R_n *are defined by the recurrence formulas of the* γ*-2-block records, i.e.:*

$$\begin{aligned} S_{n+1} = \inf\{k > S_n; Z_k > R_n\} \\ \text{and } R_{n+1} = \gamma_{R_n}(\max(R_{T_{n+1}}, R_{T_{n+1}+1})). \qquad (3.20) \end{aligned}$$

Remark 3.6. – Note that the sequence $\{(S_n, R_n)\}$ depends on the particular choice of the initial threshold u_0.

Proof. – By (Haiman (1987a), théorème 1′, p. 432), there exist $0 < u_1 < 1$, a constant C_1 and a function $\mu(u)$, $0 < \mu(u) < 1$, such that

$$\begin{aligned} \max_{n\geq 1}\left| \frac{P\{\max(Z_1, \ldots, Z_n) \leq u\}}{\mu^n(u)} - 1 + P\{Z_1 > u, Z_2 > u\} \right| \\ \leq C_1\,(P(Z_1 > u))^2. \qquad (3.21) \end{aligned}$$

Furthermore, there exist constants C_2 and C_3 such that, for $u_1 \leq u \leq 1$, we have

$$\begin{aligned} |\mu(u) - (1 - P\{Z_1 \leq u, Z_2 > u\})| \\ = |\mu(u) - u| \leq C_2(P(Z_1 > u))^2 \leq C_3(1 - u)^2. \qquad (3.22) \end{aligned}$$

Remark 3.7. – In the sequel we take for u_0 the maximum of u_0 in remark 3.2 and u_1.

For $k \geq 2$ and $u_0 \leq u < v < 1$, let

$$\mathcal{A}_k = \mathcal{A}_k(u, dv) = \{Z_1 \leq u, \ldots, Z_{k-1} \leq u, Z_k > u, \gamma_u(\max(Z_k, Z_{k+1})) \in (v; v+dv]\}. \tag{3.23}$$

It may be checked that

$$P\{\mathcal{A}_2(u, dv)\} = P\{Z_1 \leq u, Z_2 > u, \gamma_u(\max(Z_2, Z_3)) \in (v; v+dv]\} = dv. \tag{3.24}$$

Next,

$$P\{\mathcal{A}_3\} = P\{\mathcal{A}_2\} - P\{Z_1 > u, Z_2 \leq u, Z_3 > u, \gamma_u(\max(Z_3, Z_4)) \in (v; v+dv]\},$$

and, for $k \geq 4$,

$$\begin{aligned} P\{\mathcal{A}_k\} = {} & P\{Z_1 \leq u, \ldots, Z_{k-3} \leq u\} \cdot P\{\mathcal{A}_2\} \\ & - P\{Z_1 \leq u, \ldots, Z_{k-3} \leq u, Z_{k-2} > u, Z_{k-1} \leq u, \\ & Z_k > u, \gamma_u(\max(Z_k, Z_{k+1})) \in (v; v+dv]\}, \end{aligned} \tag{3.25}$$

where, by stationarity and 1-dependence, the second term is majorizated by

$$\begin{aligned} & P\{Z_1 \leq u, \ldots, Z_{k-4} \leq u\} \cdot P\{Z_1 > u\} \\ & \cdot P\{Z_1 > u, \gamma_u(\max(Z_1, Z_2)) \in (v; v+dv]\} \end{aligned} \tag{3.26}$$

(by convention, $P\{Z_0 \leq u\} = 1$). Now, by hypothesis **H2**,

$$P\{Z_1 > u, \gamma_u(\max(Z_1, Z_2)) \in (v; v+dv]\} = O_1(P\{\mathcal{A}_2\}), \quad u \to 1. \tag{3.27}$$

Thus, combining (3.21), (3.24)-(3.28), and observing that $u = 1 - (1 - u)$, we obtain, for $k \geq 2$,

$$P\{\mathcal{A}_k(u, dv)\} = u^{k-1} dv (1 + O_2((1-u)^2))^{k-1}(1 + O_3(1-u)), \tag{3.28}$$

where $|O_i(x)| \leq C_i\ |x|$ with C_i, $i = 1, 2, 3$ universal constants. We shall now construct $\{(S_n, R_n)\}$ by using recursively the following lemma.

LEMMA 3.2. (Haiman (1987a), p. 448). – *Let Y a random variable taking values in a measurable space $(\mathcal{Y}, \mathcal{F})$. Let $\varphi \in \mathcal{F}$ and let $\hat{P}$ be a probability* *measure on $(\mathcal{Y}, \mathcal{F})$ such that $0 < \hat{P}(\varphi) < 1$. Assume that on φ, the probability distribution of Y, denoted P_Y, has a Radon-Nicodym derivative $\frac{dP_Y}{d\hat{P}}$ with respect to $\hat{P}$ such that*

$$\max_{y \in \varphi} \left| \frac{dP_Y}{d\hat{P}}(y) - 1 \right| (1 - \hat{P}(\varphi))^{-1} \leq q < 1. \tag{3.29}$$

Let Q, be a Bernoulli r.v. independent of Y such that $P(Q = 0) = q$. Then there exist two random variables Y' and $\bar{Y}$, taking values respectively in $\mathcal{Y}$ and φ^c, independent of Y and Q, and such that, if we put

$$\hat{Y} = \begin{cases} Y & si\ Q = 1\ et\ Y \in \varphi \\ \bar{Y} & si\ Q = 1\ et\ Y \in \varphi^c \\ Y' & si\ Q = 0 \end{cases} \tag{3.30}$$

then, the probability distribution of $\hat{Y}$ is $\hat{P}$. (φ^c denotes the complementary event of φ.)

Let $S_0 = u_0$ and define (S_1, R_1) independently of $\{X_n\}_{n \geq 1}$. Moreover, suppose that for $1 \leq p \leq n$, (S_p, R_p) have allready been constructed and that the events of the form $\{(S_1 = s_1, R_1 \in \mathcal{A}_1), \ldots, (S_n = s_n, R_n \in \mathcal{A}_n)\}$ (where $1 \leq s_1 < \ldots < s_n$ and $\mathcal{A}_i, 1 \leq i \leq n$, are Borel sets of $\mathbb{R}$) are $\sigma\{X_1, \ldots, X_{s_n+1}\} \times \sigma'$ measurable, where σ' is a σ-field independent of $\sigma\{X_n, n \geq 1\}$.

Let $Y = Y_{s,r}$, $r \geq u_0$, be the random variable taking values in $\mathcal{Y} = \{1, 2, \ldots\} \times (r, 1)$, defined conditionaly given $S_n = s$ and $R_n = r$ as follows:

$$Y_{s,r} = (1, r'), r < r' < 1 \Leftrightarrow \{Z_{s+3} > r, \gamma_r(\max(Z_{s+3}, Z_{s+4})) = r'\} \tag{3.31}$$

and, for $t \geq 2$,

$$Y_{s,r} = (t, r'), r < r' < 1 \text{ if and only if}$$

$$\{Z_{s+3} \leq r, .., Z_{s+2+t-1} \leq r, Z_{s+2+t} > r, \gamma_r(\max(Z_{s+2+t}, Z_{s+3+t})) = r'\}. \tag{3.32}$$

Let $\hat{P}$, be the probability measure on $\mathcal{Y}$, defined conditionaly given $S_n = s$ and $R_n = r$ by

$$\hat{P}(t, [v, v + dv]) = r^{t-1} dv, \quad t \geq 1, r < v < 1. \tag{3.33}$$

Let

$$\varphi = \varphi_r = \{(t, r') \in \mathcal{Y}; 2 \leq t \leq [\psi(r)]\} \tag{3.34}$$

where

$$\psi(r) = \frac{\tau}{1-r} \ln_2\left(\frac{1}{1-r}\right), \quad u_0 < r < 1, \tag{3.35}$$

with $\tau > 1$ a fixed constant and $\ln_2(x) = \ln(\ln x)$. We then have

$$\hat{P}(\varphi_r^c) = 1 - \hat{P}(\varphi) = 1 - r + r^{[\psi(r)]} > r^{[\psi(r)]} \sim \left(\ln\left(\frac{1}{1-r}\right)\right)^{-\tau}, \quad r \to 1. \tag{3.36}$$

Next, denoting by Y_1, Y_2 the coordinates of $Y_{s,r} = Y$, observe that

$$P\{Y_1 = t, Y_2 \in [v, v + dv]\} = P\{\mathcal{A}_k(u, dv)\}. \tag{3.37}$$

Thus, combining (3.29), (3.32)-(3.38) and using the fact that $(1+x)^k = 1 + O(kx)$, $x \to 0$, we obtain

$$\max_{y \in \varphi_r} \left| \frac{dP_Y}{d\hat{P}}(y) - 1 \right| (1 - \hat{P}(\varphi))^{-1} \le c(1-r) \left(\ln\left(\frac{1}{1-r}\right)\right)^{\tau'} = q(r), \tag{3.38}$$

for some constants $C > 0$ and $\tau' > \tau$. Since $q(r) \to 0$ as $r \to 1$, for $r \ge u_2$ large enough (from now one we take for u_0 the maximum of u_0 in remark 3.7 and u_2), we may apply lemma 3.2.

Let $Q_{n+1} = Q$ with Q of the lemma (i.e. $P(Q_{n+1} = 0) = q(R_n)$).

Let L_{n+1} be a Bernoulli r.v. such that

$$P(L_{n+1} = 1) = r^2. \tag{3.39}$$

Furthermore, let Q_{n+1} and L_{n+1} be mutually independent and depend on $\{X_n\}$ and $(S_1, R_1), \ldots, (S_n, R_n)$ only through R_n.

Let $\tilde{Y} = (\tilde{Y}_1, \tilde{Y}_2)$, be a random vector taking values in $\{1, 2\} \times (R_n, 1)$ such that

$$P\{\tilde{Y}_1 = t; \tilde{Y}_2 \le v\} = r^{t-1}(v - r) \frac{1}{P(L_{n+1} = 0)}, \quad t = 1, 2, \; r < v < 1. \tag{3.40}$$

(We take $\tilde{Y}$ independent of the previously defined random elements.) We now define (S_{n+1}, R_{n+1}) as follows:
If $L_{n+1} = 0$, then

$$S_{n+1} = S_n + \tilde{Y}_1 \text{ and } R_{n+1} = \tilde{Y}_2. \tag{3.41}$$

If $L_{n+1} = 1$, then

$$S_{n+1} = S_n + 2 + \hat{Y}_1 \text{ and } R_{n+1} = \hat{Y}_2, \tag{3.42}$$

where $(\hat{Y}_1, \hat{Y}_2) = \hat{Y}$ is given by lemma 3.2.

It is not difficult to check that $\{(S_1, R_1), \ldots, (S_{n+1}, R_{n+1})\}$ satisfy (3.19) and the measurability conditions requested previously at order n also at order $n + 1$.

Now, in order to show that there exists a r.v. N such that a.s. for $n \geq N$, relation (3.20) is satisfied, it is sufficient to prove, denoting by E_n the event

$$E_n = \{L_{n+1} = 1, (\max(Z_{S_n+2}, Z_{S_n+3}) \leq R_n), Q_{n+1} = 1, Y \in \varphi_{R_n}\}, \tag{3.43}$$

that we have

$$P\{E_n^c \;\; i.o.\} = 0, \tag{3.44}$$

where E_n^c is the complementary event of E_n. In order to obtain (3.45), we will prove that

$$P\{L_{n+1} = 0 \; i.o.\} = P\{\max(Z_{S_n+2}, Z_{S_n+3}) > R_n \; i.o.\} = \\ P\{Q_{n+1} = 0 \; i.o.\} = P\{Y \in \varphi_{R_n}^c \;\; i.o.\} = 0. \tag{3.45}$$

Remark 3.8. – Observe that if $\{(S_n, R_n)\}$ satisfies (3.19), then it is possible to construct an i.i.d. uniformly on $[0, 1]$ distributed sequence $\{\hat{Z}_n\}$ such that (S_n, R_n) is the record sequence of $\{\hat{Z}_n\}$ with respect to the initial threshold u_0.

Hence, by (Haiman (1987), Lemma 3, p. 454), for any $0 < \alpha < 1$, we have

$$P\{1 - R_n > e^{-\alpha n} \;\; i.o.\} = 0, \tag{3.46}$$

and for any $\beta > 1$,

$$P\{1 - R_n < e^{-\beta n} \;\; i.o.\} = 0. \tag{3.47}$$

Next, if A_n is one of the events in the brackets in (3.46), $P\{A_n \; i.o.\} = 0$ is equivalent to $P\{A_n \cap (1 - R_n \leq e^{-\alpha n}) \; i.o.\} = 0$.

Let

$$B_n = (L_{n+1} = 0) \cap (1 - R_n \leq e^{-\alpha n}). \tag{3.48}$$

We then have from (3.39)

$$P(B_n) = \int_{1-e^{-\alpha n}}^{1} (1 - r^2) \, dP_{R_n}(r) \leq c \;\, e^{-\alpha n},$$

with c positive constant, which is the general term of a convergent series. Thus

$$P(L_{n+1} = 0 \ i.o.) = 0$$

follows by the first Borel-Cantelli lemma.

In a similar way, we obtain

$$P\{(Q_{n+1} = 0) \cap (1 - R_n \le e^{-\alpha n})\} \le c' \ e^{-\alpha n}(\alpha n)^{\tau'}, \ 1 < \tau < \tau',$$

and

$$P\{(Y \in \varphi^c_{R_n}) \cap (1 - R_n \le^{-\alpha n})\} \le c'' \ n^{-\tau},$$

from which

$$P(Q_{n+1} = 0 \ i.o.) = P\{Y \in \varphi^c_{R_n} \ i.o.\} = 0$$

follows. Let now

$$C_n = (Q_{n+1} = 1) \cap (L_{n+1} = 1) \cap (Y \in \varphi^c_{R_n}) \cap (1 - R_n \le e^{-\alpha n})$$

and

$$D_n = C_n \cap \{\max(Z_{S_n+2}, Z_{S_n+3}) > R_n\}.$$

We then similarly obtain

$$P(D_n) \le c \, e^{-\alpha n} \ln(n)$$

and thus the last equality in (3.45). This achieves the proof of lemma 3.1. ■

Proof of Proposition 3.1 *and Theorems* 3.1 *and* 3.2. – In order to prove proposition 3.1 and theorems 3.1 and 3.2, we also need the following lemmas:

LEMMA 3.3. – *Consider, with* $\{(S_n, R_n)\}$ *of lemma 3.1, the events*

$$\begin{aligned} \mathcal{A}_n &= \{\inf[m > S_{n+1} + 2; Z_m > \max(Z_{S_{n+1}}, Z_{S_{n+1}+1})] < S_{n+2}\} \\ &= \bigcup_{m \ge 2} \{[\max(Z_{S_{n+1}+2}, \ldots, Z_{S_{n+1}+m}) \le \max(Z_{S_{n+1}}, Z_{S_{n+1}+1})] \\ &\cap [Z_{S_{n+1}+m+1} \in (\max(Z_{S_{n+1}}, Z_{S_{n+1}+1}), \gamma_{R_n}(\max(Z_{S_{n+1}}, Z_{S_{n+1}+1}))]]\} \end{aligned} \tag{3.49}$$

Then

$$P\{\mathcal{A}_n \ i.o.\} = 0. \tag{3.50}$$

Remark 3.9. – Lemma 3.3 implies that there exists a r.v. M such that a.s. for $n \ge M$,

$$S_{n+1} = \inf\{k > S_n; Z_k > \max(Z_{S_n}, Z_{S_n+1})\}, \tag{3.51}$$

i.e. for $n \ge M$ the S_n are given by the recurrence formulas of the 2-block record times of $\{Z_n\}$ in (3.16).

Proof of Lemma 3.3. – Let, taking into account (3.46) and (3.47),

$$\mathcal{B}_n = \mathcal{A}_n \cap (1 - R_n \leq^{-\alpha n}) \cap (1 - R_{n+1} > e^{-\beta n}).$$

We then have

$$P(\mathcal{B}_n) = \sum_{t=2n+1}^{\infty} \int\int_{\Delta} \times P\{\mathcal{A}_n | R_n = r, S_{n+1} = t, R'_{n+1} = s\} dP_{(R_n, S_{n+1}, R'_{n+1})}(r, t, s),$$

where

$$\Delta = \{(r, s); u_0 \leq r < s < 1,\ 1 - r \leq^{-\alpha n)},\ 1 - s > e^{-\beta n}\}$$

and

$$R'_{n+1} = \max(Z_{S_{n+1}}, Z_{S_{n+1}+1}).$$

Thus

$$P\{\mathcal{A}_n R_n = r, S_{n+1} = t, R'_{n+1} = s\}$$
$$\leq P\left\{\bigcup_{m\geq 3} (\max(Z_{t+3}, \ldots, Z_{t+m}) < s) \cap (Z_{t+m+1} \in (s, \gamma_r(s)])\right\}$$
$$\leq K_1 P\{Z_1 \in (s, \gamma_r(s)]\} \sum_{k=0}^{\infty} \mu^k(s)$$

with, by (3.22),

$$\mu(s) = s + O((1 - s)^2).$$

Thus, we easily deduce that there exists a positive constant K_2 such that

$$P\{\mathcal{A}_n | R_n = r, S_{n+1} = t, R'_{n+1} = s\} \leq K_2 \frac{\gamma_r(s) - s}{1 - s}.$$

Next, by (3.9)

$$\begin{aligned}\gamma_r(s) - s &= P\{\xi_1 > r, r < \xi_2 \leq s, \xi_3 > s\} \\ &\leq P\{\xi_1 > r\} \cdot P\{\xi_3 > s\} = O((1 - r)^2). \qquad (3.52)\end{aligned}$$

Thus, there exists a constant K_3 such that on Δ

$$P\{\mathcal{A}_n | R_n = r, S_{n+1} = t, R'_{n+1} = s\} \leq K_3 \frac{e^{-2\alpha n}}{e^{-\beta n}} = K_3\, e^{-(2\alpha - \beta)n}.$$

Choose α and β such that $2\alpha - \beta > 0$. Then

$$P\{\mathcal{B}_n\} \leq K_3\, e^{-(2\alpha-\beta)n}$$

which is the general term of a convergent series and we get

$$P\{\mathcal{B}_n \ i.o.\} = 0,$$

whence (3.50). ■

Remark 3.10. – Let $\{(\tau'_n, \rho'_n)\}_{n\geq 1}$ be any sequence such that for $n \geq n_0$, (τ'_n, ρ'_n) are defined by the recurrence formulas of the γ-2-block records (3.10).

We then observe that either there exist N_1 and Q_1 such that for $n \geq N_1$

$$\tau'_{n-Q_1} = S_n \ \text{and} \ \rho'_{n-Q_1} = R_n,$$

or there exist N_2 and Q_2 such that for $n \geq N_2$

$$S_n < \tau'_{n-Q_2} < S_{n+1} \ \text{and} \ R_n < \rho'_{n-Q_2} < R_{n+1}.$$

Thus, by (3.46) and (3.47), we also have:
for any $0 < \alpha < 1$

$$P\{1 - \tau'_n > e^{-\alpha n} \ i.o.\} = 0$$

and for any $\beta > 1$

$$P\{1 - \tau'_n < e^{-\beta n} \ i.o.\} = 0.$$

Let $\mathcal{A}'_n$ be the event defined by formula (3.49), in which S_{n+1}, S_{n+2} are replaced by τ'_{n+1}, τ'_{n+2} and R_n by ρ'_n. We then may use the proof of lemma 3.3, without any other changes, to deduce that

$$P\{\mathcal{A}'_n \ i.o.\} = 0.$$

But this implies that there exists a r.v. M', such that a.s. for $n \geq M'$

$$\tau'_{n+1} = \inf\{k > \tau'_n; Z_k > \max(Z_{\tau'_n}, Z_{\tau'_n+1})\}.$$

Then, by remark 3.4, there exist r.v.'s M' and K' such that a.s. for $n \geq M'$

$$\tau'_n = S_{n-K'} \ \text{and} \ \rho'_n = R_{n-K'}. \tag{3.53}$$

This proves proposition 3.1. The proof of theorem 3.1 also readily follows by taking $\{\hat{Z}_n\}$ the i.i.d. sequence of remark 3.8, $\hat{T}_n := S_n$ and $\hat{\theta}_n := R_n$. Remark 3.9 and theorem 3.1 then imply theorem 3.2. ■

ACKNOWLEDGMENT

We wish to thank Professor J. Bretagnolle for giving us a first counterexample which inspired the one presented in remark 3.2 and Professor M.A. Lifshits for his remark leading to theorem 1.2.

REFERENCES

[1] G. Haiman, *Etude des extrêmes d'une suite stationnaire m-dépendante avec une application relative aux accroissements du processus de Wiener.*, Annales de l'institut Henri Poincaré, Vol. **23**, 1987a, N° 3, p. 425-458.

[2] G. Haiman, *Almost sure asymptotic behaviour of the record and record time sequences of a stationary gaussian process*, Mathematical Statistics and Probability Theory, vol. **A**, 1987b, p. 105-120.

[3] G. Haiman, *A strong invariance principle for the extremes of multivariate stationary m-dependent sequences*, Journal of Statistical Planning and Inference, Vol. **32**, 1992, p. 47-163.

[4] G. Haiman, and M. L. Puri, *A strong invariance principle concerning the J-upper order statistics for stationary Gaussian sequences*, Annals of Probability, Vol. **21**, N° 1, 1993, p. 86-135.

[5] G. Haiman, M. Kiki and M. L. Puri, *Extreme of Markov sequences*, Journal of Statistical Planning and Inference, Vol. **45**, 1995, p. 185-201.

[6] F. Matus, *On two-block-factor sequences and one dependence*, Preprint of the Institute of Information Theory and Automation, Academy of Sciences of the Czech Republic, 1994.

[7] N. Mayeur, *Thèse de Doctorat*, 1996.

[8] V. Nevzorov, *Records*, Theory Probab.Appl., Vol. **32**, N° 2, 1987, p. 201-228.

(Manuscript received November 29, 1996;
Revised version received March 20, 1998.)

Part IV

Appendices A, B and C

Appendix A

Volume 1

Nonparametric Methods in Statistics and Related Topics

Contents

Preface xiii

Part I. NONPARAMETRIC METHODS IN UNIVARIATE ANALYSIS

[1] Asymptotic efficiency of a class of c-sample tests
The Annals of Mathematical Statistics, 35 (1964), 102-121. 3

[2] On the estimation of contrasts in linear models
(with Subha Bhuchongkul)
The Annals of Mathematical Statistics, 36 (1965), 198-202. 23

[3] Some distribution-free k-sample rank tests of homogeneity against ordered alternatives
Communications on Pure and Applied Mathematics, 18 (1965), 51-63. 28

[4] Multi-sample analogues of some one-sample tests
(with K.L. Mehra)
The Annals of Mathematical Statistics, 38 (1967), 523-549. 41

[5] On robust estimation in incomplete block designs
(with Pranab Kumar Sen)
The Annals of Mathematical Statistics, 38 (1967), 1587-1591. 68

[6] On some optimum nonparametric procedures in two-way layouts
(with Pranab Kumar Sen)
Journal of the American Statistical Association, 62 (1967), 1214-1229. 73

[7] On Chernoff-Savage tests for ordered alternatives in randomized blocks
(with Pranab Kumar Sen)
The Annals of Mathematical Statistics, 39 (1968), 967-972. 89

[8] Multiple decision procedures based on ranks for certain problems in analysis of variance
(with Prem S. Puri)
The Annals of Mathematical Statistics, 40 (1969), 619-632. 95

[9] A class of rank order tests for a general linear hypothesis
(with Pranab Kumar Sen)
The Annals of Mathematical Statistics, 40 (1969), 1325-1343. 109

[10] Analysis of covariance based on general rank scores
(with Pranab Kumar Sen)
The Annals of Mathematical Statistics, 40 (1969), 610-618. 128

[11] On a class of rank order tests for the identity of two multiple regression surfaces
(with Pranab Kumar Sen)
Zeitschrift für Wahrscheinlichkeitstheorie und Verw. Geb., 12 (1969), 1-8. 137

[12] Statistical inference in incomplete blocks designs
(with Harold D. Shane)
Nonparametric Techniques in Statistical Inference, Cambridge University Press (1970), 131-153. 145

[13] On some selection procedures in two-way layouts
(with P.K. Sen)
Zeitschrift für Wahrscheinlichkeitstheorie und Verw. Geb., 22 (1972), 242-250. 170

[14] A note on asymptotic distribution-free tests for subhypotheses in multiple linear regression
(with P.K. Sen)
The Annals of Statistics, 1 (1973), 553-556. 179

[15] Asymptotically distribution-free aligned rank order tests for composite hypotheses for general multivariate linear models
(with Pranab K. Sen)
Zeitschrift für Wahrscheinlichkeitstheorie und Verw. Geb., 39 (1977), 175-186. 183

[16] Rank order estimators in the case of grouped data (linear regression)
(with A.R. Padmanabhan)
Sankhyā, Series B, 41 (1979), 239-259. 195

[17] Theory of nonparametric statistics for rounded-off data with applications
(with A.R. Padmanabhan)
Mathematische Operations-Forschung und Statistik, Series Statistics, 14 (1983), 301-349. 216

[18] Centering of signed rank statistics with a continuous score-generating function
(with S.S. Ralescu)
Teoria Veroyatnostei i ee Primeneniya (also in Theory of Probability and Its Applications, Translation by SIAM), 29 (1984), 580-584. 265

[19] Convergence of generalized inverses with applications to asymptotic hypothesis testing
(with Carl T. Russell and Thomas Mathew)
Sankhyā, Series A, 46 (1984), 277-286. 271

[20] Asymptotic normality of the lengths of a class of nonparametric confidence intervals for a regression parameter
(with Tiee-Jian Wu)
Canadian Journal of Statistics, 12 (1984), 217-228. 281

[21] Bhattacharyya bound of variances of unbiased estimators in nonregular cases
(with Masafumi Akahira and Kei Takeuchi)
Annals of the Institute of Statistical Mathematics, 39 (1986), 35-44. 293

[22] Almost sure linearity for signed rank statistics in the non-i.i.d. case
(with S. Ralescu)
Acta Mathematica Hungarica, 48 (1986), 273-284. 303

[23] Asymptotic behavior of multi-response permutation procedures
(with Manfred Denker)
Advances in Applied Mathematics, 9 (1988), 200-210. 315

[24] Adaptive nonparametric procedures and applications
(with N.J. Hill and A.R. Padmanabhan)
Journal of the Royal Statistical Society, Series C (Applied Statistics), 37 (1988), 205-218. 326

Part II. NONPARAMETRIC METHODS IN MULTIVARIATE ANALYSIS

[25] On the theory of rank order tests for location in the multivariate one sample problem
(with Pranab Kumar Sen)
The Annals of Mathematical Statistics, 38 (1967), 1216-1228. 343

[26] Nonparametric confidence regions for some multivariate location problems
(with Pranab Kumar Sen)
Journal of the American Statistical Association, 62 (1968), 1373-1378. 356

[27] Rank order tests for multivariate paired comparisons
(with Harold D. Shane)
The Annals of Mathematical Statistics, 40 (1969), 2101-2117. 362

[28] Asymptotic theory of likelihood ratio and rank order tests in some multivariate linear models
(with Pranab Kumar Sen)
The Annals of Mathematical Statistics, 41 (1970), 87-100. 379

[29] On the robustness of rank order tests and estimates in the generalized multivariate one-sample location problem
(with P.K. Sen)
Zeitschrift für Wahrscheinlichkeitstheorie und Verw. Geb., 22 (1972), 226-241. 393

[30] Joint asymptotic multinormality for a class of rank order statistics in multivariate paired comparisons
(with Carl T. Russell)
Journal of Multivariate Analysis, 4 (1974), 88-105. 409

Part III. NONPARAMETRIC METHODS IN DESIGN AND ANALYSIS OF EXPERIMENTS

[31] Nonparametric methods in design and analysis of experiments
(with Edgar Brunner)
In: Handbook of Statistics, volume 13, Design of Experiments. (Editors: S. Ghosh and C.R. Rao) North Holland, Amsterdam, Netherlands (1996), 631-703. 429

[32] Rank-score tests in factorial designs with repeated measures
(with Edgar Brunner and Ulrich Munzel)
Journal of Multivariate Analysis, 70 (1999), 286-317. 502

[33] Nonparametric methods in factorial designs
(with Edgar Brunner)
Statistical Papers (Springer-Verlag journal), 42 (2001), 1-52. 534

[34] A class of rank-score tests in factorial designs
(with Edgar Brunner)
Journal of Statistical Planning and Inference, 103 (2002), 329-359. 584

Part IV. MISCELLANEOUS TOPICS

[35] On estimating inter-subject variability of choice probabilities under observability constraints
(with Robert Bartoszyński)
Journal of Mathematical Psychology, 25 (1982), 175-184. 617

[36] Linear minimax estimators for estimating a function of the parameter
(with Dan A. Ralescu and Stefan S. Ralescu)
The Australian Journal of Statistics, 26 (1984), 277-283. 627

[37] The asymptotic behavior of the minimax risk for multiple decision problems
(with Olaf Krafft)
Sankhyā, Series A, 36 (1974), 1-12. 634

[38] A spherical correlation coefficient robust against scale
(with Kanti V. Mardia)
Biometrika, 65 (1978), 391-395. 646

[39] Asymptotic distributions of likelihood ratio criteria for testing latent roots and latent vectors of a covariance matrix under an elliptical population
(with Takesi Hayakawa)
Biometrika, 72 (1985), 331-338. 651

[40] On a class of multivariate multisample rank-order tests
(with Pranab Kumar Sen)
Sankhyā, Series A, 28 (1966), 353-376. 659

[41] On a class of rank order estimators of contrasts in MANOVA
(with Pranab K. Sen)
Sankhyā, Series A, 30 (1968), 31-36. 683

[42] On a class of multivariate multisample rank order tests II. Tests for homogeneity of dispersion matrices
(with Pranab K. Sen)
Sankhyā, Series A, 30 (1968), 1-22. 689

[43] On a class of rank order tests for independence in multivariate distributions
(with P.K. Sen and D.V. Gokhale)
Sankhyā, Series A, 32 (1970), 271-298. 711

[44] Nonparametric methods for stratified two-sample designs with application to multiclinic trials
(with Edgar Brunner and Shan Sun)
Journal of the American Statistical Association, 90 (1995), 1004-1014. 739

Part V. APPENDICES

A. Volume 2.
Probability Theory and Extreme Value Theory
Contents 753
B. Volume 3.
Time Series, Fuzzy Analysis, and Miscellaneous Topics
Contents 759
C. The Publications of Madan Lal Puri 765

Appendix B

Volume 3

Time Series, Fuzzy Analysis and Miscellaneous Topics

Contents

Preface xiii

Part I. TIME SERIES AND RELATED TOPICS

[1] Linear serial rank tests for randomness against ARMA alternatives
(with Marc Hallin and Jean-Francois Ingenbleek)
Annals of Statistics, 13 (1985), 1156-1181. 3

[2] Linear and quadratic serial rank tests for randomness against serial dependence
(with Marc Hallin and Jean-François Ingenbleek)
Journal of Time Series Analysis, 8 (1987), 409-424. 29

[3] Locally asymptotically optimal tests for randomness
(with Marc Hallin)
New Perspectives in Theoretical and Applied Statistics, John Wiley and Sons, New York (1987), 87-99. 45

[4] Optimal rank-based procedures for time series analysis: testing an ARMA model against other ARMA models
(with Marc Hallin)
Annals of Statistics, 16 (1988), 402-432. 58

[5] On locally asymptotically maximin tests for ARMA processes
(with Marc Hallin)
In: Statistical Theory and Data Analysis II (Editor: K. Matusita) (1988), 495-499. North Holland Publishing Company. 89

[6] Asymptotically most powerful rank tests for multivariate randomness against serial dependence
(with Marc Hallin and Jean Francois Ingenbleek)
Journal of Multivariate Analysis, 30 (1989), 34-71. 94

[7] Weak convergence of the U-statistic and weak invariance of the one-sample rank order statistic for Markov processes and ARMA models
(with Michel Harel)
Journal of Multivariate Analysis, 31 (1989), 258-265. 132

[8] Weak convergence of serial rank statistics under dependence with applications in time series and Markov processes
(with Michel Harel)
Annals of Probability, 18 (1990), 1361-1387. 140

[9] Asymptotic normality of L-statistics based on $m(n)$-decomposable time series
(with K.C. Chanda and F.H. Ruymgaart)
Journal of Multivariate Analysis, 35 (1990), 260-275. 167

[10] Time series analysis via rank order theory: signed-rank tests for ARMA models
(with Marc Hallin)
Journal of Multivariate Analysis, 39 (1991), 1-29. 183

[11] Rank tests for time series analysis. A survey
(with Marc Hallin)
In: New Directions in Time Series Analysis (the IMA Volume in Mathematics and Its Applications) 45 (1992), 111-153. (Editors: David Brillinger, Emanuel Parzen, Murray Rosenblatt et al.). Springer-Verlag, New York/Heidelberg/Berlin. 212

[12] Nonparametric prediction for random fields
(with Frits H. Ruymgaart)
Stochastic Processes and Their Applications, 48 (1993), 139-156. 255

[13] Asymptotic behavior of L-statistics for a large class of time series (with Frits H. Ruymgaart)
Annals of the Institute of Statistical Mathematics, 45 (1993), 687-701. 273

[14] Aligned rank tests for linear models with autocorrelated error terms
(with Marc Hallin)
Journal of Multivariate Analysis, 50 (1994), 175-237. 288

[15] Change curves in the presence of dependent noise
(with Frits H. Ruymgaart)
Change-Point Problems, a volume in the IMS Lecture Notes and Monograph Series, Volume 23 (1994), 242-254. 351

[16] A multivariate Wald-Wolfowitz rank test against serial dependence
(with Marc Hallin)
The Canadian Journal of Statistics, 23 (1995), 55-65. 364

[17] Nonparametric approach for non-Gaussian vector stationary processes
(with Masanobu Taniguchi and Masao Kondo)
Journal of Multivariate Analysis, 56 (1996), 259-283. 375

[18] The limiting density of unit root test statistics: A unifying technique
(with Mithat Gönen, Frits H. Ruymgaart and Martien C.A. van Zuijlen)
The Journal of Time Series Analysis, 21 (2000), 249-260. 400

Part II. FUZZY SET THEORY AND RELATED TOPICS

[19] Integration on fuzzy sets
(with Dan A. Ralescu)
Advances in Applied Mathematics, 3 (1982), 430-434. 415

[20] A possibility measure is not a fuzzy measure
(with Dan Ralescu)
Fuzzy Sets and Systems, 7 (1982), 311-313. 420

[21] Strong law of large numbers for Banach space valued random sets (with Dan A. Ralescu)
Annals of Probability, 11 (1983), 222-224. 423

[22] Strong law of large numbers with respect to a set-valued probability measure
(with Dan A. Ralescu)
Annals of Probability, 11 (1983), 1051-1054. 426

[23] Differentials of fuzzy functions
(with Dan A. Ralescu)
Journal of Mathematical Analysis and Applications, 91 (1983), 552-558. 430

[24] The concept of normality for fuzzy random variables
(with Dan A. Ralescu)
Annals of Probability, 13 (1985), 1373-1379. 437

[25] Limit theorems for random compact sets in Banach space
(with Dan A. Ralescu)
Mathematical Proceedings of the Cambridge Philosophical Society, 97 (1985), 151-158. 444

[26] Fuzzy random variables
(with Dan A. Ralescu)
Journal of Mathematical Analysis and Applications, 114 (1986), 409-422. 452

[27] Gaussian random sets in Banach space
(with D.A. Ralescu and S.S. Ralescu)
Teoria Veroyatnostei i ee Primeneniya (also in Theory of Probability and Its Applications, Translation by SIAM), 31 (1986), 598-601. 466

[28] Limit theorems for fuzzy random variables
(with E.P. Klement and D.A. Ralescu)
Proceedings of the Royal Society of London, 407 (1986), 171-182. 470

[29] Convergence theorem for fuzzy martingales
(with Dan A. Ralescu)
Journal of Mathematical Analysis and Applications, 160 (1991), 107-122. 482

[30] Limit theorems for fuzzy random variables
(with F.N. Proske)
In: Asymptotics in Statistics and Probability: Papers in Honor of George Gregory Roussas. VSP International Science Publishers, The Netherlands (2000), 331-346. 498

[31] Central limit theorem for Banach space valued fuzzy random variables
(with Frank N. Proske)
Proceedings of the American Mathematical Society, 130 (2002), 1493-1501. 515

[32] Strong law of large numbers for Banach space valued fuzzy random variables
(with Frank N. Proske)
Journal of Theoretical Probability, 15 (2002), 543-552. 524

Part III. MISCELLANEOUS TOPICS

[33] Local maxima of the sample functions of the N-parameter Bessel process
(with Lanh Tat Tran)
Stochastic Processes and Their Applications, 9 (1979), 137-145. 535

[34] Realization of l_p by spaces of random variables
(with Shey Shiung Sheu)
Applicable Analysis, 8 (1979), 337-347. 544

[35] Shorted operators and generalized inverses of matrices
(with Sujit Kumar Mitra)
Linear Algebra and its Applications, 25 (1979), 45-56. 555

[36] Complex planar splines
(with Gerhard Opfer)
Journal of Approximation Theory, 31 (1981), 383-402. 567

[37] Shorted matrices — an extended concept and some applications
(with Sujit Kumar Mitra)
Linear Algebra and its Applications, 42 (1982), 57-79. 587

[38] The fundamental bordered matrix of linear estimation and the Duffin–Morley general linear electromechanical systems
(with Sujit Kumar Mitra)
Applicable Analysis, 14 (1983), 241-258. 610

[39] Complex Chebyshev polynomials and generalizations with an application to the optimal choice of interpolating knots in complex planar splines
(with Gerhard Opfer)
Journal of Approximation Theory, 37 (1983), 89-101. 628

[40] Strong solutions of stochastic differential equations for multiparameter processes
(with Markus Dozzi)
Stochastics, 17 (1986), 19-41. 641

[41] A new semigroup technique in Poisson approximation
(with P. Deheuvels and D. Pfeifer)
Semigroup Forum, 38 (1989), 189-201. 664

[42] Information
(with H.L. Resnikoff)
Acting Under Uncertainty: Multidisciplinary Conceptions (Editor: George von Furstenberg). Kluwer Academic Publishers, Boston/Dordrecht/Lancaster (1990), 451-469. 677

[43] A local algorithm for constructing non-negative cubic splines
(with Bernd Fischer and Gerhard Opfer)
Journal of Approximation Theory, 64 (1991), 1-16. 696

[44] The time spent by the Wiener process in a narrow tube before leaving a wide tube
(with Antonia Földes)
Proceedings of the American Mathematical Society, 117 (1993), 529-536. 712

[45] Maximum likelihood estimation for stationary point processes
(with Pham D. Tuan)
Proceedings of the National Academy of Sciences, U.S.A., 83 (1986), 541-545. 720

[46] On time-reversibility and the uniqueness of moving average representations for non-Gaussian stationary time series
(with Marc Hallin and Claude Lefevre)
Biometrika, 75 (1988), 170-171. 725

[47] Locally asymptotically rank-based procedures for testing autoregressive moving average dependence
(with Marc Hallin)
Proceedings of the National Academy of Sciences, U.S.A., 85 (1988), 2031-2035. 727

[48] On some limit laws for perturbed empirical distribution functions
(with Manfred Denker)
Statistics and Probability Letters, 21 (1994), 317-321. 732

Part IV. APPENDICES

A. Volume 1.
Nonparametric Methods in Statistics and Related Topics
Contents 739

B. Volume 2.
Probability Theory and Extreme Value Theory
Contents 745

C. The Publications of Madan Lal Puri 751

Appendix C

The Publications of Madan Lal Puri

BOOKS AND MONOGRAPHS

1970

[1] Nonparametric Techniques in Statistical Inference. Cambridge University Press, England, 623 pages. (Editor).

1971

[2] Nonparametric Methods in Multivariate Analysis. John Wiley and Sons, New York, 440 pages. Co-author: P.K. Sen. Revised Reprint by Krieger Publication Company, Florida (1993).

1975

[3] Statistical Inference and Related Topics. Academic Press, New York, 352 pages. (Editor).

[4] Stochastic Processes and Related Topics. Academic Press, New York, 315 pages. (Editor).

1982

[5] Nonparametric Statistical Inference, Vol I. North Holland Publishing Company, Amsterdam, 463 pages. Co-editors: B.V. Gnedenko (Moscow State University, Moscow, U.S.S.R.) and Istvan Vincze (Hungarian Academy of Sciences, Budapest, Hungary).

[6] Nonparametric Statistical Inference, Vol. II. North Holland Publishing Company, Amsterdam, 446 pages. Co-editors: B.V. Gnedenko (Moscow State University, Moscow, U.S.S.R.) and Istvan Vincze (Hungarian Academy of Sciences, Budapest, Hungary).

1985

[7] Nonparametric Methods in General Linear Models. John Wiley and Sons, New York, 399 pages. Co-author: P.K. Sen.

1987

[8] New Perspectives in Theoretical and Applied Statistics. John Wiley and Sons, New York, 544 pages. Co-editors: J.P. Vilaplana (University of Bilbao, Spain) and Wolfgang Wertz (Technische Universität, Vienna, Austria).

[9] Mathematical Statistics and Probability Theory: Volume A, Theoretical Aspects. D. Reidel Publication Company, Holland, 326 pages. Co-editors: Pal Révész (Hungarian Academy of Sciences, Budapest, Hungary) and Wolfgang Wertz (Technische Universität, Vienna, Austria).

1993

[10] Statistical Sciences and Data Analysis: Proceedings of the Third Pacific Area Statistical Conference. VSP International Science Publishers, The Netherlands, 570 pages. Co-editors: Kameo Matusita and Takesi Hayakawa.

1994

[11] Recent Advances in Statistics and Probability: Proceedings of the 4th International Meeting of Statistics in the Basque Country, Spain. VSP International Science Publishers, Utrecht, The Netherlands, 458 pages. Co-editor: J.P. Vilaplana.

2000

[12] Asymptotics in Statistics and Probability: Papers in Honor of George Gregory Roussas. VSP International Science Publishers, The Netherlands, 429 pages. (Editor).

RESEARCH PAPERS

1964

[13] *Asymptotic efficiency of a class of c-sample tests.* The Annals of Mathematical Statistics, 35, 102-121.

1965

[14] *On the estimation of contrasts in linear models.* The Annals of Mathematical Statistics, 36, 198-202. Co-author: S. Bhuchongkul.

[15] *Some distribution-free k-sample rank tests of homogeneity against ordered alternatives.* Communications on Pure and Applied Mathematics, 18, 51-63.

[16] *On the combination of independent two-sample tests of a general class.* Review of the International Statistical Institute, 33, 229-241.

[17] *On some tests of homogeneity of variances.* Annals of the Institute of Statistical Mathematics, 17, 323-330.

[18] *On a class of testing procedures in linear models.* The Journal of Indian Statistical Association, 3, 1-8.

1966

[19] *On a class of multivariate multisample rank-order tests.* Sankhyā Series A, 28, 353-376. Co-author: P.K. Sen.

1967

[20] *Multi-sample analogues of some one-sample tests.* The Annals of Mathematical Statistics, 38, 523-549. Co-author: K.L. Mehra.

[21] *On the theory of rank order tests for location in the multivariate one sample problem.* The Annals of Mathematical Statistics, 38, 1216-1228. Co-author: P.K. Sen.

[22] *On robust estimation in incomplete block designs.* The Annals of Mathematical Statistics, 38, 1587-1591. Co-author: P.K. Sen.

[23] *On some optimum nonparametric procedures in two-way layouts.* Journal of the American Statistical Association, 62, 1214-1229. Co-author: P.K. Sen.

[24] *Combining independent one-sample tests of significance.* Annals of the Institute of Statistical Mathematics, 19, 285-300.

[25] *On some rank tests for homogeneity of scale parameters.* Trabajos de Estadistica, 28, 353-376. Co-author: P.S. Puri.

1968

[26] *On Chernoff-Savage tests for ordered alternatives in randomized blocks.* The Annals of Mathematical Statistics, 39, 967-972. Co-author: P.K. Sen.

[27] *Nonparametric confidence regions for some multivariate location problems.* Journal of the American Statistical Association, 62, 1373-1378. Co-author: P.K. Sen.

[28] *Selection procedures based on ranks: Scale parameter case.* Sankhyā, Series A, 30, 291-302. Co-author: P.S. Puri.

[29] *On a class of rank order estimators of contrasts in MANOVA.* Sankhyā, Series A, 30, 31-36. Co-author: P.K. Sen.

[30] *On a class of multivariate multisample rank order tests II. Tests for homogeneity of dispersion matrices.* Sankhyā, Series A, 30, 1-22. Co-author: P.K. Sen.

[31] *Multi-sample scale problem: Unknown location parameter.* Annals of the Institute of Statistical Mathematics, 22, 99-106.

1969

[32] *Multiple decision procedures based on ranks for certain problems in analysis of variance.* The Annals of Mathematical Statistics, 40, 619-632. Co-author: P.S. Puri.

[33] *A class of rank order tests for a general linear hypothesis.* The Annals of Mathematical Statistics, 40, 1325-1343. Co-author: P.K. Sen.

[34] *Analysis of covariance based on general rank scores.* The Annals of Mathematical Statistics, 40, 610-618. Co-author: P.K. Sen.

[35] *Rank order tests for multivariate paired comparisons.* The Annals of Mathematical Statistics, 40, 2101-2117. Co-author: H.D. Shane.

[36] *On a class of rank order tests for the identity of two multiple regression surfaces.* Zeitschrift für Wahrscheinlichkeits-theorie und verw. Geb., 12, 1-8. Co-author: P.K. Sen.

[37] *On the asymptotic normality of one-sample rank order test statistics.* Teoria Veroyatnostey i ee Primeneniya, (also in Theory of Probability and its Applications, Translation by SIAM) 14, 167-173. Co-author: P.K. Sen.

[38] *The Van Elteren W Test and non-null hypotheses.* Review of the International Statistical Institute, 37, 166-175.

[39] *On the asymptotic theory of rank order tests for experiments involving paired comparisons.* Annals of the Institute of Statistical Mathematics, 20, 163-173. Co-author: P.K. Sen.

[40] *On robust nonparametric estimation in some multivariate linear models.* Proceedings of the Second International Symposium on Multivariate Analysis, Dayton, Ohio, 33-52, Academic Press. Co-author: P.K. Sen.

[41] *Structure of distribution-free multivariate paired comparison tests.* Proceedings of the 37th Session of the International Statistical Institute, 320-322.

1970

[42] *Asymptotic theory of likelihood ratio and rank order tests in some multivariate linear models.* The Annals of Mathematical Statistics, 41, 87-100. Co-author: P.K. Sen.

[43] *On the estimation of location parameters in the multivariate one-sample and two-sample problems.* Metrika, 16, 58-73. Co-author: P.K. Sen.

[44] *On a class of rank order tests for independence in multivariate distributions.* Sankhyā, Series A, 32, 271-298. Co-authors: P.K. Sen. and D.V. Gokhale.

[45] *Statistical inference in incomplete blocks designs.* Nonparametric Techniques in Statistical Inference, Cambridge University Press, 131-153. Co-author: H.D. Shane.

1972

[46] *On the robustness of rank order tests and estimates in the generalized multivariate one-sample location problem.* Zeitschrift für Wahrscheinlichkeitstheorie und Verw. Geb., 22, 226-241. Co-author: P.K. Sen.

[47] *On some selection procedures in two-way layouts.* Zeitschrift für Wahrscheinlichkeitstheorie und Verw. Geb., 22, 242-250. Co-author: P.K. Sen.

[48] *Some aspects of nonparametric inference.* Review of the International Statistical Institute, 40, 298-327.

[49] *Rank tests for some linear hypotheses in paired comparisons designs.* Sankhyā, Series A, 34, 257-264.

1973

[50] *On parallel sum and difference of matrices.* Journal of Mathematical Analysis and Applications, 44, 92-97. Co-author: S.K. Mitra.

[51] *A note on asymptotic distribution-free tests for subhypotheses in multiple linear regression.* The Annals of Statistics, 1, 553-556. Co-author: P.K. Sen.

1974

[52] *Joint asymptotic multinormality for a class of rank order statistics in multivariate paired comparisons.* Journal of Multivariate Analysis, 4, 88-105. Co-author: C.T. Russell.

[53] *The asymptotic behavior of the minimax risk for multiple decision problems.* Sankhyā, Series A, 36, 1-12. Co-author: O. Krafft.

[54] *Asymptotic distribution of rank statistics for experiments involving incomplete block designs.* Annals of the Institute of Statistical Mathematics, 26, 421-446. Co-author: N.L. Wykoff.

[55] *Asymptotic results for rank statistics in some multivariate problems*. Bulletin of the International Statistical Institute, 35(2), 285-293.

1975

[56] *Order of normal approximation for rank test statistics distribution*. Annals of Probability, 3, 526-533. Co-author: J. Jurečková.

[57] *Distribution-free approaches to general linear models*. Proceedings of the International Symposium in Statistical Design and Linear Models, Fort Collins, Colorado; North Holland Publishing Company, 459-474. Co-author: P.K. Sen.

[58] *Miscellaneous problems of rank order theory*. Proceedings of the International Conference on Asymptotic Methods of Statistics and Their Applications in Physical and Social Sciences, Charles University, Prague, Czechoslovakia, 273-311.

1976

[59] *Augmenting Shapiro-Wilk Test for normality*. Contributions to Applied Statistics, Birkhauser Verlag, 129-139. Co-author: C.R. Rao.

[60] *On linear combination of order statistics*. Essays in Probability and Statistics, 433-449. Co-author: F. Eicker.

[61] *Remainder term estimate in the distribution of rank statistics under alternatives*. Proceedings of the Statistics Days Conference, 136-142. Co-author: H. Bergström.

1977

[62] *Convergence and remainder terms in linear rank statistics*. Annals of Statistics, 5, 671-680. Co-author: H. Bergström.

[63] *Asymptotically distribution-free aligned rank order tests for composite hypotheses for general multivariate linear models*. Zeitschrift für Wahrscheinlichkeitstheorie und verw. Geb., 39, 175-186. Co-author: P.K. Sen.

[64] *Problems of association for bivariate circular data and new test of independence*. Proceedings of the Fourth International Symposium on Multivariate Analysis, North Holland Publishing Company, 513-522. Co-author: J.S. Rao.

1978

[65] *A spherical correlation coefficient robust against scale*. Biometrika, 65, 391-395. Co-author: K.V. Mardia.

1979

[66] *Local maxima of the sample functions of the N-parameter Bessel process.* Stochastic Processes and Their Applications, 9, 137-145. Co-author: L.T. Tran.

[67] *Analysis of central place theory.* Bulletin of the International Statistical Institute, 47(2), 93-110. Co-authors: R. Edwards and K.V. Mardia.

[68] *Realization of ℓ_p by spaces of random variables.* Applicable Analysis, 8, 337-347. Co-author: S.S. Sheu.

[69] *Shorted operators and generalized inverses of matrices.* Linear Algebra and its Applications, 25, 45-56. Co-author: S.K. Mitra.

[70] *Rank order estimators in the case of grouped data (linear regression).* Sankhyā, Series B, 41, 239-259. Co-author: A.R. Padmanabhan.

[71] *Asymptotic behavior of stochastic linear rank statistics.* Bulletin of the International Statistical Institute, 47(4), 292-297. Co-author: N.S. Rajaram.

[72] *A simple test for goodness-of-fit based on spacing with some efficiency comparisons.* Contributions to Statistics — Jaroslav Hájek Memorial Volume, Academia Prague, Czechoslovak Academy of Sciences, 197-209. Co-authors: J.S. Rao and Y. Yoon.

[73] *Sample size, parameter rates and contiguity — the i.i.d. case.* Communication in Statistics — Theory and Methods, A8 (1), 71-83. Co-authors: M.G. Akritas and G.G. Roussas.

[74] *Matrices G satisfying simultaneous equations* $A * MAG = A * M$ *and* $G * NGA = G * N$. Journal of Indian Statistical Association, 17, 103-108. Co-authors: C.G. Khatri and S.K. Mitra.

1980

[75] *Empirical distribution functions and functions of order statistics for mixing random variables.* Journal of Multivariate Analysis, 10, 405-425. Co-author: L.T. Tran.

[76] *Asymptotic normality and convergence rates of linear rank statistics under alternatives.* Banach Center Publications, 6, 267-277. Co-author: N.S. Rajaram.

1981

[77] *Invariance principles for rank statistics for testing independence.* Contributions to Probability Theory. Academic Press, New York, 267-282. Co-author: L.T. Tran.

[78] *A note on predicting the results of chess championship.* Behavioral Science, 26, 85-87. Co-author: Robert Bartoszynski.

[79] *Complex planar splines.* Journal of Approximation Theory, 31, 383-402. Co-author: Gerhard Opfer.

[80] *Some remarks on strategy in playing tennis.* Behavioral Science, 26, 379-387. Co-author: Robert Bartoszynski.

[81] *Différentielle d'une fonction floue.* C.R. Acad. Sc. Paris, 293, Série I, 237-239. Co-author: Dan A. Ralescu.

[82] *A nonparametric test of ordered alternatives in the case of skewed data, with a biomedical application.* Statistics and Related Topics, North Holland Publishing Company, 279-283. Co-authors: A.R. Padmanabhan and A.K. Md. Ehsanes Saleh.

1982

[83] *Shorted matrices — an extended concept and some applications.* Linear Algebra and its Applications, 42, 57-79. Co-author: S.K. Mitra.

[84] *On estimating inter-subject variability of choice probabilities under observability constraints.* Journal of Mathematical Psychology, 25, 175-184. Co-author: Robert Bartoszyński.

[85] *Integration on fuzzy sets.* Advances in Applied Mathematics, 3, 430-434. Co-author: Dan A. Ralescu.

[86] *A possibility measure is not a fuzzy measure.* Fuzzy Sets and Systems, 7, 311-313. Co-author: Dan A. Ralescu.

[87] *On the degeneration of the variance in the asymptotic normality of signed rank statistics.* Statistics and Probability: Essays in Honor of C.R. Rao. North Holland Publishing Company, 591-608. Co-author: Stefan S. Ralescu.

[88] *Stochastic integrals and rank statistics.* Nonparametric Statistical Inference, Volume II, North Holland Publishing Company, 729-747. Co-author: N.S. Rajaram.

[89] *The asymptotic distribution theory of one sample signed rank statistic.* Statistical Decision Theory and Related Topics III, Volume 2, Academic Press, New York, 213-232. Co-author: Stefan S. Ralescu.

1983

[90] *Theory of nonparametric statistics for rounded-off data with applications.* Mathematische Operations-Forschung und Statistik, Series Statistics, 14, 301-349. Co-author: A.R. Padmanabhan.

[91] *Strong law of large numbers for Banach space valued random sets.* Annals of Probability, 11, 222-224. Co-author: Dan A. Ralescu.

[92] *On the order of magnitude of cumulants of von Mises functionals and related statistics.* Annals of Probability, 11, 346-354. Co-author: Rabi N. Bhattacharya.

[93] *Strong law of large numbers with respect to a set-valued probability measure.* Annals of Probability, 11, 1051-1054. Co-author: Dan A. Ralescu.

[94] *Differentials of fuzzy functions.* Journal of Mathematical Analysis and Applications, 91, 552-558. Co-author: Dan A. Ralescu.

[95] *Complex Chebyshev polynomials and generalizations with an application to the optimal choice of interpolating knots in complex planar splines.* Journal of Approximation Theory, 37, 89-101. Co-author: Gerhard Opfer.

[96] *The fundamental bordered matrix of linear estimation and the Duffin–Morley general linear electromechanical systems.* Applicable Analysis, 14, 241-258. Co-author: S.K. Mitra.

[97] *Improving upon the best invariant estimator in multivariate location problems.* The Australian Journal of Statistics, 25, 453-462. Co-author: Dan A. Ralescu.

1984

[98] *On Berry–Esséen rates, a law of the iterated logarithm and an invariance principle for the proportion of the sample below the sample mean.* Journal of Multivariate Analysis, 14, 231-247. Co-author: Stefan S. Ralescu.

[99] *Centering of signed rank statistics with a continuous score-generating function.* Teoria Veroyatnostei i ee Primeneniya (also in Theory of Probability and Its Applications, Translation by SIAM), 29, 580-584. Co-author: Stefan S. Ralescu.

[100] *Law of large numbers and central limit theorem for fuzzy random variables.* Cybernetics and Systems Research, 2, 525-529. Co-authors: E.P. Klement and D.A. Ralescu.

[101] *The Hausdorff α-dimensional measures of the level sets and the graph of the N-parameter Wiener process.* Metrika, 31, 275-283. Co-author: L.T. Tran.

[102] *Convergence of generalized inverses with applications to asymptotic hypothesis testing.* Sankhyā, Series A, 46, 277-286. Co-authors: Carl T. Russell and Thomas Mathew.

[103] *Rank order tests for the parallelism of several regression surfaces.* Journal of Statistical Planning and Inference, 10, 43-57. Co-author: Ching-Yuan Chiang.

[104] *Edgeworth expansions for signed linear rank statistics with regression constants.* Journal of Statistical Planning and Inference, 10, 137-149. Co-author: Munsup Seoh.

[105] *Edgeworth expansions for signed linear rank statistics under near location alternatives.* Journal of Statistical Planning and Inference, 10, 289-309. Co-author: Munsup Seoh.

[106] *Asymptotic normality of the lengths of a class of nonparametric confidence intervals for a regression parameter.* Canadian Journal of Statistics, 12, 217-228. Co-author: T.J. Wu.

[107] *Linear minimax estimators for estimating a function of the parameter.* The Australian Journal of Statistics, 26, 277-283. Co-authors: Dan A. Ralescu and Stefan S. Ralescu.

[108] *Rank procedures for testing subhypotheses in linear regression.* Annals of the Institute of Statistical Mathematics, 36, 35-50. Co-author: Ching-Yuan Chiang.

[109] *Locally most powerful rank tests for the two-sample problem with censored samples.* Journal of Organizational Behavior and Statistics, 1, 189-196. Co-author: A.K. MD. Ehsanes Saleh.

1985

[110] *Linear serial rank tests for randomness against ARMA alternatives.* Annals of Statistics, 13, 1156-1181. Co-authors: Marc Hallin and Jean-François Ingenbleek.

[111] *Cramér type large deviations for generalized rank statistics.* Annals of Probability, 13, 115-125. Co-authors: Stefan S. Ralescu and Munsup Seoh.

[112] *The concept of normality for fuzzy random variables.* Annals of Probability, 13, 1373-1379. Co-author: Dan A. Ralescu.

[113] *Limit theorems for random compact sets in Banach space.* Mathematical Proceedings of the Cambridge Philosophical Society, 97, 151-158. Co-author: Dan A. Ralescu.

[114] *On the rate of convergence in the central limit theorem for signed rank statistics.* Advances in Applied Mathematics, 6, 23-51. Co-author: Stefan S. Ralescu.

[115] *Asymptotic distributions of likelihood ratio criteria for testing latent roots and latent vectors of a covariance matrix under an elliptical population.* Biometrika, 72, 331-338. Co-author: T. Hayakawa.

[116] *Gaussian approximations of signed linear rank statistics process.* Journal of Statistical Planning and Inference, 11, 277-312. Co-author: T.J. Wu.

[117] *A sharpening of the remainder term in the higher-dimensional central limit theorem for multilinear rank statistics.* Journal of Multivariate Analysis, 17, 148-167. Co-authors: Manfred Denker and Uwe Rösler.

[118] *Asymptotic expansions of the distributions of some test statistics.* Annals of the Institute of Statistical Mathematics, 37, 95-108. Co-author: Takesi Hayakawa.

[119] *On the rate of convergence to asymptotic normality for generalized linear rank statistics.* Annals of the Institute of Statistical Mathematics, 37, 51-69. Co-author: Munsup Seoh.

[120] *Berry-Esséen theorems for signed linear rank statistics with regression constants.* Limit Theorems in Probability Statistics, (P. Révész, Editor). North Holland Publishing Company, 875-905. Co-author: Munsup Seoh.

[121] *Central limit theorem for linear rank statistics process.* Statistical Theory and Data Analysis, (K. Matusita, Editor). North Holland Publishing Company, 103-145. Co-author: Mark A. Carlson.

[122] *On the Cramér-Fréchet-Rao inequality for translation parameter in the case of finite support.* Mathematische Operationsforschung und Statistik, Series Statistics, 16, 495-506. Co-author: I. Vincze.

[123] *Tests de rangs linéaires pour une hypothèse de bruit blanc.* C.R. Acad. Sc. Paris, 301, Série I, 49-52. Co-authors: Marc Hallin and Jean-François Ingenbleek.

[124] *Solutions fortes des équations différentielles stochastiques pour des processus à indice multidimensionnel.* C.R. Acad. Sc. Paris, 301, Série I, 845-848. Co-author: Markus Dozzi.

[125] *Tests de rangs quadratiques pour une hypothèse de bruit blanc.* C.R. Acad. Sc. Paris, 301, Série I, 935-938. Co-authors: Marc Hallin and Jean-François Ingenbleek.

1986

[126] *Maximum likelihood estimation for stationary point processes.* Proceedings of the National Academy of Sciences, U.S.A., 83, 541-545. Co-author: Pham D. Tuan.

[127] *Fuzzy random variables.* Journal of Mathematical Analysis and Applications, 114, 409-422. Co-author: Dan A. Ralescu.

[128] *The order of normal approximation for signed linear rank statistics.* Teoria Veroyatnostei i ee Primeneniya (also in Theory of Probability and Its Applications, Translation by SIAM), 31, 156-163. Co-author: Tiee-Jian Wu.

[129] *Gaussian random sets in Banach space.* Teoria Veroyatnostei i ee Primeneniya (also in Theory of Probability and Its Applications, Translation by SIAM), 31, 598-601. Co-authors: Dan A. Ralescu and Stefan S. Ralescu.

[130] *Strong solutions of stochastic differential equations for multiparameter processes.* Stochastics, 17, 19-41. Co-author: Markus Dozzi.

[131] *Central limit theorem for perturbed empirical distribution functions evaluated at a random point.* Journal of Multivariate Analysis, 19, 273-279. Co-author: Stefan S. Ralescu.

[132] *Limit theorems for fuzzy random variables.* Proceedings of the Royal Society of London, 407, 171-182. Co-authors: E.P. Klement and D.A. Ralescu.

[133] *Bhattacharyya bound of variances of unbiased estimators in nonregular cases.* Annals of the Institute of Statistical Mathematics, 39, 35-44. Co-authors: M. Akahira and K. Takeuchi.

[134] *Almost sure linearity for signed rank statistics in the non-i.i.d. case.* Acta Mathematica Hungarica, 48, 273-284. Co-author: Stefan S. Ralescu.

[135] *Les tests de rangs dans l'analyse des séries chronologiques.* Cahiers du C.E.R.O., 28, 41-55. Co-authors: Marc Hallin and Jean-François Ingenbleek.

[136] *Tests de rangs localement optimaux pour une hypothèse de bruit blanc multivarié.* C.R. Academy of Sciences, Paris, 303, Série I, 901-904. Co-authors: Marc Hallin and Jean-François Ingenbleek.

1987

[137] *Linear and quadratic serial rank tests for randomness against serial dependence.* Journal of Time Series Analysis, 8, 409-424. Co-authors: Marc Hallin and Jean-François Ingenbleek.

[138] *Asymptotic normality of a class of nonparametric statistics.* Econometric Theory, 3, 313-347. Co-author: Munsup Seoh.

[139] *Linear rank statistics, bounded operators and weak convergence.* New Perspectives in Theoretical and Applied Statistics, John Wiley and Sons, New York, 171-206. Co-authors: Manfred Denker and Gerhard Keller.

[140] *Locally asymptotically optimal tests for randomness.* New Perspectives in Theoretical and Applied Statistics, John Wiley and Sons, New York, 87-99. Co-author: Marc Hallin.

[141] *Berry-Esséen theorems for signed rank statistics under near location alternatives.* Studia Scientiarum Mathematicarum Hungarica, 20, 197-211. Co-author: Munsup Seoh.

[142] *Tests of subhypotheses in linear regression based on rank order estimates.* Studia Scientiarum Mathematicarum Hungarica, 20, 237-247. Co-author: Ching-Yuan Chiang.

[143] *Limit theorems for random central order statistics.* Adaptive Statistical Procedures and Related Topics, IMS Lecture Notes and Monograph Series, 8, 447-475. Co-author: Stefan S. Ralescu.

[144] *On admissible estimation in exponential families with imprecise information.* In Statistical Decision Theory and Related Topics IV, (Editors: Shanti S. Gupta and James O. Berger), Springer Verlag, 403-408. Co-author: Stefan Ralescu.

[145] *Convergence faible de la statistique serielle linéaire de rang en condition de dépendance avec applications aux séries chronologiques et processus de Markov.* C.R. Academy of Sciences, Paris, 304, Série I, 583-586. Co-author: Michel Harel.

[146] *Convergence faible du processus empirique multidimensionnel corrigé en condition de mélange.* C.R. Academy of Sciences, Paris, 305, Série I, 93-95. Co-author: Michel Harel.

[147] *Weak convergence of weighted multivariate empirical processes under mixing conditions.* Proceedings of the Sixth Pannonian Symposium on Mathematical Statistics, Volume A: Mathematical Statistics and Probability (Editors: M.L. Puri, P. Révész and W. Wertz), Reidel Publishing Company, Holland, 121-141. Co-author: Michel Harel.

1988

[148] *Locally asymptotically rank-based procedures for testing autoregressive moving average dependence.* Proceedings of the National Academy of Sciences, U.S.A., 85, 2031-2035. Co-author: Marc Hallin.

[149] *Optimal rank-based procedures for time series analysis: testing an ARMA model against other ARMA models.* Annals of Statistics, 16, 402-432. Co-author: Marc Hallin.

[150] *On time-reversibility and the uniqueness of moving average representations for non-Gaussian stationary time series.* Biometrika, 75, 170-171. Co-authors: Marc Hallin and Claude Lefevre.

[151] *Asymptotic behavior of multi-response permutation procedures.* Advances in Applied Mathematics, 9, 200-210. Co-author: Manfred Denker.

[152] *Adaptive nonparametric procedures and applications.* Journal of the Royal Statistical Society, Series C (Applied Statistics), 37, 205-218. Co-authors: N.J. Hill and A.R. Padmanabhan.

[153] *Asymptotic properties of linear functions of order statistics.* Journal of Statistical Planning and Inference, 18, 203-223. Co-author: Cheun-Der Lea.

[154] *An invariance principle for processes indexed by two parameters and some statistical applications.* Probability and Mathematical Statistics, 9, 25-76. Co-author: Manfred Denker.

[155] *Asymptotic properties of perturbed empirical distribution functions evaluated at a random point.* Journal of Statistical Planning and Inference, 19, 201-215. Co-author: Cheun-Der Lea.

[156] *L'espace $\tilde{D}_k$ et la convergence faible du processus empirique indexé par rectangles en condition de mélange.* C.R. Academy of Sciences, Paris, 306, Série I, 207-210. Co-author: Michel Harel.

[157] *On locally asymptotically maximin tests for ARMA processes.* In Statistical Theory and Data Analysis II (Editor: K. Matusita), 495-499. North Holland Publishing Company. Co-author: Marc Hallin.

[158] *Comportement limite de la U-statistique, de la V-statistique et d'une statistique de rang à un échantillon pour des processus absolument réguliers non stationnaires.* C.R. Academy of Sciences, Paris, 306, Série I, 625-628. Co-author: Michel Harel.

[159] *Tests de rangs signés localement optimaux pour une hypothèse de dépendance ARMA.* C.R. Academy of Sciences, Paris, 307, Série I, 355-358. Co-author: Marc Hallin.

[160] *Convergence faible de la statistique sérielle linéaire de rang avec des fonctions de scores et des constantes de régression non bornées en condition de mélange.* C.R. Academy of Sciences, Paris, 307, Série I, 617-620. Co-author: Michel Harel.

1989

[161] *Asymptotic expansions for sums of nonidentically distributed Bernoulli random variables.* Journal of Multivariate Analysis, 28, 282-303. Co-authors: Paul Deheuvels and Stefan S. Ralescu.

[162] *Asymptotically most powerful rank tests for multivariate randomness against serial dependence.* Journal of Multivariate Analysis, 30, 34-71. Co-authors: Marc Hallin and Jean François Ingenbleek.

[163] *Central limit theorems under alternatives for a broad class of nonparametric statistics.* Journal of Statistical Planning and Inference, 22, 271-294. Co-author: Munsup Seoh.

[164] *Limiting behavior of U-statistics, V-statistics, and one sample rank order statistics for nonstationary absolutely regular processes.* Journal of Multivariate Analysis, 30, 181-204. Co-author: Michel Harel.

[165] *On the rate of convergence in normal approximation and large deviation probabilities for a class of statistics.* Teoria Veroyatnostei i ee Primeneniya (also in Theory of Probability and its Applications, Translation by SIAM), 33, 735-750. Co-author: Munsup Seoh.

[166] *A new semigroup technique in Poisson approximation.* Semigroup Forum, 38, 189-201. Co-authors: Paul Deheuvels and Dietmar Pfeifer.

[167] *Weak convergence of the U-statistic and weak invariance of the one-sample rank order statistic for Markov processes and ARMA models.* Journal of Multivariate Analysis, 31, 258-265. Co-author: Michel Harel.

[168] *Convergence faible de la U-statistique generalisée pour des processes nonstationnaires absolument réguliers.* C.R. Academy of Sciences, Paris, 309, Série I, 135-138. Co-author: Michel Harel.

1990

[169] *Weak convergence of serial rank statistics under dependence with applications in time series and Markov processes.* Annals of Probability, 18, 1361-1387. Co-author: Michel Harel.

[170] *Weak invariance of generalized U-statistics for nonstationary absolutely regular processes.* Stochastic Processes and Their Applications, 34, 341-360. Co-author: Michel Harel.

[171] *The space $\tilde{D}_k$ and weak convergence for the rectangle-indexed processes under mixing.* Advances in Applied Mathematics, 11, 443-474. Co-author: Michel Harel.

[172] *Asymptotic normality of L-statistics based on $m(n)$-decomposable time series.* Journal of Multivariate Analysis, 35, 260-275. Co-authors: K.C. Chanda and F.H. Ruymgaart.

[173] *A strong invariance principle concerning the J-upper order statistics for stationary m-dependent sequences.* Journal of Statistical Planning and Inference, 25, 43-51. Co-author: George Haiman.

[174] *Weak convergence of the serial linear rank statistics with unbounded scores and regression constants under mixing conditions.* Journal of Statistical Planning and Inference, 25, 163-186. Co-author: Michel Harel.

[175] *Information.* In: Acting Under Uncertainty: Multidisciplinary Conceptions (Editor: George von Furstenberg). Kluwer Academic Publishers, Boston/Dordrecht/Lancaster, 451-469. Co-author: Howard L. Resnikoff.

[176] *Measure of information and contiguity.* Statistics and Probability Letters, 9, 223-228. Co-author: Istvan Vincze.

[177] *Information and Mathematical Statistics.* Proceedings of the Fourth Prague Symposium on Asymptotic Statistics, Charles University, Prague (Editors: P. Mandl and M. Hušková) August 29, 1988 September 2, 1988, 447-456. Co-author: Istvan Vincze.

1991

[178] *A local algorithm for constructing non-negative cubic splines.* Journal of Approximation Theory, 64, 1-16. Co-authors: Bernd Fischer and Gerhard Opfer.

[179] *Time series analysis via rank order theory: signed-rank tests for ARMA models.* Journal of Multivariate Analysis, 39, 1-29. Co-author: Marc Hallin.

[180] *Convergence theorem for fuzzy martingales.* Journal of Mathematical Analysis and Applications, 160, 107-122. Co-author: Dan A. Ralescu.

[181] *Limiting behavior of one sample rank-order statistics with unbounded scores for nonstationary absolutely regular processes.* Journal of Statistical Planning and Inference, 27, 1-23. Co-author: Michel Harel.

[182] *Weak invariance of the multidimensional rank statistic for nonstationary absolutely regular processes.* Journal of Multivariate Analysis, 36, 204-221. Co-author: Michel Harel.

[183] *On Hilbert-space-valued U-statistics.* Teoria Veroyatnostei i ee Primeneniya (also in Theory of Probability and Its Applications, Translation by SIAM), 36, 604-605. Co-author: V.V. Sazonov.

[184] *Théorème central limite de l'estimateur de densité et de l'erreur quadratique intégrée pour des variables aléatoires nonstationnaires.* C.R. Academy of Sciences, Paris, 312, Série 1, No. 1, 145-148. Co-author: Michel Harel.

[185] *Invariance faible de la statistique de rang multidimensionnelle. Applications.* C.R. Academy of Sciences, Paris, 312, Série 1, No. 4, 349-352. Co-author: Michel Harel.

1992

[186] *Rank tests for time series analysis.* A survey. In: New Directions in Time Series Analysis (the IMA Volume in Mathematics and Its Applications) 45, 111-153. (Editors: David Brillinger, Emanuel Parzen, Murray Rosenblatt et al.). Springer-Verlag, New York/Heidelberg/Berlin. Co-author: Marc Hallin.

[187] *Rank-based estimation of ratio of scale parameters and applications.* Journal of Statistical Planning and Inference, 31, 23-49. Co-author: A.R. Padmanabhan.

[188] *The Neyman-Pearson probability ratio and information.* In: Data Analysis and Statistical Inference, Festschrift in Honor of Friedhelm Eicher (Editors: Siegfried Schach and Götz Trenkler), 53-64. Verlag Josef Eul, Bergisch Gladbach, Köln. Co-author: Istvan Vincze.

[189] *Asymptotic normality of two-sample linear rank statistics under U-statistic structure.* Journal of Statistical Planning and Inference, 32, 89-110. Co-author: Manfred Denker.

[190] *Some asymptotic results for a broad class of nonparametric statistics.* Journal of Statistical Planning and Inference, 32, 165-196. Co-author: Marc Hallin.

1993

[191] *A strong invariance principle concerning the J-upper upper order statistics for stationary Gaussian sequences.* Annals of Probability, 21, 86-135. Co-author: George Haiman.

[192] *The time spent by the Wiener process in a narrow tube before leaving a wide tube.* Proceedings of the American Mathematical Society, 117, 529-536. Co-author: Antonia Földes.

[193] *Weak convergence of the simple linear rank statistic under mixing conditions in the nonstationary case.* Teoria Veroyatnostei i ee Primeneniya (also in Theory of Probability and Its Applications, Translation by SIAM) 38, 579-599. Co-author: Michel Harel.

[194] *Nonparametric prediction for random fields.* Stochastic Processes and Their Applications, 48, 139-156. Co-author: Frits H. Ruymgaart.

[195] *Asymptotic behavior of L-statistics for a large class of time series.* Annals of the Institute of Statistical Mathematics, 45, 687-701. Co-author: Frits H. Ruymgaart.

[196] *On the central limit theorem in Hilbert space with application to U-statistics.* In: Statistical Sciences and Data Analysis: Proceedings of the Third Pacific Area Statistical Conference. VSP International Science Publishers, Holland. (Editors: Kameo Matusita, Madan L. Puri and Takesi Hayakawa), 407-413. Co-author: V.V. Sazonov.

[197] *Asymptotics of the perturbed sample quantile for a sequence of m-dependent stationary random process.* In: Statistical Sciences and Data Analysis: Proceedings of the Third Pacific Area Statistical Conference. VSP International Science Publishers, Holland. (Editors: Kameo Matusita, Madan L. Puri and Takesi Hayakawa), 415-426. Co-author: Shan Sun.

1994

[198] *Weak convergence of weighted empirical U-statistics processes for dependent random variables.* Journal of Multivariate Analysis, 48, 297-314. Co-authors: Colm Art O'Cinneide and Michel Harel.

[199] *Aligned rank tests for linear models with autocorrelated error terms.* Journal of Multivariate Analysis, 50, 175-237. Co-author: Marc Hallin.

[200] *Law of the iterated logarithm for perturbed empirical distribution functions evaluated at a random point for nonstationary random variables.* Journal of Theoretical Probability, 7, 831-855. Co-author: Michel Harel.

[201] *Berry–Esséen rate in asymptotic normality for perturbed sample quantiles.* Metrika, 41, 83-98. Co-author: Munsup Seoh.

[202] *On some limit laws for perturbed empirical distribution functions.* Statistics and Probability Letters, 21, 317-321. Co-author: Manfred Denker.

[203] *Change curves in the presence of dependent noise.* In: Change-Point Problems, a volume in the IMS Lecture Notes and Monograph Series, Volume 23, 242-254. Co-author: Frits H. Ruymgaart.

1995

[204] *Nonparametric methods for stratified two-sample designs with application to multiclinic trials.* Journal of the American Statistical Association, 90, 1004-1014. Co-authors: Edgar Brunner and Shan Sun.

[205] *Extremes of Markov sequences.* Journal of Statistical Planning and Inference, 45, 185-201. Co-authors: George Haiman and Maxime Kiki.

[206] *A multivariate Wald-Wolfowitz rank test against serial dependence.* The Canadian Journal of Statistics, 23, 55-65. Co-author: Marc Hallin.

[207] *Higher order asymptotic theory for normalizing transformations.* Annals of the Institute of Statistical Mathematics, 17, 581-600. Co-author: Masanobu Taniguchi.

[208] *Asymptotic expansions in statistics: A review of methods and applications.* In: Advances in Econometrics and Quantitative Economics: Essays in honor of C.R. Rao. Blackwell Publishers, Oxford, United Kingdom. (Editors: G.S. Maddala, Peter C.B. Phillips and T.N. Srinavasan), 88-122. Co-author: R.N. Bhattacharya.

[209] *Comportement asymptotique des estimateurs de la fonction de regression pour des variables aléatoires dependantes et nonstationnaires.* C.R. Academy of Sciences, Paris, 320, Série I, 75-78. Co-author: Michel Harel.

1996

[210] *Nonparametric approach for non-Gaussian vector stationary processes.* Journal of Multivariate Analysis, 56, 259-283. Co-authors: Masanobu Taniguchi and Masao Kondo.

[211] *Valid Edgeworth expansions of M-estimators in regression models with weakly dependent residuals.* Econometric Theory, 12, 331-346. Co-author: Masanobu Taniguchi.

[212] *Conditional U-statistics for dependent random variables.* Journal of Multivariate Analysis, 57, 84-100. Co-author: Michel Harel.

[213] *Asymptotic normality of nearest neighbor regression function estimates based on nonstationary dependent observations.* Special Issue on Multivariate Statistical Analysis in Honor of Professor Minoru Siotani, American Journal of Mathematical and Management Sciences, 15, 255-290. Co-author: Michel Harel.

[214] *Weak convergence of sequences of first passage processes and applications.* Stochastic Processes and Their Applications, 62, 327-345. Co-author: Stefan S. Ralescu.

[215] *Characterization of weak convergence for perturbed empirical and quantile processes under φ-mixing.* Journal of Statistical Planning and Inference, 53, 285-295. Co-authors: H.J.A. Degenhardt, Shan Sun and Martien van Zuijlen.

[216] *Nonparametric density estimators based on nonstationary absolutely regular random sequences.* Journal of Applied Mathematics and Stochastic Analysis, 9, 233-254. Co-author: Michel Harel.

[217] *Nonparametric methods in design and analysis of experiments.* In: Handbook of Statistics, volume 13, Design of Experiments. (Editors: S. Ghosh and C.R. Rao) North Holland, Amsterdam, Netherlands, 631-703. Co-author: Edgar Brunner.

1997

[218] *Normal approximation of U-statistics in Hilbert space.* Teoria Veroyatnostei i ee Primeneniya (also in Theory of Probability and its Applications, Translation by SIAM), 41, 481-504. Co-authors: Yu.V. Borovskikh and V.V. Sazonov.

1998

[219] *Autoregressive quantiles and related rank scores processes for generalized random coefficient autoregressive processes.* Journal of Statistical Planning and Inference, 68, 271-294. Co-author: Michel Harel.

[220] *Records and 2-block records of 1-dependent stationary sequences under local dependence.* Annales de l'institut Henri Poincaré, 34, 481-503. Co-authors: George Haiman, Nelly Mayeur and Valéry Nevzorov.

1999

[221] *Conditional empirical processes defined by nonstationary absolutely regular sequences.* Journal of Multivariate Analysis, 70, 250-285. Co-author: Michel Harel.

[222] *Rank-score tests in factorial designs with repeated measures.* Journal of Multivariate Analysis, 70, 286-317. Co-authors: Edgar Brunner and Ullrich Munzel.

[223] *Robust one-way ANOVA under possibly non-regular conditions.* Biometrical Journal, 41, 321-339. Co-authors: G.J. Babu and A.R. Padmanabhan.

2000

[224] *The limiting density of unit root test statistics: A unifying technique.* The Journal of Time Series Analysis, 21, 249-260. Co-authors: Mithat Gönen, Frits H. Ruymgaart and Martien C.A. van Zuijlen.

[225] *An estimation method in time series errors-in-variables models.* Journal of the Japan Statistical Society, 30, 75-87. Co-authors: Narihisa Tsuga and Masanobu Taniguchi.

[226] *Limit theorems for fuzzy random variables.* In: Asymptotics in Statistics and Probability: Papers in Honor of George Gregory Roussas. VSP International Science Publishers, The Netherlands, 331-346. Co-author: Frank Proske.

2001

[227] *Nonparametric methods in factorial designs.* Statistical Papers (Springer-Verlag journal), 42, 1-52. Co-author: Edgar Brunner.

[228] *Point and interval estimators for nonparametric treatment effects in designs with repeated measures.* In: Data Analysis from Statistical Foundations: Festschrift in Honor of Donald A.S. Fraser. (Editor: A.K.E. Saleh) Nova Science Publishers, Inc., New York, 167-178. Co-author: Edgar Brunner.

[229] *U-statistiques conditionnelles universellement consistantes pour des modèles de Markov cachés.* Comptes rendus de l'Académie des Sciences, Paris, t.333, Série I/Mathématique, 953-956. Co-author: Michel Harel.

2002

[230] *A class of rank-score tests in factorial designs.* Journal of Statistical Planning and Inference, 103, 329-359. Co-author: Edgar Brunner.

[231] *Central limit theorem for Banach space valued fuzzy random variables.* Proceedings of the American Mathematical Society 130, 1493-1501. Co-author: Frank Proske.

[232] *Strong law of large numbers for Banach space valued fuzzy random variables.* Journal of Theoretical Probability 15, 543-552. Co-author: Frank Proske.

[233] *Weighted rank statistics in factorial designs with fixed effects.* Statistica Neerlandica, 56, 179-194. Co-authors: Edgar Brunner and Sebastian Domhof.

[234] *Asymptotics of tests for a unit root in autoregression.* Journal of Statistical Planning and Inference 108, 351-364. Co-authors: Kenji Sakiyama and Masanobu Taniguchi.

[235] *The multivariate nonparametric Behrens-Fisher problem.* Journal of Statistical Planning and Inference 108, 37-53. Co-authors: Edgar Brunner and Ullrich Munzel.

[236] *LAN based asymptotic theory for time series.* In: Proceedings of the Fourth International Triennial Calcutta Symposium on Probability and Statistics, December 26-28, 2000. (Editors: Arijit Chaudhuri and Malay Ghosh), Calcutta Statistical Association Bulletin, 52, 143-170. Co-author: Masanobu Taniguchi.

2003

[237] *A strong law of large numbers for the generalized random sets from the viewpoint of empirical processes.* Proceedings of the American Mathematical Society. (To appear.) Co-author: Frank Proske.